D0990767

# Steel, Concrete, and Composite Design of Tall Buildings

Bungale S. Taranath

Second Edition

McGraw-Hill

New York San Francisco Washington, D.C. Auckland Bogotá
Caracas Lisbon London Madrid Mexico City Milan
Montreal New Delhi San Juan Singapore
Sydney Tokyo Toronto

**Library of Congress Cataloging-in-Publication Data**

Taranath, Bungale S.
Steel, concrete, and composite design of tall buildings / Bungale S. Taranath.
p. cm.
Includes index.
ISBN 0-07-062914-5 (hardcover)
1. Building, Iron and steel. 2. Concrete construction.
3. Composite construction. 4. Tall buildings—Design and construction. 5. Structural engineering.
TH1611.T37 1998
693'.71—dc21 96-49612
CIP

*McGraw-Hill*

*A Division of The **McGraw-Hill** Companies*



1 2 3 4 5 6 7 8 9 0 DOC/DOC 9 0 2 1 0 9 8 7

ISBN 0-07-062914-5



*The sponsoring editor for this book was Larry S. Hager and the production supervisor was Sherri Souffrance. It was set in Century Schoolbook by The Universities Press (Belfast) Ltd.*

*Printed and bound by R. R. Donnelley & Sons Company.*

McGraw-Hill books are available at special quantity discounts to use as premiums and sales promotions, or for use in corporate training programs. For more information, please write to the Director of Special Sales, McGraw-Hill, 11 West 19th Street, New York, NY 10011. Or contact your local bookstore.

Information contained in this work has been obtained by The McGraw-Hill Companies, Inc. ("McGraw-Hill") from sources believed to be reliable. However, neither McGraw-Hill nor its authors guarantee the accuracy or completeness of any information published herein, and neither McGraw-Hill nor its authors shall be responsible for any errors, omissions, or damages arising out of use of this information. This work is published with the understanding that McGraw-Hill and its authors are supplying information but are not attempting to render engineering or other professional services. If such services are required, the assistance of an appropriate professional should be sought.

This book is printed on acid-free paper.

*This book is dedicated to my wife Saroja*
*Daughter Anupama and Son Abhiman*

# Contents

Preface xix

Acknowledgements xxi

Chapter 1 General Considerations 1

1.1 Introduction 1
1.2 Structural concepts 3
1.3 Structural vocabulary: Case Studies 9
1.3.1 The Museum Tower, Los Angeles 10
1.3.2 Bank One Center, Indianapolis 12
1.3.3 Two Union Square, Seattle 13
1.3.4 Bank of China Tower, Hong Kong 15
1.3.5 Dallas Main Center 16
1.3.6 The Miglin-Beitler Tower, Chicago, Illinois 17
1.3.7 The NCNB Tower, North Carolina 25
1.3.8 The South Walker Tower, Chicago 29
1.3.9 AT&T Building, New York City 31
1.3.10 Trump Tower, New York City 31
1.3.11 Metro-Dade Administration Building 31
1.3.12 Jin Mao Tower, Shanghai, China 35
1.3.13 Petronas Towers, Malaysia 39
1.3.14 Tokyo City Hall 44
1.3.15 Leaning Tower, a building in Madrid, Spain 44
1.3.16 Hong Kong Central Plaza 49
1.3.17 Fox Plaza, Los Angeles 53
1.3.18 Bell Atlantic Tower, Philadelphia 53
1.3.19 Norwest Center, Minneapolis 54
1.3.20 First Bank Place, Minneapolis 55
1.3.21 Figueroa at Wilshire, Los Angeles 58
1.3.22 One Detroit Center 61
1.3.23 One Ninety One Peach Tree, Atlanta 65
1.3.24 Nations Bank Plaza, Atlanta 64
1.3.25 Allied Bank Tower, Dallas, Texas 67
1.3.26 First Interstate World Center, Los Angeles 75
1.3.27 Singapore Treasury Building, Singapore 78
1.3.28 City Spire, New York 80
1.3.29 City Corp Tower, Los Angeles 81
1.3.30 Cal Plaza, Los Angeles 82
1.3.31 MTA Headquarters, Los Angeles 84
1.3.32 The 21 Century Tower, Shanghai, China 89

Chapter 2 Wind Effects 105

2.1 Design Considerations 105
2.2 Nature of Wind 107
2.2.1 Introduction 107
2.2.2 Types of wind 109
2.3 Extreme Wind Conditions 111
2.3.1 Introduction 111
2.3.2 Thunderstorms 111
2.3.3 Hurricanes 112
2.3.4 Tornadoes 113
2.4 Characteristics of Wind 114
2.4.1 Introduction 114
2.4.2 Variation of wind velocity with height 115
2.4.3 Turbulent nature of wind 116
2.4.4 Probabilistic approach to wind load determination 119
2.4.5 Vortex-shedding phenomenon 121
2.4.6 Dynamic nature of wind 125
2.4.7 Cladding pressures 126
2.4.7.1 Introduction 126
2.4.7.2 Distribution of pressures and suctions 127
2.4.7.3 Local cladding loads and overall design loads 128
2.5 Code Wind Loads 130
2.5.1 Introduction 130
2.5.2 BOCA National Building Code (1996) 131
2.5.3 Standard Building Code (1991) 136
2.5.4 Uniform Building Code (UBC 1994) 136
2.5.4.1 General provisions 136
2.5.4.2 Wind speed map 138
2.5.4.3 Special wind regions 139
2.5.4.4 Hurricanes and Tornadoes 139
2.5.4.5 Exposure effects 139
2.5.4.6 Site exposure 140
2.5.4.7 Design wind pressures 140
2.5.4.8 The $C_e$ factor 141
2.5.4.9 Pressure coefficient, $C_q$ 143
2.5.4.10 Importance factors $I_w$ 145
2.5.4.11 Design example, UBC 1994 146
2.5.5 ANSI/ASCE 7-93 147
2.5.5.1 Overview 148
2.5.5.2 Design pressures for the main wind-force resisting system 151
2.5.5.3 Design wind pressure on components and cladding 156
2.5.5.4 External pressure coefficient $c_p$ 156
2.5.5.5 Internal pressure coefficient $c_{pi}$ 161
2.5.5.6 Design examples:
(1) Non-flexible building; Height $<$500 ft 162

(2) Non-flexible structure; Height >500 ft 168
(3) 60-storey flexible building 170
(4) 50-storey flexible building 173
2.5.5.7 Sensitivity study of gust response factor $\bar{G}$ 178
2.5.6 ASCE 7-95: Wind load provisions 179
2.5.6.1 Introduction 179
2.5.6.2 Overview 181
2.5.6.3 Wind speed-up over hills and escarpments: $K_{zt}$ factor 187
2.5.6.4 Full and partial loading 189
2.5.6.5 Gust effect factor 189
(1) Rigid structures—simplified method 190
(2) Rigid structures—complete analysis 192
(3) Flexible or dynamically sensitive buildings i93
2.5.6.6 Calculation of gust effect factor: design example 194
2.5.6.7 Sensitivity study of gust response factor 197
2.5.6.8 Calculation of wind pressures: design example 199
2.5.7 National Building Code of Canada (NBC 1990) 205
2.5.7.1 Simple procedure 205
(a) Reference pressure, $q$ 209
(b) Exposure factor, $C_e$ 209
(c) Gust effect factor, (dynamic response factor) $C_g$ 210
(d) Pressure coefficient, $C_p$ 211
2.5.7.2 Experimental procedure 211
2.5.7.3 Detailed procedure: 211
(a) Exposure factor, $C_e$ (detailed procedure) 212
(b) Gust effect factor, $C_g$ (detailed procedure) 213
2.5.7.4 Design example 215
2.5.7.5 Wind-induced building motion 217
2.6 Wind Tunnel Engineering 221
2.6.1 Introduction 221
2.6.2 Description of wind tunnels 222
2.6.3 Objective of wind tunnel tests 224
2.6.4 Rigid model studies 225
2.6.5 Aeroelastic study 232
2.6.5.1 Model requirments 235
2.6.6 High-frequency force balance model 246
2.6.7 Pedestrian wind studies 253
2.7 Field Measurements of Wind Loads 256
2.8 Motion Perception: Human Response to Building Motions 258
2.9 Comparison of Code and Wind Tunnel Test Results 259
2.10 Chapter Summary 260

Chapter 3 Seismic Design 261
3.1 Introduction 261
3.1.1 Nature of earthquakes 261
3.1.2 Some recent earthquakes 262

3.1.3 Seismograph 267
3.1.4 Earthquake magnitude and intensity 268
3.1.5 Seismic design 269
3.1.6 Uncertainties in seismic design 272
3.1.7 Design ground motion 275
3.2 Tall Building Behavior During Earthquakes 276
3.2.1 Introduction 276
3.2.2 Response of tall buildings 276
3.2.3 Influence of soil 278
3.2.4 Damping 279
3.2.5 Building motion and deflections 280
3.2.6 Seismic separation 281
3.3 Seismic Design Concept 281
3.3.1 Determination of forces 281
3.3.2 Design of the structure 281
3.3.3 Structural response 282
3.3.4 Path of forces 282
3.3.5 Demands of earthquake motion 282
3.3.6 Response of buildings 283
3.3.7 Response of elements attached to buildings 283
3.3.8 Techniques of seismic design 284
3.3.8.1 Layout 284
3.3.8.2 Structural symmetry 284
3.3.8.3 Irregular buildings 285
3.3.8.4 Lateral force-resisting systems 285
3.3.8.5 Diaphragms 289
3.3.8.6 Ductility 289
3.3.8.7 Nonstructural participation 291
3.3.8.8 Foundations 291
3.3.8.9 Damage control features 291
3.3.8.10 Redundancy 293
3.4 1994 UBC Equivalent Lateral Force Procedure (Static Method) 293
3.4.1 Design base shear 293
3.4.2 Base shear distribution along building height 299
3.4.3 Horizontal distribution 299
3.4.4 Torsion 300
3.4.5 Story shear and overturning moments 300
3.4.6 Discontinuity in lateral force-resisting elements 301
3.4.7 *p*-delta effects 302
3.4.8 Continuous load path 304
3.4.9 Redundancy 304
3.4.10 Configuration 304
3.4.11 Design example 305
3.5 Dynamic Analysis Procedure 309
3.5.1 Introduction 309
3.5.2 Response spectrum method 315
3.5.3 Development of design response spectrum 320
3.5.4 Time-history analysis 325
3.5.4.1 Introduction 325

3.5.4.2 Analysis procedure 327
3.5.5 Dynamic requirements: overview of 1994 UBC 330
3.5.6 Modal analysis: hand calculation procedure 334
3.5.6.1 Three-storey building 334
3.5.6.2 Seven-storey building 337
3.6 Seismic Vulnerability Study and Retrofit Design 345
3.6.1 Introduction 345
3.6.2 Code sponsored design 348
3.6.3 Alternative design philosophy 349
3.6.3.1 FEMA 178 method 352
3.6.3.2 Tri-services manual 360
3.6.3.3 SEAOC's Vision 2000: performance-based engineering 370
3.7 Dynamic Analysis: Theory 374
3.7.1 Introduction 374
3.7.2 Systems with single-degree-of-freedom 376
3.7.3 Multi-degree-of-freedom systems 380
3.7.4 Modal superposition method 384
3.7.4.1 Normal coordinates 385
3.7.4.2 Orthogonality 386
3.8 Summary 393

Chapter 4 Lateral Systems: Steel Buildings 397

4.1 Introduction 397
4.1.1 Steel in high-rise buildings 397
4.2 Frames with Semi-Rigid-Connections 397
4.2.1 Introduction 398
4.2.2 Review of connection behavior 398
4.2.3 Beam line concept 401
4.2.4 Type 2 wind connections 405
4.2.4.1 Design outline for type 2 wind connections 412
4.2.5 Concluding remarks 415
4.3 Rigid Frames (Moment Frames) 415
4.3.1 Introduction 415
4.3.2 Deflection characteristics 417
4.3.2.1 Cantilever bending component 418
4.3.2.2 Shear racking component 419
4.3.3 Methods of analysis 420
4.3.4 Calculation of drift 420
4.4 Braced Frames 421
4.4.1 Introduction 421
4.4.2 Behavior 421
4.4.3 Types of braces 423
4.5 Staggered Truss System 426
4.5.1 Introduction 426
4.5.2 Physical behavior 429

4.5.3 Design consideration 429
4.5.3.1 Floor system 429
4.5.3.2 Columns 433
4.5.3.3 Trusses 434
4.6 Eccentric Bracing Systems 435
4.6.1 Introduction 435
4.6.2 Ductility 436
4.6.3 Behavior of frame 436
4.6.4 Essential features of the link 438
4.6.5 Analysis and design considerations 438
4.6.6 Deflection considerations 439
4.6.7 Conclusions 439
4.7 Interacting System of Braced and Rigid Frames 440
4.7.1 Introduction 440
4.7.2 Physical behavior 443
4.8 Outrigger and Belt Truss Systems 445
4.8.1 Introduction 445
4.8.2 Physical behavior 448
4.8.3 Deflection calculations 453
4.8.4 Optimum location of single truss 457
4.8.5 Optimum location for a two outrigger system 459
4.8.5.1 Computer solution 460
4.8.5.2 Explanation of graphs 462
4.8.6 Example projects 463
4.8.7 Concluding remarks 466
4.9 Framed Tube System 466
4.9.1 Introduction 466
4.9.2 Framed tube behavior 467
4.9.3 Shear-lag phenomenon 471
4.9.4 Irregularly shaped tubes 476
4.10 Truss Tube System 480
4.11 Bundled Tube 487
4.11.1 Behavior 488
4.12 Ultimate High-Efficiency Structures 495

Chapter 5 Lateral Bracing Systems for Concrete Buildings 501

5.1 Frame Action of Column and Two-way Slab Systems 502
5.2 Flat Slab and Shear Walls 505
5.3 Flat Slab, Shear Walls, and Columns 505
5.4 Coupled Shear Walls 506
5.5 Rigid Frames 506
5.6 Widely Spaced Perimeter Tube 507
5.7 Rigid Frame with Haunch Girders 508
5.8 Core-supported Structures 509
5.9 Shear Wall-Frame Interaction 510
5.10 Frame Tube Structures 522

5.11 Exterior Diagonal Tube 523
5.12 Modular or Bundled Tube 524
5.13 Miscellaneous Systems 525

Chapter 6 Lateral Systems for Composite Construction 531

6.1 Introduction 531
6.2 Composite Elements 533
6.2.1 Composite slabs 534
6.2.2 Composite girders 534
6.2.3 Composite columns 535
6.2.4 Composite diagonals 538
6.2.5 Composite shear walls 539
6.3 Composite Building Systems 539
6.3.1 Shear wall systems 541
6.3.2 Shear wall-frame interacting systems 545
6.3.3 Tube systems 546
6.3.4 Vertically mixed systems 549
6.3.5 Mega frames with super columns 549
6.4 Example Projects 550
6.4.1 Composite steel pipe columns 550
6.4.1.1 Pacific First Center 550
6.4.1.2 Fremont Experience, Las Vegas 552
6.4.2 Formed composite columns 554
6.4.2.1 Interfirst Plaza, Dallas 554
6.4.2.2 Bank of China Tower, Hong Kong 556
6.4.2.3 The Bank of South West Tower, Houston, Texas 557
6.4.3 Composite shear walls and frames 559
6.4.3.1 First City Tower, Houston, Texas 559
6.4.4 Composite tube system 565
6.4.4.1 The America Tower, Houston, Texas 565
6.4.5 Conventional concrete system with partial steel floor framing 568
6.4.5.1 The Huntington, Houston, Texas 568
6.5 High-Efficiency Structure: Structural Concept 573

Chapter 7 Gravity Systems for Steel Buildings 577

7.1 Introduction 577
7.2 Design Loads 578
7.3 Metal Deck Systems 580
7.4 Open-Web Joist Systems 580
7.5 Wide-Flange Beams 582
7.5.1 Bending 582
7.5.2 Shear 582
7.5.3 Deflections 582
7.6 Columns 583

Chapter 8 Gravity Systems in Concrete Buildings 585

8.1 Floor Systems 586
8.1.1 Flat plates 586
8.1.2 Flat slabs 587
8.1.3 Waffle system 587
8.1.4 One-way concrete ribbed slabs 588
8.1.5 Skip joist system 588
8.1.6 Band beam system 590
8.1.7 Haunch girder and joist system 591
8.1.8 Beam and slab system 591
8.1.9 Design examples 592
8.1.9.1 One-way slab-and-beams system 592
8.1.9.2 T-beam design 600
8.1.9.3 Analysis of two-way slabs 607
8.2 Prestressed Concrete Systems 613
8.2.1 Method of prestressing 614
8.2.2 Materials 616
8.2.3 Design 618
8.2.4 Practical considerations 622
8.2.5 Building examples 623
8.2.6 Craking problems in post-tensioned floors 625
8.2.7 Preliminary design 626
8.2.7.1 Introduction 626
8.2.7.2 General step-by-step procedure 629
8.2.7.3 Simple spans 631
8.2.7.4 Continuous spans 634
Example 1 637
Example 2 645
Example 3 650
End bay design 654
8.2.7.5 Mild steel reinforcement design (strength design for flexure) 660

Chapter 9 Composite Gravity Systems 661

9.1 Composite Metal Decks 661
9.1.1 General considerations 661
9.1.2 SDI specifications 662
9.2 Composite Beams 664
9.2.1 General considerations 664
9.2.2 AISC design specifications 670
Example 682
9.3 Composite Haunch Girders 687
9.4 Composite Trusses 689
9.5 Composite Stub Girders 689
9.5.1 General considerations 689
9.5.2 Behavior and analysis 692
Example 694

9.5.3 Moment-connected stub girder 700
9.5.4 Strengthening of stub girder 700
9.6 Composite Columns 701

Chapter 10 Analysis Techniques 705
10.1 Preliminary Hand Calculations 705
10.1.1 Portal method 706
10.1.2 Cantilever method 708
10.1.3 Lateral stiffness of frames 711
10.1.4 Framed tube structures 722
10.1.5 Coupled shear walls 722
10.2 Lumping Techniques 732
10.3 Partial Computer Models 736
10.4 Torsion 738
10.4.1 Introduction 738
10.4.2 Concept of warping behavior: I-section core 749
10.4.3 Sectorial coordinate $\omega$ 754
10.4.4 Shear center 756
10.4.5 Principal sectorial coordinate $\omega$ diagram 758
10.4.6 Sectorial moment of inertia $I_{\omega}$ 758
10.4.7 Shear torsion constant $J$ 758
10.4.8 Calculation of sectorial properties: worked example 759
10.4.9 General theory of warping torsion 761
10.4.10 Torsion analysis of shear wall structures: worked examples 771
10.4.11 Torsion analysis of steel braced core: worked example 791
10.4.12 Warping torsion constants for open sections 796
10.4.13 Computer analysis 796
Modeling techniques 796
Warping stiffness of floor slab 801
Finite element analysis 804
Twisting and warping stiffness of open sections 805
Stiffness method using warping-column model 808

Chapter 11 Structural Design 813
11.1 Steel Design 814
11.1.1 Design load combinations 814
11.1.2 Tension members 815
11.1.3 Members subject to bending 817
11.1.3.1 Lateral stability 817
11.1.3.2 Compact, semicompact and noncompact sections 818
11.1.3.3 Allowable bending stresses 818
11.1.3.4 Allowable shear stress 822
11.1.4 Members subjected to compression 822
11.1.4.1 Buckling of columns 822
11.1.4.2 Column curves 824
11.1.4.3 Allowable stresses 825

11.1.4.4 Stability of frames: effective length concept 825
11.1.5 Members subjected combined axial load and bending 829
11.1.5.1 Secondary bending: $P$-$\Delta$ effects 829
11.1.5.2 Interaction equations 833
11.1.5.3 Direct analysis for $P$-$\Delta$ effects 834
11.1.6 Calculation of stress ratios 835
11.1.7 Design of continuity plates 836
11.1.8 Design of doubler plates 838
11.1.9 Additional seismic requirements (UBC 1994) 840
11.1.9.1 Ordinary moment frames 840
11.1.9.2 Special moment-resisting frames 841
11.1.9.3 Braced frames 843
11.1.9.4 Eccentrically braced frames (EBF) 843
11.1.9.5 Special concentrically braced frames 845
11.2 Concrete design 848
11.2.1 Load combinations and $\phi$ factors 848
11.2.2 Beam design 849
11.2.2.1 Design for flexture 850
11.2.2.2 Design for shear 853
Shear reinforcement in coupling beams: seismic design 855
11.2.2.3 Joint design of special moment-resisting frames 855
Determination of panel zone shear force 855
Determination of effective area of joint 857
Panel zone shear stress 858
11.2.2.4 Beam column flexural capacity ratios 858
11.2.3 Column design 859
11.2.3.1 Generation of bi-axial interaction surfaces 859
11.2.3.2 Determination of moment magnification factors 862
11.2.3.3 Additional seismic requirements 866
11.2.3.4 Design for shear 866
Determination of factored forces 866
Determination of concrete shear capacity 868
Determination of shear reinforcement 870
11.2.4 Shear wall design 870
11.2.4.1 Design for overturning moment and axial load 872
11.2.4.2 Design for shear 874
11.2.4.3 Additional seismic requirements 875
11.2.4.4 Load-moment interaction diagram 876
11.2.5 Comments on seismic details 877
Joint shear 877
Why strong column–weak beam 878
Why minimum positive reinforcement 878

Chapter 12 Special Topics 881

12.1 Differential Shortening of Columns 882
12.1.1 Calculations 884

12.1.2 Simplified approach 888
Derivation of closed-form solution 891
12.1.3 Column shortening verification during construction 895
12.1.4 Conclusions 895
12.2 Floor-Leveling Problems 897
12.3 Floor vibrations 899
12.4 Panel Zone Effects 907
12.5 Cladding Systems 913
12.5.1 Glass 916
12.5.2 Metal curtain wall 920
12.5.3 Stone cladding 921
12.5.4 Brick veneer systems 923
12.5.5 Glass fiber-reinforced concrete cladding 923
12.5.6 Curtain wall mock-up tests 923
12.6 Mechanical Damping Systems 924
12.7 Foundations 926
12.7.1 Pile foundation 927
12.7.2 Mat foundation 929
Analysis 932
Mat for a 25-storey building 936
Mat for an 85-storey building 937
12.8 Seismic Design of Diaphragms 938
12.8.1 Introduction 938
12.8.2 Diaphragm behavior 940
12.8.3 Rigid, semi-rigid, and flexible diaphragms 944
12.8.4 Metal deck diaphragms 946
12.8.5 Design criteria 948
12.9 Earthquake Hazard Mitigation Technology 949
12.9.1 Seismic base isolation 949
12.9.1.1 Introduction 949
12.9.1.2 Salient features 952
12.9.1.3 SEAOC Blue Book (1996) requirements 952
12.9.2 Energy dissipation 957
12.9.2.1 Metallic systems 957
12.9.2.2 Friction systems 957
12.9.2.3 Viscoelastic systems 960
12.9.2.4 Viscous and semi-viscous fluid systems 960
12.10 Welded Moment Connections Subjected to Large Inelastic Demands 961
12.10.1 Introduction 961
12.10.2 Overview of SAC's guidelines 965
12.11 Unit structural quantities 975
12.11.1 Introduction 975
12.11.2 Concrete floor framing quantities 975
12.11.3 Structural steel quantities 976

Selected References 987

Appendix A: Conversion Factors: U.S. Customary Units to SI Metric Units 991

Index 993

# Preface

This book has been developed to serve as a comprehensive reference for designers of tall building structures. Structural design aspects of steel, concrete and composite buildings with particular reference to wind and seismic loads are discussed. Methods of providing gravity and lateral load resistance including the state-of-the-art systems are discussed as well as the many facets of designing of structural elements.

This is itended as a *practical* book useful to engineerig students, consulting engineers, architects, engineers employed by federal, state and local governments and educators. The material has been presented in easy-to-understand form to make it useful to young engineers with their first high-rise, and to offer new approaches to those who have been involved with tall building structures in the past. Numerous examples illustrating design procedures are worked out in detail.

The book begins with a description of structural systems of tall buildings built around the world within the past two decades. The purpose is to familiarize readers with the information that currently resides in these designs, for the engineering mind constantly needs past solutions and tried formats as anchors before it can break new ground or differ markedly from *conventional wisdom.*

Chapter 2 deals with different approaches for evaluating wind loads appropriate for building design. Building code and wind tunnel procedures are discussed, including analytical methods for determining building response related to occupant comfort.

Chapter 3 outlies seismic design, highlighting the dynamic behavior of builidings. Static, dynamic, and time-history analyses are described. Seismic vulnerability study and retrofit design of buildings not meeting current building code detailing standards are also discussed.

The design of framing systems for lateral forces is the subject of the fourth, fifth and sixth chapters. Traditional and newer-type bracing systems in steel, concrete, and combinations of the two, called composite construction, are analyzed.

The seventh, eighth, and ninth chapters are dedicated to gravity design of vertical and horizintal building systems. In addition to common systems, novel techniques such as composite stub girders are also discussed.

Chapter 10 focuses on the analysis of structural systems and components. Approximate methods are discussed first, followed by computer modeling techniques for two- and three-dimensional analyses. Torsion analysis, including the author's original Ph.D. work on warping behavior of open-section shear walls, is covered in detail. This information is particularly useful in making preliminary designs and verifying three-dimensional computer models.

Chapter 11 gives an overview of code check process for verifying the design of structural elements. Design methods and equations from leading specifications including special requirements for seismic design are presented for ready reference.

The concluding chapter is devoted to the discussion of various topics unique to the design of tall buildings. Differential shortening of columns, design of curtain walls, mechanical damping systems for reducing wind-induced sway accelerations, drilled pier and mat foundations, earthquake mitigation technologies are some of the subjects covered in this chapter. Brittle fracture of welded moment connections subjected to large inelastic demands is discussed. Unit structural quantities for estimating prelimiary steel tonnage of high rise steel and composite buildings are described, followed by an update of the 1997 Uniform Building Code which is expected to serve as a source document for the entire United States begining in the year 2000.

The book attempts to achieve a number of objectives; it is intended to bridge the gap between a novice and an experienced designer while serving simultaneously as a comprehensive resource document. The first and foremost audience is the practicing structural engineer ranging from the young that have just now entered the profession to those with considerable experience. The scope of the book is intentionally broad with enough in-depth material to make it useful for practitioners of structural engineering in all stages of their careers. It is hoped that this book will also serve as a teaching tool for advanced structural courses in colleges and universities.

# Acknowledgements

The author wishes to express his sincere appreciation and thanks to many individuals and friends who helped in this endeavour: John A. Martin Sr. and Trailer A. Martin Jr., for their support and encouragement; Mysore V. Ravindra, Dr. P. V. Banavalkar, Dr. Walter P. Moore Jr., and John L. Tanner for providing information on buildings designed by their firms; Dr. Roger M. De Julio Jr., for comprehensive review of Chapter 3; Dr. Farzad Naeim, Brett W. Beekman, and Kalman V. Benuska for reviewing parts of the manuscript and offering many helpful suggestions; Mike Baltay for his assistance in gust-sensitivity study; Lupe Infante, Rima Roerish, Ivy Policar, and Betty Cooper for typing parts of the manuscript; Evita Oseguera and Andrew Besirof for supplying project photographs; and Emilio Rodriguez, Raul Oseguera, Mike Mittelstaedt, and Ben Kirton for their general assistance.

Dr. Madhu B. Kanchi offered many helpful suggestions in the preparation of the manuscript. To this I am grateful to him. I thank the entire engineering staff of John A. Martin & Associates for sharing information about their projects. Thanks to Jack Martin Sr. and Trailer Martin Jr., for giving permission to include descriptions of their projects in this book.

My daughter Anupama and son Abhiman provided a great deal of support and help to this project. My sincere thanks to both of them, especially my son who typed most of the manuscript.

Most deserving of special recognition is my wife Saroja. My source of inspiration, she helped in all phases of this endeavor—from the manuscript's inception to its final submission. Her "sunny disposition" transformed the ardous tasks of manuscript preparation into a fun-filled family project. My deepest appreciation and thanks for her unconditional love, support, sacrifice and help. Without her patience, this book could not have been written.

Chapter

# 1

# General Considerations

## 1.1 Introduction

Ancient tall structures such as the pyramids of Giza in Egypt, Mayan temples in Tikal, Guatemala, and the Kutab Minar in India are just a few examples testifying to the human aspiration to build increasingly tall structures. These buildings are primarily solid structures serving as monuments rather than space enclosures. By contrast, contemporary tall structures are human habitats, conceived in response to rapid urbanization and population growth although the sheer audacity in their vertical scale may often give them the dubious title of monuments. The difference in the usage of buildings, from solid monumental structures to space enclosures, in itself has not changed the basic stability and strength requirements; the structural issues are still the same, the materials and methods are different.

In the design of early monuments, consideration of spatial interaction between structural subsystems was relatively unimportant, because their massiveness provided for strength and stability. In comparison, the size and density of structural elements of a contemporary tall building are strikingly less, and continue to diminish motivated by the real-estate market, aesthetic principles and innovative structural solutions. Thus the trend in high-rise technology can be thought of as a progressive reduction in the quantity of structural material used to create the exterior architectural enclosure and the spaces within.

To be successful, a tall building must economically satisfy the often conflicting demands imposed by various trades such as mechanical, electrical, structural and architectural. In doing so, from a structural point of view, a building can be defined tall when its height creates different conditions in the design, construction, and use than the conditions that exist for its lower brethren. These conditions are manifest when the effects of lateral loads begin to influence its design. For example, in the design of tall buildings, in addition to the requirements of strength, stiffness and stability, the lateral deflections due to wind or seismic loads should be controlled to prevent both structural and nonstructural damage. Also the wind response of top floors in terms of their accelerations during frequent wind storms, should be kept within acceptable limits to minimize motion perception and discomfort to building occupants.

The trend in high-rise architecture is to create an overall spatial form with an intricate detailing of cladding system (Fig. 1.1). The reason is uniquely to define a tower within an urban environment, and at the same time, provide interior spaces that are highly desirable to the building tenants. More often, the resulting structural solution is complex. However, the engineer, who until the early 1970s exercised considerable influence on the building's architectural shape, no longer deems it necessary to do so. Instead, with the

**Figure 1.1** Spatial form of modern high-rise architecture. Fox Plaza, Los Angeles Johnson, Fain and Pereira; Architects John A. Martin & Asso. Inc., Structural Engineers.

immense analytical backup provided by computers, the structural engineer has freed the architect of structural restraints, especially in seismically benign regions. Needless to say, free-form architecture has demanded closer scrutiny of proven systems, challenging the engineer to either modify the proven systems or to come up with new structural solutions altogether. Although it is possible to arrive at a number of structural solutions which are equally applicable to a particular high-rise building, the final scheme more often depends on how best it meets other nonstructural requirements. Optimization of structural systems is thus a task that is studied in concert with other building disciplines.

Although the form of the building exterior plays a large role in how the building behaves under wind and seismic forces, few engineers are given the opportunity (and rightly so, otherwise all our buildings would be prismatic, and either be square or round), to influence the shape of the building. Instead, their role is confined to optimization of the structure for the particular shape which the architect and the owner provide.

## 1.2 Structural Concepts

The key idea in conceptualizing the structural system for a narrow tall building is to think of it as a beam cantilevering from the earth (Fig. 1.2). The laterally directed force generated, either due to wind blowing against the building or due to the inertia forces induced by ground shaking, tends both to snap it (shear), and push it over (bending). Therefore, the building must have a system to resist shear as well as bending. In resisting shear forces, the building must not break by shearing off (Fig. 1.3a), and must not strain beyond the limit of elastic recovery (Fig. 1.3b). Similarly, the system resisting the bending must satisfy three needs (Fig. 1.4). The building must not overturn from the combined forces of gravity and lateral loads due to wind or seismic effects; it must not break by premature failure of columns either by crushing or by excessive tensile forces; its bending deflection should not exceed the limit of elastic recovery. In addition, a building in seismically active regions must be able to resist realistic earthquake forces without losing its vertical load-carrying capacity.

In the structure's resistance to bending and shear, a tug-of-war ensues that sets the building in motion, thus creating a third engineering problem; motion perception or vibration. If the building sways too much, human comfort is sacrificed, or more importantly, non-structural elements may break resulting in expensive damage to the building contents and causing danger to the pedestrians.

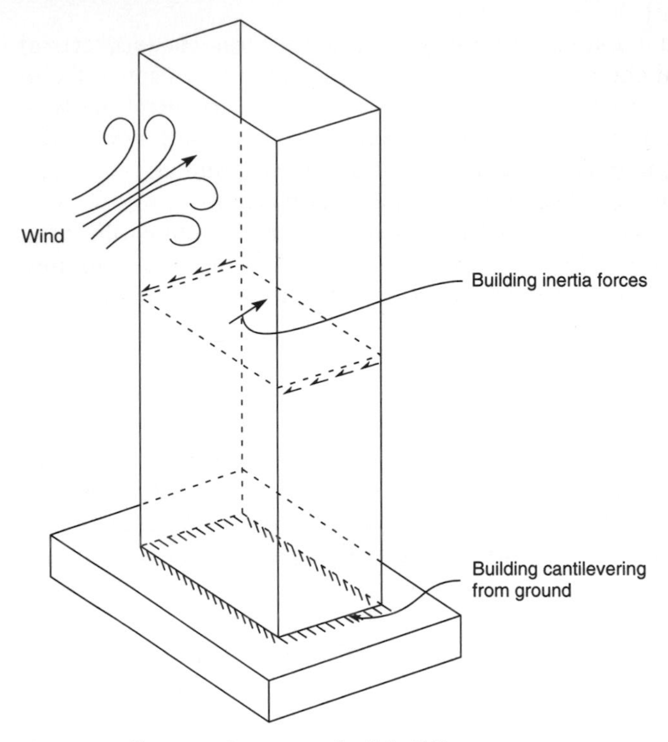


**Figure 1.2** Structural concept of tall building.

A perfect structural form to resist the effects of bending, shear and excessive vibration is a system possessing vertical continuity ideally located at the farthest extremity from the geometric center of the building. A concrete chimney is perhaps an ideal, if not an inspiring engineering model for a rational super-tall structural form. The

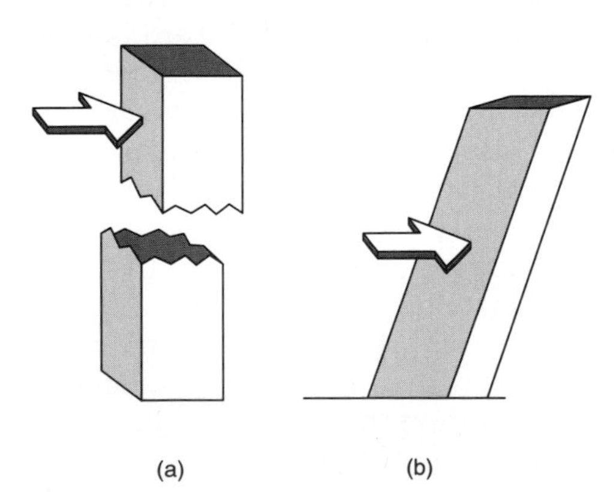


**Figure 1.3** Building shear resistance: (a) building must not break; (b) building must not deflect excessively in shear.

quest for the best solution lies in translating the ideal form of the chimney into a more practical skeletal structure.

With the proviso that a tall building is a beam cantilevering from earth, it is evident that all columns should be at the edges of the plan. Thus the plan shown in Fig. 1.5(b) would be preferred over the plan in Fig. 1.5a. Since this arrangement is not always possible, it is of interest to study how the resistance to bending is affected by the arrangement of columns in plan. We will use two parameters, Bending Rigidity Index BRI and Shear Rigidity Index SRI, first published in Progressive Architecture, to explain the efficiency of structural systems.

The ultimate possible bending efficiency would be manifest in a square building which concentrates all the building columns into four corner columns as shown in Fig. 1.6a. Since this plan has maximum efficiency it is assigned the ideal Bending Rigitidy Index (BRI) of 100. The BRI is the total moment of inertia of all the building columns about the centroidal axes participating as an integrated system.

The traditional tall building of the past, such as the Empire State Building, used all columns as part of the lateral resisting system. For columns arranged with regular bays, the BRI is 33 (Fig. 1.6b).

A modern tall building of the 1980s and 90s has closely spaced exterior columns and long clear spans to the elevator core in an arrangement called a "tube." If only the perimeter columns are used to resist the lateral loads, the BRI is 33. An example of this plan type is the World Trade Center in New York City (Fig. 1.6c).

The Sear Towers in Chicago uses all its columns as part of the lateral system in a configuration called a "bundled tube." It also has a BRI of 33 (Fig. 1.6d).

The Citicorp Tower (Fig. 1.6e), uses all of its columns as part of its

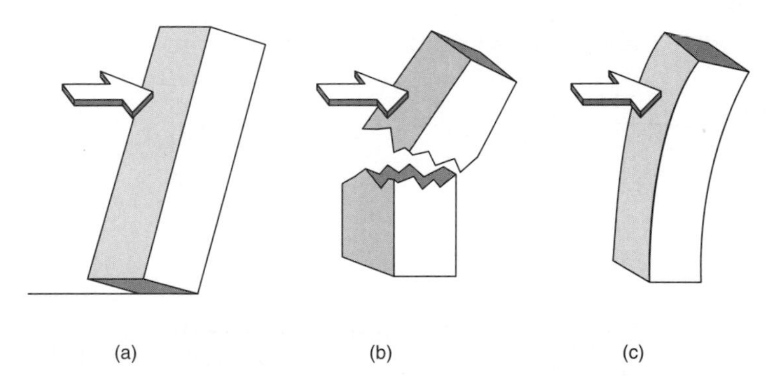


**Figure 1.4** Bending resistance of building: (a) building must not overturn; (b) columns must not fail in tension or compression; (c) bending deflection must not be excessive.

lateral system, but because columns could not be placed in the corners, its BRI is reduced to 31. If the columns were moved to the corners, the BRI would be increased to 56 (Fig. 1.6f). Because there are eight columns in the core supporting the loads, the BRI falls short of 100.

The plan of Bank of Southwest Tower, a proposed tall building in Houston, Texas, approaches the realistic ideal for bending ridigity with a BRI of 63 (Fig. 1.6g). The corner columns are split and displaced from the corners to allow generous views from office interiors.

In order for the columns to work as elements of an integrated system, it is necessary to interconnect them with an effective shear-resisting system. Let us look at some of the possible solutions and their relative Shear Rigidity Index (SRI).

The ideal shear system is a plate or wall without openings which has an ultimate Shear Rigidity Index (SRI) of 100 (Fig. 1.7a). The second-best shear system is a diagonal web system at 45 degree angles which has an SRI of 62.5 (Fig. 1.7b). A more typical bracing

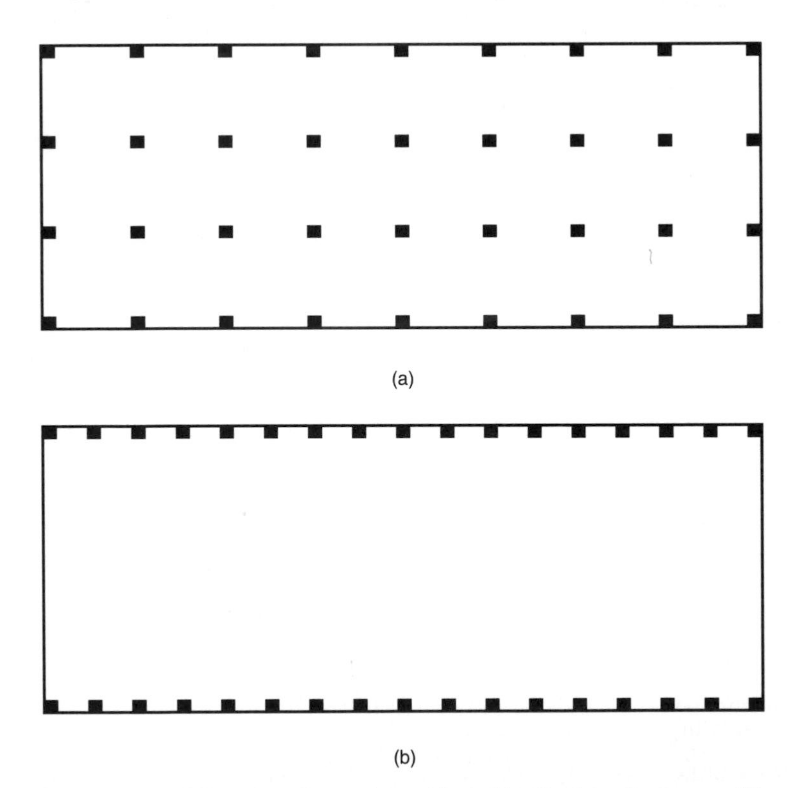


**Figure 1.5** Building plan forms: (a) uniform distribution of columns; (b) columns concentrated at the edges.

system which combines diagonals and horizontals but uses more material is shown in Fig. 1.7c. Its SRI depends on the slope of the diagonals and has a value of 31.3 for the most usual brace angle of 45 degrees.

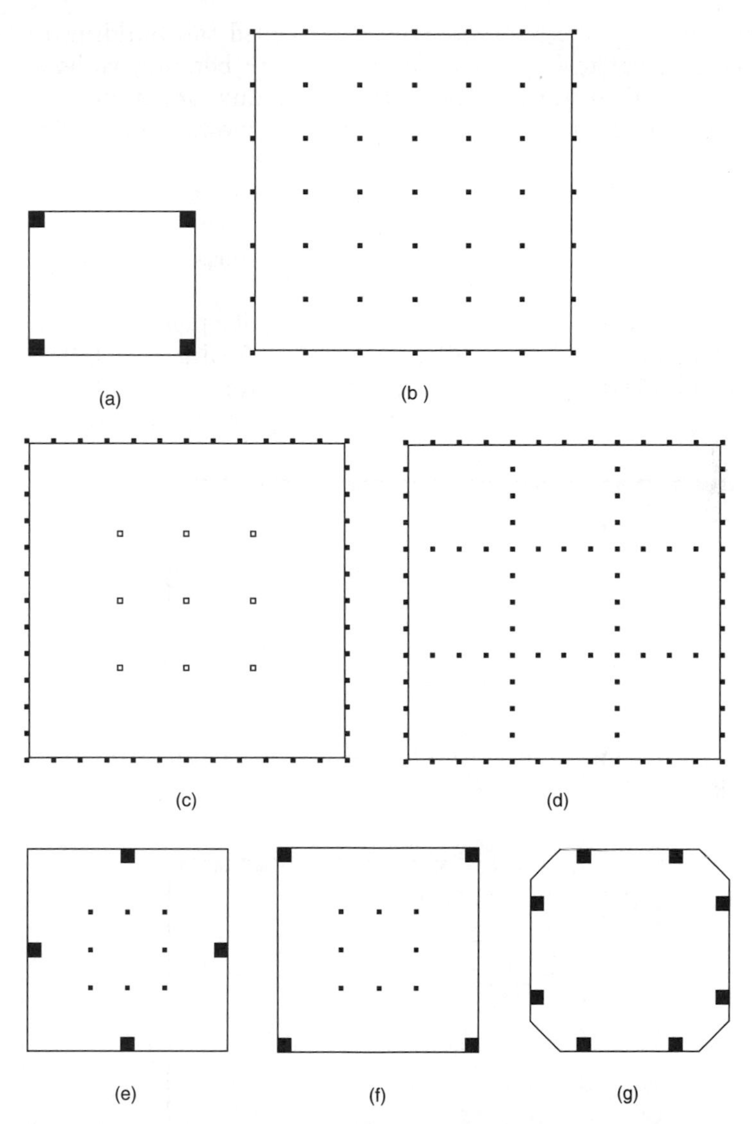


**Figure 1.6** Column layout and Bending Rigidity Index (BRI): (a) square building with corner columns: BRI = 100; (b) traditional building of the 1930s, BRI = 33; (c) modern tube building, BRI = 33; (d) Sears Towers, BRI = 33; (e) City Corp Tower: BRI = 33; (f) building with corner and core columns, BRI = 56; (g) Bank of Southwest Tower, BRI = 63.

The most common shear systems are rigidly joined frames as shown in Figs 1.7d–g. The efficiency of a frame as measured by its SRI depends on the proportions of members' lengths and depths. A frame, with closely spaced columns, like those shown in Fig. 1.7e–g,

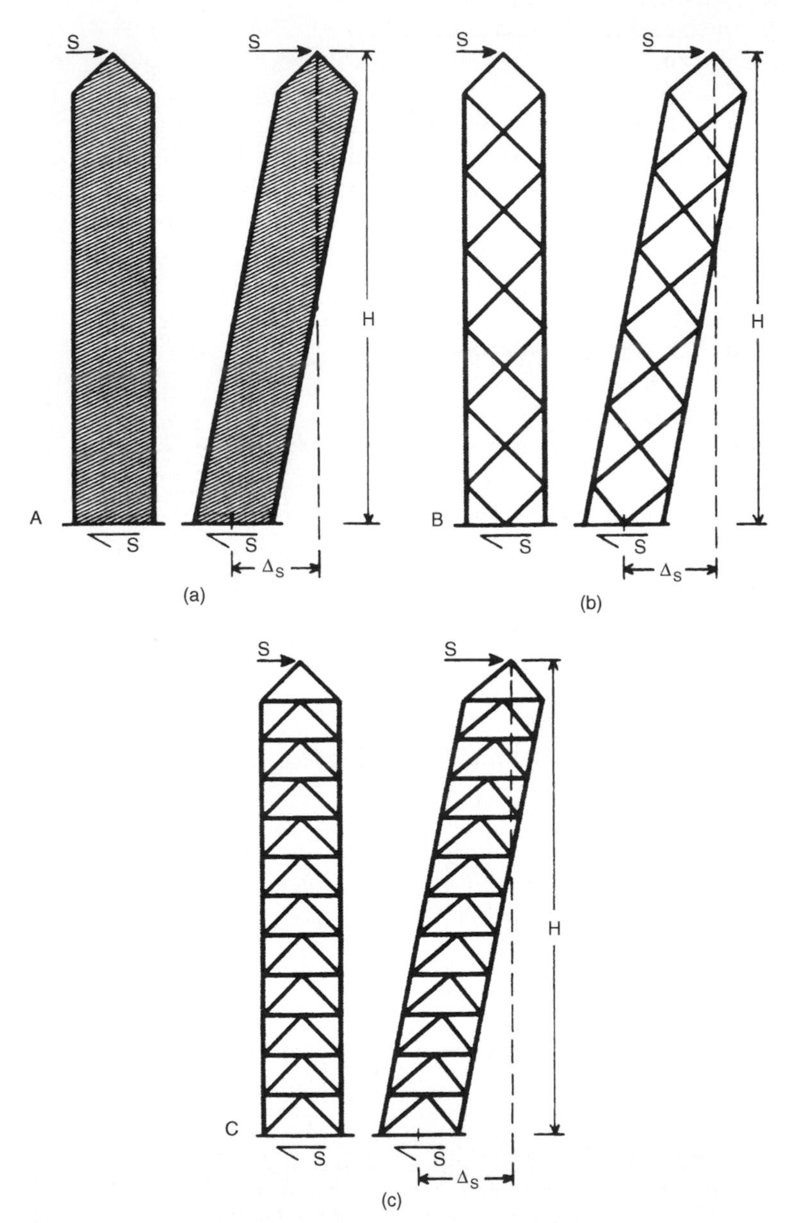


**Figure 1.7** Tall building shear systems: (a) shear wall system; (b) diagonal web system; (c) web system with diagonals and horizontals.

used in all four faces of a square building has a high shear rigidity and doubles up as an efficient bending configuration. The resulting configuration is called a "tube" and is the basis of innumerable tall buildings including the world's two most famous buildings, the Sears Tower and the World Trade Center.

In designing the lateral bracing system for buildings it is important to distinguish between a "wind design" and "seismic design". The building must be designed for horizontal forces generated by wind or seismic loads, whichever is greater, as prescribed by the building code or site-specific study accepted by the Building Official. However, since the actual seismic forces, when they occur, are likely to be significantly larger than code-prescribed forces, seismic design requires material limitations and detailing requirements in addition to strength requirements. Therefore, for buildings in high-seismic zones, even when wind forces govern the design, the detailing and proportioning requirements of seismic resistance must also be satisfied. The requirements get progressively more stringent as the zone factor for seismic risk gets progressively higher.

## 1.3 Structural Vocabulary; Case Studies

Having noted that a building must have systems to resist both bending and shear, let us visit some of the world's tall buildings to explore how prominent engineers have exploited the concept of SRI and BRI in their designs. In describing the designs, an attempt is made to present the structural scheme descriptions in a doctored form. This serves the educational purposes of this book more effectively than a prosaic recounting of the design

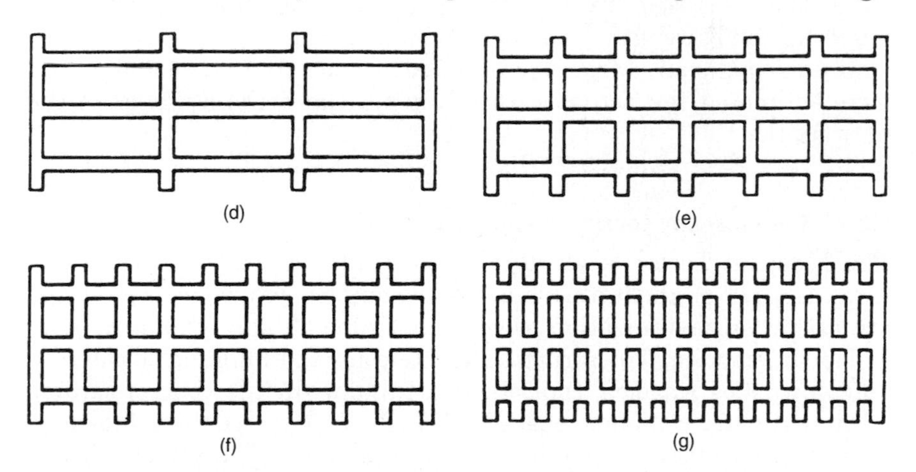


**Figure 1.7** (*Continued*). (d–g) Rigid frames.

data. Although some examples include run-of-the-mill designs that a large number of engineers have to solve on a day-to-day basis, other studies recounted are somewhat poetic, high-profile projects, even daring in their engineering solutions. Many are examples of buildings constructed or proposed in seismically benign regions requiring careful examination of their ductile behaviour and reserve strength capacity before they are applied to seismically active regions.

The main purpose of this section is to introduce the reader to the existing and new vocabulary of structural systems normally considered in the design of tall buildings. Structural design is in a period of mixing and perfecting structural systems such as megaframes, interior and exterior super diagonally braced frames, spine structures, etc., to name a few. The case studies included in this section illuminate those aspects of conceptualization and judgment that are timeless constants of the design process and can be as important and valuable for understanding structural design as are the latest computer software. The case histories are based on information contained in various technical publications and periodicals. Frequent use is made of personal information obtained from the structural engineers-of-record.

We start our world tour in New York City to pay homage to the Empire State Building which was the tallest building in the world for more than 40 years, from the day of its completion in 1931 until 1972 when the Twin Towers of the New York's World Trade Center exceeded its 1280 ft (381 m) height by almost 120 ft (37 m) (Fig. 1.7h). The structural steel frame with riveted joints, while encased in cinder concrete, was designed to carry 100% of gravity and 100% wind load imposed on the building. The encasement, although neglected in strength analysis, stiffened the frame particularly against wind load. Measured frequencies on the completed frame have estimated the actual stiffness at 4.8 times the stiffness of the bare frame.

### 1.3.1 The Museum Tower, Los Angeles

The Museum Tower, a 22-story residential complex, shown in Fig. 1.8a, is part of the California Plaza complex which is one of the largest urban revitalization projects in a zone of high seismic activity in North America. The structural system for the building, located in downtown Los Angeles, consists of a tubular ductile concrete frame with perimeter columns spaced at 13 ft (3.96 m) centers interconnected with upturned spandrel beams (Fig. 1.8b). The exterior frame is of exposed painted concrete.

The gravity system for the typical floor consists of an 8 in (203 mm) thick post-tensioned flat plate with banded and uniform tendons running in the short and long directions of the building respectively as shown in Fig. 1.8a.

Although the building is regular both in plan and elevation, and is less than 240 ft (73 m) in height, because of transfers at the base (Fig. 1.8b), a dynamic analysis using site specific spectrum was used in the seismic design. The dynamic base shear was scaled-down to a value corresponding to the 1992 UBC static base shear. To preserve the dynamic characteristics of the building, the spectral accelerations were scaled down without altering the story masses. The structural design is by John A. Martin & Associates, Inc., Los Angeles. The architecture is by Fujikawa Johnson Asso. Inc., and Barton Myers Asso. Inc.

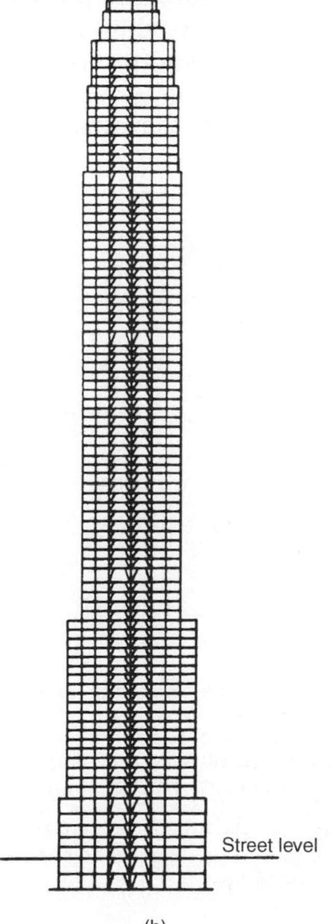


**Figure 1.7** *(Continued).* (h) Empire State Building bracing system; riveted structural steel frame encased in cinder concrete.

### 1.3.2 Bank One Center, Indianapolis

Bank One Center in Indianapolis is a 52-story steel-framed office building which rises to a height of 623 ft (190 m) above the street level. In plan, the tower is typically 190 × 120 ft (58 × 37 m) with set backs at the 10th, 13th, 23rd, 45th and 47th floors (Fig. 1.9a).

The structural system resisting the lateral forces consists of two large vertical flange trusses in the north–south direction and two smaller core braces in the east–west acting as web trusses connecting flange trusses. The flange trusses which provide maximum lever arm for resisting the overturning moments, also serve to transfer gravity loads of the core to the exterior columns. The resulting equalization of axial stresses in the truss and the nontruss perimeter columns keeps the differential shortening between them to a minimum. To assure a direct load path for the transfer of gravity load from the core to the truss columns, the core column is removed below the level of braces at every 12th level, as shown in Fig. 1.9b. In addition, the step-back corners are cantilevered to maximize the tributary area of gravity load, to compensate for the tensile force due to overturning

(a)

**Figure 1.8** The Museum Tower, Los Angeles: Architects: Fujikawa Johnson Asso. Inc. and Barton Myers Asso. Inc. Structural engineers: John A. Martin & Asso. Inc., Los Angeles. (a) building elevation; (b) lateral system; (c) typical post-tensioned floor framing plan.

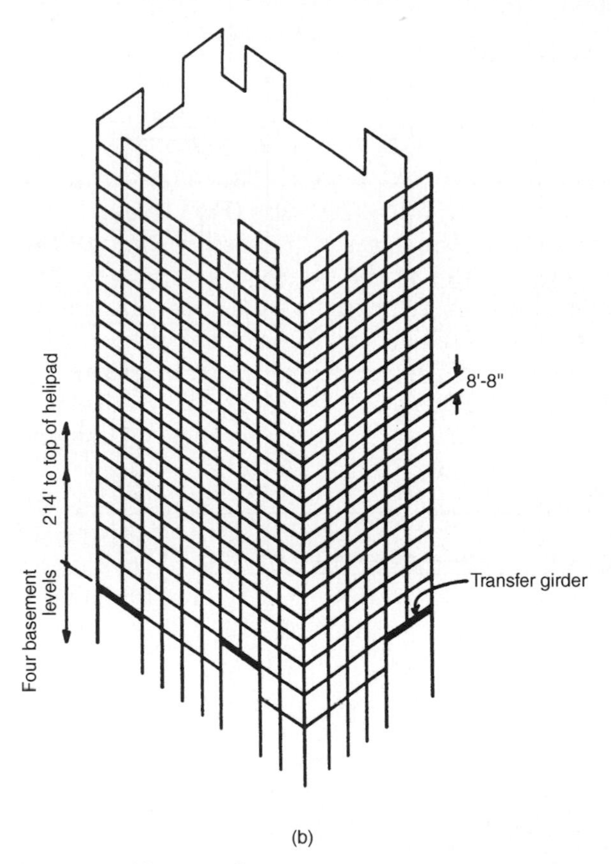


(b)

**Figure 1.8** *(Continued).*

moments. The structural design is by LeMessurier Consultants, Inc., Cambridge, Massachusetts.

### 1.3.3 Two Union Square, Seattle

This 50-story office tower (Fig. 1.10c) has a curved façade with widely spaced perimeter columns. Lateral resisting elements are placed in the interior core walls enabling the perimeter columns to be spaced approximately 44 ft (13.42 m) rather than a more typical 10 to 15 ft (3.05 to 4.58 m).

Four 10 ft diameter (3.05 m) steel pipes filled with high-strength, 19,000 psi (131 mPa) concrete are the primary lateral load-resisting elements (Fig. 10a,b). To reduce perception of lateral movement in the upper levels of the building, the building's structural system incorporates 16 dampers. Structural design is by Skilling Ward Magnusson Barkshire Inc.

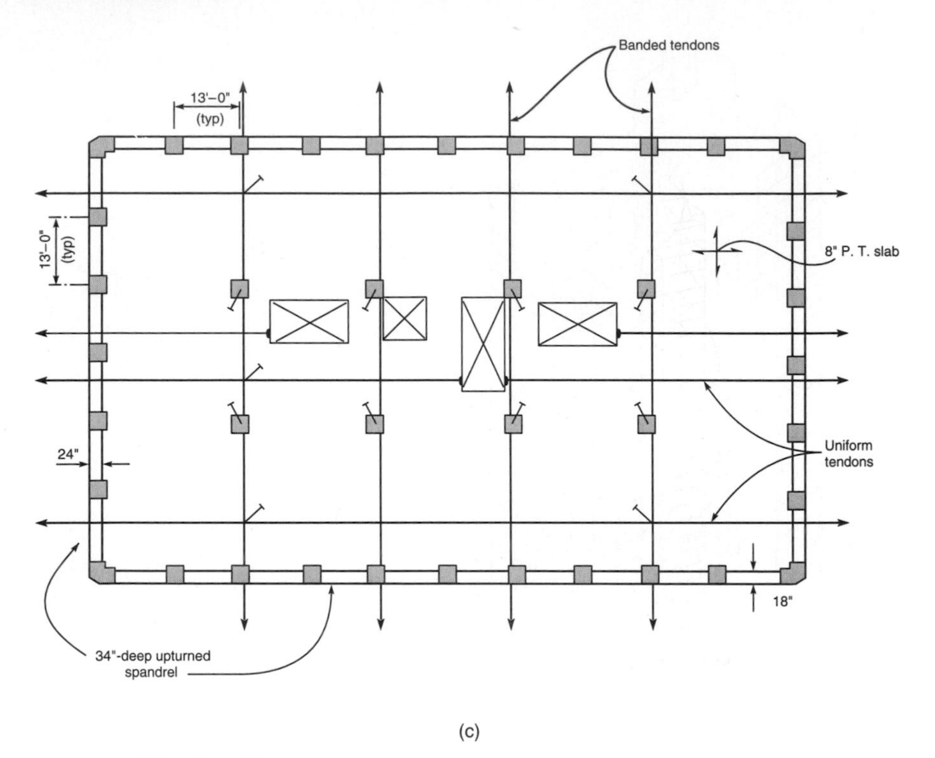


(c)

**Figure 1.8** (*Continued*).

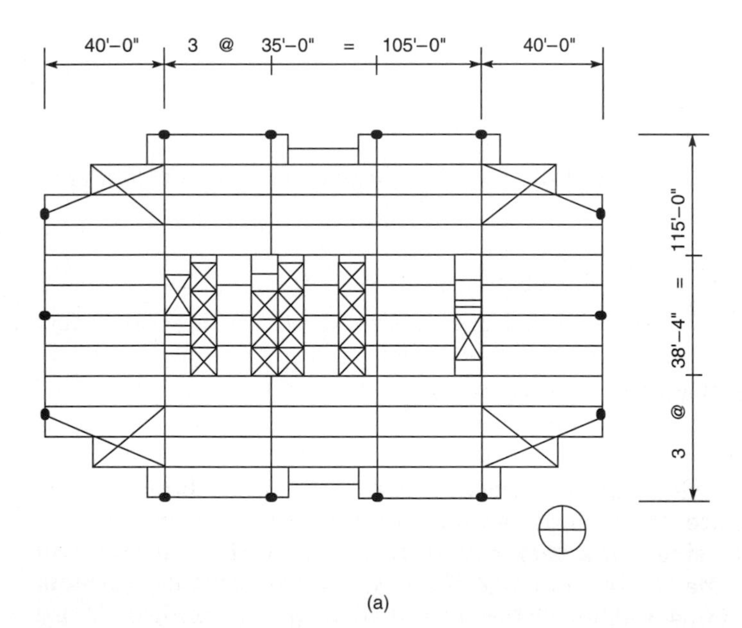


(a)

**Figure 1.9** Bank One Center, Indianapolis: (a) plan; (b) lateral system.

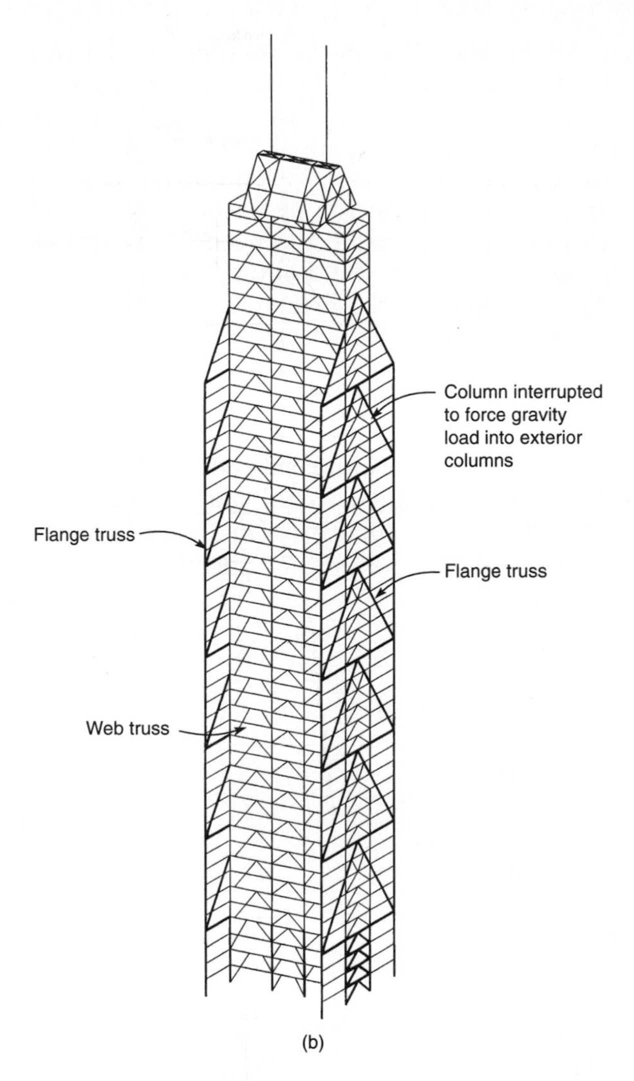


**Figure 1.9** (*Continued*).

### 1.3.4 Bank of China Tower, Hong Kong

The structural system for the 70-story, 1209 ft (368.5 m) Bank of China Tower in Hong Kong consists primarily of a cross-braced space truss. The space truss supports almost the entire weight of the building while simultaneously resisting lateral loading of typhoon winds. Both the lateral and gravity loads are carried to four composite columns at the corners of the building, allowing a 170 ft (51.82 m) clear span at the base of the building.

A fifth composite column in the center of the building begins at the

25th floor and extends to the top. The loads on this column are transferred to the corner columns at the 25th level. At the foundation level, the corner columns are 14 × 26 ft (4.3 × 7.93 m). The size of the steel section of the composite columns varies, and the concrete portion gets progressively smaller as it rises, varying by more than 10 ft (3.05 m). Compositing of frame elements by enclosing the steel members with reinforced concrete eliminated the need for expensive three-dimensional steel connections at the building corners. The structural design is by Leslie Robertson & Associates. The structural system is shown schematically in Fig. 1.11.

### 1.3.5 Dallas Main Center

The 921 ft (280.7 m) building is of composite construction consisting of 73-stories of office space. A three-dimensional moment-resisting frame made of highly repetitive 36 in. (0.30 m) rolled shapes spans

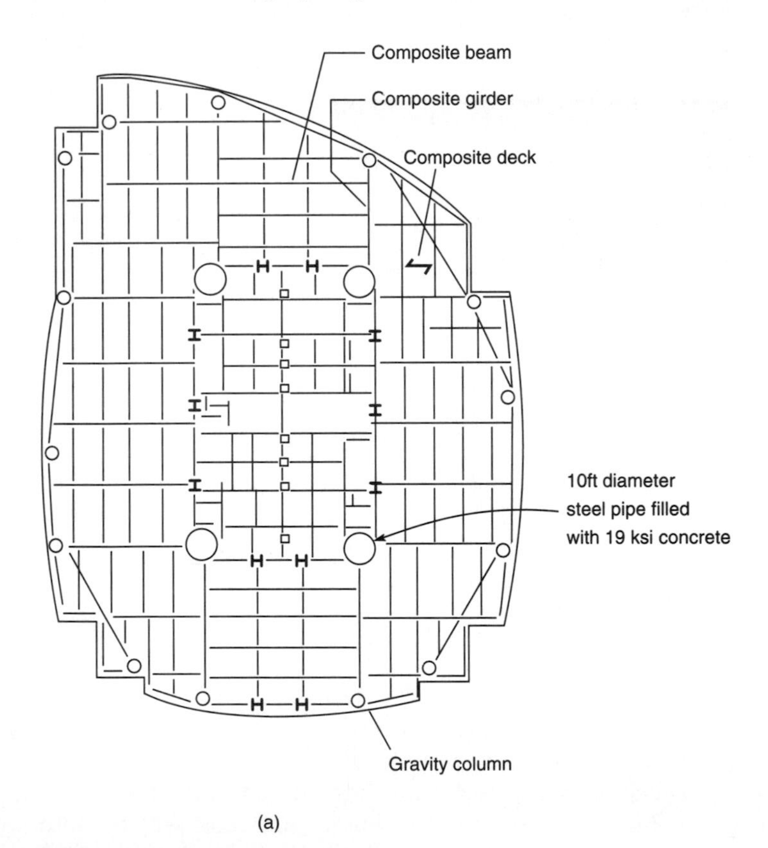


**Figure 1.10** Two Union Square, Seattle: (a) plan; (b) construction photograph; (d) building elevation.

the entire building to sixteen composite columns consisting of light steel columns encased in 10,000 psi (68.95 mPa) concrete in sizes up to 7 ft (2.14 m) square (Fig. 1.12).

The steel verticals in the core are web members of a vierendeel system and are not carried to the foundation. All gravity loads and the overturning forces from the wind are resisted by exterior concrete columns. Wind shear is resisted by the steel frame, which connects the columns across the building. The distance across the building is 127 ft (38.71 m) between columns, giving a height-to-width ratio of the frame of more than 7:1. The structural design is by LeMessurier Consultants, Inc., Cambridge, Massachusetts.

### 1.3.6 The Miglin-Beitler Tower, Chicago, Illinois

The Miglin-Beitler Tower designed by the New York Office of Thornton-Tomasetti Engineers, if built; will establish a new record

(b)

**Figure 1.10** *(Continued).*

as the world's tallest building as well as the world's tallest non-guyed structure surpassing the Sears Tower and the CN Tower which rise to heights of 1454 and 1822 ft (443.2 and 555.4 m), respectively. Rising to 1486.5 ft (453 m) at the upper sky room level, 1584.5 ft (483 m) at the top of the mechanical areas, and finally to 1999.9 ft (609.7 m) at the tip of the spire, the project will provide a regal landmark to the Chicago skyline. An elevation and schematic plan of the building are shown in Fig. 1.13a,b.

A cruciform tube structure has been developed to achieve structural efficiency, superior dynamic behavior, simplicity of construction, and unobtrusive integration of structure into leased office floor areas (Fig. 1.13f).

(c)

**Figure 1.10** (*Continued*).

The tube consists of five major components as shown in Fig. 1.13f. These components are, in order of construction sequence:

1. A 62 ft 6 in. × 62 ft 6 in. (19 × 19 m) concrete core with walls varying from a maximum thickness of 3 ft 0 in. (0.91 m) to a minimum thickness of 1 ft 6 in. (0.46 m).
2. A conventional structural steel composite floor system utilizing 18 in (0.46 m) deep rolled steel sections spaced 10 ft (3.05 m) on center with 3 in. (74 mm) deep corrugated metal deck spanning the 10 ft (3.05 m) between the beams and $3\frac{1}{2}$ in. thick (89 mm) normal weight concrete topping. The steel floor system is supported on light steel erection columns which allow the steel construction to proceed 8 to 10 floors ahead of the next concrete operation.
3. The concrete fin columns, each of which encase a pair of steel

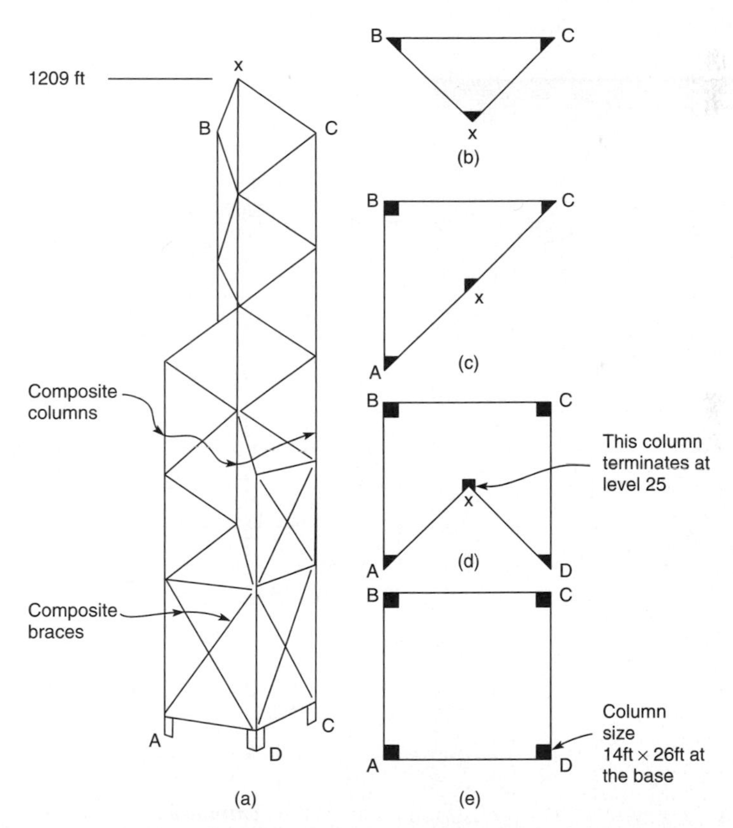


**Figure 1.11** The Bank of China Tower, Hong Kong: (a) schematic elevation; (b-e) floor plans.

erection columns, located at the face of the building. These fin columns, which extend 20 ft (6.10 m) beyond the 140 × 140 ft (42.7 × 42.7 m) footprint at the base of the building, vary in dimension from $6\frac{1}{2} \times 33$ ft. (2.0 × 10 m) at the base, $5\frac{1}{2} \times 15$ ft (1.68 × 4.6 m) at the middle, to $4\frac{1}{2} \times 13$ ft (1.38 × 4 m) near the top.

4. The next components of the cruciform tube system are the link beams which interconnect the four corners of the core to the eight fin columns at every floor. These link beams are comprised of reinforced concrete placed simultaneously with floor concrete. They become the concrete link between the fin columns and the core to make the full structural width of the building resist lateral forces. In addition to the link beams at each floor there are three two story deep outrigger walls located at the 16th story, the 56th

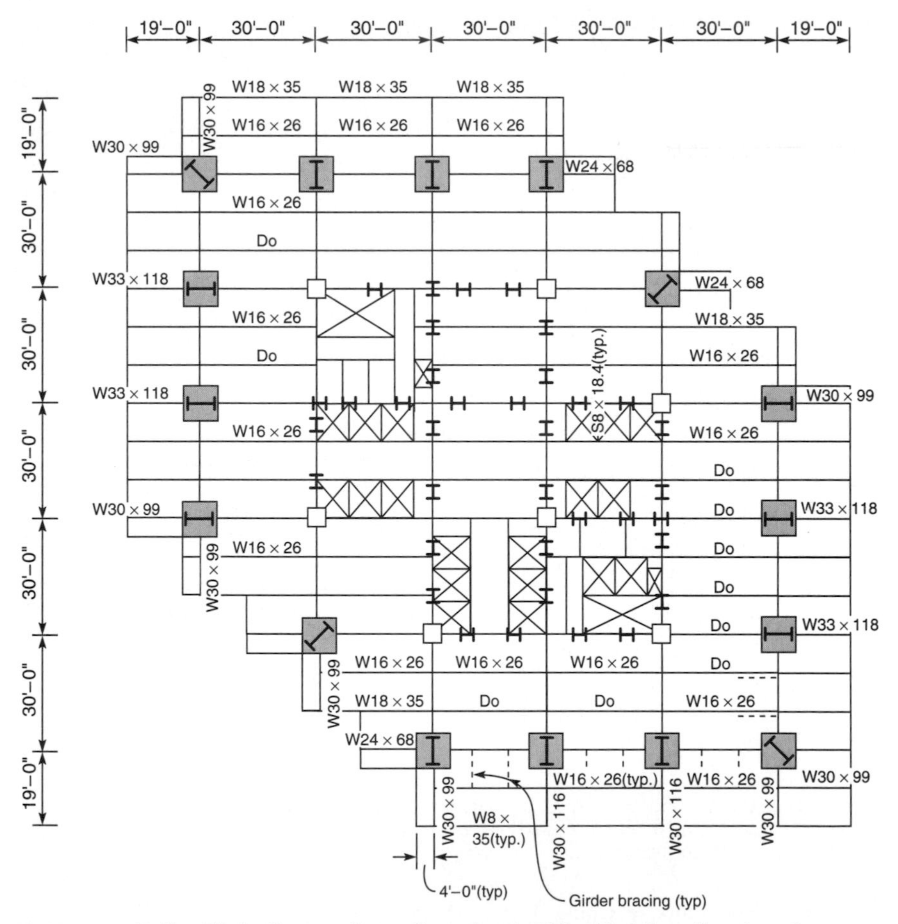


**Figure 1.12** Dallas Main Center: (Inter-first plaza), 26th–43rd floor framing plan.

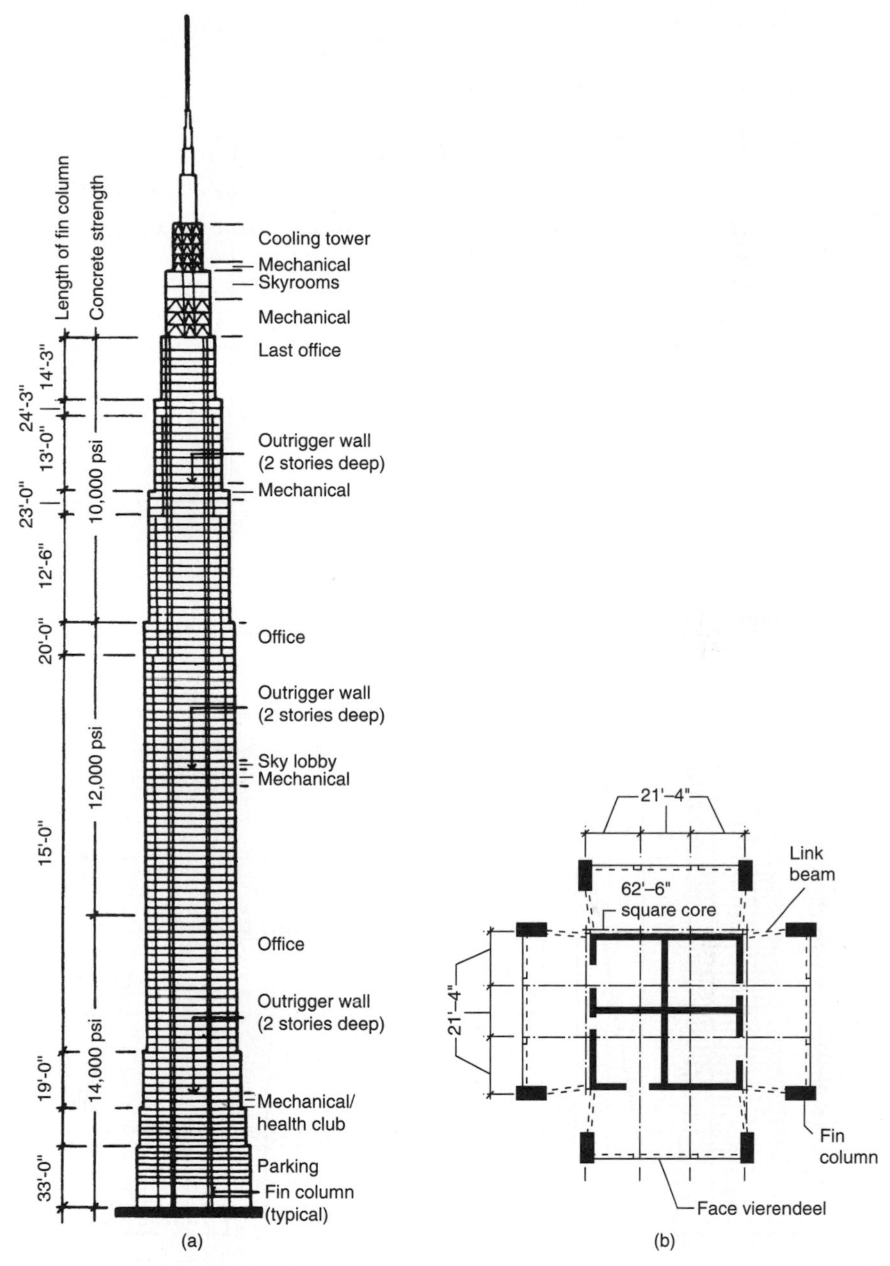


**Figure 1.13** Miglin-Beitler Tower, Chicago: (a) elevation; (b) plan; (c) wind pressures; (d) wind shear; (e) wind moment; (f) typical floor framing plan; (g) wind force distribution in fin columns.

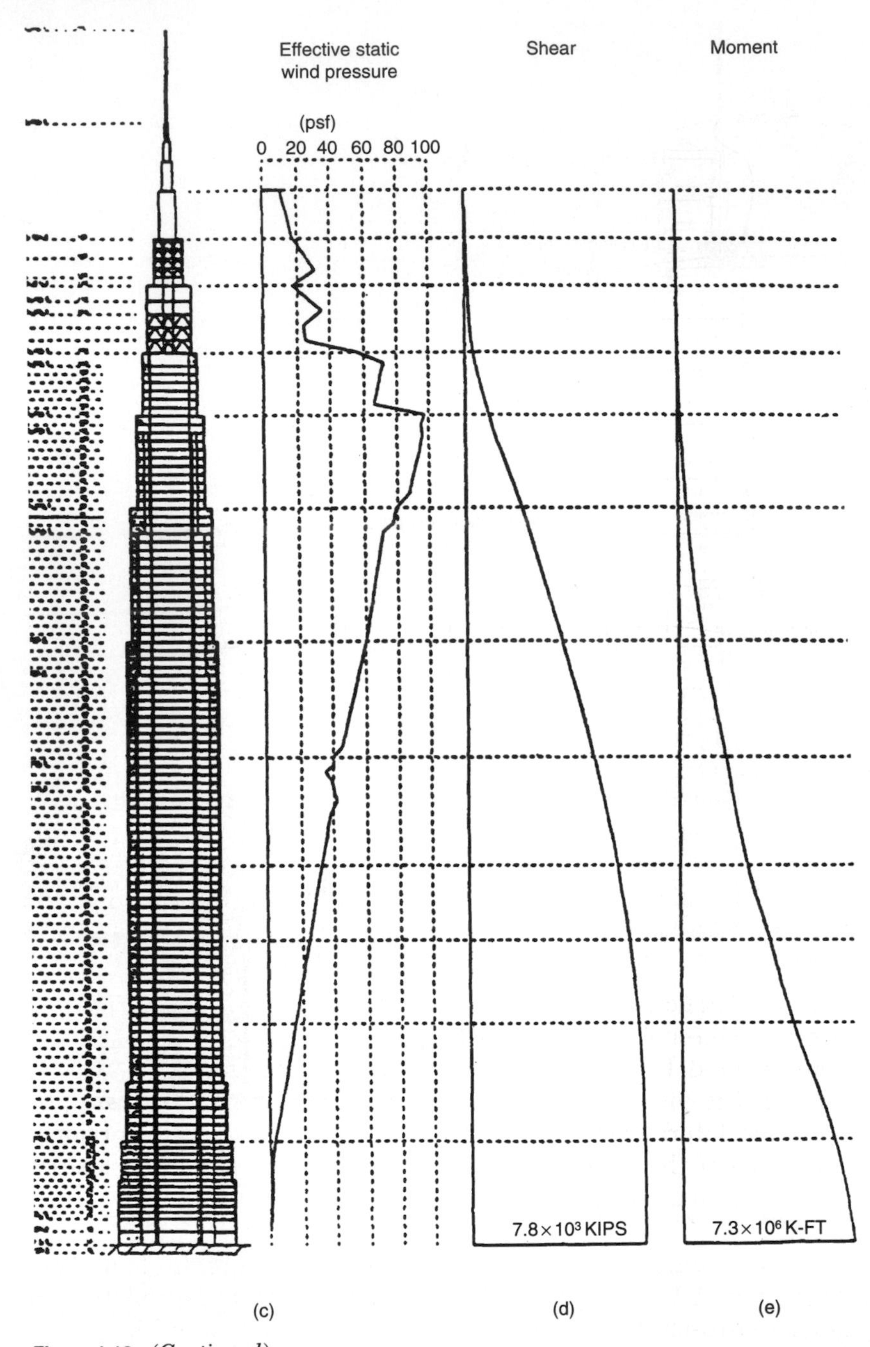


**Figure 1.13** (*Continued*).

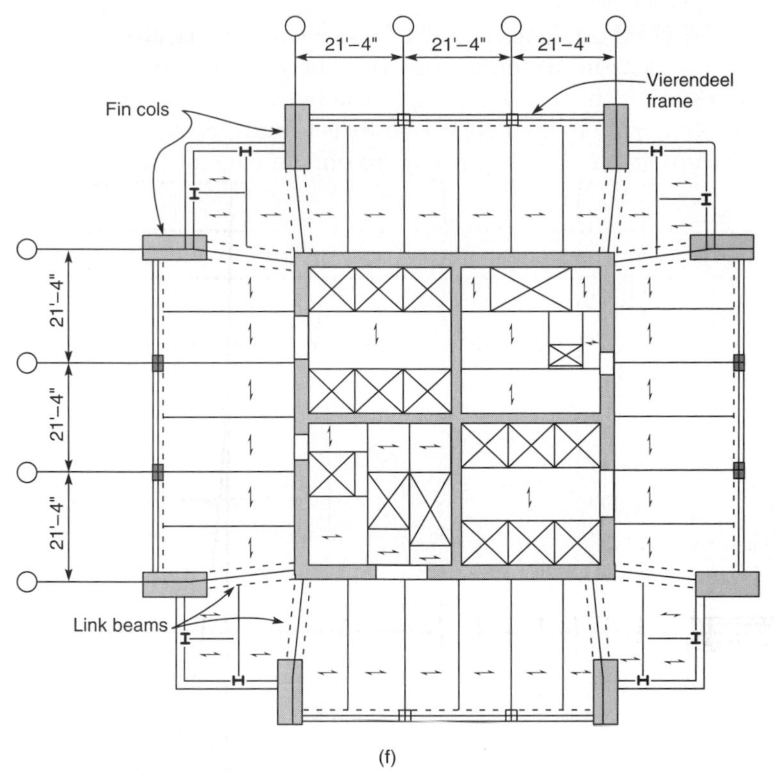


**Figure 1.13** *(Continued).*

story and the 91st story. These outrigger walls further enhance the structural rigidity by linking the exterior fin columns to the concrete core.

5. The last structural components of the cruciform tube are the exterior vierendeel trusses which are comprised of horizontal spandrel and two vertical columns at each of the 60 ft (18.3 m) faces on the four sides of the building. These vierendeels supplement the lateral force resistance and also improve the torsional resistance of the structural system. In addition, gravity loads are transferred out to the fin columns eliminating uplift forces.

The design wind pressures, shear and overturning moments are shown in Fig. 1.13c, d and e. The axial force distribution in the fin columns and the horizontal shear in the fins are shown in Fig. 1.13g.

The proposed foundation system for the project is rock caissons varying in diameter from 8 to 10 ft (2.44 to 3.0 m). The caissons will have a straight shaft, steel casing and will be embedded in rock a minimum of 6 ft (1.83 m). The length of these caissons is 95 ft (29 m). A four-foot thick (1.22 m) concrete mat ties the caissons and provides a means for resisting the shear forces at the base of the building. The bottom of the mat will be cast in a two-directional groove pattern to engage the soil in shear. Passive pressure on the edge of the mat and

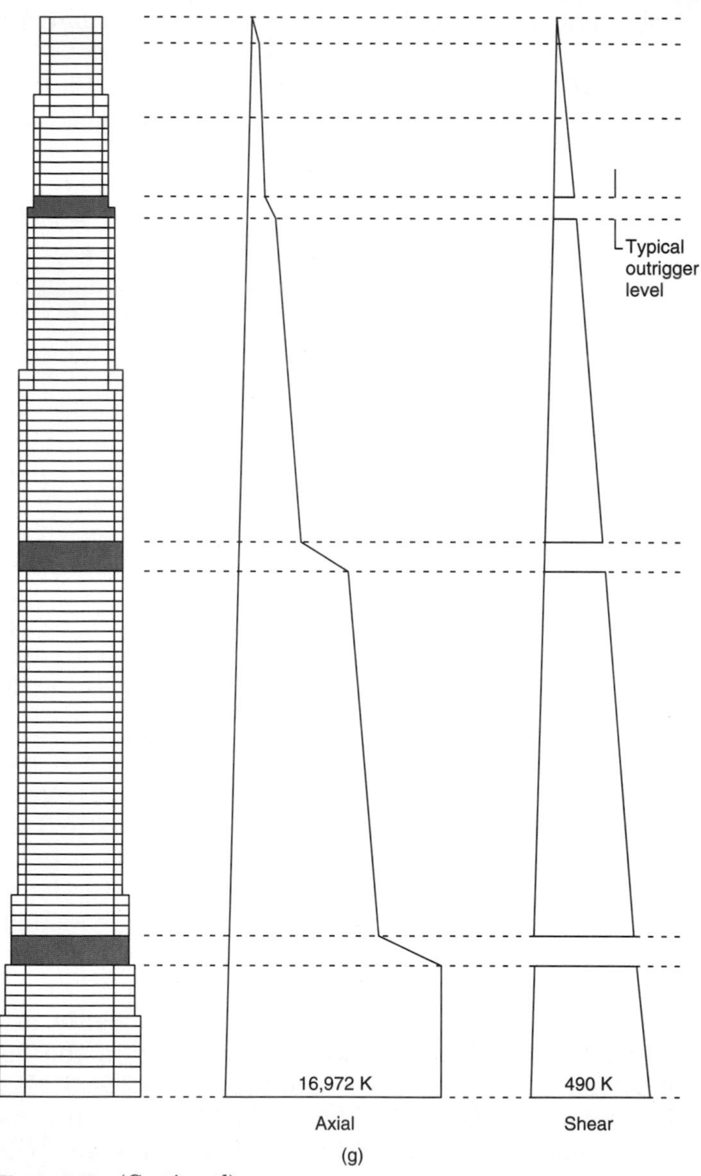


**Figure 1.13** *(Continued).*

on the projected side surface of the caisson provide additional resistance to shear at the base.

### 1.3.7 The NCNB Tower, North Carolina

The NCNB tower in Charlotte, North Carolina, is an 870 ft (265.12 m) high, concrete office building with a 100 ft (30.5 m) crown of aluminum spires (Fig. 1.14a). The building has 12 ft 8 in. (3.87 m) floor-to-floor heights and 48 ft (14.63 m) column-free spans from the perimeter to core.

The structural system for resisting lateral loads consists of a reinforced concrete perimeter tube with normal-weight concrete ranging in strength from 8000 psi (55.16 mPa) near the building's base to 6000 psi (41.37 mPa) at the top. Typical column sizes range from 24 × 38 in. (0.61 × 0.97 m) at the base to 24 × 24 in. (0.61 × 0.61 m) at the top. The floor system (Fig. 1.14b) consists of a $4\frac{5}{8}$ in. (118 mm) thick lightweight concrete slab supported on 18 in. (458 mm) deep post-tensioned beams spaced at 10 ft (3.05 m) on

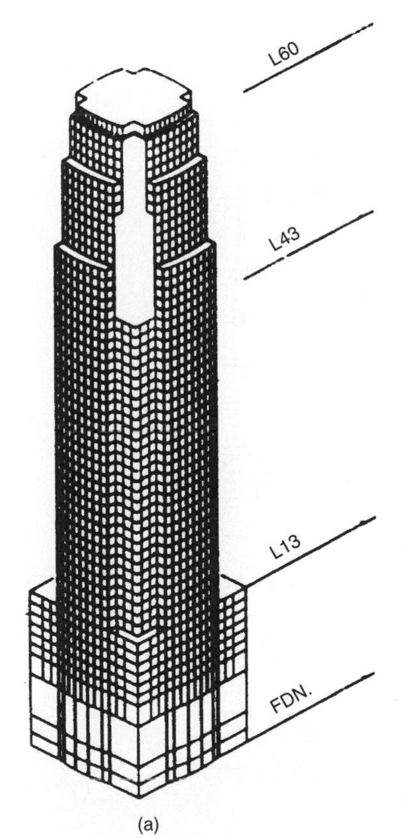


**Figure 1.14** NCNB Tower, North Carolina: (a) schematic elevation; (b) typical floor framing plan; (c) vierendeel action at perimeter frame; (d) location of ductily reinforced beams; (f) ductile diagonal reinforcement detail.

centers. Lightweight concrete was used to reduce the building weight and give the floors the required fire rating. The beams span as much as 48 ft (14.63 m) providing column-free lease space from core to the perimeter (Fig. 1.14b).

The tower's columns are spaced 10 ft (3.05 m) on center and are connected by 40 in. deep (1.01 m) spandrel beams. The building has a roughly square plan at the base, but above the 13th floor it resembles a square set over a slightly larger cross, with the building's four corners recessed and its four major faces bowed slightly outward.

To maintain tube action between the 13th and 43rd floors, engineers used L-shaped vierendeel trusses to continue the tube around the corners. Instead of using transfer girders at the building step-backs, the buildings column-and-spandrel structure is used to create multilevel vierendeel trusses on the building's main façades. These vierendeels transfer loads using another set of vierendeel trusses perpendicular to the façade at the edges of recessed corners (Fig. 1.14c).

Differential shortening between the core and perimeter columns

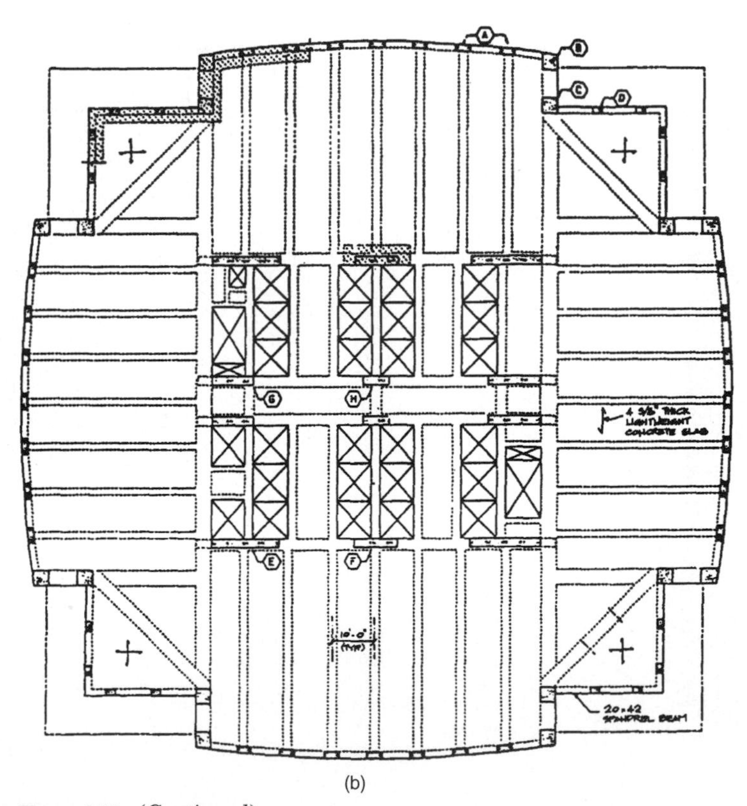


(b)

**Figure 1.14** (*Continued*).

was a concern during design because the core columns will be under significantly higher stresses than the closely spaced perimeter columns. To compensate for this, the core columns were constructed slightly longer than the perimeter columns.

Aware that Charlotte is located in an area of moderate earthquake risk, engineers considered seismic provisions applicable to moderate seismic zones. While wind loads controlled the design of the lower two-thirds of the building, seismic forces controlled the design of the upper third. To meet the seismic requirements additional ties and stirrups were added to columns and beams. The shear reinforcing for critical beams, particularly at set-back levels was designed with a diagonal configuration to provide greater ductility (Fig. 1.14d,e).

Both standard and lightweight concrete were used simultaneously. The normal-weight concrete was used for the perimeter columns, which ranged in size from 24 × 38 in. (610 × 965 mm) at the bottom to 24 × 24 in. (610 × 610 mm) at the top, as well as the core columns, ranging from 2 × 18 ft (0.61 × 3.5 m) at the base to 2 × 3 ft (0.61 × 0.92 m) at the top.

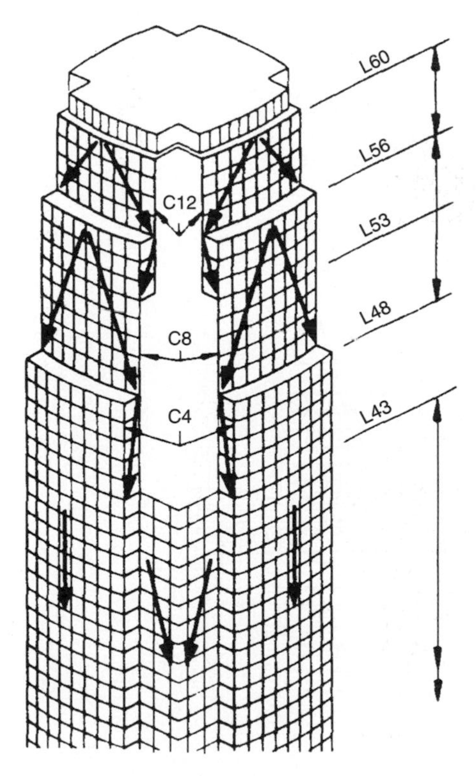


**Figure 1.14** *(Continued).*

Normal-weight concrete was also used for post-tensioned spandrels at the perimeter of each floor, but 5000 psi (34.5 mPa) lightweight concrete was used to form the $4\frac{5}{8}$ in. thick (118 m) floor slabs and the 18 in. deep (0.46 m) post-tensioned beams, spaced 10 ft (3.05 m) on center. The two types of concrete were poured in quick succession and puddled to avoid a cold joint.

The foundation system for the Tower consists of high-capacity caissons under the perimeter columns and a reinforced concrete mat for the core columns.

The high-capacity caissons were designed for a total end-bearing

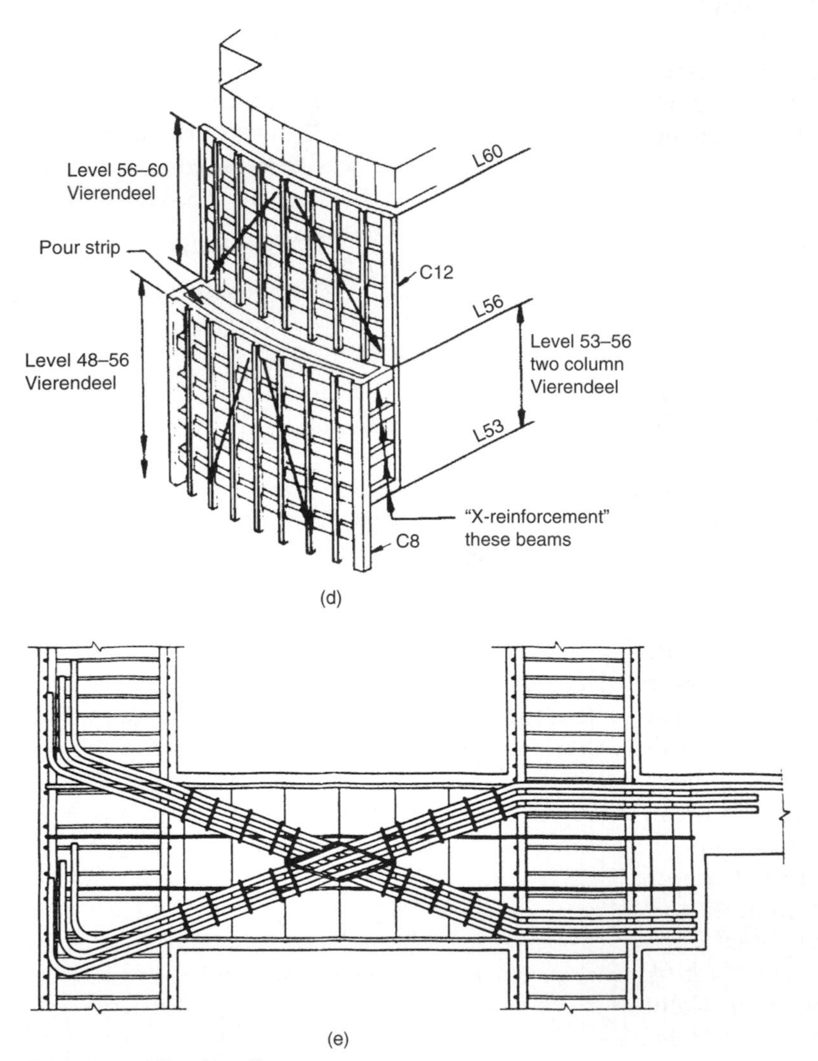


**Figure 1.14** *(Continued).*

pressure of 150 ksf (7182 kN/m$^2$) and skin friction of 5 ksf (240 kN/m$^2$). The high bearing pressure required that the caisson be advanced through the fractured and layered rock zones into high-quality bedrock. Full-length casing was provided to prevent intrusion of soil and ground water into the drilled hole and for the safety of inspectors. Each caisson excavation was inspected in the field by the project geotechnical engineer. A $1\frac{1}{2}$ in. (38 mm) diameter pilot hole was drilled to verify rock quality immediately below the caisson. Caisson diameters ranged from 54 to 72 in. (1.37 to 1.83 m), and the length ranged from 30 to 100 ft (9.15 to 30.5 m) (1.37 to 1.83 m). The concrete strength was 6000 psi. (41.37 kN/m$^2$).

The core columns were supported on a foundation mat bearing on partially weathered rock. The mat dimensions were $83 \times 93 \times 8$ ft ($25.3 \times 28.35 \times 2.44$ m). The average total sustained bearing pressure under the mat is equal to 20 ksf (958 kN/m$^2$). The mat was predicted to setle $\frac{1}{2}$ in. (12.7 mm), mostly during construction. The structural design is by Walter P. Moore and Associates, Inc., Houston, Texas.

### 1.3.8 The South Walker Tower, Chicago

This tower, 946 ft (288.4 m) in height, has a changing geometry with the east face rising in a single plane from street level to 65th floor while the other three faces change shape. To the 14th level the structure is basically a trapezoid in plan $135 \times 225$ ft ($41.15 \times 68.6$ m) overall. The building steps back at the 15th floor on three faces to provide ten corner offices on each floor. There are additional set backs at the 47th floor. At the 51st floor, the sawtooth shape is dropped and the tower becomes an octagon in plan with 70 ft long (21.4 m) sides. The slenderness ratio of the structures is 7.25 : 1. The schematic floor plans at various levels are shown in Fig. 1.15.

The core shear walls in the tower's lower floors carry much of the lateral loading with shear wall–frame interaction. There are four main shear walls—two I shapes and two C shapes—on a typical floor. These interact with the perimeter columns and perimeter spandrel beams through girders that span from core to the perimeter.

The girders have 39 in. deep (1.0 m) haunches at the columns. Spandrels are 36 in. (0.92 m) deep. Core wall concrete design strength varies from 8000 psi (55.121 mPa) at the base to 4000 psi (27.6 mPa) at the upper levels.

There is a 40 to 48 ft ($12.2 \times 14.63$ m) span between the core and the perimeter. The spacing between the perimeter columns is fairly short, about 14 ft (4.3 in.) except at two corners where the spacing is 32 ft (9.76 m). Column loads range from 12,000 kips (53,376 kN) to 30,000 kips (133,440 kN). Concrete strengths used are 12,000 and 10,000 psi (82.74 and 68.95 mPa) at low- and mid-rise areas, and columns at upper levels include both 10,000 and 4000 psi (68.95 and 27.58 mPa) concrete. The largest columns, which are 5 ft square (1.53 m), contain 52 #18, grade 75 bars, compared with the 66 #18 bars that would have been needed if the more conventional, grade 60 steel had been used.

The original floor design had 16 in deep (406.4 mm) pans with 4 in. (101.6 mm) slabs. By using a post-tensioned system, the required depth of floor joists was reduced to 10 in. deep (254 mm) with a 4.5 in. (114.3 mm) slab.

The foundation system for the tower consists of a combination of caisson and mat foundation. The caissons range in size from 36 to

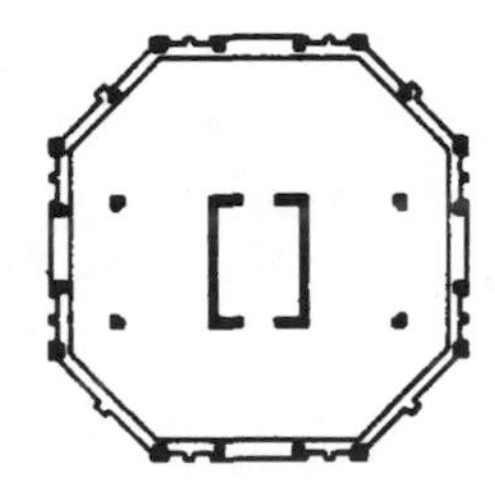

Floors 62–65

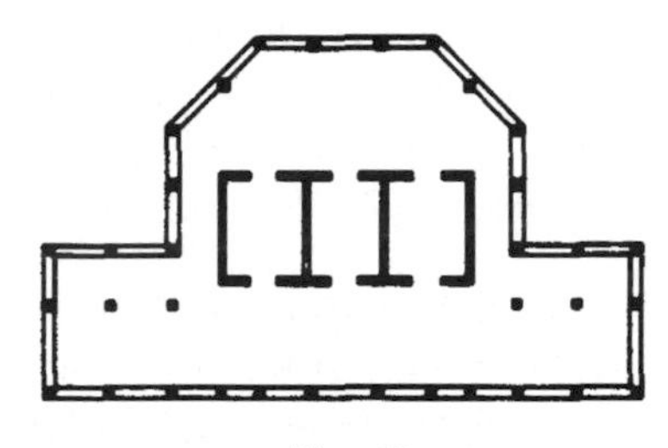

Floor 48

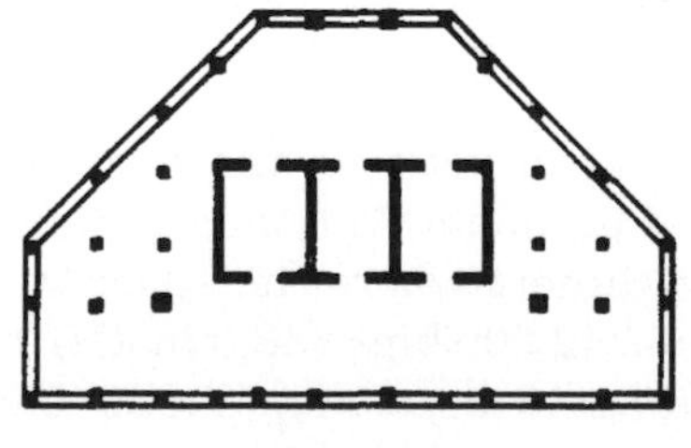

Floors 4–12

**Figure 1.15** South Walker Tower, Chicago. Schematic plans.

108 in. (0.91 to 2.75 m) in diameter. The mat under the core walls is 8 ft (2.44 m) thick. The structural design is by Brockette, Davis, Drake Inc., Dallas, Texas.

### 1.3.9 AT&T Building, New York City

The basic structural system for the building shown in Fig. 1.16a consists of a rigid-frame steel tube at the building perimeter. Additional stiffness is added along the width of the building by means of four vertical steel trusses. At every eighth floor, two I-shaped steel plate walls, with holes cut for circulation, extend from the sides of the trusses to the exterior columns on the same column line. The steel walls act as outrigger trusses mobilizing the full width of the building in resisting lateral forces. The horizontal shear at the base of the building is transferred to two giant steel plate boxes (Fig. 1.16b). Structural design is by Lesley Robertson and Associates.

### 1.3.10 Trump Tower, New York City

The 61-story Trump Tower, a reinforced concrete building shown in Fig. 1.17a, is an example of a multi-use complex which required extensive column transfers. The first seven floors are dedicated to commercial space, including a 6-story atrium, followed by eleven floors of office space and a mechanical floor, and capped at the top with apartment dwellings.

The building consists of a slender reinforced concrete core which serves as the primary lateral load-resisting element. The core is tied to the exterior columns through a large cap girder at the roof level to reduce wind over turning moments in the core and lateral deflections of the building.

The vertical loads are shared by the columns and core, but many of the columns acceptable in the apartment floors would be obstructive in the office floors. Therefore, deep concrete girders are used at the 19th floor to transfer these columns. Similarly many of the columns permitted in the offices must not drop into the commercial space. They are, therefore, transferred onto a huge post-tensioned "A" frame structure (Fig. 1.17b). The structural design is by the Office of Irwin G. Cantor, New York.

### 1.3.11 Metro-Dade Administration Building

Hurricane wind loads which may be as high as four times the U.S. Code values for inland regions played a large role in the selection of structural system for this 500 ft (152.4 m) tall building in Miami, Florida (Fig. 1.18a). The gravity loads are resisted by the concrete

(a)

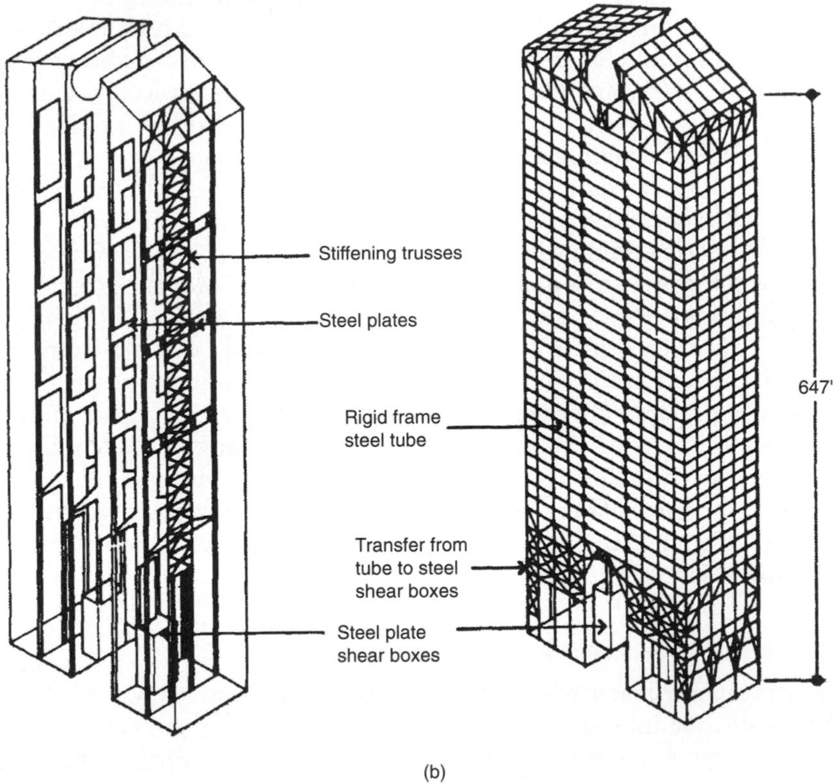


(b)

**Figure 1.16** AT&T Building, New York: (a) building elevation; (b) lateral system.

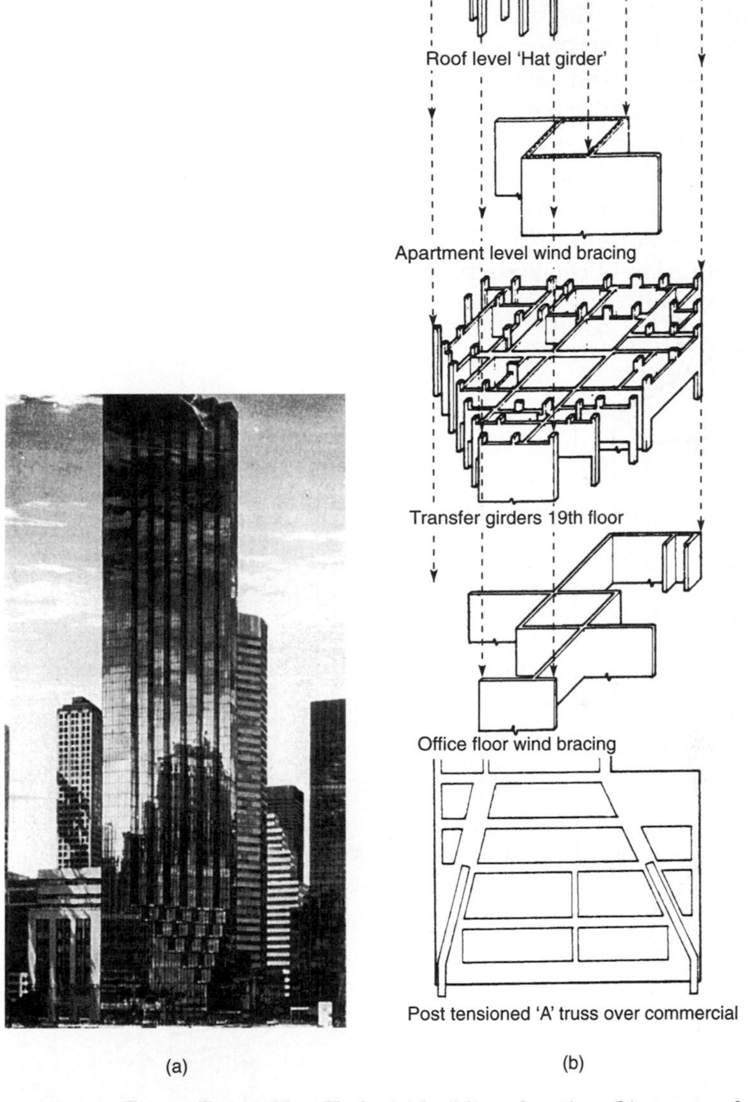


**Figure 1.17** Trump Tower, New York: (a) building elevation; (b) structural system.

columns and shear walls located at the two narrow sides of the building. The floor system consists of a 6 in. (152 mm) thick slab supported on haunch girders. The columns on the broad face of the building rest on a 60 ft deep by 2 ft thick (18.3 × 0.61 m) concrete transfer girder that spans the width of the building. When the wind strikes the broad face, the lateral load is transferred through the floor diaphragm into the end shear walls. The axial forces due to overturning in the columns on the broad face due to overturning moment are transferred to the end walls similar to the axial loads due to gravity loads. Wind on the short face of the building is resisted above the transfer girders by a combination of the end shear walls acting as cantilevers and the frame action of the exterior columns and spandrels on the broad face. The structural design is by LeMessurier Consultants, Inc., Cambridge, Massachusetts. Schematic repreresentation of the structural system are shown Figs. 1.18b,c.

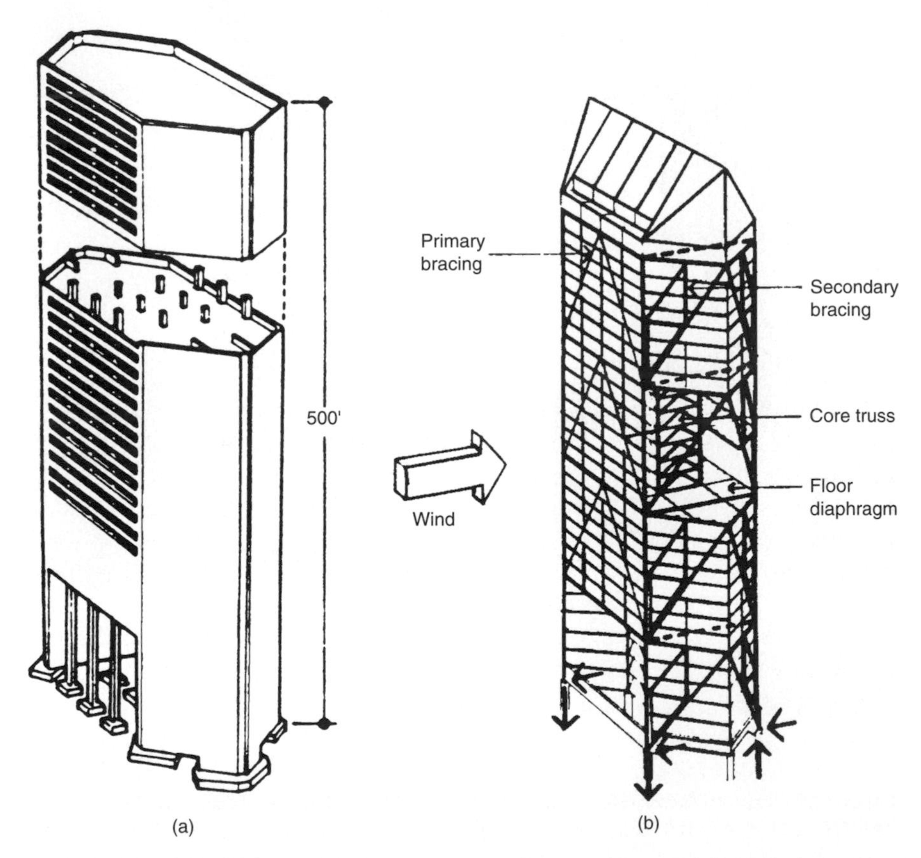


**Figure 1.18** Metro Dade County Administrative Building, Florida: (a) building elevation; (b) lateral system: transverse direction; (c) lateral system, longitudinal direction.

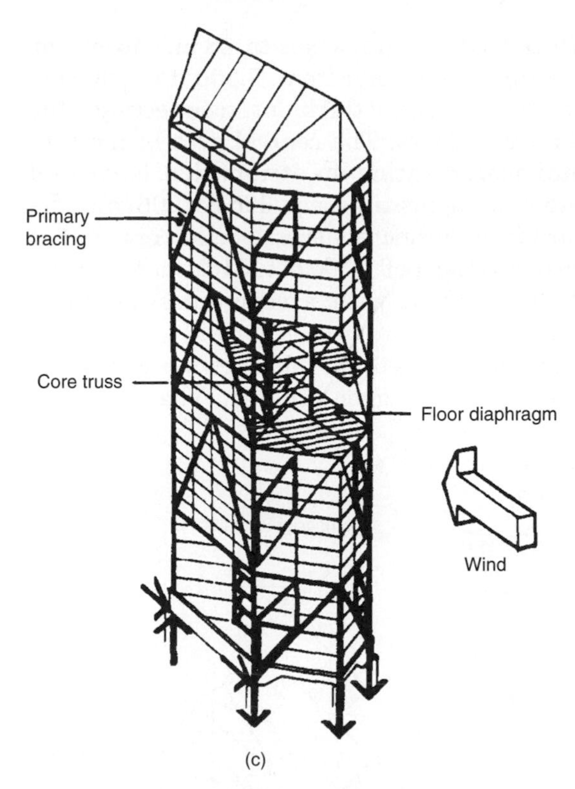


**Figure 1.18** *(Continued).*

### 1.3.12 Jin Mao Tower, Shanghai, China

Jin Mao Building consists of a 1381 ft (421 m) tower and an attached low-rise podium with a total gross building area of approximately 3 million sq ft (278 682 $m^2$). The building includes 50-stories of office space topped by 36-stories of hotel space with two additional floors for a restaurant and observation deck. Parking for automobiles and bicycles is located below grade. The podium consists of retail spaces as well as an auditorium and exposition spaces.

The superstructure is a mixed use of structural steel and reinforced concrete with many major structural members composed of both steel and concrete. The primary components of the lateral system include a central reinforced concrete core linked to exterior composite mega-columns by outrigger trusses (Fig. 1.19a,b). A central shear-wall core houses the primary building functions, including elevators, mechanical fan rooms and washrooms. The octagon-shaped core, nominally 90 ft (27.43 m) from centerline to centerline of perimeter flanges, exists from the foundation to level 87. Flanges of the core typically vary from 33 in. (84 cm) thick at the foundation to 18 in. (46 cm) at level 87 with concrete strengths varying from 7,500 to 5,000 psi (51.71 to 34.5 mPa). Four 18 in. (46 cm) thick interconnecting core wall webs exist through the office floors. The central area of the core is open throughout the hotel floor, creating an atrium that leads into the spire with a total height of approximately 675 ft (206 m). The composite megacolumns vary from a concrete cross-section of 5 × 16 ft (1.5 × 4.88 m) with a concrete strength of 7,500 psi (51.71 mPa) at the foundation, to 3 × 11 ft (0.91 × 3.53 m) with a concrete strength of 5,000 psi (34.5 mPa) at level 87.

The shear-wall core is directly linked to the exterior composite megacolumns by structural steel outrigger trusses. The outrigger

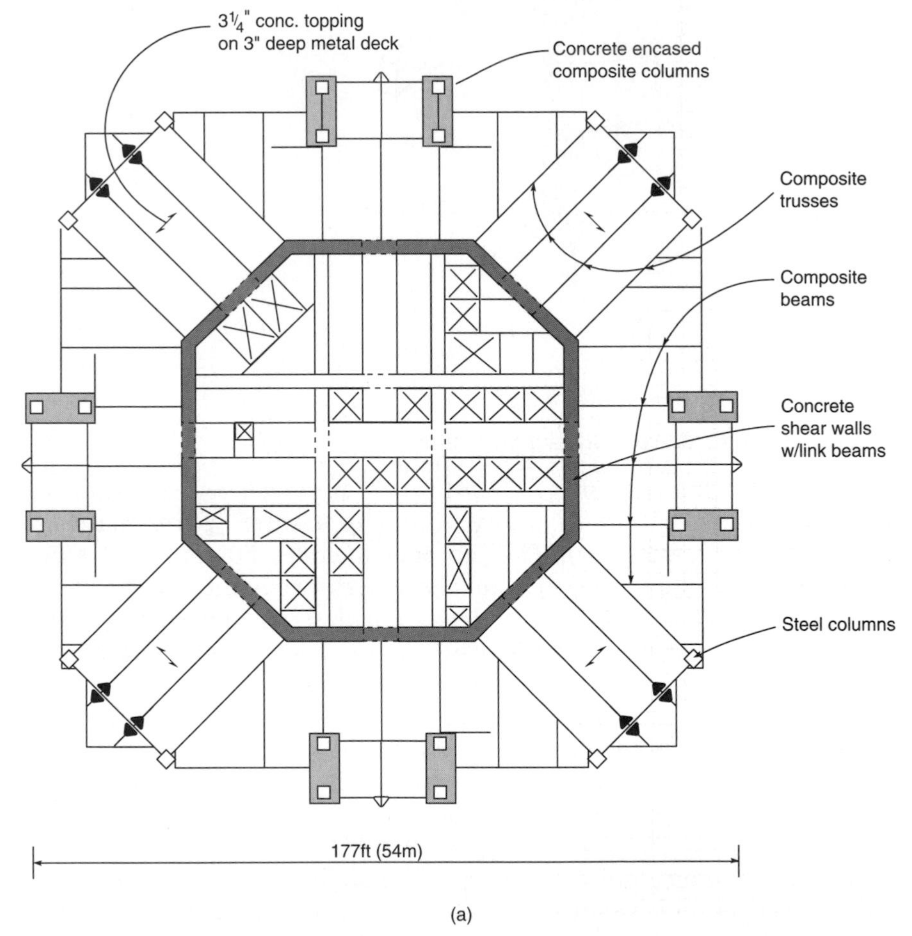


**Figure 1.19** Jin Mao Tower, Shanghai, China: (a) typical office floor framing plan; (b) structural system elevation.

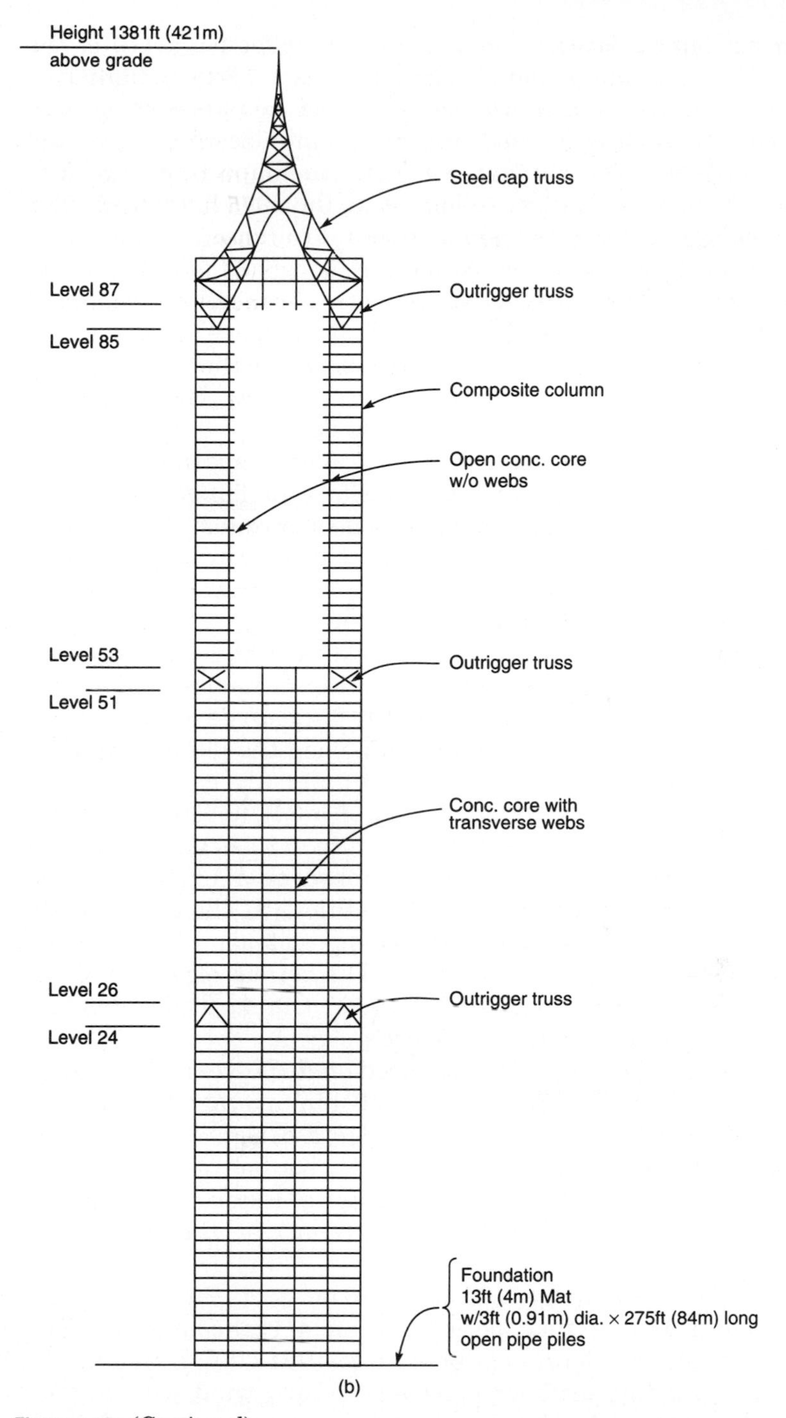


**Figure 1.19** *(Continued).*

trusses resist lateral loads by maximizing the effective depth of the structure. Under bending, the building acts as a vertical cantilever with tension in the windward columns and compression in the leeward columns. Gravity load framing minimizes uplift in the exterior composite megacolumns. The octagon-shaped core provides exceptional torsional resistance, eliminating the need for any exterior belt or frame systems to interconnect exterior columns.

The outrigger trusses are located between levels 24 and 26, 51 and 53, and 85 and 87. The outrigger truss system between levels 85 and 87 is capped with a three-dimensional steel space frame which provides for the transfer of lateral loads between the core and the exterior composite columns. It also supports gravity loads of heavy mechanical spaces located in the penthouse floors.

The structural elements for resisting gravity loads include eight structural steel built-up columns. Composite wide-flange beams and trusses are used to frame the floors. The floor-framing elements are typically 14 ft 6 in. (4.4 m) on-center with a composite 3 in. (7.6 cm) deep metal deck and a $3\frac{1}{4}$ in. (8.25 cm) normal weight concrete topping floor slab spanning between steel members.

The foundation system for the Tower consists of high-capacity piles capped with a reinforced concrete mat. High-water conditions required the use of a 3 ft 3 in. (1 m) thick, 100 ft (30 m) deep, continuous reinforced concrete slurry wall diaphragm along the 0.5 mile (805 m) perimeter of the site.

The high-capacity pile system consists of a 3 ft (0.91 m) diameter structural steel open-pipe piles with a $\frac{7}{8}$ in. (2.22 cm) thick wall typically spaced 9 ft (2.75 m) on-center capped by a 13 ft (4 m) deep reinforced concrete mat. Since the soil conditions at the upper strata are so poor, the piles were driven into a deep, stiff sand layer located approximately 275 ft (84 m) below grade. The bottom elevation of the pipe piles is the deepest ever attempted in China. The individual design-pile capacity is 1,650 kips (7340 kN).

Strength design of the structure is based on a 100-year return wind with a basic wind speed of 75 mph for a 10 min. average time for the Tower. The basic wind speed corresponds to a design wind pressure of approximately 14 psf (0.67 kN/m$^2$) at the bottom of the building and 74 psf (3.55 kN/m$^2$) at the top of the spire. Exterior wall-design pressures are in excess of 100 psf (4.8 kN/m$^2$) at the top of the building.

Wind speeds can average 125 mph (56 m/s) at the top of the building over a 10 min time period during a typhoon event. The earthquake ground accelerations compare to 1994 UBC Zone 2A.

The overall building drift for a 50-year return wind with a 2.5% structural damping is H/1142. The drift value increases to H/857 for the future developed condition in which two tall structures are proposed to be located adjacent to the Jin Mao Building. The drift based on specific Chinese code-defined winds which were eqiuvalent to a 3,000 year wind is H/575.

The structural design for the tower is governed by its dynamic behavior under wind and not by its strength, overall or interstory drift. The inherent mass, stiffness and damping characteristics of the Tower lead to achieving dynamic stability with fundamental translational periods of 5.7 sec for each principal axis and a torsional period of 2.5 sec.

Based on the results from force-balance and aeroelastic wind-tunnel study, the accelerations at the top floor of the hotel zone were evaluated using a value of 1.5 percent structural damping. The accelerations are between 9 and 13 milli g's for a 10-year return period and between 3 and 5 milli g's for a one-year return period—well within the acceptable ranges defined by international standards. Only the passive characteristics of the structural system including its inherent mass, stiffness and damping are required to control the dynamic behavior. Therefore, no mechanical damping systems are used.

Since the central core and composite megacolumns are interconnected by outrigger trusses at only three 2-story levels, the stresses in the trusses due to differential shortening of the core relative to the composite columns were of concern. Therefore, concrete stress levels in the core and megacolumns were controlled in an attempt to reduce relative movements. To further reduce the adverse effect of differential shortening, slotted connections are used in the trusses during the construction period of the building. Final bolting with hard connections is made after completion of construction to relieve the effect of differential shortening occurring during construction.

The architecture and structural engineering of the building is by the Chicago office of Skidmore, Owings and Merrill.

### 1.3.13 Petronas Towers, Malaysia

Two 1476 ft (450 m) towers, 33 ft (7 m) taller than Chicago's Sears Tower, and a sky bridge connecting the twin towers define the new record-breaking tall buildings in Kuala Lumpur, Malaysia.

The towers have 88 numbered levels but are in fact equal to 95-stories when mezzanines and extra-tall floors are considered. In addition to 6,027,800 ft$^2$ (560,000 m$^2$) of office space, the project includes 1,507,000 ft$^2$ (140,000 m$^2$) of retail and entertainment space in a 6-story structure linking the base of the towers, plus parking for 7000 vehicles in five below-ground levels.

The lateral system for the towers is of reinforced concrete consisting of a central core and perimeter columns and ring beams using concrete strengths up to 11,600 psi (80 mPa). The foundation system consists of pile and friction barrette foundations with a foundation mat.

Typical floor system consists of wide flange beams spanning from the core to the ring beams. A two-inch-deep composite metal deck system with a $4\frac{1}{4}$ in. (110 mm) concrete topping completes the floor system.

Architecturally, the towers are cylinders 152 ft (46.2 m) in diameter formed by 16 columns. The façade between columns alternates pointed projections with arcs giving unobstructured views through glass and metal curtain wall on all sides. The floor plate geometry is composed of two rotated and superimposed squares overlaid with a ring of small circles. The towers have setbacks at levels 60, 72, 82, 85 and 88 and circular appendages at level 44. Concrete perimeter framing is used up to level 84. Above this level, steel columns and ring beams support the last few floors and a pointed pinacle.

The towers are slender with an aspect ratio of 8.64 (calculated to level 88). The design wind speed in Kuala Lumpur area is based on 65 mph (35.1 m/s) peak 3 sec gust at 33 ft (10 m) above grade for a 50-year return. In terms of U.S. Standard of fastest mile wind, the corresponding wind speed is about 52 mph (28.1 m/s).

The mass and stiffness of concrete is taken advantage of in resisting lateral loads while the advantages of speed of erection and long span capability of structural steel is used in the floor framing system. The building density is about 18 lb/cu ft (290 kg/m$^3$).

As is common for buildings of high aspect ratios, the towers were wind tunnel tested to determine dynamic characteristics of the building in terms of occupant perception of wind movements and acceleration on the upper floors. The acceleration is in the range of 20 mg, well below the normally accepted criteria of 25 mg. Figure 1.20a shows a photograph of the building model.

The periods for the primary lateral modes are about 9 sec, while the torsional mode has a period of about 6 sec. The drift index for lateral displacement is of the order of 1/560.

Because the limestone bedrock lies 200 ft (60 m) to more than 330 ft (100 m) below dense silty sand formation, it was not feasible to extend the foundations to bedrock. A system of drilled friction piers were designed for the foundation but barrettes (slurry-wall concrete segments) proposed as an alternative system by the contractor were installed. A 14.8 ft (4.5 m) thick mat supports the 16 tower columns and 12 bustle columns.

The tower columns vary in size from 7.8 ft (2.4 m) diameter at the base to 4 ft (1.2 m) at the top. In the bustles, eight of the 12 columns vary from 4.6 to 3.28 ft (1.4 to 1.0 m) in diameter; the four facing the tower, being more heavily loaded, are slightly larger. Because all columns are exposed to view, they are cast with reusable steel forms, smoothed and painted.

The setbacks at floors 60, 73 and 82 are made with sloped columns over 3-story heights. The method of transfer eliminates the need for deep transfer girders that would interrupt the constant floor height necessary for double-deck elevators.

(a)

**Figure 1.20** Petronas Twin Towers, Kuala Lumpur, Malaysia: (a) elevation; (b) structural system; (c) height comparison, (1) Petronas Tower, (2) Sears Tower.

The floor corners of alternating right-angles and arcs are cantilevered from the perimeter ring beams. Haunched ring beams varying from 46 in. (1150 m) deep at columns to 31 in. (775 mm) at midspan are used to allow for ductwork in office space outside of the ring beams. A similar approach with midspan depth of 31 in. (775 m) is used in the bustles. The haunches are used primarily to increase the stiffness of ring beams.

The central core for each tower houses all elevators, exit stairs and mechanical services, while the bustles have solid walls. The core and

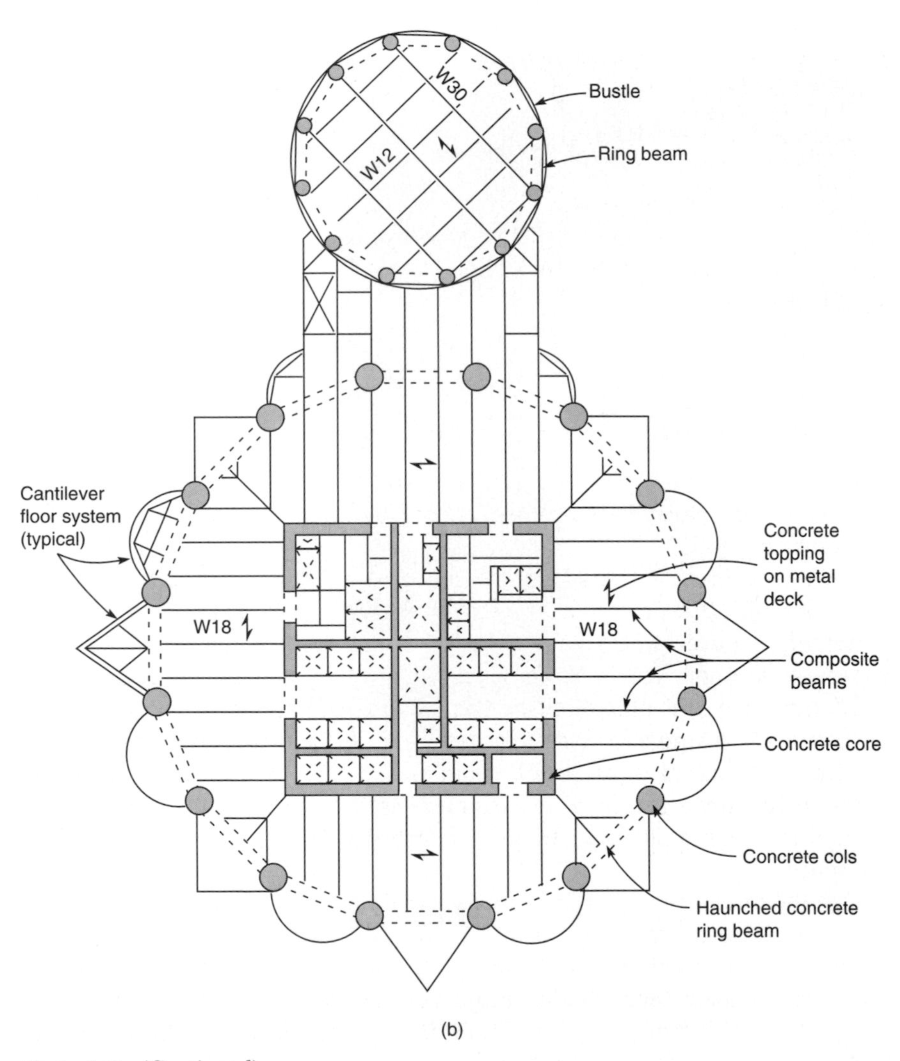


**Figure 1.20** *(Continued).*

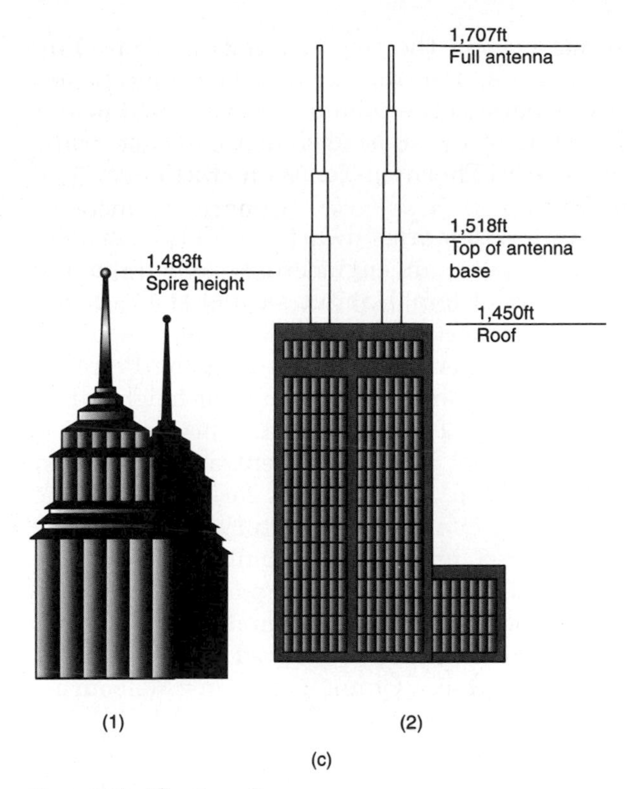


**Figure 1.20** *(Continued)*.

bustle walls carry about half the overturning moment at the foundation level.

Each core is 75 ft (23 m) square at the base, rising in four steps to 62 × 72 ft (18.8 × 22 m). Inner walls are a constant 14 in. (350 mm) thick while outer walls vary from 30 to 14 in. (750 to 350 mm). The concrete strength varies from 11,600 to 5,800 psi (80 to 40 mPa).

To increase the efficiency of the lateral system, the interior core and exterior frame are tied together by a 2-story deep outrigger truss at the mechanical equipment room (level 38). A vierendell type of truss with three levels of relatively shallow beams connected by a midpoint column is used to give flexibility in planning of building occupancy.

The tower floors Fig. 1.20b are typically composite metal deck with concrete topping varying from $4\frac{1}{2}$ in. (110 mm) in offices to 8 in. (200 mm) on mechanical floors, including a 2 in. (53 mm) deep composite metal deck. Wide-flange beams frame the floors at spans up to 42 ft (12.8 m), and on most floors are W18 or shallower to provide room for ductwork, sprinklers and lights.

Cantilevers for the points beyond the ring beams are 3.28 ft (1 m) deep prefabricated steel trusses. For the arcs, the cantilevers are beams propped with kickers back to the columns. Trusses and beams are connected to tower columns by embedded high-strength bolts. The structural engineering is by Thornton-Tomasetti Engineers, and Ranhill Bersekutu Sdn. Bhd.

Although the Sears Tower's 110-stories dwarf the Malaysian twin skyscrapers' 88 floors (Fig. 1.20c), an engineering panel from the Council on Tall Buildings and Urban Habitat says that the Sears Tower is no longer the world's tallest building. This panel which sets international building height standards, contends that the Petronas Towers' 242 foot high ornamental spires are part of their height while the radio antennas of the Sears Tower are not. This is because traditionally the measurement from ground floor entrance to highest original structural point has been the criterion for assessing the height of skyscrapers for over a quarter of a century. Executives of the Chicago skyscrapers disagree, and say their building is actually 35 feet taller if the radio bases are considered as part of the height. The debate once again confirms that the vanity and desire to build taller skyscrapers is alive and well, as evidenced by even taller skyscrapers planned in Shanghai, China, and in Melbourne, Australia.

### 1.3.14 Tokyo City Hall

Tower no. 1 is a high-rise building with a height of 800 ft (243.4 m) consisting of 48-stories above ground and 3-stories underground. The basic structural element in the vertical direction consists of a 21 ft (6.4 m) square super column. The super column is made up of four 40 × 40 in. (1020 × 1020 mm) steel box columns linked by K-braces. Eight such columns run through the building from foundation to the top. These columns are connected by an orthogonal system of beams at each floor level. The super columns are interconnected by a system of 1-story deep belt trusses at the 9th, 33rd and 44th floors.

A large and flexible column-free space of 63 × 357 ft (19.2 × 108.8 m) is established at every floor with the use of deep beams. The typical floor-to-floor height is 13.13 ft (4.0 m). Figure 1.21 shows schematic plans and sections.

### 1.3.15 Leaning tower; a building in Madrid, Spain

Beyond the needs of a typical building, rising straight up from ground, a leaning building requires an enhanced lateral-force system in the direction of the cantilever and an enhanced torsional system on account of eccentricity of transverse lateral forces. Note that although the overhang creates a substantial gravity induced overturning moment, the gravity induced shear force is zero. In addition to resisting normal lateral forces from wind and earthquakes, the structural system for a leaning building must resist the gravity induced over-turning moment—and must do so at very low levels of lateral deflection. This additional requirement is particularly severe since the lateral deflection may not be recoverable, i.e., such deflection may remain permanently in the structure. The problem is even more severe in concrete structures because the deflections may increase with time, on account of the long-term creep properties of concrete.

A triangulated structural system consisting of super diagonals offers unique advantages in terms of both stiffness and strength. Additionally, it may offer opportunities to incorporate the structural system into visual expression of the building.

Observe that in the basic triangular unit shown in Fig. 1.22a, the slopping exterior columns carry symmetrical vertical loads without bending. To carry unsymmetrical loading as is required for live loads

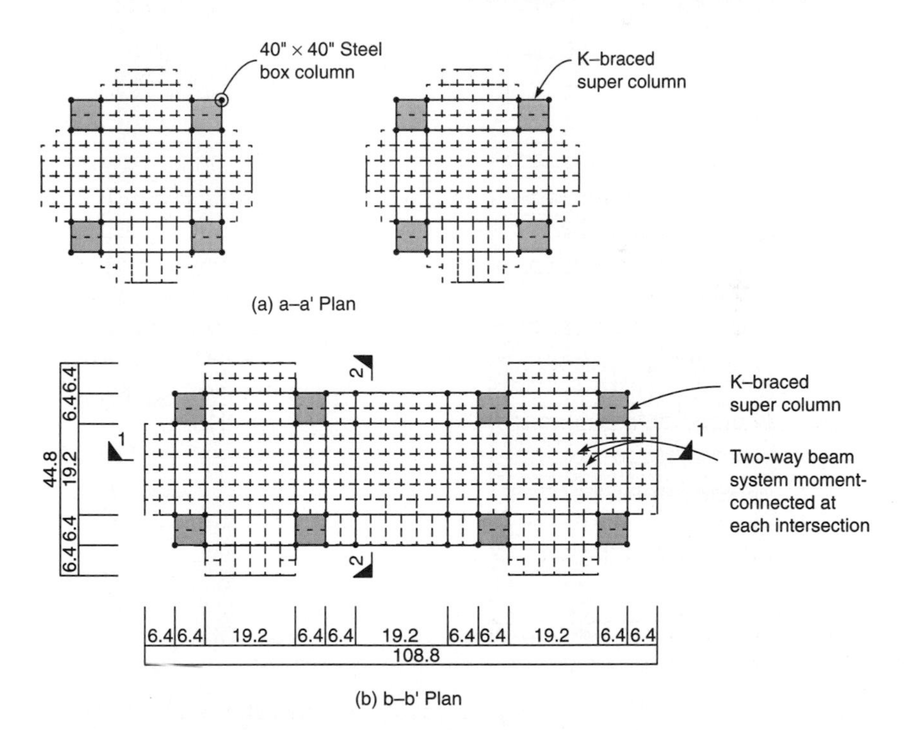


**Figure 1.21** Tokyo City Hall: (a,b) schematic plans; (c,d) schematic sections.

and lateral loads, a secondary system acting within the basic triangular unit is necessary.

A parallelogram shape consisting of two or more triangular blocks as shown in Fig. 1.22b more or less balances the dead load on one lower corner. However, under lateral forces and for unbalanced live loads the unit has to be tied down in the other lower corner. The tie-down can be composed of rock or earth anchors, tension piling or a ballast consisting of concrete mass.

To compensate for gravity-induced lateral deflection, many approaches are possible. These include cambering the structure, post-tensioning and increasing the stiffness of the structure. Any of the approaches can be combined with the other.

Post-tensioning is a proven technique for the control of gravity-induced deflections. In concept, post-tensioning is introduced into

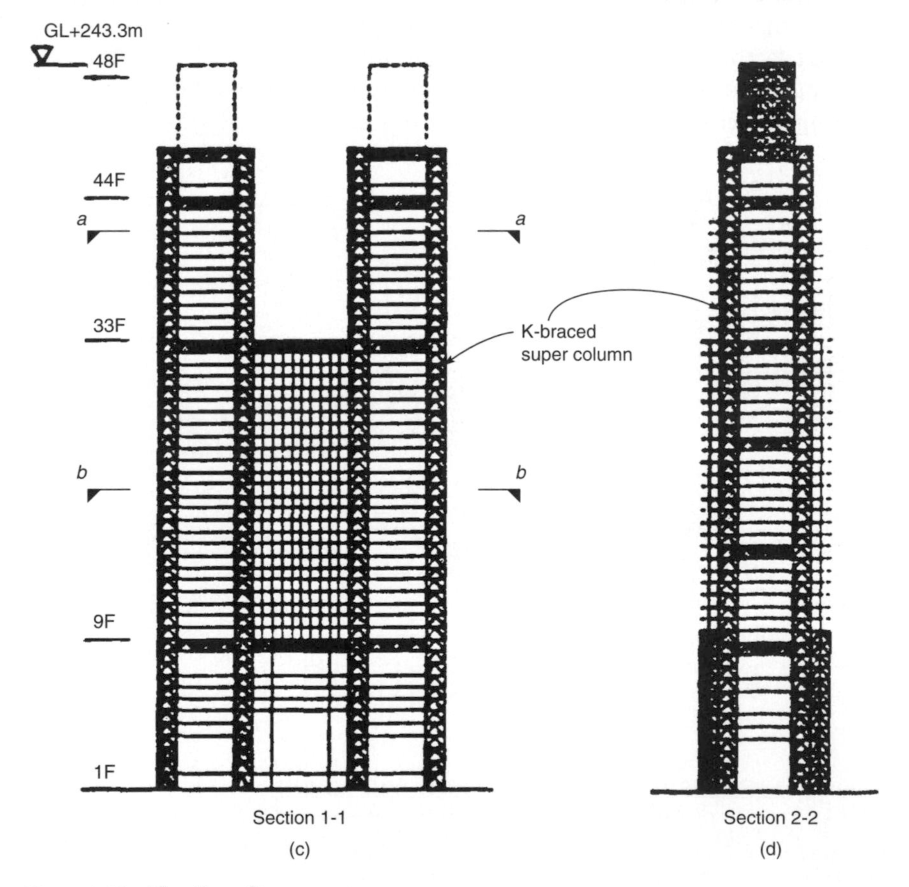


**Figure 1.21** *(Continued).*

those elements of the structural system that are under gravity-induced tension so as to place these members in compression. By post-tensioning the outer columns of a leaning building, it is possible to compensate for all or part of the lateral deflection induced by gravity loading.

As an example of a leaning tower, Fig. 1.22c shows a schematic framing system for a high-rise building in Madrid, Spain.

The building perimeter framing and the interior floor beams are of structural steel. The service core is of reinforced concrete. Tie-down ballast of reinforced concrete is used to counteract the overturning effect. The ballast is 165 ft long, 45 ft wide and 31.7 ft deep (50 × 13.7 × 9.67 m) and weighs 15,400 tons.

The post-tension used on the steel for this project is a conventional system popular in post-tensioning of concrete construction. To protect the tendons, the post-tensioning is carried in steel pipes, not post-tensioning ducts used in concrete construction. The post-tension

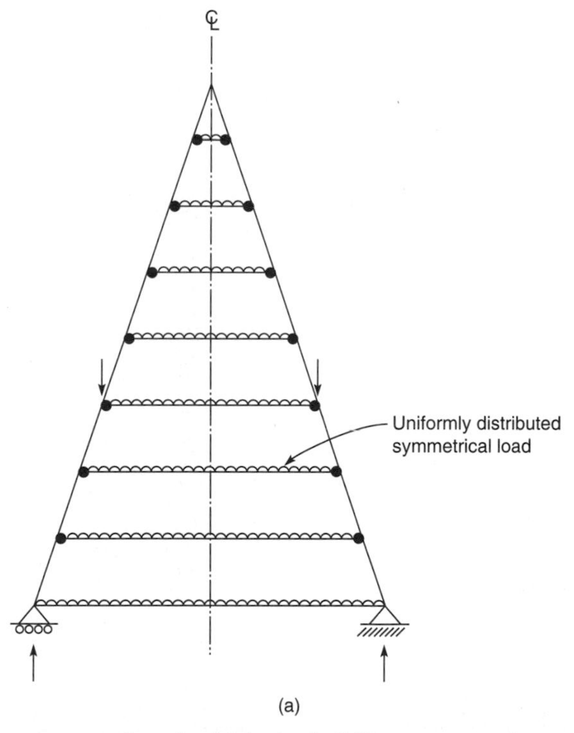


**Figure 1.22** Leaning high-rise buildings, structural concept: (a) the basic triangular building block; (b) Interconnected blocks: for symmetrical gravity loads no uplift exists at the base; (c) structural schematics, Twin Towers of Puerta de Europe.

system is anchored into the concrete ballast at the base of the building.

Transverse to the direction of the sloping faces, the lateral system is fairly straightforward. It consists of super diagonal braces running

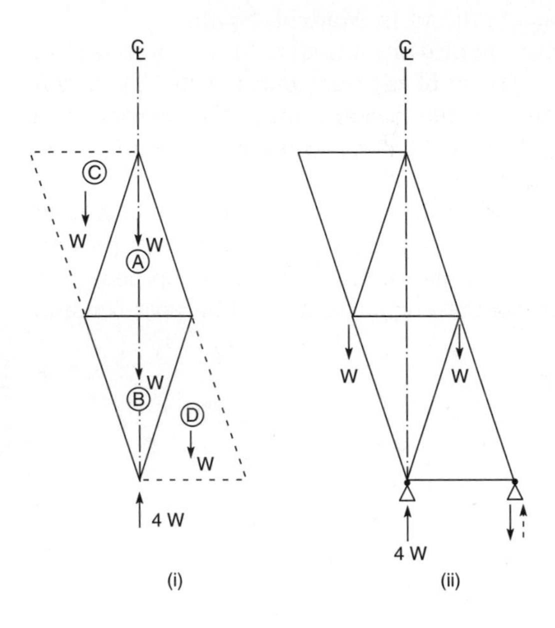


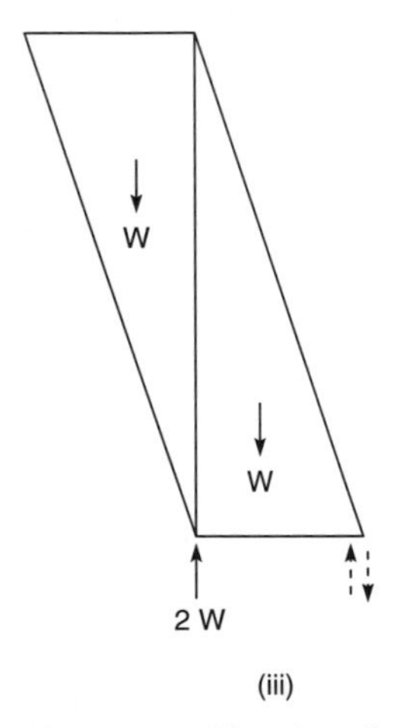


**Figure 1.21** *(Continued).*

for the full height of the building. A 25 ft deep (7.62 m) truss is used in the roof-top mechanical space to mobilize the stiffening effect of triangulated façades. The structural engineering is by the New York office of Leslie Robertson and Associates.

### 1.3.16 Hong Kong Central Plaza

The building is 78-stories with the highest office floor at 879 ft (268 m) above ground. Including the tower mast, the building is 1207.50 ft (368 m) tall (Fig. 1.23a). The building has a triangular floor plate with a sky lobby on the 46th floor.

The triangular design consisting of a typical floor area of 23,830 $ft^2$ (2214 $m^2$) was preferred over a more traditional square or rectangular plan, because the triangular shape has very few dead corners and offers more views from the building interiors.

The tower consists of three sections: (i) a 100 ft (30.5 m) tall tower

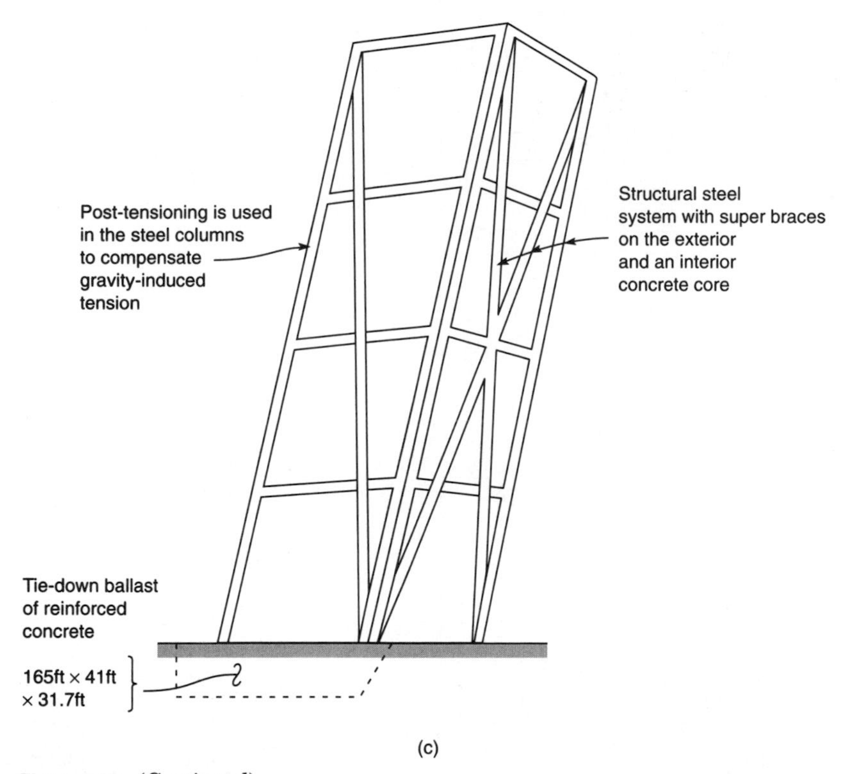


**Figure 1.22** (*Continued*).

base forming the main entrance and public circulation spaces; (ii) a 772.3 ft (235.4 m) tall tower section containing 57 office floors, a sky lobby and five mechanical floors; and (iii) a top section consisting of six mechanical floors and a 334 ft (102 m) tall tower mast.

The triangular building shape is not truly triangular because its three corners are cut off to provide better internal office layout. The building façade is clad in insulated glass. The mast is constructed of structural steel tubes with diameter up to 6.1 ft (2 m).

The triangular core design (Fig. 23b,c) provides a consistent structural and building services configuration. A column-free office space, with 30.84 to 44.3 ft (9.4 to 13.5 m) depth is provided between the core and the building perimeter.

To enhance the spatial quality of the tower at the base, the 15 ft

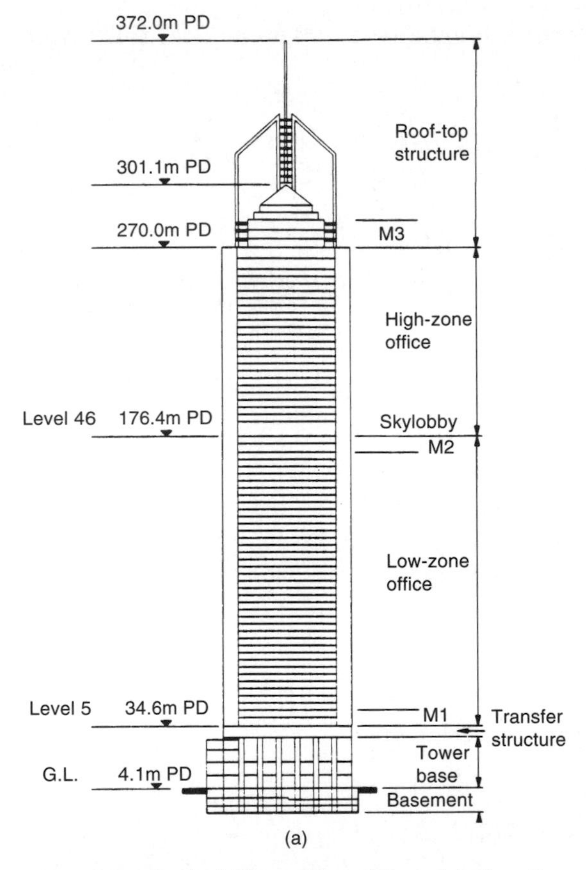


**Figure 1.23** Central Plaza, Hong Kong: (a) elevation; (b–c) floor plans; (d) lateral load transfer system.

(4.6 m) column grid of the tower is transformed to 30 ft (9.2 m) column grid by eliminating every other column. An 18 ft (5.5 m) deep transfer girder facilitates column termination.

The building site is typical of a recently reclaimed area in Hong Kong with sound bed rock lying between 82 and 132 ft (25 and 40 m) below ground level. This is overlaid by decomposed rock and marine deposits with the top 33 to 50 ft (10 to 15 m) consisting of a fill material. The allowable bearing pressure on sound rock is of the

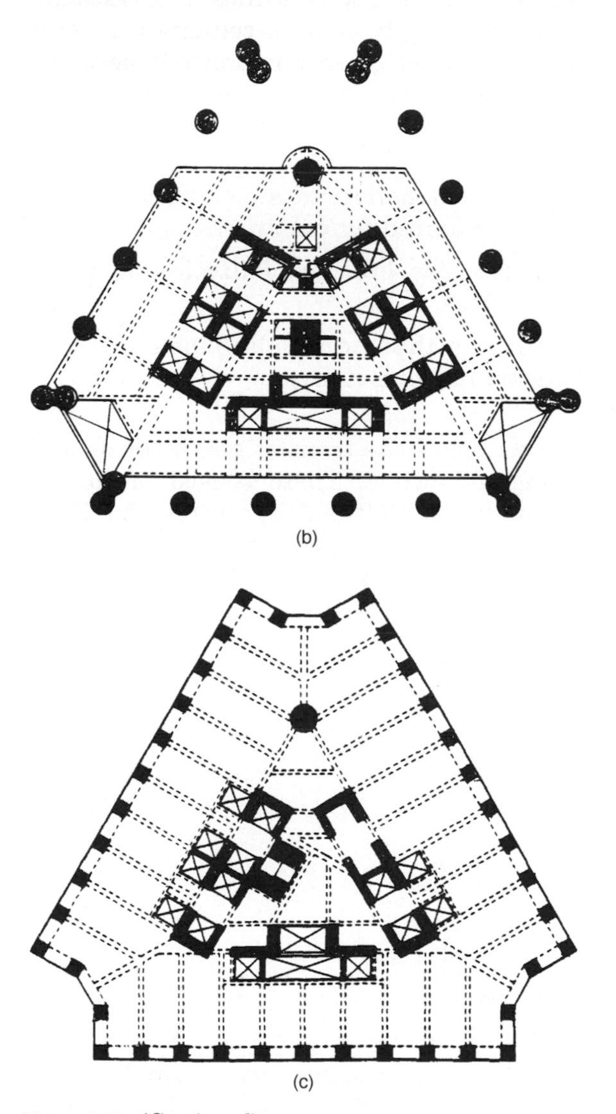


**Figure 1.23** *(Continued).*

order of 480 ton/ft$^2$ (5.0 kN/m$^2$). The maximum water table is about 6.1 ft (2 m) below ground level.

Wind loading is the major lateral load criterion in Hong Kong which is situated in an area susceptible to typhoon winds. The local wind design is based on a mean-hourly wind speed of 100 mph (44.7 m/s), a three-second gust of 158 mph (70.5 m/s) and gives rise to a lateral design pressure of 86 psf (4.1 kN/m$^2$) at 656 ft (200 m) above ground level.

The basement consisting of a diaphragm slurry wall extends around the whole site perimeter and is constructed down to and grouted to rock. The diaphragm wall design allowed for the basement to be constructed by the "top down" method. This method has three fundamental advantages:

1. It allows for simultaneous construction of superstructure and basement thus reducing time required for construction.
2. Basement floor slabs are used for bracing of diaphragm walls thereby reducing lateral tie-backs.
3. Creates a watertight box with-in the site enabling installation of hand dug caissons, traditional in Hong Kong.

The lateral system for the tower above the transfer girder consists of external façade frames acting as a tube. These consist of closely spaced 4.93 ft (1.5 m) wide columns at 15 ft (4.6 m) centers and 3.6 ft (1.1 m) deep spandrel beams. The floor-to-floor height is 11.82 ft (3.6 m). The core shear walls carry approximately 10 percent of the

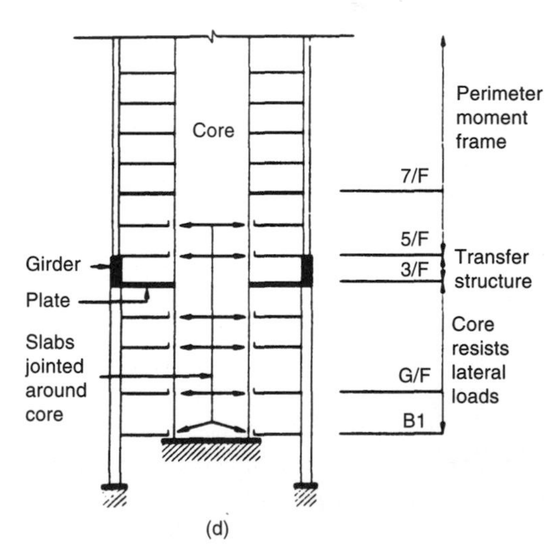


**Figure 1.23** (*Continued*).

lateral load above the transfer level. The transfer girder located at the perimeter is 18 ft (5.5 m) deep by 9.2 ft (2.8 m) wide, allowing alternate columns to be dropped from the façade, thereby opening up public area at ground level. The increased column spacing together with the elimination of spandrel beams in the tower base, results in the external frame no longer being able to carry the lateral loads acting on the building. Therefore, the wind shears are transferred to the core through the diaphragm action of 3.28 ft (1 m) thick slab located at the transfer level. The wind shear is taken out from the core at the lowest basement level, where it is transferred to the perimeter diaphragm walls. In order to reduce large shear reversals in the core walls, the floor slabs and beams are separated horizontally from the core walls (Fig. 1.23d) at certain levels. Structural engineering for the project is by Ove Arnp and Partners.

### 1.3.17 Fox Plaza, Los Angeles

The structural system for resisting lateral loads for the 35-story building, consists of special moment-resisting frames located at the building perimeter. The floor framing consists of W21 wide flange composite beams spanning 40 ft (12.2 m) between the core and the perimeters. A 2 in. (51 mm) deep 18 gauge composite metal deck with a $3\frac{1}{4}$ in. (83 mm) lightweight concrete topping is used for typical floor construction. A typical floor framing plan with sizes for typical members is shown in Fig. 1.24. The structural design is by John A. Martin & Associates, Los Angeles.

### 1.3.18 Bell Atlantic Tower, Philadelphia

This is a 53-story steel building consisting of a braced core linked to four super columns via a shear-resisting system consisting of 2-story vierendeel girders placed alternately at each side of the core. The varying configurations of the floor plates shown in Fig. 1.25a,b would have required multiple levels of perimeter column-transfer in a conventional tube system making the structural system uneconomical. The braced core system working in concert with the exterior columns, on the other hand, maintains the structural efficiency close to that of a conventional tube system, with the added benefits of providing column-free corner offices (Fig. 1.25e).

The building perimeter consists of four major built-up steel columns measuring $54 \times 30$ in. ($1.37 \times 0.76$ m). These box columns are linked by a series of vertically stacked 5-story vierendeel frames (Fig. 1.25d). Each stack of vierendeels is linked to the one below by a series of hinges (Fig. 1.25d, Detail A) designed to transfer horizontal shears only. This detail prevents the build-up of gravity loads in the vierendeel frame by systematically shedding the loads to the box columns, thereby increasing their efficiency in resisting the overturning moments. The lateral resistance in the transverse direction is provided by the braced core linked to the box columns by the 2-story vierendeel girders (Fig. 1.25e). The structural design is by CBM Engineers, Inc., Houston, Texas.

### 1.3.19 Norwest Center, Minneapolis

The structural system for the 56-story bank building (Fig. 1.26a) is similar to the Bell Atlantic Tower. The only difference is that composite super-columns are used instead of built-up all-steel box columns (Fig. 126b,c). High strength, 10,000 psi (68.94 mPa) concrete is used in the composite columns. This type of construction has been estimated in the North American construction market to be five to

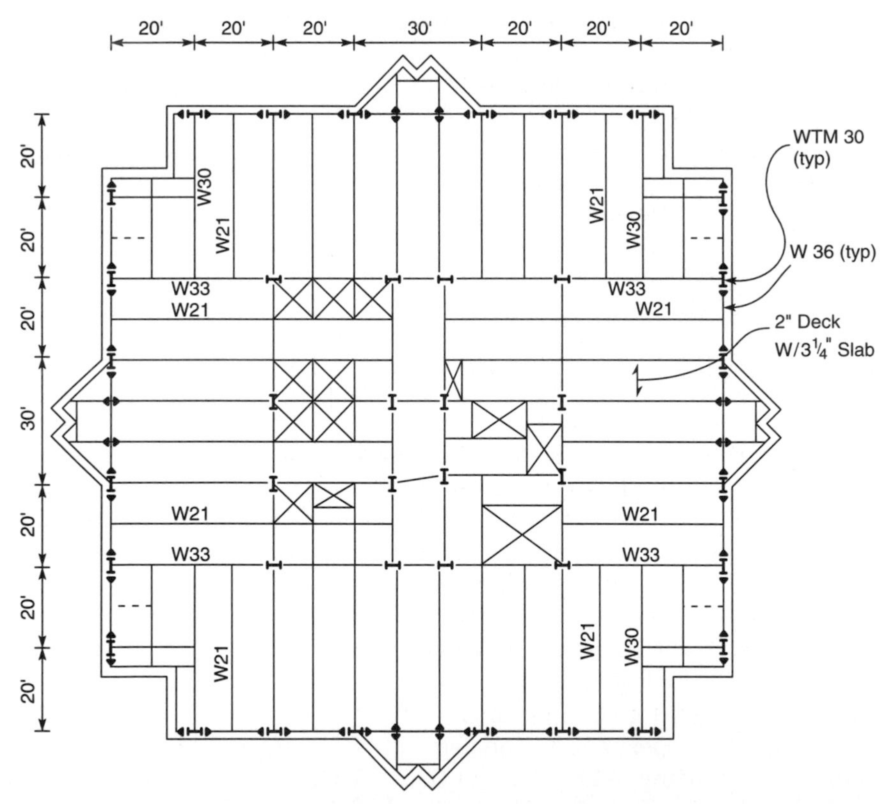


**Figure 1.24** Fox Plaza, Los Angeles. Architects: Johnson, Fain & Pereira Inc. Structural engineers: John A. Martin & Asso. Inc., Los Angeles.

six times less expensive than steel columns of equivalent strength and stiffness. A representative detail of the connection between the steel beam and composite column, and the structural actions associated with moment transfer between the two are shown in Fig. 1.26d. Details of hinge used in the vierendeel frame are shown in Fig. 1.26e. The structural design is by CBM Engineers, Inc., Houston, Texas.

### 1.3.20 First Bank Place, Minneapolis

This is a 56-story, granite-clad building (Fig. 1.27a) comprising of an array of changing floor plans (Fig. 1.27b) with an added architectural

(a)

**Figure 1.25** Bell Atlantic Tower: (a) building elevation; (b) composite floor plan; (c) floor framing plan; (d) lateral system; (e) section.

stipulation that the north-east corner of the building, line $A^1B^1$ in Fig. 1.27b, should have a minimum of structural columns to provide unobstructed views of the city. The response was to come up with an uninterrupted cruciform-shape structural spine as shown in

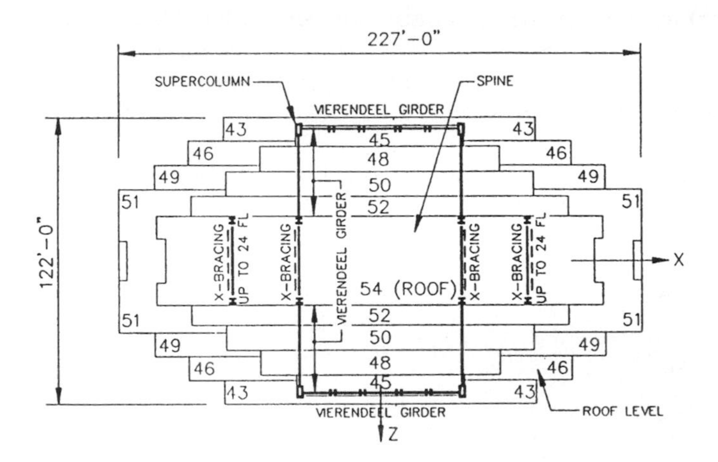


(b)

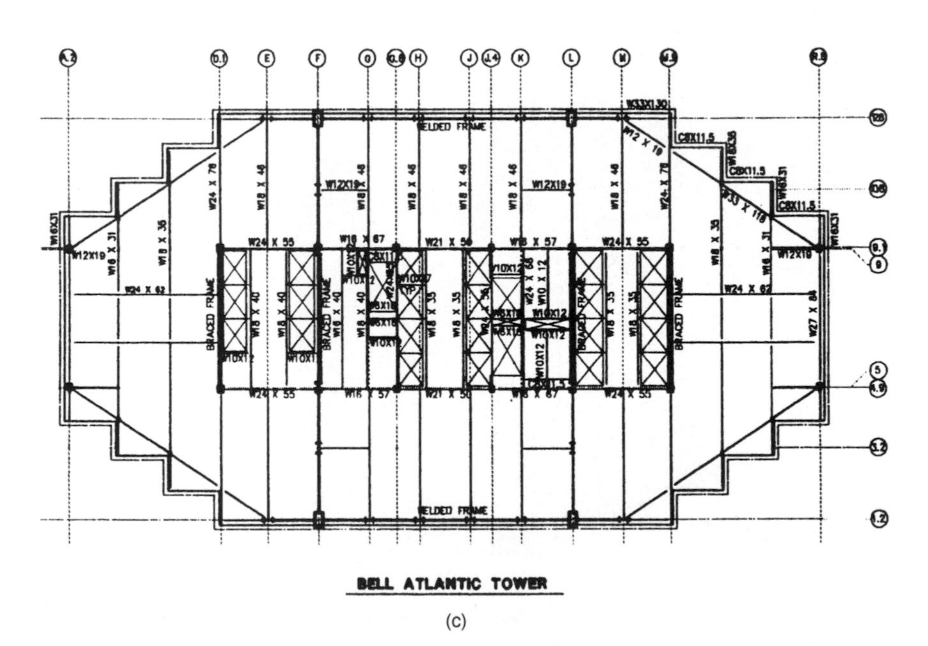


(c)

**Figure 1.25** *(Continued).*

Figs 1.27c,d,e,f. A combination of steel bracing and moment frames consistent with interior space planning is provided along the two spines, $AA^1$ and $BB^1$ (Fig. 1.27c) to act as shear membranes. Composite columns with concrete strengths up to 10,000 psi (68.44 mPa), and varying in size from 75 sq. ft (7 $m^2$) at the base, to 50 sq. ft (4.65 $m^2$) at top, are used at the spine extremities as shown in Fig. 1.27c.

Since the cruciform shape is torsionally unstable, the spine is stabilized by providing braced frames along the perimeter BC and $B^1C^1$, and by a series of 3-story tall vierendeel trusses along lines $CC^1$ and $B^1D^1$. A 12-story tall vierendeel girder along $A^1D^1$ links the

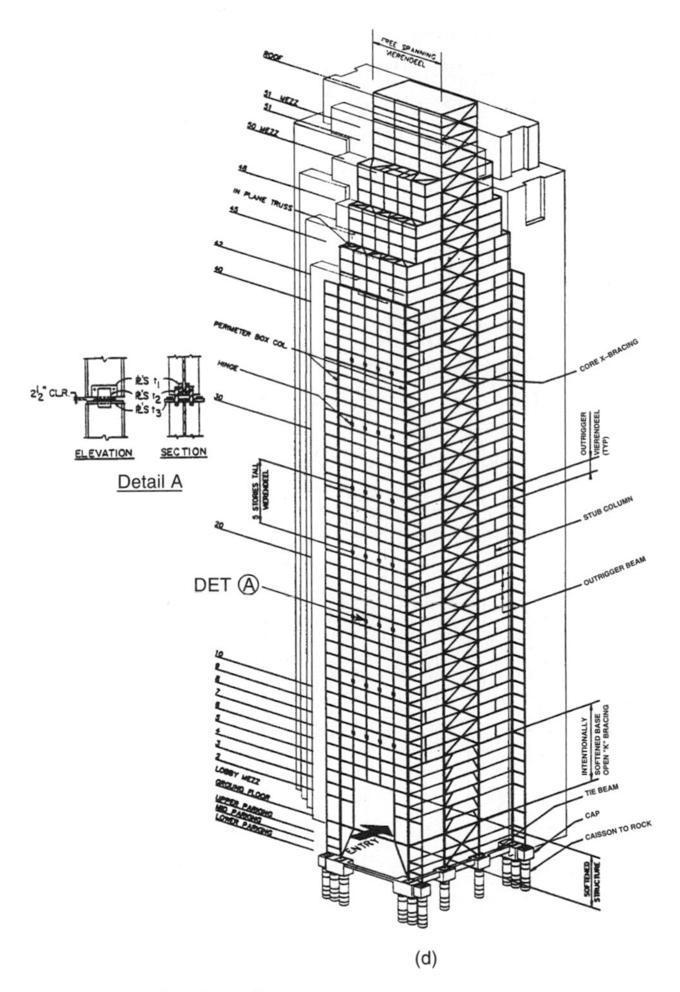


(d)

**Figure 1.25** *(Continued)*.

composite concrete super-column at $A^1$ to the perimeter steel column $D^1$ (Fig. 1.27d). It also supports the perimeter circular vierendeel along $BAB^1E$ above the 45th floor. This circular vierendeel provides both torsional and lateral resistance to the entire frame. The entire system is an outstanding example of strategic structural response to complex building geometry and leasing demands without having to pay an undue premium for the structural system. The structural design is by CBM Engineers, Inc., Houston, Texas.

### 1.3.21 Figueroa at Wilshire, Los Angeles

Floor framing plans at various step-backs and notches for the 53-story tower (Fig. 1.28a) are shown in Fig. 1.28b. The structural system, designed by CBM Engineers, Inc., Houston, Texas, consists of eight steel super-columns at the perimeter interconnected in a

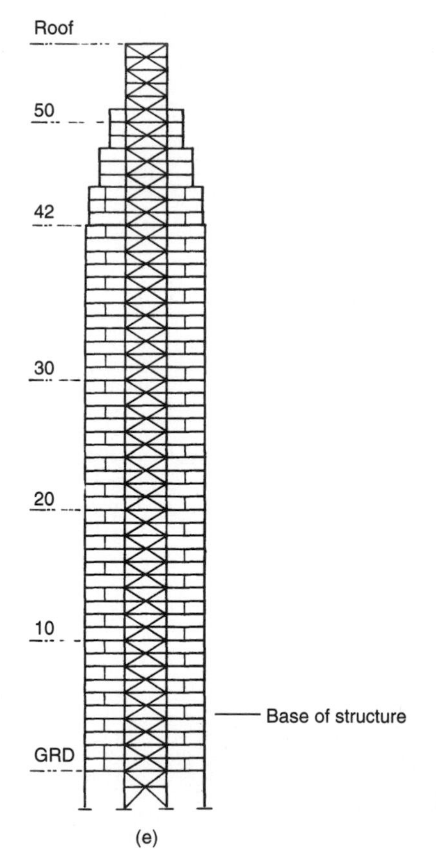


**Figure 1.25** *(Continued.)*

criss-cross manner to an interior braced core with moment connected beams acting as outriggers at each floor (Figs. 1.28c,d). The floor framing is structured such that the main columns participating in the lateral loading system are heavily loaded by gravity loads to compensate for the uplift forces due to overturning. The structural system consists of three major components.

1. Interior concentrically braced core.
2. Outrigger beams spanning approximately 40 ft from the core to the building perimeter. The beams perform three distinct functions. First, they support gravity loads. Second, they act as ductile moment-resisting beams between the core and exterior frame columns. Third, they enhance the overturning resistance of the building by engaging the perimeter columns to the core columns. To reduce the additional floor-to-floor height that might otherwise

(a)

**Figure 1.26** Norwest Center: (a) building elevation; (b) typical floor framing plan; (c) structural systems, isometric; (d) structural details, steel beam to concrete column connection; (e) hinge details.

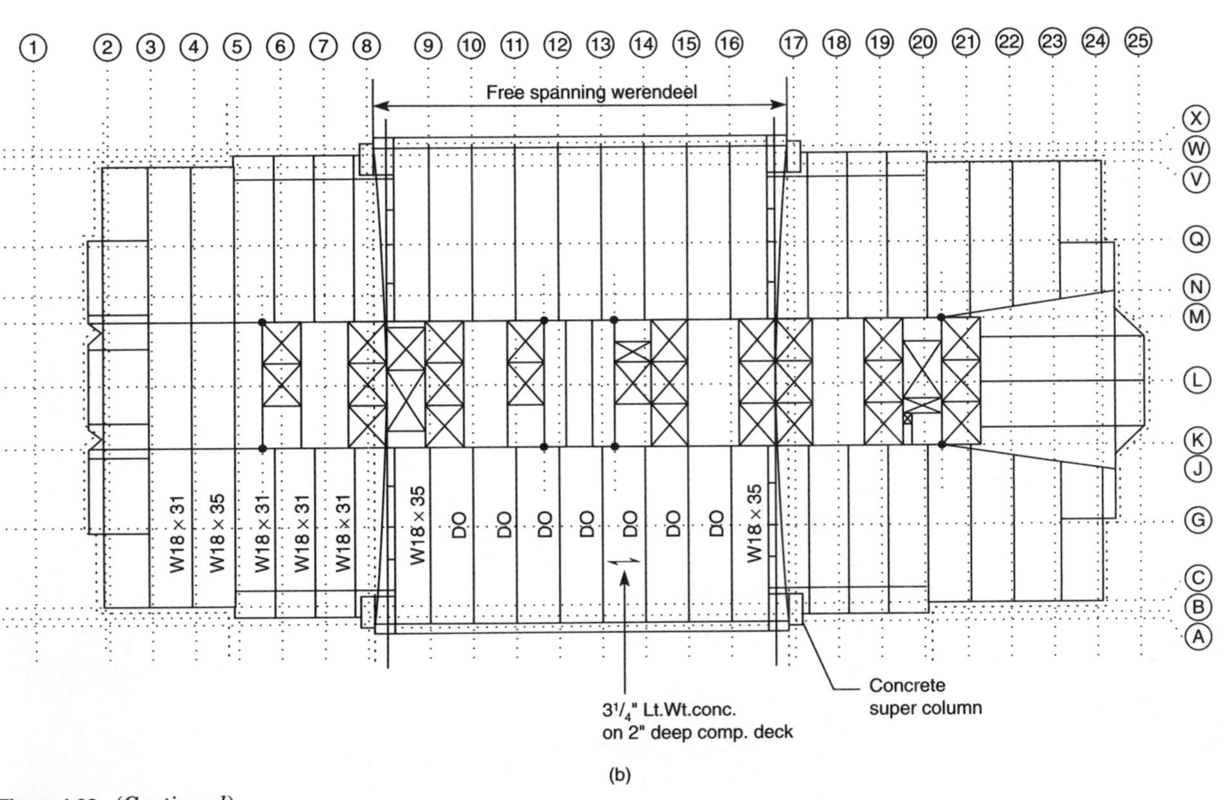


**Figure 1.26** (*Continued*).

be required, these beams are notched at the center, and offset into the floor framing as shown in Fig. 1.28e, to allow for mechanical duct work.

3. Exterior super columns loaded heavily by gravity loads to counteract the uplift effect of overturning moments.

### 1.3.22 One Detroit Center

This is a 45-story office tower with a clear 45 ft 6 in. (13.87 m) span between the core and the exterior (Fig. 1.29a). The structural system consists of eight composite concrete columns measuring 7 ft 6 in. × 4 ft 9 in. (2.28 × 1.45 m) at the base, placed 20 ft (6.1 m) away from the corners to provide column-free corner offices, and also to optimize

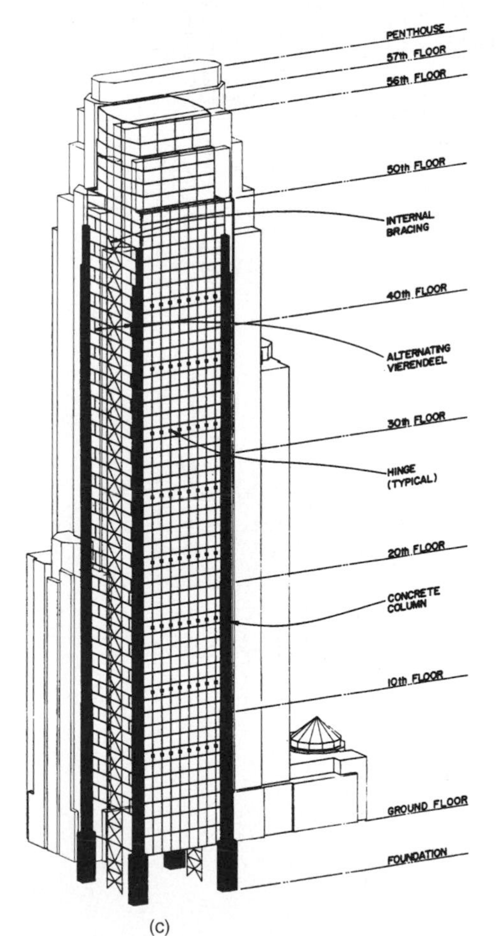


**Figure 1.26** (*Continued*).

the free-span of the vierendeel frames. The composite columns are connected at each face by a system of perimeter columns and spandrels acting as vierendeel frames. The vierendeels are stacked 4-story high and span between composite super-columns to provide column-free entrances at the base of the tower. At each fourth level, the vierendeels are linked by hinges which transfer only horizontal

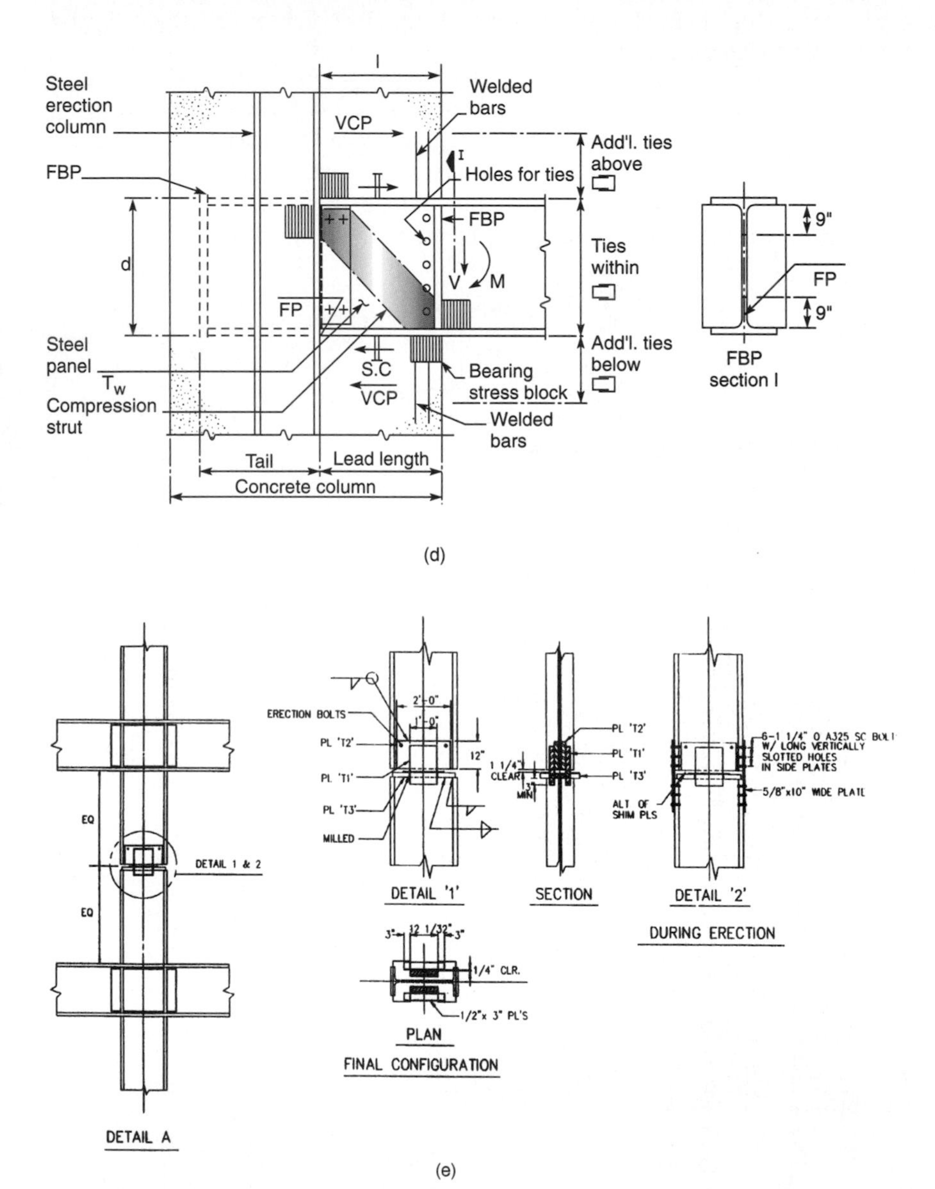


**Figure 1.26** *(Continued).*

shear between adjoining vierendeels and not gravity loads. The reason for this type of connection is to reduce: (i) the effect of creep and shrinkage of super-columns on the members and connections of the vierendeel, and (ii) gravity load transfer due to arch action of the vierendeel with associated horizontal thrusts. The 4-story vierendeel achieves uniformity in the transfer of moment and shear between horizontal steel beams and composite super-columns throughout the height of the tower.

A schematic representation of the structural system is shown in Figs 1.29b through 1.29c. Figure 1.29d shows the connection details for the vierendeel frame. The structural design is by CBM Engineers, Inc., Houston, Texas.

(a)

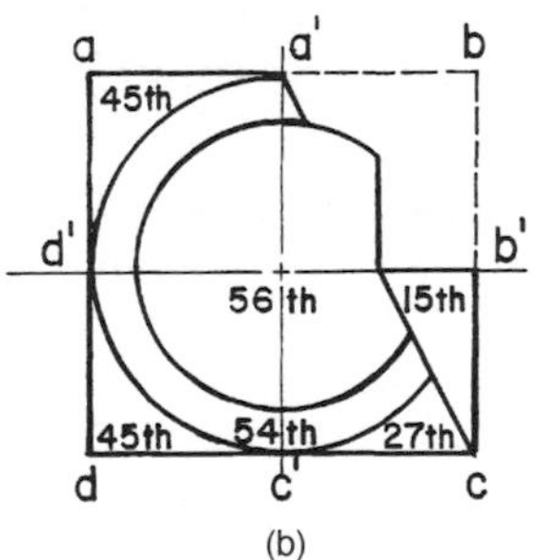


(b)

**Figure 1.27** First Bank Place: (a) building elevation; (b) composite floor plans; (c) plan of structural system; (d) isometric of structural system.

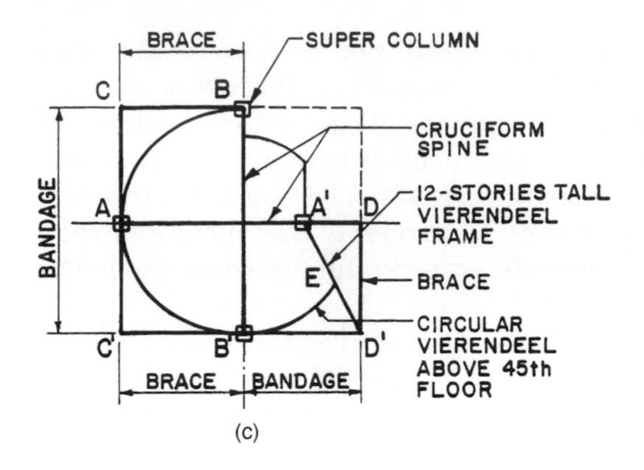


(c)

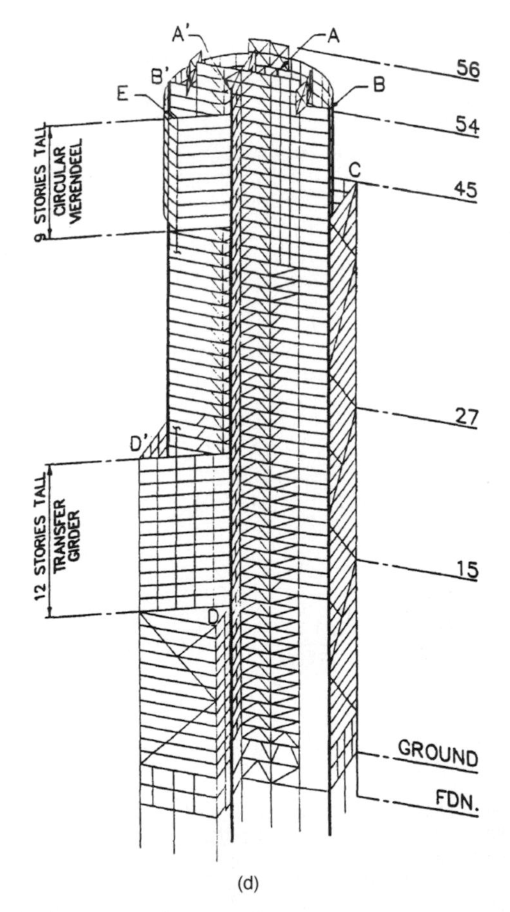


(d)

**Figure 1.27** (*Continued*).

### 1.3.23 One Ninety One Peachtree, Atlanta

This 50-story building uses the concept of composite partial tube as shown in Fig. 1.30a. The partial tubes which extend uninterrupted from foundation to the 50th floor consists of concrete columns encasing steel erection columns with cast-in-place concrete spandrels. The building interior is an all-steel structure with composite steel beams supported on steel columns (Fig. 1.30b).

Since the building does not achieve the lateral resistance until after the concrete in composite construction has reached substantial strength, a system of temporary bracing was provided in the core. The erected steel was allowed to proceed 12 floors above the completed composite frame with six floors of metal deck and six floors of concreted floors. The structural design is by CBM Engineers, Inc., Houston, Texas.

(a)

**Figure 1.28.** Figueroa at Wilshire, Los Angeles: (a) building elevation; (b) floor framing plans; (c) lateral system; (d) section; (e) reinforcement at beam notches, design concept.

### 1.3.24 Nations Bank Plaza, Atlanta

The 57-story office building has a square plan with the corners serrated to create the desired architectural appearance and to provide for more corner offices (Fig. 1.31a). The typical floor plan (Fig. 1.31b) is 162 × 162 ft (49.37 × 49.37 m) with an interior core measuring 58 ft 8 in. × 66 ft 8 in. (17.89 × 20.32 m). A five-level basement provided below the tower is of reinforced concrete construction. The foundation consists of shallow drilled piers bearing on rock.

The gravity load is primarily supported by 12 composite super-columns. Four of these are located at the corner of the core, and eight at the perimeter, as shown in Fig. 1.31b. The core columns are

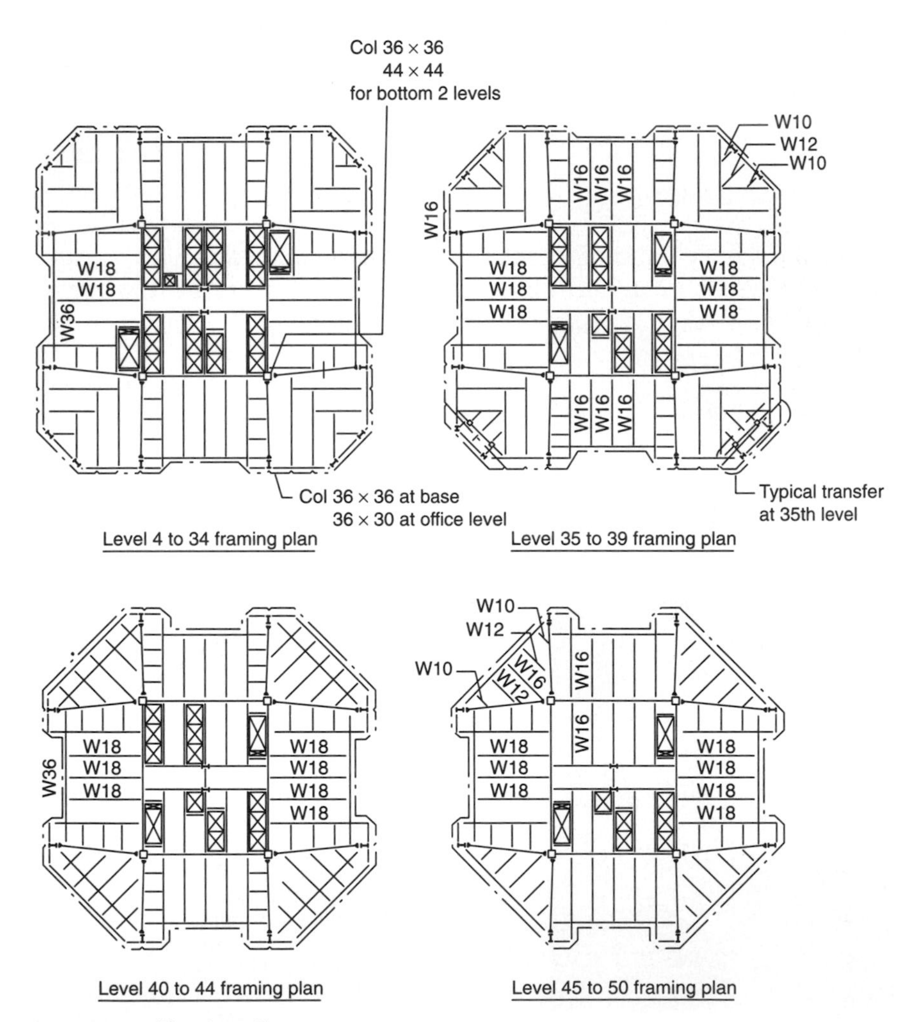


**Figure 1.28** *(Continued).*

braced on all four sides with diagonal bracing as shown schematically in Fig. 1.31c. Since the braces are arranged to clear mechanical and door openings in the core, their configuration is different on all four sides. Steel girders 36 in. deep (0.91 m) are moment-connected between the composite columns, to transfer part of the overturning moment to the exterior columns. Because the girders are deeper than other gravity beams, openings have been provided in the girders to provide for the passage of mechanical ducts and pipes. A diagonal truss is used between levels 56 and 59 to tie the core columns to the perimeter super-columns. These trusses transfer part of the overturning moment to the perimeter columns and also add considerable stiffness to the building. Above the 57th floor, the building tapers to form a 140 ft (42.68 m) tall conehead which is used to house mechanical and telecommunication equipment. The structural design is by CBM Engineers, inc., Houston, Texas.

### 1.3.25 Allied Bank Tower, Dallas, Texas

This is a 60-story building with an unconventional geometrical shape, raising to 726 ft (221.3 m) above a tree-studded plaze in downtown Dallas (Fig. 1.36a). The geometrical composition of the tower shown in Fig. 1.36b consists of: (i) a rectangular block 192 ft (58.53 m) square in plan raising to 54 ft (16.46 m) above the plaza; (ii) a 480 ft (146.3 m) tall geometric shape that wedges gradually from a square

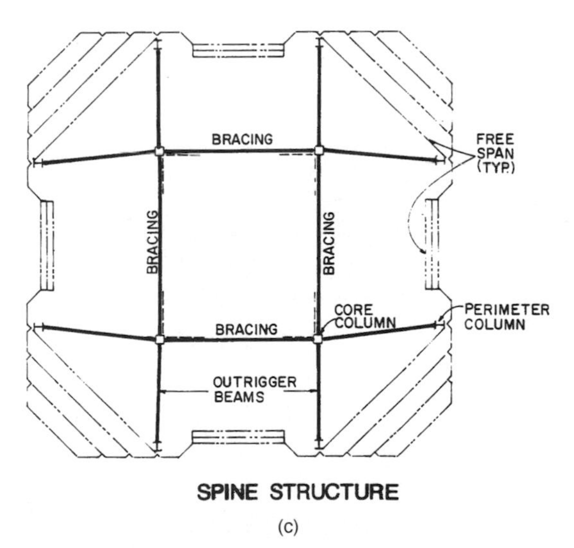


**Figure 1.28** (*Continued*).

at the bottom to a 96 × 192 ft (29.3 × 58.5 m) rhombus; (iii) a 16-story high skewed triangular prism capping the building top. The combined 672 ft (204.83 m) high building is raised on a 4-story high

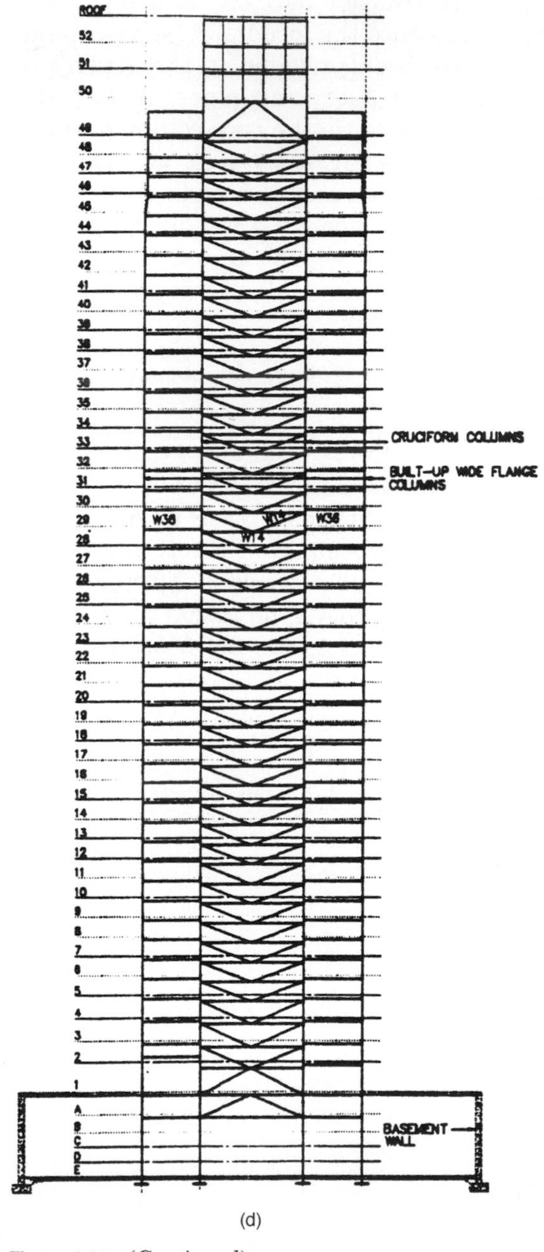


(d)

**Figure 1.28** (*Continued*).

pedestal that matches the rhombus shape at the top as shown schematically in Fig. 1.36c. The resulting unconventional shape serves as an eye-catcher with a distinct exterior façade and a unique sloping top.

The structural system shown in Fig. 1.36b,c consists of a perimeter trussed frame that performs a dual function by providing the required lateral resistance, and an architecturally desired free-span at the base. This system has a 40-story deep megatruss on each of the 156 ft (47.55 m) side, and an 8-story deep truss on each of the 96 ft side as shown in Fig. 1.36c. The trussed exterior frames are set back 3 ft (0.92 m) from the building skin to clear the curtain wall. Also,

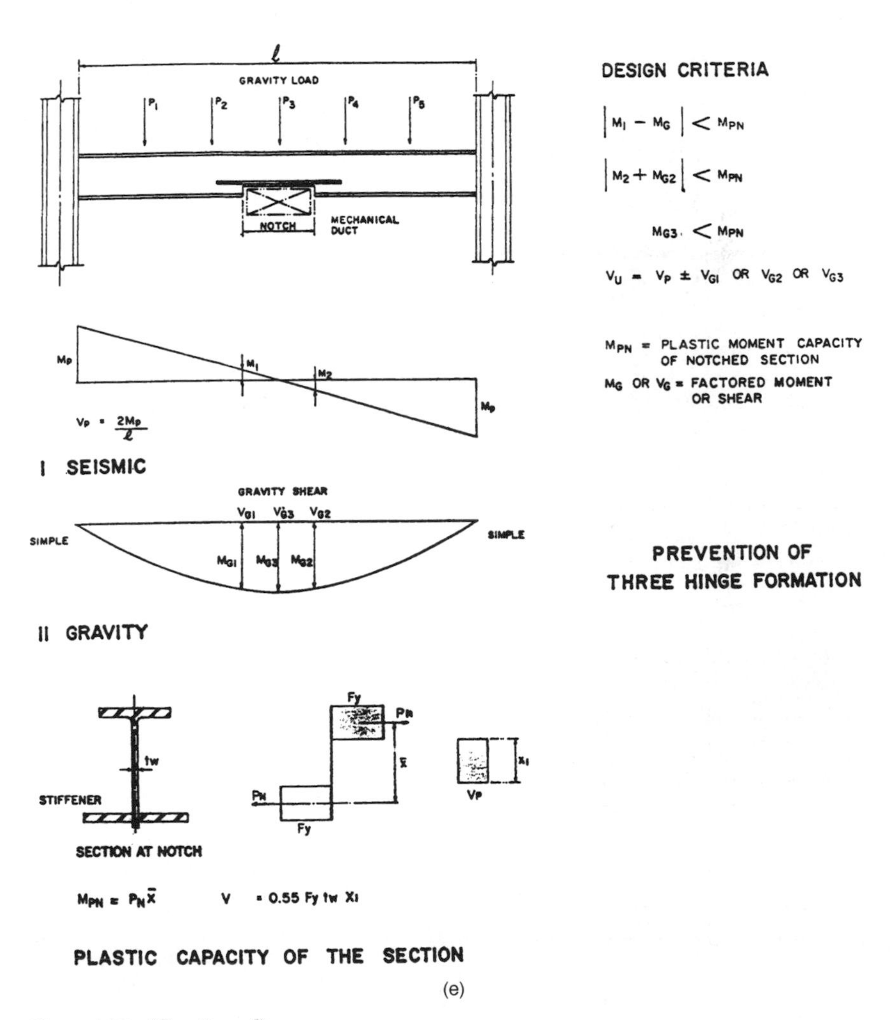


Figure 1.28 *(Continued).*

the geometry of the diagonals dictated by the architectural and leasing requirements resulted in the intersection of the diagonsl at mid-height of columns (Fig. 1.32d). Therefore, a conventional truss action in which the truss diagonal, the column, and the floor beam, all meet at a common node, could not be provided to transfer the unbalanced horizontal components of axial forces in the diagonals, directly to the floor members. As an alternative solution, a story-deep vierendeel truss was designed to resist the moments caused by the unbalanced horizontal components of the forces in the diagonals.

Above the 45th level, moment frames are used to resist the lateral loads in both directions. To facilitate a smooth flow of forces between the trussed and moment-connected regions, an overlap zone was

(a)

**Figure 1.29** One Detroit Center: (a) building elevation; (b) typical floor framing plan; (c) free-spanning vierendeel elevations; (d) structural details for vierendeel frame, (1) partial elevation, (2) detail 1, (3) detail 2.

created by extending the moment frames nine floors below the apex of the trussed floor. A schematic of the structural system is shown in Fig. 1.36c.

To eliminate the temporary shoring, otherwise required for erecting steel four levels above the open plaza, a 1-story deep truss is used all around the perimeter between the fifth and sixth levels. The truss is designed to support the erection load of eight floors above the fifth level. The 8-story deep subtruss (Fig. 1.36c) together with the vierendeel truss at the 12th level and the truss at the fifth level, supports the construction loads of the next eight floors. This sequence of supporting the construction loads is carried through, for the entire structure.

Below the fifth level, two 30 ft wide (9.15 m) pylons located on opposite sides of the building are the only elements for transferring

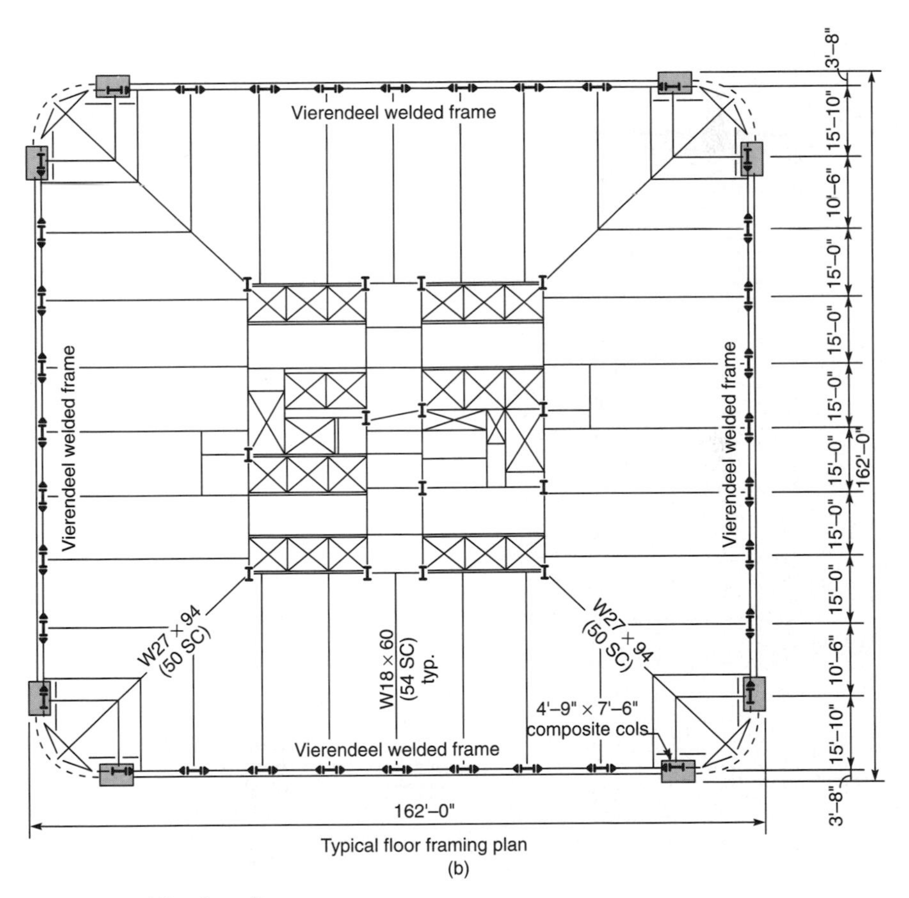


Typical floor framing plan
(b)

**Figure 1.29** *(Continued).*

the entire north-south wind shear from above. Heavy built-up columns with W14 wide flanges as cross-bracing resist shear and overturning without significant uplift in the columns.

The floor spandrels between the intersection of the diagonals are moment connected to the diagonals to serve three purposes: (i) to act as a secondary lateral system to transmit wind loads applied at each level to truss panel points; (ii) to provide additional lateral restraint to the system and (iii) to provide lateral bracing for the compression diagonals of the truss.

Both 36 and 50 ksi steel are used for the project. All diagonals of the mega truss are W14 wide flange shapes. The maximum weight of built-up trapezoidal column with 8 in. (203 mm) thick plates is 2450 lbs/ft (35 kN/m). Two-sided gusset plates are used at the intersection of diagonals with the box columns, while single-gusset plate with stiffeners matching the wide flange diagonals and

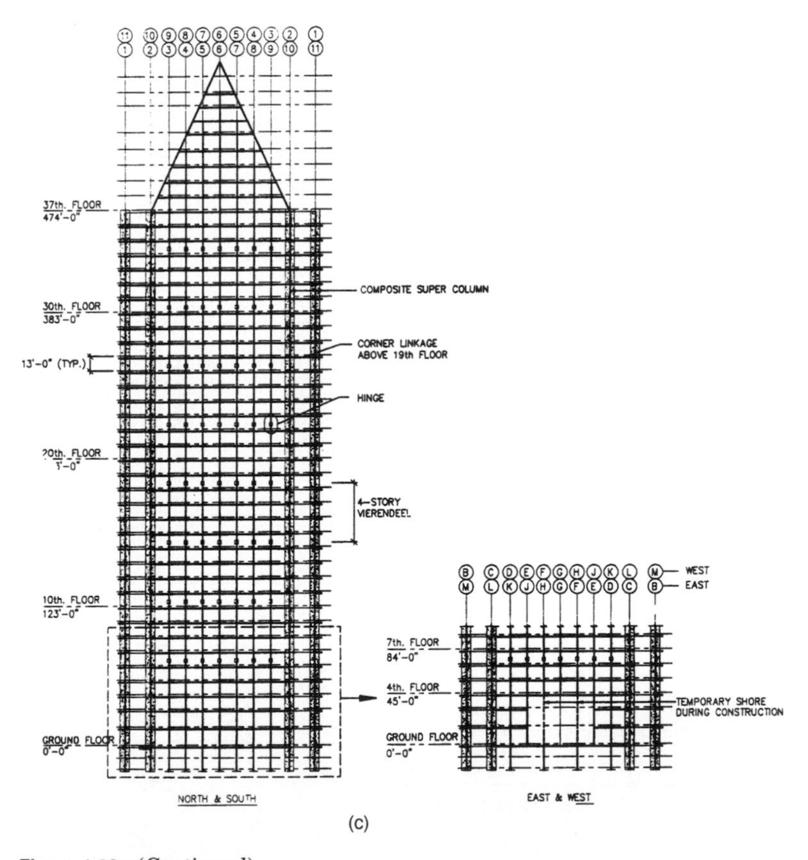


(c)

**Figure 1.29** (*Continued*).

columns are provided at the intersection of the diagonals (Figs 1.36e and 1.36f).

The typical floor construction consists of 50 ksi (344.74 mPa) composite floor beams with a 3 in. deep (76 mm) metal deck and $2\frac{1}{2}$ in. thick (63.5 mm) normal weight concrete topping. Instead of welded wire fabric, fibre reinforced concrete is used in the slab as reinforcement. Additional mild steel reinforcement is used at vierendeel floors and at levels 13 and 45, where the shape of the exterior façade changes.

The building consists of three basements below the plaza level. Shallow drilled piers bearing on 50-ton capacity rock support the building. The unit weight of the structural steel is 22 psf (1053.4 Pa). The floor-to-floor height is 12 ft (3.66 m) as compared to 13 ft (3.96 m) normally used for this type of office buildings. The exterior column spacing is 48 ft (14.63 m) offering generous views from the building interior even with the presence of diagonals. The structural design is by CBM Engineers, Inc., Houston, Texas.

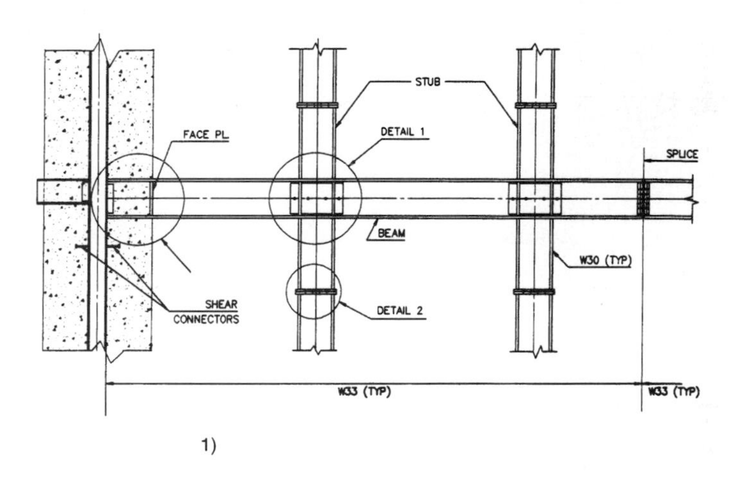


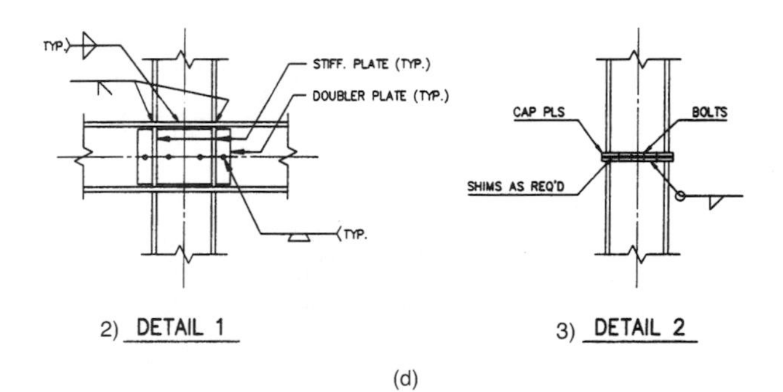


(d)

**Figure 1.29** (*Continued*).

(a)

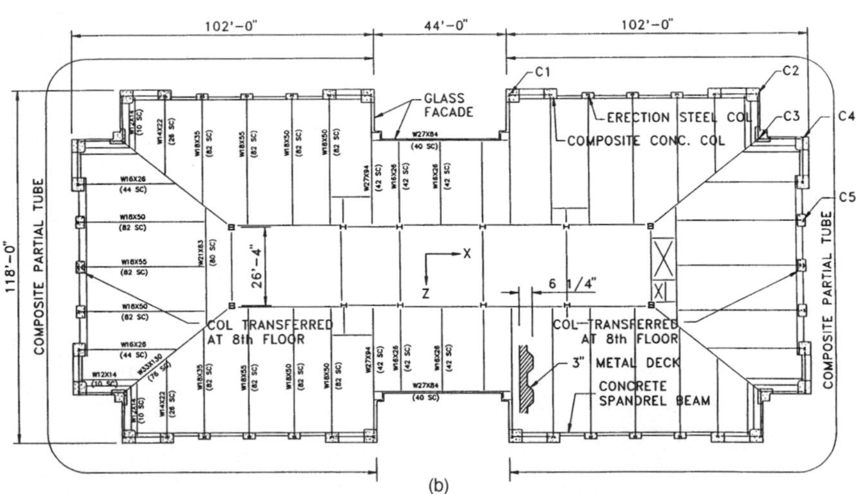


(b)

**Figure 1.30** One Ninety One Peachtree-Detroit: (a) building elevation; (b) typical floor framing plan.

### 1.3.26 First Interstate World Center, Los Angeles

This 75-story granite-clad building (Fig. 1.33a) sports multiple step-backs as shown in Figs 1.37b,c. The structural system is a dual system consisting of an uninterrupted 73 ft 10 in. (22.5 m) square braced spline interacting with a perimeter ductile moment resisting frame. The spine is a 2-story tall chevron-braced core as shown in Fig. 1.33d.

The 55 ft (16.76 m) span for the floor beams coupled with the 2-story-tall free-spanning core, loads the corner core columns in such a way that the design is primarily governed by gravity design. To study the effect of buckling of chevron-braced diagonals, two types of failure modes were investigated. In the first mode, the buckled diagonal was assumed to have lost the axial load capacity, and in the

(a)

**Figure 1.31** Nations Bank Plaza, Atlanta, Georgia: (a) building elevation; (b) typical framing plan; (c) section.

second failure mode, the lower end of the diagonal was assumed to be absent. The structural members and connections were designed for the resulting overload due to the assumed modes of failure. Also, an attempt was made to proportion the stiffness of the perimeter frame and core bracing such that ductile yielding of the frame precedes the buckling of the diagonals.

To achieve overall economy and take advantage of the increase in stresses allowed under transient lateral loads, the columns are intentionally widely spaced to collect gravity loads from large tributary areas (Fig. 1.33e). The column design is very close to an optimum solution; the design is primarily for gravity loads with the additional loads due to seismic and wind resisted by the one-third increase in allowable stresses.

The structure is designed to remain essentially elastic for an anticipated credible earthquake of magnitude 8.3 on the Richter scale at the nearby San Andreas fault. The strong column weak

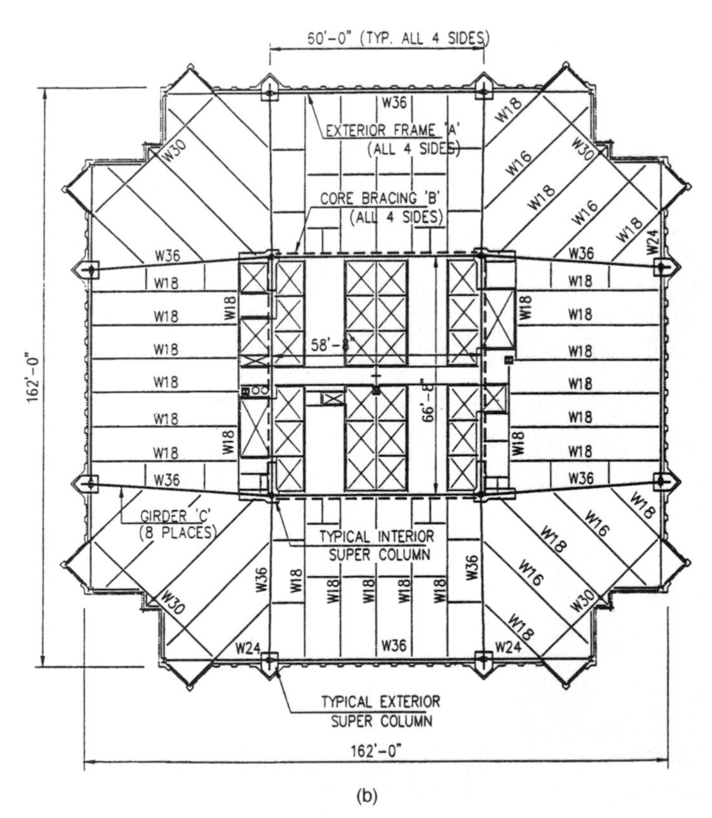


Figure 1.31 *(Continued).*

beam concept is maintained in the design of beam-column assemblies of the perimeter tube.

The sustained dead weight of the structure is 204,000 kips (927,272 kN) with the fundamental period of vibrations $T_x = 7.46$ sec, $T_y = 6.95$ sec, and $T_z = 3.57$ sec. The interaction between the interior braced core and the perimeter ductile frame is typical of dual systems; the shear resistance of core increases progressively from the top to the base of the building. Nearly 50 percent of the overturning is resisted by the core. The maximum calculated lateral deflection at the top, under a 100-year wind is 23 in. (584 mm).

Sixteen critical joints (Fig. 1.37e,f) in the braced frame were mechanically stress relieved by using the Leonard Thompson vibration method of stress relief.

The structure is founded on shale rock with an allowable bearing

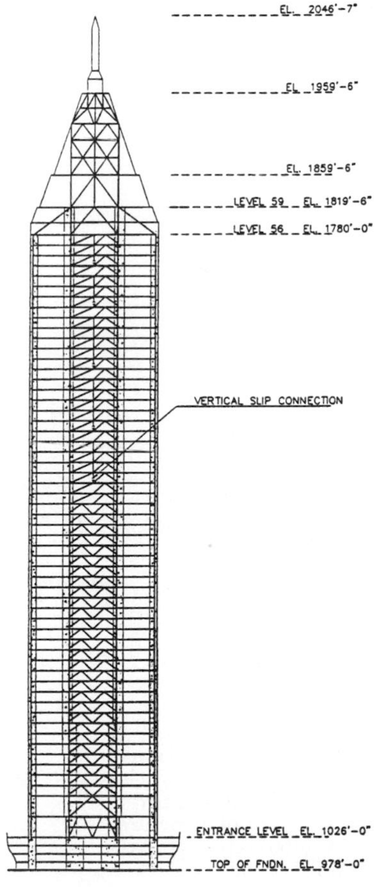


**Figure 1.31** *(Continued)*.

capacity of 7.5 tons/ft$^2$ (720 kPa). The building core is supported on an 11.5 ft (3.5 m) thick concrete mat while a perimeter ring supports the ductile frame. Typical floor framing consists of W24 wide flange composite beams spaced at 13 ft centers, spanning a maximum of 55 ft (16.76 m) from the core to the perimeter. The structural design is by CBM, Engineers, Inc., Houston, Texas. Construction photographs of chevron brace, corner columns at the base, and link beam at the lobby are shown in Figs 1.33g–i.

### 1.3.27 Singapore Treasury Building, Singapore

This 52-story office tower is unique in that every floor in the building is cantilevered from an inner cylindrical, 159 ft (48.4 m) diameter

(a)

**Figure 1.32** Allied Bank Tower, Dallas, Texas: (a) building elevation; (b) building geometry; (c) schematic view of structural system; (d) intersection of diagonals, detail 1, detail 2.

core enclosing the elevator and service areas (Fig. 1.34a). Radial beams cantilever 36 ft (11.6 m) from the reinforced concrete core wall (Fig. 1.34b). Each cantilever girder is welded to a steel erection column embedded in the core wall. To reduce relative vertical deflections of adjacent floors, the steel beams are connected at their free ends by a 1 × 4 in. (25 × 100 mm) steel tie hidden in the curtain wall. A continuous perimeter ring-truss at each floor minimizes relative deflections of adjacent cantilevers on the same floor produced by uneven distribution of live load. Additionally, the vertical ties and the ring beam provide a back-up system for the cantilever beams.

Since there are no perimeter columns, all gravity and lateral loads are resisted solely by the concrete core. The thickness of core walls

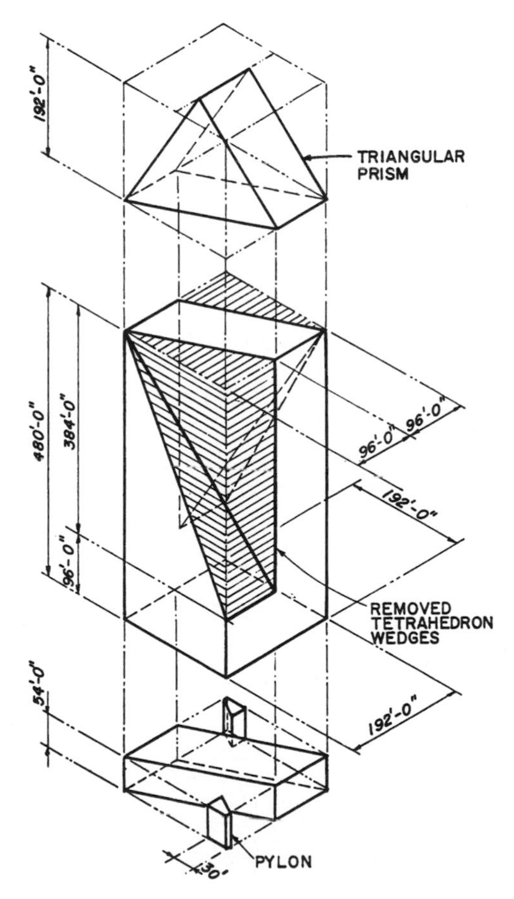


**Figure 1.32** (*Continued*).

varies from a 3.3 ft (1.0 m) at the top to 4 ft (1.2 m) at the sixteenth floor, and remains at 5.4 ft (1.65 m) below the sixteenth floor. The structural engineering is by LeMessurier Consultants, Cambridge, Massachusetts, and Ove Arup Partners, Singapore.

### 1.3.28 City Spire, New York

This 75-story, office and residential tower, with a height-to-width ratio of 10:1 is one of the most slender buildings, concrete or steel, in the world today. The critical wind direction for this building is from the west, which produces maximum across-wind response. Wind studies indicated possible problems of vortex shedding as well as

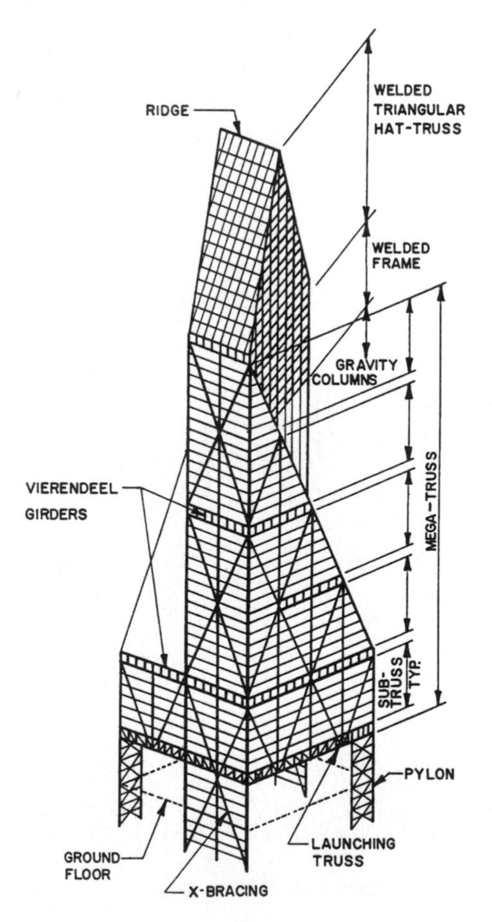


**Figure 1.32** (*Continued*).

occupant perception of acceleration. This possibility was eliminated by adding mass and stiffness to the building.

The main structural system consists of shear walls in the spine connected to exterior jumbo columns with staggered rectangular concrete panels. The structure is subdivided into nine major structural subsystems with many set backs and column transfers as evident from the plans shown in Fig. 1.35a. The structural design is by Robert Rosenwasser Associates, New York.

### 1.3.29 City Corp Tower, Los Angeles

This 54-story tower raises to a height of 720 ft (219.50 m) above the ground level and has a height-to-width ratio of 5.88:1 (Fig. 1.36a). The building has two vertical set backs of approximately 10 ft (3.05 m) at the 36th and 46th floors as shown in the composite floor plan (Fig. 1.36b). As is common to most tall buildings in seismic Zone 4, this building was designed for site-specific maximum probable and maximum credible response spectrums, which represented peak accelerations of 0.28 g and 0.35 g respectively. The corresponding critical damping ratios were 5 and 7.5 percent. The structural system consists of a steel perimeter tube with WTM24 columns

1

2

(d)

**Figure 1.32** *(Continued).*

spaced at 10 ft (3.05 m) centers, and 36 in. (0.91 m) deep spandrels. The columns at the set back levels are carried on 48 in. (1.22 m) deep transfer girders and by the vierendeel action of the perimeter frame. Typical floor plans at the set back levels are shown in Figs 1.38c,d.

The foundation for the tower consists of a 7 ft (2.14 m) deep mat with a 4-story basement for parking. The structural design is by John A. Martin & Associate, Inc., Los Angeles.

### 1.3.30 Cal Plaza, Los Angeles

The project consists of a 52-story office tower rising above a base consisting of lobby and retail levels, and six levels of subterranean parking (Fig. 1.37a). A structural steel system consisting of a ductile

(a)

**Figure 1.33** First Interstate World Center: (a) building elevation; (b) building key plan showing column transfers; (c) composite plan; (d) structural system; (e) floor framing plan; (f) location of stress-relieved joints; (g) chevron brace connection; (h) connection of corner column and braces at base; (i) link beam connection at lobby.

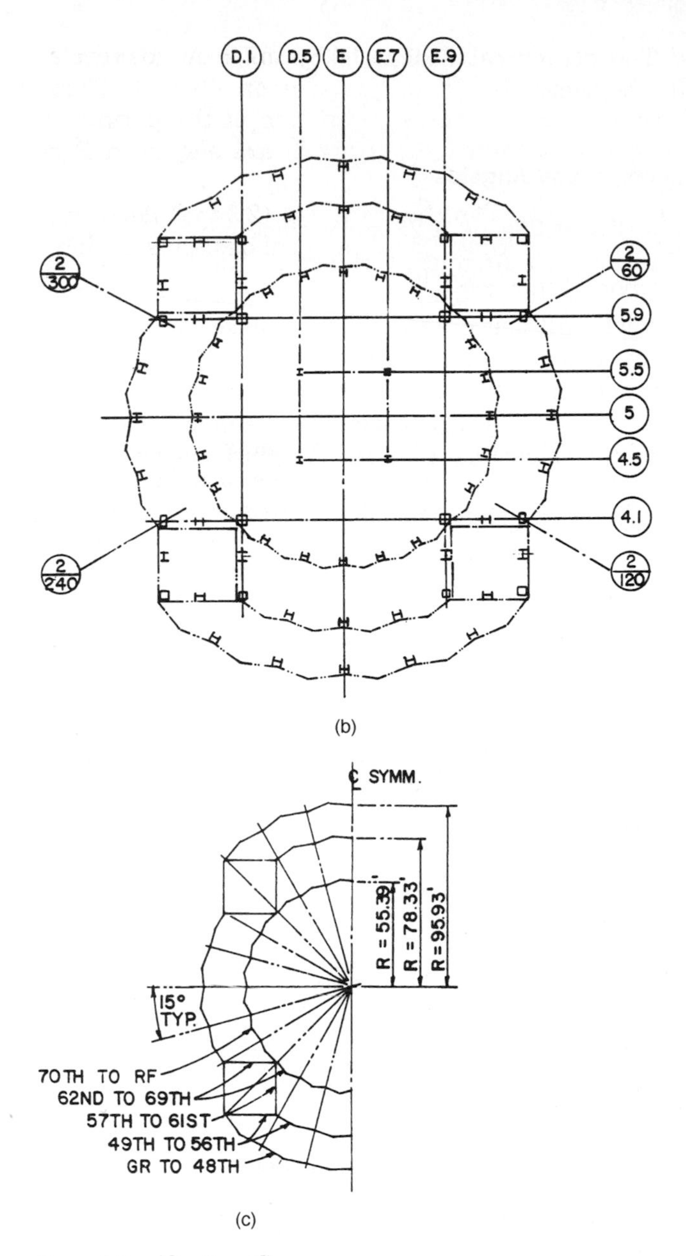


**Figure 1.33** (*Continued*).

moment-resisting frame at the perimeter resists the lateral loads. The parking areas outside of the tower consist of a cast-in-place concrete system with waffle slab and concrete columns. Figure 1.37b shows a typical midrise floor plan for the tower with sizes for typical framing elements. The structural design is by John A. Martin & Associates, Inc., Los Angeles.

### 1.3.31 MTA Headquarters, Los Angeles

MTA Head Quarters (Fig. 1.38a) is a 28-story office building located east of the Union Station, Los Angeles, California. The building has a gross area of 622,000 $ft^2$ (57,785 $m^2$). It has a four-level subterranean structure, which will serve as a common base for the MTA Tower, and two future office buildings. The basement levels, which serve as a parking garage, extend beyond the footprint of the tower. The construction for the parking structure consists of precast reinforced concrete columns and girders with cast-in-place concrete slab.

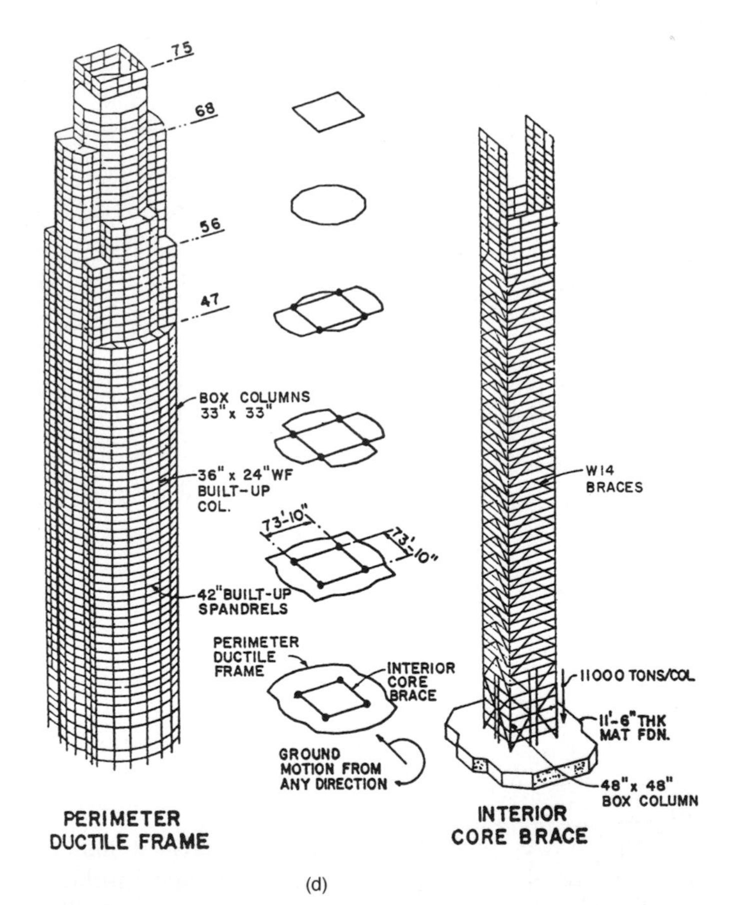


(d)

**Figure 1.33** *(Continued).*

The plaza level underneath the tower consists of a composite floor system with a $4\frac{1}{2}$ in. (114 mm) normal-weight concrete topping on a 3 in. (76 mm) deep, 18 gauge composite metal deck. The metal deck spans between composite steel beams spaced typically at 7 ft 6 in. (2.29 m) outside of the tower which has heavy landscape, and 10 ft (3.04 m) on centers within the tower footprint.

The building is essentially rectangular in plan, 118 ft × 165 ft (36 m × 50.3 m) with a slight radius on the short faces. The building height is 400 ft (122 m) with a fairly low height-to-width ratio of 3.39. Typical floor framing shown in Fig. 1.38b consists of 21 in. (0.54 m) deep composite beams spanning 41 ft (12.5 m) from the core to the exterior. A 3 in. (76 mm) deep metal deck with a $3\frac{1}{4}$ in. (83 mm) thick lightweight concrete topping completes the floor system.

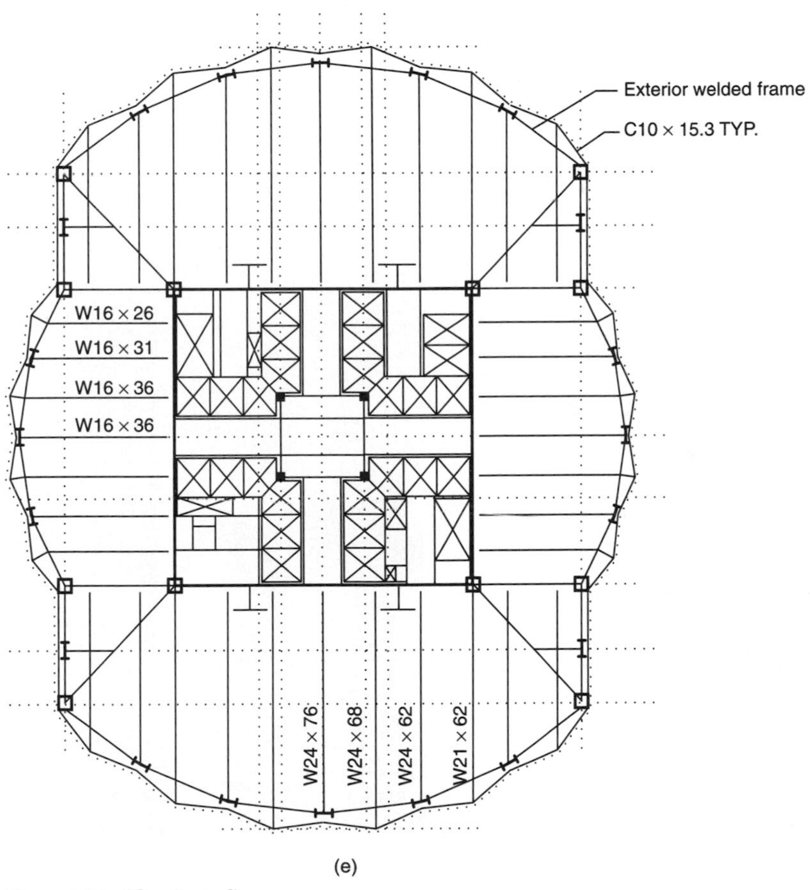


**Figure 1.33** (*Continued*).

The lateral system consists of a perimeter tube with widely spaced columns tied together with spandrel beams. The exterior columns on the broad faces vary from W30 × 526 at the plaza to W30 × 261 at the top. The spandrels vary from a WTM36 × 286 at the plaza level to W36 × 170 at the top floors. The columns on the curvilinear faces are built up, 34 × 16 in. (0.87 × 0.40 m), box columns while 24 × 24 in. (0.61 × 0.61 m) box columns are used at the corners. Plates varying in thickness from 4 in. (102 mm) at the bottom to 1 in. (25 mm) at the top are used for the built-up columns. As is common for most steel buildings in seismic zones 3 and 4, 50 ksi steel for columns, and 36 ksi steel for spandrels are used to satisfy the strong-column-weak-beam requirement. The salient seismic design parameters are as follows.

Building period $x$-direction = 4.57 sec

Building period $y$-direction = 4.40 sec

Wind shear $x$-direction = 1580 kips (7028 kN)

Wind shear $y$-direction = 2078 kips (9243 kN)

1991 UBC base shear for strength check = 2894 kips (12,873 kN)

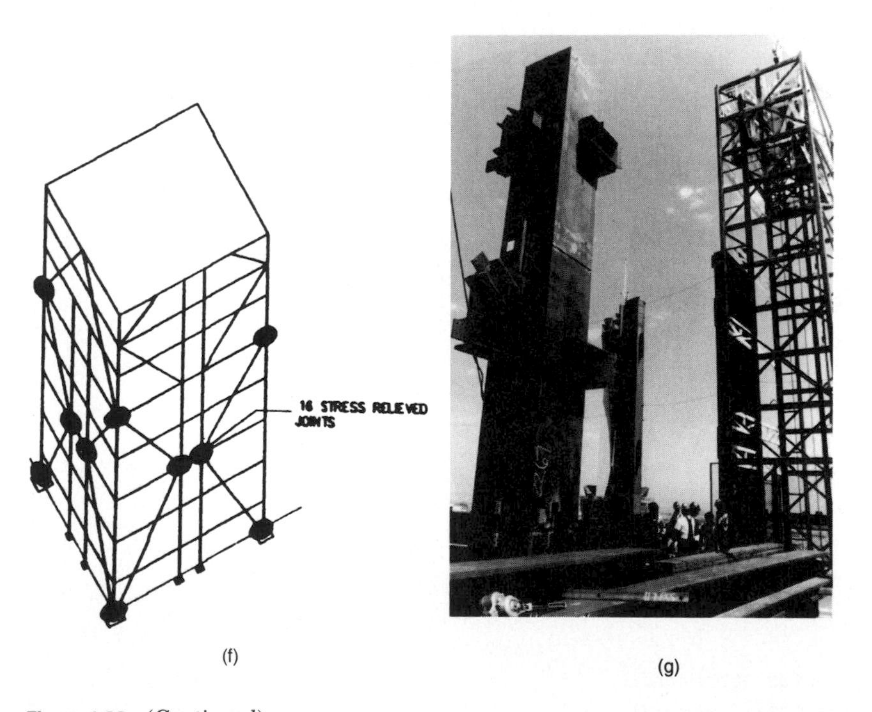


(f) (g)

**Figure 1.33** (*Continued*).

1991 UBC base shear for deflection check = 1497 kips (6659 kN)

Unscaled UBC base shear = 5163 kips (22,966 kN) $x$-direction

Unscaled base shear Maximum Credible Earthquake with 7 percent damping 6460 kips (28,735 kN) $x$-direction

Building weight above shear base (plaza level) = 77,212 kips (343,456 kN)

Building floor-to-floor height—13 ft 4 in. (4 m)

Unit density of building = 9 lbs per cubic foot (144 kg/m$^3$)

Unit weight of building = 118 psf (5650 Pa)

Steel Tonnage = Approximately 24 psf (1149 Pa)

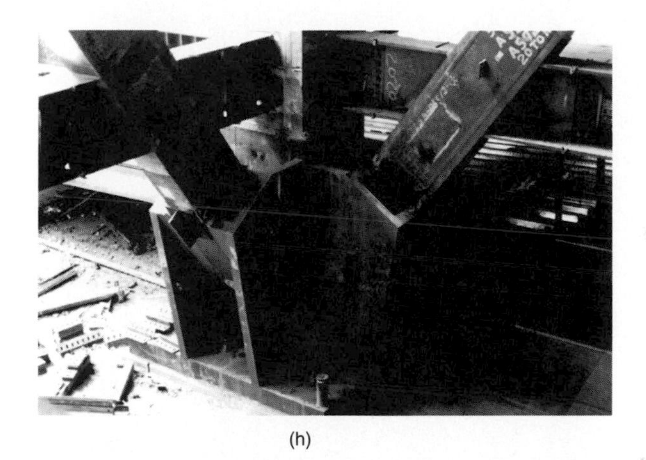

(h)

(i)

**Figure 1.33** *(Continued)*.

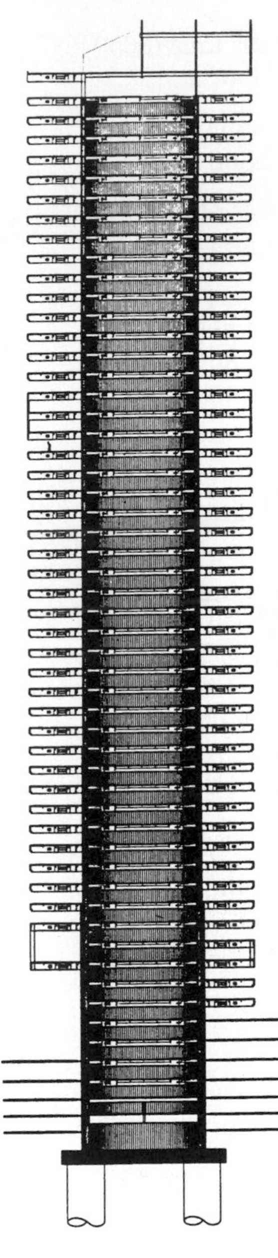

(a)

**Figure 1.34** Singapore Treasury Building: (a) schematic section; (b) typical floor framing plan.

The building architecture is by McLarand Vasquez & partners Inc., while the structural engineering is by John A. Martin & Asso. Inc., both of Los Angeles, California.

### 1.3.32 The 21 Century Tower

The 21 Century Tower (Figs 1.39a,1,2,3) when built will be a landmark 50-story, 1,000,000 sq ft (100,000 sq m) building in

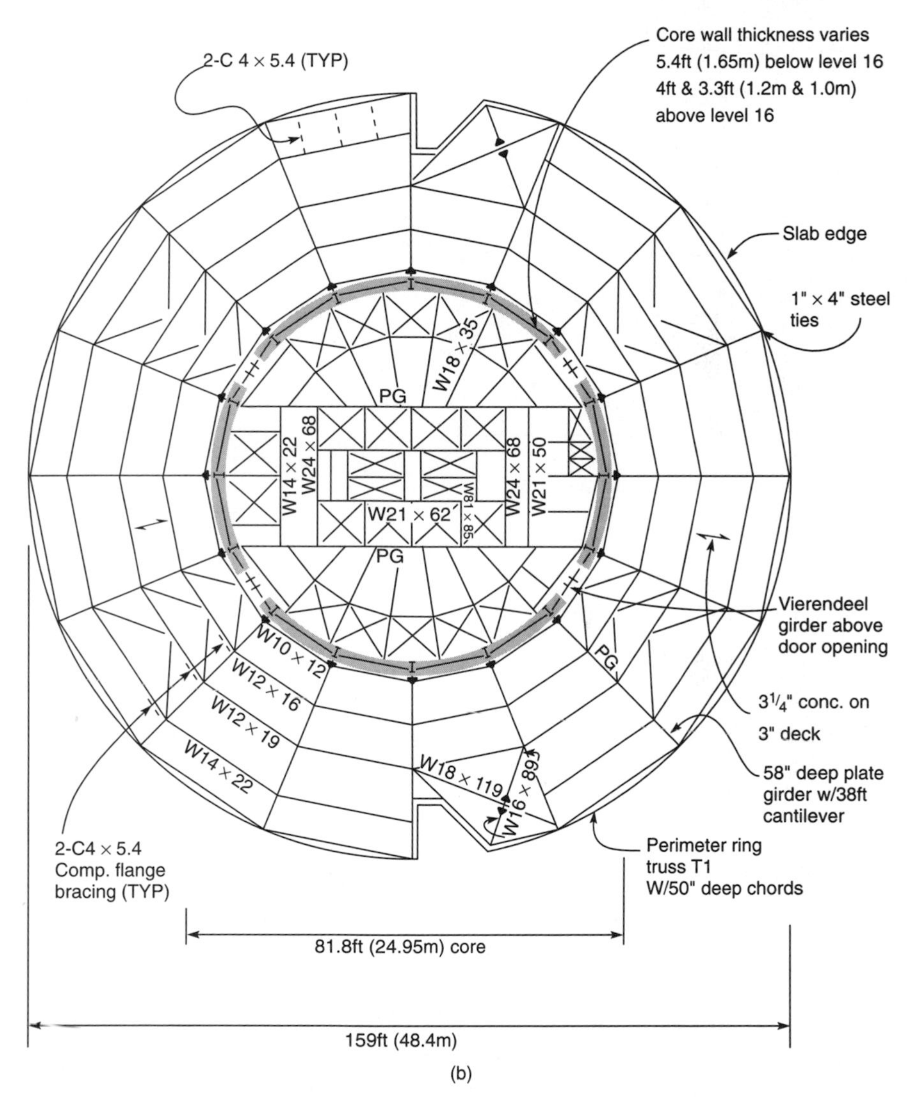


**Figure 1.34** (*Continued*).

Shanghai. Designed by the architectural firm of Murphy/Jahn Inc., Chicago, the exterior form of the building is that of a rectangular tower; however the structure is made unique by a setback base and a

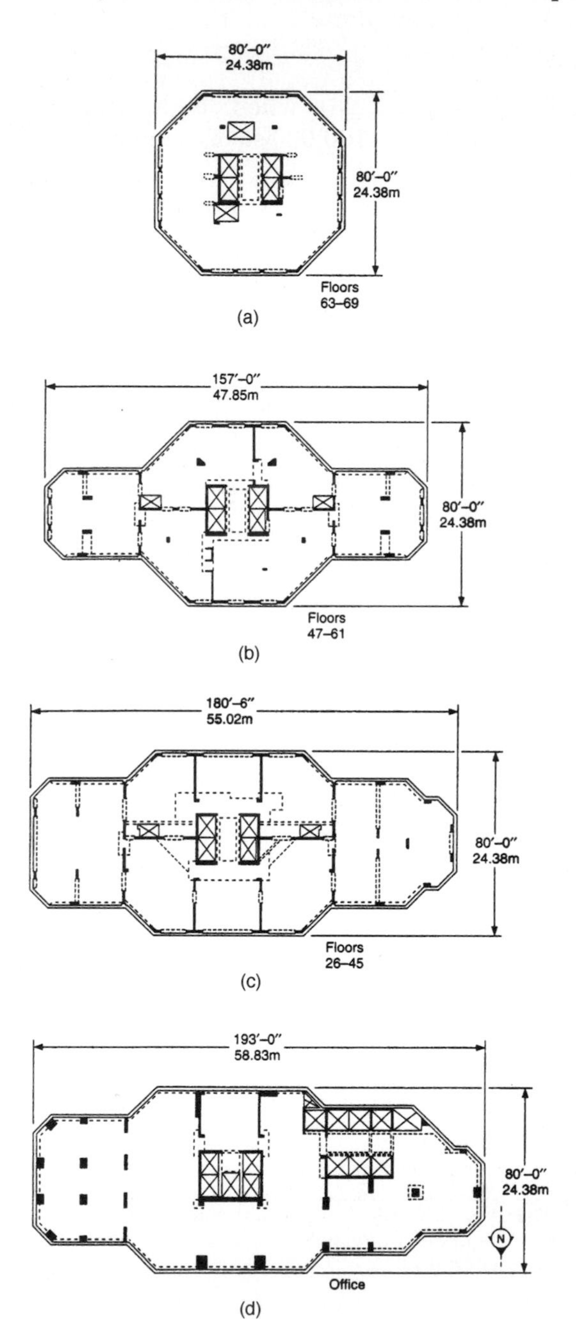


**Figure 1.35** City Spite New York.

series of 9-story high wedge-shaped atria or "winter gardens", which run the fullheight of the tower. These features effectively remove one corner column over the full height of the building, and remove the opposite column at both the top and bottom of the tower. The structure therefore takes the form of a stack of 9-story high chevrons. Other unusual architectural features include a cable-suspended skylit canopy roof over the podium, exposed rod-truss curtain wall supports at the winter gardens, and exposed truss-stringer stairs. The building façade is a futuristic expression of transparency, color and structure, each element supporting and enforcing the other so that the architecture and structure are integrated into a single entity. Structural elements, most notably the 9-story high superbraces on

(a)

**Figure 1.36** City Corp Tower, Los Angeles: (a) building elevation; (b) composite plan; (c) 36th floor framing plan; (d) 47th–52nd floor framing plan. Architects: Cesar Pelli Asso. Inc. Structural engineers: John A. Martin & Asso. Inc., Los Angeles.

each face of the tower, are boldly expressed in red, while blue and green solar glazing covers all office spaces. The winter gardens and the podium are enclosed in clear glass.

Although the building is expressed as a square, it is punctuated by a series of four 9-story high wedge-shaped winter gardens

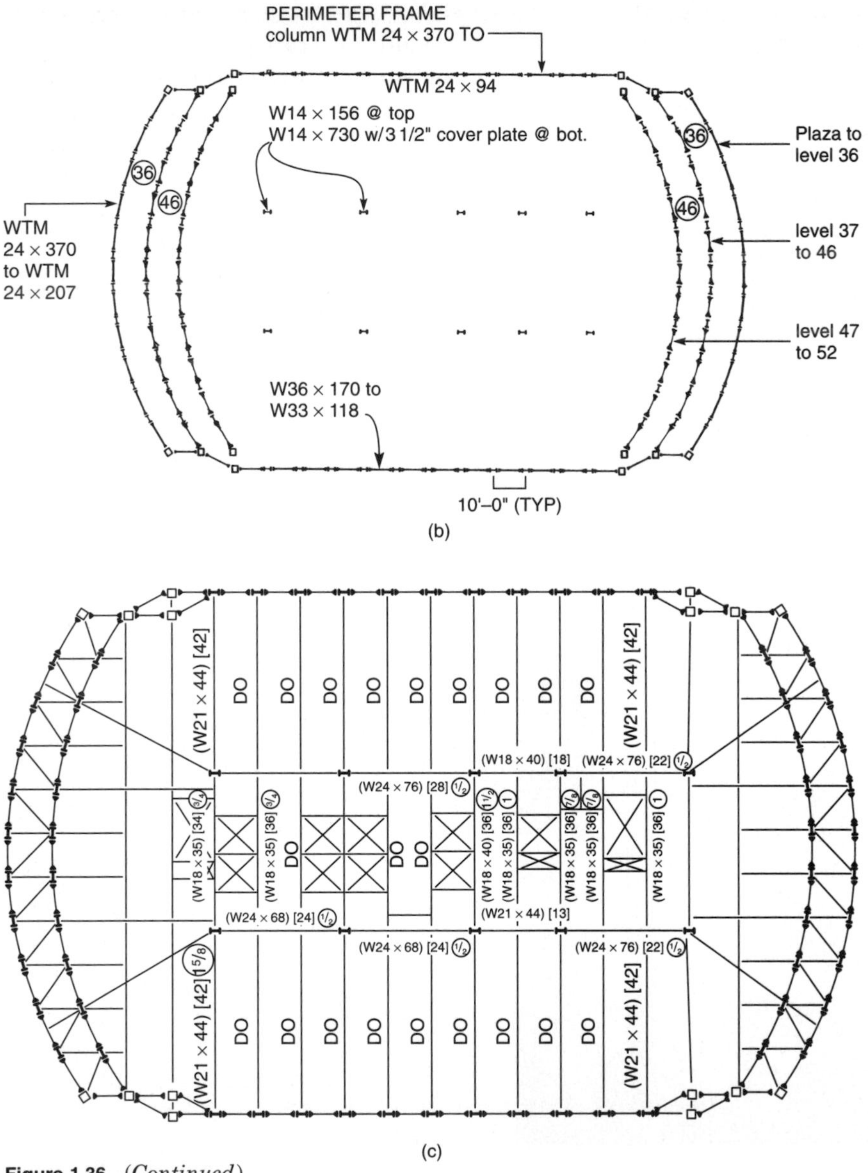


**Figure 1.36** *(Continued).*

cut into the northeast corner and two more at the southwest corner (Fig. 1.39b). The winter gardens have the effect of dividing the tower, both visually and structurally, into a stack of five modules outlined by the superbraces. At the lowest module, the northeast corner column is eliminated entirely. As a result the tower has only a single axis of symmetry, which passes at 45 degrees through the corners; in addition, nine different floor plans are requried within each module. Plans at two representative floors are shown in Figs 1.39c,d.

The building site consists primarily of sandy sedimentary soils with ground water level only 1 ft 8 in. (0.5 m) below grade. The basement substructure is a 3-story deep concrete box, roughly triangular in plan, with typical exterior walls 1 ft 8 in. (0.5 m) in thickness. A 6 ft 3 in. (1.9 m) wide zone between the basement walls and property line is provided for a slurry wall and backfill. The maximum excavation depth for the substructure is 44 ft (13.4 m), and is supported by a combination of 24 in. (600 mm) driven steel pipe

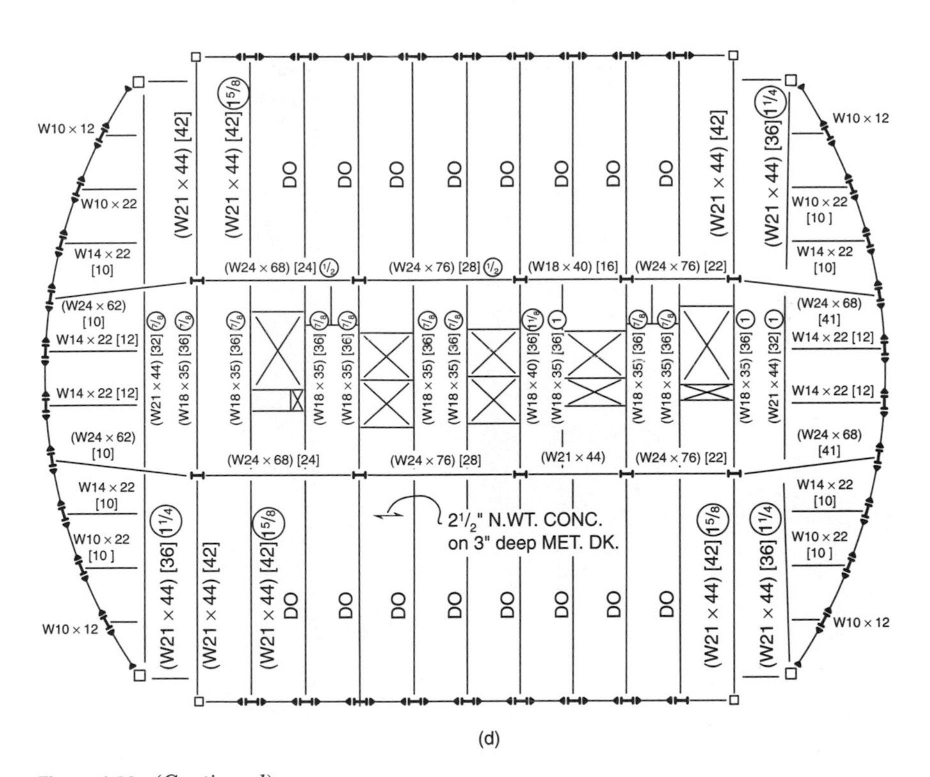


**Figure 1.36** *(Continued).*

piles below the tower and cast-in-place concrete piles of same diameter below the plaza and podium. The piles have a maximum length of 40 m, and are installed prior to the start of the basement excavation using a reusable steel driving extension. Driving from the ground level in this way eliminates the need to move and operate pile drivers on the extremely wet soils found at the bottom of the excavation; in addition the piles will pin the soil and prevent heaving of the subgrade while overburden is being removed. The piles support a 5 ft (1.5 m) thick concrete mat slab. Beneath the tower the mat is strengthened by a grid of column pedestals and inverted grade beams topped by a concrete slab: the whole system forming a complex of

(a)

**Figure 1.37** Cal Plaza, Los Angeles: (a) building elevation; (b) mid-rise floor framing plan.
Architects: Arthur Erickson Inc., Structural engineers: John A. Martin & Asso. Inc., Los Angeles.

story high cells which also serve as water tanks. This cellular system extends beneath the plaza some distance to the north and south of the tower. Beneath the remainder of the plaza the slab is covered by a 14 in. (350 mm) thick layer of crushed rock and a 6 in. (150 mm) slab at the lowest level, the rock serving as a capillary break for any water which might infiltrate through the mat. Under some portions of the substructure the hydrostatic uplift pressure exceeds the structural dead load; therefore the piles in these areas are designed as friction piles to prevent uplift.

Below the tower the framing consists of concrete-encased steel beams supporting cast-in-place concrete floor slabs. Other structural slabs are cast-in-place flat slabs with drop-panels. The design of the tower is driven by the unique architectural treatment to the building envelope, by the arrangement of the winter gardens, and by the high wind speeds of up to 115 mph (185 km/hr) resulting in design pressures as high as 136 psf (6.5 kPa).

The structural solution for the tower consists of a superbrace system on the exterior skin supplemented by an eccentrically-braced interior service core. The superbrace system makes maximum use

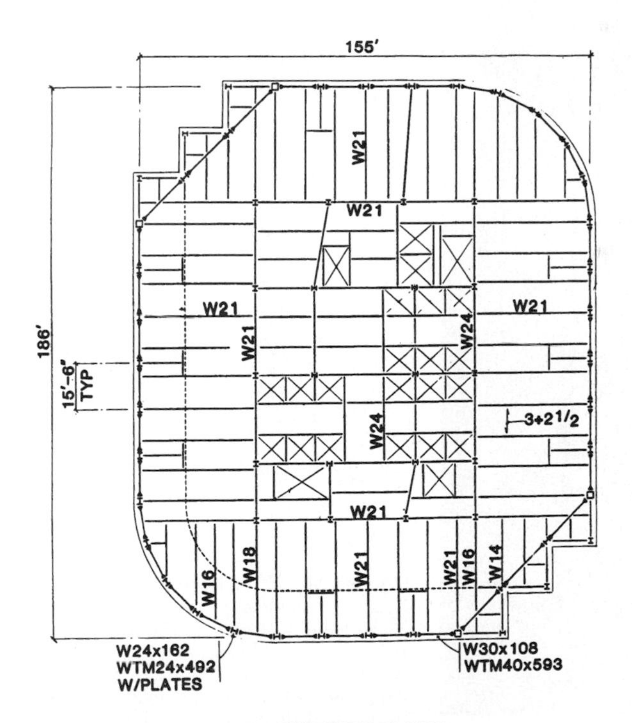


(b) **MID-RISE FLOOR**

**Figure 1.37** *(Continued).*

of the total tower width, giving optimum resistance to wind loading, maximum economy and minimum structural intrusion into interior spaces. Schematic exterior brace elevations are shown in Figs 1.39e–h. Structural action in the primary columns and braces due to lateral loads are shown in Fig. 1.39i for the lower three modules.

The braces generally consist of heavy W14 sections, field-spliced

(a)

**Figure 1.38** MTA headquarters, Los Angeles: Architects: McLarand Vasquez & Partners, Inc., Structural engineers: John A. Martin & Asso. Inc., Los Angeles. (a) building elevation; (b) typical floor framing plan.

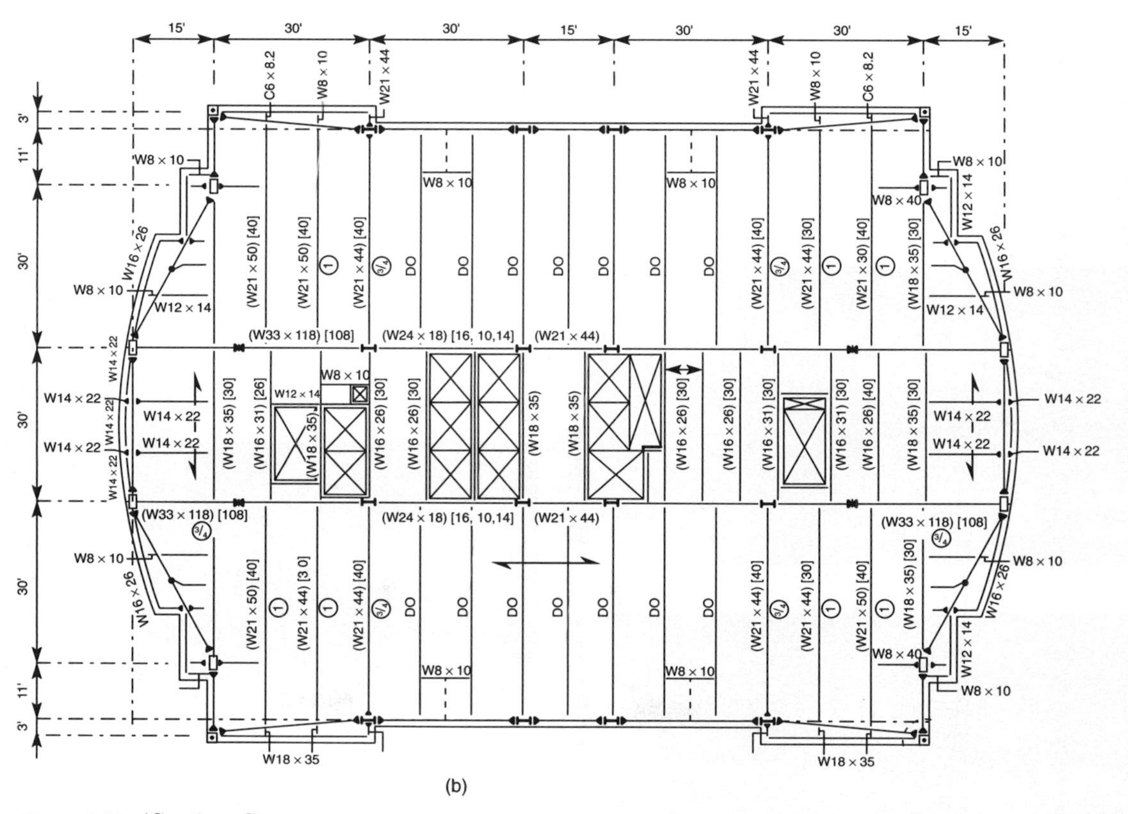


**Figure 1.38** *(Continued).*

(a)

**Figure 1.39** 21st Century Tower: Architects: Murphy/Jahn Inc., Chicago, Structural engineers: John A. Martin & Asso. Inc., Los Angeles, Martin & Huang, International, Los Angeles: (a) model photographs 1, 2 and 3; (b) bracing system; (c) framing plan; levels 19, 28 and 37; (d) framing plan, levels 20, 29 and 37; (e) exterior frame elevation, north face; (f) exterior frame elevation, west face; (g) exterior frame elevation, east face; (h) exterior frame elevation, south face; (i) structural action in primary columns and braces; (j) typical interior core bracing.

every three floors and connected at their ends to square steel box columns. The braces also act as inclined columns, and carry the vertical loading from all secondary columns above them. This arrangement maximizes the vertical loading carried by the box columns and minimizes corner uplift. Columns vary in size from 20 to 24 in. (520 to 610 mm) with plate thicknesses up to 5 in. (130 mm). The braces are arranged in five 9-story high vees with a one-bay gap in the middle of each building face; stiffness in the gap is provided by one-bay wide rigid frame. Panel points for the superbraces occur at the ground, 5th, 15th, 24th 33rd 42nd and roof levels. Horizontal members and diaphragms at these levels are stiffened to transfer horizontal brace forces. Service core bracing provides additional overall stiffness and gives lateral support to floors between the superbrace panel points. For architectural reasons, braces at the center bay of each core-face are eccentric (Fig. 1.39j). Although most

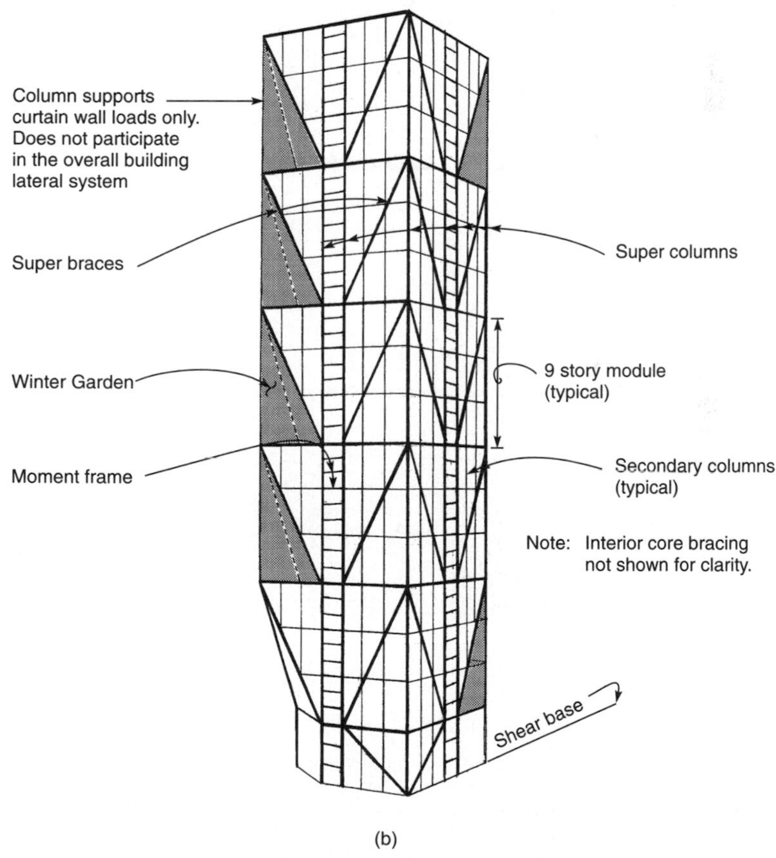


**Figure 1.39** *(Continued).*

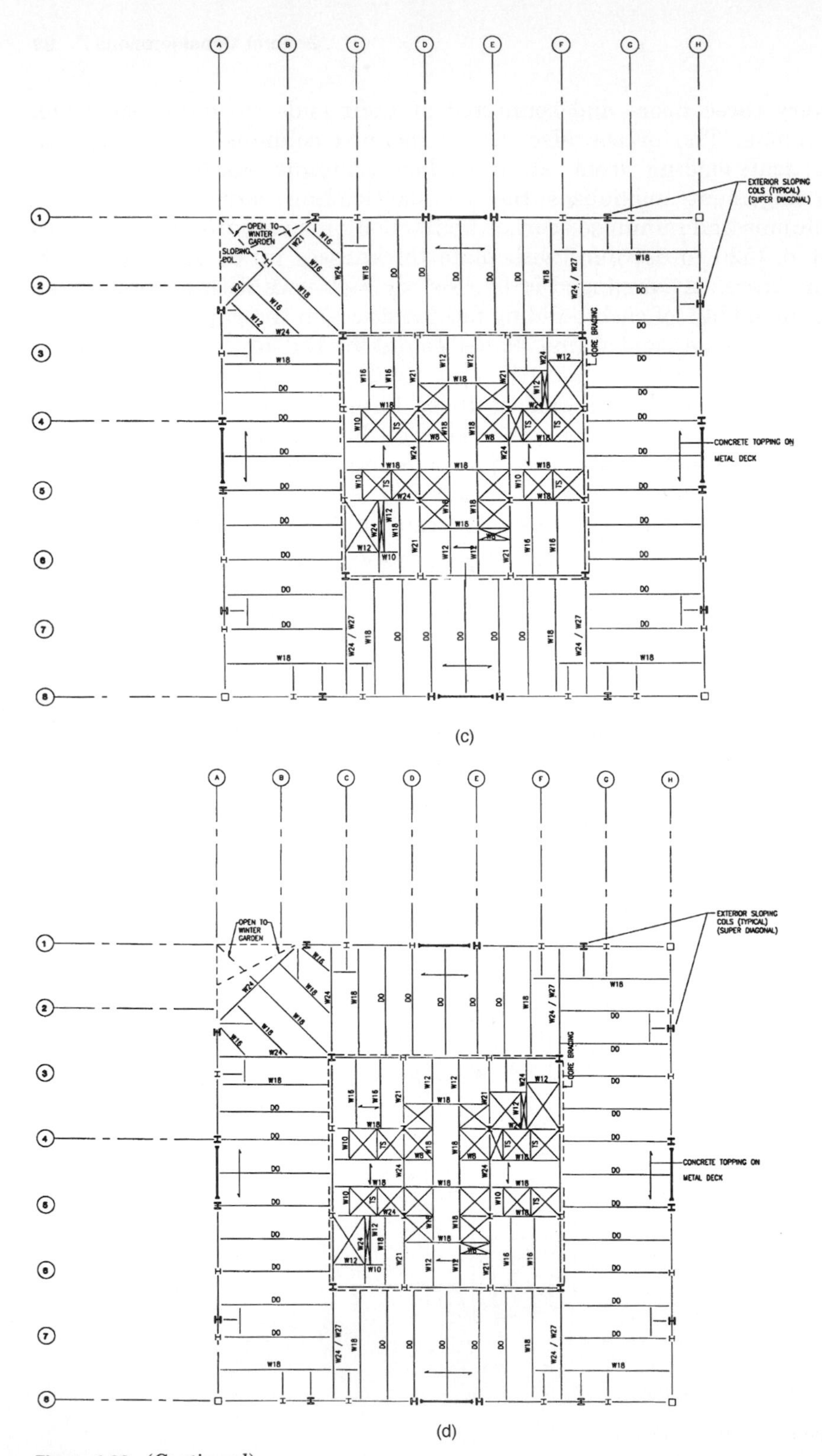


**Figure 1.39** (*Continued*).

lateral loading is transferred at the ground level to the shear walls, core bracing is extended to the foundation of the substructure. The numerous corner cutouts of the tower structure effectively rotate the principal axes of the structure by 45%, to pass through the corner columns. The lowest two modes of vibration of the tower are single-curvature bending through these axes; the third mode is torsional. Period of the first three modes are 4.93, 4.62 and 2.12 sec respectively.

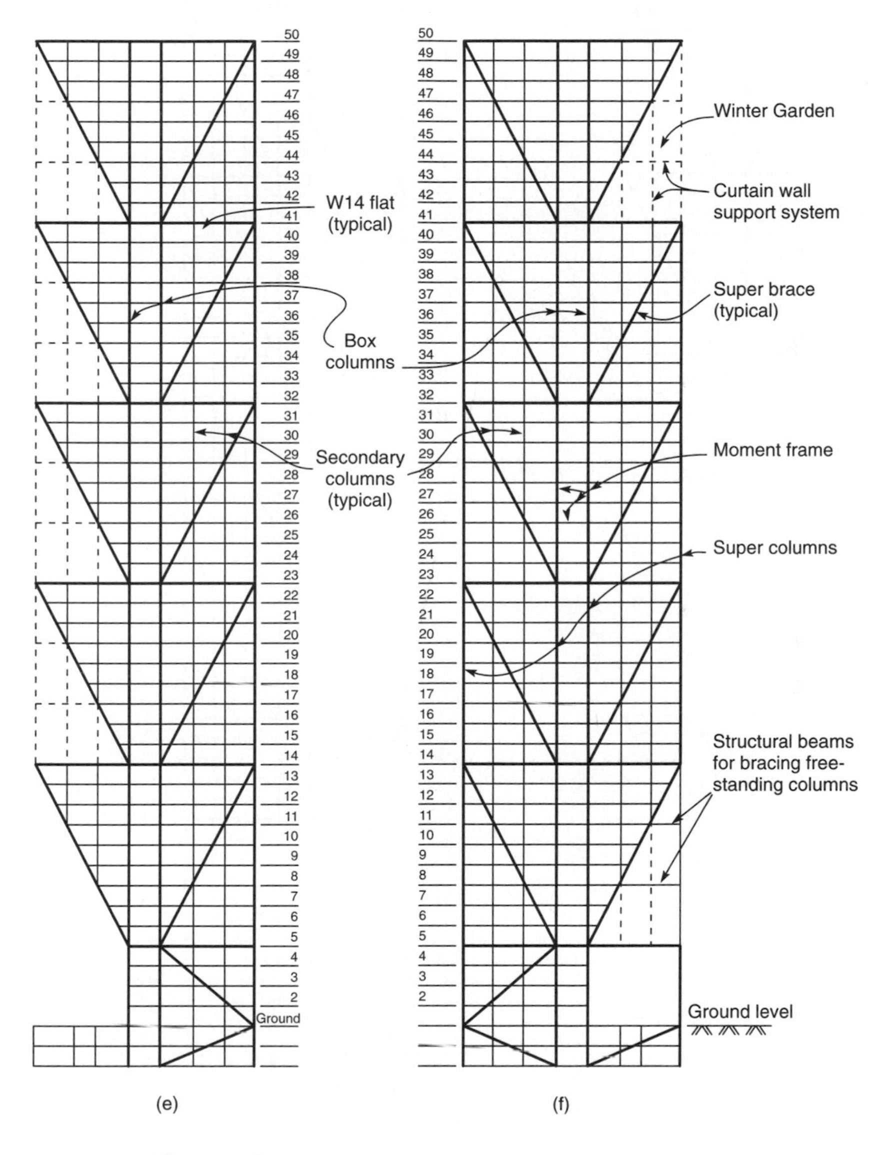


**Figure 1.39** (*Continued*).

The 21 Century Tower is an example of how the superbrace structural frame can be integrated with the architectural theme of the building to create an impressive architectural statement. By concentrating the primary lateral resistance of the building in a relatively small number of perimeter members, both high economy and maximum lateral strength are achieved. This is borne out by the

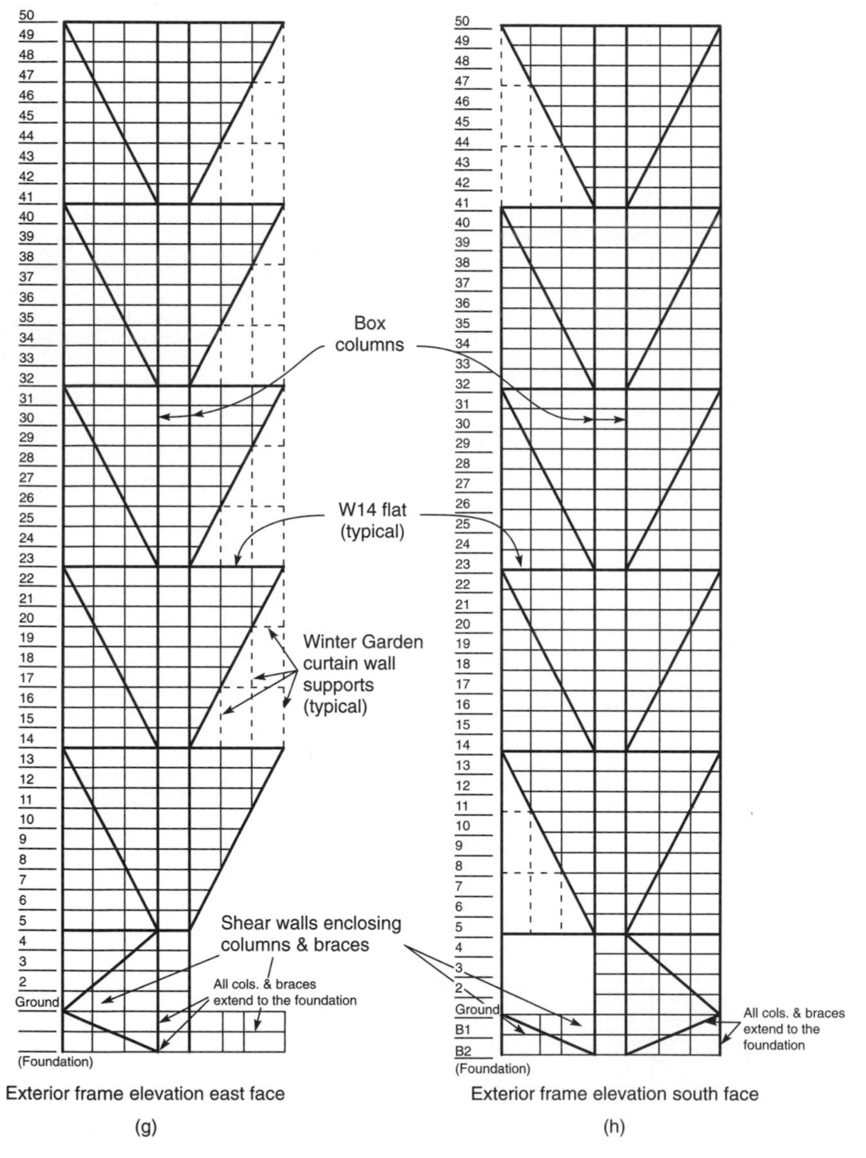


**Figure 1.39** *(Continued).*

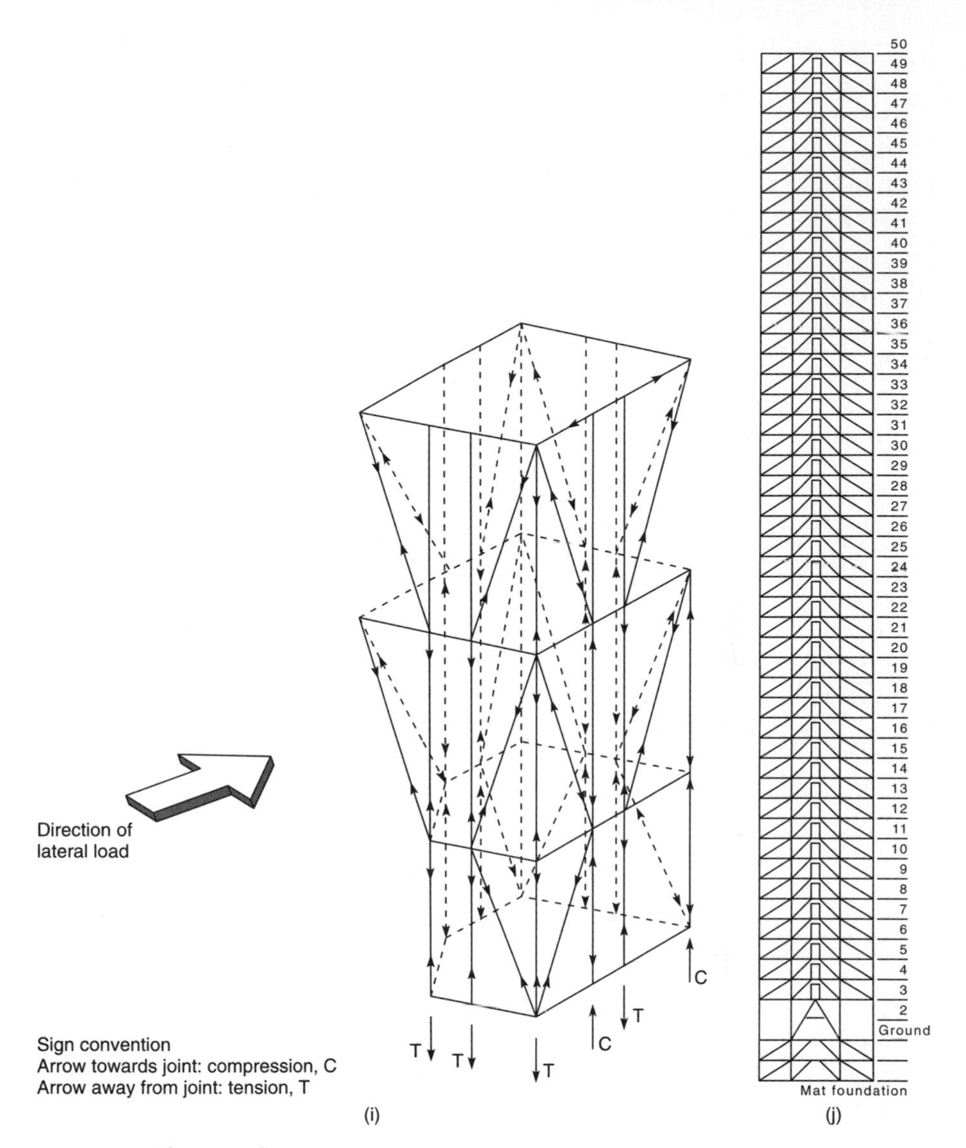


**Figure 1.39** *(Continued)*.

buildings average steel weight of 29 psf (142 kg/m$^2$). The conceptual and preliminary design is by structural engineers John A. Martin & Associates; working drawings are by Martin & Huang, International, both of Los Angeles, California.

Chapter

# 2

# Wind Effects

## 2.1 Design Considerations

Windy weather poses a variety of problems in new skyscrapers, causing concern for building owners and engineers alike. The forces exerted by winds on buildings increase dramatically with the increase in building heights. Static wind effects increase as the square of a structure's height, and the high-rise buildings of the 1980s and 1990s, which are at times 1000 ft (305 m) tall, must be 25 times stronger than the typical 200 ft (61 m) building of the 1940s. Moreover, the velocity of wind increases with height, and the wind pressures increase as the square of the velocity of wind. Thus, wind effects on a tall building are compounded as its height increases.

In designing for wind, a building cannot be considered independent of its surroundings. The influence of nearby buildings and land configuration can be substantial. The swaying at the top of a high-rise building caused by wind may not be seen by a passerby, but its effect may be of concern to those occuping the top floors. There is scant evidence that winds, except in the case of a tornado or hurricane, have caused major structural damage to new high rises. However, a modern skyscraper, which uses lightweight curtain walls, dry partitions, and high-strength materials, is more prone to wind motion problems than the early skyscrapers, which had the weight-advantage of masonry partitions, heavy stone façades, and massive structural members.

To be sure, all buildings sway during windstorms, but the motion in earlier tall buildings with locked-in gravity loads from their enormous weight is usually imperceptible and certainly has not been a cause for concern. Structural innovations and lightweight construction technology have reduced the stiffness of modern high-rise buildings. Wind action has become a major concern for the designer of today's high-rises. In buildings prone to wind motion problems, objects in a room may vibrate, catching the occupant's eye. Doors and chandeliers swing, pictures lean, and books fall sideways off shelves. If the building has a twisting action, the occupants may get an illusory sense that the world outside is moving, creating symptoms of vertigo and disorientation. In more violent storms, windows can blow out, creating safety problems to pedestrians below. Sometimes, strange and often frightening noises are heard by the occupants as the wind shakes elevators, strains floors and walls, and whistles around the sides.

Keeping the movements in the upper levels of the building to acceptable human tolerances is the goal. Exactly what this tolerance is has been very difficult to assess. Engineers today try to design structures that have inherent stiffness achieved through engineering techniques rather than depending upon dead weight to stabilize the structure. In spite of all the mathematical and engineering sophistication possible with computers, wind has still managed to dodge complete quantitative analysis, mainly because of two major problems. First, unlike dead loads which are permanent and unchanging and live loads which are tacitly assumed to change slowly, wind loads change rapidly and even abruptly, creating effects much larger than if the same loads were applied gradually. The other problem is limiting of building accelerations below human perception. The true complexity of wind and acceptable human tolerance have just begun to be understood. There is still a need for understanding the nature of wind and its interaction with a tall building, with particular reference to allowable deflections and comfort of occupants. In designing tall buildings to withstand wind forces, the following are important factors that must be considered:

1. Strength and stability requirements of structural system.
2. Fatigue in structural members and connections caused by fluctuating wind loads.
3. Excessive lateral deflection that may cause cracking of partitions and external cladding, misalignment of mechanical systems and doors, and possible permanent deformations.
4. Frequency and amplitude of sway that can cause discomfort to occupants.

5. Possible buffeting that may increase the magnitudes of wind velocities on neighboring buildings.
6. Effects on pedestrians.
7. Annoying acoustical disturbances.
8. Resonance of building oscillations with vibrations of elevator hoist ropes.

## 2.2 Nature of Wind

### 2.2.1 Introduction

*Wind* is the term used for air in motion and is usually applied to the natural horizontal motion of the atmosphere. Motion in a vertical or near vertical direction is called a *current.* Winds are produced by differences in atmospheric pressure, which are primarily attributable to differences in temperature. These temperature differences are caused largely by unequal distribution of heat from the sun, and the difference in the thermal properties of land and ocean surfaces. When temperatures of adjacent regions become unequal, the warmer and thus lighter air tends to rise and flow over the colder, heavier air. Winds initiated in this way are modified by rotation of the earth.

Movement of air near the surface of the earth is three-dimensional, with a horizontal motion much greater than the vertical motion. Motion of air is created by solar radiation, which generates pressure differences in air masses. Vertical air motion is of importance in meteorology but is of less importance near the ground surface. The surface boundary layer involving horizontal motion of wind extends upward to a certain height above which the horizontal airflow is no longer influenced by the ground effect. The wind speed at this height is called the *gradient wind speed* and generally occurs at an altitude greater than 1500 ft (458 m). In this boundary layer is precisely where most of the human activity is conducted, and therefore how the wind effects are felt within this zone is of great concern in engineering.

Although one cannot see the wind, it is a common observation that its flow is quite complex and turbulent in nature. Imagine taking a walk outside on a reasonably windy day. You no doubt experience the constant flow of wind, but intermittently you will experience sudden gusts of rushing air. This sudden variation in wind speed is called gustiness or turbulence. The up-and-down fluctuations of speed about the mean velocity that occur over long periods of time due to solar energy cycles are of little importance in engineering, but the shorter-period peaks resulting from surface-generated turbulence are of great importance for the human activity in the earth's boundary layer.

In describing global circulation and specific recurrences of certain types of wind, modern meteorology relies on wind terminology used by early long-distance sailors. For example, terms like trade winds and westerlies were used by sailors who recognized the occurrence of steady winds blowing for long periods of time in the same direction. A broad indication of the flow of wind in the lower levels resulting from the general circulation of the atmosphere can be obtained by considering the interface between the cold winds from the polar regions and the westerlies.

Near the equator, the lower atmosphere is warmed, by the sun's heat. The warm air rises, depositing much precipitation and creating a uniform low-pressure area. Into this low-pressure area, air is drawn from the relatively cold high-pressure regions from northern and southern hemispheres, giving rise to trade winds between the latitudes of 30° from the equator. The air going aloft flows counter to the trade winds to descend into these latitudes, creating a region of high pressure. Flowing northward and southward from these latitudes in the northern and southern hemispheres, respectively, are the prevailing westerlies, which meet the cold dense air flowing away from the poles in a low-pressure region characterized by stormy variable winds. It is this interface between cold, dense air and warm, moist air which is of main interest to the television meteorologists of northern Europe and North America.

Air above hot earth expands and rises. Air from cooler areas such as the oceans then floats-in to take its place. The process is called *circulation.* Two kinds of circulation produce wind: (i) general circulation extending around the earth; and (ii) smaller secondary circulations producing local wind conditions. Figure 2.1 shows a simplified theoretical model of the circulation of prevailing winds which result from the general movement of air around the earth. There are no prevailing winds within the equatorial belt, which lies roughly between latitudes 10° S and 10° N. Therefore, near the equator and up to about 700 miles (1127 km) on either side of it there lies a low-pressure belt in which the air is hot and calm. The air in this region rises above the earth instead of moving across it, creating a region of relative calm called the *doldrums.* In both hemispheres, some of the air that has risen at the equator returns to the earth's surface at about 30° latitude, producing no wind. These high-pressure areas are called *horse latitudes,* possibly because many horses died on the sailing ships that got stalled because of lack of wind. The winds that blow between the horse latitudes and the doldrums are called *trade winds* because sailors relied on them in sailing trading ships. The direction of trade winds is greatly modified by the rotation of the earth as they blow from east to west. Two other kinds of winds that result from the general circulation of the atmosphere are called the *prevailing winds* and the *polar easterlies.* The prevailing winds blow in two belts bounded by the horse latitudes and 60° north and south of the equator. The polar easterlies blow in the two belts between the poles and about 60° north and south of the equator. Thus the moving surface air produces six belts of winds around the earth as shown in Fig. 2.1.

### 2.2.2 Types of wind

Of the several types of wind that encompass the earth's surface, winds that are of interest in the design of buildings can be classified into three major types: the prevailing winds, seasonal winds, and local winds.

1. *The prevailing winds.* Surface air moving from the horse latitudes toward the low-pressure equatorial belt constitutes the prevaling winds or trade winds. In the northern hemisphere,

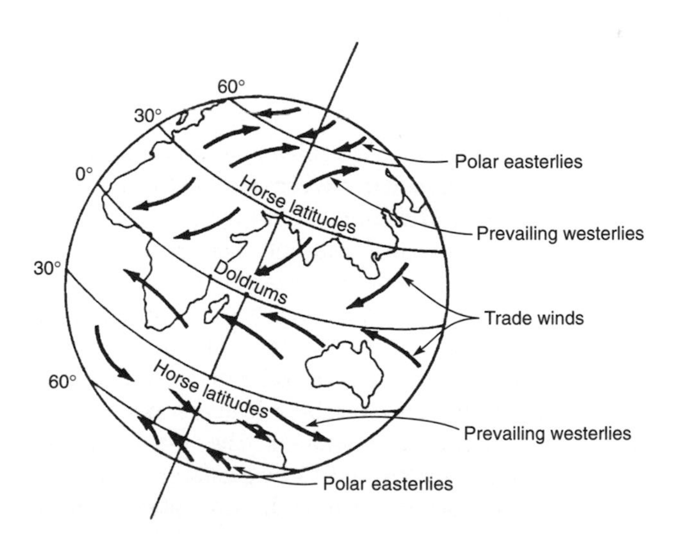


**Figure 2.1** Circulation of world's winds.

the northerly wind blowing toward the equator is deflected by the rotation of the earth to become northeasterly and is known as the northeast trade wind. The corresponding wind in the southern hemisphere is called the southeast trade wind.

On the polar regions of the horse latitudes, the atmospheric pressure diminishes and the winds moving toward the poles are deflected by the earth's rotation toward the east. Because winds are known by the direction from which they blow, these winds are known as the prevailing westerlies. The winds from the poles, particularly in the southern hemisphere, are deflected to become the polar easterlies. In comparison to the westerlies, the trade winds and the polar easterlies are shallow, and above a few thousand feet they are generally replaced by westerlies.

2. *The seasonal winds.* The air over the land is warmer in summer and colder in winter than the air adjacent to oceans during the same seasons. During summer, the continents thus become seats of low pressure, with winds blowing in from the colder oceans. In the winter, the continents are seats of high pressure with winds directed toward the warmer oceans. These seasonal winds are typified by the monsoons of the China Sea and Indian Ocean.
3. *The local winds.* Corresponding with the seasonal variation in temperature and pressure over land and water, daily changes occur which have a similar but local effect, penetrating to a distance of about 30 miles (48 km) on and off the shores. Similar daily changes in temperature occur over irregular terrain and cause mountain and valley breezes. Other winds associated with local phenomena include whirlwinds and winds associated with thunderstorms.

All three types of wind mentioned here are of equal importance in design. However, for purposes of evaluating wind loads, the characteristics of the prevailing and seasonal winds are analytically studied together, while those of local winds are studied separately. This grouping is for analytical convenience and to distinguish between the widely differing scale of fluctuations of the winds; prevailing and seasonal wind speeds fluctuate over a period of several months, whereas the local winds vary almost every minute. The variations in the speed of prevailing and seasonal winds are referred to as *fluctuations* in mean velocity. The variations in the local winds, which are of a smaller character, are referred to as *gusts*.

Flow of wind, unlike that of other fluids, is not steady and fluctuates in a random fashion. Because of this, the properties of wind are studied statistically. The roughness of the earth's surface creates frictional drag in the flow of air, causing a gradual decrease in its velocity near the earth's surface. Also, turbulence is caused by surface roughness. Before these characteristics are studied in greater detail, it is of interest to have an overview of other types of extreme wind conditions and their effect on structures.

## 2.3 Extreme Wind Conditions

### 2.3.1 Introduction

Human beings and their works are subjected to hazards arising from forces and disturbances occurring in the natural environment. One of these natural hazards is the dynamic action of wind in the shape of hurricanes and tornadoes. These winds cause direct economic damages exceeding several million dollars in any given year. The large risk to both investment and human life which occurs from wind makes necessary an understanding of the physical phenomenon involved in the wind hazard and the development of improved planning and design methods. Wind engineering is, therefore, important, both in terms of potential economic damage and life loss and in human comfort.

Extreme winds, such as thunderstorms, hurricanes, tornadoes, and typhoons, impose loads on structures that are many times more than those normally assumed in their design. Some standards, such as the ASCE 7-95 provide for hurricane wind speeds for a specified probability of occurrence but do not consider directly the effect of other types of extreme wind conditions.

### 2.3.2 Thunderstorms

Thunderstorms are one of the most familiar features of temperate summer weather, characterized by long hot spells punctuated by release of torrential rain. The essential conditions for the occurance of thunderstorms are warm, moist air in the lower atmosphere and cold, dense air at higher altitudes. Under these conditions warm air at ground level rises, building storm clouds in the upper atmosphere. Thunder and lightning accompany downpours, creating gusty winds.

Wind speeds of 20 to 70 mph (9 to 31 m/s) are typically reached in a thunderstorm and are often accompanied with swirling wind action exerting high suction forces on roof and cladding elements.

### 2.3.3 Hurricanes

Hurricanes are severe atmospheric disturbances that originate in the tropical regions of the Atlantic Ocean or Caribbean Sea. They travel north, northwest, or northeast from their point of origin and usually cause heavy rains. They originate in the doldrums and consist of high-velocity winds blowing circularly around a low-pressure center known as the eye of the storm. The low-pressure center develops when the warm saturated air prevalent in the doldrums interacts with the cooler air. From the edge of the storm toward its center the atmospheric pressure drops sharply raising the wind velocity. In a fully developed hurricane, winds reach speeds up to 70 to 80 mph (31 to 36 m/s), and in severe hurricanes can attain velocities as high as 200 mph (90 m/s). Within the eye of the storm, the winds cease abruptly, the storm clouds lift, and the seas become exceptionally violent. The most destructive hurricane in the U.S. history hit the Atlantic seaboard in June, 1972 causing at least 122 deaths and damage amounting to over $3 billion.

The maximum design wind velocity for any area of the United States specified by code is 120 mph (54 m/s), which is less than the highest wind speeds in hurricanes. Except in rare instances, such as defense installations, a structure is not normally designed for full hurricane wind speeds.

Hurricanes are one of the most spectacular forms of terrestrial disturbances that produce heaviest rains known on earth. They have two basic requirements, warmth and moisture, and consequently develop only in the tropics. Almost invariably they move in a westerly direction at first, and then swing away from the equator, either striking land with devastating results or moving out over the oceans until they encounter cool surface water to die out naturally. The region of greatest storm frequency is the northwestern Pacific, where the storms are called *typhoons*—a name of Chinese origin meaning "wind which strikes." The storms which occur in the Bay of Bengal and the seas of north Australia are called *cyclones*. Although there are some general characteristics common to all hurricanes, no two are exactly alike. However, a typical hurricane can be considered to have a 375 mile (600 km) diameter, with its circulating winds spiraling in toward the center at speeds up to 112 mph (50 m/s). The size of the eye can vary in diameter from as little as 3.7 to 25 miles (6 to 40 km). However, the typhoon which roared past the island of Guam in 1979 had a very large diameter of 1400 miles (2252 km) with the highest wind speed reaching 190 mph (85 m/s). Storms of such violence have been known to drive a plank of wood right through the trunk of a tree and blow straws end-on through a metal deck. Fortunately, storms of such magnitude are not common.

### 2.3.4 Tornadoes

Tornadoes develop within severe thunderstorms consisting of a rotating column of air usually accompanied by a funnel-shaped downward extension of a dense cloud having a vortex of several hundred feet, typically 200 to 800 ft (61 to 244 m) in diameter whirling destructively at speeds up to 300 mph (134 m/s). They contain the most destructive of all wind forces, destroying everything along their path. Tornadoes form when a cold storm-front runs over warm, moist surface air. The warm air rises through the overlaying cold storm clouds intercepted by the high-altitude winds that are even colder rapidly moving above the clouds. Warm air collides with the cooler air and begins to whirl. The pressure at the center of the spinning column of air is reduced because of the centrifugal force. This reduction in pressure causes more warm air to be sucked into it, creating a violent outlet for the warm air trapped under the storm. As the velocity increases, more warm air is drawn up to the low-pressure area created in the center of the vortex. As the vortex gains strength, the funnel begins to extend toward the ground, eventually touching it. Funnels usually form close to the leading edge of the storm. Larger tornadoes may have several vortices within a single funnel. If the bottom of the funnel can be seen, it usually means that the tornado has touched down and begun to pick up visible debris from the ground.

A typical tornado travels 20 to 30 mph (9 to 14 m/s), touches ground for 5 to 6 miles (8 to 10 km), and has a funnel 300 to 500 ft (92 to 152 m) wide. Distance from the ground to the cloud averages about 2000 ft (610 m). Tornadoes contain the most powerful of all winds, causing damage well in excess of $100 million a year in the United States. Although it is impractical to design buildings to sustain a direct hit from a tornado, it behoves the engineer to pay extra attention to anchorage of roof decks and curtain walls of buildings in areas of high tornado frequency.

Rolling plains and flat country make a natural home for tornadoes.

Statistically, flat plains get more tornadoes than other parts of the country. In North America, communities in Kansas, Nebraska, and Texas have many tornadoes and are classified as "tornado belt" areas. No accurate measurement of the inner speed of a tornado has been made because tornadoes destroy standard measuring instruments. However, photographs of tornadoes suggest that wind speeds are of the order of 167 to 224 mph (75 to 100 m/s). Although there are definite tornado seasons, tornadoes can occur at any time.

Similar to a hurricane, a tornado consists of a mass of unstable air rotating furiously and rising rapidly around the center of an area with low atmospheric pressure. The similarity ends here, because whereas a hurricane is generally of the order of 300 to 400 miles (483 to 644 km) in diameter, a large tornado is unlikely to be more than 1500 ft (458 m) across. However, in terms of destructive violence no other atmospheric disturbance compares with a tornado.

Wind alone is not the only damaging element at work in a tornado. The pressure at the center of a tornado is extremely low. As the storm passes over a building, the pressure inside the structure is far greater than the outside, causing the building to literally explode. Typically, buildings are not designed to withstand a direct hit from a tornado. However, for those which are deemed essential, such as defence installations and nuclear facilities, sufficient information is available in engineering literature to implement tornado-resistant design. This information is in the form of tornado risk probabilities, wind speeds, and forces.

## 2.4 Characteristics of Wind

### 2.4.1 Introduction

Wind is a phenomenon of great complexity because of the many flow situations arising from the interaction of wind with structures. However, in wind engineering simplifications are made to arrive at meaningful predictions of wind behavior by distinguishing the following features:

- variation of wind velocity with height (2.4.2);
- turbulent nature of wind (2.4.3);
- probabilistic approach (2.4.4);
- vortex shedding phenomenon (2.4.5);
- dynamic nature of wind-structure interaction (2.4.6).

### 2.4.2 Variation of wind velocity with height

At the interface between a moving fluid and a solid surface, viscosity manifests itself in the creation of shear forces aligned opposite to the direction of fluid motion. A similar effect occurs between the surface of the earth and the atmosphere. Viscosity reduces the air velocity adjacent to the earth's surface to almost zero. A retarding effect occurs in the layers near the ground, and these inner layers in turn successively slow down the outer layers. The slowing down is less at each layer and eventually becomes negligibly small. The velocity increase is curvilinear varying from zero at the ground surface to a maximum at some distance above the ground. The height at which the velocity ceases to increase is called the *gradient height,* and the corresponding velocity, the *gradient velocity*. The shape and size of the curve depends less on the viscosity of the air than on the type and predominance of the turbulent and random eddying motions in the wind, which in turn are affected by the type of terrain over which the wind is blowing (see Fig. 2.2). This important characteristic of variation of wind velocity with height is a well understood phenomenon as evidenced by higher design pressures specified at higher elevations in most building codes.

The variation of velocity with height can be considered as a gradual retardation of the wind nearer the ground as a result of surface friction. At heights of approximately 1200 ft (366 m) from the ground, the wind speed is virtually unaffected by surface friction and its

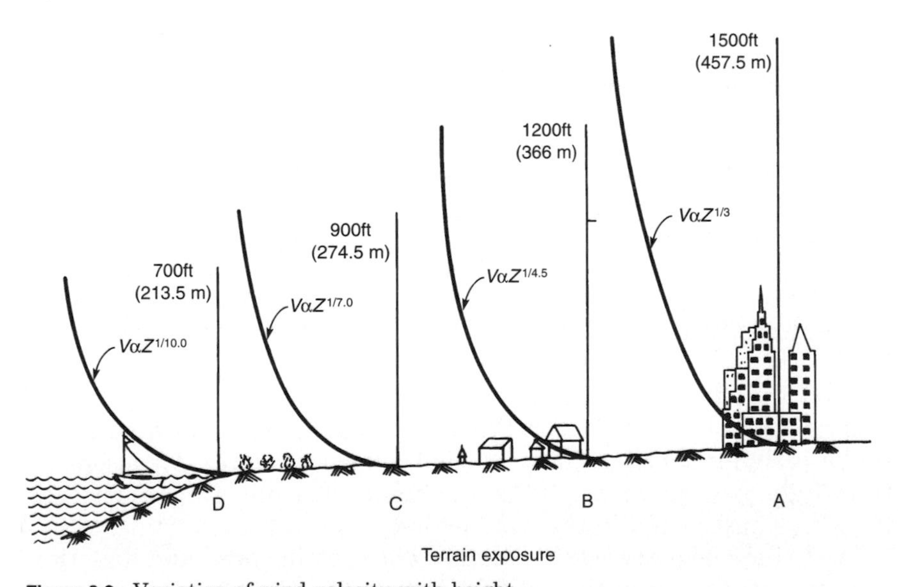


**Figure 2.2** Variation of wind velocity with height.

movement is solely dependent on the prevailing seasonal and local wind effects. The height through which the wind speed is affected by the topography is called the *atmospheric boundary layer*. The wind speed profile within this layer is in the domain of turbulent flow and can be mathematically predicted by a logarithmic equation. However, in practice wind speed variation, is given by a simpler power-law expression of the form:

$$V_Z = V_g(Z/Z_g)^\alpha \tag{2.1}$$

where $V_Z$ = the mean wind speed at height $Z$ above the ground
$V_g$ = gradient wind speed assumed constant above the boundary layer
$Z$ = height above the ground
$Z_g$ = depth of boundary layer
$\alpha$ = power law coefficient

By knowing the mean wind speed at gradient height and the value of exponent $\alpha$, the wind speeds at height $Z$ are easily calculated by using Eq. (2.1). The exponent $\alpha$ and the depth of boundary layer $Z_g$ vary with terrain roughness. The value of $\alpha$ ranges from a low of 0.14 for open country to about 0.5 for built-up urban areas, signifying that wind speed reaches its maximum value over a longer height in an urban terrain than in an open country. The pressure and suction generated by wind are a function of the wind speed, and in general increase with the building height.

### 2.4.3 Turbulent nature of wind

The motion of wind is turbulent. A concise mathematical definition of turbulence is difficult to give, except to state that it occurs in wind flow because air has a very low viscosity of about one-sixteenth that of water. Any movement of air at speeds greater than 2 to 3 mph (0.9 to 1.3 m/s) is turbulent, causing particles of air to move randomly in all directions. This is in contrast to the laminar flow of particles of heavy fluids, which move predominantly parallel to the direction of flow.

The variation of wind velocity with height describes only one aspect of wind in the boundary layer. Superimposed on the mean wind speed is the turbulence or gustiness of wind, which produces deviations in the wind speed above and below the mean, depending upon whether there is a gust or lull in the wind action.

Flow of air near the earth's surface changes in speed and direction because of the obstacles which introduce random vertical and horizontal movements at right-angles to the main direction of flow. These gusts vary over a wide range of frequencies and amplitudes, both in time and space. Shown in Fig. 2.3 are the anemometer readings of wind speeds in which $\bar{v}$ and $v'$ represent the instantaneous mean and gust speeds.

The scale and intensity of turbulence can be likened to the size and rotating speed of the eddies or vortices that make up the turbulence. It is generally found that the size of the flow affects the size of the turbulence within it. Thus, the flow of a large mass of air has a larger overall turbulence than a corresponding flow of a small mass of air. Because of its random nature, the properties of wind are studied statistically. A statistical property is the mean or the average. Because wind speed changes constantly, different averages are obtained by using different averaging times. For example, while a one-hour average of wind speed may be 30 mph (13.4 m/s), the same wind averaged for one minute may be as high as 80 mph (35.8 m/s). Therefore it is necessary to specify averaging time whenever a mean velocity is referenced. However, one of the strange phenomena that occurs when measuring wind speed is that averages taken over periods between 10 and 60 min adequately avoid the violent peaks and valleys due to gustiness. The averages taken over 10, 20, and 60 min intervals are nearly the same. This characteristic of the wind allows the comparison of mean data taken over different time periods.

For structural engineering purposes, the velocity of wind can be considered as having two components; a mean velocity component whose value increases with height and a turbulent velocity fluctuation. This can be visualized by considering the mean velocity as increasing with height as before, but subject to vortices or eddies

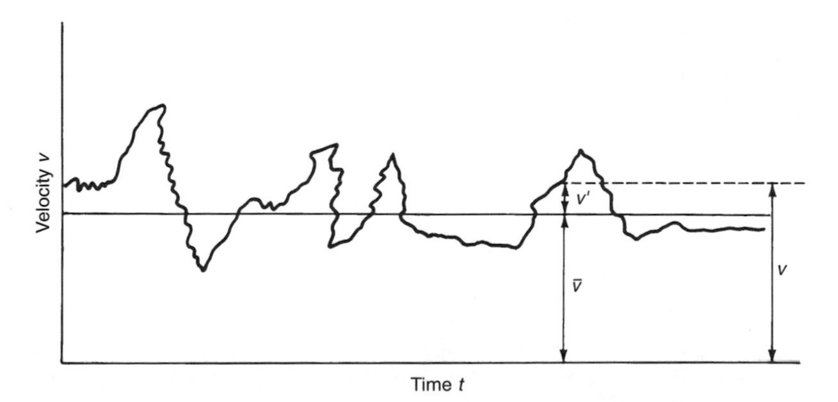


**Figure 2.3** Variation of wind velocity with time.

(small currents of air spinning in space) oriented randomly to the mean direction. By assigning different size and rates of spin to these vortices, one can equate these to the wavelength and frequency of variable components.

Spectral analysis techniques provide a convenient method for dealing with the random turbulence of wind. A complete treatment of the method is beyond the scope of the present work except to state that the method has some similarities with structural problems, wherein a required solution, for example, the deflection of a simply supported beam, can be obtained by superposition of a sufficiently large number of deflection patterns with different amplitudes and shapes.

The velocity at any instant $v_t$ can be represented as the summation of the average velocity and the instantaneous value of velocity fluctuation about the mean value as shown in Fig. 2.3. Thus

$$v_t = \bar{v} + v' \tag{2.2}$$

where $v_t$ = velocity at instant $t$
$\bar{v}$ = average or mean velocity
$v'$ = instantaneous velocity fluctuation about the mean velocity $\bar{v}$

Figure 2.4 schematically represents the fluctuation of mean and gust pressure along the height of a building.

The longest averaging time used in structural engineering practice

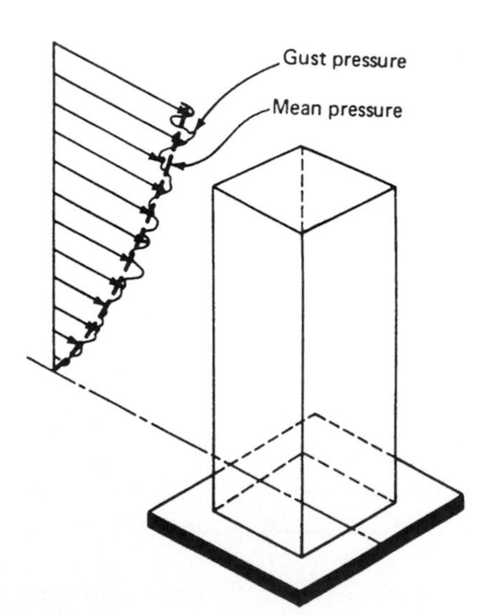


**Figure 2.4** Schematic representation of mean and gust pressure.

is one hour. As the averaging time decreases, the maximum speed of wind increases. The average or mean wind speed used in many building codes of the United States is the *fastest-mile wind,* which can be thought of as the maximum velocity measured over one mile of wind passing through an anemometer. Normally, the wind speed in structural design is in the range of 60 to 120 mph (97 to 154 m/s), giving an averaging period of 30 to 60 s.

Rapid bursts in the velocity of wind are called gusts. Tall buildings are sensitive to gusts that last about one second. Therefore, the fastest mile wind (which has the averaging period of 30 to 60 s) is inadequate for design of tall buildings. One must use the gust speed rather than the mean wind speed in the determination of wind loads. The gust speed can be obtained by multiplying the mean wind speed by a gust factor $G_v$. Thus

$$V_g = G_v V \tag{2.3}$$

where $G_v$ = the gust factor
$V_g$ = the gust speed
$V$ = the mean wind speed

Not all buildings are equally sensitive to gusts. In general, the more flexible a structure is, the more sensitive it is to gusts. The only accurate way to determine the gust factor (also called gust response factor) is to conduct a wind-tunnel test. However, attempts are made in some contemporary wind load standards, such as the National Building Code of Canada (NBC 1990) and the ASCE 7-95, to give analytical procedures for determining gust response factors. These are considered later in this chapter.

### 2.4.4 Probabilistic approach to wind load determination

In many engineering sciences the intensity of certain events is considered as a function of duration recurrence interval (return period). For example, in hydrology the intensity of rainfall expected in a region is considered in terms of return period because the rainfall expected once in 10 years is likely to be less than the one expected once in every 50 years. Similarly, in wind engineering the speed of wind is considered to vary with return periods. For example, the fastest-mile wind 33 ft (10 m) above ground in Dallas, Texas, corresponding to a 50-year return period is 67 mph (30 m/s) as compared to the value of 71 mph (31.7 m/s) for a 100-year recurrence interval.

A 50-year return period wind of 67 mph (30 m/s) means that on the average, Dallas will experience a wind faster than 67 mph within a period of 50 years. A return period of 50 years corresponds to a probability of occurrence of 1/50 = 0.02 = 2 percent. Thus the chance that a wind exceeding 67 mph (30 m/s) will occur in Dallas within a given year is 2 percent. Suppose a building is designed for a 100-year lifetime using a design wind speed of 67 mph. What is the probability that wind will exceed the design speed within the lifetime of the structure? The probability that this wind speed will not be exceeded in any year is 49/50. The probability that this speed will not be exceeded in 100 years in a row is $(49/50)^{100}$. Therefore, the probability that this wind speed will be exceeded at least once in 100 years is

$$1 - \left(\frac{49}{50}\right)^{100} = 0.87 = 87 \text{ percent}$$

This signifies that although a wind with low annual probability of occurrence such as a 50-year wind is used to design structures, there exists still a high probability of the wind being exceeded within the lifetime of the structure. However, in structural engineering practice it is believed that the actual probability of overstressing a structure is much less because of the factors of safety and the genreal conservative values used in design.

It is important for design engineers to understand the notion of probability of occurrence of design wind speeds during the service life of buildings. The general expression for probability, $P$, that the design wind speed will be exceeded at least once during the exposed period of $n$ years is given by

$$P = 1 - (1 - P_a)^n \tag{2.4}$$

where

$P_a$ = annual probabilty of being exceeded (reciprocal of the mean recurrence interval)
$n$ = exposure period in years

Consider a building in Dallas being designed for a 50-year service life instead of 100 years. The probability of exceeding the design wind speed at least once during the 50-year lifetime of the building is

$$P = 1 - (1 - 0.02)^{50} = 1 - 0.36 = 0.64 = 64 \text{ percent}$$

The probability that wind speeds of a given magnitude will be exceeded increases or decreases with exposure period of the building and the mean recurrence interval used in the design. Values of $P$ for

**TABLE 2.1 Probability of Exceeding Design Wind Speed During Design Life of Building**

| Annual probability $P_a$ | Mean recurrence interval $(1/P_a)$ years | Exposure period (design life), $n$ (years) | | | | | |
|---|---|---|---|---|---|---|---|
| | | 1 | 5 | 10 | 25 | 50 | 100 |
| 0.1 | 10 | 0.1 | 0.41 | 0.15 | 0.93 | 0.994 | 0.999 |
| 0.04 | 25 | 0.04 | 0.18 | 0.34 | 0.64 | 0.87 | 0.98 |
| 0.034 | 30 | 0.034 | 0.15 | 0.29 | 0.58 | 0.82 | 0.97 |
| 0.02 | 50 | 0.02 | 0.10 | 0.18 | 0.40 | 0.64 | 0.87 |
| 0.013 | 75 | 0.013 | 0.06 | 0.12 | 0.28 | 0.49 | 0.73 |
| 0.01 | 100 | 0.01 | 0.05 | 0.10 | 0.22 | 0.40 | 0.64 |
| 0.0067 | 150 | 0.0067 | 0.03 | 0.06 | 0.15 | 0.28 | 0.49 |
| 0.005 | 200 | 0.005 | 0.02 | 0.05 | 0.10 | 0.22 | 0.39 |

a given mean recurrence interval and a given exposure period are shown in Table 2.1.

### 2.4.5 Vortex-shedding phenomenon

In general, wind blowing past a body can be considered to be diverted in three mutually perpendicular directions, giving rise to forces and moments about the three directions. In aeronautical engineering special terminology is used to describe these forces and moments as shown in Fig. 2.5. Although all six components are significant in aeronautical work, in civil and structural engineering, the force and moment corresponding to the vertical axis (lift and yawing moment) are of little signficance. Therefore, the flow of wind is considered two-dimensional, as shown in Fig. 2.6.

*Along wind* or simply *wind* is the term used to refer to drag forces and *transverse wind* is used to refer to crosswind. The crosswind

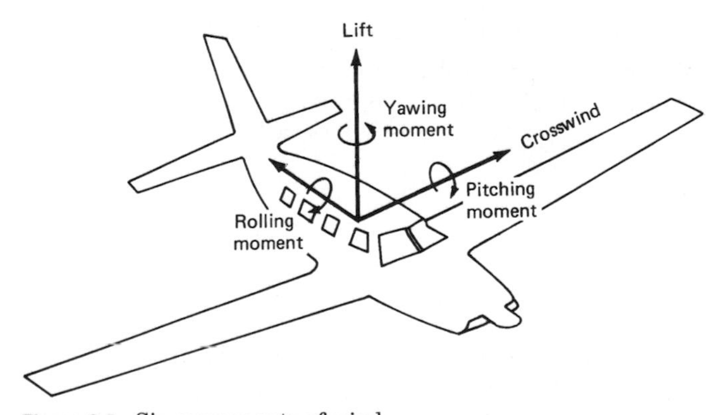


**Figure 2.5** Six components of wind.

response, that is, motion in a plane perpendicular to the direction of wind, dominates over the along-wind response for most tall buildings. This complex nature of wake-excited response is a result of interaction of turbulence, building motion, and the dynamics of wake formation.

It is perhaps baffling to the uninitiatied engineer to learn that a tall building is subject to wind excitations not only in a direction parallel to the wind but also in a direction perpendicualr to it. Yet in many instances, the major criterion for design of very tall buildings is the crosswind response. While the maximum lateral wind-loading and deflection are generally in the direction parallel with the wind (along-wind direction), the maximum acceleration of a building leading to possible human perception of motion or even discomfort may occur in a direction perpendicular to the wind (crosswind direction).

There appear to be three distinctly different reasons why a building responds in a direction at right-angles to the applied wind forces; these are: (i) the biaxial displacement induced in the structure because of either asymmetry in geometry or in applied wind loading; (ii) the turbulence of wind; and (iii) the negative-pressure wake or trail on the building sides. For tall buildings it appears that the crosswind response is caused mainly by the wake.

Consider a cylindrically shaped building subjected to a smooth wind flow. The originally parallel stream lines are displaced on

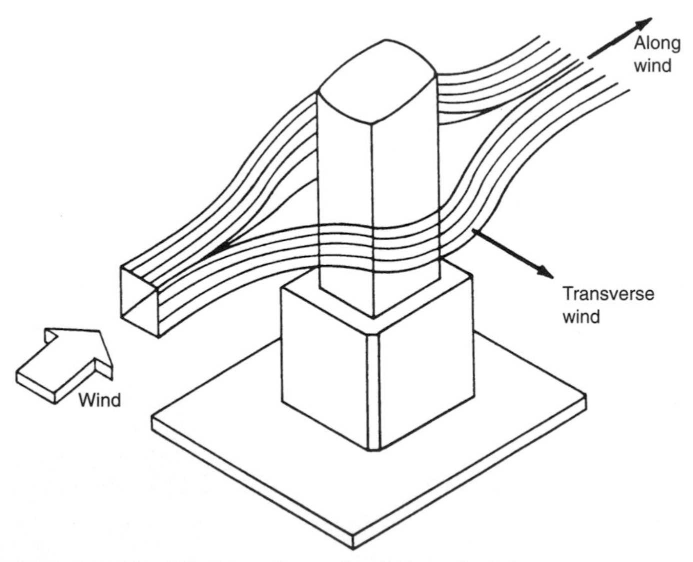


**Figure 2.6** Simplified two-dimensional flow of wind.

either side of the cylinder. This results in spiral vortices being shed periodically from the sides of the cylinder into the downstream flow of wind called the wake. At low wind speeds of say 50 to 60 mph (22.3 to 26.8 m/s), the vortices are shed symmetrically in pairs one from each side. These vortices can be thought of as imaginary projections attached to the cylinder that increase the drag force on the cylinder. When the vortices are shed, i.e., break away from the surface of the cylinder, an impulse is applied to the cylinder in the transverse direction. This phenomenon of alternate shedding of vortices for a rectangular tall building is shown schematically in Fig. 2.7.

At low wind speeds, since the shedding occurs at the same instant on either side of the building, there is no tendency for the building to vibrate in the transverse direction. It is therefore subject to along-wind oscillations parallel to the wind direction. At higher speeds, the vortices are shed alternately first from one and then from the other side. When this occurs, there is an impulse in the along-wind direction as before, but in addition, there is an impulse in the transverse direction. The transverse impulses are, however, applied alternatively to the left and then to the right. The frequency of transverse impulse is precisely half that of the along-wind impulse. This kind of shedding which gives rise to structural vibrations in the flow direction as well as in the transverse direction is called *vortex shedding* or the *Karman vortex street,* a phenomenon well known in the field of fluid mechanics.

There is a simple formula to calculate the frequency of the transverse pulsating forces caused by vortex shedding

$$f = \frac{V \times S}{D} \tag{2.5}$$

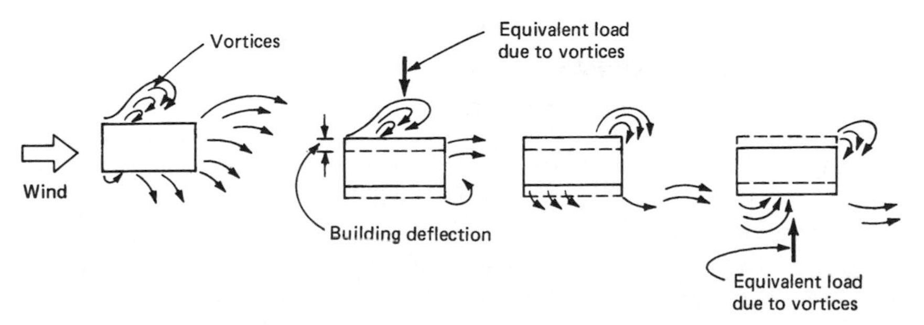


**Figure 2.7** Vortex-shedding phenomenon.

where $f$ = the frequency of vortex shedding in hertz
$V$ = the mean wind speed at the top of the building
$S$ = a dimensionless parameter called the Strouhal number for the shape
$D$ = the diameter of the building

In Eq. (2.5), the parameters $V$ and $D$ are expressed in consistent units such as ft/s and ft, respectively.

The Strouhal number is not a constant but varies irregularly with the wind velocity. At low air velocities, $S$ is also low and increases with the velocity up to a limit of 0.21 for a smooth cylinder. This limit is reached for a velocity of about 50 mph (22.4 m/s) and remains almost a constant at 0.20 for wind velocities between 50 and 115 mph (22.4 and 51 m/s).

Consider for illustration purposes, a circular prismatic-shaped high-rise building having a diameter equal to 110 ft (33.5 m) and a height-to-width ratio of 6 with a natural frequency of vibration equal to 0.16 Hz. Assuming a wind velocity of 60 mph (27 m/s), the vortex-shedding frequency is given by

$$f = \frac{V \times 0.2}{110} = 0.16 \text{ Hz}$$

where $V$ is in ft/s.

If the wind velocity increases from 0 to 60 mph (27.0 m/s), the frequency of vortex excitation will rise from 0 to a maximum of 0.16 Hz. Since this frequency happens to be very close to the natural frequency of the building, and assuming a very low damping, the structure would pulsate as if its stiffness were zero at a wind speed somewhere around 60 mph (27 m/s). Note the similarity of this phenomenon with the ringing of church bells or the shaking of a tall lamppost whereby a small impulse added to the moving mass at each end of the cycle greatly increases the kinetic energy of the system. Similarly, during vortex shedding an increase in deflection occurs at the end of each swing. If the damping characteristics are small, the vortex shedding can cause building displacements far beyond those predicted on the basis of static analysis.

When the wind speed is such that the shedding frequency becomes approximately the same as the natural frequency of the building, a resonance condition is created. After the structure has begun to resonate, further increases in wind speed by a few percent will not change the shedding frequency, because the shedding is now controlled by the natural frequency of the structure. The vortex shedding frequency has, so to speak, locked-in with the natural frequency. When wind speed increases above that causing the lock-in phenomenon, the frequency of shedding is again controlled by the speed of the wind. The structure vibrates with the resonant frequency only in the lock-in range, and for wind speeds either below or above this range, the vortex shedding will not be critical.

Vortex shedding occurs for many building shapes. The value of $S$ for different shapes is determined in wind tunnels by measuring the frequency of shedding for a range of wind velocities. One does not have to know the value of $S$ very precisely because the lock-in phenomenon occurs within a range of about 10 percent of the exact frequency of the structure.

The crosswind response mechanism is very complex, and an exact analytical method that takes into account the variables of turbulence, building shape, structure stiffness, damping, and density has not been introduced into structural engineering practice. In cases where the crosswind response may become the controlling factor in the design, the only recourse is to resort to an aeroelastic wind tunnel investigation. However, it should be recognized that even in elaborate model tests it is not possible to simultaneously scale the Reynolds number, Strouhal number, the fluctuating lift and drag coefficients, the structural stiffness and the aerodynamic damping.

### 2.4.6 Dynamic nature of wind

When wind hits a blunt object in its path, it transfers some of its energy to the object. The measure of the amount of energy transferred is called the *gust response factor*. As mentioned previously, wind turbulence (also called gustiness) is affected by terrain roughness and varies with the height above ground. A tall, slender, and flexible structure could have a significant dynamic response to wind because of buffeting. This dynamic amplification of response would depend on how the gust frequency correlates with the natural frequency of structure and also on the size of the gust in relation to the building size.

Unlike the mean flow of wind, which can be considered as static, wind loads associated with gustiness or turbulence change rapidly and even abruptly, creating effects much larger than if the same loads were applied gradually. Wind loads, therefore, need to be studied as if they were dynamic in nature. The intensity of a wind load depends on how fast it varies and also on the response of the structure. Therefore, whether the pressures on a building created by a wind gust, which may first increase and then decrease, are considered as dynamic or static depends to a large extent on the dynamic response of structure to which it is applied.

Let us consider the movement of a building 800 ft tall, designed to a drift index of $\frac{H}{400}$, acted upon by a wind gust. Under wind loads, the building bends slightly as its top moves. It first moves in the direction of wind, say with a magnitude of 2 ft (0.61 m), and then starts oscillating back and forth. After moving in the direction of wind, the top goes through its neutral position, then moves approximately 2 ft (0.61 m) in the opposite direction, and continues oscillating back and forth until it eventually stops. The time it takes a buliding to swing through a complete oscillation is known as a *period*. The period of oscillation for a tall steel building in the height range of 700 to 1400 ft (214 to 427 m) normally is in the range of 5 to 10 seconds, whereaas for a 10-story concrete or masonry building it may be in the range of 0.5 to 1 seconds. The action of a wind gust depends not only on how long it takes the gust to reach its maximum intensity and decrease again, but on the period of the building itself. If the wind gust reaches its maximum value and vanishes in a time much shorter than the period of the building, its effects are dynamic. On the other hand, the gusts can be considered as static loads if the wind load increases and vanishes in a time much longer than the period for the building. For example, a wind gust that develops to its strongest intensity and decreases to zero in 2 seconds is a dynamic load for a tall building with a period of, say 5 to 10 seconds, but the same 2 seconds gust is a static load for a low-rise building with a period of less than 2 seconds.

### 2.4.7 Cladding pressures

#### 2.4.7.1 Introduction

The design of cladding for lateral loads is of major concern to architects and engineers. Although the failure of an exterior cladding resulting in broken glass may be of less consequence than collapse of a structure in an earthquake, the expense of replacement and hazards posed to pedestrians require that careful attention be given to its design. Cladding breakage in a windstorm is a complicated phenomenon as witnessed in hurricane Alicia, which hit Galveston and downtown Houston on August 18, 1983 causing breakage of glass in several tall buildings. Wind forces play a major role in glass breakage, which is also influenced by other factors, such as solar radiation, mullion and sealant details, tempering of the glass, double or single glazing of glass, and fatigue. It is known with certainty that glass failure starts at nicks and scratches which may be caused during its manufacturing and handling operations.

There appears to be no analytical approach available for a rational design of curtain walls of all shapes and sizes. Although most codes have tried to identify regions of high wind loads around building corners, the modern trend in architecture of using nonprismatic and curvilinear shapes combined with the unique topography of each site, has necessitated experimental determination of wind loads for each building.

It has become a routine nowadays to obtain design information concerning the distribution of wind pressures over a building's surface by conducting wind-tunnel studies. In the past two decades, curtain wall has developed into an ornamental item and has emerged as a significant architectural element. Sizes of window panes have increased considerably, requiring that the glass lights be designed as structural elements for various combinations of forces due to wind, shadow effects, and temperature movement. Glass in curtain walls not only has to resist large forces, particularly in tall buildings, but must also be designed to accommodate the various distortions of the total building structure. Breaking of large panes of glass in tall buildings can cause serious damage to neighboring properties and can injure pedestrians.

#### 2.4.7.2 Distribution of pressures and suctions

It has been known for some time that when air flows around the edges of a structure, the pressures produced at the corners are much in excess of the normal pressure on the center of elevation, as evidenced by damage caused to corner windows, eave and ridge tiles, etc., in a windstorm. Wind tunnel studies conducted on scale models of buildings have shown that three distinct pressure areas are developed. These are shown schematically in Fig. 2.8 and are listed below.

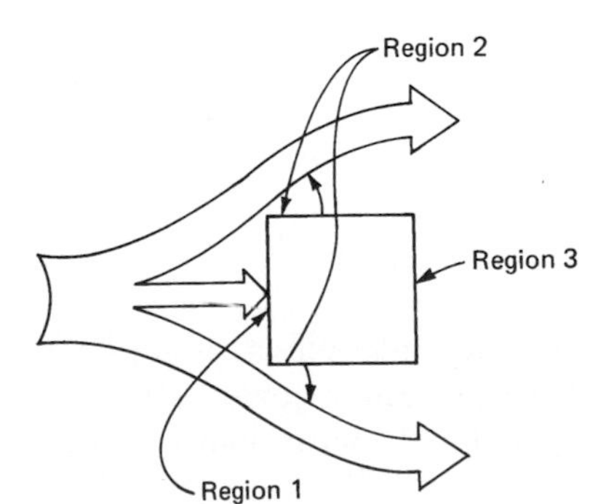


**Figure 2.8** Distribution of pressures and suctions.

1. Positive-pressure zone on the upstream face (Region 1).
2. Negative pressure zones at the upstream corners (Regions 2).
3. Negative pressure zone on the downstream face (Region 3).

Highest negative pressures are created in the upstream corners designated as Regions 2 in Fig. 2.8. These have been measured in wind-tunnel investigations and also have occurred in practice. Wind pressures on a buildings surface are not constant, but fluctuate continuously. The positive pressure on the upstream or the windward face fluctuates more than the negative pressure on the downstream or the leeward face. The negative-pressure region remains relatively steady as compared to the positive-pressure zone. The fluctuation of pressure is random and varies from point to point on the building surface. Therefore, the design of the cladding is strongly influenced by local pressures. As mentioned earlier, the design pressure can be thought of as a combination of the mean and the fluctuating velocity. As in the design of buildings, whether or not the pressure component arising from the fluctuating velocity of wind is treated as a dynamic or as a pseudo-static load is a function of the period of the cladding. The period of cladding on a building is usually on the order of 0.2 to 0.02 sec, which is very much shorter than the period it takes for wind to fluctuate from a gust velocity to a mean velocity. Therefore, it appears that it is sufficiently accurate to consider both the static and the gust components of winds as equivalent static loads in the design of cladding.

Strength of glass, and indeed any other cladding material, is not known in the same manner as for steel or concrete. For example, it is not possible to buy glass based on yield strength criteria as for concrete or steel. Therefore, the selection, testing, and acceptance criteria for glass must necessarily be based on statistical probabilities rather than on absolute strength. The glass industry has addressed this problem, and commonly uses eight failures per thousand lights of glass as an acceptable probability of failure .

#### 2.4.7.3 Local cladding loads and overall design loads

The overall wind load, consisting of the composite effect of positive and negative pressures is required to determine the required strength and stiffness of the building frame. The local wind loads which act on the various areas of the building enclosures are required for dictating the strength and stiffness of wall and roof elements and for the design of their fastenings. The two types of loads differ significantly, and it is important that these differences be understood. These are:

1. Local winds are more influenced by the configuration of the building surface on which they act than the overall loading.
2. The local load is the maximum load that may occur at any location at any time on any wall surface, whereas the overall load is the summation of all loads (with due regard to their sign) occurring simultaneously over the building surfaces.
3. The intensity and character of local loading for any given wind direction and velocity differ substantially on various parts of the building surface, whereas the overall load is considered to have a specific intensity and direction.
4. The local loading is sensitive to the momentary nature of wind, but in determining the critical overall loading, only gusts of about 2 sec or more are significant.
5. Generally, maximum local suctions are of greater intensity than the overall load.
6. Internal pressures caused by leakage of air through cladding systems have a significant effect on local cladding loads but usually are of no consequence in determining the overall load.

The relative importance of providing for these two types of wind loading is quite obvious. Although proper assessment of overall wind load is important, very few, if any, buildings have been toppled by winds. There are no classic examples of building failures comparable to the Tacoma bridge disaster. On the other hand, local failures of roofs, windows, and wall cladding are not uncommon, and in aggregate such failures continue to cost tens of millions of dollars each year.

The analytical determination of wind pressure or suction at a given surface of a building under varying wind direction and velocity is a very complex problem. Contributing to the complexity are the vagaries of wind action as influenced both by adjacent surroundings and the configuration of the wall surface itself. Much more research is needed on the microeffects of such architectural features as projecting mullions and column covers and deep window reveals. In the meantime, increasing use of model testing in boundary layer wind tunnels is providing vital information on wind loads on building surfaces.

Probably the most important fact established by these tests is that the negative or outward-acting wind loads on wall surfaces are greater and more critical than had formerly been assumed. It may be as much as twice the magnitude of positive loading. In most instances of local cladding failure, glass or panels have been blown off of the building, not into it, and the majority of such failures have occurred in areas near building corners. Therefore it is important to give careful attention to the design of both anchorage and glazing details to resist outward-acting forces, especially near the corners of high-rise buildings.

Another feature that has come to light from model testing is that wind loads, both positive and negative, on tall buildings do not vary in proportion to height above ground. Typically, the positive-pressure contours instead of being horizontal are usually found to follow a more concentric pattern as illustrated in Fig. 2.9, with the highest pressure being near the upper center of the façade and pressures at the very top being somewhat less than those a few stories below the roof.

## 2.5 Code Wind Loads

### 2.5.1 Introduction

Building codes and standards are documents which serve as compendiums for technical information and as sources for extracting minimum requirements of accepted design and construction practices. Codes and standards are, in fact, dynamic instruments that are revised periodically to reflect the state of the art. With the exception of some large cities in the United States which have adoped their own codes, about 85 percent of state and local governments have adopted or patterned their regulations on the provisions given mainly in three model codes. These are the *Basic Building Code* issued by the Building Officials and Code Administrators (BOCA) International, *Standard Building Code* (SBC) issued by Southern Building Code Congress International, and the *Uniform Building Code* (UBC) issued by the International Conference of Building Officials.

Because considerable differences exist among the model codes, the provisions of each code are discussed separately instead of collectively in the following sections. Also discussed are the provisions of the

National Building Code of Canada (NBC 1990) and the American Society of Civil Engineers Standards ASCE 7-93 and 7-95.

### 2.5.2 BOCA National Building Code (1996)

The basic wind speed map is given in Fig. 2.10. The wind pressures and suctions in pounds per square foot, to be applied simultaneously on windward and leeward walls, and roof surfaces for the design of building's main wind-force resisting system are given by:

Windward wall design pressure, $P$:

$$P = P_v I[K_z G_h C_p - K_h(GC_{pi})] \tag{2.6}$$

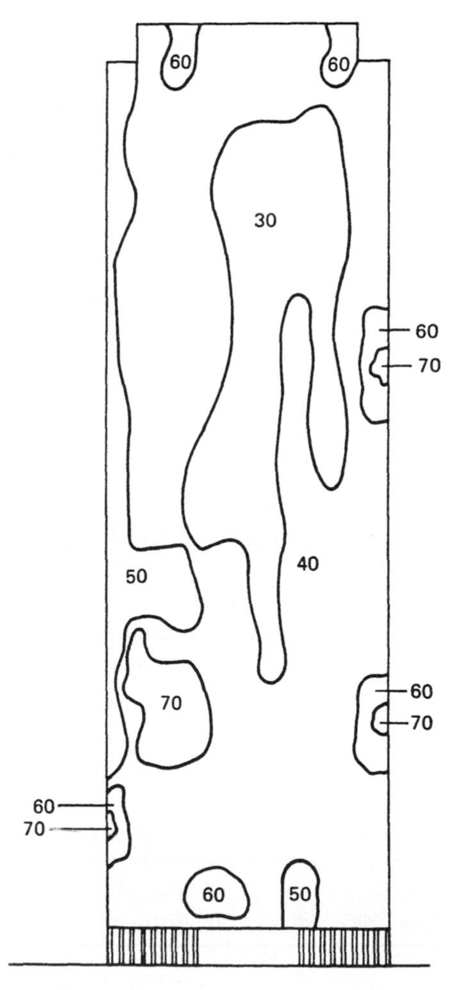


**Figure 2.9** Distribution of positive pressures in psf.

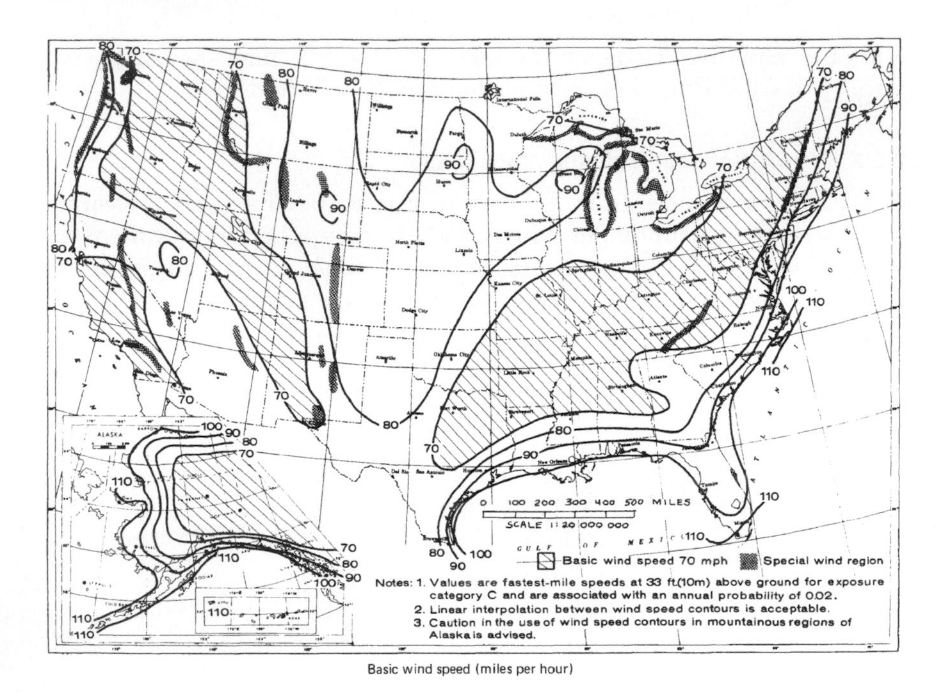


**Figure 2.10** Basic wind speed map (mph) BOCA 1996.

Leeward wall, side walls and roof design pressure, $P$:

$$P = P_v I[K_h G_h C_p - K_h(GC_{pi})] \tag{2.7}$$

where $P_v$ = basic velocity pressure (Table 2.2)
$I$ = wind load importance factor (Table 2.3)
$K_z$ = velocity pressure exposure coefficient, at the height of interest (Table 2.4)
$G_h$ = gust response factor (Table 2.5) evaluated at height $h$. The gust response factor for buildings which have a height to least horizontal dimension ratio greater

**TABLE 2.2 Basic Pressure $P_v$ (BOCA 1996)**

| Basic wind speed ($V$) (miles per hour) | 70 | 75 | 80 | 85 | 90 | 100 | 110 | 120 | 130 |
|---|---|---|---|---|---|---|---|---|---|
| Basic velocity pressure $P_V$ (lb/ft$^2$) | 12.5 | 14.4 | 16.4 | 18.5 | 20.7 | 25.6 | 31.0 | 36.9 | 43.3 |

**TABLE 2.3 Importance Factor *I* (BOCA 1996)**

| Nature of occupancy | Wind load importance factor ($I$)[a] 100 miles[c] from hurricane oceanline, and in other areas | At hurricane oceanline[b] |
|---|---|---|
| All buildings and structures except those listed below | 1.00 | 1.10 |
| Occupancies in Use Group A in which more than 300 people congregate in one area | 1.15 | 1.25 |
| Buildings and structures havnig essential facilities, including buildings containing any one or more of the indicated occupancies. 1. Fire, rescue and police stations. 2. Use Group 1-2 having surgery or emergency treatment facilities. 3. Emergency preparedness centers. 4. Designated shelters for hurricanes. 5. Power generating stations and other utilities required as emergency backup facilities. 6. Primary communication facilities. | 1.15 | 1.25 |
| Buildings and structures that represent a low hazard to human life in the event of failure, such as agricultural buildings, production greenhouses, certain temporary facilities and minor storage facilities | 0.90 | 1.00 |

[a] For regions between the hurricane oceanline and 100 miles inland, the importance factor ($I$) shall be determined by linear interpolation.
[b] Hurricane oceanlines are the Atlantic and Gulf of Mexico coastal areas.

than 5 or a fundamental frequency less than one cycle per second (period greater than 1 sec) shall be calculated by an approved rational analysis that incorporates the dynamic properties of the main windforce-resisting system

$C_p$ = external pressure coefficient (Table 2.6)

$K_h$ = velocity pressure exposure coefficient, evaluated at the mean roof height (Table 2.4)

$GC_{pi}$ = product of internal pressure coefficient and gust response factor (Table 2.7)

$h$ = mean roof height; the distance from grade to the average height of the roof

A schematic representation of external pressures for loads on main wind-force resisting system, and pressure coefficient $C_p$ are shown in

**TABLE 2.4 Velocity Pressure Exposure Coefficients $K_z$ and $K_h$ (BOCA 1996)**

| Height above ground level, $z$ (feet) | Coefficients $K_z$ and $K_h$ | | | |
|---|---|---|---|---|
| | Exposure A | Exposure B | Exposure C | Exposure D |
| 0–15 | 0.12 | 0.37 | 0.80 | 1.20 |
| 20 | 0.15 | 0.42 | 0.87 | 1.27 |
| 25 | 0.17 | 0.46 | 0.93 | 1.32 |
| 30 | 0.19 | 0.50 | 0.98 | 1.37 |
| 40 | 0.23 | 0.57 | 1.06 | 1.46 |
| 50 | 0.27 | 0.63 | 1.13 | 1.52 |
| 60 | 0.30 | 0.68 | 1.19 | 1.58 |
| 70 | 0.33 | 0.73 | 1.24 | 1.63 |
| 80 | 0.37 | 0.77 | 1.29 | 1.67 |
| 90 | 0.40 | 0.82 | 1.34 | 1.71 |
| 100 | 0.42 | 0.86 | 1.38 | 1.75 |
| 120 | 0.48 | 0.93 | 1.45 | 1.81 |
| 140 | 0.53 | 0.99 | 1.52 | 1.87 |
| 160 | 0.58 | 1.05 | 1.58 | 1.92 |
| 180 | 0.63 | 1.11 | 1.63 | 1.97 |
| 200 | 0.67 | 1.16 | 1.68 | 2.01 |
| 250 | 0.78 | 1.28 | 1.79 | 2.10 |
| 300 | 0.88 | 1.39 | 1.88 | 2.18 |
| 350 | 0.98 | 1.49 | 1.97 | 2.25 |
| 400 | 1.07 | 1.58 | 2.05 | 2.31 |
| 450 | 1.16 | 1.67 | 2.12 | 2.36 |
| 500 | 1.24 | 1.75 | 2.18 | 2.41 |
| >500 | The velocity pressure exposure coefficients shall be determined in accordance with an approved rational procedure that incorporates the dynamic properties of the main windforce-resisting system | | | |

Figs 2.11 and 2.13. The factor $GC_{pi}$ does not apply to the overall design of main wind-force resisting system. Ignoring this factor, Eqs (2.6) and (2.7) take the simpler form

Qindward wall design pressure $P$:

$$P = P_v I[K_z G_h C_p] \tag{2.8}$$

Leeward ward, side walls and roof design pressure $P$:

$$P = P_v I[K_h G_h C_p] \tag{2.9}$$

The basic velocity pressure $P_v$ given in Table 2.10 is determined using the formula $V = 0.00256V^2$

where $V$ = the basic wind speed, in miles per hour, determined from wind map Fig. 2.10. The wind speed map is based on open

**TABLE 2.5 Gust Response Factors $G_h$ and $G_z$ (BOCA 1996)**

| Height above ground level, $z$ (feet) | Factors $G_h$ and $G_z$ | | | |
|---|---|---|---|---|
| | Exposure A | Exposure B | Exposure C | Exposure D |
| 0–15 | 2.36 | 1.65 | 1.32 | 1.15 |
| 20 | 2.20 | 1.59 | 1.29 | 1.14 |
| 25 | 2.09 | 1.54 | 1.27 | 1.13 |
| 30 | 2.01 | 1.51 | 1.26 | 1.12 |
| 40 | 1.88 | 1.46 | 1.23 | 1.11 |
| 50 | 1.79 | 1.42 | 1.21 | 1.10 |
| 60 | 1.73 | 1.39 | 1.20 | 1.09 |
| 70 | 1.67 | 1.36 | 1.19 | 1.08 |
| 80 | 1.63 | 1.34 | 1.18 | 1.08 |
| 90 | 1.59 | 1.32 | 1.17 | 1.07 |
| 100 | 1.56 | 1.31 | 1.16 | 1.07 |
| 120 | 1.50 | 1.28 | 1.15 | 1.06 |
| 140 | 1.46 | 1.26 | 1.14 | 1.05 |
| 160 | 1.43 | 1.24 | 1.13 | 1.05 |
| 180 | 1.40 | 1.23 | 1.12 | 1.04 |
| 200 | 1.37 | 1.21 | 1.11 | 1.04 |
| 250 | 1.32 | 1.19 | 1.10 | 1.03 |
| 300 | 1.28 | 1.16 | 1.09 | 1.02 |
| 350 | 1.25 | 1.15 | 1.08 | 1.02 |
| 400 | 1.22 | 1.13 | 1.07 | 1.01 |
| 450 | 1.20 | 1.12 | 1.06 | 1.01 |
| 500 | 1.18 | 1.11 | 1.06 | 1.00 |
| >500 | The gust response factors shall be determined in accordance with an approved rational procedure that incorporates the dynamic properties of the main windforce-resisting system | | | |

terrain exposure category C with an annual probability of occurrence of 0.02. The descriptions of site categories, category A for large city centers, B for urban and suburban areas, C for open terrain and D for flat unobstructed areas are identical those in ASCE 7-95.

**TABLE 2.6 External Pressure Coefficient Coefficient $C_p$ (BOCA 1996)**

| Surface | $L/B$ | $C_p$ | For use with |
|---|---|---|---|
| Windward wall | All values | 0.8 | $K_z$ |
| Leeward wall | 0 to 1 | −0.5 | $K_h$ |
| | 2 | −0.3 | |
| | ≥4 | −0.2 | |
| Side walls | All values | −0.7 | $K_h$ |

**TABLE 2.7 Product of Internal Pressure Coefficient and Gust Response Factor $GC_{pi}$ (BOCA 1996)**

| | Condition | $GC_{pi}$ |
|---|---|---|
| Condition I | All conditions except as noted uncer Condition II. | +0.25<br>−0.25 |
| Condition II | Buildings which have both of the following wall opening characteristics:<br>1. The area of openings in the windward wall, and windward roof if under positive pressure, exceeds the sum of the area of openings in the remaining walls and roof surfaces by 5 percent or more; and<br>2. The openings in any one of the remaining walls or roof do not exceed 20 percent of the wall or roof area. | +0.75<br>−0.25 |

### 2.5.3 Standard Building Code (1991)

The basic wind speed map is shown in Fig. 2.12. The gust velocity pressure (which includes a gust factor) is multiplied by an appropriate pressure coefficient (Table 2.8) and by an appropriate Use factor (Table 2.9). The Use factor serves the same objective as the Importance factor specified in the UBC 1994 and ASCE 7-93 and 95. Exposure in which a specific building is sited is considered either as: (i) Standard Exposure; or (ii) Coastal Exposure applicable for a distance of 50 miles (80 km) inland from coastline.

### 2.5.4 Uniform Building Code (UBC 1994)

#### 2.5.4.1 General provisions

UBC wind provisions are based on the 1972 and 1982 editions of ANSI A58.1. One of the major philosophies of the UBC is that prevailing wind direction at the site is not considered in calculating wind forces on the structures: the direction that has the most critical

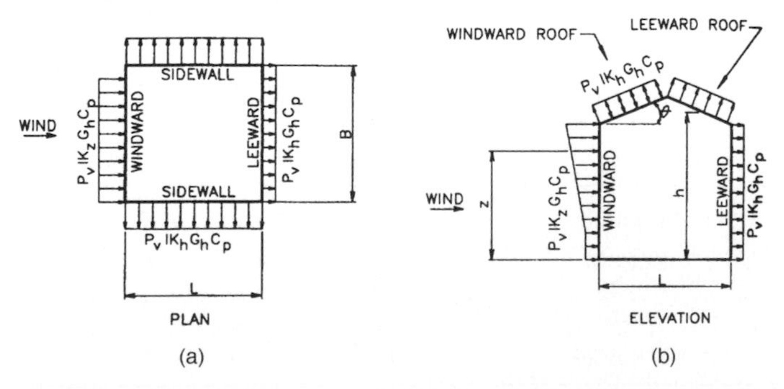


**Figure 2.11** External pressures for loads on main wind-force resisting systems (BOCA 1996): (a) plan; (b) elevation.

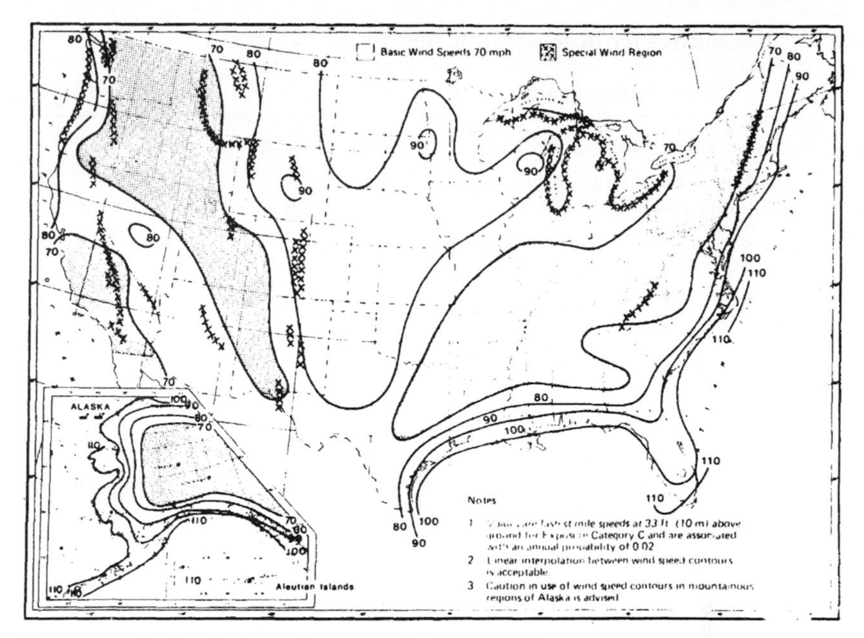


**Figure 2.12** Basic wind speed map, mph. (SBC 1991).

exposure controls the UBC design. Consideration of shielding by adjacent buildings is not allowed because firstly, studies have shown that nearby buildings can actually increase the wind speed through funnelling effects or increased turbulence. Secondly, it is also

**TABLE 2.8 Gust Velocity Pressure (SBC 1991)**

| | Fastest mile wind speed (mph) (From Fig. 2.12) | | | | | | | | | |
|---|---|---|---|---|---|---|---|---|---|---|
| | Standard exposure | | | | | Coastal exposure[c] | | | | |
| Height (ft)[a] | 70 | 80 | 90 | 100 | 110 | 70 | 80 | 90 | 100 | 110 |
| 0–30 | 8 | 11 | 14 | 17 | 21 | 14 | 18 | 23 | 29 | 35 |
| 31–50 | 10 | 14 | 17 | 21 | 26 | 16 | 21 | 27 | 33 | 40 |
| 51–100 | 13 | 17 | 21 | 26 | 32 | 19 | 25 | 32 | 39 | 47 |
| 100–200 | 16 | 21 | 26 | 33 | 39 | 22 | 29 | 36 | 45 | 54 |
| 200–300 | 19 | 25 | 31 | 39 | 47 | 25 | 33 | 41 | 50 | 61 |
| 300–400 | 21 | 28 | 35 | 43 | 53 | 27 | 35 | 45 | 55 | 66 |
| 400–500 | 23 | 31 | 39 | 48 | 58 | 29 | 37 | 47 | 58 | 70 |

[a] Incremental values of gust velocity pressure may be used to calculate windward wall design pressures on the main wind force resisting system and positive pressures for components and cladding of buildings. All other design pressures shall be based on values of gust velocity pressure at mean roof height.

[b] Linear interpolation of tabled values is permissible.

[c] Coastal exposure is applicable for a distance of 50 miles inland from the smoothed coastline.

**TABLE 2.9 Use Factors for Buildings (SBC 1991)**

| Nature of occupancy | Use factor |
|---|---|
| All buildings and structures except those listed below | 1.0 |
| Buildings and structures where the occupant load is 300 or more in any one room. | 1.15 |
| Buildings and structure designated as essential facilities, including, but not limited to:<br>1. Hospital and other medical facilities having surgery or emergency treatment areas<br>2. Fire or rescue and police stations<br>3. Primary communication facilities and disaster operation centers<br>4. Power stations and other utilities required in an emergency | 1.15 |
| Buildings and structures that represent a low hazard to human life in the event of failure, such as agricultural buildings, certain temporary facilities, and minor storage facilities | 0.9 |

possible that adjacent existing buildings may be removed during the life of the building being designed.

In order to shorten the calculation procedure, certain simplifying assumptions are made. These assumptions do not allow determination of wind loads for flexible structures that are sensitive to dynamic effects and wind-excited oscillations. The general section of the UBC directs the user to other methods and standards for these type of structures. The ASCE 7-95 and the Canadian Building Code (NBC 1990), discussed later in this chapter, are two such standards for determining the dynamic gust resposne factor required for the design of these types of buildings. The UBC provisions are not applicable to buildings taller than 400 ft (122 m) for Method 1, and 200 ft (61 m) for Method 2. Any building, including those not covered by the UBC, may be designed according to ASCE 7-95 or by using wind-tunnel data.

#### 2.5.4.2 Wind speed map

The UBC basic wind speed map (Fig. 2.14) is similar to the map in ASCE 7-95, except for the states of Washington and Oregon. The wind speeds represent fastest-mile wind speed representing exposure-C terrain at 33 ft (10 m) above grade for a 50-year mean recurrence interval. The probability of experiencing a basic wind speed faster than the indicated value in any one-year period is one in fifty or 2 percent.

#### 2.5.4.3 Special wind regions

Although basic wind speeds are surprisingly constant over hundreds of miles, some areas have local weather or topographic characteristics that affect the design wind speeds. Many of these special wind regions for which the basic wind speed may be higher than indicated on the map, are defined in the UBC map. Some jurisdictions prescribe basic wind speeds that are higher than indicated on the map, therefore it is prudent to contact the local building official before starting a project in an unfamiliar site.

#### 2.5.4.4 Hurricanes and tornadoes

The basic wind speeds shown in the UBC map come from data collected by meteorological stations throughout the Continental U.S. and the other weather-station data for Alaska, Hawaii, Puerto Rico, and the Virgin Islands. However, coastal regions did not have enough statistical measurements to predict hurricane wind speeds. Therefore data generated by computer simulations have been used to formulate basic hurricane wind speeds.

Tornado level winds are not included in the map. The mean recurrence intervals of tornadoes are in the range of 400–500 years.

#### 2.5.4.5 Exposure effects

Every building site has its unique characteristics of surface roughness and related fetch (the length of terrain upwind of the site associated with such roughness). Simplified code methods cannot account for the uniqueness of the site. Therefore the code approach is to specify broad exposure categories to which each building site is assigned.

The UBC specifies three exposure categories: Exposure B, the least severe, for urban, suburban, wooded and other terrain with numerous closely spaced surface irregularities; Exposure C for flat and generally open terrain with scattered obstructions; and the most severe, Exposure D, for unobstructed coastal areas directly exposed to large bodies of water.

Definitions of exposure categories in the UBC Code are given below.

*Exposure A* (centers of large cities where over half the buildings have a height in excess of 70 feet), included in some standards such as the NBC 1990 and ASCE 7-88, is not recognized in the UBC. The UBC considers this type of terrain as Exposure B, allowing no further decrease in wind pressure.

*Exposure B* has terrain with buildings, forest or surface irregularities, covering at least 20 percent of the ground level area extending 1 mile (1.61 km) or more from the site.

*Exposure C* has terrain which is flat and generally open, extending one-half mile (0.81 km) or more from the site in any full quadrant.

*Exposure D* represents the most severe exposure in areas of basic wind speeds of 80 mph (129 km/h) or greater and has terrain which is flat and unobstructed facing large bodies of water over one mile (1.61 km) or more in width relative to any quadrant of the building site. Exposure D extends inland from shore line $\frac{1}{4}$ mile (0.4 km) or 10 times the building height, whichever is greater.

#### 2.5.4.6 Site exposure

While a building site may appear to have different exposure categories in different directions, the most severe exposure is used for all wind-load calculations regardless of building orientation or direction of wind. This is because cladding design is governed by suction pressures acting on the side and back faces of the building which are higher than the positive design pressure values. Consequently, the UBC requires the selection of one exposure for all directions.

Exposure D is easy to determine because it is explicitly for unobstructed coastal areas directly exposed to large bodies of water. It is not easy to determine whether a site falls into Exposure B or C because the description of these categories is somewhat ambiguous. Moreover, the terrain surrounding a site is usually not uniform and can be composed of zones that would be classified as Exposure B while others would be classified as Exposure C. When such a mix is encountered, the more severe exposure governs. The UBC classifies a site as Exposure C when open terrain exists for one full 90 degree quadrant exending outward from the building for at least one-half mile. If the quadrant is less than 90 degrees or less than one-half mile, then the site is classified as Exposure B. It is essential to select the appropriate category because force levels could differ by as much as 65% between Exposure B and C. It is advisable to contact the local building official before embarking on a building design with a questionable site exposure category. If the site has a "view" of a cliff or a hill, it may be prudent to assign Exposure C to D to account for higher wind velocity effects.

#### 2.5.4.7 Design wind pressures

Design wind pressure $p$ is given as a product of combined height, exposure and gust factor coefficient $C_e$, pressure coefficient $C_q$, wind stagnation pressure $q_s$ and building Importance Factor $I_w$.

$$p = C_e C_q q_s I_w \tag{2.10}$$

The pressure $q_s$ manifesting on the surface of a building due to a mass of air with density $\rho$, moving at a velocity $v$ is given by the Bernoulli's equation:

$$q_s = \tfrac{1}{2}\rho v^2 \tag{2.11}$$

The density of air $\rho$ is 0.0765 pcf, for conditions of standard atmosphere, temperature (59 °F), and barometric pressure (29.92 in. of mercury).

Since velocity given in the wind map is in mph, Eq. (2.11) reduces to

$$\begin{aligned} q_s &= \frac{1}{2}\left[\frac{0.0765\ \text{pcf}}{32.2\ \text{ft/s}^2}\right]\left[\frac{5280\ \text{ft}}{\text{mile}} \times \frac{1\ \text{hr}}{3600\ \text{s}}\right] v^2 \\ q_s &= 0.00256V^2 \end{aligned} \tag{2.12}$$

For instance, if the wind speed is 80 mph, $q_s = 0.00256 \times 80^2 = 16.38$ psf which the UBC rounds off to 16.4 psf (Table 2.10). Note the UBC does not consider the effect of reduced air density at sites located at higher altitudes.

#### 2.5.4.8 The $C_e$ factor

The effects of height, exposure and gust factor are all lumped into one factor $C_e$ in the interest of keeping the UBC simple to use. Values of $C_e$ shown in Table 2.11 (UBC Table 16G) are essentially equal to the product of two parameters, $K_z$, the velocity pressure exposure coefficient, and $G_h$, the gust response factor. Both these parameters are defined separately in ASCE 7-95, and hence are more appropriate for "non-ordinary" buildings.

**TABLE 2.10 Wind Pressure $q_s$ at 33 ft (10 m) For Various Basic Wind Speeds (UBC 1994)**

| Basic wind speed (mph)[a] (×1.61 for km/h) | 70 | 80 | 90 | 100 | 110 | 120 | 130 |
|---|---|---|---|---|---|---|---|
| Pressure $q_S$ (psf) (×0.0479 for kN/m²) | 12.6 | 16.4 | 20.8 | 25.6 | 31.0 | 36.9 | 43.3 |

[a] Wind speed from Fig. 2.14.

The height and exposure factors account for the terrain effects on gradient heights and the lower wind speed in built-up terrain than in open terrain. The gust factor accounts for air turbulence and dynamic building behavior. For low-rise buildings with natural period of less than one second, the response is essentially static with deflection proportional to the applied force. Tall buildings, on the other hand, respond dynamically experiencing wind loads that are substantially greater than those estimated by the UBC.

The UBC uses many simplifying assumptions and lumps the dynamic effects together with height and exposure effects into the $C_e$ factor. From a structural design standpoint, the most challenging feature of a tall building is not the height *per se,* but its aspect ratio, the relation of the height to width of base. Therefore for slender buildings a procedure such as the one given in the ASCE 7-95 commentary, which takes into account the dynamic characteristic of the building, is more appropriate.

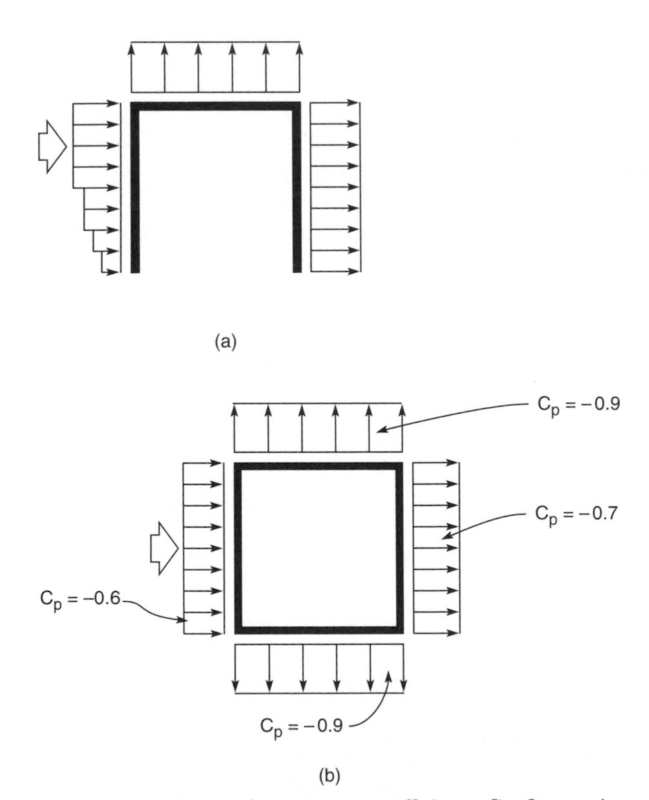


**Figure 2.13** External pressure coefficient $C_p$ for main wind-force resisting systems (BOCA 1996): (1) positive internal pressure, (a) elevation, (b) plan; (2) negative internal pressure, (a) elevation, (b) plan.

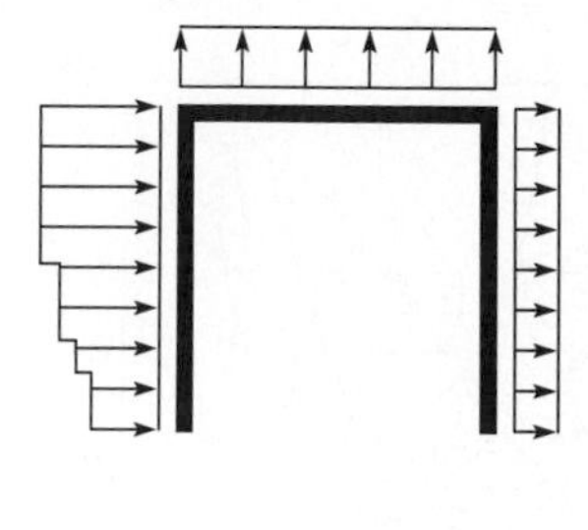

(a)

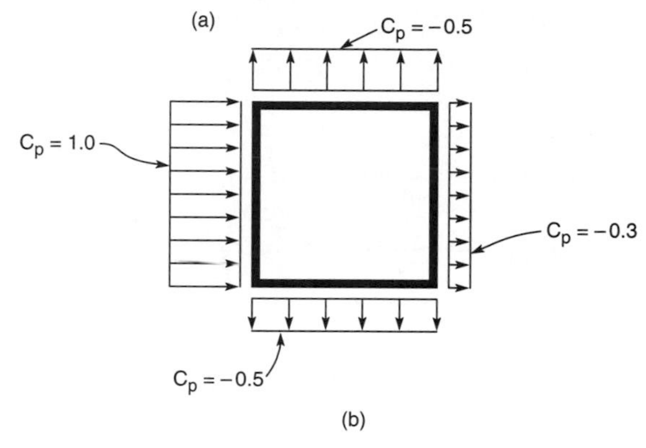


(b)

**Figure 2.13** *(Continued)*.

### 2.5.4.9 Pressure coefficient $C_q$

The coefficient $C_q$ given in Table 2.12 accounts for the shape and the location of the bulding and type of pressure loading involved. It is given as a constant for any particular shape.

**TABLE 2.11 Combined Height, Exposure and Gust Factor Coefficient $C_e$ (UBC 1994)**

| Height above average level of adjoining ground (feet) ×304.8 for mm | Exposure D | Exposure C | Exposure B |
|---|---|---|---|
| 0–15 | 1.39 | 1.06 | 0.62 |
| 20 | 1.45 | 1.13 | 0.67 |
| 25 | 1.50 | 1.19 | 0.72 |
| 30 | 1.54 | 1.23 | 0.76 |
| 40 | 1.62 | 1.31 | 0.84 |
| 60 | 1.73 | 1.43 | 0.95 |
| 80 | 1.81 | 1.53 | 1.04 |
| 100 | 1.88 | 1.61 | 1.13 |
| 120 | 1.93 | 1.67 | 1.20 |
| 160 | 2.02 | 1.79 | 1.31 |
| 200 | 2.10 | 1.87 | 1.42 |
| 300 | 2.23 | 2.05 | 1.63 |
| 400 | 2.34 | 2.19 | 1.80 |

Values for intermediate heights above 15 feet (4572 mm) may be interpolated.

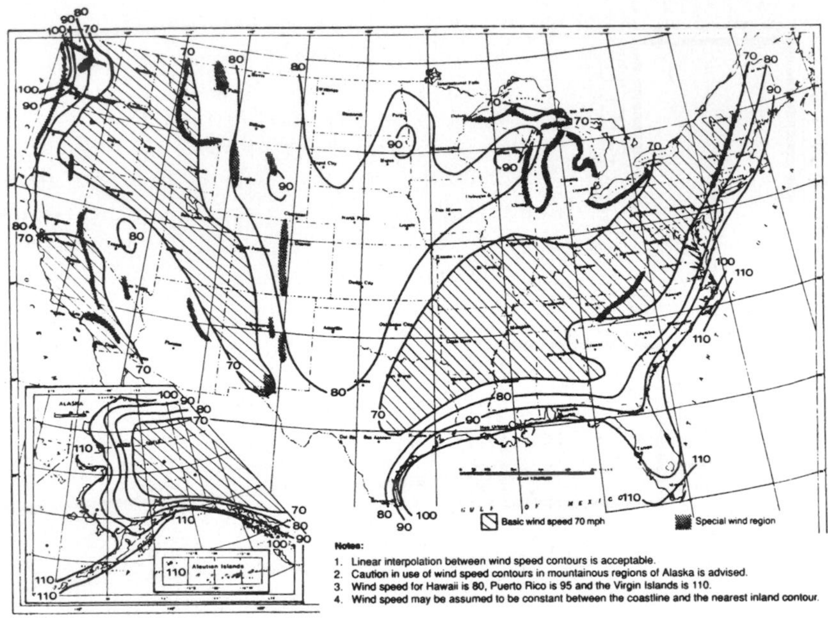


**Figure 2.14** UBC 1994: Basic wind speed map, mph.

Values of $C_q$ are specified in two parts: the first part, defined as for primary frames and systems, is for the design of an entire building. The second part is for the design of cladding systems referred to as

**TABLE 2.12 Pressure Coefficients $C_q$ for Primary Frames and Systems (UBC 1994)**

| Description | $C_q$ |
|---|---|
| Method 1 (Normal force method) | |
| Walls: | |
| Windward wall | 0.8 inward |
| Leeward wall | 0.5 outward |
| Roofs: | |
| Wind perpendicular to ridge | |
| Leeward roof or flat roof | 0.7 outward |
| Windward roof | |
| less than 2:12 (16.7%) | 0.7 owtward |
| Slope 2:12 (16.7%) to less than 9:12 (75%) | 0.9 outward or 0.3 inward |
| Slope 9:12 (75%) to 12:12 (100%) | 0.4 inward |
| Slope > (12:12 (100%) | 0.7 inward |
| Wind parallel to ridge and flat roofs | 0.7 outward |
| Method 2 (Projected area method) | 1.3 horizontal any direction |
| On vertical projected area | |
| Structures 40 feet (12,192 mm) or less in height | 1.4 horizontal any direction |
| Structures over 40 feet (12,192 mm) in height | |
| On horizontal projected area | 0.7 upward |

for "Elements and Components of Structure". Table 2.12 (UBC Table 16-H) gives pressure coefficient values $C_q$ for the first part.

Because of the gusting nature of wind, peak pressures do not occur simultaneously over large areas, therefore the wind design of primary frames and systems is governed by the average wind pressures acting over large areas. On the other hand, building components such as curtain walls and cladding are controlled by the instantaneous peak pressures and suctions acting over relatively small localized areas. For this reason, the pressure coefficients for building components are larger than those for the entire building.

The UBC wind pressures for primary systems are mainly a function of the building's height. Although the shape, roughness of the building exterior, and the aspect ratio of building plan play a significant role, these are ignored in the UBC. For example, even though wind load on a circular building is less than for a rectangular building, no reduction of forces is permitted.

**Methods 1 and 2** The UBC provides $C_q$ coefficients for two methods of determining wind pressures for primary frames (Table 2.12).

Method 1, the normal force method, is applicable to all structures, and must be applied to all gabled frames. It assumes that wind pressures act perpendicular to the surface of roof, windward and leeward walls simultaneously.

Method 2, the projected area method, is easier to use than Method 1 in that it assumes that wind pressures act on the projected horizontal and vertical areas of the structure instead of on individual surfaces of roof and walls.

Another noteworthy difference between Methods 1 and 2 is that Method 1 uses a constant value of $C_e$ based on mean roof height to calculate wind pressures on the leeward wall. Method 2 uses a $C_e$ value that varies with the height. Hence it is possible that Method 2 may underestimate the wind pressures on taller structures. For this reason, use of Method 2 is limited to structures of less than 200 ft (61 m).

### 2.5.4.10 Importance factors $I_w$

Importance factors $I_w$ are applied to increase wind loads for certain occupancy categories. The 1994 UBC gives five separate occupancy categories: Essential Facilities, Hazardous Facilities, Special Occupancy Structures, Standard Occupancy Structures and Miscellaneous Structures. Essential or Hazardous Facilities are assigned an Importance factor $I_w = 1.15$ which has the effect of increasing mean reference interval from 50 to 100 years. Special Structures, Standard

Occupancy Structures and Miscellaneous Structures have an Importance factor $I_w$ of 1.00. Office and residential buildings are typically assigned a Standard Occupancy factor of 1.00.

### 2.5.4.11 Design example, UBC 1994

In this example, wind pressures on primary wind-resisting systems will be determined for a 12 story building located in a wind zone of 80 mph.

#### Given

- Eleven-story communication building deemed necessary for post-disaster emergency communications, $I_w = 1.15$.
- Building height 120 ft (36.6 m) consisting of two bottom floors at 15 ft (4.6 m) and 9 typical floors at 10 ft (3.05 m).
- Exposure category $= C$.
- Basic wind speed $v = 83.4$ mph.
- Building width = 60 ft.

#### Required

Design wind pressures on primary wind-resisting system.

The design pressure is given by

$$p = C_e C_q q_s I_w$$

The values for combined height, exposure, and gust factor coefficient $C_e$ taken from Table 2.11 are tabulated in Table 2.13. Note for suction on the leeward face $C_e$ is at the roof height, and is constant for the full height of the building. The wind pressure $q_s$ corresponding to basic wind speed of 83.4 mph is given by

$$q_s = 0.00256v^2,$$
$$q_s = 0.00256 \times 83.4^2 = 17.80 \text{ psf}$$

The values of pressure coefficient $C_q$ (Table 2.12), using normal force method (Method 1) are 0.8 for inward pressure on the windward face, and 0.5 for the leeward face. The combined value of $0.8 + 0.5 = 1.3$ may be used throughout the height to calculate the total pressures on the primary wind resisting system. Observe Method 2 (projected area method) yields the same value of $C_q = 1.3$ for the horizontal pressures.

Design pressures and floor-by-floor wind loads are shown in Table 2.13. observe that wind pressure and suction on the lower half of the

**TABLE 2.13 Design Example (UBC 1994): Design pressures on Primary Wind-resisting System**

| Level 1 | Height above ground ft 2 | $C_e$ 3 windward | $C_e$ 3 leeward | $C_q$ 4 windward | $C_q$ 4 leeward | Windward pressure psf $p = C_eC_qq_sI_w$ 5 | Leeward suction psf $p = C_eC_qq_sI_w$ 6 | Design pressure psf 5 + 6 | Floor-by-floor load kips |
|---|---|---|---|---|---|---|---|---|---|
| Roof | 120 | 1.67 | 1.67 | +0.8 | −0.5 | 27.4 | 17.1 | 44.5 | 13.35 |
| 11 | 110 | 1.64 | 1.67 | +0.8 | −0.5 | 26.9 | 17.1 | 44.0 | 26.55 |
| 10 | 100 | 1.61 | 1.67 | +0.8 | −0.5 | 26.4 | 17.1 | 43.5 | 26.25 |
| 9 | 90 | 1.57 | 1.67 | +0.8 | −0.5 | 25.7 | 17.1 | 42.8 | 25.9 |
| 8 | 80 | 1.53 | 1.67 | +0.8 | −0.5 | 25.1 | 17.1 | 42.2 | 25.5 |
| 7 | 70 | 1.48 | 1.67 | +0.8 | −0.5 | 24.3 | 17.1 | 41.4 | 24.1 |
| 6 | 60 | 1.43 | 1.67 | +0.8 | −0.5 | 23.5 | 17.1 | 40.6 | 24.6 |
| 5 | 50 | 1.37 | 1.67 | +0.8 | −0.5 | 22.5 | 17.1 | 39.6 | 24.0 |
| 4 | 40 | 1.31 | 1.67 | +0.8 | −0.5 | 21.5 | 17.1 | 38.6 | 23.5 |
| 3 | 30 | 1.23 | 1.67 | +0.8 | −0.5 | 20.2 | 17.1 | 37.3 | 28.4 |
| 2 | 15 | 1.06 | 1.67 | +0.8 | −0.5 | 17.4 | 17.1 | 34.5 | 32.3 |

$q_s = 17.80$ psf; $I_w = 1.15$.

first story (between the ground and 7.5 ft above ground) is transmitted directly into the ground. The wind load at each level is obtained by multiplying the tributory area for the level by the average of design pressures above and below that level. For example,

$$\text{wind force at level } 10 = \frac{60 \times 10(44 + 43.5)}{1000 \times 2} = 26.25 \text{ kips.}$$

### 2.5.5 ANSI/ASCE 7-93

The full title of this standard is *American Society of Civil Engineers Minimum Design Loads for Buildings and Other Structures.* It was first published in 1972 as an ANSI Standard, and subsequently expanded and revised in 1982, 1988, 1993 and very recently in 1995 with major modifications. In this section wind provisions of ASCE 7-93 will be discussed; the provisions of ASCE 7-95 are discussed later.

In one of its nine sections, ASCE 7-93 prescribes the procedures for calculating design wind loads. It takes into consideration the differences between those wind loads which act on the building as a whole and those which dictate the design of individual structural components and cladding of buildings. Two procedures, one analytical and the other experimental, using wind tunnel tests and similar tests employing fluids other than air are given in the ANSI Standard. The analytical procedure applies to a majority of buildings, with due consideration being given to the load magnification effect caused by gusts that may be in resonance with the along-wind vibrations of the structure. This procedure does not consider the phenomenon of vortex shedding, nor does it consider buildings having unusual shapes or response characteristics. Since it is not possible to analytically define site locations which can result in channeling effects, a wind tunnel procedure is recommended for such situations.

Basic wind speeds for any location in the continental United States and Alaska are shown on a map having isotachs representing the fastest-mile velocities at 33 ft (10 m) above the ground. For Hawaii and Puerto Rico, the basic wind speeds are given in a table as 80 and 95 mph (35.7 and 42.4 m/s), respectively. The map is for a 50-year recurrence interval for Exposure B, consisting of flat, open country and grasslands with an open terrain and scattered obstructions generally less than 30 ft (9 m) in height. The minimum basic wind speed provided in the standard is 70 mph (31.3 m/s). Upgrading of minimum wind speed for special topographies such as mountain terrain, gorges, and ocean fronts is recommended.

The wind speed map for the United States and adjoining land masses is based on the data collected over a long period of time at 129 weather stations located throughout the country. The maximum velocity to be expected at any location can be found simply by reference to the wind velocity map.

Because of the relatively rare occurrence of hurricanes, records of sufficient length are not available to map the wind speed contours in the hurricane-prone regions. To alleviate this lack of available data, mathematical simulation of hurricanes has been used to analytically generate wind speed records in 58 coastal points. These have been incorporated into the wind speed map.

#### 2.5.5.1 Overview

A simple relationship between wind buffeting a building, and the corresponding pressures or suction induced on the surface of the building is given in the form of a chain equation:

$$\begin{aligned}\text{[wind pressure]} = {} & \text{[reference velocity pressure]} \\ & \times \text{[dynamic gust response factor]} \\ & \times \text{[aerodynamic shape factor]}\end{aligned}$$

$$p_z = q_z \times GRF \times C_p \tag{2.13}$$

where

$p_z$ = design wind pressure or suction, in psf, at height $z$ above ground level

$q_z$ = velocity pressure, in psf, determined at height $z$ above ground level

$GRF$ = Dynamic Response Factor, dimensionless, which magnifies the mean wind load to include the effect of: (i) random wind gusts; (ii) fluctuating forces induced by the motion of the structure itself through the wind

$C_p$ = pressure coefficient which varies with height acting as pressure (positive load) on windward face, and as suction (negative load) on non-windward faces and roof

The velocity pressure, $q_z$, is given by the equation

$$q_z = 0.00256 K_z (IV)^2 \tag{2.14}$$

$$q_h = 0.00256 K_h (IV)^2 \tag{2.15}$$

where

$K_z$ and $K_h$ = combined velocity pressure, exposure coefficients (Table 2.15) dimensionless, which take into account changes in wind speed above ground and nature of terrain, i.e., Exposure A, B, C, or D.

$I$ = Importance factor (Table 2.16) dimensionless, which modifies: (i) the 50-year wind to 100 or 25-year wind, depending on the building category and; (ii) hurricane wind speeds to probabilities that are consistent with the other wind speeds.

$V$ = Reference velocity given in wind speed map (Fig. 2.15).

The exposure categories A, B, C and D as defined below, give prescriptive definitions for classifying the terrain surrounding the building site.

*Exposure A.* Large city centers with at least 50 percent of the buildings having a height in excess of 70 ft (21.3 m). Use of this exposure category shall be limited to those areas for which terrain representative of Exposure A prevails in the upwind direction for a distance of at least one-half mile (0.8 km) or 10 times the height of the building or structure, whichever is greater. Possible channeling effects or increased velocity pressures due to the building or structure being located in the wake of adjacent buildings shall be taken into account.

*Exposure B.* Urban and suburban areas, wooded areas, or other terrain with numerously closely spaced obstructions having the size of single-family dwellings or larger. Use of this exposure category shall be limited to those areas for which terrain representative of Exposure B prevails in the upwind direction for a distance of at least 1500 ft (460 m) or 10 times the height of the building or structure, whichever is greater.

*Exposure C.* Open terrain with scattered obstructions having heights generally less than 30 ft (9.1 m). This category includes flat open country and grasslands.

*Exposure D.* Flat, unobstructed areas exposed to wind over large bodies of water. This exposure shall apply only to those buildings and other structures exposed to the wind coming from over the water. Exposure D extends inland from the shoreline a distance of 1500 ft (460 m) or 10 times the height of the building or structure, whichever is greater.

The data given in the wind speed contour, Fig. 2.15, are for wind speeds measured primarily for Exposure C, at standard height of 33 ft. Two steps are necessary to convert the standard speed to the design speed at height $z$, at the building site. First, the basic wind speed, $V_{33}$, at the standard height is converted gradient speed, $V_z$, by the relation

$$V_z = V_{33}\left(\frac{900}{33}\right)^{1/7} \tag{2.14}$$

Second, the gradient speed is converted to the required height and exposure by the relation

$$V_z = \left[V_{33}\left(\frac{900}{33}\right)^{1/7}\right]\left[\frac{z}{z_g}\right]^{1/\alpha} \tag{2.15}$$

$K_z$ is a pressure coefficient rather than a wind speed coefficient: hence, the wind speeds have to be squared to obtain the velocity pressure exposure coefficients. The equation for $K_z$ becomes

$$\begin{aligned} K_z &= \left(\frac{V_z}{V_{33}}\right)^2 \\ &= \left(\frac{900}{33}\right)^{2/7}\left(\frac{z}{z_g}\right)^{2/\alpha} \\ &= 2.58\left(\frac{z}{z_g}\right)^{2/\alpha} \quad \text{for } 15\text{ ft} \leqslant z \leqslant z_g \end{aligned} \tag{2.16}$$

Below 15 ft height, $K_z$ is taken as a constant determined at height 15 ft by the relation

$$K_z = 2.58\left(\frac{15}{z_g}\right)^{2/\alpha} \tag{2.17}$$

A step-by-step procedure for determining wind forces for main wind-force resisting system follows.

#### 2.5.5.2 Design pressure for the main wind-force resisting system

The procedure for calculating wind loads may be summarized as follows:

Step 1. Determine *I,* the importance factor, for wind from the nature of occupancy of the building and its proximity to hurricane-prone oceanlines. Tables 2.14 and 2.16 give the type of occupancy and the importance factor *I,* respectively. Buildings are classified into four categories depending upon the primary purpose of occupancy. Assembly buildings and essential facilities are treated as either category II or III, requiring design for larger wind loads. Buildings with low

**TABLE 2.14 Classification of Buildings and Other Structures for Wind, Snow, and Earthquake Loads (ANSI/ASCE 7-93)**

| Nature of occupancy | Category |
|---|---|
| All buildings and structures except those listed below | I |
| Buildings and structures where the primary occupancy is one in which more than 300 people congregate in one area | II |
| Buildings and structures designated as essential facilities, including, but not limited to:<br>1. Hospital and other medical facilities having surgery or emergency treatment areas<br>2. Fire or rescue and police stations<br>3. Primary communication facilities and disaster operation centers<br>4. Power stations and other utilities required in an emergency<br>5. Structures having critical national defense capabilities | III |
| Buildings and structures that represent a low hazard to human life in the event of failure, such as agricultural buildings, certain temporary facilities, and minor storage facilities | IV |

**TABLE 2.15 Velocity Pressure Exposure Coefficient $K_z$ (ASCE 7-93)**

| Height above ground level, $s$ (feet) | $K_z$ | | | |
|---|---|---|---|---|
| | Exposure A | Exposure B | Exposure C | Exposure D |
| 0–15 | 0.12 | 0.37 | 0.80 | 1.20 |
| 20 | 0.15 | 0.42 | 0.87 | 1.27 |
| 25 | 0.17 | 0.46 | 0.93 | 1.32 |
| 30 | 0.19 | 0.50 | 0.98 | 1.37 |
| 40 | 0.23 | 0.57 | 1.06 | 1.46 |
| 50 | 0.27 | 0.63 | 1.13 | 1.52 |
| 60 | 0.30 | 0.68 | 1.19 | 1.58 |
| 70 | 0.33 | 0.73 | 1.24 | 1.63 |
| 80 | 0.37 | 0.77 | 1.29 | 1.67 |
| 90 | 0.40 | 0.82 | 1.34 | 1.71 |
| 100 | 0.42 | 0.86 | 1.38 | 1.75 |
| 120 | 0.48 | 0.93 | 1.45 | 1.81 |
| 140 | 0.53 | 0.99 | 1.52 | 1.87 |
| 160 | 0.58 | 1.05 | 1.58 | 1.92 |
| 180 | 0.63 | 1.11 | 1.63 | 1.97 |
| 200 | 0.67 | 1.16 | 1.68 | 2.01 |
| 250 | 0.78 | 1.28 | 1.79 | 2.10 |
| 300 | 0.88 | 1.39 | 1.88 | 2.18 |
| 350 | 0.98 | 1.49 | 1.97 | 2.25 |
| 400 | 1.07 | 1.58 | 2.05 | 2.31 |
| 450 | 1.16 | 1.67 | 2.12 | 2.36 |
| 500 | 1.24 | 1.75 | 2.18 | 2.41 |

hazard to human life in the event of failure are considered as category IV, and all other buildings are classified as category I. $I$ essentially recognizes that buildings near hurricane-prone oceanlines are buffeted by stronger winds. For example, an essential building of category II located adjacent to the Atlantic Ocean or Gulf of Mexico has a value

**TABLE 2.16 Importance Factor $I$, Wind Loads (ASCE 7-93)**

| Category | $I$ | |
|---|---|---|
| | 100 miles from hurricane oceanline and in other areas | At hurricane oceanline |
| I | 1.00 | 1.05 |
| II | 1.07 | 1.11 |
| III | 1.07 | 1.11 |
| IV | 0.95 | 1.00 |

(1) The building and structure classification categories are listed in Table 2.14.
(2) For regions between the hurricane oceanline and 100 miles inland the importance factor $I$ shall be determined by linear interpolation.
(3) Hurricane oceanlines are the Atlantic and Gulf of Mexico coastal areas.

of 1.11 for $I$ as compared to 1.07 for the same building located inland 100 miles (1690 km) from the ocean.

Step 2. Select the basic wind speed $V$ for the location of the building from the 50-year mean recurrence interval map which is shown in Fig. 2.15.

Step 3. Determine the ratio of height to least horizontal dimension $H/W$, and the fundamental frequency $f$, of the building. If $H/W < 5$ and $f > 1$ Hz, the building can be considered nonflexible, permitting simpler expressions for determining the gust factor $G$. If not, the building must be considered flexible requiring a rational method for determining the gust factor, $\bar{G}$. Two such methods use empherical parameters and graphs, as given in the ASCE 7-95 commentary and in the National Building Code of Canada (NBC, 1990). These are considered later.

Step 4. Determine the external pressure coefficient $C_p$ for average loads on main wind-force-resisting systems from a knowledge of plan dimensions and building elevation.

Step 5. Determine the exposure category that adequately reflects the characteristics of the ground surface irregularities at

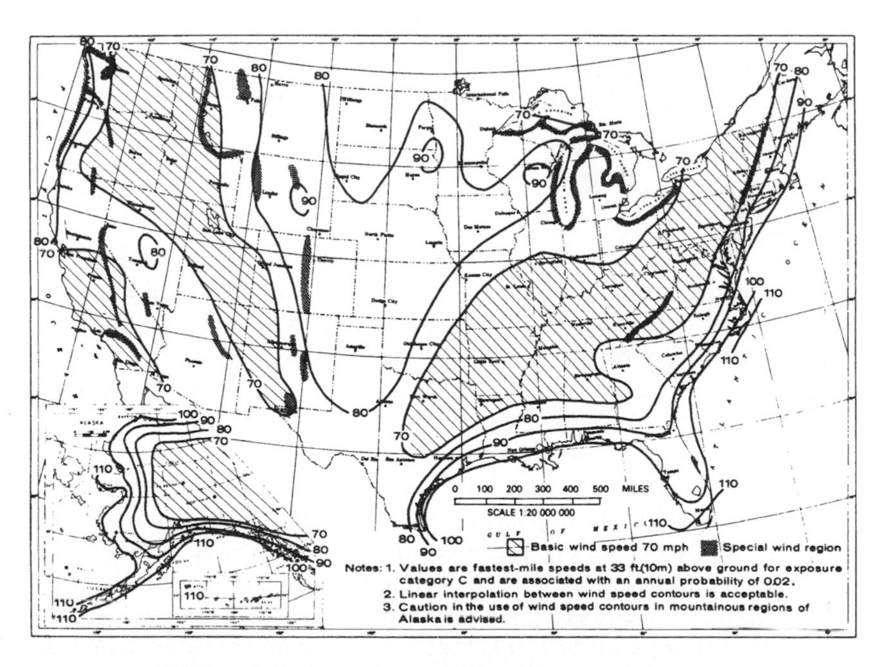


**Figure 2.15** ASCE 7-93: Basic wind speed map, mph.

the building site, taking into account both natural topography and man-made features. As mentioned previously, ASCE 7-93 has four categories of exposures to take into account the terrain effect on design wind loads for buildings. An abbreviated description of these categories follows:

*Exposure A*: Heavily built-up urban locations protected by surrounding buildings with at least 50 percent of the buildings being more than 70 ft (21.4 m) in height.

*Exposure B*: Urban and suburban areas, towns, city outskirts, wooded areas, or other terrain with obstructions having the size of single-family dwellings or larger.

*Exposure C*: Flat, open country and grasslands with scattered obstructions [generally less than 30 ft (9.14 m)].

*Exposure D*: Flat, unobstructed coastal areas directly exposed to wind flowing over large bodies of water.

The gradient height varies from a high of 1500 ft (457 m) for densely built up urban locations (Exposure A) to a low of 700 ft (213.4 m) for coastal areas (Exposure D). The corresponding values for Exposures B and C are, respectively, 1200 and 900 ft (366 and 274 m).

Step 6. Determine the velocity pressure exposure coefficient $K_z$, which depends on the type of exposure and height above ground. This is determined by using the power law expressions

$$K_z = 2.58\left(\frac{Z}{Z_g}\right)^{2/\alpha} \quad \text{for } Z \geq 15 \text{ ft} \tag{2.18}$$

$$K_z = 2.58\left(\frac{15}{Z_g}\right)^{2/\alpha} \quad \text{for } Z < 15 \text{ ft} \tag{2.19}$$

$Z_g$ is the gradient height above which the frictional effect of terrain becomes negligible. As noted earlier, this height varies with the characteristics of the ground surface irregularities of the building site, which arise as a result of natural topographic variations as well as manmade features.

The power coefficient $\alpha$ is the exponent for velocity increase in height and has values of 3.0, 4.5, 7.0, and 10.0, respectively, for Exposures A, B, C, and D. To eliminte the necessity of making these computations, ASCE 7-93 gives a table for computed values of $K_z$ for various exposures up to a height of 500 ft (152.6 m) (Table 2.15). An extended version for buildings up to 1500 ft (457 m) is given in Table 2.17 and in Fig. 2.16.

**TABLE 2.17 Velocity Pressure Exposure Coefficient $K_z$ (ASCE 7-93)**

| Height above ground level $Z$, ft | $K_z$ | | | |
|---|---|---|---|---|
| | Exposure A | Exposure B | Exposure C | Exposure D |
| 0 | 0.00 | 0.00 | 0.00 | 0.00 |
| 15 | 0.12 | 0.37 | 0.80 | 1.20 |
| 20 | 0.15 | 0.42 | 0.87 | 1.27 |
| 25 | 0.17 | 0.46 | 0.93 | 1.32 |
| 30 | 0.19 | 0.50 | 0.98 | 1.37 |
| 40 | 0.23 | 0.57 | 1.06 | 1.46 |
| 50 | 0.27 | 0.63 | 1.13 | 1.52 |
| 60 | 0.30 | 0.68 | 1.19 | 1.58 |
| 70 | 0.33 | 0.73 | 1.24 | 1.63 |
| 80 | 0.37 | 0.77 | 1.29 | 1.67 |
| 90 | 0.40 | 0.82 | 1.34 | 1.71 |
| 100 | 0.42 | 0.86 | 1.38 | 1.75 |
| 120 | 0.48 | 0.93 | 1.45 | 1.81 |
| 140 | 0.53 | 0.99 | 1.52 | 1.87 |
| 160 | 0.58 | 1.05 | 1.58 | 1.92 |
| 180 | 0.63 | 1.11 | 1.63 | 1.97 |
| 200 | 0.67 | 1.16 | 1.68 | 2.01 |
| 250 | 0.78 | 1.28 | 1.79 | 2.10 |
| 300 | 0.88 | 1.39 | 1.88 | 2.18 |
| 350 | 0.98 | 1.49 | 1.97 | 2.25 |
| 400 | 1.07 | 1.58 | 2.05 | 2.31 |
| 450 | 1.16 | 1.67 | 2.12 | 2.36 |
| 500 | 1.24 | 1.75 | 2.18 | 2.41 |
| 550 | 1.32 | 1.82 | 2.24 | 2.46 |
| 600 | 1.40 | 1.90 | 2.30 | 2.50 |
| 650 | 1.48 | 1.96 | 2.35 | 2.54 |
| 700 | 1.55 | 2.03 | 2.40 | 2.58 |
| 750 | 1.63 | 2.09 | 2.45 | 2.62 |
| 800 | 1.70 | 2.15 | 2.49 | 2.65 |
| 850 | 1.77 | 2.21 | 2.54 | 2.68 |
| 900 | 1.84 | 2.27 | 2.58 | 2.71 |
| 950 | 1.90 | 2.33 | 2.62 | 2.74 |
| 1000 | 1.97 | 2.38 | 2.66 | 2.77 |
| 1050 | 2.03 | 2.43 | 2.70 | 2.80 |
| 1100 | 2.10 | 2.48 | 2.73 | 2.82 |
| 1150 | 2.16 | 2.53 | 2.77 | 2.85 |
| 1200 | 2.22 | 2.58 | 2.80 | 2.87 |
| 1250 | 2.28 | 2.63 | 2.83 | 2.90 |
| 1300 | 2.35 | 2.67 | 2.87 | 2.92 |
| 1350 | 2.40 | 2.72 | 2.90 | 2.94 |
| 1400 | 2.46 | 2.76 | 2.93 | 2.96 |
| 1450 | 2.52 | 2.81 | 2.96 | 2.98 |
| 1500 | 2.58 | 2.85 | 2.99 | 3.00 |

Step 7. Determine the velocity pressure $q_z$ in pounds per square foot

$$q_z = 0.0025K_z(IV)^2 \tag{2.20}$$

Step 8. Determine the design wind pressure $P$ by the chain equation

$$p = q\bar{G}C_p \tag{2.21}$$

where $q = q_z$ for windward wall at height $Z$ above ground
$q = q_h$ for leeward wall at mean roof height for sloping roofs, and at roof height for flat roots
$\bar{G}$ = the gust response factor
$C_p$ = external pressure coefficient

Step 9. Finally, the level-by-level wind load required for the lateral analysis is obtained by multiplying the exposed area tributory to the level by the corresponding value of $P$ at the floor height.

#### 2.5.5.3 Design wind pressure on components and cladding

For purposes of calculating design loads, building surfaces including the roof are grouped into seven zones as shown in Fig. 2.17. Values for $GC_p$, the product of gust factor and external pressure coefficient are given for the seven zones in terms of the tributary area of the cladding element exposed to the wind. Internal pressure coefficients are significant in the design of cladding systems and depend, among other things, on the amount and disposition of openings. Therefore, values $GC_{pi}$, the product of gust response factor $G$, and internal pressure $C_{pi}$ are given for two types of opening conditions, as defined in Table 2.18. Both the positive and negative values of $GC_{pi}$ are combined with the corresponding value $GC_p$ to obtain critical design pressures and suctions. As before, the design pressure is obtained by the chain equation

$$p = q[GC_p - GC_{pi}]$$

in which $q = q_z$ for the windward, and $q = q_h$ for the leeward and side walls and roof. An example, given shortly, will clarify the procedure.

#### 2.5.5.4 External pressure coefficient $C_p$

Wind, when buffeted by buildings of different shapes, produces quite a different loading on each face of the building. For example, wind acting on a rectangular building, results in positive (inward-acting) load only on the windward face; the other three sides are subjected to negative (outward-acting) or suction loads. The pressure distribution becomes even more complicated when wind direction is angular to the building face. The distribution of positive and negative pressures is also influenced by the plan aspect ratio of the building. The pressure coefficients on the roof vary with the direction of wind and also with the pitch of the roof. The ANSI standard has assembled a table of pressure coefficient values for walls and roofs of buildings based on wind-tunnel tests, actual data collected from full-scale measurements, and data published in other standards, such as the

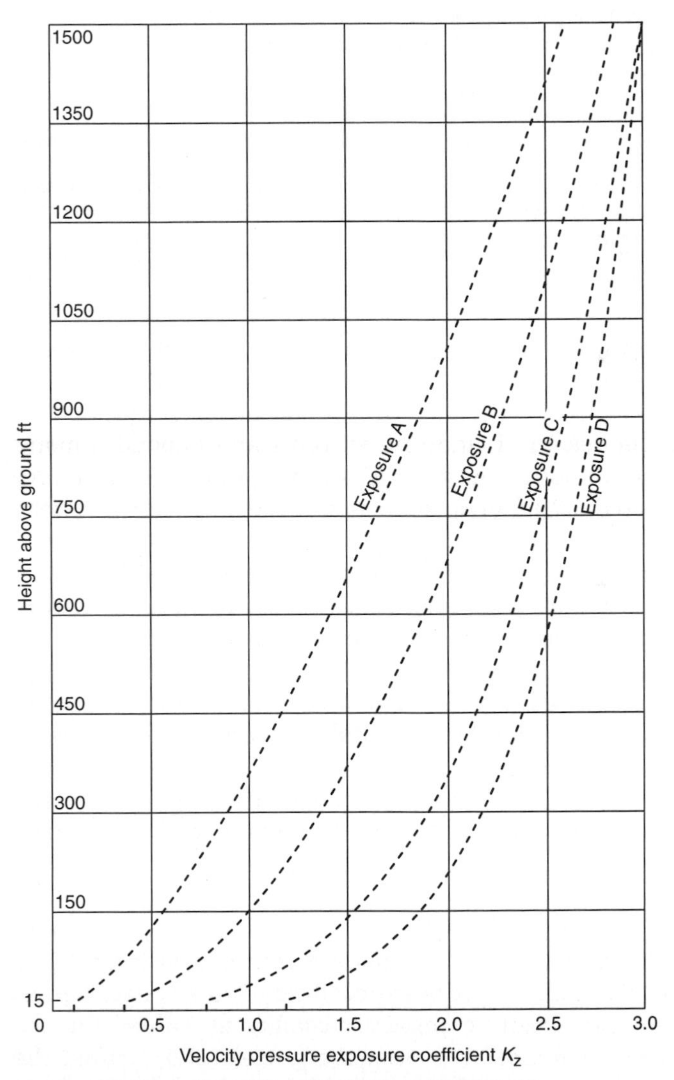


**Figure 2.16** Variation of $K_z$ with height (ASCE 7-93).

**TABLE 2.18 Internal Pressure Coefficient $GC_{pi}$ (ASCE 7-93)**

| | Condition | $GC_{pi}$ |
|---|---|---|
| Condition I | All conditions except as noted under condition II. | +0.25<br>−0.25 |
| Condition II | Buildings in which both of the following are met:<br>1. Percentage of openings in one wall exceeds the sum of the percentages of openings in the remaining walls and roof surfaces by 5% or more, and<br>2. Percentage of openings in any one of the remaining walls or roof do not exceed 20%. | +0.75<br>+0.25 |

(1) Values are to be used with $q_z$ or $q_h$.
(2) Plus and minus signs signify pressures acting toward and away from the surfaces, respectively.
(3) To ascertain the critical load requirements for the appropriate condition, two cases shall be considered a positive value of $GC_{pi}$ applied simultaneously to all surfaces, and a negative value of $GC_{pi}$ applied to all surfaces.
(4) Percentage of openings in a wall or roof surface is given by ratio of area of openings to gross area for the wall or roof surface considered.

Australian standard. The pressure coefficient values for the windward wall are referenced to the velocity pressure $q_z$. The design wind pressure, therefore, varies with the height above the ground. The coefficients for suction on the leeward wall are referenced to a single velocity pressure that is evaluated at the mean roof height. If the building has a flat roof, it is referenced to the velocity pressure at the top of the building.

Two sets of external pressure coefficients are given, one for determining the overall design forces on the wind-bracing system of the building and the other to determine the design forces on the components and cladding of the building. For the design of cladding, a distinction is made between buildings with a mean roof height of 60 ft (18.3 m) or smaller, and taller buildings. Separate values of pressure coefficients are given for each type.

The ANSI values of external pressure coefficients to be used in the design of bracing systems of the buildings are reproduced in pictorial form in Fig. 2.18. Linear interpolation is permitted for intermediate values of plan aspect ratio $L/B$.

The external pressure coefficients to be used in the design of components and cladding of tall buildings are reproduced from ANSI in Fig. 2.17. As in the case of pressure coefficients for the design of the bracing system, the positive pressure-coefficient values for the windward wall are referenced to velocity pressure $q_z$; thus the positive pressure on the cladding varies with the height above

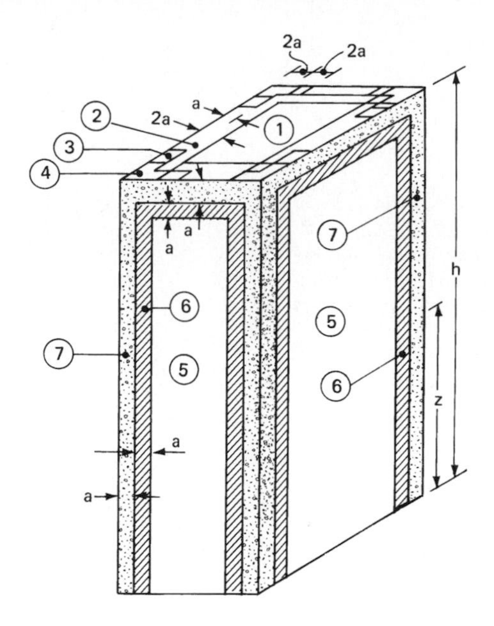


**Figure 2.17** External pressure coefficients for cladding design (ASCE 7-93). Schematic elevation identifying various zones.

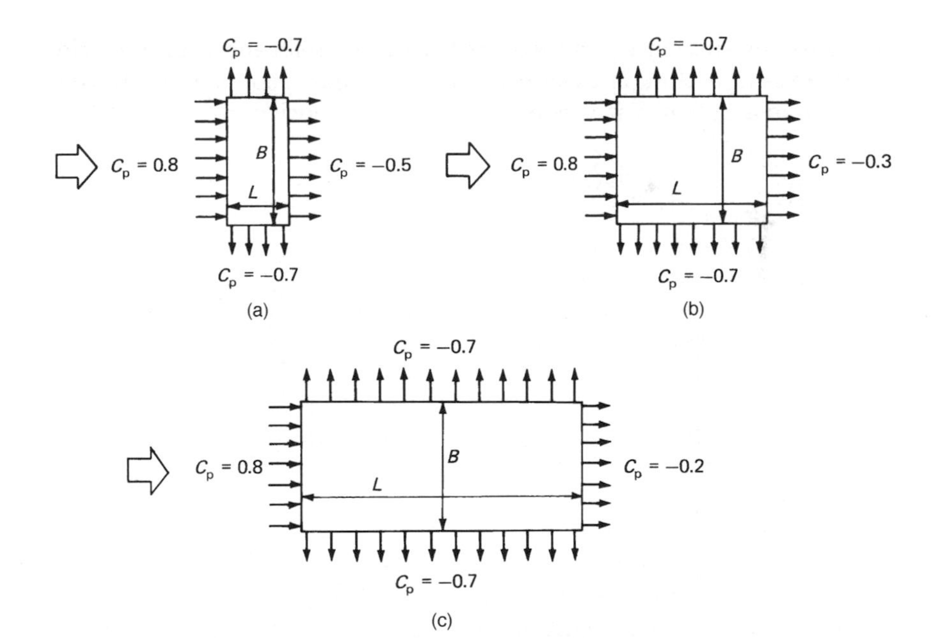


**Figure 2.18** Variation of external wind pressure coefficient $C_p$ with plan dimensions: (a) $0 \leq L/B \leq 1$; (b) $L/B = 2$; (c) $L/B > 4$, (1) plan, (2) elevations.

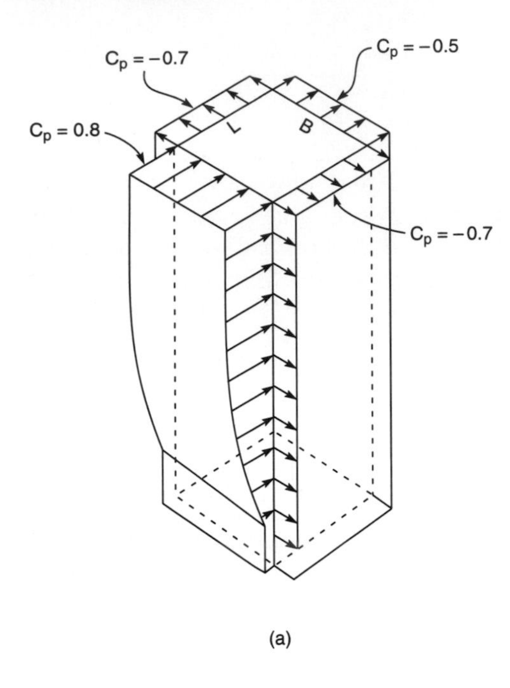


(a)

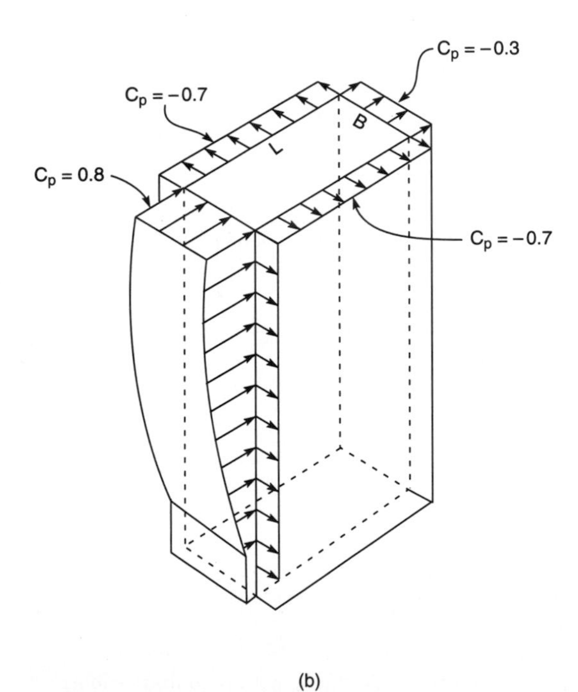


(b)

**Figure 2.18** (*Continued*).

ground. The negative pressure coefficient values for the cladding design on the leeward face and the sides of the building are referenced to the velocity pressure $q_h$ evaluated at the mean roof height for sloped roof, and at roof height for flat roof. Since wind can blow from any direction, each component and cladding member should be designed for the maximum positive and negative pressure. In determining the design forces on the cladding, due consideration should be given to its location and its tributary area. Similar to the live load reduction allowed for gravity design of members based on the tributary area of loading, some reduction in design pressure is allowed for in the lateral design of cladding components.

#### 2.5.5.5 Internal pressure coefficients $C_{pi}$

Another important consideration in the design of cladding is the pressure existing within the building. This pressure is the cumulative effect of wind seepage through the building enclosure, mechanical ventilation and air conditioning, and stack effect. The stack effect, which is also called the chimney effect, is a result of temperature difference between the interior and exterior of the building and is

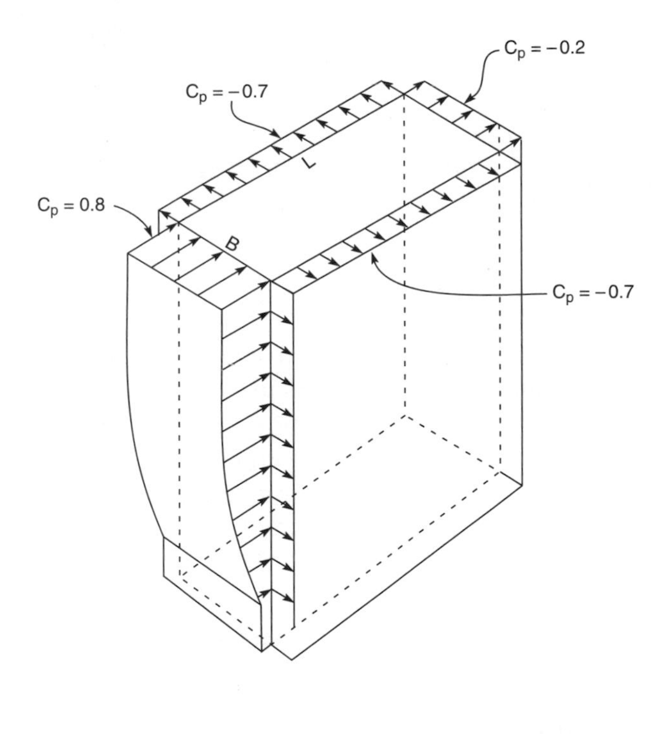


**Figure 2.18** *(Continued).*

particularly significant in tall buildings in colder climates. The internal pressure also depends on either the intentional openings in curtain walls, such as operable windows, or accidental openings likely to be caused because of glass breakage during windstorms. Because loss of window glass during high winds cannot be predicted with any certainty, the net internal pressure is variable in character and seldom can be predicted with great accuracy. On the basis of information gathered through wind-tunnel tests conducted primarily on low-rise buildings, ANSI gives a value of $\pm 0.25$ for $GC_{pi}$ the internal pressure coefficient for buildings with uniformly distributed openings in the walls. For other cases, the value of $GC_{pi}$ could vary from a positive value of 0.75 to a suction value of $-0.25$. Note the suffix $G$ in the expression for internal pressure coefficient, which signifies that the values are combined values of gust response and internal pressure coefficient.

#### 2.5.5.6 Design examples: (1) non-flexible building; height < 500 ft

In this example, design wind pressures on the main wind-force-resisting system and on components and cladding of a medium-rise building, located in an urban area of Las Vegas, Nevada will be determined.

The building data are as follows:

| | |
|---|---|
| Plan dimensions: | 60 × 120 ft (18.28 × 36.57 m) |
| Building height: | 14 floors at 10 ft floor-to-floor = 10 × 14 = 140 ft (42.67 m) |
| Fundamental frequency: | 1.1 Hertz |
| Structural framing: | Reinforced concrete flat plate for the floors. Closely spaced concrete columns and spandrels at the perimeter for resisting lateral loads |
| Exterior cladding: | Glass and mullion system with mullions at 5 ft (1.52 m) spacing, spanning 8 ft (2.43 m) between spandrels. Glazing panels are 5 × 5 ft (1.52 × 1.52 m) |

**Required**

Wind pressures for the design of: (i) primary lateral system; and (ii) components and cladding.

**Loads on primary lateral system**

Step 1. Building classification. A typical office building is not generally considered an essential facility in the aftermath of a windstorm, nor is its primary function for occupancy by more than 300 persons in one area. Therefore, the example building is judged to be in Type 1 category. However, before a building is classified into a category, it is good practice to ascertain with building owners and plan-check officials, that the category is consistent with their perception of building performance.

Results of Step 1: Category 1, $I = 1.0$

Step 2. Basic wind speed $V$. The wind speed map, Fig. 2.15, indicates that Las Vegas is between wind contours of 70 and 80 mph. An average value of 75 mph is considered appropriate for the building. Again, it is good practice to confirm the design wind speed with the local plan-check officials.

Result of Step 2: $V = 75$ mph

Step 3. Gust response factor $G$ or $\bar{G}$. The building height is less than 500 ft, the height-to-width ratio is less than 5, and its natural frequency of 1.1 Hertz is more than 1.0 Hertz. Therefore, the building may be considered a non-flexible building. Only one value of gust response factor, $G$, calculated at the building roof level, is required for determining design loads on the main lateral system. The value of $G$ is obtained from Table 2.19, for $h = 140$ ft. Thus $G = 1.26$.

Results of Step 3: $G = 1.26$

Step 4. External pressure coefficient $C_p$. From the given plan dimensions, the ratio of building width to depth ratio, $\frac{L}{B} = \frac{60}{120} = 0.5$ for wind parallel to the 60 ft face. For wind parallel to 120 ft face, the ratio $\frac{L}{B} = \frac{120}{60} = 2.0$. From Fig. 2.18 the following values of $C_p$ are obtained.

$C_p = 0.8$ for the windward wall

$C_p = 0.5$ for leeward wall, wind parallel to 60 ft face

$C_p = 0.3$ for leeward wall, wind parallel, to 120 ft face

Roof pressures and internal pressures are not relevant in determining wind loads for primary lateral systems; internal pressures and sections acting on the windward and leeward walls cancel out without adding or subtracting to the overall wind loads. Roof suction which results in uplift

**TABLE 2.19 Gust Response Factors $G_h$ and $G_z$ (ASCE 7-93)**

| Height above ground level $z$, (feet) | $G_h$ and $G_z$ | | | |
|---|---|---|---|---|
| | Exposure A | Exposure B | Exposure C | Exposure D |
| 0–15 | 2.36 | 1.65 | 1.32 | 1.15 |
| 20 | 2.20 | 1.59 | 1.29 | 1.14 |
| 25 | 2.09 | 1.54 | 1.27 | 1.13 |
| 30 | 2.01 | 1.51 | 1.26 | 1.12 |
| 40 | 1.88 | 1.46 | 1.23 | 1.11 |
| 50 | 1.79 | 1.42 | 1.21 | 1.10 |
| 60 | 1.73 | 1.39 | 1.20 | 1.09 |
| 70 | 1.67 | 1.36 | 1.19 | 1.08 |
| 80 | 1.63 | 1.34 | 1.18 | 1.08 |
| 90 | 1.59 | 1.32 | 1.17 | 1.07 |
| 100 | 1.56 | 1.31 | 1.16 | 1.07 |
| 120 | 1.50 | 1.28 | 1.15 | 1.06 |
| 140 | 1.46 | 1.26 | 1.14 | 1.05 |
| 160 | 1.43 | 1.24 | 1.13 | 1.05 |
| 180 | 1.40 | 1.23 | 1.12 | 1.04 |
| 200 | 1.37 | 1.21 | 1.11 | 1.04 |
| 250 | 1.32 | 1.19 | 1.10 | 1.03 |
| 300 | 1.28 | 1.16 | 1.09 | 1.02 |
| 350 | 1.25 | 1.15 | 1.08 | 1.02 |
| 400 | 1.22 | 1.13 | 1.07 | 1.01 |
| 450 | 1.20 | 1.12 | 1.06 | 1.01 |
| 500 | 1.18 | 1.11 | 1.06 | 1.00 |

(1) For main wind-force resisting systems, use building or structure height $h = z$.
(2) Linear interpolation is acceptable for intermediate values of $z$.
(3) For height above ground of more than 500 feet. Use Eq. $G_z = 0.65 + 3.65T_z$. See worked examples.
(4) Value of gust response factor shall be not less than 1.0.

forces is generally neglected in the design of a primary lateral system.

Results of Step 4 are $C_p = 0.8 + 0.5 = 1.3$ for wind parallel to 60 ft dimension, and $C_p = 0.8 + 0.3 = 1.1$ for wind parallel to 120 ft dimension

Step 5. Building exposure. Since the building is located in an urban terrain, the exposure category is judged to be B.

Results of Step 5: Exposure B

Step 6. Combined velocity pressure, exposure coefficient $K_z$. The gradient height, $Z_g$, and the power coefficient $\alpha$ for Exposure B are 1200 ft and 4.5 respectively. The values for $K_z$ are calculated by the relation:

$$K_z = 2.58\left(\frac{z}{1200}\right)^{2/4.5}$$

**TABLE 2.20 Design Example, Non-Flexible Building (ASCE 7-93)**

| | Values of $k_z$ and $q_z$ | | | | | | |
|---|---|---|---|---|---|---|---|
| | Windward | | Leeward | | Design pressure | | |
| (1) Height (ft) | (2) $K_z$ | (3) $q_z$ | (4) $K_h$ | (5) $q_h$ | (6) Windward | (7) Leeward | (8) Total |
| 140 (roof) | 0.99 | 14.25 | 0.99 | 14.25 | 14.4 | 9.0 | 23.4 |
| 100 | 0.86 | 12.38 | 0.99 | 14.25 | 12.5 | 9.0 | 21.5 |
| 80 | 0.77 | 11.08 | 0.99 | 14.25 | 11.2 | 9.0 | 20.2 |
| 60 | 0.68 | 9.79 | 0.99 | 14.25 | 9.9 | 9.0 | 18.9 |
| 40 | 0.57 | 8.20 | 0.99 | 14.25 | 8.3 | 9.0 | 17.3 |
| 30 | 0.50 | 7.20 | 0.99 | 14.25 | 7.3 | 9.0 | 16.3 |
| 20 | 0.42 | 6.04 | 0.99 | 14.25 | 6.0 | 9.0 | 15.0 |
| 0–15 | 0.37 | 5.32 | 0.99 | 14.25 | 5.4 | 9.0 | 14.4 |

Gust response factor GRF = 1.26; $c_p = -0.8$, windward; $c_p = -0.5$, Leeward.

Below the 15 ft height, the value of $K_z$ is taken as a constant determined at height 15 ft.

Instead of calculating the values for $K_z$ from the above equation, they can be obtained directly from Table 2.15 or Fig. 2.16.

The results of Step 6 are shown in columns 2 and 4 of Table 2.20

Step 7. Velocity pressure $q_z$. Values for $q_z$ are obtained from Eqs. (2.14) and (2.15).

The results of Step 7 are shown in columns 3 and 5 of Table 2.20

Step 8. Design pressure *P*. With the known values of *q*, *GRF* and $C_p$, the design pressure is obtained by the chain equation

$$p = q \times GRF \times C_p$$

Wind pressures and suctions on windward and leeward walls are shown in Fig. 2.19a and columns 6 and 7 of Table 2.20. Summation of the two is used in determining the pressure for the design of main wind-force-resisting system (column 8 in Table 2.30).

Step 9. Floor-by-floor wind load. This is obtained by multiplying the exposed area tributary to the level by the corresponding value of *p* at the floor height.

The procedure for wind parallel to 160 ft dimension of the building is exactly the same as in Table 2.20. The only difference is in the value of $C_p$ for the leeward face. Its value is equal to $-0.3$ instead of

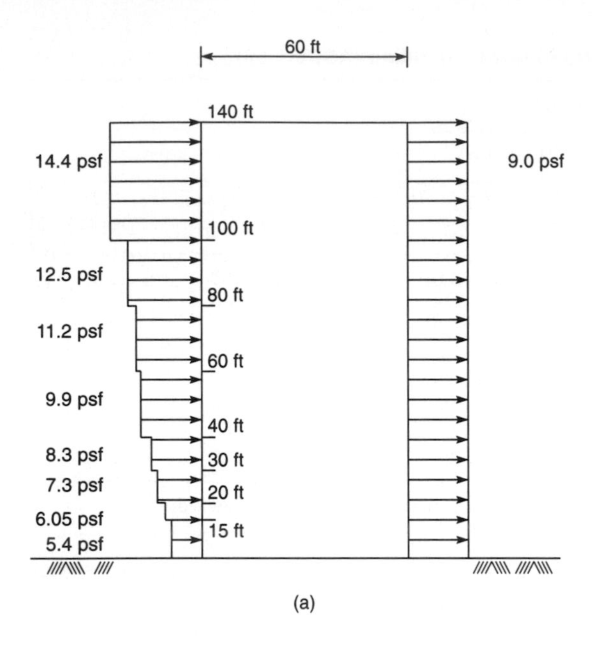


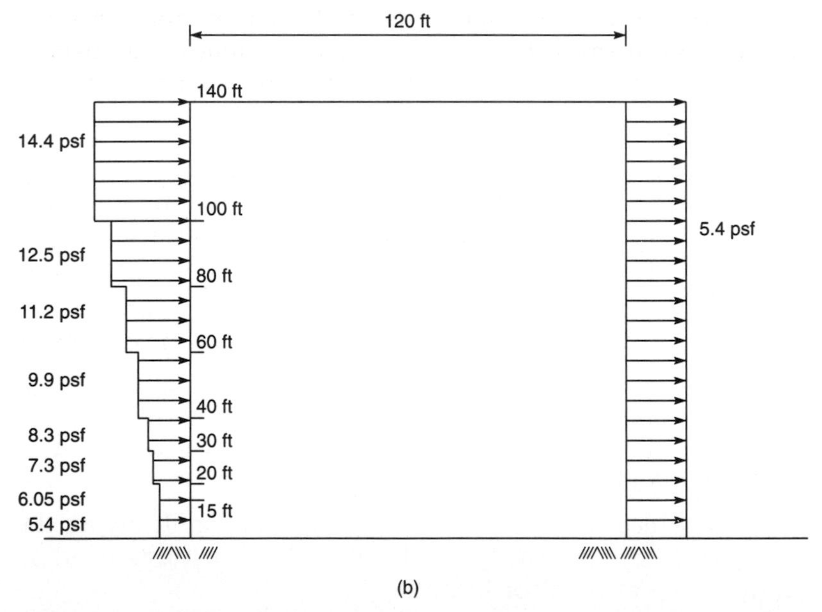


**Figure 2.19** Example problem: (ASCE 7-93) wind pressures for main wind-force resisting system (a) wind parallel to 60 ft dimension (b) wind parallel to 120 ft dimension.

−0.5. The resulting wind pressures and sections are shown in Fig. 2.19b.

### Loads on components and cladding

In the ASCE Standard, building surfaces including the roof are separated into seven zones, Zones 1 through 7, for purposes of determining design pressures and suctions on components and cladding. Zones 1 through 4 define the roof surfaces while Zones 5, 6, and 7 are used for building elevations. Values for the product of gust factor and external pressure coefficient, $GC_p$, are given for the seven zones in terms of the tributary area of the component (Fig. 2.17).

For the example problem we will calculate the design pressures and suctions for the mullions and glass located in Zones 5, 6, and 7. Design values for roof components such as skylights can be obtained in a similar manner by using the values of $GC_p$ for the roof.

The equation for the determination of wind pressures and suctions for components and cladding is given by

$$p = q[(GC_p) - (GC_{pi})] \tag{2.22}$$

which is applicable to cladding loads of both flexible and non-flexible buildings. The values for $q = q_z$ and $q_h$ for windward and leeward walls are evaluated at height $z$ and mean roof height $h$. The value of $GC_p$, is the product of gust response factor $G$ and external pressure coefficient $C_p$. Since the value of $GC_p$ already includes the effect of gust response factor, there is no need to calculate $G$ separately. The value of $GC_p$ is a function of where the cladding is located (Zones 1 through 7), and its tributary area, as shown in Table 2.21.

The value of $GC_{pi}$, the product of gust response factor $G$ and internal pressure $C_{pi}$, are given in the ASCE 7-93 for two opening conditions in the building walls and roof. However, in most buildings, openings for windows are uniform in all the faces, giving a typical value of $GC_{pi} = \pm 0.25$. Two values of $GC_{pi}$ should be considered to obtain the critical design load on the cladding: a positive value of

**TABLE 2.21 Values of $GC_p$**

| Zone | Cladding component | $+GC_p$ | $-GC_p$ |
|---|---|---|---|
| 5 | Mullion | +1.0 | −1.1 |
| | Glazing | +1.05 | −1.1 |
| 6 | Mullion | +1.0 | −1.8 |
| | Glazing | +1.05 | −1.8 |
| 7 | Mullion | +1.0 | −2.5 |
| | Glazing | +1.05 | −2.5 |

**TABLE 2.22 Example Problem: Design Pressures for Zone 5 (ASCE 7-93)**

| | Design pressures (psf) | | | |
|---|---|---|---|---|
| | Mullion | | Glazing panel | |
| Height above ground (ft) | +ve pressure | −ve pressure (suction) | +ve pressure | −ve pressure (suction) |
| 121–140 | 17.81 | 19.23 | 18.52 | 19.23 |
| 101–120 | 16.64 | 19.23 | 17.31 | 19.23 |
| 81–100 | 15.47 | 19.23 | 16.10 | 19.23 |
| 61–80 | 13.85 | 19.23 | 14.40 | 19.23 |
| 41–60 | 12.23 | 19.23 | 12.72 | 19.23 |
| 31–40 | 10.25 | 19.23 | 10.66 | 19.23 |
| 16–30 | 9.0 use 10 | 19.23 | 9.36 use 10 | 19.23 |
| 0–15 | 6.6 use 10 | 19.23 | 6.9 use 10 | 19.23 |

For example, in zone 5, design pressure for mullion located at a height between 121 and 140 ft is worked out as follows.

$$p = q[GC_p - GC_{pi}]$$

$$q = q_2 = 14.25 \text{ psf}$$

$$GC_p = 1.0, \quad -GC_p = 1.1, \quad GC_{pi} = \pm 0.25$$

$$p = 14.25[1.0 - (-0.25)] = 14.25 \times 1.25 = 17.81 \text{ psf}$$

$$p_{\text{suction}} = 14.25[-1.1 - 0.25] = 19.23 \text{ psf}$$

$GC_{pi}$ applied simultaneously to all surfaces and a negative value applied similarly. Values of design wind pressures for the specific components and cladding for the example are given in Tables 2.22–2.24, with a typical calculation shown below each table.

**(2) Non-flexible structure: height > 500 ft**

**Given**

Height = 550 ft (167.6 m)

Height-to-width ratio = 4.58

Fundamental Frequency = 1.1 Hz

Importance factor $I = 1.0$

Ratio of plan dimensions = 1.0

**Required**

Gust factor $G_h$

Since the height-to-width ratio of the building is less than 5, and the

**TABLE 2.23 Example Problem: Design Pressures for Zone 6 (ASCE 7-93)**

| | Design pressures (psf) | | | |
|---|---|---|---|---|
| | Mullion | | Glazing panel | |
| Height above ground (ft) | +ve pressure | −ve pressure (suction) | +ve pressure | −ve pressure (suction) |
| 121–140 | 17·81 | −29.2 | 17.81 | −29.2 |
| 101–120 | 16·6 | −29.2 | 16.6 | −29.2 |
| 81–100 | 15·5 | −29.2 | 15.5 | −29.2 |
| 61–80 | 13·9 | −29.2 | 13.9 | −29.2 |
| 41–60 | 12·3 | −29.2 | 12.3 | −29.2 |
| 31–40 | 10·3 | −29.2 | 10.3 | −29.2 |
| 16–30 | 9.0 use 10 | −29.2 | 9.0 use 10 | −29.2 |
| 0–15 | 6.6 use 10 | −29.2 | 6.6 use 10 | −29.2 |

For example, design pressure and suction for glazing panel located at height between 121 and 140 ft is derived as follows.

$$p = q[GC_p - GC_{pi}]$$

$$q = q_2 = 14.25 \text{ psf}$$

$$GC_p = 1.0, \quad -GC_p = -1.8, \quad GC_{pi} = \pm 0.25$$

$$+p = 14.25[1.05 - (-0.25)] = 18.52 \text{ psf}$$

$$-p = 14.25[-1.8 - 0.25] = -29.21 \text{ psf}$$

**TABLE 2.24 Example Problem: Design Pressures for Zone 7 (ASCE 7-93)**

| | Design pressures (psf) | | | |
|---|---|---|---|---|
| | Mullion | | Glazing panel | |
| Height above ground (ft) | +ve pressure | −ve pressure (suction) | +ve pressure | −ve pressure (suction) |
| 121–140 | 17.81 | 39.2 | 18.52 | 39.2 |
| 101–120 | 16.6 | 39.2 | 17.30 | 39.2 |
| 81–100 | 15.5 | 39.2 | 16.10 | 39.2 |
| 61–80 | 13.9 | 39.2 | 14.40 | 39.2 |
| 41–60 | 12.3 | 39.2 | 12.72 | 39.2 |
| 31–40 | 10.3 | 39.2 | 10.66 | 39.2 |
| 16–30 | 9.0 use 10 | 39.2 | 9.36 use 10 | 39.2 |
| 0–15 | 6.6 use 10 | 39.2 | 6.9 use 10 | 39.2 |

For example, design pressures and suctions for glazing panel located at height between 121 and 140 ft are worked out as follows.

$$p = q[GC_p - GC_{pi}]$$

$$q = q_2 = 14.25 \text{ psf}$$

$$GC_p = 1.05, \quad -GC_p = -2.5, \quad GC_{pi} = \pm 0.25$$

$$+p = 14.25[1.05 + 0.25] = 18.52 \text{ psf}$$

$$-p = 14.25[-2.5 - 0.25] = -39.2 \text{ psf}$$

**TABLE 2.25a Exposure Category Constants (ASCE 7-93)**

| Exposure category | $\alpha$ | $z_g$ | $D_0$ |
|---|---|---|---|
| A | 3.0 | 1500 | 0.025 |
| B | 4.5 | 1200 | 0.010 |
| C | 7.0 | 900 | 0.005 |
| D | 10.0 | 700 | 0.003 |

fundamental frequency is greater than 1.0 Hz, the building may be considered a non-flexible a building.

The gust factor $G_h$ is given by

$$G_h = 0.65 + 3.65T_z \tag{2.23}$$

where $T_z$ is the exposure factor at mean roof height. $T_z$ is given by the relation

$$T_z = \frac{2.35(D_0)^{1/2}}{(Z/30)^{1/\alpha}} \tag{2.24}$$

$D_0$ is a surface coefficient factor which is given as a function of type of exposure in Table 2.25a. Substituting the exposure category constants $D_0 = 0.010$ and $\alpha = 4.5$, we get

$$T_z = \frac{2.35(0.010)^{0.5}}{(550/30)^{1/4.5}} = 0.123$$

Substituting $T_z$ in Eq. (2.23), we get

$$G_h = 0.65 + 3.65 \times 0.123 = 1.099$$

**(3) 60 story flexible building**

**Given**

Basic design wind speed = 70 mph

Building height 60 floors @ 11 ft 5 in. (3.48 m) = 685 ft (208.8 m)

Fundamental Frequency = 0.143 Hz (period = 7 sec)

Damping coefficient = 0.02

Height-to-width ratio $= \dfrac{685}{120} = 5.7$

Type of exposure = $B$

Building width = 116 ft

**Required**

Gust factor, $\bar{G}$

The building is flexible on both accounts: its height-to-width ratio is greater than 5, and its natural frequency is less than 1.0 Hz.

Therefore, the gust response factor must be determined by using the expressions and graphs given in ASCE 7-93 for flexible buildings. The gust factor is given by the relation

$$\bar{G} = 0.65 + \left[\frac{P}{\beta} + \frac{(3.32T_1)^2 S}{1 + 0.0002c}\right]^{1/2} \tag{2.25}$$

where $P$ = a power factor = $\bar{f}JY$

$\bar{f} = 10.5fh/sV$ (2.26)

$J$ = pressure profile factor given as a function of parameter given in Fig. 2.20

$\gamma$ = a parameter that depends on the exposure category given in Table 2.25b.

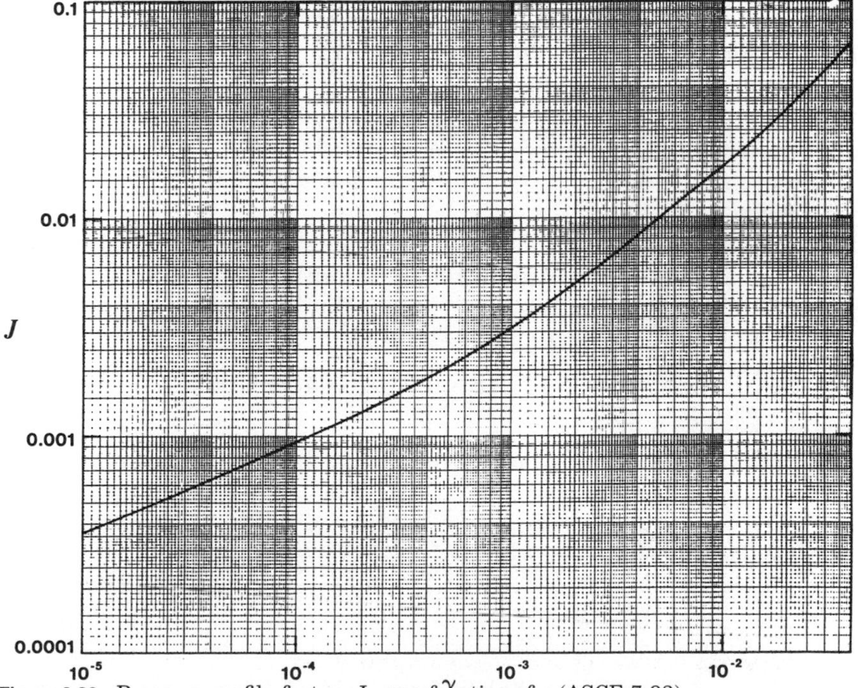


**Figure 2.20** Pressure profile factor, $J$, as a function of $\gamma$ (ASCE 7-93).

**TABLE 2.25b Parameters $S$ and $\gamma$ (ASCE 7-93)**

| Exposure category | $s$ | $\gamma$ |
| --- | --- | --- |
| A | 1.46 | 8.20/h |
| B | 1.33 | 3.28/h |
| C | 1.00 | 0.23/h |
| D | 0.85 | 0.02/h |

$Y$ = a resonance factor given as a function of width-to-height ratio $c/h$ of the building in Fig. 2.21a
$\beta$ = the damping coefficient = 0.02 for the example problem
$T_1$ = the exposure factor evaluated at two-thirds the mean roof height of the structure as shown below
$s$ = the surface friction factor given in Table 2.25b
$S$ = the structure size factor given in Fig. 2.21b
$c$ = the average horizontal dimension of the building perpendicular to wind direction = 116 ft for the example problem

We will now proceed to calculate the gust response factor $G$ for the example problem. Surface friction factor $s$ and parameter $\gamma$ for type B exposure are obtained as 1.33 and $3.28/h = 0.0048$, respectively, from Table 2.25b. From Eq. (2.26),

$$\bar{f} = \frac{10.5 \times 0.143 \times 685}{1.33 \times 70} = 11.048$$

From Fig. 2.20, $J$ is obtained as 0.01 for $\gamma = 3.28/685 = 0.0048$, and from Fig. 2.21a, resonance factor $Y$ is obtained as 0.045 for $\bar{f} = 11.048$ and $c/h = 0.17$.

$$P = \bar{f}JY = 11.048 \times 0.01 \times 0.045 = 0.005$$

$T_1$ is calculated at two-thirds the mean roof height by the relation

$$T_1 = \frac{2.35(D_0)^{0.5}}{\left(\frac{2}{3} \times \frac{z}{30}\right)^{1/\alpha}} = \frac{2.35(0.010)^{0.5}}{(2/3 \times 685/30)^{1/4.5}} = 0.1283$$

The values of $D_0$ and $\alpha$ are taken from Table 2.25a.

Substituting values for the various parameters in Eq. (2.25), we get

$$\bar{G} = 0.65 + \left\{\frac{0.005}{0.02} + \frac{[3.32(0.1283)]^2 \times 0.75}{1 + 0.002 \times 120}\right\}^{1/2} = 1.25$$

### (4) 50-story flexible building

In this example, design wind pressures on the main wind-force-resisting system of a 50-story high-rise building, located on the

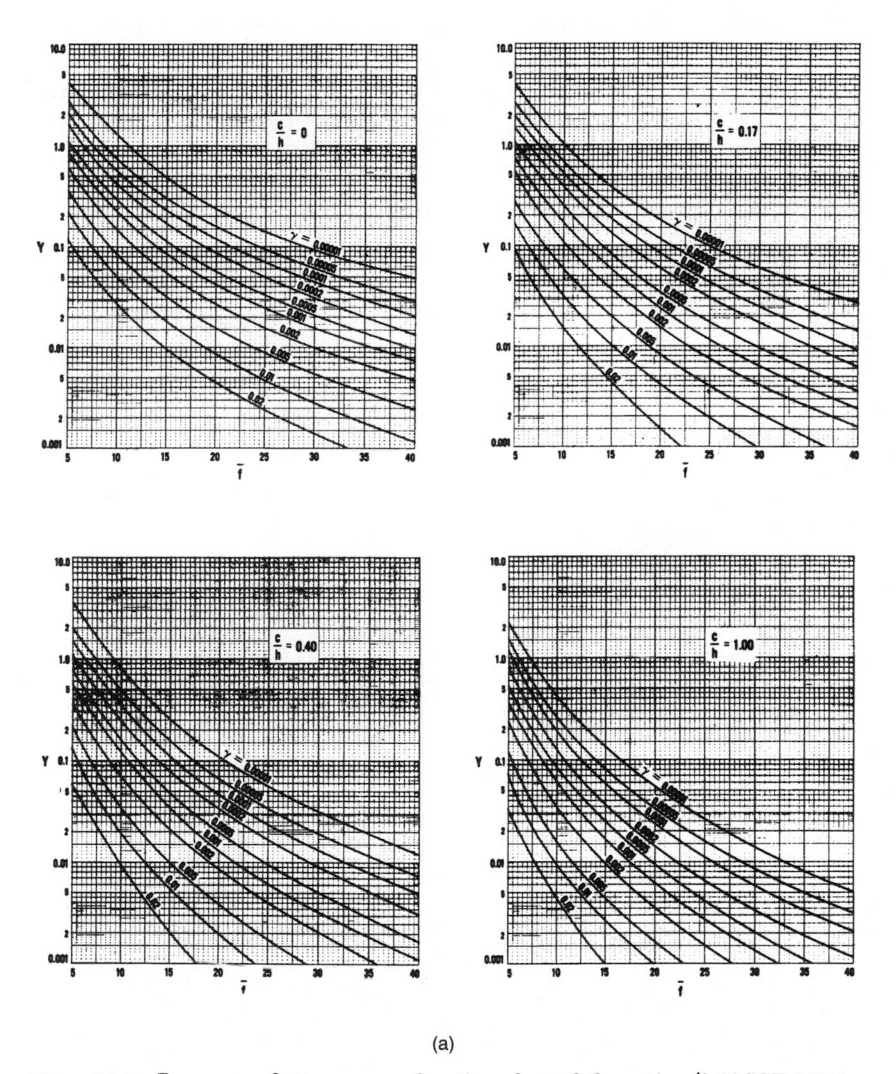


(a)

**Figure 2.21a** Resonance factor, $y$, as a function of $\gamma$ and the ratio $c/h$ (ASCE 7-93).

outskirts of Miami, Florida, are determined. The building data are as follows:

| | |
|---|---|
| Plan dimensions: | $137 \times 137$ ft |
| Building height: | 50 stories @ $12.3 = 615$ ft |
| Fundamental period: | 6 sec (Frequency $f = 0.167$ Hertz) |
| Damping coefficient: | 0.010 |

Step 1. Building classification. The building function is commercial. It it is not considered to be be an essential facility in the aftermath of a natural disaster. Hence the building classification Category 1 is judged appropriate for this building.

Results of Step 1: Category 1

Step 2. Basic Wind speed. The wind speed map indicates a basic wind speed of 110 mph. There is neither a special wind region indicated on the map near Miami, nor any reason to

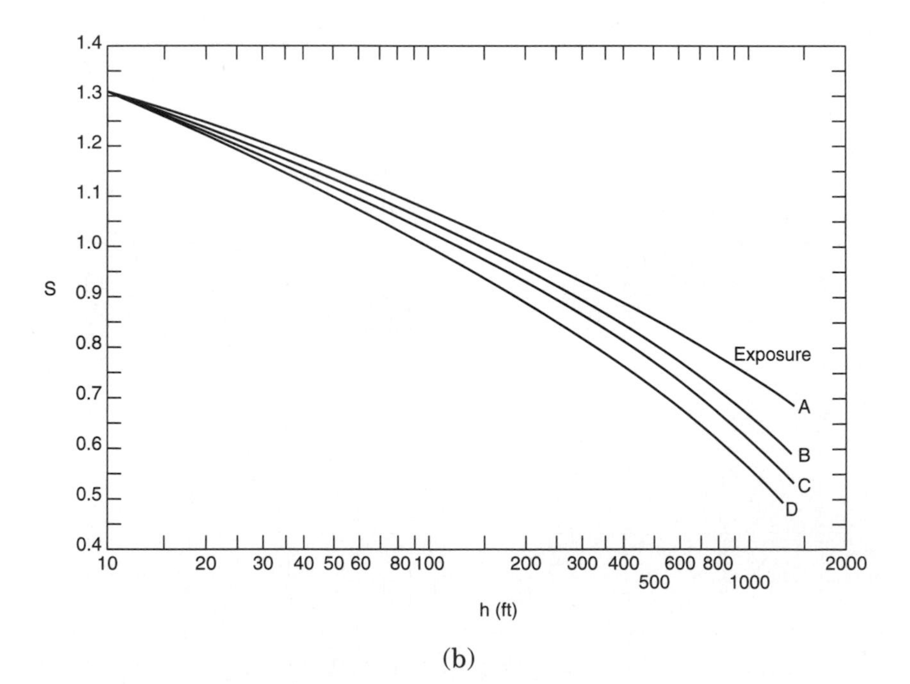


(b)

**Figure 2.21b** Structure size fractor $S$ (ASCE 7-93)

consider that winds at the site are unusual and would require additional attention.

Results of Step 2: $V = 110$ mph

Step 3. Building Exposure. The outskirts of Miami, which is the location of the example building, is assumed to be an open flat country. hence exposure Category C is judged to be appropriate.

Results of Step 3: Exposure C

Step 4. External pressure coefficient $C_p$. The ratio of the building plan dimensions is equal to 1. Therefore, from Fig. 2.18 the values of $C_p$ are as shown below.

Results of Step 4: $C_p = 0.8$ windward wall
$C_p = -0.5$ leeward wall

Step 5. Combined velocity exposure coefficient $K_z$. The gradient height, $Z_g$, and the power coefficient, $\alpha$ for Exposure C are 900 ft and 7 respectively. The values of $K_z$ are calculated by the relation

$$K_z = 2.58\left(\frac{z}{900}\right)^{2/7}$$

Below 15 ft height, the value of $K_z$ is taken as a constant determined at height 15 ft. As before, instead of calculating $K_z$ from the above equation, we can obtain the values directly from Table 2.15 or Fig. 2.16.

Results of Step 5: $K_z$ values shown in columns 2 and 4 of Table 2.26

Step 6. Velocity pressure $q_z$. This is calculated by the relation $q_z = 0.00256K_z(1 \times 110)^2$

Results of Step 6: Shown in columns 3 and 5 of Table 2.26

Step 7. Gust response factor *GRF*. The building is tall and relatively slender. Its height-to-width ratio, $\frac{615}{137} = 4.49$, is less than five. However, the fundamental frequency of vibration is 0.167 Hertz which is considerably less than 1 Hertz. Therefore, the building must be considered flexible: the *GRF* values given in Table 2.19 cannot be used. As required by the ASCE 7-93, a rational method much be used to determine the gust factor. The method used here is the one given in the ASCE 7-93 commentary. The calculations use a series of tables and equations as shown below.

**TABLE 2.26 Example 3: 50-Story Flexible Building: Design Pressures for Main Wind-force Resisting System (ASCE 7-93)**

| Height | Windward | | Leeward | | Windward pressure psf | Leeward suction psf | Design value psf |
|---|---|---|---|---|---|---|---|
| ft | $K_z$ | $q_z$ | $K_h$ | $q_h$ | $p_z = q_z\bar{G}C_p$ | $p_h = q_h\bar{G}C_p$ | $p = p_z + p_h$ |
| (1) | (2) | (3) | (4) | (5) | (6) | (7) | (8) |
| 615 | 2.31 | 71.6 | 2.31 | 71.6 | 64.2 | 40.1 | 104.3 |
| 600 | 2.30 | 71.3 | 2.31 | 71.6 | 63.9 | 40.1 | 104.0 |
| 550 | 2.24 | 69.4 | 2.31 | 71.6 | 62.2 | 40.1 | 102.3 |
| 500 | 2.18 | 67.6 | 2.31 | 71.6 | 60.6 | 40.1 | 100.7 |
| 450 | 2.12 | 65.7 | 2.31 | 71.6 | 58.9 | 40.1 | 99.0 |
| 400 | 2.05 | 63.5 | 2.31 | 71.6 | 56.9 | 40.1 | 97.0 |
| 350 | 1.97 | 61.0 | 2.31 | 71.6 | 54.7 | 40.1 | 94.8 |
| 300 | 1.88 | 58.3 | 2.31 | 71.6 | 52.2 | 40.1 | 92.3 |
| 250 | 1.79 | 55.5 | 2.31 | 71.6 | 49.7 | 40.1 | 89.8 |
| 200 | 1.68 | 52.0 | 2.31 | 71.6 | 46.6 | 40.1 | 86.7 |
| 180 | 1.63 | 50.5 | 2.31 | 71.6 | 45.2 | 40.1 | 85.3 |
| 160 | 1.58 | 49.0 | 2.31 | 71.6 | 43.9 | 40.1 | 84.0 |
| 140 | 1.52 | 47.0 | 2.31 | 71.6 | 42.1 | 40.1 | 82.2 |
| 120 | 1.45 | 45.0 | 2.31 | 71.6 | 40.3 | 40.1 | 80.4 |
| 100 | 1.38 | 42.8 | 2.31 | 71.6 | 38.3 | 40.1 | 78.4 |
| 90 | 1.34 | 41.5 | 2.31 | 71.6 | 37.2 | 40.1 | 77.3 |
| 80 | 1.29 | 40.0 | 2.31 | 71.6 | 35.8 | 40.1 | 75.9 |
| 70 | 1.24 | 38.4 | 2.31 | 71.6 | 34.4 | 40.1 | 74.5 |
| 60 | 1.19 | 36.9 | 2.31 | 71.6 | 33.1 | 40.1 | 73.2 |
| 50 | 1.13 | 35.0 | 2.31 | 71.6 | 31.4 | 40.1 | 71.5 |
| 40 | 1.06 | 32.8 | 2.31 | 71.6 | 29.4 | 40.1 | 69.5 |
| 30 | 0.98 | 30.4 | 2.31 | 71.6 | 27.2 | 40.1 | 67.3 |
| 25 | 0.93 | 28.8 | 2.31 | 71.6 | 25.8 | 40.1 | 65.9 |
| 20 | 0.87 | 27.0 | 2.31 | 71.6 | 24.2 | 40.1 | 64.3 |
| 15 | 0.80 | 24.8 | 2.31 | 71.6 | 22.2 | 40.1 | 62.3 |

Gust Response Factor $\bar{G} = 1.12$; $c_p = +0.8$ Windward face; $c_p = -0.5$ Leeward face.

**Given**

Building height, $h = 615$ ft

Building width, $c = 137$ ft

Basic wind speed, $V = 110$ mph

Building fundamental frequency, $f = 0.167$ Hz

Building damping coefficient, $\beta = 0.010$

**Gust response factor calculation.** From Table 2.25b for exposure $C$, $s = 1.00$ and $\gamma = \dfrac{0.23}{h} = \dfrac{0.23}{615} = 0.00037$. $\dfrac{c}{h} = \dfrac{137}{615} = 0.223$. This value is closer to the graph with $\frac{c}{h} = 0.17$, than for other graphs. Use graph $\frac{c}{h} = 0.017$ (Fig. 2.21a) to obtain $Y$.

From Eq. 2.26,

$$\bar{f} = \frac{10.5fh}{sV}$$

$$= \frac{10.5 \times 0.167 \times 615}{1 \times 110}$$

$$= 9.8 \text{ Hz}$$

From Fig. 2.20, $J = 0.0018$ for $\gamma = 0.00037$.

From Fig. 2.21, using graph $\frac{c}{h} = 0.17$, $Y = 0.03$ for $\gamma = 0.00037$ and $\bar{f} = 9.8$ Hz.

$$P = \bar{f}JY$$

$$= 9.80 \times 0.0018 \times 0.03$$

$$= 0.00053$$

$D_0$ for Exposure C, from Table 2.25a = 0.005

$$T_1 = \frac{2.25(D_0)^{1/2}}{\left(\frac{2}{30}\right)^{1/\alpha}}$$ Noting that $T_1$ is evaluated at $\frac{2}{3}$ the height

$$T_1 = \frac{2.25(0.005)^{1/2}}{\left(\frac{z}{3} \times \frac{615}{30}\right)^{1/7}}$$

$$= 0.109$$

$$G = 0.65 + \left[\frac{P}{\beta} + \frac{(3.32T_1)^2 S}{1.002c}\right]^{1/2}$$

$$= 0.65 + \left[\frac{0.00053}{0.01} + \frac{(3.32 \times 0.109)^2 \times 0.76}{1 + 0.002 \times 137}\right]^{1/2}$$

$$= 0.65 + 0.436 = 1.09$$

The result of Step 7: G = 1.09

since the selection of parameters $Y$, $\gamma$ etc., from the graphs are necessarily approximate, we choose to use a conservative value for gust factor $G = 1.12$.

Step 8. Design pressure $p$. The design pressure $p$ is obtained by the chain equation

$$p = q_z G C_p$$

The results for the windward and leeward faces are shown in columns 6 and 7 while the combined value is shown in column 8 of Table 2.26.

Step 9. The final step is the calculation of level-by-level wind load by multiplying the exposed area tributary to the level by the corresponding value of $p$ shown in column 8.

#### 2.5.5.7 Sensitivity study of gust response factor, $\bar{G}$

Since the graphs and tables used for evaluating the gust response factors are interdependent, it is necessary to assume specific building characteristics for studying the sensitivity of the gust response factor. The variation of gust response factor to each of the buildings characteristics such as fundamental frequency, damping ratio etc., may be assessed by varying one particular parameter at a time, while keeping the others constant.

Let us study the variation of gust response factor by changing two parameters (1) the fundamental frequency of the building and (2) the damping.

**Given**

Building height = 685 ft (208.8 m)

Building plan dimensions = 120 × 180 ft (34.67 × 54.86 m)

Fundamental frequency = 0.143 Hz

Critical damping ratio = 0.02

Basic wind speed = 70 mph, parallel to 180 ft dimension

Type of terrain = Exposure B

**Required**

Parametric study of gust response factor $\bar{G}$ by varying the building frequency and damping, one at a time. Values of frequency and

**TABLE 2.27a Sensitivity of Gust Response Factor to Fundamental Frequency (ASCE 7-93)**

| Period $T$ secs | Fundamental frequency, $f$ Hz | $\bar{f} = \frac{10.5\,fh}{SV}$ | Resonance factor $Y$ | Power factor $p = \bar{f}JY$ | Gust response factor $G$ |
|---|---|---|---|---|---|
| 3 | 0.333 | 25.75 | 0.004 | 0.0010 | 1.07 |
| 4 | 0.25 | 19.31 | 0.009 | 0.0017 | 1.12 |
| 5 | 0.20 | 15.45 | 0.022 | 0.0031 | 1.21 |
| 6 | 0.167 | 12.87 | 0.045 | 0.0050 | 1.30 |
| 7 | 0.142 | 11.05 | 0.055 | 0.0061 | 1.37 |
| 10 | 0.10 | 7.72 | 0.11 | 0.0085 | 1.47 |

damping to be used in the study are:

Frequency Hz, 0.333, 0.25, 0.20, 0.167, 0.142, 0.10
(period, seconds 3, 4, 5, 6, 7, 10)

Damping ratio 0.005, 0.01, 0.015, 0.020, 0.025, 0.030

As in the previous examples the gust response factor $\bar{G}$ is calculated by using the tables and graphs. A summary of the results is shown in Tables 2.27a and 2.27b, and in Figs 2.22a,b.

### 2.5.6 ASCE 7-95, Wind load provisions

#### 2.5.6.1 Introduction

While the wind provisions of the 1993 edition of ASCE-7 were virtually unchanged from the previous editions, the latest 1995 edition has gone through a major revision. The most visible change

**TABLE 2.27b Sensitivity of Gust Response Factor to Damping (ASCE 7-93)**

| | Gust response factor | |
|---|---|---|
| Damping ratio $\beta$ | Building period 7 sec | Building period 3 sec |
| 0.005 | 1.20 | 1.8 |
| 0.01 | 1.11 | 1.5 |
| 0.015 | 1.07 | 1.37 |
| 0.020 | 1.05 | 1.29 |
| 0.025 | 1.04 | 1.24 |
| 0.030 | 1.03 | 1.20 |

is the abandonment of the fastest mile speed in favor of a three-second-gust speed. The reasons are: (i) modern weather stations no longer measure wind speeds using the fastest-mile method; (ii) the three-second-gust speed is closer to the sensational wind speeds often quoted by news media; and (iii) it relates closely to wind conditions experienced by small buildings, and components of all buildings.

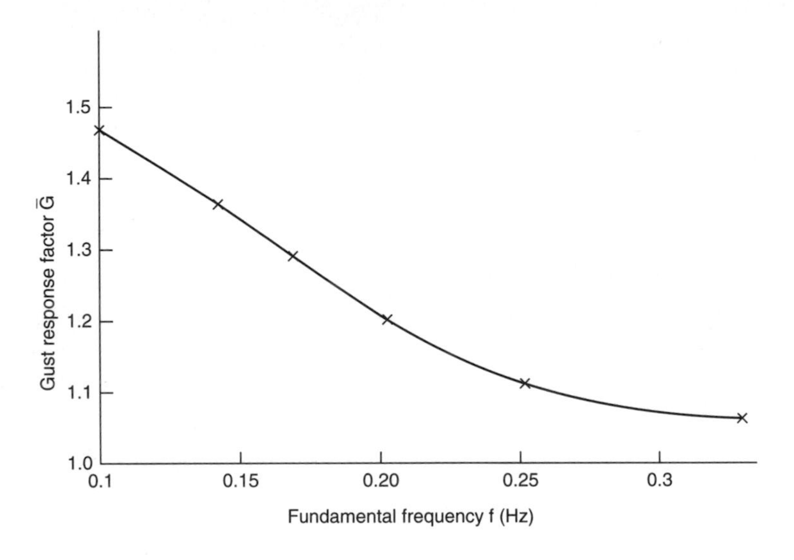


(a)

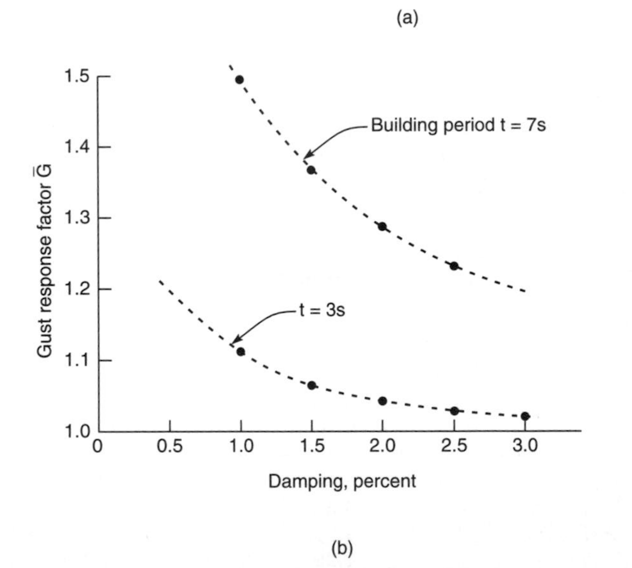


(b)

**Figure 2.22** Sensitivity Study (ASCE 7-93): (a) Gust response factor $\bar{G}$ versus fundamental frequency; (b) Gust response factor $\bar{G}$ versus damping.

Except for Exposure D, the definitions for exposure categories, A, B, and C are unchanged from ASCE 7-93. For Exposure D, large bodies of water have been defined to mean open water for a distance of at least 1 mile (1.6 km). The height and exposure factors $K_z$ and $K_h$ have been revised to reflect the change in wind speed from the fastest mile to three-second-gust speed. The exposure factors for suburban settings have been increased to reflect recent studies showing greater turbulence in Exposure B terrain than previously thought.

Two new provisions have been added. The first considers hill effects on wind speeds by introducing a non-dimensional topography factor, $K_{zt}$, and the second requires verification of torsional response of buildings taller than 60 ft (18 m).

These and other provisions, related to the calculation of gust factors, are explained in the following sections. Worked examples are given for calculating gust factors and wind loads on the main wind force resisting system.

#### 2.5.6.2 Overview

As in the previous edition, the wind pressure or section induced on the building surface is given by a chain equation: $p_z = q_z \times G \times C_p$

where $p_z$ = design wind pressure or suction, in psf, at height $z$, above ground level

$q_z$ = velocity pressure, in psf, determined at height $z$ above ground

$G$ = gust effect factor dimensionless (denoted by $G_f$ when loads for main wind force resisting system of flexible building are calculated)

$C_p$ = external pressure coefficient which varies with building height acting as pressure (positive load) on windward face, and as suction (negative load) on non-windward faces and roof. The values of $C_p$ are unchanged from the previous edition

The velocity pressure, $q$, is given by the Equations

$$q_z = 0.00256 K_z K_{zt} V^2 I \tag{2.27}$$

$$q_h = 0.00256 K_h K_{zt} V^2 I \tag{2.28}$$

where $K_h$ and $K_z$ = combined velocity pressure exposure coefficients (Table 2.29), dimensionless, which take into

**TABLE 2.28 Importance Factor *I*: Wind Loads (ASCE 7-95)**

| Category | I |
|---|---|
| I | 0.87 |
| II | 1.00 |
| III | 1.15 |
| IV | 1.15 |

account changes in wind speed above ground and the nature of terrain, exposure category A, B, C, or D. As mentioned previously, definitions for Exposures A, B, and C are unchanged from ASCE 7-93. For Exposure D, large bodies of water have been defined to mean open water for a distance of at least 1 mile (1.6 km)

$K_{zt}$ = topographic factor newly introduced in ASCE 7-95 for the first time

$I$ = importance factor (Table 2.28) dimensionless, which modifies: (i) the 50-year wind to 100 or 25-year wind, depending on the building category and, (ii) hurricane wind speeds to probabilities that are conistent with with non-hurricane wind speeds. Importance factors of 0.87 and 1.15 given in Table 2.36 correspond to mean recurrence intervals of 25 and 100 years, respectively. These values represent the square of the importance factors given in ASCE 7-93, which were applied to the basic wind speed rather than to the velocity pressure, except that the $(0.95)^2 = 0.90$ factor was changed to 0.87 based on new data analysis

$V$ = basic wind speed, Fig. 2.23, in miles per hour. The basic wind speed corresponds to a 3 sec gust speed at 33 ft (10 m) above ground in exposure category C with a 50-year mean recurrence interval

The basic wind speed is converted to the design speed at any height $z$, for a given exposure category by using the velocity exposure coefficient $K_z$, evaluated at height $z$. $K_z$ is given by

$$K_z = 2.01\left(\frac{z}{z_g}\right)^{2/\alpha} \qquad \text{for } 15\,\text{ft} < z < z_g \tag{2.29}$$

$$K_z = 2.01\left(\frac{15}{Z_g}\right)^{2/\alpha} \qquad \text{for } z < 15\,\text{ft} \tag{2.30}$$

where $z$ is the gradient above which the frictional effect of terrain becomes negligible. It varies with the characteristics of the ground surface irregularities at the building site, which arise as a result of natural topographic variations as well as manmade features.

The power coefficient $\alpha$ (Table 2.30) is the exponent for velocity increase in height and has new values of 5.0, 7.0, 9.5 and 11.5 respectively for Exposure A, B, C and D. The values of $K_z$ for various exposures up to a height of 500 ft (152.6 m) is given in ASCE 7-95. An extended version, up to a height of 1500 ft (457 m), is given in Table 2.29 and in Fig. 2.24.

The values of the gradient height, $z_g$, given in ASCE 7-95 are of course, identical to those in ASCE 7-93. This is obvious because the gradient height, $z_g$, for a given exposure does not vary with the reference wind speed. However, because the shape of wind profiles vary when the reference speed is changed from fastest mile to the 3 sec gust speed, it is necessary to modify the values of $\alpha$ in the

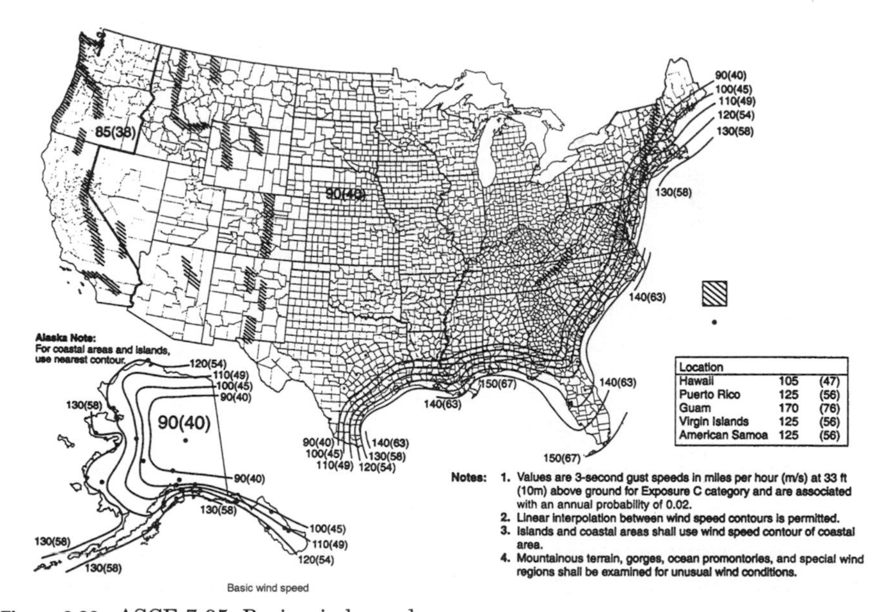


**Figure 2.23** ASCE 7-95. Basic wind speed map.

**TABLE 2.29 Velocity Pressure Exposure Coefficients, $K_h$ and $K_z$ (ASCE 7-95)**

| Height above ground level, $z$ ft | (m) | Exposure category A | B | C | D |
|---|---|---|---|---|---|
| 0–15 | (0–4.6) | 0.32 | 0.57 | 0.85 | 1.03 |
| 20 | (6.1) | 0.36 | 0.62 | 0.90 | 1.08 |
| 25 | (7.6) | 0.39 | 0.66 | 0.94 | 1.12 |
| 30 | (9.1) | 0.42 | 0.70 | 0.98 | 1.16 |
| 40 | (12.2) | 0.47 | 0.76 | 1.04 | 1.22 |
| 50 | (15.2) | 0.52 | 0.81 | 1.09 | 1.27 |
| 60 | (18) | 0.55 | 0.85 | 1.13 | 1.31 |
| 70 | (21.3) | 0.59 | 0.89 | 1.17 | 1.34 |
| 80 | (24.4) | 0.62 | 0.93 | 1.21 | 1.38 |
| 90 | (27.4) | 0.65 | 0.96 | 1.24 | 1.40 |
| 100 | (30.5) | 0.68 | 0.99 | 1.26 | 1.43 |
| 120 | (36.6) | 0.73 | 1.04 | 1.31 | 1.48 |
| 140 | (42.7) | 0.78 | 1.09 | 1.36 | 1.52 |
| 160 | (48.8) | 0.82 | 1.13 | 1.39 | 1.55 |
| 180 | (54.9) | 0.86 | 1.17 | 1.43 | 1.58 |
| 200 | (61.0) | 0.90 | 1.20 | 1.46 | 1.61 |
| 250 | (76.2) | 0.98 | 1.28 | 1.53 | 1.68 |
| 300 | (91.4) | 1.05 | 1.35 | 1.59 | 1.73 |
| 350 | (106.7) | 1.12 | 1.41 | 1.64 | 1.78 |
| 400 | (121.9) | 1.18 | 1.47 | 1.69 | 1.82 |
| 450 | (137.2) | 1.24 | 1.52 | 1.73 | 1.86 |
| 500 | (152.4) | 1.29 | 1.56 | 1.77 | 1.89 |
| 550 | (167.6) | 1.35 | 1.61 | 1.81 | 1.93 |
| 600 | (182.9) | 1.39 | 1.65 | 1.85 | 1.96 |
| 650 | (191.1) | 1.44 | 1.69 | 1.88 | 1.98 |
| 700 | (213.3) | 1.48 | 1.72 | 1.91 | 2.01 |
| 750 | (228.6) | 1.52 | 1.76 | 1.93 | 2.03 |
| 800 | (243.8) | 1.56 | 1.79 | 1.96 | 2.06 |
| 850 | (259.1) | 1.60 | 1.82 | 1.99 | 2.08 |
| 900 | (274.3) | 1.64 | 1.85 | 2.01 | 2.10 |
| 950 | (289.5) | 1.67 | 1.88 | 2.03 | 2.12 |
| 1000 | (304.8) | 1.71 | 1.91 | 2.06 | 2.14 |
| 1050 | (320) | 1.74 | 1.93 | 2.08 | 2.16 |
| 1100 | (335.3) | 1.78 | 1.96 | 2.10 | 2.17 |
| 1150 | (350.5) | 1.81 | 1.99 | 2.12 | 2.19 |
| 1200 | (365.7) | 1.84 | 2.01 | 2.14 | 2.21 |
| 1250 | (381) | 1.87 | 2.03 | 2.15 | 2.22 |
| 1300 | (396.2) | 1.90 | 2.06 | 2.17 | 2.24 |
| 1350 | (411.5) | 1.93 | 2.08 | 2.19 | 2.26 |
| 1400 | (426.7) | 1.96 | 2.10 | 2.21 | 2.27 |
| 1450 | (441.9) | 1.98 | 2.12 | 2.22 | 2.28 |
| 1500 | (457.2) | 2.01 | 2.14 | 2.24 | 2.29 |

power-law representation of the wind speed profile. The new values of $\alpha$ listed in Table 2.30 are based on a comprehensive review of existing data, and have resulted in a modification to the gust speed profiles. The multiplier in Eqs (2.29) and (2.30) for the exposure

**TABLE 2.30 Exposure Category Constants (ASCE 7-95)**

| Exposure category | $\alpha$ | $z_g$ [ft (m)] |
|---|---|---|
| A | 5.0 | 1,500 (457) |
| B | 7.0 | 1,200 (366) |
| C | 9.5 | 900 (274) |
| D | 11.5 | 700 (213) |

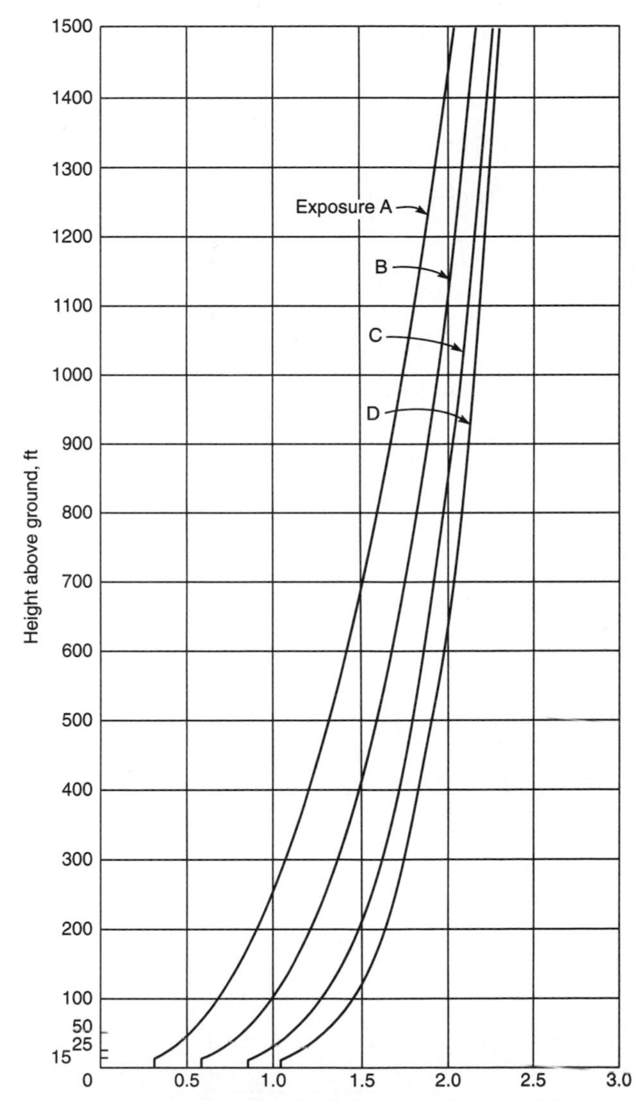


**Figure 2.24** Velocity pressure exposure coefficients, $K_h$ and $K_z$ (ASCE 7-95).

coefficient, $K_z$, has been changed from 2.58 to 2.01. As with ASCE 7-93, values of $K_z$ are assumed to be constant for heights less than 15 ft (4.6 m), and for heights greater than the gradient height, $z$.

The values of $\alpha$ in Table 2.30 increase the $K_z$ values for exposure categories A and B while decreasing the values for exposure category D as compared to previous values of $K_z$ in ASCE 7-93. In addition, differences in $K_z$ value between exposure categories are reduced.

The variation of velocity pressure, $q_z$ for exposure categories A, B, C, and D is given in Fig. 2.25.

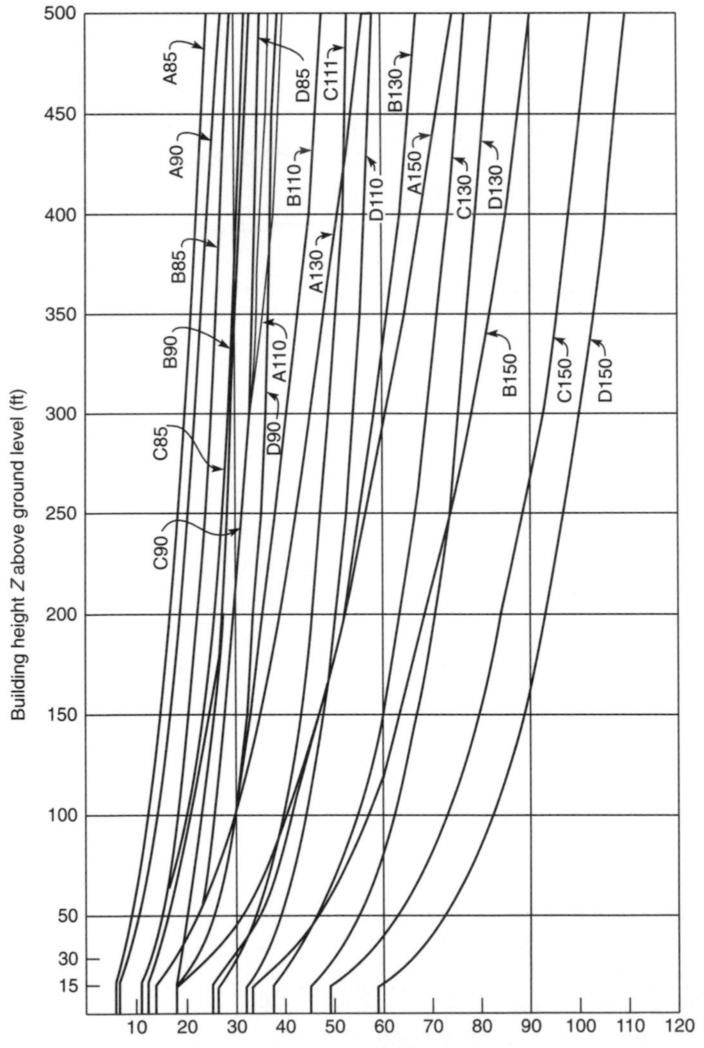


**Figure 2.25** Velocity pressure $q_z$ psf (ASCE 7-95).

#### 2.5.6.3 Wind speed-up over hills and escarpments: $K_{zt}$ factor

The topographic factor $K_{zt}$ is a new provision in ASCE 7-95 which has been added to account for the effect of isolated hills or escarpment located in Exposure B, C, or D. Buildings sited on the upper half of an isolated hill or escarpment may experience significantly higher wind speeds than buildings situated on level ground. To account for these higher wind speeds, the velocity pressure exposure coefficients are multiplied by a topographic factor, $K_{zt}$. The topographic feature is described by two parameters, $H$ and $L_h$. $H$ is the height of the hill or difference in elevation between the crest and that of the upwind terrain. $L_h$ is the distance upwind of the crest to where the ground elevation is equal to half the height of the hill. $K_{zt}$ is determined from three multipliers, $K_1$, $K_2$, and $K_3$ (Fig. 2.26). $K_1$ is related to the shape of the topographic feature and the maximum speed-up with distance upwind or downwind of the crest, $K_2$ accounts for the reduction in speed up with distance upwind or downwind of the crest, and $K_3$ accounts for the reduction in speed-up with height above the local ground surface.

The multipliers $K_1$, $K_2$ and $K_3$ are based on the assumption that the wind approaches the hill along the direction of maximum slope, causing the greatest speed-up near the crest along the direction of

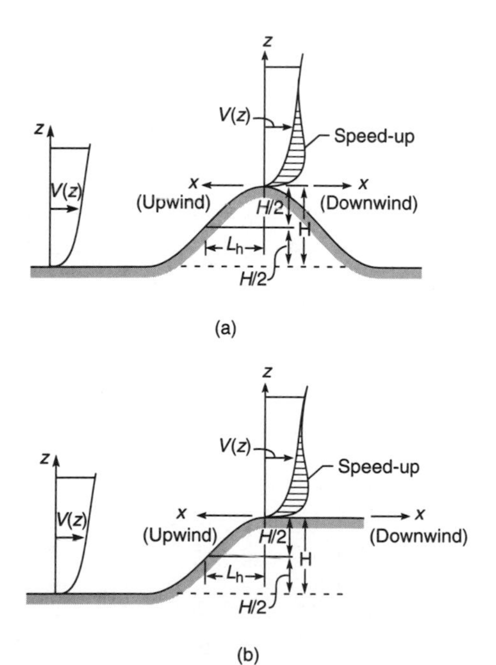


**Figure 2.26** (a) 2-D ridge or 3-D axisymmetrical hill. (b) escarpment

| $K_1$: Factor for topographic feature and maxmum speed-up effect | | | | $K_2$: Factor for reduction in speed-up with distance upwind or downwind of crest | | | $K_3$: Factor for reduction in speed-up with height above local terrain | | | |
|---|---|---|---|---|---|---|---|---|---|---|
| $H/L_h$ | 2-D Ridge | 2-D Escarpment | 3-D Axisym. hill | $x/L_h$ | 2-D Escarpment downwind of crest | All others cases | $z/L_h$ | 2-d Ridge | 3-D Escarpment | 3-D Axisym. hill |
| 0.10 | 0.14 | 0.09 | 0.11 | 0.00 | 1.00 | 1.00 | 0.00 | 1.00 | 1.00 | 1.00 |
| 0.15 | 0.22 | 0.13 | 0.16 | 0.50 | 0.88 | 0.67 | 0.10 | 0.74 | 0.78 | 0.67 |
| 0.20 | 0.29 | 0.17 | 0.21 | 1.00 | 0.75 | 0.33 | 0.20 | 0.55 | 0.61 | 0.45 |
| 0.25 | 0.36 | 0.21 | 0.26 | 1.50 | 0.63 | 0.00 | 0.30 | 0.41 | 0.47 | 0.30 |
| 0.30 | 0.43 | 0.26 | 0.32 | 2.00 | 0.50 | 0.00 | 0.40 | 0.30 | 0.37 | 0.20 |
| 0.35 | 0.51 | 0.30 | 0.37 | 2.50 | 0.38 | 0.00 | 0.50 | 0.22 | 0.29 | 0.14 |
| 0.40 | 0.58 | 0.34 | 0.42 | 3.00 | 0.25 | 0.00 | 0.60 | 0.17 | 0.22 | 0.09 |
| 0.45 | 0.65 | 0.38 | 0.47 | 3.50 | 0.13 | 0.00 | 0.70 | 0.12 | 0.17 | 0.06 |
| 0.50 | 0.72 | 0.43 | 0.53 | 4.00 | 0.00 | 0.00 | 0.80 | 0.09 | 0.14 | 0.04 |
| | | | | | | | 0.90 | 0.07 | 0.11 | 0.03 |
| | | | | | | | 1.00 | 0.05 | 0.08 | 0.02 |
| | | | | | | | 1.50 | 0.01 | 0.02 | 0.00 |
| | | | | | | | 2.00 | 0.00 | 0.00 | 0.00 |

1. For values of $H/L_h$, $x/L_h$ and $z/L_h$ other than those shown, linear interpolation is permitted.
2. For $H/L_h > 0.5$, assume $H/L_h = 0.5$, and substitute $2H$ for $L_h$ in $x/L_h$ and $z/L_h$.
3. Multipliers are based on the assumption that wind approaches the hill or escarpment along the direction of maximum slope.
4. Effect of wind speed-up shall not be required to be accounted for when $H/L_h < 0.2$ or when $H < 15$ ft (4.5 m) for Exposure D, or <30 ft (9 m) for Exposure C, or <60 ft (18 m) for all other exposures.
5. Notation:

$H$: Height of hill or escarpment relative to the upwind terrain, in feet (meters).
$L_h$: Distance upwind of crest to where the difference in ground elevation is half the height of hill or escarpment, in feet (meters).
$K_1$: Factor to account for shape of topographic feature and maximum speed-up effect.
$K_2$: Factor to account for reduction in speed-up with distance upwind or downwind of crest.
$K_3$: Factor to account for reduction in speed-up with height above local terrain.
$x$: Distance (upwind or downwind) from the crest to the building site, in feet (meters).
$z$: Height above local ground level, in feet (meters).

**Figure 2.26** (*Continued*). Multipliers $K_1$, $K_2$ and $K_3$ for evaluating topographic factor $K_{zt}$ (ASCE 7-95).

maximum slope. The average maximum upwind slope of the hill is approximately $H/L_h$ and measurements have shown that hills with slopes of less than about 0.20 are unlikely to produce significant speed-up of the wind. For values of $H/L_h > 0.5$ the speed-up effect is assumed to be independent of slope. The speed-up principally affects the mean wind speed rather than the amplitude fluctuations and this fact has been accounted for in the values of $K_1$, $K_2$, and $K_3$. Therefore, values of $K_{zt}$ are intended for use with velocity presusre exposure coefficients, $K_h$ and $K_z$ which are based on gust speeds.

The lower bounds placed on the height $H$ for exposure categories B, C, and D are intended to screen out those situations where speed-up effects can be ignored. The additional restriction that the upwind terrain be free of hills or escarpments for a distance equal to the smaller of $50H$ or 1 mile is intended to limit application where the existing information is known to apply. It should be noted that the use of $K_{zt}$ is not a substitue to the general case of wind over hilly or complex terrain for which wind tunnel test may be warranted.

#### 2.5.6.4 Full and partial loading

This is a new requirement in ASCE 7-95 for verifying torsional response of buildings taller than 60 ft (18 m). Torsion may be induced by: (i) partial wind loading, (ii) eccentricity of the elastic center with respect to the resultant wind load vector and (iii) eccentricity of the elastic center with respect to the center of mass. The load combinations described in Fig. 2.27 reflect surface pressure patterns that have been observed on tall buildings in turbulent wind. Wind tunnel tests have demonstrated that even a 25 percent selective load reduction can underestimate the wind-induced torsion in buildings with a uniform rectangular cross-section. In some structural systems, more severe effects are observed when the resultant wind load acts diagonally to the building. To account for this effect and the fact that many structures exhibit maximum response in the across-wind direction, a structure should be capable of resisting 75 percent of the design wind loads applied simultaneously along the principal axes. The full and partial load requirements are shown schematically in Fig. 2.27.

#### 2.5.6.5 Gust effect factor

The gust effect factor accounts for the loading effects in the along-wind direction due to wind turbulence and structure interaction. It also accounts for along-wind loading effects due to dynamic amplification for flexible buildings. It does not include allowances for across-wind loading effects, vortex shedding, instability due to galloping or flutter, or dynamic torsional effects. For structures susceptible to loading effects that are not accounted for in the gust effect factor, information should be obtained from wind-tunnel test.

The gust effect factor is given for three major categories namely: (1) Rigid structures-simplified method; (2) Rigid structures-complete analysis; and (3) flexible or dynamically sensitive structures.

**Rigid structures-simplified method.** No calculations are required in

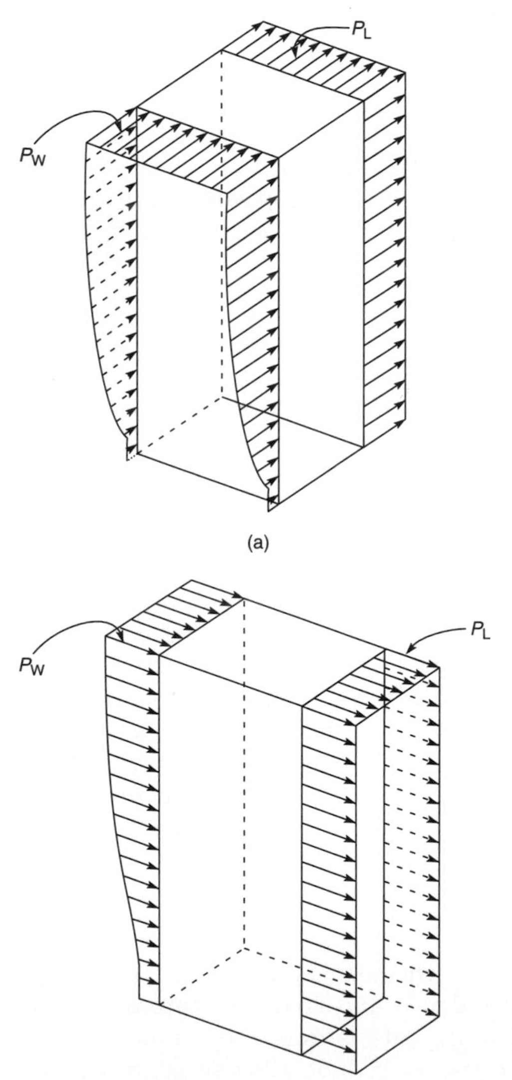


**Figure 2.27** Full and partial wind loads (ASCE 7-95). (a), (b), (c), (d), Elevations (e), (f), plans.

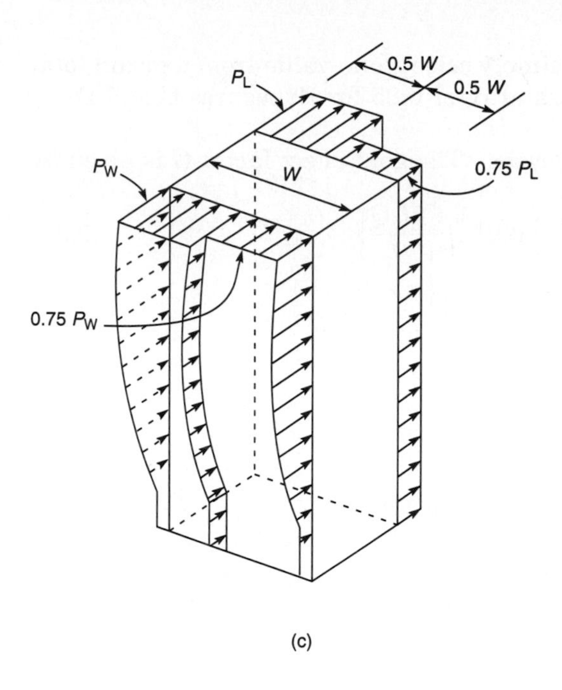

(c)

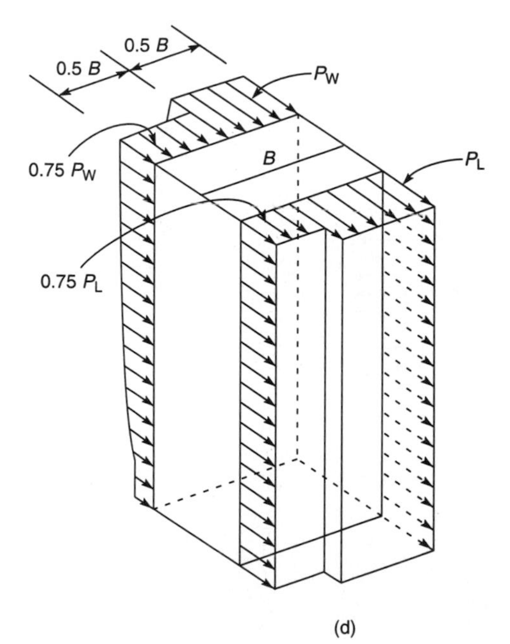

(d)

**Figure 2.27** (*Continued*).

this method. The engineer simply selects one value from a grand total of two, 0.8 for Exposures A and B, or 0.85 for Exposures C and D.

**Rigid structures-complete analysis.** The gust effect factor $G$ is given by

$$G = 0.9\left\{\frac{1 + 7I_{\bar{z}}Q}{1 + 7I_{\bar{z}}}\right\} \tag{2.31}$$

$$I_{\bar{z}} = C\left(\frac{33}{\bar{z}}\right)^{1/6} \tag{2.32}$$

where $I_{\bar{z}}$ = the intensity of turbulence at height $z$
$\bar{z}$ = the equivalent height of structure = $0.6h$

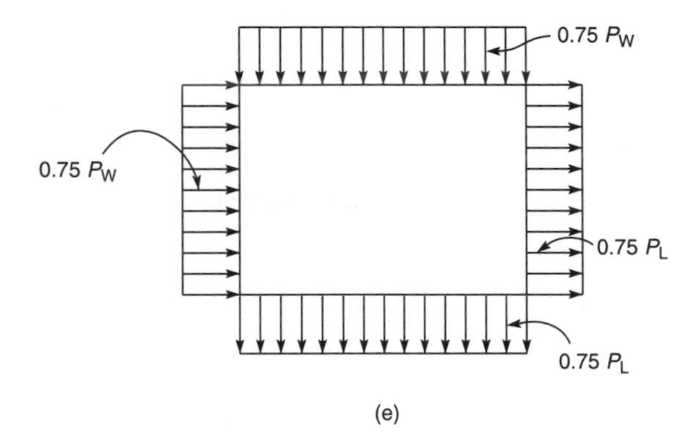


(e)

Plan view of building

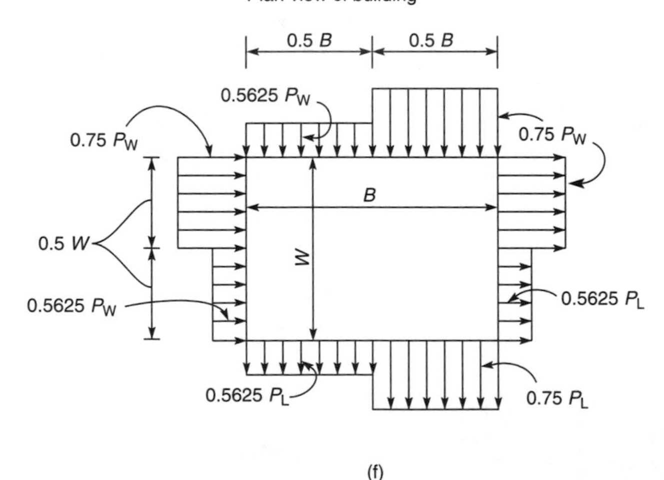


(f)

Plan view of building

**Figure 2.27** *(Continued)*.

**TABLE 2.31 Constants for Evaluating Gust Effect Factor *G* (Rigid Structures, Complete Analysis ASCE 7-95)**

| Exp | $\bar{\alpha}$ | $\bar{b}$ | $c$ | $l$ (ft) | $\varepsilon$ |
|---|---|---|---|---|---|
| A | 1/3.0 | 0.30 | 0.45 | 180 | 1/2.0 |
| B | 1/4.0 | 0.45 | 0.30 | 320 | 1/3.0 |
| C | 1/6.5 | 0.65 | 0.20 | 500 | 1/5.0 |
| D | 1/9.0 | 0.80 | 0.15 | 650 | 1/8.0 |

$c$ = the value given in Table 2.31

$Q$ = the background response given by the relation

$$Q^2 = \frac{1}{1 + 0.63\left\{\dfrac{b+h}{L_{\bar{z}}}\right\}^{0.63}} \tag{2.33}$$

where $b$ = building width parallel to wind

$h$ = building height

$L_{\bar{z}}$ = the integral length of scale of turbulence at the equivalent height given by

$$L\bar{z} = \lambda\left(\frac{\bar{z}}{33}\right)^{\varepsilon} \quad \text{in which}$$

$\lambda$ = value listed in Table 2.31

$\varepsilon$ = value listed in Table 2.31

**Flexible or dynamically sensitive buildings.** These are slender buildings that have a frequency less than 1 Hz (i.e. buildings with a fundamental period greater than 1 s). Included are buildings that have a height, $h$, in excess of four times the least horizontal dimension. The gust factor, $G_f$, accounts for the loading effects in the along-wind direction due to wind turbulence, structure interaction and dynamic amplification due to the flexibility of the building. It does not include allowances for across-wind effects, vortex shedding, instability due to galloping or flutter, or dynamic torsional effects. As mentioned previously wind tunnel test is the preferred and appropriate method. The gust factor, $G_f$ , is determined by using a series of equations as illustrated in the following.

The gust factor is given by

$$G_f = \frac{1 + 2gI_{\bar{z}}\sqrt{Q^2 + R^2}}{1 + 7I_{\bar{z}}} \tag{2.34}$$

where $R$, the resonant response factor, is given by

$$R^2 = \frac{1}{\beta} R_n R_h R_b (0.53 + 0.47 R_d)$$

$$R_n = \frac{7.465 N_1}{(1 + 10.302 N_1)^{5/3}}$$

$$N_1 = \frac{n_1 L_{\bar{z}}}{\bar{V}_{\bar{z}}}$$

$$R_l = \begin{cases} \dfrac{1}{\eta} - \dfrac{1}{2\eta^2}(1 - e^{-2\eta}) & \text{for } \eta > 0 \\ 1 & \text{for } \eta = 0 \end{cases}$$

$(l = h, b, d)$

$R_l = R_h$ setting $\eta = 4.6 n_1 h / \bar{V}_{\bar{z}}$

$R_l = R_b$ setting $\eta = 4.6 n_1 b / \bar{V}_{\bar{z}}$

$R_l = R_d$ setting $\eta = 15.4 n_1 d / \bar{V}_{\bar{z}}$

$\beta =$ damping ratio

$\bar{V}_{\bar{z}} =$ mean hourly wind speed at height $\bar{z}$, in ft/sec

$$\bar{V}_{\bar{z}} = \bar{b} \left( \frac{\bar{z}}{33} \right)^{\bar{\alpha}} \hat{V}_{\text{ref}}$$

where $\bar{b}$ and $\bar{\alpha}$ are listed in Table 2.31 and

$g = 3.5$

#### 2.5.6.6 Calculation of gust effect factor: design example

##### Given

Basic wind speed, $V$, at reference height of 33 ft in Exposure C = 140 mph

Exposure type for building terrain = $D$

Building height $h$ = 600 ft

Building width (dimension perpendicular to wind) = 100 ft

Building depth (dimension parallel to wind) = 100 ft

Building natural frequency $n_1 = 0.2$ Hz (i.e., Building period = 1/0.2 = 5.0 sec)

Damping ratio = 0.015

**Required**

Gust factor $G_f$

Gust factor $G_f$ is given by

$$G_f = \frac{1 + 2gI_{\bar{z}}\sqrt{Q^2 + R^2}}{1 + 7I_{\bar{z}}}$$

$$g = 3.5$$

$$\bar{z} = 0.6 \times 600 = 360$$

$$I_{\bar{z}} = c\left(\frac{33}{\bar{z}}\right)^{1/6} \quad [c \text{ from Table 2.30} = 0.15]$$

$$= 0.15\left(\frac{33}{360}\right)^{1/6}$$

$$= 0.10072$$

$$L_{\bar{z}} = l\left(\frac{\bar{z}}{33}\right)^{\epsilon} \quad [l = 650 \text{ ft, and } \epsilon = 1/8.0 \text{ from Table 2.30}]$$

$$= 650\left(\frac{360}{33}\right)^{1/8} = 876.27$$

$$Q^2 = \frac{1}{1 + 0.63\left\{\dfrac{b + h}{L_{\bar{z}}}\right\}^{0.63}}$$

$$= \frac{1}{1 + 0.63\left\{\dfrac{100 + 600}{876.27}\right\}^{0.63}}$$

$$= 0.64646$$

$$\bar{V}_z = \bar{b}\left\{\frac{\bar{z}}{33}\right\}^{\bar{\alpha}} \bar{v}_{\text{ref}}$$

$$= 0.80\left\{\frac{360}{33}\right\}^{1/9} 205.33$$

$$= 214.215$$

$$N_1 = \frac{n_1 L_{\bar{z}}}{\bar{V}_z}$$

$$= \frac{0.2 \times 876.27}{214.215}$$

$$= 0.818$$

$$R_n = \frac{7.465N_1}{(1 + 10.302N_1)^{5/3}}$$

$$= \frac{7.465 \times 0.818}{(1 + 10.302 \times 0.818)^{5/3}}$$

$$= 0.14514$$

$n$ for $R_b$: $$n = \frac{4.6n_1 b}{V_{\bar{z}}}$$

$$= \frac{4.6 \times 0.2 \times 100}{214.215}$$

$$= 0.42947$$

$n$ for $R_h$: $$\eta = \frac{4.6n_1 h}{\bar{V}_{\bar{z}}}$$

$$= \frac{4.6 \times 0.2 \times 600}{214.215}$$

$$= 2.57684$$

$\eta$ for $R_d$: $$\eta = \frac{15.4n_1 d}{\bar{V}_{\bar{z}}}$$

$$= \frac{15.4 \times 0.2 \times 100}{214.215}$$

$$= 1.43780$$

$$R_h = \frac{1}{\eta} - \frac{(1 - e^{-2\eta})}{2n^2}$$

$$= \frac{1}{2.576} - \frac{(1 - e^{-2 \times 2.576})}{2 \times 2.576^2}$$

$$= 0.31321$$

$$R_b = \frac{1}{\eta} - \frac{(1 - e^{-2n})}{2n^2}$$

$$= \frac{1}{0.42947} - \frac{(1 - e^{-2 \times 0.429})}{2 \times 0.429^2}$$

$$= 0.76595$$

$$R_d = \frac{1}{\eta} - \frac{(1 - e^{-2n})}{2\eta^2}$$

$$= \frac{1}{1.4378} - \frac{(1 - e^{-2 \times 1.4378})}{2 \times 1.4378^2}$$

$$= 0.46728$$

$$R^2 = \frac{1}{\beta} R_n R_h R_b (0.53 + 0.47 R_d)$$

$$= \frac{1 \times 0.14514 \times 0.31321 \times 0.76595(0.53 + 0.47 \times 0.46728)}{0.015}$$

$$= 1.74$$

$$G_f = \frac{1 + 2gI_{\bar{z}}\sqrt{Q^2 + R^2}}{1 + 7I_{\bar{z}}}$$

$$= \frac{1 + 2 \times 3.5 \times 0.10072\sqrt{0.64646 + 1.74}}{1 + 7 \times 0.10072}$$

$$= 1.225$$

#### 2.5.6.7 Sensitivity study of gust response factor

To get a feel for the magnitude of gust response factor, a sensitivity study has been performed on five buildings ranging in height from 500 ft (152 m) to 1200 ft (366 m). The buildings' characteristics used in the study are shown in Table 2.32. The gust factors have been determined for exposure categories A, B, C and D by varying the fundamental frequency, damping ratio and the basic wind speed, one at a time, for each building. The results have been summarized in a

**TABLE 2.32 Building Characteristics Used in Sensitivity Study (ASCE 7-95)**

| Height ft (m) | Width ft (m) | Depth ft (m) | Frequency values | Reference |
|---|---|---|---|---|
| 500 (152) | 120 (37) | 120 (37) | 0.25, 0.20, 0.167, 0.143, 0.125 | Fig. 2.21a,b,c |
| 600 (183) | 100 (30) | 100 (30) | 0.25, 0.20, 0.167, 0.143, 0.125 | Fig. 2.22a,b,c |
| 800 (244) | 140 (43) | 140 (43) | 0.167, 0.143, 0.125, 0.111, 0.1 | Fig. 2.23a,b,c |
| 900 (274) | 150 (46) | 150 (46) | 0.167, 0.143, 0.125, 0.111, 0.1 | Fig. 2.24a,b,c |
| 1200 (366) | 175 (53) | 175 (53) | 0.125, 0.111, 0.10, 0.091, 0.083 | Fig. 2.25a,b,c |

Wind speeds (mph): 132, 161.33, 190.66, 205.33, and 220. Damping percent: 0.015, 0.02, 0.03, 0.04, and 0.05.

set of three graphs for each building. For example, Figs 2.28a–c show the variation of $G_f$ for the 500 ft (152 m) building. In Fig. 2.28a, the building period and basic wind speeds are the same while the damping ratio is varied from 0.015 to 0.05. In Fig. 2.28b, the damping and basic wind speeds are the same while the period is varied from four to eight seconds. And finally in Fig. 2.28c, damping and periods are the same while the basic wind speed is varied from 90 mph to 150 mph (132 f/s to 220 f/s).

The other three sets of graphs in Figs 2.29–2.31 similarly show the variation of gust response factor for the 600 ft, 900 ft, and 1200 ft buildings.

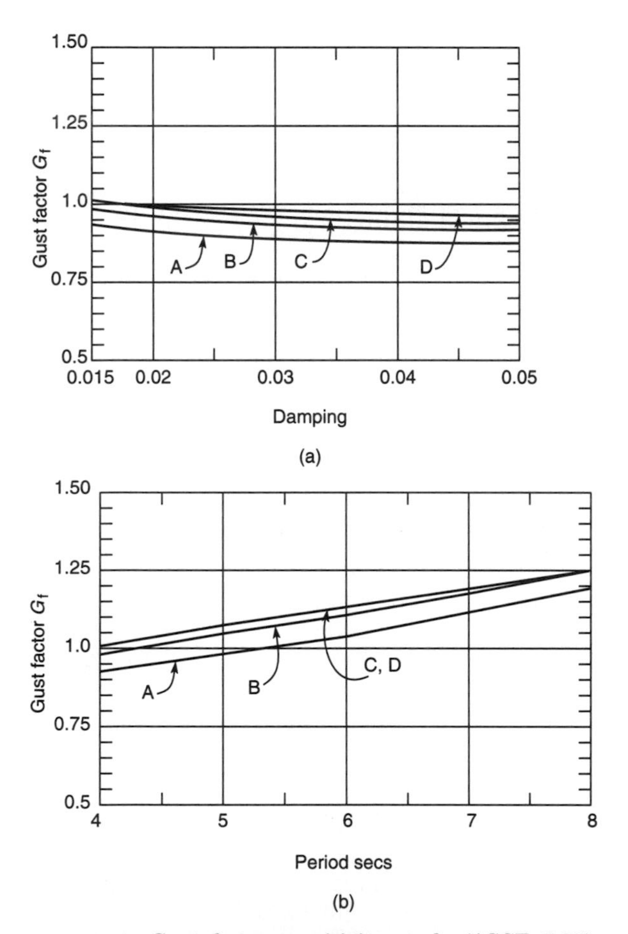


**Figure 2.28** Gust factor sensitivity study (ASCE 7-95): 500 ft building; (a) variation of $G_f$ with respect to damping; (b) variation of $G_f$ with respect to period; (c) variation of $G_f$ with respect to wind speed.

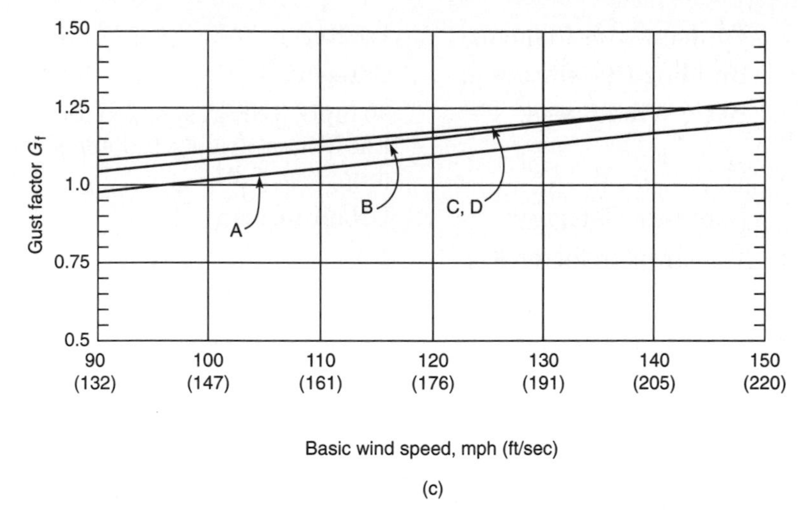


**Figure 2.28** *(Continued)*.

It should be noted that the gust factors, velocity pressure exposure coefficients, pressure coefficients and force coefficients given in ASCE 7-95 are for use with a 3 sec gust speed at 33 ft (10 m) above ground in open country. It is therefore necessary that wind speeds based on different averaging time, such as hourly mean or fastest mile, be adjusted to reflect the 3 sec gust in open country. As a guide the results of statistical studies given in Fig. 2.32a may be used for adjusting non-3 sec wind speeds to 3 sec gusts.

**Example.** Given a wind speed, averaged for a 90 sec interval is 60 mph determine the 3-second gust speed for a non-hurricane condition.

From Fig. 2.32, the ratio $\frac{V_t}{V_{3600}} = \frac{V_{90}}{V_{3600}} = 1.20$. The ratio $\frac{V_3}{V_{3600}} = 1.50$.

Therefore the 3-second gust speed $= \frac{1.50}{1.20} \times 60 = 75$ mph for a non-hurricane condition.

### 2.5.6.8 Calculation of wind pressures: design example

In this example, design wind pressures will be determined for the main force resisting of the same building used previously for illustrating the ASCE 7-93.

**Given**

| | |
|---|---|
| Plan dimensions: | $60 \times 120$ ft ($18.28 \times 36.57$ m) |
| Building Height: | 14 floors at 10 ft floor-to-floor = $10 \times 14 = 140$ ft (42.67 m) |

| | |
|---|---|
| Fundamental frequency: | 1.1 Hertz |
| Building Classification: | Category 1 |
| Basic Wind Speed: | 90 mph, 3 sec gust speed for Las Vegas, from wind speed map, Fig. 2.31 |
| Exposure Category: | Urban terrain |
| Topographic factor $K_{zt}$: | 1.0 |

**Required**

Wind pressures for the design of primary lateral system.

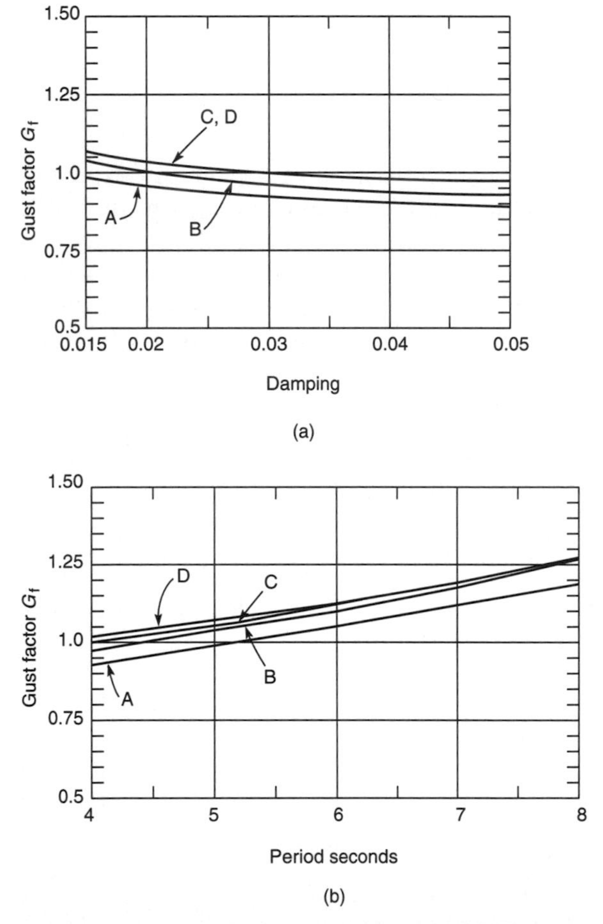


**Figure 2.29** Gust factor sensitivity study (ASCE 7-95): 600 ft building; (a) variation of $G_f$ with respect to damping; (b) variation of $G_f$ with respect to period; (c) variation of $G_f$ with respect to wind speed.

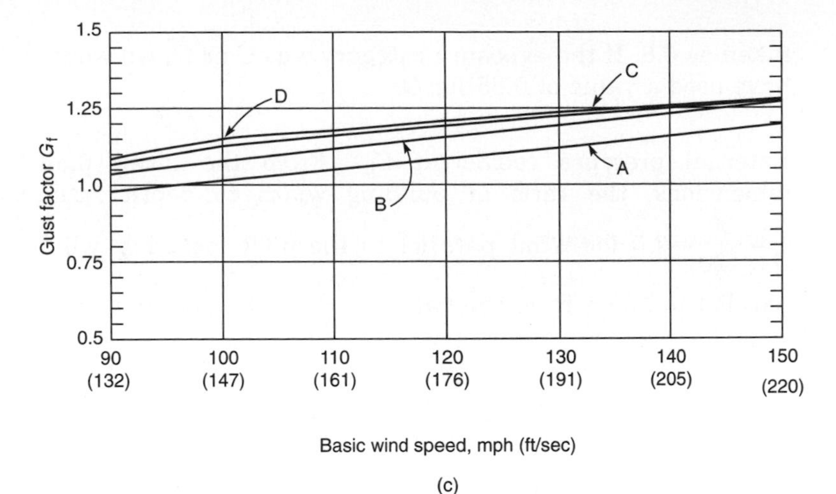


**Figure 2.29** *(Continued).*

Step 1. Building Classification. A typical office building is not generally considered an essential facility in the aftermath of windstorm, nor its primary function is for occupancy by more than 300 persons in one area. Therefore, the example building is judged to be in Type 1 Category. However, before a building is classified into a category, it is good practice to ascertain with building owners and plan-check officials, that the category is consistent with their perception of building performance.

Results of Step 1: Category 1, $I = 1.0$

Step 2. Basic Wind Speed $V$. The ASCE wind speed map, Fig. 2.23, indicates that Las Vegas is in wind contour of 90 mph. Again it is a good practice to confirm the design wind speed with the local plan check officials.

Result of Step 2: $V = 90$ mph

Step 3. Determination of Gust Response Factor $G$. The building height is less than 500 ft, the height-to-width ratio is less than 5, and its natural frequency of 1.1 Hertz is more than 1.0 Hertz. Therefore, the building may be considered a non-flexible building. Only one value of gust response factor $G$, calculated at the building roof level, is required for determining design loads on the main lateral system. The building under consideration is a non-flexible building with an exposure category B. Therefore the value of $G$ may be taken as 0.8. If the exposure category was C or D, we would have used a value of 0.85 for $G$.

Step 4. External pressure coefficient $C_p$. From the given plan dimensions, the ratio of building width to depth ratio, $\frac{L}{B} = \frac{60}{120} = 0.5$ for wind parallel to the 60 ft face. For wind parallel to 120 ft face, the ratio $= \frac{120}{60} = 2.0$. From Fig. 2.18

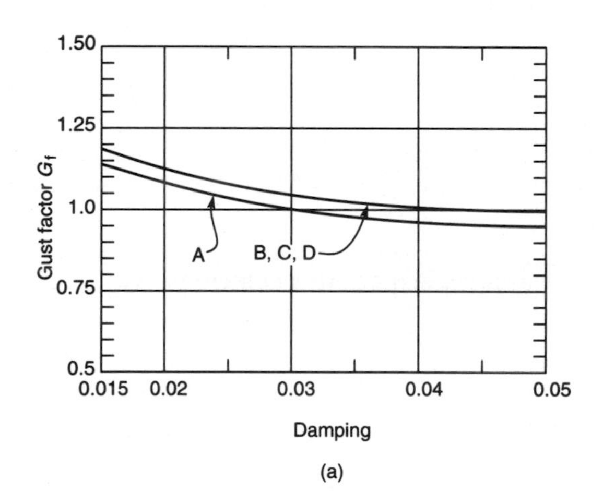


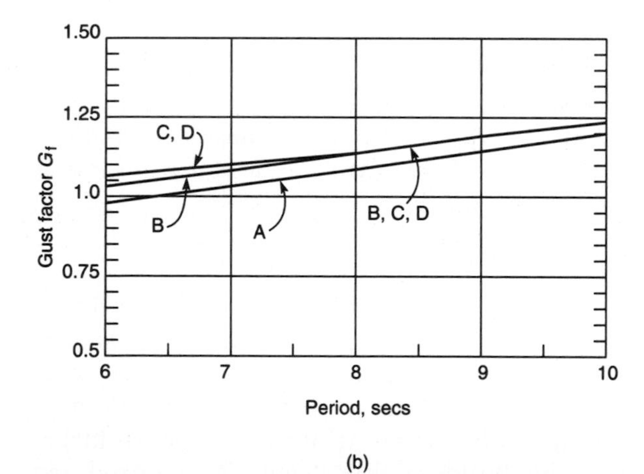


**Figure 2.30a** Gust factor sensitivity study (ASCE 7-95): 800 ft building; (a) variation of $G_f$ with respect to damping; (b) variation of $G_f$ with respect to period; (c) variation of $G_f$ with respect to wind speed.

the following values of $C_p$ are obtained.

$C_p = 0.8$ for the windward wall

$C_p = 0.5$ for leeward wall, wind parallel to 60 ft face

$C_p = 0.3$ for leeward wall, wind parallel to 120 ft face

Roof pressures and internal pressures are not relevant in determining wind loads for primary lateral system: internal pressures and suctions acting on the windward and leeward walls cancel out with adding or subtracting to the overall wind loads. Roof suction which results in uplift forces is generally neglected in the design of primary lateral system.

Results of Step 4 are shown as $C_p$ values in Table 2.33.

Step 5. Building exposure. Since the building is located in an urban terrain, the exposure category is judged to be B.

Results of Step 5: Exposure B

Step 6. Combined velocity pressure, exposure coefficient $K_z$. The gradient height, $Z_g$, and the power coefficient $\alpha$ for exposure B are 1200 ft and 7.5 respectively. The values for $K_z$ are calculated by the relation: Below the 15 ft height, the value of $K_z$ is taken as a constant determined at height 15 ft.

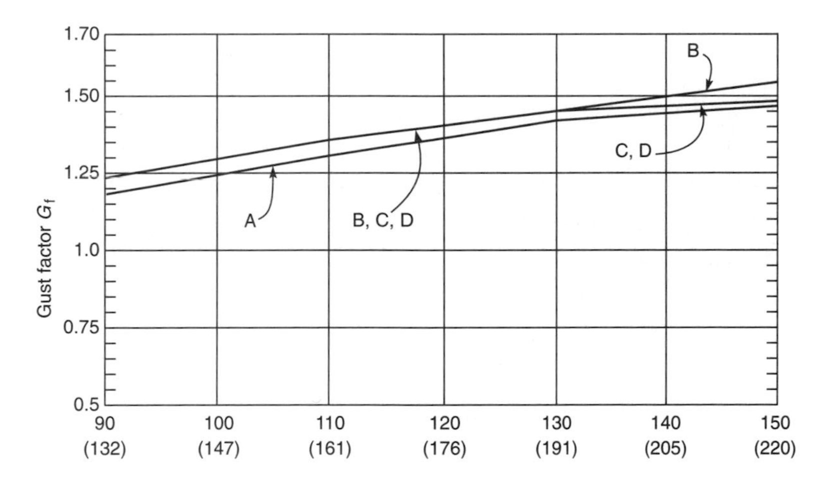


(c)

**Figure 2.30a** (*Continued*).

Instead of calculating values for $K_z$ from Eq. (2.20), they can be obtained directly from Table 2.29 or Fig. 2.24. The results of Step 6 are shown in Table 2.33.

Step 7. Velocity pressure $q_z$. Values for $q_z$ are obtained from Eqs (2.20) and (2.21).

The results of Step 7 are shown in column 3 and 5 of Table 2.33.

Step 8. Design pressure $p$. With the known values of $q$, $G$ and $C_p$, the design pressure is obtained by the chain equation

$$p = q \times G \times C_p$$

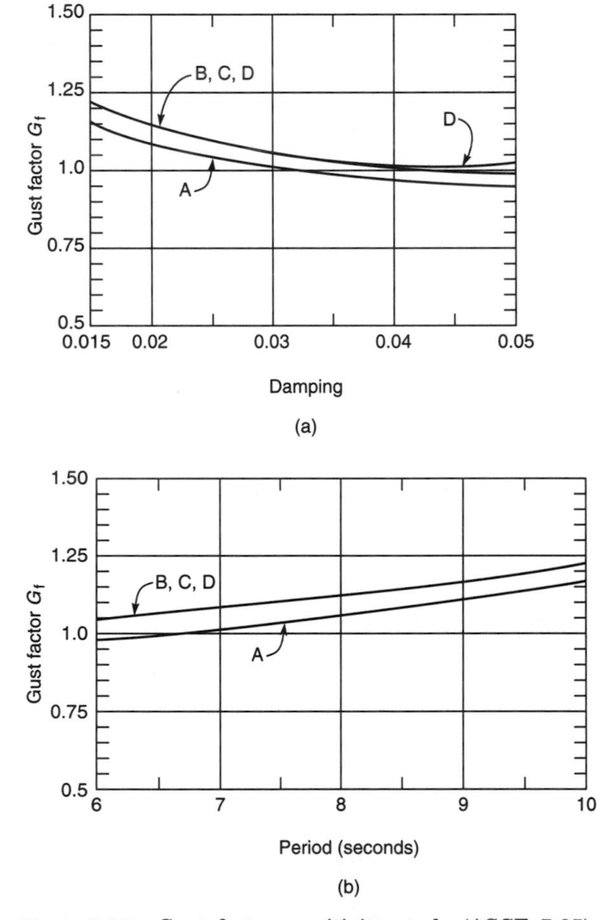


**Figure 2.30b** Gust factor sensitivity study (ASCE 7-95): 900 ft building; (a) variation of $G_f$ with respect to damping; (b) variation of $G_f$ with respect to period; (c) variation of $G_f$ with respect to wind speed.

Wind pressures and suctions on windward and leeward walls are shown in Table 2.33. Summation of the two is used in determining the pressure for the design of main wind-force resisting system.

Step 9. Floor-by-floor wind load. This is obtained by multiplying the exposed area tributary to the level by the corresponding value of design pressure at the floor height.

### 2.5.7 National Building Code of Canada (NBC 1990)

The NBC is perhaps the most exhaustive treatise to address wind loading on tall buildings. In determining the wind loads, it takes into consideration building dimensions, shape, stiffness, damping ratios, site topography, climatology, boundary layer meteorology, bluff body aerodynamics, and probability theory. Three different approaches for determining wind loads on buildings are given: (i) simple procedure; (ii) experimental procedure; and (iii) detailed procedure.

#### 2.5.7.1 Simple procedure

The simple procedure is applicable for determining the structural wind loads for a majority of low- and medium-rise buildings and also for cladding design of low-, medium-, and high-rise buildings. The method is similar to other code approaches in which the dynamic action of wind is dealt with by equivalent static loads defined independently of the dynamic properties of wind. The external

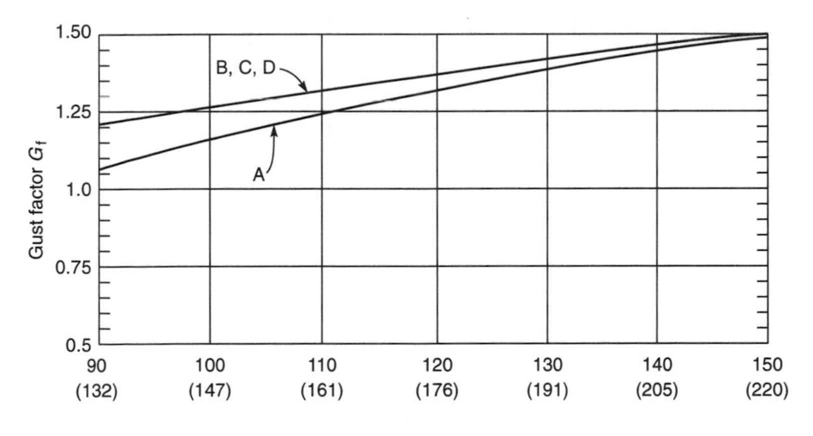


**Figure 2.30b** *(Continued).*

pressure or suction on the building surface is given by the equation

$$p = qC_eC_gC_p \tag{2.35}$$

where $p$ = design static pressure or suction, acting normal to the surface: kilo pascals
$q$ = the reference wind pressure; kilo pascals
$C_e$ = exposure factor that reflects the changes in wind speed with height and variations in the surrounding terrain: dimensionless

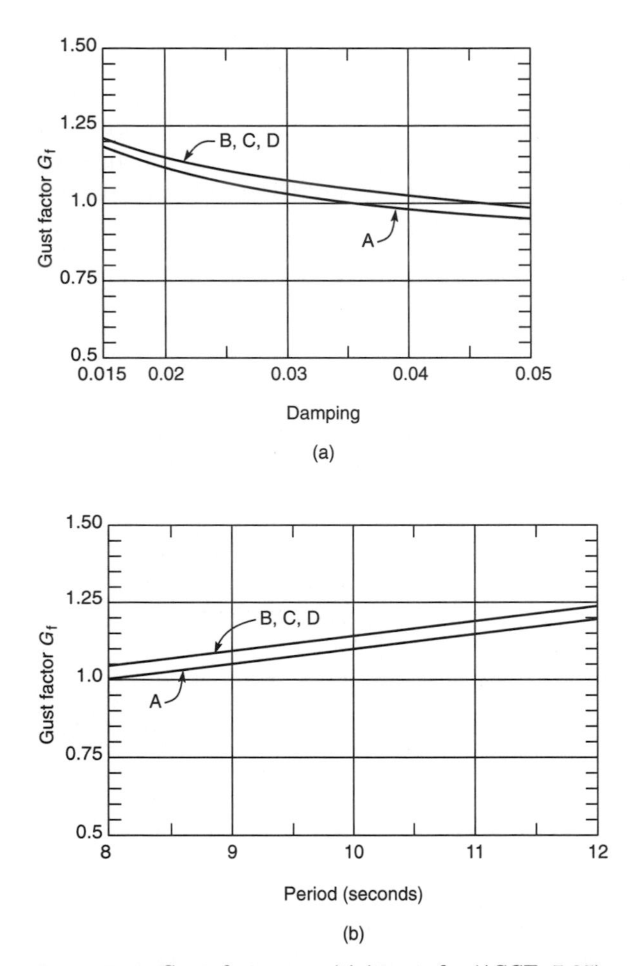


**Figure 2.31** Gust factor sensitivity study (ASCE 7-95): 1200 ft building; (a) variation of $G_f$ with respect to damping; (b) variation of $G_f$ with respect to period; (c) variation of $G_f$ with respect to wind speed.

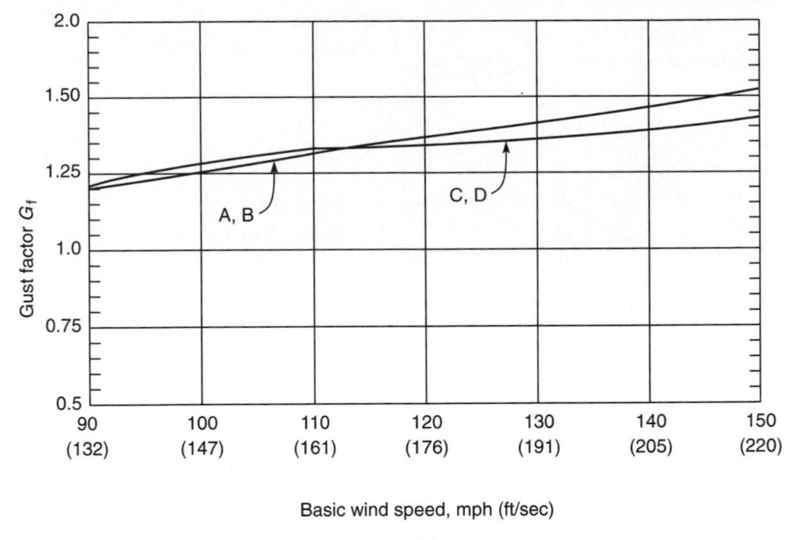


**Figure 2.31** *(Continued).*

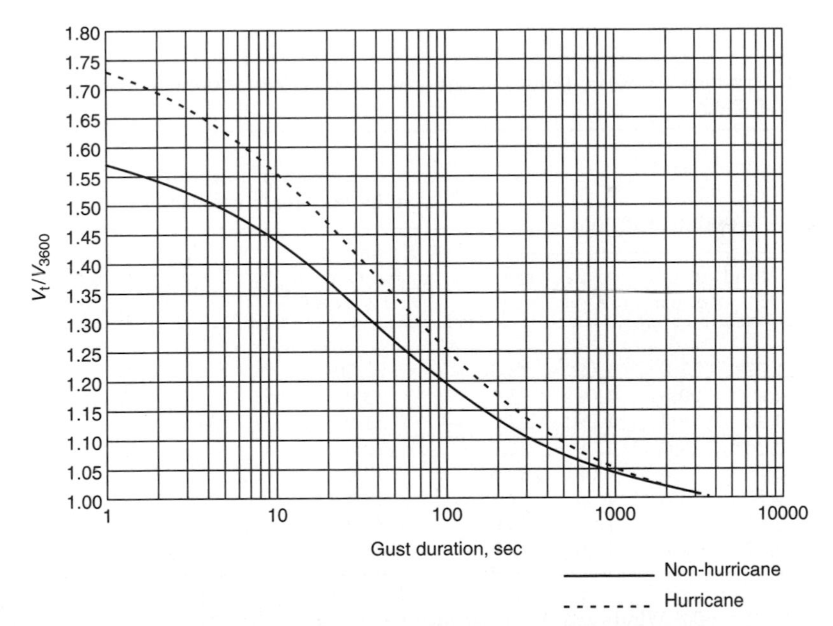


**Figure 2.32a** Ratio of probable maximum speed averaged over t-sec to hourly mean wind speed.

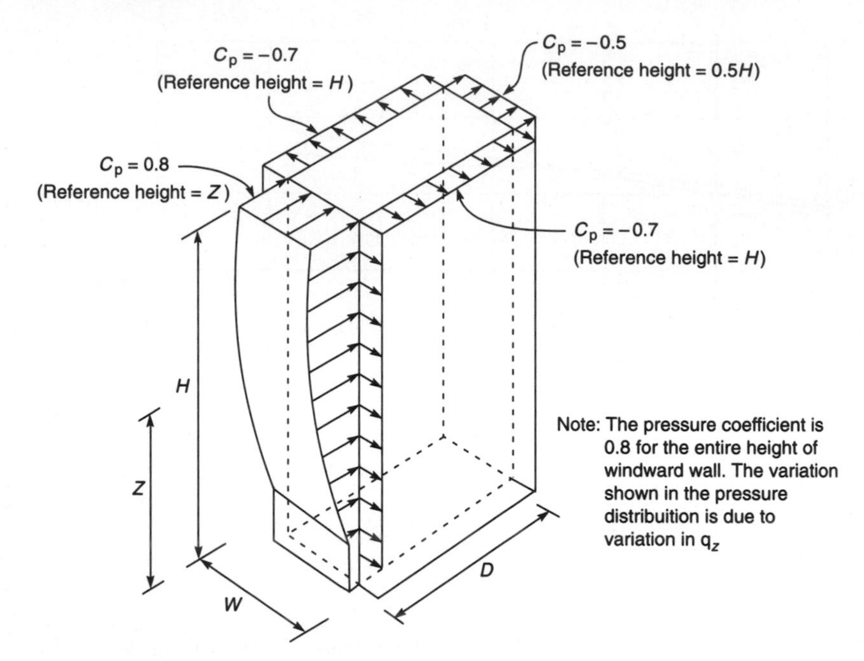


**Figure 2.32b** NBC 1990: external wind pressure coefficient, $C_p$; flat-roofed buildings, $H > W$.

**TABLE 2.33 Example Problem, Design Pressures for Main Wind-force Resisting Frame (ASCE 7-95)**

| | Value of $K_z$ and $q_z$ | | | | Design pressures for main wind-force resisting system | |
|---|---|---|---|---|---|---|
| | Windward | | Leeward | | X-wind, psf | Y-wind, psf |
| Height ft | $K_z$ | $q_z$ psf | $K_h$ | $q_h$ | $p_z = q_z G \times 0.8 + q_h G \times 0.5$ | $p_z = q_z G \times 0.8 + q_h G \times 0.3$ |
| 140 | 1.09 | 22.6 | 1.09 | 22.6 | 23.50 | 19.8 |
| 100 | 0.99 | 20.5 | 1.09 | 22.6 | 22.0 | 18.5 |
| 80 | 0.93 | 18.8 | 1.09 | 22.6 | 21.0 | 17.4 |
| 60 | 0.85 | 17.6 | 1.09 | 22.6 | 20.3 | 16.6 |
| 40 | 0.76 | 15.5 | 1.09 | 22.6 | 19.0 | 15.3 |
| 30 | 0.70 | 14.5 | 1.09 | 22.6 | 18.3 | 14.6 |
| 20 | 0.62 | 13.4 | 1.09 | 22.6 | 17.6 | 13.9 |
| 0–15 | 0.57 | 12.0 | 1.09 | 22.6 | 16.7 | 13.0 |

X-Wind: $c_p = +0.8$ Windward  Gust factor $G = 0.8$
$c_p = -0.5$ Leeward
Y-Wind: $c_p = +0.8$ Windward
$c_p = -0.3$ Leeward.

$C_g$ = gust factor, with a value of 2.0 for the structure as a whole and 2.50 for cladding: dimensionless
$C_p$ = external pressure coefficient averaged over the area of the surface considered: dimensionless

**(a) Reference pressure *q*.** The pressure $q$ in kilo pascals, is determined from referenced wind speed $\bar{V}$ by the equation:

$$q = C\bar{V}^2 \tag{2.36}$$

The factor $C$ depends on the atmospheric pressure and air temperature. If the wind speed $\bar{V}$ is in meters per second, the design pressure, in kilo pascals, is obtained by using a value of $C =$ 0.0006464. The reference wind pressure $q$, is given in the NBC supplement for three different levels of probability being exceeded per year ($\frac{1}{10}$, $\frac{1}{30}$, and $\frac{1}{100}$), that is, return periods for 10, 30, and 100 years, respectively. A 10-year recurrence pressure is used for the design of cladding and for the serviceability check of structural members for deflection and vibration. A 30-year wind pressure is used for the strength design of structural members of all buildings except those which are classified as post-disaster buildings. A 100-year wind is used for the design of post-disaster buildings such as hospitals, fire stations, etc. For example, the 10, 30 and 100-year mean hourly wind pressures in Montreal, Quebec are 0.31 kPa (6.5 psf), 0.37 kPa (7.72 psf) and 0.44 kPa (9.2 psf) respectively with corresponding wind speeds of 22 m/s (49.2 mph), 24 m/s (54 mph) and 26 m/s (58 mph).

**(b) Exposure factor $C_e$.** The exposure factor $C_e$ is based on the $\frac{1}{5}$ power law, which is appropriate for wind gust pressures in open terrain. An averaging period of 3 to 5 s is used in determining the gust factor. The gust represents a "parcel" of wind assumed to be effective over the entire building. For tall buildings, the reference height for pressures on the windward face correspond to the actual height above ground, and for suctions on the leeward face the reference height is half the height of the structure.

The exposure factor, $C_e$, reflects the changes in wind speed and height, and the effects of variations in the surrounding terrain and topography. Hills and escarpments which can significantly amplify the wind speeds are reflected in the exposure factor.

The exposure factor $C_e$ may be obtained from any of the following three methods.

1. The value shown in Table 2.34.
2. The value of the function $\left(\frac{h}{10}\right)^{1/5}$ but not less than 0.9, where $h$ is the reference height above grade, in metres.

**TABLE 2.34 Exposure Factor $C_e$ (NBC 1990)**

| Exposure factors, $C_e$ (NBC 1990) | |
|---|---|
| Height, m | Exposure factor |
| Over 0 to 6 | 0.9 |
| Over 6 to 12 | 1.0 |
| Over 12 to 20 | 1.1 |
| Over 20 to 30 | 1.2 |
| Over 30 to 44 | 1.3 |
| Over 44 to 64 | 1.4 |
| Over 64 to 85 | 1.5 |
| Over 85 to 140 | 1.6 |
| Over 140 to 240 | 1.8 |
| Over 240 to 400 | 2.0 |

3. If a dynamic approach is used, an appropriate value depending upon both height and shielding.

**(c) Gust effect factor (dynamic response factor), $C_g$.** This factor denoted by $C_g$ is defined as the ratio of the maximum to the mean effect of the loading. The increase in mean loads due to dynamic response includes the action of

- random wind gusts acting for short durations over all or part of the structure;
- fluctuating pressures induced by the wake of the structure, including vortex shedding forces;
- fluctuating forces induced by the motion of the structure itself through the wind.

All buildings are affected to some degree by these forces. The total response may be considered as a summation of the "mean component" which acts without any structural dynamic magnification, and a "resonant component" due to building vibration close to its natural frequency. For the majority of the buildings less than 120 m (394 ft) tall and height-to-width ratio less than four, the resonant component is small: the only added loading is due to gusts and can be dealt with in a simple static manner.

For buildings and components that are not particularly tall, long, slender, light-weight, flexible or lightly damped, a simplified set of dynamic gust factors is as follows.

$C_g = 2.5$ for building components and cladding

$C_g = 2.0$ for the primary structural system including anchorages to foundation.

**(d) Pressure coefficient $C_p$.** Pressure coefficient $C_p$ is a non-dimensional ratio of wind-induced pressure on a building to the velocity pressure of the wind speed at the reference height. It depends on the shape of the building, wind direction, and the profile of the wind velocity, and can be determined most reliably from wind-tunnel tests on building models. However, for the simple procedure, based on some limited measurements on full-scale buildings supplemeneted by wind-tunnel tests, the NBC gives the following values of $C_p$ for simple building shapes.

Windward wall: $C_p = +0.8$ (positive pressure)
Reference height $= Z$ above ground

Side wall and roof: $C_p = -1.0$ (negative pressure, suction)
Reference height $= H$ above ground

Leeward wall: $C_p = -0.5$ (negative pressure, suction)
Reference pressure $= 0.5H$ above ground

A schematic representation of the coefficients $C_p$ is given in Fig. 2.32 which covers the requirements for the overall design of structure having simple building geometries.

#### 2.5.7.2 Experimental procedure

The second approach is to use the results of special wind-tunnel or other experimental procedures for buildings likely to be susceptible to wind-induced vibrations. This is appropriate for tall, slender structures for which wind loading plays a major role in the structural design. A wind-tunnel test is also recommended for determining exterior pressure coefficients for cladding design of buildings whose geometry deviates markedly from more common shapes for which information is already available.

#### 2.5.7.3 Detailed procedure

The third approach consists of a series of calculations intended for a more accurate determination of the gust factor $C_g$, the exposure factor $C_e$, and the pressure coefficient $C_p$. The end product of the calculations yields a static design pressure which is expected to produce the same peak effect as the actual turbulent wind, with due considerations for the building properties such as height, width, natural frequency of vibration, and damping. This approach is primarily for determining the overall wind loading and response of tall slender structures and is not intended for determining exterior pressure coefficients for cladding design.

The code gives procedures for calculating the dynamic effects of vortex shedding for slender cylindrical towers and for tapered structures. Since the available data are limited for slender structures with cross-sections other than circular, wind-tunnel tests are recommended for estimating the likely response. To limit the cracking of masonry and interior finishes a maximum lateral deflection limitation of 1/500 of the height is specified unless detailed analysis is made and precautions are taken to permit larger movements.

The code recognizes that maximum accelerations of a building leading to a possible human perception of motion or even discomfort may occur in a direction perpendicular to the wind. A tentative acceleration limit of 1 to 3 percent of gravity for a 10-year return wind is recommended to limit the possibility of perception of motion.

**(a) Exposure factor, $C_e$ (detailed procedure).** The exposure factor, $C_e$, is based on the mean wind speed profile which varies considerably depending on the general roughness of terrain over which the wind has been blowing before it reaches the building. Three wind profile categories have been established as follows:

*Exposure A.* This is the exposure on which the reference wind speeds are based. The exposure is defined as open level terrain with only scattered buildings, trees or other obstructions, open water or shorelines.

$$C_e = \left(\frac{z}{10}\right)^{0.26}, \qquad C_e \geq 1.0 \tag{2.37}$$

*Exposure B.* Suburban and urban areas, wooded terrain or centers of large towns with terrain roughness extending in the upwind direction for at least 1.5 km.

$$C_e = 0.5\left(\frac{z}{12.7}\right)^{0.50}, \qquad C_e \geq 0.5 \tag{2.38}$$

*Exposure C.* Centers of large cities with heavy concentrations of tall buildings extending in the upwind direction for at least 1.5 km, with at least 50 percent of the buildings exceeding four stories in height.

$$C_e = 0.4\left(\frac{z}{30}\right)^{0.72}, \qquad C_e \geq 0.4 \tag{2.39}$$

Exposue factors can be calculated from Eqs (2.37)–(2.39) or obtained directly from graphs in Fig. 2.33.

**(b) Gust effect factor, $C_g$ (detailed procedure).** A general expression for the maximum or peak load effect, denoted $W_p$, is

$$W_p = \mu + g_p \sigma \tag{2.40}$$

where $\mu$ = the mean loading effect
$\sigma$ = the "root-mean square" loading effect, and
$g_p$ = peak loading

According to this expression, the dynamic gust response factor is defined as the ratio of peak loading to mean loading,

$$C_g = W_p/\mu$$
$$= 1 + g_p\left(\frac{\sigma}{\mu}\right) \tag{2.41}$$

The parameter $\frac{\sigma}{\mu}$ is given by the expression

$$\frac{\sigma}{\mu} = \sqrt{\frac{K}{C_{eH}}\left(B + \frac{sF}{\beta}\right)} \tag{2.42}$$

where $K$ = a factor related to the surface roughness coefficient of terrain
$K = 0.08$ for Exposure A
$K = 0.10$ for Exposure B
$K = 0.14$ for Exposure C

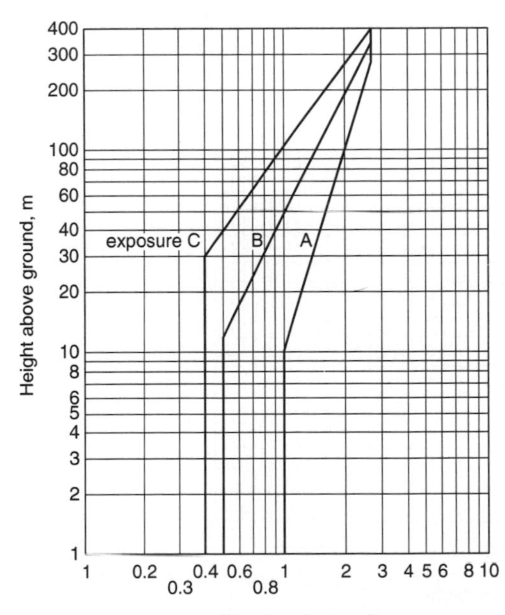


**Figure 2.33** Exposure factor $C_e$ as a function of terrain roughness and height above ground (NBC 1990).

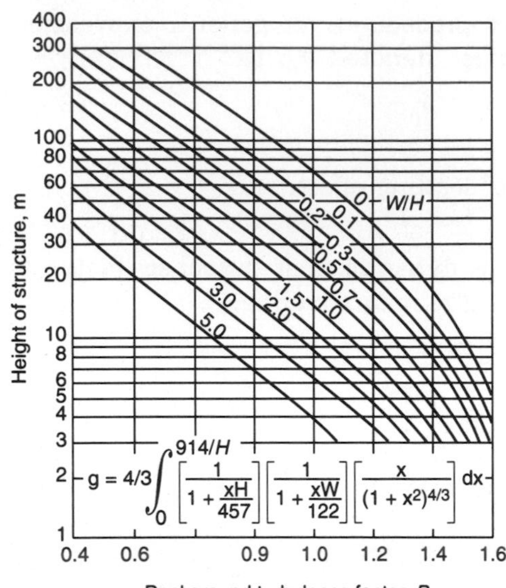


**Figure 2.34** Background turbulence factor as a function of width and height of structure (NBC 1990)

$C_{eH}$ = Exposure factor at the top of the building, $H$, evaluated by using Fig. 2.33

$B$ = background turbulence factor obtained from Fig. 2.34 as a function of building width-to-height ratio $W/H$

$H$ = height of building

$W$ = width of building

$s$ = size reduction factor obtained from Fig. 2.35 as a function of $\dfrac{W}{H}$ and reduced frequency $\dfrac{n_0 H}{V_H}$

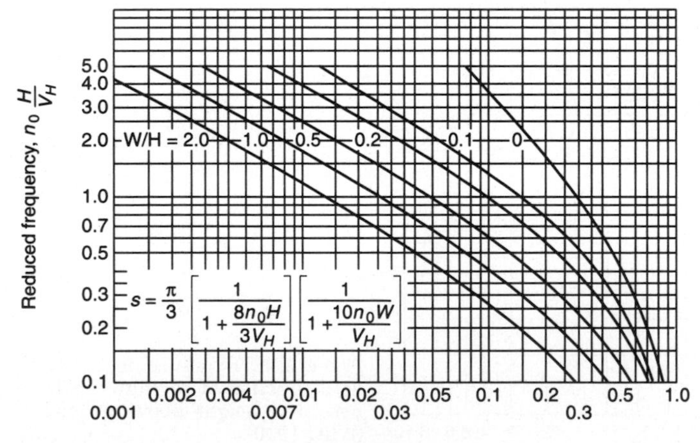


**Figure 2.35** Size reduction factor as a function of width, height and reduced frequency of structure (NBC 1990).

$n_0$ = natural frequency of vibration, Hz
$V_H$ = mean wind speed (m/s) at the top of structure, $H$
$F$ = gust energy ratio at the natural frequency of the structure obtained from Fig. 2.36 as a function of wave number $\frac{n_0}{V_H}$, and
$\beta$ = critical damping ratio, with commonly used values of 0.01 for steel, 0.015 for composite, and 0.02 for cast-in-place concrete buildings

#### 2.5.7.4 Design example

To illustrate the calculation of gust response factor, $C_g$, the following example for an office-type of occupancy will be worked out in detail. The properties of the building are given.

Height $H$ = 240 m (787.5 ft)

Width $W$ (across-wind) = 50 m (164 ft)

Depth $D$ (along-wind) = 50 m (164 ft)

Fundamental frequency $n_0$ = 0.125 Hz (period = 8 sec)

Critical damping ratio $\beta$ = 0.010

Average density of the building = 195 kg/m$^3$ (12.2 pcf)

Terrain for site = Exposure B

Maximum wind-induced lateral deflection at top of the building in along-wind direction, $\Delta$ = 0.6 m

Reference wind speed at 10 m, open terrain = 26.4 m/s (60 mph)

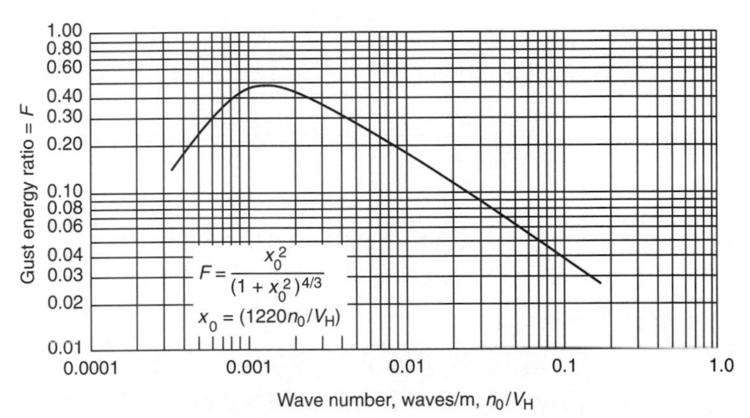


**Figure 2.36** Gust energy ratio as a function of wave number (NBC 1990).

Step 1. Calculate required parameters

$$C_{eH} = 2.17 \text{ (from Fig. 2.33)}$$

Mean wind speed at the top of the building, $V_H$, from equation

$$V_H = \bar{V}\sqrt{C_{eH}}$$
$$= 26.4 \times \sqrt{2.17}$$
$$= 36.88 \text{ m/s}$$

$$\text{Aspect ratio } W/H = \frac{50}{240} = 0.208$$

Wave number for calculation of $F$: $\dfrac{n_0}{V_H} = \dfrac{0.125}{38.88} = 0.00322$

Reduced frequency for calculation of $s$

$$n_0H/V_H = \frac{0.125 \times 240}{38.88} = 0.772$$

Step 2. Calculate $\dfrac{\sigma}{\mu}$, from Eq. (2.5)

(1) $K = 0.10$ for Exposure B

(2) $B = 0.50$ from Fig. 2.34 $\left(\text{for } \dfrac{W}{H} = \dfrac{50}{240} = 0.208\right)$

(3) $s = 0.14$ from Fig. 2.35 $\left(\text{for } \dfrac{n_0H}{V_H} = 0.772,\ W/H = 0.2\right)$

(4) $F = 0.36$ from Fig. 2.36 $\left(\text{for } \dfrac{n_0}{V_H} = 0.0032\right)$

(5) $\beta = 0.010$ (given)

(6) $$\frac{\sigma}{\mu} = \sqrt{\frac{K}{C_{eH}}\left(B + \frac{SF}{\beta}\right)}$$
$$= \sqrt{\frac{0.10}{2.17}\left(0.50 + \frac{0.14 \times 0.36}{0.010}\right)}$$
$$= 0.505$$

Step 3. Calculate $v$, from the following:

$$v = n_0\sqrt{\frac{sF}{sF + \beta B}}$$
$$= 0.125\sqrt{\frac{0.14 \times 0.36}{0.14 \times 0.36 + 0.01 \times 0.5}}$$
$$= 0.119 \text{ cycles/sec}$$

Step 4. Obtain peak factor $g_p$:

$$g_p \text{ (from Fig. 2.37)} = 3.6$$

Step 5.

$$C_g \text{ (from Eq. (2.41))} = 1 + 3.6 \times 0.505$$
$$= 2.82$$

Once the gust factor is determined, the peak dynamic forces and displacements may be determined by multiplying the mean wind loading values by $C_g$.

### 2.5.7.5 Wind-induced building motion

Although the maximum lateral wind-loading and deflection are generally in the direction parallel to the wind (along-wind direction), the maximum acceleration of a building leading to possible human perception of motion or even discomfort may occur in the direction perpendicular to the wind (across-wind direction). Across-wind accelerations are likely to exceed along-wind accelerations if the building is slender about both axes, such that the aspect ratio $\sqrt{\frac{WD}{H}}$ is less than one-third, where $W$ and $D$ are the across-wind and along-wind plan dimensions and $H$ is the height of the building.

Based on a wide range of turbulent boundary layer wind tunnel studies, the NBC gives two expressions for determining the across and along-wind accelerations.

The peak acceleration in the across-wind direction at the top is given by:

$$a_W = n_W^2 g_p \sqrt{WD}\left(\frac{a_r}{\rho_B g \sqrt{\beta_W}}\right) \tag{2.43}$$

In less slender structures or for lower wind speeds, the maximum

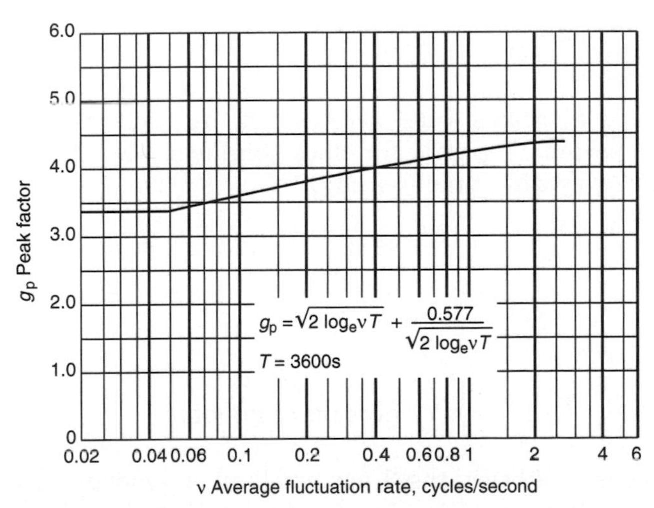


**Figure 2.37** Peak factor $g_p$ as a function of average fluctuation rate (NBC 1990).

acceleration may be in the along-wind direction. This is given by

$$a_D = 4\pi^2 n_D^2 g_p \sqrt{\frac{KsF}{c_e \beta_B} \frac{\Delta}{C_g}} \tag{2.44}$$

For the sample problem, we will assume the fundamental frequencies and the critical damping ratios in the across-wind and along-wind directions are the same.

Thus $\quad n_W = n_D = n_0 = 0.125\ \text{Hz}$

and $\quad \beta_W = \beta_D = \beta = 0.010$

The average density of the building $\rho_B = 195\ \text{kg/m}^3$ as given in the problem.

Step 6. Calculate $a_r$

$$a_r = 78.5 \times 10^{-3} \left[ \frac{V_H}{n_w \sqrt{WD}} \right]^{3.3}$$

$$= 78.5 \times 10^{-3} \left[ \frac{38.88}{0.125 \times 50} \right]^{3.3}$$

$$= 32.7\ \text{m/s}^2$$

Step 7. Calculate $a_W$ (across-wind response)
In our case, since $n_W = n_D = 0.125$ and $\beta_W = \beta_D = 0.10$

$$a_W = n_W^2 g_P \sqrt{WD} \left( \frac{a_r}{\rho_B g \sqrt{\beta_W}} \right)$$

$$= 0.125^2 \times 3.6 \times \sqrt{50 \times 50} \left( \frac{32.7}{195 \times 9.81 \sqrt{0.01}} \right)$$

$$= 0.482\ \text{m/s}^2$$

$$\frac{a_W}{g} = \frac{0.482}{9.81} \times 100 = 4.91 \text{ percent of gravity}$$

The calculated value of across-wind acceleration $a_W$ exceeds the acceptable limit of 3 percent of gravity for office buildings, warranting a detailed boundary layer wind tunnel study.

Step 8. Calculate $\frac{a_D}{g}$ (along-wind response)

$$a_D = 4\pi^2 \times 0.125^2 \times 3.6 \sqrt{\frac{0.10 \times 0.14 \times 0.36}{2.12 \times 0.10} \times \frac{0.60}{2.82}}$$

$$= 0.0729\ \text{m/s}^2$$

$$\frac{a_D}{g} = \frac{0.0729}{9.81} \times 100 = 0.74 \text{ percent of gravity}$$

The calculated value is well below the 3 percent limit. Its along-wind response is o.k.

To get a feel for the accelerations in along-wind and across-wind directions, the results of calculations for three rectangular buildings are given in Fig. 2.38. Accelerations at the top of each building are given for two cases: wind buffeting the broad face, and wind buffeting the narrow face. Observe that in Fig. 2.36a, the maximum acceleration of the building occurs in a direction perpendicular to the wind (across-wind direction) because the building is considerably more slender in the across-wind than in the along-wind direction. Across-wind accelerations control, if the building is slender about both axes,

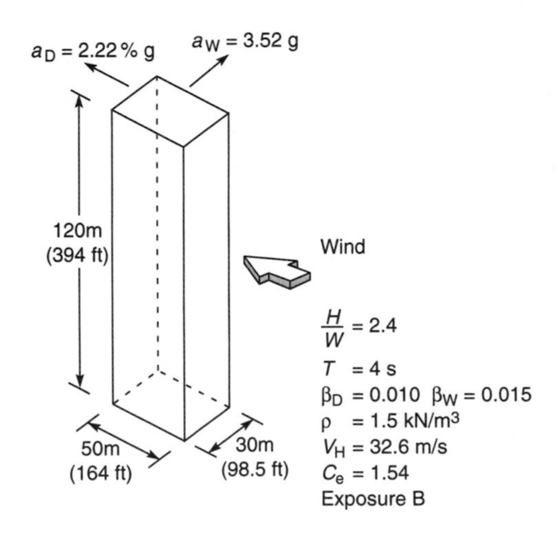


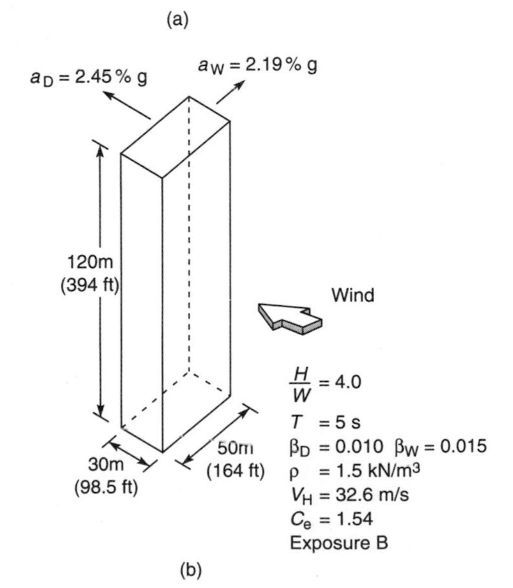


**Figure 2.38** Wind induced peak accelerations (NBC 1990): (a) $H/W = 2.4$; (b) $H/W = 4.0$; (c) $H/W = 4.8$.

that is, if $\sqrt{\frac{WD}{H}}$ is less than one-third, where $W$ and $D$ are the across-wind and along-wind plan dimensions and $H$ is the building height.

Since Eqs (2.43) and (2.44) for along-wind and across-wind accelerations are sensitive to the natural frequency of the building, and additionally Eq. (2.44) is to the building stiffness, use of approximate formulas for period calculations are not appropriate for the calculation of acceleration. Results of more rigorous methods, such as computer dynamic analyses, are recommended for use in these formulas.

In addition to the acceleration, many other factors such as visual cues, body positions and orientation and state of mind influence human perception of motion. However, research has shown that when the amplitude of acceleration is in the range of 0.5 percent to

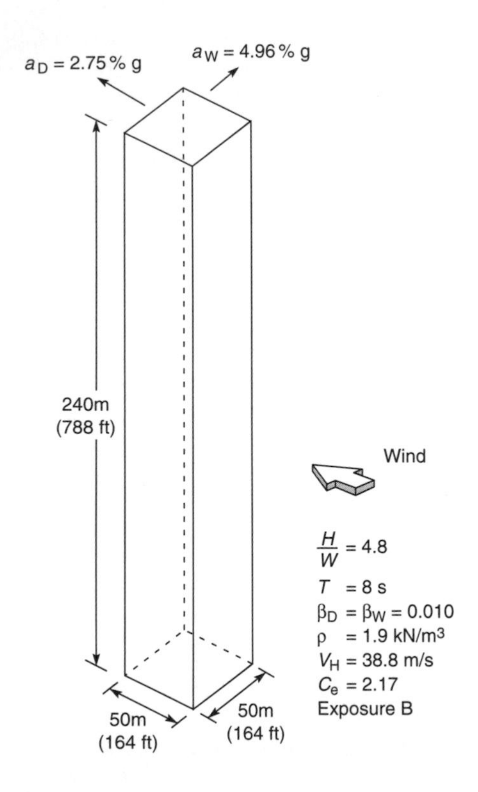


**Figure 2.38** *(Continued)*.

1.5 percent of acceleration due to gravity, movement of buildings becomes perceptible to most people. Based on this and other information, a tentative acceleration limit of 1 to 3 percent of gravity is recommended. The lower value is considered appropriate for apartment buildings, the higher values for office buildings.

## 2.6 Wind Tunnel Engineering

### 2.6.1 Introduction

It is fairly easy to visualize wind loading as an equivalent horizontal load causing shear and overturning effects on a building. But the complex action of wind, with its turbulent fluctuations about a mean value, drag, vortex shedding, and separation effects, not only results in shear and overturning effects but also induces dynamically fluctuating loads on the overall structure and its smaller components such as curtain walls. Methods of analysis which can be used to predict the complicated flow of wind and its effect on the response of the building and its components have not been developed for routine office use. In the design of very tall buildings it is recognized that use of the wind tunnel approach is a more refined method for arriving at design wind loads. Although some uncertainties still exist in the wind tunnel tests because of the complicated characteristics of natural wind, in today's engineering practice, wind tunnel results are the state of the art in tall building design.

Although wind tunnel model testing has gained wide acceptance, it is important to note that the action of wind in many situations is adequately covered in existing codes. It is therefore necessary to identify situations where wind tunnel tests are required in order to achieve reliable structural performance. Also, wind tunnel model studies generally indicate lower wind loads than prescribed in the codes and lead to more cost-effective designs. Although it is difficult to pick out which buildings need wind tunnel tests, it is prudent to include those buildings which appear to have an unusual sensitivity to the action of wind and generally fall outside existing experience. Buildings with unusual aerofoil shapes that are torsionally flexible may need to be wind-tunnel-tested even when height is not a major design consideration. Prismatic shapes, as a rule of thumb, can be considered as candidates for wind-tunnel test when the height exceeds the range of 40 to 50 stories.

Use of wind tunnels for testing buildings has been an offshoot of aeronautical engineering. From time immemorial, builders have known something about the wind factor, at least from experience, and used this more or less crudely in their designs. Gustave Eiffel, the noted engineer and constructor of the 984 ft (300 m) tower for the Paris Exposition of 1889, was the first to use an assumed wind loading on a major building-like structure. However, it was not until early this century with the coming of the airplane that the methods necessary for the understanding of aerodynamics began to appear. At first, methods were crude and the results were largely empirical. Later, with the greater complexity of engineering systems brought about by the Second World War, wind engineering analysis developed rapidly. Concern with the effects of wind on buildings and structures increased after the collapse of the Tacoma Narrows Bridge in 1940. This suspension bridge was designed to resist a steady wind of at least 100 mph (44 m/s). However, a few months after completion, the main span began to oscillate in both horizontal and torsional modes under a wind speed of only 42 mph (18.8 m/s). Within a few hours, the vibrating structure literally tore itself apart. This episode led to an extensive study of the effect of wind on structures. All the studies were conducted in aeronautical wind tunnels with relatively short sections.

In aeronautics the flow of wind is duplicated at high altitudes. At these altitudes wind is uniform with very little turbulence. The impact of turbulent flow (which is typical of the wind that blows on structures) on wind tunnel measurements was not understood until 1958, and since then the techniques for modeling the turbulent wind effects have improved considerably. Today, there are over 30 wind-tunnel facilities in North America alone, attesting to the rapid growth in the field of wind-tunnel testing of buildings and structures.

### 2.6.2 Description of wind tunnels

Aeronautical wind tunnels are designed to minimize the effects of turbulence. They do not model the natural atmospheric boundary layer because the majority of airplane flights, except for brief periods of landing and takeoff, occur at a height well above the boundary layer. Building activity, on the other hand, occurs precisely within this atmospheric boundary layer, which is characterized by the gradual retardation of wind speed and high turbulence near the surface of the earth. Therefore, aeronautical wind tunnels are not directly suitable for measuring wind effects on buildings. To overcome the disadvantage, aeronautical wind tunnels have been modified and entirely new facilities have been built to reproduce the turbulence and the natural flow of wind within the boundary layer.

Wind-tunnel tests or similar tests employing fluids other than air are considered to be properly conducted only if all of the following conditions are satisfied:

1. the natural atmospheric boundary layer has been modeled to account for the variation of wind speed with height;
2. the length scale of the longitudinal component of atmospheric turbulence is modeled to approximately the same scale as that used to model the building;
3. the modeled building and surrounding structures and topography are geometrically similar to their full-scale counterparts;
4. the projected area of the modeled building and surroundings is less than 8 percent of the test section cross-sectional area unless correction is made for blockage;
5. the longitudinal pressure gradient in the wind tunnel test section is accounted for;
6. Reynolds number effects on pressures and forces are minimized; and
7. response characteristics of the wind tunnel instrumentation are consistent with the required measurements.

Boundary-layer wind tunnels capable of developing flows that meet the conditions stipulated above, typically have test-section dimensions in the following ranges; width of 6–12 ft (2–4 m), height of 6–10 ft (2–3 m), and length of 50–100 ft (15–30 m). Maximum wind speeds are ordinarily in the range of 25–100 mph (10–45 m/s). The wind tunnel may be either an open-circuit or closed-circuit type.

Three basic types of wind-tunnel test models are commonly used. These are: (1) rigid pressure model; (2) rigid high-frequency force balance model; and (3) aeroelastic model. One or more of the models may be employed to obtain design loads for a particular building or structure. The pressure model provides local peak pressures for design of elements such as cladding and mean pressures for the determination of overall mean loads. The high-frequency model measures overall fluctuating loads for the determination of dynamic responses. When motion of a building or structure influences the wind loading, the aeroclastic model is employed for direct measurement of overall loads, deflections and accelerations.

Various techniques are used in aeronautical tunnels to generate turbulence and atmospheric boundary layer by using devices such as screens, spires, and grids. In special boundary-layer wind tunnels with long test sections thickened turbulent boundary layer is generated by installing appropriate roughness elements in the upstream flow. Another approach is to use a counterjet technique. In every case there is always some question whether the natural wind turbulence characteristic is appropriately modeled and proper gust simulation is included. The degree of scaling required to appropriately account for these may yield a very extreme scale for the building on the order of 1:500 or even more for urban environment studies.

### 2.6.3 Objective of wind tunnel tests

The objective of wind tunnel tests is to determine the design lateral loads and to predict the response of the building under the influence of wind loading. Other topics of importance are:

1. boundary layer profile and turbulence intensities;
2. intensity and duration of extreme winds;
3. influence of nearby existing and proposed building;
4. drag, vortex shedding, and wind separation from building surface;
5. dynamic response;
6. loads on cladding and glass;
7. near zone effects—stability of vehicles and pedestrians;
8. motion tolerance—occupancy discomfort;
9. buffeting of downstream structures;
10. damage to structures by flying gravel;
11. moisture penetration;
12. snow accumulation and pollution-control problems.

Although not common, occasionally model testing is done on building configurations to determine the most favourable shape of the building for wind design. However, the most often sought wind information for building design are

1. The intensity and scale of pressure fluctuations on exterior panels and glass surfaces
2. The floor-by-floor shear forces for the overall design of building.
3. The oscillation response of the building related to occupants comfort.

4. The change in the wind environment at ground resulting in uncomfortable or even dangerous conditions to pedestrians.

### 2.6.4 Rigid model studies

Within the past 20 years, the field of wind engineering has experienced tremendous growth and has become a primary tool for assessing the wind load characteristics on tall buildings. Daring architectural forms executed with lightweight building materials have created a compelling need to obtain a better description of the wind effects than described in the codes. Analytical approaches that take into account the specific design parameters created by the uniqueness of the building shape, near field turbulence characteristics of wind, etc., are not readily available for use in practice. Cladding pressure study is of great concern because of the large number of inadequately performing or failed curtain wall systems. Although some of the more advanced building codes have attempted to establish design loads with due considerations to shape factors, turbulence, and dynamic characteristics of buildings, it has become industry practice to resort to wind-tunnel tests because it is generally felt by owners and developers that the confidence in cladding wind loads obtained in wind tunnels far outweighs the cost. Many curtain wall suppliers in North America hesitate to undertake jobs if the cladding pressure studies are not available. Although the cladding results may show a larger dispersion with lower than code pressures in some areas and larger ones in others, the risk factor associated with the larger rather than the economic implications of the lower loads is the prime motivation behind the curtain wall pressure study.

Although the basic purpose of rigid model study is to obtain the local pressure fluctuations, the results can nevertheless be extrapolated to obtain the design pressure on the overall structural system for buildings that are not sensitive to aeroelastic interaction. A brief description of experimental and analytical techniques used for obtaining the cladding and overall design loads is given in the following section.

**Modeling criteria.** The underlying principle in model studies is that forces acting on the model should be proportional to the forces on the prototype. The four forces usually considered in fluid mechanics (of which wind engineering can be considered a branch) are inertia, gravity, viscosity, and surface tension. Usually it is not possible to have all forces in the model in the same proportions as they are in the prototype, and in most model studies it is sufficient to have only two forces the same without introducing serious errors. The inertial force, which is generally a predominant force, and one other force are made proportional. In fluid mechanics three nondimensional parameters are used to represent the ratios of forces of gravity, viscosity, and surface tension to the force of inertia and are called respectively, Froude number, Reynolds number, and Weber number. Depending upon the nature of the problem being investigated, equality among different numbers is sought. For example, equating the Froude numbers of the model and prototype ensures that the gravitational and inertial forces are in the same proportions. Likewise, similarity of Reynolds numbers assures that viscous and inertial forces are in the same proportion. Equating the Weber numbers results in proportionality between surface tension and inertial forces.

In wind tunnel tests of structures it is important to simulate the mean wind profile, the turbulence intensity, and the length scale of turbulence within the atmospheric boundary layer. The gradual retardation of the wind speed as it approaches the ground results in the wind profile. The wind turbulence intensity is a measure of the velocity fluctuations, which increase with the terrain roughness and are more predominant near the ground. The effect of turbulence extends to a height of 500 ft (152 m) or so in the boundary layer.

To simulate the wind profile correctly in a wind tunnel, it is necessary to have geometric and kinematic similarity of the wind velocity gradients, intensities of turbulence, and turbulence spectrums between the prototype and the model. Dynamic similarity is not required since the fluid employed for testing is the same as the prototype.

The effect of rotation of the earth on the wind profile depends upon the geographical location of the building. Rotation effects are greatest at the poles and smallest at the equator. This effect is neglected in wind tunnel tests. The wind velocity gradient governs the mean wind load; the intensity of turbulence spectrum affects the aeroelastic interaction between the wind and the building. Wind profile, as mentioned earlier, is simulated in a number of ways. Placing of screens, stacked plates, or honeycombs of varying thicknesses in the wind tunnel are some of the methods used to generate a velocity profile. Although they are quite effective in simulating the velocity profile, they cannot be used to control the turbulence characteristics. However, the boundary layer which grows along a sufficiently long stretch of wind tunnel gives a good simulation of turbulence. Therefore, when the length of the wind tunnel is limited, some artificial method of creating the turbulence is required. Barriers, mixing devices, and grids are some of the devices used to assist in the generation of turbulence.

Most commonly, pressure study models are made from methyl methacrylate sheets. This material is commonly known as Plexiglas, Lucite, and Perspex and has several advantages over wooden or aluminum alloy models in that it can be easily and accurately machined and drilled and is transparent, facilitating observation of the instrumentation inside the model. It can also easily be formed into curved shapes by heating the material to about 200 °C. Model panels can either be cemented together or joined by using flush-mounted screws.

A scale model of the prototype in a 1:300 and 1:500 range is constructed by the wind tunnel laboratory using drawings that are usually provided by the architect. In a rigid model, important building features which have significance in regard to the wind flow, such as building profile, protruding mullions, and overhangs, are simulated to the correct length scale. Because the model is rigid, measurements are obtained only for the mean and fluctuating pressures acting on the building. No attempt is made to simulate the dynamic response characteristics of the building. Characteristics, such as the mass and stiffness distribution which determine the natural frequency and mode of vibration, together with the damping characteristics, are precisely the features that require simulation in an aeroelastic model study as discussed later in this chapter.

**Measurement techniques.** The model is instrumented with a large number of pressure taps, sometimes as many as 500 to 700, and is tested in a boundary layer wind tunnel in the presence of a detailed modeling of the nearby surroundings within a radius of 1500 ft (457 m), as shown in Fig. 2.39. Flexible, transparent vinyl or polyethylene tubing of about $\frac{1}{16}$ in. (1.5 mm) internal dimaeter is used as pressure tappings and is located around the exterior surface of the model. Generally a liberal number of tappings are deployed around the model surface to obtain a good distribution of pressure. Pressure tap locations are generally more concentrated in regions of high pressure gradients such as around corners.

The pressure tappings are connected by short plastic tubing to miniature electronic pressure transducers. The length of plastic tubing is kept as short as possible to minimize damping of fluctuating pressure by the column of air trapped in the tube. Pressure transducers are used instead of monometers because of the ability of transducers to measure fluctuating loads.

It is not usually feasible to attach a separate transducer to each of the pressure tappings for reasons of economy and available space. Therefore, the pressure transducer is mounted to a pressure-scanning device such as a Scanivalue, which has the capability to automatically switch the pressure transducer to as many as 40 or 50 pressure taps one at a time. The electrical output signals from the transducers are processed through a computerized automatic data

(a)

(b)

**Figure 2.39** (a) Rigid model of high-rise buildings in wind tunnel; (b) close-up of pressure model. (Photographs courtesy of Dr. Peter Irwin of Rowan, Williams, Davis & Irwin, Inc).

acquisition system. The pressure transducer is calibrated such that the electrical output signal is converted to an equivalent pressure or velocity. Static pressure upstream of the tunnel is used as reference, since only the pressure differentials are measured by the transducer. The pressure data acquisition is done by an on-line computer system capable of sampling data in a short period of time.

The wind tunnel test is run for a duration of about 60 s, which corresponds to approximately 1 hr in real time. Sufficient numbers of readings are obtained from each port to obtain a stationary value such that fluctuations become independent of time. From the values thus obtained, the mean pressure and the root-mean-square value of the pressure fluctuations are evaluated. The rate of sampling of the pressure signal by the computer corresponds to a very small time interval, such as half a second at full scale. The computer record is divided into subintervals, and the maximum and minimum values of pressures are calculated for each subinterval. These individual maximum and minimum values are used in an extreme-value analysis to determine the most probable maximum and minimum values applicable for the whole sample period.

The boundary layer wind tunnel, by virtue of having a long working section with roughened floor and turbulence generators at the upwind end, is expected to correctly simulate the mean wind speed profile and turbulence of natural wind. The model is mounted on a turntable, thus allowing any wind direction to be simulated by rotating the model to the appropriate angle. As in other wind studies, it is necessary to simulate the features surrounding the building under study. Generally polystyrene foam is used to simulate the near field characteristics because of the ease of construction. The building model and the near field characteristics are mounted on a turntable in the test section and rotated to change the direction of the wind.

**Cladding pressures.** Measurements are taken for all wind directions spaced about 10 to 20° apart. The data are then converted into pressure coefficients that are derived from the measured dynamic pressure of the wind above the boundary layer. From the data acquired, full-scale peak exterior pressures and suctions for selected return periods at each tap location are derived by combining the wind tunnel data with a statistical model of windstorms expected at the building site. The results are given for various return periods such as 25, 50, and 100 years. A detailed account of the probability procedure used by wind engineers to analyze the meteorological data and to combine this information with the wind tunnel results to produce the peak suction and pressure values at each port is beyond the scope of this book. Suffice it to say that the method involves fitting the measured data to a probability distribution and computer simulation and analysis of hurricane events.

In order to get a feel for the amount of data that is handled by the computer, let us take a closer look at a practical example. Consider the model shown in Fig. 2.39a which is a 1:400 scale model for a 50-story building. Assume the model is instrumented with 703 pressure taps and measurements are taken for 36 wind directions at 10° intervals. If the rate of sampling is half a second at full scale, 120 readings for each port for each wind direction are required for a full-scale test duration of 1 hr. For 703 pressure taps with 36 wind directions, the total number of readings for each configuration is equal to $703 \times 120 \times 36 = 3{,}036{,}960$. This enormous amount of data is condensed into recommended cladding design loads and presented by the wind engineer in the form of block diagrams as shown in Fig. 2.40. Sometimes the information is presented in the form of pressure contours or isobars, as shown in Fig. 2.41.

In evaluating the peak wind loads on the exterior of the prototype, effects of internal pressures arising from air leakage, mechanical equipment, and stack effect should be included. In performing the internal pressure calculations, it may be necessary to take into account the possibility of window breakage caused by roof gravel scoured from roofs of adjacent buildings and other flying debris during a windstorm. As a rough guide, the resulting internal pressure can be considered to be in the range of $\pm 5$ psf (25 kg/m$^2$) at the base of the building to as much as $\pm 20$ psf (100 kg/m$^2$) at the roof for a 50-story building.

In the design of glass, a 1-minute loading is commonly used. In wind tunnel study the duration of measured peak pressures is different from the 1 min interval; usually it corresponds to 5 to 10 s or less in terms of real time. Therefore, it is necessary to reduce the peak loads obtained from the wind tunnel tests. Empirical reduction factors of 0.80, 0.94, and 0.97 have been given in glass manufacturers' recommendations for three different types of glass, namely annealed float glass, heat-strengthened glass, and tempered glass, respectively.

**Overall building loads.** The results obtained in a rigid model test are used to predict the local wind environment for the design of glass and cladding. However, for buildings that are not dynamically sensitive to wind, the results can be extrapolated to obtain the design pressure on the overall structural system. This procedure involves introducing a gust factor for converting the mean wind load to gust loads. An appropriate gust factor estimation should take into account the averaging period of the mean wind load; the terrain roughness in relation to the building height; the peak gust factor, which depends on the natural frequency of the building; the effect of turbulence; and the critical damping ratio of the building. Rigid model wind study does not take into account all of these factors, yet is considered to provide adequate design data for buildings with a height-to-weight ratio of less than 5.

The procedure for converting results of pressure model tests to obtain overall structural loads can be outlined as follows. The first step is to delineate the exterior surfaces of the building into small zonal areas tributary to each pressure tap. This is similar in procedure to assigning tributary areas for computing gravity loads in

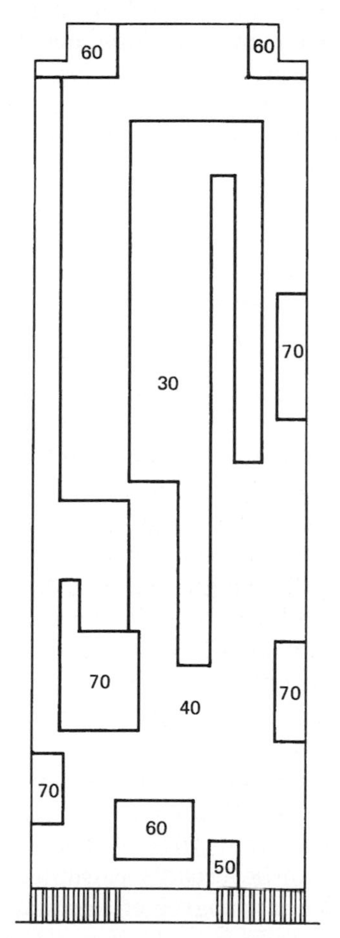


**Figure 2.40** Block pressure diagram, in psf.

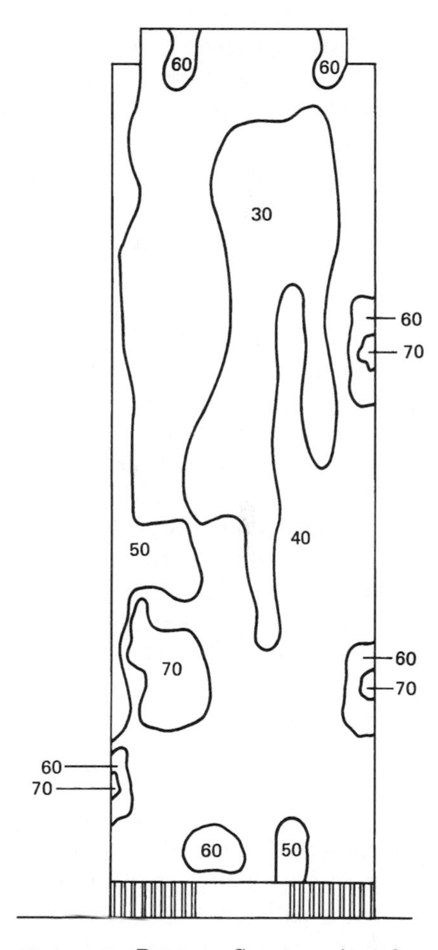


**Figure 2.41** Pressure Countours in psf.

column design. The next step is to compute the wind force components $F_x$ and $F_y$ by multiplying the tributary area corresponding to each pressure tap by the mean pressure. A gust factor appropriate to the overall loading is introduced at this stage to convert the hourly mean wind results to include effects of gusts. The components $F_x$ and $F_y$ can be appropriately summed over the building to obtain the distribution of wind loads and by statics the moments at each floor are obtained. Torsional moments are obtained by appropriate summation forces $F_x$ and $F_y$ about the vertical axis.

Although the primary purpose of pressure model study is to obtain the peak loads to be used in the design of glass and cladding, as a by-product it is becoming more and more common to use the results to design the overall structural frame. As compared to the aeroelastic and high-frequency force balance studies, the method used in rigid model study to obtain mean pressure may appear to be tedious, but it should be noted that it is the only test that provides information on the magnitude of local pressure fluctuations, which are essential to an economical and satisfactory design of cladding and glass panels.

### 2.6.5 Aeroelastic study

Although a good estimate of the mean wind load on the building as a whole can be obtained from rigid model study by integration of local pressures over the surface, the uncorrelated nature of local peak pressures precludes such an integration being carried out for the overall peak dynamic loads. A gust load factor selected for an appropriate wind gust duration, of course, can be used to increase the mean hourly load to that of a gust whose duration is sufficient for its entire effect to be fully felt by the structure. Selection of a proper gust factor that takes into account all the variables present in a wind-structure interaction is not always obvious. Although many of the interaction effects can be described qualitatively, in practice precise analytical quantification of these effects is not possible. Aeroelastic model study attempts to take the guesswork out of the gust factor computation by measuring directly the dynamic loads in the wind tunnel. The fluctuating aerodynamic loads can be measured on a variety of models ranging from very simple rigid models mounted on flexible supports to models exhibiting the multimode vibration characteristics of tall buildings. The more common types of models used in aeroelastic studies of tall buildings can be broadly classified into two categories: (1) stick models, and (2) multi-degree-of-freedom models.

In this section a brief description of each of these models is presented. Before this is undertaken, a few comments that in general apply to aeroelastic studies are necessary.

**General comments.** In tall, slender, and flexible buildings, dynamic loads are induced by the buffeting action of atmospheric turbulence. The wind-induced excitations can be significantly amplified by the dynamic response of the building, which may contribute up to 50 percent or even more to the total response of the building. Wind-induced horizontal accelerations are also significantly amplified, requiring consideration of occupant comfort and perception of motion. As in the case of rigid model studies, analytical procedures for determining the dynamic effects of wind loads are not available as a routine design procedure, thus necessitating wind tunnel tests to simulate unique characteristics of the building and its specific setting.

In addition to the similarity of the exterior geometry, the aeroelastic studies require similarity of the inertia, stiffness, and damping characteristics of the building. Although a building in reality responds dynamically to wind loads in a multimode configuration, enough evidence exists to show that the dynamic response occurs primarily in the lower modes of vibration. As a result, it is possible to study the dynamic behavior of the building by using simple dynamic models.

Aeroelastic study basically examines the wind-induced sway response, in addition to providing information on the overall wind-induced mean and dynamic loads. These tests are important for slender, flexible, and dynamically sensitive structures where aeroelastic or body-motion-induced effects are of signficance. When a tall building sways and twists under wind action, the resulting acceleration generates inertial loads, causing fluctuating stresses. At any given instant, the amplitude of twisting and swaying motion is not just a function of the magnitude of wind load at the instant but also depends on the integrated effect of the wind over the several previous minutes. Therefore, it is important to consider the building's dynamic response when predicting wind loads on the structure. In addition to providing an accurate assessments of loads for structural design, an aeroelastic model test provides one of the most reliable approaches to predicting the building response to wind which can be used by the designer to ensure that the predicted motion will not cause discomfort to the building occupants in the upper floors.

Typically, aeroelastic measurements are carried out at several wind speeds covering a range selected to provide information on both relatively common events, such as 10-year wind loads, which may influence the serviceability and occupant comfort, and relatively rare events, such as 100-year winds, which govern the strength design. The modeling of dynamic properties requires the simulation of inertial, stiffness, and damping characteristics. It is necessary, however, to simulate these properties for only those modes of vibration which are susceptible to wind excitation.

It is often difficult to determine quantitatively when an aeroelastic study is required on a building project. The following factors can be used as a guide in making a decision.

1. The building height-to-width ratio is greater than about 5; i.e., the building is slender.
2. Approximate calculations show that there is a likelihood of vortex shedding phenomenon.
3. The structure is light in density on the order of 8 to 10 lb/ft$^3$ (1.25 to 1.57 kN/m$^3$).
4. The structure has very little inherent damping, such as a welded steel construction.
5. The structural stiffness is concentrated in the interior of the building, making it torsionally flexible.
6. The calculated period of oscillation is long, on the order of 5 to 10 s.
7. Existence of unusual near field conditions that could create torsional loads and cause strong buffeting action.
8. The building is sited such that predominant winds blow from a direction most sensitive to the building oscillations.
9. The building occupancy is such that the occupants' comfort plays a more predominant role. Occupants in high-rise apartments, condominiums, and hotels are likely to experience more discomfort from building oscillations than those in office buildings.

The development during the past two decades of aeroelastic direct modeling techniques for predicting the behavior of tall buildings has been influenced by the following factors:

1. Tall buildings, although slender from structural considerations, aerodynamically speaking behave more like three-dimensional structures. Therefore, analytical formulations applicable for very slender tall structures such as chimneys cannot be successfully applied to tall buildings.

2. Gust factor methods are inadequate to describe the crosswind response of tall buildings, necessitating aeroelastic studies for slender buildings with a height-to-width ratio greater than 6 or so.
3. In most instances the surroundings have a beneficial shielding effect, but it certain situations the wake buffeting can significantly increase the dynamic response of the building.
4. Current modern architecture often results in buildings which have an erratic distribution of mass because of leaveouts, stepbacks, etc. Response of such buildings tends to be highly complex, defying analytical solutions. The complicated structural features are relatively easy to reproduce physically, thus favoring model testing.
5. Aeroelastic study provides perhaps the most sought after answer to the expected response of the building in terms of human comfort. Acceleration at the upper levels of the prototype can be predicted based on measurements on the model.

#### 2.6.5.1 Model requirements

1. Characteristics of wind simulated in the wind tunnel should represent the mean and turbulent nature of wind. This requirement is not unique to aeroelastic studies and applies equally to all wind studies of building.
2. The depth of boundary layer and the scale of turbulence intensity that can be simulated in a wind tunnel must be consistent with the scale of the model.
3. Although mismatching of Reynolds number is of little concern in turbulent flows, for building shapes that are unusually smooth, special precautions such as scoring the surface of the model may be necessary to increase the turbulence of wind.
4. The modal mass and stiffness values should correspond to the prototype values.
5. The damping device attached to the model should represent the damping characteristics of the prototype. Usual values assumed in practice for critical percentage of damping are 1.0, 1.5, and 2.0, for steel, composite, and concrete constructions, respectively.

The entire phenomenon of wind loading on the prototype cannot be modeled in the wind tunnel at a consistent geometric scale. However, it is agreed by researchers in this field that representative results can be obtained as long as the depth of boundary layer and turbulence characteristics are maintained in the wind tunnel.

**Rigid aeroelastic models.** The main objective of aeroelastic study is to obtain a more accurate prediction of design wind loads and to determine the degree of occupant sensitivity to building motion. For the results to be accurate, it is necessary that the model respond in the same way as the prototype building. Fortunately, this can be achieved without having to miniaturize all the curtain wall or structural member characteristics. It is possible to capture the essential behavior of the prototype by simpler models.

Rigid model study is based on the premise that the fundamental displacement mode as a tall building, due to shear and bending, can be approximated by a straight line. In terms of aerodynamic modeling, it is not necessary to achieve the correct density distribution along the building height as long as the mass moment of inertia about the pivot point is the same as that of the correct density distribution. It should be noted that the pivot point is chosen to obtain a mode shape which provides the best agreement with the calculated fundamental mode shapes of the prototype. For example, modal calculations for a tall building with a relatively stiff podium may show that the pivot point is located at the intersection of podium and the tower. Therefore the pivot point for the model should be at a location corresponding to this intersection point rather than at the base of the building. The springs located near the gimbals are chosen to achieve the correct frequencies of vibration in the two fundamental sway modes. An electromagnet or an oil dashpot provides the model with a damping corresponding to that of the full-scale tower. Figure 2.42 shows a rigid aeroelastic model mounted on gimbals.

An alternative method of obtaining a rigid aeroelastic model is to mount the model on a flexible steel bar attached to a vibration-free table. The width and thickness of the bar are chosen to properly simulate the tower stiffnesses in two horizontal directions. Damping is simulated by employing dashpots. Figure 2.43 shows a schematic elevation of a rigid aeroelastic model mounted to a flexible steel bar. Shown in Fig. 2.44 is a photograph of an aeroelastic model of a 62-story building. In either type of model, torsional modes are not simulated because the model effectively rotates as a rigid about the vertical axis.

In a tall building with a torsionally stiff structural system, such as a perimeter tube system, the torsional deformation from lateral loads may be of minor consequence when compared to the bending deformations. It is not necessary in such cases to simulate the torsional motions of the prototype in the wind tunnel model. In regard to the sway or the bending deformations, a fairly accurate representation can be obtained by approximating the mode shape by a straight line. A correct distribution of the mass on the model is

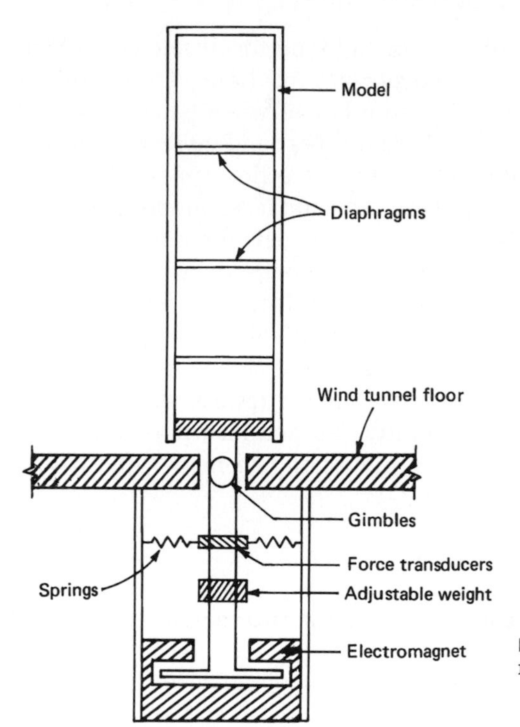


**Figure 2.42** Rigid aeroelastic model with gimbal.

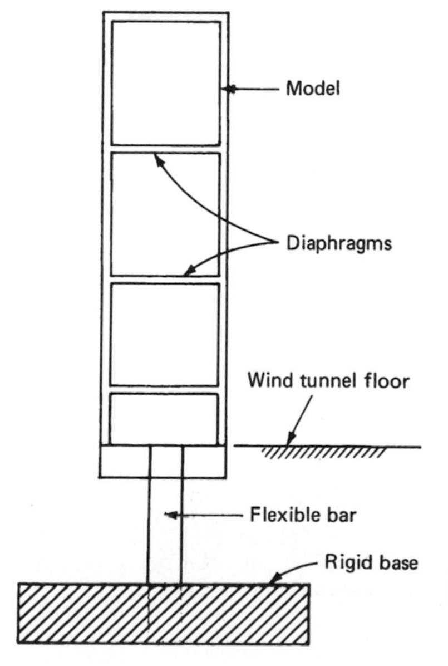


**Figure 2.43** Rigid aeroelastic model mounted to flexible steel bar.

obtained by building the model with a light material such as balsa wood and then adding weights at appropriate locations. The damping of sway motion can be achieved in a number of ways, such as using hydraulic dampers, springs in two mutually perpendicular directions, electromagnetic devices, or dashpots. As in other types of model testing, certain nondimensional parameters must be made the same on the model as in full-scale building in order for the model to experience the same nondimensional moment, shear force, and acceleration as the prototype. A complete discussion of deducing these nondimensional parameters using classical methods such as Buckingham's PI theorem is beyond the scope of this work.

The quantities that are of interest to the structural engineer are the wind pressure, shear force, moment, and acceleration that occur on the full-scale building. These are related to the model quantities

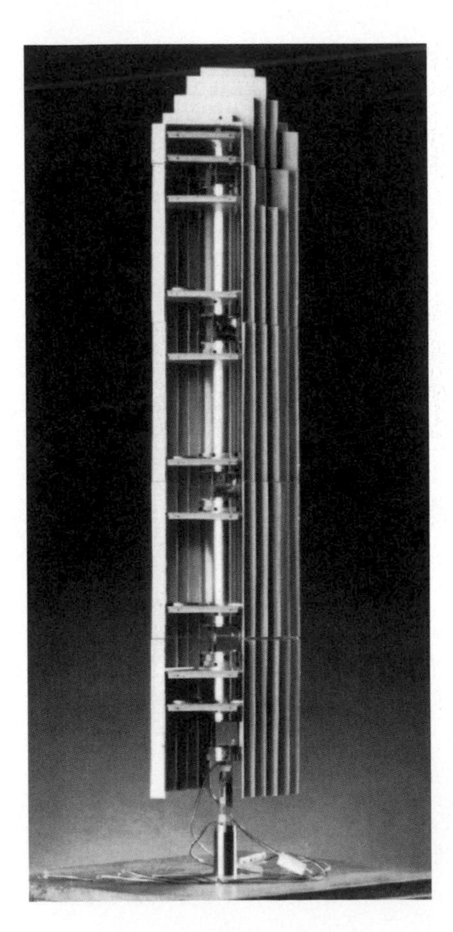

**Figure 2.44** Rigid aeroelastic model. (Photograph courtesy of Dr. Peter Irwin of Rowan, Williams, Davis & Irwin, Inc).

by the nondimensional ratios of length and the frequency scales as follows:

$$P_p = P_m\left(\frac{N_p^2 b_p^2}{N_m^2 b_m^2}\right) \tag{2.45}$$

$$F_p = F_m\left(\frac{N_p^2 b_p^4}{N_m^2 b_m^4}\right) \tag{2.46}$$

$$M_p = M_m\left(\frac{N_p^2 b_p^5}{N_m^2 b_m^5}\right) \tag{2.47}$$

$$a_p = a_m\left(\frac{N_p^2 b_p}{N_m^2 b_m}\right) \tag{2.48}$$

where $P$, $F$, $M$, and $a$ denote the pressure, shear force, moment, and acceleration, respectively. The subscripts $p$ and $m$ are used to denote the prototype and model quantities.

The model frequency $N_m$ can be any value deemed convenient for testing. Assuming it to be 10 Hz and using a value of 0.166 Hz for $N_p$, the frequency of the full-scale building which is representative of a 60-story steel building, we obtain the ratio $N_p/N_m = 0.0166$. Usually the wind tunnel model is constructed to a length scale of 1:400, giving the ratio $b_p/b_m = 400$. Using these values in Eqs (2.45)–(2.48), we get

$$P_p = 44.1 P_m$$

$$F_p = 7.05 \times 10^6 F_m$$

$$M_p = 2821.7 \times 10^6 M_p$$

$$a_p = 0.11 a_m$$

The relation between the model time $t$ and full-scale time is given by the relation $t_m = t_p(N_p/N_m)$. The model events in the example are therefore compressed to about one-sixtieth of the full time duration. Thus, 1 hr of full-scale event is typically compressed to 60 s on the model.

**Rigid models simulating torsional modes.** Torsion is a consequence of unsymmetrical distribution of building stiffness about the shear center, or it may occur because of eccentric disposition of the lateral loads with respect to the center of stiffness of the building. Centrally supported concrete-core buildings often use open section shear walls which may have their shear centers located at considerable distance from the geometrical center of the building. Unless additional lateral resisting elements such as moment frames, braces, or shear walls are used on the building perimeter, the torsional characteristics of the building may play an important role in the overall behavior of the structure. It is necessary in such instances to simulate in the aeroelastic model not only the bending characteristics of the building but also the torsional behavior. This is achieved by introducing torsional springs at appropriate locations along the building height. The torsional mode shapes of the full-scale building are simulated in a stepwise manner by the aeroelastic model. In order to allow one section of the model to rotate relative to the next, the model shell is cut around the periphery. Figure 2.45 shows a schematic representation of a model with two cuts. The model behaves as a three-degree-of-freedom system in torsion and therefore can capture essential dynamic behavior of the three lowest torsional modes.

**Flexible model.** If the building geometry is uniform for the whole height, it is reasonable to assume that the sway modes of vibration vary linearly along the height. By using only the lowest modes, it is possible to obtain acceptable results for the dynamic behavior of the building. However, for buildings of complex shapes with stepbacks and similar major variations in stiffness, this assumption may not yield acceptable results because: (1) the fundamental mode shapes may not be linear, and (2) the higher modes could contribute

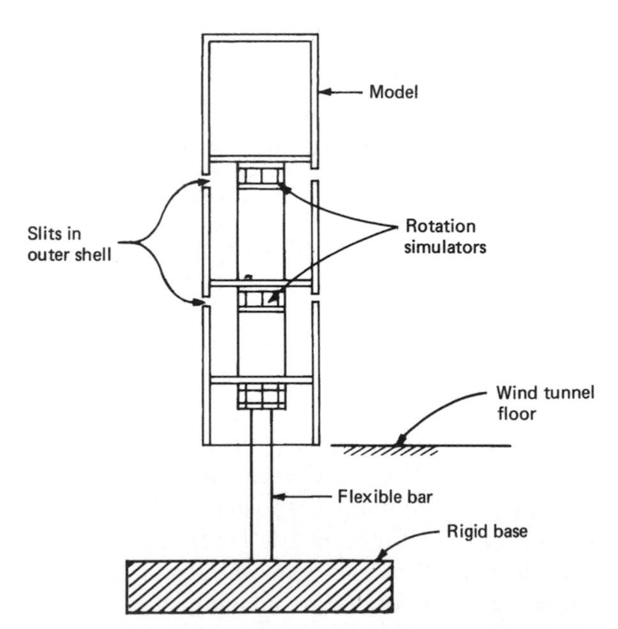


**Figure 2.45** Rigid aeroelastic model with provisions for simulating torsion.

significantly to the dynamic behavior. In such cases it is essential to capture the multimode behavior of the building. This is achieved by using a model with several lumped masses interconnected with elastic columns. A schematic representation of such a model is shown in Fig. 2.46. The building is divided into three zones, with the mass of each zone located at the center. In this model the masses are concentrated in the diaphragm representing the floor system and are interconnected by flexible columns. A lightweight shell that represents the building shape encloses the assembly of the floor system, masses, and columns. The shell is made discontinuous at the three zones to allow for relative movements between the masses. The similarity between the elastic properties of the prototype and the model can be achieved in varying degrees. For example, the behavior

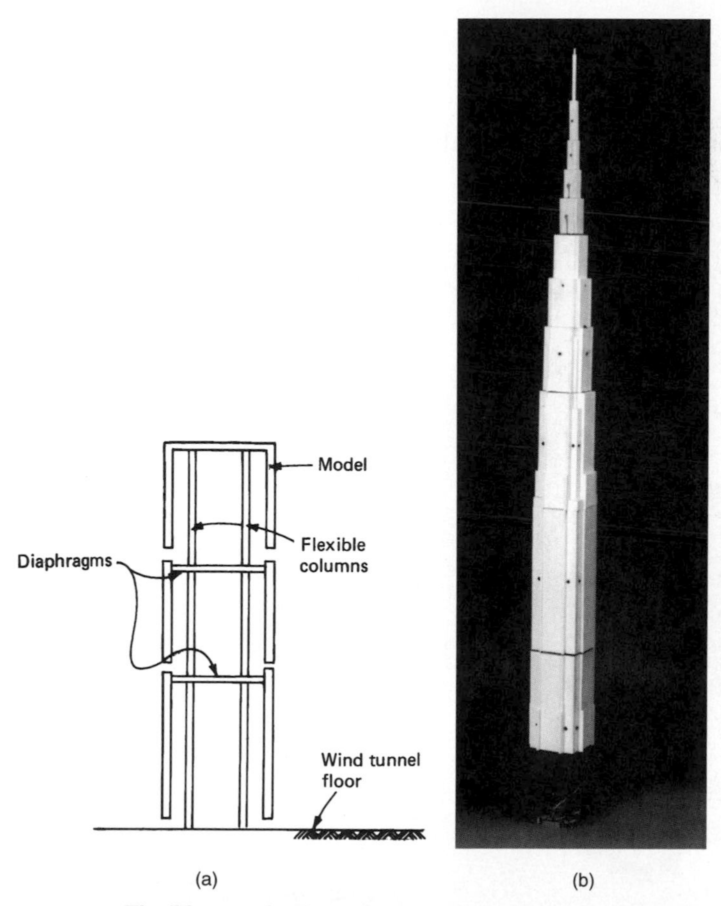


**Figure 2.46** Flexible aeroelastic model: (a) schematic cross section; (b) proposed tower in Chicago; (c) close-up of internal structure of flexible aeroelastic model. (Photograph courtesy of Dr. Peter Irwin of Rowan, Williams, Davis & Irwin, Inc).

of the so-called shear building in which the girder rotations and column axial deformations are neglected can be duplicated in the model by using a rigid diaphragm and flexible columns. The effect of girder rotations and axial deformation of the columns can, however, be simulated in the aeroelastic model at considerable fabrication effort.

**Instrumentation.** As mentioned previously, the aeroelastic study is undertaken to obtain forces and moments to be used in the design of the overall structural system and to evaluate the effect of the building motion on the occupants' comfort. Local pressures required for the design of the curtain wall and glass are not obtained, nor are the effects of wind on the pedestrian. To obtain the bending and torsional moments and shear forces, strain gauges are attached to

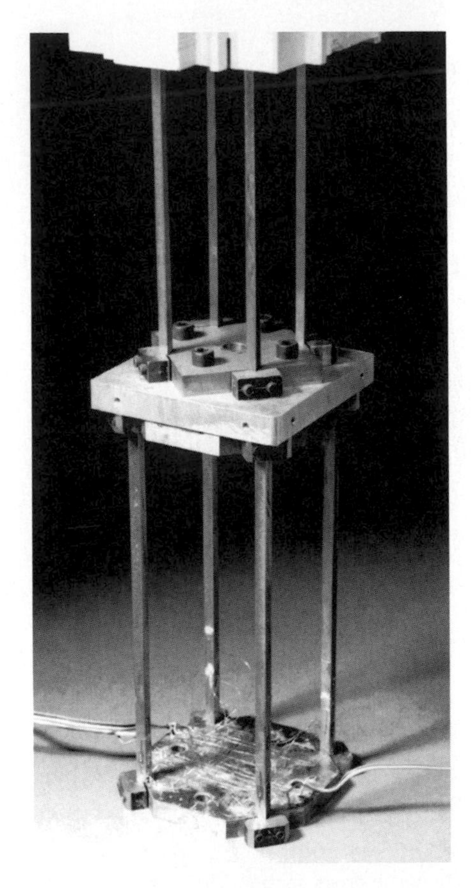

(c)

**Figure 2.46** *(Continued).*

the metal bars and calibrated by subjecting the model to known forces. Instrumentation connected to the on-line computer system gives a continuous history of these quantities for a given wind direction and speed. Figure 2.47 shows a schematic representation of measured overturning moments for a period of 3 min at full scale. These figures are for wind buffeting the broad face of the building shown schematically in Fig. 2.47. To correlate these measured moments with the physical behavior of the model, it is useful to decompose the continuously varying moments into mean values about which the maximum and minimum values oscillate.

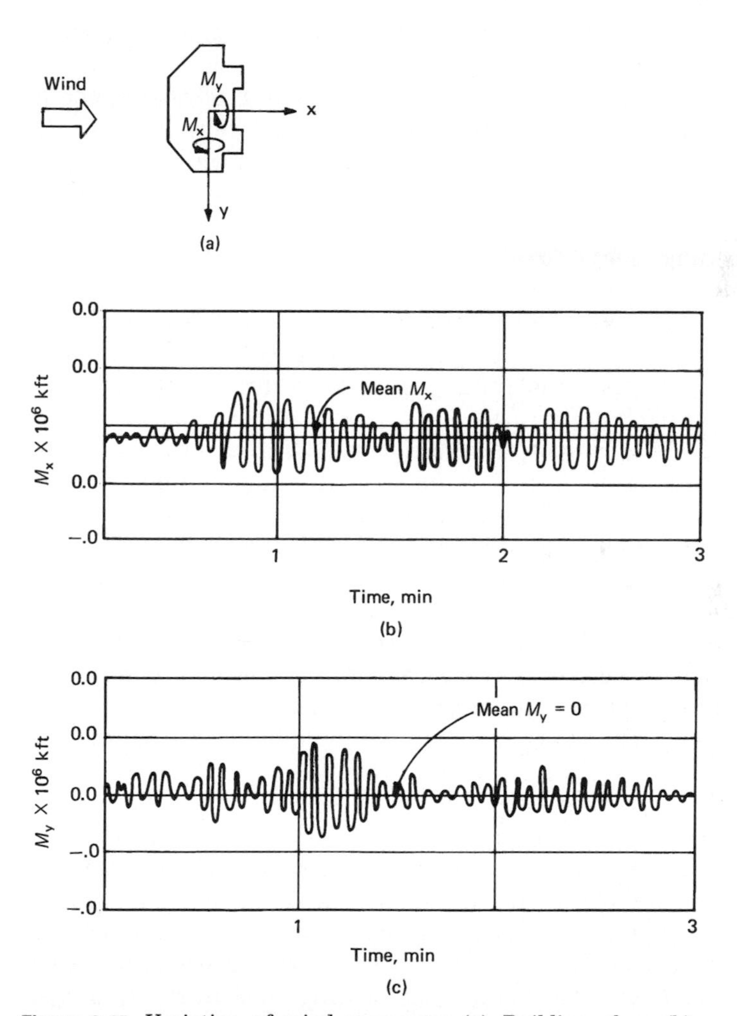


**Figure 2.47** Variation of wind moments: (a) Building plan; (b) Moment parallel to wind direction; (c) Moment perpendicular to wind direction.

It is easy to visualize tha the quasi-static component of the wind on the broad face is giving rise to the mean moment $M_x$ as shown in Fig. 2.47b. The oscillations of variable amplitudes that occur about the mean value are a consequence of the dynamic characteristics of the tower motion. They occur primarily at one frequency that corresponds to the fundamental frequency of the building and are caused by the inertial loads somewhat similar to the lateral loads generated during earthquakes. Depending upon the slenderness of the building, it is entirely possible for the dynamic component to be as large as the steady component. Hence in very tall buildings with a height-to-width ratio of greater than 5 or so, it is important to account for peak excursions of the moment rather than just the mean value.

So far, we have discussed only one component of the moment, namely $M_x$. However, a tall building oscillates not only in the direction of wind but also at right angles to it as shown in Fig. 2.47. Often the building motion is greatest in the across-wind direction, with the corresponding overturning moment due to the across-wind motion exceeding that due to alongside wind. The overturning moment due to alongside wind increases more or less proportionately with the wind speed. The across-wind moment, on the other hand, shows a distinct peak due to vortex shedding excitation. The wind speed at which vortex shedding occurs is a function of Strouls number.

If this speed corresponds to the design speed, it is very likely that the building will be subjected to dangerously high moments. One method of substantially reducing the likelihood of vortex shedding is to increase the damping of the building. An alternate approach is to increase the building's natural frequency such that the peak excitation occurs well above the design speed range. Analytical methods are not available for calculating wind speeds that correspond to the peak excitation of buildings of arbitrary shapes subjected to real wind climate, so it is almost mandatory to undertake aeroelastic wind tunnel test in cases where the vortex shedding phenomenon is suspected.

**Derivation of equivalent lateral loads.** As mentioned earlier, the response of a tall building to wind loads in general is two-dimensional. For a given wind direction, one would have to consider forces in both the $x$ and $y$ directions. The force occurring at any instant can be considered as the summation of the effects arising from the mean and dynamic components. Aeroelastic testing measaures the peak moment at the base of the structure. For structural analysis, however, a floor-by-floor distribution of lateral forces is required. The task then is to convert the measured peak overturning moment into a system of equivalent static forces acting at each floor level. This is considered next by using a specific example.

Assume that a model of a 60-story building tested aeroelastically in the wind tunnel gives peak values of 1,500,000 kft (2,033,700 kNm) and 600,000 kft (813,480 kNm) for base overturning moments $M_x$ and $M_y$ for a particular wind direction. Since these are peak values, they include the effect of dynamic components. Assume that a cladding pressure study is available for this building and the floor-by-floor loads are calculated from the results of the pressure study model. A knowledge of these forces and their distance to the base yields the static components of the moments $M_x$ and $M_y$. Let us assume that the static values for the example problem are 1,000,000 and 500,000 kft (1,356,000 and 678,000 kNm). Out of the 1,500,000 kft (2,033,700 kNm) of peak moment $M_x$ we have accounted for 1,000,000 kft (1,356,000 kNm) as the static component. The difference of 500,000 kft (678,000 kNm) that corresponds to the dynamic component remains to be distributed as an equivalent lateral load at each level. This is calculated from the mode shape of vibration and the mass distribution.

The summation of the mean force obtained from the cladding pressure study and the dynamic force obtained by the modal distribution yields the equivalent static force to be used in the structural analysis.

**Biaxial nature of wind.** Wind tunnel tests are conducted for several wind directions to account for wind's random nature. For a given wind direction, it is highly unlikely that the peak overturning moment about the $x$ axis will coincide exactly in time with the peak overturning moment about the $y$ axis. Therefore it is justifiable to reduce the peak moments when considering the biaxial behavior of the building.

Recommendations for combining the force values for key wind directions are normally given in the wind tunnel reports, and may vary from one facility to another.

**Prediction of building acceleration and human comfort.** One of the basic reasons for undertaking aeroelastic study is to ensure that the building motion will not cause discomfort to the building occupants. It is generally known that quantitative prediction of human discomfort is difficult if not impossible to define in absolute terms because the perception of sway and associated discomfort are subjective by their very nature. However, in practice certain thresholds of comfort have been established relating to the acceleration at the top floors to the frequency of windstorms. One such criterion first given by

Davenport is shown in Fig. 2.48. Accelerations at the top of the model are measured directly by installing three accelerometers, two of which measure the acceleration components in the $x$ and $y$ directions; the third is used to deduce the torsional component. The peak acceleration is evaluated from the expression:

$$\bar{a} = G_p \sqrt{a_z^2 + a_y^2 + a_z^2} \tag{2.49}$$

where $\bar{a}$ = peak acceleration

$G_p$ = a peak factor for acceleration usually in the range of 3.0 to 3.5

$a_x$ and $a_y$ = accelerations due to the sway components in the $x$ and $y$ directions

$a_z$ = acceleration due to torsional component

The peak acceleration is measured for a series of wind directions and speeds. It is then combined with the meteorological data to predict the frequency of occurrence for various levels of accelerations. A commonly accepted criterion is that for human comfort the maximum acceleration in the upper floors should not exceed 2.0 percent of gravitational acceleration for a 10-year return period storm.

### 2.6.6 High-frequency force balance model

In a tall flexible building the effect of wind load can be looked upon as created from three distinctly different contributions. First, there is

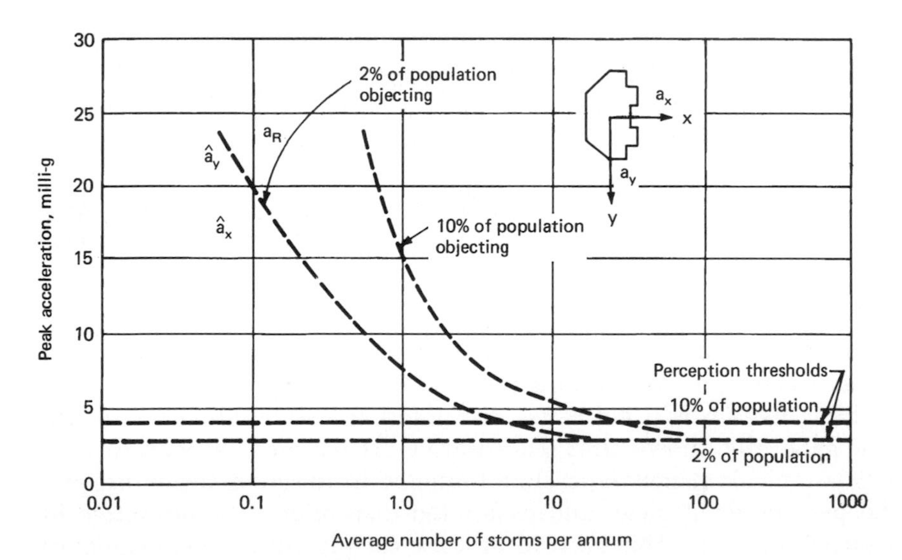


**Figure 2.48** Davenport's criterion for perception of motion.

the mean wind load, the effect of which is very easy to visualize. Since the mean wind load by definition is a constant load, it simply bends and twists a building, which returns to its normal undeflected position upon load removal. Then there is the fluctuating load resulting from the unsteady nature of the wind which brings about the oscillation of the building about the steady deflected shape. The third contribution comes from the inertia forces of the building; it is somewhat similar to the lateral forces induced in a building by earthquakes and depends at any particular instant on the deflection history of the building several minutes preceding that instant. For design purposes it is convenient to consider the inertial effects as an equivalent wind load.

A rigid model offers a convenient method of obtaining local wind pressures on the building faces, which in general consist of positive and negative pressures distributed uniquely for each building. It is possible to sum these local forces horizontally at each pressure tap level to obtain the net lateral forces in two perpendicular directions and a torsional moment at each level. By interpolating the values obtained at each pressure tap level, which in general may not correspond to the floor level of the prototype, it is possible to derive the forces and torsional moments at each floor level. The cumulative shear and overturning and torsional moments at each floor are easily obtained from simple statics, as are the base shear and overturning moments. These values calculated from the mean measurements, would have been excellent for the design of the bracing system, except for the fact that they ignore the influence of the gust factor. Therefore, when using rigid model pressure studies, it is necessary to assume a conservative gust factor to multiply the mean values. An alternate and better approach is to take the guesswork out of gust factor calculation by experimentally determining it.

A wind-tunnel-tested aeroelastic model provides comprehensive information on the dynamic loads and motions of the full-scale building because the essential structural features such as flexibility, mass, and damping of the prototype are simulated in the model. However, an aeroelastic model is quite complex to design and build and takes the best part of 10 to 12 weeks to complete the wind tunnel tests. During the last decade, the beneficial effects of reduced wind loads on the economy of the structural system were becoming known to engineers and owners alike, but the time element involved in the aeroelastic model testing often precluded the use of wind tunnel test data in the actual design. The flurry of high-rise activity and the tight schedules for design did not allow for an orderly method for incorporating the wind tunnel test results.

In response to this demand, wind tunnel engineers came up with a substitute. The high-frequency force balance approach provides an alternative, more economical, and time efficient method furnishing the desired information.

Two basic types of force balance models are in vogue. In the first type the outer shell of the model representing the architectural shape is connected to a flexible metal cantilever bar. Accelerometers and strain gauges are fitted into the model, and the aerodynamic forces are derived from the acceleration and strain measurements. In the second type, a simple foam model of the building is mounted on a five-component, high-sensitivity force balance which is used to measure bending moments and shear forces in two orthogonal directions and torsion about the vertical axis. In both methods, resulting fluctuating loads on the model as a whole are determined, and by making certain simplifying assumptions, the information of interest to the structural engineer, namely the floor-by-floor lateral loads and the expected behavior of the building in terms of acceleration at the top floor, is deduced. A brief description of the concept behind each of the two force balance models is given below.

**Flexible support model.** In this method, a lightweight rigid model is mounted on a high-frequency-response force balance and tested in a wind tunnel to obtain the overturning moments. The full-scale building loads and motions are computed from the measured power spectrum. This method is applicable in cases where building motion does not itself affect the aerodynamic forces and where torsional effects are not of prime concern. In practice, this method is applicable to many tall buildings.

It is not cost effective and may be impossible to measure the dynamic forces on the entire height of the building. Instead, in the force balance study, dynamic bending and torsional moments are measured at the base of the building and are used to calculate the dynamic response of the building. The accelerations are calculated at different heights, and from a knowledge of the weight of the building at different levels, the corresponding inertial loads are computed. The measured dynamic forces are combined with the inertial loads to obtain a peak dynamic load, which represents the gust factor contribution to the mean loading. A modal analysis is used that is similar to approaches used in aeroelastic model studies to correlate the test data to the prototype. Since the dynamic loads are obtained directly from the model without reference to the prototype structural stiffness and mass distribution, it is possible to calculate the dynamicresponse for several alternative structural systems and cladding variations.

The objectives of the technique are: (i) to provide data on wind-induced forces and moments for design of the building's structural frame and (ii) to determine the wind-induced accelerations. The high-frequency force balance model is constructed to a small scale of the order 1:500 and is tested in a wind tunnel that simulates the wind environment of the prototype in a manner similar to the pressure model tests. The balance shown in Figs 2.49 and 2.50 consists of a rectangular steel bar about 6 in. (152 mm) long. The cross-sectional dimensions of the bar are chosen to give a very high bending stiffness that results in natural frequencies of the order of 50 Hz. A set of bars, usually four, which are relatively thin compared to their width, is installed on top of the stiff bar by a flat plate to simulate the torsional stiffness of the building. These bars, because of their proportions, contribute very little to the bending stiffness of the balance, but because of their disposition about the center of rotation, they simulate the torsional mode. The bending and torsional frequencies of the balance are controlled independently of each other. The model itself is constructed out of a very lightweight material such as balsa wood and is mounted on top of the torsion spring through a relatively rigid plate. Strain gauges are attached to the bars that simulate the bending and torsional stiffness of the model. The gauges are calibrated by applying a range of known loads to the model. Tests are conducted in the wind tunnel for various wind directions, and the instantaneous base overturning and torsional moments are read from the strain gauges. Two sets of strain gauges are attached to the flexural bar to facilitate calculation of base shear by taking the difference of overturning measurements.

The instantaneous data records obtained from the strain gauge readings are analyzed to obtain the root-mean-square values for base overturning moments, torsional moments, and base shear for each of the wind directions for which the model is tested. Power spectral density functions are computed for the overturning and torsional moments and correction factors are applied to remove the effect of modal resonance. Reduction factors to the measured values are determined to account for the directional nature of wind, wind statistics for the site, and probability distribution of wind speeds. Neglecting the reduction factor amounts to making the conservative assumption that the design wind comes with equal probability from any direction. Meteorologically, it is known that strong winds generally have preferred directions of flow, which vary from region to region. Wind tunnel tests routinely take account of this phenomenon.

From the measured bending and twisting moments and known frequency and mass distribution of the prototype, it is possible to calculate the information of interest to the engineers, namely the wind forces at each floor and the expected peak accelerations.

**Five-component force balance model.** In this method, the prototype building is represented as a rigid model made out of lightweight material such as polystyrene foam. The stress components of interest at the base are the base moments, shear forces, torsional moment, and axial load. The model is attached to a measuring device consisting of a set of five highly sensitive load cells attached to a

**Figure 2.49** High-frequency force balance model. (Photograph courtesy of Dr. Peter Irwin of Rowan, Williams, Davis & Irwin, Inc).

**Figure 2.50** Detailed view of a high frequency force balance: (a) and (b) close-up of high-frequency force balance; (c) Commerz Bank, Frankfurt. (Photograph courtesy of Dr. Peter Irwin of Rowan, Williams, Davis & Irwin, Inc).

three-legged miniature frame and an interconnecting rigid beam. The arrangement is shown in Fig. 2.51 in which the load cells are schematically represented as extension springs. Horizontal forces acting in the $x$ direction produce extension of the vertical spring at 1, which can be related to the base overturning moment $M_x$ by knowing the extension of the spring and the pivotal distance $P_x$. Similarly, the base overturning moment $M_y$ can be calculated from a knowledge of extension of the spring at 2 and the pivotal distance $P_y$. The horizontal spring at 3 measures the shear force in the $x$ direction, while those at 4 and 5 measure the shear force in the $y$ direction. The difference in the measurements of springs at 4 and 5 serves to compute the torsional moment at the base about the $z$ axis. However, the results obtained for torsion are an approximation to the true response of the building because the model does not account for the relative twist present in the prototype.

From the measurements of the fluctuating aerodynamic loads at the base of the rigid model of the building, the dynamic response is calculated by modal methods. The force balance measurements are

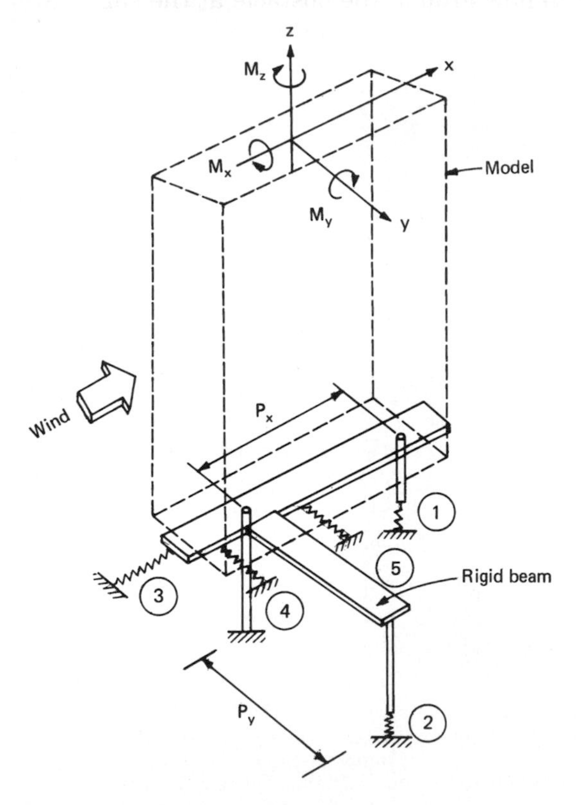


**Figure 2.51** Schematic representation of five-component force balance model.

in the form of mean and root-mean-square fluctuating aerodynamic components of the base shear and overturning moments about two orthogonal axes and a torsion for the full range of wind directions. Dynamic response calculations are based on the principle that the aerodynamic forces are proportional to the overturning moments and the torque measured in the wind tunnel. This assumption holds reasonably well for transitional modes, but not so far torsional modes because of the rigid body rotation of the model. However, this assumption normally gives conservative results for torsional forces and is considered sufficiently accurate for design purposes.

### 2.6.7 Pedestrian wind studies

A sheet of air moving over the earth's surface is reluctant to rise when it meets an obstacle such as a tall building. If the topography is suitable, it will prefer to flow around the building rather than over it. There are good physical reasons for this tendency, the predominant being that the wind will find the path of least resistance, i.e., a path that requires minimum expenditure of energy. As a rule, it requires less energy for the wind to flow around the obstacle at the same level than for it to rise, which requires more potential energy. Also, if the wind has to go up or down, additional energy has to be expended in having to compress the column of air above or below it. Generally wind will try to seek a gap at the same level. However, during high winds when the air stream is blocked by the broad side of a tall, flat building, its tendency is to drift in a vertical direction rather than to go around the building at the same level; the circuitous path around the building would require expenditure of more energy. Thus the wind is driven in two directions. Some of the wind will be deflected upward, but most of it will spiral to the ground creating a so-called standing vortex or a minitornado at the sidewalk level.

Thus tall buildings and their smooth walls are not the only victims of wind buffeting. Pedestrians who walk past tall, smooth-skinned skyscrapers may be subjected to what someone has called the "Mary Poppins syndrome," referring to the tendency of the wind to lift the pedestrian literally off his or her feet. Another effect of this phenomenon has frequently been observed and is known as the "Marilyn Monroe effect," referring to the billowing action of ladies' skirts in the turbulence of wind around and in the vicinity of the building. Whatever the popular name may be, the point is that during windy days even a simple task such as crossing the plaza or taking an afternoon stroll becomes an extremely unpleasant experience to pedestrians, especially during winter months around buildings in the cold climates. Walking may become irregular, and the only way to keep walking in the direction of wind would be to bend the upper body windward.

In skyscrapers, the downward wind along the windward faces, can cause strong ground-level winds, creating havoc for pedestrians and shops. The effect is often amplified when there are other tall buildings nearby lined up to form what is usually termed a street canyon.

A dramatic example of wind modification by buildings is arcades under tall buildings. Because of the high stagnation pressure on the upwind side of the building and the large suction on the downward side, a strong draft is generated through the arcade. Tall buildings supported by stilts have the same problem as arcades. Other examples are shown in Fig. 2.52.

For a successful design of a building at the pedestrian level, studies are therefore necessary to assist the building planners in overcoming the uncomfortable or dangerous wind conditions which often occur at the base of the buildings. Planners, designers, and developers are becoming increasingly aware of the potential for pedestrian-level problems and generally acknowledge that design assistance is required to predict the microclimate that will be adversely affected by a proposed design.

Although one can get some idea of wind flow patterns from the above examples, analytically it is impossible to estimate pedestrian-level wind conditions in outdoor areas of buildings and building complexes. There are innumerable variations to location, orientation, shapes, and topography, making it impossible to even attempt formulating an analytical solution to the problem. Based on actual field experience and results of wind tunnel studies, it is, however, possible to qualitatively recognize situations that adversely affect the pedestrian comfort within a building complex.

Model studies can provide reliable estimates of pedestrian-level wind conditions based on considerations of both safety and comfort. These studies require geometrically scaled models, including aerodynamically significant details in areas of interest. Simulation of characteristics of natural wind, including the simulation of the salient properties of the approach wind and the influence of nearby building structures and significant topographic features, is required for an accurate estimate of pedestrian wind conditions. Model studies are used to evaluate a building design and offer remedial suggestions where unacceptable pedestrian-level wind speeds are created. A model is constructed and pedestrian-level wind speed measurements are made at various locations for all the prevailing directions. The effects of adding a proposed building to an existing cluster of buildings are obtained by comparing measured wind speeds with and without the proposed building to a set of standard acceptance criteria. The acceptance criteria state how often a wind speed

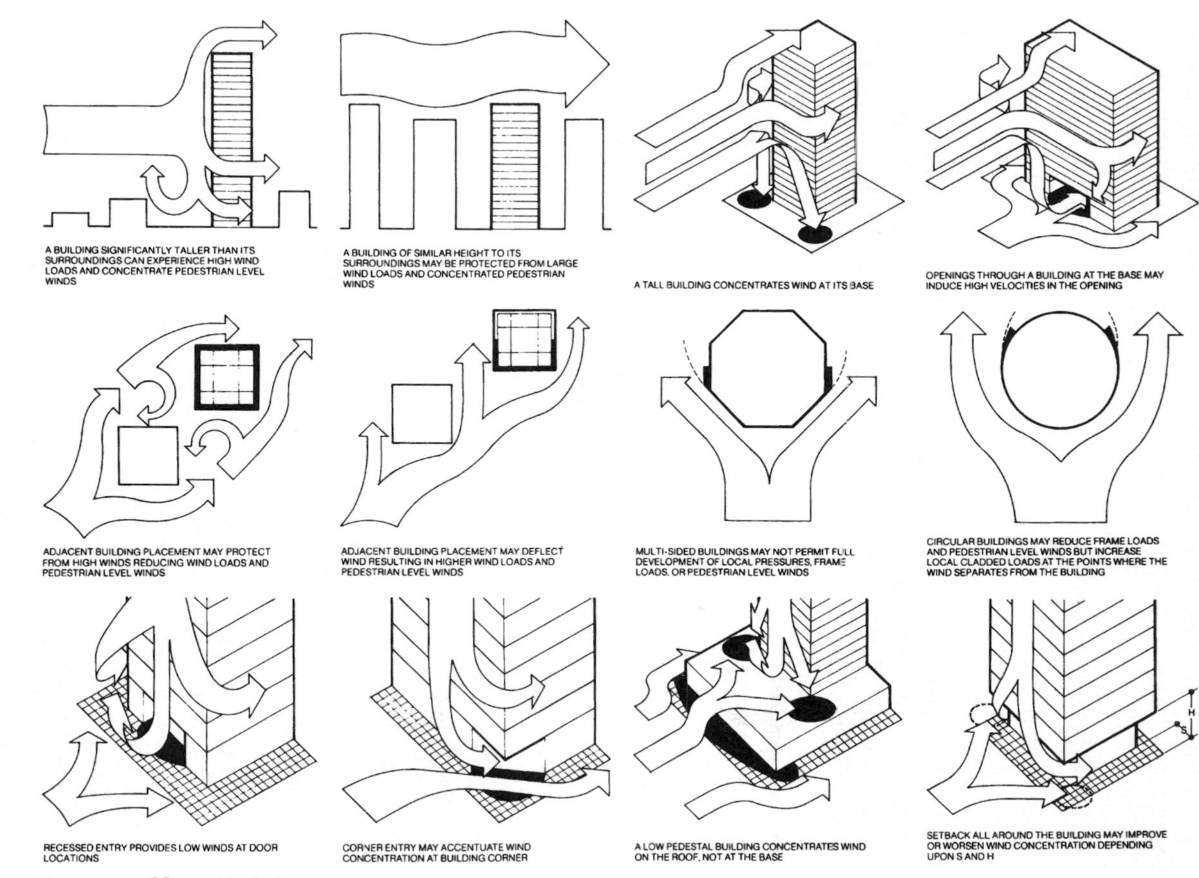


**Figure 2.52** Near wind climate.

occurrence is permitted to occur for various levels of activity. This is done for both the summer and winter seasons with acceptance criteria being more severe during the winter months. For example, an occurrence once a week of a mean speed of 15 mph (6.7 m/s) is considered acceptable for walking during the summertime, whereas only 10 mph (4.47 m/s) is considered acceptable during winter months.

## 2.7 Field Measurements of Wind Loads

Gustave Eiffel, the noted engineer for the Eiffel Tower, was, perhaps, the earliest researcher to conduct studies and make measurements of the displacements of the top of his tower under the action of wind. His observations between 1893 and 1895 showed that under the effect of wind, the displacements at the top of the tower resembled an elliptical shape. The measured values, which ranged from 2.3 to 2.75 in (60 to 70 mm), were considerably less than those predicted by him. At about the same period, a series of tests were conducted on the 16 story Monadnock Building in Chicago to measure the movements of structure. The building's vibration characteristics were checked by measuring the oscillations of plumb bobs suspended in a stair shaft from the sixteenth floor. These observations were also cross-checked by transits and showed that the building was very stiff with a measured deflection of about $1\frac{1}{2}$ in. (12.7 mm). The next set of activities was spurred during the boom years of high-rise construction in the 1940s. New York's Empire State Building was the first to be instrumented for wind load measurements. Rathbun (1940) in his ASCE paper which describes the observations on the Empire State Building, compared the building oscillations to the tines of a tuning fork. His observation showed that the building's sway motion occurs primarily in the fundamental frequency of the building. In the next two decades no similar studies involving the measurement of the wind response of tall buildings were conducted. Since 1960, however, the flurry of high-rise building activity rekindled the interest in full-scale studies in a number of countries, the most important work coming from Canada, England, Australia, Hong Kong, and Holland. In Canada, Dalghish studied three buildings in the range of 34 to 58 stories. His work encompasses multiple aspects of wind engineering that include studies of cladding pressures, meteorological data for predicting overall structural loads, and measurements of mean and fluctuating pressures.

In England, Lee has carried out full-scale studies of the structural behavior of a 157 ft (48 m) tall concrete building by subjecting the building to forced vibrations. In comparison to the number of tall buildings completed in the United States, the number of buildings instrumented to validate wind tunnel tests has been very small. Some buildings have been instrumented, but because of the nature of the tests, it will be some time before enough data are gathered for results to be published.

Many difficulties arise in the field measurement of wind loads. Stack effects can create pressure differences as large as a moderate wind. Engineers and researchers alike have been interested in comparing the results of model testing with those of the prototype in order to determine what improvements are needed in testing, and to get a better picture of the actual physical phenomenon. Full-scale measurements boost confidence in model scale tests and improve our knowledge of the effects that we do not fully understand as yet such as Reynolds number and turbulence effects. Certain scaling inequalities inherent in the wind tunnel tests appear to promote the lingering doubt about the validity of wind tunnel tests.

The field-measured mean and fluctuating loads as compared to those predicted by the wind tunnel studies appear to show a moderate amount of agreement, but more often these studies expose the shortcomings of field studies. It is agreed among researchers that there simply are not enough field data on full-scale buildings to validate wind tunnel techniques. Therefore, it is unreasonable to expect the wind tunnel results to be any more precise than any of the other parameters encountered in structural engineering practice, such as the calculation of natural frequencies and damping of tall buildings. The level of structural damping is at best an educated guess. Thus, there are many uncertainties in the wind design of tall buildings, signifying once again that structural engineering is as much an art as it is a science.

Amidst this rather pessimistic observation, structural engineer Halvorson and wind eingineer Isyumov have brought in evidence which appears to confirm the validity of wind tunnel tests. Their work consisted of a field-monitoring program to measure the dynamic response characteristics of Allied Bank Plaza Tower, a 71-story steel building in downtown Houston. The instrumentation consisted of two accelerometers located at the 71st floor of the tower. Wind measurements were made under two significant wind events, (i) an extratropical windstorm on April 1, 1983, with gusts speeds of up to 56 mph (25 m/s), and (ii) Hurricane Alicia on August 18, 1983, with fastest-mile wind speeds approaching the code-specified 50-year recurrence wind speed of 90 mph (40 m/s). Their observations showed that the sway response of the tower varied in a sinusoidal manner corresponding to the fundamental modes of vibration of the building. Full-scale measurements of accelerations showed good agreement with those predicted from the wind tunnel data. Based on the comparison between field measurements and wind tunnel results, the following conclusions are made by Halvorson and Isyumov.

1. The magnidue of structural loads experienced during Hurricane Alicia (which was determined as a 50-year recurrence event) are in good agreement with the loads predicted by the wind tunnel tests. The Houston code wind loads, on the other hand, overestimated the wind loads by approximately 100 percent.
2. The mean wind loads are only about 20 to 30 percent of the total structural loads. Dynamic effects brought about by wind gusts are 3 to 5 times as significant as the mean loads.
3. The lateral drift, which is a combination of mean and dynamic peak displacements, was evaluated from wind tunnel measurements and field records, respectively. Calculated drift indices in two directions were $2\frac{1}{2}$ and 2 ft (0.76 and 0.61 m) and compared well with the measured values of 2.58 and 1.25 ft (0.78 and 0.38 m).
4. Davenport's criterion of limiting the acceleration for a 10-year windstorm to approximately 20 mg appears to serve well the serviceability requirements of a tall building in regard to motion perception by building occupants.
5. Estimates from the field records indicate that the building exhibits a damping characteristic equivalent to 1.4 to 1.6 percent of the critical damping, which is somewhat higher than rule-of-thumb value of 1 percent used for steel buildings. The additional contribution appears to come from soil structure interaction provided by the foundation system, which consists of a $9\frac{1}{2}$ ft (289 m) thick mat founded on overconsolidated clay at a depth of 55 ft (16.77 m) below grade.

## 2.8 Motion Perception: Human Response to Building Motions

Every building or other structure must satisfy a "strength" criteria, in which each member is sized to carry the design loads without buckling, yielding, fracturing, etc. It should also satisfy the intended function (serviceability) without excessive deflection and vibration. While strength requirements are traditionally specified by the building codes, the serviceability limit states for the most part are not included within the building codes. The reasons for not codifying the serviceability requirements are several: failure to meet serviceability limits are noncatastrophic, are a matter of judgement as to their application, involve the perceptions and expectations of the user or owner, and because the benefits themselves are often subjective and difficult to quantify. The fact that serviceability limits are usually not codified should not diminish their importance. A building which is correctly designed for code standards may nontheless be too flexible for its occupants due to lack of deflection criteria. Excessive building drifts can cause safety-related frame stability problems because of excessive $P$-$\Delta$ effects. It can also cause portions of building cladding to fall, potentially injuring pedestrians below.

Perception of building motion under the action of wind is a serviceability issue. In locations where buildings are close together, the relative motion of an adjacent building may make the occupants more sensitive to otherwise imperceptible motion. Human response to building motions is a complex phenomenon involving many physiological and psychological factors. Some people are more sensitive than others to perceived building movements. Although researchers have attempted to study this problem from motion simulators, there is no firm consensus on human comfort standards. Although building motion can be described by various physical quantities including maximum values of velocity, acceleration, and rate of change of acceleration, sometimes called jerk, it is generally agreed that acceleration, especially when associated with torsional rotations, is the best standard for evaluation of motion perception in tall buildings. A commonly used criterion is to limit the acceleration of upper floors to 2.0 percent of gravity (20 mg) for a 10-year return period.

## 2.9 Comparison of Code and Wind Tunnel Test Results

Many researchers have made comparisons of structural wind loads for a number of tall buildings, determined from wind tunnel studies and code values. Twenty-four buildings located in several major North American cities were used in a comparison. The building heights ranged from just under 500 ft (152.4 m) to approximately 1,100 ft (335.3 m). The building shapes varied from simple box-like to highly irregular forms representative of post-modern architecture. On an average, wind tunnel studies indicate somewhat lower peak overturning moments, approximately 87 percent of the ASCE 7-88 values, and about 83 percent of the NBC values.

Although, in general, code values appear to be appropriate, there are situations where wind-tunnel procedures result in higher than code loads: approximately 25 percent of the buildings, as compared to ASCE 7-88, and about 15 percent as compared to NBC, exceeded the code values. Since the situations where code values are deemed insufficient cannot be predicted with certainty, it is advisable to carry out wind-tunnel tests for potentially wind-sensitive buildings.

A building sheltered by many surrounding buildings of comparable height experiences almost no mean or static wind load, and the effective pressures are almost entirely due to dynamic effects. A non-sheltered building, on the other hand, is more exposed to wind action and is subjected to a significant mean or static wind force, in addition to the dynamic component.

A building is considered wind-sensitive if it has one or more of the following characteristics:

1. has a shape that differs significantly from a uniform rectangular prism or "box-like" shape;
2. is flexible with natural frequencies below 1 Hz;
3. is subject to buffeting by the wake of upwind buildings or other structures; or
4. is subject to accelerated flow caused by channeling or local topographic features.

## 2.10 Chapter Summary

This Chapter addresses the different approaches for the evaluation of wind loading appropriate for the design of tall buildings. The reduction of mass, stiffness and damping, make contemporary tall buildings much more vulnerable to severe aerodynamic excitation requiring a careful examination of their static and dynamic wind response.

Issues affecting a tall building include: (i) The dynamic and static responses of the building as a whole due to wind acting on the primary system; (ii) Local pressures exerted on the exterior cladding, sometimes in association with internal pressures; (iii) street-level wind conditions affecting pedestrian comfort and safety.

Wind-tunnel model studies are accepted globally as alternative procedures for providing information on the action of wind as required in the design of tall buildings. Although the ASCE 7-95 and NBC attempt to quantify wind-induced building motions and related accelerations, it should be noted that the code procedures are mainly for isolated buildings located in homogeneous terrain. Such procedures may not be representative for contemporary buildings located in complex terrain or topography. For this reason, most codes alert the designer that their provisions may not be applicable to buildings with unusual properties, exterior geometry, response characteristics and siting. In addition to improved reliability of building performance there is also an economic motivation for undertaking wind-tunnel studies. Wind-tunnel model studies lead to more tailored and cost-effective designs and, on occasions, expose situations where code values are insufficient, confirming the advisability of carrying out wind-tunnel studies for potentially wind-sensitive buildings.

Chapter

# 3

# Seismic Design

## 3.1 Introduction

### 3.1.1 Nature of earthquakes

Catastrophic earthquakes appear in the headlines with discomforting frequency, causing thousands of lives to be lost and property damage running into hundreds of millions of dollars. This truly global phenomenon has just begun to be understood, and considerable emphasis is being placed on the analytical studies of earthquake response of buildings, supported by experimental studies both in the laboratory and in the field in an effort to prevent much of this destruction and loss of life.

Accounts of destructive earthquakes appear all through recorded history. Early humans, in their inability to comprehend such a strange and destructuve phenomenon, attributed the whole mechanism of earthquakes to the angry work of gods. Although in ancient times it was tempting to think of earthquakes as somehow otherworldly, they are in fact, among the most common of the earth's phenomena.

We will limit our discussion to the class of earthquakes in which the energy release is both near the earth's surface and large enough to damage structures. Such shallow-focus earthquakes are related to the forces which bring about the gradual distortions of the earth's crust. According to a well-established theory known as the elastic rebound theory, the distortions and the associated strains and stresses in the outer layers of earth build up with the passage of time until ultimately the stress at some location becomes high enough to fracture the rock or cause it to slip along some previously existing fault plane. Slippage at one location causes an increase in the stress in the adjacent rock, resulting in a rapid propagation of slippage along the fault plane. The result is the sudden rebound of the elastic strain. The strain energy that has accumulated in the rock is suddenly released and propagates in all directions from the source in a series of shock waves. If the amount of energy released is small, or if the fault slippage occurs in an uninhabited region, the shock waves travel unnoticed except by sensitive seismographs. If this energy release is great, the effect at nearby locations is chaotic. The earth may experience violent motion in all directions, lasting for a few seconds in a moderate earthquake or for a few minutes in a very large earthquake.

Although most of the earthquakes of record have occurred in well-defined earthquake belts, an examination of earthquake records for the world reveals the truly global nature of this phenomenon. Thus, even in the seismically inactive parts of the world, some measure of earthquake resistance should be built into the design of all structures in which failure will be a major catastrophe.

The earth's crust is composed of a dozen or so large plates and several smaller ones ranging in thickness from 20 to 150 miles (32 to 241 km). The plates are in constant motion, riding on the molten mantle below and normally traveling at the rate of a millimeter a week, equivalent to the growth rate of a fingernail. The plates' travels result in continental drift and the formation of mountains, volcanoes, and earthquakes. If plates carrying two continental masses grind past each other, as the Pacific and the North American plates do under California's San Andreas Fault, friction locks them together. When slippage occurs, the earth around the fault creates a so-called strike-slip earthquake. Still another kind of tectonic phenomenon results when an oceanic and a continental plate meet each other. For example, the oceanic plate which forms part of the Pacific floor off Mexico is pushing north-eastward against the North American plate, which is creeping westward. The oceanic plate dips under the continental crust, but the relative movement between the two plates is resisted by friction. When frictional forces are overcome, the stuck section of the plate lurches forward, generating shock waves of a thrust quake similar to the one responsible for the Mexican disaster of 1985.

### 3.1.2 Some recent earthquakes

Scientists estimate that over one million earthquakes occur every year. Some are very small and cause no damage, while others are violent and cause severe damage. One of the largest and most violent earthquakes to hit North America occurred near Anchorage, Alaska, in 1964. In the twentieth century there have been only three earthquakes of magnitude 8 or larger to affect a metropolitan area. The first one was the 1906 earthquake which destroyed much of San Francisco, California, estimated at 8.3 on the Richter scale ($m_L$). In 1923, Tokyo and Yokohama in Japan were badly damaged, and the third was the September 1985 quake which hit Mexico City. The most powerful earthquake ever recorded was off the coast of Chile in May, 1960. It reached 9.5 on the moment magnitude scale.

The devastating earthquake which hit Mexico City in September of 1985 measured 8.1 on the surface magnitude scale. In just 4 minutes an estimated 300 buildings collapsed in downtown Mexico City. Fifty more were later judged dangerously close to falling, and hundreds of others were regarded as unsafe. Just 36 hours after the first tremor, the second earthquake, known as the aftershock, battered Mexico City. This tremor, which lasted for about a minute and was not as powerful as the first, toppled some already weakened buildings. The estimated death toll numbered over 8000 persons with property damage estimated at $8 billion. The strength of this earthquake set the skyscrapers swinging as far north as Houston, 1100 miles (1770 km) from the epicenter. Tidal waves rolled ashore on the coast of El Salvador more than 800 miles (1287 km) to the southeast. The widespread damage is a chilling reminder that the world's well-defined quake-prone areas can be struck at any time without warning and with deadly effect. The same region in which the September 1985 earthquakes occurred had experienced six earthquakes with a magnitude of at least 7.0 since 1911. Thus, the latest shocks came as no surprise to seismologists.

Mexico City is built on soft, moist sediment of an ancient lake bed, which when jolted shakes like a bowl of jelly producing a large effect with a predominant vibration period of about 2 seconds. In addition, the city is undergoing subsidence at the rate of up to 10 in. (25.4 mm) annually, creating uneven settlement in some building foundations. The unusual severity of the quake and the resonance of the soil structure are major factors behind the extent of the damage. It is estimated that because of the soft subsoils some buildings were subjected to acceleration equal to 1.0 g. Most seriously affected by the earthquakes were structures in the 5 to 15-story range which had natural frequencies close to that of the soil, causing them to resonate with greater vigor than other buildings.

During the 1985 Mexican earthquake, coastal towns only 50 miles (80 km) from the epicenter suffered less damage than Mexico City, which was 200 miles (320 km) from the epicenter, because the shoreline is made of solid rock and thus shakes less violently than the alluvial lake bed on which Mexico City is built. Thus even though the seismic waves were diminished in intensity in their travel from the epicenter, they were amplified by the city's foundation.

At 4:31 a.m., in the pre-dawn hours of Monday January 17, 1994, the San Fernando Valley (about 30 miles north-west of central Los Angeles) was shaken by its most devastating earthquake in 60 years. The so-called Northridge earthquake had a magnitude of 6.7 and peak ground accelerations of about 1.0 g in several locations. According to news accounts, the earthquake was responsible for more than 50 deaths (of which 22 were attributed to earthquake-induced heart attacks) and at least 5000 injuries. According to the City of Los Angeles, more than 10,000 buildings were red-tagged (prohibited entry) or yellow-tagged (restricted entry), and more than 25,000 dwelling units were vacated. Some areas flooded from broken water mains. Some areas were heavily damaged by fire. Damage estimates ranged from $50 to $100 billion.

The earthquake generated a large number of strong-motion recordings over a wide variety of geologic site conditions, including free-field stations on rock and soil as well as recordings of motions from instrumented structures of varying types of construction.

Although the epicenter was located in the suburban city of Reseda in the San Fernando Valley, peak horizontal accelerations approaching 0.5 g were recorded at sites as far as 22 miles (36 km) from the epicenter in downtown Los Angeles. Recordings at two stations in the epicentral area, Sylmar County Hospital (alluvium) and Tarzana Cedar Hills Nursery (alluvium over siltstone), yielded the largest free-field accelerations on soil sites and unusually high values of peak accelerations. The Tarzana station 4.4 miles (7 km) south of the epicenter, recorded peak horizontal and vertical accelerations of 1.82 and 1.18 g, respectively.

A magnitude 7.7 earthquake occurred in the Gilan Province between the towns of Rudbar and Manjil in northern Iran on Thursday, June 21, 1990, at 12:30 a.m. local time. The earthquake, the largest ever to be recorded in that part of the Caspian Sea region, may have been amplified by two or more closely spaced earthquakes occurring in rapid succession. The event, which was exceptionally close to the surface for this region, was unusually destructive causing widespread damage in areas within a 62 mile (100 km) radius of the epicenter near the city of Rasht about 124 miles (200 km) northwest of Tehran. One hundred thousand adobe houses sustained major damage or collapsed resulting in forty thousand fatalities, and sixty thousand injured. Five hundred thousand people were left homeless.

On October 17, 1989, the 7.1 magnitude Loma Prieta earthquake occurred in the Santa Cruz mountains due to movements occurring along a 25 mile (40 km) segment of the San Andreas fault. Measurements along the surface of the earth after the earthquake showed that the Pacific plate moved 6.25 ft (1.9 m) to the northwest and 4.25 ft (1.3 m) upward over the North American plate. The upward motion resulted from deformation of the plate boundary at the bend in the San Andreas fault. At the surface the fault motion was evident as a complex series of cracks and fractures.

This earthquake, according to the geologists, was not unexpected. During the 1906 San Francisco earthquake, there was only about one metre of movement on the Santa Cruz segment of the San Andreas fault, while farther north in the San Francisco area, there was more than 8 ft (2.5 m) of movement. This indicated that all of the strain had not been released in the Santa Cruz segment in the 1906 earthquake so this segment was likely to break before the northern segment.

At the Stanford University campus, 30 miles northwest of the epicenter, 60 buildings sustained varying degrees of damage, with an estimated repair cost of $160 million.

The most deadly structural failure occurred when the upper deck of the Interstate 880 (Nimitz Freeway) in Oakland fell onto the lower roadway causing 41 deaths. Another spectacular failure occurred on the Oakland Bay Bridge. Interstate 280, the Embarcadero Freeway, and the Highway at Fell Street were also damaged.

On July 1993, at 10:17 p.m. local time, a magnitude 7.8 earthquake occurred in the Japan Sea off southwest Hokkaido. The subduction zone event and subsequent tsunami resulted in widespread damage in Northern Japan, with losses estimated at 6 billion yen ($60 million) and 196 earthquake and tsunami related deaths. The majority of the casualties and damage were tsunami caused. Tsunami run-up heights on Okushiri were as high as 100 ft (30.6 m), the highest ever recorded in Japan. Ground shaking produced relatively little damage to engineered structures, but secondary effects from liquefaction and landslides caused substantial damage at several localities.

The January 17, 1995, Kobe earthquake was the most damaging to strike Japan since the great Kanto earthquake destroyed large areas of Tokyo and Yokohama that killed 143,000 people in 1923. As of January 30, 1995, the toll from the earthquake in Kobe and adjacent cities had reached 5096 dead, 13 missing, and 26,797 injured. One-fifth of the city's 1.5 million population was left homeless and more than 103,521 buildings were destroyed with an estimated cost of restoring basic functions to be about $100 billion dollars. The total losses, including losses of privately owned property and reduction in business activity, may be twice this amount, which would be ten times higher than losses resulting from the 1994 Northridge, California earthquake. The epicenter was located about 20 km south-west of downtown Kobe. The earthquake was assigned a Japan Meteorological Agency magnitude of 7.2.

California is the most earthquake-prone state in the U.S. with a number of active faults criss-crossing the region (Fig. 3.1). Among these, the San Andreas Fault is the longest passing under the Gulf of California through the San Joquin Valley and San Francisco. It

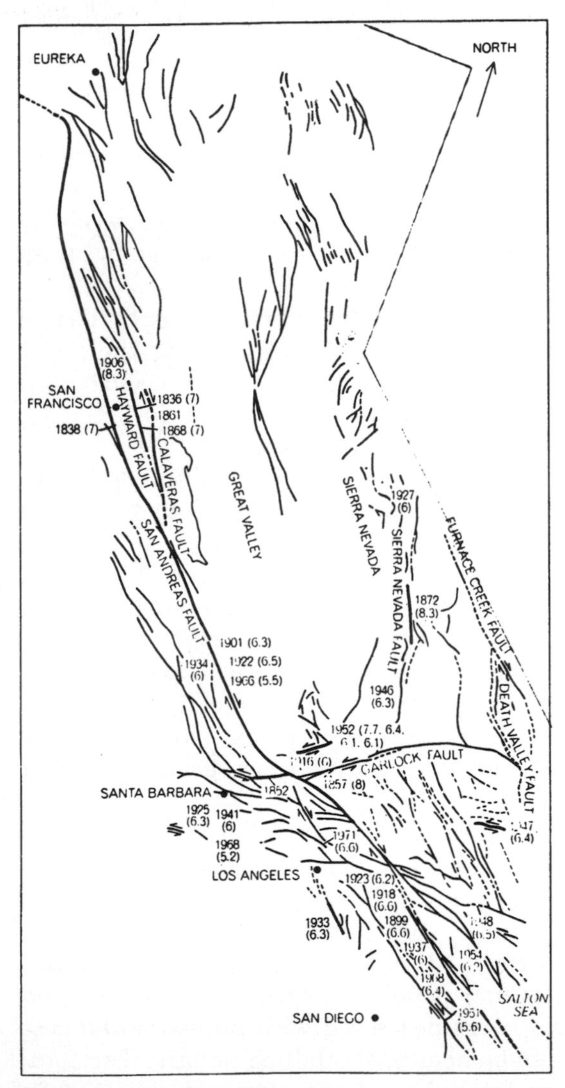


**Figure 3.1** Simplified fault map of California. (Courtesy of Dr. Farzad Naeim)

continues under the Pacific Ocean off the coast of Northern California. The land west of the fault is slowly moving north, while the land east of the fault is moving south. In 1906, movement along this fault caused the famous San Francisco earthquake.

Some faults are deep and others are close to surface of earth. The point beneath the surface where the rocks break and move is called the focus of an earthquake. Directly above the focus, on the earth's surface, is the epicenter. The most violent shaking is often found near the epicenter of an earthquake.

Three types of waves are caused during an earthquake: (i) the primary, or *P* waves, which travel the fastest, pulling and pushing rock particles; (ii) secondary shear waves, or *S* waves, which are responsible for most of earthquake damage: they are slower than *P* waves and cause the rock mass to move in a direction at right-angles to the direction of propogation; (iii) the surface waves generated by the *P* and *S* waves which move along the earth's surface, similar to the way waves travel in an ocean.

Earthquakes often occur in the same three parts of the world. One major earthquake zone extends nearly all the way around the edge of the Pacific Ocean, going through New Zealand, the Philippines, Japan, Alaska, and along the western coasts of North and South America. The San Andreas Fault, where as many as 30 major earthquakes have occurred in historic times is part of this zone. A second major zone of earthquakes is found near the Mediterranean Sea, extending across Asia, encompassing Italy, Greece, Turkey, and part of India. The third zone is located from Iceland south, through the middle of the Atlantic Ocean.

### 3.1.3 Seismograph

For purposes of seismic engineering, earthquakes are recorded on a strong-motion accelerograph. The accelerograph is normally at rest and is triggered only when the ground acceleration exceeds a preset value. Earthquake records can be made for three components of ground acceleration, two horizontal and one vertical.

A basic seismograph consists of a recording device at the end of a heavy pendulum attached to a frame in such a way that it will remain nearly still even when the earth is subjected to seismic motions. A constantly moving drum, attached to the ground through a heavy foundation mass, is placed under the writing device. When the earth is still, the writing device records a straight line, and a zig-zag or wavy line during an earthquake. The higher the wavy lines on the seismograph, the stronger the earthquake.

When an earthquake wave travels through the ground underneath such a pendulum, the motion of the ground is traced by the pen. This is because the pendulum stays stationary due to its inertia, and the trace is actually the movement of the drum. In most seismographs used today, the ground motion is made to produce a small electric signal, and this signal records the motion.

In the conceptual seismograph shown in Fig. 3.1a, the lengths of wavy lines are proportional to the ground displacement. This type of seismograph is called a displacement seismograph. It is possible, however, to record the ground acceleration by appropriately changing the period and damping of the pendulum. Such a device is called an acceleration seismograph.

The current method of recording earthquakes is by the use of accelerographs which produce an analog trace of acceleration versus time in the form of either a photographic trace on film or paper or a scratch on waxed paper.

### 3.1.4 Measures of earthquake magnitude and intensity

There are two commonly used earthquake parameters of interest to the structural engineer. They are an earthquake's magnitude and its intensity. The magnitude is a measure of the amount of energy released by an earthquake while the intensity is the apparent effect of the earthquake as experienced at a specific location. The magnitude is the easier of the two parameters to measure because, unlike the intensity which can vary with location, the magnitude of a particular earthquake is constant. The magnitude scale was first developed by Prof. Richter and then extended to cover world-wide events by others who invented the surface wave magnitude and moment wave magnitude scale. The most widely used scale to

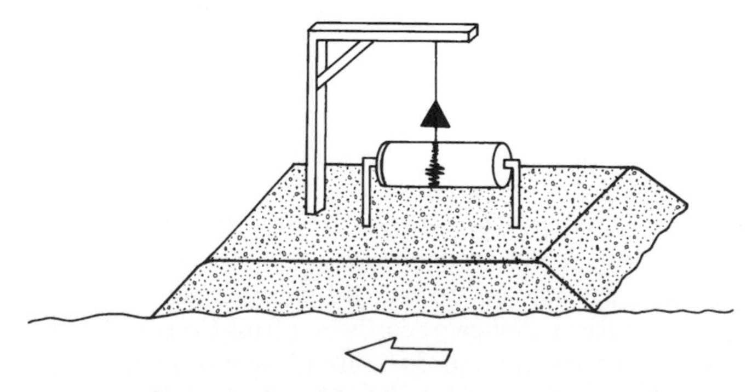

**Figure 3.1a** Conceptual model of displacement seismograph.

measure magnitude is the moment magnitude scale. Using the magnitude scale, the energy released, measured in Ergs can be estimated from the equation

$$\text{Log}\, E = 11.4 + 1.5M.$$

where $E$ is the energy released in Ergs and $M$ is the magnitude. Because the open-ended magnitude scale is logarithmic, each unit increment in the magnitude represents an increase in energy by a factor of 31.6. For example, a quake registering 6 on the magnitude is 31.6 times more powerful than one measuring 5.

Although an earthquake has only one magnitude it will have many different intensities. In the United States, intensity is measured according to the Modified Mercalli Index (MMI).

The intensity values of MMI describe the degree of shaking on a scale of 1 to 12 and are expressed in Roman numerals (I, II, ... , XII). The successive intensities are not derived from a physical measurement, but merely represent ratings given to earthquake effects upon people and the amount of damage to buildings. The intensity levels are represented by a rather long description of earthquake effects ranking earthquake intensities from I (barely felt) to XII (total destruction). Destructive earthquakes of modified Mercalli intensity IX or X at epicentral regions generally register more than 6.5 on the Richter scale.

### 3.1.5 Seismic design

For structural engineering purposes, an accelerograph record of the time-history of ground shaking gives the best measure of earthquake intensity. Figure 3.2 shows the recorded ground acceleration for the El Centro earthquake, California. From Fig. 3.2 it is seen that during a short initial period, the intensity of ground acceleration increases to strong shaking, followed by a strong acceleration phase, which is

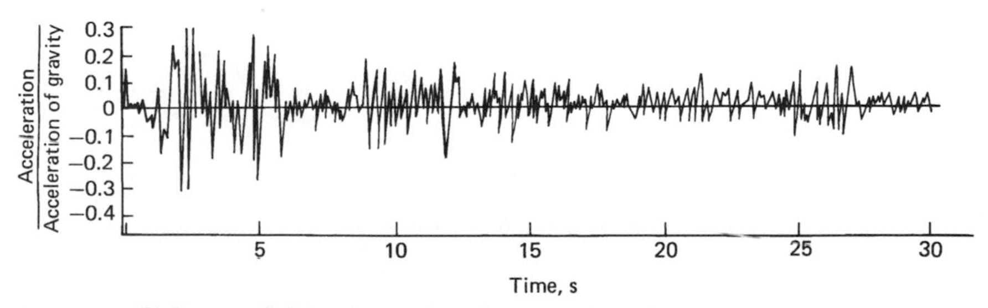


**Figure 3.2** El Centro, California, earthquake ground acceleration record.

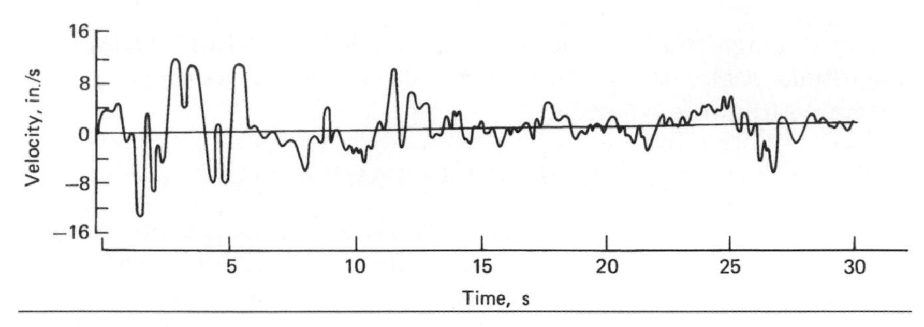


**Figure 3.3** El Centro, California, earthquake ground velocity record.

followed by gradual decreasing motion. Figures 3.3 and 3.4 show the corresponding ground velocity and displacement plotted as functions of time. The maximum recorded ground acceleration is about $0.33\,g$, where $g$ is the acceleration due to gravity of 32 ft/s$^2$ (9.75 m/s$^2$). The maximum ground velocity is 13.7 in./s (0.348 m/s), and the maximum ground displacement from the initial position is about 8.3 in (211 mm).

Shown in Fig. 3.5 are acceleration, velocity and displacement records from the more recent, 1994 Northridge earthquake, California. The peak values of ground acceleration, velocity, and displacement are $0.843\,g$, 50.39 in./s (1.28 m/s), and 12.81 in. (325.5 mm).

Although peak ground acceleration is considered most often as the single measure of damage potential, it is a combination of several characteristics of ground motion that are important to structural response. These are: (i) amplitude; (ii) duration; (iii) frequency content which are related to: (i) magnitude; (ii) site distances from the fault; (iii) site and intervening soil conditions.

It is not possible to predict with any certainty when and where earthquakes will occur, how strong they will be, and what characteristics the ground motions will have; therefore, the engineer must

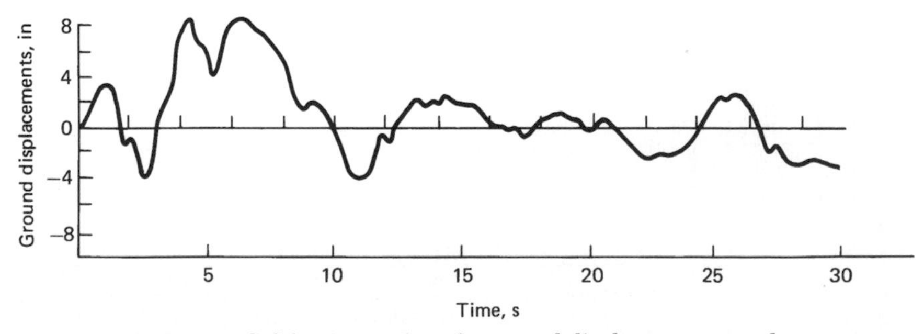


**Figure 3.4** El Centro, California, earthquake ground displacement record.

estimate the ground shaking. A simple method is to use a seismic zone map, such as those used in conjunction with building codes. The map is based on the seismic history and geological information from the area. A detailed method of estimating design earthquake is to conduct a site-specific seismic evaluation, which takes into account seismic history, active faults in the vicinity of the building site, and the stress-strain properties of materials through which the seismic waves travel.

Earthquake forces result from the vibratory motion of the ground on which the structure is supported. The vibratory motion of the ground sets up inertia forces both vertically and horizontally, but it is customary to neglect the vertical component except for cantilevers, since most members have adequate reserve strength for vertical loads because of safety factors specified in codes for gravity loads. The

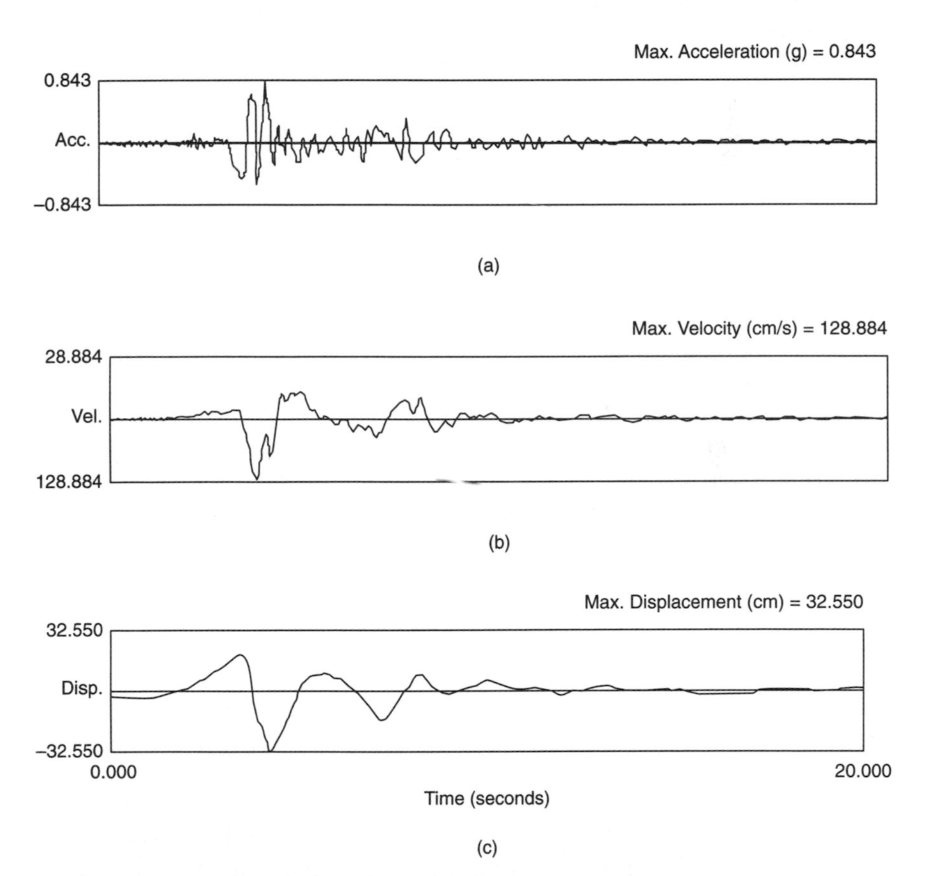


**Figure 3.5** Northridge, California earthquake ground records, 1994. Sylmar county hospital parking lot (N–S direction: (a) acceleration; (b) velocity; (c) displacement. (Courtesy of Dr. Farzad Naeim)

horizontal forces, equal to mass times acceleration, represent the inertia forces occurring at the critical instant of maximum deflection and zero velocity during the largest cycle of vibration as the structure responds to the earthquake motion. In virtually all earthquake design practice, the structure is analyzed as an elastic system, although it is acknowledged that the elastic limit of the structural members will be exceeded during the violent shaking of a major earthquake.

### 3.1.6 Uncertainties in seismic design

In most structural engineering problems, we can evaluate with a fair degree of accuracy the dead loads and the live loads which the structure must be able to support. The strength and properties of materials used in the construction of a structure are well defined by many tests. With the available theories of mechanics and digital computers we can determine to a high degree of precision the moments, shears, and forces that the members will be subjected to. From the material and geometric properties, the amount of resistance the members can provide to these forces can be assessed. With this complete and adequate information, a realistic factor of safety against collapse can be established.

Consider, for example, the lateral design of a tall building subjected to wind loads. Unless the building is very slender, with a height-to-width ratio of greater than 6 or so, the building, if designed for code-designated wind forces, is more than likely to perform adequately throughout its life span. Even when the slenderness ratio is greater than 6 and the building shape is unusual, the engineer can fairly accurately determine the design lateral loads by either using correction factors to the basic wind loads to take into account the dynamic nature of wind or by resorting to wind tunnel tests to determine the worst possible wind load that can be expected over the life of the building. In other words, engineers can get a fairly reasonable estimate of lateral loads either by analytical procedures or by wind-tunnel experiments. The measured wind loads on actual buildings, although the available data are limited, seem to indicate that the actual wind loads on buildings are somewhat smaller than those usually assumed in design, adding an additional degree of confidence.

The situation with regard to earthquake forces is entirely different. The seismic forces specified in the code are quite small, relative to the actual forces expected at least once in the life cycle of the building. It is important to understand the principles behind the code specifications and the justification for designing for lateral earthquake forces of 3 to 20 percent of gravity as compared to dynamic analysis requirements of over 50 percent of gravity. Accelerations derived from actual earthquakes are high when compared to the code forces used in design. A design based on manipulation of numbers to come up to code requirements without appreciation of the code intent will certainly not assure adequate earthquake resistance in case of a major earthquake. A better approach is to design on some reasonable basis, recognizing the uncertain nature of demands and to provide for all the reserve capacity that can be incorporated at little or no extra cost in initial construction or at only a slight sacrifice in architectural features. This in essence is the underlying philosophy of earthquake design.

The seismic loads on the structure during an earthquake result from inertia forces, which are created by ground accelerations. The magnitude of these loads is a function of the following factors:

- mass of the building;
- the dynamic properties of the building, such as its mode shapes, periods of vibration, and its damping;
- the intensity, duration and frequency content of ground motion and soil structure interaction.

From the structural viewpoint, the intensity of vibration of the earth's surface at the building site is of interest. Such intensity of vibration is a function of: (i) amount of energy released; (ii) distance from center of earthquake to the structure; and (iii) character and thickness of foundation material. The magnitude of earthquake, which is a function of the energy released, can be predicted on a regional basis from probability theories. Mathematical theories are available to predict the effect of distance for various soil conditions underlying a site using assumed bedrock vibrations. However, there are too many unknowns to be able to predict quantitatively with any degree of certainty the ground vibration for some unknown future earthquake. Qualitatively, the following are apparent:

1. Ground shaking is strongest in the vicinity of the causative fault, and the intensity diminishes with distance from the fault.
2. Deep deposits of soft soils tend to produce ground surface motions having predominantly long-period characteristics.
3. Shallow deposits of stiff soils result in ground motions having predominantly short-period characteristics.
4. The soil amplification varies with frequency and intensity of the bedrock motions.

In spite of the great strides made in earthquake engineering during the last three decades, numerous uncertainties still exist. The traditional static approach of determining the force level for a given earthquake motion and designing the structure to withstand these forces with a considerable degree of safety has very serious limitations because of the following problems:

1. There simply are not enough empirical data available at the present time to make a reliable prediction as to the intensity and nature of future earthquakes at a given time.
2. Foundation and soil interaction and geological conditions have a profound effect on the structural performance, but at present there exists no clear-cut method which can correctly incorporate these effects.
3. Analysis by elastic assumptions does not take into account the change in properties of the building materials during the progress of an earthquake.

Because of these uncertainties, it is necessary when applying the static load criteria to evaluate the capabilities of the structure to perform satisfactorily beyond the elastic-code-stipulated stresses. Ductility, which involves deformations into the inelastic range is a necessity if structures designed for the static forces are to be capable of resisting earthquakes of the intensity of those which have been actually recorded.

Recent codes require that the function of buildings be taken into account in the earthquake design. This is a direct consequence of the lessons learned from the 1971 San Fernando and later earthquakes, in which several medical facilities were rendered useless and vacated, becoming liabilities rather than maintaining emergency service. Higher loads are specified for other vital public buildings whose functioning is considered indispensable in rescue and recovery efforts.

In earthquake-resistant design, it is not necessary to consider the simultaneous action of wind and earthquake loads, since the probability of this occurrence is quite low. There is no record of an extreme wind and earthquake load hitting a building at the same time. It is expected, as in the case of wind loads, that under the action of moderate earthquake loads, the building structure will remain within the elastic range. Under code-specified earthquake forces, which represent the action of a moderately large earthquake, it is reasonable to expect the structure to maintain elastic behavior. In the case of catastrophic earthquakes, however, a different philosophy exists that permits the building to venture into the plastic range. Certain portions of the building are permitted to suffer minor damage, provided the stability of the structure as a whole is not impaired. The occasional excursion of the building into the plastic range is accepted on the premise that the peak forces produced by earthquakes are of short duration and therefore can be more readily absorbed by the movement of the building than a sustained static load.

### 3.1.7 Design ground motion

It is important to recognize that the discipline of design ground motion specification is in a state of evolution. For typical building applications, even with the use of computers for probabilistic studies, the exact prediction of ground motions at a site is not possible.

The seismic hazard at a site is usually evaluated using probabilistic methods by considering all possible earthquakes in the area, estimating the associated shaking at the site and calculating the probabilities of these occurrences. Given the current limited knowledge and understanding of the earthquake process, all assessments of earthquake hazard are inherently uncertain. Probability methods are not used by design engineers on a regular basis and therefore the methods of generating response spectra may be difficult to comprehend. However, a brief discussion of earthquake hazard analysis techniques used in determining seismic ground motions is given below.

Available procedures for assessing seismic ground motions vary from fully deterministic procedures through hybrid (partly deterministic to partly probabilistic), to fully probabilistic procedures. In a deterministic approach a single maximum earthquake is specified by magnitude and location with respect to a site of interest. By using an attenuation function, the reduction in the intensity of the seismic motion due to the distance of the site from the specified seismic source is calculated. The resulting reduced ground motion is used to design or evaluate the seismic vulnerability of a facility. The target earthquake, usually the maximum credible, also called the maximum considered earthquake, is selected by consideration of the historical seismic record and physical characteristics of the seismic source. The deterministic approach does not consider the likelihood of the occurrence of an earthquake nor does it consider the importance of the target earthquake relative to other possible seismic events, such as those due to larger but more distant earthquakes or smaller but closer earthquakes. For this reason this approach is seldom used in practice.

The probabilistic approach, on the other hand, addresses the questions of how strongly and how often the ground will shake, by considering all possible earthquakes likely to affect the site. As in the deterministic procedure, an attenuation function, together with the distance from the seismic source, is used to estimate ground motion at the site resulting from a variety of seismic events. The rate of earthquake occurrence for each seismic source is also considered. Thus the probabilistic procedure combines information on earthquake size, location, probability of occurrence and resulting ground motion to give results in terms of expected ground motion and associated annual probabilities of occurrence. The objective of the analysis is to provide an estimate of ground motion at a specific site. Typically the ground motion is expressed in terms of peak acceleration, velocity, displacement or a design spectrum at the site. Duration of strong motion is an important measure, but is not explicitly used in design criteria at the present time.

## 3.2 Tall Building Behavior During Earthquakes

### 3.2.1 Introduction

The behavior of a tall building during an earthquake is a vibration problem. The seismic motions of the ground do not damage a building by impact as does a wrecker's ball, or by externally applied pressure such as wind, but rather by internally generated inertial forces caused by vibration of the building mass. An increase in the mass has two undesirable effects on the earthquake design. First, it results in an increase in the force, and second, it can cause buckling and crushing of vertical elements such as columns and walls when the mass pushing down exerts its force on a member bent or moved out of plumb by the lateral forces. This phenomenon is known as the $p$-$\Delta$ effect. The greater the vertical force, the greater the movement due to p-$\Delta$. It is almost always the vertical load that causes buildings to collapse; in earthquakes, buildings very rarely fall over—they fall down. The seismic motions of the ground cause the structure to vibrate, and the amplitude and distribution of dynamic deformations and their duration are of concern to the engineer. Note that although duration of strong motion is an important measure, it is not explicitly used in design criteria at the present time.

### 3.2.2 Response of tall buildings

In general, tall buildings respond to seismic motion somewhat differently than low-rise buildings. The magnitude of inertia forces induced in an earthquake depends on the building mass, ground acceleration, the nature of foundation and the dynamic characteristics of the structure (Fig. 3.6). If a building and its foundation were infinitely rigid, it would have the same acceleration as the ground: the inertia force $F$ for a given ground acceleration $a$ given by Newton's Law $F = Ma$, where $M$ is the building mass. For a structure that deforms only slightly, thereby absorbing some energy, the force $F$ tends to be less than the product of mass and ground acceleration. Tall buildings are invariably more flexible than low-rise buildings, and in general experience accelerations much less than low-rise buildings. But a flexible building subjected to ground motions for a prolonged period may experience much larger forces if its natural period is near that of the ground waves. Thus, the magnitude of lateral force in a building is not a function of the acceleration of the ground alone but is influenced to a great extent by the type of response of the structure and its foundation, as well. This interrelationship of building behavior and seismic ground motion also depends on the building period and is expressed in the so-called response spectrum, explained later in this chapter.

Consider, for example, the behavior of a 30 story building during an earthquake. Although the motion of the ground is erratic and three-dimensional, the horizontal components in two mutually perpendicular directions are of importance. These components typically have varying periods and can be considered as short-period components when the period is less than 0.5 sec, and long-period components for periods in excess of 0.5 sec. The period of fundamental

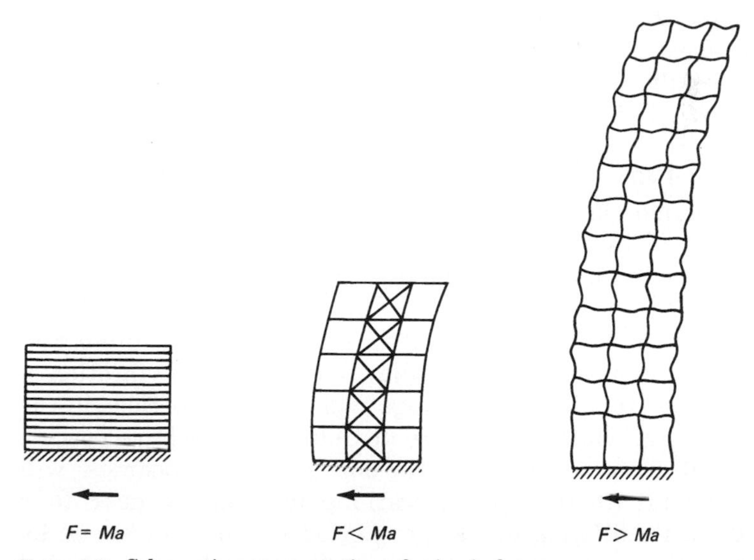


**Figure 3.6** Schematic representation of seismic forces.

frequency $T_1$ of a tall building is a function of its stiffness, mass, and damping characteristics, and can vary over a broad range anywhere from 0.05 to 0.30 times the number of stories, depending upon the materials used in the construction and the structural system employed. As a preliminary approximation for steel-framed buildings, the period $T_1$ is approximately equal to $0.15N$, where $N$ is the number of stories. A typical 30-story building would have a fundamental period of 4.5 sec, with the periods of the next two higher modes, $T_2$ and $T_3$, approximately equal to one-third and one-fifth of $T_1$.

The second and third modes of vibration for the 30-story building are thus approximately equal to 1.5 and 0.9 sec. During the first few seconds of earthquake, the acceleration of the ground reaches a peak value and is associated with relatively short-period components of the range 0 to 0.5 sec, which have little influence on the fundamental response of the building. On the other hand, the long-period components that occur at the tail-end of the earthquakes, with periods closer to the fundamental period of the building have a profound influence on its behavior.

The intensity of ground motion reduces with the distance from the epicenter of the earthquake. The reduction of ground motion intensity, called attenuation, is at a faster rate for higher-frequency components than for lower-frequency components. The cause for the change in attenuation rate is not understood, but its existence is certain. This is a significant factor in the design of tall buildings, because a tall building, although situated farther from a causative fault than a low-rise building, may experience greater seismic loads because long-period components are not attenuated as fast as the short-period components. Therefore, the area influenced by ground shaking potentially damaging to, say, a 50-story building is much greater than to a 1-story building.

### 3.2.3 Influence of soil

The seismic motion that reaches a structure on the surface of the earth is influenced by the local soil conditions. The subsurface soil layers underlying the building foundation may amplify the response of the building to earthquake motions originating in the bedrock. Although it is difficult to visualize, it is possible that a number of underlying soils can have a period similar to the period of vibration of the structure. Low- to-mid-rise buildings typically have periods in the 0.10 to 1.0 sec range while taller, more flexible buildings have periods between 1 and 5 sec or greater. Harder soils and bedrock will efficiently transmit short-period vibrations (caused by near field earthquakes) while filtering out longer-period vibrations (caused by distant earthquakes) whereas softer soils will transmit longer-period vibrations.

As a building vibrates under ground motion, its acceleration will be amplified if the fundamental period of the building coincides with the period of vibrations being transmitted through the soil. This amplified response is called resonance. Natural periods of soil are in the range of 0.5 to 1.0 sec so that it is entirely possible for the building and ground to have the same fundamental period, and therefore, for the building to approach a state of resonance. This was the case for many 5 to 10-story buildings in the September 1985 earthquake in Mexico City. An obvious design strategy, if one can predict approximately the rate at which the ground will vibrate, is to ensure that buildings have a natural period different from that of the expected ground vibration to prevent amplification.

### 3.2.4 Damping

Buildings do not resonate with the purity of a tuning fork because they are damped; the extent of damping depends upon the construction materials, type of connections and the presence of nonstructural elements. Damping is measured as a percentage of critical damping.

In a dynamic system, critical damping is defined as the minimum amount of damping necessary to prevent oscillation altogether. To get a mental picture of critical damping, imagine a tightly tensioned string immersed in a tank containing water. When the string is plucked, it oscillates about its mean position several times before coming to rest. If we replace water with a liquid of higher viscosity, the string will oscillate, but certainly not as many times as it did in water. By progressively increasing the viscosity of the liquid, it is easy to visualize that a state can be reached where the string, once plucked, will return to its neutral position without ever crossing it. The minimum viscosity of the liquid that prevents the vibration of the string altogether can be considered equivalent to critical damping.

The damping of structures under earthquake disturbances is influenced by a number of external and internal sources. Chief among them are:

1. External viscous damping caused by air surrounding the building. Since the viscosity of air is small, this effect is negligible in comparison to other types of damping.
2. Internal viscous damping associated with the material viscosity. This is proportional to the velocity and increases in proportion to the natural frequency of the structure.
3. Friction damping, also called Coulomb damping, occurring at connections and support points of the structure. It is a constant, irrespective of the velocity or amount of displacement. Steel

buildings with bolted connections have more friction damping as compared to a fully welded construction. A prestressed concrete building has less damping as compared to a mild-steel-reinforced construction because in prestressed concrete cracking of concrete is relatively less.

4. Hysteresis damping that develops when the structure is subjected to load reversals in the inelastic range. The area inside of the hysteresis loop corresponds to the energy dissipated and is referred to as hysteretic damping. It increases with the level of displacement and is independent of the velocity of the structure.
5. Radiation damping resulting from energy dissipation through the ground on which the structure is built. It is a function of the characteristics of the ground such as density, Poisson's ratio, shear and elastic modulii, and the depth to which the structure is embedded in the ground.
6. Hysteresis damping around the foundation caused by the inelastic deformation of the ground adjacent to the foundation.

It is common practice in dynamic analysis of buildings to lump all of the various sources of damping into one type and to represent it as viscous damping. To arrive at a precise value of viscous damping that can effectively take into account all the aforementioned characteristics is impractical. Representative values of damping ratios used in practice vary anywhere from 0.02 to 0.10 depending upon the material used for the building and the level of design force used in the analysis. Vibration tests of existing buildings indicate that damping ratios vary from a low of 0.005 to a high of 0.075.

In earthquake design, the idea of critical damping is used to modify the ground response spectrum by assuming certain percentages of damping, generally of the order of 2 to 15 percent of critical. Low end values are used in wind engineering, while upper-end values are used in earthquake engineering.

A level of ground acceleration generally at $0.1g$, where $g$ is the acceleration due to gravity, is sufficient to produce some damage to weak construction. An acceleration of $1.0g$, or 100 percent of gravity, is analytically equivalent, in the static sense, to a building that cantilevers horizontally from a vertical surface.

### 3.2.5 Building motion and deflections

Earthquake-induced motions, even when they are more violent than those induced by wind, evoke a totally different type of human response. First, because earthquakes occur much less frequently than windstorms, and second, because the duration of motion caused by an earthquake is generally short. People who experience earthquakes are grateful that they have survived the trauma and are less inclined to be critical of building motion. Earthquake-induced motions are, therefore, a safety rather than a human discomfort phenomenon.

Lateral deflections that occur during earthquakes should be limited to prevent distress in structural members and architectural components. Non-load-bearing in-fills, external wall panels and window glazing should be designed with sufficient clearance or with flexible supports to accommodate the anticipated movements.

### 3.2.6 Seismic separation

Drift is the lateral displacement of one floor relative to the floor below. Buildings subjected to earthquakes need drift control to limit damage to interior partitions, elevator and stair enclosures, glass, and cladding systems. Stress or strength limitations do not always provide adequate drift control especially for tall buildings with relatively flexible moment-resisting frames and narrow shear walls.

Total building drift is the absolute displacement of any point relative to the base. Adjoining buildings or adjoining sections of the same building do not have identical modes of response and, therefore, have a tendency to pound against one another. Building separations or joints must be provided to permit adjoining buildings to respond independently to earthquake ground motion.

## 3.3 Seismic Design Concept

### 3.3.1 Determination of forces

There are two general approaches to determining seismic forces: an equivalent static force procedure and a dynamic analysis procedure. In this section the equivalent static force procedure is illustrated. Dynamic analysis procedures are considered later, but some discussion of structural dynamics is included in this section in order to explain the rationale of the equivalent static force procedure.

### 3.3.2 Design of the structure

The development of an adequate earthquake-resistant design for a structure includes the following: (i) selecting a workable overall structural concept; (ii) establishing member sizes; (iii) performing a structural analysis of the members to verify that stress and displacement requirements are satisfied; and (iv) providing structural and nonstructural details so that the building will accommodate the distortions and stresses that occur in the building. Elements which cannot accommodate these stresses and distortions such as rigid stairs, partitions, and irregular wings should be isolated to reduce detrimental effects to the lateral force-resisting system.

### 3.3.3 Structural response

If the base of a structure is suddenly moved, as in the case of seismic ground motion, the upper part of the structure will not respond instantaneously but will lag because of inertial resistance and flexibility of the structure. This concept is illustrated in Figs 3.6a–c by showing the motion in one plane. The stresses and distortions in the building are the same as if the base of the structure were to remain stationary while time-varying horizontal forces are applied to the upper part of the building. These forces, called inertia forces, are equal to the product of the mass of the structure times acceleration, or $F = ma$ (The mass $m$ is equal to weight divided by the acceleration of gravity, i.e., $m = w/g$). Because the ground motion at a point on the earth's surface is three dimensional (one vertical and two horizontal), the structures affected will deform in a three-dimensional manner. Generally, however, the inertia forces generated by the horizontal components of ground motion require greater consideration for seismic design since adequate resistance to vertical seismic loads is usually provided by the member capacities required for gravity load design. In the Equivalent Static Procedure, the inertia forces are represented by equivalent static forces.

### 3.3.4 Path of forces

Buildings are composed of vertical and horizontal structural elements which resist lateral forces. The vertical elements that are used to transfer lateral forces to the ground are: (i) shear walls; (ii) braced frames; and (iii) moment-resisting frames. Horizontal elements that distribute lateral forces to vertical elements are: (i) most usually diaphragms, such as floor slabs; and (ii) horizontal bracing in special floors such as transfer floors. Horizontal forces produced by seismic motion are directly proportional to the masses of building elements and are considered to act at the center of the mass of these elements. All of the inertia forces originating from the masses on and off the structure must be transmitted to the lateral force-resisting elements, and then to the base of the structure and into the ground.

### 3.3.5 Demands of earthquake motion

The loads or forces which a structure sustains during an earthquake result directly from the distortions induced in the structure by the motion of the ground on which it rests. Base motion is characterized by displacements, velocities, and accelerations which are erratic in direction, magnitude, duration, and sequence. Earthquake loads are inertia forces related to the mass, stiffness, and energy-absorbing (e.g., damping and ductility) characteristics of the structure. During the life of a structure located in a seismically active zone, it is generally expected that the structure will be subjected to many small earthquakes, some moderate earthquakes, one or more large earthquakes, and possibly a very severe earthquake. In general, it is uneconomical or impractical to design buildings to resist the forces resulting from large or severe earthquakes within the elastic range of stress. If the earthquake motion is severe, most structures will experience yielding in some of their elements. The energy-absorption capacity of the yielding structure will limit the damage so that buildings that are properly designed and detailed can survive earthquake forces which are substantially greater than the design forces that are associated with allowable stresses in the elastic range. Seismic design concepts must consider building proportions and details for their ductility (capacity to yield) and reserve energy-absorption capacity for surviving the inelastic deformations that would result from a maximum expected earthquake. Special attention must be given to connections that hold the lateral force-resisting elements together.

### 3.3.6 Response of buildings

A building is analyzed for its response to ground motion by representing the structural properties in an idealized mathematical model as an assembly of masses interconnected by springs and dampers. The tributary weight to each floor level is lumped into a single mass, and the force-deformation characteristics of the lateral force-resisting walls or frames between floor levels are transformed into equivalent story stiffnesses. Because of the complexity of the calculations, the use of a computer program is generally necessary, even when the design is by the equivalent static force procedure.

### 3.3.7 Response of elements attached to buildings

Elements attached to the floors of buildings (e.g., mechanical equipment, ornamentation, piping, nonstructural partitions) respond to floor motion in much the same manner as the building responds to ground motion. However, the floor motion may vary substantially from the ground motion. The high-frequency components of the ground motion tend to be filtered out at the higher levels in the building while the components of ground motion that correspond to the natural periods of vibrations of the building tend to be magnified. If the elements are rigid and are rigidly attached to the structure, the forces on the elements will be in the same proportion to the mass as the forces on the structure. But elements that are flexible and have periods of vibration close to any of the predominant modes of the building vibration will experience forces in a proportion substantially greater than the forces on the structure.

### 3.3.8 Techniques of seismic design

For gravity loads, it has been a long-standing practice to design for strength and deflections within the elastic limits of the members. However, to control design within elastic behavior for the maximum expected horizontal seismic forces is impractical in high-seismicity areas. Therefore, several problems of building design should be recognized by the building owner, architect, and engineer as factors that may substantially increase the earthquake risk to their building. The solution lies in the design teams' understanding of seismic resistant design rather than in application of specific code provisions. A few of these problems are discussed below.

#### 3.3.8.1 Layout

A great deal of a building's resistance to lateral forces is determined by its plan layout. The objective in this regard is symmetry about both axes, not only of the building itself but of the arrangement of wall openings, columns, shear walls, etc. It is most desirable to consider the effect of lateral forces on the structural system from the start of the layout since this may save considerable time and money without detracting significantly from the usefulness or appearance of the building.

#### 3.3.8.2 Structural symmetry

Experience has shown that buildings which are unsymmetrical in plan have greater susceptibility to earthquake damage than symmetrical structures. The effect of asymmetry will induce torsional oscillations of the structure and stress concentrations at re-entrant corners. Asymmetry in plan can be eliminated or improved by separating *L*-, *T*-, and *U*-shaped buildings into distinct units by use of seismic joints at junctions of the individual wings. Asymmetry caused by the eccentric location of lateral force-resisting structural elements, e.g., a building that has a flexible front because of large openings and an essentially stiff (solid) rear wall, can usually be avoided by better conceptual planning, e.g., by modifying the stiffness of the rear wall, or adding rigid structural elements to bring the center of rigidity of the lateral force-resisting elements close to the center of mass.

#### 3.3.8.3 Irregular buildings

Those who have studied the performance of buildings in earthquakes generally agree that the building's form has a major influence. This is because the shape and proportions of the building have a major effect on the distribution of earthquake forces as they work their way through the building. Geometric configuration, type of structural members, details of connections, and materials of construction all have a profound effect on the structural-dynamic response of a building. When a building has irregular features, such as asymmetry in plan or vertical discontinuity, the assumptions used in developing seismic criteria for buildings with regular features may not apply. Therefore, it is best to avoid creating buildings with irregular features. For example, omitting exterior walls in the first story of a building to permit an open ground floor leaves the columns at the ground level as the only elements available to resist lateral forces, thus causing an abrupt change in rigidities at that level. This condition is undesirable. It is advisable to carry all shear walls down to the foundation. When irregular features are unavoidable, special design considerations are required to account for the unusual dynamic characteristics and the load transfer and stress concentrations that occur at abrupt changes in structural resistance. Examples of plan and elevation irregularities are illustrated in Figs 3.7 and 3.8.

#### 3.3.8.4 Lateral force-resisting systems

There are several systems that can be used effectively for providing resistance to seismic lateral forces. Some of the more common systems are shown in Fig. 3.9. All of the systems rely basically on a complete, three-dimensional space frame; a coordinated system of moment frames, shear walls or braced frames with horizontal diaphragms; or a combination of the systems.

1. In buildings where a space frame resists the earthquake forces, the columns and beams act in bending. During a large earthquake, story-to-story deflection (story drift) may be accommodated within the structural system without causing failure of

columns or beams. However, the drift may be sufficient to damage elements that are rigidly tied to the structural system such as brittle partitions, stairways, plumbing, exterior walls, and other elements that extend between floors. Therefore, buildings can have substantial interior and exterior nonstructural damage and still be considered as structurally safe. While there are excellent

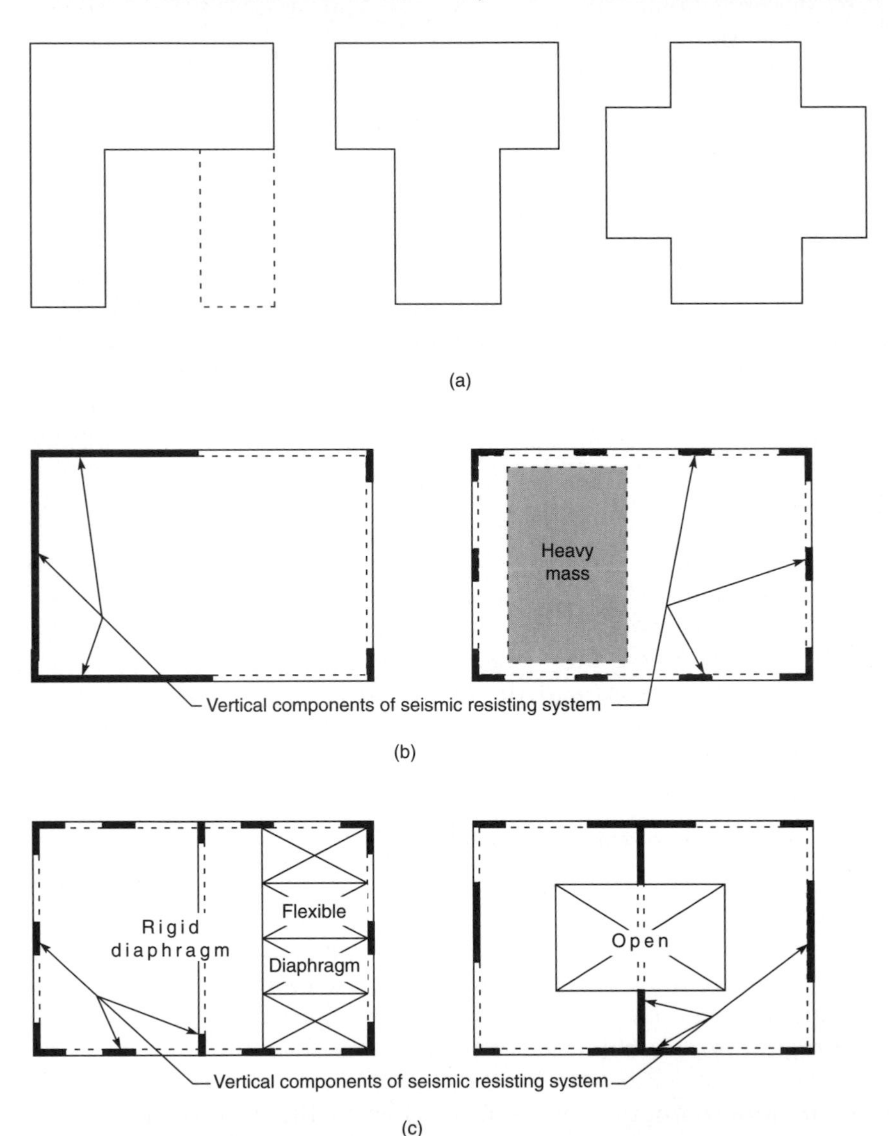


**Figure 3.7** Plan irregularities: (a) geometric irregularities; (b) irregularity due to mass-resistance eccentricity; (c) irregularity due to discontinuity in diaphragm stiffness

theoretical and economic reasons for resisting seismic forces by frame action, for particular buildings this system may be a poor economic risk unless special damage-control measures are taken.

2. A shear wall (or braced frame) building is normally more rigid than a framed structure. With low design stress limits in shear walls, deflection due to shear forces is relatively small.

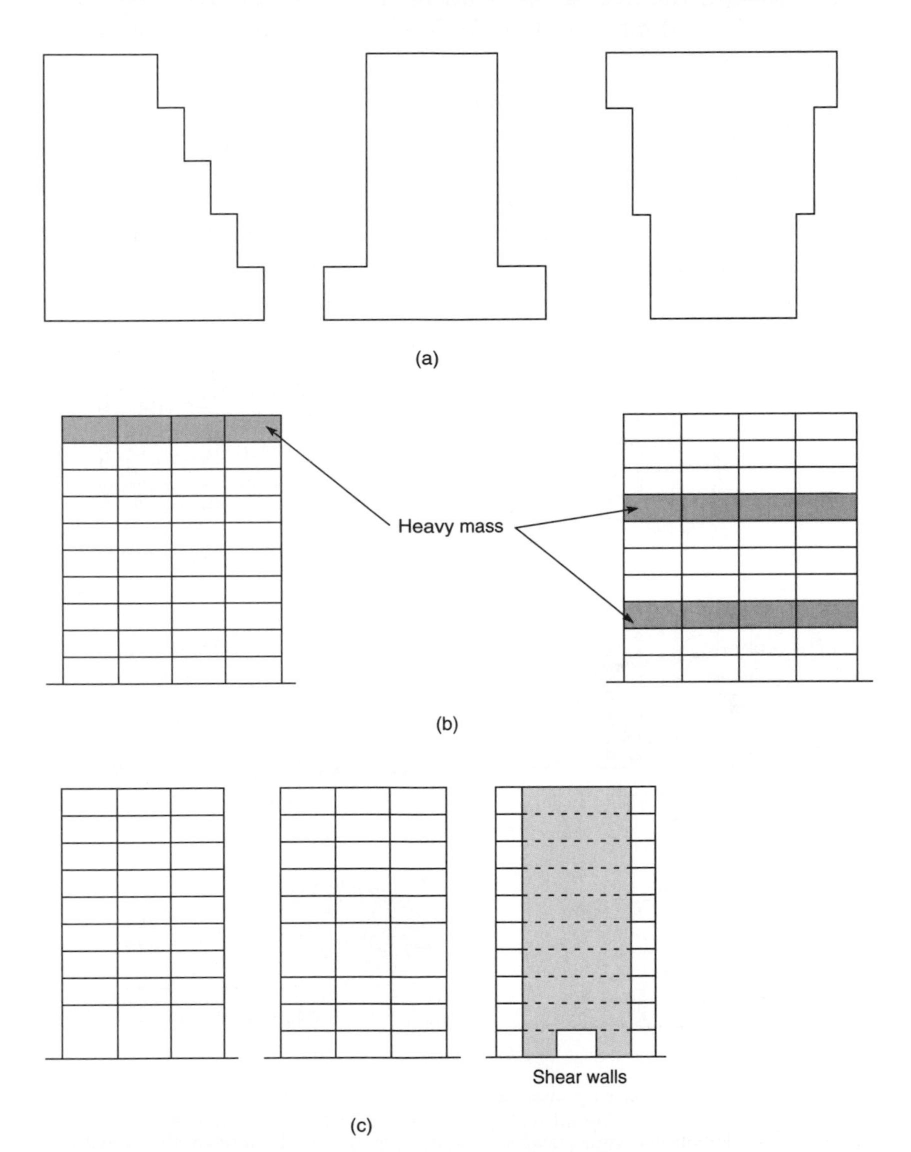


**Figure 3.8** Elevation irregularities: (a) abrupt change in geometry; (b) large difference in floor masses; (c) large difference in story stiffnesses.

Shear wall construction is an economical method of bracing buildings to limit damage, and this type of construction is normally economically feasible up to about fifteen stories. Notable exceptions to the excellent performance of shear walls occur when the height-to-width ratio becomes great enough to make overturning a problem and when there are excessive openings in the shear walls. Also, if the soil beneath its footings are relatively soft, the entire shear wall may rotate, causing localized damage around the wall.

3. The structural systems mentioned above may be used in combination. When frames and shear walls are combined, the system is called a dual bracing system. The type of structural system used, with specified details concerning the ductility and energy-absorbing capacity of its components, will establish the minimum $R_W$-value, defined later, to be used for calculating the total base shear and to distribute the lateral seismic forces.

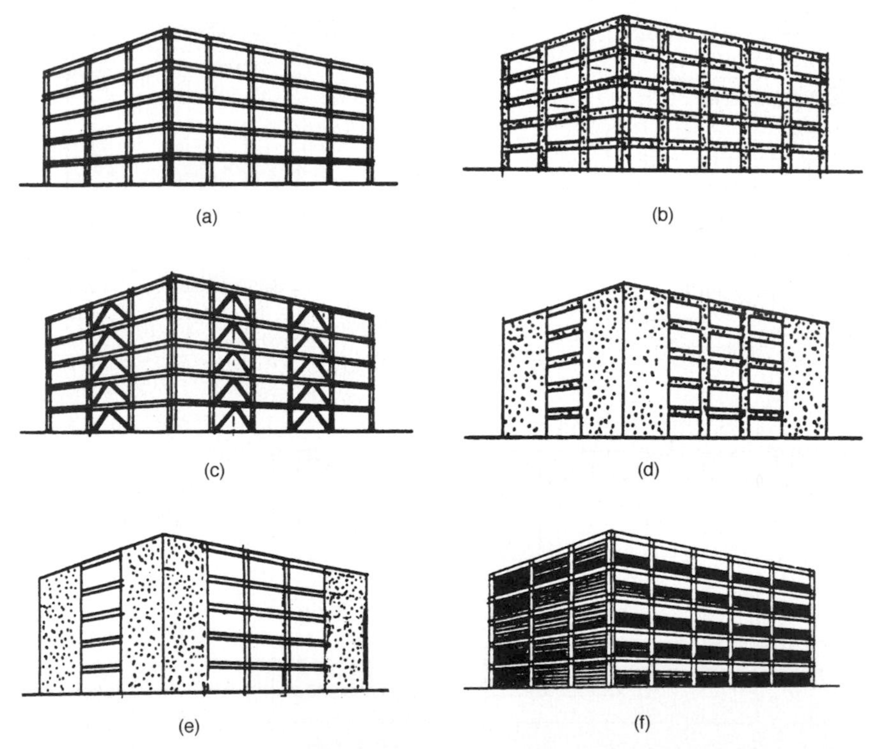


**Figure 3.9** Lateral force-resisting systems: (a) steel moment-resisting frame building; (b) reinforced concrete moment-resisting frame building; (c) braced steel frame building; (d) reinforced concrete shear wall building; (e) steel frame building with cast-in-place concrete shear walls; (f) steel frame building with in-filled walls of nonreinforced masonry.

The design engineer must be aware that a building does not merely consist of a summation of parts such as walls, columns, trusses, and similar components but is a completely integrated system or unit which has its own properties with respect to lateral force response. The designer must follow the flow of forces through the structure into the ground and make sure that every connection along the path of stress is adequate to maintain the integrity of the system. It is necessary to visualize the response of the complete structure and to keep in mind that the real forces involved are not static but dynamic, are usually erratically cyclic and repetitive, and can cause deformations well beyond those determined from the elastic design.

#### 3.3.8.5 Diaphragms

The earthquake loads at any level of a building will be distributed to the vertical structural elements through the floor and roof diaphragms. The roof/floor deck or slab responds to loads like a deep beam. The deck or slab is the web of the beam carrying the shear and the perimeter spandrel or wall is the flange of the beam resisting bending.

Three factors are important in diaphragm design.

1. The diaphragm must be adequate to resist both the bending and shear stresses and be tied together to act as one unit.
2. The collectors and drag members, see Fig. 3.10, must be adequate to transfer loads from diaphragm into the lateral load-resisting vertical elements such as shear walls and moment and braced frames.
3. Openings or re-entrant corners in the diaphragm must be properly placed and adequately reinforced.

Inappropriate location or large size openings (stair or elevator cores, atriums, skylights) create problems similar to those related to cutting a hole in the web of a beam. This reduces the natural ability of the diaphragm to transfer the forces and may cause failure (Fig. 3.11a).

#### 3.3.8.6 Ductility

Ductility is the capacity of building materials, systems, or structures to absorb energy by deforming into the inelastic range. The capability of a structure to absorb energy, with acceptable deformations and without failure, is a very desirable characteristic in any earthquake-resistant design. Brittle material such as concrete must be properly reinforced with steel to provide the ductility characteristics necessary

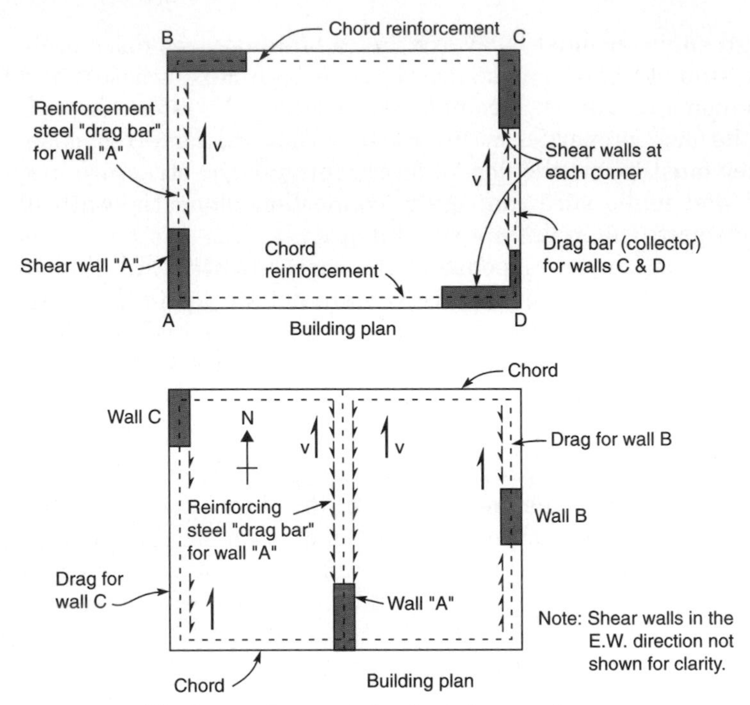


**Figure 3.10.** Diaphragm drag and chord reinforcement.

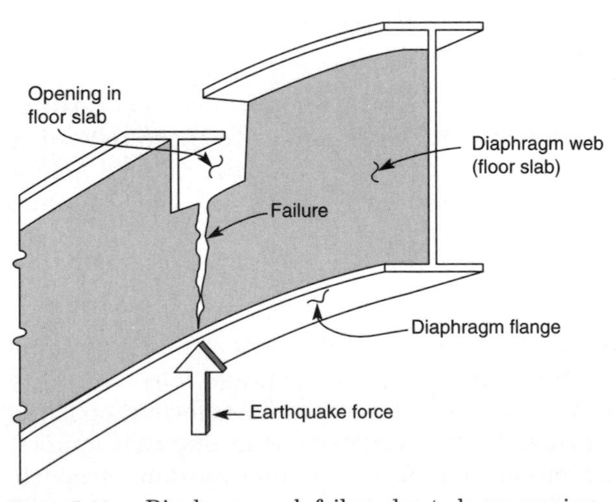


**Figure 3.11a.** Diaphragm web failure due to large opening.

to resist seismic forces. In concrete columns, for example, the combined effects of flexure (due to frame action) and compression (due to the action of the overturning moment of the structure as a whole) produces a common mode of failure; buckling of the vertical steel and spalling of the concrete cover near the floor levels. Columns must, therefore, be detailed with proper spiral reinforcing or hoops to have greater reserve strength and ductility.

Ductility is often measured by the hysteretic behavior of critical components such as a column-beam assembly of a moment frame. The hysteretic behavior is usually examined by observing the cyclic moment-rotation (or force deflection) behavior of the assembly as shown in Fig. 3.11b. The slope of the curves represent the stiffness of the structure, and the enclosed areas are sometimes full and fat, or they may be lean and pinched. Structural assemblies with curves enclosing a large area representing large dissipated energy, are regarded as superior systems for resisting seismic loading.

#### 3.3.8.7 Nonstructural participation

For both analysis and detailing, the effects of nonstructural elements such as partitions, filler walls, and stairs must be considered. The nonstructural elements that are rigidly tied to the structural system can have a substantial influence on the magnitude and distribution of earthquake forces, causing a shearwall-like response with considerably higher lateral forces and overturning moments. Any element that is not strong enough to resist the forces that it attracts will be damaged; therefore, it should be isolated from the lateral force-resisting system.

#### 3.3.8.8 Foundation

The differential movement of foundations due to seismic motions is an important cause of structural damage, especially in heavy, rigid structures that cannot accommodate these movements. Adequate design must minimize the possibility of relative displacement, both horizontal and vertical, between the various parts of the foundation and between the foundation and superstructure.

#### 3.3.8.9 Damage control features

The design of a structure in accordance with the seismic provisions will not fully ensure against earthquake damage because the horizontal deformations that can be expected during a major earthquake are several times larger than those calculated under design loads. A list of features that can be detailed to minimize earthquake damage follows.

1. Provide details which allow structural movement without damage to nonstructural elements. Damage to such items as piping, glass, plaster, veneer, and partitions may constitue a major financial loss. To minimize this type of damage, special care in detailing, either to isolate these elements or to accommodate the movement, is required.
2. Breakage of glass windows can be minimized by providing adequate clearance at edges to allow for frame distortions.

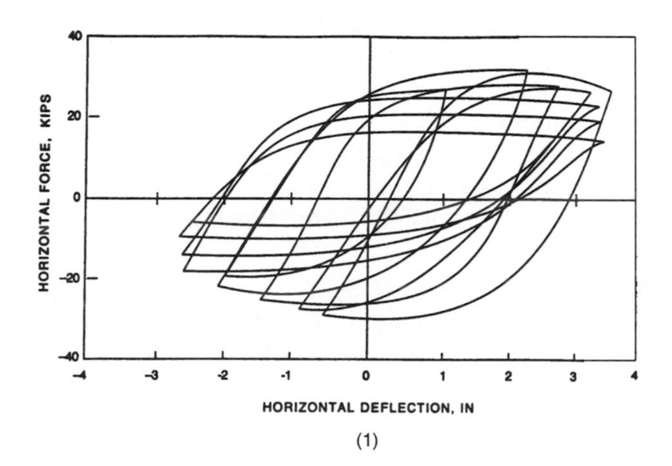


(1)

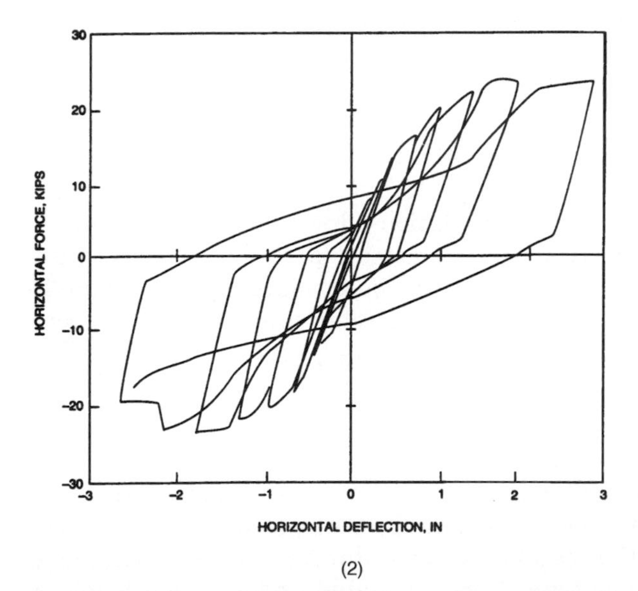


(2)

**Figure 3.11b** Hysteritic behavior: (1) curve representing large energy dissipation; (2) curve representing limited energy dissipation.

3. Damage to rigid nonstructural partitions can be largely eliminated by providing a detail at the top and sides which will permit relative movement between the partitions and the adjacent structural elements.
4. In piping installations, the expansion loops and flexible joints used to accommodate temperature movement and are often adaptable to handling the relative seismic deflections between adjacent equipment items attached to floors.
5. Fasten free-standing shelving to walls to prevent toppling.
6. Concrete stairways often suffer seismic damage due to their inhibition of drift between connected floors. This can be avoided by providing a slip joint at the lower end of each stairway to eliminate the bracing effect of the stairway or by tying stairways to stairway shear walls.

#### 3.3.8.10 Redundancy

Redundancy is a highly desirable characteristic for earthquake-resistant design. When the primary element or system yields or fails, the lateral force can be redistributed to a secondary system to prevent progressive failure.

## 3.4 1994 UBC Equivalent Lateral Force Procedure (Static Method)

### 3.4.1 Design base shear

The total seismic lateral force, also called the base shear, is determined by the relation

$$V = \frac{ZIC}{R_W} W \tag{3.1}$$

in which

$R_W$ = The response modification factor. It is a measure of the ability of the structure to withstand earthquake motions without collapse. It represents the ratio of forces in an entirely linear elastic system to the forces anticipated in a system with significant yielding. It's magnitude depends on the type and material of the structure, the possibility of failure of the vertical load system, the degree of redundancy of the system that would allow some localized failures without overall failure, and the ability of the secondary system, in the case of dual systems, to stabilize the building when the primary system suffers significant damage. The factor $R_W$ is in the denominator of the base shear equation

**TABLE 3.1 $R_W$ Values for Structural Systems in Concrete**

| | Zones 3 and 4 | | Zone 2 | | Zone 1 | |
|---|---|---|---|---|---|---|
| System | $R_W$ | $H$ | $R_W$ | $H$ | $R_W$ | $H$ |
| Bearing wall system | | | | | | |
| • Shear walls | 6 | 160 | 6 | NL | 6 | NL |
| • Braced frames | Not allowed | | 4 | NL | 4 | NL |
| Building frame system | | | | | | |
| • Shear walls | 8 | 240 | 8 | NL | 8 | NL |
| • Braced frames | Not allowed | | 8 | NL | 8 | NL |
| Moment resisting frame | | | | | | |
| • SMRF | 12 | NL | 12 | NL | 12 | NL |
| • IMRF | Not allowed | | 8 | NL | 8 | NL |
| • OMRF | Not allowed | | Not allowed | | 5 | NL |
| Dual Systems | | | | | | |
| • Shear wall + SMRF | 12 | NL | 12 | NL | 12 | NL |
| • Shear walls + IMRF | Not allowed | | 9 | NL | 9 | NL |
| • Braced frame + SMRF | Not allowed | | 9 | NL | 9 | NL |
| • Braced frame + IMRF | Not allowed | | 6 | NL | 6 | NL |

SMRF: Special Momenting Resisting Frame (Ductile Frame).
IMRF: Intermediate Moment Resisting Frame.
OMRF: Ordinary Moment Resisting Frame.
NL: No limit.
H: Height in feet.

(Eq. 3.1) so that design loads decrease for systems with large $R_W$ values, i.e., for large inelastic deformation capabilities. Numerical values for $R_W$ are shown in Tables 3.1 and 3.2.

$W$ = The total seismic dead load of the building and applicable portions of other loads. It represents the total mass of the building and includes the weights of structural slabs, beams, columns and walls, non-structural components such as floor topping, roofing, fire-proofing material, fixed electrical and mechanical equipment, partitions and ceilings. When partition locations are subject to change as in office buildings, a uniform distributed dead load of at least 10 psf of floor area is used in calculating $W$. Typical miscellaneous items such as ducts, piping and conduits can be covered by using an additional 2 to 5 psf. In storage areas, 25 percent of the design live load is included in the seismic weight $W$. In areas of heavy snow, a load of 30 psf should be used where the snow load is greater than 30 psf. However it may be reduced to as little as 7.5 psf when approved by building officials.

In addition to determining the overall weight $W$, it is necessary to evaluate tributary weight $W_x$ at each floor for both vertical and horizontal distribution of loads (Fig. 3.12).

**TABLE 3.2 $R_W$ Values for Structural Systems in Steel**

| System | Zones 3 and 4 | | Zone 2 | | Zone 1 | |
|---|---|---|---|---|---|---|
| | $R_W$ | $H$ | $R_W$ | $H$ | $R_W$ | $H$ |
| Bearing wall system | | | | | | |
| • Braced frames where bracing carries gravity loads | 6 | 160 | 6 | NL | 6 | NL |
| Building frame system | | | | | | |
| • Eccentrically braced frame (EBF) | 10 | 240 | 10 | NL | 10 | NL |
| • Ordinary braced frames | 8 | 160 | 8 | NL | 8 | NL |
| • Special concentrically braced frames | 9 | 240 | 9 | NL | 9 | NL |
| Moment resisting frame | | | | | | |
| • SMRF | 12 | NL | 12 | NL | 12 | NL |
| • OMRF | 6 | 160 | 6 | NL | 12 | NL |
| Dual Systems | | | | | | |
| • EBF + SMRF | 12 | NL | 12 | NL | 12 | NL |
| • CF + SMRF | 10 | NL | 10 | NL | 10 | NL |
| • SCBF + SMRF | 11 | NL | 11 | NL | 11 | NL |
| • SCBF + OMRF | 6 | 160 | 6 | NL | 6 | NL |

SMRF: Special Moment Resisting Frame.
EMF: Eccentric Braced Frame.
OMRF: Ordinary Moment Resisting Frame.
OBF: Ordinary Braced Frame.
SCBF: Special Concentric Braced Frame.
H: Height in feet.

Therefore, the calculations for $W$ must be done in an orderly tabular form so that overall weights as well as tributary weights can be properly accounted for.

$Z$ = Seismic zone factor with values of 0.4 in zone 4, 0.3 in zone 3, 0.2 in zone 2B, 0.15 in zone 2A, 0.075 in zone 1, and 0 in zone 0. $Z$ corresponds numerically to effective Peak Ground Acceleration (PGA) of a region, and is defined for the United States by a zoning map that is divided into regions representing five levels of ground motion (Fig. 3.13). California, being earthquake country, has the two highest PGAs of 0.4 and 0.3. Mississippi which is underlain by the New Madrid fault has a PGA of 0.3, while many midwest regions, such as Texas, have the lowest value of 0.

$C$ = a coefficient related to the fundamental period of vibration of the structure, $T$, including the site-structure response factor $S$. $C$ is given by the relation

$$C = \frac{1.25S}{T^{2/3}} \tag{3.2}$$

A lower limit $C = 0.075R_W$ is imposed because of uncertainty in predicting response for long-period structures to strong ground motions. A maximum limit of $C = 2.75$ provides a simple seismic load evaluation where it is not practical to calculate structural period and site-structure response factor $S$.

$S$ = Site coefficient with values of 1.0, 1.2, 1.5, and 2.0 as defined below:

1.0 For soil profile with either: (i) a rock-like material characterized by a shear wave velocity greater than 2500 ft/s or by other suitable means of classification; or

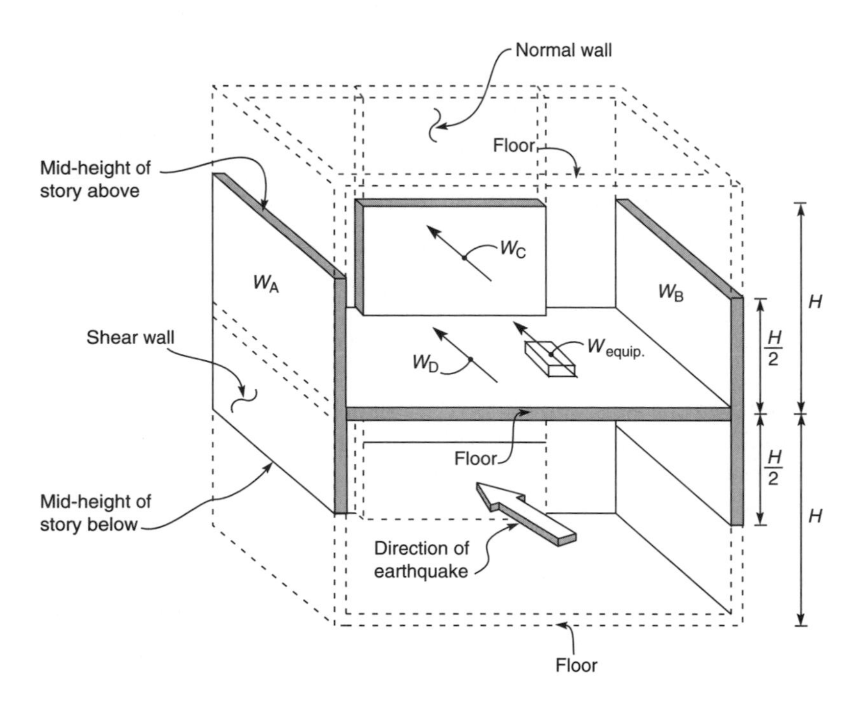


**Figure 3.12** Tributary weights for seismic load calculation.

(ii) stiff or dense soil conditions where the soil depth is less than 200 ft (soil profile type S1).

1.2 For soil profiles with dense or stiff soil conditions, where the soil depth exceeds 200 ft (soil profile type S2).

1.5 For a soil profile 70 ft or more in depth and containing more than 20 ft of soft to medium-stiff clay but not more than 40 ft of soft clay (soil profile type S3)

2.0 For a soil profile containing more than 40 ft soft clay characterized by a shear-wave velocity less than 500 ft (soil profile type S4). S was added to the 1994 UBC after the 1985 Mexico City earthquake in which the majority of damage was attributed to the resonance of the soft soil profile with building period. Buildings in the 5 to 15-story range which had natural frequencies of about 2 seconds, close to that of the soil, suffered most damage. In locations where the soil properties are not known in sufficient detail to determine the soil profile, soil profile type S3 should be used.)

$I$ = Importance Factor with four categories of occupancy: essential, hazardous, special, and standard. Essential and hazardous occupancies are assigned an Importance Factor of 1.25 while special and standard occupancies are permitted $I = 1.0$. Essential, hazardous, and special occupancies require construction observation in addition to compliance with Sections

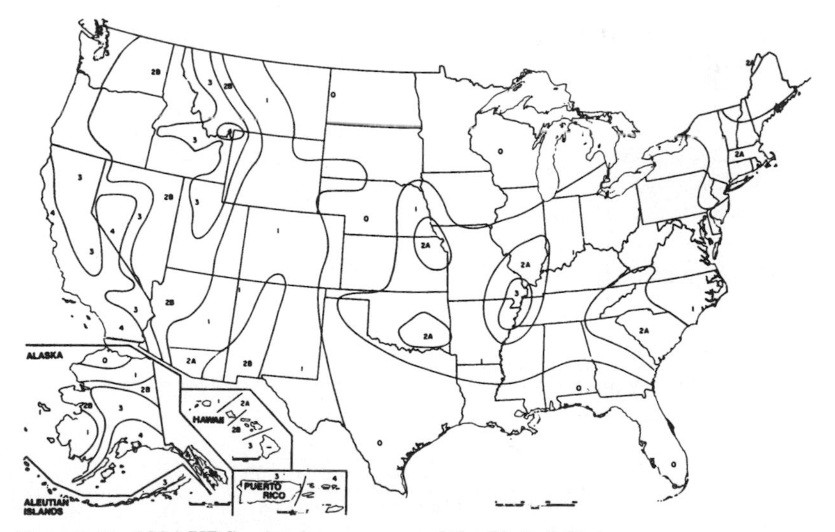


**Figure 3.13** 1994 UBC seismic zone map of the United States.

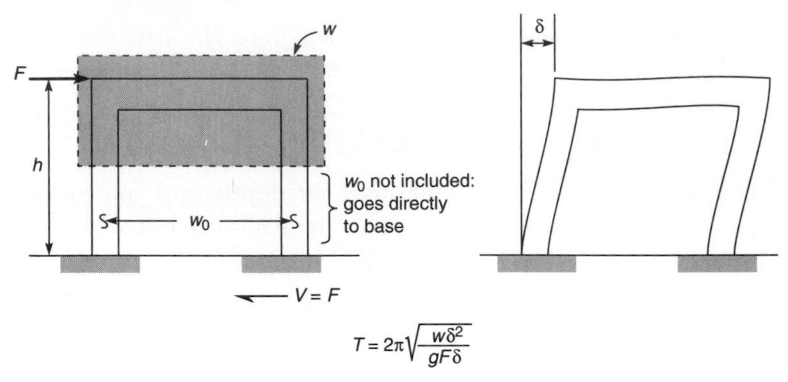


**Figure 3.14** Period calculation for a single degree-of-freedom system.

305 and 306 of the UBC, whereas standard occupancies require compliance with Sections 305 and 306 only.

$T$ = Fundamental period of vibration of the structure. Simple formulas involving only a general description of the building type and overall dimensions are given as also is a more refined formula called the Raleigh formula which can be used for all types of structures. The only restriction on the Raleigh formula is that the period used in obtaining the base shear should not exceed the period calculated by simple formulas by more than 30 percent in seismic zone 4 and 40 percent in zones 1, 2 and 3. This restriction is to prevent "excessively sharp pencil" effects which could lead to unreasonably large values for the building period $T$. Schematic representation of the Raleigh formula for period calculation are shown in Figs 3.14 and 3.15.

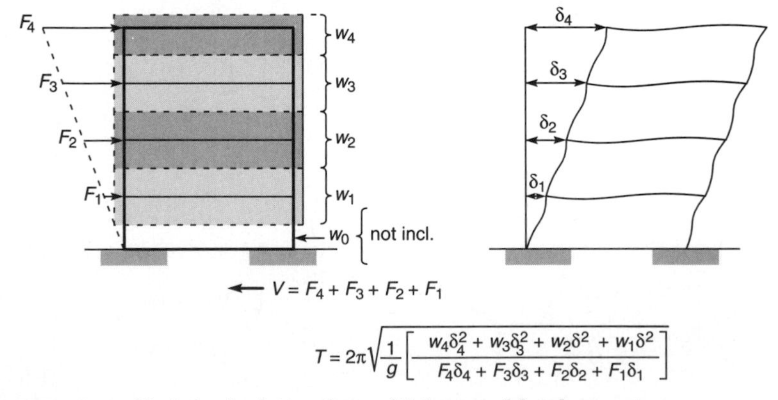


**Figure 3.15** Period calculation for multi degree-of-freedom system.

### 3.4.2 Base shear distribution along building height

The base shear Formula 3.1 does not indicate the manner in which the shear force is distributed along the building height. To proceed with the analysis, it is necessary to allocate the base shear as effective horizontal loads at various floor levels. This is given by the formula

$$F_x = \frac{(V - F_t)w_x h_x}{\sum_{i=1}^{n} w_i h_i} \tag{3.3}$$

In the above distribution, the following factors are taken into account.

1. Although the vibration of a structure at any instant during an earthquake is a combination of several modes, it is sufficiently accurate for regular buildings to consider only the fundamental mode.
2. The fundamental mode shape is assumed to be approximately linear from the base. The horizontal displacement in the first mode is, therefore, maximum at the building roof and zero at the base.
3. The maximum acceleration at any level of the structure is proportional to the horizontal displacement at that level.
4. The effective seismic load at a floor level is equal to the product of the mass assigned to that floor and the horizontal acceleration at that level.
5. For relatively stocky buildings with a uniform mass distribution over their height, the triangular distribution of seismic loads is appropriate where only the fundamental mode is significant. In slender, longer-period tall buildings, however, higher modes become significant causing a greater portion of the total base shear to act near the top. This effect, called the whip lash effect, is related to the period of the building and is reflected in the UBC, by applying a part of the total load as a concentrated horizontal force $F_t$ at the top of the building. $F_t$ is given by the relation: $F_t = 0.07\,TV$. The remainder of the base shear $(V - F_t)$ is then distributed as an inverted triangle over the height of the building.

### 3.4.3 Horizontal distribution

The horizontal distribution of weight at each floor level is required in order to calculate the eccentricity between the center of mass and the center of gravity of diaphragm forces. The weight $W$ of the diaphragm and the elements tributary thereto includes the weight of floor system, tributary weights of walls and partitions, and other elements attached to the diaphragm. The weights of shear walls and frames and items such as cladding attached thereto, that act in the same direction under consideration for the diaphragm, need not be included in the weight of the diaphragm. However, if there is a vertical discontinuity in these walls and frames, their weights should be included in the diaphragm design. The load distribution for a diaphragm generally consists of a combination of a uniform and concentrated weights.

### 3.4.4 Torsion

Horizontal torsional moments at any given story are taken as the product of the story shear and an eccentricity resulting from the calculated center of mass and center of rigidity of the story, and an accidental eccentricity of 5 percent of the plan dimension of the building perpendicular to the direction of the force being considered. The additional accidental torsion is to account for possible errors in evaluating stiffness of structural elements and the distribution of mass at floor levels. Where torsional irregularities exist, the accidental eccentricity is to be increased by an amplification factor $A_x$, relating the maximum story drift, $\delta_{\text{max}}$, at one end of the structure to the average of the story drifts, $\delta_{\text{ave}}$, of the two ends of the structure (See Fig. 3.16c)

$$A_x = \left[\frac{\delta_{\text{max}}}{1.2\delta_{\text{ave}}}\right]^2 \tag{3.4}$$

### 3.4.5 Story shear and overturning moments

The story shear at any level is the sum of all the lateral forces at and above that level. The overturning moment at any level is the sum of the moments of the seismic story shears above that level. The overturning moments and story shears are distributed to various lateral load-resisting elements in proportion to their rigidities. Usually for buildings with concrete floor slabs or metal deck and concrete topping, the distribution is made assuming the floor slabs to be completely rigid. However, in very long and narrow buildings it may be necessary to compare the deformation of floor diaphragm with the average drift of the associated story to determine whether the diaphragm is flexible or rigid. Diaphragms are considered flexible when the maximum lateral deformation of the diaphragm is more than two times the average story drift of the associated story. This is determined by comparing the computed in-plane midpoint deflection of adjoining vertical resisting elements under equivalent tributary lateral load.

### 3.4.6 Discontinuity in lateral force-resisting elements

Overturning moments on discontinuous shear-resisting elements are required to be carried down to the foundation. In zones 3 and 4,

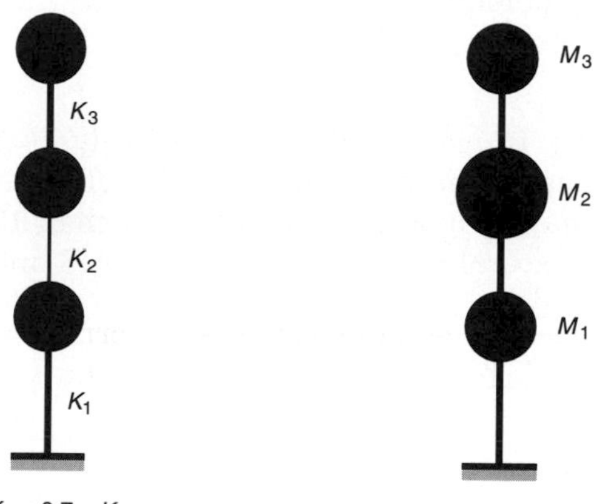


$K_2 < 0.7 \times K_3$

$K_2 < 0.8\ (K_3 + K_4 + K_5)/3$

(a)

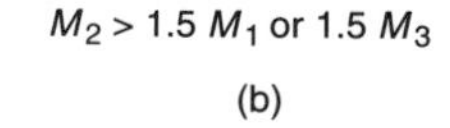


(b)

**Figure 3.16a** Stiffness irregularity.

**Figure 3.16b** Mass irregularity.

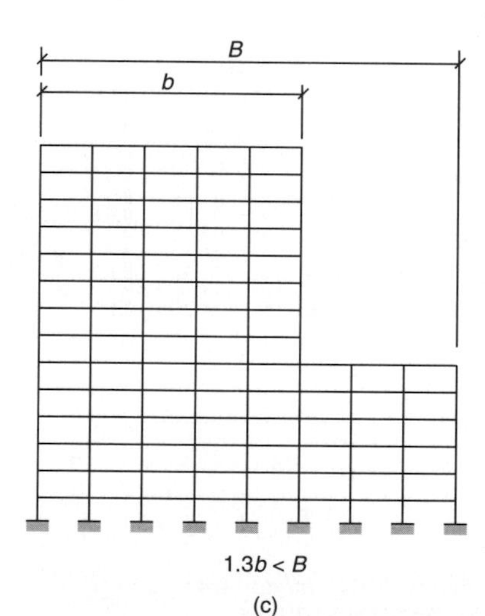


**Figure 3.16c** Vertical geometric irregularity.

columns supporting discontinuous shear walls or braces are to be designed to carry:

$$1.0DL + 0.8LL + \tfrac{3}{8}R_W E \quad (3.5)$$

$$0.85DL \mp \tfrac{3}{8}R_W E \quad (3.6)$$

These load combinations are considered as ultimate loads. Therefore, for steel columns designed using working-stress methods the ultimate capacity is determined by increasing the allowable stress by a factor of 1.7. The columns supporting discontinuous elements must also meet special detailing requirements.

### 3.4.7 *p*-delta effects

In evaluating overall structural frame stability, in general, it is necessary to consider the $p\Delta$ effects. The moment induced by the $p\Delta$ effect is a secondary effect and may be ignored when it is less than 10 percent of the primary action of lateral loads. In seismic zones 3 and

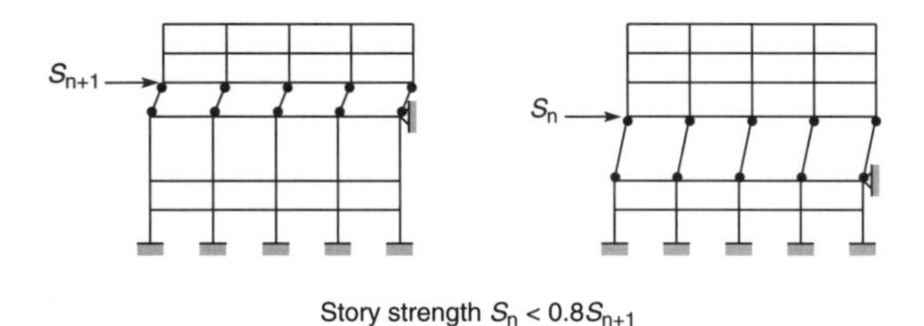


**Figure 3.16d** Discontinuity in capacity (weak story)

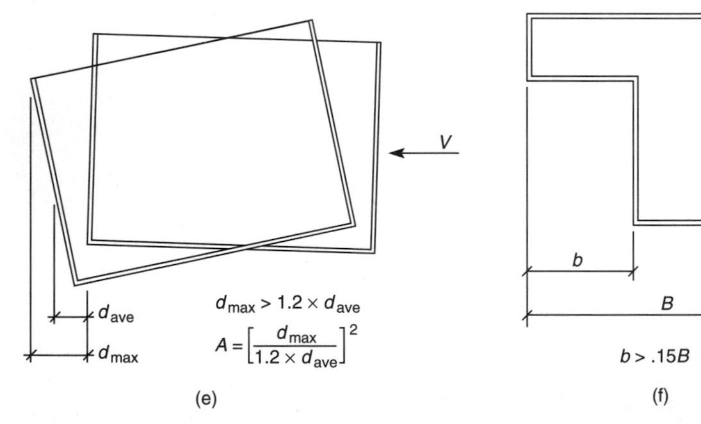


**Figure 3.16e** Torsional irregularity.

**Figure 3.16f** Irregularity due to reentrant corners.

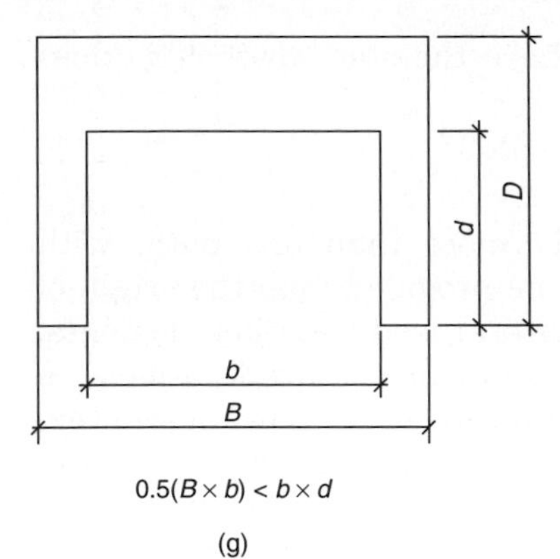


**Figure 3.16g** Irregularity due to diaphragm discontinuity.

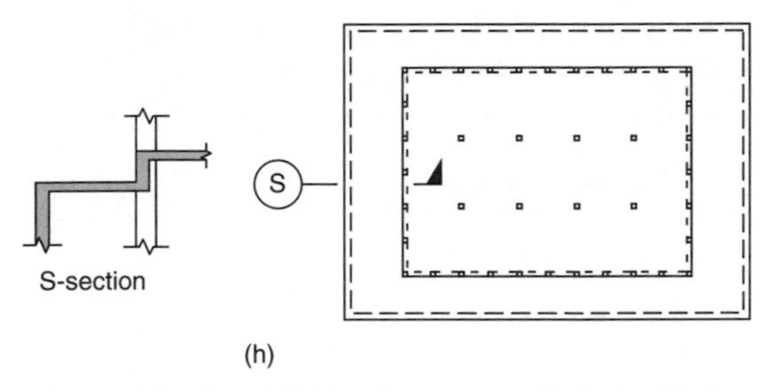


**Figure 3.16h** Irregularity due to out-of-plane offsets.

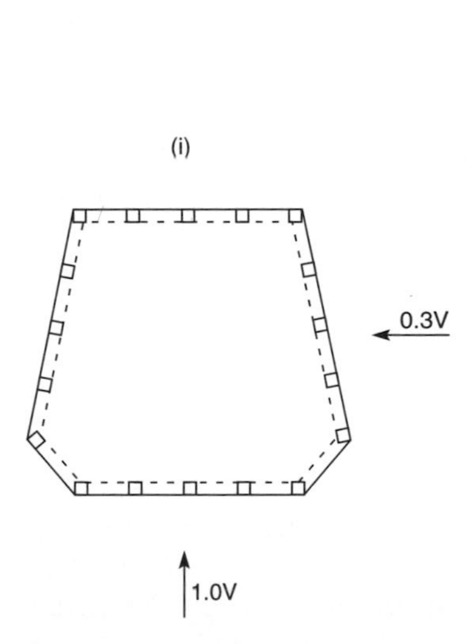


**Figure 3.16i** Irregularity due to nonparallel system.

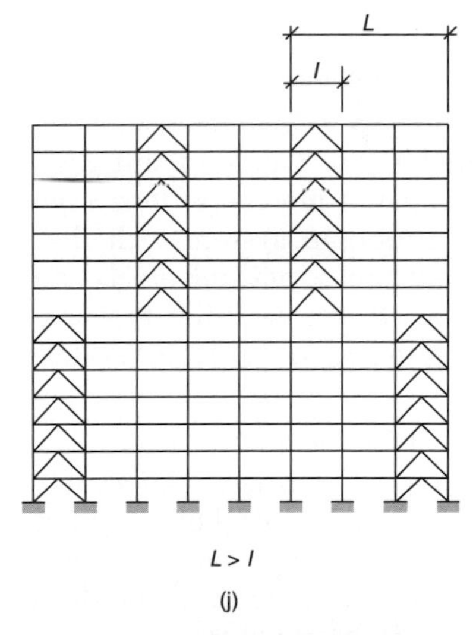


**Figure 3.16j** In-plane discontinuity.

4, $p\Delta$ effect need not be considered where the story drift ratio does not exceed $0.02/R_W$.

### 3.4.8 Continuous load path

A continuous load path, or preferably more than one path, with adequate strength and stiffness should be provided from the origin of initial load manifestation to the final lateral load resisting elements. Because it is one of the most fundamental considerations in earthquake-resistant design, this requirement is now stated explicitly at the beginning of the 1996 Blue Book, the sixth edition of Recommended Lateral Force Requirements and commentary prepared by the Seismology Committee, Structural Engineers Association of California.

The general path for load transfer, in a conceptual sense, is in reverse to the direction in which seismic loads are delivered to the structural elements. Thus the path for load transfer is as follows: inertia forces generated in an element such as a segment of exterior curtain wall, are delivered through structural connections to a horizontal diphragm (i.e., floor slab or roof); the diaphragms distribute these forces to vertical components such as moment frames, braces, and shear walls; and finally the vertical elements transfer the forces into the foundations. While providing a continuous load path is an obvious requirement, examples of common flaws in load paths are; a missing collector, or a discontinuous chord because of an opening in the floor diaphragm, or a connection that is inadequate to deliver diaphragm shear to a frame or shear wall.

### 3.4.9 Redundancy

Redundancy is a fundamental characteristic for good performance in earthquakes. It tends to mitigate high demands imposed on the performance of members. It is a good practice to provide a building with a redundant system such that failure of a single connection or component does not adversely affect the lateral stability of the structure. Otherwise, all components must remain operative for the structure to retain its lateral stability.

### 3.4.10 Configuration

A building with an irregular configuration may be designed to meet all code requirements but it will not perform as well as a building with a regular configuration. If the building has an odd shape that is not properly considered in the design, good details and construction are of a secondary value.

Two categories of structural irregularities are defined in the UBC:

**TABLE 3.3 Vertical Structural Irregularities**

| Irregularity type and definition | Reference Fig. |
|---|---|
| 1. Stiffness irregularity—soft story<br>A soft story is one in which the lateral stiffness is less than 70 percent of that in the story above or less than 80 percent of the average stiffness of the three stories above. | Fig. 3.16a |
| 2. Weight (mass) irregularity<br>Mass irregularity shall be considered to exist where the effective mass of any story is more than 150 percent of the effective mass of an adjacent story. A roof which is lighter than the floor below need not be considered. | Fig. 3.16b |
| 3. Vertical geometric irregularity<br>Vertical geometric irregularity shall be considered to exist where the horizontal dimension of the lateral force-resisting system in any story is more than 130 percent of that in an adjacent story. One-story penthouses need not be considered. | Fig. 3.16c |
| 4. In-plane discontinuity in vertical lateral-force-resisting element<br>An in-plane offset of the lateral load-resisting elements greater than the length of those elements. | Fig. 3.16j |
| 5. Discontinuity in capacity—weak story<br>A weak story is one in which the story strength is less than 80 percent of that in the story above. The story strength is the total strength of all seismic-resisting elements sharing the story shear for the direction under consideration. | Fig. 3.16d |

vertical and plan irregularities. Each of these is further defined by five different types as shown in Tables 3.3 and 3.4, and as illustrated in Figs 3.15a–j.

Vertical and plan irregularities result in building responses significantly different from those assumed in the equivalent static force procedure. Although the code gives certain recommendations for assessing the degree of irregularity, and corresponding penalties and restrictions, it is important to understand that these recommendations are not an endorsement of their design; rather, the intent is to make the designer aware of the potential detrimental effects of irregularities.

### 3.4.11 Design example

Before illustrating a design example, for convenience, the UBC 1994 provisions for determining the base shear are given again here in a summary format. $V$ is the total lateral force or shear at the base. $W$ is the total seismic dead load.

$$V = \frac{ZIC}{R_W} W.$$

$Z$ = 0.4 in zone 4, 0.3 in zone 3, 0.2 in zone 2B, 0.15 in zone 2A, 0.075 in zone 1 and 0 in zone 0.

$I$ = 1.25 for essential facilities and hazardous facilities,

= 1.0 for special occupancy structures and standard occupancy structures.

$$C = \frac{1.25S}{T^{2/3}} \geqslant 0.075.$$

$\leqslant 2.75$ (See Table 3.5 for the calculation of period $T$.)

$S$ = 1.0 for a soil profile with either: (i) a rock-like material characterized by a shear wave velocity greater than 2500 ft/s or by other suitable means of classification; or (ii) stiff or dense soil condition where the soil depth is less than 200 ft (soil profile type S1).

$S$ = 1.2 for a soil profile with dense or stiff soil conditions, where the depth of soil exceeds 200 ft (soil profile type S2).

$S$ = 1.5 for a soil profile 70 ft or more in depth and containing more than 20 ft of soft to medium stiff clay but not more than 40 ft of soft clay (soil profile type S3).

$S$ = 2.0 for a soil profile containing more than 40 ft of soft clay characterized by a shear-wave velocity less than 500 ft (soil profile type S4).

In locations where the soil properties are not known in sufficient detail to determine the soil profile, type S3 should be used.

$R_W$ = Response modification factor. See Tables 3.1 and 3.2.

### Given

A 14-story reinforced concrete moment-resisting frame building in downtown Los Angeles with the following characteristics:

Building height $h_n$ = 136 ft

The story heights are 12.5 ft for the first floor and 9 ft 6 in. for typical floors.

Fundamental period $T = 0.030(h_n)^{3/4} = 0.030(136)^{3/4} = 1.195$ sec

Seismic zone factor $Z = 0.4$

Importance factor $I = 1.0$

Response modification factor $R_W = 12.0$ (special moment-resisting frame, SMRF)

Soil factor $S = 1.2$

**TABLE 3.4 Plan Structural Irregularities**

| Irregularity type and definition | Reference Fig. |
|---|---|
| 1. Torsional irregularity—to be considered when diaphragms are not flexible<br>Torsional irregularity shall be considered to exist when the maximum story drift, computed including accidental torsion, at one end of the structure transverse to an axis is more than 1.2 times the average of the story drifts of the two ends of the structure. | Fig. 3.16e |
| 2. Reentrant corners<br>Plan configurations of a structure and its lateral force-resisting system contain reentrant corners, where both projections of the structure beyond a reentrant corner are greater than 15 percent of the plan dimension of the structure in the given direction. | Fig. 3.16f |
| 3. Diaphragm discontinuity<br>Diaphragms with abrupt discontinuities or variations in stiffness, including those having cutout or open areas greater than 50 percent of the gross enclosed area of the diaphragm, or changes in effective diaphragm stiffness of more than 50 percent from one story to the next. | Fig. 3.16g |
| 4. Out-of-plane offsets<br>Discontinuities in a lateral force path, such as out-of-plane offsets of the vertical elements. | Fig. 3.16h |
| 5. Nonparallel systems<br>The vertical lateral load-resisting elements are not parallel to or symmetric about the major orthogonal axes of the lateral force-resisting system. | Fig. 3.16i |

**TABLE 3.5 Fundamental Period of Vibration: UBC 1994**

| | |
|---|---|
| Moment-resisting frames | Steel $T_a = 0.035h_n^{3/4}$<br>Concrete $T_a = 0.030h_n^{3/4}$ |
| Other structural systems | $T_a = C_t h_n^{3/4}$, $C_t = 0.020$ or $C_t = 0.1/\sqrt{A_c}$<br>$A_c = \sum A_e\left[0.2 + \left(\dfrac{D_e}{h_n}\right)^2\right]$<br>$\dfrac{D_e}{h_n} \leqslant 0.9$ |
| All structural systems | $T = 2\pi\sqrt{\sum w_1 \delta_1^2 / g \sum w_i \delta_i}$<br>Ritz formula |

$A_c$ = combined effective area, in square feet, of shear walls in the first story.
$A_e$ = the minimum cross-sectional shear area in any horizontal plane in the first story, in square feet, of a shear wall.
$h_n$ = building height, in feet, above the shear base.
$D_e$ = the length, in feet, of shear wall in the first story in the direction parallel to the applied forces.

**Required**

Floor-by-floor seismic shears using UBC static procedure.

$$\text{Coefficient } C = \frac{1.25 \times 5}{T^{2/3}} = 1.33$$

Building seismic weight $W = 20873$ kips

$$\text{Base shear } V = \frac{ZICW}{R_W} = \frac{0.4 \times 1.33 \times 20873}{12} = 925 \text{ kips}$$

Top level additional force

$$F_t = 0.07TV = 0.07 \times 1.195 \times 925 = 77 \text{ kips}$$

$$V - F_t = 925 - 77 = 848 \text{ kips}$$

Story weight for calculation of lateral forces is the tributary weight of all elements located between two imaginary parallel planes passing the midheight of the columns below and above the story (see Fig. 3.12). It typically includes the weight of the complete floor system plus one-half the weight of story walls, columns and cladding above floor level and one-half the weight of the story walls and columns and cladding below the floor level. If partitions are laterally supported top and bottom, their weight is divided between the floor levels; however, if the partitions are free standing, the total weight is included with the supporting level.

Before using the static procedure we must verify that the structure is regular and has no significant physical discontinuity in plan or vertical configuration or in its lateral force-resisting systems. We assume the irregular features described below do not occur in this building.

**Weak story.** There are no significant strength discontinuities in any of the vertical elements in the lateral force-resisting system. The story strength at any story is not less than 80 percent of the strength of the story above.

**Soft story.** There are no significant stiffness discontinuities in any of the vertical elements in the lateral force-resisting system; the lateral stiffness of a story is not less than 70 percent of that in the story above or less than 80 percent of the average stiffness of the three stories.

**Geometry.** There are no significant geometrical irregularities. There are no setbacks, i.e., no changes in horizontal dimension of the lateral force-resisting system of more than 30 percent in a story relative to the adjacent stories.

**Mass irregularity.** There are no significant mass irregularities; there are no changes in effective mass of more than 50 percent from one story to the next, excluding light roofs. The effective mass consists of the dead weight of the floor plus the actual weights of partitions and equipment.

**Vertical discontinuities.** There are no discontinuities in the lateral force-resisting system such as shear walls, moment frames and infilled walls.

**Torsion.** We will assume that the lateral force-resisting elements form a well-balanced system that is not subject to significant torsion.

These rather optimistic assumptions permit us to perform an equivalent static analysis for determining the seismic loads for the example building. The results are summarized in Table 3.6.

## 3.5 Dynamic Analysis Procedure

### 3.5.1 Introduction

Buildings with symmetrical shape, stiffness and mass distribution and with vertical continuity and uniformity behave in a fairly predictable manner whereas when buildings are eccentric or have areas of discontinuity or irregularity, the behavioral characteristics are very complex. In such instances dynamic analysis can be helpful in determining important seismic response characteristics that may not be evident from the static procedure, such as: (i) the effects of the structure's dynamic characteristics on the vertical distribution of lateral force; (ii) increase in the dynamic loads in the structure's lateral force resisting system due to torsional motions; and (iii) the effects of higher modes that could substantially increase story shears and deformations.

The 1994 Uniform Building Code static method is based on a single mode response with approximate load distributions and corrections for higher mode response. These simplifications are appropriate for simple regular structures. However they do not consider the full range of seismic behavior in complex structures.

Therefore dynamic analysis is required for buildings with unusual or irregular geometry, since it results in distributions of seismic design forces that are in more agreement with the actual distribution of mass and stiffness of the building.

A structure is considered irregular if it has any of the characteristics given in Tables 3.3 and 3.4, and Figs 3.16.

According to the UBC, the dynamic procedure is required for all

**TABLE 3.6 Design Example: Calculation of Seismic Loads, Static Procedure UBC 1994**

| Level (1) | $h$ ft (2) | $\Delta h$ ft (3) | $w$ kips (4) | $\sum w$ kips (5) | $(2)\times(4)$ $wh$ (6) | $\frac{wh}{\sum wh}$ (7) | $\frac{(V-F_t)\times(7)}{F}$ kips (8) | $\sum(8)$ kips (9) | $(3)\times(9)$ $\Delta OTM$ $K$-ft (10) | $\sum(10)$ $OTM$ $K$-ft (11) | $(9)\div(5)$ $\frac{F_t+\sum F_i}{\sum w_i}$ (12)* |
|---|---|---|---|---|---|---|---|---|---|---|---|
| $R$ | 136.0 | 9.5 | 1424 | 1424 | 193,664 | 0.125 | $F_t = 77$ 106 | 183 | 1739 | 1739 | 0.129 |
| 14 | 126.5 | 9.5 | 1494 | 2918 | 188,991 | 0.122 | 103.5 | 286.5 | 2722 | 4461 | 0.098 |
| 13 | 117.0 | 9.5 | 1494 | 4412 | 174,798 | 0.113 | 95.8 | 382.3 | 3632 | 8093 | 0.087 |
| 12 | 107.5 | 9.5 | 1494 | 5906 | 160,605 | 0.104 | 88.2 | 470.5 | 4470 | 12,563 | 0.08 |
| 11 | 98.0 | 9.5 | 1494 | 7400 | 146,412 | 0.095 | 80.6 | 551.1 | 5235 | 17,798 | 0.074 |
| 10 | 88.5 | 9.5 | 1494 | 8894 | 132,219 | 0.086 | 72.9 | 624 | 5928 | 23,726 | 0.070 |
| 9 | 79.0 | 9.5 | 1494 | 10,388 | 118,026 | 0.076 | 66.4 | 690.4 | 6559 | 30,285 | 0.066 |
| 8 | 69.5 | 9.5 | 1494 | 11,882 | 103,833 | 0.067 | 56.8 | 747.2 | 7098 | 37,383 | 0.063 |
| 7 | 60.0 | 9.5 | 1494 | 13,376 | 89,640 | 0.058 | 49.2 | 796.4 | 7566 | 44,949 | 0.060 |
| 6 | 50.5 | 9.5 | 1494 | 14,870 | 75,447 | 0.049 | 41.6 | 838 | 7961 | 52,910 | 0.056 |
| 5 | 41.0 | 9.5 | 1494 | 16,364 | 61,254 | 0.040 | 33.9 | 871.9 | 8283 | 61,193 | 0.053 |
| 4 | 31.5 | 9.5 | 1494 | 17,858 | 47,061 | 0.030 | 25.4 | 897.3 | 8524 | 69,717 | 0.050 |
| 3 | 22.0 | 9.5 | 1494 | 19,352 | 32,868 | 0.021 | 17.8 | 915.1 | 8693 | 78,410 | 0.047 |
| 2 | 12.5 | 12.5 | 1521 | 20,873 | 19,013 | 0.012 | 10.2 | 925.3 | 11,566 | **89,976 | 0.044 |
| Ground | 0 | | | | | | | | | | |
| $\Sigma$ | | | 20,873 | | 1,543,830 | 0.998 | 925 | | **89,976 | | |

* For use in diaphragm design.
** For foundation overturning, this value may be reduced for regular buildings by 10,472 kip-ft (77 × 136) when $F_t$ is neglected. See 1994 UBC Section 1809.4.

structures not specifically listed under the static procedure including the following.

- For tall buildings over 240 ft in height. This is to determine the possible effects of higher mode response on force distribution and deformations.
- For irregular structures (Tables 3.3, 3.4, and Figs 3.16) because the cited irregularities invalidate the assumptions applicable for equivalent static analysis.
- Regular or irregular structures located on soil type S4 with fundamental periods greater than 0.7 sec. This is because there is concern about the occurrence of extensive damage to structures located on type S4 sites due to the resonance response of the type observed in Mexico City during the September 1985 earthquake.

Determining the behavior of a structure in an earthquake is basically a vibration problem. Using dynamic analysis, calculations can be made of the earthquake induced vibrations which will indicate the general nature and amplitude of deformations expected.

Two methods of dynamic analysis are permitted: (i) an elastic response spectrum analysis and (ii) an elastic or inelastic time-history analysis. A majority of engineers use the response spectrum analysis. The time-history procedure is used if it is important to represent inelastic response characteristics or to incorporate time-dependent effect when computing the structure's dynamic response.

Structures that are built into the ground and extended vertically some distance above the ground respond either as simple or complex oscillators when subjected to seismic ground motions. Simple oscillators are represented by single-degrees-of-freedom systems (SDOF), and complex oscillators are represented by multi-degree-of-freedom (MDOF) system.

A simple oscillator is represented by a single lump of mass on the upper end of a vertically cantilevered pole or by a mass supported by two columns as shown in Fig. 3.17. The idealized system represents

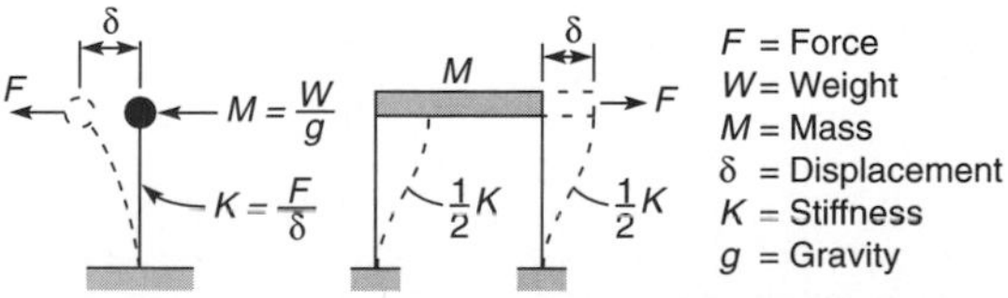

**Figure 3.17** Idealized single degree-of-freedom system.

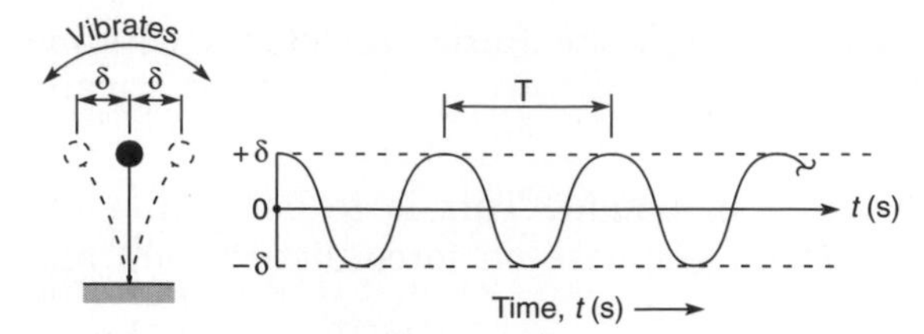


**Figure 3.18** Undamped free vibrations of a single degree-of-freedom system.

two kinds of structures: (i) a single-column structure with a relatively large mass at its top; and (ii) a single story frame with flexible columns and a rigid beam. The mass $M$ is the weight $W$ of the system divided by the acceleration of gravity $g$. i.e., $M = W/g$.

The stiffness $K$ of the system, which is a ratio equal to a horizontal force $F$ applied to the mass divided by the corresponding displacement $\Delta$. If the mass is deflected and then suddenly released, it will vibrate at a certain frequency, which is called its natural or fundamental frequency of vibration. The reciprocal of frequency is called the period of vibration. It represents the time for the mass to move through one complete cycle. The period $T$ is given by the relation:

$$T = 2\pi\sqrt{\frac{M}{K}} \tag{3.7}$$

In an ideal system having no damping ($\beta = 0$), the system would vibrate forever (Fig. 3.18). In a real system where there is always some damping, the amplitude of motion will gradually decrease for each cycle until the structure comes to a complete stop (Fig. 3.19). The system responds in a similar manner if, instead of displacing the mass at the top, a sudden impulse is applied to the base of the system.

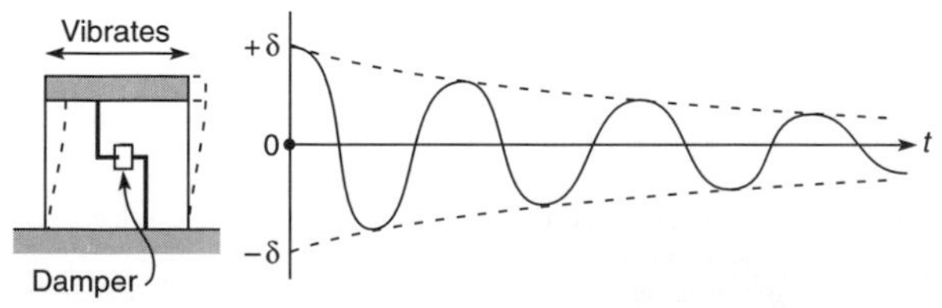


**Figure 3.19** Damped free vibration of a single degree-of-freedom system.

Tall buildings may be anlayzed as multi-degree-of-freedom systems by lumping story-masses at intervals along the length of a vertically cantilevered pole. During vibration, each mass will be deflected in one direction or another. For example, for higher modes of vibration, some masses may go to the right while others are going to the left. Or all masses may simultaneously deflect in the same direction as in the fundamental mode of vibration. An idealized MDOF system has a number of modes equal to the number of masses. Each mode has its own natural period of vibration with a unique mode shape formed by a line connecting the deflected masses. When ground motion is applied to the base of the multi-mass system, the deflected shape of the system is a combination of all mode shapes; but modes having periods near predominant periods of the base motion will be excited more than the other modes. Each mode of a multi-mass system can be represented by an equivalent single-mass system having generalized values $M$ and $K$ for mass and stiffness. The generalized values represent the equivalent combined effects of story masses $m_1, m_2, \ldots$ and stiffness $k_1, k_2 \ldots$. This concept, shown in Fig. 3.20, provides a computational basis for using response spectra based on single-mass systems for analyzing multi-storied buildings. Given the period, mode shape, and mass distribution of a multi-storied building, we can use the response spectra of single-degree-of-freedom system for computing the deflected shape, story accelerations, forces and overturning

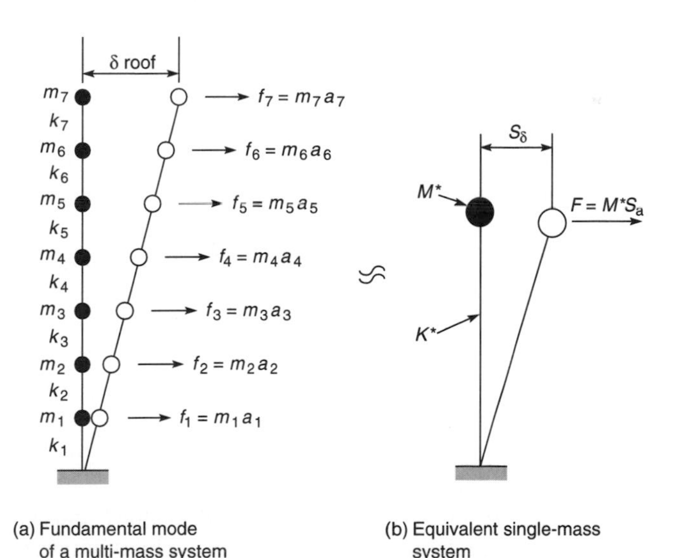


**Figure 3.20** Mathematical representation of a multi-mass system by a single-mass system.

moments. Each predominant mode is analyzed separately and the results are combined statistically to compute the multi-mode response.

Buildings with symmetrical shape, stiffness and mass distribution and with vertical continuity and uniformity behave in a fairly predictable manner whereas when buildings are eccentric or have areas of discontinuity or irregularity, the behavioral characteristics are very complex. The predominant response of the building may be skewed from the apparent principal axes of the building. The torsional response as well as the coupling or interaction of two translational directions of response must be considered. This is similar to the Mohr's circle theory of principal stresses.

Thus, three-dimensional methods of analysis are required as each mode shape is defined in three dimensions by the longitudinal and transverse displacement and the rotation about a vertical axis. Thus building irregularities complicate not only the method of dynamic analysis but also the methods used to combine modes.

For a building that is regular and essentially symmetrical, a two-dimensional model is generally sufficient for the modal analysis of the structure subject to ground motions in each of its principle axis. Note that when the aspect ratio (length-to-width) of the diaphragm is large, torsional response may be predominant thus requiring a 3-D analysis in an otherwise symmetrical and regular building.

The building is modeled as a system of masses lumped at each floor level, each mass having one-degree-of-freedom, that of lateral displacement in the direction under consideration. The weights used in computing the masses are those prescribed in the static procedure. The analysis will include, for each principal axis, all significant modes of vibration. The relative significance of higher modes will be determined by the values of modal participation factors and modal spectral accelerations.

When a structure is unsymmetrical in plan, has discontinuities in the vertical or horizontal planes, large plan aspect ratios, flexible horizontal diaphragms or other irregularities, a three-dimensional model is required. In a three-dimensional analysis, at each floor level there will be three-degrees-of-freedom. The primary displacement generally occurs in a direction parallel to the direction of the ground motion. There will also be a displacement component normal to this direction and rotation about the vertical axis of the building.

For moderate to high-rise buildings, the effects of higher modes may be significant. For a fairly uniform building, the dynamic characteristics can be approximated by using the general modal

**TABLE 3.7 General Modal Relationships**

| Mode | 1 | 2 | 3 | 4 | 5 |
|---|---|---|---|---|---|
| Ratio of period to 1st mode period | 1.000 | 0.327 | 0.186 | 0.121 | 0.083 |
| Participation factor at roof | 1.31 | −0.47 | 0.24 | −0.11 | 0.05 |
| Base shear participation factor | 0.828 | 0.120 | 0.038 | 0.010 | 0.000 |

relationship shown in Table 3.7. The fundamental period of vibration of the buildings may be estimated by using the code formulas. Approximate periods for the second through fifth modes of vibration, and the roof and base participation factors, defined presently, may be estimated by using the relationship shown in Table 3.7. For example, the second and third mode periods are equal to 0.327 and 0.186 times the fundamental mode.

For most buildings, an inelastic response can be expected to occur during a major earthquake. Although nonlinear inelastic programs are available, they are not representative of typical design practice because: (i) their proper use requires special background; (ii) results produced are difficult to interpret and apply to traditional design criteria; and (iii) the necessary computations are expensive. Therefore, analyses used in practice are essentially based on linear elastic analysis. One such method called the response spectrum method is the most widely used approach.

### 3.5.2 Response spectrum method

The word "spectrum" expresses the idea that a broad range of quantities is summarized in one graph. For a given earthquake and percentage of critical damping, the graph shows related quantities such as acceleration, velocity, or deflection for a complete range or spectrum of building periods.

The plot of a response spectrum (Figs 3.21 and 3.22), may be visualized as a response of a series of progressively longer cantilever pendulums with increasing natural periods subjected to a common lateral motion of the base. Imagine the common base being moved through a ground motion corresponding to that occuring in a given earthquake. A plot of maximum response, such as acceleration versus the period of the pendulums will provide the acceleration response spectrum as shown in Fig. 3.12. The absolute value of the peak

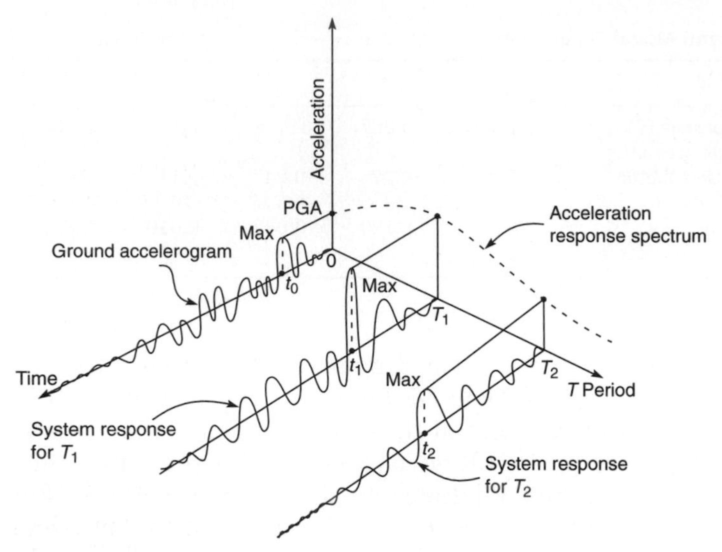


**Figure 3.21** Graphical description of response spectrum.

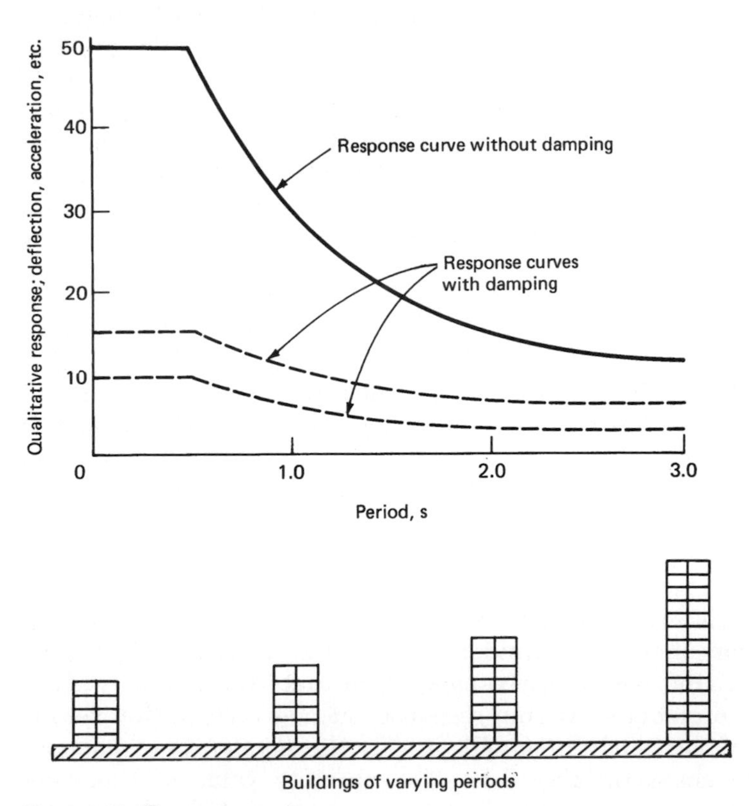


**Figure 3.22** Response spectrum.

acceleration response occurring during the excitation for each pendulum is represented by a point on the acceleration spectrum curve. In Fig. 3.23, the response spectra for the 1940 El Centro earthquake is illustrated. Using the ground acceleration as input, a family of response spectrum curves can be generated for various levels of damping, where higher values of damping results in lower spectral response.

In order to establish the concept of the response spectrum method, let us consider a single-degree-of-freedom structure such as an elevated water tank supported on columns or a revolving restaurant supported at the top of a tall concrete core. These structures can be adequately modeled as single-degree-of-freedom structures by considering the columns and the core as flexible cantilevers and the tank or the restaurant as the only mass at the tip of the cantilever; the mass of the columns or core is ignored. Let us assume that we are required to design the structures for an earthquake which will have the same characteristics as the 1940 El Centor earthquake. The recorded ground acceleration for the first crucial 30 seconds of the earthquake is shown in Fig. 3.2. The maximum acceleration recorded is 0.33 $g$, which occurred at about 2 seconds after the start of the record.

To design the building, the base of the structure is subjected to the same acceleration as the El Centro recorded acceleration. The purpose is to calculate the maximum acceleration experienced by the mass during the first 30 seconds of the earthquake. It is possible to obtain the maximum response such as displacement, velocity, and acceleration of a single-degree-of-freedom system by assuming that the ground motion corresponds to a series of impulsive loads. The maximum response can be obtained by integrating the effect of individual impulses over the duration of the earthquake. This procedure is called *Duhamel integration* and is widely used for obtaining maximum response to earthquakes. In the seismic analysis of buildings it is generally not necessary to carry out the complicated integration procedures because the maximum response values for many recorded earthquakes are already established. For example, the acceleration response for the north–south component of the El Centro earthquake is shown in Fig. 3.23.

To design the two structures mentioned earlier, assume the tank and restaurant structures weigh 720 kips (3202 kN) and 2400 kips (10,675 kN), with corresponding periods of vibration of 0.5 and 1 second, respectively. Since the response of the structure is strongly influenced by the damping factor, it is necessary to estimate damping factor for the two structures. Let us assume the damping factor for the two structures are 0.05 and 0.1, respectively. From Fig. 3.23, the acceleration for the water tank structure is 26.25 ft/s$^2$, giving a horizontal force in kips equal to the mass of the tank, $w/g$, times the acceleration

$$F = \frac{720}{32.2} \times 26.25 = 587 \text{ kips}$$

The acceleration for the restaurant structure from Fig. 3.23 is 11.25 ft/s$^2$, and the horizontal force in kips would be equal to the lumped mass at the top of the core times the acceleration, or $2400 \times 11.25/32.2 = 838.51$ kips (3730 kN). The two structures can then be designed by applying horizontal forces at the top and determining the associated forces and moments. The lateral load, which is obtained by multiplying the response spectrum acceleration by the lumped mass of the system, is referred to as the *base shear,* and its evaluation forms one of the major tasks in earthquake analysis.

In the previous example a single-degree-of-freedom example was chosen to illustrate the underlying principle of spectrum analysis. Multistory buildings, however, cannot be modeled as single-degree-of-freedom systems and their analysis is necessarily more complicated than the previous examples.

A multistory building will have as many modes of vibration as it has degrees of freedom. The use of lumped mass models to represent the actual distributed mass of a structure is a convenient tool for

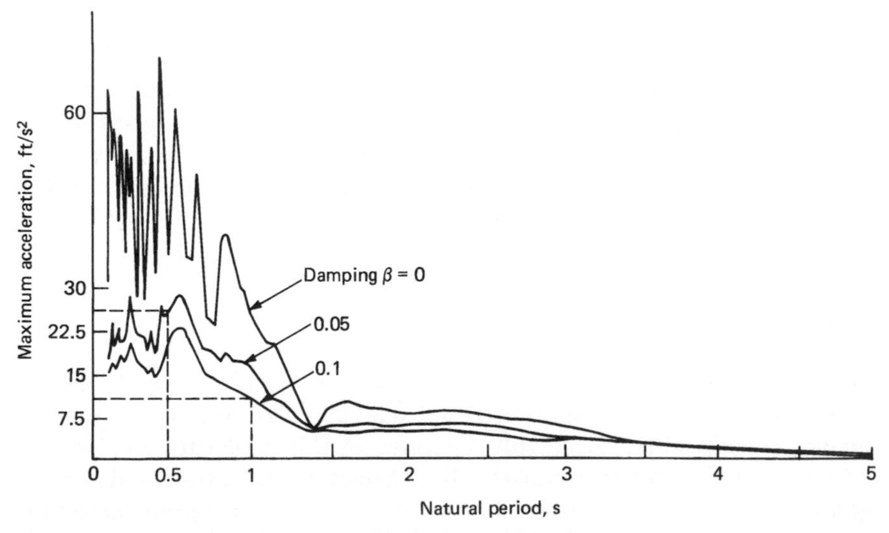


**Figure 3.23** Acceleration spectrum: El Centro earthquake.

reducing degrees of freedom to a manageable few. In multistory buildings it is generally sufficient to assume the masses as concentrated at the floor levels and to formulate the problem in terms of these masses.

Assuming that masses are concentrated at each level, for a planar analysis the number of modes of vibration corresponds to the number of levels in the multistory structure. Each mode of vibration has its own characteristic frequency or period of vibration. The actual motion of a tall building at any instant is a unique linear combination of its natural modes of vibration. During vibration, the masses of the structure vibrate in phase with displacements as measured from its initial position, always having the same relationship to each other. All masses vibrating in one of the natural modes pass the equilibrium position at the same time and reach their extreme positions at the same instant.

Using certain simplifying assumptions, it can be shown that each mode of vibration behaves as an independent single-degree-of-freedom system with a characteristic frequency. The assumptions required for and the proof of the proposition will not be attempted in this section because it is explained in subsequent sections. For now, suffice it to note that this method called the *modal superposition method,* consists of obtaining the total response of the building by appropriately combining the appropriate modes of vibration.

Since a multistory building has several degrees of freedom, in general it vibrates with as many different mode shapes and periods as it has degrees of freedom. Each mode of vibration contributes to the base shear, and for elastic action of the structure, this modal base shear can be determined by multiplying an effective mass by an acceleration read from the response spectrum for the period of that mode and for the assumed damping. Therefore, the procedure for determining the base shear for each mode of a multi-degree-of-freedom structure is the same as that for determining the base shear for a single-degree-of-freedom structure except that an effective mass is used instead of the total mass. The effective mass is a function of the actual mass at each floor and the deflection at each floor and is greatest for the fundamental mode and becomes progressively less for higher modes. The mode shape must therefore be known in order to compute the effective mass.

Since the actual deflected shape of the building consists of linear combinations of the modal shapes, higher modes of vibration also contribute, though to a lesser degree, to the structural response. These can be taken into account by using the concept of a participation factor. Further mathematical explanation of this concept is deferred to a later section, but suffice it to note that the base shear, for each mode is determined as the summation of products of effective mass and spectral acceleration. The force at each level for each mode is obtained by distributing the base shear in proportion to the product of the floor weight and displacement. The design values are then computed by using modal combination methods, such as complete quadratic combination or the square root of sum of the squares.

### 3.5.3 Development of design response spectrum

In practice it is rare that a structural engineer is called upon to develop a design response spectrum. However, it is important to understand the various assumptions used in their development. Three basic types of response spectrum are used in practice. These are:

1. Response spectrum from actual earthquake records.
2. Smoothed design response spectrum.
3. Site-specific response spectrum.

**Response spectrum from actual earthquake records.** A response spectrum curve can be generated by subjecting a series of damped single-degree-of-freedom mass-spring systems to a given ground excitation. Response spectrum graphs are generated by numerical integration of actual earthquake records to determine maximum values for each period of vibration.

Spectral curves developed from actual earthquake records are quite jagged, being characterized by sharp peaks and troughs. Because the magnitude of these troughs and peaks can vary significantly for different earthquakes, and because of the uncertainties of future earthquakes, several possible earthquake spectra are used in the evaluation of the structural response.

**Smooth response spectrum.** As an alternative to the use of several earthquake spectra, a smooth spectrum representing an upper-bound response to ground motions may be generated. The sharp peaks in earthquake records indicate the resonant behavior of the system when the natural period of the system approaches the period of forcing function, especially for systems with little or no damping. However, even a moderate amount of damping, shown as $\beta$ in Figs 3.24, 3.25 and 3.26, has a tendency to smooth out the peaks and reduce the spectral response. Because most buildings in practice have at least some degree smooth out the peaks and reduce the spectral

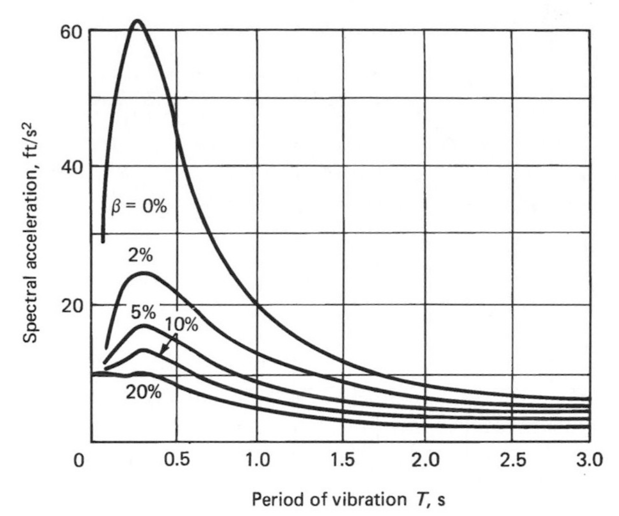


**Figure 3.24** Smoothed acceleration spectra for the El Centro earthquake.

response. Because most buildings in practice have at least some degree of damping, the peaks in response spectra are of little significance. Figure 3.24 shows the smoothed acceleration spectra for the El Centro, California, earthquake. The other two response spectra for velocity and displacement, shown in Figs 3.25 and 3.26, are obtained from the acceleration spectrum, since they are related to one another. The three spectra can be represented in one graph, as shown in Fig. 3.27, in which the horizontal axis denotes the natural period and the ordinate the spectrum velocity, both on a logarithmic

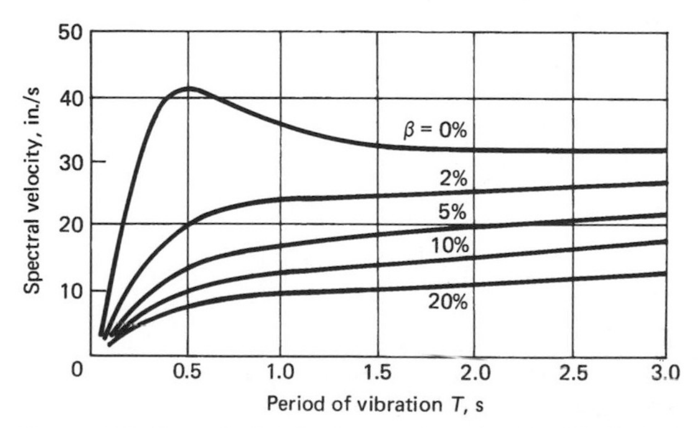


**Figure 3.25** Smoothed velocity spectra for the El Centro earthquake.

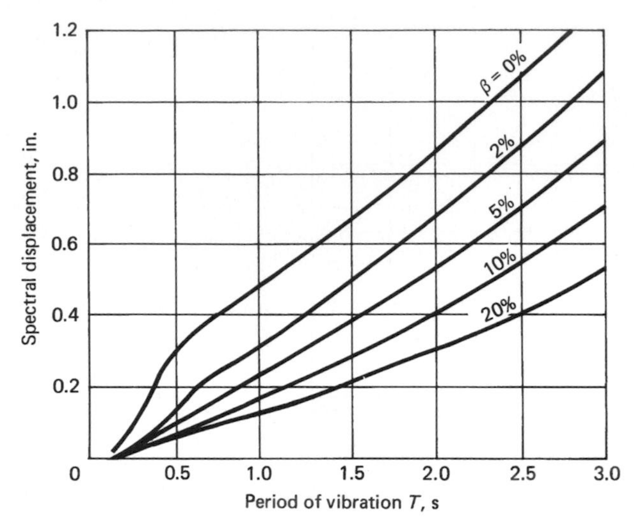


**Figure 3.26** Smoothed displacement spectra for El Centro earthquake.

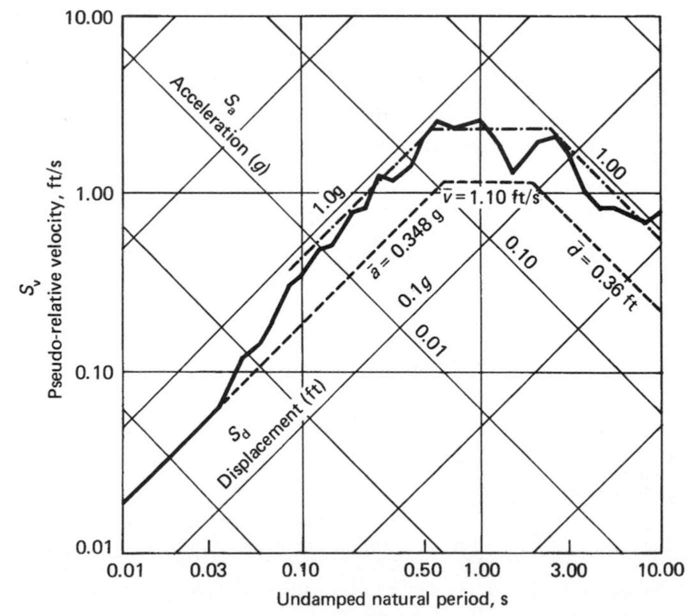


**Figure 3.27** Tripartite response spectra for El Centro earthquake (5 percent damping, north–south component).

scale. The acceleration and displacement are represented on diagonal axes inclined at 45° to the horizontal. The plot, which encompasses all the spectral parameters, is called a *tripartite response spectrum*. From this plot the following observations can be made.

1. For very stiff, structures, the spectral acceleration approaches the maximum ground acceleration. Structures in this period range would behave like rigid bodies attached to the ground.
2. For moderately short periods of the order of 0.1 to 0.3 sec, the spectral accelerations are about twice as large as the maximum ground acceleration.
3. For long period buildings, the maximum spectral displacements approach the maximum ground displacements.
4. For intermediate values of period, the maximum spectral velocity is several times the input velocity.

Thus, in the short-period range, the variation of the spectrum curve tends to show correlation with the line of maximum ground acceleration. In the medium-period range, the correlation is with maximum ground velocity while in the higher-period range, the correlation is with the displacement.

Because of the above characteristics, it is possible to represent an idealized upper-bound response spectrum by a set of three straight lines (Fig. 3.27). The values of ground acceleration ($\bar{\alpha} = 0.348\,g$), maximum velocity ($\bar{v} = 1.10$ ft/s), and displacement ($\bar{d} = 0.36$ ft) for the El Centro earthquake are also shown in Fig. 3.27.

**Unique design spectra.** For especially important structures or where local soil conditions are not amenable to simple classification, the use of recommended smooth spectrum curves is inadequate for final design purposes. In such cases, site-specific studies are performed to determine more precisely the expected intensity and character of seismic motion. The development of site-specific ground motions is generally the responsibility of geotechnical consultants working in concert with the structural engineer. However, it is important for the structural engineer to be aware of the procedure used in the generation of site specific response spectrum. This is considered next.

The seismicity of the region surrounding the site is determined from a search of an earthquake database. A list of active, potentially active, and inactive faults is compiled from the database along with their nearest distance from the site.

The predicted response of the deposits underlying the site and the influence of local soil and geologic conditions during earthquakes are determined based on statistical results of studies of site-dependent spectra developed from actual time-histories recorded by strong motion instruments.

Several postulated design earthquakes are selected for study based on the characteristics of the faults. The peak ground motions generated at the site by the selected earthquakes are estimated from empirical relationships.

The dynamic characteristics of the deposits underlying the site are estimated from the results of a nearby downhole seismic survey, from the logs of borings, static test data, and dynamic test data.

The causative faults are selected from the list of faults as the most significant faults along which earthquakes are expected to generate motions affecting the site.

Several earthquakes with different probability of occurence that may be generated along the causative faults are selected. The maximum credible earthquake (MCE) for example, constitutes the largest earthquake that appears to be reasonably likely to occur. Since the probability of such an earthquake occurring during the lifetime of the subject development is low, the ground motions associated with the MCE events are estimated to have 10% probability of being exceeded in 250 years.

Several other design basis earthquakes are also postulated from the data. For example, in Fig. 3.29d, earthquake "C" is considered to be an event having a 50% probability of being exceeded in 50 years.

The slip rates of the faults are estimated from published data. Using the slip rates, the accumulated slip over an approximate 475 year period (corresponding to 10% probability of being exceeded in 50 years) and over an approximate 72 year period (corresponding to 50% probability of being exceeded in 50 years) are determined. Using the surface displacement versus magnitude relationships, the magnitudes for each significant fault are determined.

Using a statistical analysis approach, the peak ground motion values (acceleration, velocity, and displacement) anticipated at the site are estimated. By applying structural amplification factors to these values the spectral bounds for acceleration, velocity and displacement are obtained for each desired value of structural damping, most usually 2, 5 and 10 percent of critical damping. The ground motion values vary with the magnitude of the earthquake and distance of the site from the source of energy release.

The peak ground motion values for velocity and displacement are developed by relating peak ground velocity and displacement to the peak ground acceleration for four site classifications: rock, stiff soil, deep cohesionless soil, and soft to medium soil.

The ground motion values obtained as above provide a basis by which site-dependent response spectra are computed. For each of the four site classes, spectral bounds are obtained by multiplying the ground motion values by damping-dependent amplification factors.

A schematic representation of acceleration spectra is shown in Fig. 3.28 for maximum credible and maximum probable events. Tripartite response spectra for four seismic events characterized as earthquakes A, B, C and D for a downtown Los Angeles site are shown in Figs 3.29a–d. Response spectra A is for a maximum credible earthquake of magnitude 8.25 occuring at San Andreas fault at a distance of 34 miles while B is for a magnitude 6.8 earthquake occuring at Santa Monica—Hollywood fault at a distance of 3.7 miles from the site. Response spectra C and D are for earthquakes with a 10 and 50 percent probability of being exceeded in 50 years.

### 3.5.4 Time-history analysis

#### 3.5.4.1 Introduction

The mode superposition or the spectrum method outlined in the previous section is a useful technique for the elastic analysis of structures. It is not directly transferable to inelastic analysis because the principle of superposition is no longer applicable. Also, the analysis is subject to uncertainties inherent in the modal superimposition method. The actual process of combining the different modal contributions is, after all, a probabilistic technique and in certain cases may lead to results not entirely representative of the actual behavior of the structure. Time history analysis overcomes these two uncertainties, but it requires a large computational effort. It is not normally employed as an analysis tool in practical design of

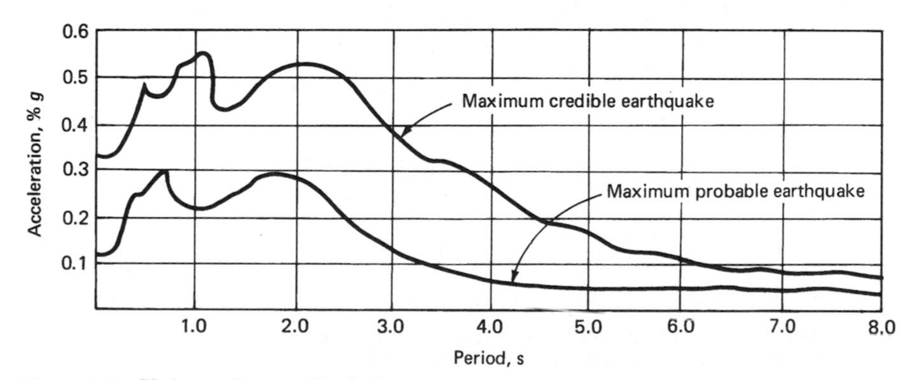


**Figure 3.28** Unique site-specific design spectra.

buildings. The method consists of a step-by-step direct integration in which the time domain is discretized into a number of small increments $\delta t$; and for each time interval the equations of motion are solved with the displacements and velocities of the previous step serving as initial functions. The method is applicable to both elastic and inelastic analyses. In elastic analysis the stiffness characteristics of the structure are assumed to be constant for the whole duration of the earthquake. In the inelastic analysis, however, the stiffness is assumed to be constant through the incremental time $\delta t$ only. Modifications to structural stiffness caused by cracking, formation of plastic hinges, etc., are incorporated between the incremental

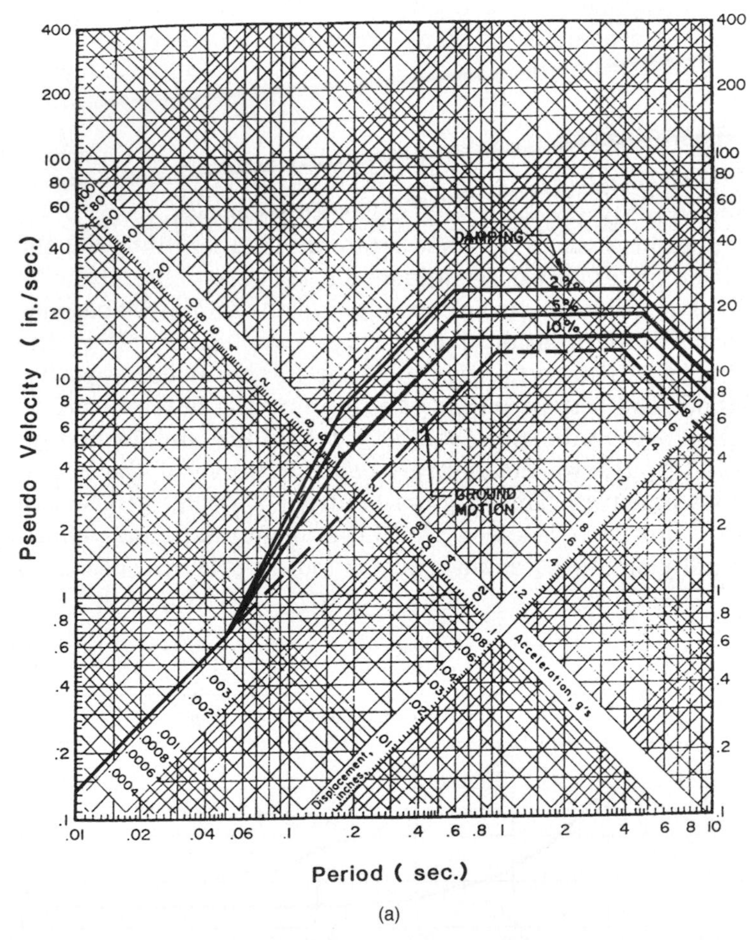


(a)

**Figure 3.29** Tripartite site-specific response spectra: (a) earthquake A; (b) earthquake B; (c) earthquake C; (d) earthquake D.

solutions. A brief outline of the method, which is thus applicable to both elastic and inelastic analysis, is given below.

### 3.5.4.2 Analysis procedure

In this method earthquake motions are applied directly to the base of the computer model of a given structure. Instantaneous stresses throughout the structure are calculated at small intervals of time for the duration of the earthquake or a significant portion of it. The maximum stresses that occur during the earthquake are found by scanning the computer output.

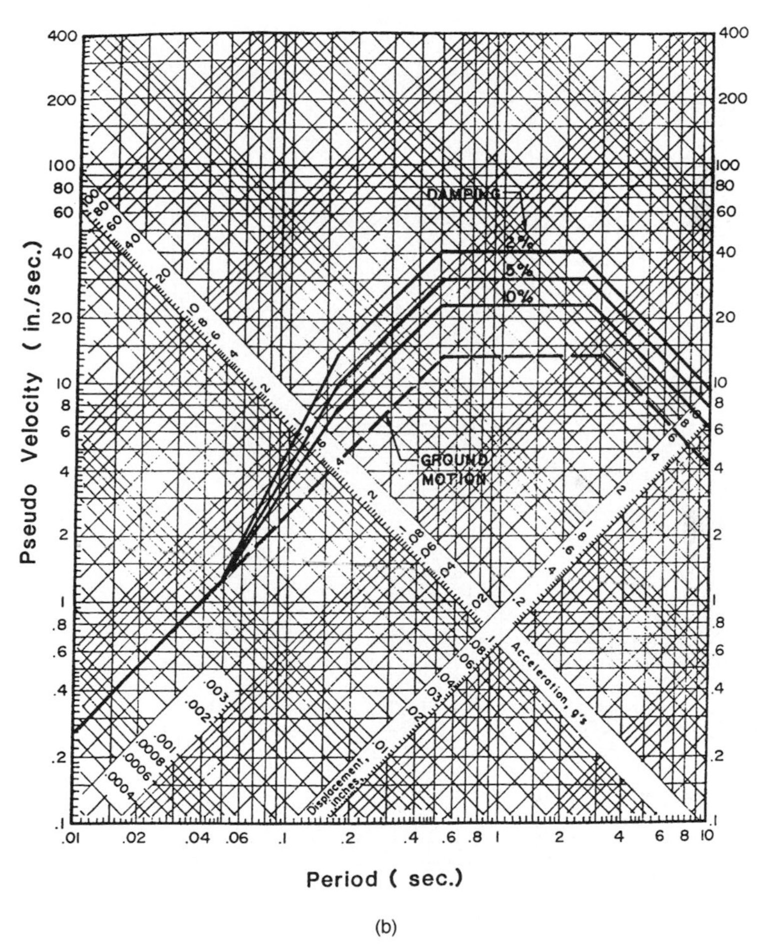


(b)

**Figure 3.29** (*Continued*).

The procedure usually includes the following steps:

1. An earthquake record representing the design earthquake is selected.
2. The record is digitized as a series of small time intervals of about 1/40 to 1/25 of a second.
3. A mathematical model of the building is set up. Usually consisting of a lumped mass at each floor. Damping is considered proportional to the velocity in the computer formulation.
4. The digitized record is applied to the model as accelerations at the base of the structure.

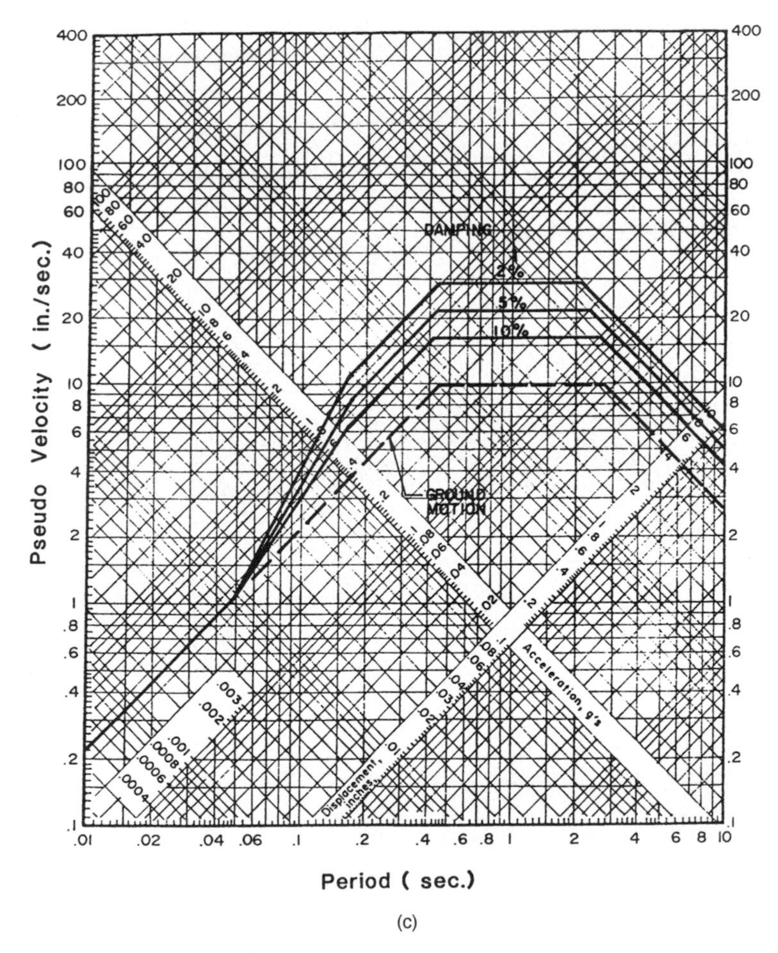


(c)

**Figure 3.29** *(Continued).*

5. The computer integrates the equations of motions and gives a complete record of the acceleration, velocity, and displacement of each lumped mass at each interval.

The accelerations and relative displacements of the lumped masses are translated into member stresses. The maximum values are found by scanning the output record.

This procedure automatically includes various modes of vibration by combining their effect as they occur, thus eliminating the uncertainties associated with modal combination methods.

The time history technique represents one of the most sophisticated method of analysis used in building design; however, it has the following sources of uncertainty:

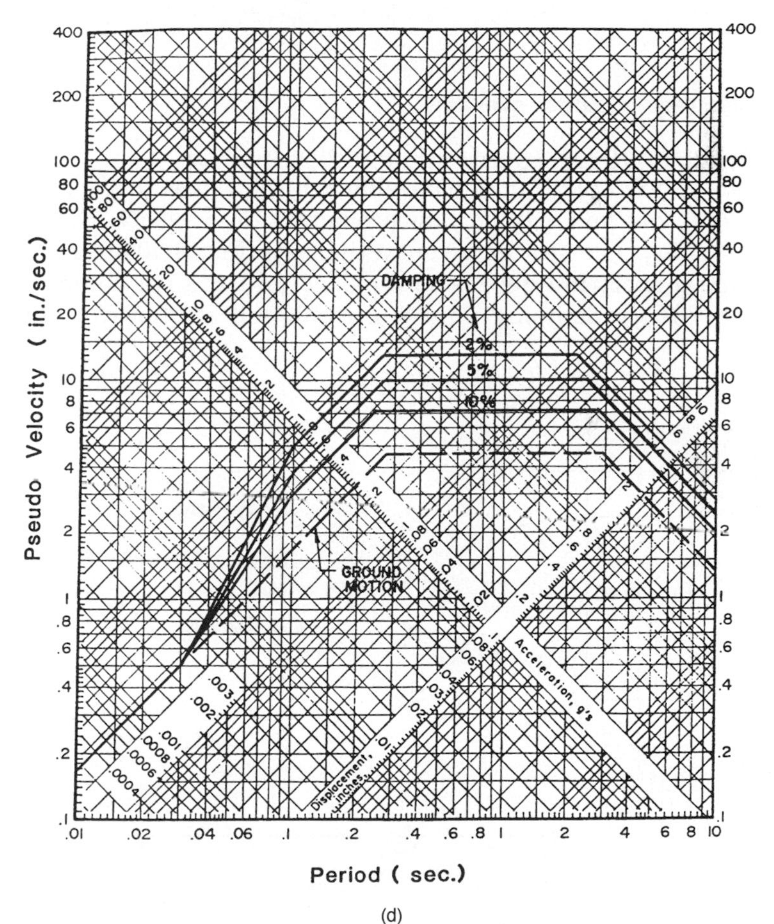


(d)

**Figure 3.29** *(Continued).*

1. The design earthquake must still be assumed.
2. If the analysis uses unchanging values for stiffness and damping, it will not reflect the cumulative effects of stiffness variation and progressive damage.
3. There are uncertainties related to the erratic nature of earthquakes. By pure coincidence, the maximum response of the calculated time history could fall at either a peak or a valley of the digitized spectrum.
4. Small inaccuracies in estimating properties of the structure will have considerable effect on the maximum response.
5. Errors latent in the magnitude of the time step chosen are difficult to assess unless the solution is repeated with several smaller time steps.

### 3.5.5 Dynamic requirements: overview of 1994 UBC

The minimum requirement for the development of a design response spectrum is one that has a 10 percent probability in 50 years. Either the normalized response spectra given in the code (Fig. 3.30) or a site specific response spectrum developed for 5 percent critical damping may be used. Other damping values are allowed if shown to be

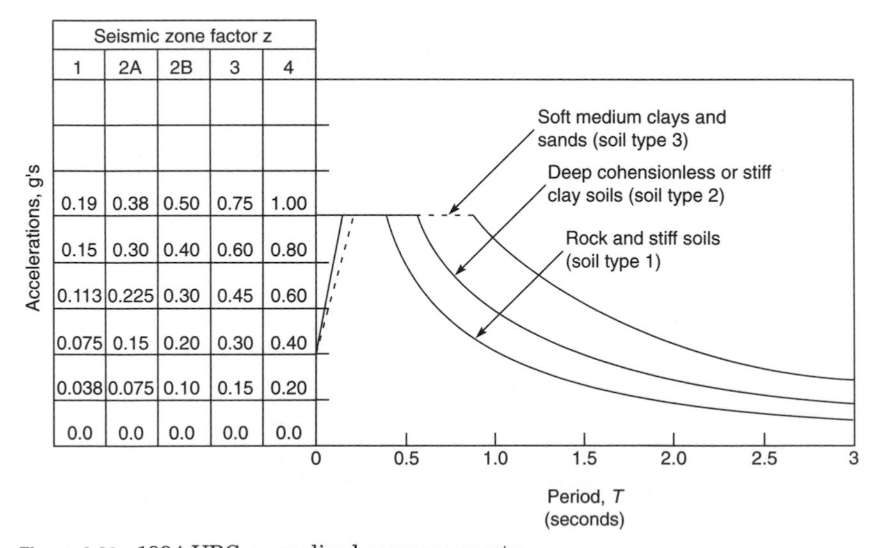


**Figure 3.30** 1994 UBC normalized response spectra.

consistent with the anticipated behavior of the structure. Since the response spectra given in the code are from actual ground motion records they are applicable for earthquakes motions characteristic of those that occur in California. A site specific spectrum is mandatory for an S4 site (soft clay soil).

For design purposes, it is permissible to use a vertical component of ground motion equal to two-thirds the peak amplitude of the horizontal motion. However, for sites closer than 6.2 miles (10 km) to the seismic source, or for unusual sites, a site specific evaluation should be performed to specify vertical motions for seismic design.

Mathematical modelling is an art requiring great care in its development. The model should adequately represent the dynamic behavior of the building without exhibiting extraneous or artificial responses and stiffnesses. Three-dimensional models are preferable to planar models because mass eccentricity of the structure can be properly retained in three-dimensional models.

An equivalent static analysis although not mandatory, is highly recommended for verifying the dynamic results. It is a good practice to plot the significant mode shapes to gain an insight into the structural performance. All modes having significant contribution to the total structural response should be included in the analysis, a requirement that is satisfied by demonstrating that 90 percent of the participating mass is included in the analysis. Although methods such as Square Root of the Sum of the Squares (SRSS) may be used to combine the effects of several modes, the preferred method is the Complete Quadratic Combination (CQC) because this method can account for modal coupling effects. Since, in the modal analysis method, the output of forces for the structural components are no longer in equilibrium, it is recommended that the predominant mode of response be used to verify equilibrium and to determine if a member is in single or double curvature.

Scaling the results of dynamic response such that the base shear is consistent with static design approach is permitted. The base shear resulting from dynamic analysis, if smaller than the static value must be increased to static value. If larger, it may be reduced. Since dynamic procedures are considered as an improvement on static procedures, using a base shear value of 90 percent of the static procedure is permitted for regular structures. No such reduction is allowed for irregular structures. Also, to prevent excessively low base shear values resulting from the use of an unreasonably long period in combination with the 90 percent static base shear, a lower limit of 80 percent of the base shear obtained by using the approximate period formulas is set.

The direction of earthquake ground motion is random and bears no relation to the principal axes of the building. In fact, earthquake motion is a composite response of concurrent motion: two translational and one torsional. However, as in the static procedure, independent design about each principal axis is deemed sufficient to provide adequate resistance for forces applied in any direction.

As in static analysis, the effects of vertical ground motion on horizontal cantilevers and prestressed elements can be accommodated by designing the cantilevers for a net upward force of 0.5 ZWp, where Wp = the weight of the horizontal element. In designing prestressed elements, no more than 50 percent of the dead load may be used to counteract the uplift effect of seismic loads. A dynamic analysis may also be used with a vertical component equal to two-thirds of the peak horizontal acceleration. However, the resulting forces may not be less than those required by the static procedure.

Torsional motions can lead to a significant increase in loads on the buildings lateral force resisting system. The building's lateral motions can be strongly coupled with torsional motions if large eccentricities exist between the centers of story resistance and centers of floor mass, or if the natural frequencies of the building's normal modes are closely spaced. A three-dimensional analysis using a Complete Quadratic Combination, CQC, for combining modal response is recommended. With the availability of computer programs, this task is no longer a heavy burden on the design engineer.

Torsional motions can occur in a building even if its centers of mass and resistance appear to be coincident and the natural frequencies of its predominant modes of vibration are well spaced. These motions can arise from several factors not typically considered in the dynamic analysis of buildings, such as: (i) spatial variation of horizontal ground accelerations; (ii) rotational components of ground motions; (iii) the effects of nonstructural elements, such as partitions, stairs, etc., on the buildings dynamic characteristics; (iv) the actual distribution of dead and live loads; and (v) uncertainties in defining the building's property for dynamic analysis.

To account for these "accidental" torsional effects, the mass in the dynamic model is displaced to alternate sides of the calculated center of mass by 5 percent of the building dimension. A three-dimensional analysis is used directly to calculate the effects of torsion.

In a dual system, such as an interacting shear wall or braced frame acting in conjunction with a specially detailed moment resisting frame, the moment frame itself should be capable of resisting at least 25 percent of the total base shear. This, in effect, provides for a "back-up" resistance to seismic forces. It is not necessary, since the frame is fully connected to the shear walls or braced frames, to perform a response spectrum analysis on the frame by itself. A static analysis of the frame subjected to 25 percent of the scaled base shear is permitted.

Time-history analysis is a numerical method of computing the dynamic response of a structure at each interval of time when the base is subjected to a specific ground motion time history. The procedure is applicable to either a linear elastic or nonlinear model of the building. It computes the time-dependent dynamic response of the building by numerical integration of the modal equations of motion. The method is more complex than response spectrum procedures and is not warranted for most structures. It is used only in situations where it is deemed necessary to represent inelastic behavior or compute time dependent response characteristics. Even then, the results may be difficult to interpret. Therefore, it is strongly recommended in code commentaries that both a static force analysis and a response spectrum analysis be carried out before any time-history analysis is performed. This may be the only method of validating the complex time-history analysis.

It is important to understand that a dynamic analysis, *per se,* will not necessarily provide response estimates consistent with actual seismic performance. It is a folly to believe that it gives all the answers or solutions to a seismic design problem. It can, however, aid in seismic design by identifying important seismic response characteristics that may not be evident from the static procedure, such as: (i) the effects of the structure's dynamic characteristics on the vertical distribution of lateral force; (ii) increase in the dynamic loads in the structure's lateral force resisting system due to torsional motions; and (iii) the effects of higher modes that could substantially increase story shears and deformations.

In general, the first three modes of vibration in each horizontal direction are sufficient for the modal analysis of low rise building. For tall buildings or for buildings with vertical irregularities, a greater number of modes may be required. A review of the participation factors for the first three modes will give an indication if more modes are required. The sum of the participation factors for all the modes at a particular story should equal unity. Also, the sum of all the modal base shear participation factors should equal unity.

The 1994 UBC states that if the sum of the modal base shear participation factors $\alpha$ for the number of modes considered in the analysis is within 10 percent of unity, it can be assumed that all the major modes have been included. It is silent regarding the modal story participation factor $PF_{xm}$ although analytically it is more exact to verify that the sum of the modal participation factors at each level is within 10 percent of unity.

### 3.5.6 Modal analysis: hand calculation procedure

Two examples are presented in the following to illustrate the modal analysis method. In the first part of each example, modal analysis is performed to determine base shear for each mode using given building characteristics and ground motion spectra. In the second part, story forces, accelerations and displacements are calculated for each mode, and are combined statistically using the SRSS combination.

The formulas for obtaining modal story participation factor, modal story lateral forces, modal base shear, modal deflections and drifts are given within the framework of the second example.

#### 3.5.6.1 Example 1: three-story building

**Given**

Masses $\frac{W}{g}$, mode shapes $\phi_s'$, periods $T_1$, $T_2$, and $T_3$, and corresponding accelerations $a_1$, $a_2$, and $a_3$. $T_1 = 0.9645$ s, $T_2 = 0.3565$ s and $T_3 = 0.1825$ s corresponding spectral accelerations from Fig. 3.31 are 0.251 g for mode 1, and 0.41 for modes 2 and 3.

**Required**

(i) Modal analysis to determine base shears.

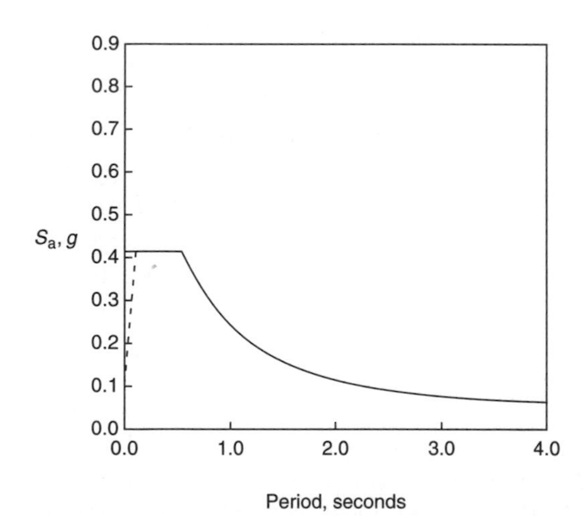


| | Period | | | | | | | |
|---|---|---|---|---|---|---|---|---|
| | 0.0 | .586 | .80 | 1.0 | 1.5 | 2.0 | 3.0 | 4.0 |
| $S_a, g$ | .14 | .41 | .300 | .240 | .160 | .120 | .080 | .060 |

**Figure 3.31** Three-story building: response spectrum.

(ii) Story forces, overturning moments, accelerations and displacements for each mode.
(iii) SRSS combinations.

### Modal analysis: influence of higher modes

The results of modal analysis for determining base shears, and story forces, accelerations and displacements are shown in Figs 3.32 and 3.33. It should be noted that higher modes of response become increasingly important for taller or irregular buildings. For this regular 3-story structure, the first mode dominates the lateral

| $T_{m_1}$, sec. | | 0.964 | | | 0.356 | | | 0.182 | | |
|---|---|---|---|---|---|---|---|---|---|---|
| | | $\phi_{R1}=0.3320$; $\phi_{31}$ 0.2044; $\phi_{21}$ 0.086̇; Mode 1 | | | $\phi_{R2}=0.2384$; −0.2201 $\phi_{32}$; −0.2075 $\phi_{22}$; Mode 2 | | | $\phi_{R3}=0.0713$; −0.2154 $\phi_{33}$; $\phi_{23}$ 0.2936; Mode 3 | | |
| Level | Mass $\left(\frac{k \cdot sec^2}{ft.}\right)$ | Mode 1 | | | Mode 2 | | | Mode 3 | | |
| | | $\phi_{x1}$ | $m_x\phi_{x1}$ | $m_x\phi_{x1}^2$ | $\phi_{x2}$ | $m_x\phi_{x2}$ | $m_x\phi_{x2}^2$ | $\phi_{x3}$ | $m_x\phi_{x3}$ | $m_x\phi_{x3}^2$ |
| R | 5.81 | 0.3320 | 1.929 | 0.640 | 0.2384 | 1.385 | 0.330 | 0.0713 | 0.4143 | 0.030 |
| 3 | 7.32 | 0.2044 | 1.496 | 0.306 | −0.2201 | −1.611 | 0.355 | −0.2154 | −1.577 | 0.340 |
| 2 | 7.32 | 0.0860 | 0.630 | 0.054 | −0.2075 | −1.519 | 0.315 | 0.2936 | 2.149 | 0.631 |
| Σ | 20.45 | | 4.055 | 1.000** | | −1.745 | 1.000 | | 0.9863 | 1.001 |
| $PF^*_{Rm}$ | | $\frac{\Sigma m\phi}{\Sigma m\phi^2}\ \phi_{R1}=$ 1.346 | | | −0.416 | | | 0.070 | | |
| $PF_{3m}$ | | 0.829 | | | 0.384 | | | −0.212 | | |
| $PF_{2m}$ | | 0.349 | | | 0.362 | | | 0.289 | | |
| $\alpha_m$ | | $\frac{(\Sigma m\phi)^2}{\Sigma m(\Sigma m\phi^2)}$ = 0.8040 | | | 0.149 | | | 0.048 | | |
| $s_a$ | | 0.251 g | | | 0.41 g | | | 0.41 g | | |
| $v=\alpha_m s_a w$ | | 132.7 Kips | | | 40.2 Kips | | | 13.0 Kips | | |

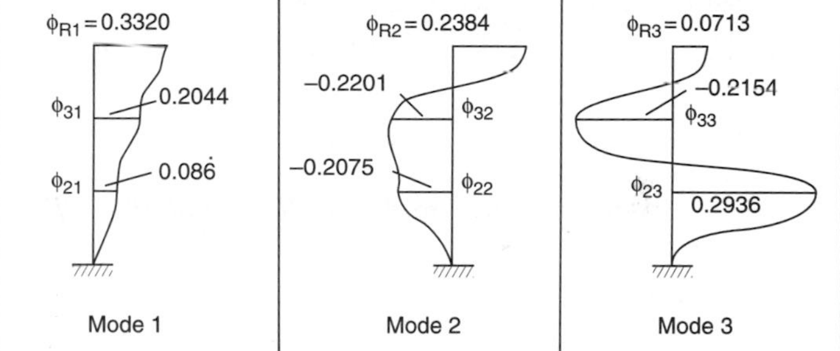

* Note that the sum of the modal participation factors $\sum_{m=1}^{3} PF_{xm}=1.0$ and the sum of modal base shear participation factors $\sum_{m=1}^{3}\alpha_m=1.0$.

** The mode shapes have been normalized by the computer program so that $\Sigma m\phi^2=1.0$.

**Figure 3.32** Three-story building: modal analysis to determine base shears.

| Level | $PF_{xm}$ | $\frac{m_x \phi_{xm}}{\sum m_x \phi_{xm}}$ | $F_{xm}$ (k) | $V_{xm}$ (k) | $\Delta OTM_{xm}$ (ft · k) | $OTM_{xm}$ (ft · k) | $a_{xm} = \frac{F_{xm}}{w_x}$ | $\delta_{xm}$ (in.) | $\Delta_{xm}$ (in.) |
|---|---|---|---|---|---|---|---|---|---|
| R | 1.346 | 0.476 | 63.2 | 63.2 | 772 | 0 | 0.337 | 3.065 | 1.182 |
| 3 | 0.829 | 0.369 | 48.9 | 112.1 | 1233 | 772 | 0.208 | 1.892 | 1.101 |
| 2 | 0.349 | 0.155 | 20.6 | 132.7 | 1416 | 2005 | 0.087 | 0.791 | 0.791 |
| | | 1.000 | | | | 3421 | | | |

(a) Mode 1

| Level | $PF_{xm}$ | $\frac{m_x \phi_{xm}}{\sum m_x \phi_{xm}}$ | $F_{xm}$ (k) | $V_{xm}$ (k) | $\Delta OTM_{xm}$ (ft · k) | $OTM_{xm}$ (ft · k) | $a_{xm}$ | $\delta_{xm}$ (in.) | $\Delta_{xm}$ (in.) |
|---|---|---|---|---|---|---|---|---|---|
| R | −0.416 | −0.793 | −31.9 | −31.9 | −389 | 0 | −0.171 | −0.212 | 0.407 |
| 3 | 0.384 | 0.923 | 37.1 | 5.2 | 57 | −389 | −0.157 | 0.195 | 0.011 |
| 2 | 0.362 | 0.870 | 35.0 | 40.2 | 429 | −332 | −0.148 | 0.184 | 0.184 |
| | | 1.000 | | | | 97 | | | |

(b) Mode 2

| Level | $PF_{xm}$ | $\frac{m_x \phi_{xm}}{\sum m_x \phi_{xm}}$ | $F_{xm}$ (k) | $V_{xm}$ (k) | $\Delta OTM_{xm}$ (ft · k) | $OTM_{xm}$ (ft · k) | $a_{xm}$ | $\delta_{xm}$ (in.) | $\Delta_{xm}$ (in.) |
|---|---|---|---|---|---|---|---|---|---|
| R | 0.070 | 0.420 | 5.5 | 5.5 | 67 | 0 | −0.029 | 0.0094 | 0.037 |
| 3 | −0.212 | −1.599 | −20.8 | −15.3 | −168 | 67 | −0.087 | −0.028 | 0.066 |
| 2 | 0.289 | 2.179 | 28.3 | 13.0 | 139 | −101 | 0.118 | 0.038 | 0.038 |
| | | 1.000 | | | | 38 | | | |

(c) Mode 3

| Level | $PF_{xm}$ | $\frac{m_x \phi_{xm}}{\sum m_x \phi_{xm}}$ | $F_{xm}$ (k) | $V_{xm}$ (k) | $\Delta OTM_{xm}$ (ft · k) | $OTM_{xm}$ (ft · k) | $a_{xm}$ | $\delta_{xm}$ (in.) | $\Delta_{xm}$ (in.) |
|---|---|---|---|---|---|---|---|---|---|
| R | | | 71.0 | 71.0 | 867 | 0 | 0.379 | 3.072 | 1.251 |
| 3 | | | 64.8 | 113.3 | 1246 | 867 | 0.275 | 1.893 | 1.094 |
| 2 | | | 49.5 | 139.3 | 1486 | 2035 | 0.208 | 0.812 | 0.813 |
| | | | | | | 3423 | | | |

(d) SRSS combination

**Figure 3.33** Three-story building: modal analysis to determine story forces, accelerations and displacements.

response. A comparison of the modal story shears and the SRSS story shears is shown in Fig. 3.34. For example, if only the first mode shears had been used for analysis, this represents 89 percent of the SRSS shear at the roof, 99 percent at the third floor and 95 percent at the second floor. While the second mode shear at the roof is 50 percent of the first mode shear, when combined on an SRSS basis the first mode accounts for 79 percent of the SRSS response with 20 percent for the second mode and 0.6 percent for the third mode. These percentages are 91 percent, 8 percent and 1 percent at the base.

| | | Mode 1 | | | Mode 2 | | Mode 3 | |
|---|---|---|---|---|---|---|---|---|
| Level | $V_{SRSS}$ | $V_1$ | $V_1/V_{SRSS}$ | $(V_1/V_{SRSS})^2$ | $V_2$ | $(V_2/V_{SRSS})^2$ | $V_1$ | $(V_3/V_{SRSS})^2$ |
| R | 71.0 | 63.2 | 0.89 | 0.79 | −31.9 | 0.202 | 5.5 | 0.006 |
| 3 | 119.3 | 112.1 | 0.989 | 0.98 | 5.2 | 0.002 | −15.3 | 0.018 |
| 2 | 139.3 | 132.7 | 0.953 | 0.91 | 40.2 | 0.083 | 13.0 | 0.009 |

**Figure 3.34** Three-story building: comparison of modal story shears and the SRSS story shears.

The effective modal weight factor, $\alpha_m$, also shows the relative importance of each mode. In this example, ($\alpha_1 = 0.804$, $\alpha_2 = 0.149$, $\alpha_3 = 0.048$). 80.4 percent of the building mass participates in the first mode, 14.9 percent in the second mode and 4.8 percent in the third mode.

### 3.5.6.2 Example 2: 7-story building

The results are shown in a format similar to the format used in the static force procedure such that a comparison of static force and dynamic analysis procedures can be made. A 7-story reinforced concrete moment resisting space frame building adapted from the Tri Services Manual, is chosen to illustrate the procedure. The modal analyses are performed on the basis of the response spectrum shown in Figs 3.35a–c. Observe that these three figures contain the same response information, only the format is different. Figure 3.35a shows spectral accelerations and periods on the vertical and horizontal axes respectively, similar to the UBC format. Figure 3.35b shows the same information on the tripartite diagram, while Fig. 3.35c is a numerical representation of the same response spectra.

The first step is to develop a mathematical model of the building. Story masses are obtained from the calculated weights of the building tributary to each story. Although three-dimensional computer models are more common in seismic design practice, we will assume, for ease of presenting the sample modal analysis, that the 7-story building is analyzed as a series of two-dimensional frames. The periods and mode shapes are determined for the first three modes of vibration by a two-dimensional computer program.

In this program, each mode is normalized for $\Sigma\,(W/g)\phi^2 = 1.0$. The mode shapes are shown in Fig. 3.36 with a normalized value of $\frac{1}{2}$ in. at the top story.

#### Modal analysis to determine total base shear and story accelerations

Figure 3.36 illustrates a hand-calculation procedure to determine the total base shear and the story accelerations using mass, mode shape, period, and response spectrum data. The following equations are used to determine the participation factors.

#### Modal story participation factor

The story modal participation factor will be calculated for each mode by using the equation

$$PF_{xm} = \left( \frac{\sum_{i=1}^{n} \frac{w_i}{g} \phi_{im}}{\sum_{i=1}^{n} \frac{w_i}{g} \phi_{im}^2} \right) \phi_{xm} \tag{3.8}$$

where $PF_{xm}$ = modal participation factor at level $x$ for mode $m$

$w_i/g$ = mass assigned to level $i$

$\phi_{im}$ = amplitude of mode $m$ at level $i$

$\phi_{xm}$ = amplitude of mode $m$ at level $x$

$n$ = level $n$ under consideration

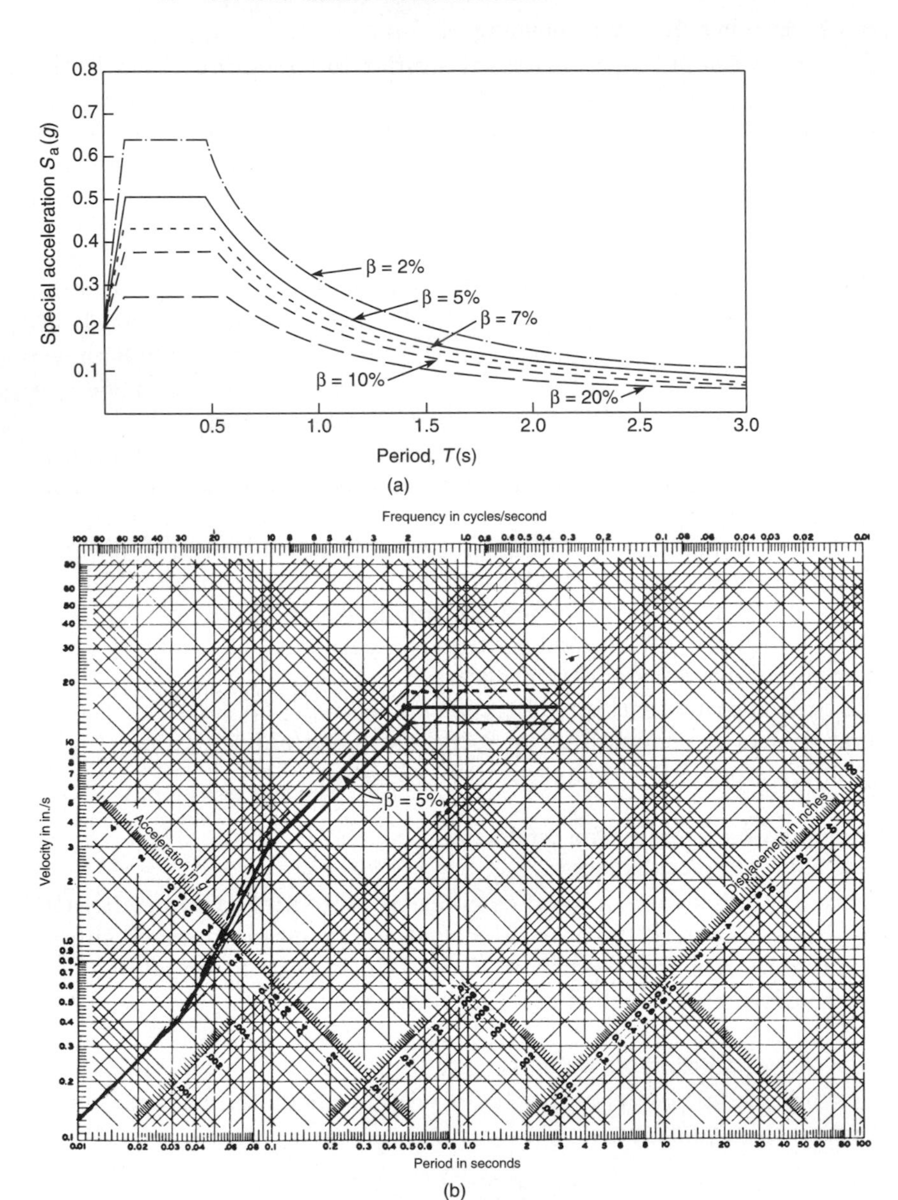


**Figure 3.35** Response spectrum for 7-story building example: (a) acceleration spectrum; (b) tripartite diagram; (c) response spectra numerical representation.

| | Spectral acceleration, $S_a(g)$ | | | | | | | | | | | |
|---|---|---|---|---|---|---|---|---|---|---|---|---|
| $\beta$ \ $T$ | 0.1 | 0.48 | 0.50 | 0.80 | 1.0 | 1.75 | 1.5 | 1.75 | 2.0 | 2.25 | 2.5 | 3.0 |
| 2% | 0.64 | 0.64 | 0.59 | 0.37 | 0.30 | 0.24 | 0.20 | 0.17 | 0.15 | 0.13 | 0.17 | 0.10 |
| 5% | 0.50 | 0.50 | 0.48 | 0.30 | 0.24 | 0.192 | 0.16 | 0.137 | 0.12 | 0.107 | 0.096 | 0.08 |
| 7% | 0.44 | 0.44 | 0.44 | 0.28 | 0.22 | 0.18 | 0.15 | 0.13 | 0.11 | 0.10 | 0.09 | 0.07 |
| 10% | 0.38 | 0.38 | 0.38 | 0.25 | 0.20 | 0.16 | 0.13 | 0.11 | 0.10 | 0.09 | 0.08 | 0.066 |
| 20% | 0.27 | 0.27 | 0.27 | 0.20 | 0.16 | 0.12 | 0.10 | 0.09 | 0.08 | 0.07 | 0.06 | 0.05 |

(c)

**Figure 3.35** *(Continued).*

It should be noted that some references define the modal participation factor as the quantity within the bracket in Eq. (3.8). Also, in some references, $\phi$ is normalized to 1.0 at the uppermost mass level.

### Modal base shear participation factor

The effective modal weight (or modal base shear participation factor) is calculated for each mode using

$$\alpha_m = \frac{\left(\sum_{i=1}^{n} \frac{w_i}{g} \phi_{im}\right)^2}{\sum_{i=1}^{n} \frac{w_i}{g} \sum_{i=1}^{n} \frac{w_i}{g} \phi_{im}^2} \tag{3.9}$$

where $\alpha_m$ = the modal base shear participation factor for mode $m$.

Next, the spectral acceleration for the period $T$ of each mode is determined from the response spectrum. The story accelerations "$a$" are determined from

$$a_{xm} = PF_{xm} S_{am} \tag{3.10}$$

where $a_{xm}$ = modal story acceleration at level $x$ for mode $m$
$PF_{xm}$ = modal participation as determined by Eq. (3.20)
$S_{am}$ = spectral acceleration for mode $m$

Next, the base shears "$V$" are determined from

$$V_m = \alpha_m S_{am} W \tag{3.11}$$

| Level | $\frac{w}{g}$ $\left(\frac{\text{k-s}^2}{\text{ft}}\right)$ | Mode 1 | | | | Mode 2 | | | | Mode 3 | | | | SRSS |
|---|---|---|---|---|---|---|---|---|---|---|---|---|---|---|
| | | $\phi_1$ | $\frac{w}{g}\phi_1$ | $\frac{w}{g}\phi_1^2$ | $a_1$ $(g)$ | $\phi_2$ | $\frac{w}{g}\phi_2$ | $\frac{w}{g}\phi_2^2$ | $a_2$ $(g)$ | $\phi_3$ | $\frac{w}{g}\phi_3$ | $\frac{w}{g}\phi_3^2$ | $a_3$ $(g)$ | $a_x$ $(g)$ |
| Roof | 43.78 | 0.0794 | 3.48 | 0.276 | 0.362 | 0.0747 | 3.27 | 0.744 | −0.235 | 0.0684 | 2.99 | 0.205 | 0.120 | 0.448 |
| 7 | 45.34 | 0.0745 | 3.38 | 0.252 | 0.340 | 0.0411 | 1.86 | 0.076 | −0.129 | −0.0040 | −0.18 | 0.001 | −0.007 | 0.364 |
| 6 | 45.34 | 0.0666 | 3.02 | 0.201 | 0.304 | −0.0042 | −0.19 | 0.001 | 0.013 | −0.0644 | −2.92 | 0.188 | −0.113 | 0.325 |
| 5 | 45.34 | 0.0558 | 2.53 | 0.141 | 0.254 | −0.0471 | −2.14 | 0.101 | 0.148 | −0.0630 | −2.86 | 0.180 | −0.111 | 0.314 |
| 4 | 45.34 | 0.0425 | 1.93 | 0.082 | 0.194 | −0.0718 | −3.26 | 0.234 | 0.226 | −0.0023 | −0.10 | 0.000 | −0.004 | 0.298 |
| 3 | 45.34 | 0.0279 | 1.27 | 0.035 | 0.127 | −0.0697 | −3.16 | 0.220 | 0.219 | 0.0604 | 2.74 | 0.166 | 0.106 | 0.275 |
| 2 | 56.83 | 0.0149 | 0.85 | 0.013 | 0.068 | −0.0467 | −2.65 | 0.124 | 0.147 | 0.0677 | 3.85 | 0.261 | 0.119 | 0.201 |
| 1 | — | 0 | 0 | 0 | 0 | 0 | | 0 | 0 | 0 | 0 | 0 | 0 | 0 |
| $\Sigma$ | 327.31 | | 16.46 | 1.000 | | | −6.27 | 1.000 | | | 3.52 | 1.001 | | |
| $PF_{\text{roof}}$ | Eq. (3.8) | $\frac{16.46}{1.000}(0.0794) = 1.31$ | | | | $\frac{-6.37}{1.000}(0.0747) = -0.47$ | | | | $\frac{3.52}{1.001}(0.0684) = 0.24$ | | | | $\Sigma = 1.08$ |
| $\alpha$ | Eq. (3.9) | $\frac{(16.46)^2}{(327.31)(1.000)} = 0.828$ | | | | $\frac{(-6.27)^2}{(327.31)(1.000)} = 0.120$ | | | | $\frac{(3.52)^2}{(927.31)(1.001)} = 0.038$ | | | | $\Sigma = 0.986$ |
| $T$ | | 0.880 sec | | | | 0.288 sec | | | | 0.164 sec | | | | |
| $S_a$ | | $0.276\,g$ | | | | $0.500\,g$ | | | | $0.500\,g$ | | | | |
| $a_{\text{roof}}$ | Eq. (3.10) | $(1.31)(0.276) = 0.362\,g$ | | | | $(-0.47)(0.500) = -0.235\,g$ | | | | $(0.24)(0.500) = 0.120\,g$ | | | | 0.448 |
| $V$ | Eq. (3.11) | $(0.828)(0.276)(10{,}539) = 2408$ kips | | | | $(0.12)(0.500)(10{,}539) = 632$ kips | | | | $(0.038)(0.500)(10{,}539) = 200$ kips | | | | 2498 kips |
| $V/W$ | | 0.229 | | | | 0.060 | | | | 0.019 | | | | 0.237 |

$W = \Sigma\left(\frac{w}{g}\right) \times g = 327.31 \times 32.2 = 10{,}539$ kips = Building Weight.

$A_G = 0.20\,g$ Site PGA.

$\beta = 0.05$ Damping Factor.

**Figure 3.36** Seven-story building: modal analysis to determine base shears.

where $V_m$ = total lateral force for mode $m$

$W$ = total seismic dead load of the building which includes the dead load plus applicable portions of other loads

For the example problem, the sum of the participation factors, $PF_{xm}$ and $\alpha_m$, add up to 1.08 and 0.986, respectively. These values being close to 1.0 indicate that most of the modal participation is included in the three modes considered in the example. The story accelerations and the base shears are combined by the square-root-of-the-sum-of-the-squares (SRSS). The modal base shears are 2408 kips, 632 kips, and 200 kips for the first, second, and third modes respectively. These are used in Fig. 3.40 to determine story forces. The SRSS base shear is 2498 kips.

### Story forces, accelerations, and displacements

Figures 3.32–3.39 are set up in a manner similar to the static design procedure of Section 3.4. In the static lateral procedure, $\frac{Wh}{\sum Wh}$ is used to distribute the force on the assumption of a straight line mode

$T = 0.880$ sec

Modal base shear $V = 2408$ kips

| (1) Story | (2) $\phi$ | (3) $h$ ft | (4) $\Delta h$ ft | (5) $w$ kips | (6) $\frac{w\phi}{\sum w\phi}$ | (7) $F$ kips ($V_1$) $\times$(6) | (8) $V$ kips $\sum$(7) | (9) $\Delta$OTM $K$-ft (4)–(8) | (10) OTM $K$-ft $\sum$(9) | (11) Accel. $g$ (7) ÷ (5) | $\delta$* ft | $\Delta\delta$ ft |
|---|---|---|---|---|---|---|---|---|---|---|---|---|
| Roof | 0.0794 | 65.7 | | 1410 | 0.211 | 508 | | | 0 | 0.360 | 0.228 | |
| | | | 8.7 | | | | 508 | 4420 | | | | 0.014 |
| 7 | 0.7450 | 57.0 | | 1460 | 0.205 | 494 | | | 4420 | 0.338 | 0.214 | |
| | | | 8.7 | | | | 1002 | 8717 | | | | 0.022 |
| 6 | 0.0666 | 48.3 | | 1460 | 0.184 | 443 | | | 13,137 | 0.303 | 0.192 | |
| | | | 8.7 | | | | 1445 | 12,572 | | | | 0.031 |
| 5 | 0.0558 | 59.6 | | 1460 | 0.154 | 371 | | | 25,709 | 0.254 | 0.161 | |
| | | | 8.7 | | | | 1816 | 15,799 | | | | 0.039 |
| 4 | 0.0425 | 30.9 | | 1460 | 0.117 | 282 | | | 41,508 | 0.193 | 0.122 | |
| | | | 8.7 | | | | 2098 | 10,253 | | | | 0.042 |
| 3 | 0.0279 | 22.2 | | 1460 | 0.077 | 185 | | | 59,761 | 0.127 | 0.080 | |
| | | | 8.7 | | | | 2283 | 19,862 | | | | 0.057 |
| 2 | 0.0149 | 13.5 | | 1830 | 0.052 | 125 | | | 79,623 | 0.068 | 0.043 | |
| | | | 13.5 | | | | 2408 | 32,508 | | | | 0.043 |
| Grd. | 0 | 0 | | 0 | 0 | 0 | | | 112,131 | 0 | 0 | |
| | | | | $\sum$ | 1.000 | 2408 | | 112,191 | | | | |

$$\text{*Displacement } \delta_{x1} = \frac{g}{4\pi^2} \times T_1^2 \times \frac{F}{W}$$

$$= \frac{32}{4\pi^2} \times 0.88^2 \times \text{acceleration}$$

$$= 0.632 \times \text{acceleration}$$

**Figure 3.37** Seven-story building: first mode forces and displacements.

$T_2 = 0.288$ sec

Modal base shear $V_2 = 632$ kips

| (1) | (2) | (3) | (4) | (5) | (6) | (7) | (8) | (9) | (10) | (11) | | |
|---|---|---|---|---|---|---|---|---|---|---|---|---|
| Story | $\phi$ | $h$ ft | $\Delta h$ ft | $w$ kips | $\frac{w\phi}{\sum w\phi}$ | $F$ kips $(V_2) \times (6)$ | $V$ kips $\sum (7)$ | $\Delta$OTM $K$-ft $(4) \times (8)$ | OTM $k$-ft $Z(9)$ | Accel. $g$ $(7) \div (5)$ | $\delta^*$ ft | $\Delta\delta$ ft |
| Roof | 0.0747 | 65.7 | | 1410 | 0.522 | −330 | | | 0 | −0.234 | −0.016 | |
| | | | 8.7 | | | | −330 | −2871 | | | | 0.007 |
| 7 | 0.0411 | 57.0 | | 1460 | 0.297 | −188 | | | −2871 | −0.129 | −0.009 | |
| | | | 8.7 | | | | −518 | −4507 | | | | 0.010 |
| 6 | −0.0042 | 48.3 | | 1460 | 0.030 | 19 | | | −7378 | 0.013 | 0.001 | |
| | | | 8.7 | | | | −499 | −4341 | | | | 0.009 |
| 5 | −0.0471 | 39.6 | | 1460 | 0.341 | 216 | | | −11,719 | 0.148 | 0.010 | |
| | | | 8.7 | | | | −283 | −2462 | | | | 0.005 |
| 4 | −0.0718 | 30.9 | | 1460 | 0.520 | 329 | | | −14,181 | 0.225 | 0.015 | |
| | | | 8.7 | | | | 46 | 400 | | | | 0.000 |
| 3 | −0.0697 | 22.2 | | 1460 | 0.504 | 319 | | | −13,781 | 0.219 | 0.015 | |
| | | | 8.7 | | | | 365 | 3176 | | | | 0.005 |
| 2 | −0.0467 | 13.5 | | 1830 | 0.423 | 267 | | | −10,605 | 0.146 | 0.010 | |
| | | | 13.5 | | | | 632 | 8532 | | | | 0.010 |
| Grd. | 0 | 0 | | | | | | | −2073 | 0 | 0 | |
| | | | | $\sum$ | 0.999 | 632 | | −2073 | | | | |

$$^*\text{Displacement } \delta_{x2} = \frac{g}{4\pi^2} \times T_2^2 \times \frac{F}{w}$$

$$= \frac{32}{4\pi^2} \times 0.288^2 \times \text{acceleration}$$

$$= 0.068 \times \text{acceleration}$$

**Figure 3.38** Seven-story building: second mode forces and displacements.

$T_3 = 0.164$ sec

Modal base shear $V_3 = 200$ kips

| (1) | (2) | (3) | (4) | (5) | (6) | (7) | (8) | (9) | (10) | (11) | | |
|---|---|---|---|---|---|---|---|---|---|---|---|---|
| Story | $\phi$ | $h$ ft | $\Delta h$ ft | $w$ kips | $\frac{w\phi}{\sum w\phi}$ | $F$ kips $(V_3) \times (6)$ | $V$ kips $\sum (7)$ | $\Delta$OTM $K$-ft $(4) \times (8)$ | OTM $K$-ft $\sum (9)$ | Accel. $g$ $(7) \div (5)$ | $\delta^*$ ft | $\Delta\delta$ ft |
| Roof | 0.0684 | 65.7 | | 1410 | 0.849 | 170 | | | 0 | 0.121 | 0.003 | |
| | | | 8.7 | | | | 170 | 1479 | | | | 0.003 |
| 7 | −0.0040 | 57.0 | | 1460 | −0.051 | −10 | | | 1479 | −0.007 | 0.000 | |
| | | | 8.7 | | | | 160 | 1392 | | | | 0.003 |
| 6 | −0.0644 | 48.3 | | 1460 | −0.830 | −166 | | | 2871 | −0.114 | −0.003 | |
| | | | 8.7 | | | | −6 | −52 | | | | 0.000 |
| 5 | −0.0630 | 39.6 | | 1460 | −0.813 | −163 | | | 2819 | −0.112 | −0.003 | |
| | | | 8.7 | | | | −169 | −1470 | | | | 0.003 |
| 4 | −0.0023 | 30.9 | | 1460 | −0.028 | −6 | | | 1349 | −0.004 | 0.000 | |
| | | | 8.7 | | | | −175 | −1523 | | | | 0.002 |
| 3 | 0.0604 | 22.2 | | 1460 | 0.778 | 156 | | | −174 | 0.107 | 0.002 | |
| | | | 8.7 | | | | −19 | −165 | | | | 0.001 |
| 2 | 0.0677 | 13.5 | | 1830 | 1.094 | 219 | | | −339 | 0.120 | 0.003 | |
| | | | 13.5 | | | | 200 | 2700 | | | | 0.003 |
| Grd. | 0 | 0 | | | | | | | 2361 | 0 | 0 | |
| | | | | $\sum$ | 0.999 | 200 | | 2361 | | | | |

$$^*\text{Displacement } \delta_{x3} = \frac{g}{4\pi^2} \times T_3^2 \times \frac{F}{W}$$

*d*

$$= \frac{32}{4\pi^2} \times 0.64^2 \times \text{acceleration}$$

d

$$= 0.022 \times \text{acceleration}$$

**Figure 3.39** Seven-story building: third mode forces and displacements.

shape. In the dynamic analysis, the more representative $\frac{W\phi}{\sum W\phi}$ is used to distribute the forces for each mode. Story shears and overturning moments are determined in the same manner for each method. Modal story accelerations are determined by dividing the story force by the story weight. Modal story displacements are calculated from the accelerations and the period by using the following equations:

$$\delta_{xm} = PF_{xm}S_{am} = PF_{xm}S_{am}\left(\frac{T_m}{2\pi}\right)^2 g \tag{3.12}$$

where $\delta_{xm}$ = lateral displacement at level $x$ for mode $m$
$S_{am}$ = spectral displacement for mode $m$ calculated from response spectrum
$T_m$ = modal period of vibration

Modal interstory drifts $\Delta\delta$ are calculated by taking the difference between the $\delta$ values of adjacent stories. The values shown in Figs 3.37–3.39 are summarized in Fig. 3.40.

The fundamental period of vibration as determined from a computer analysis is 0.88 sec. The periods of the second and third modes of vibration are 0.288 sec and 0.164 sec respectively. From Figs 3.35a–c, using a response curve with 5 percent of critical damping ($\beta = 0.05$), it is determined that the second and third mode spectral accelerations (0.500$g$) are 80 percent greater than the first mode spectral acceleration (0.276$g$). On the basis of mode shapes and modal participation factors, modal story forces, shears, overturning moments, acceleration and displacements are determined.

Figure 3.40a shows story forces obtained by multiplying the story acceleration by the story mass. The shapes of story force curves, (Fig. 3.40a) are quite similar to the shapes of the acceleration curves (Fig. 3.40d), because the building mass is essentially uniform.

Figure 3.40b shows story shears which are a summation of the modal story forces in Fig. 3.40a. The higher modes become less significant in relation to the first mode because the forces tend to cancel each other due to the reversal of direction. The SRSS values do not differ substantially from the first mode values.

Figure 3.40c shows the building overturning moments. Again, the higher modes become somewhat less significant because of the reversal of force direction. The SRSS curve is essentially equal to the first mode curve.

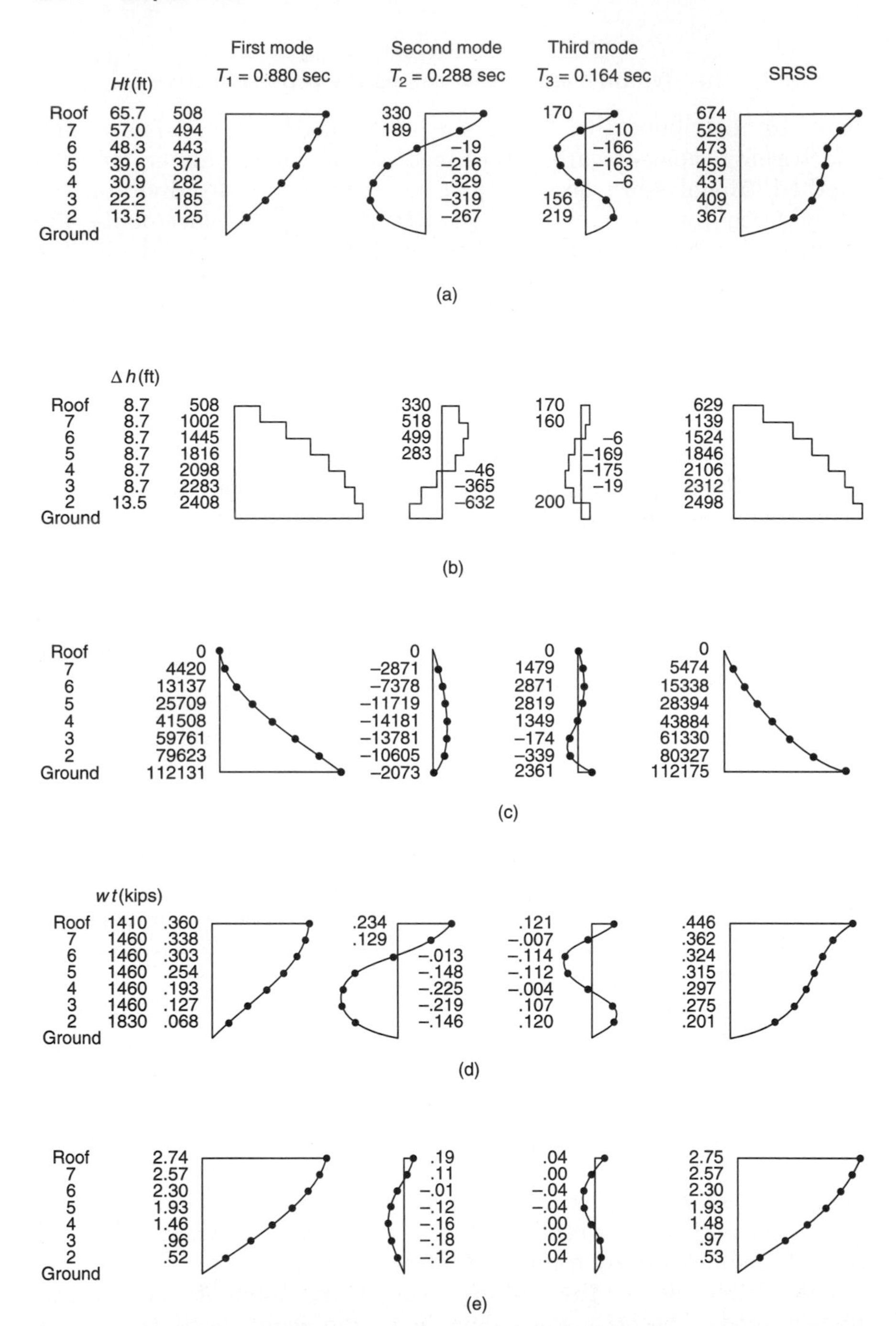


**Figure 3.40** Seven story building. Modal analysis summary: (a) modal story forces (kips); (b) modal story shears (kips); (c) modal story overturning moments (kip-ft); (d) modal story accelerations ($g$'s); (e) modal lateral displacements (in.).

Figure 3.40d shows story accelerations. Observe that the second and third modes do play a significant role in the structure's maximum response. While the shape of an individual mode is the same for displacements and accelerations, accelerations are proportional to displacements divided by the squared value of the modal period, which accounts for the greater accelerations in the higher modes. The shape of the SRSS combination of the accelerations is substantially different from the shapes of any of the individual modes because it accounts for the predominance of the various modes at different story levels.

Figure 3.40e shows the modal displacements. Observe that the fundamental mode predominates, while the second and third mode displacements are relatively insignificant. The SRSS combination does not differ greatly from the fundamental mode. It should be noted that for taller and irregular buildings the influence of the higher modes becomes larger. Consideraton of higher modes of vibration may become necessary to demonstrate that at least 90 percent of the participating mass of the structure is included in the analysis.

## 3.6 Seismic Vulnerability Study and Retrofit Design

### 3.6.1 Introduction

Earthquakes create havoc. Even minor earthquakes which take no lives can cause severe disruptions to utilities, business, transportation, and individuals whose homes and livelihooods are drastically changed within minutes. They happen without warning, and they can happen in many places, from Maine to Alaska, along the New Madrid fault in the Midwest, in areas we usually don't think of as vulnerable.

The most visible damage from an earthquake is to man-made structures; to buildings, bridges, and roads. Older buildings in particular suffer damage from earthquakes. Many buildings in almost all parts of the world were constructed before earthquake-resistant technology was developed. And building damage, whether it is total or partial building collapse, is generally considered the most dangerous aspect of an earthquake, because that's how most lives are lost.

Most Americans are aware of California's vulnerability to earthquakes. But California is not the only state facing earthquake risks. At least 40 states are seismically vulnerable to some degree. The vulnerability is obviously less severe, in terms of people's safety, in some geographic regions than others. Yet the possibility of functional disruption and significant property damage from an earthquake, even if no lives are lost, should be a matter of concern for every community.

Seismic rehabilitation entails costs as well as disruption. In fact, the effects of a rehabilitation program are similar to those of an earthquake because strengthening, in terms of cost and the need to vacate while strengthening is under way, is analogous to building repair after an earthquake. The crucial difference is that strengthening occurs at a specified time and no deaths or injuries will occur during the process.

Before selecting an approach to seismic rehabilitation, building owners must first get the facts on the nature of its seismic hazard, and on the risk to buildings likely to be affected by an earthquake and the people who live and work in the affected buildings. Seismic rehabilitation involves six basic steps.

Step one is seismic hazard assessment, usually done by geologists and seismologists. This is needed to identify the level of risk associated with the seismic vulnerability of buildings.

Step two is to develop an overall damage estimate by determining the probable extent of damage to various types of buildings and structures, as well as the disruptions to the use of the facilities.

Step three is determining building priorities, i.e., categorize the buildings to be strengthened and, within each category, set the priorities for attention. Again, building owners will exercise their judgement on what needs to be done in what sequence.

Step four involves decisions on engineering methods, construction costs, indirect costs and overall effectiveness of the methods employed.

Step five covers the economic impact of rehabilitation. Once the data have been gathered on vulnerability, loss, damage and impact, the emphasis in the seismic rehabilitation program development is on the process of analyzing data, assessing impacts, considering trade-offs and evaluating options. The intention is to reach consensus on acceptable levels of risk, given the economic resources available.

The final step is implementing the seismic strengthening program appropriate for the building and its usage.

In a seismic vulnerability study it is convenient to classify the damage within a building into two categories, structural and nonstructural.

Structural damage refers to degradation of the building's support system, such as frames and walls while nonstructural damage is any damage that does not affect the integrity of the building's physical support system. Examples of nonstructural damage are chimnies that collapse, broken windows or ornamental features, and a collapsed ceiling. The expected type of damage depends on the building's structural characteristics and age, its configuration, construction materials, the site conditions, the proximity of the building to neighboring buildings, and the type of nonstructural elements.

An earthquake can cause a building to experience four types of damage:

1. the entire building collapses;
2. portions of the building collapse;
3. components of the building fail and fall;
4. entry-exit routes are blocked, preventing evacuation and rescue.

Any of the above, may result in unacceptable risk to human lives. It can also mean loss of property, and interruptions of use and normal function.

Another type of damage that should be included in the vulnerability study is the structural damage from the "pounding" action that results when two buildings, insufficiently separated, collide. This condition is particularly severe when the floor levels of the two buildings do not match, and the stiff floor framing of one building can badly damage the more fragile walls or columns of its neighbor.

There are two basic ways to improve a building's structural performance in an earthquake: strengthening load-resisting components and decreasing demand on load-resisting components.

The strengthening of load-resisting members may range from upgrading a single element to replacing the entire system including the diaphragms and chord members because after all, a seismic strengthening program is a response to a unique set of demands.

The second method of improving a building's earthquake resistance is by decreasing demand on existing systems. There are four basic ways to reduce the demand:

1. reduce the weight of the building;
2. increase the fundamental period of vibration;
3. increase the response reduction factor, $R_w$;
4. provide alternative methods such as base isolation and supplemental damping techniques.

Nonstructural architectural elements can also create life-threatening hazards. For example, windows may break or architectural cladding such as granite veneer with insufficient anchorage can cause damage. Consequently, a seismic retrofit program should explore techniques for dealing with nonstructural components, such as veneers, lighting fixtures, glass doors and windows, raised computer access floors, as well as ceilings. Similarly, because damage to mechanical and electrical components can impair building functions that may be essential to life safety, seismic strengthening should be considered for components such as mechanical and electrical equipment, ductwork and piping, elevators, emergency power systems, communication systems and computer equipment.

### 3.6.2 Code sponsored design

It is well known that the level of forces experienced by a structure during a major earthquake is much larger than the actual design forces. Usually, it is neither practical nor economically feasible, to design a building to remain elastic during a major seismic event. Instead, the structure is designed to remain elastic for a reduced force level. And, by prescribing "detailing requirements" the structure is relied upon to sustain post-yield displacements without collapse when subjected to higher levels of earthquake ground motion. The rationale of designing for lower forces is based on the premise that special detailing requirements specified in the code are expected to allow for additional deformation during a major event without collapse. Historically, this approach has produced buildings with a strength capacity up to the scaled-down seismic forces, and more importantly with adequate performance characteristics beyond the elastic range. It is the consensus of the structural engineering profession that a building properly designed to both the code-specified forces and the detailing requirements, will have an acceptable level of seismic safety.

The ability of a member to undergo large deformations beyond the elastic range is termed ductility. In a building, the property that allows it to absorb earthquake-induced damage and yet remain stable, may be considered, in a conceptual sense similar to ductility. Ductile structures may deform excessively under load, but they remain, by and large, intact. This ability prevents total structural collapse and provides protection to occupants of buildings. Therefore, providing capacity for displacement beyond the elastic range without collapse is the primary goal in code-sponsored seismic design.

Aside from this philosophy implicit in the code, there are no explicit earthquake performance objectives stated in the building code. However, they can be found in the first few pages of the 1996 Blue Book reproduced here with some editorial changes.

Building structures designed in conformance with the UBC regulations are expected, in general, to do the following:

1. Resist a minor level of earthquake ground motion without damage.
2. Resist a moderate level of earthquake ground motion without structural damage, but possibly experience some non-structural damage.
3. Resist a major level of earthquake ground motion of an intensity equal to the strongest either experienced or forecast for the building site, without collapse but possibly with some structural as well as non-structural damage.

It is expected that structural damage, even in a major earthquake, will be limited to a repairable level for structures that meet these requirements. However, conformance to these provisions is not an assurance that significant structural damage will not occur in the event of maximum level of earthquake ground motion. Therefore, additional requirements are given in the code to provide for structural stability in the event of extreme structural deformations. The protection of life rather than prevention/repairability of damage is the primary purpose of the code; the protection of life is reasonably provided for but not with complete assurance.

The first two criteria, which are commonly referred to as the serviceability limit state, can be achieved by: (i) specifying service (or moderate) design earthquake levels; (ii) limiting the maximum stresses or internal forces in critical members; and (iii) limiting the story drift ratio. The third criterion, which is prevention of building collapse is achieved not by limiting maximum stresses or story drift, but by providing sufficient strength and ductility to ensure that the structure does not collapse in a severe earthquake.

### 3.6.3 Alternative design philosophy

Although earthquake performance objectives are stated in the Blue Book, significant questions about these objectives still linger on. Does the commentary adequately define the expected earthquake performance? Can the performance be actually delivered? Should the earthquake response objectives be explicitly stated in the code rather than relegated to the commentary in the Blue Book? Is it feasible to make an existing non-ductile building conform to the current detailing/ductility provisions? If not, what level of upgrade will provide "minimum life safety"? Should the upgrade conform to the code?

Explicit answers to these and other similar questions cannot be found in the current building codes because the code approach is mainly empirical and for a new construction. Although a set of minimum design loads are prescribed, the loads may not be appropriate for seismic verification and upgrade design because:

1. The code provisions do not provide a dependable or an established method to evaluate the performance of non-code compliant structures.
2. They are not readily adaptable to a modified criterion, such as one that attempts to limit damage.
3. Since the primary purpose is protection of life safety, the code does not address building owners' business concerns such as protection of property, environment or business.

To overcome the aforementioned shortcomings, a procedure that uses a two-phase design/analysis approach has been in use for some time. The technique explicitly requires verification of serviceability and survival limit states by using two distinct design earthquakes; one that defines the threshold of damage and the other the collapse. The serviceability level earthquake is normally characterized as an earthquake, which has a maximum likelihood of occurring once during the life of the structure. The collapse threshold (i.e., survival limit state) is typically associated with the maximum earthquake that can occur at the building site in the presently known tectonic framework. These characterizations can vary, however, to suit the specifics of the project such as the nature of the facility, associated risk levels, and the threshold of damageability.

The principle behind the two-phase approach may be explained by recalling the primary goal in seismic design, which is to provide capacity for displacement beyond the elastic range. Any number of elastic and post-elastic deformation capabilities may be invoked to attain this goal. For example, at one end of the spectrum we could have a system that would remain elastic throughout the displacement range. This system would have high elastic strength but low ductility. Conversely, it is entirely possible to have a system with a relatively low elastic strength but high ductility, meeting the same design objectives. It is, perhaps, easier to understand the methodology if it is recognized that a specific earthquake excitation causes about the same displacement in a structure whether it responds elastically or with any degree of inelasticity.

In Fig. 3.41, the behavior of an idealized structure subjected to three distinct levels of earthquake forces $F_L$, $F_U$, and $F_C$ corresponding to lower-level, upper-level, and collapse-level earthquakes are compared. Also shown is the earthquake force $F_E$ experienced by the structure if the structure remained completely elastic during the earthquake. The structure designed to the lower-level earthquake force $F_U$ needs to deform elastically from $O$ to $E$ and inelastically from $E$ to $U'$ while the same structure if designed for a substantially higher elastic force of $F_E$ needs to deform $O$ to $U$, responding elastically all through the displacement range. Both systems are capable of attaining the anticipated deformation. Clearly, the $F_L$ criterion will require a more ductile system than the

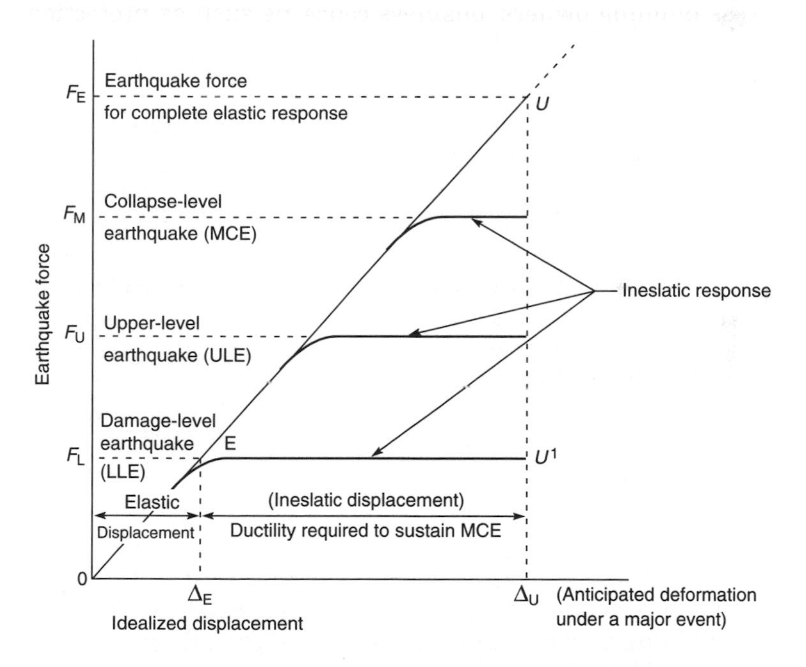


**Figure 3.41** Idealized earthquake force-displacement relationships.

$F_E$ building, and more importantly it will suffer more damage should the postulated event occur. Nevertheless, both systems achieve the primary goal, namely attainment of displacement $C$ without collapse. As can be seen in Fig. 3.10, it is possible to design the structure for any level of force between $F_L$ and $F_E$. For example, a structure designed for force levels, $F_U$ and $F_C$ (upper- and collapse-level earthquake loads) shown in Fig. 3.39 requires higher strength but less ductility than if it is designed for force level $F_L$. It is a matter of choice as to how much strength can be traded off in part for deformability, and conversely, deformability for strength. Expressed in another way, structural systems of limited ductility may be considered valid in seismic design, provided they can resist correspondingly higher seismic forces.

This is the basis of approach used in the seismic vulnerability study and retrofit design of existing buildings. Since buildings of pre-1970 vintage do not have the required ductile detailing, the purpose of the analysis is qualitatively to establish the strength levels that can be traded off in part for lack of required ductility.

As in other disciplines, there are a large number of differing philosophies and theories regarding seismic study. This is especially so with respect to the level of seismic risk in a given locale as well as the behavior of materials, members, and structural systems. As a consequence a number of different methodologies are being developed for seismic vulnerability study and retrofit design. Two of these approaches as given in: (i) FEMA Publication 178, NEHRP Handbook for Seismic Evaluation of Existing Buildings and; (ii) Tri-Services Manual's Demand-Capacity Comparison and Capacity Spectrum method (TM 809-10-1 and 10-2) are considered next, followed by a brief description of performance-based engineering as given in SEAOC's Vision 2000 document.

#### 3.6.3.1 FEMA-178 method

**Seismic demand.** In dealing with seismic evaluation, the ultimate capacity of all components in the building are checked against the following demand due to the effects of gravity and seismic forces.

$$Q = 1.1Q_D + Q_L + Q_S \pm Q_E \tag{3.13}$$

or

$$Q = 0.9Q_D \pm Q_E \tag{3.14}$$

where $Q$ = the effect of the combined loads
$Q_D$ = the effect of dead load
$Q_L$ = the effective live load equal to 25 percent of the unreduced design live load but not less than the actual live load
$Q_S$ = the effective snow load equal to either 70 percent of the full design snow load or, where conditions warrant and approved by the regulatory agency, not less than 20 percent of the full design snow load except that, where the design snow load is less than 30 pounds per square foot, no part of the load need be included in seismic loading; and
$Q_E$ = the effect of seismic forces i.e., the element forces and deflections obtained from a lateral analysis. Where elements of a lateral force resisting system are believed to have less ductility than the system as a whole, the term $Q_E$ is modified to account for their premature brittle failure. For example, if the column splice details of a moment resisting steel frame do not include connections of both flanges and web of columns, the seismic load on the column $Q_E$ is amplified by a factor $C_d/2 = 5.5/2 = 2.75$. The term $C_d$ is a deflection modification factor with a unique value for each structural system as given in Table 8.

**Analysis.** The base shear $V$, for the determination of seismic demand is given by:

$$V = C_S W \tag{3.15}$$

where $C_S$ = the seismic design coefficient determined below and
$W$ = the total dead load and applicable portions of the following

(1) in storage and warehouse occupancies, a minimum of 25 percent of floor live load,

(2) where an allowance for partition load is included in the floor load design, the actual partition weight or a minimum weight of 10 psf of floor area, whichever is greater;

(3) total operating weight of all permanent equipment; and

(4) the snow load as defined above.

There are two equations for determining the seismic coefficient, $C_S$. Equation (3.16) depends on the building period and Eq. (3.17), which gives an upper limit to the value of $C_S$, is applicable to buildings with short periods.

The seismic coefficient, $C_S$, for existing buildings is taken as 67 percent of the value for new buildings as given in the 1988 *NEHRP Recommended Provisions* as follows:

$$C_S = 0.67\left(\frac{1.2A_V S}{RT^{2/3}}\right)$$

$$= \frac{0.80A_V S}{RT^{2/3}} \qquad (3.16)$$

where $A_V$ = the peak velocity-related acceleration coefficient given in Fig. 3.42

$S$ = the site coefficient given in Table 3.28 (in locations where the soil properties are not known in sufficient detail to determine the soil profile type or where the profile does not fit any of the four types, the value of $S$ should be taken as 1.5.)

$R$ = a response modification coefficient from Table 3.8 and

$T$ = the fundamental period of the building

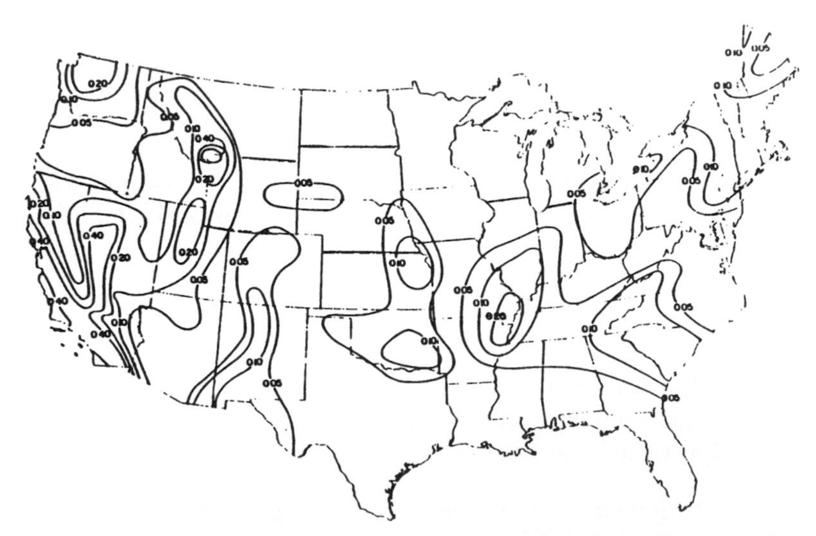

**Figure 3.42** Contour map for the effective peak acceleration coefficient $A_a$ for the continental United States.

**TABLE 3.8 Response Coefficients**

| $R$ | $C_d$ | System |
|---|---|---|
| | | **Bearing Wall Systems** |
| 6.5 | 4 | Light-framed walls with shear panels |
| 4.5 | 4 | Reinforced concrete shear walls |
| 3.5 | 3 | Reinforced masonry shear walls |
| 4 | 3.5 | Concentrically braced frames |
| 1.25 | 1.25 | Unreinforced masonry shear walls |
| | | **Building Frame Systems** |
| 8 | 4 | Eccentrically braced frames, moment resisting connections at columns away from link |
| 7 | 4 | Eccentrically braced frames, non-moment resisting connections at columns away from link |
| 7 | 4.5 | Light-framed walls with shear panels |
| 5 | 4.5 | Concentrically braced frames |
| 5.5 | 5 | Reinforced concrete shear walls |
| 4.5 | 4 | Reinforced masonry shear walls |
| 3.5 | 3 | Tension-only braced frames |
| 1.5 | 1.5 | Unreinforced masonry shear walls |
| | | **Moment Resisting Frame System** |
| 8 | 5.5 | Special moment frames of steel |
| 8 | 5.5 | Special moment frames of reinforced concrete |
| 4 | 3.5 | Intermediate moment frames of reinforced concrete |
| 4.5 | 4 | Ordinary moment frames of steel |
| 2 | 2 | Ordinary moment frames of reinforced concrete |
| | | **Dual System with a Special Moment Frame Capable of Resisting at Least 25% of Prescribed Seismic Forces** |
| | | *Complementary seismic resisting elements* |
| 8 | 4 | Eccentrically braced frames, moment resisting connections at columns away from link |
| 7 | 4 | Eccentrically braced frames, non-moment resisting connections at columns away from link |
| 6 | 5 | Concentrically braced frames |
| 8 | 6.5 | Reinforced concrete shear walls |
| 6.5 | 5.5 | Reinforced masonry shear walls |
| 8 | 5 | Wood sheathed shear panels |
| | | **Dual System with an Intermediate Moment Frame of Reinforced Concrete or an Ordinary Moment Frame of Steel Capable of Resisting at Least 25% of Prescribed Seismic Forces** |
| | | *Complementary seismic resisting elements* |
| 5 | 4.5 | Concentrically braced frames |
| 6 | 5 | Reinforced concrete shear walls |
| 5 | 4.5 | Reinforced masonry shear walls |
| 7 | 4.5 | Wood sheathed shear panels |
| | | **Inverted Pendulum Structures** |
| 2.5 | 2.5 | Special moment frames of structural steel |
| 2.5 | 2.5 | Special moment frames of reinforced concrete |
| 1.25 | 1.25 | Ordinary moment frames of structural steel |

The value of $C_S$ need not be greater than 85 percent of the limiting value given in the 1988 NEHRP provisions as follows;

$$C_S = 0.85\left(\frac{2.5A_a}{R}\right)$$

$$= \frac{2.12A_a}{R} \tag{3.17}$$

where $A_a$ = the effective peak acceleration coefficient given in Fig. 3.43

**Period calculation.** The building period is calculated using either Method 1 or Method 2.

**Method 1.** The value of $T$ may be taken to be equal to the approximate fundamental period of the building, $T$, determined as follows:

a. For buildings in which the lateral-force-resisting system consists of moment resisting frames capable of resisting 100 percent of the required lateral force and such frames are not enclosed or adjoined by more rigid components tending to prevent the frames

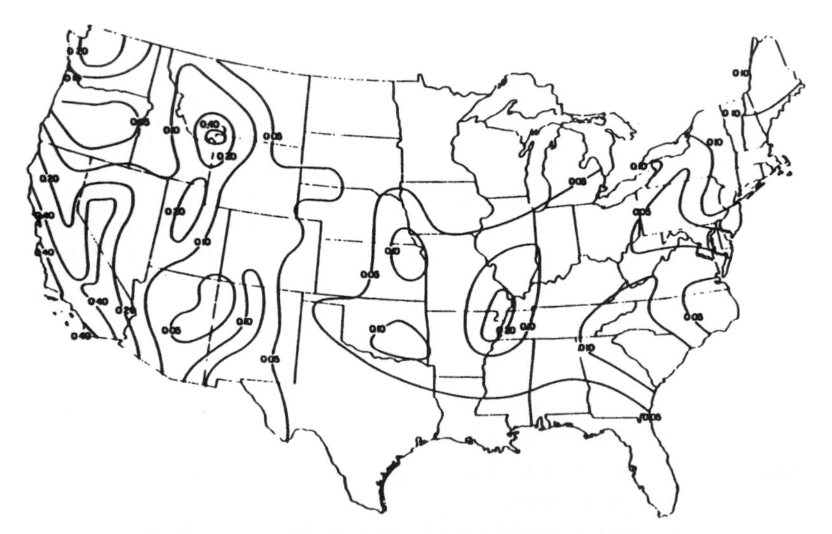

**Figure 3.43** Contour map for the effective peak velocity-related acceleration coefficient $A_V$ for the continental United States.

from deflecting when subjected to seismic forces:

$$T_a = C_T h_n^{3/4} \tag{3.18}$$

where $C_T = 0.035$ for steel frames
$C_T = 0.030$ for concrete frames, and
$h_n =$ the height above the base to the highest level of the building

b. As an alternate for concrete and steel frame buildings of 12 stories or fewer with a minimum story height of 10 feet, the equation $T = 0.10N$, where $N =$ number of stories, may be used in lieu of above equation.

c. For all other buildings,

$$T_a = \frac{0.05h_n}{\sqrt{L}} \tag{3.19}$$

where $L =$ the overall length (in feet) of the building at the base in the direction under consideration

**Method 2.** The fundamental period $T$ may be estimated using the structural properties and deformational characteristics of the resisting elements in a properly substantiated analysis. This requirement may be satisfied by using the following equation:

$$T = 2\pi\sqrt{\frac{\Sigma\,(w_i d_i^2)}{g\,\Sigma\,(f_i d_i)}} \tag{3.20}$$

The values of $f_i$ represent any lateral force, associated weights $w$. The elastic deflections, $d_i$, should be calculated using the applied lateral forces, $f_i$. The period used for computation of $C_S$ shall not exceed $C_a T_a$ where $C_a$ varies with $A_V$ as follows: $C_a = 1.2, 1.3, 1.4, 1.5, 1.7, 1.7$; $A_V = 0.4, 0.3, 0.2, 0.15, 0.1, 0.05$. For example, the period $T$ for a building in Los Angeles with $A_v = 0.4$, should not exceed 1.2 times the value obtained from the simpler analysis, i.e. $T \leqslant 1.2T_a$. And for a building in Las Vegas with $A_v = 0.2$, $T \leqslant 1.4T_a$.

**Vertical distribution of forces.** The lateral force, $F$, induced at any level is determined as follows

$$F_x = C_{vx} V \tag{3.21}$$

and

$$C_{vx} = \frac{w_x h_x^k}{\sum_{i=1}^{n} w_i h_i^k} \tag{3.22}$$

where $C_{vx}$ = vertical distribution factor

$V$ = total design lateral force at the base of the building

$w_i$ and $w_x$ = the portion of the total gravity load of the building $W$ located or assigned to Level $i$ or $x$

$h_i$ and $h_x$ = the height in feet from the base to Level $i$ or $x$, and

$k$ = an exponent related to the building period as follows
For buildings having a period of 0.5 sec or less, $k = 1$
For buildings having a period of 2.5 sec or more, $k = 2$
For buildings having a period between 0.5 and 2.5 sec, $k$ may be taken as 2 or may be determined by linear interpolation between 1 and 2

For checking deformations and drift limits, the elastic deformations caused by the seismic forces are computed and then multiplied by the factor $C_d$ to determine the total deformation. Interstory drifts are compared to 1.33 times the limits for new buildings. For purposes of this drift analysis only, it is permissible to use the computed fundamental period $T$ of the building without the upper bound limitations.

**Diaphragms.** Floor and roof diaphragms should be designed to resist the seismic forces determined as follows: a minimum force equal to $0.5A$ times the weight of the diaphragm and other elements of the building attached thereto. The portion of the seismic shear force at that level, $V$, required to be transferred to the components of the vertical seismic resisting system should also be added to the design seismic shear. These additional forces are caused by offsets or changes in stiffness of the vertical components above and below the diaphragm.

**Element capacities.** The element capacities are calculated on the ultimate-strength basis. When calculating capacities of deteriorated elements, appropriate reductions should be made in the material strength, the section properties, and any other aspects of the capacity affected by the deterioration.

**Dynamic analysis.** The use of a dynamic analysis procedure is required for tall buildings, for buildings with vertical irregularities caused by significant mass or geometric irregularities, and for other buildings where the distribution of the lateral forces departs from that assumed in the equivalent lateral force procedure. Scaling of results, either up or down to correspond to that required by the equivalent lateral force procedure, is permitted.

**Acceptance criteria.** The basic acceptance criterion is:

$$Q \leqslant C \tag{3.23}$$

Where elements or portions of a lateral force resisting system are expected to behave in a less ductile manner than the system as a whole, the term $Q_E$ in Eqs (3.13) or (3.14) should be modified or special calculations be made to account for the different failure modes of the various elements. This is because less ductile, i.e. brittle, elements may fail prematurely unless they are strong enough to remain elastic.

The following are some instances in which the earthquake demand $Q_E$ in Eqs (3.13) and (3.14) should be modified.

1. Buildings with significant strength discontinuity in the lateral force resisting system; the story strength at any level is less than 65 percent of the strength of story above. To compensate for this condition the design forces should be amplified by the factor $C_d/2$ but not less than 1.5.
2. Buildings with vertical discontinuity in the lateral force resisting system such as with a discontinuous shear wall. The overturning forces in the columns supporting the discontinuous element should be designed for seismic forces amplified by the factor $C_d/2$ but not less than 1.5, or for the shear capacity of the wall if this is greater.
3. Splice details in a steel frame column does not have connections to both flanges and web of columns. The seismic force on the column is amplified by a factor $C_d/2 = \dfrac{5.5}{2} = 2.75$.

If all significant elements meet the basic acceptance criteria, no further analysis is needed. Elements that do not meet the acceptance criteria are the deficiencies that should be addressed in a further study or a rehabilitation program.

The criterion $Q \leqslant C$ is an indication of whether an element meets the requirements of FEMA-178; however, because $Q$ involves gravity effects, the ratio of $Q$ to $C$ for an element may not necessarily be a good indicator of the seriousness of the earthquake hazard. However, the seriousness of the deficiencies can be assessed by listing the $D_E/C_E$ ratios in descending order where $D_E$ and $C_E$ are the earthquake portion of the seismic demand and capacity. The element with the largest value is the weakest link in the building. If the element can fail without jeopardizing the building or can be fixed easily, then attention should be focused on the element with the next lower ratio, and so on.

**Final evaluation.** The final evaluation should be made based on a review of the qualitative and quantitative results.

The evaluating engineer is urged to dwell on the issues carefully, to refrain from penalizing the building over fine technical points beyond those contained in this evaluation methodology, and to visualize the building in its ultimate condition in the earthquake, being aware of the risks of brittle failure and buckling. Due consideration should be given to the mitigating influence of good workmanship, structural integrity, and the strengths and redundancies that are not explicitly considered to be part of the lateral-force-resisting system. Most importantly, engineering judgment based on sound seismic design principles should be exercised before pronouncing the building as unsafe.

### 3.6.3.2 TriServices manual

**(1) Demand-capacity method.** The method given here with a slight revision in format is based on the concept of Inelastic Demand Ratios, IDR. An elastic dynamic analysis is made using the design earthquake response spectrum without the use of $R_W$ factor. A higher damping value, for example, a 10 percent instead of 5 percent, is allowed to account for the effects of yielding and high stress levels in structural components. Reduced stiffnesses are used to allow for lengthening of the building periods due to inelastic response. For each component of the structure, the ratio of the calculated demand to the capacity of each component are compared to allowable consensus values of IDR's.

**Seismic forces.** Two earthquakes, EQ-1 and EQ-2, are defined. EQ-1 has a 50 percent probability of being exceeded in 50 years

**TABLE 3.9 Inelastic Demand Ratios**

| Building system | Element | Essential | High risk | Others |
|---|---|---|---|---|
| Steel SMRSF | Beams | 2.0 | 2.5 | 3.0 |
| | Columns* | 1.25 | 1.5 | 1.75 |
| Braced Frames | Beams | 1.5 | 1.75 | 2.0 |
| | Columns* | 1.25 | 1.5 | 1.75 |
| | Diag. Braces† | 1.25 | 1.5 | 1.5 |
| | K-Braces‡ | 1.0 | 1.25 | 1.25 |
| | Connections | 1.0 | 1.25 | 1.25 |
| Concrete SMRSF | Beams | 2.0 | 2.5 | 3.0 |
| | Columns* | 1.25 | 1.5 | 1.75 |
| Concrete Walls | Shear | 1.25 | 1.5 | 1.75 |
| | Flexure | 2.0 | 2.5 | 3.0 |
| Masonry Walls | Shear | 1.1 | 1.25 | 1.5 |
| | Flexure | 1.5 | 1.75 | 2.0 |
| Wood | Trusses | 1.5 | 1.75 | 2.0 |
| | Columns* | 1.25 | 1.5 | 1.75 |
| | Shear Walls | 2.0 | 2.50 | 3.0 |
| | Connections (other than nails) | 1.25 | 1.50 | 2.0 |

* In no case shall axial loads exceed the elastic buckling capacity.

† Full panel diagonal braces with equal member acting in tension and compression for applied lateral loads.

‡ *K*-bracing and other concentric bracing systems that depend on compression diagonal to provide vertical reaction for tension diagonal.

(return period of 72 years) while EQ-2 has a 10 percent probability in 100 years (return period of 950 years). The structure is evaluated for its ability to resist EQ-1 by elastic behavior, and EQ-2 by post-elastic behavior, both with ductility limitations given in Table 3.9.

Buildings are classified into four categories: hazardous, essential, high-risk, and others, as defined below.

1. Hazardous facilities: These facilities, which include nuclear power plants, dams, and LNG facilities, are beyond the scope of design normally performed by building design engineers. Therefore, they are not discussed further here.
2. Essential facilities: These are structures housing facilities that are necessary for post-disaster recovery and require continuous operation during and after an earthquake, e.g., hospitals, fire stations, rescue stations, and garages for emergency vehicles, etc.
3. High-risk: This classification includes those structures where primary occupancy is for the assembly of a large number of people: e.g., Auditorium, Dining Hall, Confinement Facilities such as Prisons, etc.

4. All others: All other buildings not covered by the above descriptions.

Permissible, Inelastic Demand Ratio, IDR, for various systems for the three building categories are given in Table 3.9.

Essential facilities are evaluated for their ability to resist the two levels of earthquake motion, EQ-1 and EQ-2 without any decrease in the forces predicted by the analyses.

High-risk buildings are evaluated similar to essential facilities using the same two-level approach, except that the forces resulting from EQ-1 are reduced by 15 percent.

All other buildings (buildings that are not classified as essential or high-risk facilities) are evaluated similar to essential facilities, except that forces from EQ-1 are reduced by 30 percent.

**Gravity loads.** The magnitude of lateral loads experienced by a building is a function of the total dead load and applicable portions of other loads which contribute to the total mass of the building. The dead load includes weight of structural slabs, beams, columns, and walls as well as non-structural components such as partitions, ceilings, floor topping, roofing, fireproofing material, and fixed electrical and mechanical equipment. When partition locations are subject to change, a uniformly distributed load of 20 pounds per square foot of floor space is used as additional dead load. It is customary to include a certain fraction of this intensity as an equivalent mass in calculating the seismic lateral loads (50 percent per UBC 1994).

Live loads result from expected usage, and in most cases they are simulated by uniformly distributed live loads placed over the entire area of the floor or roof. The effect of live loads needs to be considered when an estimate is made for the mass of a building.

**Demand capacity comparison.** In a broad sense, demand is conceptually similar to a load applied to a structure while capacity may be considered similar to available resistance. The demands for verifying the seismic behavior of the building under EQ-1 and EQ-2 are obtained from corresponding response spectra. The loads obtained from response spectra are considered as factored ultimate loads, that is, no additional load factors are applied over and above the loads obtained from the spectra. The following two combinations are used in evaluating the demand, $D$ under various earthquakes.

$$D = DL + \frac{LL}{2} + E_i \quad (3.24)$$

$$D = 0.9Dl \pm E_i \quad (3.25)$$

where $DL$ = dead load
$LL$ = design live load
$E_i$ = earthquake load corresponding to EQ-1 and EQ-2
$D$ = demand due to combined earthquake and gravity loads

The capacity, i.e., the available resistance in a member is determined by using an equation of the following form:

$$C = \alpha \times EC \times \eta_i \quad (3.26)$$

where $C$ = capacity of the member
$EC$ = nominal elastic capacity obtained by upgrading the working strength capacity to yield strength criteria using appropriate interaction relations with gravity loads
$\alpha$ = reduction factor, set essentially by engineering judgment, to modify the nominal elastic capacity of members and connections. The reason for reducing the capacity is to allow for reduced energy absorption capacity in existing structures which fail to meet the design/detailing criteria of the current UBC
$\eta_i$ = magnification factor that displays the amount of ductility that can be permitted

This factor, also referred to as the Inelastic Demand Ratio, IDR, is an indication of the relative extent of post-yield behavior that will be permitted for each type of element. The permissible IDR values are shown in Tables 3.9.

A comparison of demand versus the capacity for each member is made by using the following equations:

$$DL + \frac{LL}{2} + E_i \leqslant \alpha \times EX \times \eta_i \quad (3.27)$$

$$0.9DL + E_i \leqslant \alpha \times EC \times \eta_i \quad (3.28)$$

**Material properties and design parameters**

1. Concrete
   (a) Since most in-situ concretes possess compressive strengths greater than those specified due to strength gains over time, it

is rational to assume strengths of existing concrete 15 percent above the specified concrete strengths. If deemed necessary, the assumed strengths may be verified by a testing program.
(b) Ultimate strains of 0.007 may be used in-lieu of 0.003 normally used in design, the rationale being that a large body of test results support the use of ultimate strains greater than 0.003.

2. Reinforcing steel
Since most reinforcing steels exhibit yield strengths greater than the specified values, using an over-strength of 25 percent appears to be reasonable. If a sensitivity study indicates that higher values are critical for the survival of the building, it may be necessary to confirm strengths by tests.

3. Capacity reduction factor
Buildings may have less strength in members and connections due to imperfect workmanship, e.g., undersized members, bars placed out of position or voids in concrete. In new designs this is accounted for by using a capacity reduction factor $\phi$ varying from 0.7 to 0.9. In seismic evaluations use of a reduction factor of $\phi = 1.0$ (i.e., no reduction in capacity) may be justified. This may be revised to an appropriate value depending upon the results of a sensitivity study.

4. Member properties
Consideration should be given to using moments of inertia less than the gross moments of inertia. However, bench-mark analyses should use gross properties. In evaluating the ultimate deformation of the inelastic structure, use of properties substantially less than gross properties are recommended.

**Presentation of results.** At the conclusion of the analysis the results should be assembled and reviewed to establish a list of deficiencies. The evaluation should be enhanced if required by further investigation of the elements that do not meet the basic acceptance criteria. The elements with the highest $D/C$ ratios are the ones of most concern. The importance of these elements should be assessed by addressing questions such as: How many have an excessive $D/C$? How high are the ratios? What would be the consequences of failure of these elements? The assessment should include descriptions of probable cracking, spalling or partial collapses.

**(2) Capacity spectrum method.** In this method, given in the Triservices Manual, a global force-displacement capacity curve obtained from a pushover analysis is graphically compared to the response spectrum earthquake demands.

The demand curve is represented by a composite of two spectra: one for the structure acting at service level, and the other for the structure acting in the post-yield (inelastic) range. For example, a 5 percent damped spectrum may be used to represent demand when the structure is responding elastically, and a 10 or 20 percent damped spectrum may be used to represent reduced demand in the inelastic range.

In a pushover analysis the lateral forces are applied to a structure until some components reach their elastic limit or yield point. Components that are about to yield or develop plastic moments are hinged to allow them to yield without taking any additional force. As additional lateral forces are applied, the loads are redistributed to the remaining elastic components until additional components reach their elastic limit. The procedure is repeated until the lateral-force resisting system can no longer take additional force or until a target displacement associated with a specified performance level is reached. The former condition is reached when a plastic mechanism forms, excessive displacement occurs that causes vertical instability, components degrade, or brittle failure occurs. This limit state, at which the structure can resist no further increase in load is the ultimate capacity state. On the other hand, the target displacement generally defined as the roof displacement at the center of mass of structure, serves as an estimate of the global displacement experienced by the structure in a design earthquake.

The pushover represents the action of the whole building. At each increment of base shear $V$, the story forces are applied in proportion to the product of story weight times the story mode shape factor $\left(\text{i.e., } F_x = \left(\frac{w_x \phi_x}{\sum w_x \phi_x}\right) V\right)$. Cumulative $V$ is plotted against cumu lative roof displacement $\Delta_R$ at each increment. The roof displacement and base shear coefficient coordinates are converted to spectral displacements $S_d$ and spectral acceleration $S_a$, respectively, by using modal participation factors and effective modal weights as determined from dynamic characteristics of the fundamental mode of the structure.

If the capacity curve can extend through the envelope of the demand curve, the building survives the earthquake. The intersection of the capacity curve and the appropriate damped demand curve represents the inelastic force and displacement of the structure.

Assume that a building has the following characteristics as determined by a dynamic analysis.

### Given

Building weight $W = 12{,}000$ kips
Building height, 6-stories @ 11 ft = 66 ft
Modal roof participation factor $PF_n$, effective modal weight $\alpha$, and fundamental period $T$ corresponding to points A, B, C and D on the pushover curve are as shown in Table 3.10.

### Required

An estimate of the elastic and inelastic response of the building to two earthquakes represented by a 5 and 10 percent damped spectra (Fig. 3.45) is required. Only the fundamental mode of response is sought.

### Procedure

Step 1. Determine the elastic capacity EC of each structural element. The capacities are the strengths of the elements at the point of yielding (e.g., negative and positive moment capacities at each end of girder, interaction diagrams at $\phi = 1.0$ for each column, shear and moment capacities of shear walls at various key locations).

Step 2. Determine the net capacity available for earthquake loading in each element using the load combination criteria given in Eqs (3.29) and (3.30). Note that the net earthquake capacity is reduced by gravity loads $D$ and $L$ when they are in the same sense as seismic loads and increased by dead loads $D$ when they are in the opposite sense.

$$\text{Net earthquake capacity} = EC - D - L \tag{3.29}$$

$$\text{Net earthquake capacity} = EC + D \tag{3.30}$$

Step 3. Perform a modal analysis using the response spectrum method. Obtain member forces (i.e., moments, shears, axial forces).

Step 4. Divide the net earthquake capacities for each element by the corresponding earthquake forces. This gives the local elastic capacity ratio for each element. Find the element with the lowest ratio, or the group of elements whose ratios fall within 10 percent of the lowest. These elements will yield first: they define the global elastic capacity ratio for the structure.

Step 5. Establish the point of initial major yielding, the first point on the capacity curve, by multiplying the base shear and lateral roof displacement by the global elastic capacity ratio for the structure. This point is represented as point $A$ by $V = 2200$ kips and $\delta_n = 2.3$ in. for the example building.

Step 6. Determine the first post-yield segment of the capacity curve. The structure is essentially frozen at the point of initial major yielding. The balance of net capacity in each element still available for additional earthquake loading is determined. Elements that are at or near (e.g., within 10 percent) their yield capacities are modeled as plastic hinges (e.g., beam elements might have their moments of inertia reduced to near zero values). Lateral forces proportional to the fundamental mode shape are applied to the revised mathematical model. For the example building, the base shear is assumed equal to 1000 kips and the corresponding roof displacement equal to 2.0 in. The resulting forces on the elements are compared to the balance of net earthquake capacities, and lateral displacements are calculated. It is determined that 40 percent of the applied loads will form a new group of yielding elements. A second point on the capacity curve is determined at $V = 2600$ kips and $\delta_n =$ 3.1 in. (2200 kips at point A plus 40 percent of 1000 kips, and 2.3 in. at point A plus 40 percent of 2.0 in.), represented by point B.

Step 7. Determine sequential post-yield segments on the capacity curve by repeating the procedure in Step 6 above (e.g., points C and D in Fig. 3.44 using revised mode shapes and mathematical models).

Step 8. The procedure is repeated until a failure mechanism, instability, or excessive deformations occur. Rotational ductility demands can be approximated by using M/EI diagrams for the yielding girders, taking into account the reduced EI's used in the yielding mathematical model. Ductility demands for flexure should not exceed 2 times the Inelastic Demand Ratios of Table 3.9, and for all other conditions they should not exceed the values shown in Table 3.9. Interstory displacements are determined by superposition of the lateral story displacements of the sequential models. For the example building, the ultimate global capacity of the structure is represented by point D at $V = 3000$ kips and $\delta_n = 8.7$ in. in Fig. 3.44.

Step 9. Determine lateral displacements and drift demands. The capacity curve is converted to $S_a$ and $T$ coordinates and superimposed on the response spectrum curve. If the curves do not intersect, irreparable damage or collapse of the structure is anticipated. If the curves do cross, the intersection can be used to approximate the response of the structure. For the example building, the force-deflection curve of Fig. 3.44 is converted into a spectral curve shown in Fig. 3.45. A table of calculations is set up as shown in Table 3.10. Spectral values are calculated for each of the yield points, A, B, C, and D. Base shear $V$ and roof deflection $\delta_n$ are entered from Fig. 3.44. The quantity $V/W$ is calculated. The spectral quantities are determined by the following formulas. The modal roof participation factor $PF_N$ and the effective modal weight $\alpha$ are calculated from Formulas given in Eqs (3.8) and (3.39).

$$S_a = \frac{V/W}{\alpha} \tag{3.31}$$

$$S_d = \frac{\delta_n}{PF_n} \tag{3.32}$$

$$T = 2\pi\sqrt{\frac{S_d}{S_a g}} \tag{3.33}$$

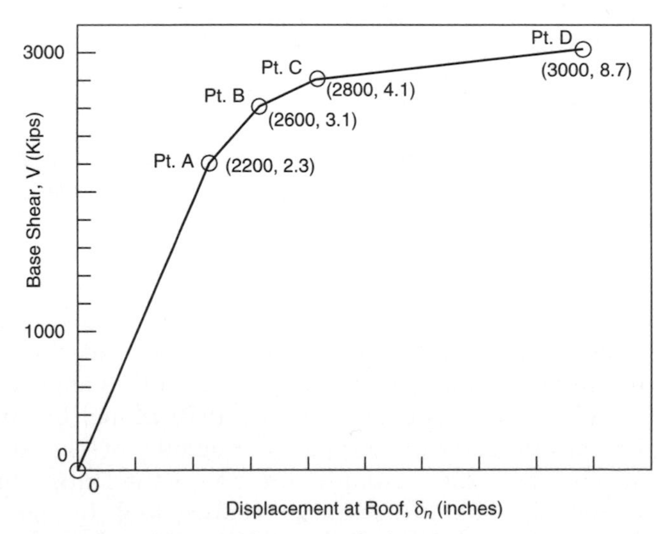


**Figure 3.44** Example problem: push over analysis.

The values of $S_a$ and $T$ are used to construct the spectral capacity curve of Fig. 3.45. Use the 5 percent-damped demand curve for the elastic capacity ($T < 0.80$ sec) and the 10 percent-damped demand curve for the ultimate capacity ($T > 1.4$ sec). A transition curve is drawn between $T =$ 0.80 sec and $T = 1.4$ sec. The intersection of the capacity and demand curves occurs at a point about $S = 0.35g$ and $T = 1.0$ sec. The lateral story displacements at this intersection are calculated from the following Formula

$$\delta_n = PF_n S_a \left(\frac{T}{2\pi}\right)^2 g$$

$$= 1.28 \times 0.35 \times \left(\frac{1.0}{2\pi}\right)^2 \times 386$$

$$= 4.38 \text{ in.} \qquad (3.34)$$

The roof displacement equals about 4.4 in. for a 6-story building, 66 ft high. Maximum interstory displacements can be obtained from a composite deflected shape estimated from the sequential incremental analysis done above. For the example building, the average interstory drift is 0.73 in. The maximum interstory drift, which is at the second story,

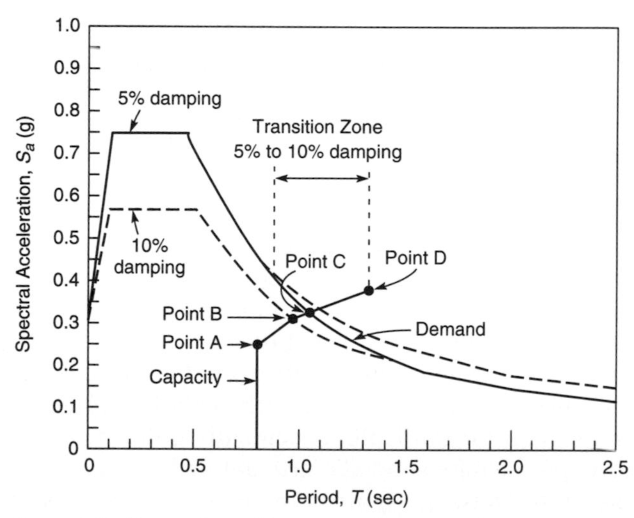


**Figure 3.45** Example problem: capacity spectrum.

**TABLE 3.10 Conversion of Base Shear $V$ and roof Deflection $\delta_n$ to Spectral Acceleration $S_n$, Spectral Displacement $S_d$ and Spectral Period $T$**

| Point on pushover curve | Results from pushover analysis | | $\frac{V}{w}$ | Results from modal analysis | | Spectral values | | |
|---|---|---|---|---|---|---|---|---|
| | Base shear $V$ (kips) | Roof deflection $\delta_n$ (in.) | | Modal roof participation factor | Effective modal weight $\alpha$ | Spect2al acceleration $S_a$ (g) | Spectral displacement (in.) | Fundamental period of vibration $T$ (sec) |
| | (1) | (2) | (3) | $PF_n$ (4) | (5) | (6) $= \frac{3}{5}$ | (7) $= \frac{2}{4}$ | (8) |
| A | 2200 | 2.3 | 0.22 | 1.30 | 0.78 | 0.280 | 1.77 | 0.80 |
| B | 2600 | 3.1 | 0.26 | 1.28 | 0.80 | 0.325 | 2.42 | 0.87 |
| C | 2800 | 4.1 | 0.28 | 1.28 | 0.80 | 0.350 | 3.2 | 0.97 |
| D | 3000 | 8.7 | 0.33 | 1.26 | 0.83 | 0.361 | 6.9 | 1.40 |

equals 1.1 in. or 0.0083 times the story height. Thus, it satisfies the generally specified requirements of drift (i.e., less than 0.015).

Step 10. The results of this procedure give an estimate of the inelastic response of a building to a severe earthquake. In general, it will result in lower force levels and larger displacements than the results of Demand-Capacity method. Neither procedure is necessarily more accurate than the other; however, an evaluation of both procedures should give the designer enough insight to determine the weak links in the structural system, evaluate the potential for instability, and suggest possible structural modifications.

#### 3.6.3.3 SEAOC's Vision 2000: performance-based engineering

Vision 2000 is a SAEOC project specially devoted to the development of a framework for performance-based engineering of buildings—buildings that are engineered to avoid economic losses associated with damage and post-earthquake loss of function.

It is a radical departure from current practice in that it seeks to provide the structural engineering profession with tools to explicitly, rather than implicitly, design for multiple, specifically defined, levels of performance. These performance levels are defined in terms of specific limiting damage states, against which a structure's performance can be objectively measured. Recommendations are developed as to which performance levels should be attained, by buildings of different occupancy and use, under several levels of earthquake loading. This tiered specification of performance levels to be achieved at predetermined earthquake hazard levels becomes the design performance objective and a basis for design. It recognizes the importance of the performance of all the various component systems to overall building performance and defines both responsibility and methodology for design to obtain the desired performance.

Performance-based engineering is defined as "selection of design criteria, appropriate structural systems, layout, proportioning, and detailing for a structure and its nonstructural components and contents and the assurance of construction quality control such that at specified levels of ground motion and with defined levels of reliability, the structure will not be damaged beyond certain limiting states."

Vision 2000 has identified five performance levels (Fig. 3.46). These are: Fully Operational, Operational, Life Safe, Near Collapse, and Collapse. Each of these performance levels has assocaited with it defined levels of damage to structural, architectural, mechanical, and electrical building components as well as tenant furnishings. Figure 3.46 provides a broad overview of where each performance level falls within the overall spectrum of possible damage states.

As with performance levels, there are infinite possible hazard levels that could be used in the development of design performance objectives. Among these four hazard levels have been selected. As shown in Fig. 3.47 these are: Frequent earthquakes, having a 50 percent chance of exceedance in 30 years (43 year mean return period); Occasional earthquakes, having a 50 percent chance of exceedance in 50 years (72 year mean return period); Rare earthquakes, having a 10 percent change of exceedance in 50 years (475 year mean return period) and Very Rare earthquakes, having a 10 percent chance of exceedance in 100 years (950 year return period).

In order to execute performance-based engineering, it is necessary to have a series of design parameters and acceptance criteria for each performance level for the various structural and nonstructural components which comprise the building. Design parameters are calculable response measures such as element forces, interstory drifts, and plastic rotations, etc. that can be derived from a structural analysis of building response to a particular design earthquake. Acceptance criteria are the limiting values for design parameters in order to attain a given performance level. As an example, if interstory drift ratio is a design parameter used for certain classes of building, acceptance criteria would be certain defined drift ratios for each performance level such as 0.020 for the Near Collapse level, 0.015 for the Life Safe level, 0.01 for the Operational level, and 0.005 for the Fully Operational level. A wide variety of potential design parameters may need to be defined including deformation, strength, and energy-based parameters.

Vision 2000 also notes that since current seismic resistant design

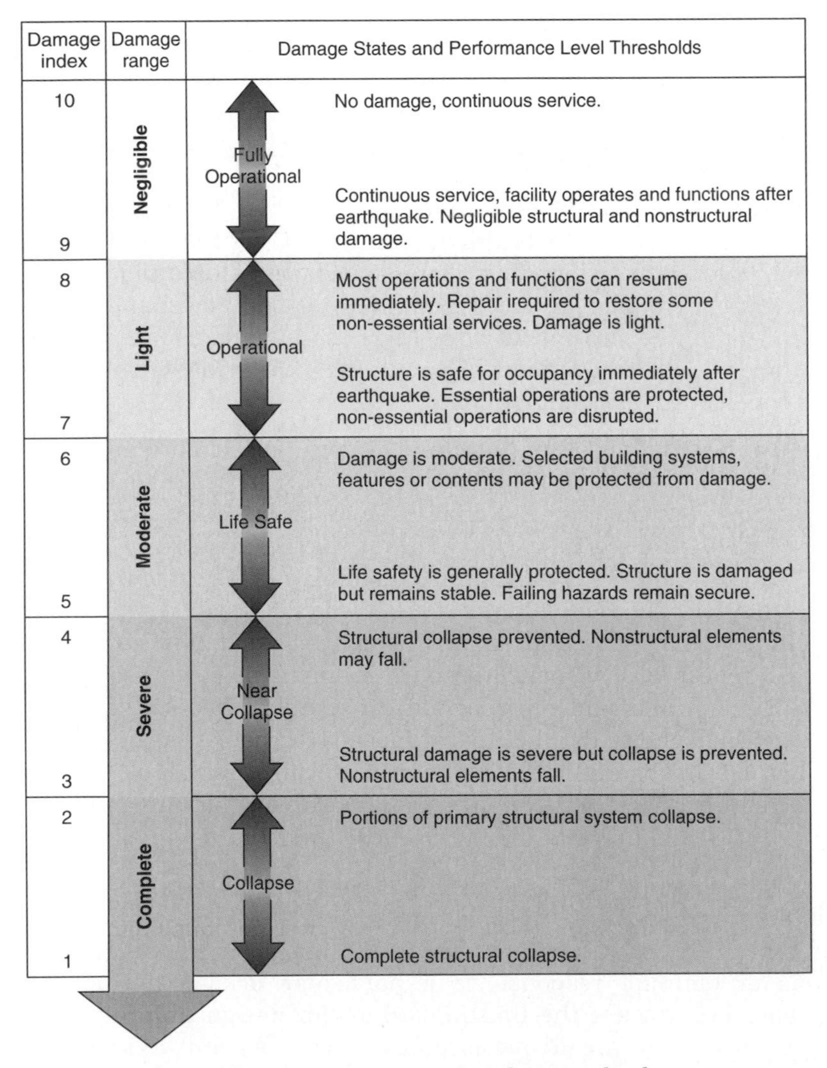


**Figure 3.46** Vision 2000: damage states and performance levels.

relies on the inelastic response of the structure to dissipate most of the input energy of the earthquake, it is imperative that the post-elastic behavior of the structure be addressed at the conceptual state of design. Application of capacity design principals is recommended to provide for the inelastic response by designating the ductile links or "fuses" in the lateral force resisting system. The designated "fuses" will be counted upon to yield and to dissipate the input energy of the earthquake while the rest of the system remains elastic. This concept gives a clear understanding of the inelastic response of the structure such that the design and quality assurance programs may be focused on the critical links in the system. At the final design and detailing stage, these critical links will be rigorously detailed to provide the ductility required.

At each stage of design, an acceptability check is performed to verify that the selected performance objectives are being met. The specific extent and methodology of the analysis will vary with the performance objectives and the design approach used, however, the general concept of the acceptability analysis remains the same. Several acceptability analysis approaches are summarized in the Vision 2000 report including general elastic procedures, component based elastic analysis procedures, the capacity spectrum procedure, pushover analysis methods, dynamic nonlinear time history procedures, and the drift demand method.

Vision 2000 design methodology is included in Appendix B of the 1996 Blue Book, and is expected to go through a consensus review and trial design process. It is anticipated that in about ten years all

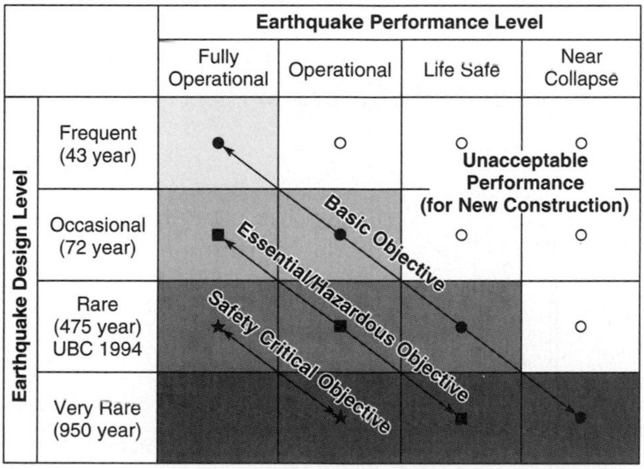

| Earthquake Design Level | Earthquake Performance Level: Fully Operational | Operational | Life Safe | Near Collapse |
|---|---|---|---|---|
| Frequent (43 year) | | | | |
| Occasional (72 year) | | | | |
| Rare (475 year) UBC 1994 | | | | |
| Very Rare (950 year) | | | | |

Unacceptable Performance (for New Construction)
Basic Objective
Essential/Hazardous Objective
Safety Critical Objective

**Figure 3.47** Vision 2000: performance objectives for building design.

design approaches and acceptability checks will become part of the code.

## 3.7 Dynamic Analysis: Theory

### 3.7.1 Introduction

A good portion of the loads that occur in multistory buildings can be considered as static loads requiring static analysis only. Although almost all loads except dead loads are transient, it is customary in most designs to treat these loads as static. For example, lateral loads imposed by transient wind pulses are usually treated as static loads. Even in earthquake design, which is clearly a dynamic problem, one of the acceptable methods of design is to use the so-called equivalent force system that is supposed to represent the static equivalent of dynamic forces. Under such assumptions, the analysis of a multistory building reduces to a single solution of the problem under static loads. The magnitude, nature, and origin of loads can be clearly visualized, and with the availability of computer programs the analysis can be performed without undue complexities. For example, once the wind load, building geometry, and member properties are known, the analysis for forces and deflections becomes a trivial computing task.

Although the equivalent static load approach is a recognized method of earthquake analysis, the state-of-the-art method for high-rise buildings uses a dynamic solution. Indeed, most building codes make the dynamic analysis mandatory for buildings whose configurations violate the assumptions made in the derivation of code equivalent forces. In today's architectural environment, it is more than likely that a tall building will not meet the spirit of the code assumptions either because the building has vertical steps, has a

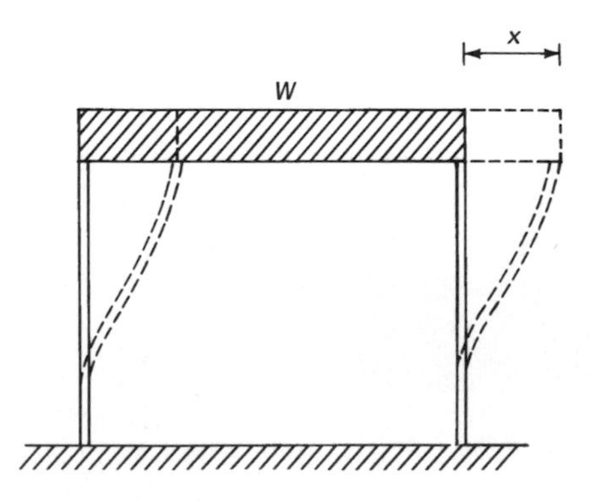


**Figure 3.48** Single-bay single-story portal frame.

nonuniform configuration, or other idiosyncrasies. It is, therefore, necessary to have a thorough understanding and mastery of the concepts and methods of solutions used in dynamic analysis.

Consider a multistory, multibay plane frame structure subjected to lateral wind loads. Although wind load is dynamic in nature, normal design practice, except in the case of very slender buildings, is to treat the wind load as an equivalent static load. The variation of wind load with respect to time is not considered: the load is assumed to be applied gradually to the structure. Under these conditions, the analysis of the structure becomes trivial. For a given set of loads, properties of the members and boundary conditions, there is but one unique solution.

Assume that the structure, instead of being subjected to wind, is subjected to seismic loads. During an earthquake the structure is subjected to rapid ground displacements and experiences a number of different forces that include inertia forces, damping forces, and elastic forces. Although there is no external force per se, the mass of the structure generates an equivalent forcing function when subjected to acceleration.

It is easy to visualize this force by considering a simplified response of the structure during an earthquake. Before the start of an earthquake, the structure is in static equilibrium and would remain so if the movement of the ground due to the earthquake takes place very slowly; the structure would simply ride to the new displaced position. When the ground moves suddenly, the inertia of mass distributed in the structure attempts to prevent the displacement of the structure, thus creating seismic loads analogous to an externally applied lateral force.

Earthquake forces are considered dynamic, because they vary with time. Since the load is time-varying, the response of the structure, including deflections, forces, and bending moments, are also time-dependent. Instead of a single solution as in a static case, a separate solution is required at each instant of time for the entire duration of earthquake. The resulting inertia forces are a function of deflections, which are themselves related to the inertia forces. It is therefore necessary to formulate the problem in terms of differential equations by relating the inertia forces to the second derivative of structural displacements.

In the following sections a brief mathematical treatment is presented for single-degree-of-freedom systems with and without damping forces, followed by analysis of multi-degree-of-freedom systems. The modal superposition method and orthogonality conditions which form the backbone of dynamic analysis are explained with reference to a system with two-degrees-of-freedom to show how coupled equations of motions for a multi-degree-of-freedom system are transformed into a set of independent single-degree-of-freedom systems. The section concludes with a summary highlighting the practical aspects of dynamic analysis.

### 3.7.2 Systems with single-degree-of-freedom

Consider a portal frame shown in Fig. 3.48 consisting of an infinitely stiff beam supported by flexible columns. Assuming that the beam is completely rigid and that columns have neglible mass as compared to the beam, the structure can be visualized as a spring-supported mass for the horizontal motion of the beam.

An analytical model for the system is shown in Fig. 3.49. Under the action of gravity force $W$, the spring will be extended by a certain amount. If the spring is very stiff, the extension is small, and vice versa.

The extension $x$ experienced by the spring can be related to the stiffness of the spring $k$ by the relation

$$x = \frac{W}{k} \tag{3.19}$$

$k$ is called the spring constant or spring stiffness and denotes the load required to produce unit extension of the spring. If $W$ is measured in kips and the extension in inches, the spring stiffness will have a dimension of kips per inch. The weight $W$ comes to rest after the spring has extended by the length $x$. Equation (3.19) expresses the

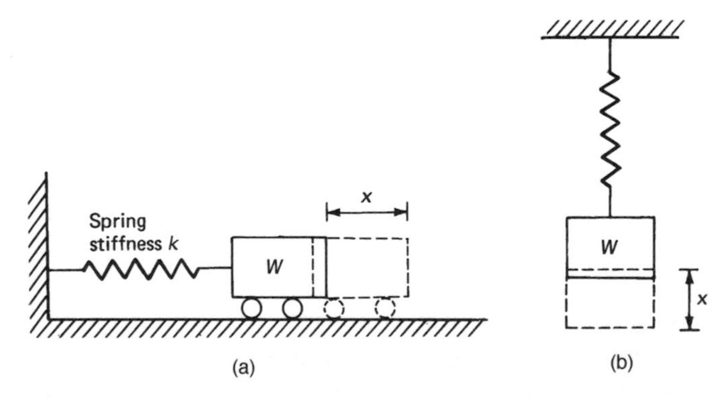


**Figure 3.49** Analytical models for single degree-of-freedom system: (a) model in horizontal position; (b) model in vertical position.

familiar static equilibrium condition between the internal force in the spring and the externally applied force $W$.

If a vertical force is applied or removed suddenly, vibrations of the system are produced. Such vibrations, maintained by the elastic force in the spring alone, are called free or natural vibrations. The weight moves up and down, and therefore is subjected to an acceleration given by the second derivative of displacement $x$, with respect to time, $\ddot{x}$. At any instant $t$, there are three forces acting on the body: the dynamic force equal to the product of the body mass and its acceleration, the gravity force $W$ acting downward, and the force in the spring equal to $W + kx$ for the position of weight shown in Fig. 3.50. These are in a state of dynamic equilibrium given by the relation

$$\frac{W}{g}\ddot{x} = W - (W + kx) = -kx \tag{3.20}$$

The above equation of motion it is called Newton's law of motion and is governed by the equilibrium of inertia force that is a product of the

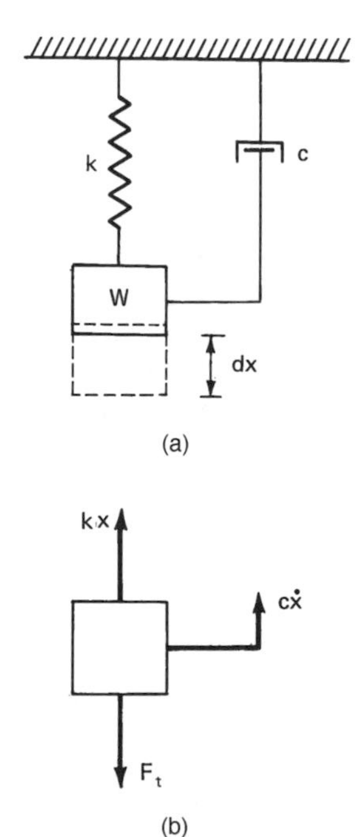


**Figure 3.50** Damped oscillator: (a) analytical model; (b) forces in equilibrium.

mass $W/g$, and acceleration $\ddot{x}$, and the resisting forces that are a function of the stiffness of the spring.

The principle of virtual work can be used as an alternative to Newton's law of motion. Although the method was first developed for static problems, it can readily be applied to dynamic problems by using D'Alembert's principle. The method establishes dynamic equilibrium by the including inertial forces in the system.

The principle of virtual work can be stated as follows: For a system that is in equilibrium, the work done by all the forces during a virtual displacement is equal to zero. Consider a damped oscillator subjected to a time-dependent force $F_{(t)}$ as shown in Fig. 3.50*a*. The free-body diagram of the oscillator subjected to various forces is shown in Fig. 3.50*a*.

Let $\delta x$ be the virtual displacement. The total work done by the system is zero and is given by

$$m\ddot{x}\ \delta\dot{x} + c\dot{x}\ \delta x + kx\ \delta x - F_{(t)}\ \delta x = 0 \tag{3.21}$$

$$(m\ddot{x} + c\dot{x} + kx - F_{(t)})\ \delta x = 0 \tag{3.22}$$

since $\delta x$ is arbitrarily selected,

$$m\ddot{x} + c\dot{x} + kx - F_{(t)} = 0 \tag{3.23}$$

This is the differential equation of motion of the damped oscillator.

The equation of motion for an undamped system can also be obtained from the principle of conservation of energy, which states that if no external forces are acting on the system and there is no dissipation of energy due to damping, then the total energy of the system must remain constant during motion and consequently its derivative with respect to time must be equal to zero.

Consider again the oscillator shown in Fig. 3.50*a* without the damper. The two energies associated with this system are the kinetic energy of the mass and the potential energy of the spring.

The kinetic energy of the spring

$$T = \tfrac{1}{2}m\dot{x}^2 \tag{3.24}$$

where $\dot{x}$ is the instantaneous velocity of the mass.

The force in the spring is $kx$; work done by the spring is $kx\ \delta x$. The potential energy is the work done by this force and is given by

$$V = \int_0^x kx\, \delta x = \tfrac{1}{2}kx^2 \tag{3.25}$$

The total energy in the system is a constant. Thus

$$\tfrac{1}{2}m\dot{x}^2 + \tfrac{1}{2}kx^2 = \text{constant } c_0 \tag{3.26}$$

Differentiating with respect to $x$, we get

$$m\dot{x}\ddot{x} + kx\dot{x} = 0 \tag{3.27}$$

Since $\dot{x}$ cannot be zero for all values of $t$, we get

$$m\ddot{x} + kx = 0 \tag{3.28}$$

which has the same form as Eq. (3.20). This differential equation has a solution of the form:

$$x = A \sin(\omega t + \alpha) \tag{3.29}$$

$$\dot{x} = \omega A \cos(\omega t + \alpha) \tag{3.30}$$

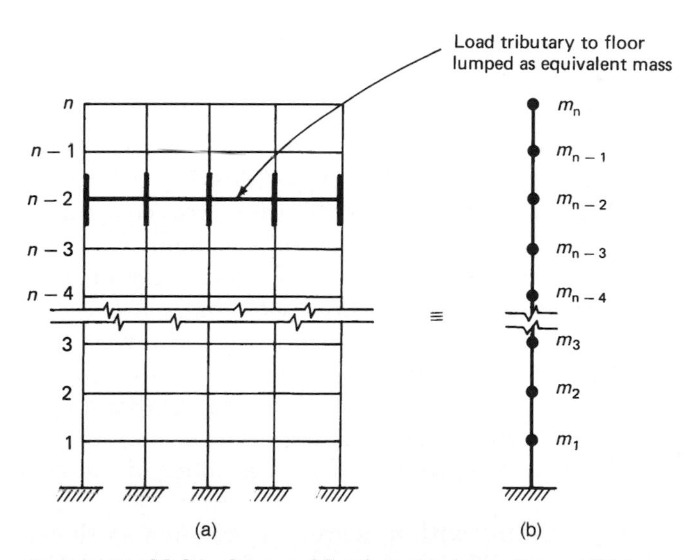


**Figure 3.51** Multi degree-of-freedom system: (a) multistory frame; (b) analytical model with lumped masses.

where $A$ is the maximum displacement and $\omega A$ is the maximum velocity. Maximum kinetic energy is given by

$$T_{\max} = \tfrac{1}{2}m(\omega A)^2 \tag{3.31}$$

Maximum potential energy is

$$V_{\max} = \tfrac{1}{2}kA^2 \tag{3.32}$$

Since $T = V$,

$$\tfrac{1}{2}m(\omega A)^2 = \tfrac{1}{2}kA^2$$

or

$$\omega = \sqrt{\frac{k}{m}} \tag{3.33}$$

which is the natural frequency of the simple oscillator. This method, in which the natural frequency is obtained by equating maximum kinetic energy and maximum potential energy, is known as *Rayleigh's method.*

### 3.7.3 Multi-degree-of-freedom systems

In these systems, the displacement configuration is determined by a finite number of displacement coordinates. The true response of a multi-degree system can be determined only by evaluating the inertia effects at each mass particle because structures are continuous systems with an infinite number of degrees-of-freedom. Although analytical methods are available to describe the behavior of such systems, the methods are limited to structures with uniform material properties and regular geometry. The methods are complex, requiring formulation of partial differential equations. The analysis is greatly simplified by replacing the entire displacement of the structure by a limited number of displacement components, and assuming the entire mass of the structure is concentrated in a number of discrete points.

Consider a multistory building with $n$ degrees-of-freedom as shown in Fig. 3.51. The dynamic equilibrium equations for undamped free vibration can be written in the general form

$$\begin{bmatrix} m_{11} & m_{12} & m_{13} & \cdots & m_{1m} \\ m_{21} & m_{22} & m_{23} & \cdots & m_{2m} \\ m_{31} & m_{32} & m_{33} & \cdots & m_{3m} \\ \cdots & \cdots & \cdots & \cdots & \cdots \\ m_{n1} & m_{n2} & m_{n3} & \cdots & m_{nm} \end{bmatrix} \begin{bmatrix} \ddot{x}_1 \\ \ddot{x}_2 \\ \ddot{x}_3 \\ \vdots \\ \ddot{x}_n \end{bmatrix} + \begin{bmatrix} k_{11} & k_{12} & k_{13} & \cdots & k_{1n} \\ k_{21} & k_{22} & k_{23} & \cdots & k_{2n} \\ k_{31} & k_{32} & k_{33} & \cdots & k_{3n} \\ \cdots & \cdots & \cdots & \cdots & \cdots \\ k_{n1} & k_{42} & k_{n3} & \cdots & k_{nm} \end{bmatrix} \begin{bmatrix} x_1 \\ x_2 \\ x_3 \\ \vdots \\ x_n \end{bmatrix} = 0$$

The above system of equations can be written in matrix form:

$$[M]\{\ddot{x}\} + [K]\{x\} = 0 \tag{3.34}$$

where $[M]$ = the mass or inertia matrix
$\{\ddot{x}\}$ = the column vector of accelerations
$[K]$ = the structure stiffness matrix
$\{x\}$ = the column vector of displacements of the structure

If the effect of damping is included, the equations of motion would be of the form

$$[M]\{\ddot{x}\} + [C]\{\dot{x}\} + [K]\{x\} = \{P\} \tag{3.34a}$$

where $[C]$ = the damping matrix
$\{\dot{x}\}$ = the column vector of velocity
$\{P\}$ = the column vector of external forces

General methods of solutions of these equations are available, but tend to be cumbersome. Therefore, in solving seismic problems simplified methods are used; the problem is first solved by neglecting damping. Its effects are later included by modifying the design spectrum to account for damping. The absence of precise data on damping does not usually justify a more rigorous treatment. Neglecting damping results in dropping the second term, and limiting the problem to free-vibrations results in dropping the right-hand side of Eq. (3.34a). The resulting equations of motion will become identical to Eq. (3.34).

During free vibration the motions of the system are simple harmonic, which means that the system oscillates about the stationary position in a sinusoidal manner; all masses follow the same harmonic function, having similar angular frequency $\omega$. Thus

$$x_1 = a_1 \sin \omega_1 t$$

$$x_2 = a_2 \sin \omega_2 t$$

$$\vdots$$

$$x_n = a_n \sin \omega_n t$$

or in matrix notation

$$\{x\} = \{a_n\} \sin \omega_n t$$

where $\{a_n\}$ represents the column vector of modal amplitudes for the $n$th mode, and $\omega_n$ the corresponding frequency. Substituting for $\{x\}$

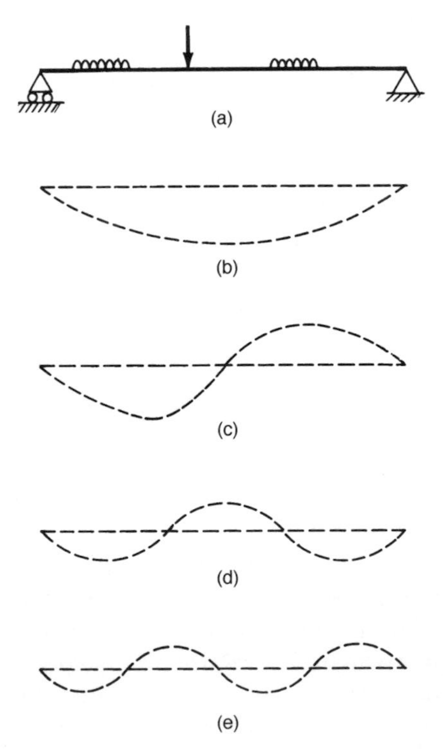


**Figure 3.52** Generalized displacement of simply supported beam: (a) loading; (b) full sine curve; (c) half-sine curve; (d) one-third sine curve; (e) one-fourth sine curve

and its second derivative $\{\ddot{x}\}$ in Eq. (3.34) results in a set of algebraic expressions.

$$-\omega_n^2[M]\{a_n\} + [K]\{a_n\} = 0 \tag{3.35}$$

Using a procedure known as Cramer's rule, the above expressions can be solved for determining the frequencies of vibrations and relative values of amplitudes of motion $a_{11}, a_{12}, \ldots, a_n$. The rule states that nontrivial values of amplitudes exist only if the determinant of the coefficients of $a$ is equal to zero because the equations are homogeneous, meaning that the right-hand side of Eq. (3.35) is zero. Setting the determinant of Eq. (3.35) equal to zero, we get

$$\begin{bmatrix} k_{11} - \omega_1{}^2 m_{11} & k_{12} - \omega_1{}^2 m_{12} & k_{13} - \omega_1{}^2 m_{13} & \cdots & k_{1n} - \omega_n m_{1n} \\ k_{21} - \omega_2{}^2 m_{21} & k_{22} - \omega_2{}^2 m_{12} & k_{23} - \omega_2{}^2 m_{23} & \cdots & k_{2n} - \omega_n m_{2n} \\ k_{31} - \omega_3{}^2 m_{31} & k_{32} - \omega_3{}^2 m_{32} & k_{33} - \omega_3{}^2 m_{33} & \cdots & k_{3n} - \omega_n m_{3n} \\ \cdots - \cdots & \cdots - \cdots & \cdots - \cdots & \cdots & \cdots - \cdots \\ k_{n1} - \omega_n{}^2 m_{n1} & k_{n2} - \omega_n{}^2 m_{n2} & k_{n3} - \omega_n{}^2 m_{n3} & \cdots & k_{nn} - \omega_n m_{nn} \end{bmatrix} = 0 \tag{3.36}$$

With the understanding that the values for all the stiffness coefficients $k_{11}$, $k_{12}$, etc., and the masses $m_1$, $m_2$, etc., are known, the determinant of the equation can be expanded, leading to a polynomial expression in $\omega^2$. Solution of the polynomial gives one real root for each mode of vibration. Hence for a system with $n$ degrees of freedom, $n$ natural frequencies are obtained. The smallest of the values obtained is called the fundamental frequency and the corresponding mode the fundamental or first mode.

In mathematical terms the vibration problem is similar to those encountered in stability analyses. The determination of frequency of vibrations can be considered similar to the determination of critical loads, while the modes of vibration can be likened to evaluation of buckling modes. Such types of problems are known as *eigenvalue* or *characteristic value* problems. The quantities $\omega^2$ which are analogous to critical loads are called eigenvalues or characteristic values, and in a broad sense can be looked upon as unique properties of the structure similar to geometric properties such as area or moment of inertia of individual elements.

Unique values for characteristic shapes, on the other hand, cannot be determined because substitution of $\omega^2$ for a particular mode into the dynamic equilibrium equation [Eq. (3.35)] results in exactly $n$ unknowns for the characteristic amplitudes $x_1 \cdots x_n$ for that mode. However, it is possible to obtain relative values for all amplitudes in terms of any particular amplitude. We are, therefore, able to obtain the pattern or the shape of the vibrating mode but not its absolute magnitude. The set of modal amplitudes that describe the vibrating pattern is called *eigenvector* or *characteristic vector.*

### 3.7.4 Modal superposition method

Modal superposition is a method in which the equations of motions are transformed from a set of $n$ simultaneous differential equations to a set of $n$ independent equations by the use of so-called normal coordinates. These equations are solved for the response of each mode, and the total response of the system is obtained by superposing individual solutions. Two concepts which are necessary for the understanding of the modal superposition method are: (i) normal coordinates and; (ii) property of orthogonality. These will be explained first, followed by application of the method to a 2-story structure. The treatments are necessarily incomplete in the mathematical sense but are sufficiently thorough to provide a sound conceptual understanding.

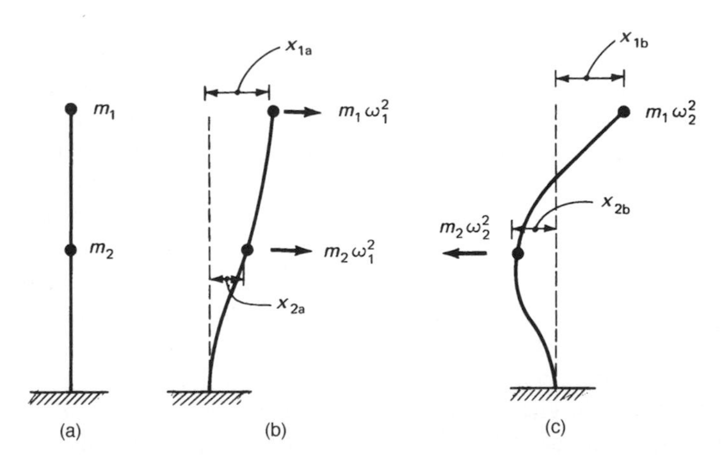


**Figure 3.53** Two-story lumped-mass system illustrating Betti's reciprocal theorem: (a) lumped model; (b) forces acting during first mode of vibration; (c) forces acting during second mode of vibration.

#### 3.7.4.1 Normal coordinates

In a static analysis it is common and convenient to represent the displacements of a structure by a system of geometric coordinates such as the cartesian system of coordinates that indicate the linear and angular positions of the elements with respect to a static position. For example, in a planar system, coordinates $x$ and $y$ and rotation $\theta$ are used to describe the position of the displaced structure. If the structure is restrained to move only in the horizontal direction and if rotations are of no consequence, only one coordinate is sufficient to describe the displacement. The displacements in general can also be obtained indirectly by any independent system of coordinates which are sufficient in number to specify the deflected position of all elements of the system. These coordinates are called generalized coordinates and their number is equal to the number of degrees-of-freedom of the system. In dynamic analysis, however, it is more advantageous to use free-vibration mode shapes to represent the displacements because of their orthogonality properties, which will be explained shortly, and because a close approximation of the displacement can be made by considering only the first few modes.

The dynamic analysis of multi-degree-of-freedom systems becomes extremely difficult if a system of direct coordinates is employed to describe their motion. A static analysis, on the other hand, can be handled without undue complications by using any set of consistent coordinates, and the resulting forces and displacements can be converted from one set of coordinates to another without much difficulty. To avoid the computational problems, in structural dynamics the normal modes of vibration are employed as generalized coordinates to describe the motion. Using this system, the undamped motion equations become uncoupled, greatly simplifying their solution. While the mathematical description of normal modes and their properties may be intriguing, there is nothing complicated about the concept. Let us indulge in some analogies to bring home the concept. For example, we can consider a set of normal modes as being similar to the primary colors, red, blue, and yellow. None of the primary colors can be constructed as a combination of the others, but any secondary color such as green or pink can be created by combining the primary colors, each with a distinct proportion. The proportions can be looked upon as scale factors, while the primary colors themselves can be considered similar to normal modes. To further reinforce the concept of generalized coordinates, it is helpful to recall its application in the solution of beam bending problems in which the deflection curve of the beam is represented in the form of trigonometric series. Considering the case of a simply supported beam subjected to vertical loads as shown in Fig. 3.52a, the deflection at any point can be represented by the following series:

$$y = a_1 \frac{\sin \pi x}{l} + a_2 \frac{\sin 2\pi x}{l} + a_3 \frac{\sin 3\pi x}{l} \tag{3.37}$$

Geometrically, this means that the deflection curve can be obtained by superposing simple sinusoidal curves such as shown in Fig. 3.52b–e.

The first term in Eq. (3.37) represents the full sine curve, the second term, the half-sine curve, etc. The coefficients $a_1$, $a_2$, $a_3$, etc., represent the maximum ordinates of the sine curves, and the numbers 1, 2, and 3 the number of waves or mode shapes. By determining the coefficients $a_1$, $a_2$, $a_3$, etc., the trigionometric series can be made to represent any deflection curve with a degree of accuracy that depends on the number of terms considered in the series.

### 3.7.4.2 Orthogonality

An important property of force-displacement relationship rarely used in static problems but very important in structural dynamics is the so-called orthogonal property. This property is best explained with reference to an example shown in Fig. 3.53.

Consider a 2-story lumped-mass system subjected to free-vibrations. The two modes of vibration can thus be considered as elastic displacements of the system due to two different loading conditions. A theorem known in structural mechanics as Betti's reciprocal theorem is used to derive the orthogonality conditions. This law states that the work done by one set of loads on the deflection due to a second set of loads is equal to the work done by the second set of loads acting on the deflections due to the first. Using this theorem with reference to Fig. 3.53, we get

$$\omega_1^2 m_1 x_{1b} + \omega_1^2 m_2 x_{2b} = \omega_2^2 m_1 x_{1a} + \omega_1^2 m_2 x_{2a} \tag{3.38}$$

This can be written in matrix form

$$\omega_1^2 \begin{bmatrix} m_1 & 0 \\ 0 & m_2 \end{bmatrix} \begin{bmatrix} x_{1b} \\ x_{2b} \end{bmatrix} = \omega_2^2 \begin{bmatrix} m_1 & 0 \\ 0 & m_2 \end{bmatrix} \begin{bmatrix} x_{1a} \\ x_{2a} \end{bmatrix}$$

or

$$(\omega_1^2 - \omega_2^2)\{x_b\}^T [M]\{x_a\} = 0 \tag{3.39}$$

If the two frequencies are not the same, i.e., $\omega_1 \neq \omega_2$, we get

$$\{x_b\}^T [M]\{x_a\} = 0 \tag{3.40}$$

This condition is called the orthogonality condition, and the vibrating shapes $\{x_a\}$ and $\{x_b\}$ are said to be orthogonal with respect to the mass matrix $[M]$. By using a similar procedure it can be shown that

$$\{x_a\}^T[k]\{x_b\} = 0 \tag{3.41}$$

The vibrating shapes are therefore orthogonal with respect to stiffness matrix as they are with respect to the mass matrix. In the general case of the structures with damping, it is necessary to make a further assumption in the modal analysis that the orthogonality condition also applies for the damping matrix. This is for mathematical convenience only and has no theoretical basis. Therefore, in addition to the two orthogonality conditions mentioned previously, a third orthogonality condition of the form

$$\{x_a^T\}c\{x_b\} = 0 \tag{3.42}$$

is used in the modal analysis.

To bring out the essentials of the normal mode method, it is convenient to consider the dynamic analysis of a two-degree-of-freedom system. We will first analyze the system by a direct method and then show how the analysis can be simplified by the modal superposition method.

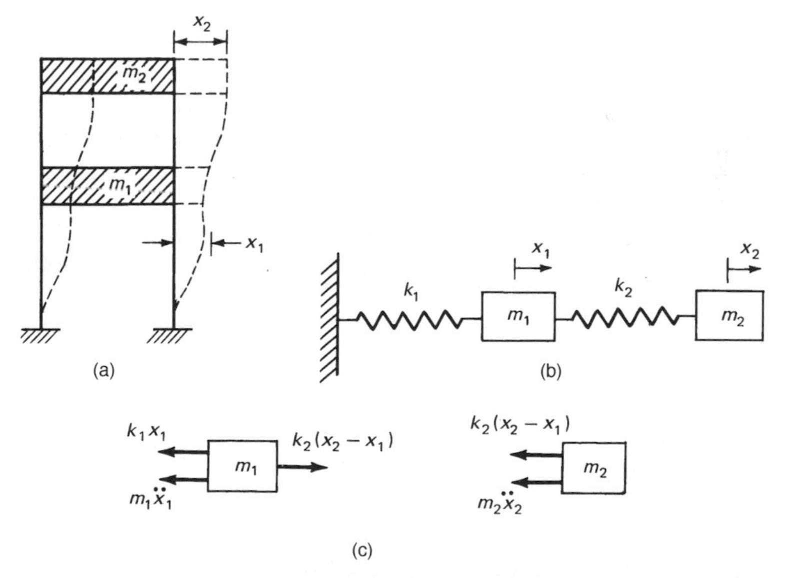


**Figure 3.54** Two-story shear building: free vibrations: (a) building with lumped masses; (b) mathematical model; (c) free-body diagram with lumped masses.

Consider a 2-story dynamic model of a shear building shown in Fig. 3.54 subject to free vibrations. The masses $m_1$ and $m_2$ at levels 1 and 2 can be considered connected to each other and to the ground by two springs having stiffnesses $k_1$ and $k_2$. The stiffness coefficients are mathematically equivalent to the forces required at levels 1 and 2 to produce unit horizontal displacements relative to each level.

It is assumed that the floors and therefore the masses $m_1$ and $m_2$ are restrained to move in the direction $x$ and that there is no damping in the system. Using Newton's second law of motion, the equations of dynamic equilibrium for masses $m_1$ and $m_2$ are given by

$$m_1\ddot{x}_1 = -k_1 x + k_2(x_2 - x_1) \tag{3.43}$$

$$m_2\ddot{x}_2 = -k_2(x_2 - x_1) \tag{3.44}$$

Rearranging terms in these equations gives

$$m_1\ddot{x}_1 + (k_1 + k_2)x_1 - k_2 x_2 = 0 \tag{3.45}$$

$$m_2\ddot{x}_2 - k_2 x_1 + k_2 x_2 = 0 \tag{3.46}$$

The solutions for the displacements $x_1$ and $x_2$ can be assumed to be of the form

$$x_1 = A\sin(\omega t + \alpha) \tag{3.47}$$

$$x_2 = B\sin(\omega t + \alpha) \tag{3.48}$$

where $\omega$ represents the angular frequency and $\alpha$ represents the phase angle of the harmonic motion of the two masses. $A$ and $B$ represent the maximum amplitudes of the vibratory motion. Substitution of Eqs (3.47) and (3.48) into Eqs (3.45) and (3.46) gives the following equations

$$(k_1 + k_2 - \omega^2 m_1)A - k_2 B = 0 \tag{3.49}$$

$$k_2 A + (k_2 - \omega^2 m_2)B = 0 \tag{3.50}$$

To obtain solution for the nontrivial case of $A$ and $B \neq 0$, the determinant of the coefficients of $A$ and $B$ must be equal to zero, thus

$$\begin{bmatrix} (k_1 + k_2 - \omega^2 m_1) & -k_2 \\ -k_2 & (k_2 - \omega^2 m_2) \end{bmatrix} = 0 \tag{3.51}$$

Expansion of the determinant gives the relation;

$$(k_1 + k_2 - \omega^2 m_1)(k_2 - \omega^2 m_2) - k_2^2 = 0 \tag{3.52}$$

or

$$m_1 m_2 \omega^4 - m_1 k_2 + m_2(k_1 + k_2)\omega^2 + k_1 k_2 = 0 \tag{3.53}$$

Solution of this quadratic equation yields two values for $\omega^2$ of the form

$$\omega_1^2 = \frac{-b + \sqrt{b^2 - 4ac}}{2a} \tag{3.54}$$

$$\omega_2^2 = \frac{-b - \sqrt{b^2 - 4ac}}{2a} \tag{3.55}$$

where $a = m_1 m_2$
$b = -[m_1 k_2 + m_2(k_1 + k_2)]$
$c = k_1 k_2$

As mentioned previously, the two frequencies $\omega_1$ and $\omega_2$ which can be considered as intrinsic properties of the system are uniquely determined.

The magnitudes of the amplitudes $A$ and $B$ cannot be determined uniquely but can be obtained in terms of ratios $r_1 = A_1/B_1$ and $r_2 = A_2/B_2$ corresponding to $\omega_1^2$ and $\omega_2^2$, respectively. Thus

$$r_1 = \frac{A_1}{B_1} = \frac{k_2}{k_1 + k_2 - \omega_1^2 m_1} \tag{3.56}$$

$$r_2 = \frac{A_2}{B_2} = \frac{k_2}{k_1 + k_2 - \omega_2^2 m_1} \tag{3.57}$$

The ratios $r_1$ and $r_2$ are called the amplitude ratios and represent the shapes of the two natural modes of vibration of the system.

Substituting the angular frequency $\omega_1$ and the corresponding ratio $r_1$ in Eqs (3.47) and (3.48), we get

$$x_1' = r_1 B_1 \sin(\omega_1 t + \alpha_1) \tag{3.58}$$

$$x_2' = B_1 \sin(\omega_1 t + \alpha_1) \tag{3.59}$$

These expressions describe the first mode of vibration, also called the fundamental mode. Substituting the larger angular frequency $\omega_2$ and the corresponding ratio $r_2$ in Eqs (3.47) and (3.48), we get

$$x_1'' = r_2 B_2 \sin(\omega_2 t + \alpha_2) \tag{3.60}$$

$$x_2'' = B_2 \sin(\omega_2 t + \alpha_2) \tag{3.61}$$

The displacements $x_1''$ and $x_2''$ describe the second mode of vibration. The general displacement of the system is obtained by summing the modal displacements, thus

$$x_1 = x_1' + x_1''$$

$$x_2 = x_2' + x_2''$$

Thus for systems having two degrees of freedom, we are able to determine the frequencies and mode shapes without undue mathematical difficulties. Although the equations of motions for multi-degree systems have similar mathematical form, solutions for modal amplitudes in terms of geometrical coordinates become unwieldy. Use of orthogonal properties of mode shapes makes this laborious process unnecessary. We will demonstrate how the analysis can be simplified by using the modal superposition method. Consider again the equations of motion for the idealized 2-story building discussed in the previous section. As before, damping is neglected, but instead of free vibrations we will consider the analysis of the system subject to time-varying force functions $F_1$ and $F_2$ at levels 1 and 2. The dynamic equilibrium for masses $m_1$ and $m_2$ is given by

$$m_1\ddot{x}_1 + (k_1 + k_2)x_1 - k_2x_2 = F_1 \tag{3.62}$$

$$m_2\ddot{x}_2 - k_2x_1 + k_2x_2 = F_2 \tag{3.63}$$

These two equations are interdependent because they contain both the unknowns $x_1$ and $x_2$. These can be solved simultaneously to get the response of the system, which was indeed the method used in the previous section to obtain the values for frequencies and mode shapes. Modal superposition method offers an alternate procedure for solving such problems. Instead of requiring simultaneous solution of the equations, we seek to transform the system of interdependent or coupled equations into a system of independent or uncoupled equations. Since the resulting equations contain only one unknown function of time, solutions are greatly simplified. Let us assume that solution for the above dynamic equations is of the form:

$$x_1 = a_{11}z_1 + a_{12}z_2 \tag{3.64}$$

$$x_2 = a_{21}z_1 + a_{22}z_2 \tag{3.65}$$

What we have done in the above equations is to express displacement $x_1$ and $x_2$ at levels 1 and 2 as a linear combination of properly scaled values of two independent modes. For example, $a_{11}$ and $a_{12}$, which are the mode shapes at level 1, are combined linearly to give the displacement $x_1$. $z_1$, and $z_2$ can be looked upon as scaling functions.

*Substituting for* $x_1$ and $x_2$ and their derivatives $\ddot{x}_1$ and $\ddot{x}_2$ in the equilibrium Eqs (3.62) and (3.63) we get

$$m_1a_{11}\ddot{z}_1 + (k_1 + k_2)a_{11}z_1 - k_2a_{21}z_1 + m_1a_{12}\ddot{z}_2 + (k_1 + k_2)a_{12}z_2 - k_2a_{22}z_2 = F_1 \quad (3.66)$$

$$m_2a_{21}\ddot{z}_1 - k_2a_{11}z_1 + k_2a_{21}z_1 + m_2a_{24}z_2 - k_2a_{12}z_2 + k_2a_{22}z_2 = F_2 \quad (3.67)$$

We seek to uncouple Eqs (3.66) and (3.67) by using the orthogonality conditions. Multiplying Eqs (3.66) by $a_{11}$ and Eqs (3.67) by $a_{21}$ we get

$$m_1a_{11}^2\ddot{z}_1 + (k_1 + k_2)a_{11}^2z_1 - k_2a_{11}a_{21}z_1 + m_1a_{11}a_{12}\ddot{z}_2 + (k_1 + k_2)a_{11}a_{12}z_2 - k_2a_{11}a_{22}z_2 = a_{11}F_1 \quad (3.68)$$

$$m_1a_{21}^2\ddot{z}_1 - k_2a_{11}a_{21}z_1 + k_2a_{21}^2z_1 + m_2a_{21}a_{22}\ddot{z}_2 - k_2a_{12}a_{21}z_2 + k_2a_{21}a_{22}z_2 = a_{21}F_2 \quad (3.69)$$

Adding the above two equations, we get

$$(m_1a_{11}^2 + m_2a_{21}^2)\ddot{z}_1 + \omega_1^2(m_1a_{11}^2 + m_2a_{21}^2)z_1 = a_{11}F_1 + a_{21}F_2 \quad (3.70)$$

Similarly, multiplying Eqs (3.66) and (3.67) by $a_{12}$ and $a_{22}$ and adding we obtain

$$(m_1a_{12}^2 + m_2a_{22}^2)\ddot{z}_2 + \omega_2^2(m_1a_{12}^2 + m_2a_{22}^2)z_2 = a_{12}F_1 - a_{22}F_2 \quad (3.71)$$

Equations (3.70) and (3.71) are independent of each other and are the uncoupled form of the original system of coupled differential equations. These can be further written in a simplified form by making use of the following abbreviations:

$$\begin{aligned} M_1 &= m_1a_{11}^2 + m_2a_{21}^2 \\ M_2 &= m_1a_{12}^2 + m_2a_{22}^2 \end{aligned} \quad (3.72)$$

$$\begin{aligned} K_1 &= \omega_1^2M_1 \\ K_2 &= \omega_2^2M_2 \end{aligned} \quad (3.73)$$

$$\begin{aligned} P_1 &= a_{11}F_1 + a_{21}F_2 \\ P_2 &= a_{12}F_1 + a_{22}F_2 \end{aligned} \quad (3.74)$$

$M_1$ and $M_2$ are called the generalized masses, $K_1$ and $K_2$ the generalized stiffnesses, and $P_1$ and $P_2$ the generalized forces.

Using these notations, each of the Eqs (3.70) and (3.71) takes the form similar to the equations of motion of a single-degree-of-freedom system, thus

$$M_1\ddot{z}_1 + k_1z_1 = P_1 \quad (3.75)$$

$$M_2\ddot{z}_2 + k_2z_2 = P_2 \quad (3.76)$$

The solution of these uncoupled differential equations can be found by any of the standard procedures given in textbooks on vibration analysis. In particular, Duhamel's integral provides a general method of solving these equations irrespective of the complexity of the loading function. However, in seismic analysis usually a response spectrum is available for the forcing function. Therefore, the maximum values of the response corresponding to each modal equation is obtained from the response spectrum. Direct superposition of modal maximum would, however, give only an upper limit for the total system which, in many engineering problems, would be too conservative. To alleviate this problem approximations based on probability considerations are generally employed. One method employs the so-called root mean square procedure, also called the square root of sum of the squares (SRSS) method. As the name implies, a probable maximum value is obtained by combining the square root of the sum of the squares of the modal quantities. Although this method is simple and widely used, it is not always a conservative predictor of earthquake response because more severe combinations of modal quantities can occur, for example, when two modes have nearly the same natural period. In such cases a more appropriate combination of modal quantities, such as the Complete Quadratic Combination (CQC) is more appropriate.

An attempt has been made in this section to bring out the essentials of structural dynamics as related to seismic design of buildings. A certain amount of mathematical presentation has been unavoidable. Lest the reader lose the physical meaning of the various steps, it is worthwhile to summarize the essential features of dynamic analysis.

Dynamic analysis of high-rise buildings is accomplished by idealizing them as systems with multiple degrees-of-freedom. The dead load of the building together with a percentage of live load (estimated to be present during an earthquake) is modeled as a system of masses lumped at floor levels. In a planar analysis, each mass has one degree-of-freedom corresponding to lateral displacement in the direction under consideration, while in a three-dimensional analysis it has three degrees-of-freedom corresponding to two translational and one torsional displacement. Free-vibrations of the buildings are evaluated, without including the effect of damping. Damping is taken into account by modifying the design response spectrum. The dynamic model representing the building has a number of mode shapes equal to the number of degrees-of-freedom of the model. Mode shapes have the property of orthogonality, which means that no given mode shape can be constructed as a combination of others, yet any deformation of the dynamic model can be described as a combination of its mode shapes, each multiplied by a scale factor. Each mode shape has a natural frequency of vibration. The mode shapes and frequencies are determined by solving an eigenvalue problem. The total response of the building to a given response spectrum is obtained by summing a number of modal responses. The number of modes required to adequately determine the forces for design is a function of the dynamic characteristics of the building. Generally for tall, regular buildings, six to ten modes in each direction are considered sufficient. Since each mass responds to earthquakes in more than one mode, it is necessary to evaluate effective modal mass values. These values indicate the percentage of the total mass that is mobilized in each mode. The acceleration experienced by each mass undergoing various modal deformations is determined from the response spectrum, that has been adjusted for damping. Product of acceleration for a particular frequency multiplied by the effective modal mass gives the static equivalent of forces at each discrete level. Since these forces do not reach their maximum values simultaneously during an earthquake event, statistical methods are used to achieve the combinations. The resulting forces are used as design static forces.

## 3.8 Summary

Since earthquakes can occur almost anywhere, some measure of earthquake resistance in the form of reserve ductility and redundancy should be built into the design of all structures to prevent catastrophic failures. The magnitude of inertial forces induced by earthquakes depends on the building mass, ground acceleration and the dynamic response of the structure. The shape and proportion of a building have a major effect on the distribution of earthquake forces as they work their way through the building. If irregular features are unavoidable, special design considerations are required to account for load transfer at abrupt changes in structural resistance.

Two approaches are recognized in modern codes for estimating the magnitude of seismic loads. The first approach, termed the equivalent lateral force procedure uses a simple method to take into account the properties of the structure and the foundation material. The second is a dynamic analysis procedure in which the modal responses are combined in a statistical manner to find the maximum values of the building response. Note that the level of force experienced by a structure during a major earthquake is much larger than the forces usually employed in the design. By prescribing detailing requirements the structure is relied upon to sustain post-yield displacements without collapse.

The complex random nature of an accelerogram makes it necessary to employ a more general characterization of ground motion. The most practical representation is by earthquake response spectra to postulate the intensity and vibration content future ground motion at a given site. Duration of ground motion, although important, is not used explicitly in design criteria at the present time (1997).

Multistory buildings are analyzed as a multi-degree-of-freedom system. They are represented by lumped masses attached at story intervals along the height of a vertically cantilevered pole. Each mode of the building system is represented by an equivalent single-degree-of-freedom system using the concept of generalized mass and stiffness. With the known period, mode shape, mass distribution and response spectrum, one can compute the deflected shape, story accelerations, forces and overturning moments. Each predominant mode is analyzed separately, and by using either the SRSS or CQC method, the peak modal responses are combined to give a reasonable value between an upper bound as the absolute sum of the modes and the lower bound as the maximum value of a single mode.

The time-history analysis technique represents the most sophisticated method of dynamic analysis for buildings. In this method, the mathematical model of the building is subjected to accelerations from earthquake records that represent the expected earthquake at the base of the structure. The equations of motion are integrated by using computers to obtain a complete record of acceleration, velocity, and displacement of each lumped mass. The maximum value is found by scanning the output record. Even with the availability of sophisticated computers, use of this method is restricted to the design of special structures such as nuclear facilities, military installations, and base-isolated structures.

In seismic design, nearly elastic behavior is interpreted as allowing some structural elements to slightly exceed specified yield stress on the condition that the elastic linear behavior of the overall structure is not substantially altered. For a structure with a multiplicity of structural elements forming the lateral-force-resisting system, the yielding of a small number of elements will generally not affect the overall elastic behavior of the structure if excess load can be distributed to other structural elements that have not exceeded their yield strength. For ductile framing systems the maximum number of moment-frame-beams with flexural overstresses should be limited to 20 percent of the beams in the direction of force on any story. The number of frame-columns with flexural overstress should be limited to 10 percent of the frame-columns in any story.

Although for new buildings, the ductile design approach is quite routine, seismic retrofitting of existing non-ductile buildings with poor confinement details is generally extremely expensive. Therefore, it is necessary to formulate an alternative method which attempts at a realistic assessment of damage resistance of the building. The method is based on the concept of "trade-off" between ductility and strength. In other words, structural systems of limited ductility may be considered valid in seismic design, provided they can resist correspondingly higher forces. The concept of an "inelastic demand ratio" is used to describe the ability of the structural elements to resist stresses beyond yield stress.

Chapter

# 4

# Lateral Systems: Steel Buildings

## 4.1 Introduction

### 4.1.1 Steel in high-rise buildings

Although use of steel in structures can be traced back to 1856 when Bessemer's steel-making process was first introduced, its application to tall buildings received stimulus from the 984 ft (300 m) Eifel Tower, which was constructed in 1889. After the turn of the nineteenth century, several tall buildings, from the 286 ft (87 m) Flatiron Building in 1902 to the 1046 ft (319 m) Chrysler Building in 1929, where constructed in downtown Chicago and Manhattan. The height record was broken by the 1250 ft (381 m) Empire State Building in 1931, the twin towers of the World Trade Center buildings at 1350 ft (412 m) in 1972, followed almost immediately by the 1450 ft (442 m) Sears Tower in Chicago in 1974.

The role of steel members, which in the early structures were relegated to carrying gravity loads only, has been compltely upgraded to include wind and seismic resistance in systems ranging from the modest portal frame at one end of the spectrum, to innovative systems involving outrigger systems, mega frames, interior super-diagonally braced frames etc, at the other.

Today there are innumerable structural steel systems that can be used for the lateral bracing of tall buildings. It would be an exercise in futility to try to classify all these systems into distinct categories because there is no single criterion that can be used for a comprehensive cataloging of all systems. However, for purposes of presentation, the different structural systems currently used in the design of tall steel buildings are broadly divided into the following categories roughly based on their relative effectiveness in resisting the lateral loads.

- Frames with Semi-Rigid Connections (4.2).
- Rigid frames (4.3).
- Braced frames (4.4).
- Staggered truss system (4.5).
- Eccentric bracing systems (4.6).
- Interacting system of braced and rigid frames (4.7).
- Outrigger and belt truss systems (4.8).
- Framed tube systems (4.9).
- Trussed tube system (4.10).
- The bundled tube (4.11).

## 4.2 Frames with Semi-Rigid Connections

### 4.2.1 Introduction

*Semirigid connections,* as the name implies, are those with rotational characteristics intermediate in degree between fully rigid and simple connections. These connections offer known rotational restraint at the beam ends resulting in significant reduction in midspan gravity moments. However, they are not sufficiently rigid to prevent the entire rotation between beam and intersecting column.

Although several specifications such as the AISC, the British, and the Australian codes permit semi-rigid connections it has rarely been used because of the difficulty in predicting the rather complex response of these connections. However, reasonable success has been obtained by another type of partially rigid connection which the AISC designates as Type 2 wind connection, with similar provisions found in the British and Australian codes.

In the following sub-sections a brief description of the behavior of three basic types of connections is given with particular emphasis on the design of Type 2 wind connection.

### 4.2.2 Review of connection behavior

Three basic types of construction and associated assumptions are permissible under the AISC specifications, and each governs in a specific manner the size of members and the types and strength of their connections.

Type 1. Commonly designated as "rigid-frame" (continuous frame), assumes that beam-to-column connections have sufficient rigidity to hold virtually unchanged the original angles between intersecting members.

Type 2. Commonly designated as "simple framing" (unrestrained, free-ended), assumes that, insofar as gravity loading is concerned, ends of beams and girders are connected for shear only and are free to rotate under gravity load.

Type 3. Commonly designated as "semi-rigid framing" (partially restrained), assumes that the connections of beams and girders possess a dependable and known moment capacity intermediate in degree between the rigidity of Type 1 and the flexibility of Type 2.

Connections in simple frames are typically designed to transfer vertical shear only. They are also designed for axial loads if they transfer chord and drag forces due to seismic loads. In either case it is assumed that there is no bending moment at the connection. Connections in fully rigid frames on the other hand, are called upon to develop resistance to both shear and bending moment. They are assumed to have sufficient rigidity to hold virtually unchanged the original angles between connecting members. Semi-rigid connection behavior is intermediate between the simple and rigid condition, its fixity varying anywhere from a low 5 percent to a high 90 percent of full-fifty.

Completely simple and completely rigid behavior are, of course, ideal conditions. Practically, it is necessary to accept something less than ideal, since real frames perform in the broad range between fully a rigid and simple support condition. For example, consider the typical beam-to-column connection consisting of a double-angle web connection as shown in Fig. 4.1. The angles fastened to the beam web are usually considered completely flexible and are assumed to carry only shear. Actually, they offer some restraint to moment and thus oppose the rotation at the beam end. The relationship between the applied moment and rotation is complex and can only be determined by experiment. When rotation under gravity load takes place, the upper part of the connection is in tension while the lower part is compressed against the column. The rotation is accommodated by deformation of the angles. Therefore, to minimize the rotational restraint, the angles are made as thin as possible.

Unstiffened seated beam connection shown in Fig. 4.2 is an example of Type 2 connection. The behavior of the seat angle is shown schematically in Fig. 4.3. The bottom angle bends as a cantilever, except its bending is partially restrained by the bottom flange of the beam. The moment-rotation characteristics of a seat angle connection primarily depend on the beam depth, thickness of top angle, diameter of bolts, and thickness of column flanges. The connection is typically stiffer than web angle connection but still considered to be a simple flexible connection. By adding top and bottom angles to the double angle web connection, it is possible to develop a connection that has greater moment resistance than either of the previously described connections.

The top and bottom angles are assumed to carry the moment, and web angles the shear. Although the load distribution may appear to be arbitrary, such a division of function produces adequately proportioned connections. Structural tees used in place of top and bottom angles increase the rotational restraint considerably. The increase occurs because the top tee is loaded in tension without any eccentricity, whereas the top angles are loaded eccentrically, resulting in large deformations.

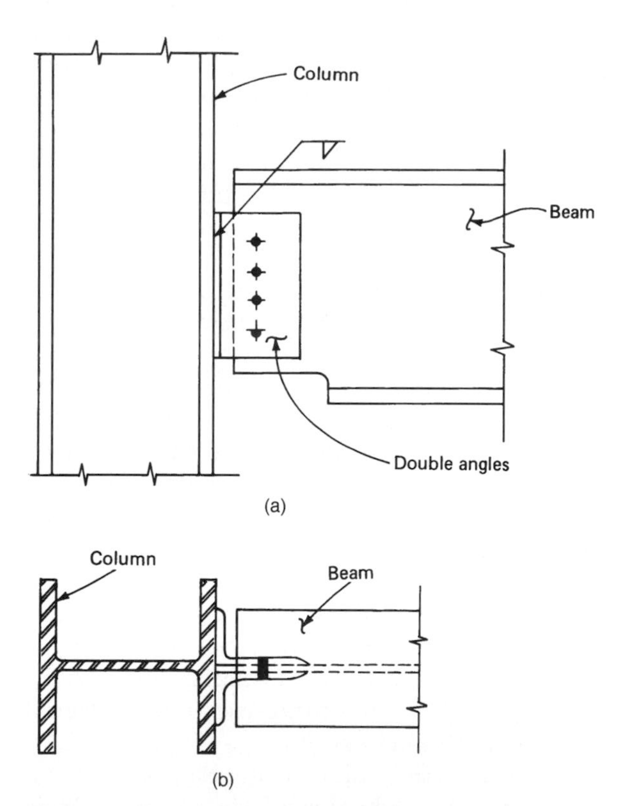


**Figure 4.1** Beam-column bolted shear connection: (a) elevation; (b) plan.

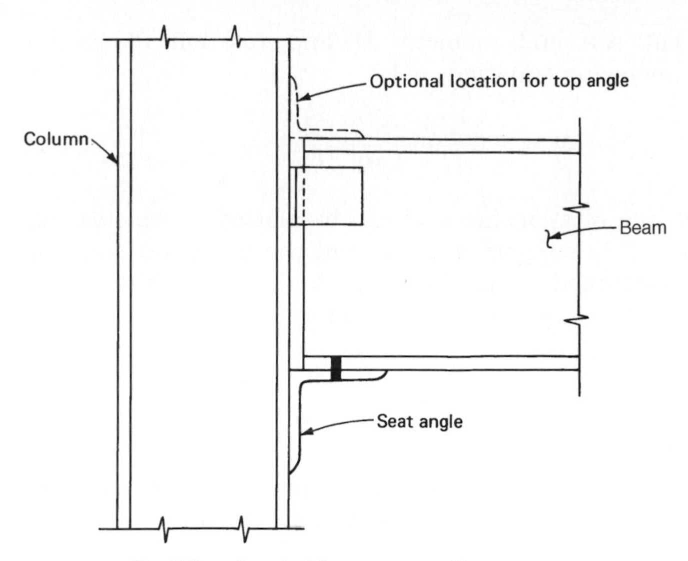


Figure 4.2 Unstiffened seated beam connection.

### 4.2.3 Beam line concept

One of the methods for understanding the behavior of beam-to-column connection is to study a plot of moment-rotation characteristics, as shown in Fig. 4.4. The vertical axis shows the end moment. The resulting rotation is plotted along the horizontal axis in radians. Superimposed upon this plot is the so-called beam line, which expresses the resulting end moment $M$ and rotation $\theta$ for a uniformly loaded beam for any end restraint, ranging from full fixity to simply supported condition.

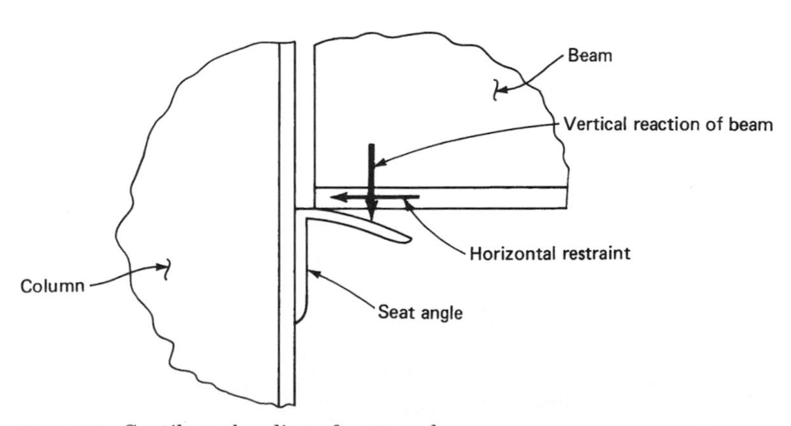


Figure 4.3 Cantilever bending of seat angle.

The relation between end moment $M$ and rotation $\theta$ can be expressed by the following equation

$$M = -\frac{2EI\theta}{L} - \frac{WL}{12} \tag{4.1}$$

This is a straight-line relationship and can be plotted by considering the rotation of a simply supported beam and the fixed end moment of a completely restrained beam. Point $a$ on the beam line is the end moment when the connection is completely restrained. Thus, in Eq. (4.1), rotation $\theta = 0$, giving

$$M = -\frac{WL}{12} \tag{4.2}$$

Point $b$ is the rotation at the end of beam when the beam has zero restraint at the ends. In other words, the beam behaves as a simply supported beam. Substituting $M = 0$ in Eq. (4.1), we get

$$\theta = -\frac{WL^2}{24EI} \tag{4.3}$$

The point at which the beam line intersects the connection line gives the resulting end moment and rotation under the given load. The

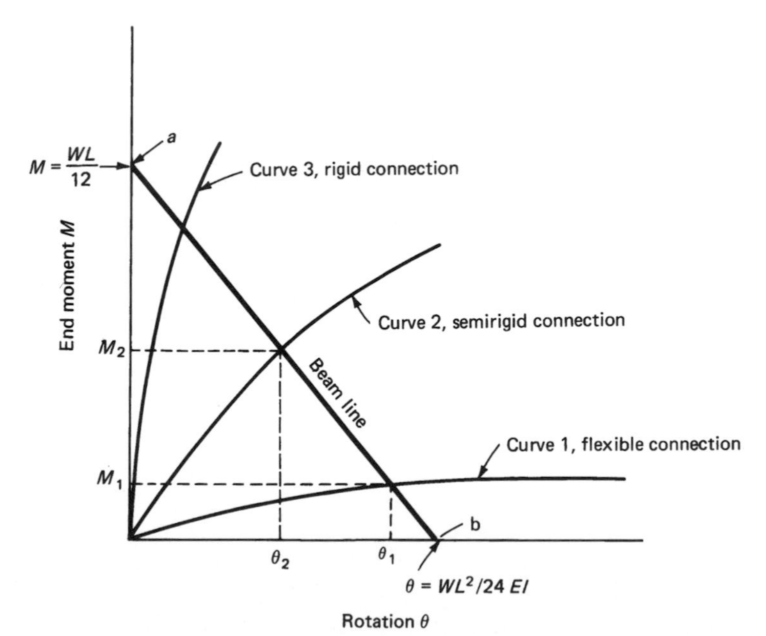


**Figure 4.4** Beam line concept: moment–rotation ($M$–$\theta$) curves.

dependence of the beam behavior on the rigidity of the connection can be studied by using this diagram. In developing the $M$-$\theta$ relationship, it is assumed that the behavior of the two end connections is the same and that the beam is subjected to loads placed symmetrically on the beam. The behavior of all the three types of connections, namely the flexible, the semirigid, and rigid connection, can be studied by using the beam line diagram. Curve 1 represents a flexible connection which is typical of a double-angle web connection. Under a uniform load $W$, the beam ends rotate through an angle $\theta_1$, which is very nearly equal to the rotation $\theta$ of a completely unrestrained beam. Corresponding to this rotation, a moment $M$ is generated at the ends signifying that even with the so-called flexible connection, some end moment is generated. Normally the bending moment developed is about 5 to 20 percent of the fully fixed moment.

Curve 2 represents a semirigid connection such as an end plate connection, Fig. 4.5, detailed in such a way that under working loads it elastically yields to provide the necessary rotation of the connection. Although the beam is detailed to undergo a rotation equal to $\theta_2$, significant moment $M_2$ corresponding to the rotation $\theta_2$ develops at the beam ends. The restraint offered by this type of connection can vary anywhere from a low of 20 percent to a high of 90 percent of the full fixity. The resulting end moments could be 20 to 90 percent of the moment generated in a fully fixed beam.

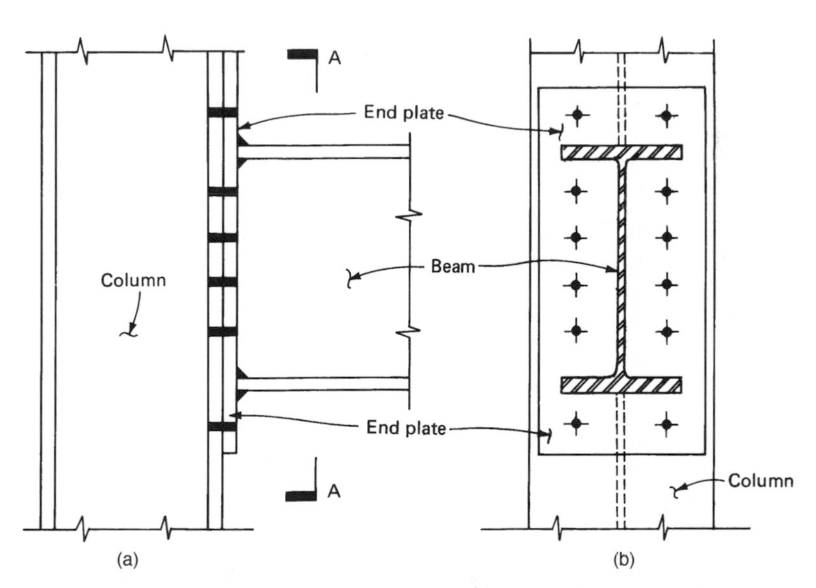


**Figure 4.5** Beam-column end-plate connection: (a) elevation; (b) section A-A, side view.

Curve 3 represents the moment-rotation characteristics of a rigid connection such as a welded beam-to-column connection shown in Fig. 4.6. The beam develops end moments which are about 90 to 95 percent of fully fixed condition, especially when column flange stiffeners are used as in Fig. 4.6.

Wind moments applied to flexible and semirigid connections present some very intriguing problems because some means of transferring these moments must be provided in these connections, which are supposed to be flexible. Additional restraint provided for carrying the

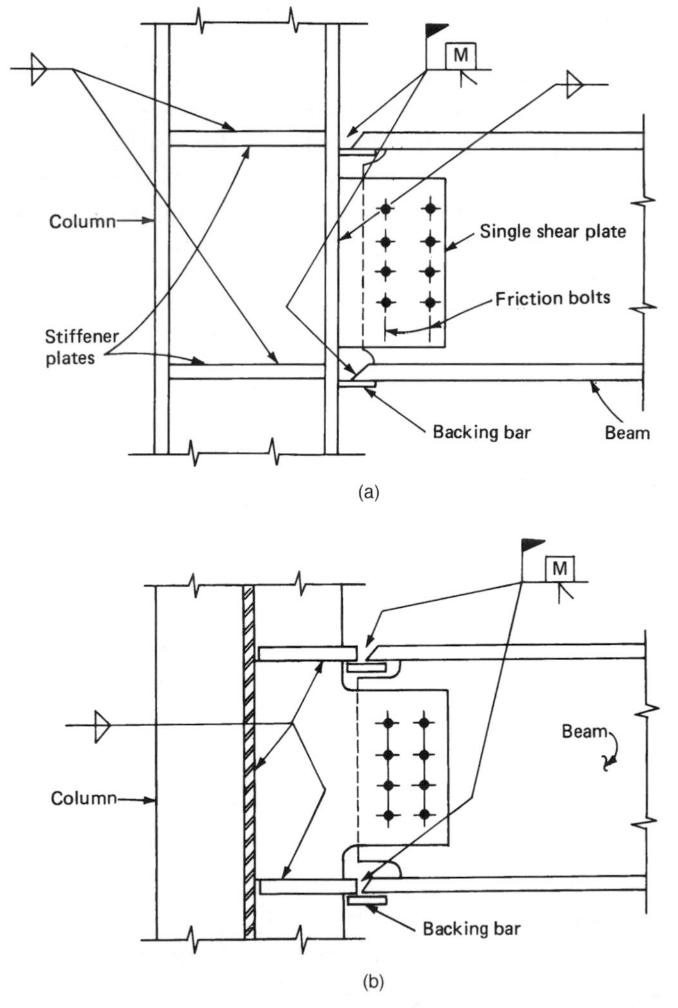


**Figure 4.6** Beam-column welded moment connections: (a) column strong axis connection; (b) column weak axis connection.

wind load will result in an increase in the end moment due to gravity loads. A rigorous mathematical solution of flexible and semirigid connections is not possible, but based on the performance of buildings using these types of connections, AISC provides for two approximate solutions. In the first method the connection is designed for the moment caused by the combination of gravity and wind loads using a one-third increase in the allowable stresses. In the second method, the connection is designed for the moment induced by wind loads only, using a one-third increase in the stress allowances. The connection must, however, be designed to yield plastically for any combination of gravity and wind moments. Any additional moment that could occur at the ends beyond the wind moments is relieved because of the yielding of the connection. This type of connection necessitates some inelastic but self-limiting deformation of the connection components without overstressing the fasteners. The *self-limiting deformation criteria* is imposed to prevent the use of semirigid connections for cantilever beams. This is because cantilever beam deflections are not limited by the rotation of the connections but may continue to progress, resulting in failure of the connection, and even the beam itself.

Although the AISC specification permits the designer to take advantage of reduction in the midspan moment of a beam with semirigid connections, in practice this procedure has not found wide acceptance primarily because of lack of reliable analytical techniques. The Type 2 wind connection, which basically ignores the beam restraint for gravity loads, has found relatively greater acceptance. The behavior of the Type 2 wind connection is considered next.

### 4.2.4 Type 2 wind connections

Although the design of Type 2 wind connections is empirical in its approach, many significant tall buildings have been built with this technique, the most notable example being the Empire State Building, for many years the world's tallest. It must be pointed out, however, that significant stability and stiffness are incorporated into the structure by the exterior stone cladding and interior braces. Other major buildings that have used Type 2 construction are the United Nations Secretariat Building and the Chrysler Building, both in New York, and the Alcoa Building in Pittsburgh.

Design of Type 2 wind connections for tall buildings is based on the practice of ignoring beam end moments generated by a connection's resistance to gravity load while counting on the same connection to resist wind moments calculated on the assumption of fully rigid behavior. This is a time-tested procedure and it is safe provided the actual end moment, which can be higher than the design moment, does not overstress the fasteners. Connections designed under this procedure are generally semirigid with components which, by deforming inelastically, prevent fastener distress. Since it is difficult to calculate the true combined moment at the connections, reliance must be placed on joint configurations of demonstrable ductility. Thus the Type 2 wind connection can be defined as a type of connection that develops lateral resistance through special wind connections which provide some restraint to the ends of a simple beam designed for gravity loads only. Basic to the Type 2 wind design is the requirement that the connection should have adequate inelastic deformation capacity to avoid connector overstress under full loading.

To understand the Type 2 wind connection, it is instructive to trace its behavior through a complete loading sequence, starting from the gravity loads to reversible wind loadings. Figure 4.4 shows a moment-rotation curve for a beam with a known moment-rotation characteristic. The shape of the curve depends on the design of the connection and may range from almost full fixity of Type 1 rigid frame to Type 2 simple framing. In Fig. 4.4, point 1 at the intersection of the connection curve and the beam line *ab* represents the gravity moment $M_1$ and the corresponding rotation $\theta_1$ at the ends due to a uniformly distributed load $w$. The end moment at the left-hand side of the girder is counterclockwise, while the end moment on the right-hand side is clockwise as shown in the free-body diagram of the beam in Fig. 4.7. The beam can be considered as a

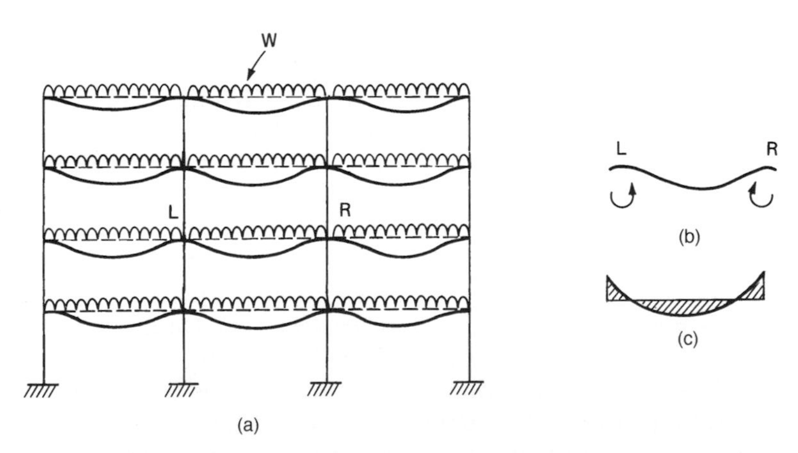


**Figure 4.7** (a) Portal frame subjected to gravity loads: (b) end moments; (c) moment diagram.

typical beam of a moment frame as shown in Fig. 4.7. The intersection points $L_1$ and $R_1$ in Fig. 4.9 represent the application of vertical load only to the beam; there is no wind moment acting on the connection or the beam. Assume that the frame is subject to wind loads acting from left to right, as represented in Fig. 4.8a. The beam is subjected to end moments which act in a clockwise direction as shown in the free-body diagram of the beam in Fig. 4.8b. The windward or the left-end moment acts in a direction opposite to the gravity moment, while the right-end or the leeward moment acts in the same sense as the gravity moment. The gravity moment at the left end is relieved by the wind action, while the right-end gravity moment becomes additive to the wind moment. Because of the additional moment, the right-end connection moves from the original point $R_1$ to $R_2$ along the connection curve. The left-end moment moves downward from $L_1$ to point $L_2$ because of the reduction in gravity moment. The windward moment does not retrace its path along the $M$-$\theta$ curve but travels on a line parallel to the slope of the curve. Recall that this characteristic is similar to the inelastic stress-strain diagram of a material such as steel subjected to load reversals. The decrease in moment occurs along a straight line because the moment is entirely elastic in this region. The rotations at the ends of the beam are, however, the same. The beam rotates by the same amount until the entire wind moment is developed between points $R$ and $L$ as illustrated in Fig. 4.9.

Since wind is a transient load, we have to consider the condition when it stops acting on the structure. The only loads on the beam are gravity loads similar to the condition that we started with, except the moment at the ends will not revert back to $M_1$ because during the

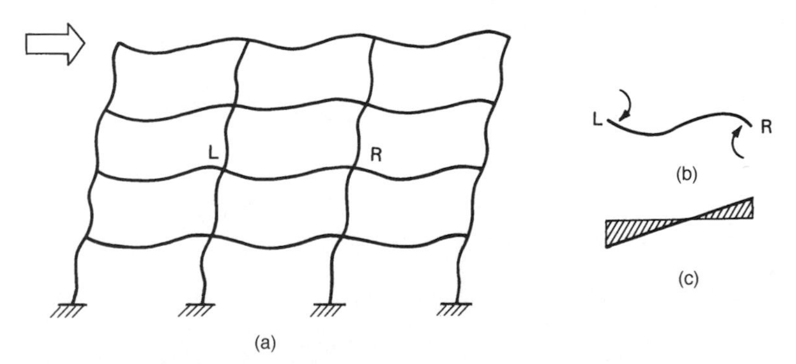


**Figure 4.8** (a) Portal frame subjected to lateral loads from left: (b) end moments; (c) moment diagram.

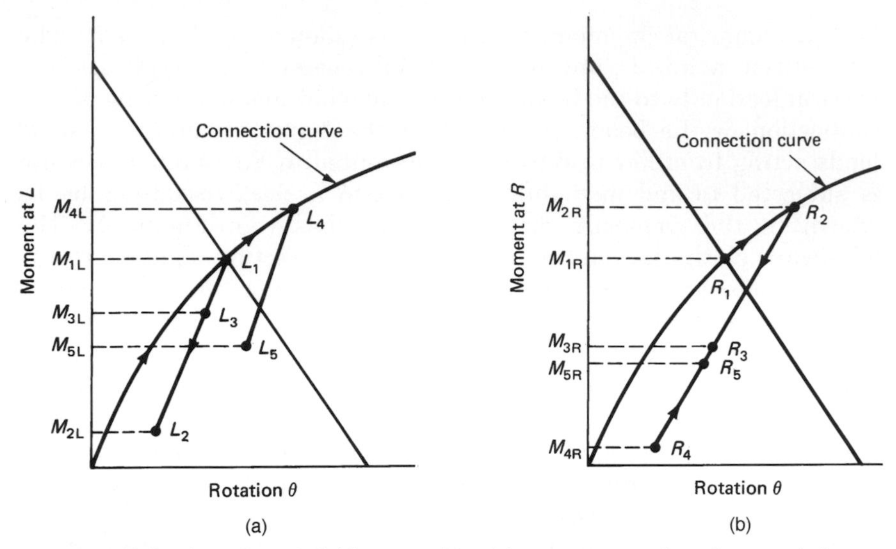


**Figure 4.9** Behavior of semi-rigid connection: (a) moment-rotation characteristics at left support; (b) moment-rotation characteristics at right support.

loading cycle the connection has undergone inelastic rotations. The left-end connection goes from $L_2$ to $L_3$ and the right-end connection goes from $R_2$ to $R_3$. Both the connections move on an elastic line parallel to the initial slope. Static equilibrium requires that the moment at each end be the same. The resulting moment is, however, less than $M_1$.

Consider the wind now acting in the right-to-left direction as shown in Fig. 4.10a. The left-end connection receives additional

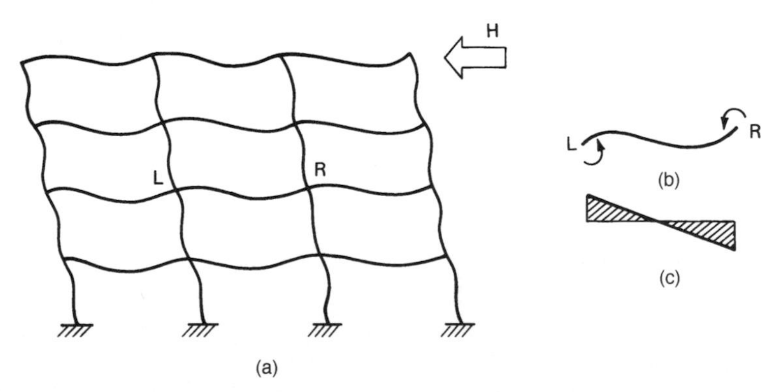


**Figure 4.10** (a) Portal frame subjected to lateral loads from the right: (b) end moments; (c) moment diagram.

moment and travels from $L_3$ to $L_4$. Part of this rotation is inelastic and part is elastic as shown in Fig. 4.9a. The right-end connection is partially relieved of the moment and moves down to point $R_4$ as shown in Fig. 4.9b. Both the connections rotate through the same angle until the entire moment is developed. Figure 4.9a,b shows the behavior when the wind stops and the only loads are gravity loads. The connections move from $R_4$ to $R_5$ and $L_4$ to $L_5$. Both ends have the same moment in order to satisfy static equilibrium. The moments corresponding to $R_5$ and $L_5$ are considerably less than the gravity moment at points $R_1$ and $L_1$ under gravity loads only. This phenomenon of reduction in moment due to cyclic loading is referred to as "shakedown." The gravity moment, therefore, has shaken down considerably. From this point on, the connection behaves entirely as an elastic connection. Regardless of the direction of wind, the maximum moment on the connection can never exceed points $L_5$ and $R_5$. The connection is said to have undergone a complete shakedown. After shakedown the maximum moment occurring at the connection is considerably smaller than the original gravity movement. A brief summary of the connection behavior is given in Table 4.1.

One of the drawbacks for the not-so-wide use of semi-rigid frames is the lack of sufficient information available to the designers about the moment-rotation relationship. The moment-rotation characteristics of a connection depend upon many physical parameters such as type of connection, the size of angles, end plates, top and bottom angles, and gauge for bolt location. In short, an exact relationship can only be obtained by conducting experiments. However, Lothers in his text *Advanced Design in Structural Steel,* has presented formulas for a parameter $Z$ to define the relationship between moment and rotation for the connections.

The factor $Z$ for a connection is analogous to the flexibility factor $L/4EI$ which corresponds to the rotation at the end of a beam upon the application of a unit moment. The slope of the moment-rotation relation defines the parameter $Z$. Although the moment-rotation characteristics for most types of connections are nonlinear for the full spectrum of elastic and inelastic deformations, their behavior in the design range can be considered elastic. The reciprocal of the initial tangent to the moment-rotation curve can be considered as sufficiently accurate for determining the value of $Z$.

Approximate values of $Z$ are given for four types of connections by Defalco and Marino in the 1960 AISC Journal. These are reproduced here, by permission, in Figs 4.11–4.14. Type A connection consists of a double-angle connection as shown in Fig. 4.11, while Type B, shown in Fig. 4.12, consists of a top and bottom clip angle connection. Type

**TABLE 4.1 Summary of Connection Behavior**

| Load case | | Moment at $L$ | Moment at $R$ | Connection behavior |
|---|---|---|---|---|
| 1. Gravity | $W$ | $M_{1L}$ | $M_{1R}$ | Equal gravity moments at each end |
| 2. Gravity plus wind from left | $W + H$ | $M_{2L}$ | $M_{2R}$ | $L$ unloads elastically while $R$ loads along connection curve |
| 3. Remove wind from left | $W$ | $M_{3L}$ | $M_{3R}$ | Elastic recovery of $M_{H/2}$ at both connections |
| 4. Gravity plus wind from right | $W - H$ | $M_{4L}$ | $M_{4R}$ | $L$ loads elastically up to and then along the connection curve. $R$ unloads elastically |
| 5. Remove wind from right | $W$ | $M_{5L}$ | $M_{5R}$ | Elastic recovery of $M_{H/2}$ at both connections |
| 6. Gravity plus wind from left | $W + H$ | $M_{6L} = M_{5L} - \frac{M_H}{2}$ | $M_{6R} = M_{5R} + \frac{M_H}{2}$ | $L$ unloads elastically by $M_{H/2}$ while $R$ loads elastically by $M_{H/2}$ |
| 7. Remove wind from left | $W$ | $M_{5L}$ | $M_{5R}$ | Elastic response at both ends. The connections have "shaken down" with the gravity moments considerably smaller than the initial gravity moments |
| 8. Gravity plus wind from right | $W - H$ | $M_{5L} + \frac{M_H}{2}$ | $M_{5R} - \frac{M_H}{2}$ | |
| 9. Gravity plus wind from left | $W + H$ | $M_{5L} - \frac{M_H}{2}$ | $M_{5R} + \frac{M_H}{2}$ | |

C connection in Fig. 4.13 is similar to type B with the exception that the shear capacity of the beam is augmented by bolting two angles to the beam web. Type D connection is a seat angle connection with a top plate connection as shown in Fig. 4.14. More important than the ultimate moment capacity is the initial stiffness of the curve, which is represented by the slope of the $M$-$\theta$ curve at the origin. The connection rotational stiffness, designated as $Z$, represents a zero value for a perfectly pin-ended connection and infinity for a fully fixed beam.

Analyses of frames that incorporate Type 2 wind and semi-rigid (Type 3) connections must include considerations of:

1. Connection ductility.
2. Evaluation of the drift characteristics of frames with less than fully rigid connections.
3. Effect of partial restraints on column and frame stability.

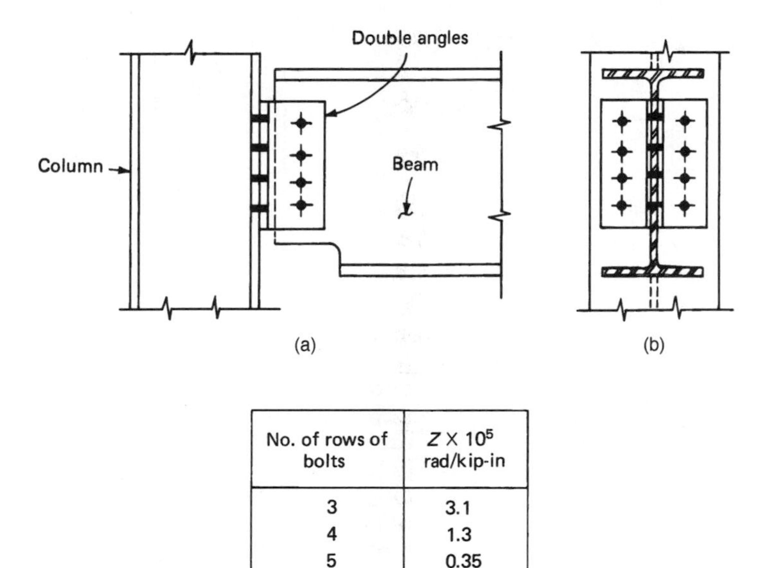


| No. of rows of bolts | $Z \times 10^5$ rad/kip-in |
|---|---|
| 3 | 3.1 |
| 4 | 1.3 |
| 5 | 0.35 |
| 6 | 0.20 |
| 7 | 0.11 |
| 8 | 0.075 |
| 9 | 0.052 |
| 10 | 0.035 |

(c)

**Figure 4.11** Double-angle connection (Type A): (a) elevation; (b) side view; (c) Values of stiffness factor $Z$.

### 4.2.4.1 Design outline for Type 2 wind connections

1. Calculate end reaction due to gravity loads.
2. Determine the magnitude of wind moment by an elastic analysis, assuming rigid connections.
3. Select the type of semirigid connection from an inventory of standard connections, such as double-angle, single-angle, shear tab, top-and-bottom angle, or header plate connection. Make an initial guess on the various dimensions and thicknesses for connection components.

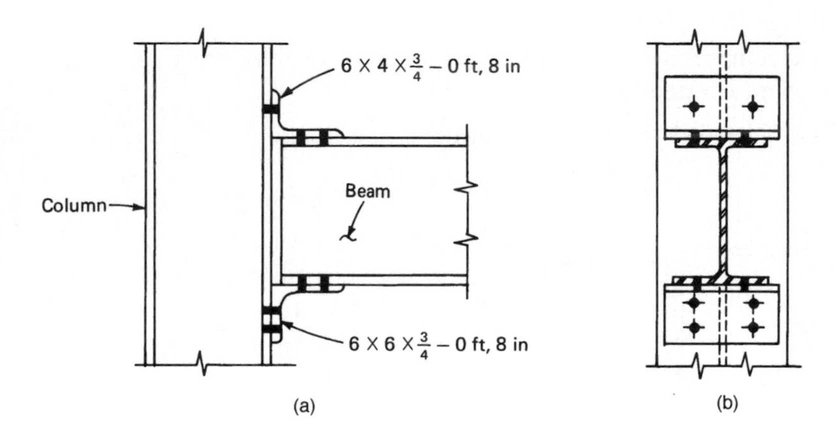


| Depth of beam, in. | $Z \times 10^5$, rad/kip-in |
|---|---|
| 8 | 0.046 |
| 10 | 0.036 |
| 12 | 0.028 |
| 14 | 0.023 |
| 16 | 0.018 |
| 18 | 0.014 |
| 21 | 0.012 |
| 24 | 0.010 |
| 27 | 0.0078 |
| 30 | 0.0066 |
| 33 | 0.0055 |
| 36 | 0.0046 |

(c)

**Figure 4.12** Top and bottom clip angle connection (Type B): (a) elevation; (b) side view; (c) Values of stiffness factor $Z$.

4. Determine the moment-rotation characteristics of the connection to evaluate the $Z$ value. Since the available data do not cover all types of connections, it is advisable to restrict connection designs to those for which $M$-$\theta$ curves have been well established. The AISC, however, allows use of any connection as long as its behavior can be demonstrated by either tests or by a rational analysis.
5. With the known value of $Z$ for the connection, plot the moment-rotation curve on the beam line.
6. Check the connection for wind moment by calculating the design

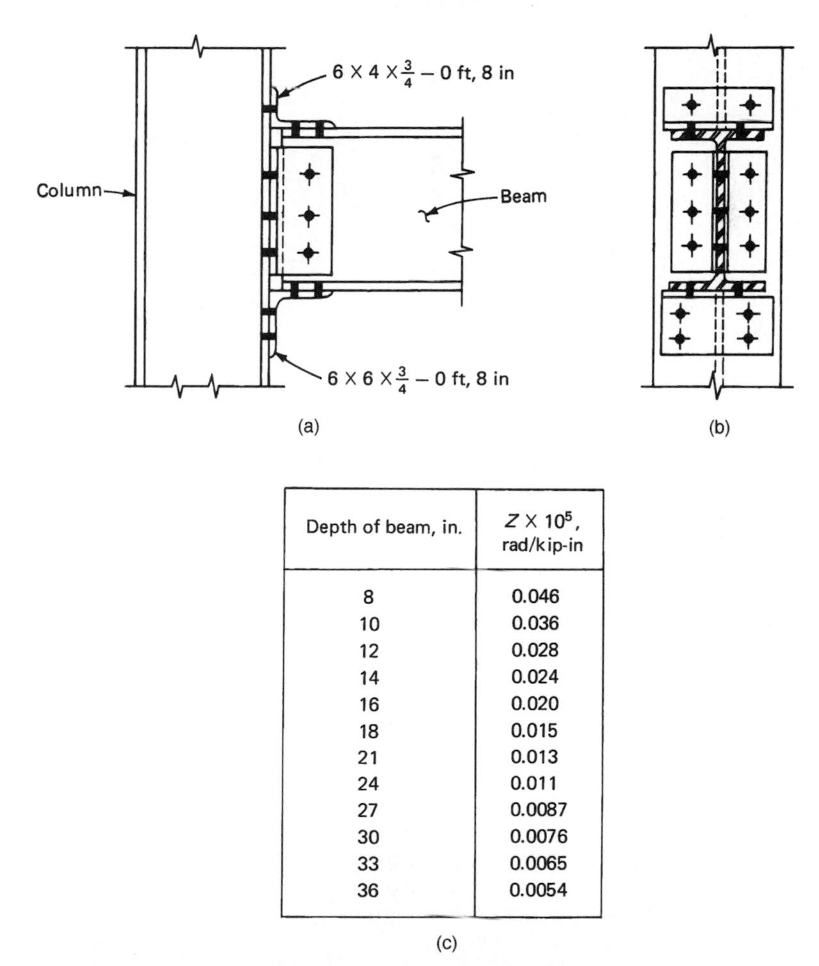


| Depth of beam, in. | $Z \times 10^5$, rad/kip-in |
|---|---|
| 8 | 0.046 |
| 10 | 0.036 |
| 12 | 0.028 |
| 14 | 0.024 |
| 16 | 0.020 |
| 18 | 0.015 |
| 21 | 0.013 |
| 24 | 0.011 |
| 27 | 0.0087 |
| 30 | 0.0076 |
| 33 | 0.0065 |
| 36 | 0.0054 |

(c)

**Figure 4.13** Type C top and bottom clip angle connection with shear plate: (a) elevation; (b) side view; (c) values of stiffness factor $Z$.

values for each component such as bolts, connecting angles, welds, etc.

7. Check the connection for ductility by verifying that all connection materials such as bolts, welds, and plates are not stressed beyond their ultimate strengths (with proper safety factors) under the simultaneous action of gravity and wind moments.
8. Since the connections were assumed fully rigid in the wind analysis, Steps 1–7 yield a conservative design. If required, the connection design can be modified by incorporating the effect of nonrigid connections in the wind analysis. This can be done by using a reduced bending rigidity for the beam to account for less than 100 percent rigidity of the connection.

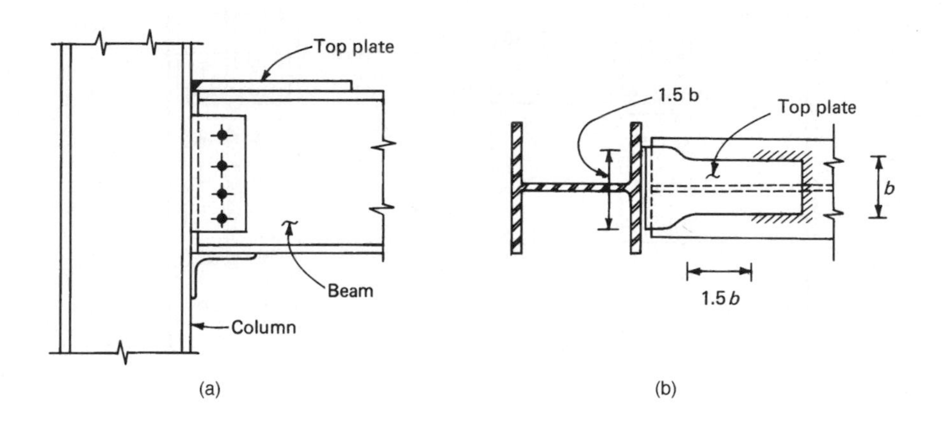


| Depth of beam, in. | $Z \times 10^5$ rad/kip-in |
|---|---|
| 8 | 0.21 |
| 10 | 0.13 |
| 12 | 0.093 |
| 14 | 0.068 |
| 16 | 0.052 |
| 18 | 0.041 |
| 21 | 0.030 |
| 24 | 0.023 |
| 27 | 0.018 |
| 30 | 0.015 |
| 33 | 0.012 |
| 36 | 0.010 |

(c)

**Figure 4.14** Type D seat angle with top plate connection: (a) elevation; (b) plan; (c) values of stiffness factor $Z$.

Iteration of Steps 2–8 gives results that will converge to an optimum solution. Since semirigid connections provide less frame stiffness than fully rigid connections, it is necessary to use a modified girder stiffness, $k_r$ which incorporates the $Z$ factor as follows:

$$K_r = \frac{3(I/L)}{4(L'/L) - (L/L')} \tag{4.4}$$

where $K_r$ = modified relative stiffness of beam

$L' = L + \dfrac{3EI}{Z}$

$L$ = beam span

$I$ = moment of inertia of the beam

Although the use of reduced stiffness of the beam in the determination of wind moments is optional, it is important that reduction in stiffness be accounted for in determining the lateral drift and $p$-$\Delta$ effects.

### 4.2.5 Concluding remarks

In spite of reported success of many buildings built by using Type 2 wind connections, there exists little unanimity of opinion about its applicability for buildings taller than 5-stories or so. Engineers who design buildings using Type 2 wind connections are automatically put in a defensive position of explaining the paradox of the joints acting as rigid for wind, and as pins for gravity loading. A straightforward application of the method for frames can result in structures in which the columns are overstressed and exhibit sway deflections much in excess of calculated values. Other cautious approaches that take into consideration the relative softness of connections have been proposed by several investigators but have not found general application in the design office.

## 4.3 Rigid Frames (Moment Frames)

### 4.3.1 Introduction

A frame is considered rigid when its beam-to-column connections have sufficient rigidity to hold virtually unchanged the original angles between intersecting members. A rigid-frame high-rise structure typically comprises of parallel or orthogonally arranged bents consisting of columns and girders with moment resistant joints. Resistance to horizontal loading is provided by the bending resistance of the columns, girders, and joints. The continuity of the frame also assists in resisting gravity loading more efficiently by reducing the positive moments in the center span of girders.

Typical deformations of a moment-resisting frame under lateral load are indicated in Fig. 4.15. The point of contraflexure is normally located near the midheight of the columns and midspan of the beams. The lateral deformation of a frame as will be seen shortly, is due partly to frame racking, which might be called shear sway, and partly to column shortening. The shear-sway component constitutes approximately 80 to 90 percent of the overall lateral deformation of the frame. The remaining component of deformation is due to column shortening, also called cantilever or chord drift component.

Moment-resisting frames have advantages in high-rise construction due to their flexibility in architectural planning. They may be placed on the exterior, or throughout the interior of the building with minimal constraint on the architectural planning.

The size of members in a moment-resisting frame is often controlled

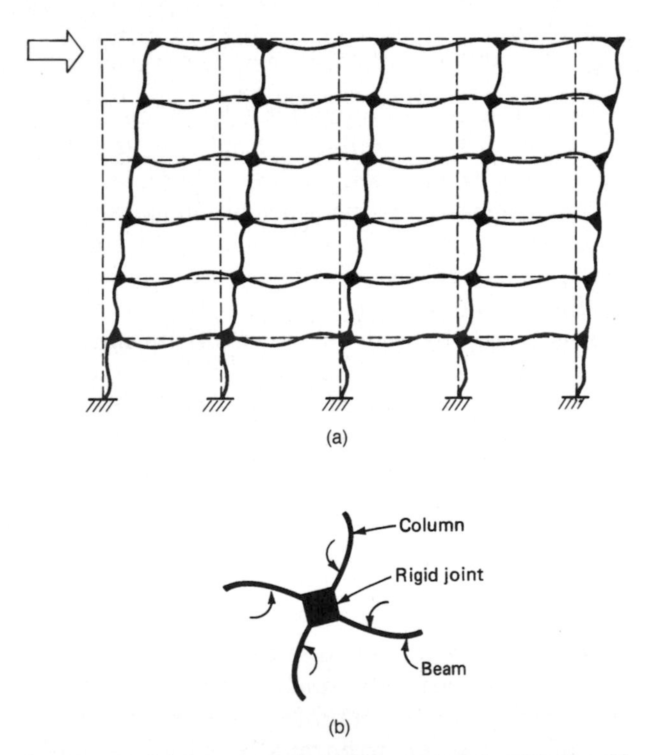


**Figure 4.15** (a) Response of rigid frame to lateral loads: (b) flexural deformation of beams and columns due to nondeformability of connections.

by stiffness rather than strength to control drift under lateral loads. The lateral drift is a function of both the column stiffness and beam stiffness. In a typical application, the beam spans are 20 ft (6 m) to 30 ft (9 m) while the story heights are usually between 12 ft (3.65 m) to 14 ft (4.27 m). Since the beam spans are greater than the floor heights, the beam moment of inertia needs to be greater than the column inertia by the ratio of beam span to story height for an effective moment-resisting frame.

Moment-resisting frames are normally efficient for buildings up to about 30-stories in height. The lack of efficiency for taller buildings is due to the moment resistance derived primarily through flexure of its members.

The connections in steel moment resisting frames are important design elements. Joint rotation can account for a significant portion of the lateral sway. The strength and ductility of the connection are also important considerations especially for frames designed to resist seismic loads.

Steel moment-resisting frames with welded connections have been regarded up until the January 1994 Northridge earthquake, as one of the safest for having the required strength, ductility and reliability.

The Northridge, magnitude 6.7 earthquake which caused damage to over 200 steel moment resisting frame buildings has shaken engineers' confidence in its use for seismic design. Almost without exception, the connections that failed were of the type with full penetration field weld of top and bottom flanges, and a high strength bolted shear tab connection. The majority of the damage consists of fractures of the bottom flange weld between the column and girder flanges. There were also a large number of instances where top flange fractures occurred. Although many factors may have contributed to the poor performance it is believed that the basic joint configuration is not conducive to ductile behavior.

### 4.3.2 Deflection characteristics

The lateral deflection components of a rigid frame can be thought of as being caused by two components similar to the deflection components of a prismatic cantilever beam. One component can be likened to the bending deflection and the other to the shear deflection. Normally for prismatic members when the span-to-depth ratio is greater than 10 or so, the bending deflection is by far the more predominant component. Shear deflections contribute a small portion to the overall deflection and are therefore generally neglected in calculating deflections. The deflection characteristics of a rigid frame, on the other hand, are just the opposite; the component analogous to the beam shear deflection dominates the deflection picture and may amount to as much as 80 percent of the total deflection, while the remaining 20 percent comes from the bending component. The bending and the shear components of deflection are usually referred to as the cantilever bending and frame racking, each with its own distinct deflection mode.

#### 4.3.2.1 Cantilever bending component

This phenomenon is also known as *chord drift*. The wind load acting on the vertical face of the building causes an overall bending moment on any horizontal cross-section of the building. This moment, which reaches its maximum value at the base of the building, causes the building to rotate about the leeward column and is called the *overturning moment*. In resisting the overturning moment, the frame behaves as a vertical cantilever responding to bending through the axial deformation of columns resulting in compression in the leeward columns and tension or uplift in the windward columns. The columns lengthen on the windward face of the building and shorten on the leeward face. This column length change causes the building to rotate and results in the chord drift component of the lateral deflection, as shown in Fig. 4.16a.

Because of the cumulative rotation up the height, the story drift due to overall bending increases with height, while that due to racking tends to decrease. Consequently the contribution to story drift from overall bending may, in the uppermost stories, exceed that from racking. The contribution of overall bending to the total drift, however, will usually not exceed 10 to 20 percent of that of racking, except in very tall, slender, rigid frames. Therefore the overall deflected shape of medium-rise frame usually has a shear configuration.

For a normally proportioned rigid frame, as a first approximation, the total lateral deflection can be thought of as a combination of three factors:

1. Deflection due to axial deformation of columns (15 to 20 percent).
2. Frame racking due to beam rotations (50 to 60 percent).
3. Frame racking due to column rotations (15 to 20 percent).

In addition to the above, there is a fourth component that contributes to the deflection of the frame which is due to deformation of the joint. In a rigid frame, since the sizes of joints are relatively small compared to column and beam lengths, it is a common practice to ignore the effect of joint deformation. However, its contribution to building drift in very tall buildings consisting of closely spaced columns and deep spandrels could be substantial, warranting a closer study. This effect is called *panel zone deformation* and is discussed at length in Chap. 11.

#### 4.3.2.2 Shear racking component

This phenomenon is analogous to the shear deflection in a beam and is caused in a rigid frame by the bending of beams and columns. The accumulated horizontal shear above any story of a rigid frame is resisted by shear in the columns of that story (Fig. 4.16b). The shear causes the story-height columns to bend in double curvature with points of contraflexure at approximately mid-story-height levels. The moments applied to a joint from the columns above and below are resisted by the attached girders, which also bend in double curvature, with points of contraflexure at approximately mid-span. These

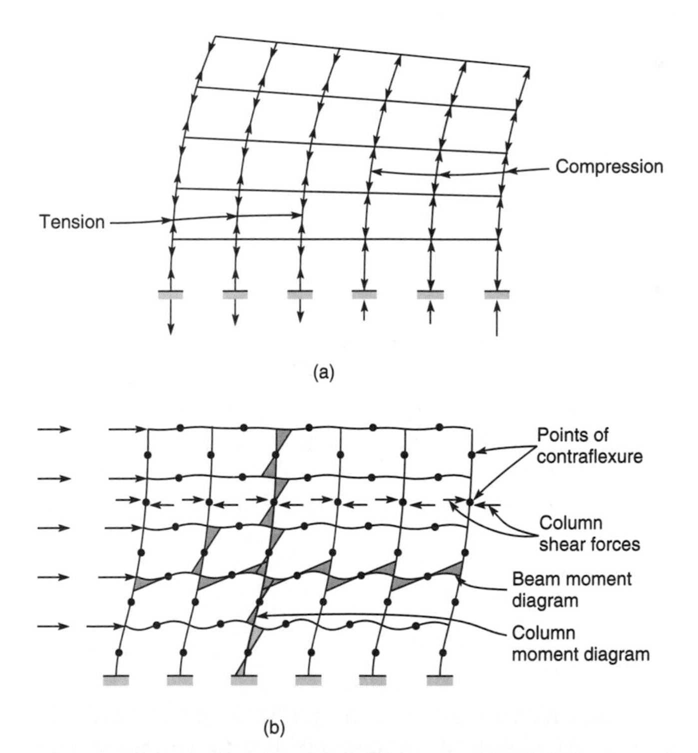


**Figure 4.16** Rigid frame deflections: (a) forces and deformations caused by external overturning moment; (b) forces and deformations caused by external shear.

deformations of the columns and girders allow racking of the frame and horizontal deflection in each story. The overall deflected shape of a rigid frame structure due to racking has a shear configuration with concavity upwind, a maximum inclination near the base, and a minimum inclination at the top, as shown in Fig. 4.16b. This mode of deformation accounts for about 80 percent of the total sway of the structure. In a normally proportioned rigid-frame building with column spacing at about 35 to 40 ft (10.6 to 12.2 m) and a story height of 12 to 13 ft (3.65 to 4.0 m), beam flexure contributes about 50 to 65 percent of the total sway. The column rotation, on the other hand, contributes about 10 to 20 percent of the total deflection. This is because in most unbraced frames the ratio of column stiffness to girder stiffness is very high, resulting in larger joint rotations of girders. So generally when it is desired to reduce the deflection of unbraced frame, the place to start adding stiffness is in the girders. However, in nontypical frames, such as those that occur in framed tubes with column spacing approaching floor-to-floor height, it is necessary to study the relative girder and column stiffness before making adjustments in the member properties.

### 4.3.3 Methods of analysis

Because of the large-scale availability of computers, the analysis of rigid-frame buildings, even in preliminary design stages, is accomplished most effectively by using stiffness analysis programs. Hand calculations are rarely undertaken except for very preliminary purposes because of the approximate nature of analysis and the longer time it takes to do hand computations.

Among the better known approximate methods of analysis, the cantilever and portal methods are perhaps the most popular and considered by some engineers to be sufficiently accurate for use in the final analysis of buildings of intermediate height range. The portal method is considered reasonably valid for buildings less than 25-stories, and the cantilever method is assumed to be valid for buildings in the 25- to 30-story range. A brief description of each method is given in Chap. 10.

### 4.3.4 Calculation of drift

Calculation of drift due to lateral loads is a major task in the analysis of tall building frames. Although it is convenient to consider the lateral displacements to be composed of two distinct components, whether or not the cantilever or the racking component dominates the deflection is dependent on factors such as height-to-width ratio of the building and the relative rigidity of the column to girder connection. Unless the building is very tall or very slender, it is usually the racking component that dominates the deflection picture. A simple method for determining the deflection of a tall building is to assume that the entire structure acts as a vertical cantiliver in which the axial stress in each column is proportional to its distance from the centroidal axis of the frame. This approach assumes that the frame is infinitely stiff with respect to longitudinal shear and hence underestimates the deflection. Methods of calculation which take into account the shear racking component are given in Chap. 10.

## 4.4 Braced Frames

### 4.4.1 Introduction

Rigid frame systems are not efficient for buildings taller than about 30-stories because the shear racking component of deflection due to the bending of columns and girders causes the drift to be too large. A braced frame attempts to improve upon the efficiency of a rigid frame by virtually eliminating the bending of columns and girders. This is achieved by adding web members such as diagonals or chevron braces. The horizontal shear is now primarily absorbed by the web and not by the columns. The webs carry the lateral shear predominantly by the horizontal component of axial action allowing for nearly a pure cantilever behavior.

### 4.4.2 Behavior

In simple terms, braced frames may be considered as cantilevered vertical trusses resisting lateral loads primarily through the axial stiffness of columns and braces. The columns act as the chords in resisting the overturning moment, with tension in the windward column and compression in the leeward column. The diagonals and girders work as the web members in resisting the horizontal shear, with diagonals in axial compression or tension depending upon their direction of inclination. The girders act axially, when the system is a fully triangulated truss. They undergo bending also when the braces are eccentrically connected to them. Because the lateral load on the building is reversible, braces are subjeced in turn, to both compression and tension; consequently, they are most often designed for the more stringent case of compression.

The effect of the chords' axial deformations on the lateral deflection of the frame is to tend to cause a "flexural" configuration of the structure, that is, with concavity downwind and a maximum slope at the top (Fig. 4.17a). The effect of the web member deformations, however, is to tend to cause a “shear” configuration of the structure (i.e., with concavity upwind, a maximum slope at the base, and a zero slope at the top; Fig. 4.17b). The resulting deflected shape (Fig. 4.17c) is a combination of the effects of the flexural and shear curves with a resultant configuration depending on their relative magnitudes, as determined mainly by the type of bracing. Nevertheless, it is the flextural deflection that most often dominates the deflection characteristics.

The role of web members in resisting shear can be demonstrated by following the path of the horizontal shear down the braced bent. Consider the typical braced frames, shown in Fig. 4.18a–e, subjected to an external shear force at the top level. In Fig. 4.18a, the diagonal in each story is in compression, causing the beams to be in axial tension; therefore, the shortening of the diagonal and extension of the beams gives rise to the shear deformation of the bent. In Fig. 4.18b, the forces in the braces connecting to each beam end are in equilibrium horizontally with the beam carrying insignificant axial load.

In Fig. 4.18c, half of each beam is in compression while the other half is in tension. In Fig. 4.18d, the braces are alternately in

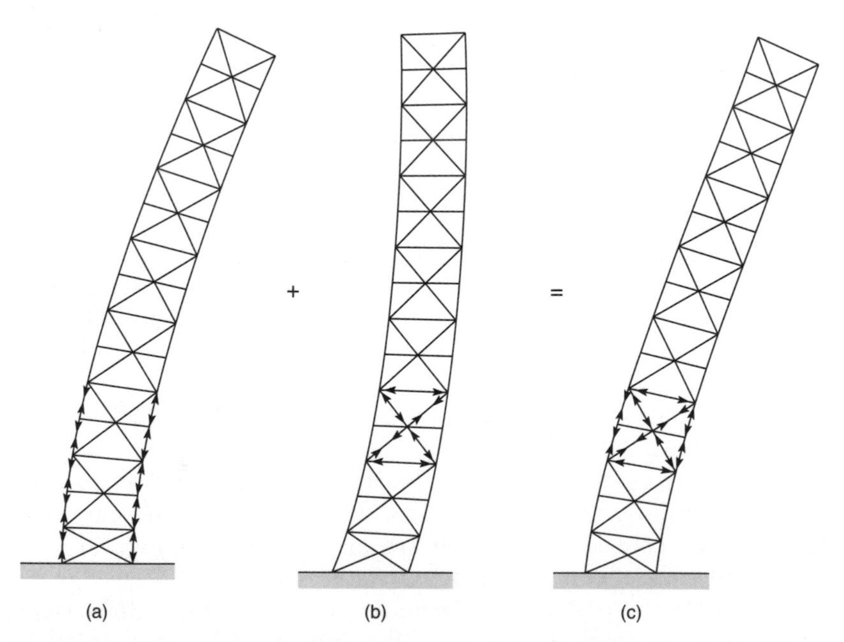


**Figure 4.17** Braced frame deformation: (a) flexural deformation; (b) shear deformation; (c) combined configuration.

compression and tension while the beams remain basically unstressed. And finally in Fig. 4.18e, the end parts of the beam are in compression and tension with the entire beam subjected to double curvature bending. Observe that with a reversal in the direction of horizontal load, all actions and deformations in each member will also be reversed.

In a braced frame the principal function of web members is to resist the horizontal shear forces. However, depending upon the configuration of the bracing, the web members may pick up substantial compressive forces as the columns shorten vertically under gravity loads. Consider, for example, the typical bracing configurations shown in Fig. 4.19. As the columns in Fig. 4.19a,b shorten, the diagonals are subjected to compression forces because the beams at each end of the braces are effective in resisting the horizontal component of the compressive forces in the diagonal. At a first glance, this may appear to be the case for the frame shown in Fig. 4.19c. However, the diagonals shown in Fig. 4.19c will not attract significant gravity forces because there is no triangulation at the ends of beams where the diagonals are not connected (nodes A and D, in Fig. 4.19c). The only horioznetal restraint at the beam end is by the bending resistance of columns, which usually is of minor significance in the overall behavior. Similarly, in Fig. 4.19d the vertical restraint from the bending stiffness of the beam is not large; therefore as in the previous case, the diagonals experience only negligible gravity forces.

### 4.4.3 Types of braces

Braced frames may be grouped into two categories as either concentric braced frames (CBF) or eccentric braced frames (EBF)

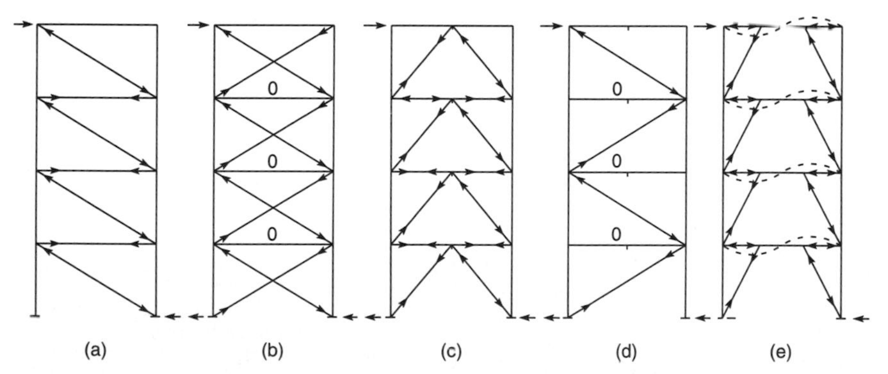


**Figure 4.18** Load path for horizontal shear through web numbers: (a) single diagonal bracing; (b) $X$-bracing; (c) chevron bracing; (d) single-diagonal, alternate direction bracing; (e) knee bracing.

depending upon their ductility characteristics. In CBFs the axes of all members, i.e., columns, beams and braces intersect at a common point such that the member forces are axial. EBFs, which will be discussed later in this chapter, utilize axis offsets to deliberately introduce flexure and shear into framing beams to increase ductility.

The CBFs can take many forms some of which are shown in Fig. 4.20. Depending upon the diagonal force, length, required stiffness and clearances, the diagonal member can be made of double angles, channels, tees, tubes or wide flange shapes. Figure 4.16 shows an inverted K bracing also called chevron bracing consisting of double angles. Besides performance, the shape of the diagonal is often based on connection considerations.

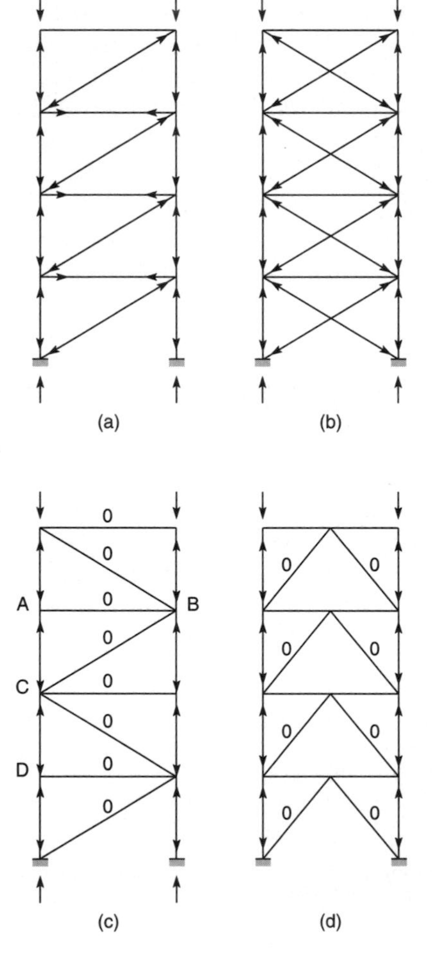


**Figure 4.19** Gravity load path: (a) single diagonal single direction bracing; (b) $X$-bracing; (c) single diagonal alternate direction bracing; (d) chevron bracing.

The least objectionable locations for braces are around service cores and elevators, where frame diagonals may be enclosed within permanent walls. The braces can be joined together to form a closed or partially closed three-dimensional cell for effectively resisting torsional loads.

Common types of interior bracing are shown in Fig. 4.20a–n. Figure 4.20e–n shows bracings across single bays in 1-story increments. Figure 4.20a shows diagonal bracing in 2-story increments. Shown in Fig. 4.20b,c is a K-braced frame, while Fig. 4.20d shows bracing for a three-bay frame. Any reasonable pattern of braces with single or multiple braced bays can be designed for resisting the lateral loads.

Finding an efficient and economical bracing system for a tall

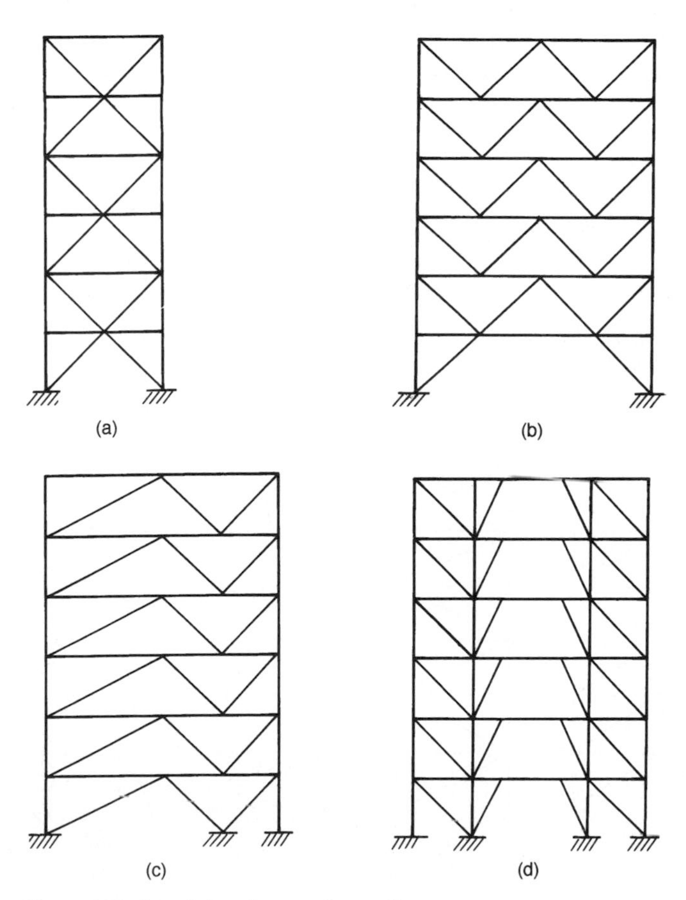


**Figure 4.20** (a–n): bracing configurations.

building presents the structural engineer with an excellent opportunity to use innovative design concepts. However, availability of proper depth for bracing is often an overriding consideration. As a preliminary guide, a height-to-width ratio of 8 to 10 is considered proper for a reasonably efficient bracing system.

## 4.5 Staggered Truss System

### 4.5.1 Introduction

Most high-rise residential-type structures such as apartments and hotels are generally in the neighborhood of 60 by 150 ft (18.3 × 45.75 m) to 200 ft (61 m) long. Their floor plans normally lend themselves to central double-loaded corridors of about 6 to 8 ft (1.83 to 2.43 m) in width. A study was done at the Massachusetts Institute of Technology in the mid-1960s under the sponsorship of U.S. Steel Corporation for the purpose of developing an economical framing system for such tall, narrow structures. The staggered truss system

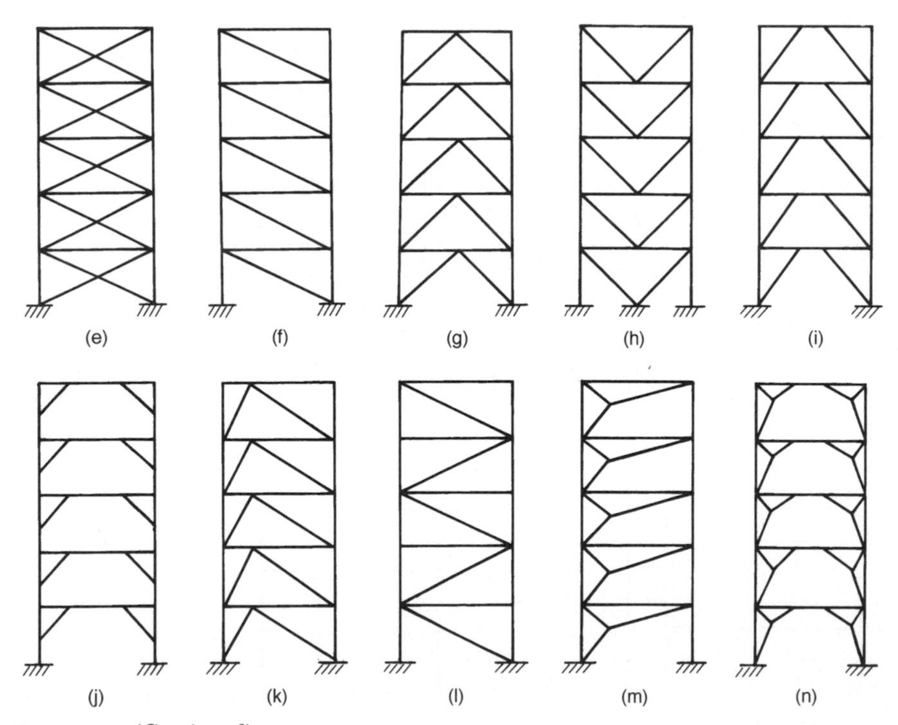


**Figure 4.20** *(Continued).*

evolved as an out-growth of the research done by the departments of architecture and civil engineering at MIT. In this system story-high trusses span in the transverse direction between the columns at the exterior of the building. The required flexibility in residential unit layouts is achieved by arranging the trusses in a staggered plan at alternate floors, as shown schematically in Fig. 4.21. The floor system acts as a diaphragm transferring lateral loads in the short direction to the trusses. Lateral loads are thereby resisted by truss diagonals and are transferred into direct loads in the columns. The columns therefore receive no bending moments.

The truss diagonals are eliminated at the corridor location to allow for openings. Since the diagonal is eliminated, the shear is carried by the bending action of the top and bottom chord members. Similarly, other openings can be provided for in the truss to allow for additional openings at a slight structural premium when required by architectural layout. The system was first used for a housing project for the elderly in St. Paul, Minnesota, completed in 1967. Since then, a number of long, narrow, high-rise buildings for apartment houses, hotels, and in some cases for office buildings have been built using this concept.

Because the staggered truss resists major gravity and lateral loads in direct stresses, the system is quite stiff. In general, no material needs to be added for drift control, and high-strength steels are conveniently used throughout the entire frame. The system has been used for 35- to 40-story buildings. Spans must be long enough to make the trusses efficient, with 45 ft (13.72 m) being considered as the minimum practical limit. In a typical hotel or residential building, a staggered truss system will normally reduce the steel requirement by as much as 30 to 40 percent as compared to a conventional moment-connected framing. Since the trusses are supported only by the perimeter columns, the need for interior columns and associated foundations is eliminated, contributing to the economy of the system.

An added advantage of the system is that it allows for public spaces free of interior columns on the lower levels. The most economical use of staggered trusses is achieved by placing the trusses between units, since in a normal hotel or housing these units are spaced uniformly across the length of the building. It is possible to extend these units through trusses by providing for additional openings. However, varying the spacings could create a variety of unit sizes that can be accommodated within the trusses. Thus, one-, two-, or three-bedroom apartments can be arranged on a single floor merely by varying the column spacings. The system is not limited to simple rectangular

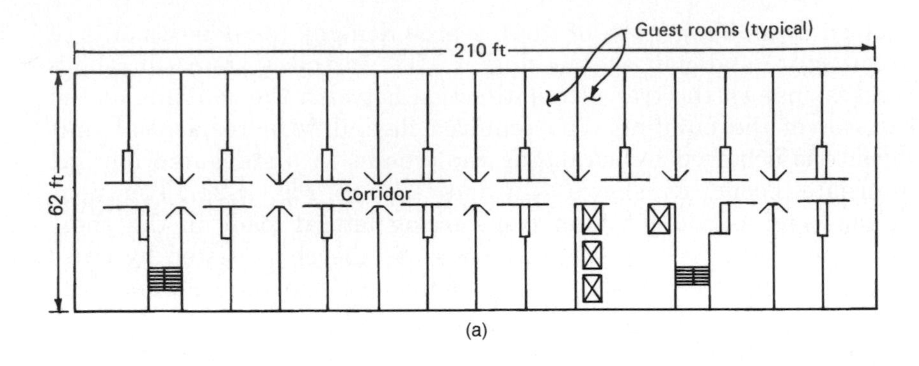


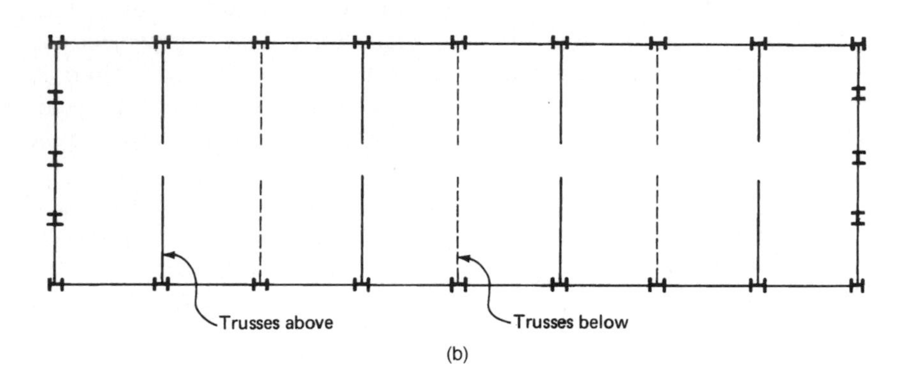


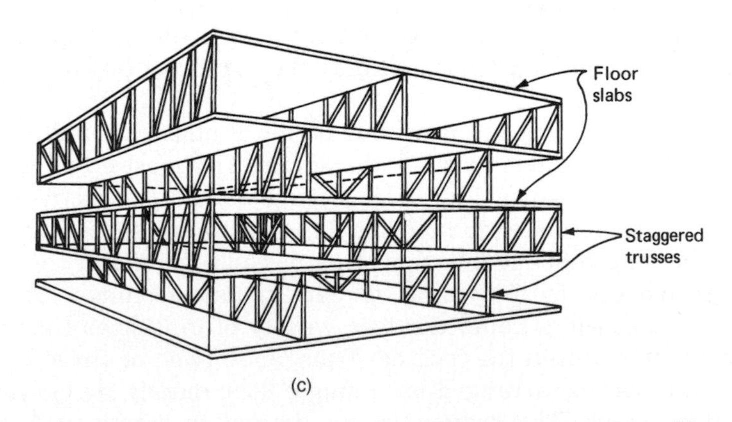


**Figure 4.21** Staggered truss system: (a) hotel plan showing guest rooms; (b) arrangement of staggered trusses; (c) perspective view of truss arrangement.

plans. It can be effectively used in curvilinear plans as shown in Fig. 4.22.

### 4.5.2 Physical behavior

Consider the three-dimensional interaction between vertical bracings of a building interconnected through floor diaphragms, as shown in Fig. 4.23. Assume that for architectural reasons it is required to eliminate the bracing at column lines A and B below level 2. If there is no other bracing below level 2, the columns at the extremities of the bracing must resist both the overturning moment and shear forces below level 2.

The overturning moment manifests itself as compressive and tensile forces in the columns, while the shear forces introduce bending moments in the columns resulting in a rather inefficient structural system. Assume that architecturally it is permissible to introduce a bracing at the center of the building below level 2 as shown in Fig. 4.23a,b. For purposes of lateral analysis, generally it is sufficiently accurate to assume that the slab diaphragm is rigid in its own plane. As a consequence of this assumption, most of the shear at the exterior braces at level 2 is transferred to the interior bracings through the diaphragm action of the floor slab. The columns under the exterior braces are therefore subjected to axial stresses only, while the shear is resisted by the interior bracing. This in essence is the structural action in a staggered truss system in which the lateral force is transmitted across the floor to the truss on the adjacent column line and continues down on the truss line across the next floor down the next truss, etc., as shown schematically in Figs 4.24 and 4.25. Thus, between the floors, lateral forces are resisted by the truss diagonals, and at each floor these forces are transferred to the truss below by the floor system acting as a diaphragm. The columns between the floors receive no bending moments, resulting in a very efficient and stiff structure. Since the trusses are placed at alternate levels on adjacent column lines, two-bay-wide column-free interior floor space is created in the longitudinal direction.

### 4.5.3 Design considerations

#### 4.5.3.1 Floor system

As in other structural systems, the floor system in a staggered truss scheme needs to fulfill two primary requirements: (i) collect and transmit the gravity loads to the vertical elements; and (ii) resist the

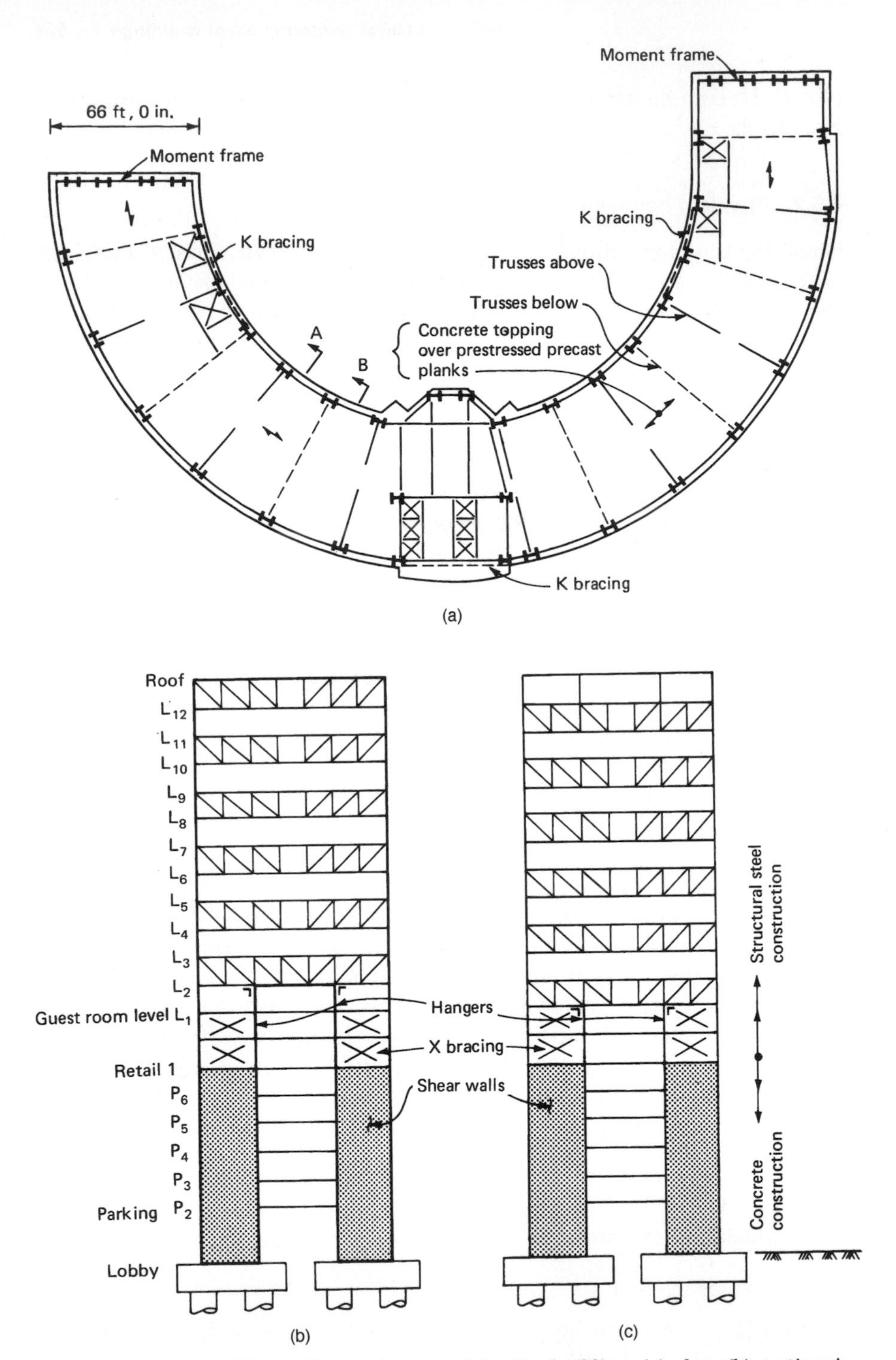


**Figure 4.22** Staggered truss system for a semicircular building: (a) plan; (b) section A (c) section B.

lateral loads as a shear diaphragm and provide a continuous path for transferring the lateral loads from the bottom chord of one truss to the top chord of the adjacent truss down through the structure. In addition to these structural requirements, the floor system must permit flexibility for apartment size and location, must provide fireproofing and an acceptable ceiling, and should be usable as temporary bracing during steel erection. Thus, one could use precast concrete planks, long-span composite steel decks, open-web joist system, or any other system consistent with the structural and

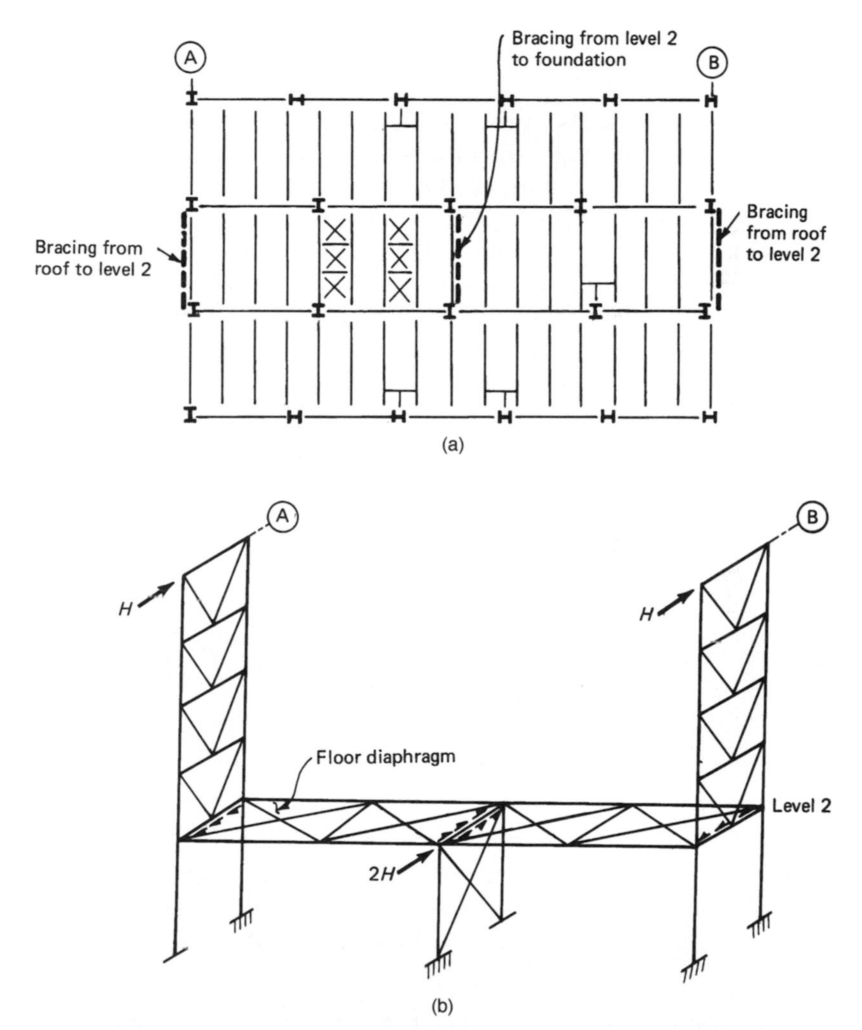


**Figure 4.23** Conceptual model for staggered truss system: (a) building plan; (b) lateral load transfer through diaphragm action.

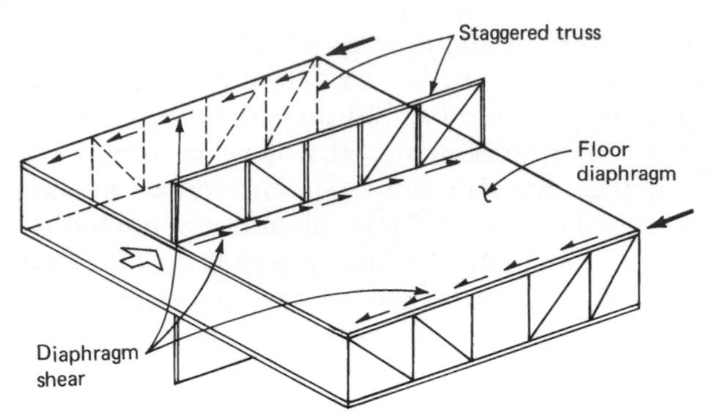


**Figure 4.24** Load path in staggered truss system.

architectural requirements. Precast planks and flat-bottomed steel decks are often used as exposed ceilings with minimum of finish. For spans up to 30 ft (9.15 m), 8 in. (203 mm) planks are required, while for spans less than 24 ft (7.3 m), 6 in. (152 mm) planks are adequate. In a composite steel deck system, a $7\frac{1}{2}$ in. (190 mm) deck is required for spans up to 30 ft (9.15 m), and for spans up to 24 ft (7.3 m), 6 in. (152 mm) deep steel deck is adequate. When precast planks are used, shear transfer is achieved by the use of welded plates cast in the planks or by welding shear connectors on the truss chord.

In the case of a metal deck system, generally adequate shear transfer is obtained by the connection of the steel deck to the trusses. Planks used for erection purposes should have connection weld plates, even when shear connectors are provided. The choice of the floor system depends on the geographical location as well as local

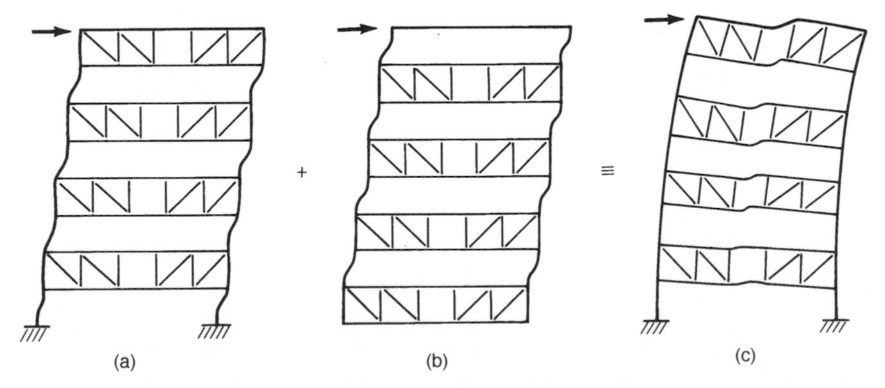


**Figure 4.25** Conceptual two-dimensional model for staggered truss system: (a,b) lateral deformation of adjacent bays; (c) overall behavior. Note the absence of local bending of columns.

conditions. In earthquake zones the lighter floor produces smaller seismic forces. In cold climates the cost of grouting between precast planks in winter is increased by the necessity for heating. The floor system may consist of either a series of simple or continuous spans over the chords of the trusses. Because of the large spacing between the trusses, the continuous spans are usually limited to a maximum of two bays. Generally one end of each span is supported on the lower chord, while the other end is made continuous by simply running the floor slab across the top chord of the truss.

Since the trusses are staggered at alternate floors, the equivalent lateral load on each truss is equal to the lateral load acting on two bays. Hence floor panels on each side of the truss must transmit half of that load to the adjacent truss in the story immediately below. The floor system acts as a deep beam resisting both the in-plane shears and bending moments. Adequate provision should be made to transfer longitudinal shear between each span so that the whole floor system acts as a single unit.

Although any floor system that is capable of carrying gravity loads and diaphragm shear can be used in staggered truss systems, economic considerations generally favor the use of either precast concrete planks or long span composite metal deck, both with a topping of concrete reinforced with welded-wire fabric.

The design of the floor components for gravity loads is identical to that of a conventional high-rise building. Although continuity may be developed at the support connections to the truss chords, simple span behavior is assumed for convenience. Because the staggered truss system depends upon the diaphragm action of the entire floor to transfer lateral loads from one truss to another, the floor is assumed to have chords to resist the in-plane shears and moments. The bending can be resisted by the floor slab or by flange action at the exterior walls. The floor system of two adjacent bays is considered to be a continuous beam for the in-plane forces. Since there are no published criteria for general use, for any specific project it is advisable to study the diaphragm behavior in somewhat greater detail to develop and decide upon design procedures and criteria.

#### 4.5.3.2 Columns

In the staggered truss system the lateral loads are taken out of the system by the floor, which acts as a diaphragm, and the truss diagonals, which act as vertical braces. Therefore, lateral loads are transferred into direct loads in the columns. No shear exists in the columns to create bending in the transverse direction of the building. Thus the columns can be designed as braced members to resist minor axis buckling, and the strong direction of the column can be used to resist lateral loads in the longitudinal direction. Another aspect of column design for staggered truss systems that should be considered is the effect of truss deflections in causing excessive weak-axis bending moment. The axial compression of the top chord and elongation of the bottom chort introduce bending in columns. This problem can be solved by introducing a camber in the truss by deliberately making the truss bottom chord smaller than the top chord. As an alternate, the connection between the bottom chord and column can be designed to slip under dead load conditions. Torquing the bolts after application of dead loads to the truss will limit the bending moments in the columns.

As a second alternative, the connection of the bottom chord can be designed to remain loose, increasing the effective length of the column in the weak axis to 2-stories between the top chords of the trusses at two levels. If none of the above procedures is applicable, then these moments should be provided for in the design of the columns.

#### 4.5.3.3 Trusses

The design of the staggered truss system is quite conventional. Loading conditions and design methods are similar in principle to other framing systems. The floor system spans only from the bottom chord of one truss to the top chord of the next, and the resulting large floor area supported by each truss allows maximum live load reduction.

In the transverse direction the lateral loads are transferred from the bottom chord of a truss to the top chord of an adjacent truss through floor diaphragm action down through the truss diagonals to the bottom chord. The sequence of events starts over, thus causing the entire transverse lateral load to be transferred through the floor system by diaphragm action and through the truss by direct stresses. The only additional bending occurs in the truss chords at the corridor openings or in other places where diagonals are eliminated.

The span-to-depth ratio of trusses is usually in the range of 6:1, giving adequate depth for the efficient design of top and bottom chords. Usually the panel width of trusses is not a governing criterion. Larger panel lengths with fewer web members decrease the fabrication costs and may work out to be more economical.

For maximum efficiency, just as in any other structural system, it is preferable to maintain a uniform spacing of trusses. This allows for maximization of typical truss units and reduces fabrication costs. However, when required by architectural arrangement, it is possible to vary the column and thus the truss spacing. Vierendeel openings other than those required for corridors should be avoided in the interest of economy. Truss design is based on continuous chord and pinned members, preferably using a computer analysis. Generally W or S shapes are selected for the chord members since angles are not efficient in resisting the secondary bending. Also, when planks are used wide flanges offer good bearing areas. Since the staggered truss system resists loads primarily by direct stresses, deflections are generally not a problem and therefore high-strength steels can be economically employed. The truss as a member supports a very large area and it is likely that if reduced live loads are used, the maximum live load reduction will be permissible. If any chord member is considered to support an area equal to the truss panel length times the bay spacing, it would be prudent to base the chord moment design on a reduced live load based on the smaller tributary area.

The simplest method of stacking trusses is a configuration called the checkerboard pattern, in which the trusses are placed at alternate columns and floors. It is possible, however, to obtain greater variety of spaces by using different layouts on alternate levels.

Longitudinally, the lateral forces in a staggered truss structure can be resisted by any conventional bracing system such as braced frames and core shear walls. However, many projects lend themselves to the design of rigid frames on the two broad faces: because (i) the main columns are oriented with webs parallel to the spandrels and; (ii) a large number of columns are generally available to participate in the moment frame. In some cases deep precast fascia beams used for architectural reasons can be directly bolted to the columns to serve as stiff structural spandrels.

In the transverse direction, at the roof and at the bottom floors it is normally not possible to carry the rhythm of the staggered truss for the full height of the building. Posts and hangers are usually required to support these levels. At the bottom level the lateral loads may need to be transferred to the foundation by diagonal bracing. These are shown schematically in Fig. 4.22b,c with respect to the curvilinear layout of a staggered truss system shown in Fig. 4.22a.

## 4.6 Eccentric Bracing Systems

### 4.6.1 Introduction

Concentric braced frames are excellent from strength and stiffness considerations and are therefore used widely either by themselves or in conjunction with moment frames when the lateral loads are caused by wind. However, they are of questionable value in seismic regions because of their poor inelastic behavior. Moment-resistant frames possess considerable energy dissipation characteristics but are relatively flexible when sized from strength considerations alone. Eccentric bracing is a unique structural system that attempts to combine the strength and stiffness of a braced frame with the inelastic behavior and energy dissipation characteristics of a moment frame. The system is called eccentric because deliberate eccentricities are employed between beam-to-column and beam-to-brace connections. The eccentric beam element acts as a fuse by limiting large forces from entering and causing buckling of braces. The eccentric segment of the beam, called the link, undergoes flexural or shear yielding prior to formation of plastic hinges in the other bending members and well before buckling of any compression members. Thus the system maintains stability even under large inelastic deformations. The required stiffness during wind or minor earthquakes is maintained because no plastic hinges are formed under these loads and all behavior is elastic. Although the deformation is larger than in a concentrically braced frame because of bending and shear deformation of the "fuse," its contribution to deflection is not significant because of the relatively small length of the fuse. Thus the elastic stiffness of the eccentrically braced frames can be considered the same as the concentrically braced frame for all practical purposes.

### 4.6.2 Ductility

The ductile behavior is highly desirable when the structure is called upon to absorb energy such as when subjected to strong ground motions. Steel's capacity for deformation without fracture combined with its high strength makes it an ideal material for use in eccentric bracing systems. In a properly designed and executed connection, steel continues to resist loads even after the maximum load is reached. This property by virtue of which steel sustains the load without fracture is called ductility. A brittle material, on the other hand, does not undergo large deformations at the onset of yielding. It fractures prior to, or just when it reaches the maximum load.

### 4.6.3 Behavior of frame

Eccentrically braced frames can be configured in various forms as long as the brace is connected to at least one link. The underlying principle is to prevent buckling of the brace from large overloads that may occur during major earthquakes. This is achieved by designing the link to yield.

The shear yielding of beams is a relatively well defined phenomenon; the load required for shear yielding of a beam of given dimensions can be calculated fairly accurately. The corresponding axial load and moments in columns and braces connected to the link can also be assessed fairly accurately. Using certain overload factors, which will be explained shortly, the braces and columns are designed to carry more load than could be imposed by the shear yielding of link. This assures that in the event of a large earthquake, it is the link that blows the fuse and not the columns or braces connected to it.

Consider the bracing shown in Fig. 4.26 subjected to horizontal

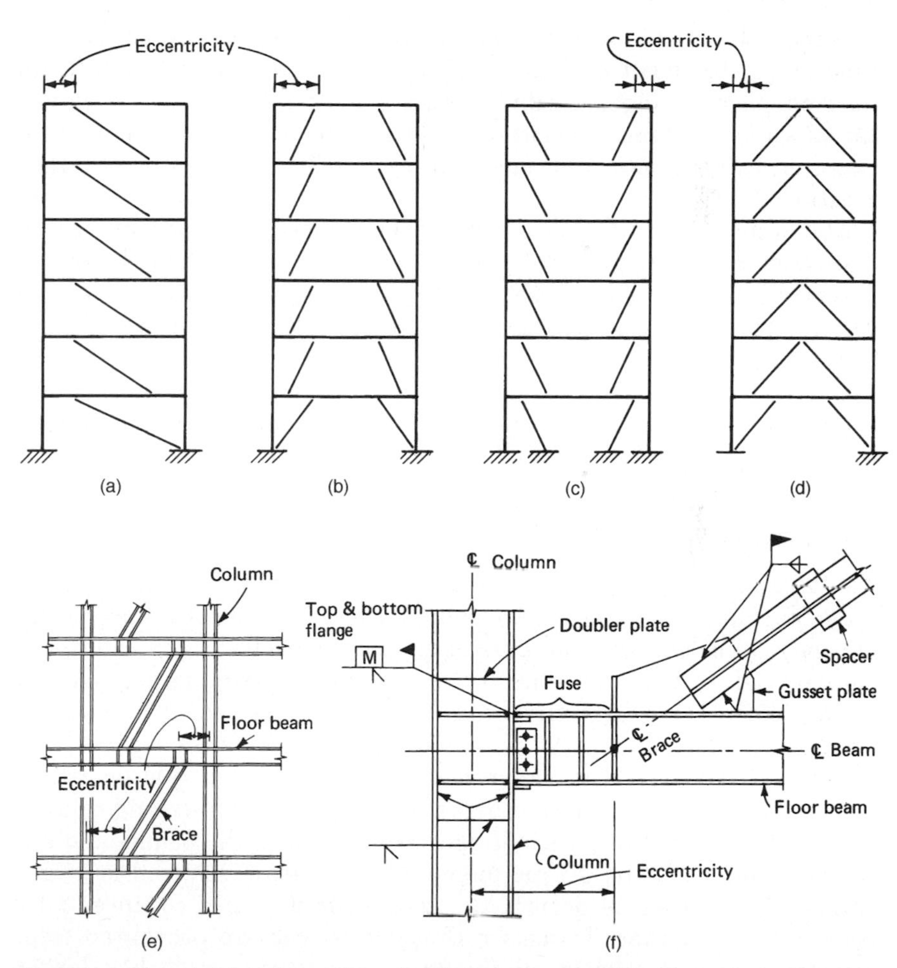


**Figure 4.26** Eccentric bracing system: (a–d) common types of bracing; (e) elevation; (f) detail.

loads. Note that the connections between the column and beams are moment-connected to achieve brace action. The force in the brace is transmitted to the beam as a horizontal force inducing axial stresses, and as a vertical force inducing shear stresses in the beam web. Of more concern in the design of the link are the cyclic shear forces induced in the beam. Assuming the link and its moment connection to the column are adequate in bending, the mechanism of failure is by shear yielding of the beam web provided web buckling is prevented. This is achieved by providing adequate stiffeners in the link.

### 4.6.4 Essential features of the link

Whether the link develops plastic hinges or yields in shear is a function of its length. Links longer than twice the depth tend to develop plastic hinges while shorter links tend to yield in shear. Links can be identified either as short or long. The short link experiences moderate rotation, while the long one a relatively larger rotation.

The cyclic shear yielding is an excellent energy dissipation mechanism because large cyclic deflections can take place without failure or deterioration in the hysteretic behavior. This is because yielding occurs over a large segment of the beam web and is followed by a cyclic diagonal field. The web buckles after yielding in shear, but the tension field takes over the load-carrying mechanism to prevent failure, resulting in a hysteretic loop having a large area representing good energy dissipation.

### 4.6.5 Analysis and design considerations

To force the formation of a hinge in the beam web, the plastic moment capacity of the beam should exceed the beam shear yield capacity. In calculating the plastic moment capacity of the beam, the contribution of the web is neglected because the web is assumed to have yielded. In design the beam is first selected for the required shear capacity and then the plastic moment capacity is checked to be slightly larger than the shear yield capacity. As in ductile frame design, the column is selected by using the weak beam-strong column concept to assure that plastic hinges are formed in the beams and not in the columns. If the plastic moment of the beam selected is larger than that required by design, the column is designed in an equally conservative manner. To assure that the braces are prevented from buckling, they are designed to withstand forces somewhat larger than those given by the analysis.

This conservatism is necessary to take into account the fact that the actual beam designed is likely to have additional capacity due to factors such as: (i) beam strain hardening; (ii) actual yield stress being more than the theoretical value; (iii) interaction of floor slabs with beams tending to increase the plastic moment capacity. The brace-to-beam connection can be designed either as a welded or bolted connection. The bolts are deisgned as friction bolts and checked for bearing capacity because of the likelihood of slippage in the event of a large earthquake. The beam-to-column connection is designed as a moment connection by welding the beam flanges to the column with full-penetration welds. Single-side shear plate connection with fillet welds is used to develop the high shear forces in the link. Lateral support is provided at the top and bottom flanges of the beam to prevent lateral torsional buckling and weak axis bending.

### 4.6.6 Deflection considerations

The lateral deflection of an eccentrically braced frame can be estimated as the sum of three components: (i) deflection due to elongation of the brace; (ii) deflection due to axial strain in the columns, usually referred to as the chord drift; and (iii) the deflection due to deformation of the eccentric element. Because the braces and columns are designed to remain elastic even under a severe earthquake, their deflection contributions are very nearly constant even after the shear yielding of the link. The beams in eccentric bracing are much heavier than in a corresponding concentrically braced frame, therefore are likely to contribute little to the deflection under elastic condition. Therefore, an eccentrically braced frame is not an unreasonably flexible system as compared to a concentric frame.

### 4.6.7 Conclusions

Buildings using eccentric bracing are lighter than moment-resisting frames and, while retaining the elastic stiffness of concentrically braced frames, are more ductile. The eccentric bracing system has the following characteristics:

1. It provides a stiff structural system without imposing undue penalty on the steel tonnage.

2. Eccentric beam elements yielding in shear, act as fuses to dissipate excess energy during severe earthquakes.
3. Premature failure of the link does not cause the structure to collapse because the structure continues to retain its vertical load-carrying capacity and stiffness.

## 4.7 Interacting System of Braced and Rigid Frames

### 4.7.1 Introduction

Even for buildings in the range of 10 to 15-stories, unreasonably heavy columns result if lateral bracing is confined to the building service core because the available depth for bracing is usually limited. In addition, high uplift forces that may occur at the bottom of core columns can present foundation problems. In such instances an economical structural solution can be arrived at by using rigid frames in conjunction with the core bracing system. Although deep girders and moment connections are required for frame action, rigid frames are often preferred because they are least objectionable from the interior space planning considerations. Often times, architecturally, it may be permissible to use deep spandrels and closely spaced columns on the building façade because usually the columns will not interfere with the space planning and the depth of spandrels need not be shallow for passage of air conditioning ducts. A schematic floor plan of a building using this concept is shown in Fig. 4.27a.

As an alternative to perimeter frames, a set of interior frames can be used with the core bracing. Such an arrangement is shown in Fig. 4.27b in which frames on grid lines 1, 2, 6, and 7 participate with core bracing on lines 3, 4 and 5. Yet another option is to moment-connect the girders between the braced core and perimeter columns as shown in Fig. 4.27c in which the frame beams act as outriggers by engaging the exterior columns to resist the bending moments.

For slender buildings with height-to-width ratios in excess of 5, an interacting system of moment frames and braces becomes uneconomical if braces are placed only within the building core. In such situations, a good structural solution is to spread the bracing to the full width of the building along the façades if such a system does not compromise the architecture of the building. If it does, then a possible solution is to move the full-depth bracing to the interior of the building. Such a bracing concept is shown in Fig. 4.27d, in which moment frames located at the building façade interact with two interior-braced bents. These bents stretched out for the full width of the building form giant K braces, resisting overturning and shear forces by developing predominantly axial forces. A transverse cross section of the building is shown in Fig. 4.27e, wherein a secondary system of braces required to transfer the lateral loads to the panel points of the K braces is also indicated. The diagonals of the K braces

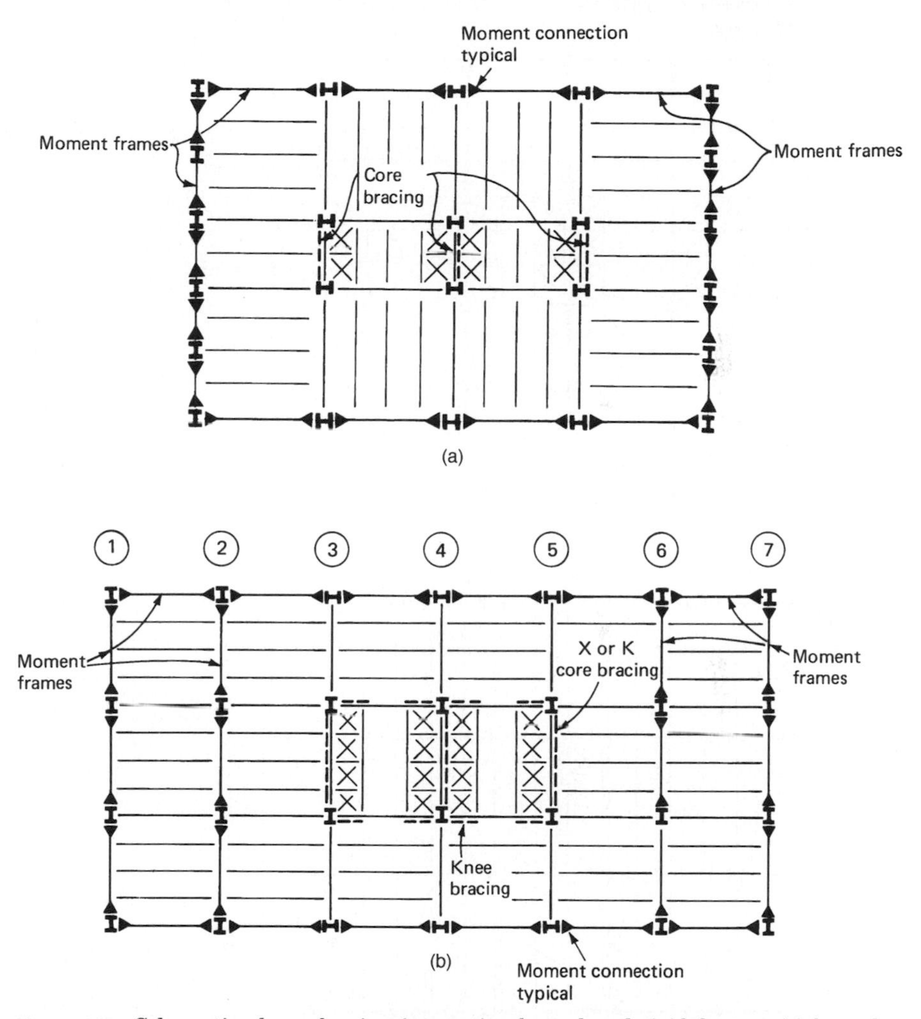


**Figure 4.27** Schematic plans showing interacting braced and rigid frames: (a) braced core and perimeter frames; (b) braced core and interior and exterior frames; (c) braced core and interior frames; (d) full depth interior bracing and exterior frames; (e) transverse cross-section showing primary interior bracing, secondary bracing, and basement construction.

running through the interior of the building result in sloping columns whose presence has to be acknowledged architecturally as a trade-off for structural efficiency.

All of the above bracing systems and any number of their variations can be used either singly or in combination and can be made to interact with moment-connected frames. The magnitude of

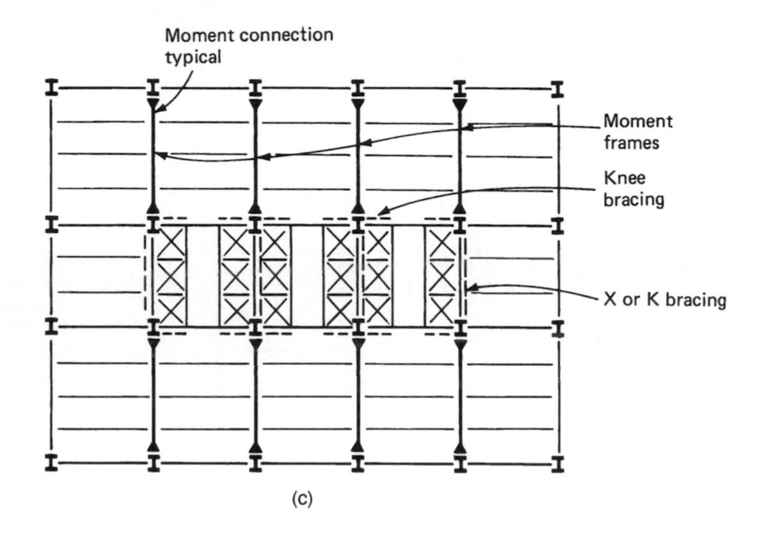


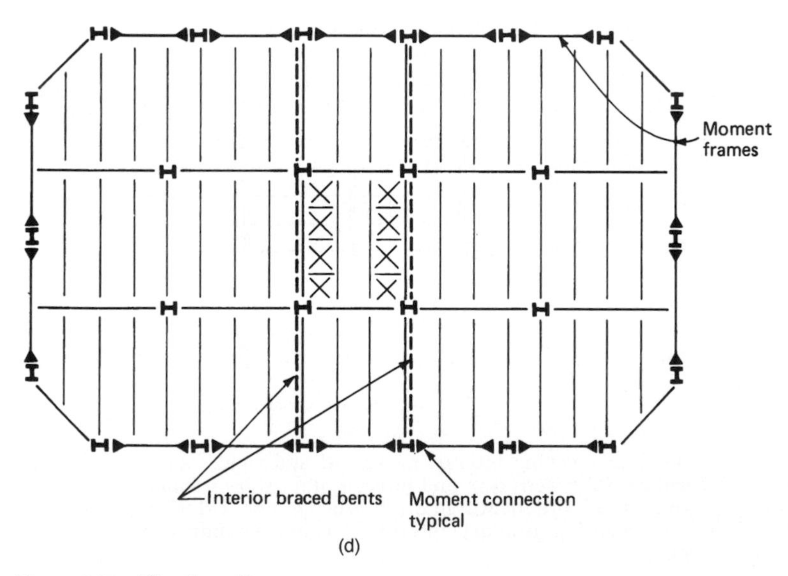


**Figure 4.27** (*Continued*).

their interaction can be controlled by varying the relative stiffness of various structural elements to achieve an economical structural system.

### 4.7.2 Physical behavior

If the lateral deflection patterns of braced and unbraced frames are similar, the lateral loads can be distributed between the two systems according to their relative stiffness. However, in normally proportioned buildings the unbraced and braced frames deform with their own characteristic shapes, necessitating that we study their behavior as a unit.

Insofar as the lateral-load-resistance is concerned, rigid and braced frames can be considered as two distinct units. The basis of classifiction is the mode of deformation of the unit when subjected to lateral loading. The deflection characteristics of a braced frame are similar to those of a cantilever beam. Near the bottom the vertical truss is very stiff, and therefore the floor-to-floor deflections will be less than half the values near the top. Near the top the floor-to-floor

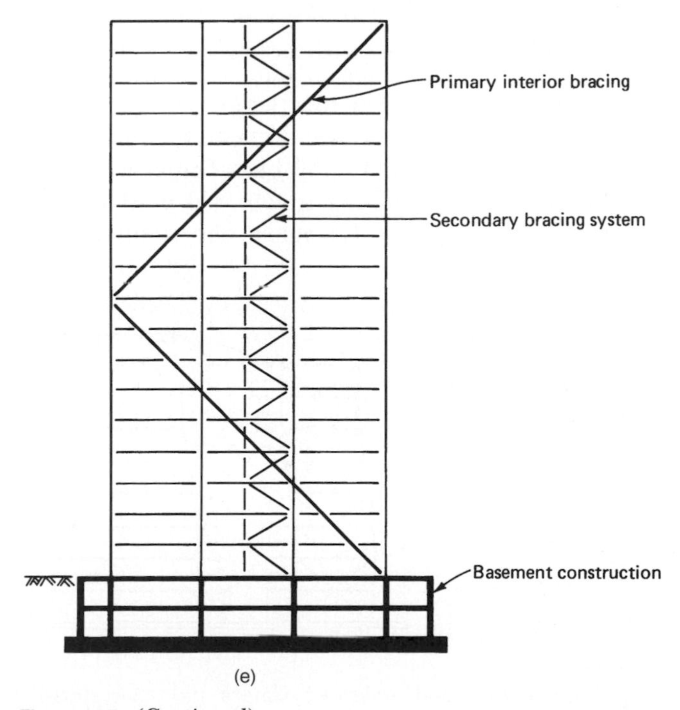


**Figure 4.27** *(Continued).*

deflections increase rapidly mainly due to the cumulative effect of chord drift. The column strains at the bottom of the building produce a deflection at the top; and since this same effect occurs at every floor, the resulting drift at the top is cumulative. This type of deflection often referred to chord drift is difficult to control requiring material quantities well in excess required for gravity needs.

Rigid frames deform predominantly in a shear mode. The relative story deflections depend primarily on the magnitude of shear applied at each story level. Although near the bottom the deflections are larger, and near the top smaller as compared to the braced frames, the floor-to-floor deflections can be considered more nearly uniform. When the two systems, the braced and rigid frames are connected by rigid floor diaphragms, a nonuniform shear force develops between the two. The resulting interaction helps in extending the range of application of the two systems to buildings up to about 40-stories in height.

Figure 4.28 shows the individual deformation patterns of a braced and unbraced frames subjected to lateral loads. Also shown are the horizontal shear forces between the two frames connected by rigid floor slabs. Observe that the braced frame acts as a vertical cantilever beam, with the slope of the deflection greatest at the top of the building, indicating that in this region the braced frame contributes the least to the lateral stiffness.

The rigid frame has a shear mode deformation, with the slope of deformation greater at the base of the structure where the maximum shear is acting. Because of the different lateral deflection characteristics of the two elements, the frame tends to pull back the brace in the upper portion of the building while pushing it forward in the lower portion. As a result, the frame participates more effectively in the

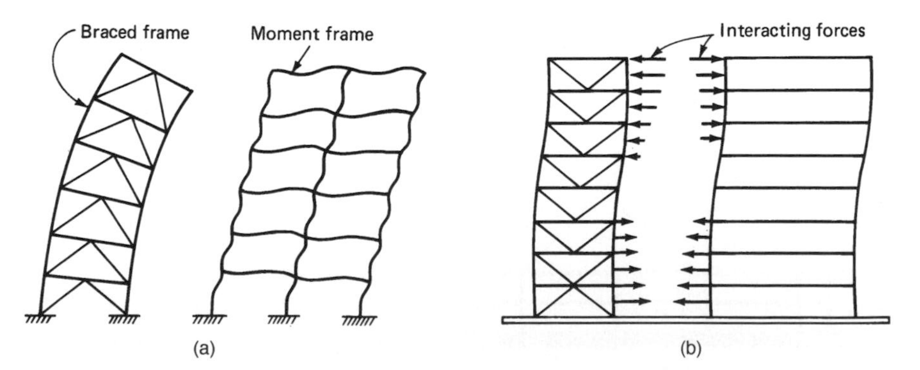


**Figure 4.28** Interaction between braced and unbraced frames: (a) characteristic deformation shapes; (b) variation of shear forces resulting from the interaction.

upper portion of the building where lateral shears are relatively less. The braced frame carries most of the shear in the lower portion of the building. Thus, because of the distinct difference in the deflection characteristics, the two systems help each other a great deal. The frame tends to reduce the lateral deflection of the trussed core at the top, while the trussed core supports the frame near the base. A typical variation of horizontal shear carried by each frame is shown in Fig. 4.28b in which the length of arrows conceptually indicates the magnitude of interacting shear forces.

Although the framed part of a high-rise structure is usually more flexible in comparison to the braced part, as the number of stories increases, its interaction with the braced frame becomes more significant, contributing greatly to the lateral resistance of the building. Therefore, when the frame part is fairly rigid by itself, its interaction with the braced portion of the building can result in a considerably more rigid and efficient design.

## 4.8 Outrigger and Belt Truss Systems

### 4.8.1 Introduction

Innovative structural schemes are continuously being sought in the design of high-rise structures with the intention of limiting the wind drift to acceptable limits without paying a high premium in steel tonnage. The savings in steel tonnage and cost can be dramatic if certain techniques are employed to utilize the full capacities of the structural elements. Various wind-bracing techniques have been developed to this end; this section deals with one such system, namely, the belt truss system, also known as the core-outrigger system in which the axial stiffness of the perimeter columns is invoked for increasing the resistance to overturning moments.

This efficient structural form consists of a central core, comprising either braced frames of shear walls, with horizontal cantilever "outrigger" trusses or girders connecting the core to the outer columns. The core may be centrally located with outriggers extending on both sides (Fig. 4.29a) or it may be located on one side of the building with outriggers extending to the building columns on one side (Fig. 4.29b)

When horizontal loading acts on the building, the column-restrained outriggers resist the rotation of the core, causing the lateral deflections and moments in the core to be smaller than if the free-standing core alone resisted the loading. The result is to increase the effective depth of the structure when it flexes as a vertical cantilever, by inducing tension in the windward columns and compression in the leeward columns.

In addition to those columns located at the ends of the outriggers, it is usual to also mobilize other peripheral columns to assist in restraining the outriggers. This is achieved by including a deep spandrel girder, or a "belt truss", around the structure at the levels of the outriggers.

To make the outriggers and belt truss adequately stiff in flexure and shear, they are made at lease one, and often 2-stories-deep. It is also possible to use diagonals extending through several floors to act as outriggers as shown in Fig. 4.29c. And finally, girders at each floor

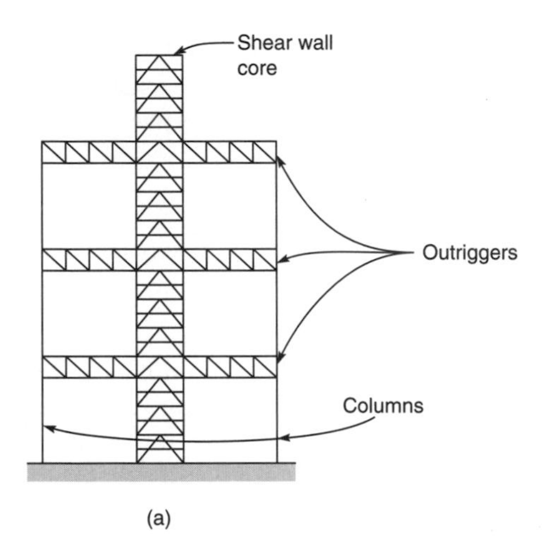


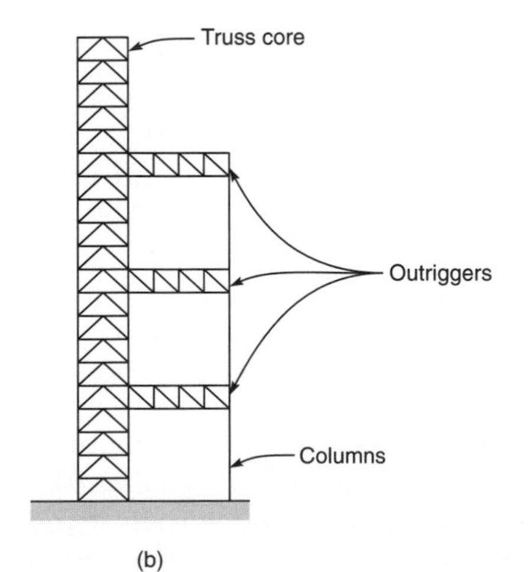


**Figure 4.29** (a) Outrigger structure with central core: (b) outrigger structure with offset core; (c) diagonals acting as outriggers; (d) floor girders acting as outriggers; (e) plan with cap truss; (f,g) behavior of tied cantilever; (h) restraining spring at $X = 0$, (i) spring at $X = 0.25L$; (j) spring at $X = 0.50L$; (k) spring at $X = 0.75L$; (l) simplified analytical model for single outrigger system.

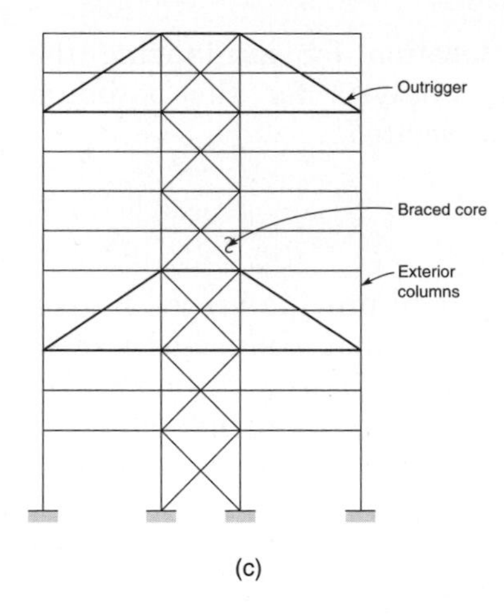


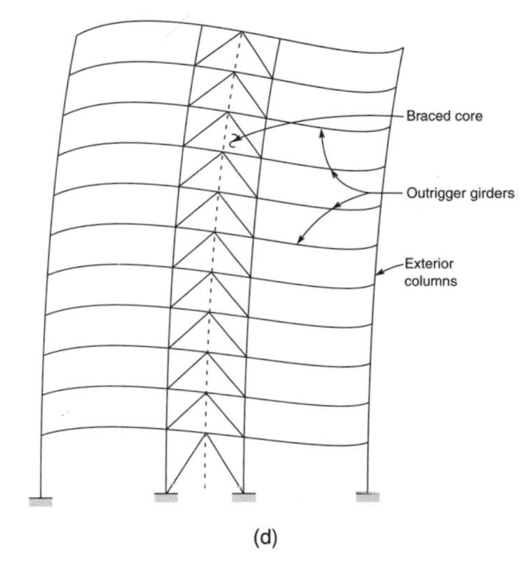


**Figure 4.29** (*Continued*).

may be transformed into outriggers by moment connections to the core and, if desired, to the exterior columns as well (Fig. 4.29d). In all these cases it should be noted that while the outrigger system is very effective in increasing the structure's flexural stiffness, it does not increase its resistance to shear, which has to be carried mainly by the core.

In the following sub-section the stiffening effect of a single outrigger located at the top of the structure is examined first. Next, the effect of lowering the truss along the height is studied with the object of finding the most optimum location for minimizing the building drift. Finally a compatibility analysis for the optimum location of a two-outrigger structure is presented.

### 4.8.2 Physical behavaior

A traditional approach to lateral bracing is to provide braced frames around the building core with a system of moment connected frames at the exterior. However, for buildings taller than 40-stories or so, the core, if kept consistent with the vertical transportation, does not provide adequate stiffness to keep the drift down to acceptable limits.

A method of increasing the efficiency of the system is to use a "cap" or "hat" truss to tie the braced core to the exterior columns. The tied columns, in addition to the usual function of supporting the gravity loads, also assist in resisting the overturning moments. The tie-down

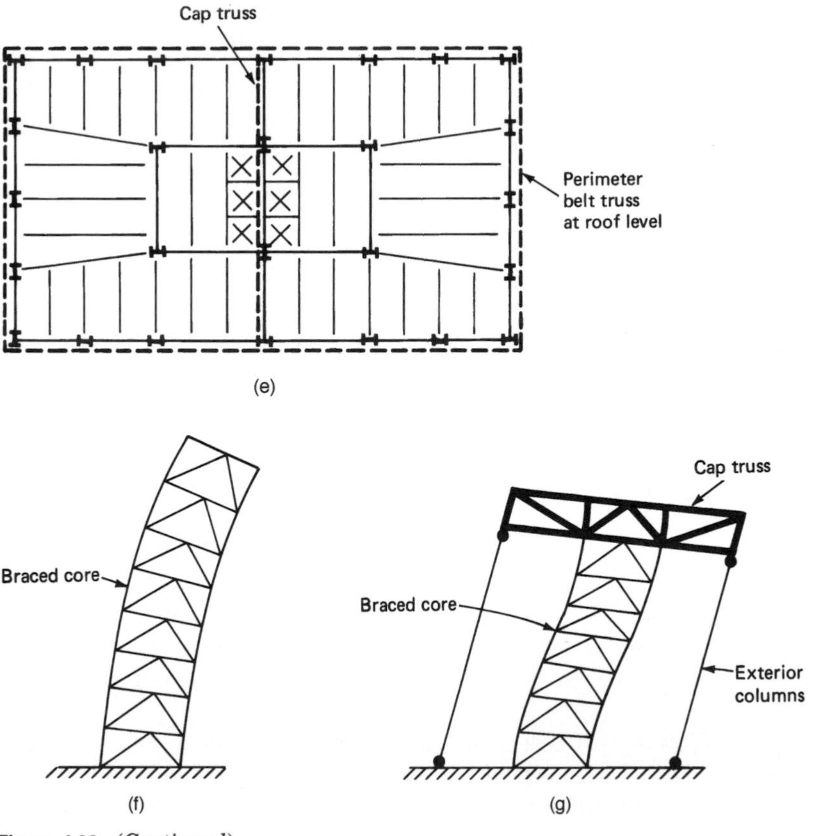


**Figure 4.29** (*Continued*).

action of the cap truss creates a restoring couple resulting in a point of inflection in its deflection curve. This reversal in curvature reduces the bending moment in the core, and hence the building drift at the top. The belt truss functioning as a horizontal fascia stiffener mobilizes other exterior columns to take part in restraining the rotation of the cap truss. A general improvement of up to 25 to 30 percent in stiffness can be realized by using a sufficiently stiff belt truss to tie all the peripheral columns.

The cap and belt truss system is also beneficial in equalizing the

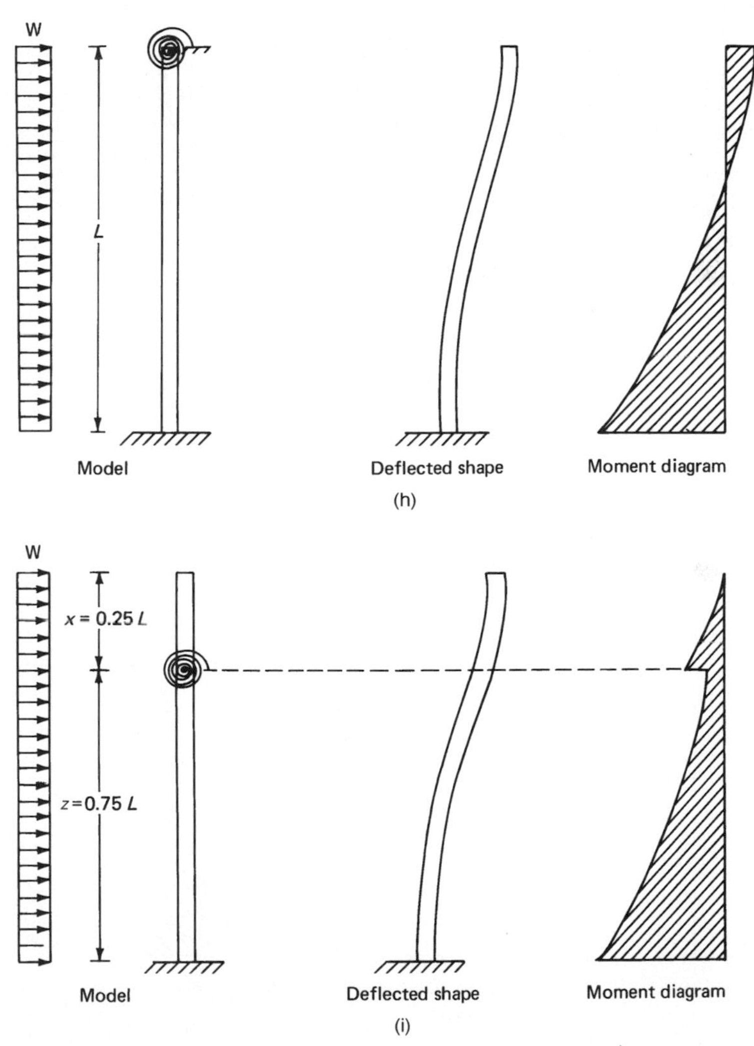


**Figure 4.29** (*Continued*).

differential movement between interior and exterior columns resulting from temperature effects, and unequal shortening due to axial load imbalance.

The behavior of the system is explained with reference to Fig. 4.29e which shows a schematic plan of a core braced building. Although the building has a belt truss connecting all the exterior columns, we will assume that their effect may be represented by two equivalent columns tied to each end of the cap truss (Fig. 4.29f,g). This

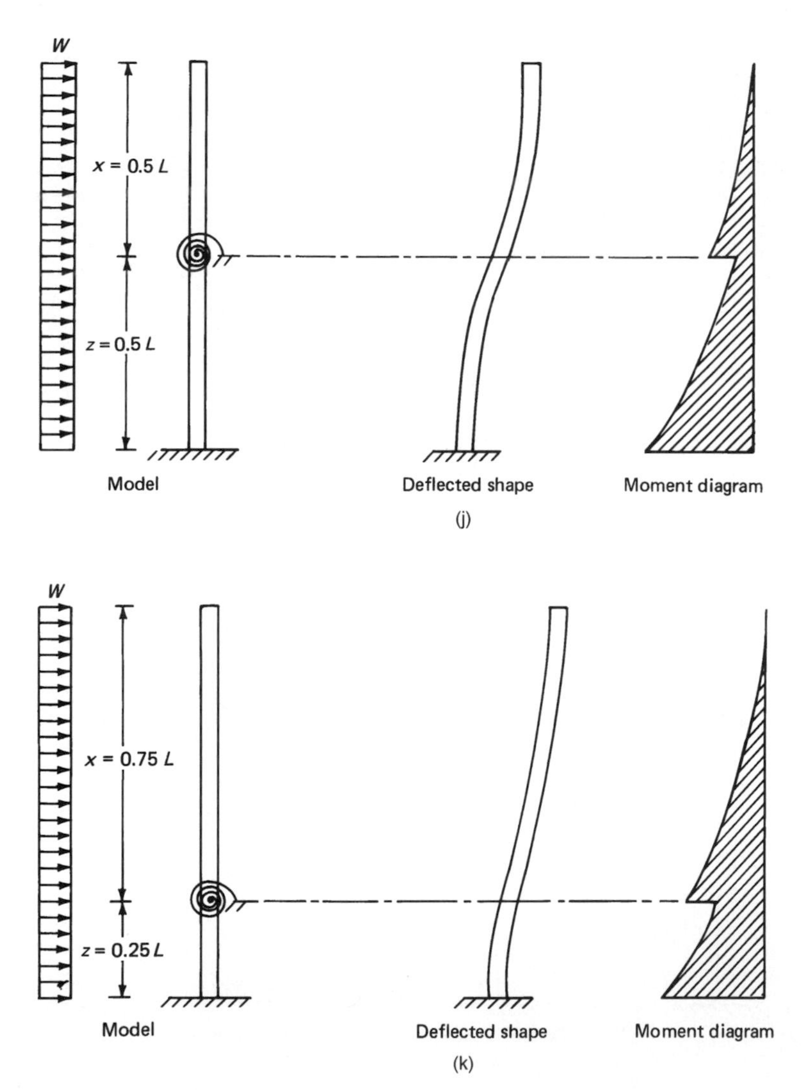


**Figure 4.29** *(Continued).*

idealization is not necessary for evaluating the restraining effect of columns, but keeps the explanation simple.

The structure in Fig. 4.29f,g is basically a cantilever with a single degree of redundancy because its rotation is restrained only at the top by the stretching and shortening of windward and leeward columns. The resultant of the tensile and compressive forces is equivalent to a restoring couple opposing the rotation of the core.

In terms of its restraining action, the cap truss may be conceptualized as an equivalent spring located at the top. Its stiffness, defined as the restoring couple due to a unit rotation of the core, may be calculated as follows.

Since the cap truss is assumed rigid, the exterior columns under go axial compression and tension equal in magnitude to the product of the rotation of core and their distance from the center of the core. Assuming that the columns are located at a distance of $d/2$ from the center of the core, the axial deformation of the columns is equal to $\theta \times \dfrac{d}{2}$ where $\theta$ is the rotation of the core. Since by our definition,

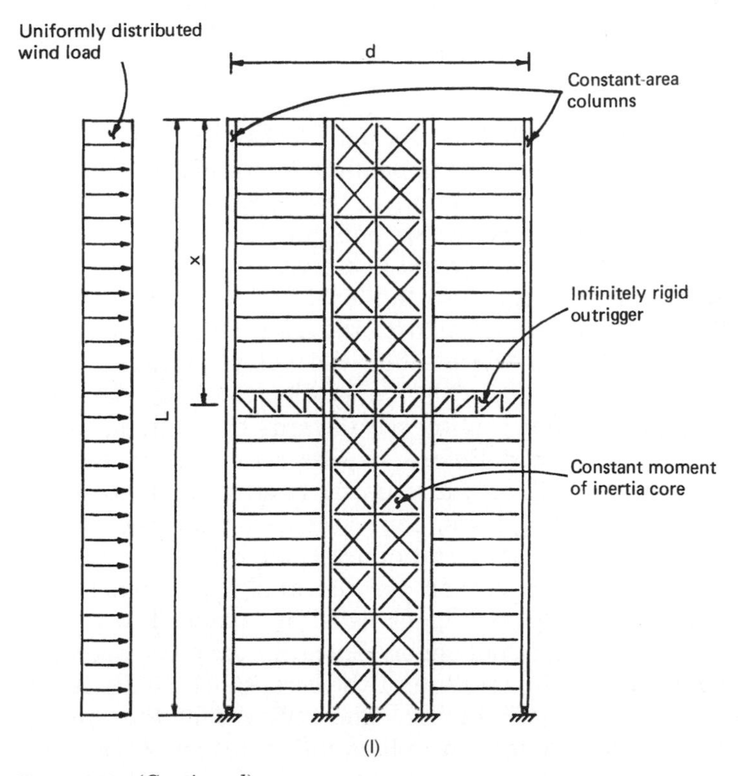


**Figure 4.29** *(Continued).*

$\theta = 1$, the axial deformation is equal to $d/2$ units. The axial load $P$ in the columns corresponding to this deformation is given by

$$P = \frac{AEd}{2L}$$

where $P$ = axial load in the columns
$A$ = area of columns
$E$ = modulus of elasticity
$d$ = distance between the exterior columns
$L$ = height of the building

The restoring couple which is the rotational stiffness of the cap truss is given by the product of the axial load in the columns and the distance from the center of core. Using the notation $K$ for the stiffness we get

$$K = \sum P_i d_i$$

$$= P \times \frac{d}{2} \times 2$$

$$= Pd$$

The equivalent spring located at the top reduces the deflection of the core by inducing a reversal in its bending curvature as shown in Fig. 4.29. The magnitude of reduction in drift due to the cap truss depends on the given geometry of the building and the area of the columns. The stiffnesses of the cap and belt truss which in a practical building are not infinitely large, also influence the reduction in drift.

Now let us consider the interaction of the core with the outrigger located not at top but somewhere along the height. How dces the location influence its effectiveness? Is top location the best? What are the structural implications if it is moved towards the bottom, say to the mid-height of the building? Before we seek a general solution for these rather intriguing questions, it is useful to study its restraining action for a few specific locations, for example at the top, three-quarters of the height, midheight and quarter height.

A compatibility method is chosen for the analysis in which the rotation of the core at the outrigger level is matched with the corresponding rotation of the outrigger. Since the primary action of the outrigger is to reduce the rotation of the core, conceptually it can be replaced by an equivalent spring. The stiffness of the spring, defined as the moment required to induce unit rotation of the core, however, it not a constant but depends on its location. Its derivation will be explained shortly but for now we will proceed with the idea that the outrigger is mathemtically equivalent to a spring with a stiffness inversely proportional to its distance from the bottom. In other words, its stiffness is minimum when at the top, and maximum at the bottom.

### 4.8.3 Deflection calculations

**Case 1: Outrigger located at top, $x = 0$, $z = L$** (Fig 4.29h)

The rotation compatibility condition at $Z = L$ can be written as

$$\theta_W - \theta_S = \theta_L \tag{4.5}$$

where $\theta_W$ = rotation of the cantilever at $Z = L$ due to a uniform lateral load $W$, in radians

$\theta_S$ = rotation due to spring restraint located at $Z = L$, in radians. The negative sign indicates that the rotation of the cantilever due to the spring stiffness acts in a direction opposite to the rotation due to external load

$\theta_L$ = final rotation of the cantilever at $Z = L$, in radians

For a cantilever with uniform moment of inertia $I$ and modulus of elasticity $E$ subjected to uniform horizontal load $W$,

$$\theta_W = \frac{WL^3}{6EI} \tag{4.6}$$

If $M_1$ and $K_1$ represent the moment and stiffness of the spring located at $Z = L$, Eq (4.5) can be rewritten thus;

$$\frac{WL^3}{6EI} - \frac{M_1 L}{EI} = \frac{M_1}{K_1} \tag{4.7}$$

and

$$M_1 = \frac{WL^3}{1/K_1 + L/EI} \tag{4.8}$$

The resulting drift $\Delta_1$ at the building top can be obtained by superposing the deflection of the cantilever due to external uniform load $W$ and the deflection due to the moment induced by the spring, thus

$$\Delta_1 = \Delta_{\text{load}} - \Delta_{\text{spring}}$$

$$= \frac{WL^4}{8EI} - \frac{M_1 L^2}{2EI}. \tag{4.9}$$

$$= \frac{L^2}{2EI}\left(\frac{WL^2}{4} - M_1\right) \tag{4.10}$$

**Case 2: Outrigger located at $z = \frac{3L}{4}$** (Fig. 4.29i)

The general expression for lateral deflection $y$ for a cantilever subjected to a uniform lateral load is given by

$$y = \frac{W}{24EI}(x^4 - 4L^3 x + 3L^4) \tag{4.11}$$

Note that $x$ is measured from the top.

Differentiating with respect to $x$, the general expression for slope of the cantilever is given by

$$\frac{dy}{dx} = \frac{W}{6EI}(x^3 - L^3) \tag{4.12}$$

The slope at the spring location is given by substituting $Z = 3L/4$, i.e., $x = L/4$ in Eq. (4.12). Thus

$$\frac{dy}{dx}\left(\text{at } z = \frac{3L}{4}\right) = \frac{W}{6EI}\left(\frac{L^3}{64} - L^3\right)$$

$$= \frac{WL^3}{6EI} \times \frac{63}{64} \tag{4.13}$$

Using the notation $M_2$ and $K_2$ to represent the moment and stiffness of spring at $Z = 3L/4$, the compatibility equation at location 2 can be written thus

$$\frac{WL^3}{6EI}\left(\frac{63}{64}\right) - \frac{M_2}{EI}\left(\frac{3L}{4}\right) = \frac{M_2}{K_2} \tag{4.14}$$

Noting that $K_2 = 4K_1/3$, the expression for $M_2$ can be written thus

$$M_2 = \left(\frac{WL^3/6EI}{1/K_1 + L/EI}\right)\frac{63/64}{3/4} = \left(\frac{WL^3/6EI}{1/K_1 + L/EI}\right)1.31 \tag{4.15}$$

Noting that the terms in the parenthesis represent $M_1$, Eq. (4.15) can be expressed in terms of $M_1$

$$M_2 = 1.31M_1$$

The drift is given by the relation

$$\Delta_2 = \frac{WL^4}{8EI} - \frac{M_2 3L}{4EI}\left(L - \frac{3L}{8}\right) \tag{4.16}$$

or

$$\Delta_2 = \frac{L^2}{2EI}\left(\frac{WL^2}{4} - 1.23M_1\right) \tag{4.17}$$

**Case 3: Outrigger at mid-height, $z = \frac{L}{2}$** (Fig. 4.29j)

The rotation at $Z = L/2$ due to external load $W$ can be shown to be equal to $7WL^3/48EI$, giving the rotation compatibility equation

$$\frac{7WL^3}{48EI} - \frac{M_3 L}{2EI} = \frac{M_3}{K_3} \tag{4.18}$$

where $M_3$ and $K_3$ represent the moment and stiffness of the spring at $Z = L/2$. Noting that $K_3 = 2K_1$, the expression for $M_3$ works out as

$$M_3 = \left(\frac{WL^3/6EI}{1/K_1 + L/EI}\right) \times \frac{7}{4} \tag{4.19}$$

Since the expression in the parentheses is equal to $M_1$, $M_3$ can be expressed in terms of $M_1$

$$M_3 = 1.75M_1 \tag{4.20}$$

The drift is given by the equation

$$\Delta_3 = \frac{WL^4}{8EI} - \frac{M_3 L}{2EI}\left(L - \frac{L}{4}\right) \tag{4.21}$$

or

$$\Delta_3 = \frac{L^2}{2EI}\left(\frac{WL^2}{4} - 1.31M_1\right) \tag{4.22}$$

**Case 4: Outriggers at quarter-height, $z = \frac{L}{4}$** (Fig. 4.29k)

The rotation at $Z = L/4$ due to uniform lateral load can be shown to be equal to $WL^3/6EI[(37/64)]$, giving the rotation compatibility equation

$$\frac{WL^3}{6EI}\left(\frac{37}{64}\right) - \frac{M_4L}{4EI} = \frac{M_4}{K_4} \tag{4.23}$$

where $M_4$ and $K_4$ represent the moment and stiffness of the spring at $Z = L/4$. Noting that $K_4 = 4K_1$, $M_4$ in Eq. (4.23) can be expressed in terms of $M_1$.

$$M_4 = 2.3M_1 \tag{4.24}$$

The drift for this case is given by the expression

$$\Delta_4 = \frac{WL^4}{8EI} - \frac{M_4L}{4EI}\left(L - \frac{L}{8}\right) \tag{4.25}$$

or

$$\Delta_4 = \frac{L^2}{2EI}\left(\frac{WL^2}{4} - M_1\right) \tag{4.26}$$

Equations (4.10), (4.14), (4.22), and (4.26) give the building drift for four different locations of the belt and outrigger trusses.

The value of $K_1$ which corresponds to stiffness of the spring when it is located at $Z = L$ can be derived as follows. A unit rotation given to the core at the top results in extension and compression of all perimeter columns, the magnitudes of which are given by their respective distances from the center of gravity of the core. The resulting force multiplied by the lever arm gives the value for stiffness $K_1$. Thus, if $p_1$ is measured in kips and the distance in feet, $K_1$ has units of kip feet. The force $p$ in each exterior column is given by the relation $p = AE\,\delta/L$; since by definition $\delta$ corresponds to column extension or compression due to unit rotation of the core, $\delta = d/2$, where $d$ is the distance between the exterior columns. Therefore,

$$p = \frac{AE}{L}\left(\frac{d}{2}\right) \tag{4.27}$$

and its contribution to the stiffness $K_1$ is given by the relation

$$\begin{aligned} K_i &= p_i d \\ &= \frac{A_iE}{L}\frac{d^2}{2} \end{aligned} \tag{4.28}$$

The total contribution of all exterior columns on the long faces is given by the summation relation

$$K_1 = \sum_{i=1}^{n} K_i = \frac{d^2E}{2L} \sum_{i=1}^{n} A_i \tag{4.29}$$

The contribution of the columns on the short faces can be worked out in a similar manner.

### 4.8.4 Optimum location of single truss

The preceding analysis has indicated that the beneficial action of the outrigger is a function of two distinct characteristics, the stiffness of the equivalent spring and the magnitude of the rotation of cantilever at the spring location due to external loads. The stiffness varies inversely as the distance of the outrigger from the base. For example, it is at a minimum when located at the top and a maximum when at the bottom. The rotation of the free cantilever for a uniformly distributed horizontal load varies parabolically from a maximum value at the top to zero at the bottom. Therefore, from the point of view of spring stiffness it is desirable to locate the outrigger at the bottom, whereas from a consideration of rotation, the converse is true. It is obvious that the optimum location is somewhere in between.

The method of analysis is based on the following assumptions.

1. The area of the perimeter columns and the moment of inertia of the core are uniform throughout the height.
2. The outrigger and the belt trusses are flexurally rigid and induce only axial forces in the columns.
3. The lateral resistance is provided only by the bending resistance of the core and the tiedown action of the exterior columns.
4. The core is rigidly fixed at the base.
5. The rotation of the core due to shear deformation is negligible.
6. The intensity of lateral load remains constant for the whole height.
7. The structure is linearly elastic.

Figure 4.29l shows the analytical model for the single out-rigger truss located at a distance $x$ from the top. As before, a compatibility method is used by matching the rotations. From the compatibility relation, the restoring moment $M_x$ at the location of the outrigger is evaluated. Next, the deflection of the core at top due to the restoring couple is calculated and maximized using the principles of calculus. Solution of the resulting third degree polynomial yields an optimum value of $x$ for which the deflection of the core due to external load is minimum. The detail calculations are as follows.

The rotation $\theta$ of the cantilever at a distance $x$ from the top, due to a uniformly distributed load $w$ is given by the relation:

$$\theta = \frac{W}{EI}(x^3 - L^3)$$

The rotation at top due to the restoring couple $M_x$ is given by the relation.

$$\theta = \frac{M_x}{EI}(L - x)$$

The compatibility relation at $x$ is given by

$$\frac{W}{6EI}(x^3 - L^3) - \frac{M_x}{EI}(L - x) = \frac{M_x}{K_x} \tag{4.30}$$

where $W$ = intensity of the wind load per unit height of the structure

$M_x$ = the restoring moment due to outrigger restraint

$K_x$ = spring stiffness at $x$ equal to $\frac{AE}{(L-x)} \times \frac{d^2}{2}$

$E$ = modulus of elasticity of the core

$I$ = moment of inertia of the core

$A$ = area of the perimeter columns

$L$ = height of the building

$x$ = location of truss measured from the top

$d$ = distance out-to-out of columns

Next, obtain the deflection at the top of the structure due to $M_x$:

$$Y_M = \frac{M_x(L - x)(L + x)}{2EI} \tag{4.31}$$

From our definition, the optimum location of the belt truss is that location for which the deflection $Y_M$ is a maximum. This is obtained by differentiating Eq. (4.31) with respect to $x$ and equating to zero. Thus,

$$\frac{d}{dx}\left[\frac{W(x^3 - L^3)(L + x)}{12(EI)^2(1/AE + 1/EI)}\right] = 0 \tag{4.32}$$

$$4x^2 + 3x^2L - L^3 = 0 \tag{4.33}$$

giving the optimum location at $x = 0.455L$. If the flexibility of the outrigger is taken into account, even for the overly simplified model, the corresponding equation for the solution of $x$ becomes too involved for hand calculations. Extension of the solution to two or more outrigger trusses further complicates the solution, thus necessitating a formulation suitable for a computer. This is considered next.

### 4.8.5 Optimum location for a two-outrigger system

In the preceding analyses only one compatibility equation was necessary because the one-outrigger structure is once redundant. On the other hand a two-outrigger structure is twice redundant requiring two compatibility equations corresponding to the degree of redundancy.

The analytical model is shown schematically in Fig. 4.30a. Since

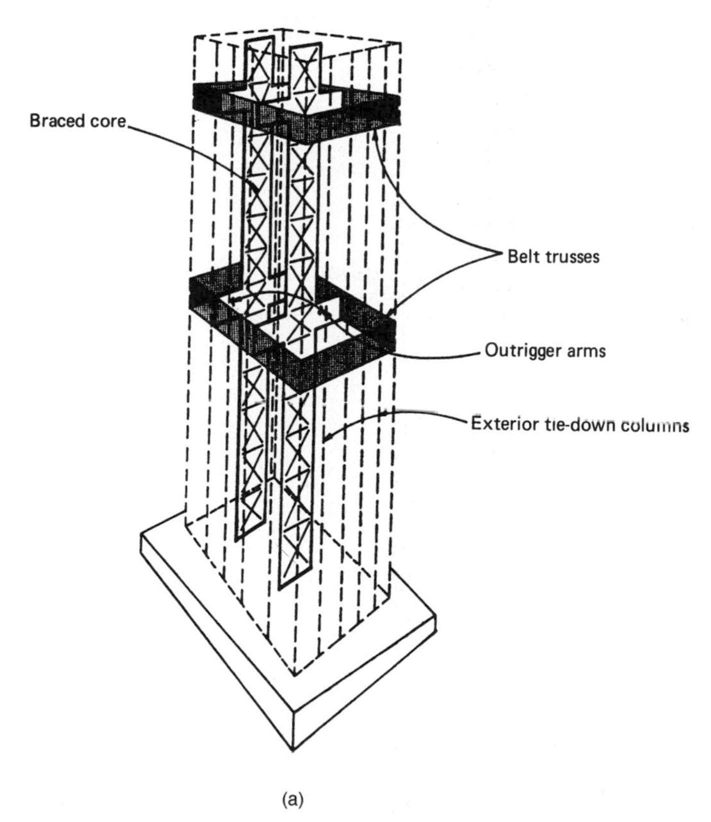


**Figure 4.30** (a) Conceptual model for two outriggers: (b) analytical model.

the formulation is for computer analysis, the column and core properties, and the distribution of lateral load need not be uniform. A linear variation which is a reasonable representation of a real structure is used along the building height as shown in Fig. 4.30b.

#### 4.8.5.1 Computer solution

The analytical model is shown in Fig. 4.31. A flexibility approach has been employed for the solution. The method is briefly explained with reference to the example problem. The moments at the outrigger locations are chosen as the unknown arbitrary constants $M_1$ and $M_2$, and the structure is released by removing the rotational restraints, making it statically determinate, so that the effect of any loading can be easily calculated. The flexibility coefficients $f_{11}$ and $f_{22}$ are calculated by using integrals of the form

$$\int \frac{m_i m_j}{EI} ds + k \int \frac{s_i s_j}{GA} ds + \int \frac{n_i n_j \, ds}{EA} \tag{4.34}$$

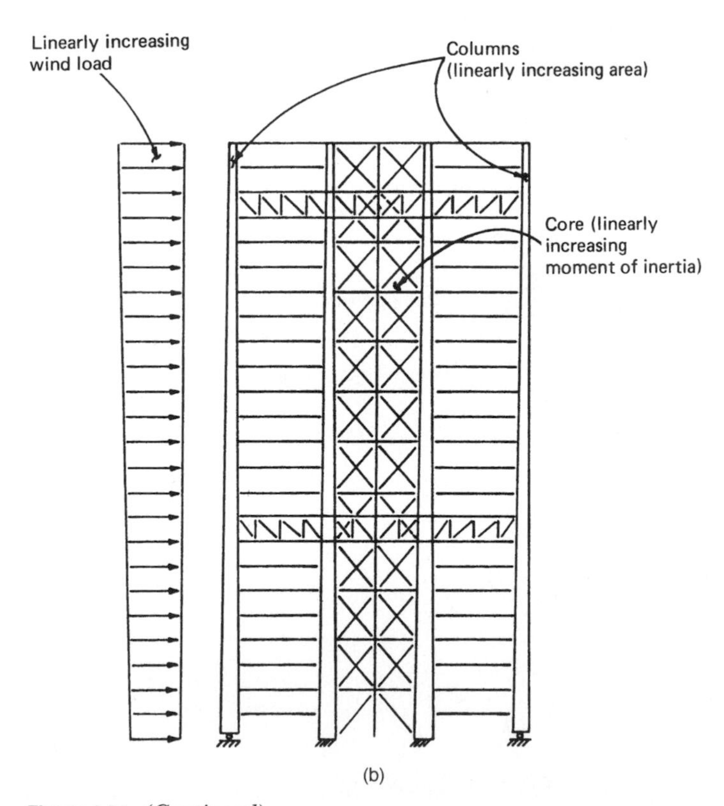


**Figure 4.30** *(Continued).*

where $m$, $s$, and $n$ represent the moment, shear force, and the axial load distribution on the statically determinate system due to the application of a unit moment at the location and in the direction of the arbitrary constants. $E$, $G$, $I$, and $A$ are the familiar notations for material and member properties of the element of the structure for which the integral is being calculated. Note that different forms of energy are significant in different members.

Next, the compatibility equations for the rotations at the truss locations are set up and the magnitudes of the arbitrary constants $M_1$ and $M_2$ obtained. The tip deflection for the structure is obtained by superposition of the solutions for the external load and for the moments $M_1$ and $M_2$. A single solution to the problem is trivial and may easily be carried out by hand calculations. A computer solution is necessary, however, since the object of the exercise is to seek an optimum combination of the truss locations to minimize lateral drift, requiring many solutions for different truss locations. A computer

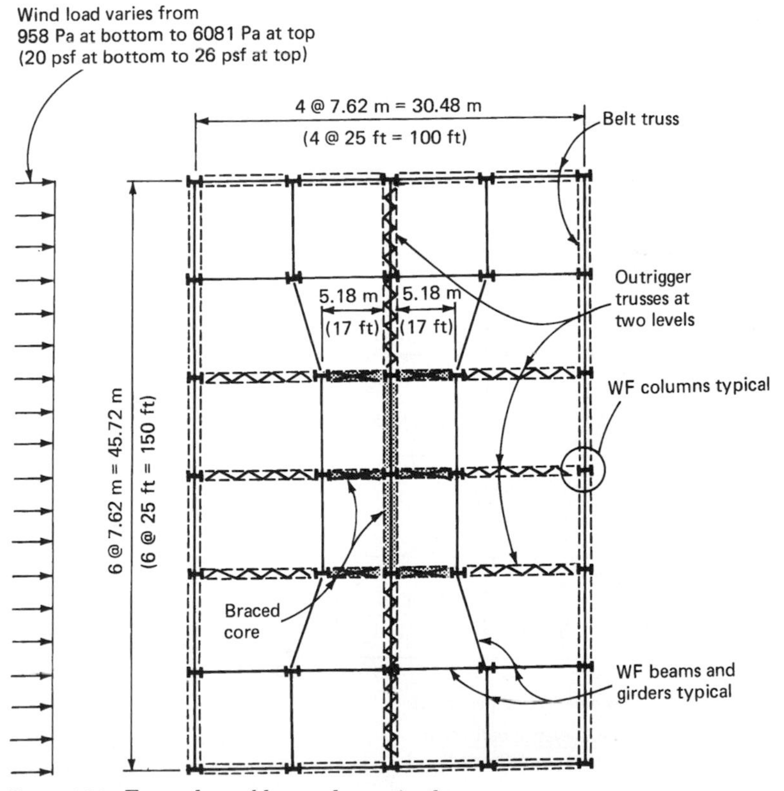


**Figure 4.31** Example problem: schematic plan.

program was written for this purpose and computations were carried out for a 46-story example structure shown in Fig. 4.31. The results of the analysis are given in the form of graphs in Fig. 4.32.

#### 4.8.5.2 Explanation of graphs

The magnitudes of the top floor deflection of the structure for three assumed modes of resistance have been presented in a nondimensional form in Fig. 4.32. The vertical ordinate with the value of the deflection parameter equal to 1 represents the top floor deflection, obtained by assuming that there are no belt trusses. The resistance is provided by the cantilever action of the braced core alone. The curve designated as S represents the deflection, assuming that a single belt truss located anywhere along the height of the structure is acting in conjunction with the braced core. The deflection for a particular location of the truss is obtained by the horizontal distance between the curve S and the vertical axis measured at the floor level (e.g.,

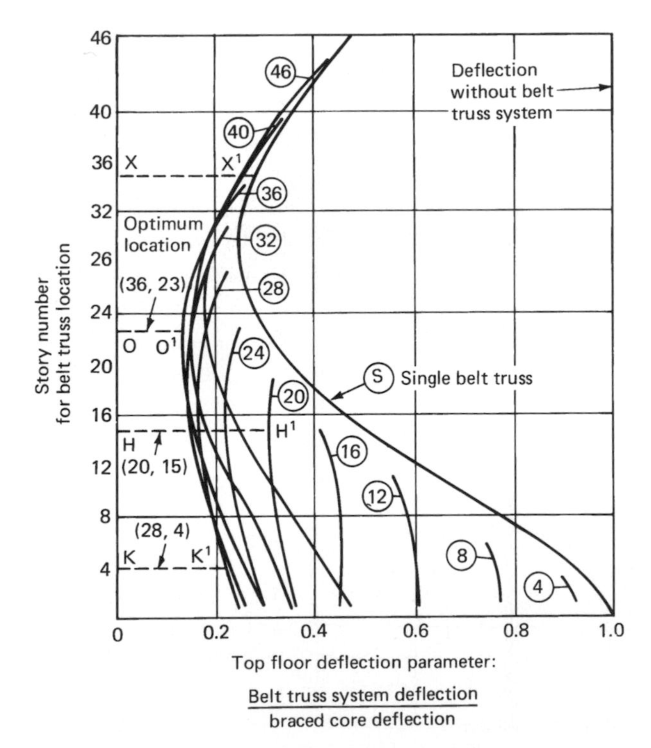


**Figure 4.32** Graph for optimum belt truss location.

distance $XX'$ multiplied by the cantilever deflection gives the top floor deflection for the location of the belt truss at floor 35). It is seen that the drift is quite sensitive to the truss location. The most favorable location is at floor 27; the resulting deflection is reduced to less than a third of the pure cantilever deflection.

The curves designated as $4, 8, \ldots, 46$ represent the top floor deflections obtained by assuming that there are two belt trusses located anywhere along the height of the structure. To obtain each curve, the location of the upper outrigger was considered fixed in relation to the building height, while the location of the lower outrigger was moved in single-story increments, starting from the first floor to the floor immediately below the top outrigger.

The number designations of the curves represent the floor number at which the upper outrigger is located. The second outrigger location is given on the vertical axis. The horizontal distance between the curve and the vertical axis is the top deflection parameter for the particular combination of truss locations given by the curve designation and the vertical ordinate. For example, let us assume that the deflection at top is desired for the combination (20, 15), the numbers 20 and 15 being the floors at which the upper and lower outriggers are located. The procedure is to select the curve with the designation 20 and to draw a horizontal line from the vertical ordinate at 15 to this curve. The requried top deflection parameter is the horizontal distance between the vertical axis and the curve 20 (distance $HH'$ in Fig. 4.32). Similarly, distance $KK'$ gives the deflection parameter for the combination (28, 4). It is seen from Fig. 4.32 that the relative location of the trusses has a significant effect on controlling the drift. Furthermore, it is evident that a deflection very nearly equal to the optimum solution can be obtained for a number of combinations. For the example problem, a deflection parameter of 0.15, which differs negligibly from the optimum value of 0.13, is achieved by the combinations (40, 23), (32, 23), etc. The effectiveness of the belt truss system is self-evident from the figure.

### 4.8.6 Example projects

Figures 4.33a,b shows photographs of The First Wisconsin Center, a 42-story, 1.3 million square foot (120,770 $m^2$) bank and office building which utilizes the concept of belt and outrigger system of lateral bracing. The building rises 601 ft (183 m) from a two-level glass-enclosed plaza and sports three belt trusses located at bottom, middle, and top of the building. The belt truss at the bottom serves

(a)

(b)

**Figure 4.33a,b** First Wisconsin Center, Milwaukee.

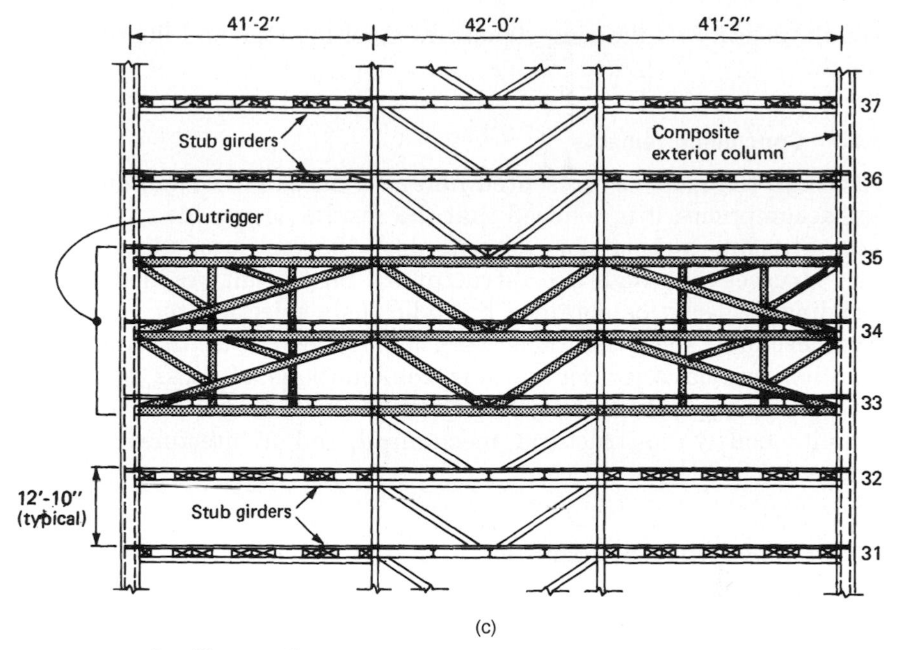


**Figure 4.33c** One Houston Center.

as a transfer truss, eliminating every other exterior column. Outrigger trusses are used at the top and middle of the building to engage the braced core to the exterior columns. Mechanical levels are located at the trussed floors. The architecture and engineering for the project is by the Chicago Office of Skidmore, Owings & Merrill.

As a second example Fig. 4.33c shows a schematic partial elevation of wind bracing system for a building in Houston, Texas called One Houston Center. The building, consisting of 48-stories, rises to a height of 681 ft (207.5 m) above grade. To minimize the building drift, 2-story-deep outrigger trusses tie the perimeter columns to the K-braced core between the 33rd and 35th levels. Three outrigger trusses on each side of the core run through the building space. Use of these outrigger trusses helped in reducing the drift to less than 1/460 of the building height. The structural engineering for the project was by the Houston office of Walter P. Moore & Associates. A novel system of floor framing, called a stub girder system, was used to frame typical bays measuring 41 ft 2 in. by 30 ft (12.54 m by 9.14 m). This unique system of floor framing, which is discussed in Chap. 9, attempts to reduce the building cost by simultaneously minimizing structural steel tonnage and floor-to-floor height.

### 4.8.7 Concluding remarks

Although the analysis presented herein is based on certain simplifying assumptions, it is believed that the results do provide sufficiently accurate information for determining the optimum location of belt trusses in high-rise structures. Significant reductions in building drift can be obtained by judiciously selecting the locations. Furthermore, since solutions very nearly equal to the optimum solution are obtained for various combinations of truss locations, it should be relatively easy to choose a combination that satisfies simultaneously the structural, mechanical, and architectural requirements.

## 4.9 Framed Tube System

### 4.9.1 Introduction

In its simplest terms the tube system can be defined as a fully three-dimensional system that utilizes the entire building perimeter to resist lateral loads.

At present four of the five world's tallest buildings are tubular systems. They are the 110-story Sears Tower, the 100-story John Hancock Building, and the 83-story Standard Oil Building, all in Chicago, and the 110-story World Trade Center towers in New York. The earliest application of the tubular concept is credited to the late Dr. Fazlur Khan of the architectural engineering firm of Skidmore, Owings & Merill, who first introduced the system in a 43-story apartment building in Chicago.

The introduction of the tubular system for resisting lateral loads has brought about a revolution in the design of high-rise buildings. All recent high-rise buildings in excess of 50- to 60-stories employ the tubular concept in one form or another. In essence the system strives to create a three-dimensional wall-like structure around the building exterior. In a framed tube this is achieved by arranging closely spaced columns and deep spandrels around the entire perimeter of the building. Because the entire lateral load is resisted by the perimeter frame, the interior floor plan is kept relatively free of core bracing and large columns, thus increasing the net leasable area of the building. As a trade-off, views from the interior of the building may be hindered by closely spaced exterior columns.

The necessary requirement to create a wall-like structure is to place columns on the exterior relatively close to each other and to use deep spandrel beams to tie the columns. The structural optimization reduces to examining different column spacings and member proportions. In practice the framed tubular behavior is achieved by placing columns at 10 ft (3.05 m) to as much as 20 ft (6.1 m) apart, with spandrel depth varying from 3 to 5 ft (0.90 to 1.52 m).

The tube system can be constructed of reinforced concrete, structural steel, or a combination of the two, termed composite construction.

The method of achieving the tubular behavior by using columns on close centers connected by a deep spandrel is by far the most used system because rectangular windows can be accommodated in this design. A somewhat different system for steel buildings that permits larger spacing of columns is called the braced tube which has diagonal or K-type of bracing at the building exterior. A simiiar concept for concrete buildings is to infill window penetrations in a systematic pattern to achieve the same effect as a diagonal or K-type bracing. Yet another is to use two or more tubes tied together to form a bundled tube. In this sub-section a description of the framed tube system is given; other systems are explained in later sections.

### 4.9.2 Framed tube behavior

To understand the behavior of a framed tube, consider a square-shaped 50-story building as shown in Fig. 4.34a consisting of closely

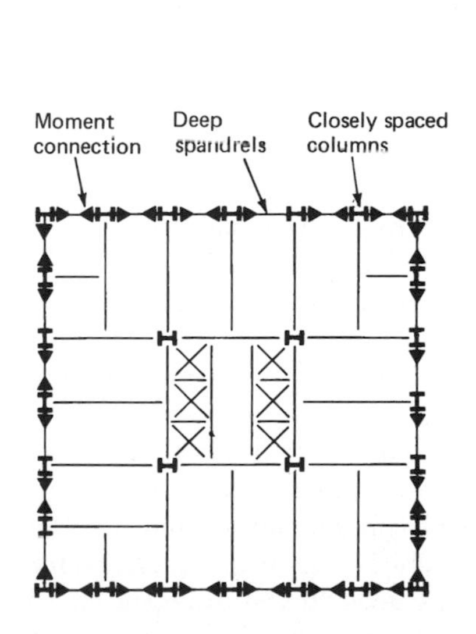


**Figure 4.34a** Schematic plan of framed tube.

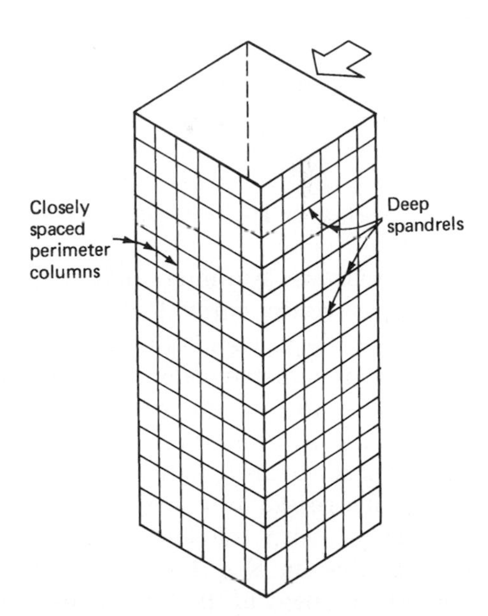


**Figure 4.34b** Isometric view of framed tube.

spaced exterior columns and deep spandrel beams. Assuming that the interior columns are designed for gravity loading only, their contribution to lateral load resistance is negligible. The floor system, as in other types of lateral bracing systems, is considered a rigid diaphragm and is assumed to distribute the wind load to various elements according to their stiffness. Its contribution to lateral resistance in terms of its out-of-plane action is considered negligible. The system resisting the lateral load thus comprises of four orthogonal rigidly jointed frame panels forming a tube in plan as shown in Fig. 4.34b.

The frame panels are formed by closely spaced perimeter columns that are connected by deep spandrel beams at each floor level. In such structures, the "strong" bending direction of the columns is aligned along the face of the building, in contrast to the typical rigid frame bent structure where it is aligned perpendicular to the face. The basic requirement has been to place as much of the lateral load-carrying material at the extreme edges of the building to maximize the inertia of the building's cross-section. Consequently, in many structures of this form, the external tube is designed to resist the entire lateral loading. The frames parallel to the lateral load act as "web" of the perforated tube, while the frames normal to the loads act as "flanges." Vertical gravity forces are resisted partly by the exterior frames and partly by some interior columns or an interior core. When subjected to bending under the action of lateral forces, the primary mode of action is that of a conventional vertical cantilevered tube, in which the columns on opposite sides of the neutral axis are subjected to tensile and compressive forces. In addition, the frames parallel to the direction of the lateral load are subjected to the usual inplane bending, and the shearing or racking action associated with an independent rigid frame.

The discrete columns and spandrels may be considered, in a conceptual sense, equivalent to a continuous three-dimensional wall. The model becomes a hollow tube cantilevering from the ground with a basic stress distribution as shown in Fig. 4.35.

Although the structure has a tube-like form, its behavior is much more complex than that of a solid tube; unlike a solid tube it is subjected to shear lag effects. The influence of shear lag is to increase the axial stresses in the corner columns and reduce those in the inner columns of both the flange and the web panels as shown by the dotted lines, in Fig. 4.35. Ignoring the shear lag consequences for now, the analogy of the hollow tube can be used to visualize the axial stress distribution in buildings with other plan forms such as rectangular, circular, and triangular as shown in Fig. 4.36a–c. This philosophy of creating a fully three-dimensional structural system utilizing the entire building foot-print to resist lateral loads has allowed for considerable freedom in manipulating building plans. The rigorous organization of orthogonal bay spacing required with the previous types of bracing is no longer necessary. The only requirements are for the structure to be continuous around the exterior to invoke a three-dimensional response, and be of a closed-cell form, to resist torsional loads. Depending upon the height and dimensions of the building, the exterior column spacing is usually of the order of 10 to 15 ft (3 to 4.6 m), although a spacing as close as 3.8 ft (1.0 m) has been used for the 110-story World Trade Center twin towers, New York (Fig. 4.37a). The efficiency of the system is directly related to building height-to-width ratio, plan dimensions, spacing, and size of columns and spandrels.

Figure 4.37 shows examples of free-form tubular configurations. Although in simplistic terms the tube is similar to a hollow cantilever, in reality its response to lateral loads is in a combined bending and shear mode. The overall bending of the tube is due to axial shortening and elongation of the columns while the shear

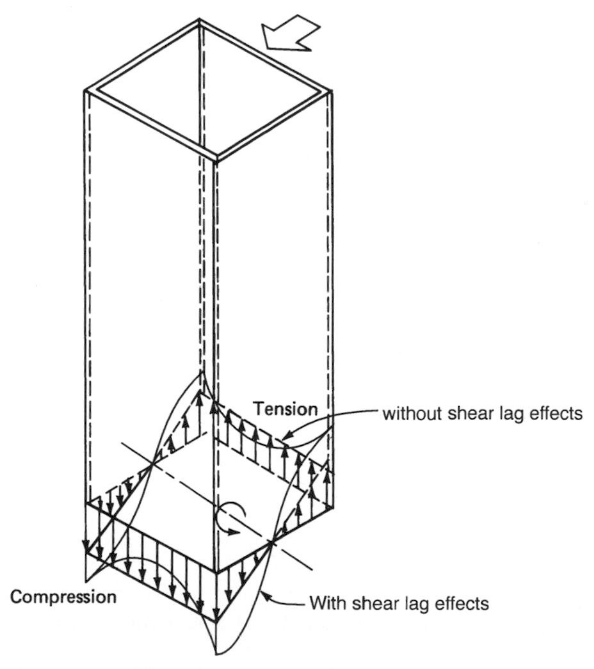


**Figure 4.35** Axial stress distribution in square hollow tube with and without shear lag.

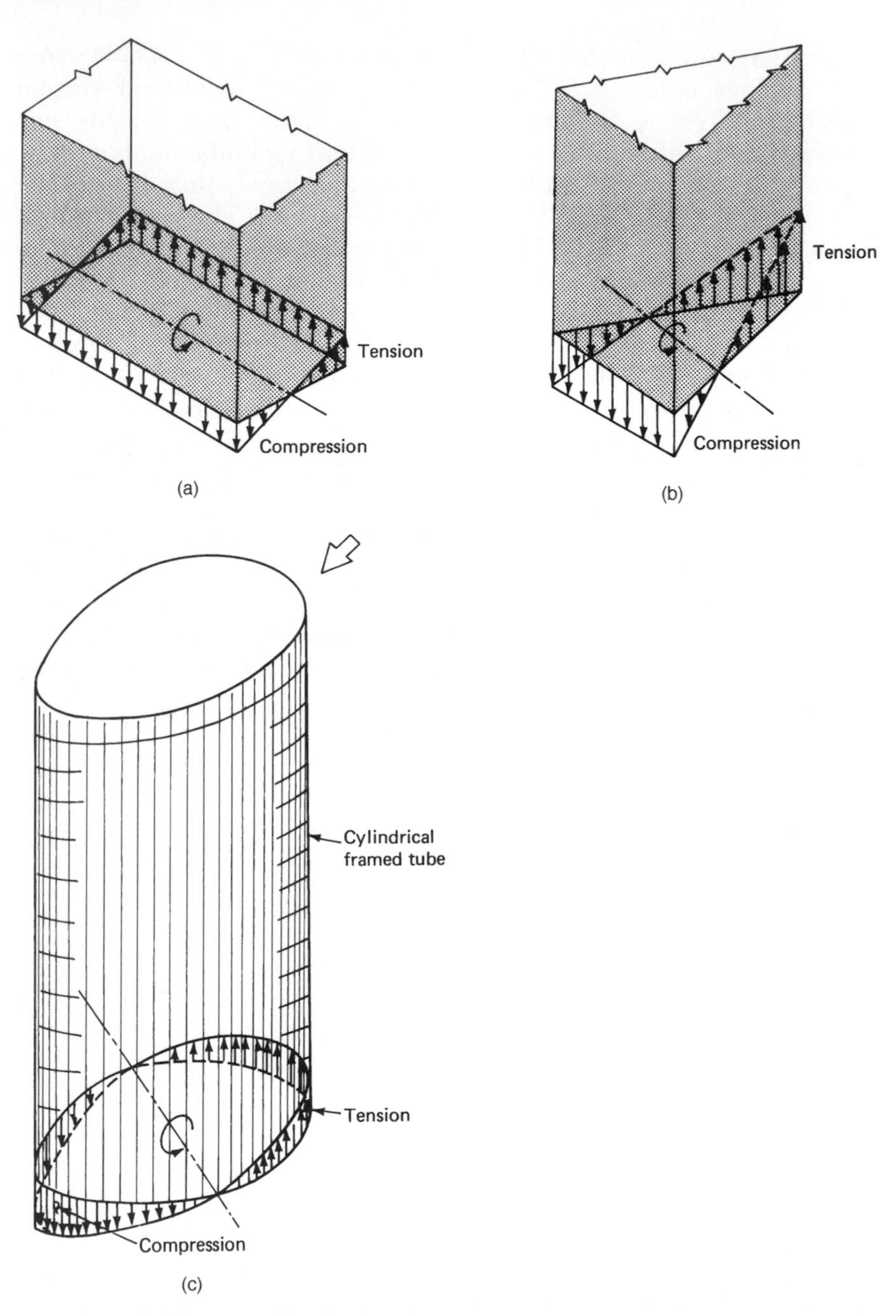


**Figure 4.36** Axial stress distribution in tube structures: (a) rectangular tube; (b) triangular tube; (c) circular tube.

deformation is due to bending of individual columns and spandrels. The underlying principle for an efficient design is to eliminate or minimize the shear deformation so that the tower as a whole bends essentially as a cantilever.

### 4.9.3 Shear lag phenomenon

The primary action of the hollow tube as discussed previously is complicated by the fact that the flexibility of the spandrel beams increases the stresses in the corner columns and reduces the same in the inner columns of both the flange and the web panels. Contrary to what one may expect, even for a solid-wall tube, the distribution of

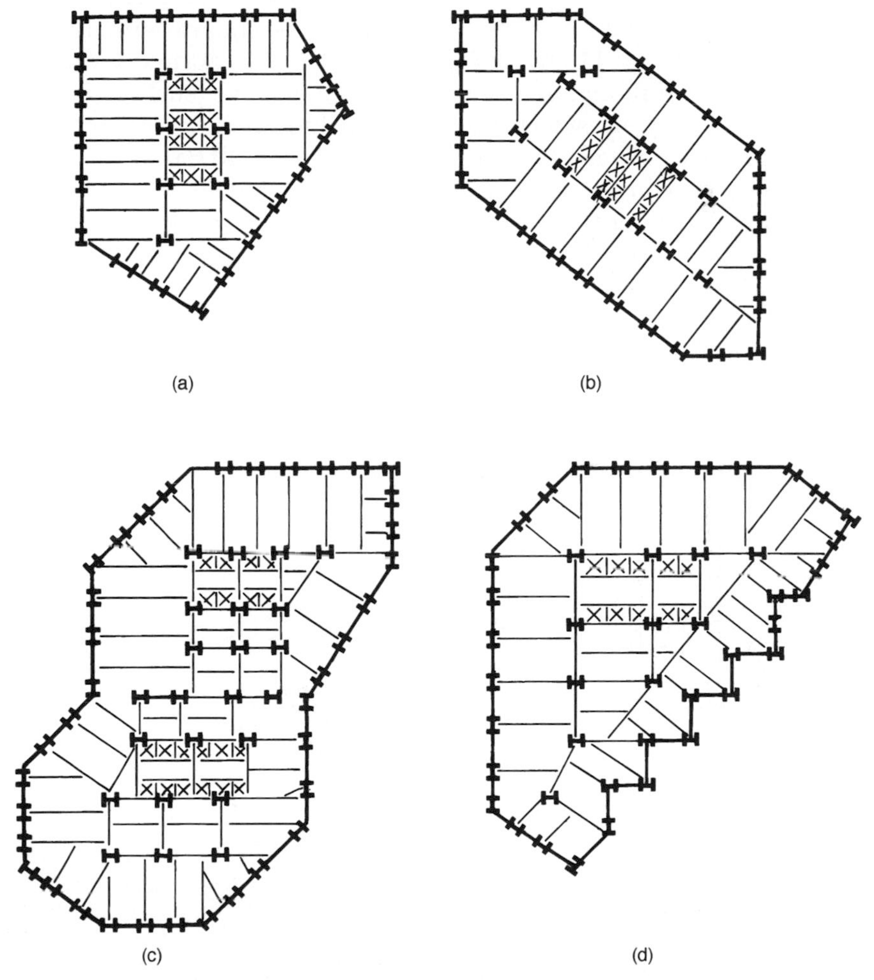


**Figure 4.37** Free-form tubular configurations.

axial forces is not uniform over the windward and leeward walls and linear over the side walls. The behavior can be readily appreciated by considering the shear deformations of the tube walls which are relatively "thin" as compared to the height and plan dimensions of the building. The extent to which the actual axial load distribution departs from the ideal is termed "shear lag effect". An understanding of this phenomenon is essential in developing optimal tubular systems.

The closest structure to a perfect tube is a system of continuous perimeter walls without any discontinuities. A hollow box represents such a system, as shown in Fig. 4.38a. In order to fix ideas, assume the hollow box represents a 50-story steel building with a typical floor-to-floor height of 13 ft (3.94 m), giving a total height of 650 ft (198 m) for 50-stories. Assume the building is square with a plan dimension of 110 by 110 ft (33.5 by 33.3 m). For a normally proportioned 50-story building, in a seismically benign zone, such as zone 1, the unit quantity of structural steel, including that required for gravity, is in the range of 22 to 24 psf (1053 to 1149 Pa) of the gross area of the building. Conservatively, assume that all of the 24 psf (1149 Psa) of structural steel is available for the lateral bracing of the building. The most efficient manner of using this material, in an

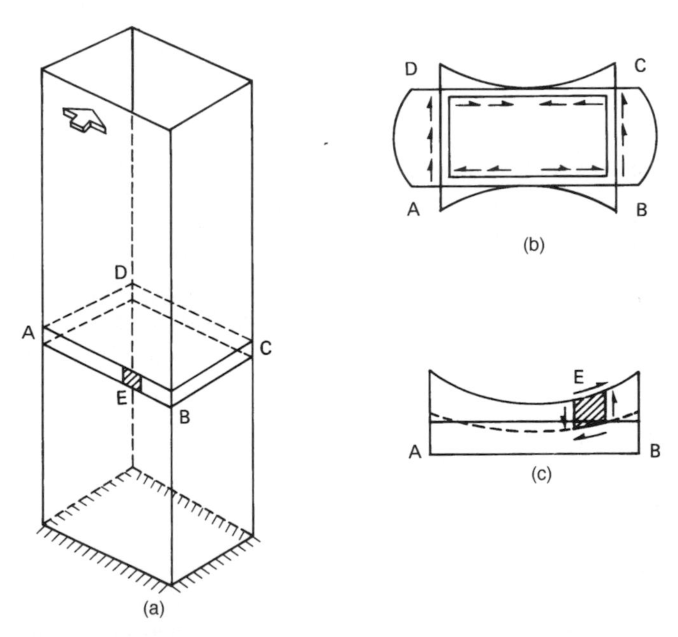


**Figure 4.38** Shear lag effects in tube structures: (a) cantilever tube subjected to lateral loads; (b) shear stress distribution; (c) distortion of flange element caused by shear stresses.

academic sense, is to convert the 24 psf of steel into an equivalent wall located at the perimeter of the tube as shown below.

Total steel available for bracing

$$= 24 \times 110 \times 1150 \times 50 = 14{,}520{,}000 \text{ lb } (64{,}585 \text{ kN})$$

$$\text{Area of perimeter wall} = 4 \times 110 \times 650 = 286{,}000 \text{ ft}^2 \; (26{,}569 \text{ m}^2)$$

Equivalent thickness of wall

$$= 14{,}520{,}000/286{,}000 \times 3.4 = 15 \text{ in } (381 \text{ mm})$$

In comparison to the plan dimensions of the building, the calculated wall thickness of 15 in, (381 mm) is relatively small, giving a length-to-thickness ratio of 1:88. Because of this characteristic, the structure has a tendency to behave like a thin-walled beam. In a thin-walled beam the shear stresses and strains are much larger than those in a solid beam and often result in large shearing deformations with a significant effect on the distribution of bending stresses. Because of the resulting large shear strains, the usual assumption used in engineers' bending theory is violated. This assumption, which states that plane sections before bending remain plane after bending, is known as the *Bernoulli hypothesis* and forms the basis for mathematical relations used in engineering mechanics. However, in thin-walled structures the large shearing strains cause the plane of bending to distort. For the hollow box structure, the element E on the flange face distorts as shown in Fig. 4.38c. The final outcome due to the cumulative effect of distortion of all such elements is that under lateral load, the originally flat plane of the cross section distorts as shown in Fig. 4.39. Because of these distortions, the simple stress distribution given by the engineers' theory of bending is no longer applicable. The bending stresses will not be proportional to the distance from the neutral axis of the section; the stress at the center of the flanges "lags" behind the stress near the web because of the lack of shear stiffness of the wall panel. This phenomenon is

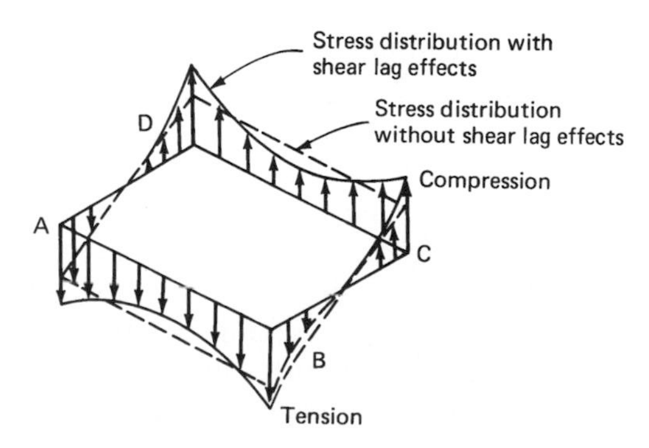


**Figure 4.39** Axial stress distribution.

known as shear lag and plays an important role in the design of tubular high-rise structures. The bending stresses in the webs are also affected in a similar manner.

To better appreciate the shear lag phenomenon, let us take a closer look at our hypothetical high-rise building. In a normally proportioned steel building, the usual method of framing for gravity loads necessitates that interior columns be located in and around the core to maintain the span of floor beams in the economical range of 35 to 45 ft (10.6 to 13.64 m). The interior gravity columns and floor beams amount to about one-third of the total steel required for the building. For the example, the effective unit quantity of steel available for the bearing wall is therefore equal to 16 psf (766 Pa), effectively reducing the equivalent thickness of the tube wall to 10 in (254 mm), with a proportional increase in the shear strain. The departure of bending stress distribution from those predicted on the basis of plane sections becomes even more severe. A high-rise building in practice has to accommodate penetrations in the exterior wall for obvious reasons, which means that the bending efficiency of the tube is further reduced because of these penetrations. Therefore, even with the most efficient distribution of material, shear lag is still present. It cannot be completely eliminated but can be minimized. Thus the structural optimization in a framed tube design reduces to an examination of different column spacings, and corresponding sizes with spandrels, which result in least shear lag effects.

The shear lag effects in tubular buildings consisting of discrete columns and spandrels may readily be appreciated by considering the basic mode of action involved in resisting lateral forces. The primary resistance comes from the web frames which deform so that the columns T are in tension and C are in compression (Fig. 4.40). The web frames are subjected to the usual in-plane bending and racking action associated with an independent rigid frame. The primary action is modified by the flexibility of the spandrel beams which causes the axial stresses in the corner columns to increase and those in the interior columns to decrease.

The principal interaction between the web and flange frames occur through the axial displacements of the corner columns. When column C, for example, is under compression, it will tend to compress the adjacent column C1 (Fig. 4.40) since the two are connected by the spandrel beams. The compressive deformations of C1 will not be identical to that of corner column C since the flexible connecting spandrel beam will bend. The axial deformation of C1 will be less, by an amount depending on the stiffness of the connecting beam. The deformation of column C1 will in turn induce compressive deformations of the next inner column C2, but the deformation will again be less. Thus each successive interior column will experience a smaller deformation and hence a lower stress than the outer ones. The stresses in the corner column will be greater than those from pure tubular action, and those in the inner columns will be less. The stresses in the inner columns lag behind those in the corner columns. Hence the term shear lag is used to describe this phenomenon.

The difference between stress distribution, as predicted by ordinary engineer's beam theory and the actual distribution is illustrated in Fig. 4.40. Because the column stresses are distributed less effectively than in an ideal tube, the moment resistance and the

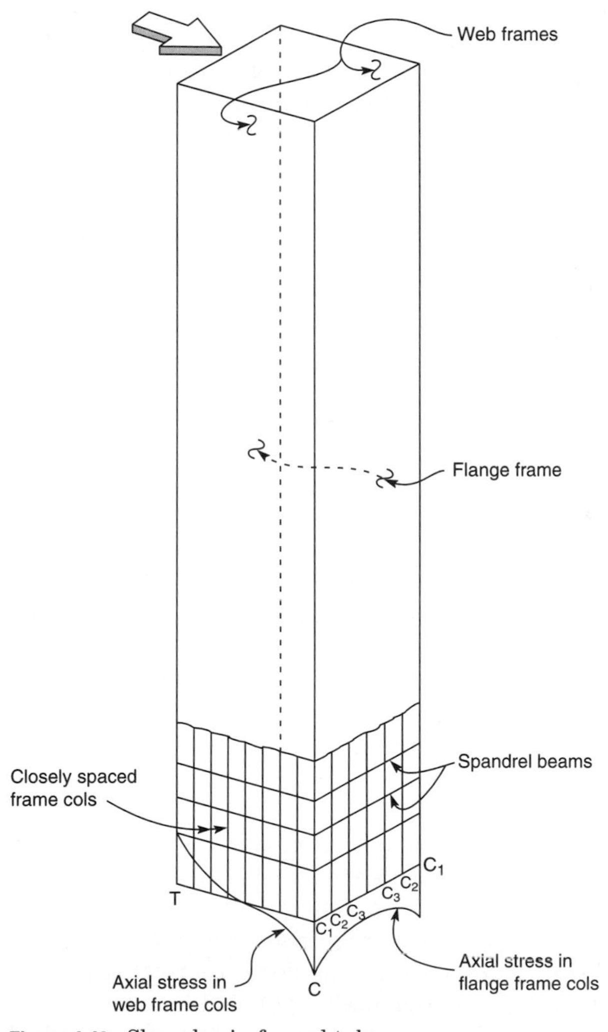


**Figure 4.40** Shear lag in framed tube.

flexural rigidity are reduced. Thus, although a framed tube is highly efficient for tall buildings, it does not fully utilize the potential stiffness and strength of the structure because of the effects of shear lag in the perimeter frames.

### 4.9.4 Irregularly shaped tubes

The framed tube concept can be executed with any reasonable arrangement of column and spandrels around the building perimeter. However, non-compact plans, and plans with re-entrant corners considerably reduce the efficiency of the system. For framed tubes, a compact plan may be defined as one with an aspect ratio of not greater than 1.5 or so. Elongated plans with longer aspect ratios impose considerable premium on the system because: (i) in wind controlled design, the elongated building elevation acts like a sail collecting large wind loads; (ii) the resulting shear forces most usually require closer spacing and or larger size columns and spandrels parallel to the wind; (iii) shear leg effects are more pronounced especially for columns orientated perpendicular to the direction of wind.

In a similar manner a sharp change in the tubular form results in a less efficient system because the shear flow must pass around the corners solely through axial shortening of the columns. Also, a secondary frame action at these locations alters the load distribution in the framed tube columns as explained below.

Consider the framed tubes shown in Figs 4.41a,b. To keep the explanation simple, let us assume that shear lag effects are negligible. For lateral loads in the N–S direction the leeward columns A and B are subjected to compression while the windward columns C and B to tension. In addition to the primary action, the frames AB and CD experience a secondary bending action about their own local axes 1-1 and 2-2, as shown in Fig. 4.41. The resulting axial forces in columns A and B are compression and tension respectively. The effect of local bending is to increase the compressive force in A while decreasing the same for column B. The final force is the summation of the two; the primary tube action and the secondary local frame action.

In framed tubes, a limited number of columns can be transferred with little, if any, structural premium because the vierendeel action of the façade frame is generally sufficient to transfer the load. However, if the transfer is too severe requiring removal of a large number of columns, a 1- or 2-story deep transfer girder of truss may be necessary. Temporary shoring is usually required to support the dead and construction loads until a sufficient number of vierendeel

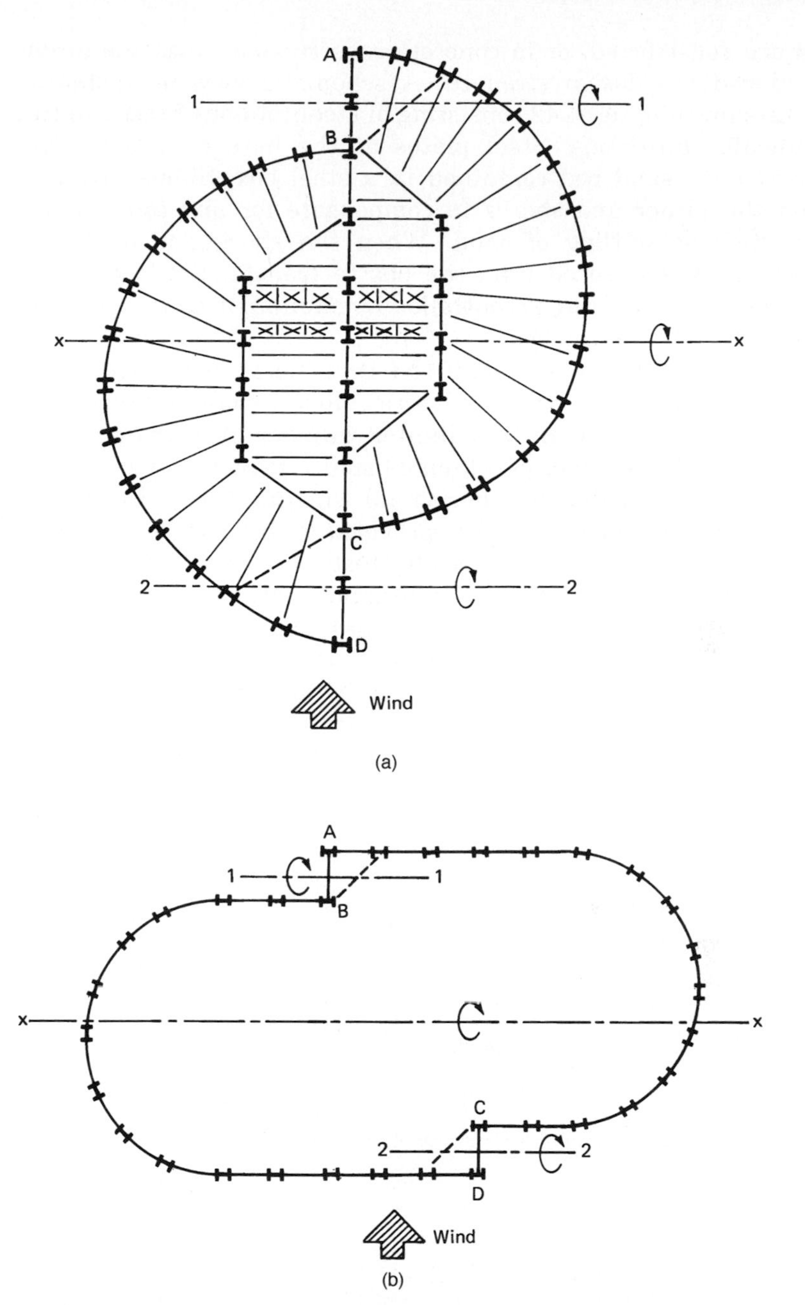


**Figure 4.41** Offset tubes: (a) semicircular tubes; (b) rectangular tube with semicircular sides.

frames are constructed, or in concrete construction, until the girder has achieved the design strength. A schematic view of a shoring system is shown in Fig. 4.42 consisting of steel columns braced in two perpendicular directions. Steel plates at the base of columns are supported on a sand bed contained in a steel box. Shims are used between the girder and shores to compensate for any construction irregularities or settling of sand. When the shoring is no longer required, sand is released from the box to transfer the load to the girder. The rate of loading is controlled by manipulating the quantity of sand removed from the box.

Generally in a transfer system, be it a vierendeel frame, steel truss, plate or concrete girder, the vertical load is collected from tube columns, and channeled in to a limited number of columns below. Removal of a large number of columns usually requires an additional lateral system to carry the horizontal shears. Generally braced frames or concrete shear walls are provided within the building core to resist the shear forces. Occasionally steel plates with welded shear studs are used to increase the shear resistance of concrete walls. The shear forces from the frame columns are transferred to the core through the diaphragm action of the slab at the transfer level. In a steel building this is sometimes achieved by providing diagonal steel bracing under the transfer floor and, in concrete building by increasing the thickness of the concrete slab and adding additional reinforcement.

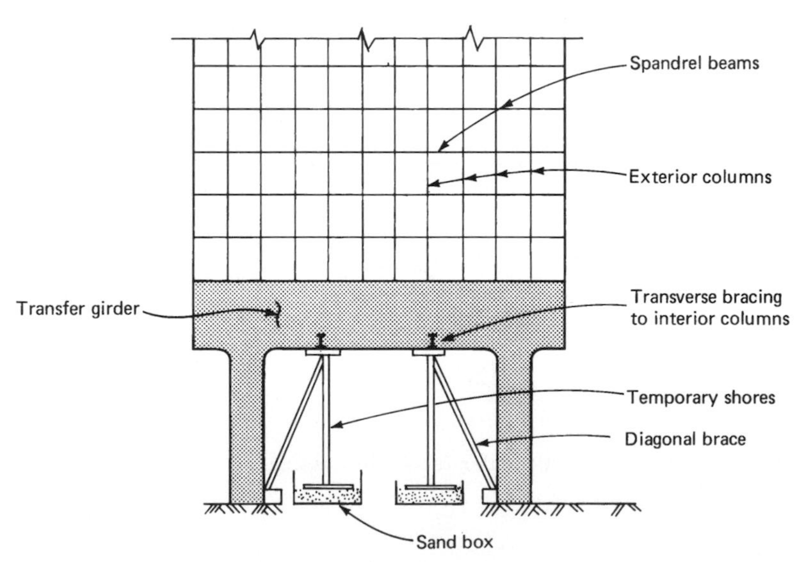


**Figure 4.42** Shoring system for a tube structure.

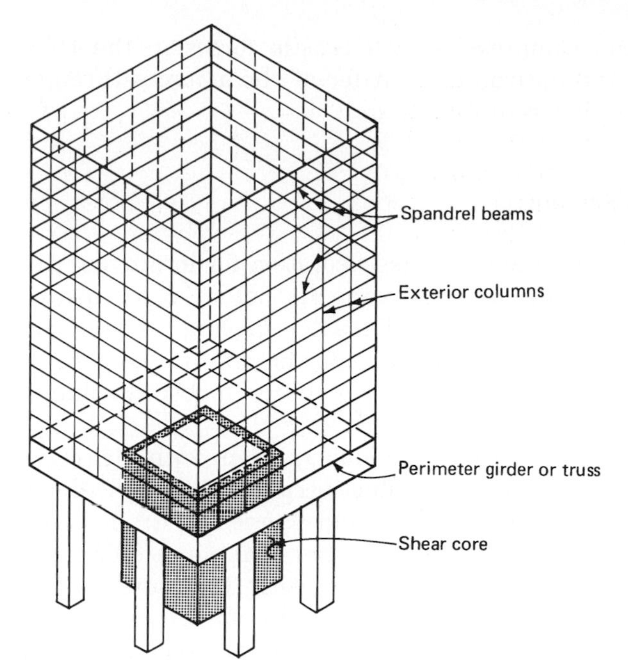


**Figure 4.43** Vertical transfer.

Figures 4.43–4.45 show conceptually the ideas discussed above. Figure 4.3 shows the vertical transfer of a frame tube to a set of eight columns. A transfer girder spanning between the columns supports the column loads. Figure 4.44 shows the transfer of overturning

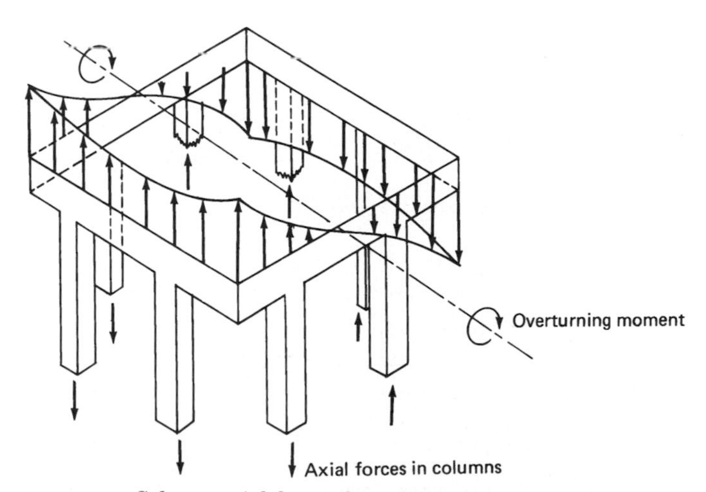


**Figure 4.44** Column axial forces due to overturning moment.

moment, i.e., the axial compressive and tensile forces in the tube columns. And finally the horizontal transfer of shear forces through the floor diaphragm is shown in Fig. 4.45.

## 4.10 Trussed Tube System

A trussed tube system represents a classic solution for improving the efficiency of the framed tube by increasing its potential for use to even greater heights as well as allowing greater spacing between the columns. This is achieved by adding diagonal bracing to the face of the tube to virtually eliminate the shear lag effects in both the flange and web frames.

The framed tube, as discussed previously, even with its close spacing of columns is somewhat flexible because the high shear stresses in the frames parallel to the wind cannot be transferred effectively around the corners of the tube. For maximum efficiency, the tube should respond to lateral loads with the purity of a cantilever with compression and tension forces spread uniformly across the windward and leeward faces. The framed tube, however, behaves like a thin-walled tube with openings. The axial forces tend to diminish as they travel around the corners, with the result that the columns in the middle of the windward and leeward faces may not sustain their fair share of compressive and tensile forces. This effect referred to previously as the shear lag effect limits the framed tube application to 50- or 60-story buildings unless the column spacing is very small as with the 109-story World Trade Center Towers, New York which has columns at 3.8 ft (1.0 m). For taller

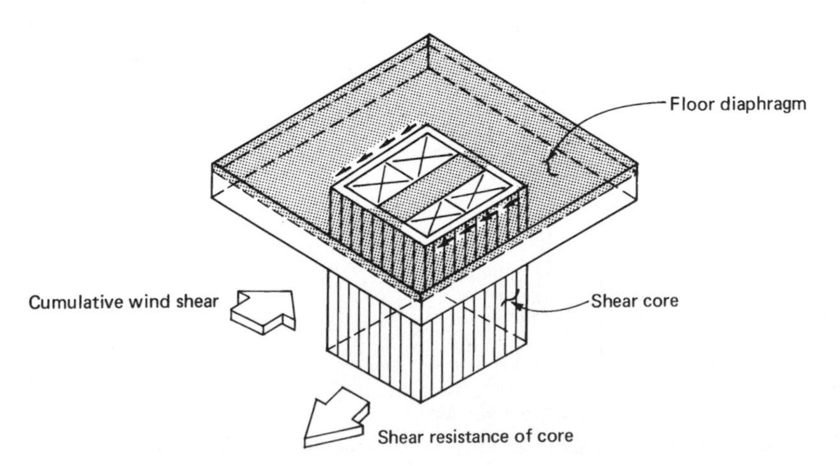


**Figure 4.45** Shear transfer through floor diaphragm.

buildings with their usual column spacing of 10 to 15 ft (3.0 to 4.58 m), the frames parallel to the lateral loads act as multibay rigid frames. Consequently, column and beam designs are controlled by their bending action rather than axial, resulting in unacceptably large sizes. Furthermore, out of the total sway, only about 25 percent is from the cantilever component, while the remainder is from the frame shear racking component. Because of the shear racking, the corner columns take more than their share of axial load, while those in between less, as compared to an ideal tube. A trussed tube overcomes this problem by stiffening the exterior frames.

The most effective trussed tube action may be obtained by replacing vertical columns with closely spaced diagonals in both directions (Fig. 4.46). However, this system is not popular because it creates problems in the detailing of curtain walls.

The diagonally braced tube, shown in Fig. 4.47a,b is by far the most usual method of increasing the efficiency of the framed tube. It represents an elegant solution by introducing a minimum number of diagonals on each façade that intersect at the same point as the corner columns. The system is tubular in that the fascia diagonals interact with the trusses on the perpendicular faces to achieve three-dimensional behavior.

The diagonals of a braced tube connected to the columns at each intersection, virtually eliminate the effects of shear lag in both the flange and web frames. As a result, the structure behaves under lateral loading more like a braced frame, with greatly diminished bending of frames. Consequently, the spacing of the columns can be larger and the size of the columns and spandrels less, thereby allowing larger size windows than in the conventional tube structure. In the braced-tube structure the bracing contributes also to the improved performance of the tube in carrying gravity loading. Differences between gravity load stresses in the columns are evened out by the braces transferring axial loading from the more highly to the less stressed columns.

The principle of façade diagonalization can readily be used for partial tubular concepts. For example, in long rectangular buildings, the end frames along the short face may be diagonalized with moment-resisting frames on the long faces. The end diagonal frame may be in the form of a channel or C shape to provide lateral resistance in both directions. Many variations are possible, each having an impact on the exterior architecture.

To use the idea of a trussed tubular system in reinforced concrete construction, a diagonal pattern of window perforations in an otherwise framed tube construction is filled in between adjacent columns and girders. The result is a reduction in shear lag for the system under lateral loads. As with the steel-framed trussed tube, the façade diagonalization offers the additional benefit of equalizing the gravity loads in the exterior columns.

## Examples

### 1. First International Plaza, Dallas

An example of a trussed tube structural system is shown in Fig. 4.48a, a 56-story office building in downtown Dallas. Designed by structural engineers Ellisor & Tanner, Inc., the building consists of 1.9 million square feet (174,437 $m^2$) of office space and rises to a height of 710 ft (216 m) above grade. Exterior columns are spaced at 25 ft (7.62 m) on center with a floor-to-floor height of 12 ft, 6 in. (3.81 m) to facilitate intersection of diagonals with columns and spandrels. There are two X braces, each 28-stories tall, on each of the

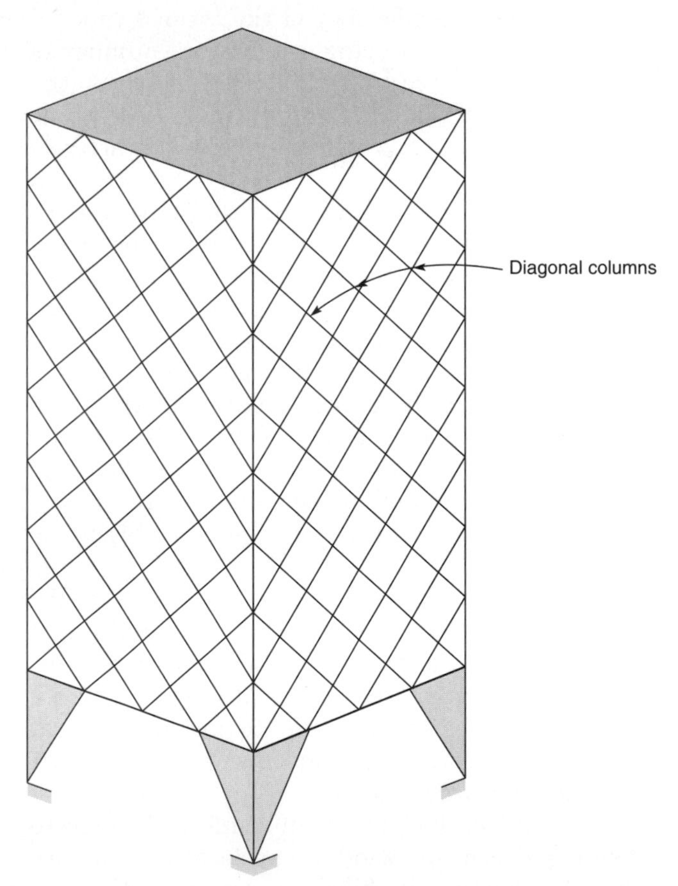


**Figure 4.46** Tube building with closely spaced diagonal columns.

four sides, as shown in Fig. 4.48b. The diagonal bracing, in addition to carrying the wind loads, helps to distribute gravity loads along the exterior columns on each side of the building. The vertical bracing is located within the glass line to minimize problems associated with fireproofing the exterior structure and the effects of temperature on exposed steel. All members in the primary exterior framing—beams,

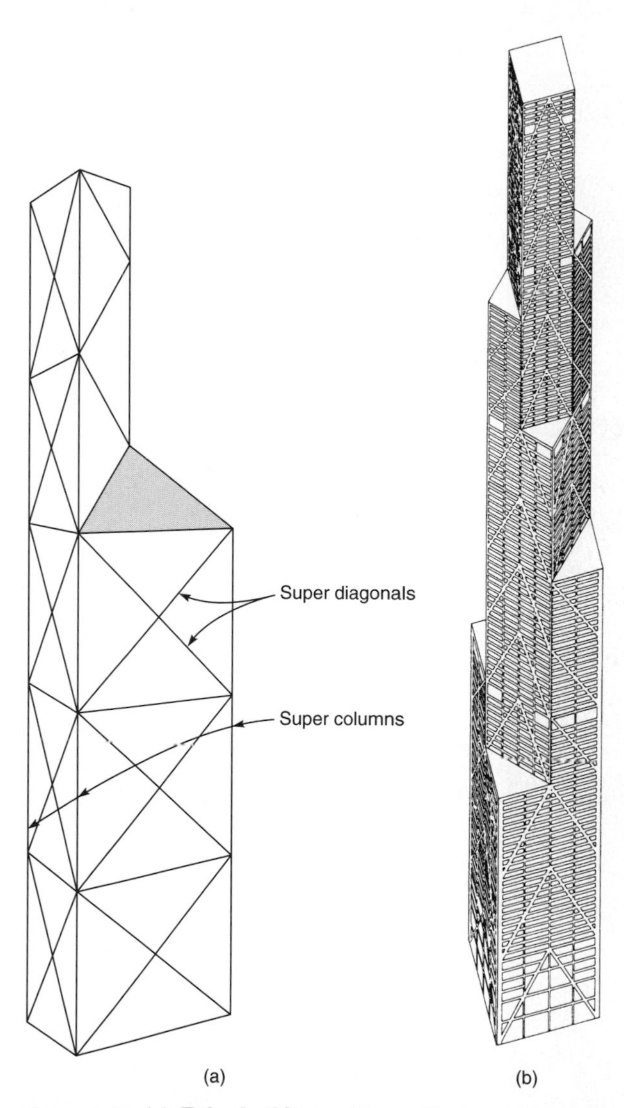


**Figure 4.47** (a) Tube building with multi-story diagonal bracing: (b) rotated square tube with super diagonals. (Adapted from an article by Mahjoub Elnimeiri, in Civil Engineering Journal)

columns, and diagonals—were fabricated from high-strnegth W14 shapes except for a few near the bottom which were built-up shapes. W14 shapes were selected because of the wide range of available sections and because all shapes had the same nominal inside

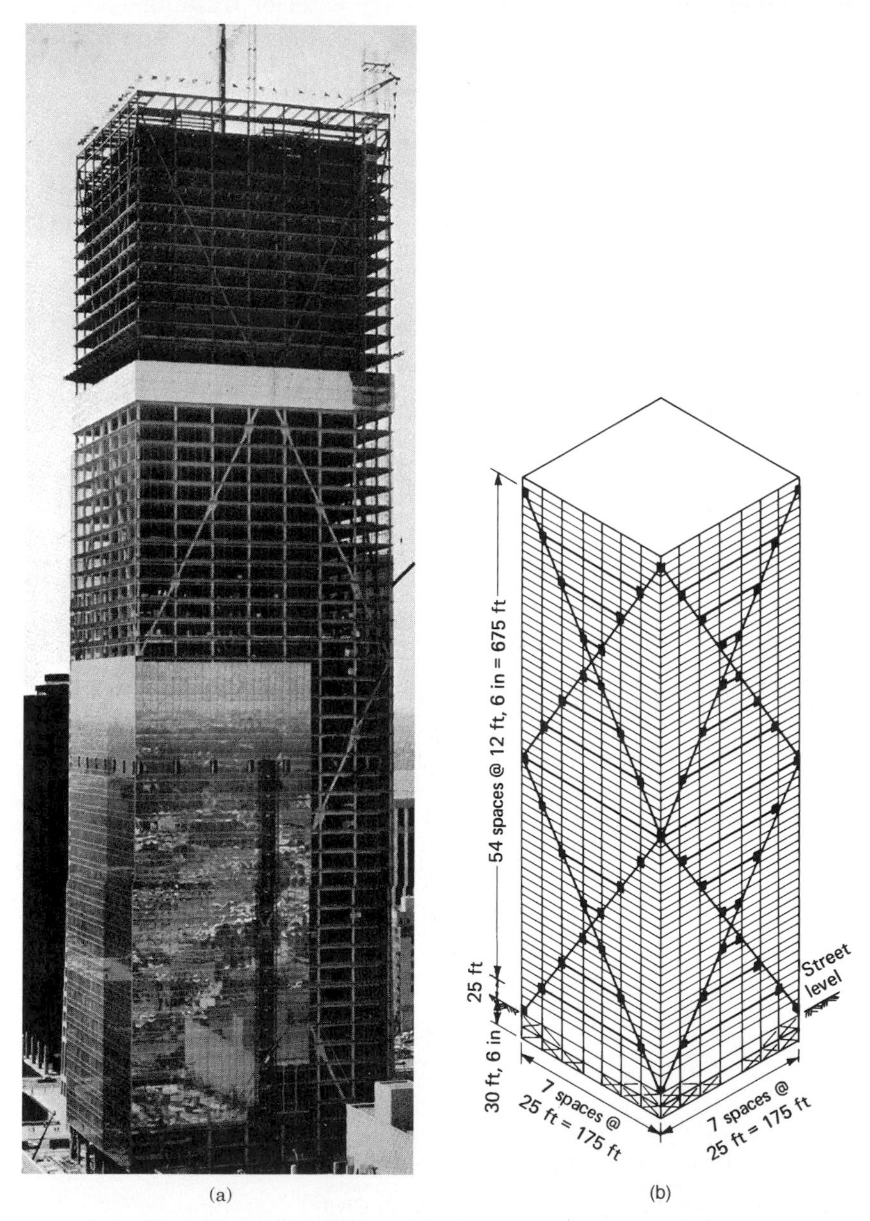


**Figure 4.48** First International Plaza: (a) photograph; (b) schematic bracing.

dimension between flanges, thus facilitating connection details. Gusset plates used to connect the diagonals, columns, and beams are approximately 10 ft (3.0 m) wide and 12 ft (3.65 m) tall. The diagonals, which were fabricated 4-stories in length, were field-welded with full-penetration welds at one end and were high-strength-bolted through splice plates at the other end. The welded connection at one end reduced gusset and splice material, while the bolted connection at the other end provided the necessary tolerances for erection. Corner gusset assemblies were required where diagonal bracing met at the corners of the building. These were fabricated from four plates, two in each of the two directions. The four plates were joined by electroslag welds and were stress-relieved after fabrication.

## 2 John Hancock Center, Chicago

Perhaps the most notable example of this structural system is the John Hancock Center in Chicago (Fig. 4.49) designed by the Chicago office of Skidmore, Owings & Merrill. The building is 100-stories with a rectangular plan that tapers from the ground level to the top. The diagonals are placed at 45° angles to each other, forming enormous X braces on each side. The diagonals serve multiple functions, acting as inclined columns to resist some of the gravity load, absorbing most of the wind shear, and stiffening the tube so that it mimics the behavior of a solid tube. This unique design, combined with the use of high-strength steel, enabled the engineers to achieve a 100-story building with only 29.7 psf (1422 Pa) of steel as compared to 42.2 psf (2020.5 Pa) for the 102-story Empire State Building.

The building is for multi-use involving commercial, parking, office and apartment-type space in one building. The ground floor plan measures 164 ft (50 m) by 262 ft (80 m) and the clear span from the central core is approximately 60 ft (18 m). The building is tapered to the top to a dimension of 100 ft (30 m) by 160 ft (49 m), and the clear span reduces to 30 ft (9 m). The floor height is 12 ft 6 in. (3.8 m) in the office sector and 9 ft 4 in. (2.8 m) in the apartment sector. The structural system consists of diagonally braced exterior frames which act together as a tube. The floors are 5 in. (127 mm) thick including lightweight concrete topping and metal deck. The columns, diagonals and ties are I-sections fabricated from three plates with maximum thickness of 6 in. (150 mm). The maximum column dimension is 36 in. (920 mm). Floor framing, fabricated from rolled beams with simple connections, are designed for gravity loading only. The interior columns are designed for gravity loads using rolled and built-up sections. Almost all the steel is ASTM A-36. Connections were shop welded and field bolted except that field welding was used in spandrels, main ties and column splices. The building was completed in 1968 reaching 1127 ft (344 m).

### 3 CityCorp Center, New York

For the CityCorp Center in New York, structural engineer William J. LeMessurier incorporated giant triangular trusses into the exterior façade of the building. These façade trusses collect about half the gravity loads and resist the entire wind loads on the building. The loads collected on the façade are channeled into four massive

**Figure 4.49** (a) John Hancock Tower, Chicago.

columns at the base. Because the shear resistance of the giant trusses is no longer available below the transfer level, a central core is designed to resist wind shears. Diagonal bracing is employed in the plane of the transfer floor to achieve shear load transmission from the tubular frame to the core. The structural system is shown schematically in Fig. 4.50.

## 4.11 The Bundled Tube

The previous sections discussed tubular systems which are generally applicable to prismatic vertical profiles including a variety of nonrectilinear, closed-plan forms, such as circular, hexagonal, triangular, and other polygonal shapes. The most efficient shape is a square, whereas a triangular shape has the least inherent efficiency. The high torsional stiffness characteristic of the exterior tubular system has advantages in structuring unsymmetrical shapes. However, for buildings with significant vertical offsets, the discontinuity in the tubular frame introduces some serious inefficiencies. A bundled tube

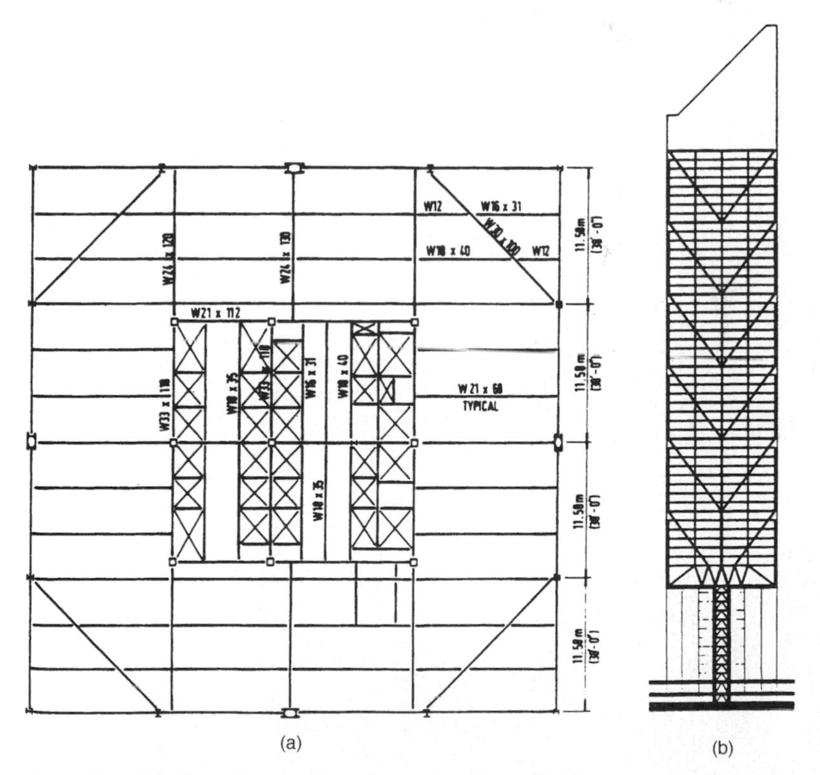


**Figure 4.50** CitiCorp Center (Structural Engineers, Le Messurier consultants Inc.): (a) typical floor framing plan; (b) elevation; (c) lateral bracing system.

configured with many cells, on the other hand has the ability to offer vertical offsets in buildings without loss in efficiency.

It allows for wider column spacings in the tubular walls than would be possible with only an exterior framed tube. It is this spacing which makes it possible to place interior frame lines without seriously compromising interior space planning. In principle, any closed-form shape may be used to create the bundled form (Fig. 4.51).

### 4.11.1 Behavior

The bundled tube structure may be regarded as a set of tubes that are interconnected with common interior panels to form a perforated multicell tube, in which the frames in the lateral load direction resist

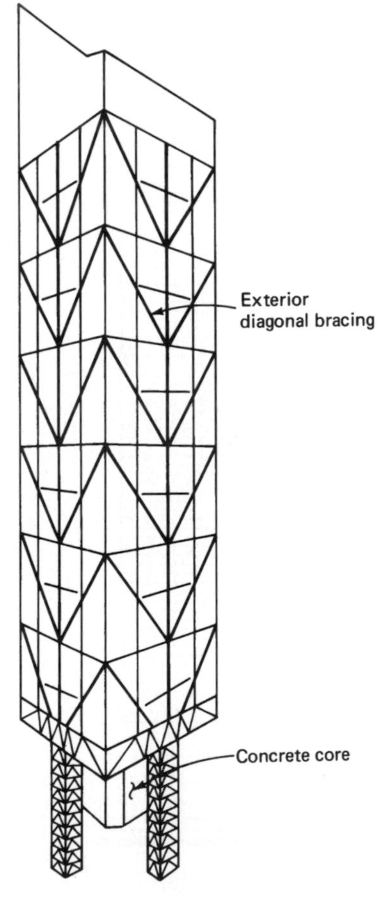


**Figure 4.50** (*Continued*).

the shears, while the flange frames carry most of the overturning moments. The cells can be curtailed at different heights without deminishing structural integrity. The torsional loads are readily resisted by the closed form of the modules. The greater spacing of the columns, and shallower spandrels, permitted by the more efficient bundled tube structure, provides for larger window openings than are allowed in the single-tube structure.

The shear lag experienced by conventional framed tubes is greatly reduced by the addition of interior framed "web" panels across the entire width of the building. When the building is subjected to bending under the action of lateral forces, in high in-plane rigidity of the floor slabs constrains the interior web frames to deflect equally with external-web frames, and the shears carried by each are proportional to their lateral stiffness. Since the end columns of the interior webs are activated directly by the webs, they are more highly stressed than in a single tube where they are activated indirectly by the exterior web through the flange frame spandrels. Consequently, the presence of the interior webs reduces substantially the nonuniformity of column forces caused by shear lag. The vertical stresses in the normal panels are more nearly uniform, and the structural behavior is much closer to the proper tube than the framed tube. Any interior transverse frame panels will act as flanges in a similar manner to the external normal frames.

Because the bundled tube design is derived from the layout of individual tubes, it is possible to achieve a variety of floor configurations by simply terminating a tube at any desired level. In the simple case of two tubes, the corresponding plan shapes are as shown schematically in Fig. 4.52. Figure 4.53 shows the diversity in plans that can be achieved in a three-cell bundled tube. For a structure to

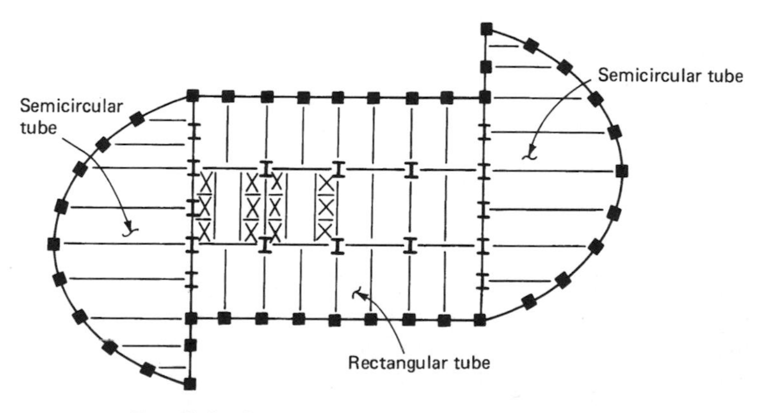


**Figure 4.51** Bundled tube.

behave as a bundled tube it is not necessary that the adjacent tubes be of similar shape. A bundled tube consisting of a square and a triangular cell would respond, in a conceptual sense, in a manner similar to two square or two triangular cells.

A distinct advantage of the bundled tube is that the individual tubes can be assembled in any configuration and terminated at any level without loss of structural integrity. This feature enables creation of setbacks with a variety of shapes and sizes. The disadvantage, however, is that the floors are divided into tight cells by a series of columns that run across the building width.

The structural principle behind the modular concept is that the interior rows of columns and spandrels act as internal webs of a huge tubular cantilever in resisting shear forces, thus minimizing the shear lag effects. Without the beneficial effect of these internal diaphragms, most of the exterior columns in a framed tube toward the center of the building would be of little use in resisting the overturning moment. The modular system can be seen as an extension of the perimeter tubular system with stiffened interior frames in both directions. The interior frames of the cell parallel to the wind resist shear forces, in a manner similar to the two end frames generating peak axial stresses at points of intersection of the

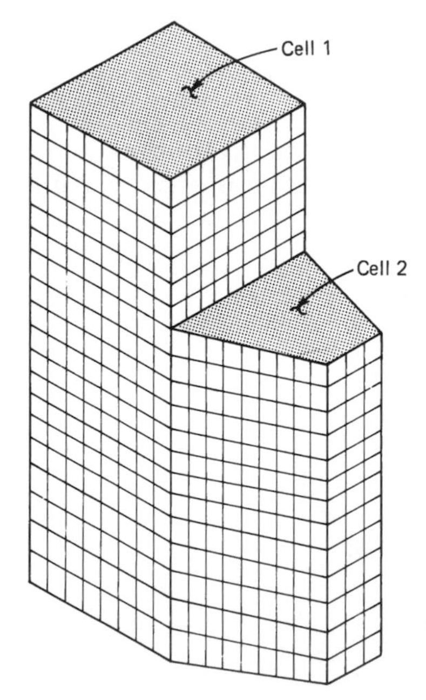


**Figure 4.52** Two-celled tube.

web frames with the flange frames. The internal diaphragms tend to distribute the axial stresses equally along the flange frames by minimizing shear lag effects.

## Examples

### 1 The Sears Tower, Chicago

Reaching to a height of 1454 ft (443 m) above street level, the Sears Tower contains 109 floors and encloses 3.9 million gross sq. ft (362,000 $m^2$) of office space. The basic shape is composed of nine areas 75 ft (22.9 m) square, for an overall floor dimension 225 ft

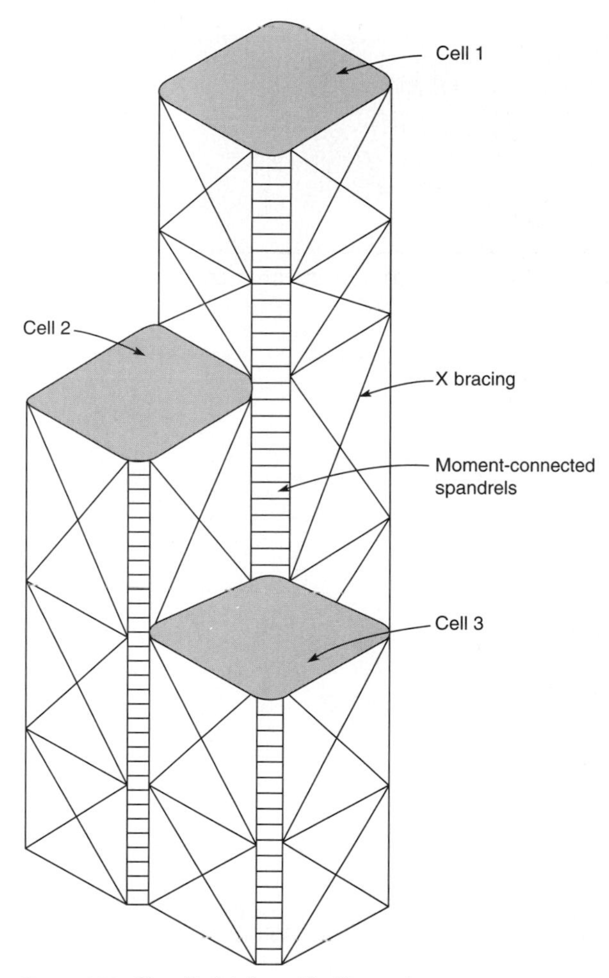


**Figure 4.53** Bundled tube with diagonals.

(68.6 m) square. Two of the nine constituent tubes are terminated at the 50th floor, two more at the 66th and three at the 90th, creating a variety of floor shapes ranging from 41,000 sq. ft (3,800 m$^2$) to 12,000 sq. ft (1,100 m$^2$) in gross area (Fig. 4.54a). The structure acts as a vertical cantilever fixed at the ground in resisting wind loads. The walls of the nine tubes bundled together are composed of columns at 15 ft (4.6 m) on centers and deep beams at each floor. Two adjacent tubes share one set of columns and beams. All beam-column connections are welded. Trussed levels with diagonals between columns are provided at three intermediate mechanical levels, two of them immediately below the setbacks at the 66th and 90th floors.

(a)

**Figure 4.54** (a) Sears Tower, Chicago: (b) Four Allen Center, Houston, plan; (c) building section.

The beams are 42 in, (1.07 m) and columns 39 in. (0.99 m) deep built-up I-sections with the flange width and thickness decreasing with increasing height. Except for column splices, all field connections are bolted. The floors are supported on one-way 40 in (1.0 m) deep trusses spanning 75 ft (23 m) and spaced at 15 ft (4.6 m). Each truss frames directly into a column. The span direction is alternated every six floors to equalize the gravity loads on columns. A floor slab of 2.5 in. (63 mm) lightweight concrete cast on a 3 in. (76 mm) steel deck spans 15 ft (4.6 m) between the trusses.

Beams and columns of built-up I sections are 42 and 39 in. (1070 and 490 mm) depth, respectively. Column flanges vary from 24 by 4 in. (609 by 102 mm) at the bottom to 12 by 2.75 in. (305 by 19 mm) at the top, and beam flanges from 16 by 2.75 in. (16 by 70 mm) to 10 by 1 in. (254 by 25 mm). A total of 76,000 tonnes (69,000 tonnes) of structural steel was used in the project, consisting of grades A588, A572, and A36.

The steel-tube structure was shop-fabricated into units of 2-story-high columns and half-span beams each side, typically weighing 14 tonnes (15 tons). The shop fabrication eliminated 95 percent of field welding. Automated electroslag welding was used for the butt welds of beams to columns. The continuity plates across columns at the joints were fillet-welded by innershield welding process.

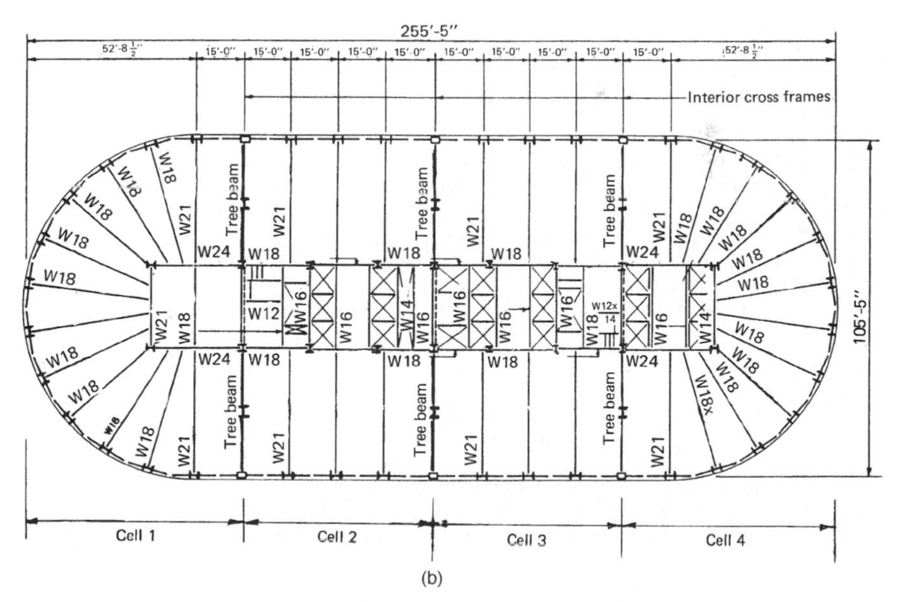


**Figure 4.54** *(Continued).*

Because site storage space was unavailable, the frame units were delivered exactly when needed and lifted off the truck into place. Except for column splices, all field connections were grade A490 high-strength friction-grip bolts in shear connections. The building designed by the Chicago Office of Skidmore, Owings & Merrill was completed in 1974.

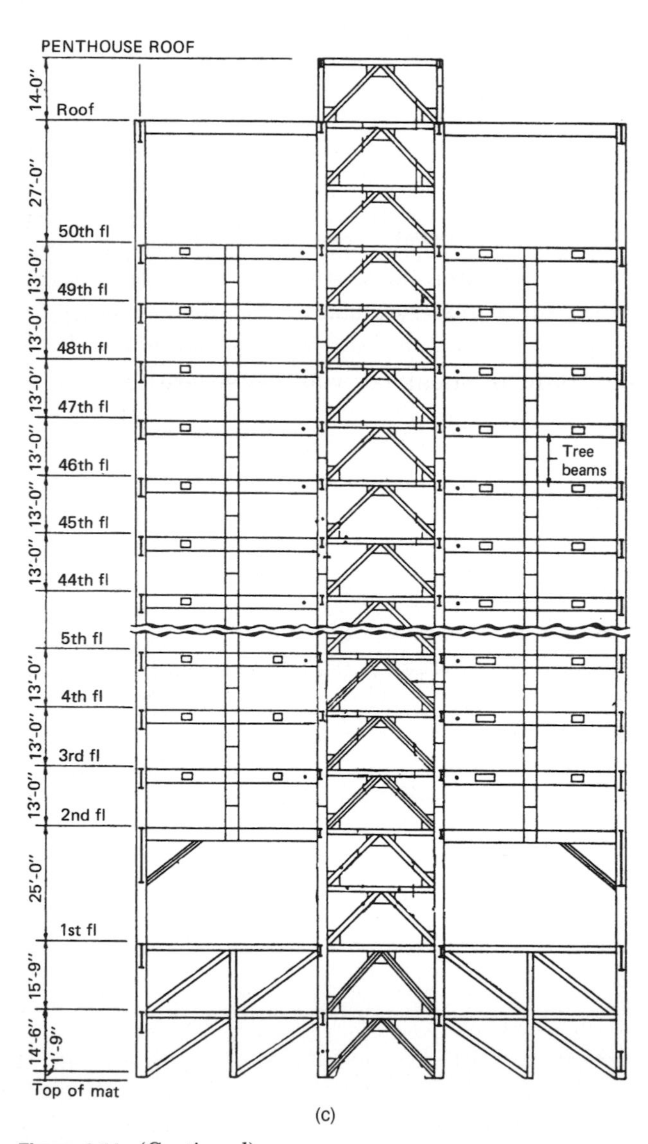


**Figure 4.54** *(Continued).*

### 2 Four Allen Center, Houston, Texas

To achieve the structural action of a bundled tube it is not necessary to have closely spaced columns dividing the building plan into secondary cells. It is possible to achieve an equivalent action, though not as efficiently as that with closely spaced columns, by inserting a minimum number of columns between the windward and leeward faces of the tube to reduce the shear lag. An example of this application can be seen in the 695 ft (211.83 m) office building called Four Allen Center in downtown Houston. In plan, the building is an elongated rectangle with semicircular ends with overall dimensions of approximately 110 by 260 ft (33.5 by 79.25 m) as shown in Fig. 4.54b,c. The tower is remarkably slender, with a height-to-width ratio in excess of 6.0. The elongated shape of the tower creates a striking knife-edged silhouette on the skyline. Although the columns are spaced at a reasonably close interval of 15 ft (4.57 m) along the entire perimeter, which is normally sufficient to achieve a framed tube solution for a building of this height, it was clear from the analysis that a pure framed tube was impractical because of the high plan-aspect ratio. The extent of shear lag was too severe to allow a single perimeter tube solution to be economical. Structural engineers Ellisor & Tanner devised a modified bundled tube by introducing interior cross frames, effectively subdividing the plan into a four-celled grid as shown in Fig. 4.54b. These cross frames were formed by horizontal tree beams interacting with diagonal trusses. A tree-column element consists of short vertical stub columns attached to the girders at middistance between the exterior and core columns. This in effect creates a vierendeel truss action between the core and exterior columns, greatly increasing the shear stiffness of the system. Connections between the stub columns are designed to carry no axial forces and therefore the columns are not required to go all the way to the foundation. A schematic elevation of the framing is shown in Fig. 4.54c.

## 4.12 Ultimate High-Efficiency Structures

It is interesting to recall that there have been no quantum leaps in the height of tall buildings that are being built today as compared to the buildings built during the 1930s and 1940s. For example, the Empire State Building scraped the sky at 1250 ft (381 m), which is not a small feat by any stretch of the imagination. The success of this awe-inspiring building and other similar bulidings of the era is credited to the holding-down power of the heavy exterior stone cladding and the masonry partitions that were in vogue during that period. Modern high-rise technology has largely replaced the heavy cladding and interior masonry systems with relatively lightweight counterparts. The holding-down power of these systems is no longer present in modern construction. Let us examine how we can employ the relatively lightweight materials to help in providing resistance to the lateral loads. The tube system, with its characteristic deployment of the columns at the building perimeter, certainly provides the much-required separation between the windward and leeward faces of the building for resisting the overturning moments. However, since the exterior columns, especially in a framed tube, are placed relatively close to each other, their tributary areas for collection of gravity loads are rather small. Therefore, the beneficial effect of gravity load in counteracting the tensile forces of the columns is somewhat limited, first because of the relatively light materials used in current construction practice, and second by the limited tributary area for the exterior columns. If somehow we could induce more gravity loads into these columns, would not the efficiency of the system be improved? This is precisely the idea behind the ultimate high-efficiency structures first envisioned by the master builder Dr Fazlur Khan. The main premise behind the ultimate high-efficiency structure is to transfer as much gravity load as practicable to the columns resisting the overturning moments. Note that this idea is routinely used by engineers when they are confronted with high uplift forces while designing interior-core-braced buildings with limited separation between the leeward and windward columns. They would normally rearrange the floor framing to achieve more flow of gravity loads into the wind columns or may choose to eliminate certain interior columns altogether in order to collect more gravity loads on the wind columns. A similar approach is possible in a tube building—eliminate as many interior columns as possible, perhaps all the interior columns. This way the holding-down power of gravity loads is put to use in the most efficient manner. It must be recognized that there is a certain amount of trade-off in the floor framing system because it is economically prohibitive to clear-span the floor members using traditional approaches. Accommodation has to be made to achieve the transfer of entire building load into the exterior columns without paying a significant premium. This is considered next.

Consider a tube building with closely spaced exterior columns and deep spandrels. Imagine that all the interior columns are eliminated completely, leaving a column-free volume inside. Within this basic configuration it is possible to provide a system of transfer floor trusses at approximately every 15th floor, corresponding to the levels at which the low, low-mid, high-mid, and high-rise elevators terminate. The trusses could be designed as one- to two-story-deep vierendeel trusses spanning in two directions for the full width and length of the building. Where appropriate, the transfer levels could be made into skylobbies or other forms of common areas. The trusses then can support interior columns within the zones between two trusses. Any type of conventional structural steel framing such as composite rolled beams, haunch girders, or stub girders can be employed to span the distance from the core columns and the exterior of the building. Since the interior columns carry gravity loads from a limited number of floors, their sizes will be substantially smaller than in a conventional system in which they would rise from foundation level to the building top. Another advantage is that the columns within any particular zone bounded by two transfer levels may be located at will to suit the interior space planning desired for specific occupancies. This in itself may be reason enough to consider this system because of the tremendous flexibility offered in the layout of columns for mixed development use. For instance, in the space planned for office use, the interior columns may be located at 60 ft (18.28 m), and so on. For major tenants willing to lease the full block of floors between two transfer levels, columns can be arranged to suit their particular needs. Column transfers within the zones can be achieved with little structural difficulty because: (i) the loads to be transferred are relatively small; and (ii) it is possible to hang the upper six floors or so in one zone from the transfer girder, creating opportunities to have some floors entirely column free. In fact, the floor framing options are limited only by imagination of the designer. The principal structural advantage is, of course, that total dead and live load from every floor is transferred only to the exterior columns, thereby increasing the structural capacity of the system to withstand lateral loading. Undoubtedly there would be some premium in the tonnage of steel for the trusses and their associated fabrication and erection costs. However, the premium spread over the total square footage of the building is likely to be small.

It is, of course, necessary to tie the windward and the leeward columns of the tube with a structural system capable of resisting the shear forces caused by the lateral loads. This can be achieved with a system of deep spandrel beams when the perimeter columns are closely spaced, or with a system of diagonal bracing when the columns are spaced apart as in a trussed tube system. Dr Fazlur Khan has shown that by progressively shifting the exterior columns to the corners of a rectangular trussed tube, the efficiency of the system can be greatly improved. An ultimate structure for a rectangular building, then, will have just four corner columns interconnected by massive diagonals as shown in Fig. 4.55.

The efficiency of a building to resist lateral loads can be increased

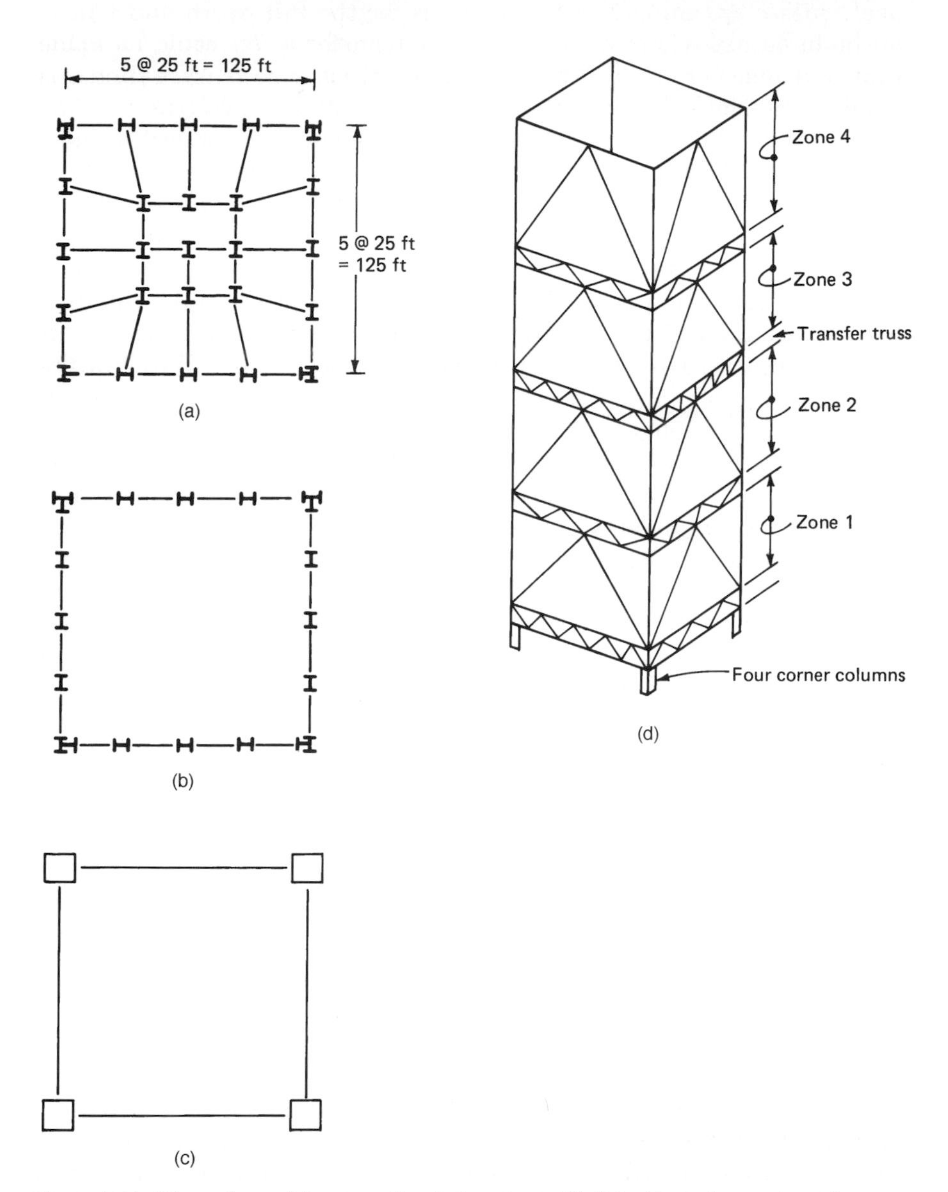


**Figure 4.55** Floor plans: (a) conventional framing with interior and exterior columns extending to the foundation; (b) exterior columns extend to foundation; (c) interior columns are supported by transfer trusses; (d) Khan's concept of the ultimate high rise building.

further by using interior bracing with a structural system in which the total gravity load of the building is made to bear on limited number of exterior columns. To increase the uplift capacity, interior columns are completely eliminated within the building envelope (Figs 4.55a–d). However, to achieve an economical floor system, it is necessary to use columns in the core area without unduly interfering with the leasing requirements. Therefore, a structural system is needed for transferring the loads from the interior columns to the building envelope. If this system can simultaneously work as a shear-resisting element, then the need for closely spaced columns or diagonal bracing on the perimeter can be eliminated. In other words a system of interior bracing that performs the dual function of channeling the loads from the interior columns to the exterior while at the same time functions as a shear element between the windward and leeward columns is likely to be the optimum system. From a pure structural point of view, such a system is highly desirable. Examples of buildings using this concept are given in Chap. 6.

Chapter

# 5

# Lateral Bracing Systems for Concrete Buildings

Analogous to steel or composite construction, concrete offers a wide range of structural systems suitable for high-rise buildings. There are perhaps as many structural concepts as there are engineers, making it awkward if not impossible to classify all the concepts into distinct categories. However, for purposes of presentation, it is convenient to group the most common systems into separate categoreis, each with an applicable height range as shown in Fig. 5.1. Although the height range for each group is logical for normally proportioned buildings, the appropriateness of each system to a particular building can only be judged when all other factors influencing the lateral load behaviour are taken into account. Such factors include building geometry, severity of exposure to wind, seismicity of the region, ductility of the frame, and limits imposed on the size of the structural members. Oftentimes, systems combining the characteristics of two or more can be employed to fulfill the specific project requirement. The multitude of systems available presents an opportunity for an experienced engineer to come up with a structural system that will serve its optimum function in the overall sense of the project. Although the selection of a system requires knowledge of both horizontal and lateral systems, the material presented in this chapter emphasizes the requirements of lateral systems only. Gravity systems are covered separately in Chap. 8.

Figure 5.1 shows 14 different categories of structural systems, starting with the most elementary system consisting of floor slabs and columns. At the other end of the spectrum is the bundled tube system, which is appropriate for very tall buildings and for buildings with a large plan aspect ratio. Almost all of the systems described in

Chap. 4, and shortly in Chap. 6, are equally applicable to concrete buildings. Therefore, proper selection can only be made when the engineer has become familiar with the systems described not only in this chapter, but elsewhere throughout this work.

## 5.1 Frame Action of Column and Two-way Slab Systems

Concrete floors in tall buildings often consist of a two-way floor system such as a flat plate, flat slab, or a waffle system. In a flat-plate system the floor consists of a concrete slab of uniform thickness which frames directly into columns. The flat slab system makes use of either column capitals, drop panels or both to increase the shear and moment resistance of the system at the columns where the shears and moments are greatest. The waffle slab consists of two rows of joists at right angles to each other; commonly formed by using square domes. The domes are omitted around the columns to increase the moment and shear capacity of the slab. Any of the three systems can be used to function as an integral part of the wind-resisting systems for buildings in the 10-story range.

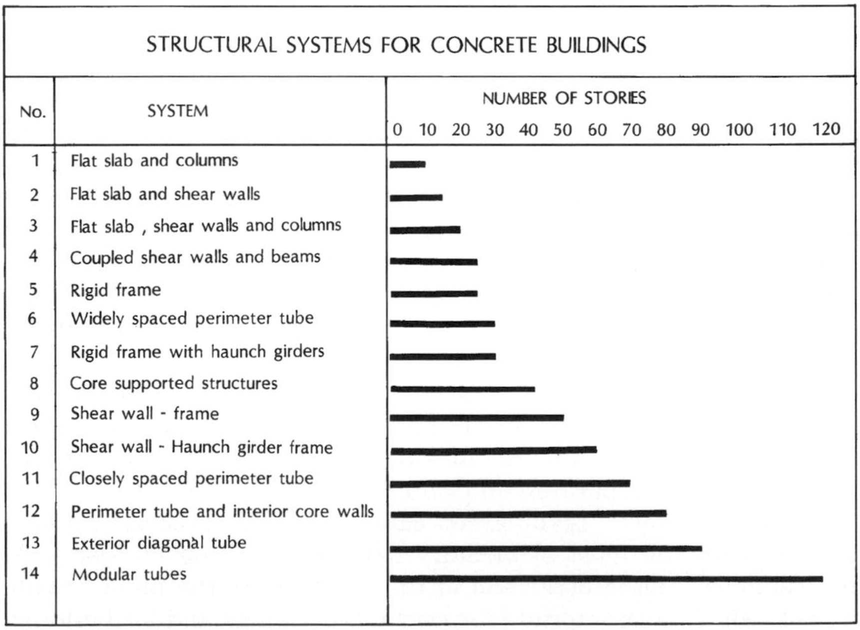

STRUCTURAL SYSTEMS FOR CONCRETE BUILDINGS

| No. | SYSTEM | NUMBER OF STORIES<br>0 10 20 30 40 50 60 70 80 90 100 110 120 |
|---|---|---|
| 1 | Flat slab and columns | |
| 2 | Flat slab and shear walls | |
| 3 | Flat slab , shear walls and columns | |
| 4 | Coupled shear walls and beams | |
| 5 | Rigid frame | |
| 6 | Widely spaced perimeter tube | |
| 7 | Rigid frame with haunch girders | |
| 8 | Core supported structures | |
| 9 | Shear wall - frame | |
| 10 | Shear wall - Haunch girder frame | |
| 11 | Closely spaced perimeter tube | |
| 12 | Perimeter tube and interior core walls | |
| 13 | Exterior diagonal tube | |
| 14 | Modular tubes | |

**Figure 5.1** Structural systems for concrete buildings.

The slab system shown in Fig. 5.2 has two distinct actions in resisting lateral loads. First, because of its high in-plane stiffness, it distributes the lateral loads to various vertical elements in proportion to their bending stiffness. Second, because of its significant out-of-plane stiffness, it restrains the vertical displacements and rotations of the columns as if they were interconnected by a shallow wide beam.

The concept of “effective width” as explained below can be used to determine the equivalent width of slab. Although physically no beam exists between the columns, for analytical purposes it is convenient to consider a certain width of slab as a beam framing between the columns. The effective width is however, dependent on various parameters, such as column aspect ratios, distance between the columns, thickness of the slab, etc. Research has shown that values less than, equal to, and greater than full width are all valid depending upon the parameters mentioned above.

Note that the American Concrete Institute, ACI code permits the full width of slab between adjacent panel center lines for both gravity and lateral loads. The only stipulation is that the two-way systems analysis should take into account the effect of slab cracking in evaluating the stiffness of frame members. Use of full width is explicit for gravity analysis, and implicit (because it is not specifically prohibited) for the lateral loads.

However, engineers generally agree that using full width of the slab gives unconservative results for lateral load analysis. The method tends to overestimate the slab stiffness and underestimate the column stiffness, compounding the error in estimating the distribution of moments due to lateral loads.

The shortcomings of using full width in lateral load analysis can be overcome by determining the equivalent stiffness on the basis of a finite-element analysis. The stiffness thus obtained is appropriate for both gravity and lateral analysis.

Of particular concern in the detailing of two-way systems is the problem of stress concentration at the column-slab joint where, especially under lateral load, nonlinear behavior is initiated through concrete cracking and steel yielding. Shear reinforcement at the column-slab joint is necessary to improve the joint behavior and avoid early stiffness deterioration under lateral cyclic loading. Note that two-way slab systems without beams are not permitted by the ACI code in regions of high seismic risk (zones 3 and 4). Their use in regions of moderate seismic risk (zone 2) is permitted subject to certain requirements, mainly relating to reinforcement placement in the column strip. Since the requirements are too detailed to

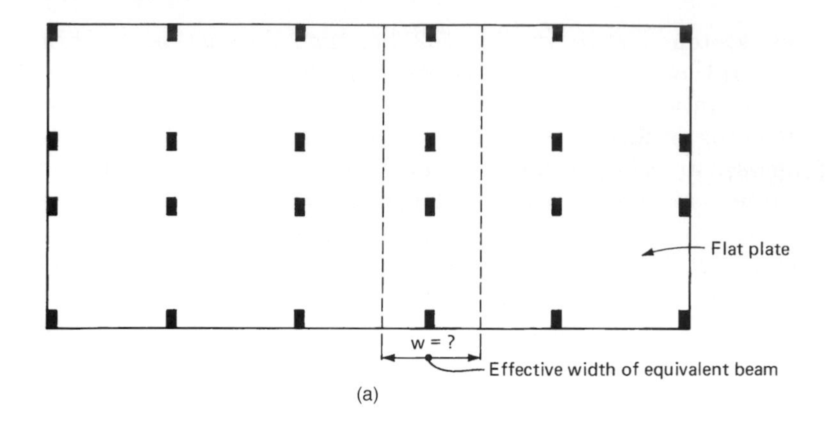


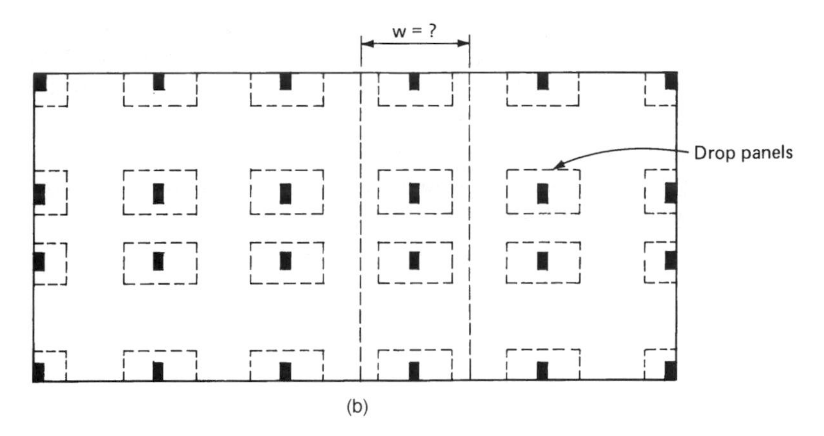


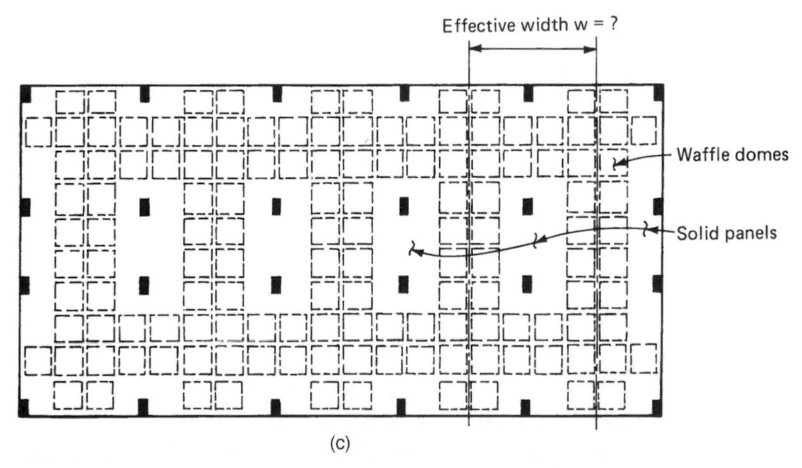


**Figure 5.2** Lateral systems using slab and columns: (a) flat plate; (b) flat slab with drop panels; (c) two-way waffle system.

be the reader is referred to the UBC and ACI codes for further details.

## 5.2 Flat Slab and Shear Walls

Frame action obtained by the interaction of slabs and columns is not adequate to give the required lateral stiffness for buildings taller than about 10-stories. The advantages of beamless flat ceilings could still be achieved in taller buildings by strategically locating shear walls to function as the main lateral-load-resisting element. The walls can be either planar, open, closed, or any combination of the three. Figure 5.3 shows an example in which planar shear walls are located in a simple orthogonal orientation. Skewed or irregular layouts require a three-dimensional analysis to include the effect of torsional loads. Planar shear walls in essence behave as slender cantilevers. When no major openings are present, stresses and deflections can be determined using simple bending theory. Complicated open-section shapes need special modelling and analyses techniques as outlined in Chap. 10.

## 5.3 Flat Slab, Shear Walls, and Columns

The applicable height range of slab and shear wall systems can be increased marginally by including the frame action of column and slabs. The system is best suited for apartments, condominiums, and hotels, and is identical to the system described in the previous section. The difference is only in the analysis in that the frame action of column and slabs is also taken into account in the lateral load

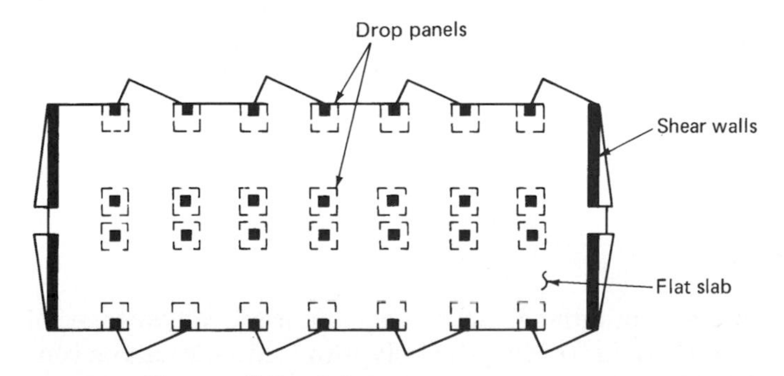


**Figure 5.3** Shear wall flat slab system.

analysis. Whether this action is significant or not is a function of relative stiffness of various elements. In most apartment or hotel layouts, the frame resistance to overturning moments is no more than 10 to 20 percent of the resistance offered by the shear walls. Many engineers, therefore, ignore the frame action altogether by designing the shear walls to carry the total lateral loads. However, in keeping with the current trend of taking advantage of all available structural actions, it is advisable to include the frame action in the analysis.

## 5.4 Coupled Shear Walls

When two or more shear walls are interconnected by a system of beams or slabs, the total stiffness of the system exceeds the summation of the individual wall stiffnesses because, the connecting slab or beam restrains the individual cantilever action by forcing the system to work as composite unit.

Such an interacting shear wall system can be used economically to resist lateral loads in buildings up to about 40-stories. However, planar shear walls are efficient lateral load carriers only in thier plane. Therefore, it is necessary to provide walls in two orthogonal directions. However, in long and narrow buildings sometimes it may be possible to resist wind loads in the long direction by the frame action of columns and slabs because first, the area of the building exposed to the wind is small, and second, the number of columns available for frame action in this direction is usually large. The layout of walls and columns should take into consideration the torsional effects.

Walls around elevators, stairs, and utility shafts offer an excellent means of resisting both lateral and gravity loads without requiring undue compromises in the leasability of buildings. Closed- and partially closed-section shear walls are efficient in resisting torsion, bending moments, and shear forces in all directions, especially when sufficient strength and stiffness are provided around door openings and other penetrations. This is discussed in greater detail in Chap. 10.

## 5.5 Rigid Frame

Cast-in-place concrete buildings have the inherent advantage of continuity at joints. Girders framing directly into columns, can be considered rigid with the columns; such a girder-column arrangement can be thought of as a portal frame. However, girders that carry shear and bending moments due to lateral loads often require additional construction depth, necessitating increases in the overall height of the building.

The design and detailing of joints where girders frame into building columns should be given particular attention, especially when buildings are designed to resist seismic forces. The column region within the depth of the girder is subjected to large shear forces. Horizontal ties must be included to avoid uncontrolled diagonal cracking and disintegration of concrete. Specific detailing provisions are given in the UBC and ACI codes to promote ductile behavior in high-risk seismic zones 3 and 4, and somewhat less stringent requirements in moderate-risk seismic zone 2. The underlying philosophy is to design a system that can respond to overloads without loss in gravity-load carrying capacity.

Rigid-frame systems for resisting lateral and vertical loads have long been accepted as a standard means of designing buildings because they make use of the stiffness in the beams and columns that are required in any case to carry the gravity loads. In general, rigid frames are not as stiff as shear wall construction and are considered more ductile and less susceptible to catastrophic earthquake failures when compared to shear wall structures.

As discussed in Chap. 4, a rigid frame is characterized by its flexibility due to flexure of individual beams and columns and rotation at their joints. The strength and stiffness of the frame is proportional to the beam and column size and inversely proportional to the column spacing. Internally located frames are not very popular in tall buildings because the leasing requirements of most buildings limit the number of interior columns available for frame action. The floor beams are generally of long spans and are of limited depth. However, frames located at the building exterior do not necessarily have these disadvantages. An efficient frame action can thus be developed by providing closely spaced columns and deep beams at the building exterior.

## 5.6 Widely Spaced Perimeter Tube

The term tube, in the usual building terminology suggests a system of closely spaced columns 8 to 15 ft on center (2.43 to 4.57 m) tied together with a relatively deep spandrel. However, for buildings with compact plans it is possible to achieve tube action with relatively widely spaced columns interconnected with deep spandrels. An example of such a layout for a 28-story building is shown in Fig. 5.4. Lateral resistance is provided by the perimeter frame, consisting of 5 ft (1.5 m) wide columns, spaced at 25 ft (7.62 m) centers, and tied to a 5 ft (1.53 m) deep spandrel.

## 5.7 Rigid Frame with Haunch Girders

One of the drawbacks of a rigid frame system is the excessive depth of girder required to make the rigid frame economical. Rigid frames, when located at the perimeter, often can use deep spandrels without adversely affecting the floor-to-floor height of the building. Also it may be architecturally acceptable to use relatively closely spaced columns at the exterior.

Office buildings usually are laid out with a lease depth of about 40 ft (12.19 m) without any interior columns between the core and the exterior. To economically frame such a large span would require a girder depth of about 3 ft (0.91 m) unless the beams are post-tensioned. This requirement clearly impacts the floor-to-floor height and often is unacceptable because of additional cost of curtain wall

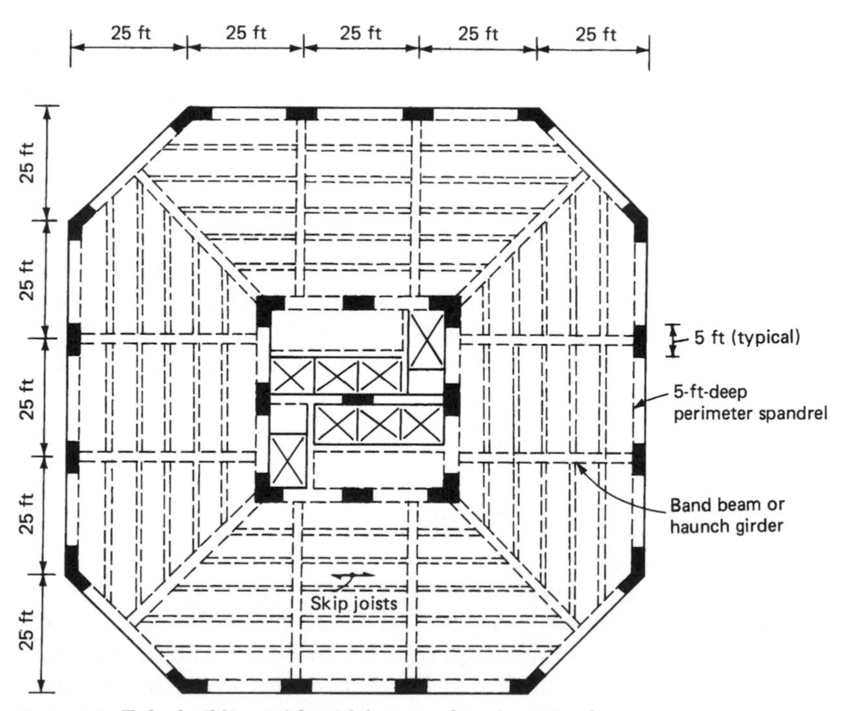


**Figure 5.4** Tube building with widely spaced perimeter columns.

and the extra heating and cooling loads due to increased volume of the building. A haunch girder, which consists of a girder of variable depth, gives the required stiffness for lateral loads without having to increase the floor-to-floor height. This is achieved by making the mid-section of the girder flush with the floor system, thus providing ample beamless space for passage of mechanical ducts.

Girders with haunches on either end are ideal for resisting lateral loads, but certain types of column layouts may limit the haunches to one end of the girder. Such an arrangement with single-end haunch girders is shown for a 28-story building in Fig. 5.5. To keep the skip-joist framing simple, the girders are framed into the spandrel without haunches at the exterior end. Girders that match haunch girder depth are used between the core columns, as shown in Fig. 5.5. The depth of these girders did not affect the mechanical distribution and therefore did not require increase in floor-to-floor height. A 9 ft (2.74 m) high ceiling was accomplished with a 12 ft 10 in (3.91 m) floor-to-floor height.

## 5.8 Core-supported Structures

One of the most frequent uses of shear walls is in the form of box-shaped cores around stairs and elevators, because this arrangement makes structural use of vertical enclosures required around the cores. Arrangement of internal cores is especially suitable for office buildings because it frees the lease space outside of the core from massive vertical elements. The walls around the core can be considered as a spatial system capable of transmitting lateral loads in both directions. Additional advantage of core structures is that being spatial structures, they have the ability to resist all types of loads: vertical loads, shear forces and bending moments in two directions, as well as torsion, especially when adequate stiffness and strength are provided between flanges of open sections. The shape of the core to a large extent is governed by the elevator and stair requirements. Variations could occur from a single rectangular core to complicated arrangements of planar shear walls. Other structural elements surrounding the core may consist of either a cast-in-place or precast concrete or structural steel construction. In a precast or steel surround system, it is more than likely that the stiffness of the core will overwhelm the stiffness of other vertical elements. Even in a cast-in-place system, unless the exterior frame consists of relatively closely spaced columns and a deep spandrel, it is justifiable to ignore the resistance of other vertical members and to design the core system for the entire lateral load. Figures 5.6a–c show some examples of core arrangements.

## 5.9 Shear Wall-Frame Interaction

Without question this system is one of the most, if not the most, popular system for resisting lateral loads. The system has a broad range of application and has been used for buildings as low as 10-stories to as high as 50-story or even taller buildings. With the advent of haunch girders, the applicability of the system is easily extended to buildings in the 70- to 80-story range.

The interaction of frame and shear walls has been understood for quite some time; the classical mode of the interaction between a prismatic shear wall and a moment frame is shown in Fig. 5.7; the frame basically deflects in a so-called shear mode while the shear wall predominantly responds by bending as a cantilever. Compatibility of horizontal deflection produces interaction between the two. The linear sway of the moment frame, when combined with the parabolic sway of the shear wall results in an enhanced stiffness because the wall is restrained by the frame at the upper levels while at the lower levels the shear wall is restrained by the frame. However, it is not always easy to differentiate between the two modes because a frame consisting of closely spaced columns and deep

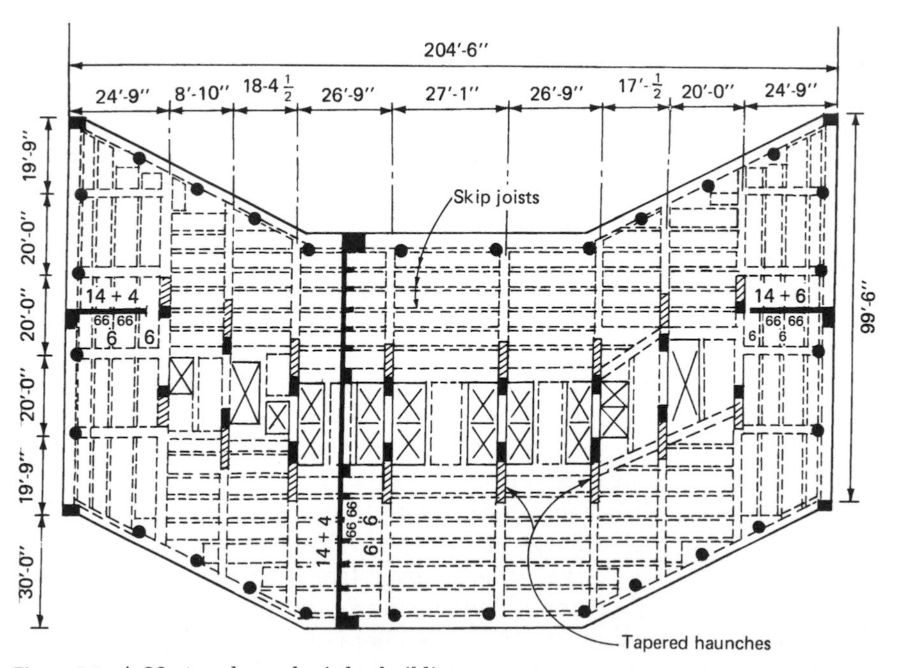


Figure 5.5 A 28-story haunch girder building.

beams tends to behave more like a shear wall responding predominantly in a bending mode. And similarly, a shear wall weakened by large openings may tend to act more like a frame by deflecting in a shear mode. The combined structural action, therefore, depends on the relative rigidity of the two, and their modes of deformation. Furthermore, the simple interaction diagram given in Fig. 5.7 is valid only if

- the shear wall and frame have constant stiffness throughout the height or;
- if stiffnesses vary, the relative stiffness of the wall and frame remains unchanged throughout the height.

Since architectural and other functional requirements frequently influence the configuration of structural elements, the above conditions are rarely met in a practical building. In a contemporary

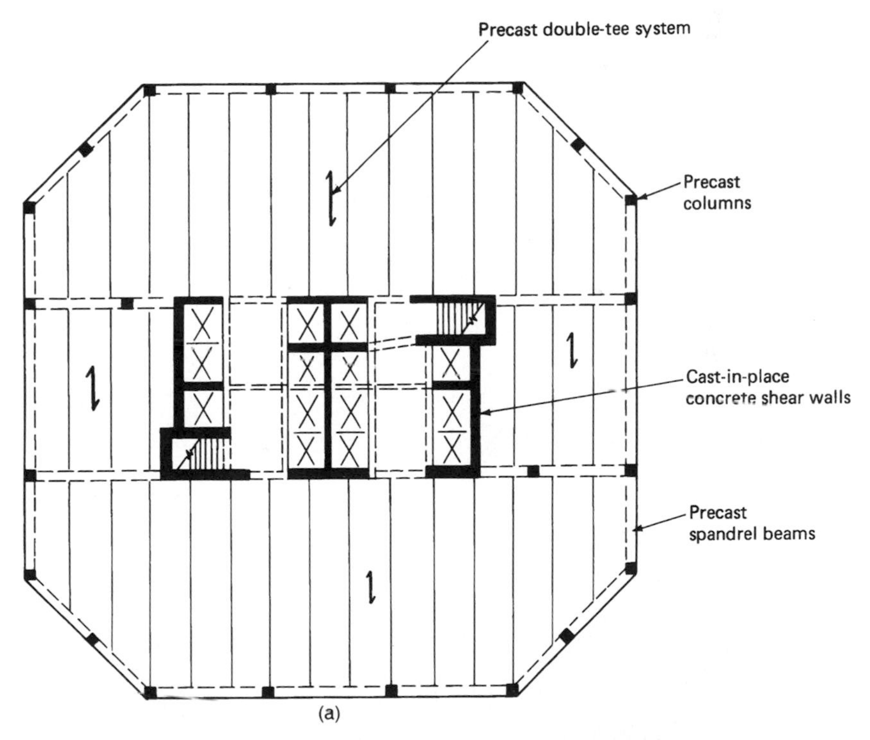


**Figure 5.6** Examples of shear core buildings: (a) cast-in-place shear walls with precast surround; (b) shear walls with post-tensioned flat plate; (c) shear walls with one-way joist system.

high-rise building, very rarely can the geometry of walls and frames be the same over the full height. For example, walls around the elevators are routinely stopped at levels corresponding to the elevator dropoffs, columns are made smaller as they go up, and the building geometry is very rarely the same for the full height. Because of the abrupt changes in the stiffness of walls and frames combined with the variation in the geometry of the building, the simple interaction shown in Fig. 5.7 does not even come close to predicting the actual behavior of the building structures. However, with the availability of two- and three-dimensional computer programs,

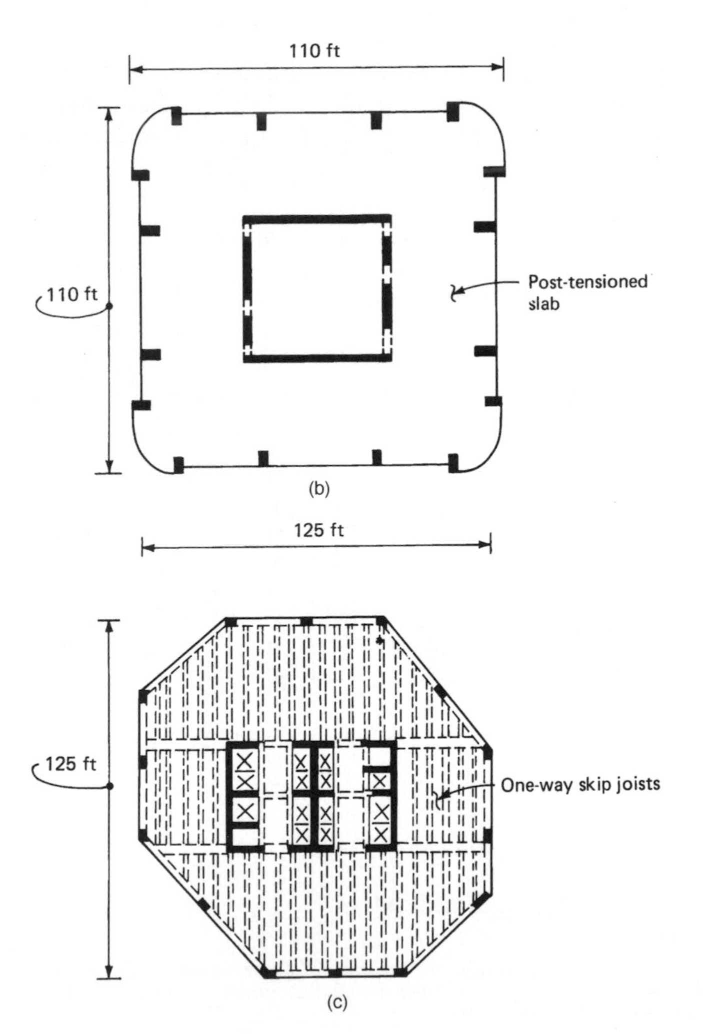


**Figure 5.6** *(Continued).*

capturing the essential behavior of the shear wall frame system is within the reach of everyday engineering practice.

To understand qualitatively the nature of interaction between shear walls and frames, consider the framing plan shown in Fig. 5.8a. The building is unusual in that it exhibits almost perfect symmetry in two directions and maintains a reasonably constant stiffness throughout its height. Therefore, in a qualitative sense, interaction between the frames and shear walls is expected to be similar to that shown in Fig. 5.7.

The building is 25-stories and consists of four levels of basement below grade. The floor framing consists of 6 in. wide by 20 in. deep (152.4 by 508 mm) skip joist framing between haunch girders which span the distance of 35 ft 6 in. (10.82 m) between the shear walls and the exterior of the building. The girders are 42 in. wide by 20 in. deep (1.06 by 0.5 m) for the exterior 28 ft 6 in. (8.67 m) length, with a haunch at the interior tapering from a pan depth of 20 to 33 in. (0.5 to 0.84 m). Four shear walls of dimensions 1 ft 6 in. by 19 ft 6 in. (0.45 by 5.96 m) rise for the full height from a 5 ft (1.52 m) deep mat foundation. The exterior columns vary from 38 by 34 in. (965 by 864 mm) at the bottom to 38 by 24 in. (965 by 610 mm) at the top. Note that the haunch girder is made deliberately wider than the exterior column to simplify removal of flying forms. However, by using "hinges", it is also possible to simplify removal of formwork. Therefore, width of haunch girders can be consistent with the structural requirements; they need not be wider than exterior columns.

The lateral load resistance in the short direction of the building is provided by a combination of three types of frames: (i) two exterior

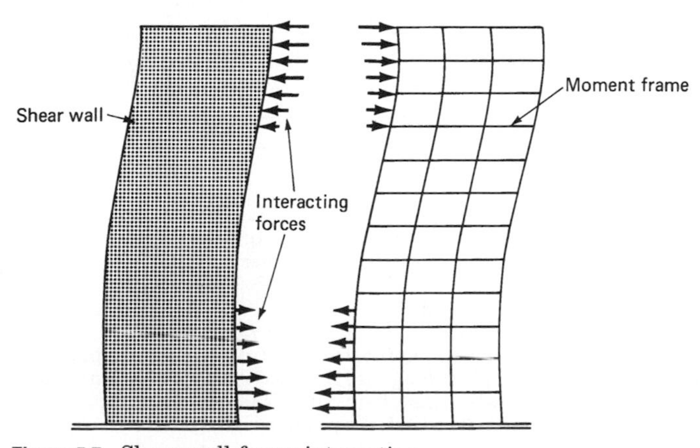


**Figure 5.7** Shear wall-frame interaction.

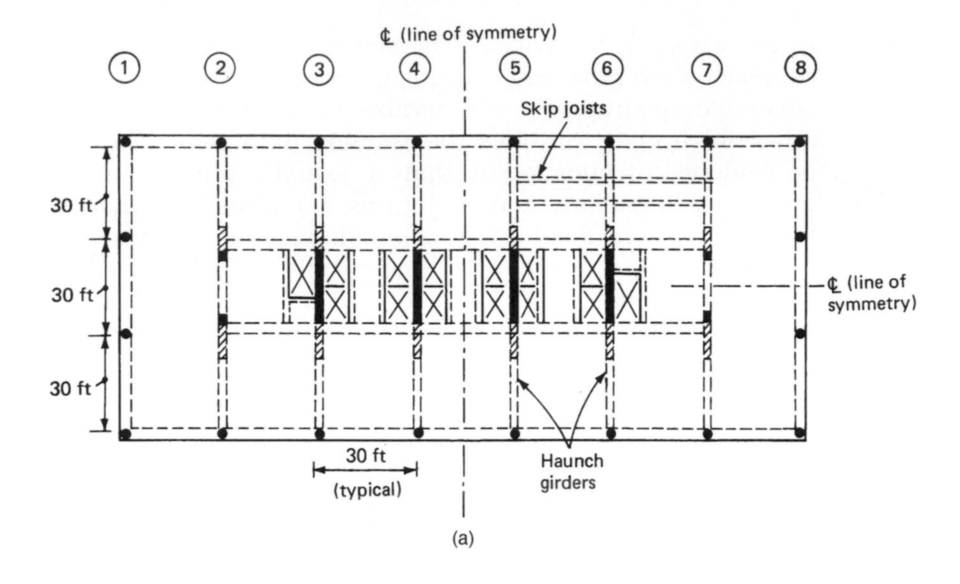


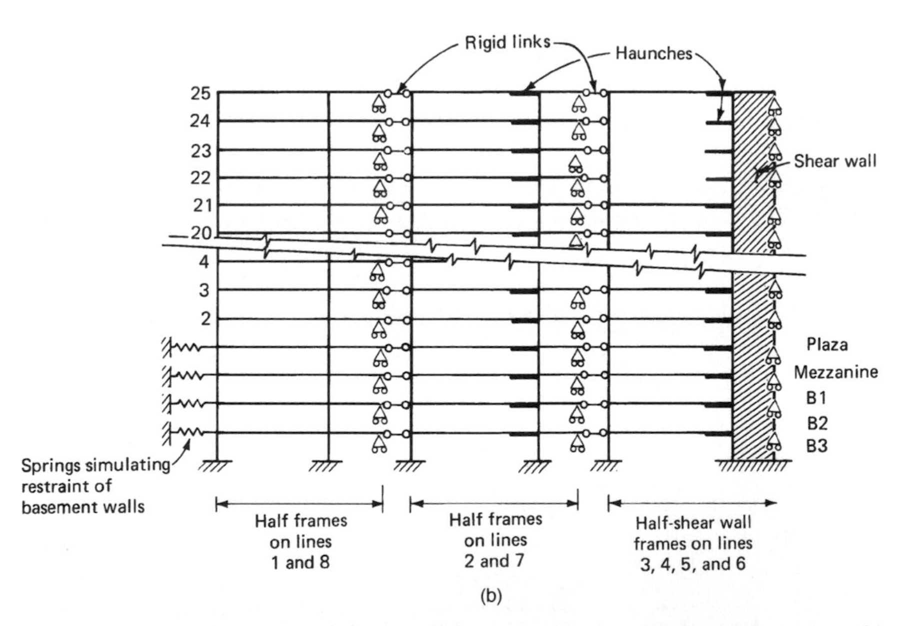


**Figure 5.8** Practical example of shear wall-frame interaction: (a) typical floor plan; (b) analytical two-dimensional model for wind on long face; (c) distribution of lateral loads, (i) applied lateral loads, (ii) lateral loads resisted by end frames, (iii) lateral loads resisted by interior frames, (iv) lateral loads resisted by shear wall frames.

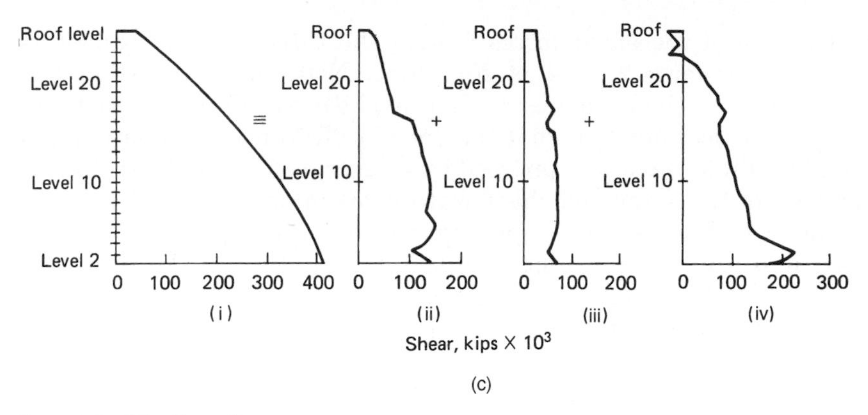


**Figure 5.8** *(Continued).*

frames along grids 1 and 8; (ii) two haunch girder frames along grids 2 and 7; and (iii) four shear wall-haunch girder frames along grids 3 through 6. The lateral load resistance in the long direction is provided primarily by frame action of the exterior columns and spandrels along the broad faces.

For purposes of structural analysis, the building is considered symmetrical about the two centerlines as shown in Fig. 5.8a. The lateral load analysis can be carried out by lumping together similar frames and using only one-half the building in the computer model. This is shown in Fig. 5.8a for a two-dimensional computer model for analysis of wind on the broad face. In the model only three equivalent frames are used to represent the lateral load resistance of eight frames, and only one-half of each frame is used to simulate the structural action of a full frame. The latter simplification is achieved by restraining the vertical displacement at the end of each frame as shown in Fig. 5.8b. Note spring restraints at the basement levels B1, B2, B3, and at the plaza are used to simulate the lateral restraint of the basement walls and soil structure interaction. The rigid links shown in Fig. 5.8b between the individual frames simulate the diaphragm action of the floor slab by maintaining the lateral displacements of each frame the same at each level.

The purpose of the example is to show qualitatively the nature of interaction that exists between shear walls and frames in a building without abrupt changes in the stiffness of either element. We will not burden the reader with an avalanche of computer results, but will limit the presentation to the distribution of horizontal shear in various frames. Part 1 of Fig. 5.8c shows the diagram of cumulative shear forces applied along the broad face of the building. The distribution of the shear forces among the three types of frames is indicated in parts 2 to 4 of Fig. 5.8c. Note the reversal in the direction of the shear force at the top of shear wall-frame in part 4 of Fig. 5.8c, which indicates that the shear wall-frame combination has a tendency to behave as a propped cantilever, not unlike the behavior observed in the simplified shear wall-frame interaction model shown in Fig. 5.7. This is not surprising because although the practical example consists of a combination of three different types of frames, in essence the structural system can be considered as a single shear wall acting in combination with a frame. There are no sudden changes in the stiffness of walls or frames along the building height contributing to its departure from the fundamental behavior.

The example shown in Fig. 5.8 (although taken from an actual building in Houston, Texas) has very little in common with the usual types of structural systems that structural engineers are called upon to design. It is a rare event indeed to be commissioned to design a building that is symmetrical and has no significant structural discontinuity over its height. More usually, there is asymmetry because either some shear walls or frames drop off, or the building shape is architecturally modulated. The reality of present-day architecture precludes generalization of even qualitative comments regarding the interaction between shear walls and frames. The structural engineer has very little choice, other than using computer analysis for determining the distribution of loads to various elements. Standard interaction diagrams that were helpful in assigning the lateral loads to simple shear wall and frame systems have limited application and are therefore generally not used in practice.

As a second example, let us consider a contemporary high-rise building which has asymmetrical floor plans and abrupt variation in stiffness of shear walls and frames throughout its height. Figure 5.9 shows a perspective of a twin-tower high-rise office development called The Lone Star Towers, scheduled for construction in Dallas, Texas. The first phase of the complex consists of a four-level, 1600-car below-grade parking structure and a 1,000,000 $\text{ft}^2$ (92,903 $\text{m}^2$) office space. The building is 655 ft tall (200 m) and consists of 50-stories above grade. A variety of floor plans are accommodated between the second and the roof levels, resulting in a number of setbacks as shown in Fig. 5.10, with a major transfer of columns at level 40 (Fig. 5.10d). The floor plan, which is essentially rectangular at the second floor, progressively transforms into a circular shape at the upper levels. Figure 5.10a–f shows the framing plans at various levels. The resistance to lateral load is provided by a system of I- and C-shaped shear walls interacting with haunch girder frames. The floor framing is of lightweight concrete consisting of

6 in. (152.4 mm) wide skip joists at 6 ft 6 in. (1.98 m) centers. The 20 in. (508 mm) depth of haunch girders at midspan matches that of pan joist construction. Tapered haunches are used at both ends of girders and vary from a depth of 20 in. (508 mm) at midbay to a depth of 2 ft 9 in. (0.84 m) at the face of columns and shear walls. High-strength normal-weight concrete of up to 10 ksi (68.95 mPa) is used in the design of columns and shear walls. The shear walls around low, mid, mid-low, mid-high, and high-rise elevators are either terminated or made smaller in their dimensions at various levels corresponding to the zoning of elevators. The analytical model necessarily resulted in a system of shear walls and frames which were asymmetrical not only with respect to the plan dimensions but also varied in stiffness over the building height. The final lateral load analysis was accomplished by using a three-dimensional computer

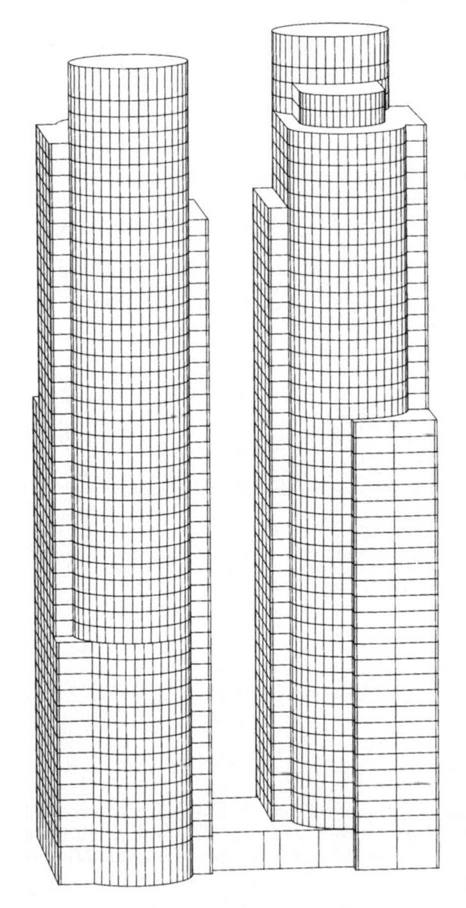

**Figure 5.9** Lone Star Towers.

model which included each and every structural member. The results were, however, verified by comparing the results with those of: (i) a relatively simpler three-dimensional model in which every other floor was lumped; and (ii) an equivalent shear wall and frame which represented the lateral stiffness of the building in the short direction. The agreement between the various computer results was within acceptable limits.

Figure 5.10g shows the distribution of horizontal shear forces among various lateral-load-resisting elements for wind loads acting on the broad face of the building. The lateral loads in the shear wall frames are shown by the curves designated as 12, 13, 14, and 15, which correspond to their locations on the grid lines shown in

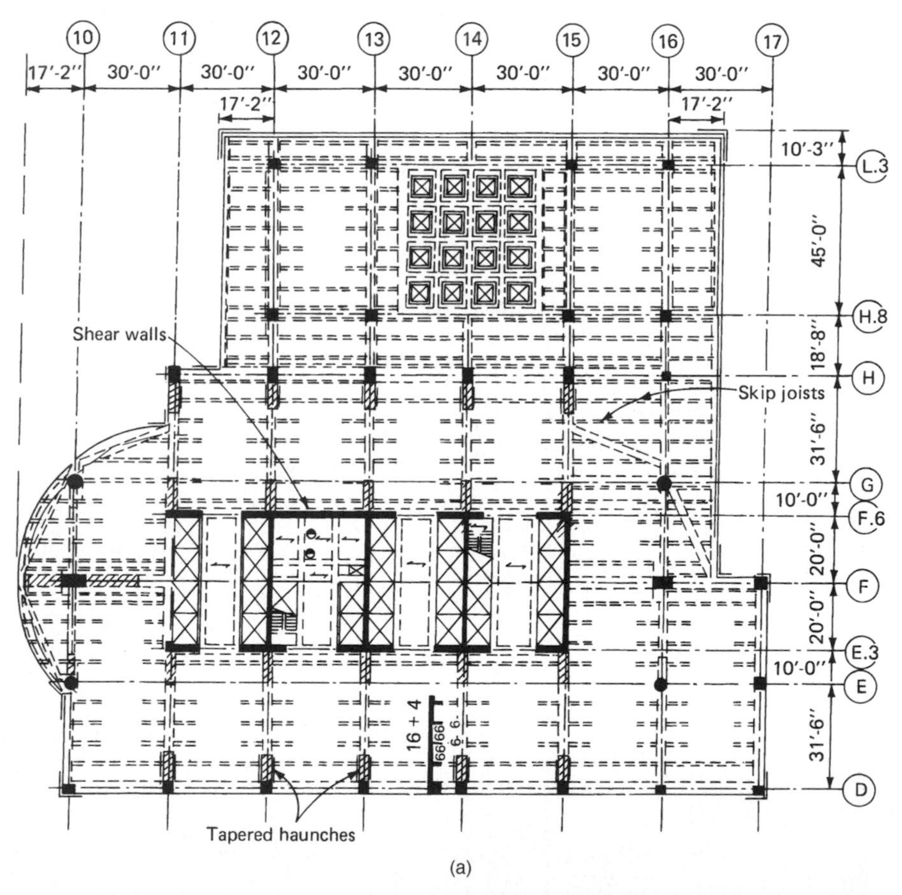


**Figure 5.10** Example of shear wall-frame interaction in a 50-story building: (a) levels 2 through 14 floor plan; (b) levels 15 through 26 floor plan; (c) levels 27 through 39 floor plan; (d) levels 40 through 47 floor plan; (e) levels 48 and 49 floor plan; (f) levels 50 and 51 floor plan; (g) distribution of lateral loads.

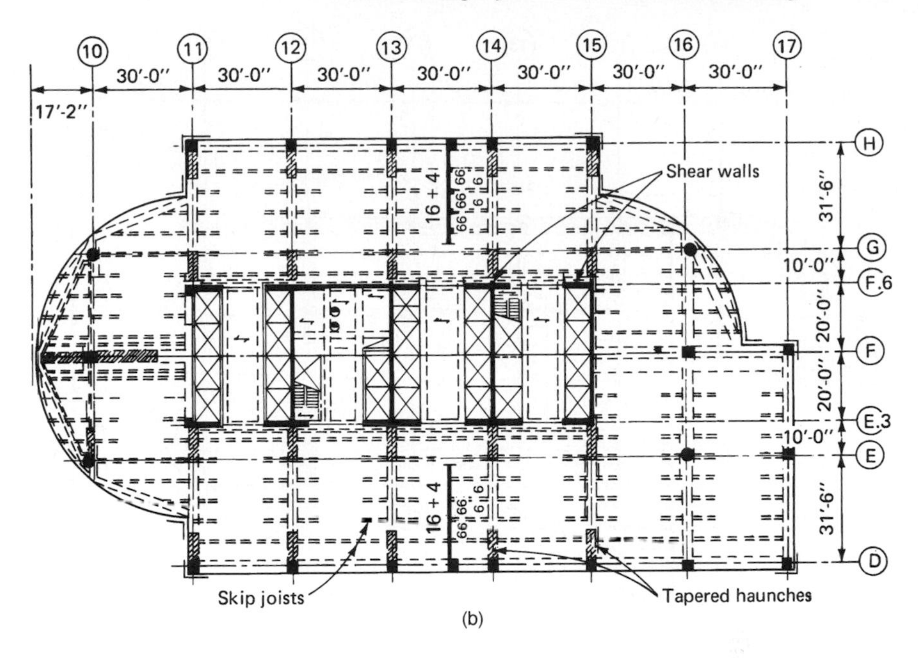


(b)

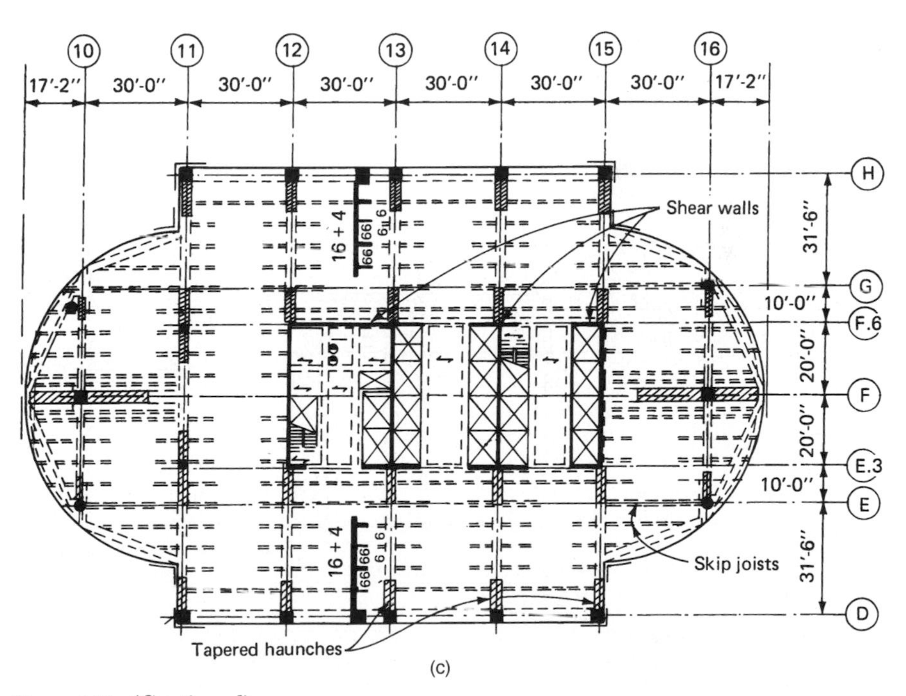


(c)

**Figure 5.10** *(Continued).*

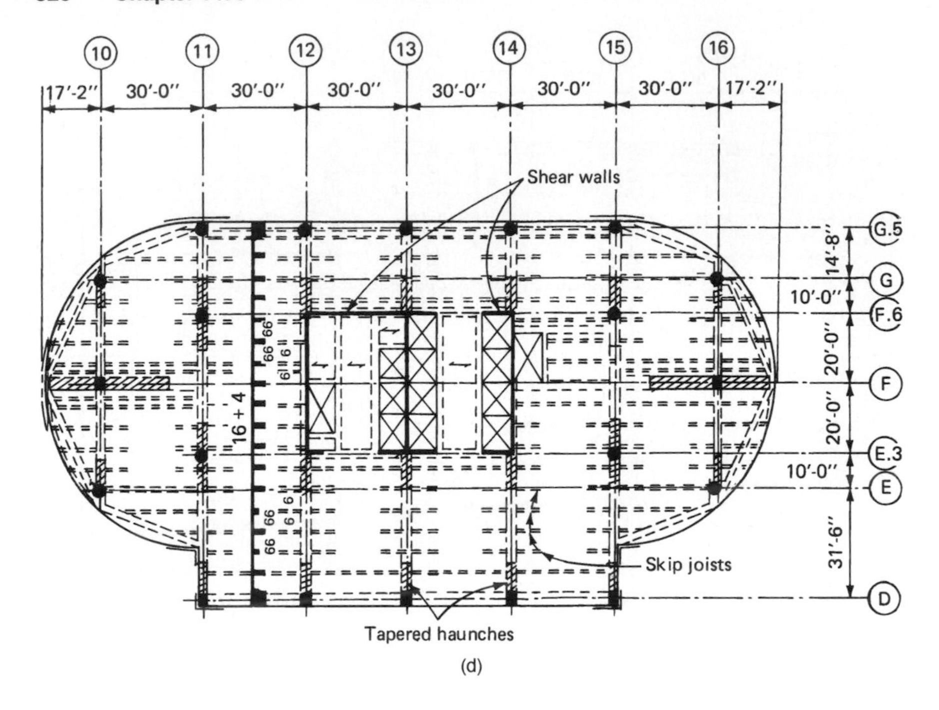


(d)

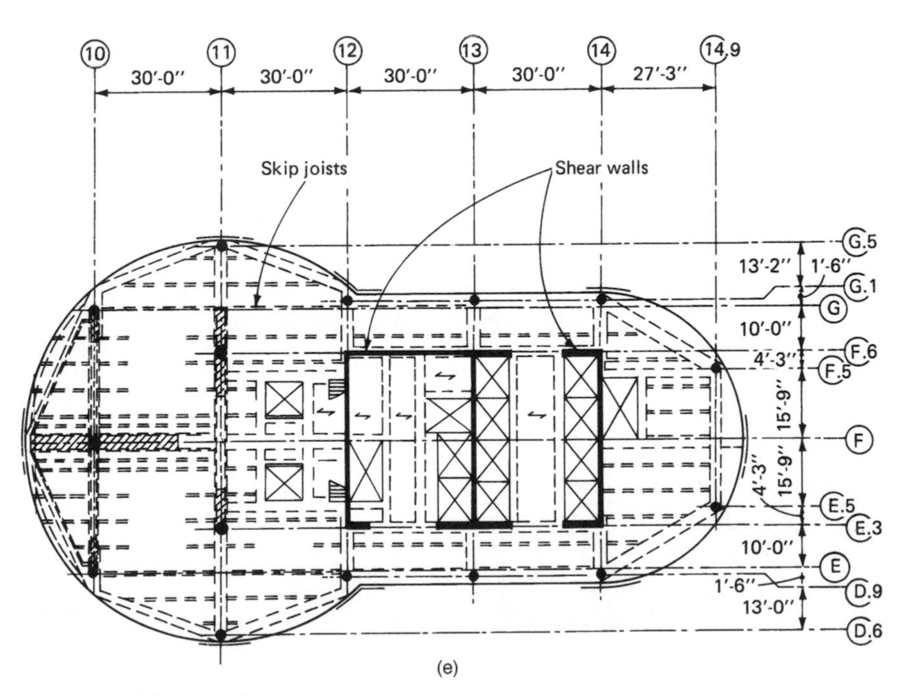


(e)

**Figure 5.10** *(Continued).*

Fig. 5.10a–f. The shear forces in the frames are shown by curves designated as 10, 11, 16, 17, and 18. The results shown are from an earlier version of the tower, which consisted of 42 floors with slight modifications to the floor plans shown in Fig. 5.10.

The purpose of presenting limited results of the analysis is to show qualitatively how the distribution of transverse shear in a practical building is considerably different than in a structure with regularly placed, full-height shear walls and frames of fairly uniform stiffness. The large difference in the pattern of transverse shear distribution occurs for two reasons. First, the structure is complex, with stiffness varying significantly over the height. Second, the mathematical assumption of a rigid diaphragm that is commonly used in the

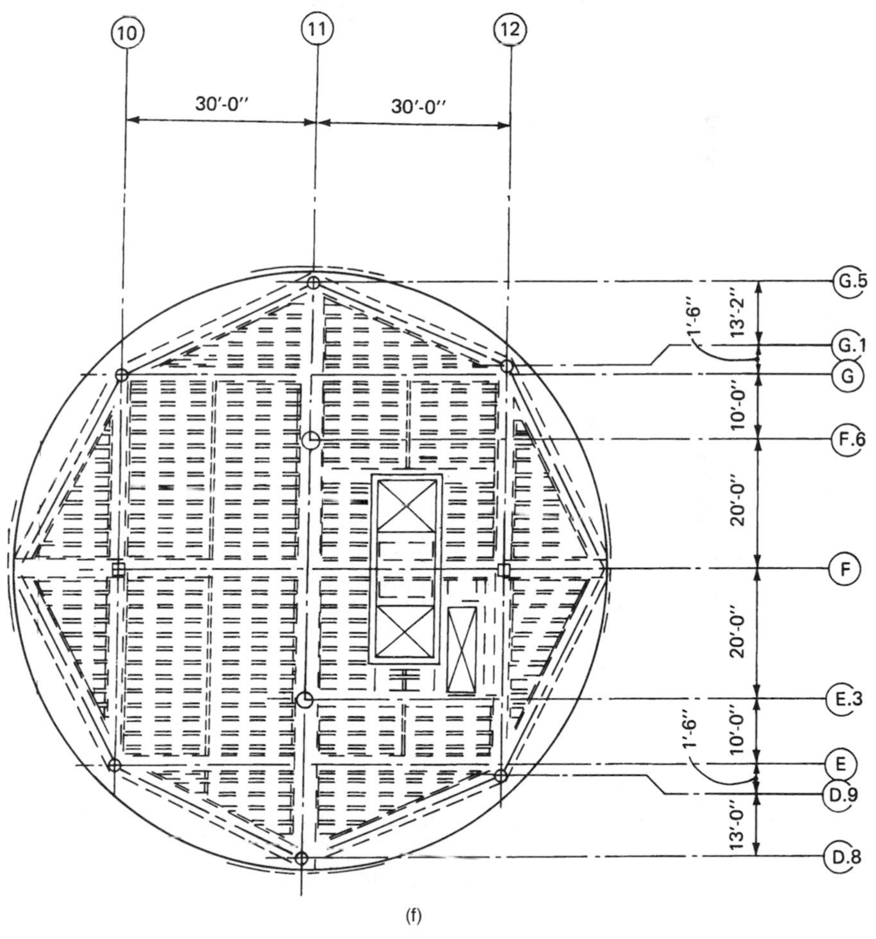


**Figure 5.10** (*Continued*).

modeling of the floor slab tends to bring about sharp shear transfers at levels where the stiffness of the building changes abruptly. Although it is possible to smooth out the harsh distribution and sudden reversals of transverse shear by modeling the floor slab as a flexible diaphragm by using a finite element model, such a complex modeling technique will have little effect on the final shear wall design.

## 5.10 Frame Tube Structures

The tube concept is an efficient framing system for tall slender buildings. In this system, the perimeter of the building consists of closely spaced columns connected by a relatively deep spandrel. The resulting system works as a giant vertical cantilever and is very efficient because of the large separation between the windward and leeward columns. The tube concept in itself does not guarantee that the system satisfies stiffness and vibration limitations. The "chord" drift caused by the axial displacement of the columns and the "web" drift brought about by the shear and bending deformations of the spandrels and columns may vary considerably depending upon the geometric and elastic properties of the tube. For example, if the plan aspect ratio is large, say much in excess of 1:1.5, it is likely that a

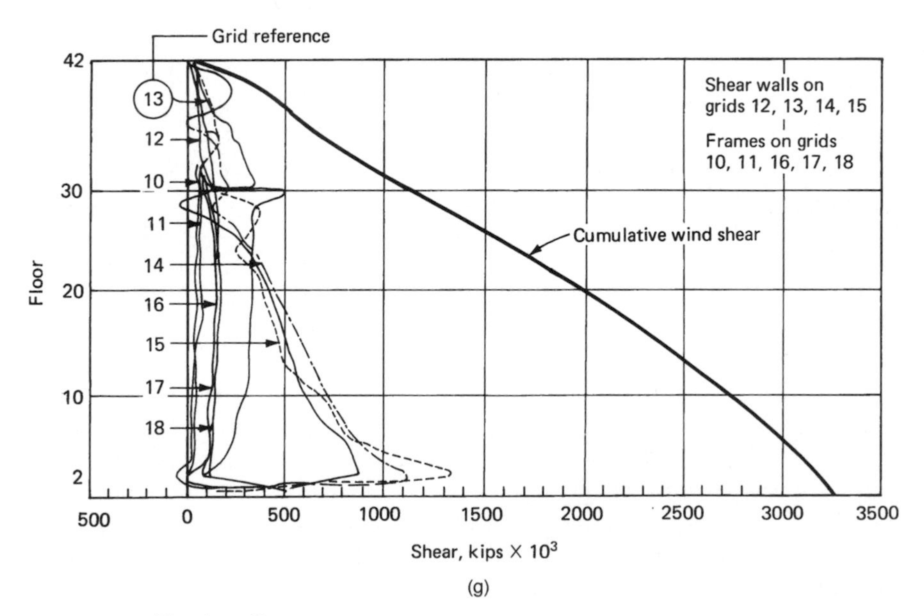


(g)

**Figure 5.10** (*Continued*).

supplemental lateral bracing is necessary to satisfy drift limitations. The number of stories that can be achieved economically by using the tube system depends on a number of factors such as spacing and size of columns, depth of perimter spandrels, and plan aspect ratio of the building. The system should be given serious consideration for buildings taller than about 40-stories.

## 5.11 Exterior Diagonal Tube

Master builder Fazlur Khan of Skidmore, Owings & Merrill envisioned as early as 1972 that it was possible to build high rises in concrete rivaling those in structural steel. His quest to find a structural solution for eliminating the shear lag phenomenon led him to the diagonal tube concept. A brilliant manifestation of this principle in steel construction is seen in the John Hancock Tower in Chicago. Applying similar principles, Khan visualized a concrete version of the diagonal truss tube consisting of exterior columns spaced at about 10 ft (3.04 m) centers with blocked out windows at each floor to create a diagonal pattern on the façade. The diagonals could then be designed to carry the shear forces, thus eliminating bending in the tube columns and girders. Although Khan enunciated the principle in the 1970s, the idea had to wait almost 15 years to find its way to a real building. Currently, two high rises have been built using this approach. The first is a 50-story office structure located on Third Ave. in New York and the second is a mixed-use building located on Michigan Ave. in Chicago. The structural system for the building in New York consists of a combination of a framed and trussed tube interacting with a system of interior core walls. All the three subsystems, namely, the framed tube, trussed tube, and shear walls, are designed to carry both lateral and vertical loads. The building is 570 ft (173.73 m) high with an unusually high height-to-width ratio of 8:1. The diagonals created by filling in the windows serve a dual function. First, they increase the efficiency of the tube by diminishing the shear lag, and second they reduce the differential shortening of exterior columns by redistributing the gravity loads. A stiffer, much more efficient structure is realized with the addition of diagonals. The idea of diagonally bracing this structure was suggested by Fazlur Khan to the firm of Robert Rosenwasser Associates, who executed the structural design for the building. Schematic elevation and floor plan of the building are shown in Fig. 5.11.

The Chicago version of the braced concrete tube is a 60-story multiuse project. The building rises in two tubular segments above a flared base. According to the designers, diagonal bracing was used primarily to allow maximum flexibility in the interior layout needed for mixed uses. In contrast to the building in New York, which has polished granite as cladding, the Chicago building sports exposed concrete framing and bracing.

## 5.12 Modular or Bundled Tube

The concept of bundled tube in concrete high rises is similar to that discussed in Chap. 4. The underlying principle is to connect two or more individual tubes in a bundle with the object of decreasing the shear lag effects. Figure 5.12a shows a schematic plan of a bundled tube structure. Two versions are possible using either framed or diagonally braced tubes as shown in Fig. 5.12b,c. A mixture of the two is, of course, possible.

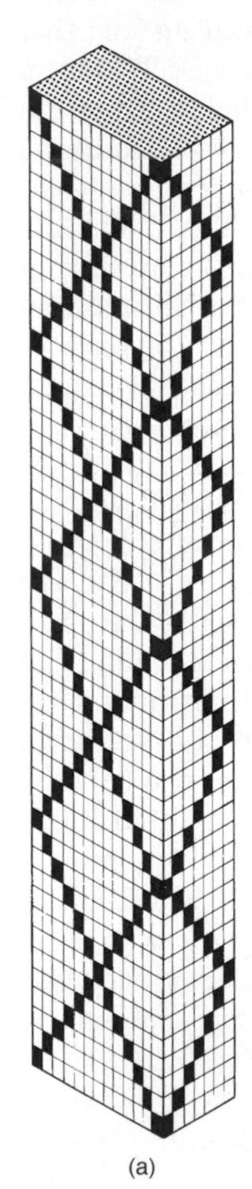

(a)

**Figure 5.11** Exterior braced tube: (a) schematic elevation; (b) plan.

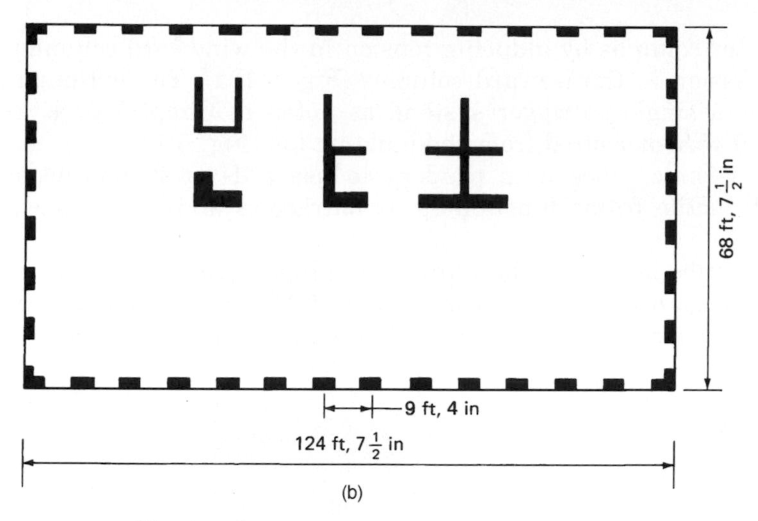


**Figure 5.11** *(Continued).*

## 5.13 Miscellaneous Systems

Figure 5.13 shows a schematic plan of a building with a cap truss. The cap truss takes on the form of a 1- or 2-story-high outrigger wall connecting the core to the perimeter columns. A 1- or 2-story wall at the perimeter acting as a belt truss may be used to tie the exterior columns together. As in steel systems, the introduction of cap truss results in a reversal of curvature in the bending mode of the shear core. A substantial portion of moment in the core is transferred to

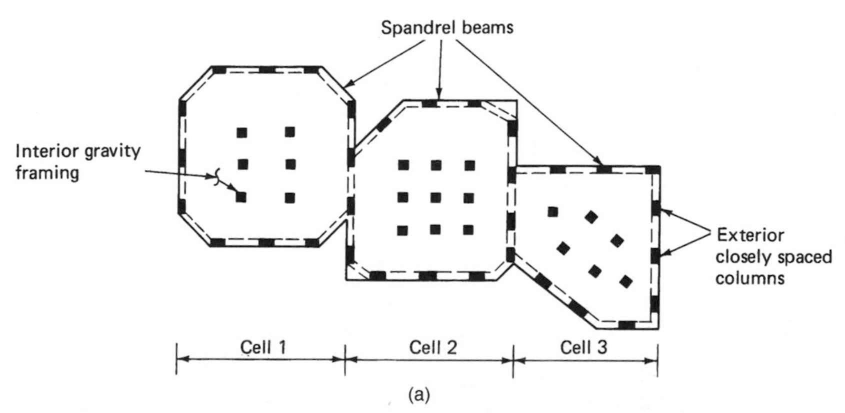


**Figure 5.12** Bundled tube: (a) schematic plan; (b) framed bundled tube; (c) diagonally braced bundled tube.

the perimeter columns by inducing tension in the windward columns and compression in the leeward columns (Fig. 5.13a). The optimum location for a single outrigger system, as noted in Chap. 3 is at a height $x = 0.45H$ measured from the building top (Fig. 5.14).

In high seismic zones it is prudent to use a 1- or 2-story-deep vierendeel ductile frame functioning as outriggers and belt trusses (Fig. 5.15).

A cellular tube in which a buliding with a high plan aspect ratio is divided into four cells is shown in Fig. 5.16. By introducing a minimum number of interior columns, three at every other floor in

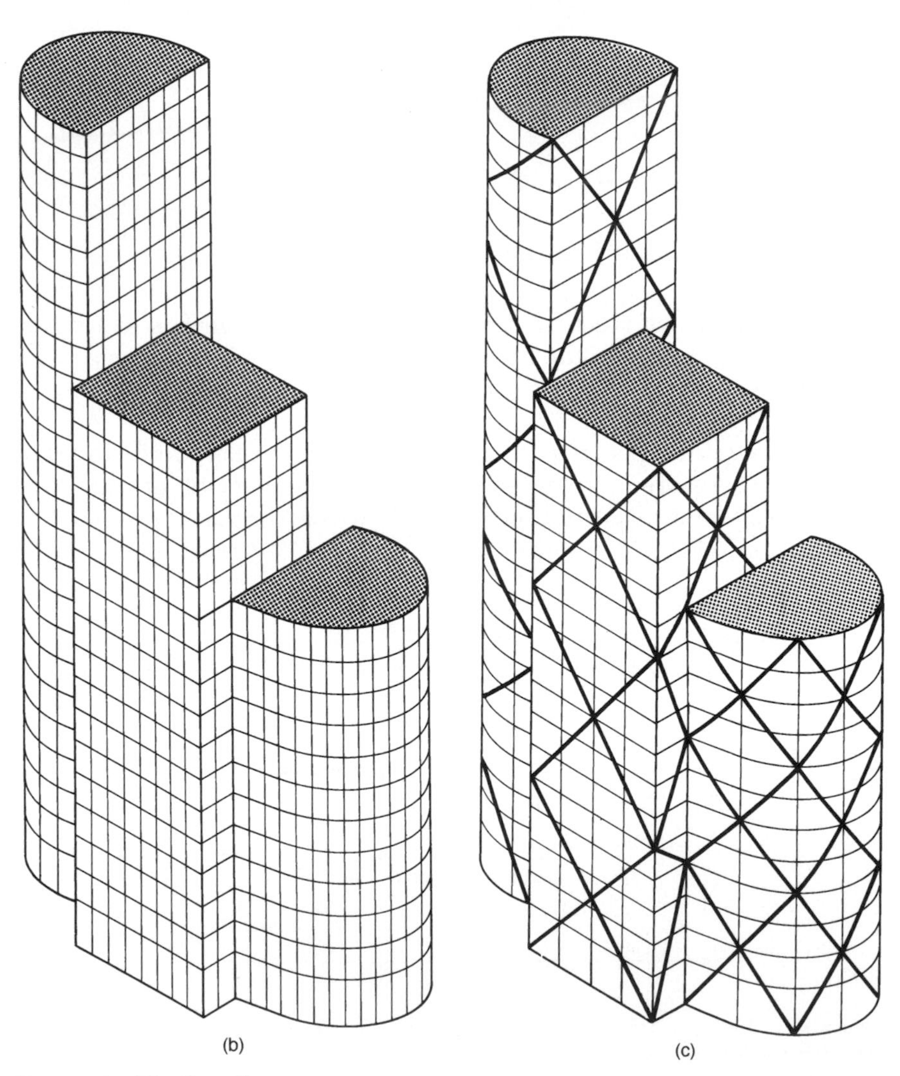


**Figure 5.12** *(Continued).*

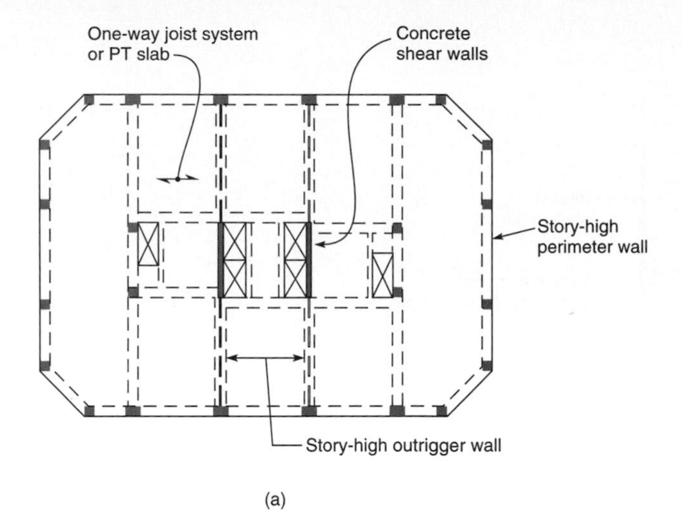


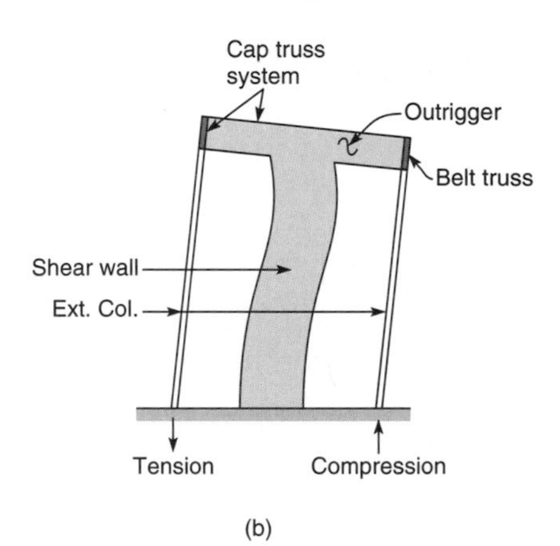


**Figure 5.13** Building with cap truss: (a) schematic plan; (b) structural behavior.

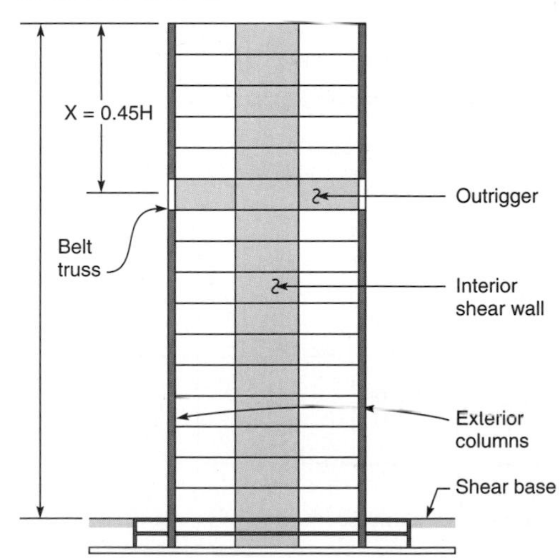


**Figure 5.14** Single outrigger system: optimum location.

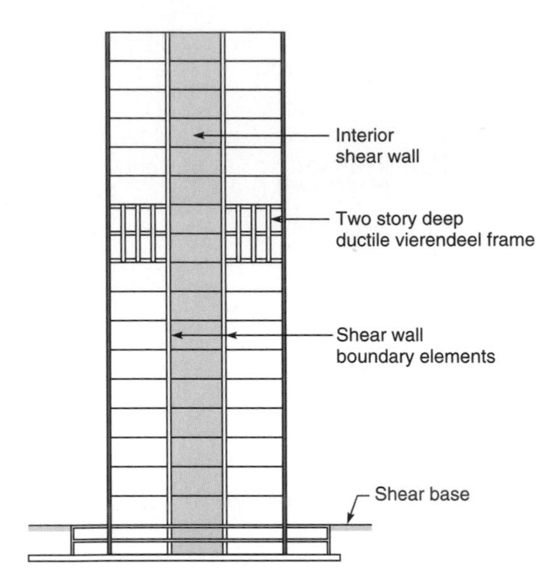


**Figure 5.15** Outrigger system: seismic version.

the example building, it is possible to reduce the effect of shear lag on the long faces of the building. A 2-story haunch girder vierendeel frame effectively ties the building exterior columns to the interior shear walls thus mobilizing the entire flange frame in resisting the overturning moments.

The concept of full-depth interior bracing interacting with the building perimeter frame is shown in Fig. 5.17. The interior diagonal

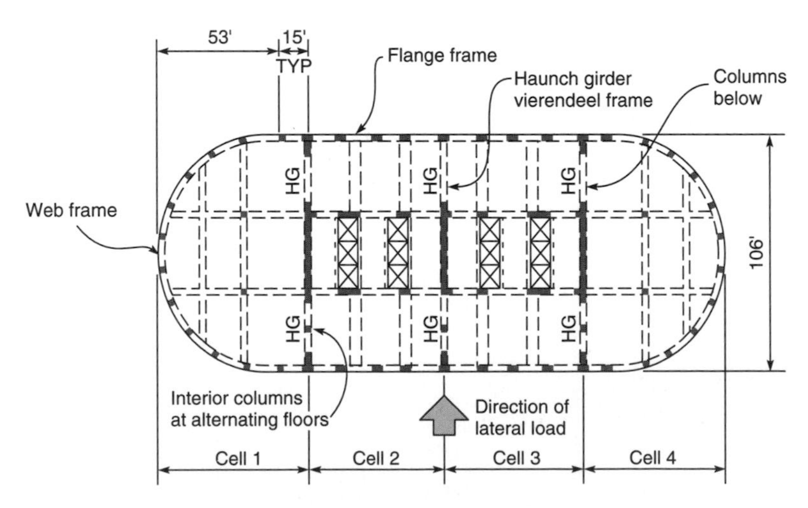


**Figure 5.16** Cellular tube with interior virendeel frames.

bracing consists of a series of wall panels interconnected between interior columns to form a giant K brace stretched out for the full width of the building (Fig. 5.17b).

A system suitable for super tall buildings, consisting of a service core located at each corner of the building and interconnected by a

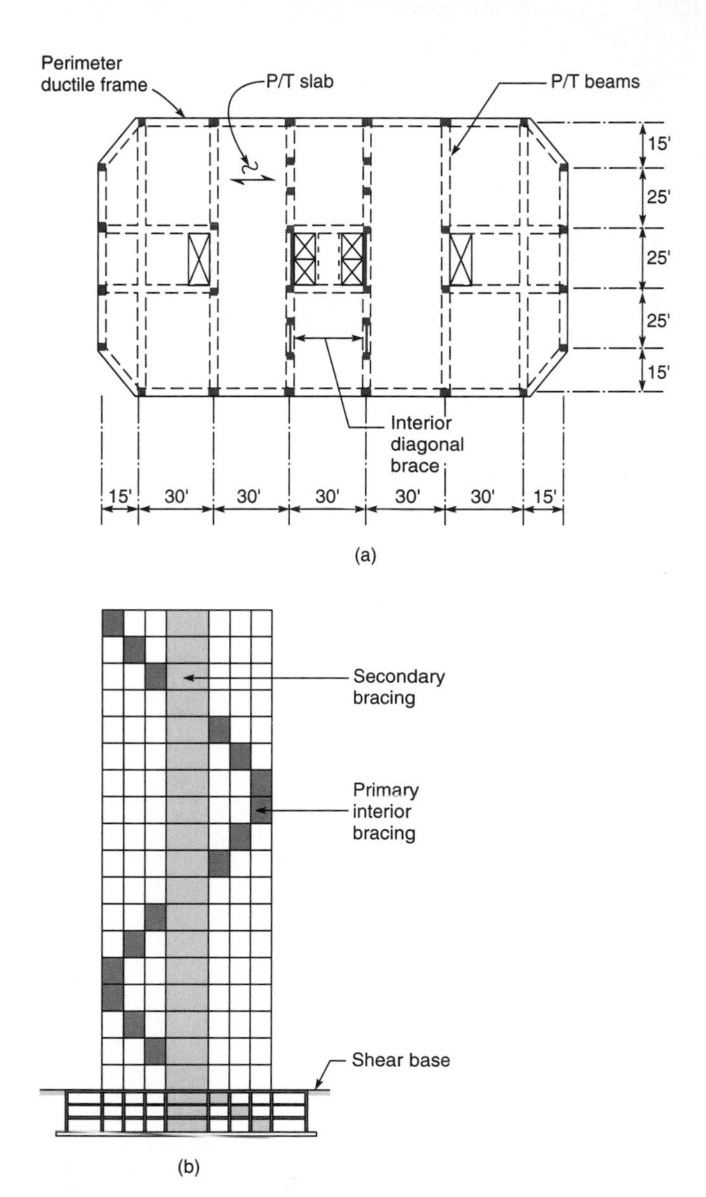


**Figure 5.17** Full-depth interior brace: (a) plan; (b) schematic section.

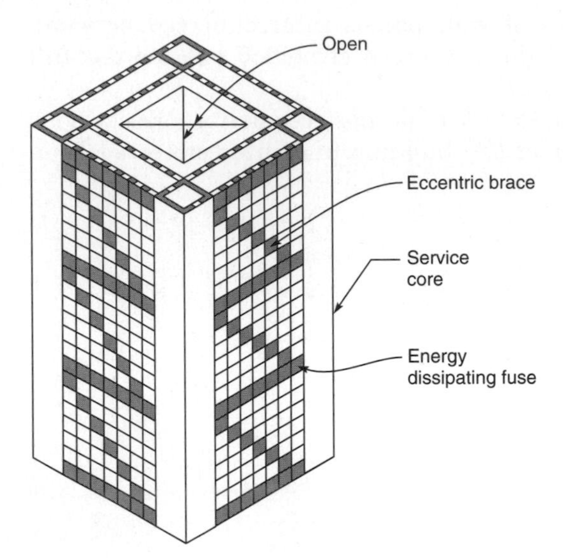


**Figure 5.18** Eccentric bracing system for super tall buildings.

diagonal in-fill bracing system, is shown in Fig. 5.18. The service core at each corner acts as a giant column carrying most of the gravity load and overturning moments. The diagonal braces at the building exterior resist the shear forces. Because of the deliberate eccentricity at one end, the resulting fuse helps in dissipating energy due to high seismic loads.

Chapter

# 6

# Lateral Systems for Composite Construction

## 6.1 Introduction

In a broad sense, all high-rises are composite buildings because a functional building cannot be built by using only steel or concrete. For example, concrete is invariably used for floor slabs in an otherwise all steel building. Similarly, in a critical sense, use of mild steel reinforcement transforms a concrete building into a composite building. Without being too pragmatic, we will settle for the definition of composite buildings as those with a blend of structural steel and reinforced concrete.

The earliest composite construction consisted of structural steel beams and reinforced concrete slabs, with shear connectors in-between. The system, called the composite floor system, first developed for bridge construction, was readily adopted to buildings. Its phenomenal success inspired engineers to develop composite building systems by combining structural steel and reinforced concrete in a variety of vertical building systems. Since 1960, with the advent of high- and ultra-high-strength concrete, i.e., concretes with compressive strengths in excess of 7000 psi to 19,000 psi (48.3 to 131 MPa), there has been a growing realization that a concrete column is more economical than a pure steel column. In fact, studies in North America indicate that concrete or composite columns are four to five times less expensive than all-steel columns. The favorable economics of concrete combined with its high stiffness and fireproofing characteristics has led to its marriage with steel, which has its own advantages namely strength, speed of construction, long span capability, and lightness.

Depending upon regional preferences, especially in areas of low seismic risk, the economics of concrete versus steel could go either way, except perhaps, for apartments, condominiums, and hotels. Concrete is favored for these buildings because the underside of floor slab is often used as finished ceiling. Unlike office buildings, the air conditioning duct work required in these buildings is relatively simple and thus there is no need for hung ceilings. On the other hand, an office building can be of either steel or reinforced concrete; the choice, although, depends to a great extent on the in-place cost of the frame, is nevertheless influenced by the speed of construction. If the building can be constructed faster and can bring a return on the investment sooner, the speed of construction necessarily enters the cost equation. The material choice, in other words, is based not on cost alone but on the speed of construction as well. Both steel and concrete possess advantages and disadvantages. It follows, therefore, that an ideal structural system is one that overcomes the disadvantages and exploits the advantages of both materials in a unified structural system.

Structural steel is well suited for providing generally column-free lease spaces required in contemporary high-rise office buildings. Because of its light weight, it imposes less severe foundation requirements and goes up faster. And its lightness is quite often a major consideration in seismic design. Another advantage of steel construction is the use of steel decking for floor construction which offers a more flexible system for wiring a building than a solid concrete slab construction. Also, a steel frame is simpler to modify, to meet the changing needs of building tenants. With steel framing it is less expensive to increase the load capacity of the floor, or cut holes in the floor to install stairways, atriums, etc., which may be required by changes in tenancy. Therefore, the adaptability to renovation and rehabilitation is an important factor in the selection process. Similarly, concrete buildings have their own advantages too. The advent of superplasticizers and high-strength concrete have made possible construction of reinforced concrete buildings without columns becoming cumbersomely large. Floor-framing techniques have progressed from flat slab construction to skip-joist and haunch girder systems, resulting in an increase in the spanibility of concrete floor systems. Lateral bracing systems have been developed that rival those of steel systems. The advantages of concrete framing are low material costs, moldability, insulating and fire-resisting quality, and most of all, inherent stiffness. However, in relation to steel, concrete construction is generally slow.

The two building systems, concrete and steel, evolved independently of each other, and up until the 1960s, engineers were trained to think of tall buildings either in steel or concrete. Dr Fazlur Khan, of Skidmore, Owings & Merrill, broke this barrier in 1969 by blending steel and concrete into a composite system for use in a relatively short, 20-story building, in which the exterior columns and spandrels were encased in concrete to provide the required lateral resistance. The system was basically a steel frame stabilized by reinforced concrete. However, today the advent of high-strength concrete has ushered in the era of super columns and mega frames where the economy, stiffness and damping characteristics of large concrete elements are combined with the lightness and constructability of steel frames. Without this type of framing, many of our contemporary tall buildings may never have been built in their present form.

The term composite system has taken on numerous meanings in recent years to describe many combinations of steel and concrete. As used here, the term means any and all combinations of steel and reinforced concrete elements and is considered synonymous with other definitions such as mixed systems, hybrid systems, etc. The term is used to encompass both gravity- and lateral-load resisting elements.

Since the advent of the first composite system, engineers have not hesitated to use a whole range of combinations to capitalize on the advantages of each material. This has resulted in numerous systems making distinct categorization a nearly impossible task. The systems can be best described by citing examples from the buildings built within the last two decades.

## 6.2 Composite Elements

To get an insight into different composite building schemes, it is instructive to study the various techniques of compositing both the horizontal and vertical elements. These are:

1. Composite slabs.
2. Composite girders.
3. Composite columns.
4. Composite diagonals.
5. Composite shear walls.

The behavior of horizontal members such as slab, beam, girder, and spandrels subjected to gravity loads is covered in Chap. 9. In this section their behavior only under lateral loads is considered.

### 6.2.1 Composite slabs

In high-rise steel buildings and the use of high-strength, light-gauge (16 to 20 gauge) metal deck with concrete topping has become the standard floor-framing method. The metal deck usually has embossments pressed into the sheet metal to achieve composite action with the concrete topping. Once the concrete hardens, the metal deck acts as bottom tension reinforcement while the concrete acts as the compression component. The resulting composite slab acts as a horizontal diaphragm interconnecting all vertical elements at each level providing for the horizontal transfer of shear forces to bracing elements. Furthermore, it acts as a stability bracing for the compression flange of steel beams.

The shear stresses induced due to diaphragm shear forces is mostly in the concrete topping because the in-plane stiffness of concrete slab is significantly more than that of the metal deck. Thus the horizontal forces must transfer from the slab to the beam top flange through the welded studs. In addition to supporting the gravity loads, the slab serves as a load path for transferring tributary floor lateral loads to the lateral load-resisting elements. Diaphragm behavior, an important aspect of seismic design, is discussed later in Chap. 12.

### 6.2.2 Composite girders

Consider a typical steel moment frame consisting of steel beams rigidly connected to columns. As discussed previously in Chap. 4, the stiffness of the frame most usually depends on the stiffness of the girder. This is because in a frame with its typical column spacing of 25 to 35 ft (7.6 to 10.67 m), and a floor-to-floor height of $12\frac{1}{2}$ to $13\frac{1}{2}$ ft (3.81 to 4.12 m), the columns are much stiffer than the beams. To limit sway under lateral loads it is more prudent to increase the girder stiffness rather than the column stiffness. Although frame beams are designed as non-composite beams, it is usual practice to use shear connectors at a nominal spacing of say, 12 in., especially in areas of high seismic risk. The shear connectors primarily provided for the transfer of diaphragm shear also increase the moment of inertia of the girder. However, the moment of inertia does not increase for the full length of the girder because the girder responds by bending in a reversed curvature. Since concrete is ineffective in tension, the composite moment of inertia can be counted on only in the positive moment region. Although design rules are not well established, a rational method may be used to take advantage of the increased moment of inertia. Occasionally engineers have used a dual approach in wind design by using bare steel beam properties for strength calculations, and composite properties in the positive regions for drift calculations.

### 6.2.3 Composite columns

Two types of columns are used in composite building systems. The first type consists of a steel core surrounded by a high-strength reinforced concrete envelope. The second type consists of a large-steel pipe or tube filled with high-strength concrete. In the first type the steel section, most usually a wide flange section, is either a light section designed to carry construction dead loads only, or may be of a substantial weight to part take in resisting axial, shear and bending moments together with the mild steel reinforcement and high strength concrete envelope. Conceptually the behavior of a composite column is similar to a reinforced concrete column, if the steel section is replaced with an equivalent mild steel reinforcement. In fact, this concept provides the basis for generating the interaction diagram for the axial load and moment capacities of composite columns.

Compositing of only exterior columns by encircling steel sections with concrete is by far the most frequent application of composite columns. The reasons are entirely economic, because concrete forming around interior columns is quite involved and is not readily applicable to jump forms. Exterior columns, on the other hand, are relatively open-faced: concrete forms can be "folded" around the steel columns for placement of concrete, then unfolded and jumped to the next floor repeating the cycle without having to dismantle the entire framework.

However, in Japanese construction it is common practice to composite the interior columns as well. Their construction makes extensive use of welding for vertical as well as transverse reinforcement (Fig. 6.1).

Prior to 1986, the design of composite columns was covered only by the ACI code. The ACI design rules are based on the same principles as the design of reinforced concrete columns. The strength of the column approaches that of concrete columns as the percentage of steel column decreases. The ACI code permits the use of any shape inside of a composite column, with a stipulation that the yield strength of steel cannot be greater than 50 ksi. For tied columns, the minimum reinforcement ratio is one percent while the maximum ratio is limited to 8 and 6 percent for wind and seismic designs respectively. The lateral spacing of longitudinal bars cannot exceed

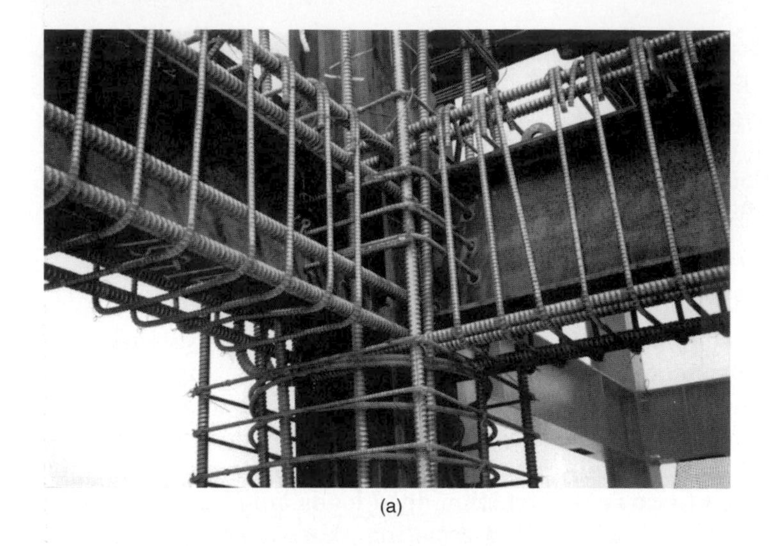

(a)

(b)

**Figure 6.1** Japanese composite construction details: (a) (i) beam column intersection; (b, c) composite column with welded ties; (d) general view.

half the length of the shorter side of the section, 16 diameters of the longitudinal bars or 48 diameters of lateral ties.

On the other hand, the AISC rules are similar to steel column design and requires that the cross-sectional area of the steel shape be at least 4 percent of the total composite cross-section. The strength of composite columns approaches that of steel columns as the percentage of steel section increases.

Both longitudinal and transverse reinforcing bars are required in

(c)

(d)

**Figure 6.1** (*Continued*).

the concrete section that encases composite steel shapes. For non-seismic applications the minimum reinforcements requirements of the ACI effectively serve the purpose of maintaining the concrete encasement around the steel shapes. For seismic applications, however, more confinement of the concrete is required to achieve ductility under cyclic lateral loading. In encased composite columns the confinement can be provided by closed hoops of transverse reinforcement as in non-composite concrete construction.

The second type of composite column consists of a steel pipe or tube filled with high-strength concrete. Typically neither vertical nor transverse reinforcement is used in the columns. However, as in the previous type, shear connectors are welded to the inner face of the pipe to provide for the interaction between the concrete and outer shell. Since the compositing of column does not require any form work this type is applicable to both the interior and exterior building columns.

The novel structural concept of concrete-filled steel columns for multistory construction was first proposed by Dr A. G. Tarics of Reid & Tarics Associates, San Francisco, California, in 1972. This idea has been used in several buildings by the Seattle engineers Skilling, Ward, Magnisson, Berkshire Inc. Two Union Square building is one such application of huge pipe columns filled with concrete in composite construction. The 58-story building includes a core that carries 40 percent of the gravity loads and all of the lateral loads. The core framing consists of four 10 ft (3 m) diameter pipe columns filled with 19,000 psi (130 MPa) concrete.

The usual method of attaching steel members to pipe columns is to use a welded connection to the outside of the pipes in which case the flange forces are carried directly by the pipe. Sometimes plates are welded to the inside of the pipes acting as stiffeners to minimize local stresses in the thin-walled pipes.

Although the concept of concrete-filled steel columns has been used in several tall buildings, including some in high seismic zones, there are still some nagging questions about the system. The post-yield behavior, the performance of bond between concrete and steel under cyclic loading, the potential for local buckling of the pipe, and the heat of hydration of the concrete are some of the unanswered questions which need further investigation.

### 6.2.4 Composite diagonals

Diagonals in a braced frame resist lateral forces primarily through axial stresses acting as part of a vertical truss. As a result, braced frames are more economical than moment-resisting frames on a material quantity basis. However, their use is often limited, because of potential interference of braces with architectural planning concerns.

The majority of braced frame applications is in structural steel. However, braced frames of composite construction have started finding application since the mid 1980s. The majority of these have been applied in concert with composite columns, in the form of "super columns" composed of large diameter circular pipes filled with high-strength concrete.

### 6.2.5 Composite shear walls

In this system, as in conventional concrete systems, concrete shear walls are placed around the building core. The walls are usually C or I shapes with webs placed parallel to the elevator banks as shown in Fig. 6.2. The flanges of adjacent shear walls are invariably connected with link beams to increase bending stiffness about an axis parallel to the web. The high shear forces in the link beams can result in brittle fracture unless the beam is properly detailed with diagonal reinforcement. In one version of composite shear walls, mostly used in areas of low-seismic risk, structural steel beams are used to link the shear walls. In this type of construction steel columns with-in the shear walls are erected with the steel frame including floor beams.

The most usual method of connecting the steel beam is to extend it through to steel columns within the shear walls. The moment capacity is achieved by welding shear connectors to the top and bottom flanges of the beam as shown in Fig. 6.3. This type of construction has not been used in areas of high-seismic activity. Much research work needs to be done before seismic design guidelines are established. For resisting high in-plane shear forces, a full length steel plate compositely attached to a concrete shear wall may be used. An example of such a construction is the core wall of Bank of China Building, Hong Kong. In this building all the lateral forces are transferred to the core at the base. To resist the high shear forces steel plates are attached to the concrete core through shear studs welded to the steel plates as shown in Fig. 6.4.

## 6.3 Composite Building Systems

Composite building systems in use today may be conveniently classified into the following categories.

1. Shear wall systems.
2. Shear wall–frame interacting systems.

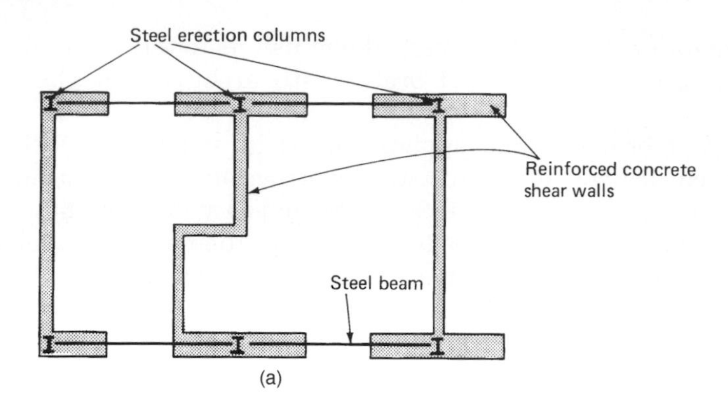


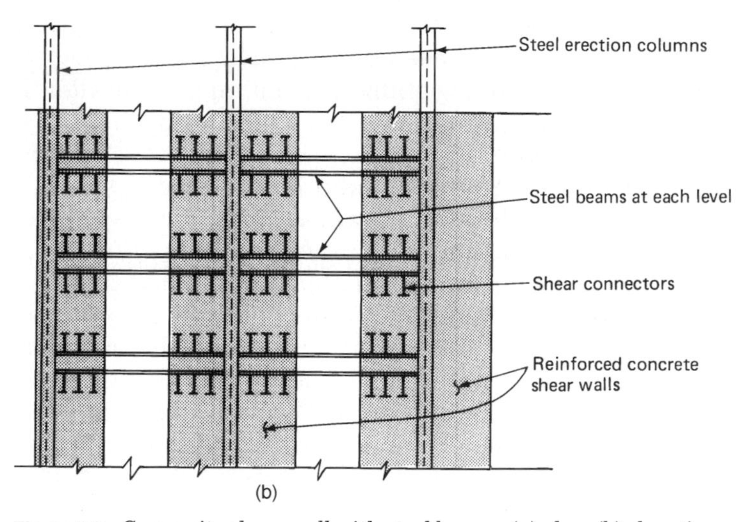


**Figure 6.2** Composite shear wall with steel beams: (a) plan; (b) elevation.

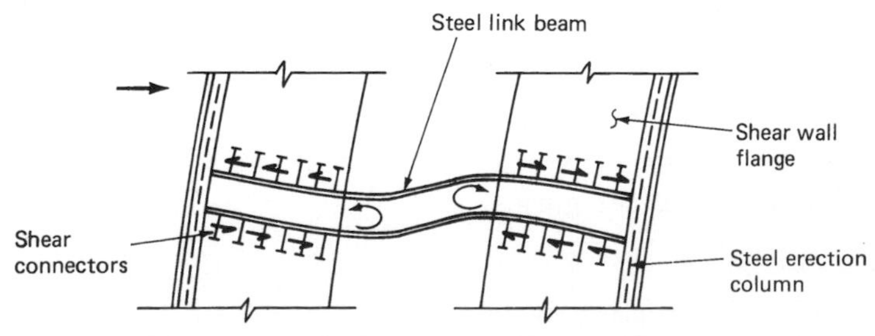


**Figure 6.3** Moment transfer between steel beam and concrete wall.

3. Tube systems.
4. Vertically mixed systems.
5. Mega frames with super columns.

### 6.3.1 Shear wall systems

Core walls enclosing building services such as elevators, mechanical and electric rooms, and stairs have been used extensively to resist lateral loads in tall concrete buildings. Simple forms such as C and I shapes around elevators interconnected with coupling or link beams are used extensively. Their popularity as a lateral-load-resisting element has once again come to the forefront because of the current trend in architecture which appears to favor lean vertical elements around the perimeter. Earlier applications of this system were limited to buildings in the 30- to 40-story range, but with the advent of superplasticizers and high-strength concrete, it is now possible to use this system for taller buildings in the 50- to 60-story range. The range of application is mostly a function of available depth of core. Buildings using four-deep elevators provide a core depth of approximately 40 ft (12.2 m), which is sufficiently large to give an economical system for buildings in the 50- to 60-story range. The shear core basically responds to lateral loads in a bending mode and does not present undue complications in its analysis.

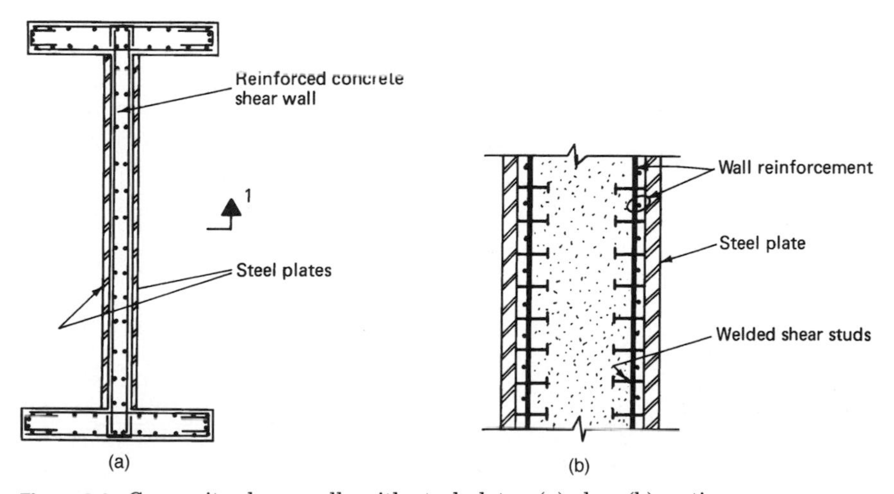


**Figure 6.4** Composite shear walls with steel plates: (a) plan; (b) section.

In the core-only system the total lateral load is resisted by the shear walls and hence the remainder of the building can be conveniently framed in structural steel. Whether or not concrete or steel comes first in the construction is not always a known priority and is often influenced by the choice of construction method. In one version, concrete core is cast first by using jump or slip forms, followed by erection of the steel surround as shown in Fig. 6.5. Although the structural steel framing may not proceed as quickly as in a conventional steel building, the overall construction time is likely to be reduced because elevators, mechanical and electrical services can be installed rapidly in the core while construction outside the core proceeds. In another version, steel erection columns are used within the shear walls to serve as erection columns and steel erection proceeds as in a conventional steel building. After the steel erection

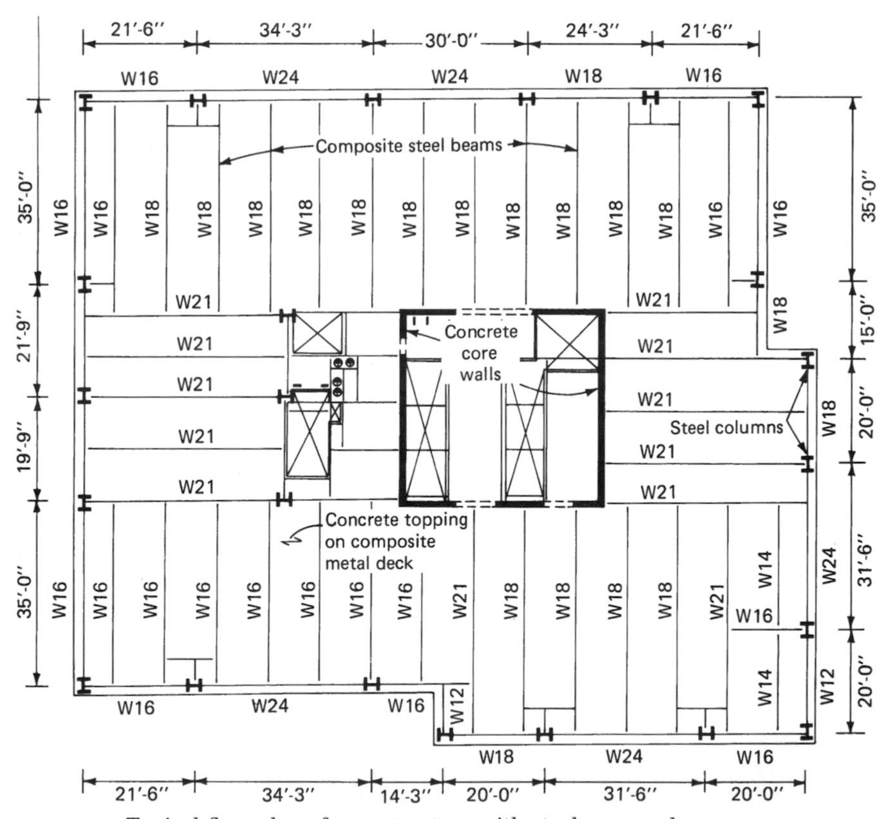


**Figure 6.5** Typical floor plan of core structure with steel surround.

has reached a reasonable level, concreting of core starts using conventional forming techniques. In order to facilitate faster jumping of forms to the next higher level, temporary openings are left in the floor slab around the shear walls.

The structural behavior of this system is no different than a concrete building with shear walls designed to take all the lateral forces. However, it behooves the engineer to recognize the absence of torsional stiffness. It is advisable to provide moment frames or other types of bracing around the building perimeter to counteract the torsional effects.

If all the lateral loads are resisted by concrete shear walls, the steel surround is designed as a simple framing for gravity loads only. Since there are no moment connections with welding or heavy bolting, the erection of steel proceeds much faster. The only nonstandard connection is between the shear walls and floor beams. Various techniques have been developed for this connection, chief among them, the embedded plate and pocket details, as shown in Fig. 6.6a. The floor construction invariably consists of composite metal deck with a structural concrete topping. This system has the advantage of keeping the steel fabrication and erection simple. Since columns carry only gravity loads, high-strength steel can be employed with the attendant savings. Interior as well as exterior columns can be made small, increasing the space-planning potential.

The floor within the core can be constructed either with cast-in-place beams and slabs or structural steel framing, steel decking, and concrete fill. At the connection between the floor slabs, both within and outside of the core shear keys are provided in the core walls to transmit lateral diaphragm forces from the floor system to the core. Although several options are available for connecting the steel beams to the concrete core, a weld plate detail shown in Fig. 6.6a is most popular, especially so in a slip-formed construction. During slip-form operation, the weld plates are set at the required locations, with the outer surface of the plate set flush with the wall surface. Anchorage of the weld plate is achieved by shear connectors welded to the inner surfaces, sometimes supplemented with a top-bent steel bar to resist high tensile forces. The shear connectors and steel bar ultimately become embedded in the core wall. Experience with slip-formed composite buildings indicates that it is prudent to overdesign these connections to compensate for misalignment of cores. Subsequent to the installation of weld plates, structural tee or shear tab connections with slotted holes are field-welded to the plate. Slotted holes are used for bolting of steel beams to provide additional tolerance for erection.

Slip forming is a special construction technique that uses a mechanized moving platform system. The process of slip forming is similar to an extrusion process. The difference is that whereas in an extrusion process the extrusion moves; in a slip-forming process the die moves while the extrusion remains fixed.

Although traditionally the core has been used to resist lateral forces, its contribution in supporting the vertical load is limited because relatively small floor area is supported by the core. For a very tall building, this can result in an unfavorable stability condition. A method of overcoming this limitation is to apply external prestressing forces to the core to relieve the tensile stresses. An equivalent passive prestressing effect can be obtained by channeling

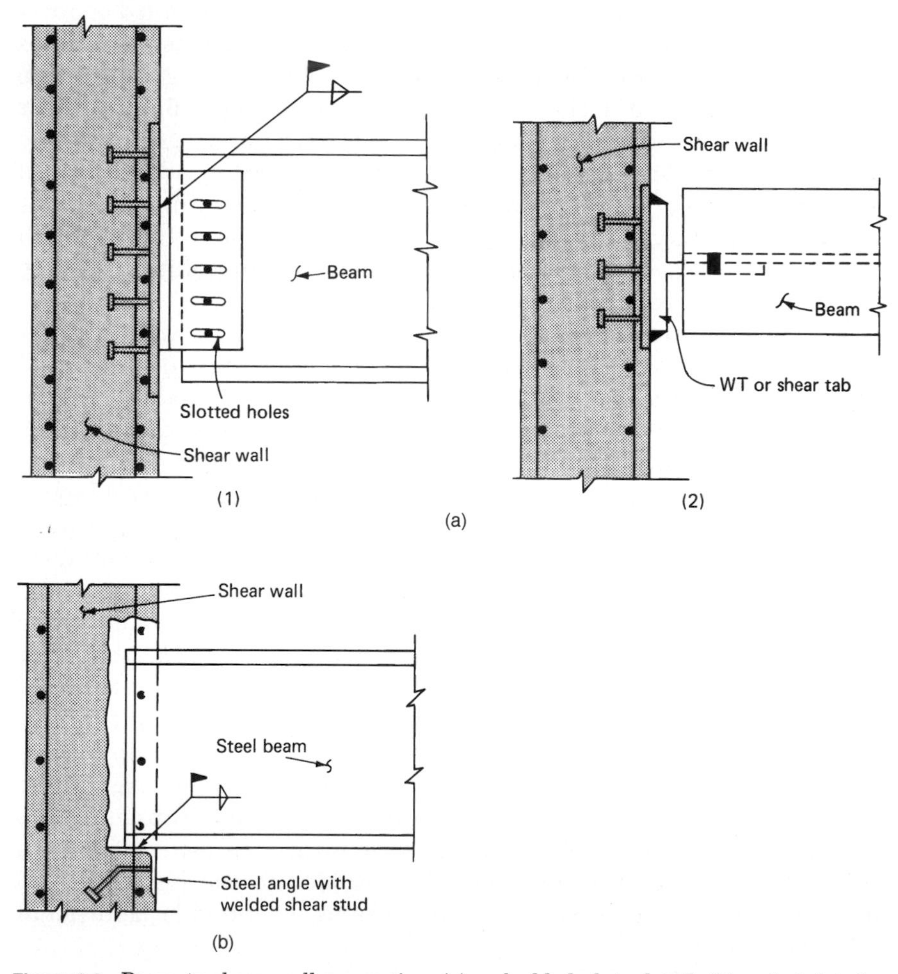


**Figure 6.6** Beam to shear wall connection: (a) embedded plate detail; (b) pocket detail.

increased vertical load to the core. Extending this fundamental idea to its limit results in the concept of a building totally supported on a single-core element. In practice, depending upon the floor area and the number of levels, several options are available for supporting the floors from the core. For example: (i) floors can be hung from the top of the center core; (ii) story-deep cantilever trusses located at one or two intermediate levels, such as at top and midheight of the building, can be used, the advantage being reduced length of hangers with fewer floor-leveling problems; or (iii) the floor system can be cantilevered at each level. The selection of a suitable system depends on the economic consequence of each method. In addition to providing views unobstructed by exterior columns at each floor, the absence of columns provides for the commonly sought column-free space at the building entrances. Also, the undulations on the building exterior common in today's architecture are easy to accommodate in a core-only structural scheme. Galvanized bridge-strand cables can be used as hangers to support the structural steel framing which normally consists of composite beams, metal deck, and concrete topping. The floor beams are attached to the hanger with simple supports, while at the core pockets or anchor plates cast into the core walls provide for the support. It is common practice to slip-form the center core with an average concrete growth rate of 6 to 18 in./h (152 to 457 mm/h). After completion of the core, the second stage of construction in the hung-floor system is the erection of roof girders and draping of the floor-supporting cables. Erection of typical floor members between the core and the perimeter cables proceeds similarly to any other building. Erection of steel floor decks and welding of beams for composite action follows by placement of concrete topping. Because the elongation of the cable due to cumulative floor loads can be substantial, it is necessary to compensate for this effect properly.

### 6.3.2 Shear wall-frame interacting systems

This system has applications in buildings that do not have cores sufficiently large to resist the total lateral loads; interaction of shear walls with other moment frames located in the interior or at the exterior is called upon to supplement the lateral stiffness of the shear walls. When frames are located on the interior, usually the columns and girders are of steel because the cost of form work for enclosing interior columns and girders for composite construction far outstrips the advantages gained by additional stiffness of the frame. Interior columns typically have beams framing in four directions, making placement of mild steel reinforcement and form work around them extremely cumbersome. On the other hand, it is relatively easy to form around exterior columns, and if desired, even the exterior spandrels can be made out of concrete without creating undue complexity. If steel erection precedes concrete construction, it is usually more cost effective to use steel beams as interconnecting link beams between the shear walls. A schematic plan of a building using this system is shown in Fig. 6.7.

### 6.3.3 Tube systems

A framing system used extensively in Louisiana and Texas is the composite concrete tube. It makes use of the well-known virtues of the tube system with the speed of steel construction. As in concrete and steel systems, closely spaced columns around the perimeter and deep spandrels form the backbone of the system. Two versions are currently popular: one system uses composite columns and concrete spandrels and the other uses structural steel spandrels

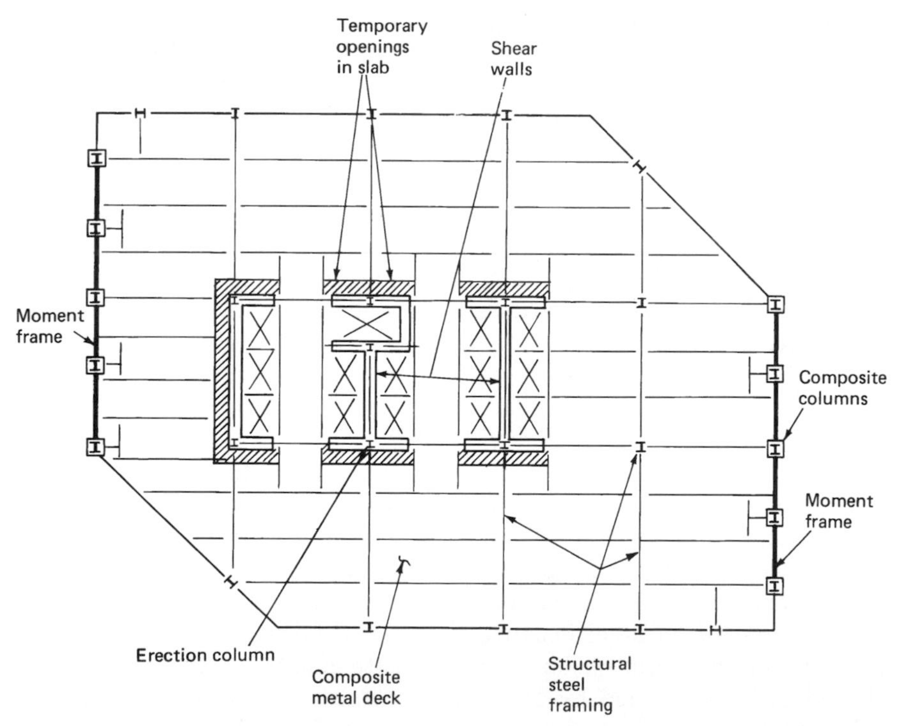


**Figure 6.7** Typical floor plan of a composite building using shear wall frame interaction.

in place of concrete spandrels. A small steel section can be used as a steel spandrel in the former scheme to stabilize the steel columns. However, in the design of the concrete spandrel, its strength and stiffness contribution is generally neglected because of its relatively small size. Schematic plan and sections for the two versions of tubular system are shown in Figs 6.8 and 6.9.

In either system, the speed of construction, rivaling that of an all-steel building, is maintained by erecting a steel skeleton first

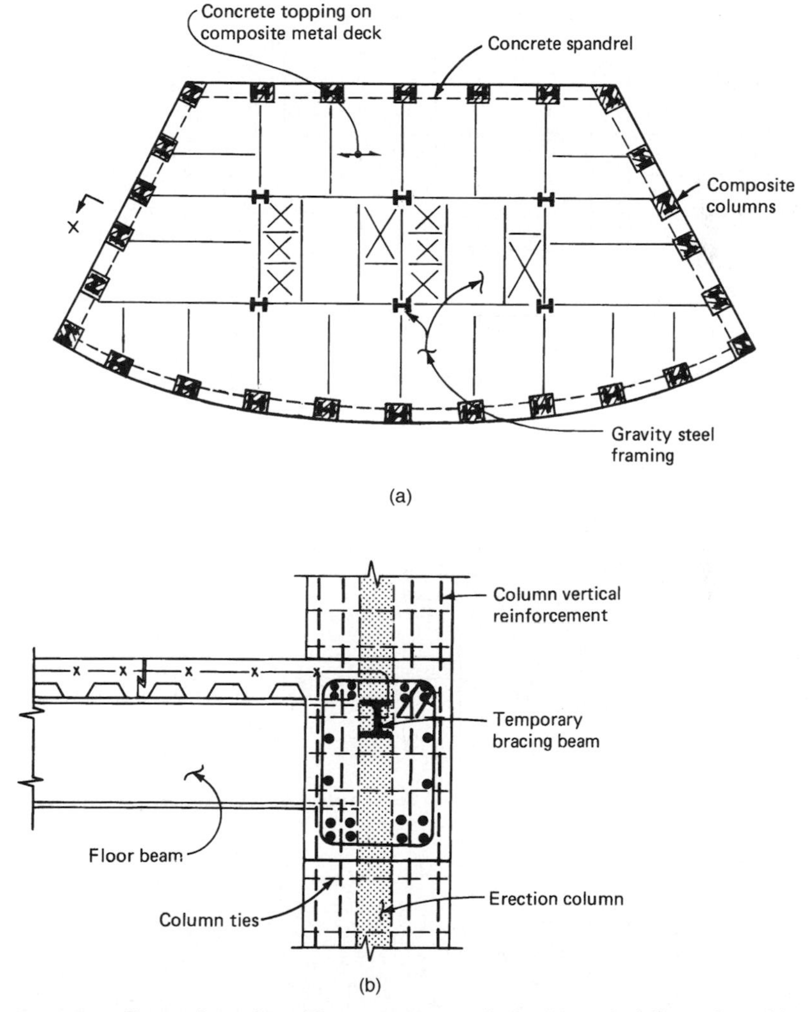


**Figure 6.8** Composite tube with concrete spandrels: (a) typical floor plan; (b) typical exterior cross section.

with interior steel columns, steel floor framing, and light exterior columns. Usually the steel frame is erected some 10- to 12-stories ahead of the perimeter concrete tube. The key to the success of this type of construction lies in the rigidity of closely spaced exterior columns, which, together with deep spandrels, results in an exterior façade that behaves more like a bearing wall with punched windows than a moment frame.

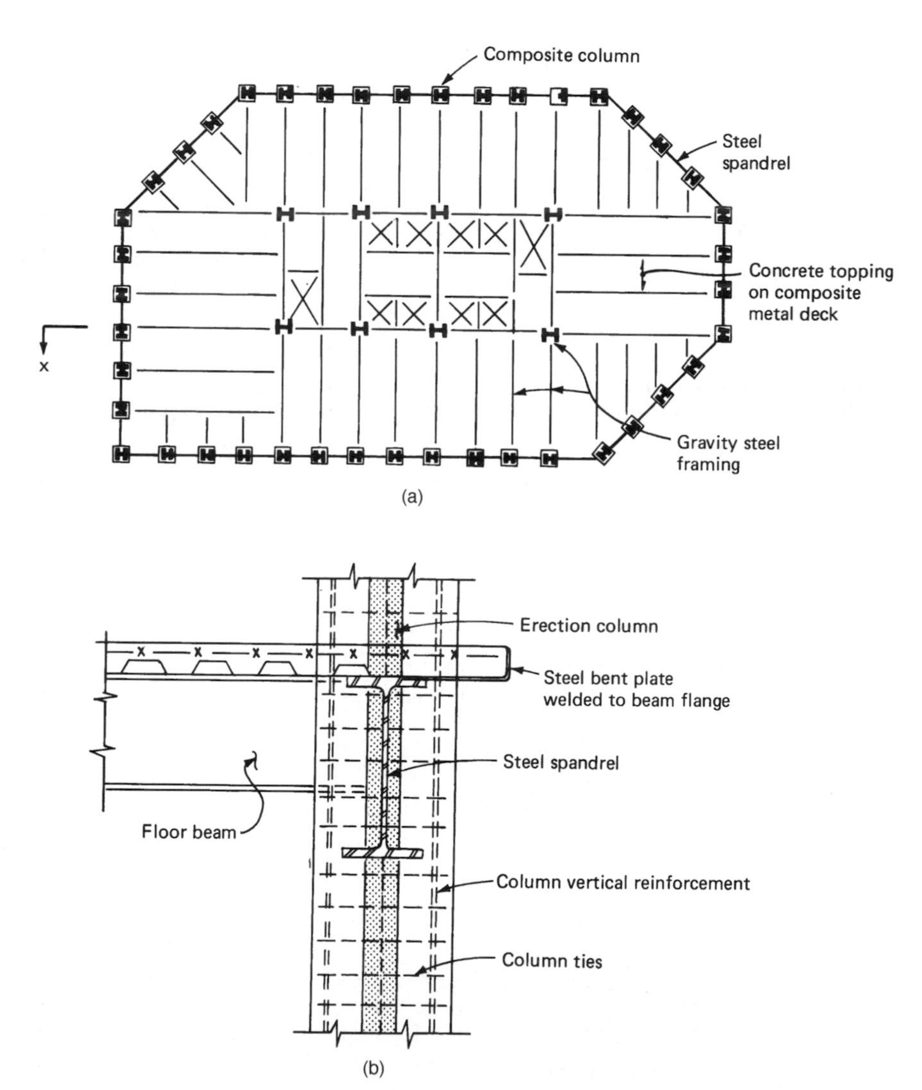


**Figure 6.9** Composite tube with steel spandrels: (a) typical floor plan; (b) typical exterior cross section.

### 6.3.4 Vertically mixed systems

Mixed-use buildings provide for two or more types of occupancies in a single building by vertically stacking different amenities. For example, lower levels of a building may house parking; middle levels, office floors; and top levels, residential units, such as apartments and hotel rooms. Since different types of occupancies economically favor different types of construction, it seems logical to mix construction vertically up the building height. As mentioned earlier, beamless flat ceilings are preferrred in residential occupancies because of minimum finish required underneath the slab. Also, large spans of the order of 40 ft (12.2 m) required for optimum lease space for office buildings are too large for apartments. Additional columns can be introduced without unduly affecting the architectural layout of residential units. The decrease in span combined with the requirement of flat-plate construction gives the engineer an opportunity to use concrete in the upper levels for apartments and hotels.

In certain types of buildings, use of concrete for the lower levels and structural steel for the upper levels may provide an optimum solution as shown in Fig. 6.10. The bracing for the concrete portion of the building is provided by the shear walls, while a braced steel core provides the lateral stability for the upper levels. A suggested technique of transferring steel columns onto a concrete wall consists of embedding steel columns for one or two levels below the transfer level. Shear studs shop-welded to the embedded steel column provide for the transfer of axial loads from the steel column to the concrete walls.

### 6.3.5 Mega frames with super columns

As mentioned in Chap. 4, the most efficient method of resisting lateral loads in tall buildings is to provide "super columns", placed as far apart as possible with-in the foot print of the building. The columns are connected with a shear-resisting system such as welded steel girders, vierendeel frames, or diagonals. This idea has given rise to a whole new category of composite systems characterized by their use of super columns.

The construction of super columns can take on many forms. One system uses large-diameter thin-walled pipes or tubes filled with high-strength concrete generally in the range of 6 to 20 ksi (41 to 138 MPa) compressive strength. Generally, neither longitudinal nor transverse reinforcement is used thereby simplifying construction. Another method is to form the composite column using conventional forming techniques; the only difference is that steel columns embedded in the super column are used for the erection of steel framing and also for additional axial and shear strength.

## 6.4 Example Projects

### 6.4.1 Composite steel pipe columns

The 44-story Pacific First Center designed by Skilling Ward Magnusson, Inc., is an example of a building with large-diameter composite pipes. It has eight 7.5 ft (2.3 m) diameter pipe columns at the core,

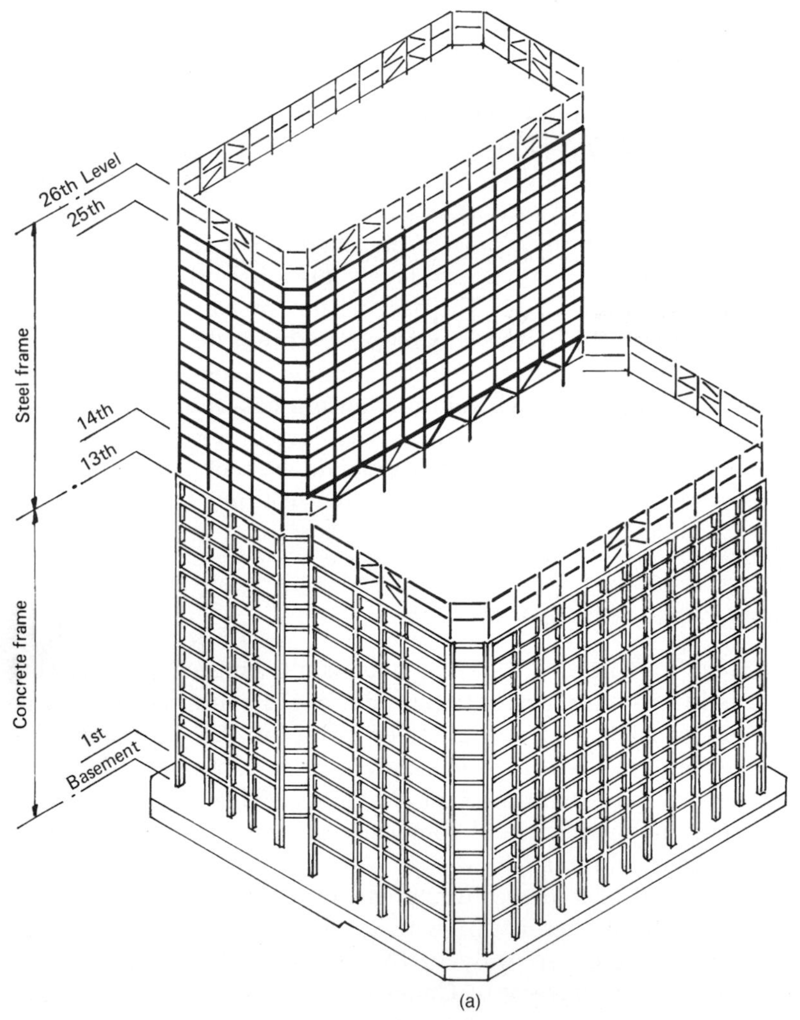


**Figure 6.10** Vertically mixed system: (a) schematic perimeter framing; (b) schematic bracing concept.

and perimeter columns with a maximum diameter of 2.5 ft (0.76 m), both filled with 19 ksi (131 MPa) concrete. Another example is the 62-story Gateway Tower in which 9 ft (2.7 m) diameter pipe columns exposed at the inner square of the hexagon are tied together with 10-story high X-braces.

The Union Square designed by the same firm is a 58-story building with four 10 ft (3 m) diameter pipe columns in the core. The pipes are filled with 19,000 psi (131 MPa) concrete. The building has fourteen more composite steel pipe columns of smaller diameter placed along

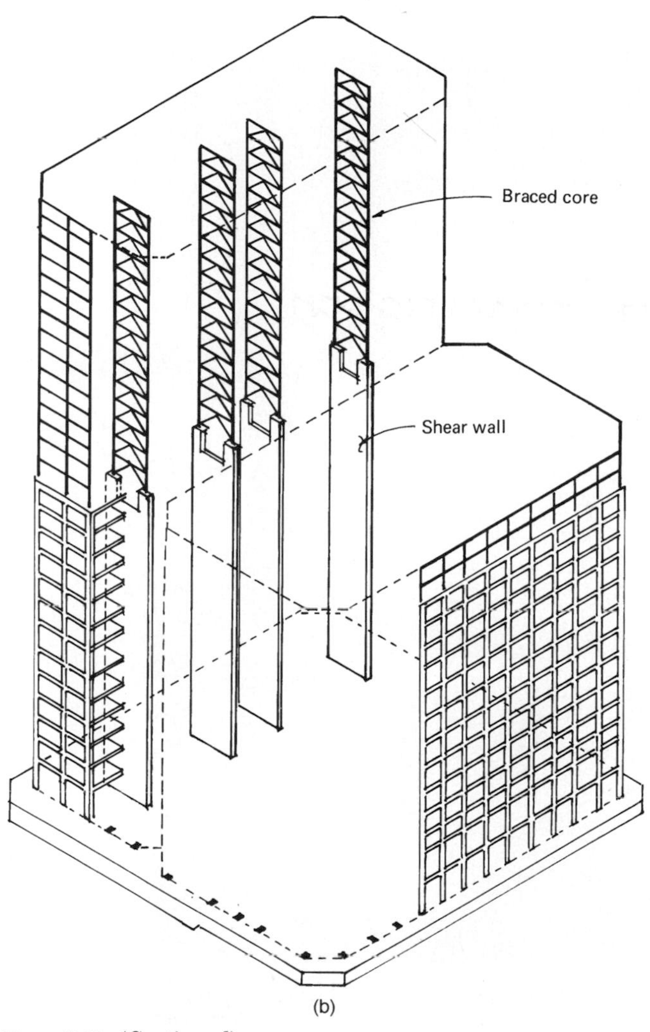


**Figure 6.10** *(Continued)*.

the building perimeter to carry gravity loads. The steel pipes provided erection steel and replaced forms as well as vertical bars and horizontal ties for the high-strength concrete. There are no reinforcing bars in the pipe columns. The pipes are connected to the concrete with studs welded to the pipe's interior surfaces.

As a diversion from high-rise buildings, it may be interesting to look at composite columns in a non-high-rise application. Figure 6.11 is a photograph of a project called The Fremont Street Experience in Las Vegas, Nevada. By covering the street with a space frame vault, five downtown blocks have been transformed into a pedestrian mall. The space frame provides partial shade from the sun and supports a

**Figure 6.11** Fremont Street Experience: (a) general view; (b) typical section through vault; (c) composite column; (d) tie-beam reinforcement detail.

graphic display system to entertain visitors to Fremont Street. The overall dimensions of the frame are 1387 ft long (422.75 m) and 100 ft (30.5 m) wide, with a 50 ft (15.25 m) radius forming a semi-circle in cross-section. The space frame with a depth of 5.77 ft (1.76 m) is supported on composite columns at intervals of between 180 and 200 ft (54.87 and 61 m) along the lengthwise span of the vault. Typical space frame steel components are 3 in. (76 mm) diameter round steel tubing, with a wall thickness of 0.120 in. (3 mm). Struts with

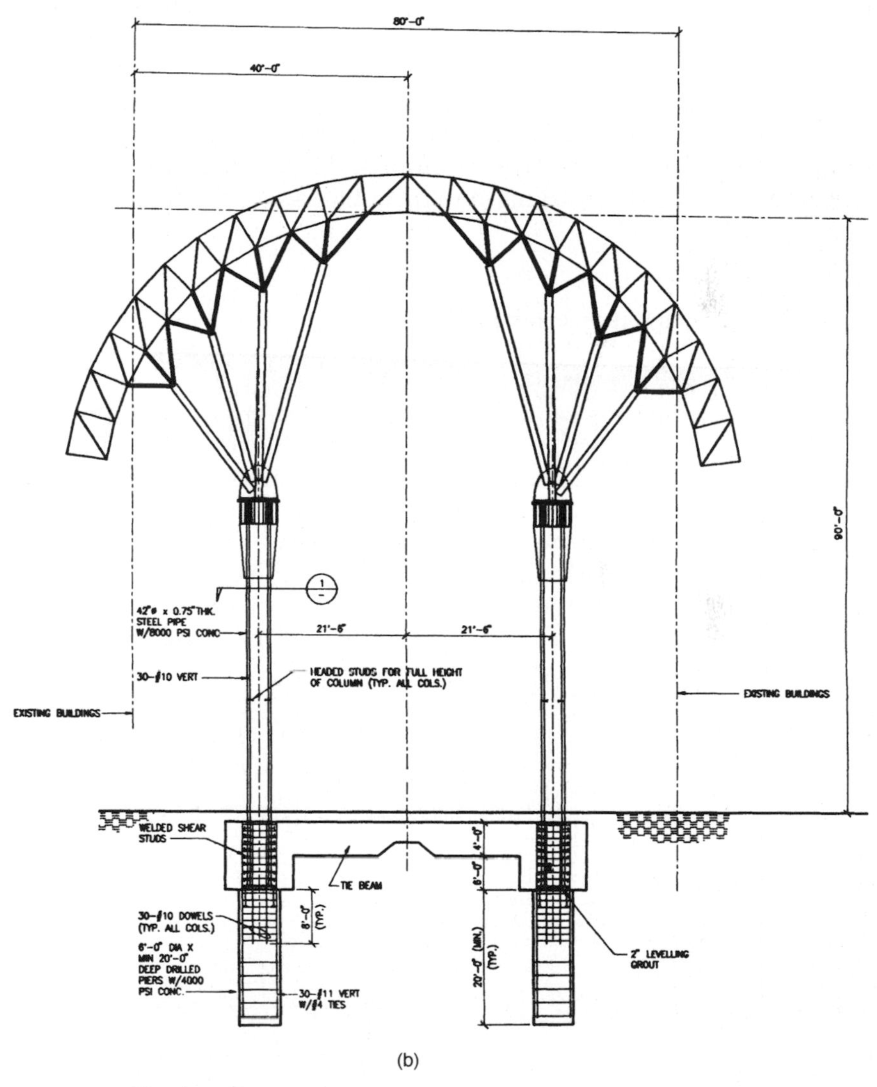

(b)

**Figure 6.11** *(Continued).*

heavier wall thickness are also used where additional strength is required.

The composite columns consist of 42 in. diameter by 0.75 in. thick (1067 mm × 19 mm) steel pipes with 8000 psi (55.16 MPa) concrete. Headed studs $\frac{1}{2}$ in. (12.7 mm) diameter by 8 in. (203 mm) are welded to the inside face of the tube at a vertical spacing of 12 in. (305 mm) and a radial spacing of 9 in. (228 mm). The bending capacity of the pipe column is developed below the street level by using: (i) welded shear studs around the outer surface of the column imbedded in the foundation; and (ii) by extending mild steel reinforcement inside of the column into the foundation. Figures 6.11b–d show schematic section through vault, composite column and grade beam reinforcement details. The architectural design is by the Jerde Partnership Inc., Venice, California while the structural engineering of the support system for the vault is by John A. Martin & Associates, Inc., Los Angeles. The space frame design is by Pearce Systems International, Inc.

### 6.4.2 Formed composite columns

#### 6.4.2.1 Interfirst Plaza, Dallas

A variation of the tube concept using formed composite columns was developed by LeMessurier Consultants, Inc for the 73-story 921 ft (281 m) tall Interfirst Plaza, Dallas, completed in 1985. To satisfy the request for offices with uninterrupted views, the weight of the entire building is placed on sixteen composite columns located up to 20 ft (6 m) in-board from the building exterior. The 20 ft distance between the glass and columns allowed for a continuous band of offices with uninterrupted views. To compensate for loss of bending ridigity, all

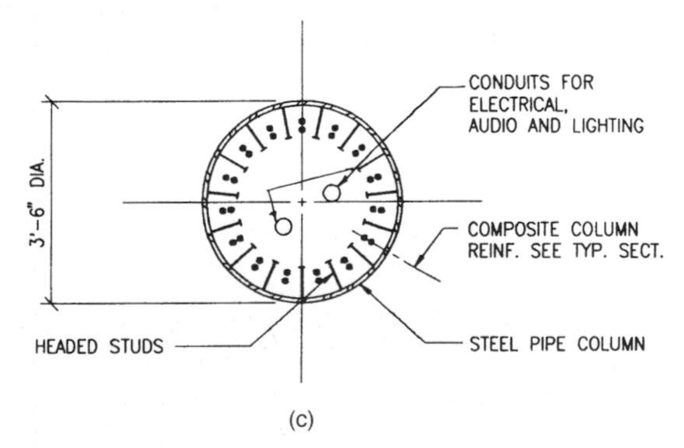


**Figure 6.11** *(Continued).*

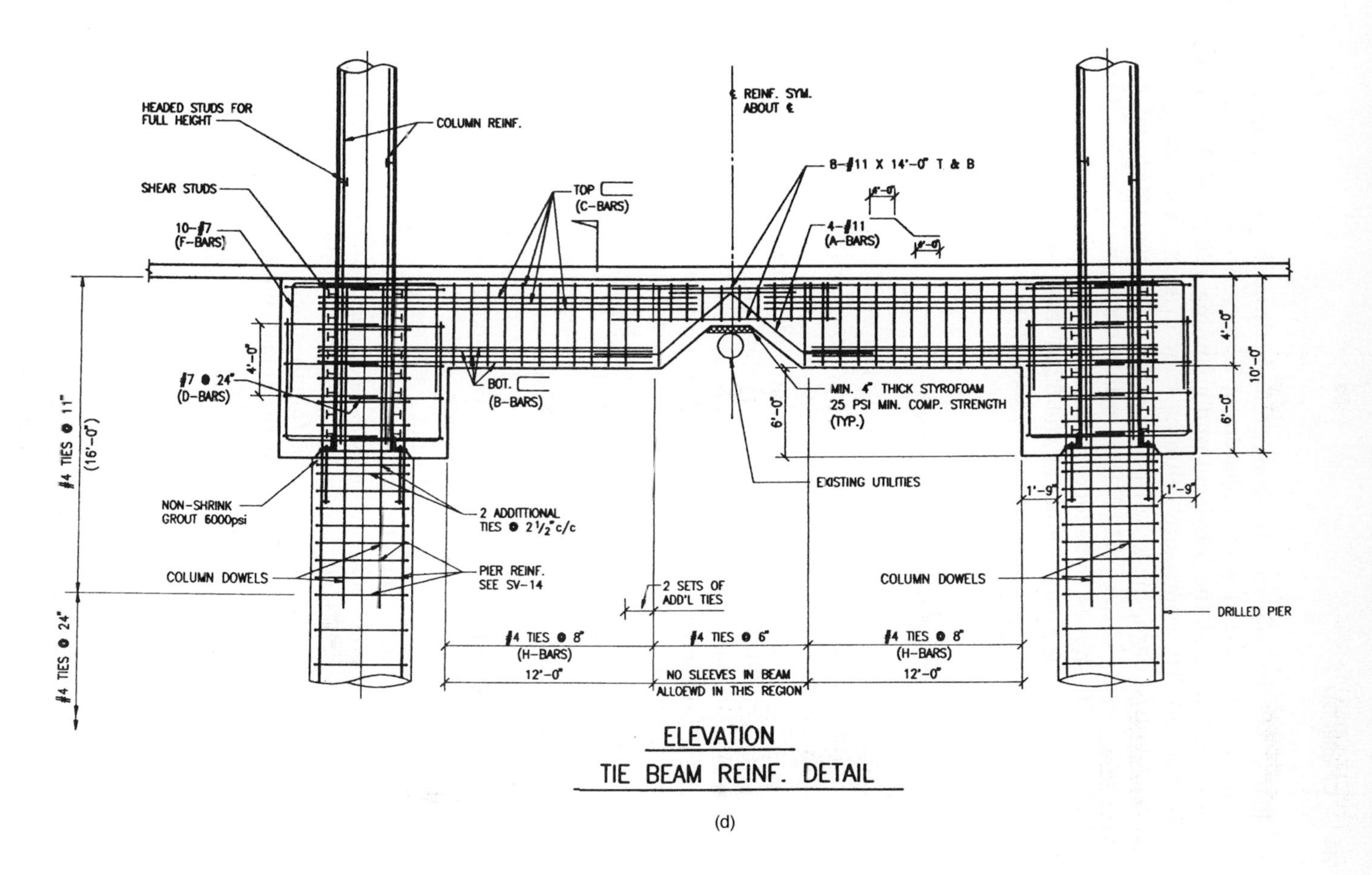
HEADED STUDS FOR
FULL HEIGHT
COLUMN REINF.
℄ REINF. SYM.
ABOUT ℄
8-#11 X 14'-0" T & B
SHEAR STUDS
TOP
(C-BARS)
10-#7
(F-BARS)
4-#11
(A-BARS)
4'-0"
#7 @ 24"
(D-BARS)
BOT.
(B-BARS)
MIN. 4" THICK STYROFOAM
25 PSI MIN. COMP. STRENGTH
(TYP.)
6'-0"
4'-0"
10'-0"
6'-0"
#4 TIES @ 11"
(16'-0")
EXISTING UTILITIES
1'-9"
1'-9"
NON-SHRINK
GROUT 6000psi
2 ADDITIONAL
TIES @ 2 1/2" c/c
COLUMN DOWELS
PIER REINF.
SEE SV-14
2 SETS OF
ADD'L TIES
COLUMN DOWELS
DRILLED PIER
#4 TIES @ 24"
#4 TIES @ 8"
(H-BARS)
12'-0"
#4 TIES @ 6"
NO SLEEVES IN BEAM
ALLOEWD IN THIS REGION
#4 TIES @ 8"
(H-BARS)
12'-0"
ELEVATION
TIE BEAM REINF. DETAIL
(d)

all loads are transferred to the ground through the composite columns interconnected with a system of 7-story two-way vierendeel trusses, beginning at the 5th level and spanning 120 ft (36.6 m) and 150 ft (45.7 m). The composite columns vary in size from 5 to 7 feet (1.5 to 2.1 m) square, made with 10 ksi (69 MPa) concrete, and are reinforced with 75 ksi (517 MPa) reinforcing bars and 50 ksi (345 MPa) W36 shapes imbedded in concrete. The concrete encasement of wide flange shapes ends at the 62nd level. A schematic floor framing plan and discussion of structural system is given in Chap. 1.

#### 6.4.2.2 Bank of China Tower, Hong Kong

The prism-shaped building, designed by the architectural firm of I. M. Pei and Partners, and structural engineer Leslie E. Robertson, rises to a height of 76-stories. Each of the four quadrants of the building rises to a different height, and only one out of the four reaches the full 76-stories. The bracing system uses a system of space trusses to resist both lateral loads and almost the entire weight of the building. From the top quadrant down, the gravity load is systematically transferred out to the building corner columns. Transverse trusses wrap around the building at various levels and help in transferring the load to the corner columns. At the 25th floor, the column at the center of four quadrants is transferred to the four corners by the space truss system, providing an uninterrupted 158 ft (48 m) clear span for the banking lobby. To achieve continuity between different truss members of the space frame, instead of complex three-dimensional connections requiring expensive weldments, the members are made to act as a single unit by encasing them in reinforced concrete columns. The concrete encasement surrounding the steel columns acts as a shear transfer mechanism and also counteracts eccentricities in the truss system. The lateral loads are thus carried down to the fourth story through the space truss system and corner columns. At the fourth floor, shear forces are transferred to a system of interior composite core walls through $\frac{1}{2}$ in. (12 mm) thick steel plate diaphragms acting compositely with the concrete slab. Although much of the shear force collected by the interior core walls is transferred to the 3 ft (0.9 m) thick slurry walls at the perimeter, the core walls are continued to the foundation to serve the dual function of resisting shear and of forming walls for the bank vault. The corner columns, which continue to the foundation, resist the overturning moments. The foundation for the building consists of caissons at bedrock. Some of the caissons are as large as 30 ft (9.1 m) in diameter. A schematic representation of floor plans, bracing concept, and photograph of the building are shown in Figs 6.12 and 6.13.

#### 6.4.2.3 Bank of Southwest Tower, Houston, Texas

The Bank of Southwest Tower, an 82-story, 1220 ft (372 m) building, proposed for downtown Houston, Texas uses the unique concept of

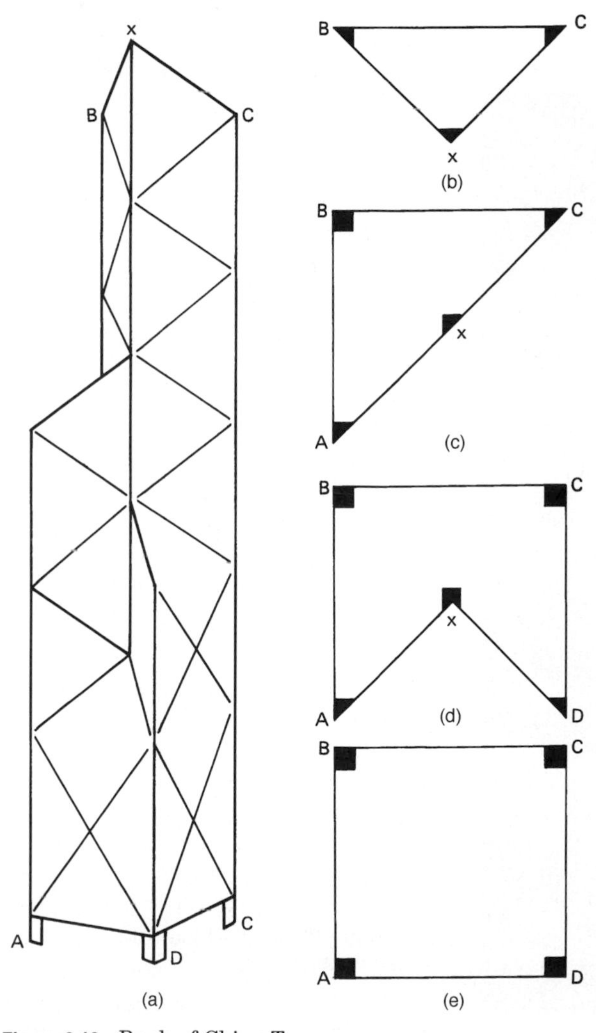


**Figure 6.12** Bank of China Tower.

composite columns with interior steel diagonal bracing. The diagonals are used to transfer both the gravity and lateral loads into eight composite super columns. The building has a base of only 165 ft (50.32 m), giving it a height-to-width ratio of 7.4. The characteristic feature of the design consists of a system of internal braces that extend through the service core and span the entire width of the building in two directions. A typical bracing consists of an inverted K-type brace rising for nine floors, there being two such braces in each direction. Eight of these 9-story trusses are assembled one on top of another within the tower. All the gravity loads are transferred to eight massive composite columns located at the building perimeter.

**Figure 6.13** Bank of China Tower: photograph.

Note that because the transverse diagonals are interconnected, a three-dimensional behavior is invoked; all eight columns participate in resisting the overturning moment. As compared to a framed tube, the eight-column structural system frees the perimeter from view-obstructing columns, especially at the building corners. The structural engineering is by LeMessurier Consultants, Inc., and Walter P. Moore & Associates, Inc.

### 6.4.3 Composite shear walls and frames

#### 6.4.3.1 First City Tower

The building designed by structural engineers Walter P. Moore & Associates, Inc., consists of a base structure extending one level below the plaza, covering an entire downtown Houston block and tower extending through 49-stories giving 1.4 million square feet (130,000 $m^2$) of office space. The building is a parallelogram in plan and is positioned on the site to create views (Fig. 6.14). Among the architecturally distinctive features of First City Tower are the four 11-story-high indentations on the façade where the glass panels open the building for viewing. These "vision strips" rise in staggered formation along the height of the two long faces.

**Structural anatomy.** In common with most other projects, First City Tower went through a design metamorphosis. The final scheme

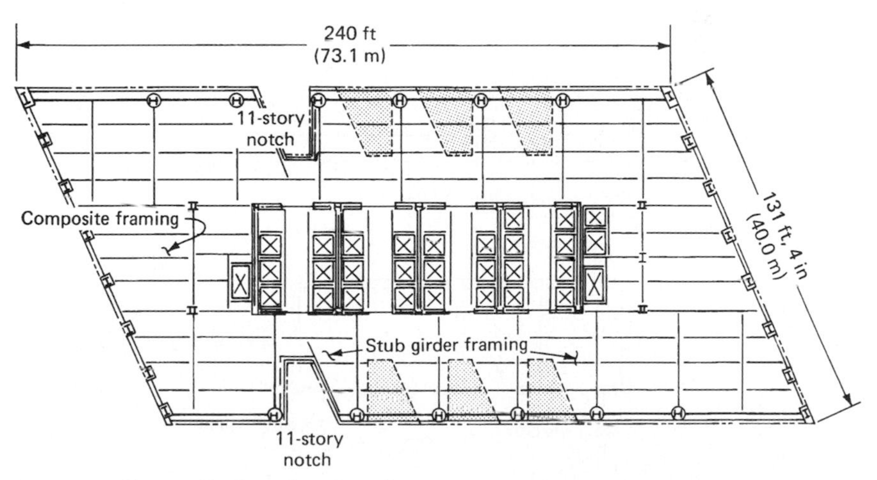


**Figure 6.14** Composite floor-framing plan.

adopted has the following composite components, as shown schematically in Fig. 6.15.

1. Composite floor-framing system, consisting of steel beams and concrete slab on formed steel deck interconnected with shear studs.
2. Composite stub girder system, consisting of rolled steel beams connected to floor slab with a series of stubs welded to the beam. This system discussed in Chap. 9, minimizes floor-to-floor height by combining the space required for mechanical ductwork within the design of the structural system.
3. Composite columns consisting of steel rolled shapes embedded in reinforced concrete columns (Fig. 6.16).
4. Composite frame consisting of composite columns and moment-connected steel beams (Fig. 6.16).
5. Composite shear walls consisting of a series of I- and C-shaped reinforced concrete shear walls interconnected with steel link beams (Fig. 6.16).
6. Composite construction that allows the initial growth of steel followed by placing of concrete to form composite columns and shear walls as shown schematically in Fig. 6.18.

**Design.** Figure 6.16 shows the arrangement of vertical reinforcement and structural steel core column for composite vertical framing

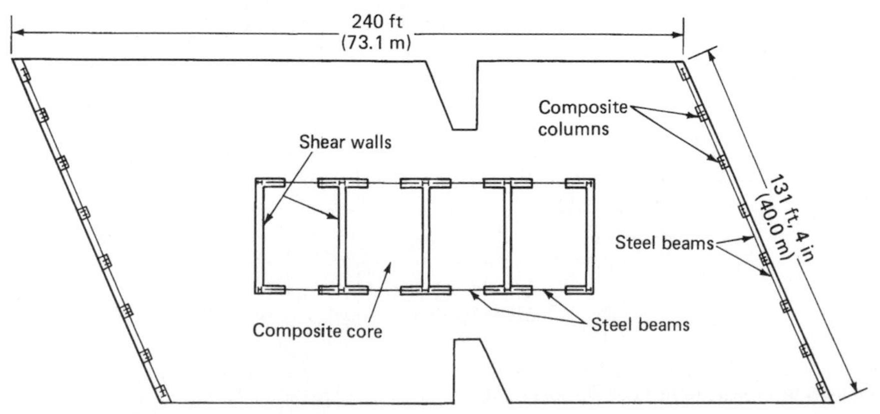


**Figure 6.15** Composite elements.

elements. Typically, the embedded steel columns vary from a W14 by 370 lb/ft (455 by 418 mm, 551 kg/m) at the bottom to a W14 by 68 lb/ft (356 by 254 mm, 101 kg/m) member at the top. For rectangular columns, width of the concrete envelope is maintained for full height, while the depth of column into the building is reduced in the upper floors.

The vertical reinforcement in the columns varied from #18 bars (57 mm in diameter) at the bottom to #7 bars (22 mm in diameter) at

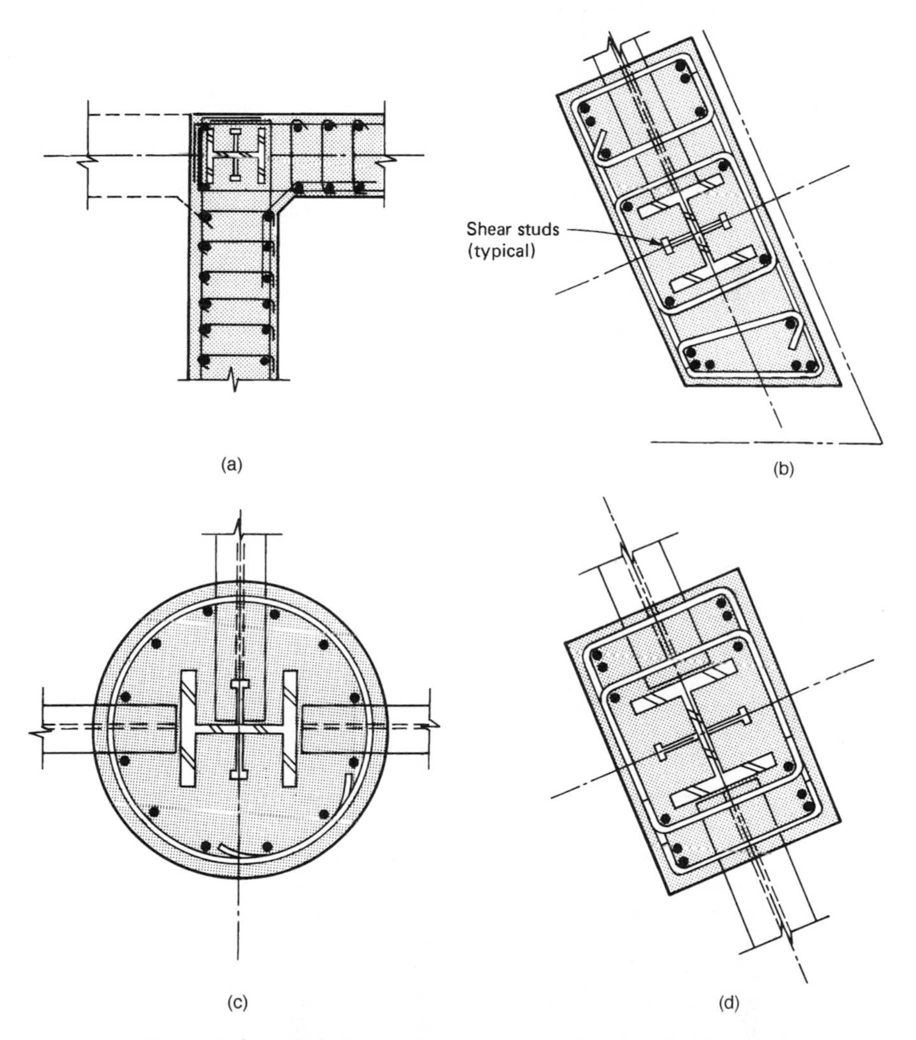


**Figure 6.16** Composite vertical elements: (a) composite shear wall; (b) composite corner column; (c) typical circular column on long faces; (d) typical exterior column on short faces.

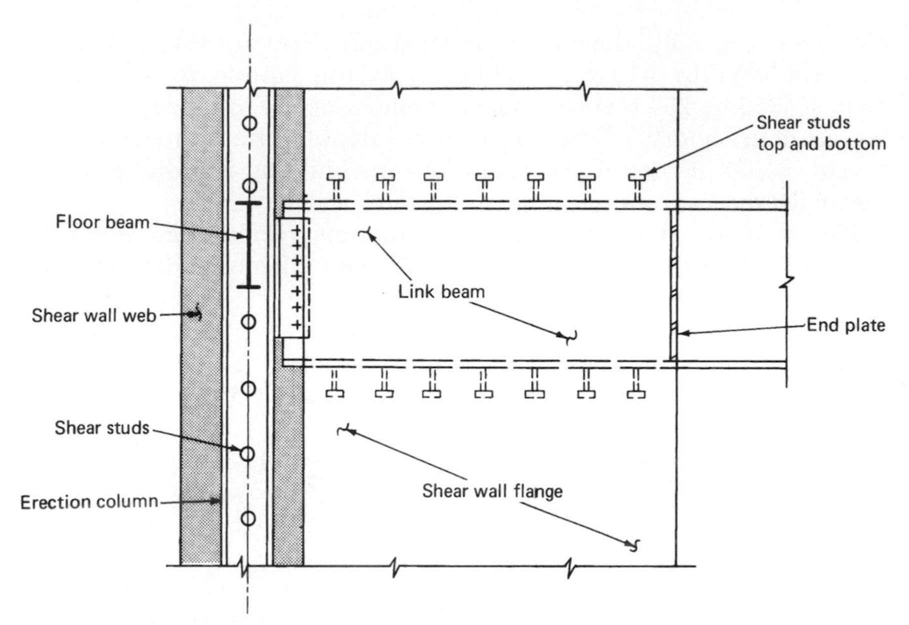


**Figure 6.17** Arrangement of link beam in shear wall.

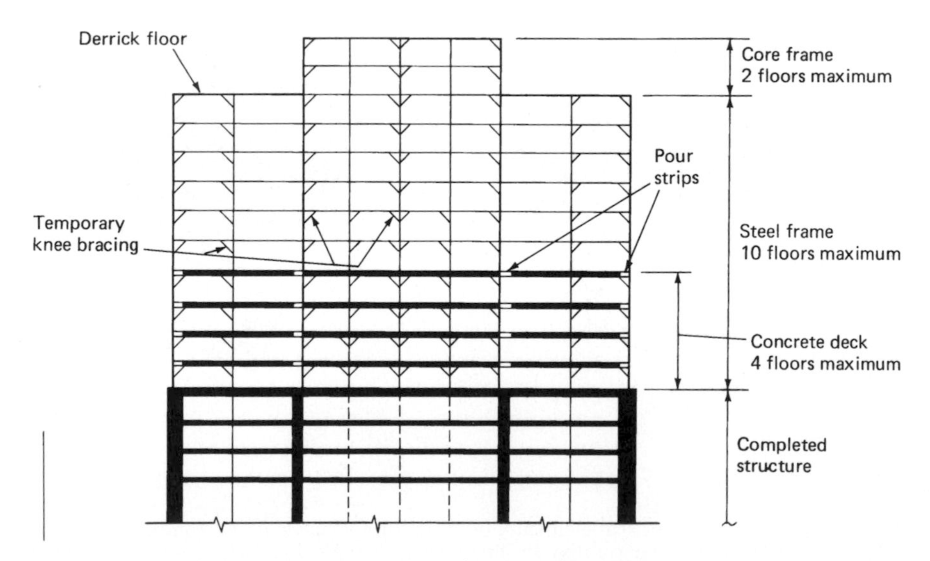


**Figure 6.18** General construction sequence in composite structures.

the top. Open ties permitted in low seismic zones are used throughout to facilitate placement of column reinforcement. Mechanical tension splices are used for frame columns on the short faces, and compression mechanical splices are used for the gravity columns on the long faces. Shear studs shop-welded to the webs of steel columns are used to achieve load transfer from concrete to steel.

Figure 6.16a shows the arrangement of reinforcement and erection column embedded in the shear wall. Typically, a W10 by 72 lb/ft (267 by 257 mm, 107 kg/m) steel column is embedded at each shear wall corner. The ties enclosing the vertical reinforcement are needed if 1 percent or more reinforcement is required for compression in the walls. Typically, the ties are used in the shear walls up to level 6 only.

Figure 6.17 shows a typical connection detail between concrete shear wall and a typical interconnecting beam. The beam to embedded steel column connection is a typical bolted shear connection. Moment capacity required at the face of the shear wall is developed by means of shear transfer mechanism between the concrete and shop-welded studs at the top and bottom flanges of the beam. The stiffener plate, set flush with the wall face, has no structural purpose but helps in simplifying shear wall forming around the beam. A conventional moment connection was used between the core columns and beams at levels where the shear walls were dropped.

**Composite construction.** The basic idea of composite construction is to let the construction of the steel frame advance to a predetermined number of stories first and then to envelop the column with concrete, as shown schematically in Fig. 6.18. The step-by-step process is as follows. After completion of the foundation system, the steel frame erection is started using standard procedures and AISC tolerances. Since the general idea is not to wait for the concreting of shear walls before steel erection, small steel columns are utilized for vertical support at the intersection of shear wall web and flanges. In order to limit the size, heavy steel sections are embedded in the exterior columns. The erection of exterior spandrels, stub girder, and purlins proceeds as in a conventional steel frame.

The size of the interior steel columns embedded within the shear walls depends on the maximum separation of the derrick floor from the level to which concrete shear walls have been completed. For this project, the criterion was set at a maximum of 10 floors. This was established during the early phase of design with proper coordination from the steel erector and the general contractor. Metal deck installation and welding of spandrels on the short faces follow closely behind steel erection. Diaphragm action of the deck is established by welding metal deck to beam flanges and by the installation of shear studs welded through the deck. Concrete topping is placed on metal deck several floors above the concreted portion, with the exception of pour strips around the shear walls and exterior columns.

The completed floor slab serves as a platform to transport concrete from a material hosit to shear walls and exterior columns. The reinforcing steel for the exterior columns and shear walls is tied into position several floors high with story-high bar lengths. The form work is then placed around the reinforcing cage.

Concrete is hoisted through the material hoist at the exterior of the building and then buggied over the concrete slab to the required location. Columns and walls below the floor are concreted by pouring concrete through the pour strips left around the shear walls and columns. Metal deck in the pour strip around the shear walls is cut to provide room for hoisting of shear wall forms. The forms are hoisted to the floor above through these temporary openings.

As a next step, the shear walls above the floors are poured and integration between floor slab diaphragm and shear walls is completed by concreting the pour strips. The various stages of the shear wall construction sequence are shown schematically in Fig. 6.19.

With proper sequencing of different trades, the construction of concrete shear walls and columns can proceed at a pace equal to the

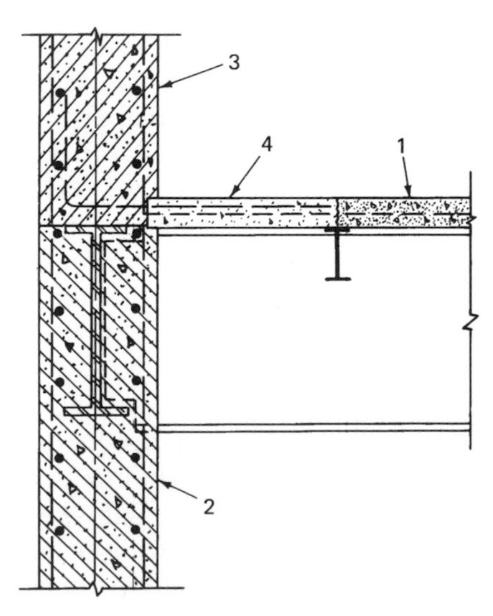


**Figure 6.19** Shear wall construction sequence.

overall speed of a conventional steel building. Typically for First City Tower, steel erection proceeded at the rate of two floors per week, with concreting of shear walls and columns following closely behind at the same rate.

**Temporary bracing.** In a pure steel or concrete building, there is no measurable lag between the levels at which the construction is proceeding and the level at which the building lateral resistance is established. For example, in a steel building moment connections or braces are connected between steel columns immediately behind the erection of the steel frame in order to provide resistance and stability to lateral loads. Monolithic casting of concrete beams and columns and other elements, such as shear walls, automatically provides the required stability. However, for a composite building, stability is not achieved until after the concrete is placed and cured. Since the level at which the steel framing is erected is deliberately made sufficiently high above the concreted level, there exists in composite construction a distinct need for considering bracing during the erection process. Steel cables traditionally used in steel buildings to plumb and stabilize the structure are often inadequate, requiring a more positive method of lateral stability. This was provided for the First City Tower by supplementary knee bracing around the building perimeter and also in the building core. The braces at the perimeter were later removed, while those in the core were left in place to become embedded in the composite shear walls since they did not interfere with the architecture of the building.

### 6.4.4 Composite tube system

#### 6.4.4.1 The America Tower, Houston, Texas

In this system a concrete exterior tube consisting of closely spaced columns and deep spandrels forms the essential wind-bracing system. Fast erection of the steel frame is achieved by using light structural columns around the perimeter. Interior columns and floor framing are identical to gravity-formed steel construction.

A somewhat new technique, as mentioned earlier, adapts a variation of this concept. This system does away with the concreting of the spandrels; instead, steel spandrels are employed in conjunction with composite columns to resist the lateral loads. This is the system used on the 42-story America Tower building, designed by structural engineers Walter P. Moore and Associates, Inc., Houston, Texas.

The America Tower project consists of 1 million square feet

($92{,}858\ m^2$) of office space and is elliptical in plan with offsets on the long faces to highlight the verticality of the building and provide more corner offices. The exterior façade is designed with a curtain wall of alternating glass and precast panels. The architectural and structural framing plans are shown in Figs 6.20 and 6.21.

**Structural anatomy.** The salient structural features are given in Table 6.1. A framed tube consisting of closely spaced exterior columns and spandrels was chosen as the proper structural solution.

**Construction.** As is common with most tube buildings, the system used for America Tower is the so-called prefabricated tree column system (Fig. 6.22). A typical tree column consists of a 2-story column and two levels of spandrels. The major difference in the mixed tube construction is that the column sizes are small compared to an all-steel scheme, resulting in savings in the fabrication and erection of steel members.

Another feature that has resulted in considerable savings is the use of a hybrid tubular frame on the perimeter of the building. Composite columns used in the typical levels (floors 3 through roof) are integrated into pure structural steel columns at the second level. This marriage of steel columns to composite columns simplified the construction of the exterior columns at nontypical lower levels.

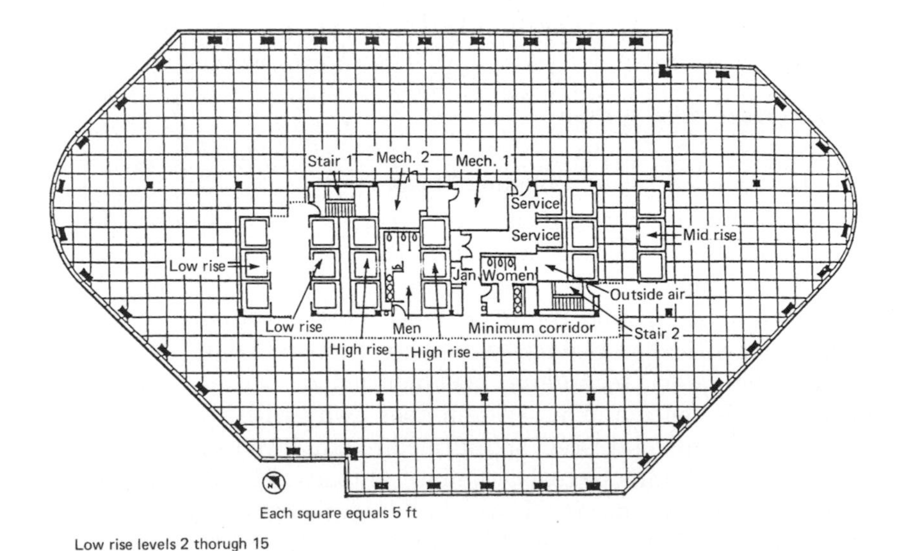


**Figure 6.20** America Tower architectural floor plan.

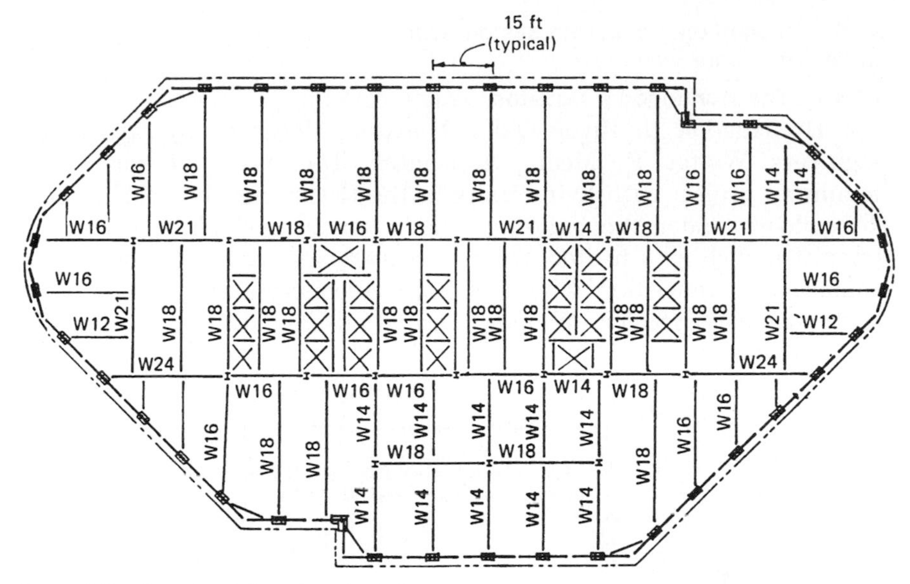


**Figure 6.21** Structural floor framing plan.

Although the savings in the structural cost for this scheme appeared to be only marginal from the comparative study, the contractor elected to use this scheme because his recent job experience on another similar high-rise project had proved that a hybrid frame is more economical than a full-height composite scheme. Figure 6.23 shows a detail of intersection of composite column with a steel column. The resistance to lateral loads during the erection of the tower was provided by the tube action of the erection columns and spandrel beams. The columns, which were sized for 12 floors of construction gravity loads, were found to be adequate for providing resistance to lateral loads. Additional welds between the erection column and the spandrel and stiffeners were necessary in order to develop the bending capacity of the columns.

**TABLE 6.1 America Tower: Structural Features**

| | |
|---|---|
| Height above mat | 619 ft 7 in. (188.85 m) |
| Width | 125 ft 4 in. (38.2 m) |
| Height/width ratio | 4.94 |
| Interior columns | W14 rolled steel sections |
| Floor framing | W18 composite beams span 41 ft (12.45 m) |
| Exterior columns | 42 by 27 in. (1.07 by 0.88 m) composite columns |
| Spandrels | W36, W33, W30 steel shapes |
| Mat | 6 ft (1.83 m) thick |

## 6.4.5 Conventional concrete system with partial steel floor framing

### 6.4.5.1 The Huntingdon, Houston, Texas

The Huntingdon in River Oaks, Houston, designed by structural engineers Walter P. Moore Associates, Inc. is a 34-story condominium project with living units in the tower, resident parking on three floors under the building, and a landscaped plaza featuring amenities such as a heated swimming pool, spa, pool-side room, etc. Construction consists of a poured-in-place concrete structure with a white precast exterior. Windows are dual-pane tinted glass with operable vents.

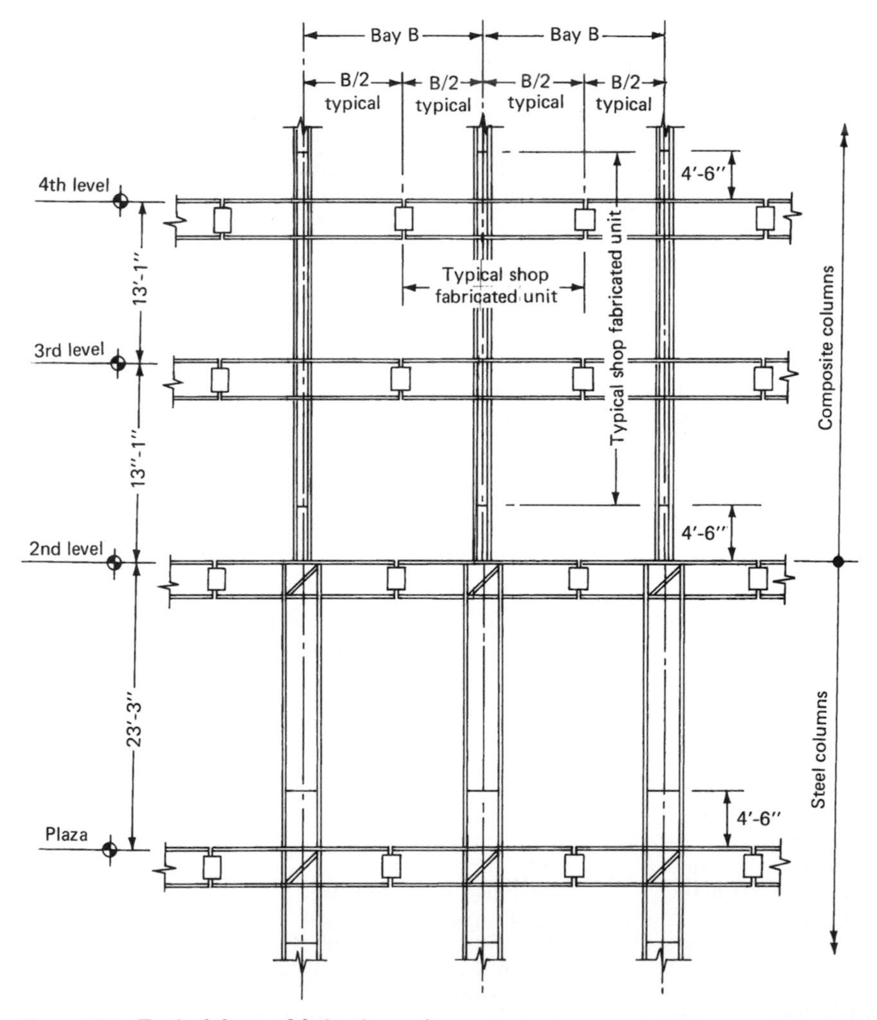


**Figure 6.22** Typical frame fabrication unit.

**Structural anatomy.** The building is essentially symmetrical in plan with a rectangular dimension of 142 ft 6 in. by 106 ft 6 in. (43.5 by 32.5 m) and is 490 ft (149.5 m) tall. The architectural design called for column-free space between the core and exterior of the building. The desired elevation dictated that exterior columns be spaced at 20 ft (6.1 m) centers around the building except at the corners. Columns were not permissible at the corners because each home has a balcony offering views in two directions. With this layout of exterior columns it was not possible to develop the tube action to resist the

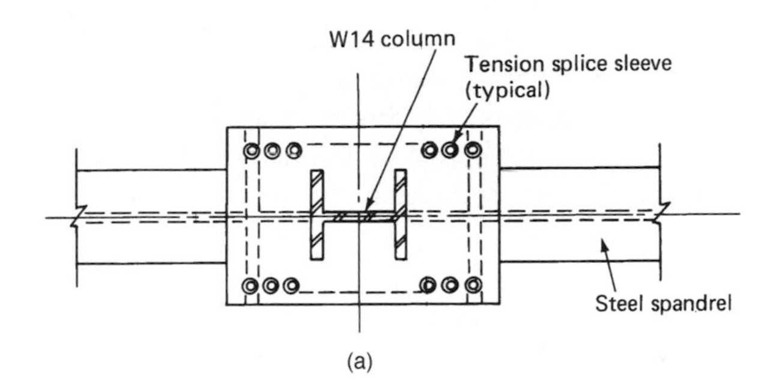


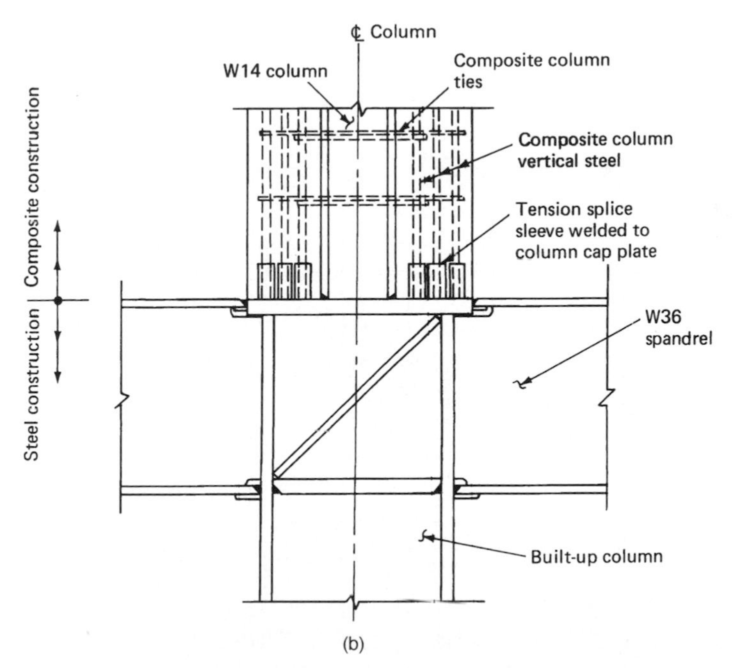


**Figure 6.23** Composite column to steel column connection detail: (a) plan; (b) elevation.

lateral loads because of lack of continuity of structural resistance around the corners. A preliminary analysis was performed with 24 in. (610 mm) deep pan joists for floor system and an interactig system of shear wall and end frames for lateral loads. The analysis indicated that the system could be designed for strength without too much of a premium in material quantities, but the calculated lateral deflection under Houston hurricane wind code was too large for the occupancy of the building. Addition of stiffening elements without introducing more columns was necessary to maintain reasonable size of structural members and to reduce deflection.

The addition of girders between the core and exterior at each column line as shown in Fig. 6.24 was the answer. In comparison to the pan joist floor system, these girders are very stiff and help in mobilizing the exterior columns to resist the wind loads. The resulting structural system consisting of the core, end frames, and exterior columns reduced the deflection to well within the drift limitations.

**Haunch girders.** A framing system employing girders of constant depth criss-crossing the interior space between the core and the exterior often presents nonstructural problems because it limits the space available for passage of air conditioning ducts. A system

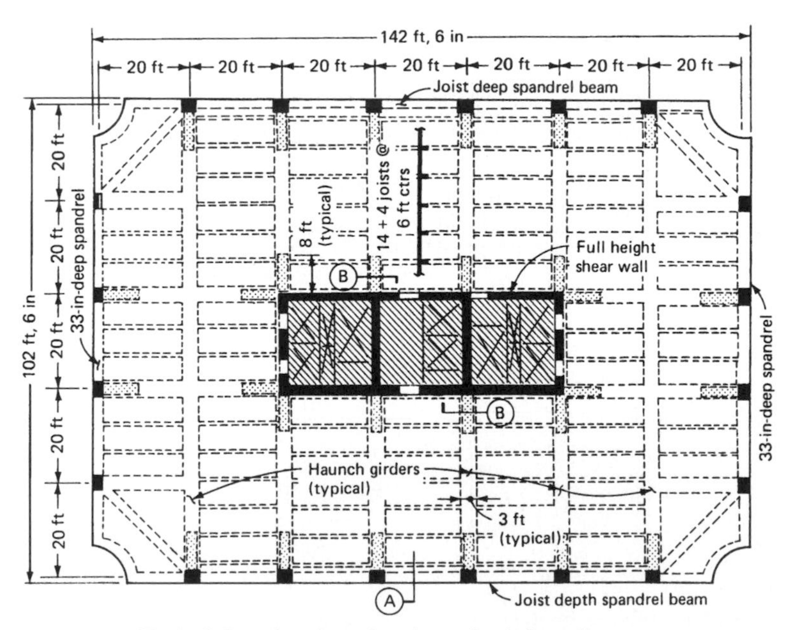


**Figure 6.24** Typical floor framing plan: haunch girder scheme.

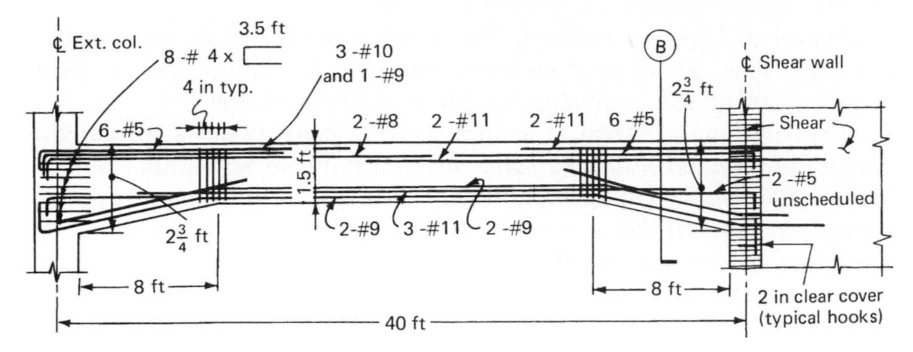


**Figure 6.25** Haunch girder elevation and reinforcement.

known as the haunch girder system, with wide acceptance in many parts of North America, alleviates nonstructural problems without making undue compromises in the structure. The basic system consists of a girder of variable depth as shown in Figs 6.25 and 6.26. The use of shallow depth at the center facilitates the passage of mechanical ducts and eliminates the need to raise the floor-to-floor height. Material quantity comparisons for gravity loads indicated that haunch girder framing is very economical. Additional comparison for lateral load design indicated even more savings from the reduced floor-to-floor height.

**Flying forms.** For the *in situ* construction to be competitive with other forms of construction, such as precast concrete or steel, it is imperative that the forming for the floor system should lend itself to repeated uses in a multistory frame. One-way joist system using flange-type steel pans is a reusable system that is currently popular

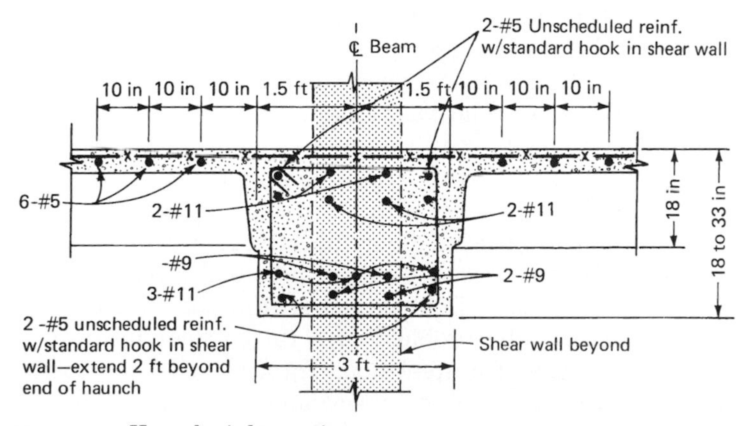


**Figure 6.26** Haunch girder section.

in almost all high-rise buildings in North America. Wide pans in two common sizes of 53 and 66 in. (1346 and 1676 mm) widths have become a standard, replacing the 30 in. (762 mm) pans.

A conventional method of concrete joist construction is to nail the flanges of pans on a formed solid decking built on adjustable shores. When concrete has gained sufficient strength, shores are removed and pans stripped and stored in an area to await reuse until shoring and wooden decking are erected for the next floor. This method of forming is labor-intensive and also time consuming, particularly when joists of different lengths are used.

A method of forming that fully utilizes the opportunity for repetition offered in multistory construction is a system known as flying form system. In this system joist pan forms are attached to slab, creating a system of deck surface, adjustable jack, and supporting framework. For stripping, the form is lowered by jack, slipped out of the slab, and lifted to the next floor for reuse as one rigid structure, resulting in economy of handling and placing.

The floor framing used in the Huntingdon project lent itself to the use of flying forms. To further simplify the form work, the width of haunch girders was kept slightly larger than the maximum width of the exterior column.

**Mixed-floor construction.** The architectural requirement for the project dictated that typical floors be constructed at two different elevations with a 2 ft 6 in. (762 mm) drop occurring at the outline of the interior core. This design allows flexibility to the buyers in locating wet areas, whereas the common areas such as the service and passenger elevator lobbies and stair landings are kept higher because of predetermined finish. Forming of a poured-in-place concrete slab at two different elevations on either side of the core would have been somewhat complicated. This required taking a fresh look at the framing of slab inside the core.

Compared to an office building, the vertical transportation requirements for a condominium are relatively small. The core size dictated by the elevator and mechanical requirements for this project worked out to 50 by 20 ft (15.2 by 6.1 m). The elevator lobbies normally required in an office building are eliminated in the core arrangement because the elevators open on each floor directly to a private foyer serving a maximum of two homes, making it possible to have a very small core. The floor area that is enveloped within the core is very small. This, combined with the problem of a raised floor inside the core, made possible the use of a steel floor framing with composite deck, unwittingly creating yet another variation to the many types of mixed construction currently in vogue for tall buildings.

Pockets were provided in the shear wall to receive the steel beams in the core. Shelf angles were quick-bolted to the walls in areas where supports for the metal deck were required. This idea of using conventional steel framing in the core facilitated the construction of shear walls and the floor framing independent of the framing inside the core. Typically, the construction of shear walls and concrete framing proceeded at the rate of one floor a week. The steel framing in the core was deliberately left to lag behind the concrete framing so that the ironworkers could work without interfering with other trades.

## 6.5 High-Efficiency Structure: Structural Concept

A super-tall building is generally defined in architectural terms as a skyscraper when its silhouette has a slender form with a height-to-width ratio well in excess of 6. The slender proportion imposes engineering demands far greater than those of gravity, requiring the engineer to come up with innovative structural framing schemes. The ideal structural form is one that can at once resist the effect of bending, torsion, shear and vibration in a unified manner. A perfect form is a chimney with its walls located at the farthest extremity from the horizontal center but as an architectural form it is less than inspiring for a buliding application. The next best and a more practical form is a skeletal structure with vertical stiffness, i.e., columns located at the farthest extremity from the building center. Two additional requirements need to be incorporated within this basic concept to achieve high efficiency: (i) transfer as much gravity load, preferably all the gravity load into these columns to enhance their capacity for resisting overturning effects due to lateral loads; and (ii) connect columns on opposite faces of the building with a system capable of resisting the external shear forces. The ultimate structure for a rectangular building, then, will just have four corner columns interconnected with a shear resisting system. Such a concept, proposed by the author for a super tall building, is shown in Fig. 6.27, in which the total gravity loads and overturning moments are resisted by four composite columns. The columns are located in-board from the corners to allow for architectural freedom in modulating the short sides of the building. The shear in the transverse direction is resisted by a system of 12-story-high chevron braces while in the longitudinal direction the shear resistance is primarily provided by the full height vierendeel frames located on the long faces. The story-high longitudinal trusses located at every 12th floor permit cantilevering of the floor system. The primary function of

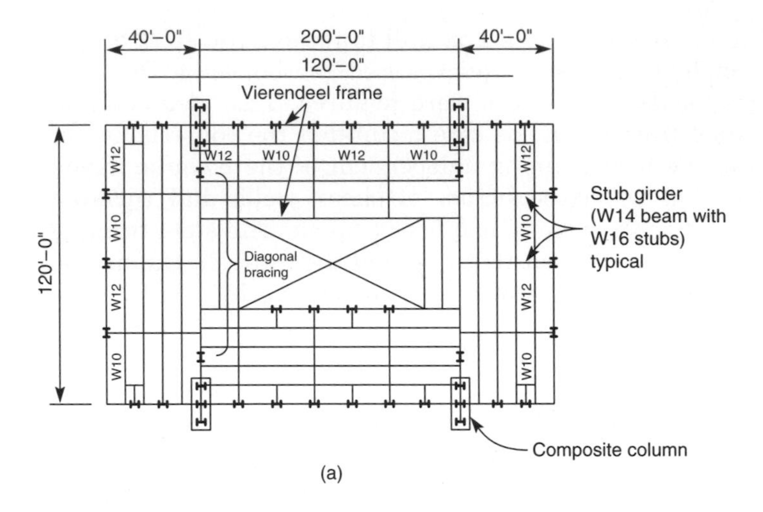


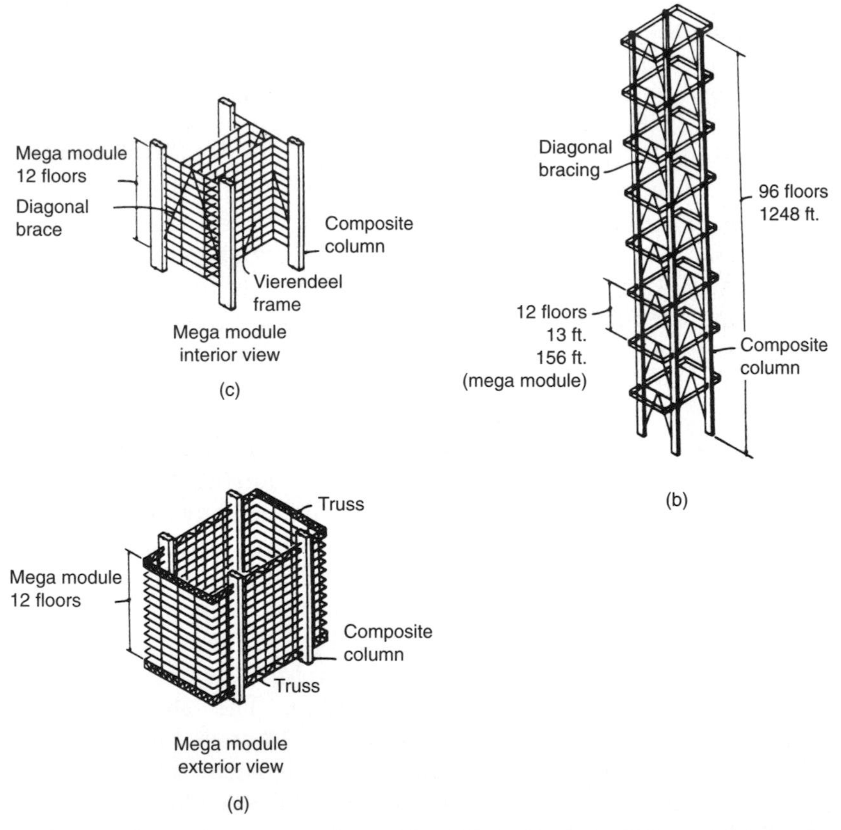


**Figure 6.27** Structural concept for a super tall building: (a) Plan; (b) schematic elevation; (c) interior view of mega module; (d) exterior view of mega module.

the interior vierendeel frame is to transfer the gravity loads of the interior columns to the composite columns via chevron braces. However, because of its geometry, it also resists external shear forces in the long direction.

Any number of conventional framing using composite beams, haunch girders, or stub girders may be used to span the distance from the core to the building perimeter. The scheme in Fig. 6.27 shows stub girders consisting of a W14 wide flange beam with W16 stubs welded at intervals to the top flange of W14 beam. This system discussed in Chap. 9 has the advantage of reducing simultaneously the floor-to-floor height and the unit quantity of steel required for floor framing.

The author believes that the scheme shown in Fig. 6.27 can be modified readily for a variety of architectural building shapes. Any desired slicing and dicing of the building on the short faces may be accommodated without inflicting an undue penalty on systems efficiency. The structural concept is complete. Like all other schemes it has to await its time before finding an application to a real building.

Chapter

# 7

# Gravity Systems for Steel Buildings

## 7.1 Introduction

There are basically three groups of structural steel available for use in bridges and buildings.

1. Carbon steel: ASTM-A36 and A529.
2. High-strength, low-alloy stels: ASTM-A440, A441, A572, and A588.
3. High-treated low-alloy steels: ASTM-A514.

In the A572 category six grades of steel: 42, 45, 50, 60, and 65 are available for structural use. The grade numbers correspond to the minimum yield point in kilopounds per square inch of the specified steel. Carbon steel is available in 36 and 42 grades, while the A514 group consists of tempered steel with specified yield points ranging from 90 to 100 ksi (620.5 to 689.5 MPa).

The most commonly available type and grade of structural steel shapes in stock is grade 36, comprising approximately 75 percent of U.S. production of structural shapes. Until about a decade ago, it was a routine and economical choice because of availability for early delivery and maximum competition among bidders. Although low-alloy, high-strength steels are available with yield points ranging from 40 to 65 ksi (276 to 448 MPa), the most common choice for high-strength steel is ASTM-A572, grade 50.

In the north American construction market, it is becoming more common for steel mills to sell dual certification ASTM A36, A572 steel meeting the requirements for both A36 and A572, grade 50. Since the cost is typically competitive with the cost of A36 steel, designers are able to take advantage of higher strength with little or no cost premium.

Steel buildings in the United States are designed as per the American Institute of Steel Construction (AISC) specifications, which were first published in 1923. The specifications are revised periodically to keep pace with new research findings and availability of new materials. Steel construction for buildings is commonly referred to as steel skeleton framing, signifying that a majority of the members consist of linear structural elements such as beams and columns.

Skeleton framing is normally erected in 2-story increments, each increment being called a tier. Light-gauge steel decking serving as a permanent form and as positive reinforcement for concrete topping is the most common method of slab construction.

The rules for the design of structural steel members subject to any one, or combination of stress conditions due to bending, shearing, axial tension, axial compression, and web crippling are given in the AISC specifications. Members can be designed by plastic, elastic or LRFD theories. Members subjected to gravity and lateral loads are designed with a 33 percent increase in allowable stresses.

The functional needs of occupancy invariably dictate that floors be relatively flat. In a steel building this is most often achieved by horizontal subsystems consisting of beams, girders, spandrels, and trusses over which spans a light-gauge metal deck. Concrete topping over the metal deck completes the floor system.

In this chapter a brief description of some of the elements normally employed in the framing of steel buildings is given. We first describe gravity loads, followed by a description of metal deck, steel beams, joists, and finally, columns. The description of metal deck is somewhat limited in this chapter because most usually metal deck is employed as a composite member, which we cover in greater detail in Chap. 9.

## 7.2 Design Loads

Gravity loads on buildings are of two kinds: (i) static; and (ii) dynamic. Static load is considered permanent, whereas the dynamic load is time-dependent. The weight of every element within the structure is a static load and includes weights of load-bearing elements, beams, slabs, columns, walls, ceiling, floor and wall finishes, sprinkler systems, light fixtures, sheet metal ducts, permanent partitions, exterior cladding, cooling towers, central plants, pump rooms, thermal storage tanks, and other mechanical equipment.

Although their effect is similar to dead loads and essentially static, live loads are less accurately predictable because they are subject to greater variation. Most often in North America a minimum value of 50 psf (2394 Pa) for live load plus 20 psf (958 Pa) for partition allowance is used in the design of speculative office buildings. Sometimes if tenants' requirements are known prior to the design of the building, allowance is made for expected usage of heavier loads such as book stacks, filing cabinets, computers and business machines. It is becoming increasingly common in certain cities of the United States to design an area 20 ft (6.0 m) deep adjacent to the exterior of the building for the minimum 50 plus 20 psf (2394 plus 958 Pa) partition load, while the interior space is designed for a heavier load of 100 psf (4788 Pa). The rationale is that the 20 ft (60 m) deep space adjacent to the building exterior invariably consists of office space with light furniture, whereas heavily loaded areas such as storage and computer rooms are tucked away from the exterior closer to the building core.

The loads given in the codes have built-in empirical safety factors to account for maximum possible loading conditions. They are given in the form of equivalent uniform and concentrated loads. Although there is a common understanding among engineers that the values given in the codes are rather conservative, invariably the design is done to the standards given in the codes applicable to the particular occupancy. However, when the actual layout and height of partitions are known, an attempt is made to justify a lower partition load. This is done more to justify the load-carrying capacity of an existing framing than to take advantage of load reduction in the design of a new facility.

Concentrated loads indicate possible single-load action at critical locations and are in addition to the uniform distributed load. It is perhaps obvious that the chances of having the full occupancy load simultaneously on every square foot of a large area in a building are next to zero. The larger the area under consideration, the less is the potential for having full occupancy load. This probability aspect is taken into consideration in the codes by allowing the use of live load reduction factors. However recent failures of hung, long-span structures have brought to light the necessity of taking extra precautions, especially in the detailing of connections to account for extraordinary situations, such as people crowding because of ceremonies, parties, fire drills, etc.

Construction loads are caused by building construction activities requiring stockpiling of materials on relativity small areas. Construction materials, such as dry walls and glass lights for curtain wall are usually stacked on each floor.

## 7.3 Metal Deck Systems

The primary function of a floor system is to collect and distribute gravity loads to vertical elements such as columns and walls and occasionally to tension members such as hangers. This is accomplished by the out-of-plane bending action of the floor system. Another structural function of the floor system is to transmit the lateral loads to various lateral-load-resisting systems. This action is called the diaphragm action. In addition to these, the nonstructural functions of the floor system are: (i) to support nonstructural components such as finish materials in ceiling and floor, piping, ducts, wiring, lighting, and sprinklers; (ii) to provide protection from damage that may be caused by fire; and (iii) to provide resistance to transmission of sound.

## 7.4 Open-Web Joist System

Open-web joists have been in use as floor and roof framing members since the early 1920s. The first joist used in 1923 was a warren-truss type with top and bottom chords of round bars and a web formed from single continuous bent bar. Since then, many types of joists have been developed, primarily to provide an economical floor and roof system. Their capacities and sizes are standardized and they are delivered to the site completely fabricated. They are made in standard depths from 8 to 30 in. (0.2 to 0.76 m) for the H series and 18 to 48 in. (0.46 to 1.2 m) for LH series, and 52 to 72 in. (1.32 to 1.84 m) for the DLH series. Recently K series have been introduced which are more economical than H series. Open-web joists are manufactured utilizing hot-rolled or cold-formed steel. They are designed according to the standards set by Steel Joist Institute (SJI), primarily as simply supported uniformly loaded trusses. The top chord is assumed to be continuously braced by a floor or roof deck. The bottom chord is designed as an axially loaded tension member. The web members are designed to resist both the vertical and horizontal shear. Either a diagonal or horizontal bridging is used to stabilize the joist. The number of rows of bridging required is a function of the clear span of the joist and the chord size. Typically three to five rows may be required for office buildings.

The reasons for providing bridging in open-web joist construction are: (i) to provide stability for the joist during construction; (ii) to maintain alignment of the joist at the specified locations; (iii) to control the slenderness ratio of tension bottom chord to below 240. Although this limitation is not essential for the structural integrity of tension members, it is adhered to in practice to maintain a certain minimum stiffness of the members to prevent undesirable lateral movements; and (iv) to provide lateral stability for compression diagonals.

Although other types of floor decks, such as precast concrete, gypsum, wood, or other materials capable of supporting the load can be used, in high-rise buildings concrete topping on metal deck is typical. The load tables published by the SJI are based on uniform load conditions. They can be used in selecting joists for gravity loads that can be expressed in terms of load per unit length of joist. Partitions, heavy pipes, and other elements running perpendicular to the joist, and mechanical units mounted on the joists should be treated as concentrated loads. In such cases the joist should be designed for the full combination uniform load and concentrated load.

In everyday engineering practice the engineer, relieved of the burden of designing the joists per se, selects the standard type and size required for the span and load conditions. Joists are manufactured with 50 ksi (344.75 MPa) steel for the chords and 36- or 50 ksi (248.2 and 344.75 MPa) steel for web members at the supplier's option. Chords of joists used for floor members are essentially parallel. Top chords are designed for uniform loads assuming either simple or continuous supports over the panel points.

It is a common practice to design the compression diagonals of an open-web joist with an effective length factor of $K = 1$ on the premise that the transverse flexural stiffness of the bottom tension chord coupled with the resistance of tension diagonals and bottom tension chord braces provides adequate stiffness. It behooves the engineer to verify that adequate bracing exists in unusual conditions with high stresses in the compression diagonals.

The roof joists should be checked for uplift capacity due to vertical suction forces induced by the wind loads. The net uplift force on roof joists is the gross uplift force due to wind, less the dead load of the roof system tributary to the joist, including the weight of the joist. If an adequate dead load exists in excess of the gross uplift load, no further consideration is necessary. If not, the design of the joist should be reviewed for reversal of stresses, such as compression of the bottom chord. Lateral bracing required to stabilize the bottom chord can be provided by: (i) additional bridging; (ii) rearranging the bridging provided for normal loading; or (iii) increasing the size of the bottom chord. The web system may also undergo stress reversal. Therefore, all components of the joist must be checked for stress reversal. As in other wind design, a 33 percent increase in allowable stresses is allowed, as long as the required section is not less than that required for dead and live load combination.

## 7.5 Wide-Flange Beams

Wide-flange rolled structural shapes constitute the most common type of flexural members used in steel buildings to support transverse loads. These are used as beams, girders, and spandrels and can be designed as simply supported, fixed-ended, or partially fixed. They can be proportioned for shears and moments determined by elastic analysis using allowable stresses. Designs using plastic and LRFD theories, although are allowed in the AISC code, have not come about as a general office procedure and therefore, will not be discussed here.

### 7.5.1 Bending

Under the usual bracing conditions the allowable bending stress $F_b = 0.66F_y$ where $F_y$ is the yield strength of steel. Almost all beams are designed to bend about their major axis and are required to be laterally braced within certain intervals to resist lateral torsional buckling. Lateral supports for flexural members are required because the compression flange behaves in a manner similar to a column and tends to buckle in the absence of lateral supports. Metal decking welded to the beam top flange at sufficiently close intervals constitutes lateral bracing, as do cross beams framing into the sides of the beam if adequate connection is made to the beam top flange. With adequate lateral bracing, the design of a steel beam boils down to the selection of a bending member having a section modules equal to or slightly larger than the calculated value.

### 7.5.2 Shear

Steel beams used in floor framing are most often wide-flange shapes with an axis of symmetry about the plane of bending. They are designed for shear assuming: (i) the contribution of flanges to the shear capacity is negligible; and (ii) the parabolic variation of shear stress in the web is replaced by an average stress on the gross area of the web. With these assumptions, for purposes of calculating the shear stresses, symmetrical shapes can be reduced to an equivalent rectangle with dimensions $t_w d$ where $t_w$ is the thickness of web and $d$ is the beam depth. The calculated shear stress $F_v = V/t_w d$ should not exceed the allowable shear stress $F_v = 0.40F_y$.

### 7.5.3 Deflections

Steel beams are usually cambered for dead loads when the calculated deflections exceed 0.75 in. (19 mm). The maximum camber should be limited to about 2.5 in. (63.5 mm) because of floor-leveling problems during concreting operations. The allowable live-load deflection is a function of the type of ceiling suspended from the beams, and for plastered ceilings it should not exceed 1/360 of the span. The AISC commentary recommends that the depth-to-span ratio of fully stressed beams in floors be not less than $F_y/800{,}000$, and if subject to shock and vibration, not less than $F_y/650{,}000$.

## 7.6 Columns

The design of a gravity column primarily carrying vertical loads reverts to selecting a steel section with calculated compressive stresses below or equal to the allowable stresses given in the AISC

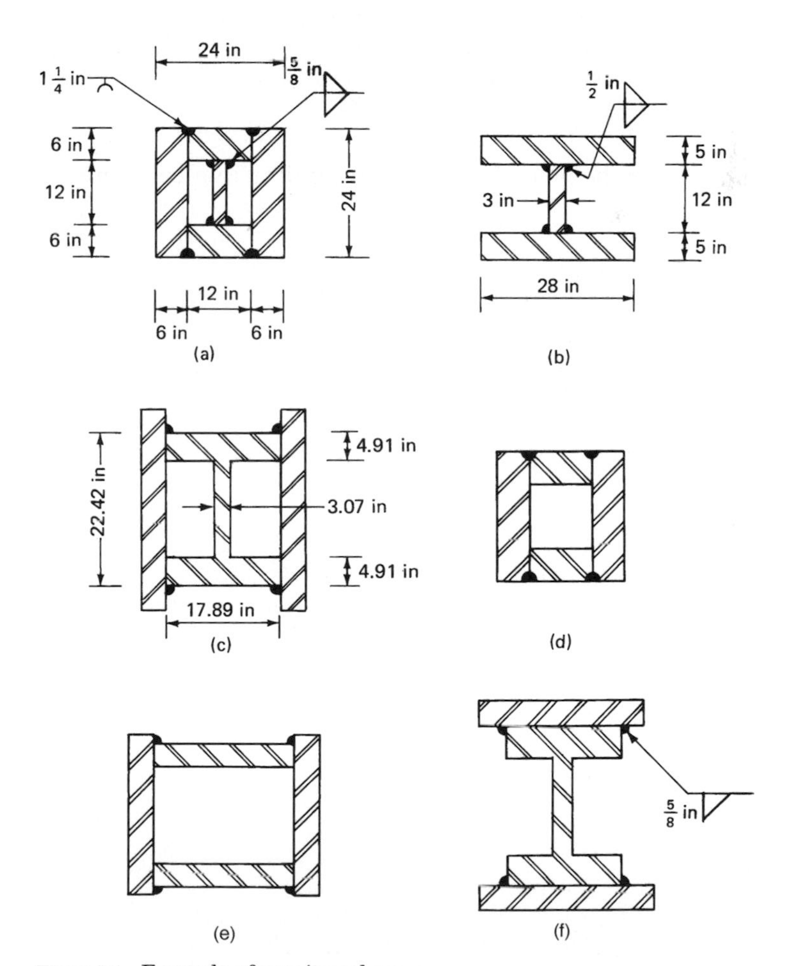


**Figure 7.1** Example of gravity columns.

specifications. An effective length factor of $K = 1$ is used in the design. For buildings in excess of 40-stories or so, it becomes necessary to use built-up columns or cover plated rolled-shapes. A few examples of these are given in Fig. 7.1.

Chapter

# 8

# Gravity Systems in Concrete Buildings

Concrete floor systems are cast on temporary form work or centering of lumber, plywood, or metal panels that are removed when the concrete has reached sufficient strength to support its own weight and construction loads. This procedure dictates that form work be simple to erect and remove and be repetitive to achieve maximum economy. Although in general floor systems for high-rise buildings are the same as for their lower brethern, there are several characteristics which are unique to high-rise buildings. Floor systems in high-rise buildings are duplicated many times over, necessitating optimum solutions in their design because:

1. Savings that might otherwise be insignificant for a single floor may add up to a considerable sum because of large number of floors
2. Dead load of floor system has a major impact on the design of vertical-load-bearing elements such as walls and columns

The desire to minimize dead loads is not unique to concrete floor systems but is of greater significance because the weight of concrete floor system tends to be heavier than steel floors and therefore has a greater impact on the design of vertical elements and foundation system. Another consideration is the impact of floor depth on the floor-to-floor height. Thus it is important to design a floor system that is relatively lightweight without being too deep.

One of the necessary features of cast-in-place concrete construction is the large demand for job-site labor. Form work, reinforcing steel, and the placing of concrete are the three aspects that demand most labor. Repetition of form work is a necessity for economical construction of cast-in-place high-rise buildings. Forms which can be used repetitively are "ganged" together and carried forward, or up the building, in large units, often combining column, beam, and slab elements in a large piece of form work. Where the layout of the building frame is maintained constant for several stories, these result in economy of handling and placing costs. "Flying" form for flat work is another type that is used to place concrete for large floor areas. In a conventional construction method sometimes called the stick method, plywood sheets are nailed to a formed solid decking built on adjustable shores. When concrete has attained sufficient strength, the shores are removed and the plywood sheets are stripped, and if undamaged, they are stored for reuse for the next floor. This method of forming is labor-intensive and also time consuming. If the floor-to-floor cycle is delayed one day because of form work, the construction time of a tall building can be lengthened significantly. The flying form system typically shortens the construction time. In this system, floor forms are attached to a unit consisting of deck surface, adjustable jack, and supporting frame work. For stripping, the form is lowered, slipped out of the slab, and shifted to the next floor as one rigid structure, resulting in economy of handling and placing.

## 8.1 Floor Systems

### 8.1.1 Flat plates

Concrete slabs are often used to carry vertical loads directly to walls and columns without the use of beams or girders. Such a system called a flat-plate (Fig. 8.1) is used where spans are not large and loads are not heavy as in apartment and hotel buildings.

*Flat plate* is the term used for a slab system without any column

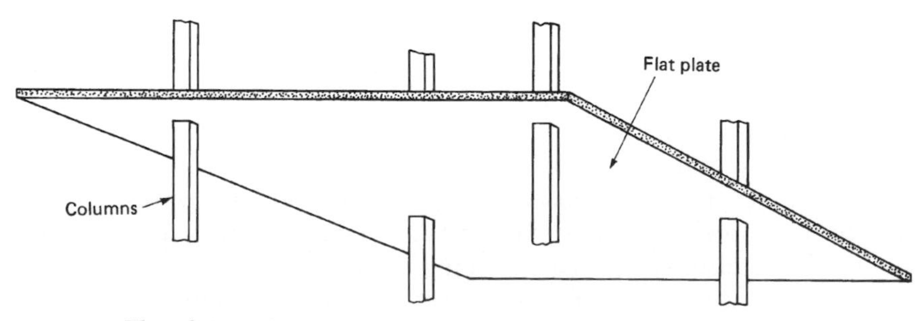


**Figure 8.1** Flat plate system.

flares or drop panels. Although column patterns are usually on a rectangular grid, flat plates can be used with irregularly spaced column layouts. They have been successfully built using columns on triangular grids and other variations.

### 8.1.2 Flat slabs

Flat slab (Fig. 8.2) is also a two-way system of beamless construction but incorporates a thickened slab in the region of columns and walls. In addition to the thickened slab, the system can have flared columns. The thickened slab and column flares, referred to as drop panels and column capitals, reduce shear and negative bending stresses around the columns.

A flat slab system with a beamless ceiling has minimum structural depth, and allows for maximum flexibility in the arrangement of air-conditioning ducts and light fixtures. For apartments and hotels, the slab can serve as a finished ceiling for the floor below and therefore is more economical. Since there are no beams, the slab itself replaces the action of the beams by bending in two orthogonal directions. Therefore, the slab is designed to transmit the full load in each direction, carrying the entire load in shear and in bending.

The limitations of span are dependent upon the use of column capitals or drop panels. The criterion for thickness of the slab is usually the punching shear around columns and long term deflection of the slab. In high-rise builidings the slabs are generally 5 to 10 in. (127 to 254 mm) thick for spans of 15 to 25 ft (4.56 to 7.6 m).

### 8.1.3 Waffle system

This system also called a two-way joist system (Fig. 8.3) is closely related to the flat slab system. To reduce the dead load of a solid slab construction, metal or fiberglass domes are used in the form work in

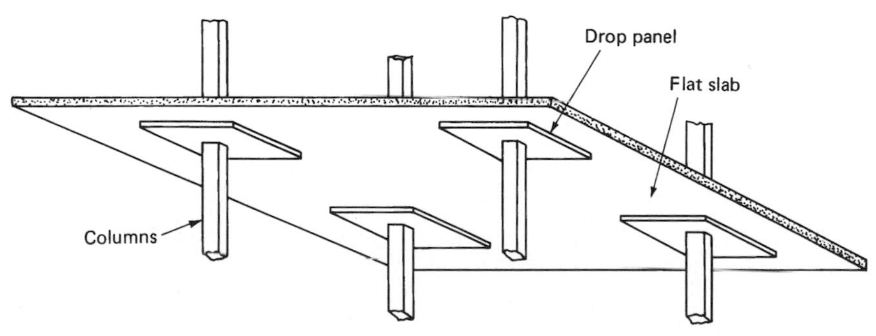


**Figure 8.2** Flat slab system.

a rectilinear pattern as shown in Fig. 8.3. Domes are omitted near columns resulting in solid slabs to resist the high bending and shear stresses in these critical areas.

In contrast to a joist which carries loads in a one-way action, a waffle system carries the loads simultaneously in two directions. The system is therefore more suitable for square bays than rectangular bays. The overall behavior of the system is similar to a flat slab. However, the waffle is more efficient for spans in the 30 to 40 ft (9.1 to 12.2 m) range because it has greater overall depth than a fat slab without the penalty of added dead weight.

### 8.1.4 One-way concrete ribbed slabs

This system also referred to as a one-way joist system is one of the most popular systems for high-rise building construction in North America. The system is based on the well-founded premise that since concrete in a solid slab below the neutral axis is well in excess of that required for shear, much of it can be eliminated by forming voids. The resulting system shown in Fig. 8.4, has voids between the joists made with removable forms of steel, wood, plastic, or other material. The joists are designed as one-way T beams for the full-moment tributary to its width. However, in calculating the shear capacity, the ACI code allows for a 10 percent increase in the allowable shear stress of concrete. It is standard practice to use distribution ribs at approximately 10 ft (3.0 m) centers for spans greater than 20 ft (6 m). For maximum economy of form work, the depth of beams and girders should be made the same as for joists.

### 8.1.5 Skip joist system

In this system instead of a standard 3 ft (0.91 m) spacing, joists are spaced at 5 ft or 6 ft 6 in. (1.52 and 1.98 m) spacings using 53 and 66 in. (1346 and 1676 mm) wide pans. The joists are designed as

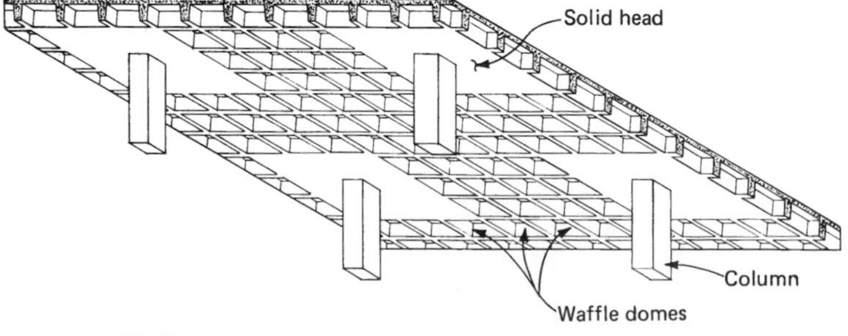


**Figure 8.3** Waffle system.

beams without using 10 percent increase in the shear capacity allowed for standard joists. Also the system is designed without distribution ribs thus requiring even less concrete. The spacing of vertical shores can be larger than for standard pan layout and consequently the form work is more economical. Figure 8.5 shows a typical layout.

The fire rating requirement for floor systems is normally specified in the governing building codes. The most usual method of obtaining the rating is to provide a slab that will meet the code requirement without the use of sprayed-on fireproofing. In the United States building codes, normally the slab thickness required for 2 hour fire rating is 4 in. (101.6 mm) for lightweight concrete and $4\frac{1}{2}$ to 5 in. (114.3 to 127 mm) for normal-weight concrete, depending upon the

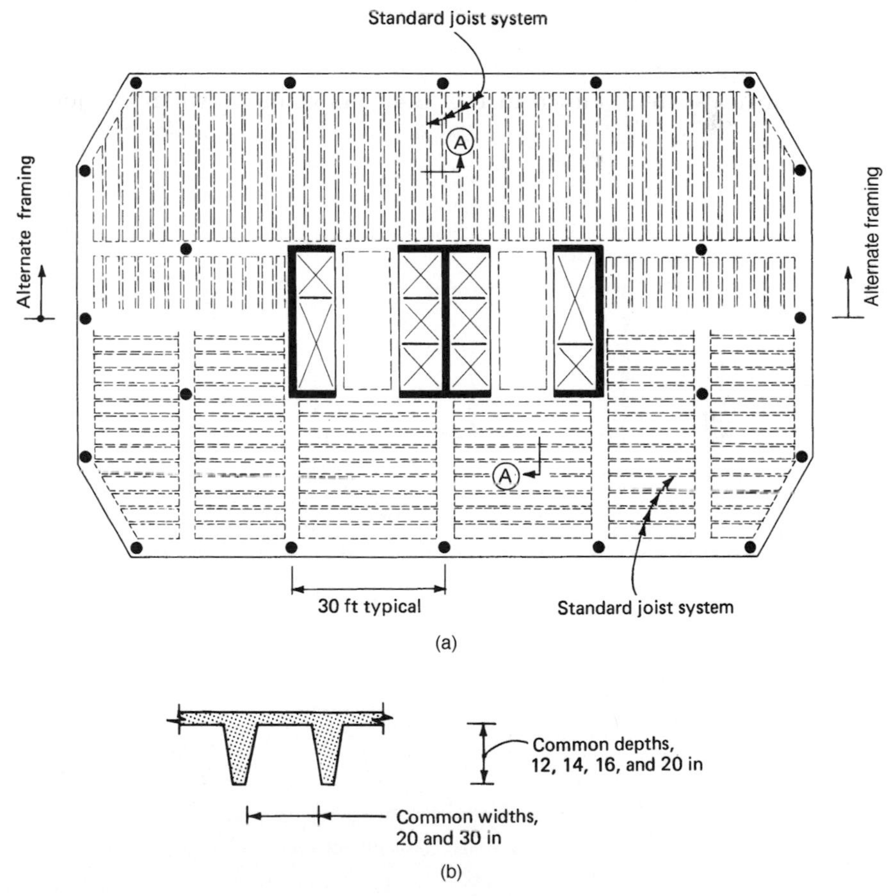


**Figure 8.4** One-way joist system: (a) building plan; (b) section A.

type of aggregate. The thickness of slab required for fire rating is much in excess of that required by structural design. Therefore use of special pan forms with joists at 8 to 10 ft centers (2.43 to 3.04 m) should be investigated for large projects.

### 8.1.6 Band beam system

This system shown in Fig. 8.6 uses wide shallow beams and should be investigated for buildings in which the floor-to-floor height is critical. Note that if lateral loads are resisted by parameter framing, it is not necessary to line band beams with either the exterior or interior columns. The slab in between the band beams is usually designed as a bending-member with varying moment of inertia to take into account the increased thickness at the beams. A variation

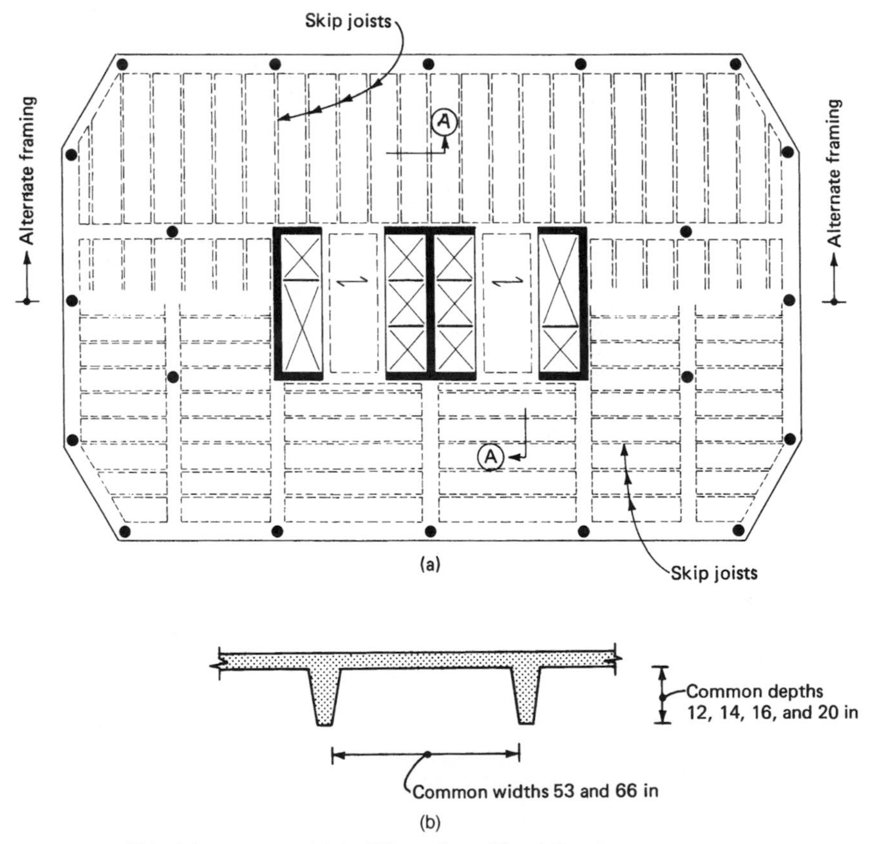


**Figure 8.5** Skip joist system: (a) building plan; (b) section A.

of the scheme uses standard or skip joists to span between the band beams.

### 8.1.7 Haunch girder and joist system

A floor-framing system with girders of constant depth criss-crossing the interior space between the core and the exterior often presents nonstructural problems because it limits the space available for the passage of air conditioning ducts. The haunch girder system widely accepted in certain parts of North America, achieves more headroom without making undue compromises in the structure. The basic system shown in Fig. 8.7 consists of a girder of variable depth. The shallow depth at the center facilitates the passage of mechanical ducts and reduces the need to raise the floor-to-floor height. Two types of haunch girders are in vogue. One uses a tapered haunch (Fig. 8.7b) and the other a square haunch (Fig. 8.7c).

### 8.1.8 Beam and slab system

This system consists of a continuous slab supported by beams generally spaced at 10 to 20 ft (3.04 to 6.08 m) on center. The thickness of the slab is selected from structural considerations and is usually much in excess of that required for fire rating. The system has broad application and is generally limited by the depth available in the ceiling space for the beam stem. This system considered a "heavy-duty" system is often used for framing nontypical floors such as ground floor and plaza levels, which are typically subjected to

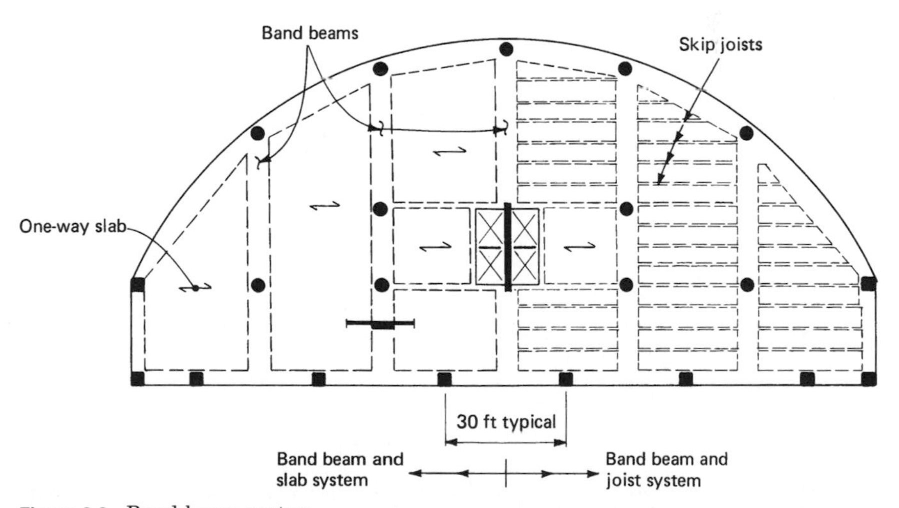


**Figure 8.6** Band beam system.

heavier superimposed loads due to landscape and other architectural features.

### 8.1.9 Design examples

#### 8.1.9.1 One-way slab-and-beam system

The analysis of the one-way slab system will be discussed to illustrate the simplifications normally made in a design office in analyzing the slabs and beams for gravity loads.

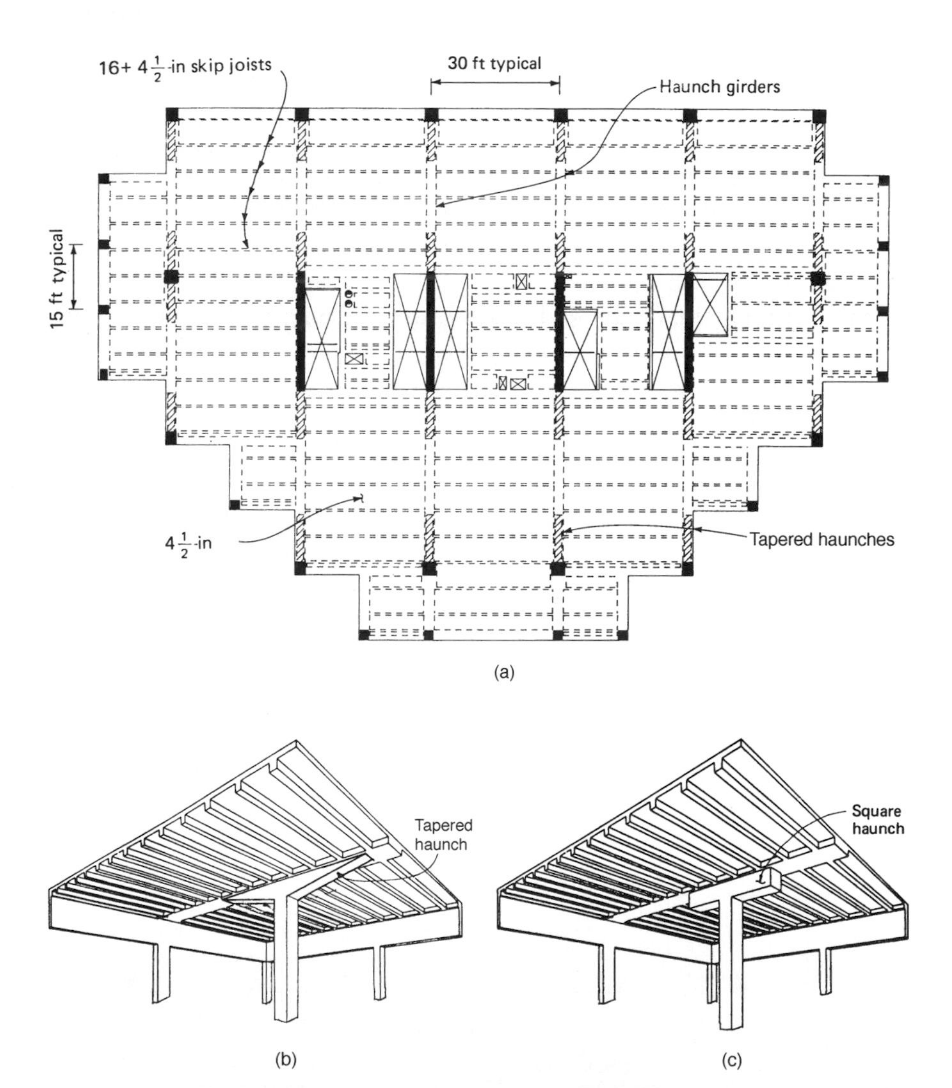


**Figure 8.7** (a) Haunch girder and joist system. (b) Tapered haunch girder with skip joist system; (c) square haunch girder with skip joist system.

Figure 8.8 shows a uniformly loaded floor slab where the intermediate beams divide the floor slab into a series of one-way slabs. If a typical 1 ft width of slab is cut out as a free-body in the longitudinal direction, it is evident that the slab will bend with a positive curvature between the beam stems, and a negative curvature at the supports. The deflected shape is similar to that of a continuous beam spanning between transverse girders, which act as simple supports. The assumpgion of simple support neglects the torsional stiffness of the beams supporting the slab. If the distance between the beams is the same, and if the slabs carry approximately the same load, the torsional stiffness of the beams has little influence on the moments in the slab.

However, the exterior beams loaded from one side only, are twisted by the slab. The resistance to the end rotation of the slab offered by the exterior beam is dependent on the torsional stiffness of the beam. If the beam is small and its torsional stiffness low, a pin support may be assumed at the exterior edge of slab. On the other hand, if the exterior beam is large with a high torsional rigidity, it will apply a restraining moment to the slab. The beam, in turn will be subjected to a torque requiring design for torsion.

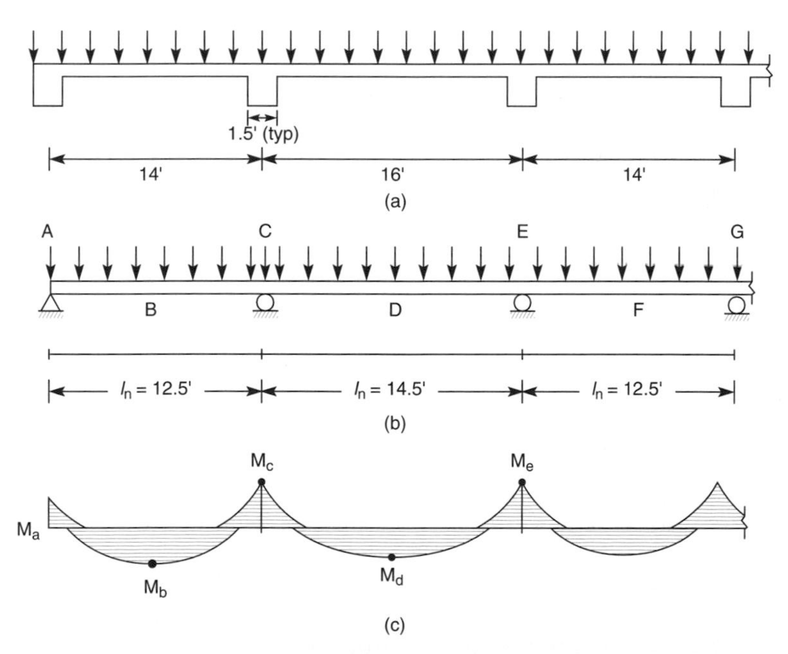


**Figure 8.8** One-way slab example: (a) typical 1 ft strip; (b) slab modelled as a continuous beam; (c) design moments.

### Analysis by ACI coefficients

Analysis by this method is limited to structures in which span lengths are approximately the same (with the maximum span difference between adjacent spans no more than 20 percent), the loads are uniformly distributed, and the live load does not exceed three times the dead load.

ACI values for positive and negative design moments are illustrated in Figs 8.9, 8.10 and 8.11. In all expressions, $l_n$ equals clear span for positive moment and shear, and the average of adjacent clear spans for negative moment.

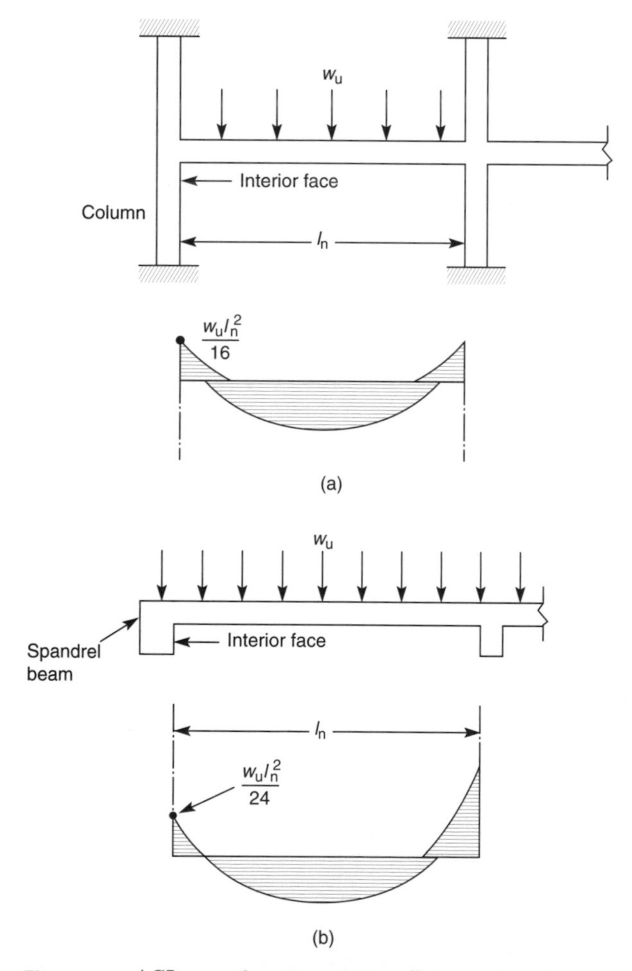


**Figure 8.9** ACI negative moment coefficients at exterior supports: (a) slab built integrally with column; (b) slab built integrally with spandrel beam.

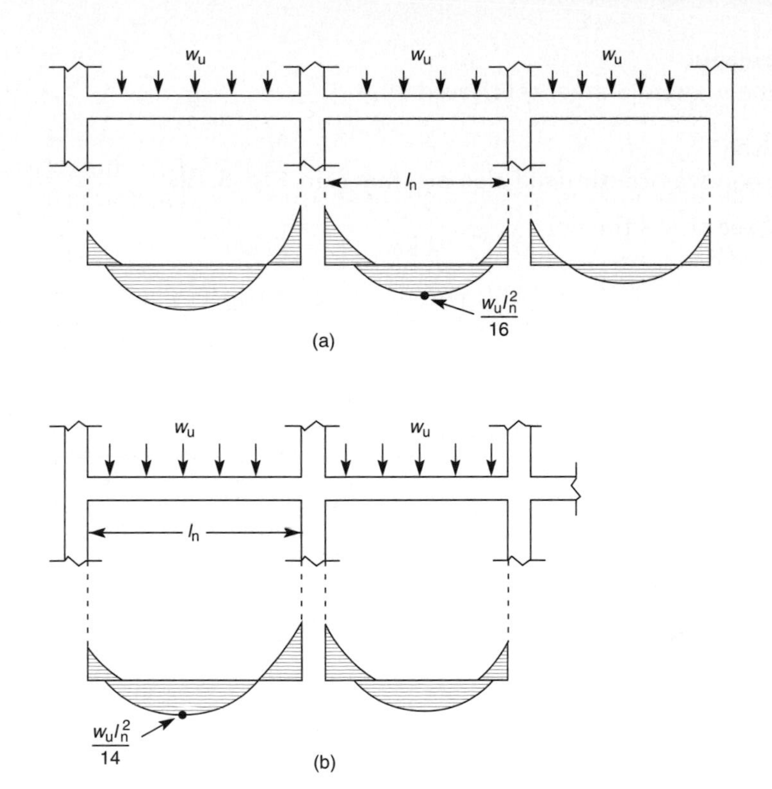


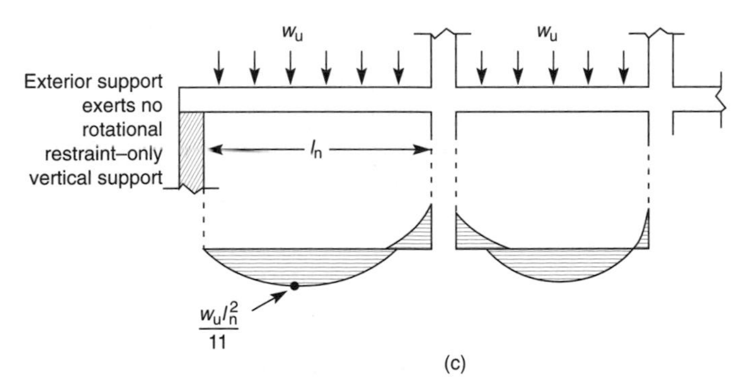


**Figure 8.10** ACI positive moment coefficients: (a) interior span; (b) exterior span, discontinuous end integral with supports: (c) exterior span, discontinuous end unrestrained.

**Example**
One-way mild steel reinforced slab.

**Given**
A one-way continuous slab as shown in Fig. 8.12.

$$f'_c = 4\text{ ksi}, \quad f_y = 60\text{ ksi}$$

$$\text{Ultimate load} = 0.32\text{ kip/ft}$$

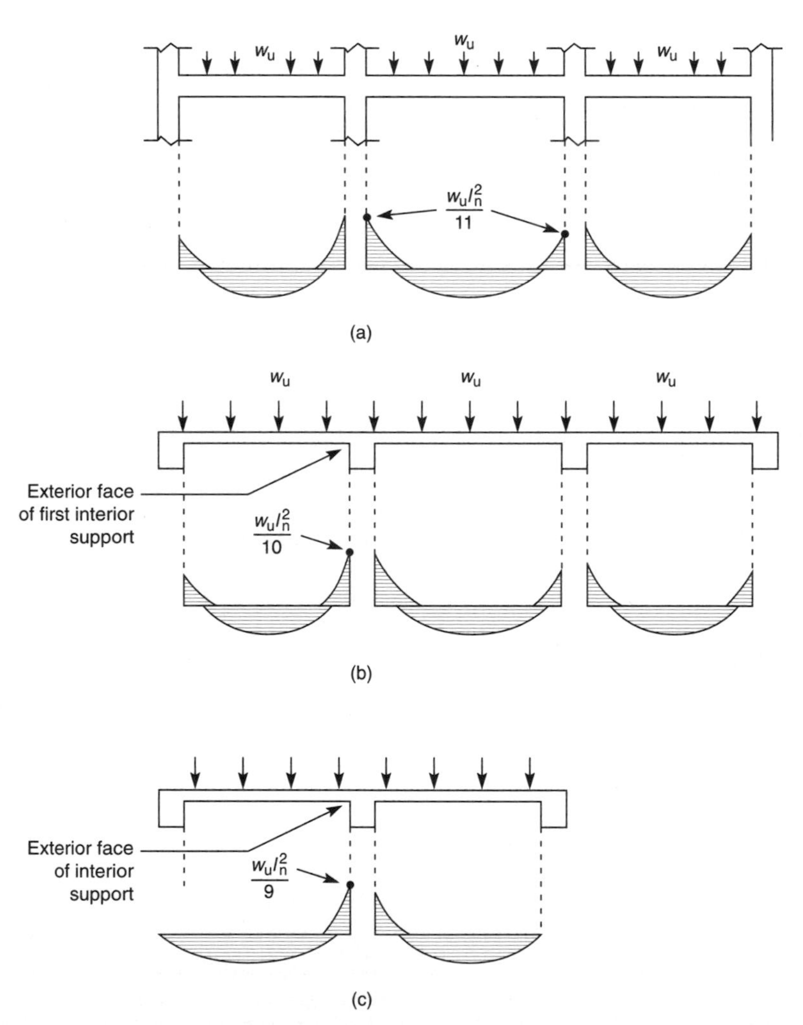


**Figure 8.11** ACI coefficients for negative moments: (a) at interior supports; (b) at exterior face of first interior support, more than two spans; (c) at exterior face of first interior support, two spans.

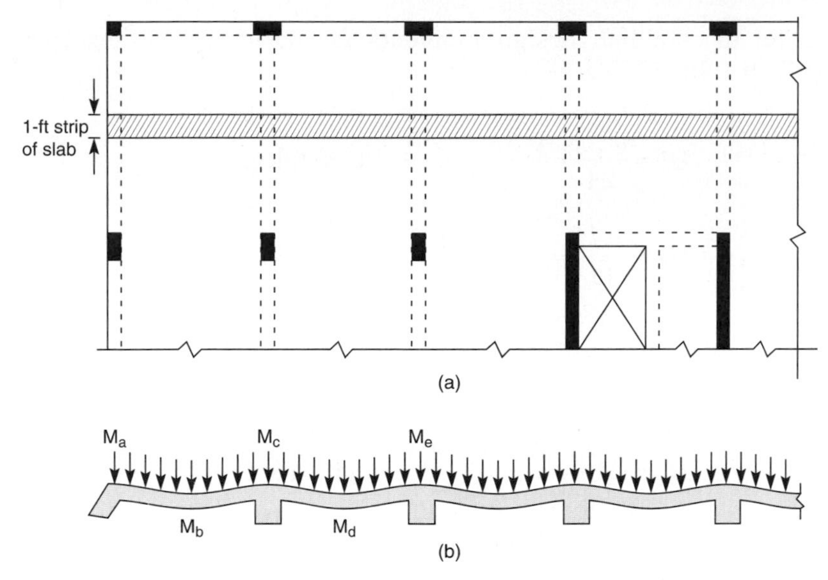


**Figure 8.12** Design example, one-way slab: (a) partial floor plan; (b) section.

### Solution

Use Table 8.1 to determine the minimum thickness of slab required to satisfy deflection limitations.

$$h_{\min} = \frac{l}{28} = \frac{12 \times 16}{28} = 6.86 \text{ in.} \quad \text{Use } 6\tfrac{1}{2} \text{ in. (165 mm)}$$

Analyze a 1 ft width of slab as a continuous beam using ACI

**TABLE 8.1 Minimum Thickness of Beams or One-way Slabs Unless Deflections are Computed (ACI Code Table 9.5*a*)†**

Members not supporting or attached to partitions or other construction likely to be damaged by large deflections

| | Minimum thickness $h$ | | | |
|---|---|---|---|---|
| | Simply supported | One end continuous | Both ends continuous | Cantilever |
| Solid one-way slabs | $l/20$ | $l/24$ | $l/28$ | $l/10$ |
| Beams or ribbed one-way slabs | $l/16$ | $l/18.5$ | $l/21$ | $l/8$ |

† Span length $l$ in inches. Values in the table apply to normal-weight concrete reinforced with steel of $f_y = 60{,}000$ lb/in.$^2$. For structural lightweight concrete with a unit weight between 90 and 120 lb/ft$^3$ multiply the table values by $1.65 - 0.005w$ respectively, but not less than 1.09; the unit weight $w$ is in lb/ft$^3$. For reinforcement having a yield point other than 60,000 lb/in.$^2$, multiply the table values by $0.4 + f_y/100{,}000$ with $f_y$ in lb/in.$^2$

coefficients to establish design moments for positive and negative steel (Fig. 8.12b).

$$M_a = \frac{w_u l_n^2}{24} = \frac{0.32 \times 12.5^2}{24} = 2.08 \text{ ft kips}$$

$$M_b = \frac{w_u l_n^2}{11} = \frac{0.32 \times 12.5^2}{11} = 4.55 \text{ ft kips}$$

At $C$, for negative moment, $l_n$ is the average of adjacent clear spans: $l_n = (12.5 + 14.5)/2 = 13.5$ ft

$$M_c = \frac{w_u l_n^2}{10} = \frac{0.32 \times 13.5^2}{10} = 5.83 \text{ ft kips}$$

$$M_d = \frac{w_u l_n^2}{16} = \frac{0.32 \times 14.5^2}{16} = 4.21 \text{ ft kips}$$

$$M_e = \frac{w_u l_n^2}{11} = \frac{0.32 \times 14.5}{11} = 6.12 \text{ ft kips}$$

Compute reinforcement $A_s$ per foot width of slab at critical sections. For example, at the second interior support, top steel must carry $M_e = 6.12$ kip ft. Note ACI code requires a minimum of $\frac{3}{4}$ in. cover for slab steel not exposed to weather or in contact with the ground.

We will use the "trial method" for determining the area of steel. In this method, the length of arm between the internal couple is estimated. Next the tension force $T$ is evaluated by using the basic relationship that applied moment equals the design strength; i.e.,

$$M_u = \phi T \times \text{arm}$$

$$T = \frac{M_u}{\phi \times \text{arm}}$$

where $\phi = 0.9$ for flexure, and $M_u$ = factored moment.

To start the procedure, the arm is estimated as $d - a/2$ by giving a value of $a = 0.15d$ where $d$ is the effective depth. The appropriate area of steel $A_s$ is computed by dividing $T$ by $f_y$.

To get a more accurate value of $A_s$, the components of the internal couple are equated to provide a close estimate of the area $A_c$ of the stress block. The compressive force $C$ in the stress block is equated to the tension force $T$.

$$C = T$$

$$0.85f'_cA_c = T$$

$$A_c = \frac{T}{0.85f_c}$$

Once $A_c$ has been evaluated, locate the position of $C$, which is the centroid of $A_c$ and recompute the arm between $C$ and $T$. Using the improved value, find the second estimates of $T$ and $A_s$. Regardless of the initial assumption for the arm, two cycles should be adequate for determining the required steel area.

For the example problem, the effective depth $d$ for the slab is given by:

$$d = h - \left(0.75 - \frac{d_b}{2}\right) = 6.5 - (0.75 + 0.25) = 5.5 \text{ in.}$$

$$M_u = \phi T(d - a/2)$$

As a first trial, guess $a = 0.15d = 0.15 \times 5.5 = 0.83$ in.

$$6.12 \times 12 = 0.9T\left(5.5 - \frac{0.82}{2}\right) = 4.58T$$

$$T = 16.03 \text{ kips}$$

$$A_s = \frac{T}{f_y} = \frac{16.03}{60} = 0.27 \text{ in.}^2/\text{ft}$$

Repeat the procedure using an arm based on improved value of $a$. Equate $T = C$

$$16.03 = 0.85f'_cA_c = 0.85 \times 4 \times a \times 12$$

$$a = 0.39 \text{ in.}$$

$$\text{Arm} = d - \frac{a}{2} = 5.5 - \frac{0.39}{2} = 5.31 \text{ in.}$$

$$T = \frac{M_u}{\phi\left(d - \frac{a}{2}\right)} = \frac{6.12 \times 12}{0.9 \times 5.31} = 15.37 \text{ kips}$$

$$A_s = \frac{15.37}{60} = 0.26 \text{ in.}^2$$

Check for temperature steel $= 0.0018A_g$

$$= 0.0018 \times 6.5 \times 12 = 0.14\ \text{in.}^2/\text{ft}$$

Determine spacing of slab reinforcement to supply 0.26 in.$^2$/ft.

$$\text{Using \#4 rebars, } s = \frac{0.20}{0.26} \times 12 = 9.23\ \text{in.} \quad \text{Say 9 in.}$$

$$\text{Using \#5 rebars, } s = \frac{0.31}{0.2} \times 12 = 14.31\ \text{in.} \quad \text{Say 14 in.}$$

Use #4 @ 9 top at support *e*. Also by ACI code, maximum spacing of flexural reinforcement should not exceed 18 in. or 3 times the slab thickness.

$$9\ \text{in.} < 3\ (6.5\ \text{in.}) = 19.5\ \text{in.} \quad 9'' \text{ spacing is OK.}$$

A schematic placement diagram of top steel is shown in Fig. 8.13.

### 8.1.9.2 T-beam design

**Design for flexure.** Design reinforcement for a simply supported T beam spanning 30 ft as shown in Fig. 8.14.

$$W_u = 0.32 \times \frac{(16 \times 14)}{2} = 4.80\ \text{kip/ft}$$

Use $W_u = 5.0$ k/ft including the self weight of beam. The minimum depth of beam to control deflections from Table 8.1 is

$$h_{\min} = \frac{l}{16} = \frac{30 \times 12}{16} = 22.5\ \text{in.} \quad \text{Use 22.5 in.}$$

Try $b_w = 18$ in. The width must be adequate to carry shear and allow for proper spacing between reinforcing bars.

The effective width of T beam $b_{\text{eff}}$ is the smallest of:

1. One-fourth the beam span:

$$\frac{30}{4} = 7.5\ \text{ft} = 90\ \text{in.} \quad \text{(controls)}$$

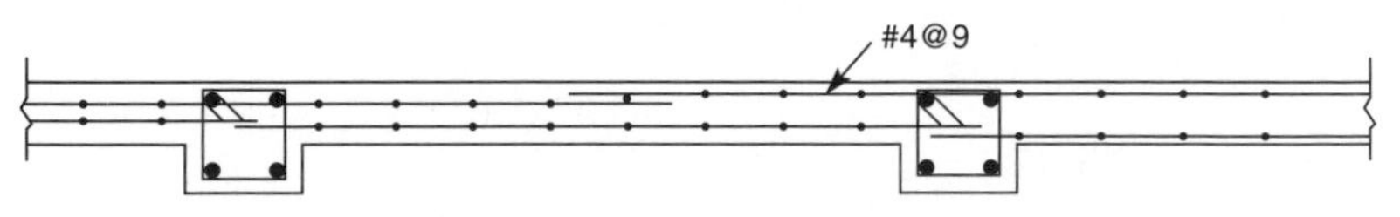


**Figure 8.13** Slab reinforcement.

2. Eight times the slab thickness on each side of stem plus the stem thickness:

$$8 \times 6.5 \times 2 + 18 = 122 \text{ in.}$$

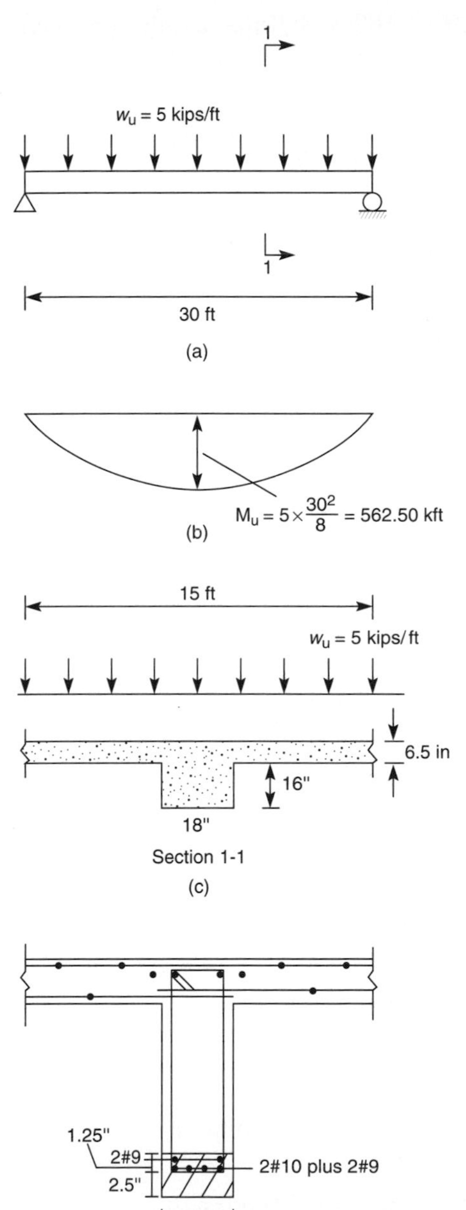


**Figure 8.14** Design example, simple beam.

3. Center-to-center spacing of panel:

$$\frac{(16 \times 14)}{2} \times 12 = 180 \text{ in.}$$

Select the flexural steel $A_s$ for $M_u = 562.50$ ft/kips using the trial method.

$$\text{Estimate } d = h - 2.6 = 22.5 - 2.6 = 19.9 \text{ in.}$$

$$M_u = \phi T\left(d - \frac{a}{2}\right) \quad \text{Guess } a = 0.8 \text{ in.}$$

$$562.50 \times 12 = 0.9T\left(19.9 - \frac{0.8}{2}\right) = 17.557$$

$$T = 384.62$$

$$A_s = \frac{T}{f_y} = \frac{384.62}{60} = 6.41 \text{ in.}^2$$

Check value of $a$

$$384.62 = T = C = ab_{\text{eff}}(0.85f_c)$$
$$= a(90)(0.85)4$$
$$a = 1.26$$

Repeat the procedure using an arm based on improved value of $a$.

$$M_u = 562.50 \times 12 = 0.9T\left(19.9 - \frac{1.26}{2}\right) = 17.34T$$

$$T = \frac{562.50 \times 12}{17.34} = 389.2 \text{ kips}$$

Check value of $a$

$$389.2 = T = C = a \times 90 \times 0.85 \times 4$$

$$a = 1.27 \text{ in.}$$

$$A_s = \frac{T}{f_y} = \frac{389.2}{60} = 6.49 \text{ in.}^2$$

$$A_{s,\min} = \frac{200b_w d}{f_y} = \frac{200 \times 18 \times 19.9}{60{,}000} = 1.19 \text{ in.}^2$$

Since 6.49 in.$^2$ controls, use two #10 and four #9 bars

$$A_{s,\text{supplied}} = 6.54 \text{ in.}^2$$

**Check reinforcement pattern for crack width**

The crack width is limited by the ACI code to a maximum of 0.013 in. (0.33 mm) and 0.016 in. (0.41 mm) respectively for interior and exterior exposures. The corresponding value for the parameter $Z$ given by the equation

$$z = f_s\sqrt[3]{d_c A}$$

is not to exceed 145 kip/in. (25.4 MN/m) and 175 kip/in. (30.6 MN/m). In this equation, $f_s$ is the steel stress and may be taken as $0.6f_y$ in kips per square inch, and $d_c$ is the distance from tension surface to the center of the row of reinforcing bars closest to outside surface, and $A$ is the effective tension area of concrete divided by the number of reinforcing bars.

We now proceed with the example problem to verify if the reinforcement pattern satisfies the ACI code requirements of crack control for exterior exposure.

Locate the center of gravity of steel by summing moments of bar areas about an axis through the base of beam stem.

$$A_{st}\bar{Y} = \sum A_n \bar{Y}_n$$

$$(6.54\ \text{in.}^2)\bar{Y} = (2\ \text{in.}^2)(2.5\ \text{in.}) + (2\ \text{in.}^2)(3.75\ \text{in.}) + (2.54\ \text{in.}^2)(2.63\ \text{in.})$$

$$\bar{Y} = 2.93\ \text{in.}$$

The effective tension area of concrete (Fig. 8.14d) is the product of beam stem width and a height of web equal to twice the distance between centroid of steel and tension face. When the reinforcement consists of more than one bar size, as in the example, the number of bars is expressed by the size of the largest bar.

$$\text{Number of bars} = \frac{\text{total area of steel}}{\text{area of largest bar}}$$

For the example, number of bars $= \dfrac{6.54}{1.27} = 5.15$. Therefore

$$A = \frac{18 \times 2.93 \times 2}{5.15} = 20.48$$

$$\begin{aligned} Z &= f_s\sqrt[3]{d_c A} \\ &= 0.6 \times 60\sqrt[3]{2.15 \times 20.48} \\ &= 127 < 145\ \text{kip/in. Therefore, O.K.} \end{aligned}$$

**T-beam design for shear.** The ACI procedure for shear design is an empirical method based on the assumption that a shear failure occurs on a vertical plane when shear force at that section due to factored service loads exceeds the concrete's fictitious vertical shear strength. The shear stress equation by strength of materials is given by

$$v = \frac{VQ}{Ib}$$

where $v$ = shear stress at a cross-section under consideration
$V$ = shear force on the member
$I$ = moment of inertia of the cross-section about centroidal axis
$b$ = thickness of member at which $v$ is computed
$Q$ = moment about centroidal axis of area between section at which $v$ is computed and outside surface of member

This expression is not directly applicable to reinforced concrete beams. The ACI, therefore, uses a simple equation to calculate the average stress on the cross section

$$v_c = \frac{V}{b_w d}$$

where $v_c$ = nominal shear stress
$V$ = shear force
$b_w$ = width of beam web
$d$ = distance between centroid of tension steel and compression surface

To emphasize that $v_c$ is not an actual stress but merely a measure of the shear stress intensity, it is termed a nominal shear stress.

For nonseismic design, the ACI code assumes that concrete can carry some shear regardless of the magnitude of the external shearing force and that shear reinforcement must carry the remainder. Thus

$$V_u = \phi V_n = \phi(V_c + V_s)$$

where $V_u$ = factored or ultimate shear force
$V_n$ = nominal shear strength provided by concrete and reinforcement
$V_c$ = nominal shear strength provided by concrete
$V_s$ = nominal reinforcement provided by shear reinforcement
$\phi$ = strength reduction factor = 0.85 for shear and torsion

Shear design computations can be made in terms of shear force $V$ or in terms of unit shear stress $v$. Stress is easier to compare with allowable values, and gives engineers a better frame of reference reducing chances of error.

The shear strength equation in terms of shear stress is given by

$$v_u \leqslant \phi v_n = v_c + v_s$$

A conservative value for $v_c$ often used in design because of its simplicity is $v_c = 2\sqrt{f'_c}$.

The nominal shear stress $v_u$ can be calculated from

$$v_u = \frac{V_u}{\phi b_w d}$$

For vertical stirrups:

$$v_u = v_c + \frac{A_v f_y}{b_w s}$$

where $A_v$ = area of vertical shear reinforcement
$f_y$ = yield strength of shear reinforcement
$s$ = spacing of shear reinforcement

Returning to the example problem, we have

$$V_u = 5.0 \times 15 = 75 \text{ kips}$$
$$d = 19.9 \text{ in.}, \quad b_w = 18 \text{ in.}$$

$$V_u \text{ at distance } d \text{ from the support} = 5.0\left(15 - \frac{19.4}{12}\right) = 70 \text{ kips}$$

$$v_u = \frac{70}{0.85 \times 18 \times 19.9} = 0.230 \text{ ksi}$$

$$v_c = 2\sqrt{f'_c} = 2\sqrt{4000} = 126 \text{ psi}$$

Shear stress to be carried by reinforcement

$$\begin{aligned} v_s &= v_u - v_c \\ &= 0.230 - 0.126 \\ &= 0.110 \end{aligned}$$

$$s = \frac{A_v f_y}{v_s}$$

For two-legged #4 stirrups $s = \dfrac{2 \times 0.2 \times 60}{0.104 \times 18} = 12.8$ in. This should be checked for maximum spacing as will be done presently.

We now calculate the stirrups using strength equation in terms of shear forces.

$$V_u = \phi(V_c + V_s)$$

For the example problem,

$$V_u = 70 \text{ kips}$$

$$V_c = 2\sqrt{f'_c}\, b_w d$$

$$= 2\sqrt{4000}\; 18 \times 19.9$$

$$= 45^k$$

$$\phi\frac{V_c}{2} = 0.85 \times \frac{45}{2} = 19.1^k$$

Since $V_u = 70$ kips exceeds $\phi\frac{V_c}{2}$, stirrups are required.

$$V_s = \frac{V_u}{\phi} - V_c$$

$$= \frac{70}{0.85} - 45 = 37.35^k$$

Spacing for two-legged #4 stirrups,

$$s = \frac{A_v f_y d}{V_s}$$

$$= \frac{2 \times 0.2 \times 60 \times 19.9}{37.35^k}$$

$$= 12.8 \text{ in.}$$

Since $V_s$ is less than $4\sqrt{f'_c}\, b_w d = 90$ kips,

$$s = \frac{d}{2} = \frac{19.9}{2} = 9.9 \text{ say } 10 \text{ in.}$$

If $V_s \geqslant 4\sqrt{f'_c}\, b_w d$, the maximum spacing would have been $\frac{d}{4}$ but not to exceed 12 in.

The spacing $s$ should not be less than $\frac{d}{2} = \frac{19.9}{2} = 9.95$ say 9 in., and the minimum area

$$A_{v,\min} = \frac{50 b_w s}{f_y}$$

$$= \frac{50 \times 18 \times 9}{60{,}000}$$

$$= 0.135 \text{ in.}^2$$

$$A_{v,\text{provided}} = 0.4 \text{ in.}^2 > 0.135 \text{ in.}^2$$

Use #4 stirrups at 9 in. near the supports. A reduced spacing of stirrups equal to $d$ may be used within the span where the calculated shear stress $v_u \leqslant \frac{v_c}{2}$.

#### 8.1.9.3 Analysis of two-way slabs

Although two-way slabs may be designed by any method that satisfies the strength and serviceability requirements of the ACI code, most usually they are designed by the "equivalent-frame method" using computers. In this section however only the direct design method is discussed.

In this method the simple beam moment in each span of a two-way system is distributed as positive and negative moments at midspan and at supports. Since stiffness considerations, except at the exterior supports, are not required, computations are simple and can be carried out rapidly.

Three steps are required for the determination of positive and negative design moments.

1. Determine simple beam moment:

$$M_0 = \frac{w_u l_2 l_n^2}{8}$$

where $M_o$ = simple beam moment
$w_u$ = ultimate uniform load
$l_2$ = slab width between columns transverse to the span under consideration
$l_n$ = clear span between face of columns or capitals

2. For interior spans divide $M_o$ into $M_c$ and $M_s$, midspan and support moments as shown in Fig. 8.15. For exterior spans use Fig. 8.16 to divide $M_0$ into moments $M_1$, $M_2$, and $M_3$.
3. Distribute $M_c$ and $M_s$ in the transverse direction across the width between column and middle strips by using Tables 8.2 and 8.3 which give percentage of moment in the column strips. The remainder is assaigned to the middle strip.

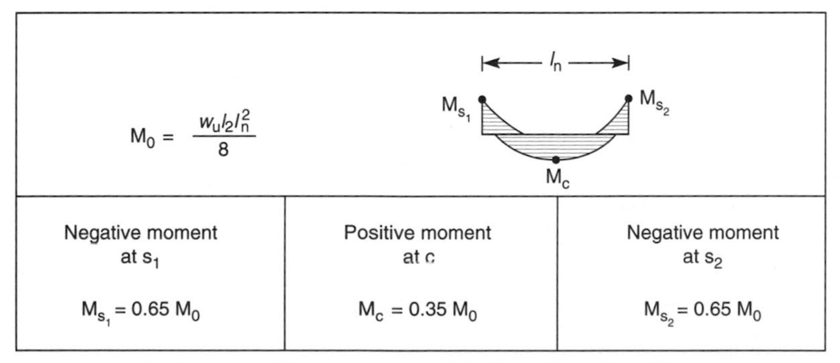


**Figure 8.15** Assignment of moments at critical sections: interior span.

| $M_0 = \dfrac{w_u l_2 l_n^2}{8}$ | | | |
|---|---|---|---|
| Edge restraint condition | Exterior negative moment at 1 $M_1$ | Positive moment at 2 $M_2$ | interior negative moment at 3 $M_3$ |
| 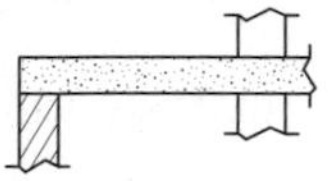 (a) | 0 | $0.63\ M_0$ | $0.75\ M_0$ |
| 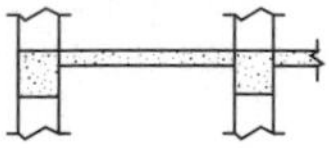 (b) | $0.16\ M_0$ | $0.57\ M_0$ | $0.70\ M_0$ |
| 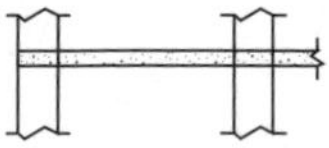 (c) | $0.26\ M_0$ | $0.52\ M_0$ | $0.70\ M_0$ |
| 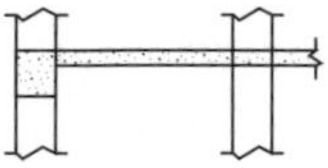 (d) | $0.30\ M_0$ | $0.50\ M_0$ | $0.70\ M_0$ |
| 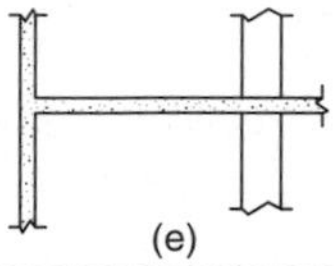 (e) | $0.65\ M_0$ | $0.35\ M_0$ | $0.65\ M_0$ |

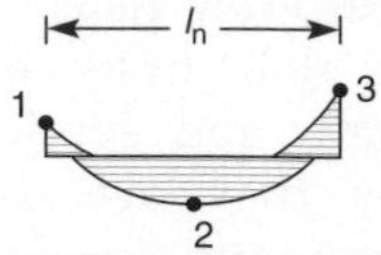

**Figure 8.16** Assignment of moments to critical sections: exterior span.

**TABLE 8.2 Percentage of Positive Moment to Column Strip, Interior Span**

| $\alpha_1 \dfrac{l_2}{l_1}$ | $l_2/l_1$ | | |
|---|---|---|---|
| | 0.5 | 1.0 | 2.0 |
| 0 | 60 | 60 | 60 |
| ≥1 | 90 | 75 | 45 |

**TABLE 8.3 Percentage of Negative Moment to Column Strip at an Interior Support**

| $\alpha_1 \frac{l_2}{l_1}$ | $l_2/l_1$ 0.5 | 1.0 | 2.0 |
|---|---|---|---|
| 0 | 75 | 75 | 75 |
| ≥1 | 90 | 75 | 45 |

Observe in Fig. 8.15, that for an interior span, the positive moment $M_c$ at midspan equals $0.35M_0$, and the negative moment $M_s$ at each support equal $0.65M_0$, values which are approximately the same as for a uniformly loaded fixed-end beam. These values are based on the assumption that an interior joint undergoes no significant rotation, a condition that is assured by the ACI code restrictions that limit: (i) the difference between adjacent span lengths to one-third of the longer span and; (ii) the ratio of live load to the dead load to 3.

The final step is to distribute the positive and negative moments in the transverse direction between column strip and middle strips. The percentage of distribution factors are tabulated (Tables 8.2 and 8.3) for three values (0.5, 1, 2) of panel dimensions $\frac{l_2}{l_1}$, and two values (0 and 1) of $\alpha_1 \frac{l_1}{l_2}$. For intermediate values linear interpolation may be used. Table 8.2 is for interior spans while Table 8.3 is for exterior spans. For exterior spans the distribution of moment is influenced by the torsional stiffness of spandrel beam. Therefore an additional parameter $\beta_t$, the ratio of the torsional stiffness of spandrel beam to flexural stiffness of slab is given in Table 8.4.

For exterior spans the distribution of total negative and positive moments between columns strips and middle strips is given in terms of the ratio $\frac{l_2}{l_1}$, and the relative stiffness of the beam and slab, and

**TABLE 8.4 Percentage of Negative Moment to Column Strip at an Exterior Support**

| $\alpha_1 \frac{l_2}{l_1}$ | $\beta_t$ | $l_2/l_1$ 0.5 | 1.0 | 2.0 |
|---|---|---|---|---|
| 0 | 0 | 100 | 100 | 100 |
| 0 | ≥2.5 | 75 | 75 | 75 |
| ≥1 | 0 | 100 | 100 | 100 |
| ≥1 | ≥2.5 | 90 | 75 | 45 |

the degree of torsional restraint provided by the edge beam. The parameter $\alpha = \dfrac{E_{cb}I_b}{E_{cs}I_s}$ is used to define the relative stiffness of the beam and slab spanning in either direction. $E_{cb}$ and $E_{cs}$ are the moduli of elasticity of the beam and slab and $I_b$ and $I_s$ are the moments of inertia. Subscripted parameters $\alpha_1$ and $\alpha_2$ are used to identify $\alpha$ for the directions of $l_1$ and $l_2$ respectively.

The parameter $\beta_t$ in Table 8.4 defines the torsional restraint of edge beam. If there is no edge beam, i.e. $\beta = 0$, all of the exterior moment at 1 (Fig. 8.16) is apportioned to the column strip. For $\beta_t \geqslant 2.5$ i.e. for very stiff edge beams 75 percent of moment at 1 is assigned to the column strip. For values in-between, linear interpolation is permitted. In most practical designs, distributing 100 percent of the moment at 1 to the column strip while using minimum slab reinforcement in the middle strip yields acceptable results.

### Design example

#### Given

A two-way slab system as shown in Fig. 8.17.

$$w_d = 150\ \text{psf}, \quad w_l = 80\ \text{psf}$$

Determine the slab depth and design moments by the direct method

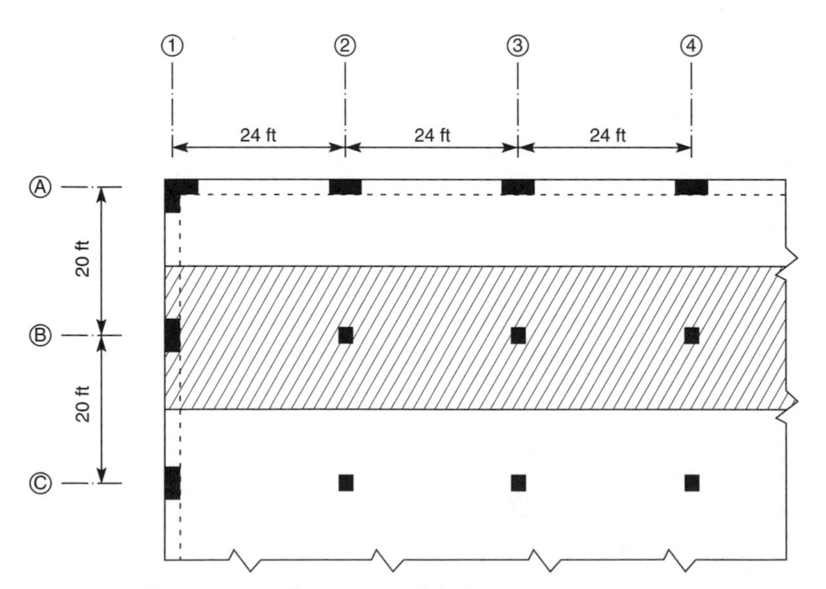


**Figure 8.17** Design example, two-way slab.

**TABLE 8.5 Minimum Thickness* of Slabs Without Interior Beams (ACI Code, Table 9.5*c*)**

| Yield stress $f_y$, psi† | Without drop panels | | | With drop panels | | |
|---|---|---|---|---|---|---|
| | Exterior panels | | | Exterior panels | | |
| | Without edge beams | With edge beams‡ | Interior panels | Without edge beams | With edge beams‡ | Interior panels |
| 40,000 | $\frac{l_n}{33}$ | $\frac{l_n}{36}$ | $\frac{l_n}{36}$ | $\frac{l_n}{36}$ | $\frac{l_n}{40}$ | $\frac{l_n}{40}$ |
| 60,000 | $\frac{l_n}{30}$ | $\frac{l_n}{33}$ | $\frac{l_n}{33}$ | $\frac{l_n}{33}$ | $\frac{l_n}{36}$ | $\frac{l_n}{36}$ |

* Minimum thickness for slabs without drop panels is 5 in. Minimum thickness for slabs with drop panels is 4 in.

† For values of reinforcement yield stress between 40,000 and 60,000 psi minimum thickness shall be obtained by linear interpolation.

‡ Slabs with beams between columns along exterior edges. The value of $\alpha$ for the edge beam shall not be less than 0.8.

at all critical sections in the exterior and interior span along column line B.

**Solution.**
From Table 8.5, for $f_y = 60$ ksi, and for slabs without drop panels, the minimum thickness of slab is determined to be $\frac{l_n}{33}$ for the interior panels. The same thickness is used for the exterior panels since the system has beams between the columns along the exterior edges.

For the example, $l_n$ the clear span in the long direction $= 24 - 2 = 22$ ft. The minimum thickness $h = \frac{2.2 \times 12}{33} = 8$ in.

**Interior span**

$$w_u = 1.4(0.15) + 1.7(0.08) = 0.346 \text{ ksf}$$

$$M_o = \frac{w_u l_2 l_n^2}{8}$$

$$= \frac{0.346 \times 20 \times 22^2}{8}$$

$$= 418.7 \text{ ft kips}$$

Divide $M_o$ between sections of positive and negative moments.

At mid span:

$$M_c = 0.35M_o$$

$$= 0.35 \times 418.7 = 146.5 \text{ ft kips}$$

At supports:

$$M_s = 0.65M_o$$

$$= 0.65 \times 418.7 = 272.2 \text{ ft kips}$$

For the distribution of the midspan moment $M_c$ between column and middle strips use Table 8.12. The value for $\alpha_1$, the ratio of beam stiffness to slab stiffness for the example problem, is zero since there are no beams in the direction of spans under consideration. The ratio $\frac{l_2}{l_1} = \frac{20}{24} = 0.833$. From Table 8.2 column strip moment is 60 percent of total moment.

$$\text{Moment to column strip} = 0.60 \times 146.5 = 87.9 \text{ ft kips}$$

$$\text{Moment to middle strip} = 0.40 \times 146.5 = 58.6 \text{ ft kips}$$

For the distribution of support moment $M_s$ between column and middle strips use Table 8.3. Since $\alpha_1 = 0$, and $\frac{l_2}{l_1} = 0.833$, from Table 8.3 column strip moment is 75 percent of total moment.

$$\text{Moment in column strip} = 0.75 \times 272.2 = 204 \text{ ft kips}$$

$$\text{Moment in middle strip} = 0.25 \times 272.2 = 68 \text{ ft kips}$$

**Exterior span.** The magnitude of the moments at critical sections in the exterior span is a function of both $M_o$, the simple beam moment, and $\alpha_{ec}$, the ratio of stiffness of exterior equivalent column to the sum of the stiffness of the slab and beam framing into the exterior joint. Instead computing $\alpha_{ec}$, we use edge condition (d) given in Fig. 8.16 to evaluate the design moments at critical sections.

At the exterior column face

$$M_1 = 0.30 \times M_o = 0.30 \times 418.7 = 125.6 \text{ ft kips}$$

At mid span

$$M_2 = 0.50 \times M_o = 0.50 \times 418.7 = 209.4 \text{ ft kips}$$

At the interior column face

$$M_3 = 0.7 \times M_o = 0.7 \times 418.7 = 293 \text{ ft kips}$$

At the exterior edge of slab, the transverse distribution of the design moment to column strip is given in Table 8.4. Instead of calculating the value of $\beta_t$, we conservatively assign 100 percent of exterior moment to the column strip.

The moment to the column strip $= 1 \times 125.6$ ft kips. The middle strip is assumed to be controlled by the minimum steel requirements, an assumption which is satisfactory in almost all practical designs.

## 8.2 Prestressed Concrete Systems

Although mild-steel-reinforced concrete is well suited for the construction of high-rise floor systems, it requires cosiderable depth to work efficiently, especially in the 35 to 40 ft (10.66 to 12.2 m) span range normally required in modern office buildings. For example, to keep the long-term creep deflection to within acceptable limits, span-to-depth ratios of two-way flat slab and one-way beam or joist systems are limited to about $\frac{1}{30}$ to $\frac{1}{35}$ and $\frac{1}{15}$ to $\frac{1}{20}$, respectively. The relatively short span capability of flat slab system limits its application to high-rise multiple-unit residential buildings which can accommodate relatively closed spaced columns within fixed and frequent partition layouts. When used for office layouts, the flat slab system becomes too heavy, imposing undue structural penalty, especially in difficult foundation conditions. The one-way joist system, although relatively lightweight, requires about 20 to 24 in. (508 to 610 mm) of structural depth for a 35 to 40 ft (10.66 to 12.2 m) span range. By using prestressing it is possible to overcome the aforementioned shortcomings. Prestressing can thus be looked upon as an aid to boost the span range of conventionally reinforced floor systems by about 30 to 40 percent. This is the primary reason for the increase in the use of prestressed concrete. Some of the other advantages of prestressed concrete are:

1. Prestressed concrete is generally crack-free and therefore, more durable.
2. Shallower sections can be used because a larger depth of compression block is available in flexure.
3. Prestress concrete is resilient. Cracks due to overloading completely close and deformations are recovered soon after the removal of overload.
4. Fatigue strength (though not a design consideration in building design) is considerably more than that of conventionally reinforced concrete because tendons are subjected to smaller variations in stress due to repeated loadings.
5. Prestressed concrete members are generally crack-free and, therefore, are stiffer than conventional concrete members.
6. The structural members are self-tested for material and workmanship during stressing operations, thereby safeguarding against unexpected poor performance in service.
7. Prestress design can be controlled more since a predetermined force is introduced in the system; the magnitude, location, and technique of introduction of such an additional force is left to the designer, who can tailor the design according to project requirements.

A major motivation for the use of prestressed concrete comes from the reduced structural depth, which translates into lower floor-to-floor height, reduction in the area of curtain wall and building volume, with a consequent reduction in heating and cooling loads.

In prestressed concrete there is no savings in using high-strength strands instead of mild steel. This is because the savings in mild steel reinforcement quantities are just about offset by the higher unit cost of prestressing steel. The cost savings, however, come from the reduction in the quantity of concrete combined with indirect nonstructural savings resulting from reduced floor-to-floor height. Although from an initial cost consideration prestress concrete may be the least expensive, other costs associated with future tenant improvements such as providing large openings in slabs must be considered before selecting the final scheme.

### 8.2.1 Methods of prestressing

Centuries ago humans discovered the principle of prestressing by using metal bands or ropes around wooden planks to form barrels. They may not have had the dubious pleasure of figuring out the exact nature and magnitude of stresses, but they knew intuitively that the stronger the bands, the better the chance of containing liquids in the wooden barrels. It was not until the 1880s that a similar principle was applied to a reinforced concrete slab with the idea of counterbalancing the tensile stresses in concrete. Intentional compressive stress in concrete was induced to overcome the tensile stresses developed from external loads. These early attempts at prestressing did not meet with great success because the amount of prestressing that could be imposed via conventional steel was limited by the strength of the steel itself. Even at low stresses, it was difficult to maintain the prestress because creep and shrinkage of concrete would destroy the prestress in the course of time.

The eminent French engineer Eugene Freyssinet is credited as the forefather of prestressing as we know it today. He established the use of high-tensile-strength steel to assure that even after creep and shrinkage losses, there remained adequate prestress to counteract the external loads.

Current methods of prestressing can be studied under two groups: (i) pretensioning and; (ii) post-tensioning. In pretensioning, the tendons are stressed first and then, concrete is placed around the tendons. After the concrete has hardened, the tendons are released to impart prestress into the concrete member.

In post-tensioning, the tendons, which may consist of steel wires, strands, or bars, are tensioned and anchored against the concrete after it has hardened. The tensioning is accomplished by using hydraulic jacks. The tendons usually remain permanently unbonded to concrete and are placed directly in the forms, and stressed after the concrete has reached a minimum 75 percent of the design strength. The measured elongations are compared against the calculated values, and if satisfactory, the tendons projecting beyond the concrete are cut off. Form work is removed after post-tensioning of tendons. However the floor is back-shored to support shoring and construction loads from the floors above.

Post-tensioning is accomplished by using high-strength strands, wires, or bars as tendons. In North America the use of strands by far leads the other two types. The strands are either bonded or unbonded depending upon the project requirements. In bonded construction the ducts are filled with a mortar grout after stressing the tendons while there is no grouting in unbonded construction.

For high-rise construction, unbonded construction is preferred because it eliminates the need for grouting. Post-tensioned members in multistory construction consist of slabs, joists, beams, and girders, with a large number of small tendons. Grouting each of the multitude of tendons is a time consuming and expensive operation. Therefore, unbounded construction using strands is popular in the North American high-rise construction.

### 8.2.2 Materials

**Post-tensioning steel.** The basic requirement is the loss of tension in the steel due to shrinkage and creep of concrete and the effects of stress relaxation should be a relatively small portion of the total prestress. In practice, the loss of prestress generally varies from a low of 15 ksi (103.4 MPa) to a high of 50 ksi (344.7 MPa). If mild steel having a yield of 60 ksi (413.7 MPa) is employed with an initial prestress of, say, 40 ksi, it is very likely that most of the prestress, if not the entire prestress, is lost because of shrinkage and creep losses. To limit the prestress losses to a small percentage of say, about 20 percent, the initial stress in the steel must be in excess of 200 ksi (1379 MPa). Therefore, high-strength steel is invariably used in prestressed concrete construction.

Although in general high-strength steel is produced by using alloys such as carbon, manganese, and silicon, prestressing steel achieves its high tensile strength by virtue of the process of cold-drawing, in which high-strength steel bars are drawn through a series of progressively smaller dyes. During this process, the crystallography of the steel is improved, because cold-drawing tends to realign the crystals.

High-strength steel used in North America is available in three basic forms: (i) uncoated stress-relieved wires; (ii) uncoated stress-relieved strands; and (iii) uncoated high-strength steel bars. Stress-relieved wires and high-strength steel bars are not generally used for post-tensioning and therefore are not considered here.

High-strength strands are fabricated in factories by helically twisting a group of six wires around a slightly larger center wire by a mechanical process called stranding. The resulting seven-wire strands are stress-relieved by a continuous heat treatment process to produce the required mechanical properties.

ASTM specification A416 specifies two grades of steel, 250 and 270 ksi (1724 and 1862 MPa), the higher-strength being more common in the building industry. A modulus of elasticity of 27,500 ksi (189,610 MPa) is used for calculating the elongation of strands. To prevent the use of brittle steel which would result in a failure pattern similar to that of an overreinforced beam, ASTM A-416 specifies a minimum elongation of 3.5 percent at rupture.

A special type of strand called low-relaxation strand is increasingly used because it has a very low loss due to relaxation, usually about 20 to 25 percent of that for stress-relieved strand. With this strand less post-tensioning steel is required, but the cost is more because of the special process used in its manufacture.

Corrosion of unbonded strand is a possibility and can be prevented by using galvanized strands. This is not, however, popular in North America because: (i) various anchorage devices in use for post-tensioned systems are not suitable for use with galvanized strand because of low coefficient of friction; (ii) damage can result to the strand because the heavy bite of the anchoring system can ruin the galvanizing; and (iii) galvanized strands are more expensive.

A little-understood, and infrequent occurrence of great concern in engineering is the so-called stress corrosion which occurs in highly stressed strands. The reason for the phenomenon is little known, but chemicals such as chlorides, sulfides, and nitrates are known to start this type of corrosion under certain conditions. It is also known that high-strength steels exposed to hydrogen ions are susceptible to failure because of loss in ductility and tensile strength. This phenomenon is called *hydrogen embrittlement* and is best counteracted by confining the strands in an environment having a pH value greater than 8.

**Concrete.** Concrete with compressive strengths of 5000 to 6000 psi (34 to 41 MPa) is commonly employed in the prestress industry. This relatively high-strength is desirable, for the following reasons. First, commercial anchorages are designed on the basis of high-strength concrete to prevent failure of concrete during the application of prestressing. Second, high-strength concrete has higher resistance in tension, shear, bond, and bearing and is desirable for prestressed structures which are typically under higher stresses than ordinary reinforced concrete. Third, its shrinkage is less and its higher modulus of elasticity and smaller creep result in smaller loss of prestress.

Post-tensioned concrete is considered as a self-testing system because if the concrete is not crushed under the application of prestress it should withstand subsequent loadings in view of the strength gain with age. In practice it is not the 28-day strength that dictates the mix design but rather the strength of concrete at the transfer of prestress. Construction schedules on high-rise projects require post-tensioning as early as possible to facilitate early removal of forms for reuse in higher floors. Typically the minimum strength of concrete at transfer is 70 to 75 percent of the 28-day strength. Assuming that stressing operation is on the fourth day or so, it is more than likely that the actual 28-day strength is much more than the specified strength. For example, assume that the design specifies a 28-day compressive strength of 5000 psi (34.47 MPa). The minimum strength required at transfer of prestress is 70 to 75 percent of 5000 psi, approximately equal to 4000 psi (27.6 MPa) at 4 days. This requirement would normally yield a concrete of 28-day strength of about 6000 psi (41.37 MPa), which is well in excess of the specified design strength. This rather wasted strength can be avoided by using the higher strength in the actual design.

Although high early strength (Type III) portland cement is well-suited for post-tension work because of its ability to gain the required strength for stressing relatively early, it is not generally used because of higher cost. Invariably, Type I cement conforming to ASTM C-150 is employed in buildings.

The use of admixtures and flyash is considered a good practice. However, use of calcium chlorides or other chlorides is prohibited since the chloride ion may result in stress corrosion of prestressing tendons.

A slump of between 3 to 5 in. (76 to 127 mm) gives good results. The aggregate used in the normal production of concrete is usually satisfactory in prestressed concrete, including lightweight aggregates. However, care must be exercised in estimating the volumetric changes so that a reasonable prestress loss is calculated. Lightweight aggregates manufactured by using expanded clay or shale have been used in post-tensioned buildings. Lightweight aggregates that are not crushed after burning maintain their coating and therefore absorb less water. Such aggregates have drying and shrinkage characteristics similar to the normal-weight aggregates, although the available test reports are somewhat conflicting. The size of aggregate, whether lightweight or normal weight, has a more profound effect on shrinkage. Larger aggregates offer more resistance to shrinkage and also require less water to achieve the same consistency, resulting in as much as 40 percent reduction in shrinkage when the aggregate size is increased from say $\frac{3}{4}$ to $1\frac{1}{2}$ in. (19 to 38 mm). It is generally agreed that both shrinkage and creep are more a function of the cement paste than the type of aggregate. Lightweight aggregate has been gaining acceptance in prestressed construction since about 1955 and has a good track record.

### 8.2.3 Design

The design involves the following steps:

1. Determination of the size of concrete member.
2. Establishing the tendon profile and prestressing force.
3. Verifying the section for ultimate bending and shear capacity.

**TABLE 8.6 Approximate Span Depth Ratios for Post-tensioned Systems**

| Floor system | Simple spans | Continuous spans | Cantilever spans |
|---|---|---|---|
| One-way solid slabs | 40–48 | 42–50 | 14–16 |
| Two-way flat slabs | 36–45 | 40–48 | 13–15 |
| Wide band beams | 26–30 | 30–35 | 10–12 |
| One-way joists | 20–28 | 24–30 | 8–10 |
| Beams | 18–22 | 20–25 | 7–8 |
| Girders | 14–20 | 16–24 | 5–8 |

The above values are intended as a preliminary guide for the design of building floors subjected to a uniformly distributed superimposed live load of 50 to 100 psf (2394 to 4788 Pa). For the final design, it is necessary to investigate for possible effects of camber, deflections, vibrations, and damping. The designer should verify that adequate clearance exists for proper placement of post-tensioning anchors.

4. Verifying the serviceability characteristics, primarily in terms of stresses and long-term deflections.

It is well known that the depth of a member subjected to bending depends on many variables such as the magnitude of the design loads, shape of the cross section, available clearance, span length, and allowable deflections. The deflections of prestressed members tend to be small because under service loads they are usually uncracked and are much stiffer than nonprestressed members of the same cross section. Also, the prestressing force induces deflections in an opposite direction to those produced by external loads. The final deflection, therefore, is a function of tendon profile and the magnitude of prestress. Appreciating this fact, the ACI code does not specify minimum depth requirements for prestressed members. However, as a rough guide, the suggested span-to-depth ratios given in Table 8.6 can be used to establish the depth of continuous flexural members. Another way of looking at the suggested span-to-depth ratios is to consider, in effect, that prestressing increases the span range by about 30 to 40 percent over and above the values normally used in nonprestressed concrete construction.

The tendon profile is established based on the type and distribution of load with due regard to clear cover required for fire resistance and corrosion protection. Clear spacing between tendons must be sufficient to permit easy placing of concrete. For maximum economy, the tendon should be located eccentric to the center of gravity of the concrete section to produce maximum counteracting effect to the external loads. For members subjected to uniformly distributed loads a simple parabolic profile is ideal, but in continuous structures parabolic segments forming a smooth reversed curve at the support are more practical. The effect is to shift the point of contraflecture away from the supports. This reverse curvature modifies the load imposed by post-tensioning from those assumed using a parabolic profile between tendon high points.

The design of a simple span is rather trivial and can be accomplished with hand calculations. In continuous and indeterminate structures, the induced moments are not directly proportional to the tendon eccentricity because the deflection due to post-tensioning is resisted at the supports. The support restraint introduces moments called the secondary moments. The name is a misnomer because it does not mean that its values are negligible or necessarily smaller than the primary moments.

The initial post-tension force immediately after transfer is less than the jacking force because of: (i) slippage of anchors; (ii) frictional losses along tendon profile; and (iii) elastic shortening of concrete. The force is reduced further over a period of months or even years due to change in the length of concrete member resulting from shrinkage and creep of concrete and relaxation of the highly stressed steel. The effective prestress is the force in the tendon after all the losses have taken place. For routine designs, empirical expressions for estimating prestress losses yield sufficiently accurate results, but in cases with unusual member geometry, tendon profile, and construction methods it may be necessary to make refined calculations.

In North American practice it is usually sufficient to specify effective force and tendon profile. The post-tension contractor submits calculations of prestress losses for the engineer's review. Therefore, the engineer is spared the drudgery of calculating the prestress losses. The post-tension design of a statically determinate structure is trivial and can be accomplished with little difficulty by hand. The floor framing systems normally encountered in practice are invariably statically indeterminate and are most usually designed by using computer programmes. Most programs use the concept of load balancing.

In this concept, prestressing is seen as a method to balance a certain portion of the external loads by inducing a counteracting load. This method, first developed by T. Y. Lin is very popular. Its application to statically indeterminate systems could be visualized just as easily as for statically determinate structures. Also, the procedure gives a simple method of calculating deflections by considering only that portion of the external load not balanced by the prestress. If the effective prestress completely balances the sustained loading, the post-tensioned member will not undergo any deflection and will remain horizontal irrespective of the modulus of rigidity or flexural creep of concrete.

In the load-balancing approach, the analysis of a prestressed member is reduced to the analysis of a nonprestressed member subjected to the load differential between externally applied loads and internally applied prestress. Since the analysis is performed with only the unbalanced portion of the external load, the inaccuracies in the method of analysis become relatively insignificant. Often, approximate method is all that is necessary for the final design. The load balancing method can be conveniently applied to multiple-span beams and slabs. The prestressing force need not be the same in all the spans. The load in each span can be balanced by choosing a suitable prestress and profile. For spans requiring higher prestressing, additional tendons can be added.

A question that usually arises is how much of the external load is to be balanced. The answer, however, is not simple. Balancing all the dead load often results in too much prestressing, leading to uneconomical design. On the other hand, there are situations in which the live load is significantly heavier than the dead load, making it more economical to prestress not only for full dead loads but also for a significant portion of the live load. However, in the design of typical floor framing systems, the prestressing force is normally selected to balance about 75 to 95 percent of the dead load and, occasionally, a small portion of the live load. This leads to an ideal condition with the structure having little or no deflection under dead loads.

Limiting the maximum tensile and compressive stresses permitted in concrete does not in itself assure that the prestressed member has an adequate factor of safety against flexural failure. Therefore, its nominal bending strength is computed in a procedure similar to that of a reinforced concrete beam. Underreinforced beams are assumed to have reached the failure load when the concrete strain reaches a value of 0.003. Since the yield point of prestressing steel is not well defined, empirical relations based on tests are used in evaluating the strain and hence the stress in tendons.

The shear reinforcement in post-tensioned members is designed in a manner almost identical to that of nonprestressed concrete members, with due consideration for the longitudinal stresses induced by the post-tensioned tendons. Another feature unique to the design of post-tensioned members is the high stresses in the vicinity of anchors. Prestressing force is transferred to the concrete by anchoring the tendons with the aid of anchorages. Large stresses are developed at the anchorages, which have to be dealt with properly by providing well-positioned reinforcement in the region of high stresses. At a cross section of a beam sufficiently far away (usually 2 to 3 times of the larger cross-sectional dimension of the beam) from the end zones, the axial and bending stresses in the beam due to an eccentric prestressing force are given by the usual $P/A$ and $MC/I$

relations. But in the vicinity of stress application, the stresses are distributed in a complex manner. Of importance are the transverse tensile forces generated at the end blocks for which reinforcement is to be provided. The bursting tensile stress has a maximum value along the axis of the force. Its distribution depends on the location of bearing area and its relative proportion with respect to the areas of the end face.

Because of the indeterminate nature and intensity of the stresses, the design of reinforcement for the end block is primarily based on empirical expressions. Reinforcement is designed to carry the tensile stresses created in the end block by the tendon reactions and usually consists of closely spaced stirrups tied together with horizontal bars.

### 8.2.4 Practical considerations

Condominiums and apartment buildings are economical if a structural system with a flat ceiling is used for floor construction. A post-tensioned flat plate is one such system for column spacings in the range of 20 to 30 ft (6.09 to 9.14 m). The form work is simple and lends itself to quick construction. The resulting flat plate system has good acoustical characteristics while maintaining a minimum floor-to-floor height. A flat slab with drop panels, which can be considered as an extension of the flat plate system, is suitable for office buildings with clear spans in the range of 35 to 45 ft (10.6 to 13.7 m). In such systems it may be economical to use long and narrow shear heads to accommodate flying forms. Post-tensioned joists clear spanning between the core and building exterior offer an alternative method for framing office buildings.

As in reinforced concrete and structural steel construction, the use of post-tensioned concrete is only limited by the imagination and ingenuity of the engineer and the relative economics of various construction materials and labor at the bid time. Certain rules of thumb for span-to-depth ratios and the average value of post-tensioning stresses in structural members are useful in conceptual design. The depth for slabs usually works out between $L/40$ and $L/50$, while for joists it is between $L/25$ and $L/35$. Beams can be much shallower, with a depth in the range of $L/20$ and $L/30$. Band beams offer perhaps the least depth without using as much concrete as flat slab construction. Although a span-to-depth ratio approaching 35 is adequate from strength and serviceability points of view, it is necessary to make sure that adequate space exists for proper detailing of anchorages. Detailing of beam-column intersection of shallow band beams should be carefully developed to avoid conflicts between the post-tensioning tendon anchorage, and main vertical column reinforcement. Bundling of column bars may be required to relieve congestion. Adequate clearance must be provided to permit access to stressing equipment.

Another thumb rule used in preliminaty design is the compression stress level in the members due to post-tensioning. A minimum compression level of 125 and 150 psi (862 and 1034 KPa) is a practical and economical range for slabs while a range of 250 to 300 psi (1724 to 2068 KPa) has been found to be adequate for beams. Compression stresses as high as 500 psi (3447 KPa) have been used successfully in band-beam systems.

### 8.2.5 Building examples

For the first example refer to the two-way flat plate framing plan of the Museum Tower shown in Fig. 1.8c. The primary tendons are $\frac{1}{2}$ in. diameter (12.7 mm) strands which are banded in the north–south direction. Unbanded tendons run from left to right across the building width. Additional tendons are used in the end panels to allow for the increased moments due to lack of continuity at one end.

As a second example, Fig. 8.18 shows the framing plan for a post-tensioned band-beam-slab system. Shallow beams only 16 in. (0.40 m) deep span across two exterior bays of 40 ft (12.19 m) and an interior bay of 21 ft (6.38 m). Post-tensioned slabs 8 in. (203 mm) deep, span between the band beams, typically spaced at 30 ft (9.14 m) on center. In the design of the slab additional beam depth is considered as a haunch at each end. Primary tendons for the slab

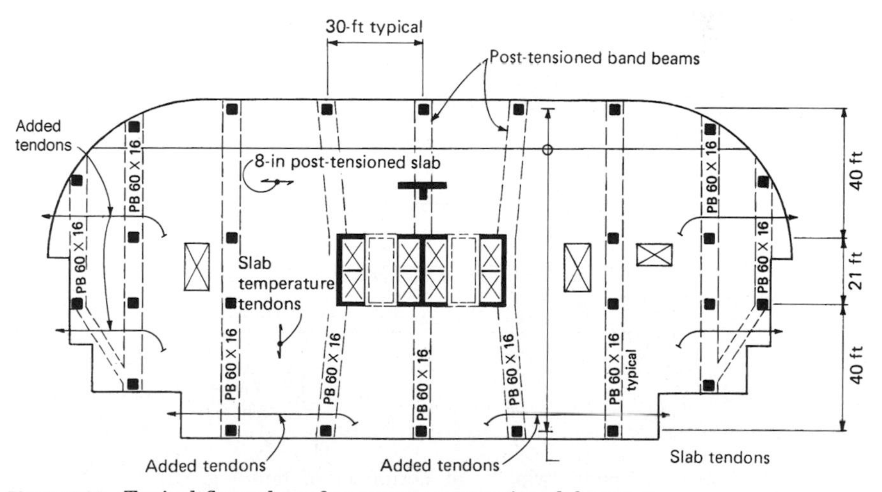


**Figure 8.18** Typical floor plan of one-way post-tension slab system.

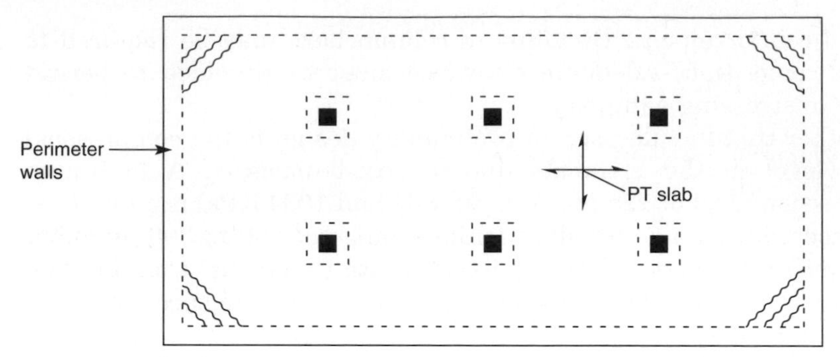


**Figure 8.19a** Cracking in post-tensioned slab caused by restraint of perimeter walls.

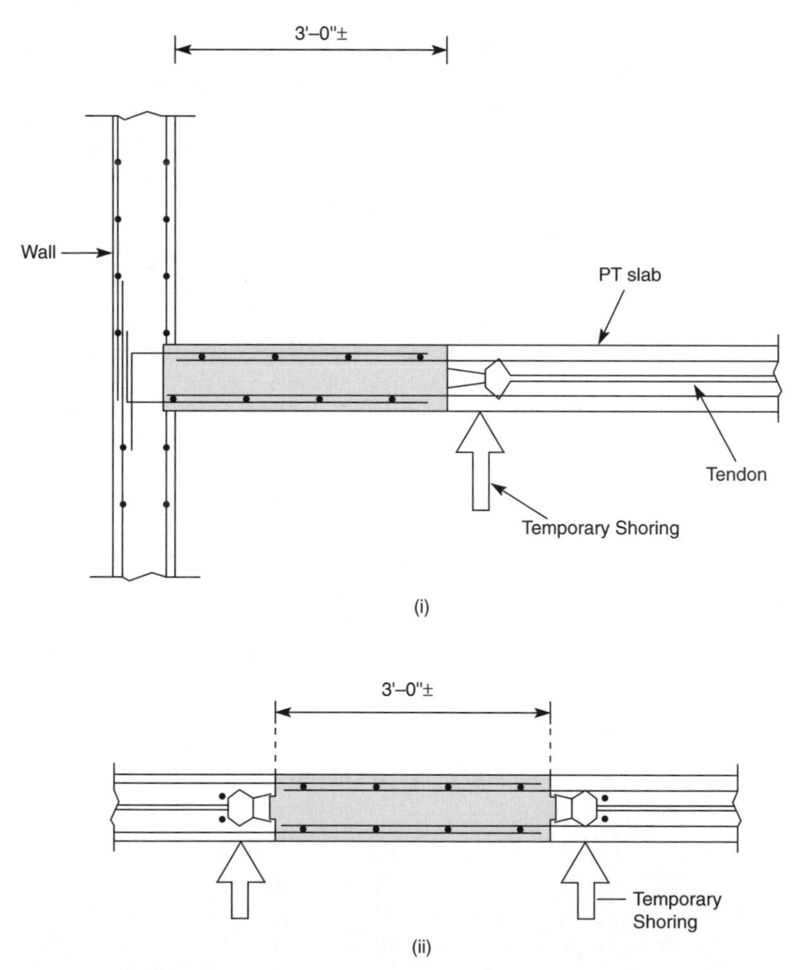


**Figure 8.19b** Temporary pour strip: (i) at perimeter of building; (ii) at interior of slab.

run across the building width, while the temperature tendons are placed in the north–south direction between the band beams.

### 8.2.6 Cracking problems in post-tensioned floors

Cracking caused by restraint to shortening is the biggest problem associated with post-tensioned floor systems. The reason for this is that the restraint to shortening is a time-dependent complex phenomenon with only subjective empirical solution. Exact numerical solutions which prevent cracking altogether have not been developed yet.

Shrinkage of concrete is the biggest contributor to shortening in both prestressed and non-prestressed concrete. In prestressed concrete, out of the total shortening, only about 15 percent is due to elastic shortening and creep. Therefore the problem is not that post-tensioned floors shorten that much more than non-prestressed concrete but it is the manner in which they shorten.

As a non-prestressed concrete slab tries to shorten, its movement is resisted internally by the bonded mild steel reinforcement. The reinforcement is put into compression and hence the concrete in tension. As the concrete tension builds up the slab cracks at fairly regular intervals allowing the ends of the slab to remain in the same position in which they were cast. In a manner of speaking, the concrete has shortened by about the same magnitude as a post-tensioned system, but not in overall dimensions. Instead of the total shortening occuring at the ends, the combined widths of many cracks which occur across the slab make-up for the total shortening. The reinforcement distributes the shortening throughout the length of the slab in the form of numerous cracks. Thus reinforced concrete tends to take care of its own shortening problems internally by the formation of numerous small cracks, each small enough to be considered acceptable. Restraints provided by stiff vertical elements such as walls and columns tend to be of minor significance, since provision for total moment has been provided by the cracks in concrete.

This is not the case with post-tensioned systems in which shrinkage cracks, which would have formed otherwise, are closed by the post-tensioning force. Much less mild steel is present and consequently the restraint to the shortening provided is much less. The slab tends to shorten at each end generating large restraining forces in the walls and columns particularly at the ends where the movement is greatest (Fig. 8.19a). These restraining forces can produce severe cracking in the slab, walls, or columns at the slab extremities causing problems to engineers and building owners alike. The most serious consequence is perhaps water leakage through the cracks.

The solution to the problem lies in eliminating the restraint by separating the slab from the restraining vertical elements. If a permanent separation is not feasible, cracking can be minimized by using temporary separations to allow enough of the shortening to occur prior to making the connection.

Cracking in post-tensioned slab also tends to be proportional to initial pour size. Some general guidelines that have evolved over the years are as follows: (i) the maximum length between temporary pour strips (Fig. 8.19b) is 150 ft (200 ft if restraint due to vertical elements is minimal); (ii) the maximum length of post-tensioned slab irrespective of the number of pour strips provided, is 300 ft. The length of time for leaving the pour strips open is critical and can range anywhere from 30 to 60 days. A 30-day period is considered adequate for average restraint conditions with relatively centered, modest length walls while a 60-day period is more the norm for severe shortening conditions with large pour sizes and stiff walls at the ends.

To minimize cracking caused by restraint to shortening, it is a good idea to provide a continuous mat of reinforcing steel in both directions of the slab. As a minimum #4 bars at 36 in. on centers both ways, is recommended for typical conditions. For slab-pours in excess of 150 ft in length with relatively stiff walls at the ends, the minimum reinforcement should be increased to #4 bars at 24 in. on centers both ways.

### 8.2.7 Preliminary design

#### 8.2.7.1 Introduction

The aim of post-tensioned design is to determine the required prestressing force and hence the number, size, and profile of the tendons for behavior at service loads. The ultimate capacity must then be checked at critical sections to assure that prestressed members have an adequate factor of safety against failure.

In statically determinate structures, as in simple beams, the moments induced by the post-tensioning are directly proportional to the eccentricity of the tendons with respect to the neutral axis of the beam. In indeterminate structures, as in continuous beams, the moments due to post-tensioning are usually not proportional to the tendon eccentricity. The difference is due to the restraint imposed by the supports to post-tensioning deformations. The moments resulting from the restraints to prestressing deformations are called "secondary moments".

Primary and secondary moments, as well as the total moments due to post-tensioning are shown in Fig. 8.20 for a two span continuous

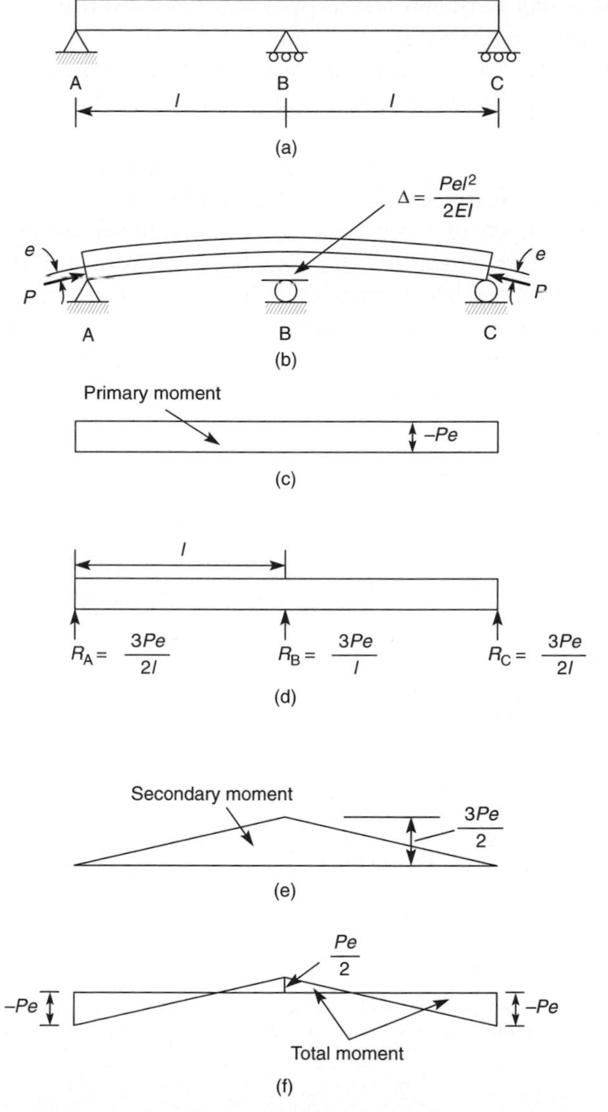


**Figure 8.20** Concept of secondary moment: (a) two-span continuous beam; (b) vertical upward displacement due to PT; (c) primary moment; (d) reactions due to PT; (e) secondary moment; (f) final moments.

beam. The beam has a post-tensioning force $P$, acting at a constant eccentricity $e$. Hence the primary moment in the beam is $Pe$ as shown in Fig. 8.20c. The primary moment will cause a theoretical upward deflection of $Pel^2/2EI$ at the center support. The reactions necessary to retain the support at its original position are shown in Fig. 8.20d. Observe the secondary moments are functions of the reactions, and for this reason, vary linearly between the supports. Also note that for this case, the secondary moment is 150 percent of the primary moment. The total moment due to post-tensioning may be expressed as the superposition of the primary and secondary moments as shown in Fig. 8.20f.

The preliminary design method presented in this section is based on an article published in Ref. 25. It uses the technique of load balancing in which the effect of prestressing is considered as an equivalent external load. For example, the parabolic profile in Fig. 8.21a exerts a horizontal force $P_1$ at the ends along with vertical components equal to $P \sin \theta$. The vertical component is neglected in design because it occurs directly over the supports. In addition to these loads, the parabolic tendon exterts a continuous upward force

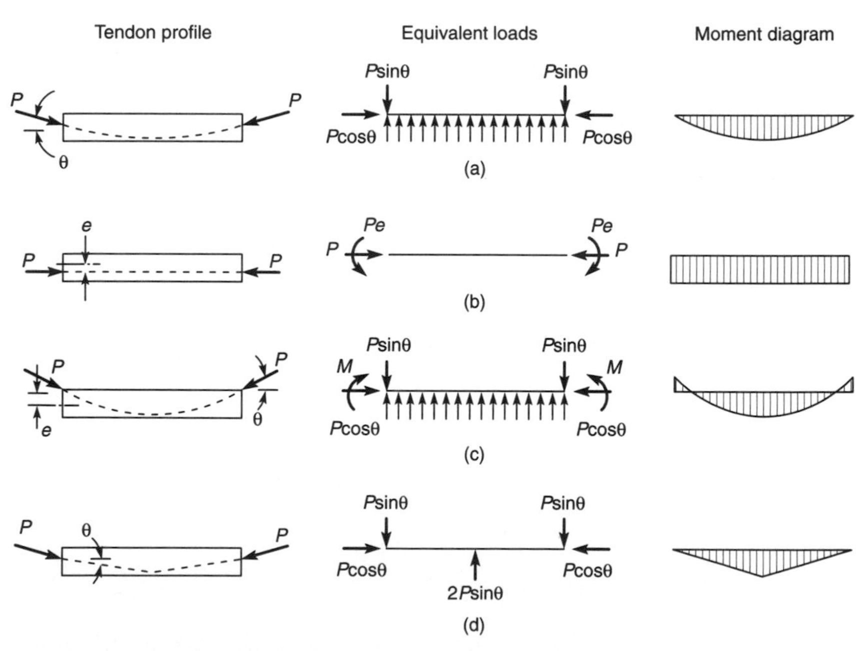


**Figure 8.21** Equivalent loads and moments produced by prestressed tendons: (a) upward uniform load due to parabolic tendon; (b) constant moment due to straight tendon; (c) upward uniform load and end-moments due to parabolic tendon not passing through the centroid at the ends; (d) vertical point-load due to sloped tendon.

on the beam along its entire length. By neglecting friction between the tendon and concrete, we can assume that: (i) the upward pressure exerted is normal to the plane of contact; and (ii) tension in tendon is constant. The upward pressure exerted is equal to the tension in the tendon divided by the radius of curvature. Due to the shallow nature of post-tensioned structures, the vertical component of the tendon force may be assumed constant. Considering one-half of the beam as a free-body Fig. 8.22a, the vertical load exerted by the tendon may be derived by summing moments about support $A$. Thus $w_p = \frac{8pe}{L^2}$. Equivalent loads and moments produced by other types of tendon profile are shown in Fig. 8.21.

#### 8.2.7.2 General step-by-step procedure

1. Determine preliminary size of members using the values given in Table 8.8 as a guide.
2. Determine section properties of the member such as the area $A$, moment of inertia $I$, and section modulus $S_t$ and $S_b$.
3. Determine tendon profile with due regard to cover and location of mild steel reinforcement.
4. Determine effective span $L_e$ by assuming $L_1 = \frac{1}{16}$ to $\frac{1}{19}$ of the span

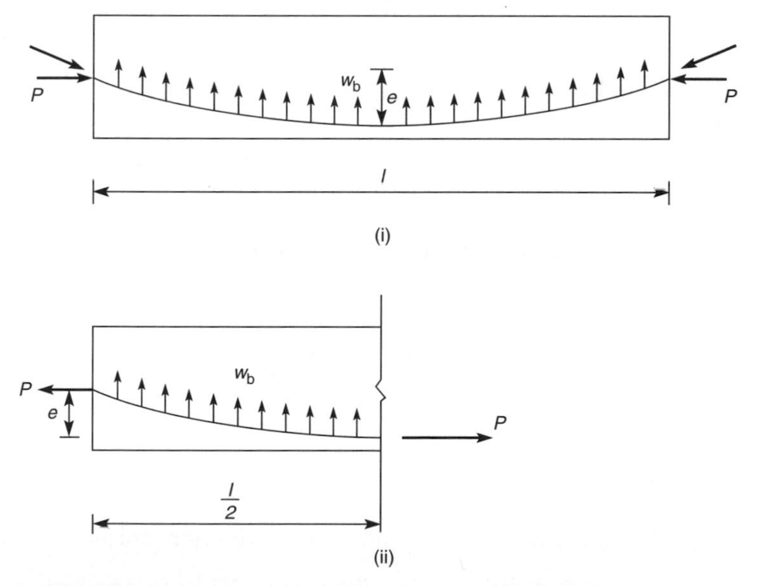


**Figure 8.22a** Load balancing concept: (i) beam with parabolic tendon; (ii) free-body diagram.

length for slabs, and $L_e = \frac{1}{10}$ to $\frac{1}{12}$ of the span length for beams (see Fig. 8.23).

5. Start with an assumed value for balanced load $w_p$ equal to 0.7 to 0.9 times the total dead load.
6. Determine the elastic moments for the total dead plus live loads (working loads). For continuous beams and slabs use computer plane-frame analysis programs, moment distribution method or ACI coefficients if applicable, in the decreasing order of preference.
7. Reduce negative moments to the face of supports.
8. By proportioning the unbalanced load to the total load, determine the unbalanced moments at $M_{ub}$ at critical sections such as at the supports and at the center of spans.
9. Calculate stress at bottom and top, $f_b$ and $f_t$ at critical sections. Typically at supports the stresses $f_t$ and $f_b$ are in tension and compression. At center of spans the stresses are in compression and tension.
10. Calculate the minimum required post-tension stress $f_p$ by using the following equations.

    For negative zones of one-way slabs and beams: $f_p = f_t - 6\sqrt{f'_c}$

    For positive moments in two-way slabs: $f_p = f_t - 2\sqrt{f'_c}$

11. Find the post tension force $P$ by the relation $P = f_p \times A$ where $A$ is the area of the section.
12. Calculate the balanced load $w_p$ due to $P$ by the relation

$$w_p = \frac{8 \times Pe}{L_e^2}$$

    where

    $e$ = drape of the tendon

    $L_e$ = effective length of tendon between inflection points.

13. Compare the calculated value of $w_p$ from step 12 with the value assumed in step 5. If they are about the same, the selection of

post-tension force for the given loads and tendon profile is complete. If not, repeat steps 9–13 with a revised value of $w_p = 0.75w_{p_1} + 0.25w_{p_2}$. $w_{p_1}$ is the value of $w_p$ assumed at the beginning of step 5 and $w_{p_2}$ is the derived value of $w_p$ at the end of step 12. Convergence is fast requiring no more than three cycles in most cases.

### 8.2.7.3 Simple spans

The concept of preliminary design discussed in this section is illustrated in Fig. 8.22b where a parabolic profile with an eccentricity of 12 in. is selected to counteract part of the imposed load consisting of a uniformly distributed dead load of 1.5 kip/ft and a live load of 0.5 kip/ft.

In practice, it is rarely necessary to provide a prestress force to fully balance the imposed loads. A value of prestress, often used for building system is 75 to 95 percent of the dead load. For the illustrative problem, we begin with an assumed 80 percent of the dead load as the unbalanced load.

**First cycle.** The load being balanced is equal to $0.80 \times 1.5 = 1.20$ kip/ft. The total service dead plus live load $= 1.5 + 0.5 = 2.0$ kip/ft of which 1.20 kip/ft is assumed in the first cycle to be balanced by the prestressing force in the tendon. The remainder of the load equal to $2.0 - 1.20 = 0.80$ kip/ft acts vertically downward

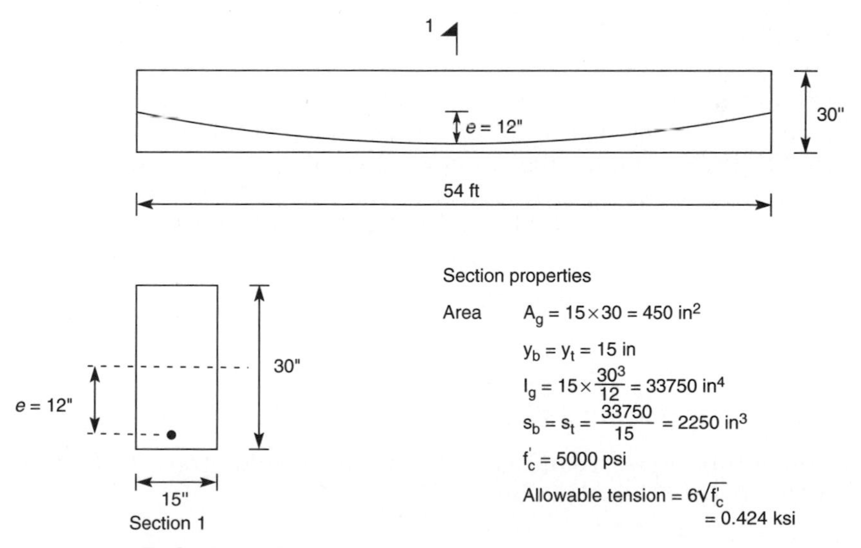


**Figure 8.22b** Preliminary design: simple span beam.

producing a maximum unbalanced moment $M_{ub}$ at center span given by

$$M_{ub} = 0.80 \times \frac{54^2}{8}$$

$$= 291.6 \text{ kip ft}$$

The tension and compression in the section due to $M_{ub}$ is given by

$$f_c = f_b = \frac{291.6 \times 12}{2250}$$

$$= 1.55 \text{ ksi}$$

The minimum prestress required to limit the tensile stress to $6\sqrt{f'_c} = 0.424$ is given by

$$f_p = 1.55 - 0.424 = 1.13 \text{ ksi}$$

Therefore the required minimum prestressing force $P$ = Area of beam $\times$ 1.13 = 450 $\times$ 1.13 = 509 kips. The load balanced by this force is given by

$$w_p \times \frac{54^2}{8} = Pe = 509 \times 1$$

$\therefore$ $w_p$ = 1.396 kip/ft compared to the value of 1.20 used in the first cycle. Since these two values are not close to each other, we repeat the above calculations starting with a more precise value for $w_p$ in the second cycle.

**Second cycle.** We start with a new value of $w_p$ by assuming the new value as 75 percent of the initial value + 25 percent of the derived value. For this example problem new value of

$$w_p = 0.75 \times 1.20 + 0.25 \times 1.396 = 1.25 \text{ kip/ft}$$

$$M_{ub} = (2 - 1.25) \times \frac{54^2}{8} = 273.3 \text{ kip ft}$$

$$f_b = f_t = \frac{273.3 \times 12}{2250} = 1.458 \text{ ksi}$$

The minimum stress required to limit the tensile stress to $6\sqrt{f'_c} = 6\sqrt{5000} = 0.424$ ksi is given by

$$f_p = 1.458 - 0.424 = 1.03 \text{ ksi}$$

Minimum prestressing force $P = 1.03 \times 450 = 465$ kips. The balanced load corresponding to the prestress value of 465 is given by

$$w_p = \frac{8Pe}{L^2} = \frac{8 \times 465 \times 1}{54^2}$$

$\therefore w_p = 1.27$ kip/ft which is nearly equal to the value assumed in the second cycle. Thus the minimum prestress required to limit the tensile stress in concrete to $6\sqrt{f_c}$ is 465 kips.

To demonstrate how rapidly the method converges to the desired answer, we will rework the problem by assuming an initial value of $w_p = 1.0$ kip/ft in the first cycle.

**First cycle**

$$w_p = 1.0 \text{ kip/ft}$$

$$M_{ub} = (2 - 1) \times \frac{54^2}{8} = 364.5 \text{ kip ft}$$

$$f_b = f_t = \frac{364.5 \times 12}{2250} = 1.944 \text{ ksi}$$

$$f_p = 1.944 - 0.454 = 1.49 \text{ ksi}$$

$$P = 1.49 \times 450 = 670.5 \text{ kips}$$

$$w_p \times \frac{54^2}{8} = 670.5 \times 1$$

$$w_p = 1.84 \text{ kip/ft}$$

compared to 1.0 kip/ft used at the beginning of first cycle.

**Second cycle**

$$w_p = 0.75 \times 1 + 0.25 \times 1.84 = 1.21 \text{ kip/ft}$$

$$M_{ub} = (2 - 1.21) \times \frac{54^2}{8} = 288 \text{ kip ft}$$

$$f_b = f_c = \frac{288 \times 12}{2250} = 1.536 \text{ ksi}$$

$$f_p = 1.536 - 0.454 = 1.082 \text{ ksi}$$

$$P = 1.082 \times 450 = 486.8 \text{ kips}$$

$$w_p = \frac{486.8 \times 1 \times 8}{54^2}$$

$$= 1.336 \text{ kip/ft}$$

compared to the value of 1.21 used at the beginning of second cycle.

**Third cycle**

$$w_p = 0.75 \times 1.21 - 1.21 \times 0.25 \times 1.336 = 1.24 \text{ kip/ft}$$

$$M_{ub} = (2 - 1.24) \times \frac{54^2}{8} = 276.67 \text{ kip ft}$$

$$f_b = f_c = \frac{276.47 \times 12}{2250} = 1.475 \text{ ksi}$$

$$f_p = 1.475 - 0.454 = 1.021 \text{ ksi}$$

$$P = 1.021 \times 450 = 459.3 \text{ kips}$$

$$w_p = 459.3 \times \frac{1 \times 8}{54^2} = 1.26 \text{ kip/ft}$$

compared to 1.24 assumed at the beginning of third cycle. The value of 1.26 kip/ft is considered close enough for design purposes.

#### 8.2.7.4 Continuous spans

The above example illustrates the salient features of load balancing. These are that generally the prestressing force is selected to counteract or balance a portion of dead load, and under this loading condition the net stress in the tension fibers is limited to a value $= 6\sqrt{f'_c}$. If it is desired to design the member for zero stress at the bottom fiber at center span (or any other value less than the code allowed maximum value of $6\sqrt{f'_c}$) it is only necessary to adjust the amount of post-tensioning provided in the member.

There are some qualifications to the above procedure that should be kept in mind when using the technique to continuous beams. Chief among them is the fact that it is not usually practical to install tendons with sharp break in curvature over supports as shown in Fig. 8.23a. The stiffness of tendons requires a reverse curvature (Fig. 8.23b) in the tendon profile with a point of contraflexure some distance from the supports. Although this reverse curvature modifies the equivalent loads imposed by post-tensioning from those assumed for a pure parabolic profile between the supports, a simple revision to the effective length of tendon as will be seen shortly, yields results sufficiently accurate for preliminary designs.

Consider the tendon profiles shown in Figs 8.24 for a typical exterior and an interior span. Observe three important features.

1. The effective span $L_e$, the distance between the inflection points which is considerably shorter than the actual span.
2. The sag or drape of the tendon numerically equal to average height of inflection points, less the height of the tendon midway between the inflection points.
3. The point midway between the inflection points is not necessarily the lowest point on the profile.

The upward equilalent uniform load produced by the tendon is given by

$$w_p = \frac{8Pe}{L_e^2}$$

where $w_p$ = equivalent upward uniform load due to prestress
$P$ = prestress force
$e$ = cable drape between inflection points
$L_e$ = effective length between inflection points

Note that relatively high loads acting downward over the supports result from the sharply curved tendon profiles located within these regions (Fig. 8.25).

Since the large downward loads are confined to a small region, typically $\frac{1}{10}$ to $\frac{1}{8}$ of the span, their effect is secondary as compared to the upward loads. Slight differences occur in the negative moment regions between the external load moments and the moment due to

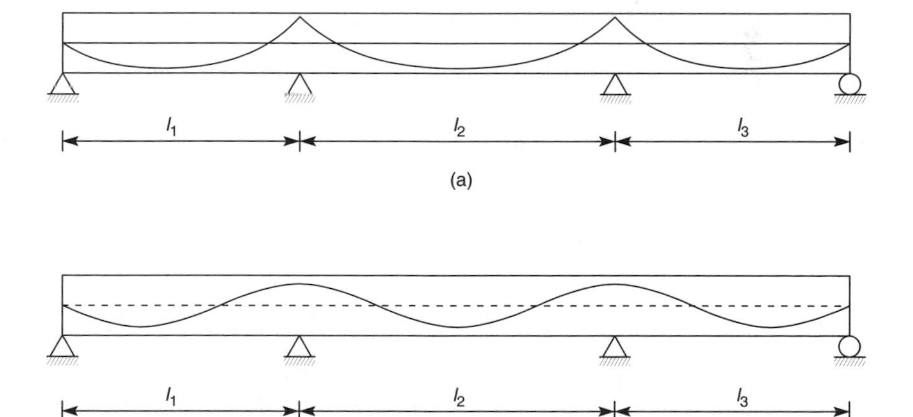


**Figure 8.23** Tendon profile in continuous beams: (a) simple parabolic profile; (b) reverse curvature in tendon profile.

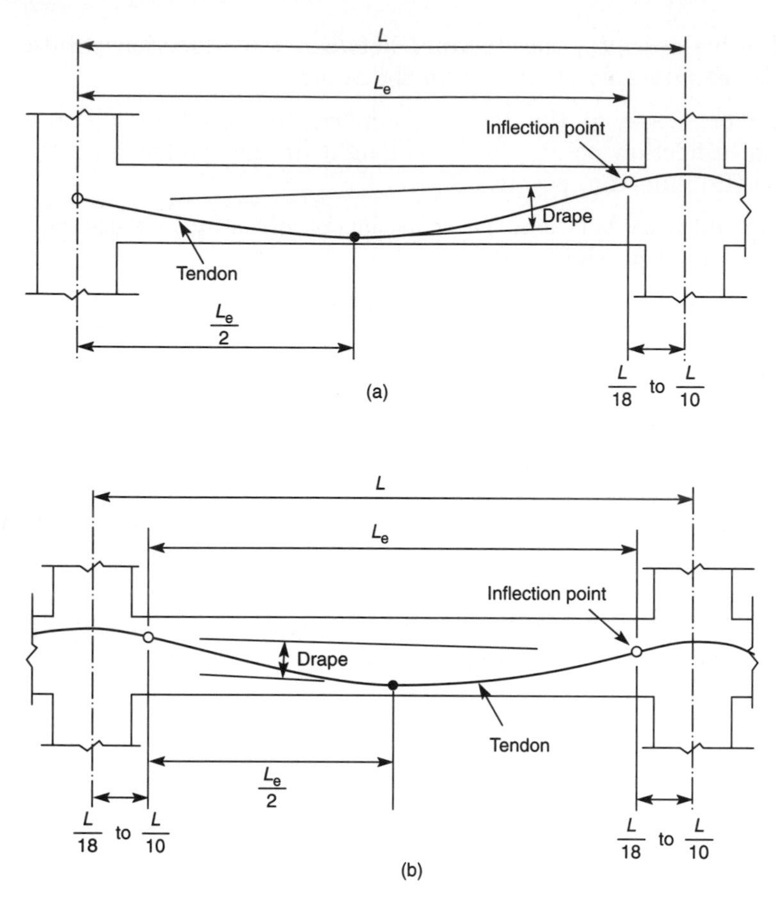


**Figure 8.24** Tendon profile: (a) typical exterior span; (b) typical interior span.

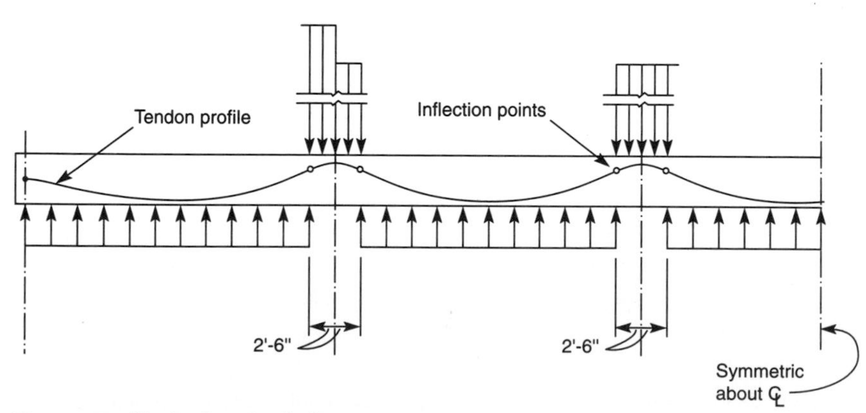


**Figure 8.25** Equivalent loads due to prestress.

prestressing force. The differences are of minor significance and can be neglected in the design without losing meaningful accuracy.

As in simple spans the moments caused by the equivalent loads are subtratced from those due to external loads, to obtain the net unbalanced moment which produces the flexural stresses. To the flexural stresses, the axial compressive stresses from the prestress are added to obtain the final stress distribution in the members. The maximum compressive and tensile stresses are compared to the allowable values. If the comparisons are favorable, an acceptable design has been found. If not either the tendon profile or force (and very rarely the cross-sectional shape of the structure) is revised to arrive at an acceptable solution.

In this method, since the moments due to equivalent loads are linearly related to the moments due to external loads, the designer can by-pass the usual requirement of determining the primary and secondary moments.

**Example 1: one-way slab.** Given a 30′ 0″ column grid layout design a one-way slab spanning between beams shown in Fig. 8.26.

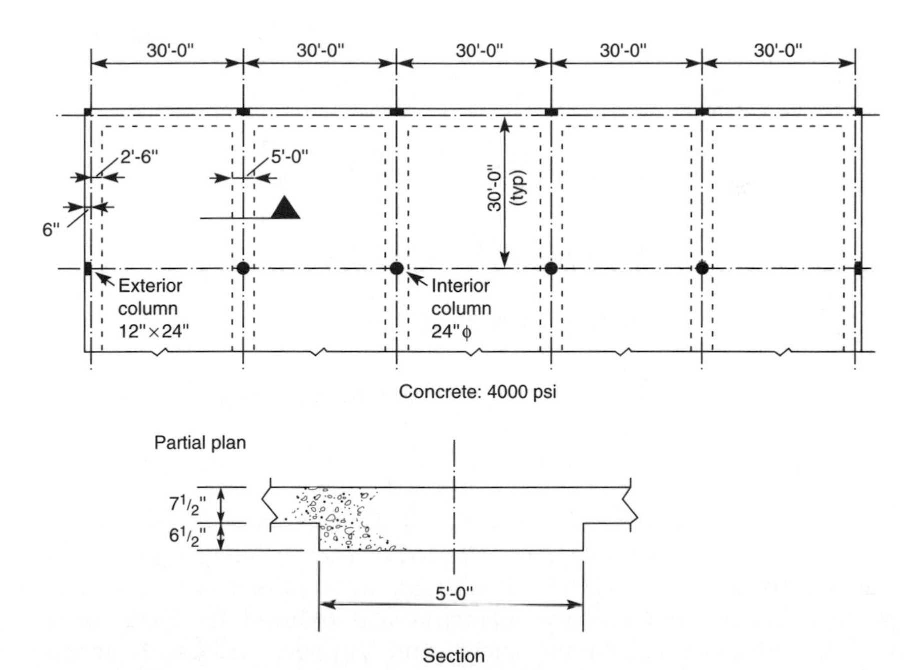


**Figure 8.26** Example 1: one-way post-tensioned slab.

Slab and beam depths:

Clear span of slab = 30 − 5 = 25 ft

$$\text{Recommended slab depth} = \frac{\text{span}}{40} = \frac{25 \times 12}{40} = 7.5 \text{ in.}$$

Clear span for beams = 30 ft center-to-center span, less 2′ 0″ for column width = 30 − 2 = 28 ft

$$\text{Recommended beam depth} = \frac{\text{span}}{25} = \frac{28 \times 12}{25} = 13.44 \text{ in. use } 14 \text{ in.}$$

Loading:

| | | |
|---|---|---|
| Dead load: | 7.5″ slab | = 94 psf |
| | Mech. & lights | = 6 psf |
| | Ceiling | = 6 psf |
| | Partitions | = 20 psf |
| | Total dead load | = 126 psf |
| Live load: | Office load<br>Code minimum is 50 psf<br>Use 100 psf per owner's request | = 100 psf |
| | Total $D + L$ | = 226 psf |

Slab design: Slab properties for 1′-0″ wide strip

$$I = \frac{bd^3}{12} = 12 \times \frac{7.5^3}{12} = 422 \text{ in.}^4$$

$$S_{\text{top}} = S_{\text{bot}} = \frac{422}{3.75} = 112.5 \text{ in.}^3$$

$$\text{Area} = 12 \text{ in.} \times 7.5 = 90 \text{ in.}^2$$

A 1 ft width of slab is analyzed as a continuous beam. The effect of column stiffness is ignored.

The moment diagram for a service load of 226 plf is shown in Fig. 8.27.

Moments at the face of supports have been used in the design instead of center line moments. Negative center line moments are reduced by a "$Va/3$" factor ($V$ = shear at that support, $a$ = total support width), and positive moments are reduced by $Va/6$ using average adjacent values for shear and support widths. A frame analysis may of course be used to obtain more accurate results.

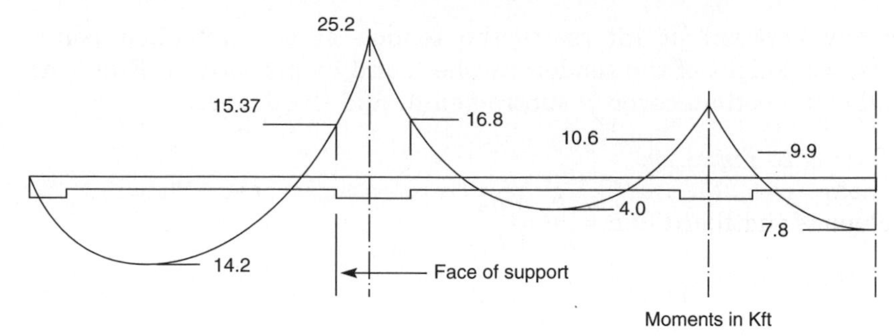


**Figure 8.27** Example 1: one-way post-tensioned slab service load, $(D + L)$, moment diagram.

The design of continuous strands will be based on the negative moment of 10.6 kip/ft. The additional prestressing required for the negative moment of 16.8 kip/ft will be provided by additional tendons in the end bays only.

**Determination of tendon profile.** Maximum tendon efficiency is obtained when the cable drape is as large as the structure will allow. Typically, the high points of the tendon over the supports and the low point within the span are dictated by concrete cover requirements and the placement of mild steel.

The high and low points of tendon in the interior bay of the example problem are shown in Fig. 8.28. Next, the location of inflection points are determined. For slabs, the inflection points usually range within $\frac{1}{16}$ to $\frac{1}{19}$ of the span. The fraction of span length used is a matter of judgement, and is based on the type of structure. For this example, we choose $\frac{1}{16}$ of span which works out to $1'\ 10\frac{1}{2}''$.

An interesting property useful in determining tendon profile shown in Fig. 8.29, is that, if a straight line (chord) is drawn connecting the tendon high point over the support, and the low point

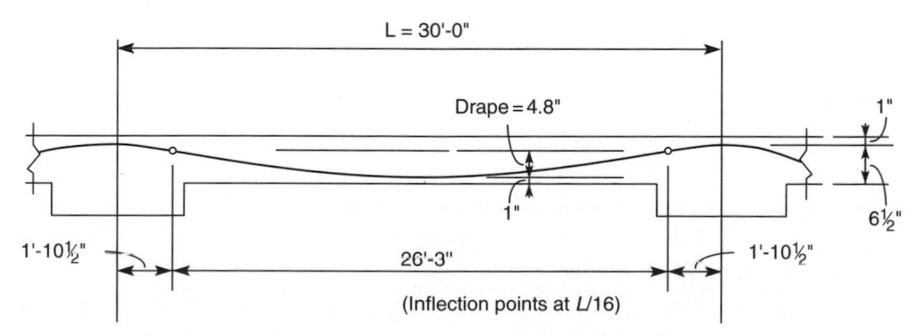


**Figure 8.28** Example 1: one-way post-tensional slab tendon profile, interior bay.

midway between, it intersects the tendon at the inflection point. Thus, the height of the tendon can be found by proportion. From the height, the bottom cover is subtracted to find the drape.

Referring to Fig. 8.29:

$$\text{Slope of the chord line} = \frac{h_1 - h_2}{(L_1 + L_2)}$$

$$\begin{aligned} h_3 &= h_2 + L_2 \times (\text{Slope}) \\ &= h_2 + \frac{L_2(h_1 - h_2)}{(L_1 + L_2)} \end{aligned}$$

$$\text{This simplifies to } h_3 = \frac{(h_1 L_2 + h_2 L_1)}{(L_1 + L_2)}$$

The drape $h_d$ is obtained by subtracting $h_2$ from the above equation. Note, that notation $e$ is also used in these examples to denote drape $h_d$.

The height of the inflection point as given above is exact for symmetrical layout of the tendon about the center span. If the tendon is not symmetrical, the value is approximate but sufficiently accurate for preliminary design.

Returning to our example problem we have $h_1 = 6.5''$, $h_2 = 1''$, $L_1 = 1.875'$ and $L_2 = 13.125'$.

Height of tendon at the inflection point:

$$h_3 = \frac{(h_1 L_2 + h_2 L_1)}{(L_1 + L_2)}$$

$$h_3 = \frac{6.5 \times 13.125 + 1 \times 1.875}{(1.875 + 13.125)} = 5.813''$$

Drape $h_d = e = 5.813 - 1 = 4.813''$ use 4.8″

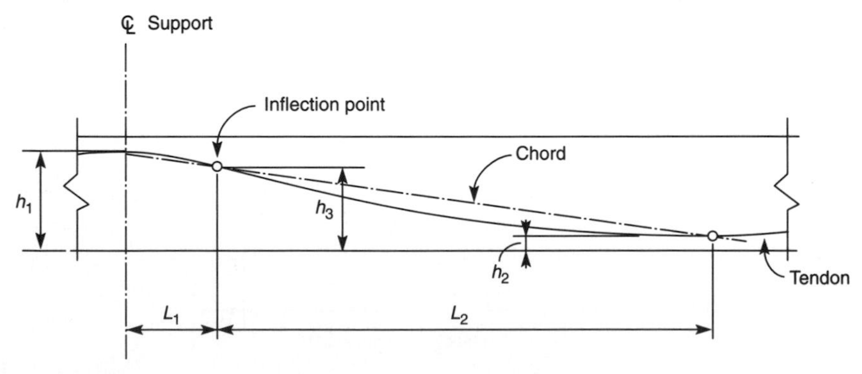


**Figure 8.29** Dimensions for determining tendon drape.

Allowable stresses from UBC 94 are as follows.

$$f_t = \text{Tensile stress} = 6\sqrt{f'_c}$$
$$f_c = \text{Compressive stress} = 0.45f'_c$$

For 4000 psi concrete:

$$f_t = 6\sqrt{4000} = 380 \text{ psi}$$
$$f_c = 0.45 \times 4000 = 1800 \text{ psi}$$

**Design of through strands.** The design procedure is started by making an initial assumption of the equivalent load produced by the prestress. A first value of 65% of the total dead load is used.

**First cycle.** Assume

$$w_p = 0.65w_d$$

where $w_p$ = Equivalent upward load due to post tensioning
$w_d$ = Total dead load
$\therefore w_p = 0.65 \times 126 = 82$ plf

The balancing moment caused by the equivalent load is calculated from

$$M_{pt} = M_s \frac{w_{pt}}{w_s}$$

where $M_{pt}$ = balancing moment due to equivalent load (also indicated by notation $M_b$).
$M_s$ = moment due to service load, $D + L$

In our example, $M_s = 10.6$ kip ft for the interior span.

$$M_{pt} = 10.6 \times \frac{82}{226} = 3.85 \text{ kip ft}$$

Next, $M_{pt}$ is subtracted from $M_s$ to give the unbalanced moment $M_{ub}$. The flexural stresses are then obtained by dividing $M_{ub}$ by the section modulii of the structure's cross-section at the point where $M_s$ is determined: Thus

$$f_t = \frac{M_{ub}}{S_t} \tag{8.1}$$

$$f_b = \frac{M_{ub}}{S_b} \tag{8.2}$$

In our case, $M_{ub} = 10.6 - 3.85 = 6.75$ kip ft. The flexural stress at the top of the section is found by

$$f_t = \frac{M_{ub}}{S_t} = \frac{6.75 \times 12}{112.5} = 0.72 \text{ ksi}$$

The minimum required compressive prestress is found by subtracting the maximum allowable tensile stress $f_a$ given below, from the tensile stresses from Eqs (8.1) and (8.2). Thus the smallest required compressive stress is:

$$f_p = f_{ts} - f_a$$

where $f_{ts}$ = the tensile stress found from Eqs. (8.1) to (8.2) depending upon the sign of the moment

$f_a = 6\sqrt{f'_c}$ for one-way slabs or beams for the negative zones

$f_a = 2\sqrt{f'_c}$ for positive moments in two-way slabs

In our case,

$$f_p = 0.720 - 0.380 = 0.34 \text{ ksi}$$

and

$$P = 0.34 \times 7.5 \times 12 = 30.60 \text{ kip/ft}$$

Use the following equation to find the equivalent load due to prestress

$$w_p = \frac{8Pe}{L_e^2}$$

$$= 8 \times \frac{30.6 \times 4.81}{12 \times (26.25)^2} = 0.142 \text{ klf} = 142 \text{ plf}$$

This is more than 82 plf. N.G.

Since the derived value of $w_p$ is not equal to the initial assumed value, the procedure is repeated until convergence is achieved. Convergence is rapid by using a new initial value for the subsequent cycle, equal to 75 percent of the previous initial value $w_{p1}$, plus 25 percent of the derived value $w_{p2}$, for that cycle.

**Second cycle.** Use the above criteria to find the new value of $w_p$ for the second cycle

$$w_p = 0.75w_{p1} + 0.25w_{p2} = 0.75 \times 82 + 0.25 \times 142 = 97 \text{ plf}$$

$$M_b = \frac{97}{226} \times 10.6 = 4.55 \text{ kip ft}$$

$$M_{ub} = 10.6 - 4.55 = 6.05 \text{ kip ft}$$

$$f_t = f_b = \frac{6.05 \times 12}{112.5} = 0.645 \text{ ksi}$$

$$f_p = 0.645 - 0.380 = 0.265 \text{ ksi}$$

$$P = 0.265 \times 90 = 23.89 \text{ kips}$$

$$w_p = \frac{8 \times 23.89 \times 4.81}{12 \times (26.25)^2} = 0.111 \text{ klf} = 111 \text{ plf}$$

This is more than 97 psf. N.G.

**Third cycle**

$$w_p = 0.75 \times 97 + 0.25 \times 111 = 100.5 \text{ plf}$$

$$M_b = \frac{100.5}{226} \times 10.6 = 4.71 \text{ kip ft}$$

$$M_{ub} = 10.6 - 4.71 = 5.89 \text{ kip ft}$$

$$f_t = f_b = \frac{5.89 \times 12}{112.5} = 0.629 \text{ ksi}$$

$$f_p = 0.629 - 0.380 = 0.248 \text{ ksi}$$

$$P = 0.248 \times 90 = 22.3 \text{ kips}$$

$$w_p = \frac{8 \times 22.3 \times 4.81}{12 \times (26.25)^2} = 0.104 \text{ klf} = 104 \text{ plf}$$

This is nearly equal to 100.5 plf, therefore, satisfactory.
Check compressive stress at the section.

Bottom flexural stress = 0.629 ksi. Direct axial stress due to prestress $= \dfrac{22.3}{90} = 0.246$ ksi

Total compressive stress = 0.629 + 0.246 = 0.876 ksi is less than $0.45f'_c = 1.8$ ksi. Therefore, satisfactory.

**End bay design.** Design end bay prestressing using the same procedure for a negative moment of 15.37 kip ft.

Assume that at left support the tendon is anchored at the center of gravity of slab with a reversed curvature as shown in Fig. 8.33, profile 1. Assume further that the center of gravity of tendon is at a distance of 1.75 in. from the bottom of slab. With these assumptions we have: $h_1 = 3.75''$, $h_2 = 1.75''$, $L_1 = 1.875'$ and $L_2 = 13.125'$.

The height of the tendon inflection point at left end:

$$h_3 = \frac{3.75 \times 13.125 + 1.75 \times 1.875}{15} = 3.25 \text{ in.}$$

The height of the right end:

$$h_3 = \frac{6.5 \times 13.125 + 1.75 \times 1.875}{15} = 5.906 \text{ in.}$$

$$\text{Average height of tendon} = \frac{3.25 + 5.906}{2} = 4.578'' \text{ use 4.6 in.}$$

Drape $h_d = e = 4.6 - 1.75 = 2.85$ in.

**First cycle.** We start with the first cycle, as for the interior span, by assuming $w_{pt} = 82$ plf.

$$M_{pt} = 15.37 \times \frac{82}{226} = 5.58 \text{ kip ft}$$
$$M_{ub} = 15.37 - 5.58 = 9.79 \text{ kip ft}$$
$$f_t = f_b = \frac{9.79 \times 12}{112.5} = 1.04 \text{ ksi}$$
$$f_p = 1.04 - 0.380 = 0.664 \text{ ksi}$$
$$P = 0.664 \times 90 = 59.7 \text{ kip/ft}$$
$$w_p = \frac{8 \times 59.7 \times 2.85}{12 \times (26.25)^2} = 0.165 \text{ klf} = 165 \text{ plf}$$

This is more than 82 plf. N.G.

**Second cycle**

$$w_p = 0.75 \times 82 + 0.25 \times 165 = 103 \text{ plf}$$
$$M_{pt} = 15.37 \times \frac{103}{226} = 7.0 \text{ kip ft}$$
$$M_{ub} = 15.37 - 7.0 = 8.37 \text{ kip ft}$$
$$f_t = f_b = \frac{8.37 \times 12}{112.5} = 0.893 \text{ ksi}$$
$$f_p = 0.893 - 0.380 = 0.513 \text{ ksi}$$
$$P = 0.513 \times 90 = 46.1 \text{ kips}$$
$$w_p = \frac{8 \times 46.1 \times 2.85}{12 \times (26.25)^2} = 0.127 \text{ klf} = 127 \text{ plf}$$

This is more than 103 plf. N.G.

**Third cycle**

$$w_p = 0.75 \times 103 + 0.25 \times 127 = 109 \text{ plf}$$
$$M_{pt} = 15.37 \times \frac{109}{226} = 7.41 \text{ kip ft}$$
$$M_{ub} = 15.37 - 7.41 = 7.96 \text{ kip ft}$$
$$f_t = f_b = \frac{7.96 \times 12}{112.5} = 0.849 \text{ ksi}$$
$$f_p = 0.849 - 0.380 = 0.469 \text{ ksi}$$
$$P = 0.469 \times 90 = 42.21 \text{ kips}$$
$$w_p = \frac{8 \times 42.21 \times 2.85}{12 \times (26.25)^2} = 0.116 \text{ klf} = 116 \text{ plf}$$

This is nearly equal to 109 plf used at the start of third cycle. Therefore, satisfactory.

Check compressive stress at the section:

$$f_b = 0.849 \text{ ksi}$$

$$\text{Axial stress due to prestress} = \frac{42.21}{90} = 0.469 \text{ ksi}$$

$$\text{Total compressive stress} = 0.849 + 0.469 = 1.318 \text{ ksi}$$

This is less than 1.8 ksi. Therefore, design is O.K.

Check the design against positive moment of 14.33 kip ft

$$w_p = 116 \text{ plf}$$

$$M_b = 14.33 \times \frac{116}{226} = 7.36 \text{ kip ft}$$

$$M_{ub} = 14.33 - 7.36 = 6.97 \text{ kip ft}$$

$$\text{Bottom flexural stress} = \frac{6.97 \times 12}{112.5} = 0.744 \text{ ksi (tension)}$$

$$\text{Axial compression due to prestress} = \frac{42.21}{12 \times 7.5} = 0.469 \text{ ksi}$$

$$\text{Tensile stress at bottom} = 0.744 - 0.469 = 0.275 \text{ ksi}$$

This is less than 0.380 ksi. Therefore, end bay design is O.K.

**Example 2: Beam design.** Refer to Fig. 8.30 for dimensions and loading. Determine flange width of beam using the criteria given in the 1994 UBC.

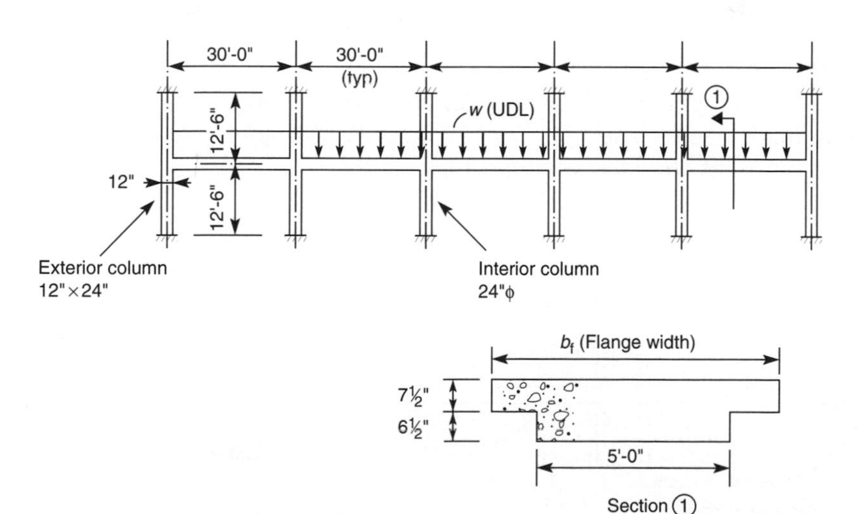


**Figure 8.30** Example 2: post-tensioned beam, dimensions and loading.

The flange width $b_f$ is the least of

1. Span/4
2. Web width + 16 × (flange thickness)
3. Web width + $\frac{1}{2}$ clear distance to next web

Therefore

$$b_f = \frac{30}{4} = 7.5 \text{ ft (controls)}$$

$$= 5 + 16 \times \frac{7.5}{12} = 15 \text{ ft}$$

$$= 5 + \frac{25}{2} = 17.5 \text{ ft}$$

Section properties

$$I = 16{,}650 \text{ in.}^4 \quad Y = 7.69 \text{ in.}$$
$$S_t = 2637 \text{ in.}^3$$
$$S_b = 2166 \text{ in.}^3$$
$$A = 1065 \text{ in.}^2$$

Loading

Dead load of $7\frac{1}{2}$ in slab = 94 psf
Mech. & Elec. = 6 psf
Ceiling = 6 psf
Partitions = 20 psf

$$\text{Additional dead load due to beam self wt} = \frac{615 \times 60 \times 150}{144 \times 30} = 13.5 = 14 \text{ psf}$$

Total dead load = 140 psf
Live load at owner's request = 80 psf
$D + L$ = 220 psf

Uniform load per ft of beam = 0.220 × 30 = 6.6 klf. The resulting service load moments are shown in Fig. 8.31. As before we design for the moments at the face of supports.

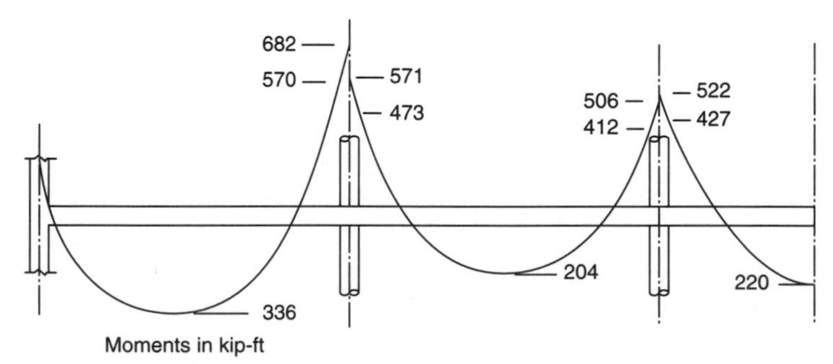


**Figure 8.31** Example 2: post-tensioned beam service load moments.

**Interior span.** Calculate through tendons by using interior span moment of 427 kip ft at the inside face of third column (Fig. 8.31).

Assume $h_1 = 11.5$ in., $h_2 = 2.5$ in., $L_1 = 2.5$ ft, $L_2 = 12.5$ ft. Refer to Fig. 8.28 for notations.

The height of inflection point

$$h_3 = \frac{11.5 \times 12.5 + 2.5 \times 2.5}{15} = 10 \text{ in.}$$

$$h_d = e = 10 - 2.5 = 7.5 \text{ in.}$$

**First cycle.** Assume

$$w_p = 3.5 \text{ klf}$$

$$M_p = \frac{3.5}{6.6} \times 427 = 226 \text{ kip ft}$$

$$M_{ub} = 427 - 226 = 201 \text{ kip ft}$$

$$f_t = \frac{201 \times 12}{2637} = 0.915 \text{ ksi}$$

$$f_p = 0.915 - 0.380 = 0.535 \text{ ksi}$$

$$P = 0.535 \times 1065 = 570 \text{ kips}$$

$$w_p = \frac{8 \times 570 \times 7.5}{12 \times (27.5)^2} = 3.77 \text{ klf}$$

which is greater than 3.5 klf. N.G.

**Second cycle.** New value of

$$w_p = 0.75 \times 3.5 + 0.25 \times 3.77 = 3.57 \text{ klf}$$

$$M_p = \frac{3.57}{6.6} \times 427 = 231 \text{ kip ft}$$

$$M_{ub} = 427 - 231 = 196 \text{ kip ft}$$

$$f_t = \frac{196 \times 12}{2637} = 0.892 \text{ ksi}$$

$$f_p = 0.892 - 0.380 = 0.512 \text{ ksi}$$

$$P = 0.512 \times 1065 = 545 \text{ kips}$$

$$w_p = \frac{8 \times 545 \times 7.5}{12 \times (27.5)^2} = 3.60 \text{ klf}$$

which is nearly equal to 3.57 klf. Therefore design is satisfactory.

Check design against positive moment of 220 kip/ft.

$$M_p = \frac{3.6 \times 220}{6.6} = 120 \text{ kip ft}$$

$$M_{ub} = 220 - 120 = 100 \text{ kip ft}$$

$$F_{\text{bot}} = \frac{100 \times 12}{2166} = 0.554 \text{ ksi (tension)}$$

$$\text{Axial comp. stress} = \frac{545}{1065} = 0.512 \text{ ksi (comp.)}$$

$$f_{\text{total}} = 0.554 - 0.512 = 0.042 \text{ ksi (tension)}$$

This is less than the allowable tensile stress of 0.380 ksi. Therefore, the design is satisfactory.

**End span.** Determine end bay prestressing for a negative moment of 570 kip ft at the face of first interior column (Fig. 8.31).

**First cycle.** As before, assume

$$w_p = 3.5 \text{ klf}$$

$$M_p = \frac{3.5}{6.6} \times 570 = 302 \text{ kip ft}$$

$$M_{ub} = 570 - 302 = 268 \text{ kip ft}$$

$$f_t = \frac{268 \times 12}{2637} = 1.22 \text{ ksi}$$

$$f_p = 1.22 - 0.380 = 0.84 \text{ ksi}$$

$$P = 0.84 \times 1065 = 894 \text{ kips}$$

$$w_p = \frac{8 \times 894 \times 7.5}{12 \times (27.5)^2} = 5.912 \text{ klf}$$

which is greater than 3.5 klf. N.G.

**Second cycle.** New value of

$$w_p = 0.74 \times 3.5 + 0.25 \times 5.912 = 4.1\,\text{klf}$$

$$M_p = \frac{4.1}{6.6} \times 570 = 354\,\text{kip ft}$$

$$M_{ub} = 570 - 354 = 216\,\text{kip ft}$$

$$f_t = \frac{216 \times 12}{2637} = 0.983\,\text{ksi}$$

$$f_p = 0.983 - 0.380 = 0.603\,\text{ksi}$$

$$P = 0.603 \times 1065 = 642\,\text{kips}$$

$$w_p = \frac{8 \times 642 \times 7.5}{12 \times (27.5)^2} = 4.24\,\text{klf}$$

This is nearly equal to 4.1 klf. However, a more accurate value is calculated as follows:

$$w_p = 0.75 \times 4.1 + 0.25 \times 4.24 = 4.13\,\text{klf}$$

Check the design against positive moment of 336 kip ft

$$M_p = \frac{4.13}{6.6} \times 336 = 210\,\text{kip ft}$$

$$M_{ub} = 336 - 210 = 126\,\text{kip ft}$$

$$\text{Bottom flexural stress} = \frac{126 \times 12}{2166} = 0.698\,\text{ksi (tension)}$$

$$\text{Axial compressive stress due to post-tension} = \frac{642}{1065} = 0.603\,\text{ksi (comp.)}$$

$$f_{\text{total}} = 0.698 - 0.603 = 0.095\,\text{ksi}$$

This is less than the allowable tensile stress of 0.380 ksi. Therefore the design is O.K.

**Example 3: Flat plate system.** Figure 8.32 shows a schematic section of a two-way flat plate system. Design of post-tension slab for an office-type loading is required.

**Given**

Specified compressive strength of concrete, $f'_c = 4000$ psi
Modulus of elasticity of concrete, $E_c = 3834$ ksi

Allowable tensile stress in precompressed tensile zone $= 6\sqrt{f'_c} = 380$ psi

Allowable fiber stress in compression $= 0.45f'_c = 0.45 \times 4000 = 1800$ psi

Tendon cover: Interior spans Top 0.75 in.
Bot. 0.75 in.

Exterior spans Top 0.75 in.
Bot. 1.50 in.

Tendon diameter $= \frac{1}{2}''$

Minimum area of bonded reinforcement:

In negative moment areas at column supports:

$$A_s = 0.00075\,\text{hl}$$

In positive moment areas where computed concrete stress in tension exceeds $2\sqrt{f'_c}$:

$$A_s = \frac{N_c}{0.5 f_y}$$

Rebar yield: 60 ksi. Max bar size = #5

Rebar cover 1.63″ at top and bottom

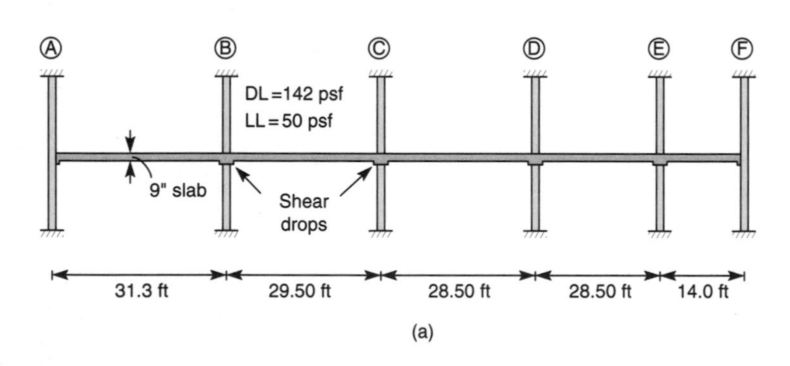


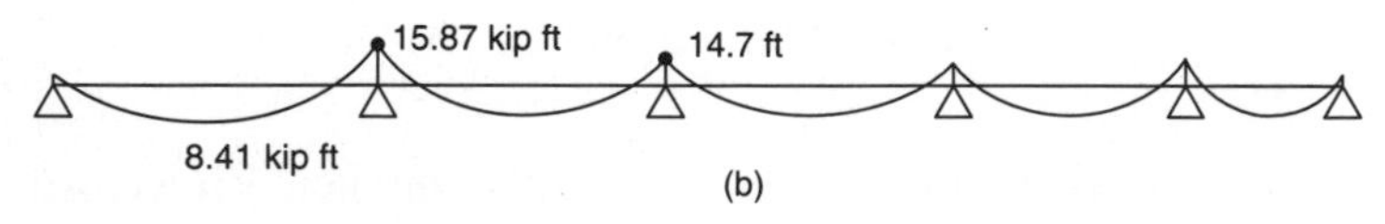


**Figure 8.32** Example 3: Flat plate: (a) span and loading conditions; (b) elastic moments due to dead plus live loads.

Post-tension requirements

Minimum post tensioned stress = 125 psi

Minimum balanced load = 65 percent of total dead load

**Design**

The flat plate is sized using the span-depth ratios given in Table 8.6. The maximum span is 31 ft 4 in. between grids A and B. Using a span-depth ratio of 40, the slab thickness is $\frac{31.33 \times 12}{40} = 9.4$ in., rounded to 9 in.

The flat plate has "shear drops" intended to increase only the shear strength and flexural support width. The shear heads are smaller than a regular drop panel as defined in the ACI code. Therefore shear heads cannot be included in calculating the bending resistance.

| Loading: | Dead load of 9″ slab | 112 psf |
|---|---|---|
| | Partitions | 20 psf |
| | Ceiling and mechanical | 10 psf |
| | Reduced live load | 50 psf |

$$\text{Total service load} = 112 + 20 + 10 + 50 = 192 \text{ psf}$$

$$\text{Ultimate load} = 1.4 \times 142 + 1.7 \times 50 = 285 \text{ psf}$$

Slab properties (for a 1 ft-wide strip)

$$I = \frac{bh^3}{12} = 12 \times \frac{9^3}{12} = 729 \text{ in.}^4$$

$$S_{\text{top}} = S_{\text{bot}} = \frac{729}{4.5} = 162 \text{ in.}^3$$

$$\text{Area} = 12 \times 9 = 108 \text{ in.}^2$$

The moment diagram for a 1 ft-wide strip of slab subjected to a service load of 192 psf is shown in Fig. 8.92b.

The design of continuous strands will be based on a negative moment of 14.7 kip ft at the second interior span. The end bay prestressing will be based on a negative moment of 15.87 kip ft.

**Interior span.** Calculate the drape of tendon using procedure given for the previous problem. See Figs 8.29 and 8.35a.

$$h_3 = \frac{h_1 L_2 + h_2 L_1}{L_1 + L_2}$$

$$L_1 = 1.84\ \text{ft} \qquad h_1 = 8\ \text{in.}$$

$$L_2 = 12.90\ \text{ft} \qquad h_2 = 1.25\ \text{in.}$$

$$L_e = 12.9 \times 2 = 25.8\ \text{ft}$$

$$h_3 = \frac{8 \times 12.90 + 1.25 \times 1.84}{14.75} = 7.153$$

$$\text{Tendon drape} = 7.153 - 1.25 = 5.90\ \text{in.}$$

**First cycle**

$$\text{Minimum balanced load} = 0.65 \times (\text{total DL})$$

$$= 0.65(112 + 10 + 20) = 92\ \text{psf}$$

$$\text{Moment due to balanced load} = \frac{92}{192} \times 14.7$$

$$= 7.04\ \text{kip ft}$$

This is subtracted from the total service load moment of 14.7 kip ft to obtain the unbalanced moment $M_{ub}$.

$$M_{ub} = 14.7 - 7.04 = 7.66\ \text{kip ft}$$

The flexural stresses at top and bottom are obtained by dividing $M_{ub}$ by the section modulii of the structure's cross-section.

$$f_t = \frac{7.66 \times 12}{162} = 0.567\ \text{ksi}$$

$$f_b = \frac{7.66 \times 12}{162} = 0.567\ \text{ksi}$$

The minimum required compressive prestress $f_p$ is found by subtracting the maximum allowable tensile stress $f_a = 6\sqrt{f'_c}$ from the calculated tensile stress. Thus the smallest required compressive stress is:

$$f_p = f_t - f_a$$

$$f_p = 0.567 - 380 = 0.187 \text{ ksi}$$

The prestress force is calculated by multiplying $f_p$ by the cross-sectional area:

$$P = 0.187 \times 9 \times 12 = 20.20 \text{ kip/ft}$$

Determine the equivalent load due to prestress force $P$ by the relation

$$w_p = \frac{8Pe}{L_e^2}$$

For the example problem,

$$P = 20.20 \text{ kip/ft}, \quad e = 5.90''$$

$$L_e = 2 \times 12.90 = 25.8 \text{ ft}$$

$$\therefore \quad w_p = \frac{8 \times 20.20 \times 5.90}{25.8^2 \times 12} = 0.120 \text{ klf} = 120 \text{ plf}$$

Comparing this with the value of 92 plf assumed at the beginning of first cycle, we find the two value are not equal. Therefore, we assume a new value and repeat the procedure until convergence is obtained.

**Second cycle**

$$w_p = 0.75 \times 92 + 0.25\,(120) = 99 \text{ plf}$$

$$M_b = \frac{99}{192} \times 14.7 = 7.58 \text{ kip ft}$$

$$M_{ub} = 14.7 - 7.58 = 7.12 \text{ kip ft}$$

$$f_t = f_b = \frac{7.12 \times 12}{162} = 0.527 \text{ ksi}$$

$$f_p = 0.527 - 0.380 = 0.147 \text{ ksi}$$

$$P = 0.147 \times 9 \times 12 = 15.92 \text{ kip/ft}$$

$$w_p = \frac{8 \times 15.92 \times 5.90}{25.8^2 \times 12} = 0.094 \text{ klf} = 94 \text{ plf}$$

This is less than 99 plf assumed at the beginning of second cycle. Therefore, we assume a new value and repeat the procedure.

**Third cycle**

$$w_p = 0.75 \times 99 + 0.25\ (94) = 97.7 \text{ plf}$$

$$M_b = \frac{99.7 \times 14.7}{192} = 7.48 \text{ kip ft}$$

$$M_{ub} = 14.7 - 7.48 = 7.22 \text{ kip ft}$$

$$f_t = f_b = \frac{7.22 \times 12}{162} = 0.535 \text{ ksi}$$

$$f_p = 0.535 - 0.380 = 0.155 \text{ ksi}$$

$$P = 0.155 \times 9 \times 12 = 16.74 \text{ kip/ft}$$

$$w_p = \frac{8 \times 16.74 \times 5.90}{25.8^2 \times 12} = 0.99 \text{ klf} = 99 \text{ plf}$$

This is nearly equal to 97.7 plf assumed at the beginning of the third cycle. Therefore O.K.

Check compressive stress at the support:

$$M_p = \frac{99 \times 14.7}{192} = 7.58 \text{ kip ft}$$

$$M_{ub} = 14.7 - 7.58 = 7.12 \text{ kip ft}$$

$$f_b = \frac{7.12 \times 12}{162} = 0.527 \text{ ksi} = 527 \text{ psi}$$

$$\text{Axial compressive stress due to post-tension} = \frac{16.74 \times 1000}{9 \times 12} = 155 \text{ psi}$$

$$\text{Total compressive stress} = 527 + 155 = 682 \text{ psi}$$

This is less than the allowable compressive stress of 1800 psi. Therefore, the design is satisfactory.

**End bay design.** The placement of tendon within the end bay presents a few problems. The first problem is in determining the location of the tendon over the exterior support. Placing the tendon above the neutral axis of the member results in an increase in the total tendon drape allowing the designer to use less prestress than would otherwise be required. Raising the tendon, however, introduces an extra moment that effectively cancels out some of the benefits from the increased drape. For this reason, the tendon is usually placed at neutral axis at exterior supports.

The second problem is in making a choice in the tendon profile: whether to use a profile with a reverse curvature over each support (profile 1), or over the first interior support only (profile 2). See Fig. 8.33. A profile with the reversed curvature over the first interior support only, gives a greater cable drape than the first profile suggesting a larger equivalent load with the same amount of prestress. On the other hand, the effective length $L_e$ between inflection points of profile 1 is less than that of profile 2 which suggests the opposite. To determine which profile is in fact more efficient, it is necessary to evaluate the amount of prestress for both profiles. More usually, a tendon profile with reverse curvature over both supports is 5 to 10 percent more efficient since the equivalent load produced is a function of the square of the effective length.

The last item addresses the extra end bay prestressing required in most situations. The exterior span, in an equal span structure has the greatest moments due to support rotations. Because of this, extra prestressing is commonly added to end bays to allow efficient design of end spans. Although for design purposes, the extra end bay prestressing is considered to act within the end bay only, these tendons actually extend well into the adjacent span for anchorage as shown in Fig. 8.34. Advantage can be taken of this condition by designing the through tendons using the largest moment found within the interior spans including the moment at the interior face of the first support. The end bay prestress force is determined using the largest moment within the exterior span. The stress at the inside face of the first support is checked using the equivalent loads produced by the through tendons and the axial compression provided by both the through and added tendons. If the calculated stresses are less than the allowable values, the design is complete. If not, more stress is provided either by through tendons or added tendons or both.

The design of end bay using profiles 1 and 2 follows.

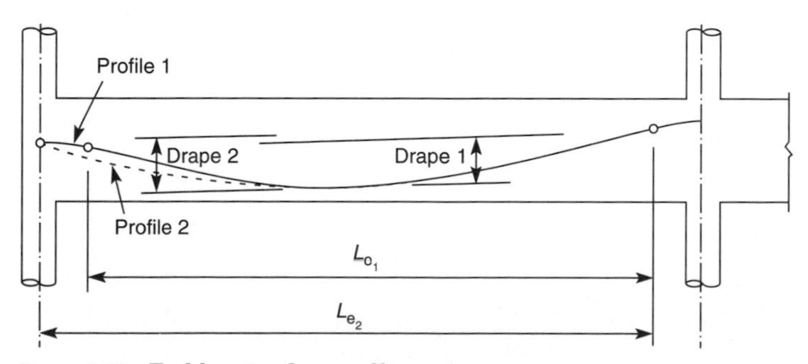


**Figure 8.33** End bay tendon profiles.

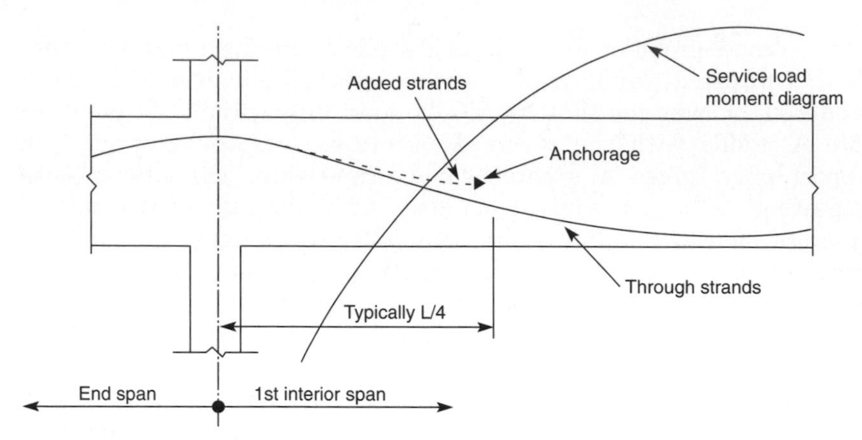


**Figure 8.34** Anchorage of added tendons.

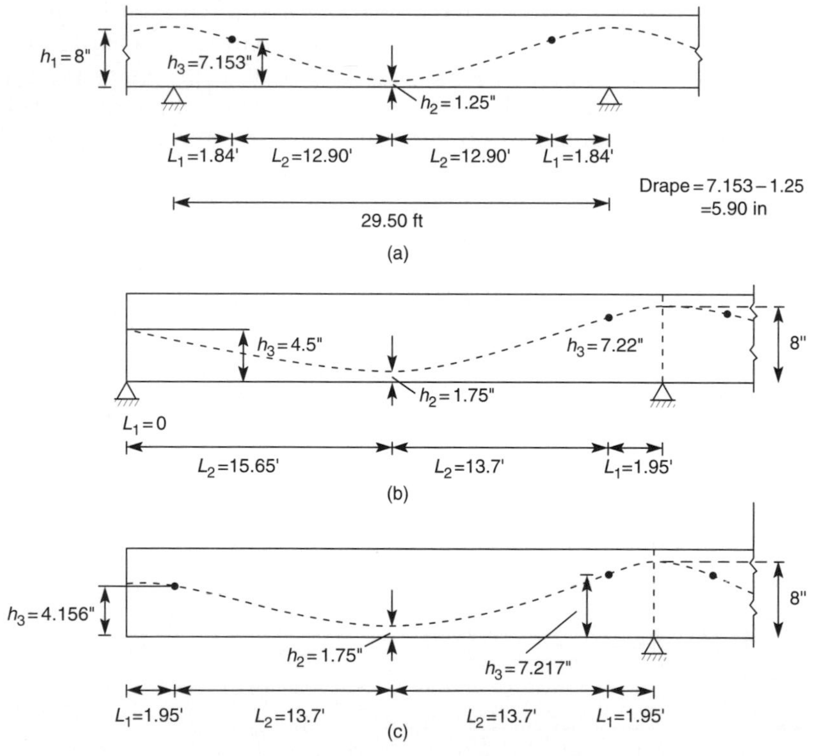


**Figure 8.35** Example problem 3: flat plate, tendon profiles: (a) interior span; (b) exterior span, reverse curvature at right support; (c) exterior span, reverse curvature at both supports.

**Profile 1: Reverse curvature at the right support only (Fig. 8.35b).** Observe the height of inflection point is exact if the tendon profile is symmetrical about the center of span. If it is not, as in span 1 of the example problem, sufficiently accurate value can be obtained by taking the average of the tendon inflection point at each end as follows.

Left End

$$h_3 = \frac{4.5 \times 15.65 + 1.75 \times 0}{0 + 15.67} = 4.5 \text{ in.}$$

Right End

$$h_3 = \frac{8 \times 13.7 + 1.75 \times 1.95}{(1.95 + 13.70)} = 7.22 \text{ in.}$$

$$\text{Average } h_3 = \frac{4.5 + 7.22}{2} = 5.86 \text{ in.}$$

$$\text{Drape} = 5.86 - 1.75 = 4.11 \text{ in.}$$

**First cycle.** To show the quick convergence of the procedure, we start with a rather high value of

$$w_p = 0.75 \text{ DL} = 0.75 \times 142 = 106 \text{ plf}$$

$$M_b = \frac{106}{192} \times 15.87 = 8.76 \text{ kip ft}$$

$$M_{ub} = 15.87 - 8.76 = 7.11 \text{ kip ft}$$

$$f_t = f_b = \frac{7.11 \times 12}{162} = 0.527 \text{ ksi}$$

$$f_p = 0.527 - 0.380 = 0.147 \text{ ksi}$$

$$P = 0.147 \times 9 \times 12 = 15.87 \text{ kips}$$

$$w_p = \frac{8Pe}{L_e^2}$$

$$= \frac{8 \times 15.87 \times 4.11}{29.35^2 \times 12}$$

$$= 0.050 \text{ klf} = 50.0 \text{ plf}$$

This is less than 106 plf. N.G.

**Second cycle**

$$w_p = 0.75\ (106) + 0.25\ (50.0) = 92\ \text{plf}$$

$$M_b = \frac{92}{192} \times 15.87 = 7.60\ \text{kip ft}$$

$$M_{ub} = 15.87 - 7.60 = 8.27\ \text{kip ft}$$

$$f_t = f_b = \frac{8.27 \times 12}{162} = 0.612\ \text{ksi}$$

$$f_p = 0.612 - 0.380 = 0.233\ \text{ksi}$$

$$P = 0.233 \times 12 \times 9 = 25.16\ \text{kip/ft}$$

$$w_p = \frac{8 \times 25.16 \times 4.11}{29.35^2 \times 12}$$

$$= 0.080\ \text{klf} = 80\ \text{plf}$$

This is less than 91.5 psi used at the beginning of second cycle N.G.

**Third cycle**

$$w_p = 0.75 \times 92 + 0.25 \times 80 = 89\ \text{plf}$$

$$M_b = \frac{89}{192} \times 15.87 = 7.356\ \text{kip ft}$$

$$M_{ub} = 15.87 - 7.356 = 8.5\ \text{kip ft}$$

$$f_t = f_b = \frac{8.5 \times 12}{162} = 0.631\ \text{ksi}$$

$$f_p = 0.631 - 0.380 = 0.251\ \text{ksi}$$

$$P = 0.251 \times 12 \times 9 = 27.10\ \text{kips}$$

$$w_p = \frac{8 \times 27.10 \times 4.11}{29.35^2 \times 12} = 0.086\ \text{klf} = 86\ \text{plf}$$

This is nearly equal to 89 plf used at the beginning of third cycle. Therefore O.K.

**Profile 2. Reverse curvature over each support (Fig. 8.35c)**

Left end

$$h_3 = \frac{4.5 \times 13.70 + 1.75 \times 1.95}{(13.70 + 1.95)} = 4.156\ \text{in.}$$

Right end

$$h_3 = \frac{8 \times 13.70 + 1.75 \times 1.95}{(13.70 + 1.95)} = 7.221 \text{ in.}$$

$$\text{Average } h_3 = \frac{4.156 + 7.221}{2} = 5.689 \text{ in.}$$

$$e = h_d = 5.689 - 1.75 = 3.939 \text{ in.}$$

**First cycle.** We start with an assumed balanced load of 0.65 DL = 92 plf

$$\text{Balanced moment } M_b = 15.87 \times \frac{92}{192} = 7.60 \text{ klp ft}$$

$$M_{ub} = 15.87 - 7.6 = 8.27 \text{ kip ft}$$

$$f_t = f_b = \frac{8.27 \times 12}{162} = 0.613$$

$$f_p = 0.613 - 0.380 = 0.233 \text{ ksi}$$

$$P = 0.233 \times 9 \times 12 = 25.12 \text{ kips}$$

$$w_p = \frac{8 \times 25.12 \times 3.937}{(27.38)^2 \times 12} = 0.088 \text{ klf} = 88 \text{ plf}$$

This is less than 92 plf. N.G.

**Second cycle**

$$w_p = 0.75 \times 92 + 0.25 \times 88 = 91 \text{ plf}$$

$$M_b = 15.87 \times \frac{91}{192} = 7.52 \text{ kip ft}$$

$$M_{ub} = 15.87 - 7.52 = 8.348 \text{ kip ft}$$

$$f_t = f_b = \frac{8.348}{162} \times 12 = 0.618 \text{ ksi}$$

$$f_p = 0.618 - 0.380 = 0.238 \text{ ksi}$$

$$P = 0.238 \times 9 \times 12 - 25.75 \text{ kips.}$$

$$w_p = \frac{8 \times 25.75 \times 3.937}{(27.38)^2 \times 12} = 0.090 \text{ klf} = 90 \text{ plf}$$

This is nearly equal to the value at the beginning of second cycle. Therefore O.K.

Check the design against positive moment of 8.41 kip ft.

$$w_p = 0.090 \text{ klf}$$

$$M_b = 8.41 \times \frac{0.090}{0.142} = 5.33 \text{ kip ft}$$

$$M_{ub} = 8.41 - 5.33 = 3.08 \text{ kip ft}$$

Bottom flexural stress $= \dfrac{3.08 \times 12}{162} = 0.228$ ksi (tension). Axial compression due to post-tension $= \dfrac{25.75}{12 \times 9} = 0.238$ ksi. Total stress at bottom $= 0.228 - 0.238 = -0.10$ ksi (compression). This is less than allowable tension of 0.380 ksi. Therefore, design O.K.

#### 8.2.7.5 Mild steel reinforcement design (strength design for flexure)

In the design of prestress members it is not enough to limit the maximum values of tensile and compressive stresses within the permitted values at various loading stages. This is because although such a design may limit deflections, control cracking, and prevent crushing of concrete, an elastic analysis offers no control over the ultimate behavior or the factor-of-safety of a prestressed member. To assure that prestressed members will be designed with an adequate factor-of-safety against failure, the ACI code requires that $M_u$, the moment due to factored service loads, not exceed $\phi M_n$, the flexural design strength of the member.

The nominal bending strength is computed in nearly the same manner as that of a reinforced concrete beam. The only difference is in the method of stress calculation in the tendon at failure. This is because the stress-strain curves of high-yield-point steels used as tendons do not develop a horizontal yield range once the yield strength is reached. It continues to slope upward at a reduced slope. Therefore, the final stress in the tendon at failure $f_{ps}$, must be predicted by an empirical relationship. The reader is referred to standard reinforced concrete design textbooks for further discussion of strength design of prestressed members.

Chapter

# 9

# Composite Gravity Systems

Gravity systems in composite construction can be broadly classified into composite floor systems and composite columns. Composite floor systems can consist of simply supported prismatic or haunched structural steel beams, trusses or stub girders linked via shear connectors to a concrete floor slab to form an effective T-beam flexural member. Formed metal deck supporting a concrete topping slab is an integral component in these floor systems used nearly exclusively in steel framed buildings in North America.

## 9.1 Composite Metal Decks

### 9.1.1 General considerations

Metal deck is manufactered from steel sheets by a fully mechanized, high-speed cold-rolling process. Although it is possible to produce shapes up to $\frac{1}{2}$ in. (12.7 mm) and even $\frac{3}{4}$ in. (19 mm) thick by cold forming, cold-formed steel construction is generally restricted to plates and sheets weighing from 0.5 psf (24 Pa) to a maximum of 9 psf (431 Pa).

Composite metal deck is manufactured with deformations specifically designed-in to produce composite action under flexure between the metal deck and concrete. The shear connection between the two is provided through lugs, corrogations, ridges or embossments formed in the profile of the sheet to increase the chemical bond between the two materials. The steel deck profile is typically trapezoidal with relatively wide flutes suitable for through-deck welding of shear studs. Metal deck may also include closed cells to accommodate floor electrification system. Non-celular deck panels may be blended with cellular panels as part of the total floor system. Metal deck is commonly available in $1\frac{1}{2}$, 2, and 3 in. (38, 51, and 76 mm) depths with rib spacings of 6, $7\frac{1}{2}$, 8, 9 and 12 in. (152, 190, 208, 228 and 305 mm).

A composite slab is usually designed as a simply supported reinforced concrete slab with the steel deck acting as positive reinforcement. Typical mesh used for control of temperature cracking does not provide enough negative reinforcement for typical beam spacing of 8 to 15 ft (2.44 to 4.57 m). Although the slab is designed as a simple span, it is a good practice to provide a nominal reinforcement of say #4 @ 18″ c-c at the top to control excessive cracking of slab. It is generally believed that cracking of slab in the negative regions does not materially impair the composite beam strength.

The Steel Deck Institute (SDI), regarded as the industry standard by the metal deck manufacturers, has published a manual which encompasses the design of composite decks, form decks, and roof decks. A brief description of the SDI specifications is given in the following section.

### 9.1.2 SDI specifications

The steel used for the fabrication of composite metal deck shall have a minimum yield point of 33 ksi (227.5 MPa). The specified yield point is the primary criterion for strength under static loading. The tensile strength is of secondary importance because fatigue strength and brittle fracture, which relate to tensile strength rather than yield point, are rarely of consequence; metal deck is rarely subjected to repetitive loads and the characteristic thinness invariably precludes the development of brittle fracture.

Considerable variations in thickness of metal deck may occur because of rolling tolerances. Therefore, SDI stipulates that the delivered thickness of bare steel without the finish such as phosphotising and galvanizing shall be not less than 95 percent of the specified thickness. The increase in the stiffness of deck due to galvanizing is not relied upon in the design of metal deck.

Opinions differ among engineers whether the metal deck used inside of a building which is not directly exposed to weather needs galvanizing or not. SDI does not mandate any particular type of finish. The appropriate finish is left to the discretion of the engineer with the recommendation that due consideration be given to the effects of environment to which the structure is subjected. However, SDI in its commentary recommends a galvanized coating conforming to ASTM A-525 G.60 requiring a minimum galvanizing of 0.75 ounce per square feet (2.24 Pa) of metal deck. Other salient features of SDI specifications are as follows:

1. Minimum compressive strength of concrete $f'_c$ shall be 3.0 ksi (20.68 MPa). The compressive stress in concrete is limited to $0.4f'_c$ under the applied load for unshored construction and under the total dead and live loads for shored construction. The flexural or shear bond is to be based on ultimate strength analysis with a minimum safety factor of 2. The minimum temperature and shrinkage reinforcement in a composite slab is a function of the area of concrete, as in ordinary reinforced concrete slab, but only the concrete area above the metal deck need be considered in calculating the area of concrete.

2. The use of admixtures containing chloride salts is prohibited because salts can corrode the steel deck.

3. When designing the section as a form in bending, the section properties are to be calculated as per AISC *Specification for the Design of Cold-formed Steel Structural Members.*

4. Bending stress is limited to 0.6 times the yield strength of steel. An upper limit of 36.0 ksi (248.2 MPa) is imposed on the allowable stress. In addition to the weight of wet concrete and deck, allowance should be made for construction live loads of 20 psf (958 Pa) of uniform load or a 150 lb (667 N) concentrated load. This is to account for the weight of one person working on a 1 ft (305 mm) width of deck. It is a common practice to allow for a 200 lb (890 N) point load as the equivalent load. This is because the loading is considered temporary with a 33 percent increase in the stress, which is equivalent to reducing the 200 lb (890 N) load by 25 percent. Clear spans are to be used in the moment calculations.

5. For calculating deflections, it is not necessary to consider the construction loads since the deck, which is designed to remain elastic, will rebound after the removal of construction loads. The calculated deflection based on the weight of concrete is limited to the smaller value $L/180$ or $\frac{3}{4}$ in. (19 mm), in which $L$ is the clear span of the deck. Deflections of composite slabs due to live loads of 50 to 80 psf (2394 to 3830 Pa) are seldom a design concern because the deflections are usually less than $L/360$, where $L$ is the span of deck. Because the slab is assumed to have cracked at the supports, the deflections are best predicted by using the average of the cracked and uncracked moment of inertia using transformed sections. Note that when slabs are cast level to compensate for the deflection of metal deck, a 10 to 15 percent of additional concrete is required for slab construction.

6. A minimum bearing of $1\frac{1}{2}$ in. (38 mm) is specified for proper deck seating on supports.
7. A maximum average spacing of 12 in. (305 mm) for puddle welds is specified to obtain proper anchorage to supporting members. The maximum spacing between adjacent welds is limited to 18 in. (457 mm). Welding of decks with thickness less than 0.028 in. (0.71 mm) is not practical because of the likelihood of burning off the sheet. Therefore, SDI stipulates use of welding washers for floor decks of less than 0.028 in. (0.71 mm) in thickness. Stud welding through the metal deck to the steel top flange can be used instead of puddle welds to satisfy the minimum spacing requirements. However, since it is possible to get uplift forces during wind storms, puddle welds should be used to prevent metal decks from blowing off buildings during construction.
8. Mechanical fasteners which satisfy the anchorage criteria can be used in lieu of puddle welds.
9. Side laps with proper fasteners are required between two longitudinal pieces of deck to: (i) prevent differential deflection; (ii) provide sufficient diaphragm strength; and (iii) sustain local construction loads without distortion or separation. The edges of metal deck shall be connected with $\frac{3}{4}$ in. (19 mm) diameter fusion welds at a maximum spacing of 3 ft (0.9 m) throughout for simply supported spans. Button punching at 2 ft (0.61 m) on centers is an acceptable alternative to minimum fusion welding.

To function as form work, decks supporting cantilevers should be proportioned to satisfy the following criteria: (i) dead load deflection should be limited to $L/90$ of overhang or $\frac{3}{8}$ in. (9.5 mm), whichever is smaller; (ii) steel stress should be limited to 26.7 ksi (184 MPa) for dead load plus 200 lb (890 N) concentrated load at the outer edge of overhang, or steel stresses limited to 20.0 ksi (138 MPa) for dead load plus 20 psf (958 Pa) of additional load, whichever is more severe; (iii) the deck should receive one fusion weld at the cantilever end, and the spacing of welds throughout the cantilever span should not exceed 12 in. (0.30 m). Button punching can be used as an acceptable alternative to fusion welding.

## 9.2 Composite Beams

### 9.2.1 General considerations

Two types of composite construction are recognized by the AISC specifications: (i) fully encased steel beams; and (ii) steel beams with shear connectors. In fully encased steel beams, the natural bond between concrete and steel interface is considered sufficient to provide the resistance to horizontal shear provided that: (i) the concrete thickness is 2 in. (50.8 mm) or more on the beam sides and soffit with the top of the beam at least $1\frac{1}{2}$ in. (38 mm) below the top and 2 in. (50.8 mm) above the bottom of the slab; (ii) the encasement is cast integrally with the slab and has adequate mesh or other reinforcing steel throughout the depth and across the soffit of the beam to prevent spalling of concrete.

Design of encased beams can be accomplished by two methods. In the first method for unshored construction, the stresses are computed by assuming that the steel beam alone resists all the dead load applied prior to hardening of concrete. The superimposed dead and live loads applied after hardening of concrete are assumed to be resisted by composite action. In addition to providing the composite action, the concrete encasement is assumed to restrain the steel beam from both local and lateral torsional buckling. Therefore an allowable of stress of $0.66F_y$, instead of $0.60F_y$ can be used when the analysis is based on the properties of transformed section. Thus for positive bending moments we get

$$f_b = \frac{M_D}{S_s} + \frac{M_L}{S_{tr}} \leq 0.66F_y$$

where $f_b$ = computed stress in the bottom flange for positive bending moment
$M_D$ = dead-load bending moment
$M_L$ = superimposed dead- and live-load bending moment
$S_s$ = section modulus of the steel section referred to its bottom flange
$S_{tr}$ = section modulus of the transformed section referred to its bottom flange

The second method of design of encased beams is a recognition of a common engineering practice where it is desired to eliminate the calculation of composite section properties. This provision permits a higher stress of $0.76F_y$ in steel when the steel beam alone is designed to resist all loads. Thus

$$f_b = \frac{M_D + M_L}{S_s} \leq 0.76F_y$$

The second type of composite steel beam, namely, composite beams with shear connectors, is by far the more popular in the construction of buildings in North America. Invariably composite action is achieved by providing shear connectors between steel top flange and concrete topping because encasing beams with concrete requires expensive form work.

Composite sections have greater stiffness than the summation of the individual stiffness of slab and beam and, therefore, can carry larger loads or similar loads with appreciably smaller deflection and are less prone to transient vibrations. Composite action results in an overall reduction of floor depth, and for high-rise buildings, the cumulative savings in curtain walls, electrical wiring, mechanical ductwork, interior walls, plumbing, etc., can be considerable.

Composite beams can be designed either for shored or unshored construction. For shored construction, the cost of shoring should be evaluated in relation to the savings achieved by the use of lighter beams. For unshored construction, steel is designed to support by itself the wet weight of concrete and construction loads. The steel section, therefore, is heavier than in shored construction.

In composite floor construction, the top flanges of the steel beams are attached to the concrete by the use of suitable shear connectors. The concrete slab becomes part of the compression flange. As a result, the neutral axis of the section shifts upward, making the bottom flange of the beam more effective in tension.

Since concrete already serves as part of the floor system, the only additional cost is that of the shear connectors. In addition to transmitting horizontal shear forces from the slab into beam, the shear connector prevents any tendency for the slab to rotate independently of the beam.

The stud shear connector is a short length of round steel bar welded to the steel beam at one end and having an anchorage provided in the form of a round head at the other end. The most common diameters are $\frac{1}{2}$, $\frac{5}{8}$, and $\frac{3}{4}$ in. (12, 16, and 19 mm). The length is dependent on the depth of metal deck and should extend at least $1\frac{1}{2}$ in. (38 mm) above the top of the deck. The welding process typically reduces their length by about $\frac{3}{16}$ in. (5 mm). The upset head thickness of the studs is usually $\frac{3}{8}$ or $\frac{1}{2}$ in. (9 to 12 mm), and the diameter $\frac{1}{2}$ in. (12 mm) larger than the stud diameter. The studs are normally welded to the beam with an automatic welding gun, and when properly executed, the welds are stronger than the steel studs. Studs located on the side of the trough toward the beam support are more effective than studs located toward the beam centerline. The larger volume of concrete between the stud and the pushing side of the trough helps in the development of a larger failure cone in concrete, thus increasing its horizontal shear resistance.

The length of stud has a definite effect on the shear resisted by it. As the length increases, so does the size of the shear cone, with a

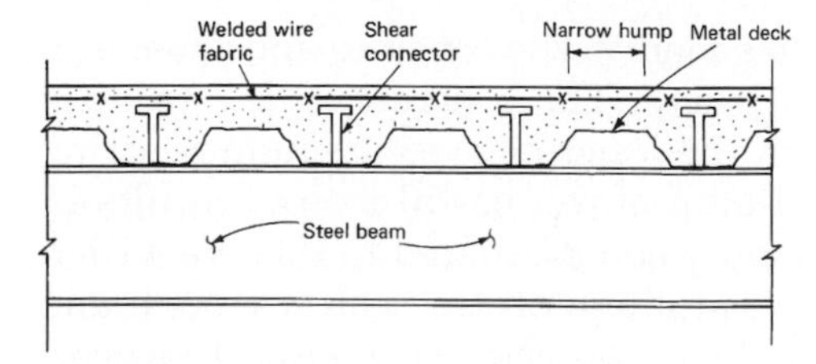


**Figure 9.1** Composite beam with narrow hump metal deck.

consequent increase in the shear value. The shear capacity of the stud also depends on the profile of the metal deck. To get a qualitative idea, consider the two types of metal decks shown in Figs 9.1 and 9.2. The deck in Fig. 9.1 has a narrow hump compared to the

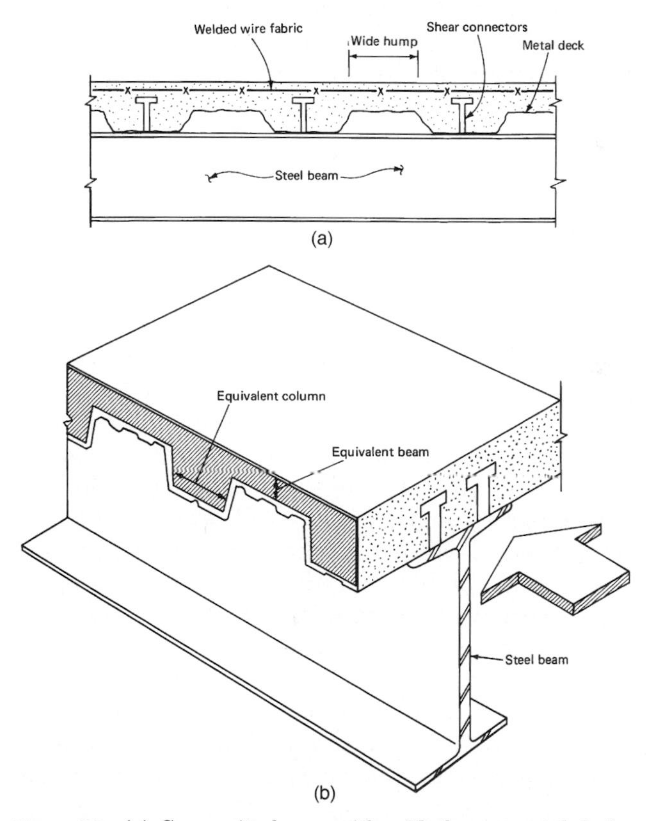


**Figure 9.2** (a) Composite beam with wide hump metal deck: (b) simplified analytical model of composite metal deck subjected to horizontal shear.

one in Fig. 9.2. When subjected to a load $V$, the concrete and the metal deck tend to behave as a portal frame. The concrete in the troughs can be thought of as columns with the concrete over the humps acting as beams (Fig. 9.2b). A narrow hump of the portal frame, results in an equivalent beam of smaller span when compared to the one with a wider hump, meaning that a deck profile with the widest trough and narrowest hump will yield the highest connector strengths. However, other considerations such as volume of concrete, section modulus, and the stiffness of deck also influence the shear strength of the connector.

Metal decks for composite construction are available in the United States in three depths, $1\frac{1}{2}$ in. (38 mm), 2 in. (51 mm), and 3 in. (76 mm). The earlier types of metal deck did not have embossments, and the interlocking between concrete and metal deck was achieved by welding reinforcement transverse to the beam. Later developments of metal deck introduced embossments to engage concrete and metal deck and dispensed with the transverse-welded reinforcement.

The spans utilizing composite metal deck are generally in the range of 8 to 15 ft (2.4 to 4.6 m).

In floor systems using $1\frac{1}{2}$ in. (38 mm) decks, the electrical and telephone services are generally provided by the so-called poke-through system, which is simply punching through the slab at various locations and passing the under-floor ducts through them. A deeper deck is required if the power distribution system is integrated as part of the structural slab; as a result, 2 and 3 in. (51 and 76 mm) metal decks were developed. Experiments have shown that there is very little loss of composite beam stiffness due to the ribbed confiugration of the metal deck in the depth range of $1\frac{1}{2}$ to 3 in. (38 to 76 mm). As long as the ratio of width to depth of the metal deck is at least 1.75, the entire capacity of the shear stud can be developed similar to beams with solid slabs. However, with deeper deck a substantial decrease in shear strength of the stud occurs, which is attributed to a different type of failure mechanism. Instead of the failure of shear stud, the mode of failure is initiated by cracking of the concrete in the rib corners. Eventual failure takes place by separation of concrete from the metal deck. When more than one stud is used in a metal deck flute, a failure cone can develop over the shear stud group, resulting in lesser shear capacity per each stud. The shear stud strength is therefore closely related to the metal deck configuration and factors related to the surface area of the shear cone.

Often special details are required in composite design to achieve the optimum result. Openings interrupt slab continuity, affecting capacity of a composite beam. For example, beams adjacent to elevator and stair openings may have full effective width for part of their length and perhaps half that value adjacent to the openings. Elevator sill details normally require a recess in the slab for door installations, rendering the slab ineffective for part of the beam length. A similar problem occurs in the case of trench header ducts, which require elimination of concrete, as opposed to the standard header duct, which is completely encased in concrete. When the trench is parallel to the composite beam, its effect can easily be incorporated in the design by suitably modifying the effective width of compression flange. The effect of the trench oriented perpendicular to the composite beam could range from negligible to severe depending upon its location. If the trench can be located in the region of minimum bending moment, such as near the supports in a simply supported beam, and if the required number of connectors could be placed between the trench and the point of maximum bending moment, its effect on the composite beam design is minimum. If, on the other hand, the trench must be placed in an area of high bending moment, its effect may be so severe as to require that the beam be designed as a noncomposite beam.

The slab thickness normally employed in high-rise construction with composite metal deck is usually governed by fire-rating-requirements rather than the thickness required by the bending capacity of the slab. In certain parts of the United States it may be economical to use the minimum thickness required for strength and to use sprayed-on or some other method of fireproofing to obtain the required ratings. Some major projects have used $2\frac{1}{2}$ in. (63.5 mm) thick concrete on 3 in. (76.2 mm) deep metal deck spanning as much as 15 ft (4.57 m). A comparative study which takes into consideration the vibration characteristics of the floor is necessary to zero in on the most economical scheme.

In continuous composite beams the negative moment regions can be designed such that: (i) the steel beam alone resists the negative moment; or (ii) it acts compositely with mild steel reinforcement placed in the slab parallel to the beam. In the latter case, shear connectors must be provided through the negative moment region.

Careful attention should be paid to the deflection characteristics of composite construction because the slender not-yet-composite shape deflects as wet concrete is placed on it. There are three ways to handle the deflection problem.

1. Use relatively heavy steel beams to limit the dead-load deflection and pour lens-shaped tapering slabs to obtain a nearly flat top. Although a reasonably flat surface results from this construction,

the economic restraints of speculative office buildings do not usually permit the luxury of the added cost of additional concrete and heavier steel beams.

2. Camber the steel beam to compensate for the weight of steel beam and concrete. Place a constant thickness of slab by finishing the concrete to screeds set from the cambered steel. Continuous lateral bracing as provided by the metal deck is required to prevent the lateral torsion buckling of beam. If steel deck is not used, this system requires a substantial temporary bracing system to stabilize the beam during construction.
3. Camber and shore the steel beam. The beam is fabricated with a camber calculated to compensate for the deflection of the final cured composite section. Shores are placed to hold the steel at its curved position while the concrete is being poured. As in method 2, slab is finished to screeds set from cambered steel. Although methods 1 and 3 are occasionally used, the trend is to use method 2 because it is the least expensive.

### 9.2.2 AISC design specifications

Including provisions for solid slab, there are three categories of composite beams in the AISC specifications each with a differing effective concrete area.

1. *Solid slab.* The total slab depth is effective in compression unless the neutral axis is above the top of the steel beam. In high-rise floor systems with relatively thin slabs the neutral axis of steel beams is invariably below the slab, rendering the total slab depth effective in compression.
2. *Deck perpendicular to beam* (Fig. 9.3).
   (a) Concrete below the top of steel decking shall be neglected in computations of section properties and in calculating the number of shear studs, but the concrete below the top flange of deck may be included for calculating the effective width.
   (b) The maximum spacing of shear connectors shall not exceed 32 in (813 mm) along the beam length.
   (c) The steel deck shall be anchored to the beam either by welding or by other means at a spacing not exceeding 16 in. (406 mm).
   (d) A reduction factor as given by the AISC formula I5-1

$$\left(\frac{0.85}{\sqrt{N_r}}\right)\left(\frac{w_r}{h_r}\right)\left(\frac{H_s}{h_r}-1\right)\leq 1.0$$

should be used for reducing the allowable horizontal shear capacity of stud connectors. In the above formula $h_r$ is the nominal rib height, in inches. $H_s$ is length of stud connector after welding, in inches. An upper limit of $(h_r + 3)$ is placed on the length of shear connectors used in computations even when longer studs are installed in metal decks. $N_r$ is the number of studs in one rib. A maximum value of 3 can be used in computations although more than three studs may be installed. $w_r$ is average width of concrete rib.

3. *Deck ribs parallel to beam* (Fig. 9.4).
   (a) The major difference between perpendicular and parallel orientation of deck ribs is that when deck is parallel to beam, the concrete below the top of the decking can be included in the calculations of section properties and must be included when calculating the number of shear studs.

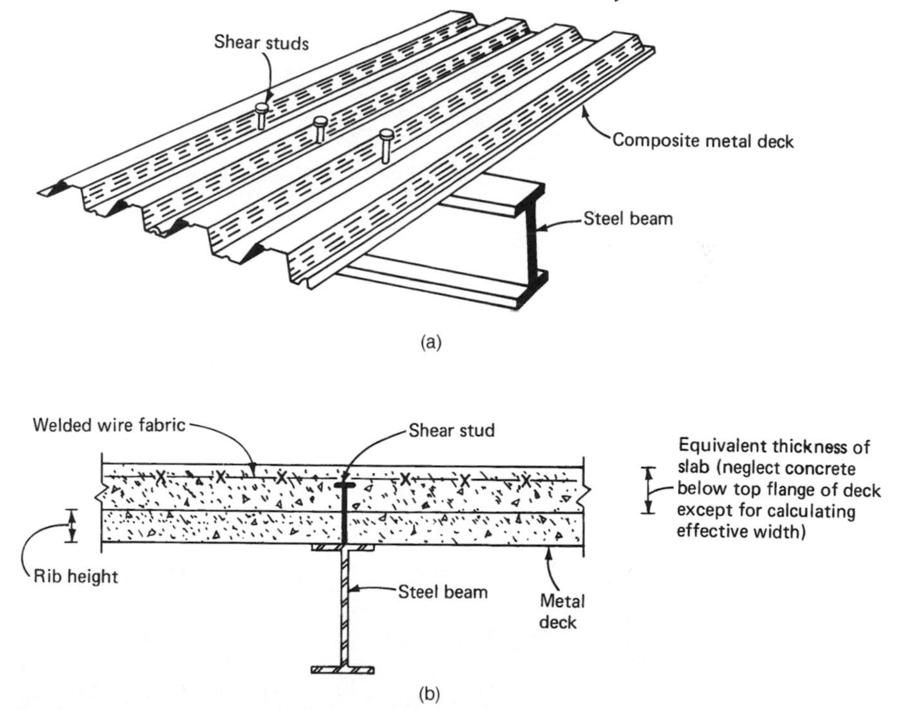


**Figure 9.3** Composite beam with deck perpendicular to beam: (a) schematic view; (b) section A showing equivalent thickness of slab.

(b) If steel deck ribs occur on supporting beam flanges, it is permissible to cut high-hat to form a concrete haunch.
(c) When nominal rib height is $1\frac{1}{2}$ in. (38.1 mm) or greater, minimum average width of deck flute should not be less than 2 in. for the first stud in the transverse row plus four stud diameters for each additional stud. This gives minimum average widths of 2 in. (51 mm) for one stud, 2 in. plus $4d$ for two studs, 2 in. plus $8d$ for three studs, etc., where $d$ is the diameter of stud. Note that if a metal deck cannot accommodate this width requirement, the deck can be split over the girder to form a haunch.
(d) A reduction factor as given by AISC formula I5-2

$$0.6\left(\frac{w_r}{h_r}\right)\left(\frac{H_s}{H_r} - 1.0\right) \leq 1.0$$

shall be used for reducing the allowable horizontal shear capacity of stud connectors.

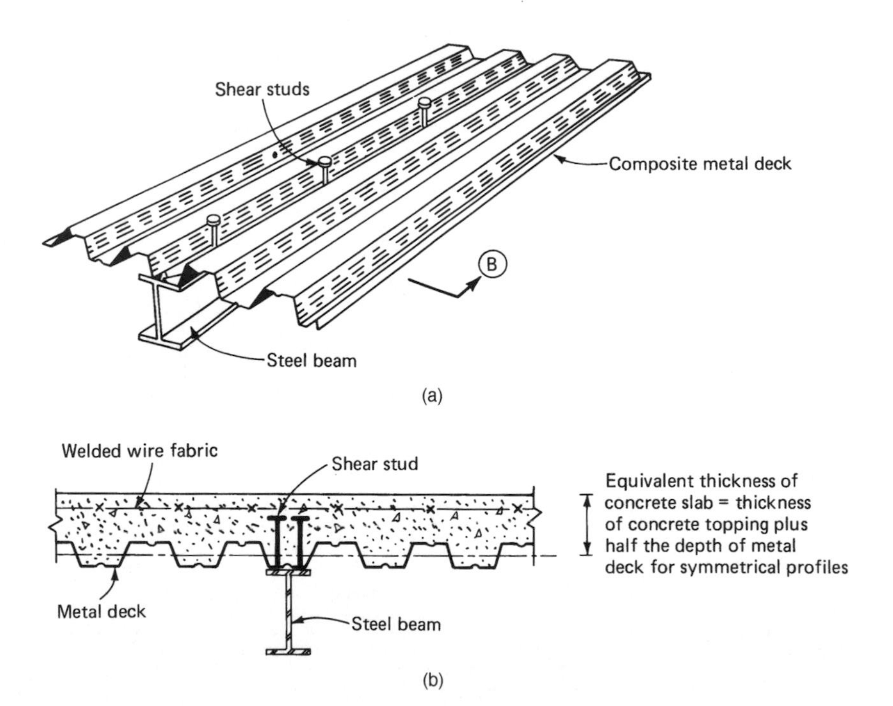


**Figure 9.4** Composite beam with deck parallel to beam: (a) schematic view; (b) section A showing equivalent thickness of slab.

4. Certain specific ASCE requirements applicable to formed steel deck construction are shown schematically in Fig. 9.5. More general comments are as follows:
   (a) The deck rib height shall not exceed 3 in. (76.5 mm).
   (b) Rib average width shall not be less than 2 in. (51 mm). If the deck profile is such that the width at the top of the steel deck is less than 2 in. (51 mm), this minimum clear width shall be used in the calculation.
   (c) The section properties do not change a great deal from deck running perpendicular or parallel to the beam, but the change in the number of studs can be significant.

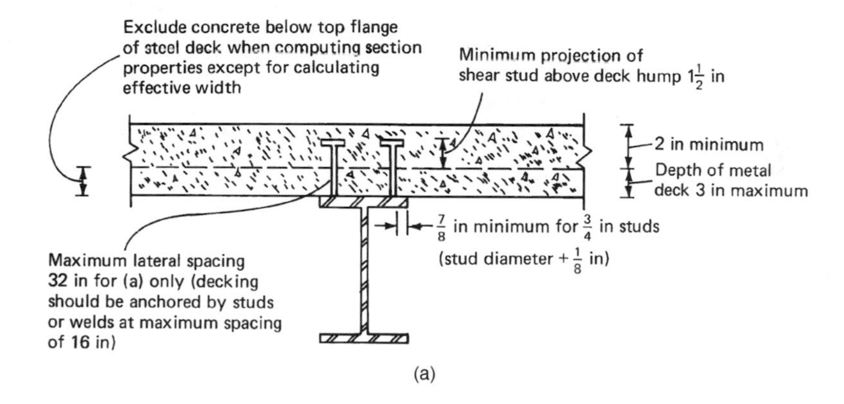


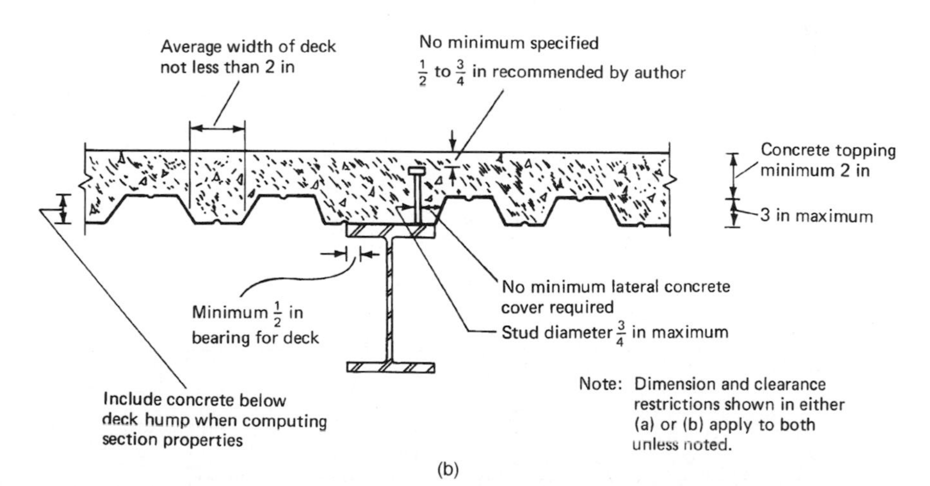


**Figure 9.5** Composite beam, AISC requirements: (a) deck perpendicular to beam; (b) deck parallel to beam.

(d) The reduction formula for stud length is based on rib geometry, number of studs per rib, and embedment length of the studs.
(e) The equation for calculating the partial section modulus makes the choice of heavier, stiffer beams with fewer studs economically more attractive.
(f) Higher shear values can be used in longer shear studs. Concrete cover over the top of the stud is not limited by the AISC specifications, but for practical reasons the author recommends a minimum of $\frac{1}{2}$ in. (12.7 mm).
(g) Studs can be placed as close to the web of deck as needed for installation and to maintain the necessary spacing.
(h) Deck anchorages can be provided by the stud welds.
(i) Maximum diameter of shear connectors is limited to $\frac{3}{4}$ in. (19 mm).
(j) After installation, the studs should extend a minimum of $1\frac{1}{2}$ in. (38 mm) above the steel deck.
(k) Total slab thickness including the ribs is used in determining the effective width without regard to the orientation of the deck with respect to the beam axis.
(l) The slab thickness above the steel deck shall not be less than 2 in. (51 mm).

For design purposes, a composite floor system is assumed to consist of a series of T beams, each made up of one steel beam and a portion of the concrete slab. The AISC limits on the width of slab that can be considered effective in the composite action are shown in Fig. 9.6. When slab extends on one side of the beam only, as in spandrel beams and beams adjacent to floor openings, the effective width naturally is less than when the slab extends on both sides of the beam. For slabs extending on both sides of the beam, the maximum

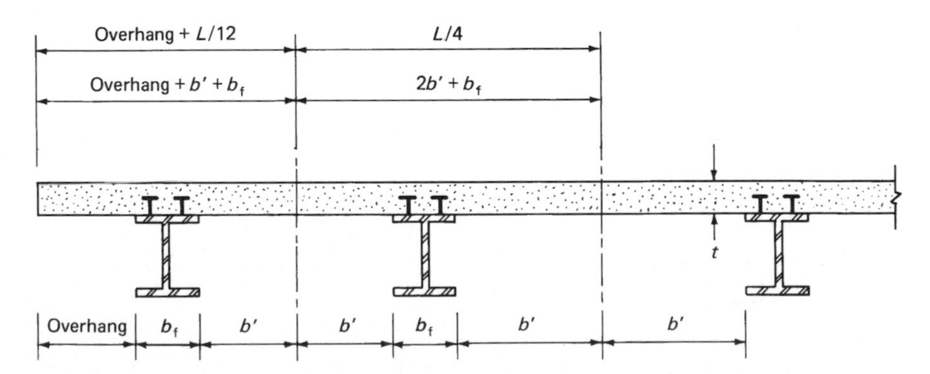


**Figure 9.6** Effective width concept as defined in the AISC specifications.

effective flange width $b$ may not exceed: (i) one-fourth of the beam span $L$; or (ii) one-half the clear distances to adjacent beams on both sides plus $b_f$, the width of steel beam flange. When the slab extends on only one side of the beam, the maximum effective width $b$ may not exceed: (i) one-twelfth of the beam span $L$; or (ii) one-half the clear distance to the adjacent beam plus $b_f$. Furthermore, the out-board effective width may not exceed the actual width of overhang, and the in-board effective width must not extend beyond the centerline between the edge beam and the adjacent interior span.

The design of composite beams is usually achieved by the transformed area method, in which the concrete effective area of the composite beam is transformed into an equivalent steel area. It is equally admissible to transform the steel area into an equivalent concrete area, but calculations are somewhat simplified by the former method. The method assumes transverse compatibility at the concrete and steel interface. The unit stress in each material is equal to the strain times its modulus of elasticity. Because of strain compatibility, the stress in steel is $n$ times the stress in concrete, where $n$ is the modular ratio $E_s/E_c$. A unit area of steel is, therefore, mathematically equivalent to $n$ times the concrete area. Therefore, the effective area of concrete $A_c = bt$ can be replaced by an equivalent steel area $A_c/n$.

Concrete is neither linearly elastic nor ductile and its stress-strain curve exhibits a constantly changing slope with a sudden brittle failure. In spite of these characteristics, concrete is considered elastic within a stress strain range of up to $0.50f'_c$ and the modulus of elasticity in pounds per square inch can be approximated by the relation $E_c = W_c^{1.5} \times 33\sqrt{f'_c}$, where $W_c$ is the unit weight of concrete in pounds per cubic foot and $f'_c$ is the compressive strength of concrete in pounds per square inch. The compressive strength $f'_c$ of concrete normally used in floor construction is in the range of 3000 to 5000 psi (20.7 to 34.4 MPa) giving a value of $E_c$ for normal weight concrete of $3.12 \times 10^6 < E_c < 4.03 \times 10^6$ psi $(21{,}512 < E_c < 27{,}787$ MPa), compared to $E_s$ of steel at $29 \times 10^6$ psi (199,955 MPa). The value of $n = E_s/E_c$, therefore, lies between 9.3 and 7.2 and is usually approximated to the whole number in recognition of the great amount of error in the formula for $E_c$ when compared to actual performance.

For strength calculations, the AISC specification uses the value of $n$ for normal-weight concrete of the specified strength. For deflection computations, $n$ depends not only on the specified strength but also the weight. Therefore, in computing deflections, especially for beams subjected to heavy sustained loads, it is necessary to account for the effects of creep. This is even more important in shored construction when the dead load of the concrete is resisted by the composite action. Creep effect is accounted for in computing deflections by using a higher modular ratio, $n$. A multiplication factor of 2 for creep effects appears to be adequate in building designs. Live loads are always resisted by the composite section. If they are of short duration, the deflections are computed using the short-term modular ratio.

The transformed steel section can be conveniently considered as the original steel beam with an added cover plate to the top flange of thickness $t$ equal to slab thickness, and an equivalent width $b/n$. The properties of the transformed section are calculated by locating the neutral axis and the transformed moment of inertia $I_{tr}$. The maximum stress in steel is given by

$$f_{bs} = \frac{MY_{tr}}{I_{tr}}$$

where $M$ is the bending moment and $Y_{tr}$ is the distance of the extreme bottom steel fibers from the neutral axis. The maximum bending stress for concrete is given by

$$f_{bc} = \frac{MC_t}{nI_{tr}}$$

where $C_t$ is the distance from the neutral axis to the extreme concrete fibers and $n$ is the modular ratio. The value

$$S_{tr} = \frac{I_{tr}}{Y_{tr}}$$

is called the transformed section modulus of the beam referred to the bottom flange.

For construction without temporary shores, the concrete stress is based upon the load applied after the concrete has reached 75 percent of the required strength. This stress is limited to $0.45f'_c$, just as in working stress design of reinforced concrete beams.

The total horizontal shear to be resisted between the point of maximum positive moment and point of zero moment is the smaller of the two values as determined by

$$V_h = \frac{0.85f'_cA_c}{2}$$

$$V_h = \frac{A_sF_y}{2}$$

where $f'_c$ = specified compressive strength of concrete
$A_c$ = actual area of effective concrete flange
$A_s$ = area of steel beam
$F_y$ = specified yield stress of steel beam

Note that the formula $V_h = 0.85f'_cA_c/2$ assumes that there is no longitudinal reinforcing steel in the compression zone of composite beam. If the compressive zone is designed with mild steel reinforcement, the formula for horizontal shear is to be modified as follows:

$$V_h = \frac{0.85f'_cA_c}{2} + \frac{A'_sF_{yr}}{2}$$

where $A'_s$ = area of the longitudinal compressive steel
$F_{yr}$ = yield stress of the reinforcing steel

AISC permits averaging of horizontal shear flow; that is, the total number of connectors between the point of maximum moment and point of zero moment must be sufficient to satisfy the total shear flow within that length. The shear connector formulas represent the horizontal shear at ultimate load divided by 2 to approximate conditions at working loads.

The number of shear connectors required for full composite action is determined by dividing the smaller value of $V_h$ by the shear capacity of one connector. The number of connectors obtained represents the shear connectors required between the point of maximum positive moment and point of zero moment. For example, in a simply supported, uniformly loaded beam this represents half the span; and in a simply supported beam with two equidistant concentrated loads, this represents the distance between the point load to the support point. The total number of connectors required for the entire span is thus double the number obtained as above.

A composite beam subject to negative bending moment experiences tensile stresses in the concrete zone and loses much of its advantage. However, when reinforcement is placed parallel to the beam within the effective width of slab, and is anchored adequately to develop the tensile forces, the advantage of continuous construction is restored. The steel used in the tensile zone is included in computing the property of the composite section. Similarly, when the compressive stress in concrete subject to positive moment exceeds the allowable stress, it is permissible to use compressive steel in the effective width zone to reduce stresses.

Consider a continuous composite beam shown in Fig. 9.7. The total horizontal shear to be resisted by shear connectors between an interior support and each adjacent point of contraflexure (regions a, b, and c in Fig. 9.7c) is given as

$$V_h = \frac{A_{sr}F_{yr}}{2}$$

where $A_{sr}$ = area of reinforcing steel provided at the interior support within the effective flange width

$F_{yr}$ = yield stress of the reinforcing steel

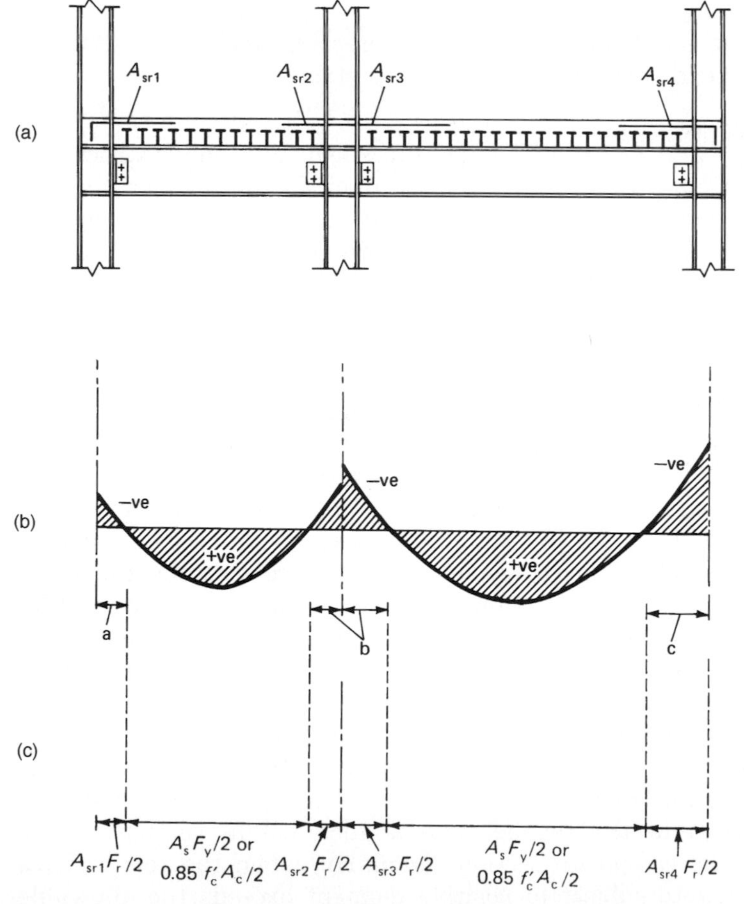


**Figure 9.7** Continuous composite beam subjected to uniformly distributed load: (a) elevation; (b) moment diagram; (c) horizontal shear resisted by studs in the positive and negative moment regions.

The AISC permits uniform spacing of connectors between the points of maximum positive moment and the point of zero moment. Also, the connectors required in the region of negative bending can be uniformly distributed between the point of maximum moment and each point of zero moment. For concentrated loads, the numbers of shear connectors $N_2$ required between any concentrated load and the nearest point of zero moment is determined by AISC formula

$$N_2 = \frac{N_1[(M\beta/M_{\max}) - 1]}{\beta - 1}$$

where $M$ = moment at concentrated load point (less than the maximum moment)

$N_1$ = number of connectors required between point of maximum moment and point of zero moment

$\beta$ = ratio of transformed section modulus to steel section modulus

This relation is schematically shown in Fig. 9.8.

In the design of composite beams for high-rise buildings, it is often

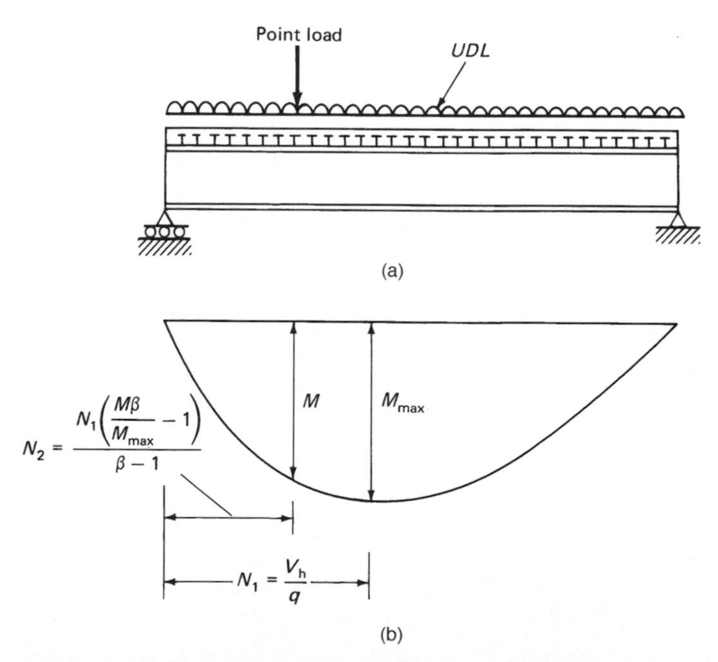


**Figure 9.8** Shear connector requirements for composite beams subjected to concentrated loads: (a) schematic loading diagram; (b) shear connector requirements; (c) composite beam design example; (1) plan; (2) section.

unnecessary to develop the full composite action. A partial composite action with fewer studs is all that may be necessary to achieve the required composite action. AISC permits the design of less than 100 percent composite beams by introducing the concept of effective section modulus as determined by the relation

$$S_{\text{eff}} = S_s + \left(\frac{V'_h}{V_h}\right)^{1/2} (S_{tr} - S_s)$$

where $S_s$ is the section modulus of steel beam, and $V'_h$ is the shear capacity provided by the shear connectors, obtained by multiplying the number of connectors used and the shear capacity of one connector. Transposing the above equation we get

$$V'_h = V_h \left(\frac{S_{\text{reqd}} - S_s}{S_{\text{avail}} - S_s}\right)^2$$

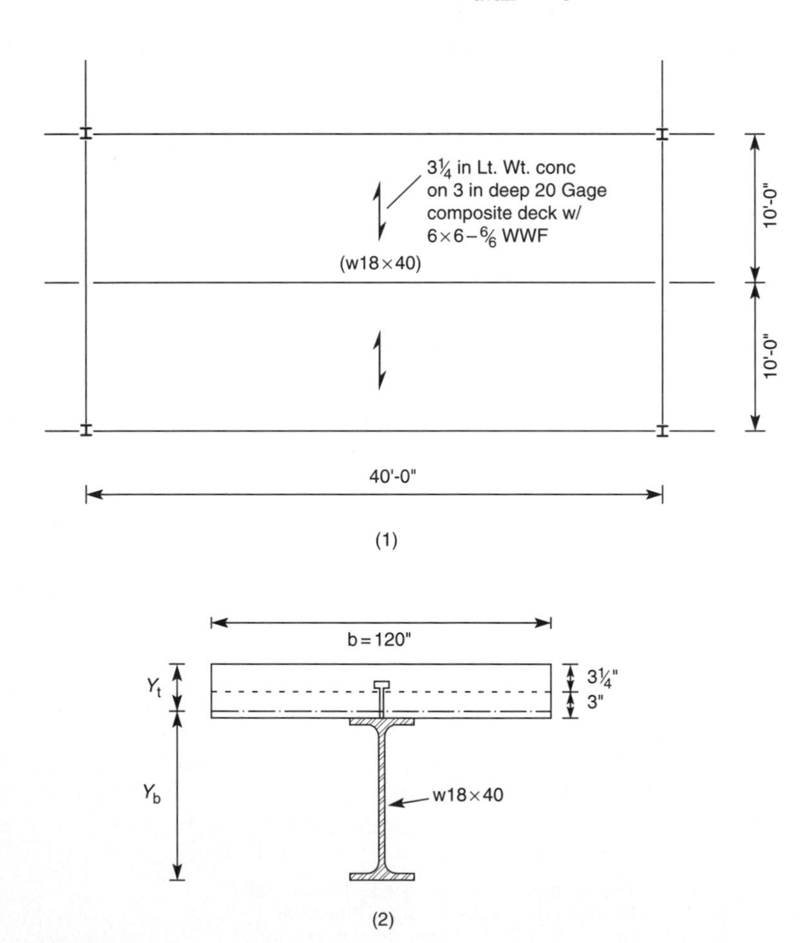


**Figure 9.8** *(Continued).*

AISC stipulates that the minimum requirement of $V'_h$ be not less than $0.25V_h$ to prevent excessive slip and loss in beam stiffness. This minimum requirement does not apply if shear studs are used for reasons other than increasing the flexural capacity, such as increase in stiffness of beams or to increase the diaphragm action.

The AISC specification gives three criteria for stud placement: (i) a minimum center-to-center spacing of six stud diameters between the studs in the longitudinal direction; (ii) minimum spacing of four stud diameters in the transverse direction; and (iii) maximum spacing in the longitudinal direction of 32 in. (813 mm). Note that if stud spacing exceeds 16 in. (406.4 mm), a plug weld between the studs is required to resist uplift forces.

If the required bending capacity is provided by the steel beam alone without depending upon the composite action, the maximum spacing requirement of 32 in. (813 mm) need not be met.

The recommended sequence for installing studs when deck is perpendicular to beam is as follows:

- Deck ribs at 6 in. (153 mm) on center. Start at beam ends and place a single stud at every fourth flute, working toward the center of beam. If studs remain, fill in empty ribs, again starting at beam ends and working toward the center without exceeding 30 in. (762 mm) for stud spacing.
- Deck ribs at 12 in. (305 mm) on center. Start at beam ends and place a single stud in every other flute working toward the center of beam. If studs remain, fill in empty ribs, again starting at beam ends and working toward the center of beam without exceeding 24 in. (610 mm) for stud spacing.
- If the number of studs is more than the number of ribs, place a double or triple row as needed, always starting from beam ends and working toward the beam center. In general, if studs cannot be uniformly spaced, the greatest number of studs should occur at the ends.

The recommended sequence for installing studs when deck is parallel to the girder is as follows. Start at the girder ends by placing the first stud at approximately 12 in. (305 mm) from the centerline of support and work toward the center of girder with uniform spaces between the studs. If a double row of studs is required, it is a good practice to place them in a staggered pattern rather than side by side.

The allowable shear for stud connectors is influenced by several factors when used in metal deck construction. As in solid slabs, the strength and type of concrete, whether regular or lightweight, determines the allowable horizontal loads. The rib geometry of metal deck and the height of the stud above metal deck (when deck is parallel to the girder) are other factors influencing the allowable horizontal loads. For girders, the wider the rib opening and the greater the penetration of the stud above the deck, the more closely the allowable horizontal shear load will approach the published AISC value for studs in solid concrete slab.

### Example

#### Given

(W18 × 40), 50 ksi steel beam, Beam span = 40

Beam spacing = 10 ft (Fig. 9.8c)

Tributary width for dead and live loads = 10 ft

Composite floor construction: $3\frac{1}{4}$ in. of 115 pcf concrete slab over 3 in. metal deck. Rib width = 6.0 in.

Compressive strength of concrete, $f'_c = 3.0$ ksi

#### Loading

(1) Slab with allowance for steel beam = 50 lb/ft$^2$.

(2) Additional precomposite dead load due to extra concrete required for compensating beam deflection = 5 lb/ft$^2$.

(3) Additional composite dead loads:
—partitions = 20 lb/ft$^2$;
—ceiling plus miscellaneous = 10 lb/ft$^2$.

(4) Live load 50 lb/ft$^2$.

#### Required

Verification of (W18 × 40) for final design. AISC specifications (AISCS) 9th edition.

#### Solution

The ASD design is based on elastic analysis using the transformed section properties for composite beams. Because the compression flange is continuously braced, the allowable stress in the steel section is $0.66F_y$. Lateral-torsional buckling is not a concern for the completed structure, but it must be guarded against during construction.

The allowable stress in the concrete slab is $0.45f'_c$. In building design, typically the neutral axis is close to the top of the section. Therefore, stress in the steel is usually the controlling factor.

The section properties for unshored construction are computed by the elastic theory. The bending stress in the steel beam is taken as the sum of: (i) the stress based on the assumption that steel section alone resists all loads applied prior to concrete reaching 75 percent of its specified strength; (ii) the stress based on the assumption that subsequent loads are resisted by the composite section.

For the example problem we have:

$$\text{Uniform precomposite load} = (50 + 5) \times 10 = 550 \text{ plf}$$

$$\text{Uniform postcomposite load} = (30 + 50) \times 10 = 800 \text{ plf}$$

$$\text{Total} = 1350 \text{ plf}$$

$$\text{Maximum moment} = \frac{(1350)(40)^2}{8 \times 1000} = 270 \text{ kip ft}$$

$$\begin{array}{l}\text{Required section modulus} \\ \text{for 50 ksi steel}\end{array} = \frac{270 \times 12}{(0.66)(50)} = 98.18 \text{ in.}^3$$

### Modulus of elasticity of concrete $E_c$

For stress check, AISCS permits the use of normal weight concrete properties even for light weight concrete topping. For deflection calculations, however, use of actual properties are required.

$$E_c = 33W^{1.5}\sqrt{f'_c}$$

$$E_c \text{ for normal weight} = \frac{33 \times 145^{1.5}}{1000}\sqrt{3000} = 2229 \text{ ksi}$$

$$\text{Modular ratio} = \frac{29000}{2229} = 13.06$$

$$E_c \text{ for light weight} = \frac{33 \times 115^{1.5}}{1000}\sqrt{3000} = 3156 \text{ ksi}$$

$$\text{Modular ratio} = \frac{29000}{3156} = 9.19$$

**Composite beam properties**

The effective flange width $b$, of the composite section is the smaller of

(1) $L/4 = \frac{40}{4} \times 12 = 120$ in.

(2) spacing of beams $= 10 \times 12 = 120$ in.

For 115 pcf concrete, $b/n = 120/13 = 9.22$ in. The composite beam properties of the transferrred section are calculated by normal procedures. The resulting values are

$$Y_t = 5.46 \text{ in.}, \quad Y_b = 18.69 \text{ in.}$$
$$I = 2199 \text{ in.}^4$$

Section modulus for tension $S_{tr} = \dfrac{2199}{18.69} = 117.64$ in.$^3$

Section modulus for compression at top $S_t = \dfrac{2199}{5.46} = 402.75$ in.$^3$

For normal weight concrete, $b/n = 120/9.19 = 13.06$ in.

$$Y_t = 4.58 \text{ in.}, \quad Y_b = 19.57 \text{ in.}$$
$$I = 2351 \text{ in.}^4$$
$$S_{tr} = 2351/19.57 = 120.1 \text{ in.}^3$$
$$S_t = 2351/4.58 = 513.5 \text{ in.}^3$$

**Stress check**

The allowable stress $f_b$ in steel for unshored construction is verified for two conditions

$$f_b = \frac{M_{D1+D2}}{S_s} + \frac{M_{D3+L}}{S_{tr}} \leqslant 0.9F_y \quad \text{and}$$

$$f_b = \frac{M_{D1+D2} + M_{D_3+L}}{S_{tr}} \leqslant 0.66F_y$$

where $S_s$ = section modulus of steel section
$S_{tr}$ = section modulus of composite section
$M_{D_1+D_2}$ = moment resisted by steel beam prior to composite action
$M_{D_3+L}$ = moment resisted by composite section

In our case

$$f_b = \frac{110 \times 12}{68.38} + \frac{160 \times 12}{12 \times 120.1} = 35.12 \text{ ksi} \leqslant 0.9Fy = 45 \text{ ksi} \quad \text{O.K.}$$

$$f_b = \frac{(110 + 160) \times 12}{120.1} = 26.9 \text{ ksi} \leqslant 0.6\,F_y = 33 \text{ ksi} \quad \text{O.K.}$$

Allowable concrete compressive stress is determined for the composite section based on the load applied after concrete has attained 75 percent of its required strength.

Moment due to postcomposite load $= \dfrac{800 \times 40^2}{8 \times 1000} = 160$ kip ft

$$f'_c = \frac{160 \times 12}{9.19 \times 513.5} = 0.41 \text{ ksi} < 0.45\sqrt{f'_c}$$

$$0.45\sqrt{3000} = 1.35 \text{ ksi} \quad \text{O.K.}$$

**Horizontal shear $V_h$**

It is the minimum of

(i) $V_h = 0.85 \times f'_c \times \dfrac{A_c}{2} = 0.85 \times 3 \times 120 \times 3.25/2 = 497.25$ kips

(ii) $V_h = \dfrac{A_s F_y}{2} = 11.80 \times 50/2 = 295.00$ kips

$$V_h = 295.00 \text{ kips controls the design}$$

**Shear connectors, partial composite action**

Usually economy is achieved by using fewer shear connectors than required for full composite action.

The required section modulus $= \dfrac{(110 + 160) \times 12}{0.66 \times 50} = 98.18$ in.$^3$

The modulus $S_{tr}$ furnished $= 120.1$ in.$^3$

$$V'_h = V_h \left(\frac{S_{\text{eff}} - S_s}{S_{tr} - S_s}\right)^{1/2}$$

$$= 295\left(\frac{98.18 - 68.38}{120.1 - 68.38}\right)^{1/2}$$

$$= 224 \text{ kips}$$

The percent of composition action is given by $\dfrac{V'_h}{V_h} \times 100$. In our case this is equal to $\dfrac{224}{295} \times 100 = 76$ percent. This is greater than 25% stipulated in the AISCS. Therefore O.K.

Using $\frac{3}{4}$ in. studs the allowable shear in normal weight concrete, per connector is 11.5 kips. The reduction factor for 115 pcf concrete from AISCS Table I4.2 is 0.86.

$$\text{Number of shear studs required} = \frac{2 \times 224}{0.86 \times 11.5} = 46$$

It is assumed that there are no reduction factors associated with deck geometry and stud layout.

### Deflections

$$\text{Effective moment of inertia } I_{\text{eff}} = I_s + (I_{tr} - I_s) \times (\text{percent comp.})^{1/2}$$

$$I = 612 + (2199 - 612) \times \left(\frac{76}{100}\right)^{1/2} = 1995 \text{ in.}^4$$

For unshored construction,

$$\text{Deflection under pre-composite loads} = \frac{5WL^4 \times 12^3}{384EI_s}$$

$$= \frac{5 \times 0.550 \times 40^4 \times 1728}{384 \times 29{,}000 \times 612} = 1.78 \text{ in.}$$

Camber beam for 75 percent of calculated unshored condition.

$$\therefore \text{ Camber specified} = 1.78 \times 0.75 = 1.34 \quad \text{say } 1.25 \text{ in.}$$

Deflection under superimposed dead and live loads

$$W = 850 \text{ lbs/ft}, \quad I = 2122 \text{ in.}^4$$

$$= \frac{5 \times 0.85 \times 40^4 \times 1728}{384 \times 29{,}000 \times 1995} = 0.84 \text{ in.}$$

Compared to $L/360 = 40 \times 12/360 = 1.33$ in. the calculated deflection is small. Therfore design is O.K.

It should be noted that the lowest percentage of partial composite allowed by the AISC specifications is 25 percent. Some designers however will not allow partial composite action below 50 percent.

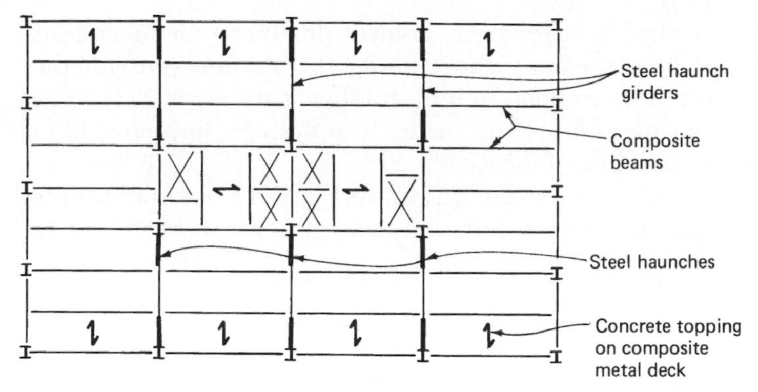


**Figure 9.9** Schematic floor plan showing haunch girders.

## 9.3 Composite Haunch Girders

Composite haunch girders, although not often used as a floor framing system, merit mention because they minimize the floor-to-floor height without requiring complicated fabrication. Figure 9.9 shows a schematic floor plan in which composite haunch girders frame between exterior columns and interior core framing. The haunch girder typically consists of a shallow steel beam, 10 to 12 in. (254 to 305 mm) deep for spans in the 35 to 40 ft (10.6 to 12.19 m) range. At each end of the beam a triangular haunch is formed by welding a diagonally cut wide-flange beam usually 24 to 27 in. (610 or 686 mm) deep (Fig. 9.10). The haunch is welded to the shallow beam and to the columns at each end of the girder. In this manner the last 8 or 9 ft (2.4 or 2.7 m) of the haunch girder at either end flares out toward the column with a depth varying from about 10 or 12 in. (254 or 305 mm) at the center to about 27 in. (686 mm) at the ends. The

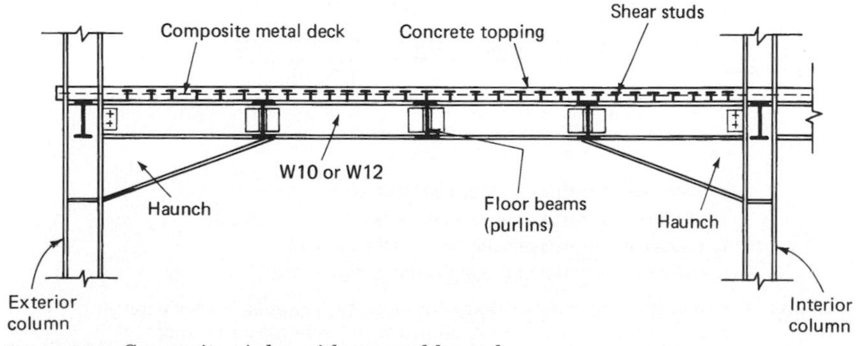


**Figure 9.10** Composite girder with tapered haunch.

system uses less steel and provides greater flexibility for mechanical ducts, which can be placed anywhere under the shallow central span. The reduction in floor-to-floor height further cuts costs of exterior cladding and heating and cooling loads. The system, however, is not common because of higher fabrication costs.

A variation of the same concept shown in Fig. 9.11 uses nontapered haunches at each end. The square haunch girder can be fabricated

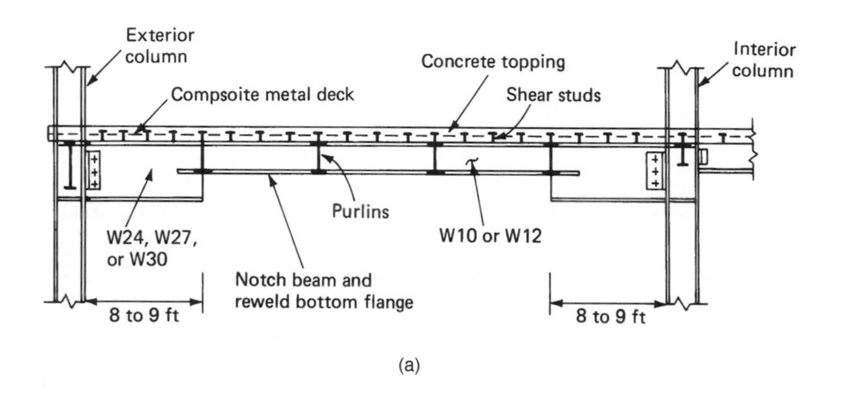


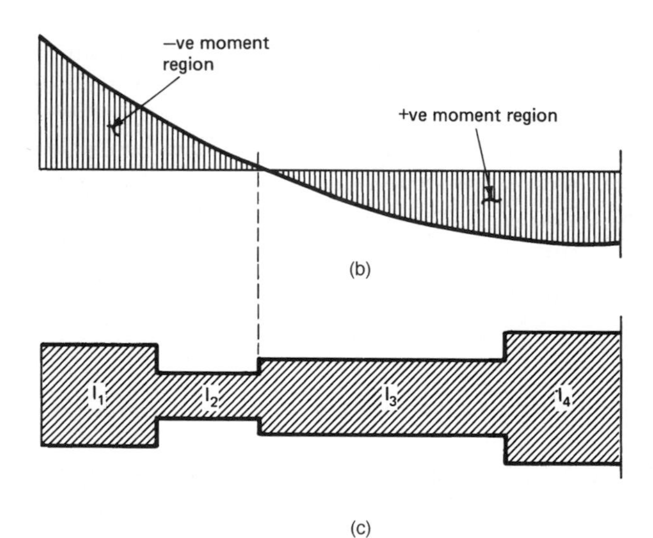


$I_1$ = Moment of inertia of unnotched steel section
$I_2$ = Moment of inertia of notched steel section
$I_3$ = Moment of inertia of composite notched section
$I_4$ = Moment of inertia of composite unnotched section

**Figure 9.11** Composite girder with square haunch: (a) schematic elevation; (b) combined gravity and wind moment diagram; (c) schematic moment of inertia diagram.

using a shallow-rolled section in the center and two deep-rolled sections, one at each end. Another method of fabricating the girder is to notch the bottom portion of the girder at midspan and reweld the flange to the web. The method requires more steel but comparatively less fabrication work.

In comparison to a shallow girder of constant depth, a haunch girder is significantly stiffer. Figure 9.11 shows the moment diagram in a haunch girder subjected to combined gravity and lateral loads. The corresponding stiffness properties including the effect of composite action in the positive moment regions is also shown in Fig. 9.11.

## 9.4 Composite Trusses

Figure 9.12 shows a typical floor-framing plan with composite trusses. To keep the fabrication simple, the top and bottom chords consist of T sections to which double-angle web members are welded directly without the use of gusset plates. The top chord is made to act compositely with the floor system by using welded shear studs. The space between the diagonals is used for the passage of mechanical and air-conditioning ducts. When the space between the diagonals is not sufficient, vertical members may be welded between the chords to form a vierendeel panel.

## 9.5 Composite Stub Girders

### 9.5.1 General considerations

In high rise design, maximum flexibility is achieved if structural, mechanical, electrical, and plumbing trades have their own designated space in the ceiling. This is achieved in a conventional system by placing HVAC ducts, lights, and other fixtures under the beams. Where deep girders are used, penetrations are made in the girder webs to accommodate the ducts. In an office building the typical span between the core and the exterior is about 40 ft (12.2 m), requiring 18 to 21 in. (457 to 533 mm) deep beams. Usual requirements of HVAC ducts, lights, sprinklers, and ceiling construction result in depths of 4 to 4.25 ft (1.21 to 1.3 m) between the ceiling and top of floor slab. The depth can, however, be decreased at a substantial penalty either by providing penetrations in relatively deep beams or by using shallower, less economical beam depths.

The stub girder system, Fig. 9.13, invented by engineer Dr Joseph Caloco, attempts to eliminate some of these shortcomings while at the same time reduces the floor steel weight. The key components of

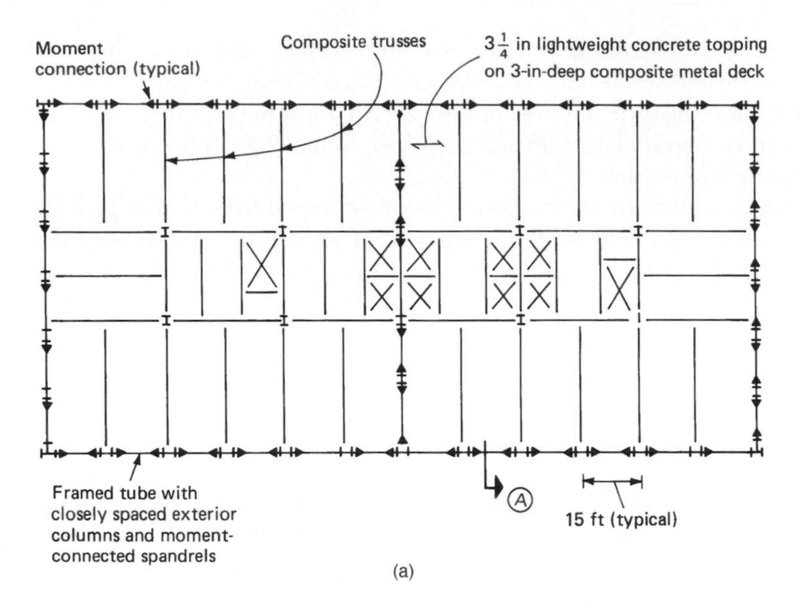


(a)

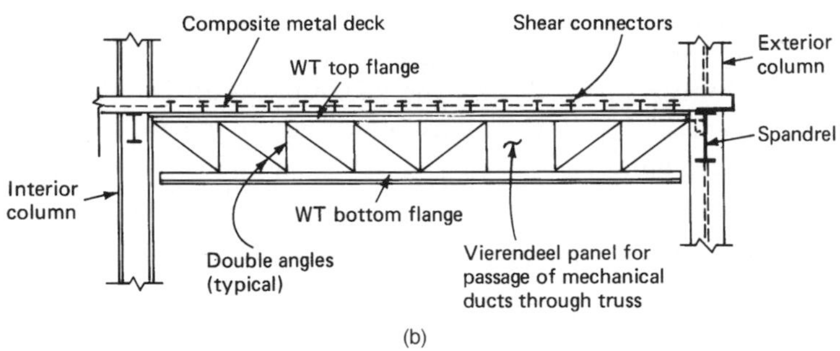


(b)

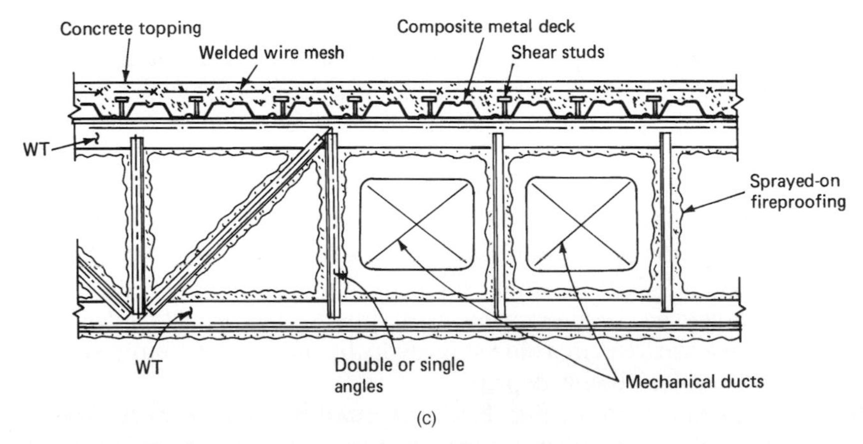


(c)

**Figure 9.12** Composite truss: (a) framing plan; (b) elevation of truss, section A; (c) detail of truss.

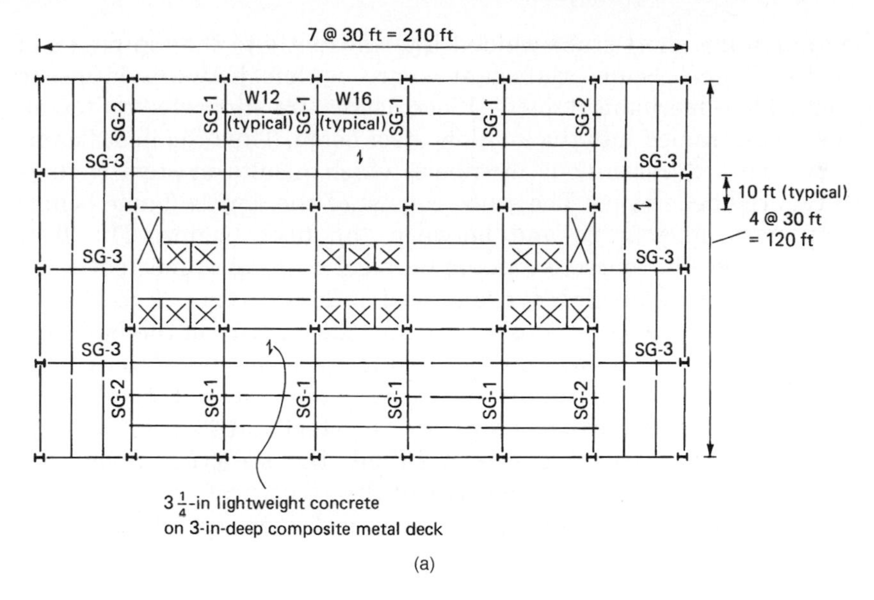


(a)

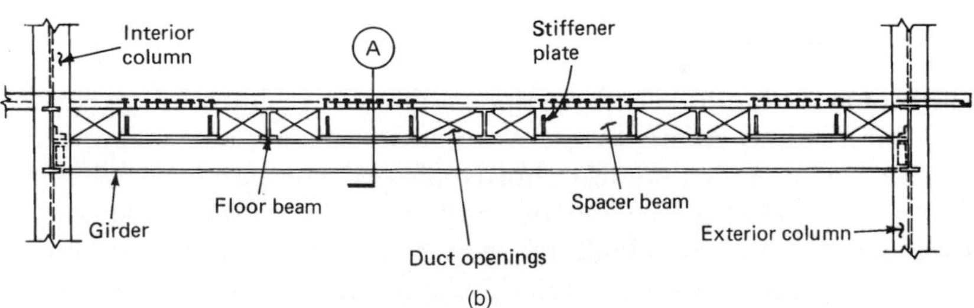


(b)

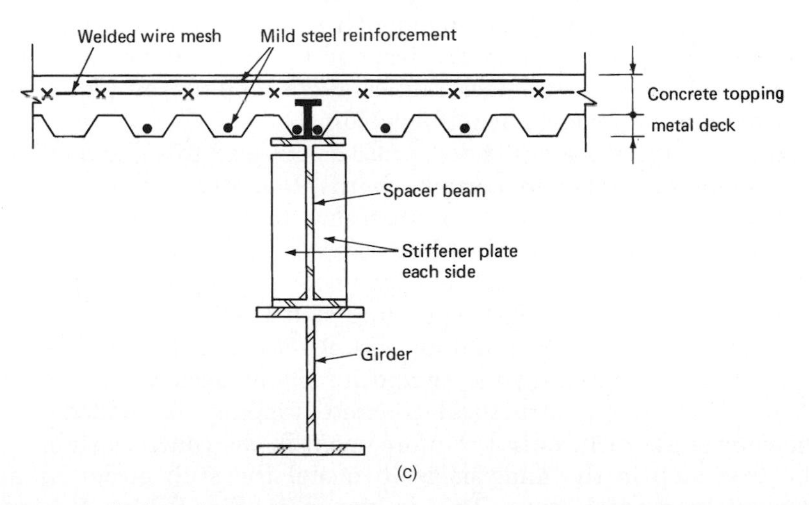


(c)

**Figure 9.13** Stub girder framing: (a) framing plan; (b) elevation of stub girder SG-1; (c) section A through stub girder.

the system are short stubs welded intermittently to the top flange of a shallow steel beam. Sufficient space is left between stubs to accommodate mechanical ducts. Floor beams are supported on top of, rather than framed into the shallow steel beam. Thus the floor beams are designed as continuous members which results in steel savings and reduced deflections. The stubs consist of short wide flange beams placed perpendicular to and between the floor beams. The floor system consists of concrete topping on steel decking connected to the top of stubs. The stub girders are spaced at 25 to 35 ft (7.62 to 10.7 m) on center, spanning between the core and the exterior of the building.

The behaviour of a stub girder is akin to a vierendeel truss; the concrete slab serves as the compression chord, the full-length steel beam as the bottom tension chord, and the steel stubs as vertical web members. From an overall consideration, the structure allows installation of mechanical system within the structural envelope, thus reducing floor-to-floor height; the mechanical ducts run through and not under the floor.

### 9.5.2 Behavior and analysis

The primary action of a stub girder is similar to that of a vierendeel truss; the bending moments are resisted by tension and compression forces in the bottom and top chords of the truss and the shearing stresses by the stub pieces welded to the top of wide-flange beam. The bottom chord is the steel wide-flange which resists the tensile forces. The compression forces are carried by the concrete slab. The effective width of concrete slab varies from 6 to 7 ft (1.83 to 2.13 m) requiring additional reinforcement to supplement the compression capacity of the concrete. The shear forces are resisted by the stub pieces, which are connected to the metal deck and concrete topping through shear connectors and to the steel beam by welding.

Because the truss is a vierendeel truss as opposed to a diagonalized truss, bending of top and bottom chords constitutes a significant structural action. Therefore, it is necessary to consider the interaction between axial loads and bending stresses in the design.

Figure 9.13a shows a typical floor plan with stub girders SG1, SG2, etc. Consider stub girder SG1, spanning 40 ft (12.19 m) between the exterior and interior of the building (Fig. 9.13b). The deck consists of a 2 in. (51 mm) deep 19-gauge composite metal deck with a $3\frac{1}{4}$ in. (82.5 mm) lightweight structural concrete topping. A welded wire fabric is used as crack control reinforcement in the concrete slab.

The first step in the analysis is to model the stub girder as an equivalent vierendeel truss. This is shown in Fig. 9.14a. A 14 in. (356 mm) wide flange beam is assumed as the continuous bottom chord of the truss. The slab and the steel beam are modeled as equivalent top and bottom chords. Note the beam elements representing these members are at the neutral axes of the slab and beam, as shown in Fig. 9.14b.

The stub pieces are modeled as a series of vertical beam elements between the top and the bottom chords of the truss with rigid panel zones at the top and bottom.

The various steps of modeling of stub girder are summarized as follows:

1. *Top chord of vierendeel truss.* As shown in Fig. 9.15, the top chord consists of an equivalent transformed area of the concrete topping which is obtained by dividing the effective width of concrete slab by the modular ration $n = E_s/E_c$. In calculating the transformed properties, advantage can be taken by including the mild steel reinforcement used in the slab. Although for strength calculations, modulus of elasticity of normal-weight concrete is used even for

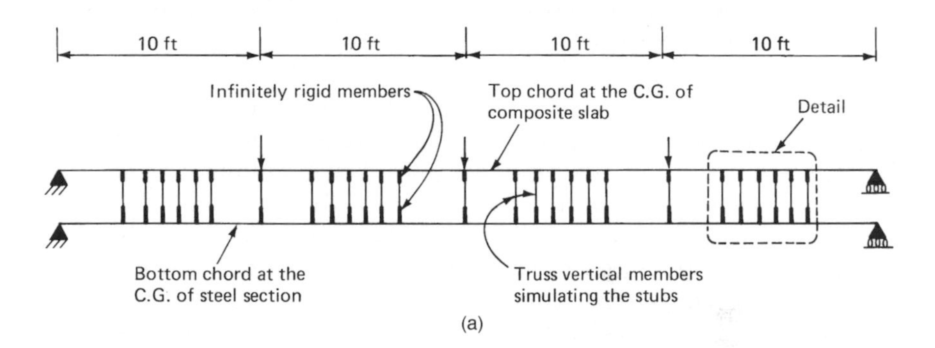


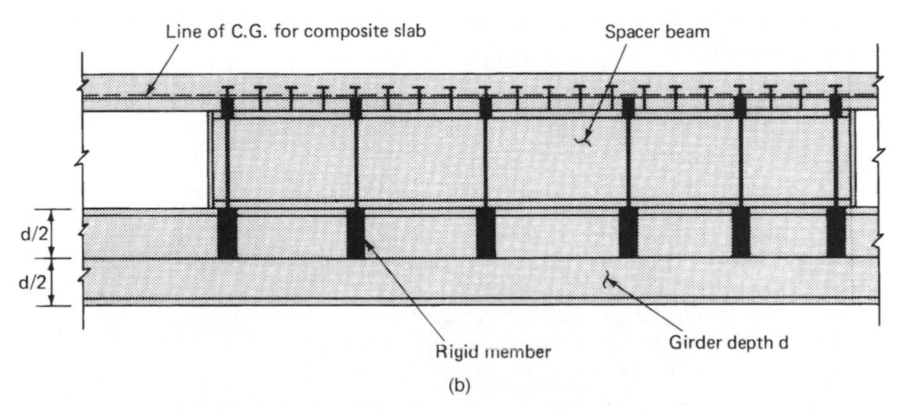


**Figure 9.14** (a) Elevation of vierendeel truss analytical model: (b) partial detail of analytical model.

lightweight concrete slabs in composite beam design, in stub girder the lower value of $n$ for lightweight concrete is used both for deflection and strength calculations. The moment of inertia $I_t$ of the top chord is obtained by multiplying the unit value of $I$ of composite slab, given in deck catalogs, by the effective width of slab.

2. *Bottom chord.* The properties of the steel section are directly used for the bottom chord properties.
3. *Stub pieces.* The web area and moment of inertia of the stub in the plane of bending of the stub girder are calculated and apportioned to a finite number of vertical beam elements representing the stubs. The more elements employed to represent the stub pieces, the better will be the accuracy of the solution. As a minimum, the author recommends one vertical element for 1 ft (0.3 m) of stub width. The vertical segments between the top and bottom flanges of the stub and the neutral axes of the top and bottom chords is treated as an infinitely rigid member. Stiffener plates used at the ends of stubs can be incorporated in calculating the moment of inertia of the stubs.

**Example.** A $3\frac{1}{4}$ in. (82.55 mm) lightweight, 3 ksi (20.7 MPa) concrete topping is to be used on 3 in. (76.2 mm) deep 18-gauge composite metal deck with nominal welded wire fabric reinforcement, with spans between steel purlins at 10 ft (3.05 m) centers. W14 × 53 (356 mm × 773 N/m), 50 ksi (344.75 MPa) structural steel beam is used for the bottom continuous chord of stub girder. Exterior and interior stub pieces consist of W16 × 26 (406 mm × 379 N/m), 36 ksi (248.3 MPa) steel beams with $6\frac{1}{2}$ and $3\frac{1}{2}$ ft (1.98 and 1.07 m) lengths, respectively. It is required to model the stub girder as a vierendeel truss to design and check various elements under the AISC and ACI specifications.

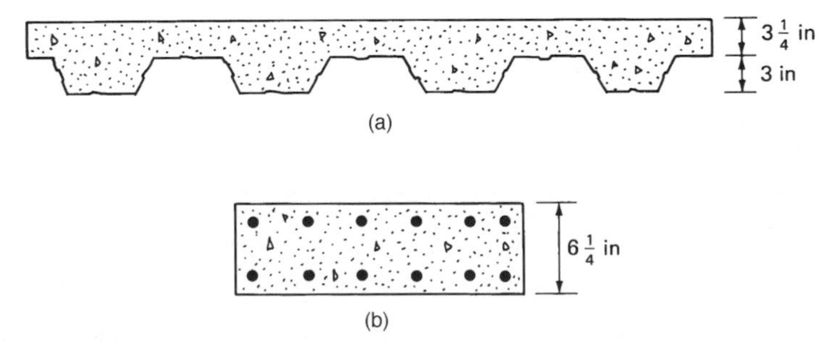


**Figure 9.15** Equivalent slab section.

As a first step, we compute the equivalent properties of: (i) top chord; (ii) bottom chord; and (iii) stub pieces for setting up the computer model.

1. *Top chord.* The concrete slab extends on both sides of the stub girder. The effective width of concrete flange is determined by considering three values:

$$b = 16t + b$$

$$b = L/4$$

$$b = \text{distance between stub girders}$$

For the example, the least value of effective width for the top chord is

$$16t + b = 16(1.5 + 3.25) + 5 = 81 \text{ in.} = 6.75 \text{ ft } (2.06 \text{ m})$$

Area $A$ of transformed section using $n = 14$

$$A = (3.25 + 1.5) \times \frac{81}{4} = 27.48 \text{ in.}^2 \ (17{,}732 \text{ mm}^3)$$

Equivalent moment of inertia $I_e$ is the value of $I$ for the particular metal deck and slab thickness given in the product catalog, multiplied by the effective width of compression chord. Assuming that the moment of inertia of $3\frac{1}{4}$ in. (82.55 mm) slab on 3 in. (76.2 mm) composite deck is 5.82 in.$^4$/ft, we get

$$I = 5.82 \times 6.75 = 39.3 \text{ in.}^4 \ (0.0164 \times 10^{-1} \text{ m}^4)$$

2. *Bottom chord.* W14 × 53 steel wide-flange beam has the following section properties:

$$A = 15.3 \text{ in.}^2 \ (9872 \text{ mm}^2), \quad I = 541 \text{ in.}^4 \ (0.225 \times 10^{-3} \text{ m}^4)$$

$$\text{Shear area } A_v = \text{depth of web} \times \text{web thickness}$$

$$= 13.92 \times 0.37 = 5.15 \text{ in.}^2 \ (3323 \text{ mm}^2)$$

3. (a) Exterior stub piece $W16 \times 26$, $6\frac{1}{2}$ ft (1.98 m) long

$$\text{Area} = 78 \times 0.25 = 19.5 \text{ in.}^2 \ (12581 \text{ mm}^3)$$

Moment of inertia in the plane of stub girder

$$I = \frac{bh^3}{12} = \frac{0.25 \times 78^3}{12} = 9886 \text{ in.}^4 \ (4.12 \times 10^{-3} \text{ m}^4)$$

We divide the area and moment of inertia of the stub piece into the six elements. These are used in the computer model to represent the stub piece. Therefore, the area of each vertical element is $19.5/6 = 3.25 \text{ in.}^2$ ($2097 \text{ mm}^3$), and the moment of inertia $I$ is $9886/6 = 1648 \text{ in.}^4$ ($0.686 \times 10^{-3} \text{ m}^4$).

(b) Interior stub piece W16 × 26, $3\frac{1}{2}$ ft (1.07 m) long

$$A = 42 \times 0.25 = 10.5 \text{ in.}^2 \ (6775 \text{ mm}^2)$$

$$I = \frac{0.25 \times 42^3}{12} = 1543.5 \text{ in.}^4 \ (0.642 \times 10^{-3} \text{ m}^4)$$

Since three vertical members are used to represent the interior stub piece, we divide the area and moment of inertia values by 3 to get the equivalent values for the computer model.

$$A = \frac{10.5}{3} = 3.5 \text{ in.}^2 \ (2258 \text{ mm}^2)$$

$$I = \frac{1548.5}{3} = 514.3 \text{ in.}^4 \ (0.214 \times 10^{-3} \text{ m}^4)$$

The next step is to set up the model of the equivalent vierendeel truss to obtain a computer solution for axial load, bending moment, and shear forces in all the members. The adequacy of each member under the action of combined forces is checked using the ACI and AISC procedures. A brief description of the procedure follows.

*Bottom chord.* Assume that the maximum axial tension and bending moment obtained from the computer run are $T = 265$ kips

(1178.8 kN) and $M = 90$ kip/ft (122.0 kN · m), respectively. Check W14 × 53 for combined tension and moment thus

$$f_a = \frac{265}{15.3} = 17.2 \text{ ksi } (118.6 \text{ MPa})$$

$$f_b = \frac{90 \times 12}{77.8} = 13.75 \text{ ksi } (94.8 \text{ MPa})$$

$$F_a = 0.6F_y = 0.6 \times 50 = 30 \text{ ksi } (206.8 \text{ MPa})$$

$$\frac{d}{t} = \frac{13.92}{0.37} = 37.62 > \frac{257}{\sqrt{F_y}} = 36.3$$

$$\therefore\ F_b = 0.60F_y = 30 \text{ ksi } (206.8 \text{ MPa})$$

$$\frac{f_a}{F_a} + \frac{f_b}{F_b} = \frac{17.2}{30} + \frac{13.75}{30} = 1.03$$

This is very nearly equal to 1.0, and therefore is OK.

*Top chord.* The top chord of the vierendeel truss is subjected to compression and bending moment and therefore is designed as a reinforced concrete column subjected to compressive forces and bending. In the opinion of the author any rational method that does not violate the spirit of the ACI code can be used in the design. One procedure is to neglect the contribution of metal deck and design the slab section as an equivalent column. For purposes of calculation of moment magnification factor, the column can be conservatively assumed to have an effective length of 10 ft (3.04 m), which is equal to the distance between the purlins.

For the example problem, assume that the computer results for axial compression and bending moment at critical sections are 250 kips (1112.0 kN) and 10 kip ft (13.56 kN · m), respectively.

We now proceed to design the equivalent section of the compression chord shown in Fig. 9.15 as a reinforced concrete column subject to axial loads and bending moment. First calculate the slenderness ratio $Kl_u/r$ by conservatively ignoring the restraint offered to the slab at the interface of stub pieces. The assumption that the equivalent column is hinged at the purlins gives a value of 1.0 for effective length factor $K$. The unsupported length $l_u$ of the equivalent column can be considered equal to 10 ft (3.05 m), which is the distance between the purlins. The radius of gyration for the equivalent rectangular column is 0.3 times the overall dimension in the direction of bending, i.e., 0.3 × 6.25 = 1.875 in. (47.62 mm).

The slenderness ratio

$$\frac{Kl_u}{r} = \frac{1 \times 10 \times 12}{1.875} = 64.0$$

Since this ratio is greater than 22, it is necessary to consider slenderness effects in the design of column. The moment magnification procedure will be used to take into account the slenderness effects. A conservative approximation will be made by assuming that the value for the coefficient $C_m$ (which relates the actual moment diagram to an equivalent moment diagram) is 1.0.

Since the axial load $P$ and moment $M$ obtained from the computer analysis are working stress values, these are converted to ultimate values by multiplying them with an average load factor of 1.5. Therefore,

$$P_u = 1.5 \times 260 = 390 \text{ kips } (1735 \text{ kN})$$

$$M_u = 1.5 \times 10 = 15 \text{ kip ft } (20.34 \text{ kN} \cdot \text{m})$$

The critical load $P_c$ is given by the relation

$$P_C = \frac{\pi^2 E_C I}{(Kl_u)^2}$$

For the example problem, we have

$$E_C = w_C^{1.5} \times 33\sqrt{f'_C} = 110^{1.5} \times 33\sqrt{3000} = 2085 \text{ ksi } (14378 \text{ MPa})$$

$$I = \frac{40.5 \times 6.25^3}{12} = 824 \text{ in.}^4$$

Substituting, we get

$$P_C = \frac{3.14^2 \times 2085 \times 824}{(10 \times 12)^2} = 1176 \text{ kips } (5230 \text{ kN})$$

The moment magnification factor is given by

$$\delta_1 = \frac{C_m}{1 - (P_u/\phi P_C)}$$

$$= \frac{1}{1 - (390/0.7 + 1176)} = 1.90$$

Therefore, design $M_u = 15 \times 1.9 = 28.5$ kip ft (38.65 kN · m).

The equivalent column is designed for $P_u = 390$ kips (1735 kN) and

$M_u = 28.5$ kip ft (38.65 kN · m). The required reinforcement is obtained by using a procedure conforming to the ACI code. For the present example, ten #5 longitudinal reinforcement is found to be adequate to carry the design axial load and bending moment.

*Computation of number of shear studs.* The shear studs between the stub pieces and the concrete slab form the backbone of composite stub girders. Their design is similar to composite beam design for which the shear connector formulas represent the horizontal shear at ultimate load divided by 2 to approximate conditions at working loads. The total horizontal shear resisted by the connectors between the point of maximum moment and each end of stub girder is the smaller of the values obtained from the following equations:

$$V_h = \frac{0.85 f'_C A_C}{2}$$

or

$$V_h = \frac{A_S F_y}{2}$$

For the example problem, it is found that a value of $V_h = 458$ kips (2037 kN) obtained from the first equation governs the design. Using a value of 9.5 kips (42.26 kN) as the allowable shear load, the number of shear studs $N = 458/9.5 = 48.2 \approx 50$, giving 32 and 18 shear connectors at the exterior and interior stubs.

*Check exterior stub W16 × 26, 6½ ft (1.98 m) long.* The design check is performed for shear and bending stresses per the AISC specifications. The summation of shear forces in the six elements used in the computer model to represent the exterior stub gives the design shear. The design moment for the stub is obtained by multiplying the accumulated shear by the height stub. Assume for the example problem that the accumulated shear = 210 kips (934 kN). The design moment then is $210 \times \frac{16}{12}$ or 280 kip ft (380 kN · m). The shear stress is $\frac{210}{78} \times 25$ or 10.76 ksi (74.25 MPa). The allowable shear stress is $0.4 \times F_v = 0.4 \times 36 = 14.4$ ksi (99.3 MPa). Therefore, the stub is okay for shear force.

To check the bending stresses, we calculate the moment of inertia and section modulus of the stub by including the contribution of the stiffener plates at the ends of stub. Without burdening the presentation with trivial calculations, let us assume that the section modulus of the stub piece and stiffener plate is equal to 300 in.$^3$ ($4.92 \times 10^6$ mm$^3$). The bending stress

$$f_b = \frac{280 \times 12}{300} = 11.2 \text{ ksi (77.3 MPa)}$$

This stress is checked against the allowable stresses per the AISC specifications.

A similar procedure is used to check the bending and shear stresses in the interior stub.

### 9.5.3 Moment-connected stub girder

The stub girder system, due to its large overall depth of approximately 3 ft (0.92 m), has a very large moment of inertia and can be used as part of a wind-resisting system. The model used for analysis is a vierendeel truss, where the concrete slab and the bottom steel beam are simulated as linear elements and each stub piece is divided into sev ements. The gravity and wind load shear forces and moments are introduced as additional load cases in the computer analysis, and the combined axial forces and moments in each section of the stub girder are obtained. All parts of the stub girder are checked for combined axial forces, shear, and moments as shown above. The controlling section for the slab is generally at the end of the first stub piece furthermost from the column. Particular care is required to transfer the moment at the column girder interfaces. If wind moments are small, moment transfer can take place between the slab and the bottom steel beam. The slab needs to be attached to the column either by long deformed wire anchors or by welding reinforcing bars to the column. For relatively large moments, the solution for moment transfer is to extend the first stub piece to the column face. The top flange of the stub piece and the bottom flange of the W14 girder are welded to the column as in a typical moment connection. The design of the connection is, therefore, identical to welded beam-column moment connection. The girder should be checked along its full length for the critical combination of gravity and wind forces. Depending upon the magnitude of reversal of stresses due to wind load, bracing of the bottom chord may be necessary.

### 9.5.4 Strengthening of stub girder

Strengthening of existing stub girders for tenant-imposed higher loads is more expensive than in conventional composite construction. A speculative type of investment building is usually designed for imposed loads of 50 psf (2.4 kN/m$^2$) plus 20 psf (0.96 kN/m$^2$) as partition allowance. For heavier loads strengthening of local framing is required. The bottom girder, which is in tension and bending, is relatively easy to reinforce by welding additional plates or angles to the existing steel member. Reinforcing the top chord of the stub girder, which is in compression and bending, is somewhat tricky. Addition of structural steel angles by using expansion anchors to the underside of metal deck and welding of additional stub pieces to reduce the effective length of compression chord, which acts like a column, have been used in practice with good results. From the point of view of ultimate load behavior, it is acceptable to strengthen the bottom chord to resist to total load without the truss action. However, it is important to check the lateral bracing requirements for the top flange of bottom chord.

## 9.6 Composite Columns

The term "composite column" in the building industry is taken to represent a unique form of construction in which structural steel is made to interact compositely with concrete. The structural steel section can be a tubular section filled with structural concrete or it could be a steel wide-flange section used as a core with a reinforced concrete surround.

Historically, composite columns evolved from the concrete encasement of structural steel shapes primarily intended as fire protection. Although the increase in strength and stiffness of the steel members due to concrete used as fireproofing was intuitively known, it was not until the 1940s that methods to actually incorporate the increases were developed. In fact, in the earlier days the design of the steel column was penalized by considering the weight of concrete as an additional dead load on the steel column. Later developments took account of the increased radius of gyration of the column because of the concrete encasement, and allowed for some reduction in the amount of structural steel. In some earlier high rise designs the concrete encasement was ignored from the viewpoint of strength considerations, but the additional stiffness of concrete was included in calculating the lateral deflections.

After the development of sprayed-on contact fireproofing in the 1950s and 1960s, use of concrete for fireproofing of structural steel was no longer an economical proposition. The high form-work cost of concrete could not be justified for fireproofing.

Over the last 20 years or so, the use of encased structural steel columns has found applications in buildings varying from as low as 10-stories to as high as 70-story or even taller buildings. These columns have been incorporated in an overall construction known as the composite system, which has successfully captured the essential advantages associated with steel and concrete construction: the speed of steel with the stiffness and moldability of concrete. Concrete columns with small steel-core columns used as erection columns were perhaps the earliest applications. Later much heavier columns were used, serving the dual purpose for both steel erection and load resistance. The heavier steel columns were used essentially to limit the size of composite vertical elements.

Another version consists of exterior concrete columns acting compositely with steel plate or precast cladding. Yet another version popular in some countries uses laced columns fabricated from light structural shapes such as angles, T sections, and channels. The concrete enclosure provides both fire-proofing qualities and also imparts additional stiffness to the light structural shapes, inhibiting their local buckling tendencies. Additional conventional reinforcement can be accommodated in the concrete encasement, as in conventionally reinforced concrete columns.

The ACI building code encompasses the design of all types of composite columns under one unified method using the same general principles as for conventionally reinforced concrete columns.

The ACI procedure is based on an ultimate concrete strain of 0.3 percent. As in conventionally reinforced concrete design, the tensile stress in concrete is disregarded. Either a parabolic or an equivalent uniform concrete strain can be assumed in the compression zone. The axial load assigned to the concrete portion of the composite column is required to be developed by direct bearing through studs, lugs, plates, or reinforcing bars welded to the structural steel plate prior to casting of concrete. In other words, the code requires a positive method for the transfer of axial load between the steel core and the concrete surround for strength calculations. For calculation of stiffness, however, merely wrapping the concrete around the steel core would suffice. Axial loads induced in the concrete section of the composite column due to bending of columns need not be transferred in direct bearing.

Tied composite columns are required by the ACI code to have more lateral ties than ordinary reinforced concrete columns. In fact, the ACI code stipulates twice as many ties, but this is based on somewhat questionable assumptions. First, it assumes that concrete that is laterally contained by the ties is thin. Second, it assumes that the concrete has a tendency to spall out from the smooth faces of the steel core. To prevent this separation, the lateral ties are specified to be vertically spaced no more than half the least dimension of the composite member. The ACI code does not permit the use of longitudinal bars in the evaluation of stiffness of columns on the premise that the longitudinal bars are rendered ineffective because of separation of concrete at high strains. They may, however, be included in the calculation of strength. Finally, the yield strength of

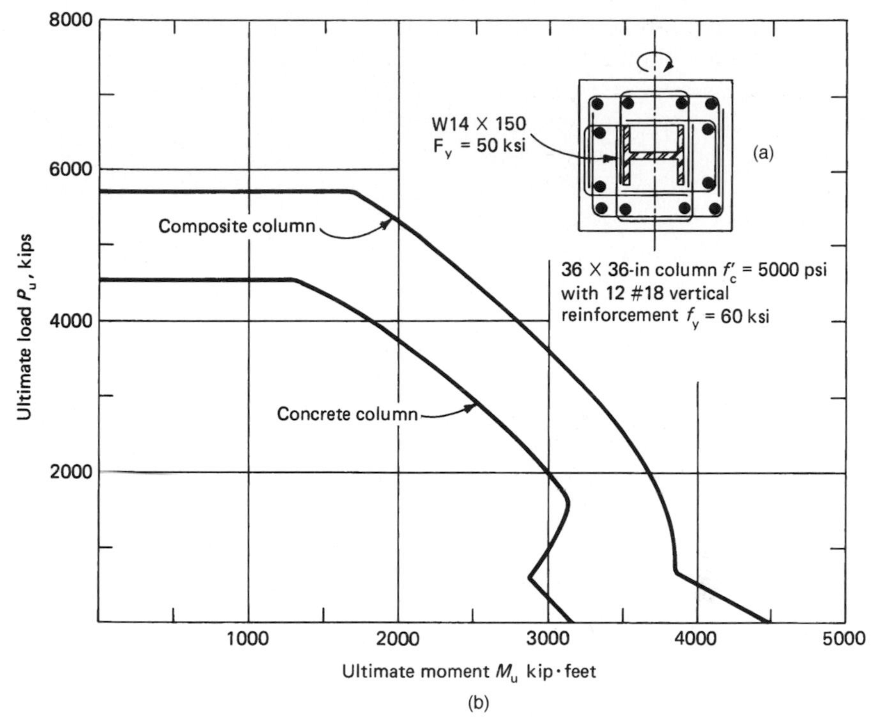


**Figure 9.16** Comparison of interaction diagrams: (a) column detail; (b) load moment interaction diagram.

the steel core is limited to 52 ksi (359 MPa) to correspond to the yielding strain of concrete of 0.0018.

The practical approach to the design of composite columns is to assume that the steel wide-flange section behaves as a reinforcing steel. With this assumption, interaction diagrams can be generated for various combinations of concrete column size, structural steel shape, and reinforcing steel. Figure 9.16 shows an interaction diagram generated for a 36 × 36 in. (915 × 915 mm) column with twelve #18 (57 mm diameter) reinforcing bars and a W14 × 150 (378 × 394 mm × 2188 N/m) structural steel shape. For comparison purposes, the interaction diagram for the same concrete column without the embedded structural steel shape is given. It can be seen that large increases in column capacity occur when structural steel shapes are included within the concrete envelope.

Chapter

# 10

# Analysis Techniques

## 10.1 Preliminary Hand Calculations

Even in today's high-tech computer-oriented world with all its sophisticated analysis capability, there still is a need for approximate analysis of structures. First, it provides a basis for selecting preliminary member sizes because the design of a structure, no matter how simple or complex, begins with a tentative selection of members. With the preliminary sizes, an analysis is made to determine if design criteria are met. If not, an analysis of the modified structure is made to improve its agreement with the requirements, and the process is continued until a design is obtained within the limits of acceptability. Starting the process with the best possible selection of members results in a rapid convergence of the iterative process to the desired solution.

Second, it is almost always necessary to compare several designs before choosing the one most likely to be the best from the points of view of structural economy and how well it fits in with other disciplines. Of the myriad structural systems only two or three schemes may be worthy of further refinement. Approximate methods are all that may be required to sort out the few final contenders from among the innumerable possibilities. Preliminary designs are therefore very useful in weeding out the weak solutions.

In the lateral load analysis of buildings, wind and earthquake forces are treated as equivalent loads and are reduced to a series of horizontal concentrated loads applied to the building at each floor. Portal and cantilever methods offer quick ways of analysis of rigid frames with unknown sizes. Both these methods are based on the well-observed characteristic of portal frames, namely that the points of contraflexure in beams and columns tend to form near the center of each column and girder segment. For purposes of analysis, the inflection points are assumed to occur exactly at the center of each member.

In the portal method, a rigid frame is treated as a series of consecutive single-bay portal frames. Interior columns are considered as part of two such portals, and the direct compression arising from the overturning effect on the leeward column of one portal is offset by the direct tension arising from the overturning effect on the windward column of the adjacent portal. If the widths of portals are unequal, the distribution of wind shear resisted by each portal can be assumed proportional to the aisle widths to maintain the interior column free of direct stress. Alternately, the column shears can be assumed to be unaffected by aisle widths resulting in axial stresses in the interior columns. With the shears in each column known and the points of contraflexure preestablished, the moments in beams and columns are determined. Simple statics yields axial and shear forces in beams and columns.

In the cantilever method the building is analyzed as a cantilever standing on end fixed at the ground level. The overturning moment is assumed to be resisted by the axial compresion of columns on the leeward side of the neutral axis and tension of columns on the windward side. The neutral axis for the frame is determined as the centroid of the areas of the columns in the bent. The axial forces in the columns due to overturning are assumed to be proportional to their distances from the neutral axis. As in the portal method, the points of inflection are assumed to occur at midheight of columns and midspan of girders. From the known axial forces in columns and the locations of the points of contraflexure, moments in columns and girders are obtained.

### 10.1.1 Portal method

Consider the application of the portal method to a 30-story frame shown in Fig. 10.1, consisting of two equal exterior bays and a smaller interior bay. Table 10.1 lists the lateral loads assumed in the analysis. The procedure is as follows. Distribute the accumulated story shears to each column in proportion to the aisle widths such that there are no direct stresses in the interior columns. Calculate the moments in the top and bottom of each column from the known shear in the column and the story height. Next, starting at the upper left corner of the frame, determine the girder moments where the column and girder moments are the same. Since the points of contraflexure are assumed at the center of the girder, the moments at each end are equal but opposite in sign. Determine the girder shears by the relation that shear multiplied by half of span length equals girder end moment. Next the axial stresses at the exterior columns are determined directly from girder shears. The results for the example frame are shown in Fig. 10.2.

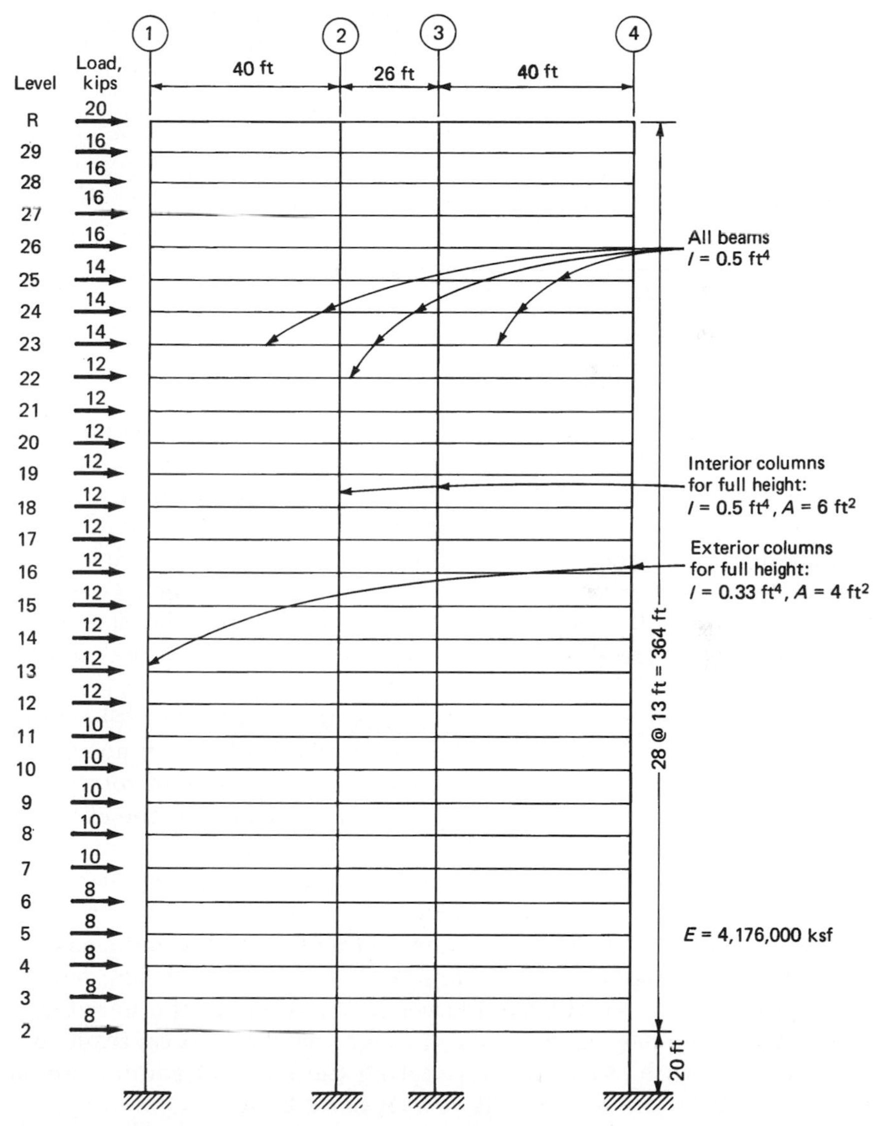


**Figure 10.1** Example frame: dimensions and properties.

**TABLE 10.1 Lateral Loads for 30-story Building Shown in Fig. 10.1**

| Level | Story shear, kips | Accumulated shear, kips | Level | Story shear, kips | Accumulated shear, kips |
|---|---|---|---|---|---|
| *R* | 20 | 20 | 15 | 12 | 222 |
| 29 | 16 | 36 | 14 | 12 | 234 |
| 28 | 16 | 52 | 13 | 12 | 246 |
| 27 | 16 | 68 | 12 | 12 | 258 |
| 26 | 16 | 84 | 11 | 10 | 268 |
| 25 | 14 | 98 | 10 | 10 | 278 |
| 24 | 14 | 112 | 9 | 10 | 288 |
| 23 | 14 | 126 | 8 | 10 | 298 |
| 22 | 12 | 138 | 7 | 10 | 308 |
| 21 | 12 | 150 | 6 | 8 | 316 |
| 20 | 12 | 162 | 5 | 8 | 324 |
| 19 | 12 | 174 | 4 | 8 | 332 |
| 18 | 12 | 186 | 3 | 8 | 340 |
| 17 | 12 | 198 | 2 | 8 | 348 |
| 16 | 12 | 210 | | | |

*Note*: 1 kip = 4.448 kN.

### 10.1.2 Cantilever method

Two assumptions are made in the analysis: (i) inflection points form at midspan of each beam and at midheight of each column; and (ii) the axial stresses in the columns vary as the distance from the frame centroidal axis. It is usually further assumed that all columns in a given story are of equal area resulting in column forces which vary as the distance from the center of gravity of the frame. To get a comparison with the portal method, we shall apply the cantilever method to the three-bay portal frame (Fig. 10.1) analyzed in the previous section.

The first assumption locates the points of contraflexure. Shown in Fig. 10.3a, part 1 is a free-body diagram of the top story above the points of contraflexure in the columns. The frame axis of rotation is located at the center of gravity of the columns, which for the example problem coincides with the line of symmetry of the frame. The column axial forces for the top story are obtained by equating the moment of the column reactions about the frame axis to the moment of the lateral forces taken about a horizontal plane through the assumed hinges of the top floor. These are also shown in Fig. 10.3a, part 1.

In a similar manner the axial forces in the columns of other stories are computed by passing a section through the points of contraflexure of columns of each story and considering the moment equilibrium of the frame above the section (Fig. 10.3a, parts 2–4).

After the column axial forces are found, the girder shears are determined at once. For example, in Fig. 10.3a, part 3, the tension in the exterior windward column at the fifteenth level is 210.72 kips (937.28 kN). Tension in the same column at the fourteenth level is 187.08 kips (832.13 kN). Therefore, by the relation that the summation of the axial forces in the columns and the girder shear is equal to 0 at the joint where the fifteenth-story girder joins the exterior windward column, the girder shear is 210.72 − 187.08 = 23.64 kips (105.15 kN). Figure 10.3a, part 3, shows the method of obtaining this and the remaining shears for the fifteenth-level girder.

With the girder shears known, the girder moments follow directly. These equal the shear in the girder times one-half the span length.

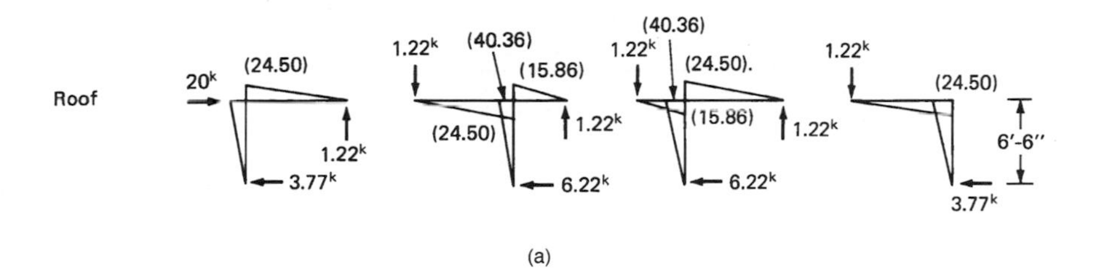


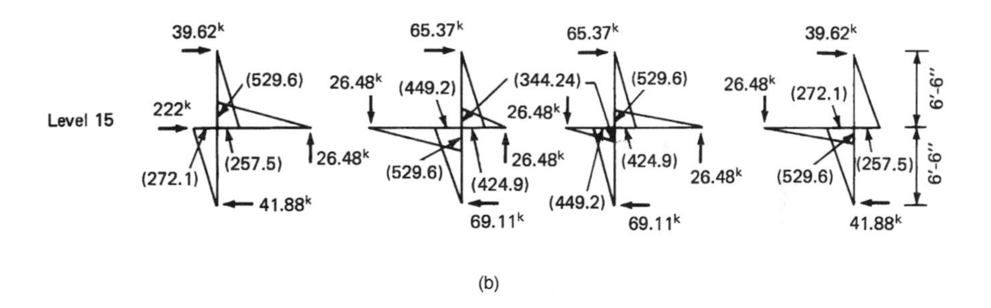


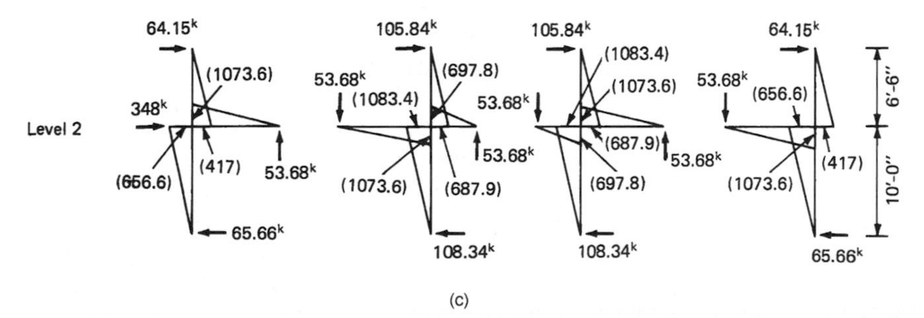


**Figure 10.2** Portal method: (a) moment and forces at roof level; (b) moment and forces at level 15; (c) moment and forces at level 2. Note: all moments are in kip/ft and forces in kips.

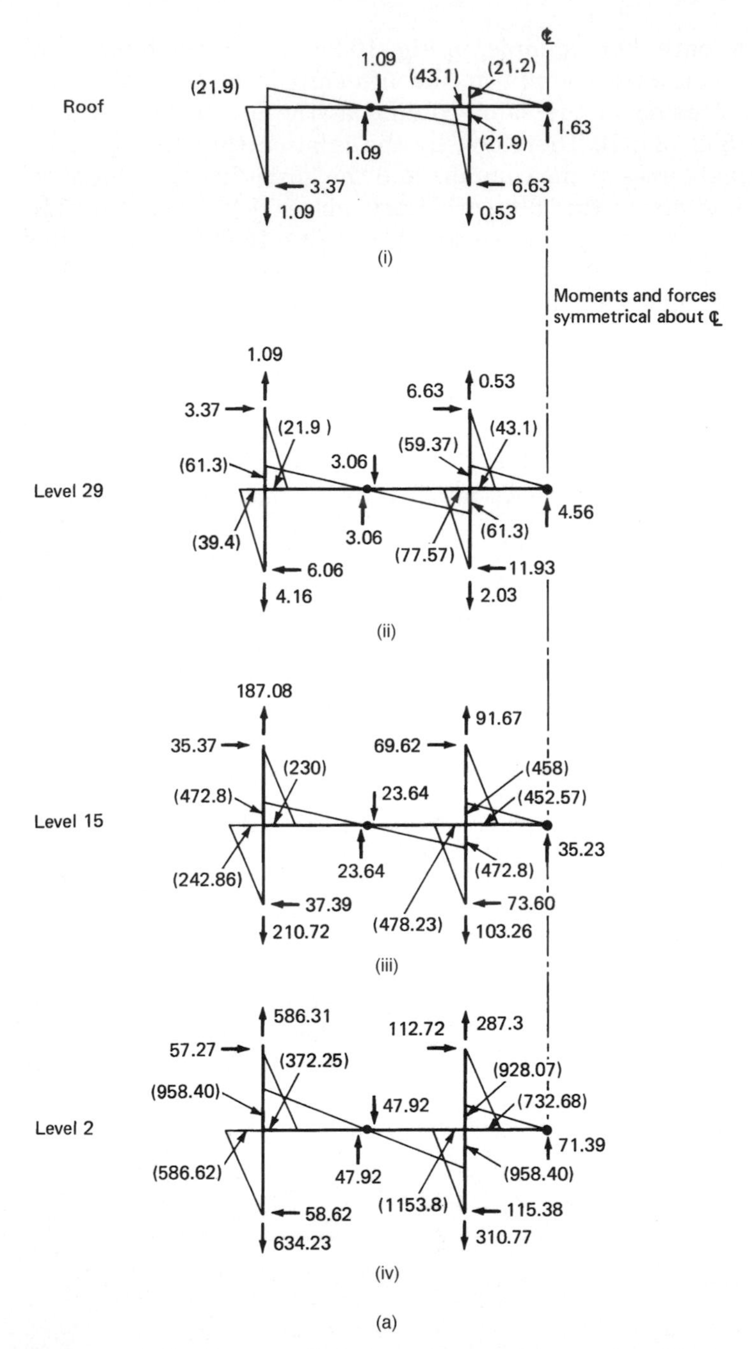


**Figure 10.3a** Cantilever method: (a) moment and forces at roof level; (b) moment and forces at level 29; (c) moment and forces at level 15; (d) moment and forces at level 2. Note: All moments are in kip/ft and forces in kips.

The study of the various joints will show that from the relation that $\sum M = 0$ at any joint, the sum of column moments must equal the sum of girder moments. Using this principle, the moments in the columns at the roof are obtained from roof girder moments (Fig. 10.3a, part 1). Since the points of contraflexure in the columns are at midheight, the column moments above the 29th level have the same value as at the roof level (Fig. 10.3a, part 2). Moments in the columns below the 29th level are obtained from the relation $\sum M = 0$, and in a similar manner column moments in other floors are found. The column shears are obtained by dividing column moments by half the height of columns. As a check, observe that the shear in the columns of any level equals the sum of the horizontal external loads above that level. The moments and forces obtained by using the above procedure for the example problem are shown in Fig. 10.3a.

To get a feel for the accuracy of the procedures, the frame in Fig. 10.1 has been analyzed by a plane-frame computer analysis (Fig. 10.3b). The computer results vary considerably from either of the two methods. Chief among the reasons for the discrepancy are: (i) points of contraflexure in the lower stories are not at the midpoints; and (ii) the shears are greater in exterior girders than in the interior girders of that floor.

### 10.1.3 Lateral stiffness of frames

The lateral displacement of one floor relative to the floor below results from a combination of bending and shear deformation of the bent. The bending deformation or the chord drift, as it is sometimes called, is a consequence of axial deformation of the columns and is independent of the size, type, location, and arrangement of the web system. The shear deformation is due to the rotation of the joints in the frame, which causes bending of columns and girders of the frame. For relatively short frames with height-to-width ratios less than 3, the deflection due to axial shortening of columns can be neglected and the deflection of the frame can be assumed to be entirely due to joint rotations. Its contribution to deflection can, however, be obtained by considering the frame as a cantilever with an equivalent moment of inertial $I = 2ad^2$ where $a$ is the area of exterior column and $d$ is half the base of the portal frame. For taller frames, it is prudent to consider the axial deformation of the interior columns; the equivalent moment of inertia is determined by the relation $I = \sum_1^n a_1 d_1^2$, where $a_1, a_2, \ldots, a_n$ represent the areas of the columns and $d_1, d_2, \ldots, d_n$ represent their corresponding distances from the natural axis of the frame. To derive the shear deformation equations, consider the

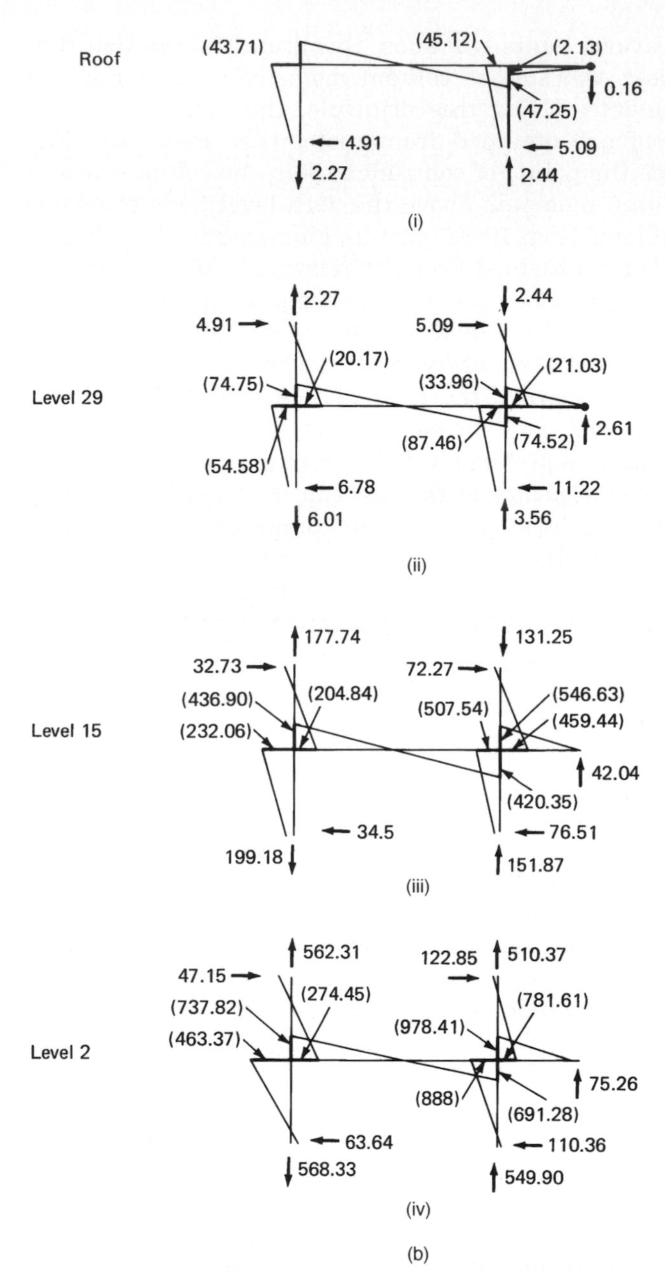


**Figure 10.3b** Cantilever method, moment and forces at: (a) roof level; (b) level 29; (c) level 15; (d) level 2. Note: all moments are in kip ft and forces in kips.

frame shown in Fig. 10.4. Isolate a typical floor and column segment between the points of contraflexure above and below the floor as shown in the figure.

**Deflection due to column rotations.** Consider the free-body diagram of a typical story bounded between the points of contraflexure in the

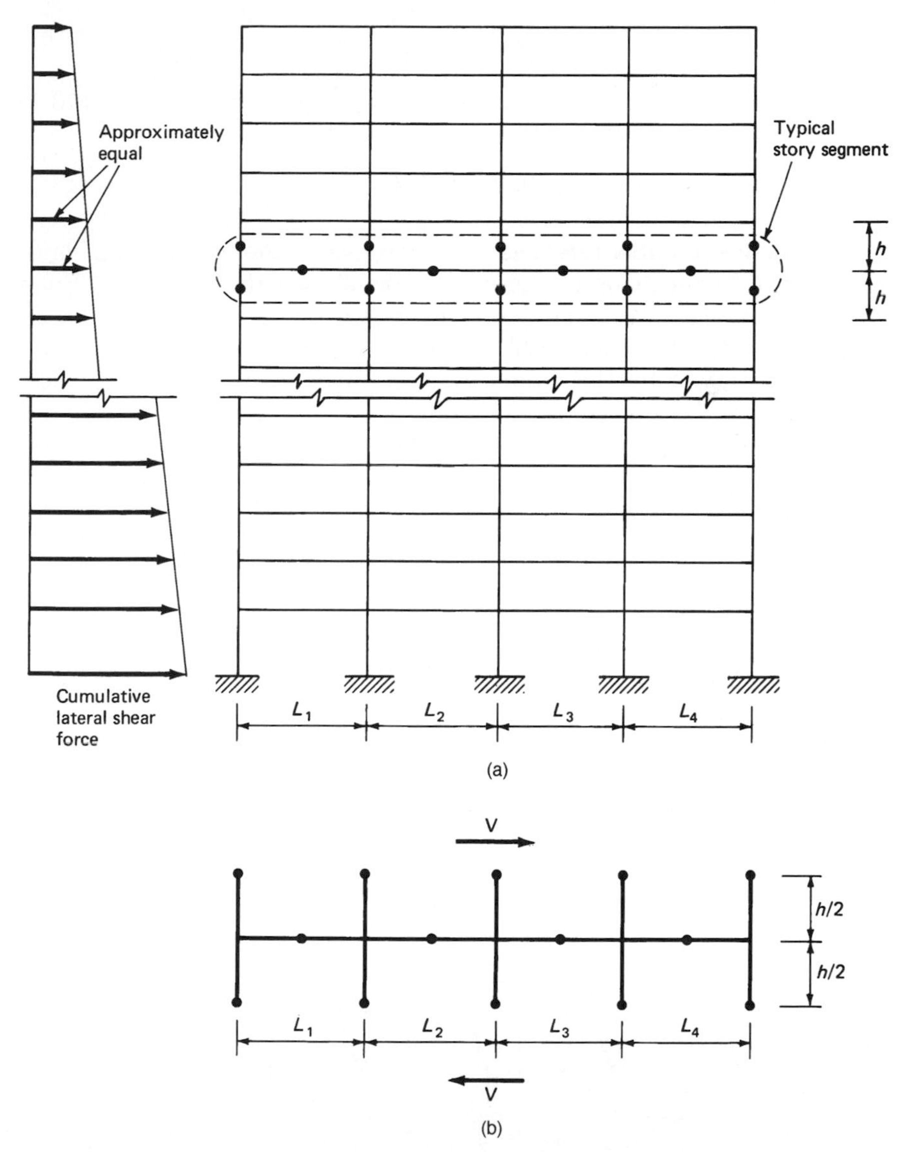


**Figure 10.4** Deflections of portal frame: (a) frame subjected to lateral loads; (b) typical story segment.

columns above and below the $i$th level as shown in Fig. 10.5. When the number of stories is large, it is reasonable to assume that the shears in the columns above and below the floor do not differ appreciably. If the floor girders are rigid, the lateral deflection $\Delta_1/2$ of each column would be equal to the sum of the deflections of the two cantilevers of length $h/2$ under the action of wind shears $V$ (Fig. 10.5).

$$\frac{\Delta_1}{2} = \frac{V\left(\frac{h}{2}\right)^3}{3EI_c} \quad \text{or} \quad \Delta_1 = \frac{Vh^3}{12EI_c} \tag{10.1}$$

giving for all columns $\Delta_1 = Vh^3/12E\sum I_c$.

**Deflection due to girder rotations.** Next consider the columns as rigid, giving rise to rotations of the girders as shown in Fig. 10.6a. Each girder undergoes a rotation equal to $\theta$ at each end giving rise to an internal moment of $12EI\theta/L$ for each girder. The total internal moment is given by the summation of such terms for each girder. Thus the total internal moment due to girder rotation is $12E\theta\sum(I_{bi}/L_i)$. The external moment due to wind shears $V$ is given by $V \times h$. Equating external moment to internal moment and noting that $\theta$ produces a displacement $\Delta_2 = \theta h$, we get

$$\Delta_2 = \frac{Vh^2}{12E\sum(I_{bi}/L_i)} \tag{10.2}$$

The total frame shear deflection $\Delta_s$ is given by

$$\Delta_s = \Delta_1 + \Delta_2 = \frac{Vh^2}{12}\left\{\frac{h}{(\sum EI)_{\text{col}}} + \frac{1}{\sum(EI/L)_{\text{beam}}}\right\} \tag{10.3}$$

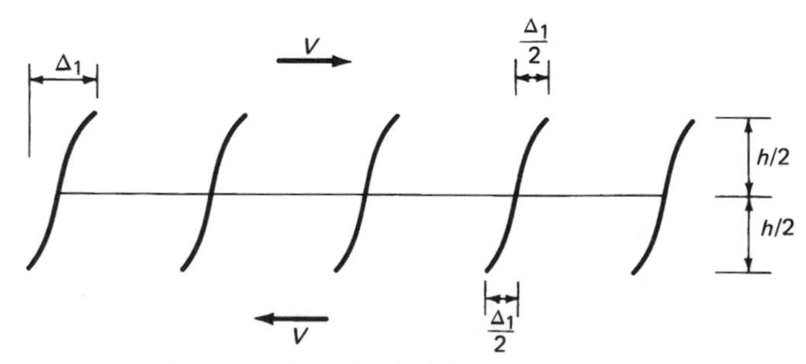


**Figure 10.5** Lateral deflection of typical story due to bending of columns.

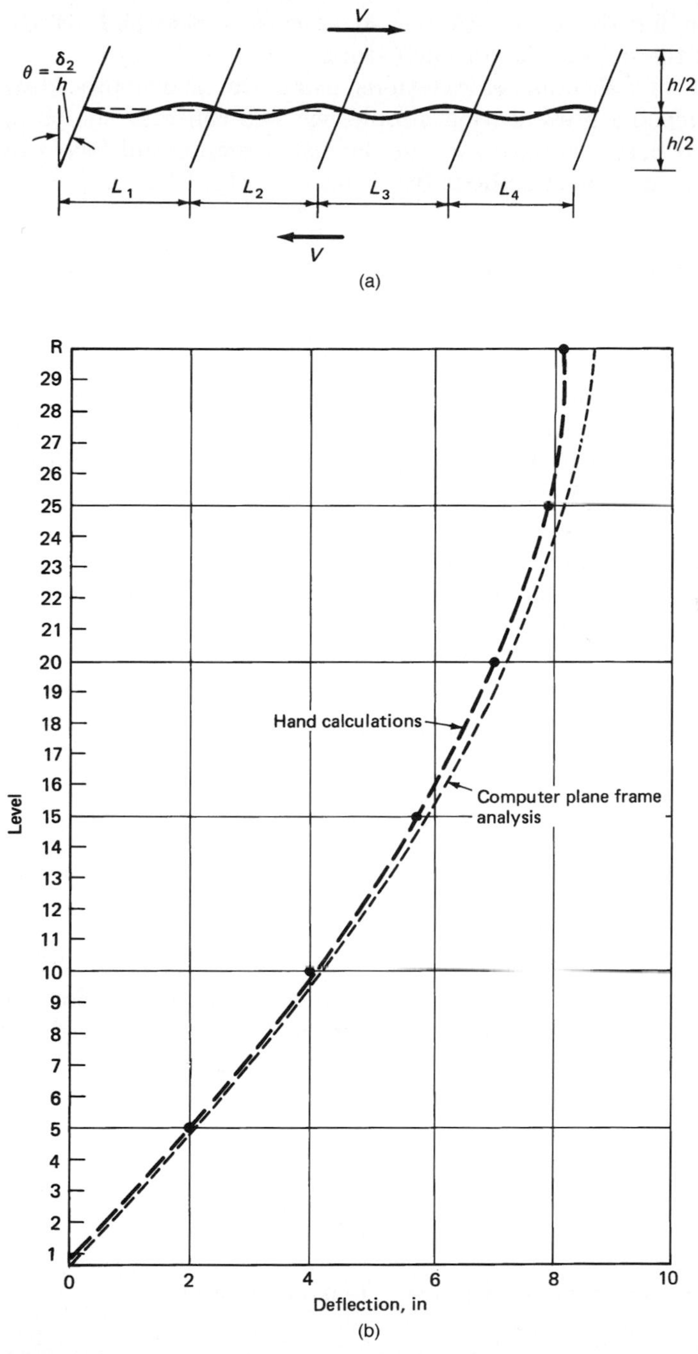


**Figure 10.6** (a) Lateral deflection of typical story due to bending of girders: (b) deflection comparison (30-story frame).

The deflection for the total number of stories is obtained by the summation of the deflections for each story.

An example of deflection calculations using the above procedure follows. To keep the presentation simple, we will consider the same example frame that was used for calculating moments and forces by the portal and cantilever methods (refer back to Fig. 10.1).

### Deflection calculations for frame shown in Fig. 10.1

**Cantilever deflection.** The neutral axis for the frame lies on the line of symmetry. The moment of inertia of the frame about the neutral axis is given by $I = 2(a_1 d_1^2 + a_2 d_2^2)$ where $a_1$ and $a_2$ are the areas of the exterior and interior columns and $d_1$ and $d_2$ their distance from the neutral axis. Substituting $a_1 = 4\text{ ft}^2$ and $a_2 = 6\text{ft}^2$, $d_1 = 53\text{ ft}$ and $d_2 = 13\text{ ft}$, we get $I = 2(4 \times 53^2 + 6 \times 13^2) = 24{,}500\text{ ft}^4$ ($211.46\text{ m}^4$).

For purposes of deflection calculation, we can assume that the frame is subjected to a uniformly distributed horizontal load $= \frac{12}{13} =$ 0.9231 kip/ft. The cantilever deflection at the top is given by

$$\Delta_{\text{cant}} = \frac{wl^4}{8EI} = \frac{0.9231 \times 384^4}{8 \times 4{,}176{,}000 \times 24{,}500} = 0.0245\text{ ft (7.47 mm)}$$

**Shear deflection due to column rotations.** This is given by

$$\Delta_1 = \frac{Vh^3}{12E \sum I_c}$$

For the example problem, the moments of inertia for the exterior and interior columns are, respectively, equal to $0.33\text{ ft}^4$ and $0.5\text{ ft}^4$, giving $\sum I_c = 2 \times 0.33 + 2 \times 0.5 = 1.66\text{ ft}^4$. Using an average cumulative shear value of $V = 210$ kips and $h = 13$ ft,

$$\Delta_1 = \frac{210 \times 13^3}{12 \times 4{,}176{,}000 \times 1.66} = 0.0056\text{ ft (1.70 mm)}$$

**Shear deflection due to girder rotations.** This is given by

$$\Delta_1 = \frac{Vh^2}{12E \sum (I/L)}$$

For the example problem, $\sum I/L$ of girders = 0.5/40 + 0.5/26 + 0.5/40 = 0.0442 ft, giving

$$\Delta_2 = \frac{210 \times 13^2}{12 \times 4{,}176{,}000 \times 0.0442} = 0.016 \text{ ft/floor (4.87 mm/floor)}$$

The total shear deflection $\Delta_s = \Delta_1 + \Delta_2 = 0.0056 + 0.016 = 0.0216$ ft/floor (6.58 mm/floor). The shear deflection at top of 30-stories is given by $30 \times 0.0216 = 0.648$ ft. Therefore total deflection at top due to chord drift and shear deformation is $0.0245 + 0.648 = 0.6725$ ft (204.97 mm). A comparison of floor-by-floor deflections obtained by using the above approach with those of a computer plane frame analysis is given in Fig. 10.6b. The appropriateness of the method for preliminary design is obvious.

Another method of calculation of frame deflection consists of representing the columns and beams as a single cantilever column with an equivalent flexural stiffness of $I_e$ and shear stiffness of $A_e$ to simulate the cantilever and shear modes of bending of the frame. The method is best explained with reference to Fig. 10.7 which shows a 19-story, three-bay unsymmetrical portal-frame with columns of varying moments of inertia. We first locate $x$, the distance of frame axis of bending from the windward column by equating moments of individual column areas to the moment of total area about the windward column. Using the values given in Fig. 10.7, we get,

$$4 \times 30 + 6 \times 50 + 6 \times 90 = (4 + 6 + 6 + 4)x$$

giving

$$x = 48 \text{ ft (14.63 m)}$$

from the windward column.

Calculate the moment of inertia of the frame about its axis of bending by the relation $I = \sum Ax^2$. Since the areas of the columns change at four locations, the corresponding four values of frame moment of inertia from the top work out equal to 21,120 ft$^4$, 42,240 ft$^4$, 63,360 ft$^4$, and 84,480$^4$ ft, respectively (182.3 m$^4$, 364.6 m$^4$, 546.86 m$^4$, 729.15 m$^4$).

Figure 10.8 shows the equivalent cantilever with varying moments of inertia. If the beams were infinitely rigid, the deflection calculated for the cantilever would have represented the total lateral deflection of the frame. Since in reality the beams are flexible, the deflection of the cantilever is increased by the racking component, which is equivalent to the shear deformation of the cantilever. This was shown equal to

$$\Delta_s = \frac{Vh^2}{12}\left[\frac{h}{(EI/h)_{\text{col}}} + \frac{1}{(EI/L)_{\text{beam}}}\right] \tag{10.4}$$

Defining story stiffness as the deflection per unit of horizontal shear, the equivalent story stiffness is given by the relation

$$\frac{V}{\Delta_1} = \frac{12}{h^2\{1/(\sum EI)_{\text{col}} + 1/[\sum (EI/L)]_{\text{beam}}\}} \tag{10.5}$$

An equivalent shear area for the cantilever is worked out as follows. Consider the shear deformation of the cantilever per unit height $h$ subjected to horizontal forces $V$ as shown in Fig. 10.9. The shear deflection $\Delta_s$ is given by

$$\Delta_s = \frac{Vh}{GA_v} \tag{10.6}$$

The story stiffness $\Delta_s/h$ works out equal to $0.4EA_v/h$ in which it is assumed that $G = 0.4E$. Equating story stiffness relations of Eqs

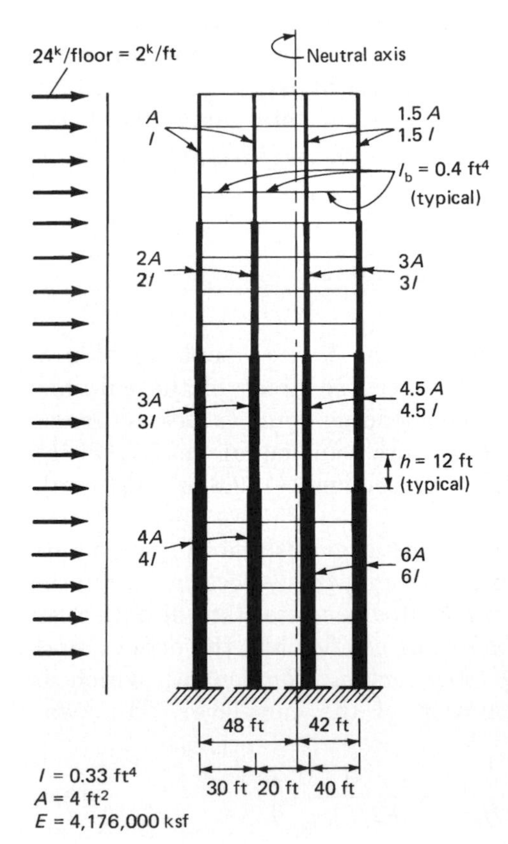


**Figure 10.7** Example portal frame for deflection calculations. Note the variation of column areas and moments of inertia at four locations.

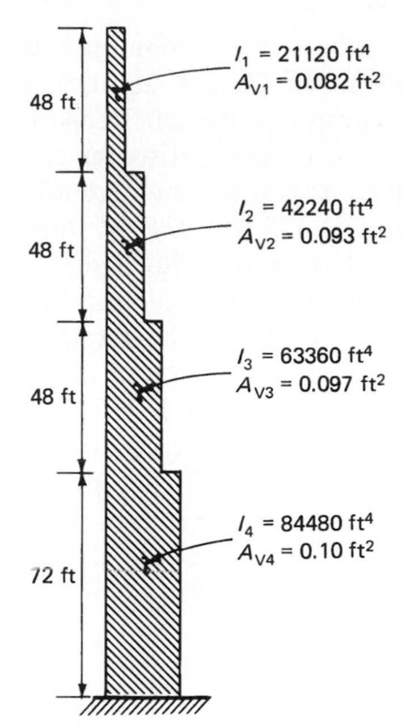


**Figure 10.8** Equivalent cantilever representing the portal frame.

(10.5) and (10.6), we get

$$\frac{0.4EA_v}{h} = \frac{12}{h^2\{1/(\sum E_cI)_{\text{col}} + 1/[\sum (E_bI/L)]_{\text{beam}}\}}$$

Assuming $E$ is constant for beams and columns, i.e., $E_c = E_b = E$, we get

$$A_v = \frac{30}{h\{1/(\sum I)_{\text{col}} + 1/[(IL)]_{\text{beam}}\}} \tag{10.7}$$

Using the numerical values shown in Fig. 10.7, the equivalent shear areas at four vertical locations work out, respectively, equal to 0.082 ft², 0.093 ft², 0.097 ft², and 0.1 ft² (0.0076 m², 0.0086 m², 0.0090 m², 0.0093 m²) from the top. These values are shown schematically in Fig. 10.8.

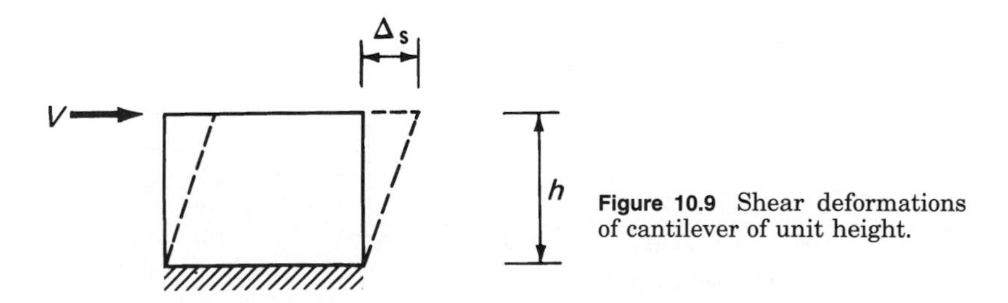


**Figure 10.9** Shear deformations of cantilever of unit height.

The deflection of the equivalent cantilever of varying moments of inertia can be obtained either by long-hand methods such as virtual work or by using a relatively simple stick computer model. Reasonable results can be obtained by assuming average properties for the equivalent cantilever. The average values for $I$ and $A$ for the example problem work out equal to 56,320 ft$^4$ and 0.093 ft$^2$ (486 m$^4$ and 0.0086 m$^2$), respectively. Using a value of 216 kips for the average cumulative shear $V$, we get a total top deflection of 0.319 ft (94 mm) as compared to a value of 0.28 ft (82.3 mm) obtained from a stick computer model and a value of 0.24 ft (73 mm) as obtained from a plane frame analysis. Comparison of deflections are shown in Fig. 10.10.

The analysis presented thus far is based on the centerline dimensions, which in general overestimate the deflection. Although all structural members have finite widths, it is unnecessary, especially in view of the approximate nature of the analysis, to be overly concerned about the effect of joint widths on the stiffness of the structure. However, in those cases in which the dimensions of the members are large in comparison to story height and girder spans, it is possible to incorporate the effect of joints by assuming that no member deformation occurs within the joint. An approximate expression for the equivalent shear area for the equivalent column can be

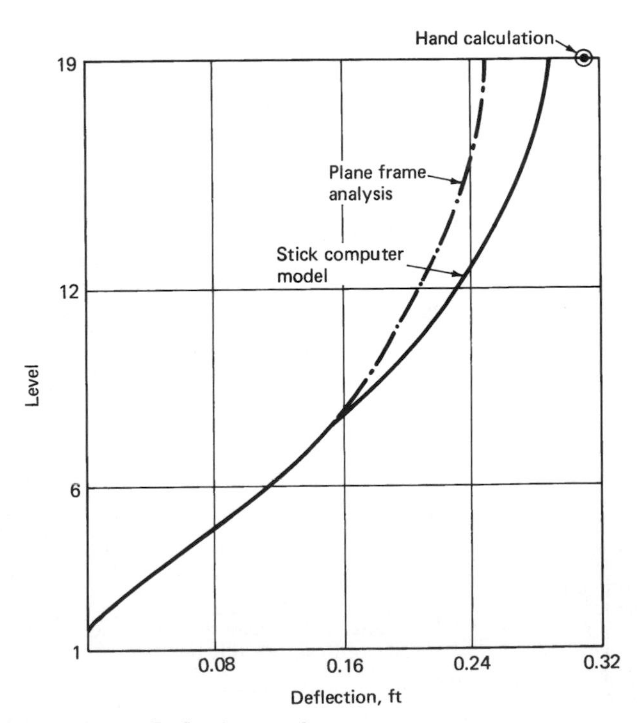


**Figure 10.10** Deflection results.

shown to be:

$$A_v = \frac{30}{h^2\{h\alpha_1^3/(\sum I)_{col} + \alpha_2^3/(\sum I/L)]_{beam}\}} \tag{10.8}$$

where $\alpha_1$ = the average ratio of clear height to center to center heights of columns (Fig. 10.11)

$\alpha_2$ = the average of the ratio of the clear span to the centerline spans of girders (Fig. 10.11)

Analytical and experimental investigations have shown that an analysis based on rigid offset lengths to the outer face of supports overestimates the stiffness of the structure. The analysis should therefore include some method for compensating the deformations that do exist in the panel zones. A rigid zone reduction factor can be used to reduce the lengths of rigid offsets—a method similar to that employed in many commercial computer programs. Arbitrary reductions are assigned to joint sizes in an effort to compensate for the joint deformation.

The underlying principle in both the portal and cantilever methods is the assumption that the point of contraflexure is located at midheight and midspan of columns and girders. Rigorous computer analyses show that this assumption is violated in various degrees, especially at the top and bottom floors of a tall building. It is possible, however, to improve the results of the approximate analyses by refining the locations for points of contraflexure.

For example, the points of contraflexure at the lower floors, especially at the first floor may be assumed to, occur at a location closer to about $h/3$ below the second floor. Equivalent shear stiffness for the first story can be shown to be:

$$A_v = \frac{20}{h\{1/(\sum EI)_{col} + 1/5[\sum (EI/L)_{beam}]\}} \tag{10.9}$$

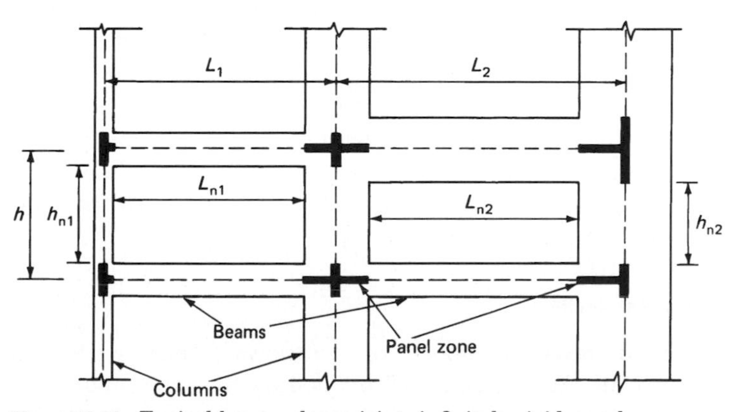


**Figure 10.11** Typical beam-column joint; infinitely rigid panel zone.

Further refinement of the analysis is generally considered unnecessary in view of the approximate nature of the analysis.

### 10.1.4 Framed tube structures

As mentioned earlier, the framed tube system in its simplest form consists of closely spaced exterior columns tied at each floor level by relatively deep spandrels. The behavior of the tube is in essence similar to that of a hollow perforated tube. The overturning moment under lateral load is resisted by compression and tension in the columns while the shear is resisted by bending of columns and beams primarily in the two sides of the building parallel to the direction of the lateral load. The bending moments in the beams and columns of these frames, called the web frames, can be evaluated using either the portal or the cantilever method. It is perhaps more accurate to use the cantilever method because tube systems are predominantly used for very tall buildings in the 40- to 80-story range in which the axial forces in the columns play a dominant role.

As mentioned earlier, because of the continuity of closely spaced columns and spandrels around the corners of the building, the flange frames are coaxed into resisting the overturning moment. Whether or not all the flange columns, or only a portion thereof, contribute to the bending resistance is a function of shear rigidity of the tube. A method for approximating the shear lag effects in a rectangular tube is to model it as two equivalent channels as shown in Fig. 10.12. The determination of width of the channel flange is subjected to engineering judgment and is usually taken as 15 to 20 percent of the width of the building.

Shown in Fig. 10.13 is the plan of a framed tube building delineating a limited number of columns in the leeward and windward faces as part of equivalent channel flanges. The axial forces obtained on the basis of equivalent channels are, as shown in Fig. 10.13. Shown in Fig. 10.14 are the axial forces obtained from a three-dimensional computer analysis.

An equivalent column approach, as shown in the previous section, can be used to obtain approximate deflection values. In calculating the moment of inertia of the frame it is only necessary to include the contribution of equivalent flange columns on the windward and leeward sides of the tube.

### 10.1.5 Coupled shear walls

Frequently, vertical rows of doors or windows occur within a continuous shear wall. When coupled by beams at each floor, the wall is usually referred to as a coupled shear wall. Another system popular in 20- to 30-story apartment and hotel buildings is the cross-wall system. This system (Fig. 10.21) consists of a continuous one-way slab spanning between load-bearing reinforced concrete walls, which resist both horizontal and vertical loads. The shear walls are either staggered in plan or placed, in line with each other, as shown in Fig. 10.15. From the point of view of structural analysis, the behaviour of both walls is very similar with a high degree of interaction between the horizontal and the vertical elements. Take, for example, the rotation $\theta$ at the center of gravity of the shear wall, as shown in Fig. 10.16. In having to comply with the deformed shape of the wall, the floor system undergoes not only a rotation, but a corresponding vertical displacement at all locations except at the center of gravity of the wall.

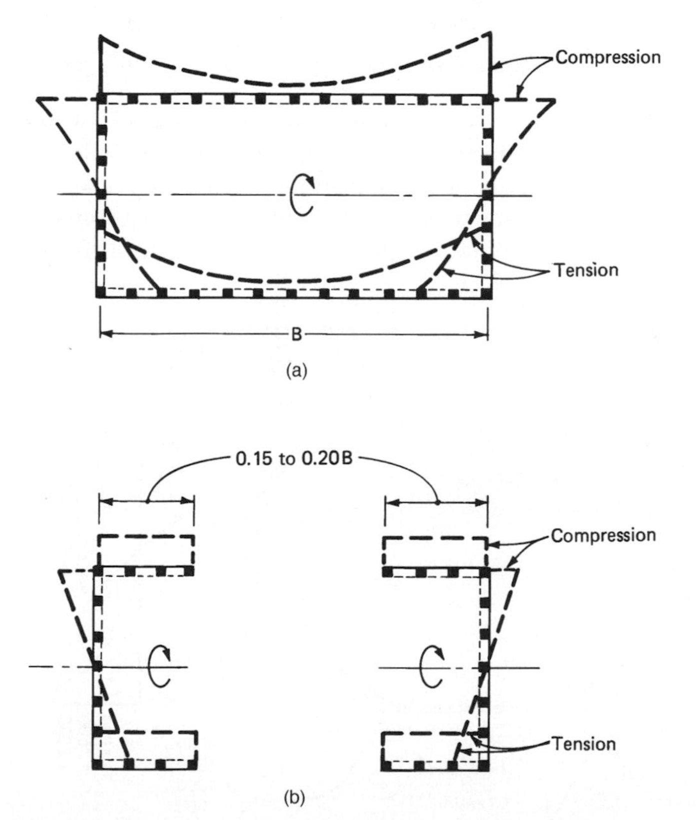


**Figure 10.12** Framed tube: (a) axial stress distribution with shear lag effects; (b) axial stress distribution in equivalent channels without shear lag effects.

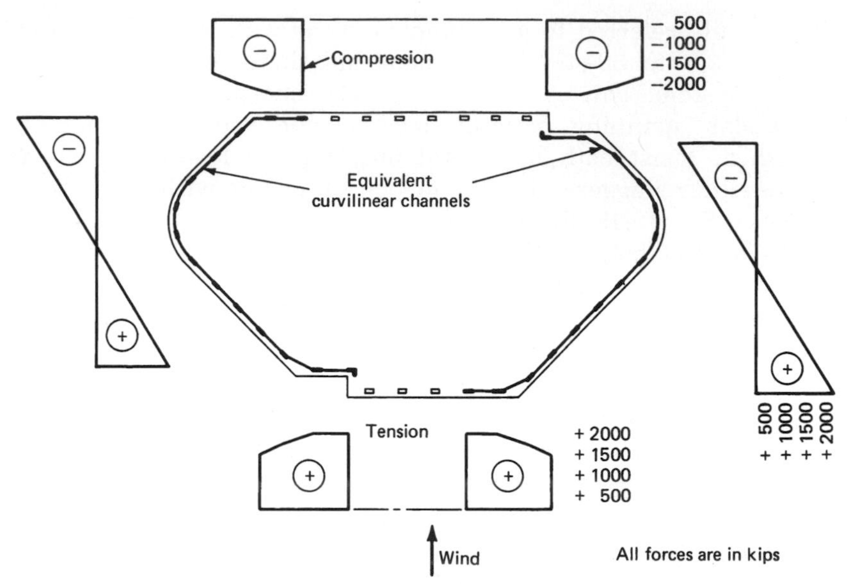


**Figure 10.13** Axial stresses in columns assuming two equivalent curvilinear channels.

**Continuous medium method.** Briefly, the analysis procedure is follows. The individual connecting beams of finite stiffness $I_b$ are replaced by an imaginary continuous connection or laminae. The equivalent stiffness of the laminae for a story height $h = I_b/h$ giving a stiffness

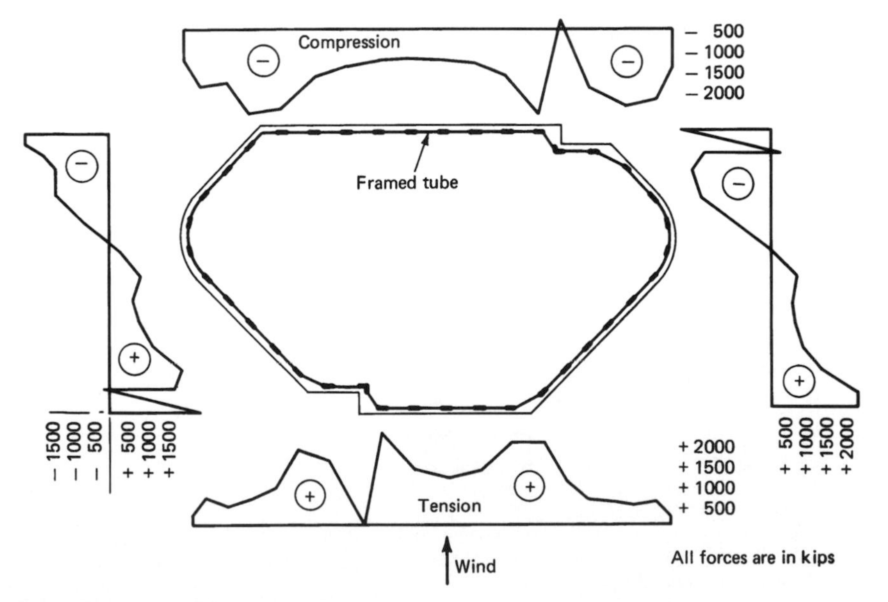


**Figure 10.14** Axial forces in columns from three-dimensional analysis of framed tube.

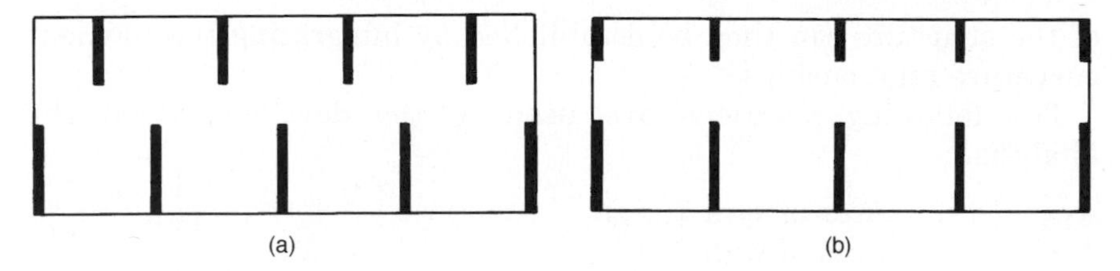


**Figure 10.15** Coupled shear walls: (a) staggered shear walls; (b) walls in line with each other.

of $I_b\,dx/h$ for a height $dx$. When the wall is subjected to horizontal loading, the walls deflect, inducing vertical shear forces in the laminae. The system is made statically determinate by introducing a cut along the center of beams which is assumed to lie on the points of contraflexure. The displacement at each wall is determined and, by considering the compatibility of deformation of the laminae, a second-order differential equation with the vertical shear force as a variable is established. The solution of this differential equation for the appropriate boundary conditions, of which fixed base is most common, leads to an equation for the integral shear $T$ from which the moments and axial loads in the walls can be established. Once the distribution of $T$ has been established, the shear force in the connecting beam at any level is obtained as the difference between the values of integral shear $T$ at levels $h/2$ above and below the beam. At any level $x$ the bending moment in each wall can be established by superposition of the moment due to external lateral loads and a counteracting moment due to eccentricity of integral shear force from the center of gravity of the wall. The deflected form

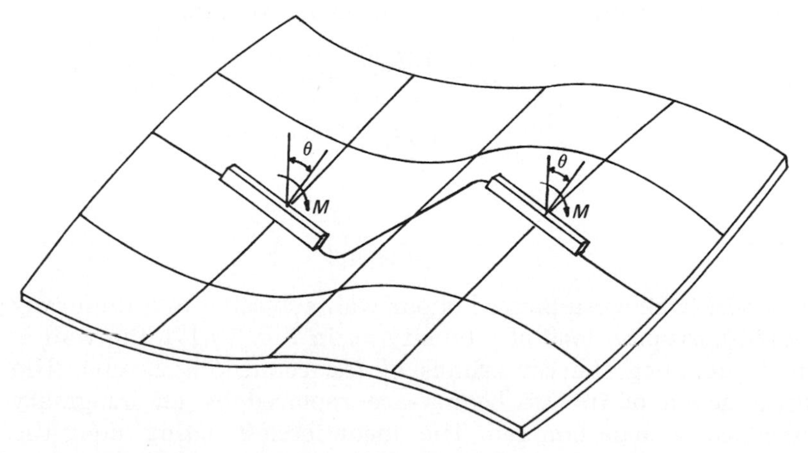


**Figure 10.16** Displacement compatibility between slab and walls.

of the structure can then be established by integrating the moment curvature relationships.

The following notations are used in the development of the analysis.

$A_1$ Area of wall 1
$A_2$ Area of wall 2
$I_2$ Moment of inertia of wall 1
$I_2$ Moment of inertia of wall 2
$H$ Total height of wall
$h$ Story height
$I'_b$ Moment of inertia of interconnecting beam
$l$ Distance between centroids of walls 1 and 2
$T$ Integral shear force or sum of the laminae shears above a given level
$q_n$ Shear force in the laminae
$w$ Uniformly distributed lateral load
$E$ Modulus of elasticity assumed constant for the system
$b$ Width of opening
$A_b$ Area of connecting beam
$I$ $I_1 + I_2$
$G$ Shear modulus
$I_b$ Moment of inertia of connecting beam reduced to take into account the effect of shear deformation in the beam. This is given by the relation:

$$I_b = \frac{I'_b}{1 + 2.4(d/b)^3(1 + \nu)}$$

$d$ depth of interconnecting beam
$\nu$ Poisson's ratio
$\alpha$, $\beta$, and $\mu$ Parameters given by the following relations

$$\alpha^2 = \frac{12I_b}{hb^3}\left[\frac{l^2}{I} + \frac{A}{A_1A_2}\right]$$

$$\beta = \frac{6wlI_b}{Ib^3h}$$

$$\mu = 1 + \frac{AI}{A_1A_2l^2}$$

Figure 10.17a shows a pierced shear wall subjected to a uniformly distributed horizontal load of intensity $w$. In Fig. 10.17b the wall is imagined cut along the centerlines of the connecting beams. The structural action of the tie beams are replaced by an imaginary equivalent continuous laminae. The shear force $T$ acting along the vertical axis of the wall is determined by considering the relative displacement of each wall.

Figure 10.17c shows an experimental setup for testing a coupled shear wall subjected to horizontal load.

Under lateral loads the two ends of beam at the cuts experience a vertical displacement consisting of contributions $\delta$, $\delta_2$, $\delta_3$, and $\delta_4$ as shown in Fig. 10.18.

The relative displacement $\delta_1$ due to bending of each wall element (Fig. 10.18a) is given by

$$\delta_1 = l\frac{dy}{dx}$$

The shear force $T = qh$ acting at each floor level at the center of connecting beams will cause a relative displacement $\delta_2$ (Fig. 10.18b)

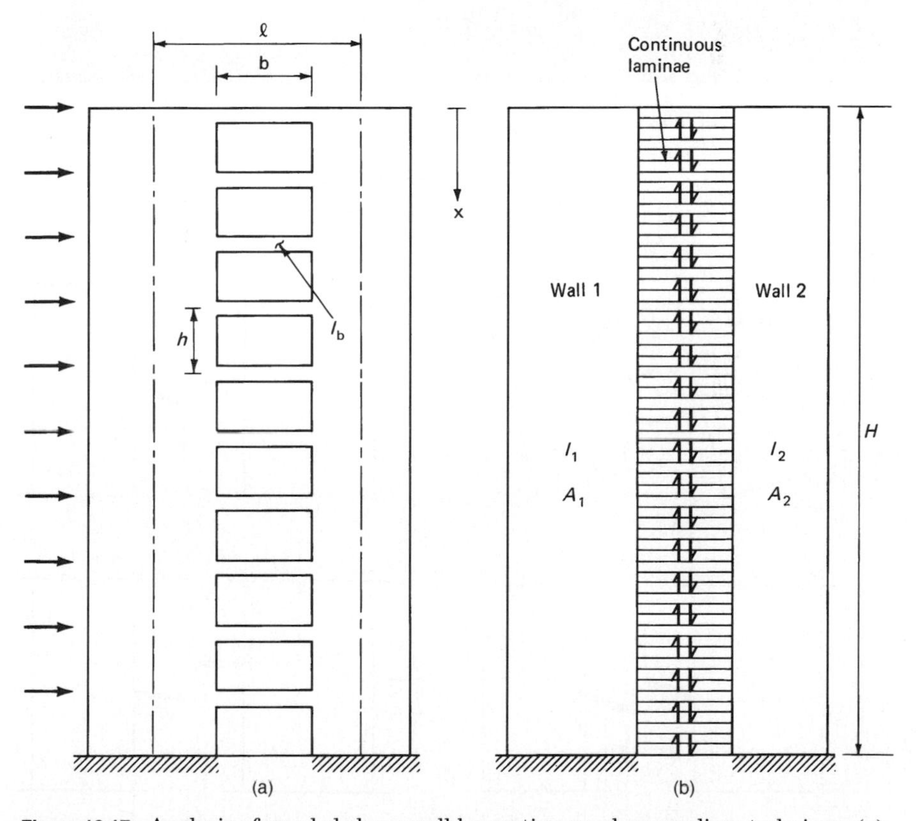


**Figure 10.17** Analysis of coupled shear wall by continuous shear medium technique: (a) interconnected shear walls; (b) representation of discrete beams by a continuous shear medium; (c) experimental set-up for a coupled shear wall model.

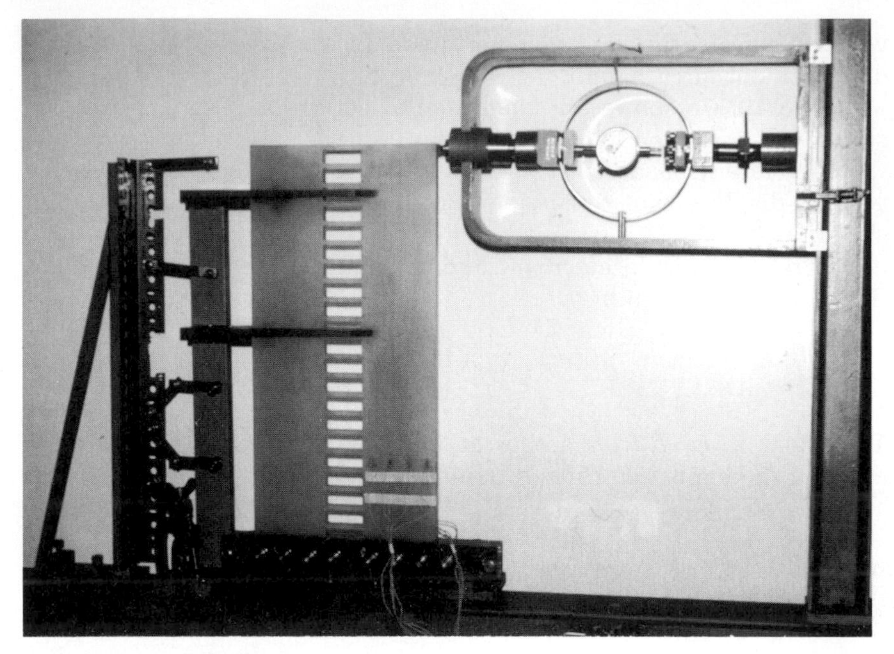

**Figure 10.17** *(Continued).*

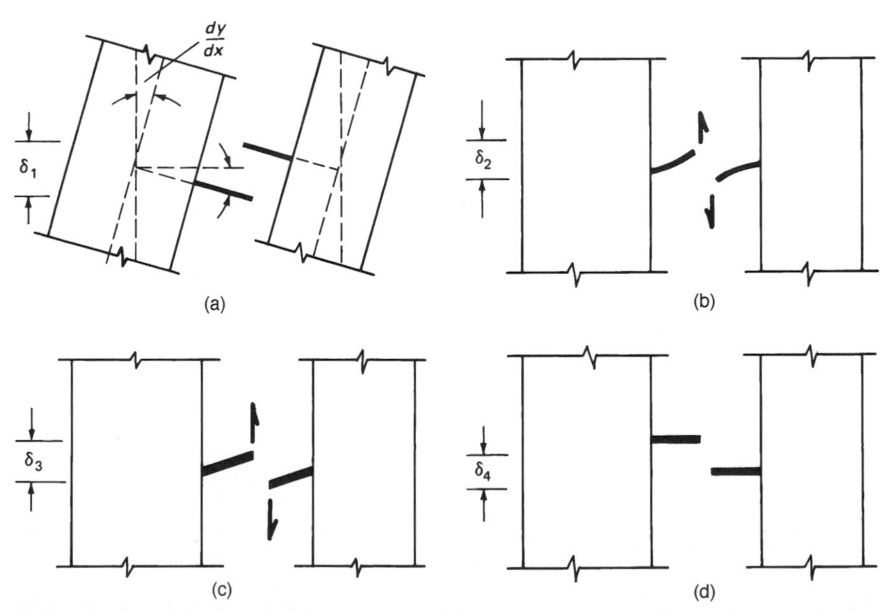


**Figure 10.18** Displacement of connecting beam due to: (a) wall rotation; (b) beam bending; (c) beam shear deflection; (d) axial tension and compression in wall segment.

due to bending of these beams. This is given by

$$\delta_2 = \frac{qb^3h}{12EI'_b}$$

The same shear force causes a shear deformation $\delta_3$ (Fig. 10.18c) in the beam given by

$$\delta_3 = \frac{qbh}{GA'_b}$$

For rectangular sections effective cross-sectional shear area $A'_b$ can be considered $= A_b/1.2$. Therefore,

$$\delta_3 = \frac{1.2qbh}{GA_b}$$

The displacement $\delta_4$ (Fig. 10.18d) is the relative displacement of the two wall elements due to the axial deformation of the walls caused by $T$ acting as a vertical load on the wall elements. This is determined as follows. The axial force in wall at any height is

$$T = \int_0^x q\,dx$$

and therefore

$$q = \frac{dT}{dx}$$

The strain in wall 1 due to axial loads is equal to

$$\frac{\text{stress}}{E} = \frac{\text{force}}{E \times \text{area}} = \frac{T}{EA_1}$$

Therefore total increase in length for wall 1 is

$$\int_x^H \frac{T}{A_1E}\,dx$$

Similarly, total axial shortening for wall 2 is

$$\int_x^H \frac{T}{A_2E}\,dx$$

Therefore, relative displacement is

$$\delta_4 = \frac{1}{E}\left(\frac{1}{A_1} + \frac{1}{A_2}\right)\int_x^H T\,dx$$

Since the two walls are connected, the compatibility condition stipulates that the relative displacements must vanish, i.e., $\delta_1 + \delta_2 + \delta_3 + \delta_4 = 0$. Substituting the above-derived expressions for $\delta_1$, $\delta_2$, $\delta_3$, and $\delta_4$, we get

$$l\frac{dy}{dx} + \frac{b^3hq}{12EI'_b} + \frac{qhb}{1.2GA_b} + \frac{1}{E}\left(\frac{1}{A_1} + \frac{1}{A_2}\right)\int_x^H T\,dx = 0$$

Substituting $q = dT/dx$ and $A = A_1 + A_2$, we get

$$\frac{dy}{dx} = -\frac{b^3h}{12lEI'_b}\frac{dT}{dx} - \frac{bh}{1.2GA_bl}\frac{dT}{dx} - \frac{A}{lEA_1A_2}\int_x^H T\,dx$$

Differentiating with respect to $x$, we get

$$EI\frac{d^2y}{dx^2} = -\frac{b^3hI}{12lI'_b}\frac{d^2T}{dx^2} - \frac{2bhI}{1.2lA_b}\frac{d^2T}{dx^2} - \frac{IAT}{lA_1A_2} \quad (10.13a)$$

The first two terms on the right-hand side of the above equation, which pertain to the bending and shear deflection of the beam, can be combined into a single term by reducing the moment of inertia of the beam to include the effect of shear deformation. The reduced moment of inertia $I_b$ is given by the relation

$$I_b = \frac{I'_b}{1 + 2.4(d/b)^3(1 + \nu)}$$

Using this relation Eq. (10.3b) reduces to

$$EI\frac{d^2y}{dx^2} = -\frac{b^3hI}{12I_bl}\frac{d^2T}{dx^2} - \frac{AI}{lA_1A_2}T \quad (10.13b)$$

The total applied moment $M_x$ at any point $x$ is given by

$$M_x = \frac{wx^2}{2}$$

Hence equation of statistical equilibrium is arrived at as follows. The applied moment less the moment due to $T$ acting at an eccentricity $l$ is

$$EI\frac{d^2y}{dx^2} = \frac{wx^2}{2} - Tl \tag{10.13c}$$

From Eqs (10.13b and c) the following governing second-order differential equation is derived:

$$\frac{d^2T}{dx^2} - \alpha^2 T = -\beta^2 x^2 \tag{10.13d}$$

where

$$\alpha^2 = \frac{12I_b}{hb^3}\left[\frac{l^2}{I} + \frac{A}{A_1A_2}\right]$$

$$\beta = \frac{6wlI_b}{hb^3I}$$

Ths solution of Eq. (10.13d) gives

$$T = c_1 \sinh \alpha x + c_2 \cosh \alpha x + \frac{\beta}{\alpha^2}\left(x^2 + \frac{2}{\alpha^2}\right) \tag{10.14a}$$

where $c_1$ and $c_2$ are constants of integration.

To eliminate these constants we introduce boundary conditions. Most commonly, the boundary condition at the base of shear walls where $x = H$ is to assume a rigid foundation which permits no deformation. The deformation at the cut ends of the laminae is zero and hence $q = 0$ or $dT/dx - 0$ at $x = H$. At top of walls where $x = 0$ the wall is free; therefore, the integral of shear force must vanish, i.e., $T = 0$ at $x = 0$. Substituting the boundary conditions in Eq. (10.14a) we get

$$T = \frac{2\beta}{\alpha^4}\left\{1 + \frac{\sinh \alpha - \alpha H}{\cosh \alpha H}\sinh \alpha x - \cosh \alpha x + \frac{\alpha^2 x^2}{2}\right\} \tag{10.14b}$$

Once the distribution force $T$ has been obtained, the shear force in the coupling beam may be determined as the difference in values of $T$ at levels $h/2$ above and below that level. The bending moment in the beam is obtained by the product of shear force and half the clear span of beam. Since the walls are assumed to deflect equally, the bending moments are proportional to their stiffness. Therefore, the bending moments in walls 1 and 2 are given by

$$M_1 = \left(\frac{wx^2}{2} - Tl\right)\frac{I_1}{T}$$

$$M_2 = \left(\frac{wx^2}{2} - Tl\right)\frac{I_2}{I}$$

The general expression for deflection $y$ at any point $x$ can be obtained by integrating Eq. (10.13c) twice and substituting appropriate boundary conditions. Assuming the foundation for the walls to be rigid, the boundary conditions are

$$y = 0 \quad \text{and} \quad \frac{dy}{dx} = 0 \quad \text{at } x = H$$

Although interstory drifts are important, most usually in preliminary analysis the maximum deflection at top is of prime interest. This is given by the following expression:

$$y_{\text{max}} = \frac{wH^4}{2EI}\left[0.25\left(1 - \frac{1}{\mu}\right) - \frac{2}{\mu}\left\{\frac{\alpha H \sinh \alpha H - \cosh \alpha H + 1}{(\alpha H)^4 \cosh \alpha H} - \frac{1}{2(\alpha H)^2}\right\}\right] \tag{10.15}$$

where

$$\mu = 1 + \frac{AI}{A_1 A_2 l^2}$$

To analyze a system of coupled shear walls by this method requires laborious calculations. Several researchers have proposed simplifications of this procedure. Of particular interest is the one proposed by Coull and Choudhury (ref. 56). In this method, the stress distribution in coupled shear walls is obtained as a combination of two distinct actions: (i) walls acting together as a single composite cantilever with the neutral axis located at the centroid of the two elements; and (ii) walls acting as independent cantilevers bending about their own neutral axes. Semigraphical methods are presented for rapidly evaluating maximum stresses and deflections in coupled shear walls subjected to a variety of loading cases. The interested reader is referred to ref. 56 for further details of this method.

## 10.2 Lumping Techniques

Analysis of tall buildings is a highly complex and indeterminate problem. Even with the availability of large-capacity high-speed computers, certain simplifying assumptions are necessary for all but the simplest of structures. One such assumption commonly made in rigid frame analysis is that the points of contraflexure occur at the midspan and midheight of beams and columns. This assumption makes it possible to lump a number of typical floors into a single floor. This is explained with reference to Fig. 10.19. The prototype frame, consisting of a 30-story, three-bay frame, is to be modeled as a lumped frame. Generally, in a high-rise building the floor-to-floor heights at the bottom and top few levels are different from typical floor-to-floor heights. Also as noted before, the points of contraflexure

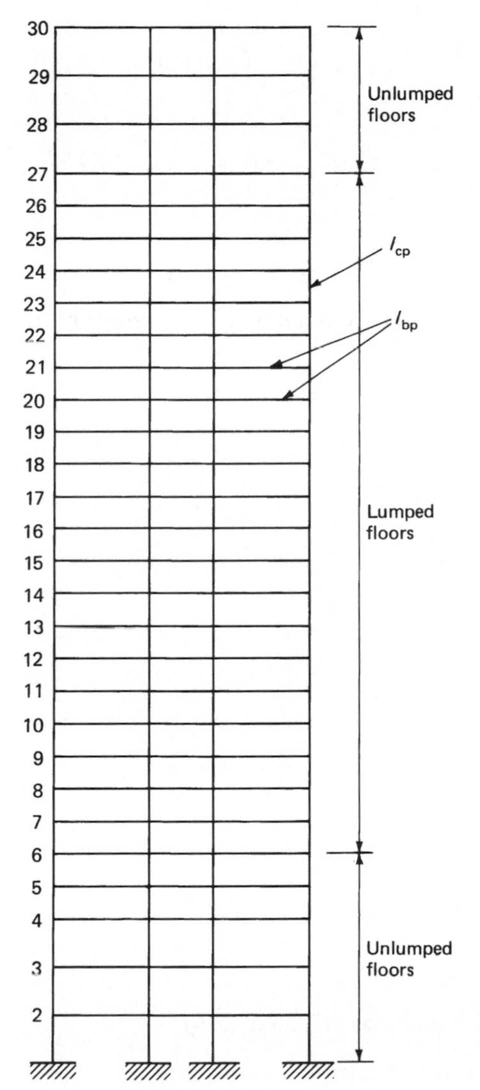


**Figure 10.19** Prototype of unlumped frame.

for beams and columns at these floors are not at their center. Therefore it is appropriate to limit lumping to typical floors only, as shown in Fig. 10.20. For purposes of illustration let us assume that two floors of the prototype are considered equivalent to a single floor of the lumped model.

The behavior of a rigid frame, as mentioned many times over, is a combination of the cantilever and shear racking modes. The cantilever behavior is a function of the location and axial stiffness of the columns only and does not depend on the arrangement of beams.

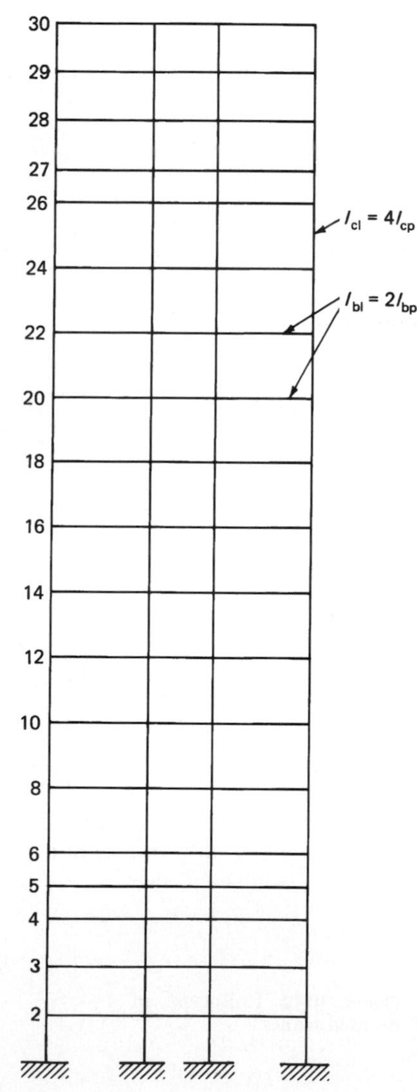


**Figure 10.20** Lumped model.

Therefore, to maintain the cantilever behavior intact between the prototype and the lumped model, the areas and location of the columns are kept the same in both unlumped and lumped models; the lumped model column occupies the equivalent position of the prototype column with the actual prototype column areas.

The equivalence of the raking component between the two models is maintained by keeping the ratio of column and girder stiffness factors the same between the two. Since two floors of the actual model are lumped into one floor, the moment of inertia and area of the girder in the lumped model should be twice their values in the prototype model. If $n$ floors are lumped into one floor, the corresponding properties will be $n$ times the prototype values. To keep the explanation simple, it is useful to introduce the following notations:

$I_{cp}$ = moment of inertia of clumn in the unlumped model (prototype)
$I_{cl}$ = moment of inertia of the column in the lumped model
$L$ = length of girder which is the same in both models
$h_{cp}$ = height of column in the unlumped model (prototype)
$h_{cl}$ = height of column in the lumped model

In the present example, two stories are lumped together. Therefore, ratio of $h_{cl}/h_{cp} = 2.0$. In general, this ratio can be considered as $n$, where $n$ is the lumping ratio. Equating stiffness ratios of column and beams between the two models gives

$$\frac{I_{cp}/h_{cp}}{I_{cp}/L} = \frac{I_{cl}/h_{cl}}{I_{bl}/L} \tag{10.16}$$

which simplifies to

$$I_{cl} = I_{cp}\left(\frac{h_{cl}}{h_{cp}}\right)\left(\frac{I_{bl}}{I_{bp}}\right) \tag{10.17}$$

Since in the example the ratio of the heights of prototype and model columns is 2.0 and the moment of inertia of the model beam is twice that of the prototype, we get

$$I_{cl} = I_{cp}(2)^2 \tag{10.18}$$

In the general case, the moment of inertia of the lumped model column works out to be $n^2$ times the prototype value. Lumping of nontypical floors can also be accomplished by assuming locations of point of contraflexure at, say, one-third the height for the lower stories and by using the principle of virtual work to equate the deflection properties. However, since the mathematical expressions get unwieldy, this procedure is not discussed here.

Discrepancies always exist in all but the simplest of structures between the unlumped and lumped models, especially at regions of abrupt change in stiffnesses and geometry. Although such deviations may be high locally, in general the overall behavior is kept unaltered.

## 10.3 Partial Computer Models

Consider a plane frame with an even number of bays subjected to lateral loads as shown in Fig. 10.21. The frame is symmetrical about column 4, and therefore computationally it is more efficient to analyze only one-half of the frame. It is only necessary to reduce the geometric properties of column 4 by 50 percent and to introduce appropriate kinematic boundary conditions at the centerline of the column. For the cantilever bending action of the frame due to lateral loads, the neutral axis can be considered to pass through the centerline of column 4. To reproduce this effect in the half model, it is

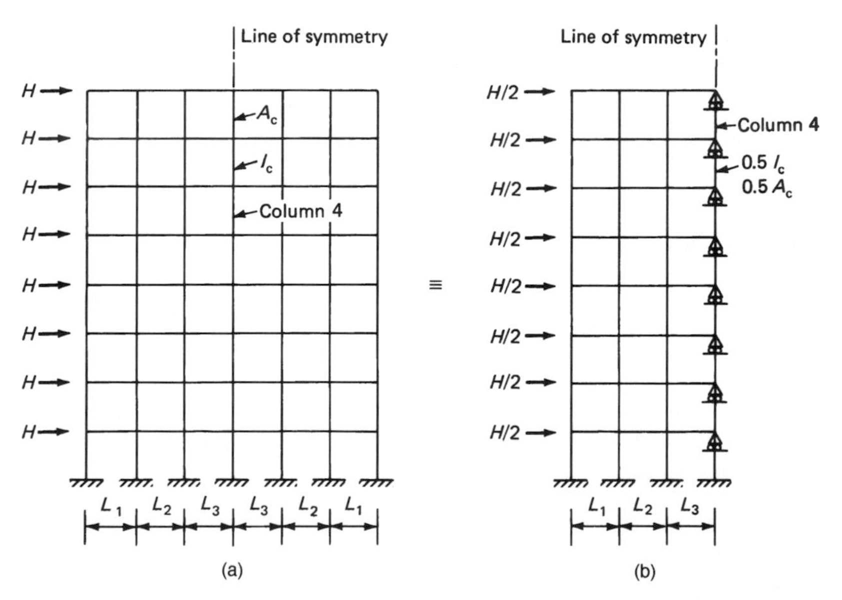


**Figure 10.21** Symmetrical frame with even numbers of bays: (a) full model; (b) partial model.

only necessary to restrain the axial deformation of column 4 at each floor. Since only one-half the frame is analyzed, the model is subjected to half the horizontal loads as shown schematically in Fig. 10.21b.

The procedure for analyzing a symmetrical structure with an odd number of bays is similar to the previous procedure as shown in Figs 10.22 and 10.23. The only difference is that the neutral axis for frame bending action passes through the center of beam spans. This affect is duplicated in the half model by introducing fictitious vertical supports at the midspan of beams. Only one-half of the lateral loads is applied to the model as before.

A symmetrical tube can be analyzed by considering only a quarter of the model. In addition to the aforementioned boundary conditions, it is necessary to assure that the three-dimensional model does not twist due to the application of horizontal loads.

A tube is a three-dimensional structure, and as such responds by bending about both its principle axes and rotation about a vertical axis. In analyzing a quarter or half model, it is necessary to restrain the transverse bending and rotation of the tube. The kinematic restraints that preclude transverse movement and rotation of the model are shown in Fig. 10.23.

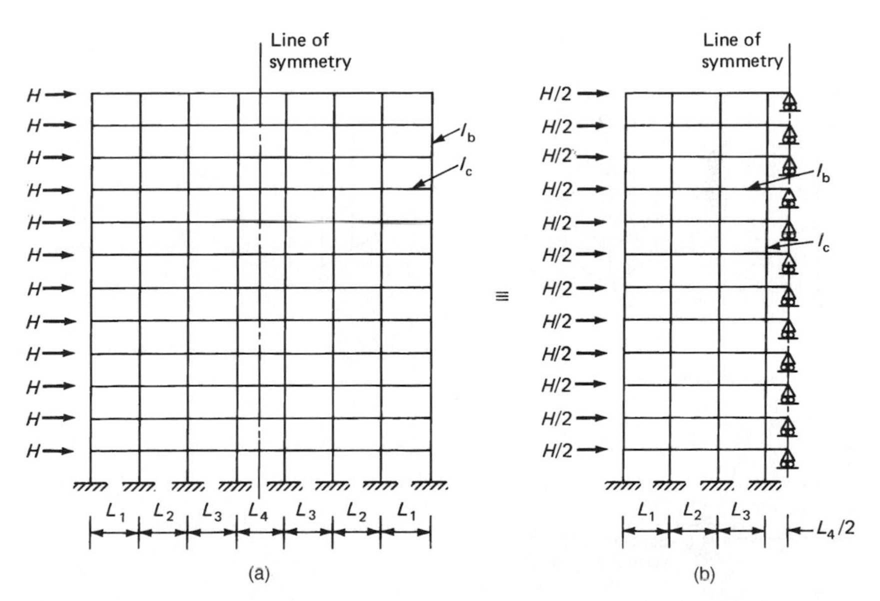


**Figure 10.22** symmetrical frame with odd number of bays: (a) full model; (b) partial model.

## 10.4 Torsion

### 10.4.1 Introduction

This section gives an overview of analysis of structural systems for torsion with particular emphasis on the torsion analysis of open-section cores. At first, we take a cursory look at the classical methods

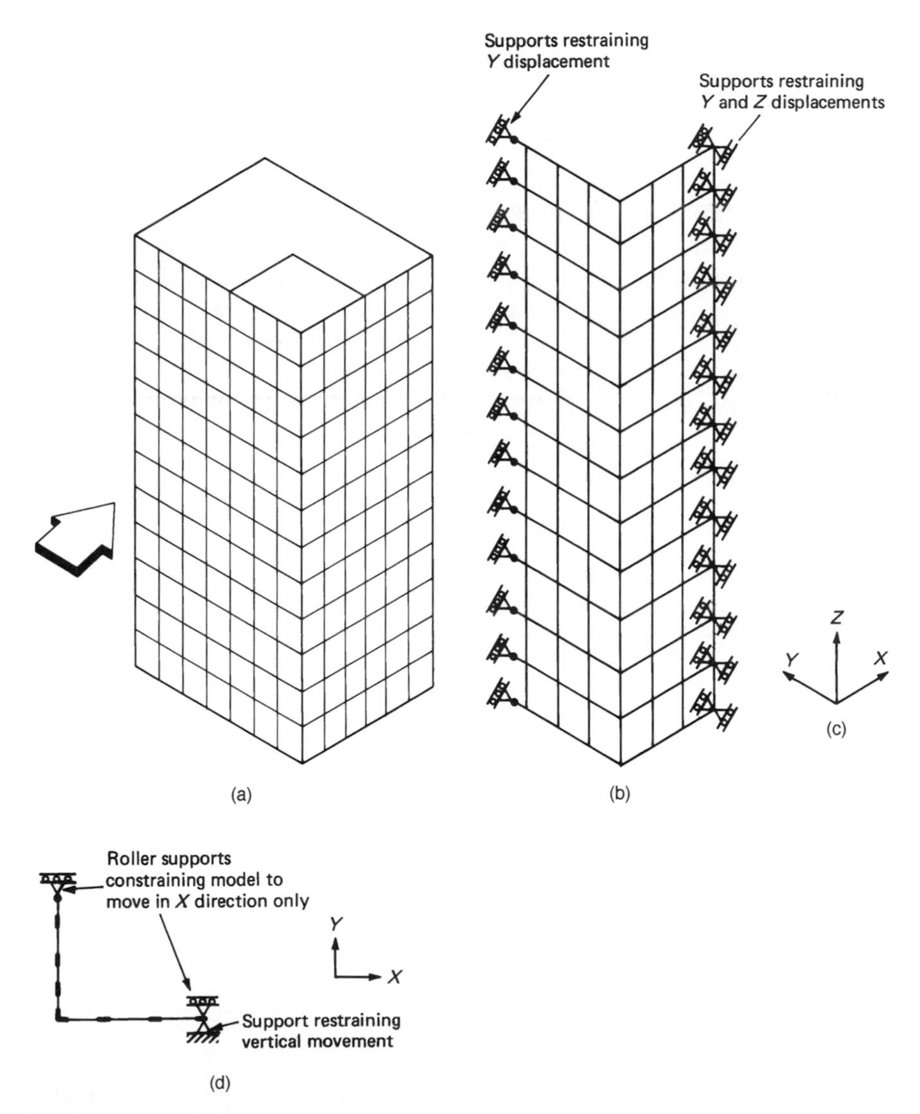


**Figure 10.23** Partial analytical model for framed tube: (a) full model; (b) partial model; (c) coordinate axes system; (d) plan of partial model.

**TABLE 10.2(a) Torsion Terminology**

| Uniform (St Venant) torsion | Warping torsion |
|---|---|
| • Torsional shear stress | • Shear center |
| • Twist | • Open section |
| • Polar moment of inertia | • Warping deformation |
| • Membrane analogy | • Sectorial coordinate |
| • Shear flow | • Warping moment of inertia |
| • Cellular sections | • Bimoment |
| | • Normal stress |
| | • Tangential stress |

of torsion of elements such as circular, non-circular, and cellular sections, and later on discuss warping torsion of structural systems consisting of open cores.

The terminology used in torsion analysis may be conveniently grouped under two headings: uniform or St Venant's torsion and warping torsion, often times referred to as constrained torsion or torsion-bending. The terms for uniform torsion are well established and given in most textbooks on structural mechanics. The purpose of recalling them here is to show how they relate to the warping theory.

The terms shown on the right-hand-side of Table 10.2 relating to warping torsion have, in the past, been given little attention. Consequently, designers are generally not at ease with neither the concepts of warping behavior nor with its methods of analysis. The aim here is to introduce the concept of warping without indulging in an abundance of mathematics and to show, by numerical examples, the importance of considering warping in practical cases.

**TABLE 10.2(b) Analogy Between Bending and Warping Torsion**

| Elementary bending theory | Warping theory |
|---|---|
| Plane sections remain plane | Profile warps |
| $I_x = \int_A y^2\, dA$ | $I_\omega = \int_A \omega^2\, dA$ |
| $\Delta_x$ | $\theta_z$ |
| $M_x = -EI_x \dfrac{d^2x}{dz^2}$ | $B = -EI_\omega \dfrac{d^2\theta}{dz^2}$ |
| $\sigma_x = \dfrac{MC}{I_x}$ | $\sigma_\omega = \dfrac{B\omega}{I_\omega}$ |
| $\tau_x = -\dfrac{VQ_x}{I_x t}$ | $\tau_\omega = -\dfrac{HQ_\omega}{I_\omega t}$ |

Terms such as sectorial coordinates, sectorial moment of inertia, bimoment, etc., will be introduced as and when the concepts are discussed instead of defining all of them at once at this stage.

Torsion, at an elemental level, occurs in various practical situations. One of the most common examples is a heavy curtain wall supported from a spandrel beam (Fig. 10.24a). This may not be much of a problem when the spandrel beam is part of the seismic or wind frame. However, when the building perimeter is not part of a lateral frame, the simple connections at the ends of spandrel do not offer adequate torsional restraint resulting in excessive rotations. Another example, with torsion in the reverse direction, is shown in Fig. 10.24b. In this example, taken from author's experience, light gage metal studs were welded to a steel wide flange spandrel before concrete topping was placed on the metal deck. Subsequent concreting of deck resulted in an inward rotation of the spandrel with a consequent misalignment of the studs. A stair-support beam (Fig. 10.24c) located between stair and mechanical shafts is another example because the slab which prevents torsion is absent.

Torsional effects on buildings as a whole are enhanced when the center of twist is eccentric from the center of gravity for inertial loading, or from the center of area for wind loading. Minimum

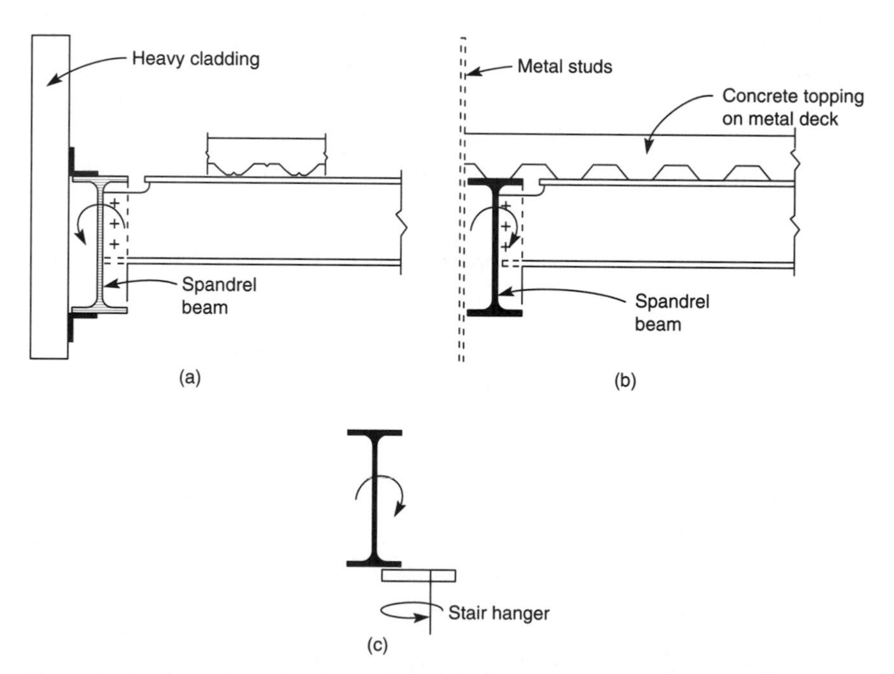


**Figure 10.24** Examples of torsion: (a) anti-clockwise rotation due to heavy cladding; (b) clockwise rotation due to weight of concrete; (c) beam rotation due to eccentric hanger.

eccentricities are prescribed by building codes to account for accidental seismic torsion. And, to reflect the observed torsional behavior of buildings in turbulent wind, some building codes such as the NBC and the Houston building code require that all buildings be designed for partial as well as full wind loading.

Consider the twisting of a circular shaft as shown in Fig. 10.25a. The twisting of the shaft does not produce any longitudinal stress, i.e., axial compression or tension, but only pure shear stresses. The shear stresses vary from zero at the center of the shaft, to maximum value at the perimeter. Because of the absence of axial deformation, a cylindrical layer peeled off of the shaft, changes its shape under the action of twist, from a rectangle to a parallelogram (Fig. 10.25b). The absence of longitudinal stresses indicates that the surfaces at the ends of the shafts remain plane. In otherwords, no warping will take place. The work done by the twisting moment is expended in developing shear stresses and only shear stresses as shown in Fig. 10.26.

Consider a rectangular section subjected to the action of a vertical load at the center of gravity of the section (Fig. 10.27). To find shear stress at any horizontal section, we introduce an imaginary horizontal cut at that section and obtain the shear stress by the relation $VQ/It$. By inspection, the resultant of the vertical shear stresses is at the center of gravity of the beam.

Next we take a look at the torsional behavior of a thin-walled section. The main reason why a thin-walled section must be given special consideration is, the shear stresses and strains in it are much larger than those in solid sections. An examination of distribution of

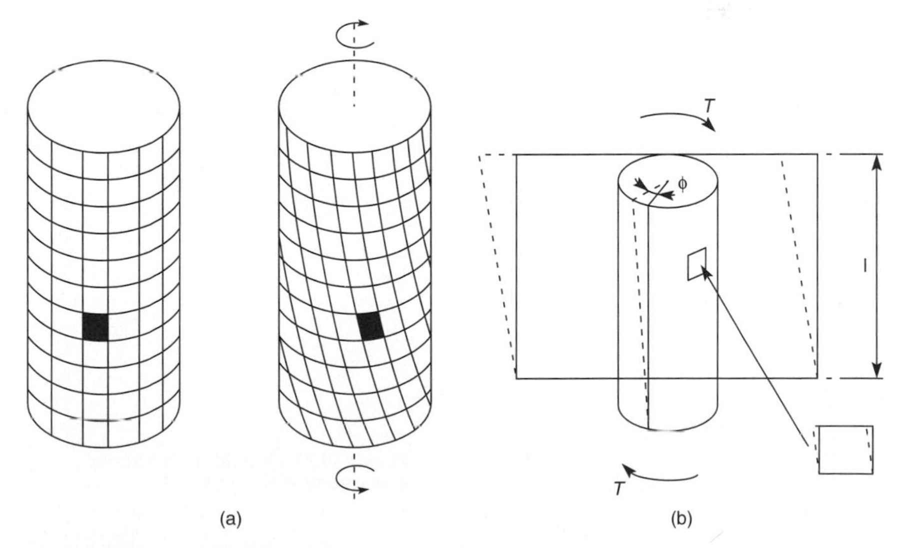


**Figure 10.25** (a, b) Twisting of circular shaft.

shear stresses through the cross-section shows that the shear stresses flow through the cross-section as if they were a fluid: hence the name, shear flow (Fig. 10.28).

Now consider a flanged section such as a C-shaped shear wall (Fig. 10.29). To find the shear flow, we abandon the idea of the horizontal cut. Instead, we consider a cut perpendicular to the profile and find the shear along the profile. The shear $R_2$ in web is in equilibrium with the vertical load $V$ and while the horizontal shears $R_1$ and $R_3$ in the webs result in no net horizontal load, the resulting moment requires offsetting of the vertical load to a location left of the web. The resultant forces from shear stresses

$$R_1 = R_3 = \int_0^a \tau t\, dx_1$$

$$= \frac{Vbta^2}{4I_y}$$

$$= \frac{3Va^2}{b^2\left(1 + \dfrac{6a}{b}\right)}$$

For vertical equilibrium $\qquad R_2 = V,$

For zero rotational effect, $\qquad R_1 b = R_2 e$

Hence
$$e = \frac{R_1 b}{R_2} = \frac{3a^2}{b\left(1 + \dfrac{6a}{b}\right)} \tag{10.19}$$

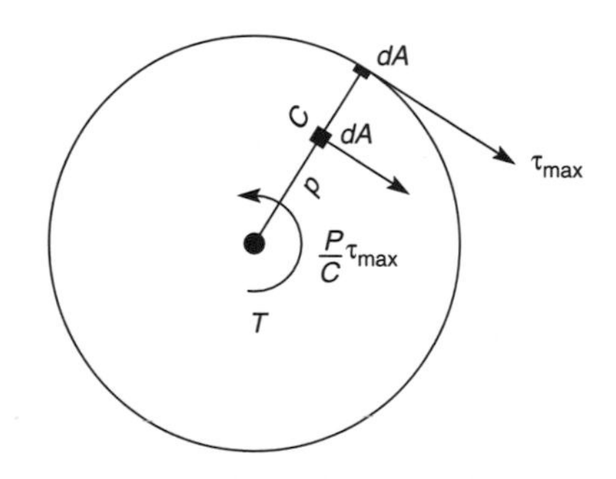


$$T = \int_A \frac{P}{C}\tau_{max}\, dA\; p$$


$$T = \tau_{max}\frac{J}{C}$$

$$J = \int_A p^2 dA$$


**Figure 10.26** Variation of shear stresses in circular shaft.

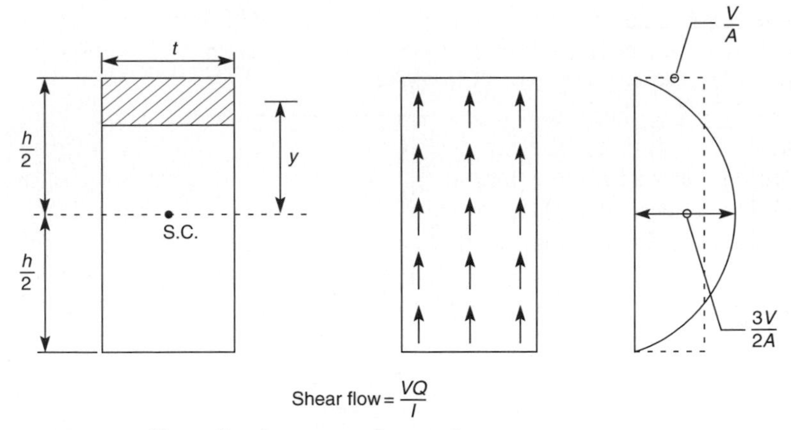


**Figure 10.27** Shear flow in rectangular section.

To find shear stresses in a cellular section, Fig. 10.30, a two-step approach is required because the problem is statically indeterminate. First, the section is rendered statically determinate by inserting a horizontal cut along the length of the section and the shear flow in the section is evaluated by the relation $VQ/I$. Next, the shear flow required to close the gap is evaluated. The final shear stress is evaluated by combining the two.

As an example, Fig. 10.31 shows schematically the final shear stresses in a hollow rectangular section. The section consists of webs of unequal thickness and is subjected to a vertical load at its shear center.

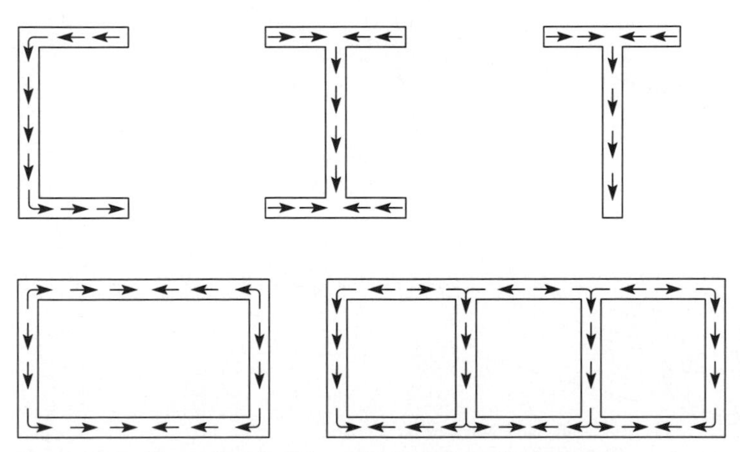

**Figure 10.28** Shear flow in thin walled sections: load at shear center.

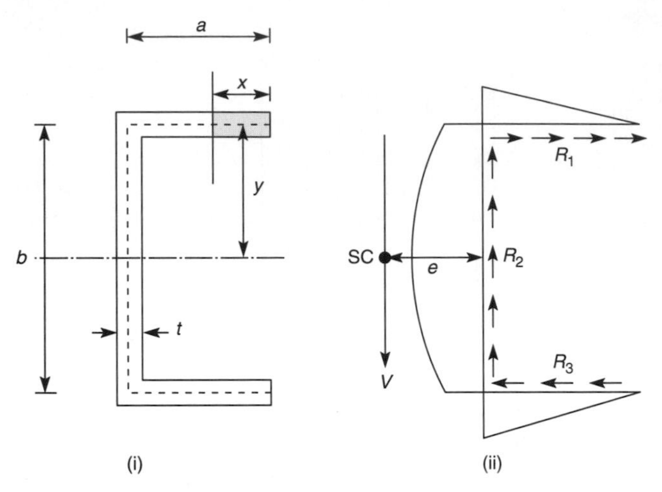


**Figure 10.29** Shear center in $C$-section.

If we have a multiple cellular-section, the procedure is similar to that for a single-cell section. The only difference is the problem is statically indeterminate to the $n$th degree where $n$ represents the number of cells. The example in Fig. 10.32 has two cells, hence, $n = 2$. Two cuts are made at A and B to render the section open. The shear flows $q_1$ and $q_2$ are evaluated by solving two simultaneous equations, and the final shear stress obtained by superposition.

The theory of torsion and related formulas discussed above are

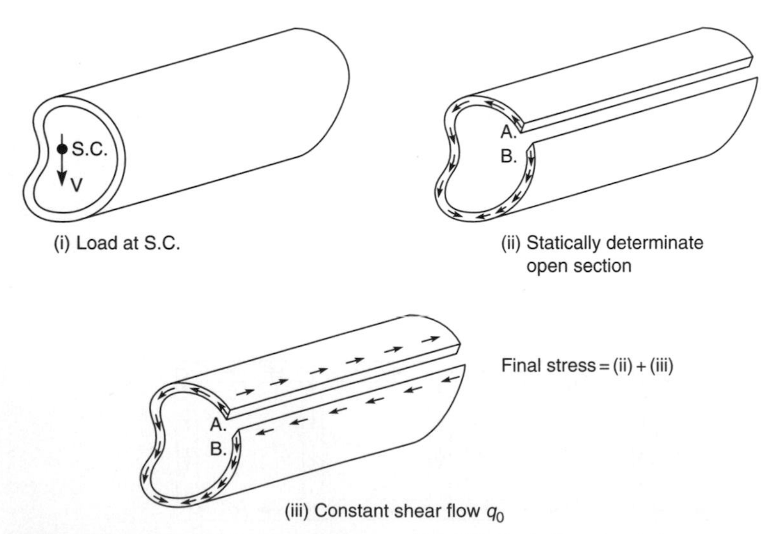


**Figure 10.30** Shear stress in hallow section: load at shear center.

commonly referred to as St Venant torsion formulas, and are valid for beams of circular cross-sections. His formula can be accepted for non-circular sections only when the additional stress caused by warping deformation is ignored. Consider, for example a rectangular section shown in Fig. 10.33a. The vertical fibers of the section are moving up and down from their initial position in space due to torsion. The top and bottom of the beam do not remain plane, but become warped. However, no additional stresses are induced because the warping deformations are not restrained either at the ends or at any section along its length.

Let us examine the case when the bottom of the beam is fixed. The warping of the bottom surface of the beam is restrained resulting in longitudinal strains and stresses. If we separate an imaginary elemental beam, as shown in Fig. 10.33b, it can be seen that the deflected shape is similar to that of a laterally loaded cantilever. It is obvious that bending stresses manifest at the fixed end of the beam.

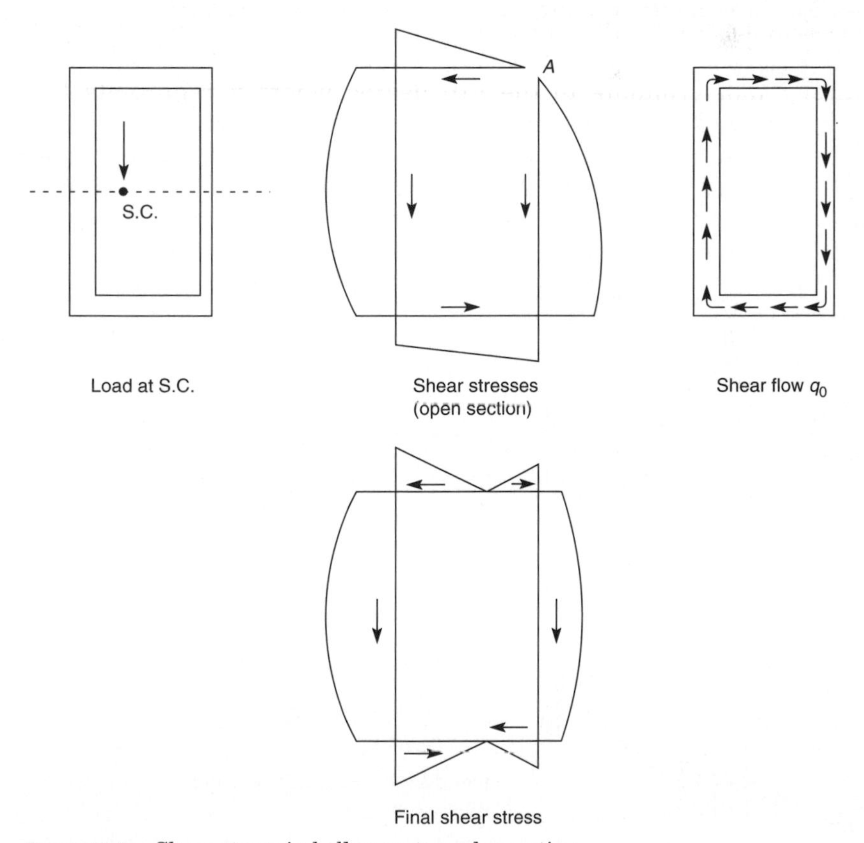


**Figure 10.31** Shear stress in hollow rectangular section.

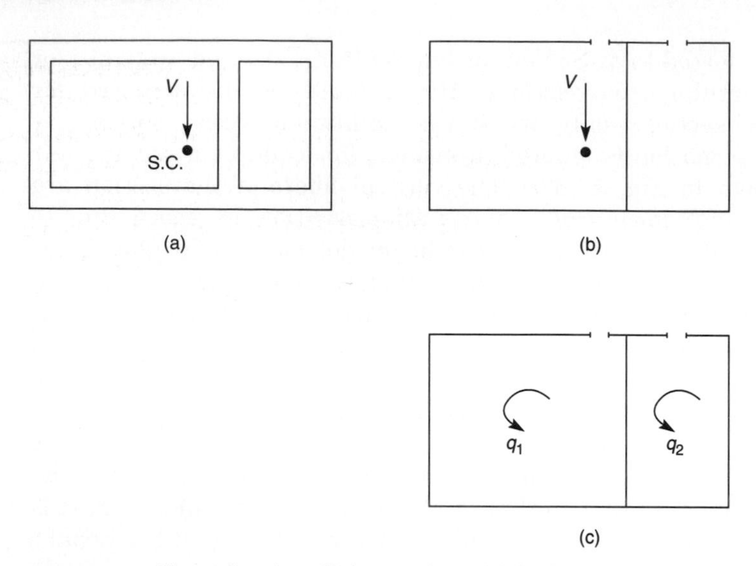


**Figure 10.32** Shear flow in cellular sections: (a) load at shear center; (b) section rendered open with two cuts; (c) shear flows required for compatibility; (d) final shear flow $= b + c$.

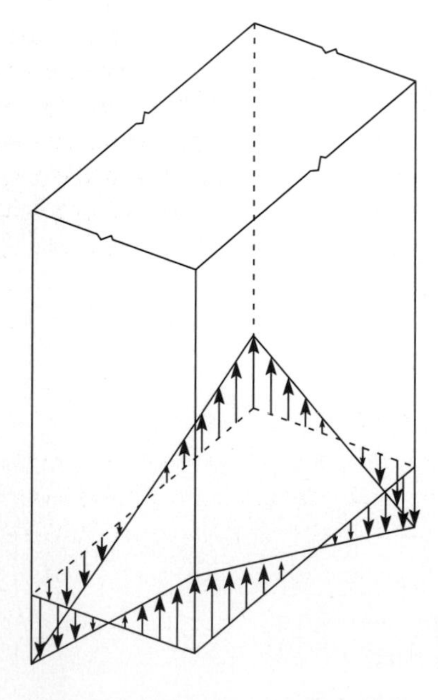

**Figure 10.33a** Warping of solid beams.

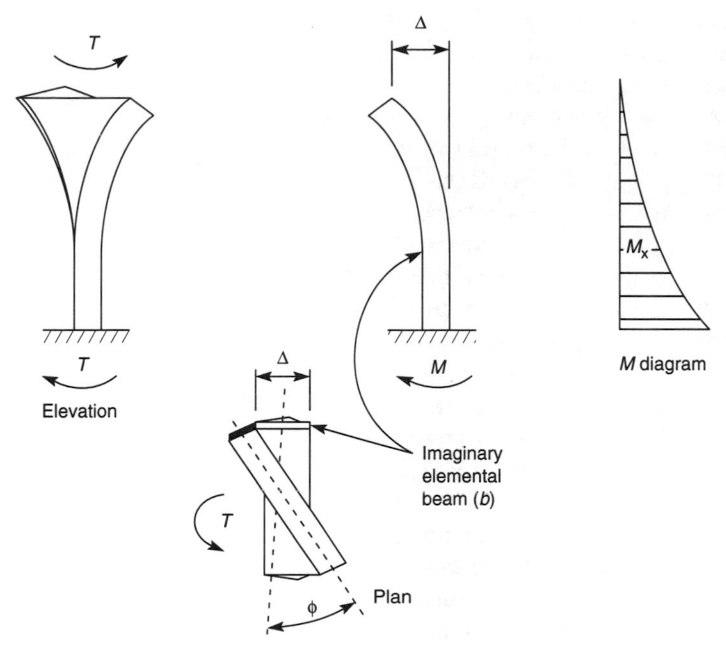


**Figure 10.33b** Thin rectangular beam: bending moment due to warping restraint.

The presence of bending stresses implies that part of the work done by the twisting moment is used up in bending the beam and only the remainder will develop shear stresses associated with the St Venant twist. Hence the resistance to external twisting moment is offered as the sum of pure torsion plus some additional torsion which causes bending of the section. This second part is called "warping torsion, non-uniform torsion or flexural twist."

For the thin rectangular beam shown in Fig. 10.33b, very little energy is expended to cause elemental bending about the weak axis. For such beams we can safely neglect the warping component of the twisting moment because the effect of constraining warping is usually restricted to the vicinity of the restraint. This phenomenon is valid, to a lesser extent for thin-walled closed sections. On the other hand, the effect of constraining warping of thin-walled open sections does not diminish rapidly and has a considerable influence on the stress distribution over a greater portion of the section.

Flexural twist causes a pair of moments. Such a pair of moments, called "bimoment" although is a mathematical function, can be visualized in most practical cases. For example, consider a two-span continuous beam supported by an interior column as shown in

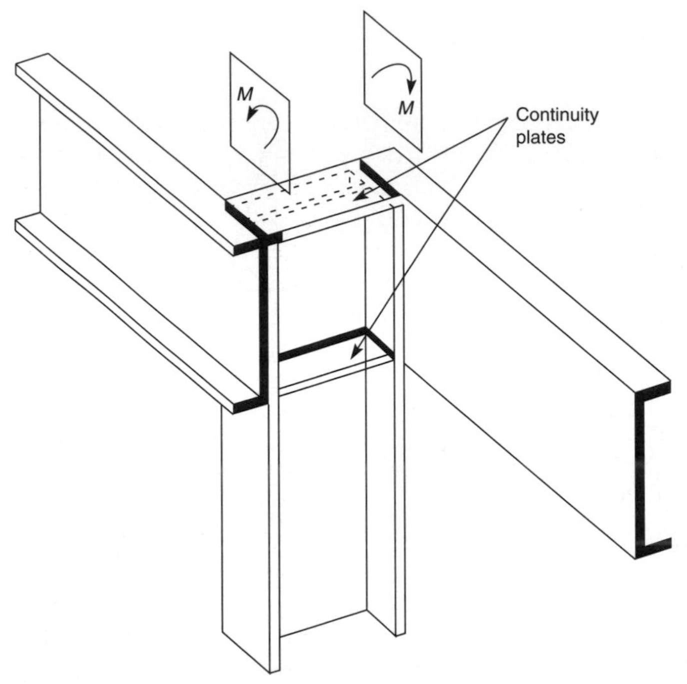


**Figure 10.34** Bimoment in wide flange column.

Fig. 10.34. Since the two channels frame into opposite flanges of the column, a bimoment is introduced at the top of the column.

Restrained warping behavior involves a set of so-called sectorial parameters each of which has counterpart in the theory of bending of

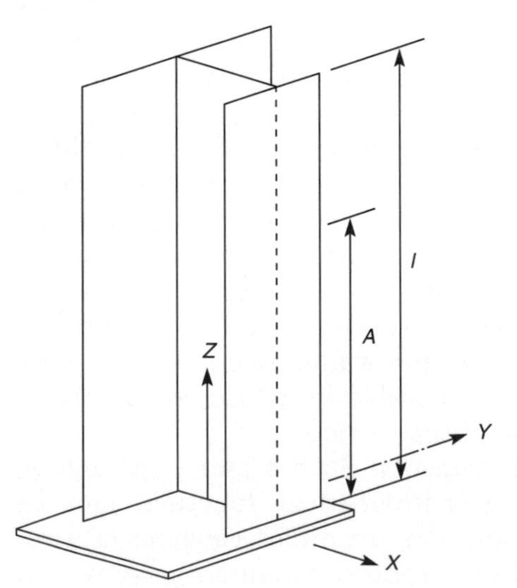


**Figure 10.35** *I*-section core.

beams. Since the sectorial parameters are generally unfamiliar to practising engineers it is perhaps appropriate to review them briefly here.

The sectorial coordinate, $\omega$, at a point on the profile of a warping core is the parmeter that expresses the axial response such as axial stress and strain at that point relative to other points around the core. The $\omega$ diagram can be constructed with the known location of the shear center and a point of zero warping deflection as an origin. The principal sectorial coordinate in warping theory is analogous to the distance $c$ of a point from the neutral axis of a section in bending. Just as the parameter $c$ is used in developing the well-known bending theory, the parameter $\omega$ is used in developing the warping theory.

A great advantage of the theory of bimoment is that internal strains and stresses can be found from formulas as simple as those used in the engineer's theory of bending. The bimoment and flexural twist can be used in a manner similar to bending moment and shearing forces. The procedure differs in that we use the sectorial coordinate $\omega$ instead of the linear coordinate $c$, to calculate the physical properties related to warping torsion.

To a beginner, the thin-walled beam theory with its differential equations presented later in this chapter, may look too academic for use in a down-to-earth practical design. In reality, once the idea of bimoment is assimilated, its use is not much more difficult than the use of bending moments or shear forces. It provides the engineer with a means for verifying the behavior tall shear wall buildings subjected to torsion.

### 10.4.2 Concept of warping behavior: I-section core

Perhaps the easiest model to describe the warping theory is an I-shaped shear wall with unequal flanges as shown in Figs 10.35 and 36. In most shear wall buildings the core around elevators and stairs consists of a series of I and C-shaped shear walls. Therefore, the model chosen has practical significance. Since torsion is the subject of discussion, the location of shear center of the cross-section is of importance. It's location is determined in a manner similar to the location of the center of gravity of the section. The only difference is that instead of dealing with the areas of the segments, we use their moments of inertia.

If an axial force is applied to the center of area, only axial deformations and stresses will occur. If, however, the axial force is applied through a point other than the c.g., bending about the transverse axes, and possibly warping, can also occur. Neglecting the web, the position of the center of gravity also called the center of area is given by

$$\bar{y}_1 = \frac{A_2 L}{A_1 + A_2} \quad \text{and} \quad \bar{y}_2 = \frac{A_1 L}{A_1 + A_2} \tag{10.20}$$

The location of c.g. is important in relation to vertical axial forces. The shear center $s$ on the other hand is important in relation to transverse forces. If a transverse force acts through $s$, the member will only bend. If, however, a transverse force acts elsewhere than through $s$, the member will twist and warp as well as bend. The shear center in this case is located along the $y$ axis by

$$y_1 = \frac{I_2}{I_1 + I_2} L \quad \text{and} \quad y_2 = \frac{I_1}{I_1 + I_2} \tag{10.21}$$

An inspection of Eqs (10.20) and (10.21) indicates that the center of the area and the shear center generally will not coincide unless the section is doubly symmetric, in which case both points lie at the center of symmetry.

When a torque $T$ is applied to the top of the member shown in Fig. 10.37a, it twists about the shear center axis causing the flanges to: (i) bend in opposite directions, about the $y$ axis; and (ii) twist about their vertical axes. The effect of the flange-bending is to cause the flange sections to rotate in opppsite directions about their $y$ axes so that initially plane sections through the member become nonplanar, or warped. Diagonally opposite corners 1 and 4, in Fig. 10.37a

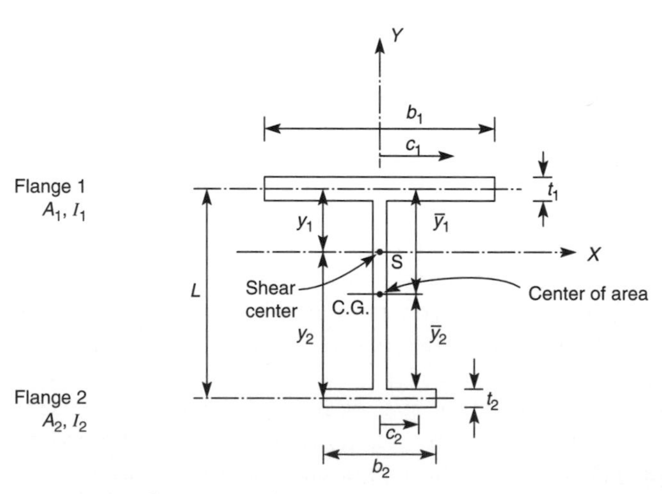


**Figure 10.36** Core properties.

displace downwards while 2 and 3 displace upward. At any level $z$ up the height of the core, the torque $T = T_z$ is resisted internally by a couple $T_\omega(z)$ resulting from the shears in the flanges and associated

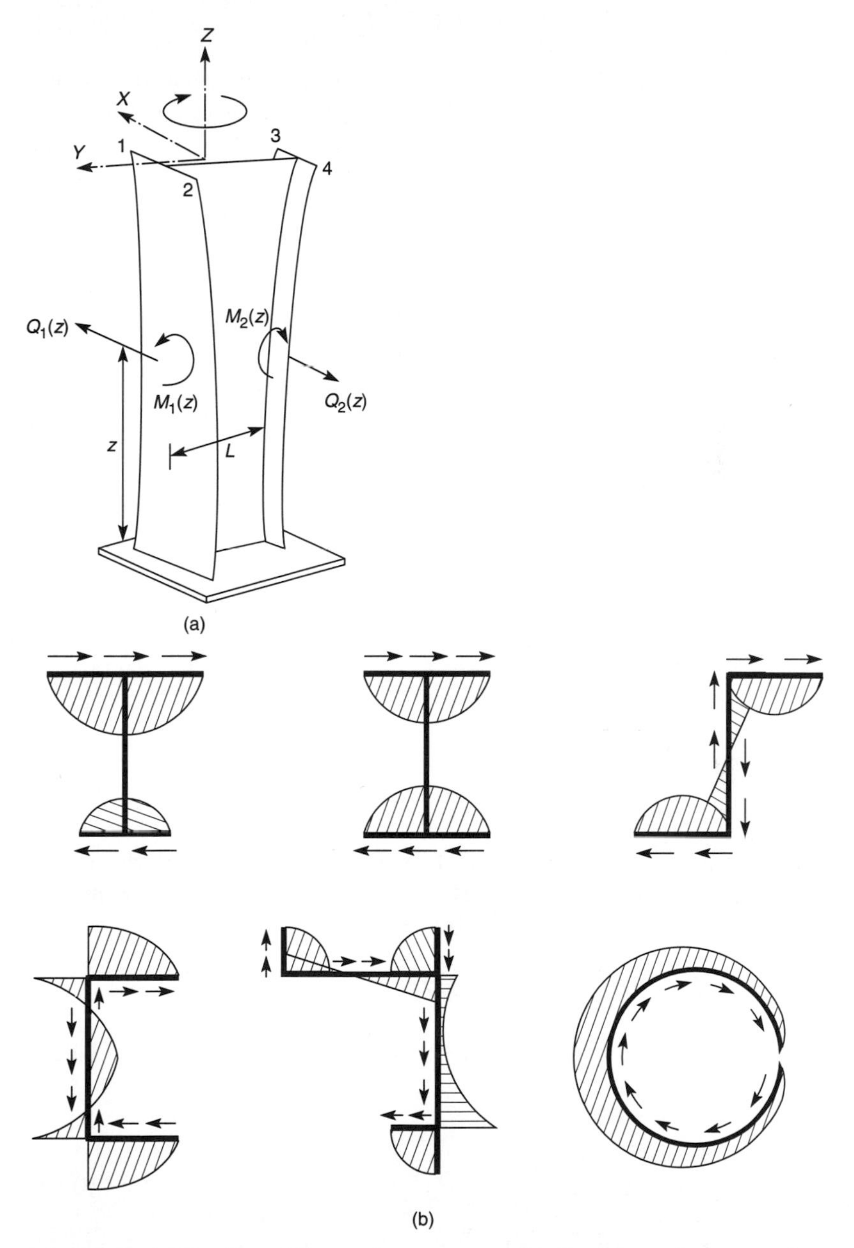


**Figure 10.37** (a) Bending of flanges due to torque: (b) shear forces due to warping torsion.

with their inplane bending, and a couple $T_v(z)$ resulting from shear stresses circulating within the section and associated with the twisting of the flanges and the web. Then

$$T_\omega(z) + T_v(z) = T_z \tag{10.22}$$

The rotation of the member about its shear center axis at a height $z$ from the base is $\theta_z$ hence the horizontal displacement of flange #1 at that level is

$$x_1(z) = y_1 \theta_{(z)} \tag{10.23}$$

and its derivatives are

$$\begin{aligned} \frac{dx_1}{dz}(z) &= y_1 \frac{d\theta}{dz}(z) \\ \frac{d^2x_1}{dz^2}(z) &= y_1 \frac{d^2\theta}{dz^2}(z) \\ \frac{d^3x_1}{dz^2}(z) &= y_1 \frac{d^3\theta}{dz^3}(z) \end{aligned} \tag{10.24}$$

Similar expressions may be written for flange #2.

The shear associated with the bending in flanges #1 and #2 can be expressed by

$$Q_1(z) = -EI_1 \frac{d^3x}{dz^3}(z) = -EI_1 y_1 \frac{d^3\theta}{dz^3}(z) \tag{10.25}$$

and

$$Q_2(z) = -EI_2 \frac{d^3x}{dz^3}(z) = -EI_2 y_2 \frac{d^3\theta}{dz^3}(z) \tag{10.26}$$

Multiplying the shear forces $Q_1$ and $Q_2$ by their respective distances from the shear center we obtain the torque resisted by these forces. Therefore, the torque contributed by these shear forces is

$$\begin{aligned} T_\omega(z) &= Q_1 y_2 + Q_2 y_2 \\ &= -(EI_1 y_1^2 + EI_2 y_2^2) \frac{d^3\theta}{dz^3}(z) \end{aligned} \tag{10.27}$$

or

$$T_\omega(z) = -EI_\omega \frac{d^3\theta}{dz^3}(z) \tag{10.28}$$

where

$$I_\omega = I_1 y_1^2 + I_2 y_2^2 \tag{10.29}$$

$I_\omega$ is a geometric property of the section similar to the moments of inertia $I_x$ and $I_y$, and is called the warping moment of inertia or warping constant. It expresses the capacity of the section to resist warping torsion. Neglecting the web, the torque resisted by the twisting of the section is

$$T_v(z) = GJ_1 \frac{d\theta}{dz}(z) \tag{10.30}$$

where $J_1$ is the torsion constant of the section given by

$$J_1 = \frac{b_1 t_1^3}{3} + \frac{b_2 t_2^3}{3} \tag{10.31}$$

in which $b_1$ and $b_2$ are the widths, and $t_1$ and $t_2$ are the thicknesses, of flanges #1, #2, respectively.

Summing the two internal torques, Eqs (10.28) and (10.30), and equating the sum to external torques as in Eq. (10.22)

$$-EI_\omega \frac{d^3\theta}{dz^3}(z) + GJ_1 \frac{d\theta}{dz}(z) = T \tag{10.32}$$

Equation (10.32) is the fundamental equation for restrained warping torsion. It simply states that an external torque applied to an open core is resisted by a combination of internal torque due to St Venant shear stresses and a couple due to equal and opposite shear forces in the flanges. The distribution of shear forces due to torsion in typical shear wall profiles is shown in Fig. 10.37b.

Considering the stresses in the flanges due to bending, the compressive stress in flange #1 at $c_1$ from the $y$ axis and $z$ from the base is

$$\sigma_1(c_1, z) = \frac{M_1(z)c_1}{I_1} \tag{10.33}$$

The tensile stress in flange #2 at $c_2$ from the $y$ axis is

$$\sigma_2(c_2, z) = \frac{M_2(z)c_2}{I_2} \tag{10.34}$$

Multiplying the right-hand side of Eq. (10.33), by the expression

$$\frac{L}{(y_1 + y_2)} \frac{y_1}{y_1}$$

which is equal to unity, and noting since $Q_1 = Q_2$, and the flange moments $M_1 = M_2 = M$, gives

$$\sigma_1(c_1, z) = \frac{M_{(z)} L y_1 c_1}{I_1 y_1^2 + I_1 y_1 y_2} \tag{10.35}$$

and since, from Eq. (10.21)

$$I_1 y_1 y_2 = I_2 y_2^2 \tag{10.36}$$

Substituting Eq. (10.36) in Eq (10.35)

$$\sigma_1(c_1, z) = \frac{M_{(z)} L y_1 c_1}{I_1 y_1^2 + I_2 y_2^2} \tag{10.37}$$

or

$$\sigma_1(c_1, z) = \frac{B_{(z)} \omega(c_1)}{I_\omega} \tag{10.38}$$

in which $B_{(z)} = M_{(z)} L$ is an action termed a bimoment, and $\omega_{(c)} = y_1 c_1$, is a coordinate termed the sectorial area, or principal sectorial coordinate, for that point of the section. In its simplest form, as considered here, a bimoment consists of a pair of equal and opposite couples acting in parallel planes. Its magnitude is the product of the couple and the perpendicular distance between the planes.

The above simple treatment of torsion of an I section explains the concept of warping and how the equations of torsion bending, also called restrained warping, are related to simple bending theory. The analogy is perhaps even more obvious by comparing the terms given in Table 10.2b.

### 10.4.3 Sectorial coordinate $\omega'$

The sectorial coordinate also called the warping function at a point on the profile of a warping core is the parameter that expresses the axial response (i.e., displacement, strain, and stress) at that point, relative to the response at other points around the section. Conceptually this is similar to the distance $c$ we use in bending formula $f = \dfrac{Mc}{I}$ to find the bending stress $f$ at a point in the cross section located at a distance $c$ from the neutral axis.

The warping coordinate is defined in relation to two points: a pole $0'$ at an arbitrary position in the plane of the section, and an origin $P_0$ at an arbitrary location on the profile of the section (Fig. 10.38a,b). The value of the sectorial coordinate at any point $P$ on the profile is then given by the area

$$\omega'_{(s)} = \int_0^s h \, ds \tag{10.39}$$

where $h$ is the perpendicular distance from the pole $0'$ to the tangent to the profile at $P$ and $s$ is the distance of $P$ along the profile $P_0$. It is evident that the warping function is an area and its magnitude depends on the location of the pole and of the point in the profile from which the integration is started.

In effect, the sectorial coordinate $\omega'$ is equal to twice the area swept out by the radius vector $0'P$ in moving from $P_0$ to $P$. The

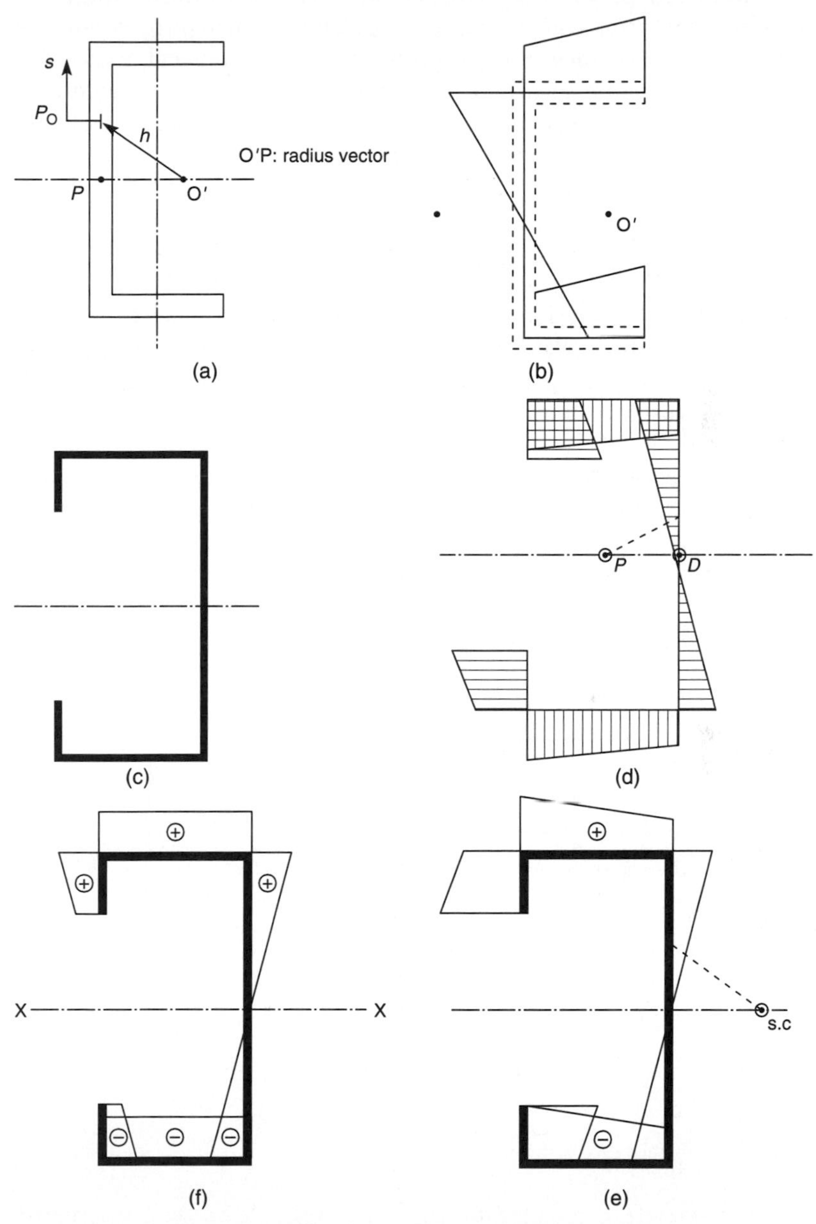


**Figure 10.38** (a) Section profile: (b) sectorial coordinate $\omega_s$ diagram; (c) singly symmetric core; (d) $\omega'$ diagram; (e) $Y$ coordinate diagram; (f) principal sectional coordinates; (g) sectorial coordinates for common profiles.

sectorial coordinate diagram (Fig. 10.38b) indicates the values of $\omega'$ around the profile. When the sectorial coordinates are related to the shear center as a pole, and to the origin of known zero warping displacement, Eq. (10.39) gives the principal sectorial coordinate values, $\omega$ and their plot is the principal sectorial coordinate diagram. The principal sectorial coordinate of a section in warping theory is analogous to the distance $c$ of a point from the neutral axis of a section in bending. The parameters $\omega$ and $c$ are used in developing the corresponding warping and bending stiffness properties of the sections, and in determining the axial displacements and stresses. Sectorial coordinates for common profiles are shown in Fig. 10.38g.

### 10.4.4 Shear center

The shear center of a section is a point in its plane through which a load transverse to the section must pass to avoid causing torque and twist. It is also the point to which warping properties of a section are related, in the way that bending properties of a section are related to the neutral axis.

Tall building cores are often singly or doubly symmetric in plan, which simplifies the location of the shear center. In doubly symmetric sections, the shear center lies at the center of symmetry while, in singly symmetric sections, it lies on the axis of symmetry.

The procedure for determining the location of shear for a singly symmetric section (Fig. 10.38c) is as follows.

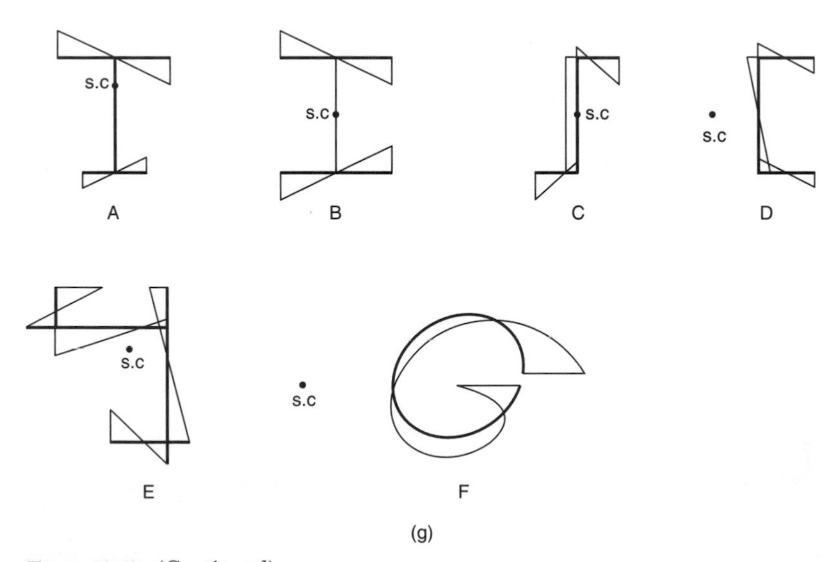


**Figure 10.38** *(Continued).*

**TABLE 10.3 Product Integral Tables**

| | Linear M diagrams | | | | Parabolic M diagrams | | |
|---|---|---|---|---|---|---|---|
| M / m | M, L | $M_0$, L | $M_1$, L | $M_0$, $M_1$, L | Origin, $M_1$, L/2 L/2 | Origin, $M_1$, L | Origin, $M_0$, $M_1$, L |
| m, L | $mML$ | $\frac{1}{2}m_0ML$ | $\frac{1}{2}mM_1L$ | $\frac{1}{2}mL(M_0+M_1)$ | $\frac{2}{3}mM_1L$ | $\frac{1}{3}mM_1L$ | $\frac{1}{3}mL(2M_0-M_1)$ |
| $m_0$, L | $\frac{1}{2}m_0ML$ | $\frac{1}{3}m_0M_0L$ | $\frac{1}{6}m_0M_1L$ | $\frac{1}{6}m_0L(2M_0+M_1)$ | $\frac{1}{3}m_0M_1L$ | $\frac{1}{12}m_0M_1L$ | $\frac{1}{12}m_0L(5M_0-M_1)$ |
| $m_1$, L | $\frac{1}{2}m_1ML$ | $\frac{1}{6}m_1M_0L$ | $\frac{1}{3}m_1M_1L$ | $\frac{1}{6}m_1L(2M_1+M_0)$ | $\frac{1}{3}m_1M_1L$ | $\frac{1}{4}m_1M_1L$ | $\frac{1}{4}m_1L(M_0-M_1)$ |
| $m_0$, $m_1$, L | $\frac{1}{2}ML(m_0+m_1)$ | $\frac{1}{6}M_0L(2m_0+m_1)$ | $\frac{1}{6}M_1L(m_0+2m_1)$ | $\frac{L}{6}[m_0(2M_0+M_1)$ $+m_1(2M_0+M_1)]$ | $\frac{1}{3}M_1L(m_0+m_1)$ | $\frac{1}{12}M_1L(m_0+3m_1)$ | $\frac{L}{12}[m_0(5M_0-M_1)$ $+3m_1(M_0-M_1)]$ |

1. Construct the $\omega_p$ diagram (Fig. 10.38d) by taking an arbitrary pole $P$ on the line of symmetry, an origin $D$ where the line of symmetry intersects the section, and by sweeping the ray $PD$ around the profile.
2. Using the $\omega_p$ and the $y$ diagrams for the section, Figs 10.38d and 10.34e, respectively, calculate the product of inertia of the $\omega_p$ diagram about the $X$ axis $I_{\omega_p x}$ using

$$I_{\omega_p x} = \int^{A} \omega_p y \, dA \tag{10.40}$$

   in which $dA = t\,ds$, the area of the segment of the profile of thickness $t$ and lengths $ds$. The integral in Eq. (10.40), may be evaluated simply by using the product integral table, Table 10.3.
3. Calculate $I_{xx}$, the second moment of area of the section about the axis of symmetry.
4. Finally, calculate the distance $\alpha_x$ of the shear center 0 from 0′, along the axis of symmetry, using

$$\alpha_x = \frac{I_{\omega_p x}}{I_{xx}} \tag{10.41}$$

### 10.4.5 Principal sectorial coordinate $\omega$ diagram

The $\omega$ diagram is related to the shear center 0 as its pole and a point of zero warping deflection as an origin. In a symmetrical section the intersection of the axis of symmetry with the profile at $D$ defines a point of antisymmetrical behavior, and hence of zero warping, therefore it may be used as the origin.

Values of $\omega$ can be found by sweeping the ray $OD$ around the profile and taking twice the values of the swept areas.

For the section of Fig. 10.38c, the principal sectorial coordinate diagram is shown in Fig. 10.38f.

### 10.4.6 Sectorial moment of inertia $I_\omega$

This geometric parameter expresses the warping torsional resistance of the core's sectional shape. It is analogous to the moment of inertia in bending.

The sectorial moment of inertia is derived from the principal sectorial coordinate distribution using the relation

$$I_\omega = \int_0^A \omega^2 \, dA \tag{10.42}$$

Note the similarity with the expression for the moment of inertia

$$I_{yy} = \int_0 x^2 \, dA$$

### 10.4.7 Shear torsion constant $J$

When a beam is twisted its fibres must undergo a shear strain to accommodate the twist. Associated with the strain are the shear stresses called St Venant shear stress. When an open-section core is subjected to torque (Fig. 10.38a) each wall twists developing St Venant shear stresses within the thickness of the wall. The stresses are distributed linearly across the thickness of the wall, acting in opposite directions on opposite sides of the wall's middle line. As the effective lever arm of these stresses is equal to only two-thirds of the wall thickness, the torsional resistance of these stresses is low. The torsion constant for this plate twisting action is

$$J = \tfrac{1}{3}k \sum^{n} bt^3 \tag{10.43}$$

in which $b$ is the width and $t$ the thickness of a wall. The summation includes the $n$ walls that comprise the section. The plate twisting rigidity of an open section core is given by $GJ$.

$k$ is a factor which makes allowance for small fillets within the cross section. If there are no fillets its value is equal to 1.00.

### 10.4.8 Calculation of sectorial properties: worked example

Consider again the shear core with unequal flanges as shown in Fig. 10.39. It is required to determine for the core,

1. The location of the shear center.
2. The principal sectorial coordinate, $\omega_s$ diagram.
3. The sectorial moment of inertia $I_\omega$.
4. The St Venant torsion constant $J$.

**1. Location of shear center.** The axis of symmetry of the section is $OY$, therefore the shear center lies on the $OY$ axis. We select an arbitrary pole $P$ at the junction of the web and the upper flange of

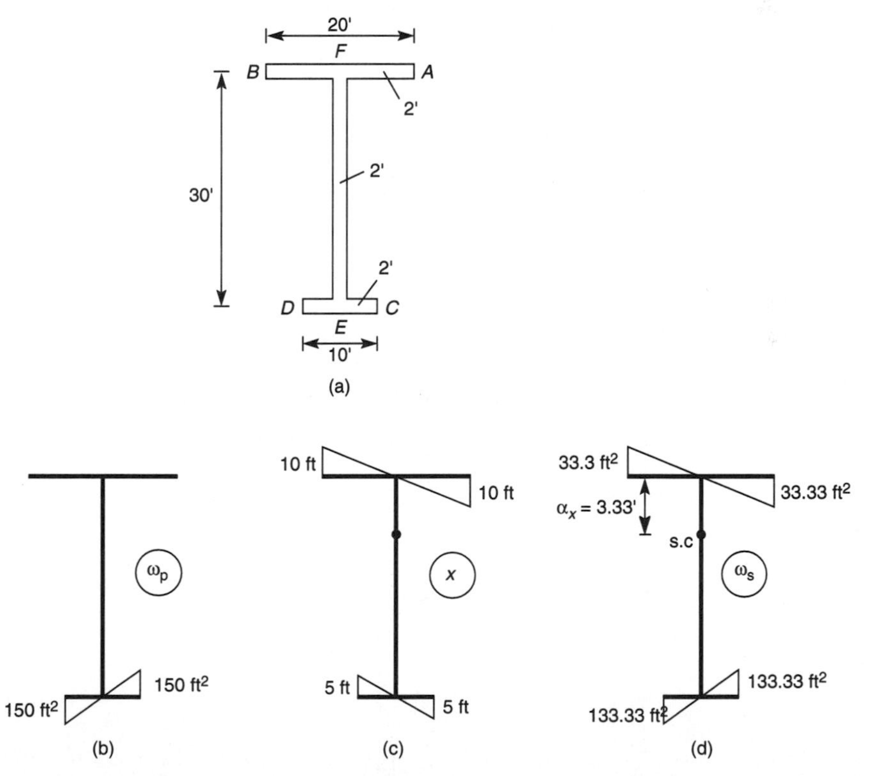


**Figure 10.39** Calculation of sectorial properties: (a) cross-section; (b) $\omega_p$ diagram; (c) $x$-coordinate diagram; (d) principal sectorial coordinate $\omega_s$ diagram.

the core. The $\omega_p$ diagram is constructed as shown in Fig. 10.39 by taking an arbitrary point on the web as the sectorial origin. The sectorial areas for the section of the upper flange and the web are equal to zero while they are distributed skew symmetrically for the lower flange.

Using the $\omega_p$ and the $Y$ coordinate diagrams, Fig. 10.39b and 10.39c, we calculate the integral $\omega_p\,dA$ by using the product integrals given in Table 10.3.

A summary of the calculations is given in the following Table 10.4.

For the whole section, $I_{\omega p} = 2500 \times 2 = 5000\ \text{ft}^5$. The moment of inertia of the section about $y$ axis

$$I_{yy} = \tfrac{1}{12}(2 \times 10^3 + 2 \times 20^3) = 1500\ \text{ft}^4$$

From Eq. (10.41), the distance of the shear center from the center of web is

$$\alpha_x = \frac{I_{\omega_p x}}{I_{yy}} = \frac{5000\ \text{ft}^5}{1500\ \text{ft}^4} = 3.33\ \text{ft}$$

**2. Principal sectorial coordinate diagram.** This is constructed by using the shear center *s.c.* as the pole and sweeping the ray from the middle of the web, around the profile (Fig. 10.39d).

**3. Sectorial moment of inertia $I_\omega$.** From Eq. (10.50)

$$I_\omega = \int \omega^2\,dA = \int^S \omega^2\,t\,ds$$

**TABLE 10.4 Calculations for Integral $\omega_p\,dA$**

| Segment | $\omega_p$ | $x$ | $\int_0^s \omega_p x\,t\,ds$ |
|---|---|---|---|
| *DE* | 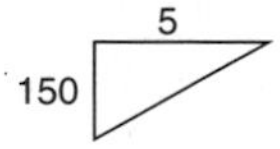  | 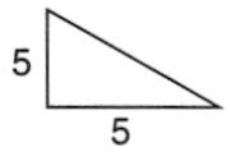  | $\frac{1}{3} \times 5 \times 150 \times 5 \times 2 = 2500\ \text{ft}^5$ |
| *EC* | 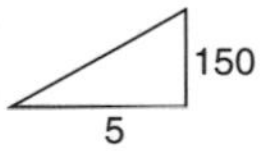  | 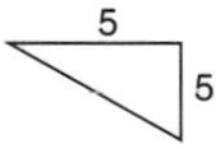  | $\frac{1}{3} \times 5 \times 150 \times 5 \times 2 = 2500\ \text{ft}^5$ |
| *AF* | 0 | 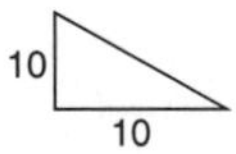  | 0 |
| *BF* | 0 | 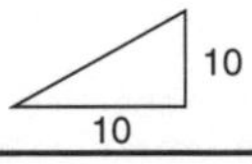  | 0 |

**TABLE 10.5 Calculations for Sectorial Moment of Inertia**

| Segment | Variation of $\omega$ | $\int \omega^2 t\, ds$ |
|---|---|---|
| $DE$ | 5; 133.33 | $\frac{1}{3} \times 5 \times 133.33^2 \times 2 = 59{,}260 \text{ ft}^6$ |
| $EC$ | 133.33; 5 | $\frac{1}{3} \times 5 \times 133.33^2 \times 2 = 59{,}260 \text{ ft}^6$ |
| $BF$ | 33.33; 10 | $\frac{1}{3} \times 10 \times 33.33^2 \times 2 = 7406 \text{ ft}^6$ |
| $AF$ | 10; 33.33 | $\frac{1}{3} \times 10 \times 33.33^2 \times 2 = 7406 \text{ ft}^6$ |

$\therefore I_\omega$ for the whole section $59260 \times 2 + 7406 \times 2 = 133{,}332 \text{ ft}^6$

Using the $\omega$ diagram (Fig. 10.39) and the product integral table, Table 10.3, the calculations for evaluating $I_\omega$ are as shown in the following Table 10.5.

$\therefore I_\omega$ for the whole section $59260 \times 2 + 7406 \times 2 = 133{,}332 \text{ ft}^6$

**4. Torsion constant *J*.** For the $I$-section core, using Eq. (10.43)

$$J = \frac{1}{3} \sum^{n} bt^3 = \frac{1}{3} \times 2^3(20 + 10 + 30) = 160 \text{ ft}^4$$

### 10.4.9 General theory of warping torsion

Before derivation of general warping torsion equations, it is instructive to consider qualitatively the difference between the behavior of thin-walled open sections and solid sections. A major difference lies in the manner in which the stresses attenuate along their length. Consider a square cantilever column loaded at top corner by a vertical load $P$ as shown in Fig. 10.40a. The load can be replaced by four sets of loads acting at each corner, which together constitute a system of loads statically equivalent to the applied force $P$. The first set represents axial loading, the second and third sets represent bending about the $x$ and $y$ axes. The resulting axial and bending stresses can be computed by the usual engineer's theory of bending, which assumes that Bernoulli's hypothesis is valid. In the last loading case, the cross sections do not remain plane because the two pairs of loads on opposite faces of the column tend to twist the cross section in opposing directions. This equal and opposite twisting results in warping of the cross-section. The last set of loads is, however, statically equivalent to zero and can be ignored by invoking St Venant's principle, which states that the perturbations imposed on a structure by a set of self-equilibrating system of forces affect the structure locally and will not appreciably affect parts of the structure away from the immediate region of applied forces. This statement simply means that the effect of self-equilibrating system of forces can be neglected in the analysis. The stresses caused by these forces attenuate rapidly toward zero and just about vanish at a distance equal to the characteristic dimension of the cross section. The stresses due to the self-equilibrating system of forces can be ignored throughout the whole length of the cantilever except at the very top region.

Now consider an I-shaped shear wall as shown in Fig. 10.41 which has the same overall dimensions as the column with the exception that it is composed of thin plates of thickness $t$. The first three sets of loads result in stress distributions which can be obtained as before by using the Bernoulli hypothesis. Although the fourth loading is self-equilibrating as before, its effect is far from local. The flanges, which are bending in opposite directions, do so as though they were independent of each other. The web acts as a decoupler separating the self-equilibrating load into two subsets one in each flange. Each subset is not self-equilibrating and causes bending in each flange. The bending action of the flanges can be thought of as being brought about by equal and opposite horizontal forces parallel to the flanges. The compatibility condition between the web and flanges results in a

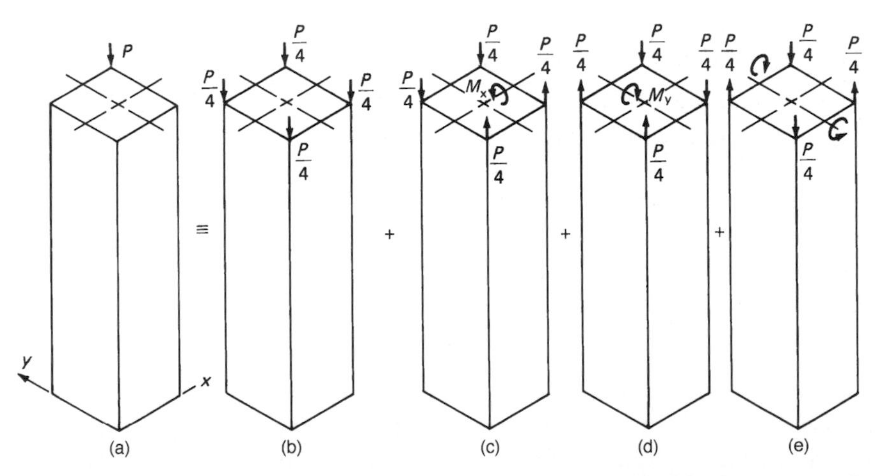


**Figure 10.40** Cantilever column of solid cross section: (a) vertical load at corner; (b) symmetrical axial loading; (c) bending about $x$-axis; (d) bending about $y$-axis; (e) self-equilibrating loading producing bimoment. Note: the resulting warping stresses are negligible.

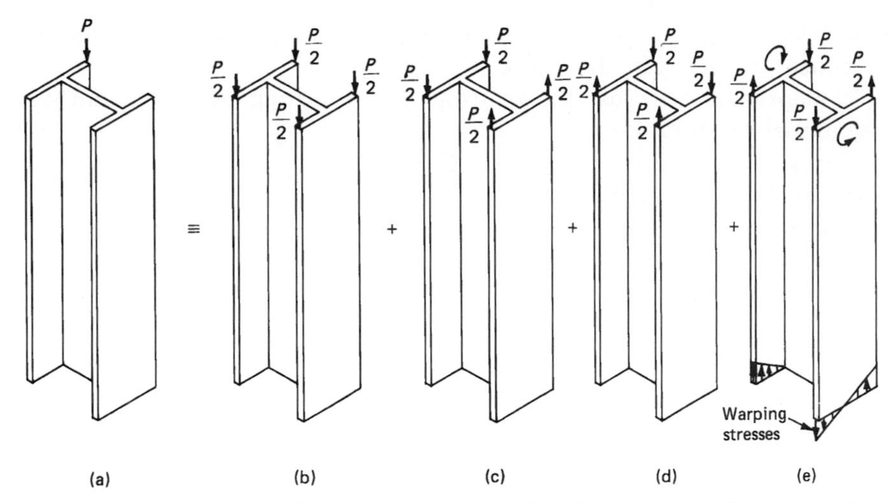


**Figure 10.41** $I$-shaped cantilever beam: (a) vertical load at a corner; (b) symmetrical axial loading; (c) bending about $x$-axis; (d) bending about $y$-axis; (e) self-equilibrating loading producing bimoment. Note: the flanges bend in opposite directions producing significant warpings stresses.

twisting of the cross section as shown in Fig. 10.42. Although the cross-section of each of the flanges remains plane, the wall as a whole is subjected to warping deformations. The restraint at the foundation prevents free warping at this end and sets up warping stresses.

The system of skew-symmetric loads which is equivalent to an internally balanced force system arising out of warping of cross section is termed a *bimoment* in thin-walled beam theory. Mathematically it can be construed as a generalized force corresponding to the warping displacement, just as moment and torsion are associated with rotation and twisting deformation, respectively. In the present example bimoment can be visualized as a pair of equal and opposite moments acting at a distance $e$ from each other. Its magnitude is

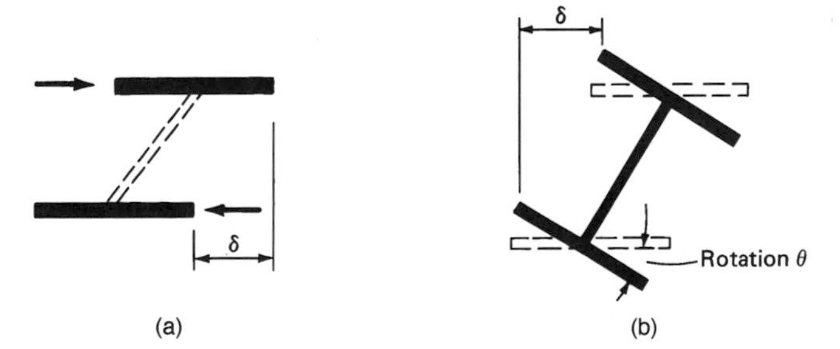


**Figure 10.42** Plan section of $I$-shaped column: (a) displacement of flanges due to bimoment load; (b) rotation due to geometric compatibility between flanges and web.

equal to $M$ times $e$ and has units of force times the square of the distance (lb · in.$^2$, kip · ft$^2$, etc.).

Presently it will be shown that the warping stresses can be calculated by the relation

$$\sigma_\omega = \frac{B_\omega \omega_s}{I_\omega} \tag{10.44}$$

where $B_\omega$ = the bimoment, a term that represents the action of a set of self-equilibrating forces
$\omega_s$ = the warping function
$I_\omega$ = the warping moment of inertia

The three terms $B_\omega$, $\omega_s$, and $I_\omega$ are conceptaully equivalent to moment $M$, linear coordinate or $x$ or $y$, and moment of inertia $I$ encountered in bending problems. Note the similarity between the bending stress as calculated by the familiar relation $\sigma_b = My/I$ and the warping stress formula given in Eq. (10.44).

In open section cores the magnitude of warping stresses can be very large and may even exceed the value of bending stress, depending upon the aspect ratio between the height, width, breadth, and thickness of the member.

A general theory which can account for warping was developed by Vlasov; in what follows a brief derivation of the fundamental equation and a description of the method of solution is given. The following notation is used in the derivation of warping torsion equation.

| | |
|---|---|
| $M_1$ | Torsional moment due to constrained torsion shear stresses |
| $M_2$ | Torsional moment due to St Venant torsion shear stresses |
| $M$ | Total torsional moment |
| $z, s$ | Orthogonal system of coordinates |
| $u$ | Longitudinal displacement due to warping along the $z$ axis |
| $v$ | Transverse displacement along the tangent to the profile |
| $\theta_z$ | Torsional rotation at $z$ |
| $\omega_s$ | Sectorial area |
| $\sigma_{z,s}$ | Longitudinal stress due to warping |
| $\varepsilon$ | Strain |
| $E$ | Modulus of elasticity of the material |
| $G$ | Modulus of rigidity of the material |
| $\mu$ | Poisson's ratio of the material |
| $\gamma$ | Shear strain |

| | |
|---|---|
| $\tau$ | Shear stress |
| $t_s$ | Thickness of thin-walled beam |
| $I_\omega$ | Warping moment of inertia |
| $m_z$ | Intensity of applied torque |
| $B_z$ | Bimoment |
| $l$ | Length of beam |
| $J$ | St Venant torsion constant |
| $k$ | Characteristic parameter, $l\sqrt{(1-\mu^2)GJ/EI_\omega}$ |
| $Z_z$ | Action matrix |
| $G_0$ | Distribution matrix |
| $Z_0$ | Initial boundary restraint matrix |
| $C_1, C_2, C_3, C_4$ | Constants of integration |

In Vlasov theory two fundamental assumptions are made:

1. The cross-section is completely rigid in the transverse direction.
2. The shear strain of the middle surface is negligible.

These two assumptions are almost completely satisfied in a practical core structure. The high in-plane stiffness of the floor slabs practically prevents distortion of the core at frequent intervals along its length, and the second assumption is valid for all but very low buildings.

Consider an open tube shown in Fig. 10.43. An orthogonal system of coordinates $z$, $s$ is chosen, consisting of a generator and the middle line of the profile (Fig. 10.43a). The origin for the coordinate $z$ is

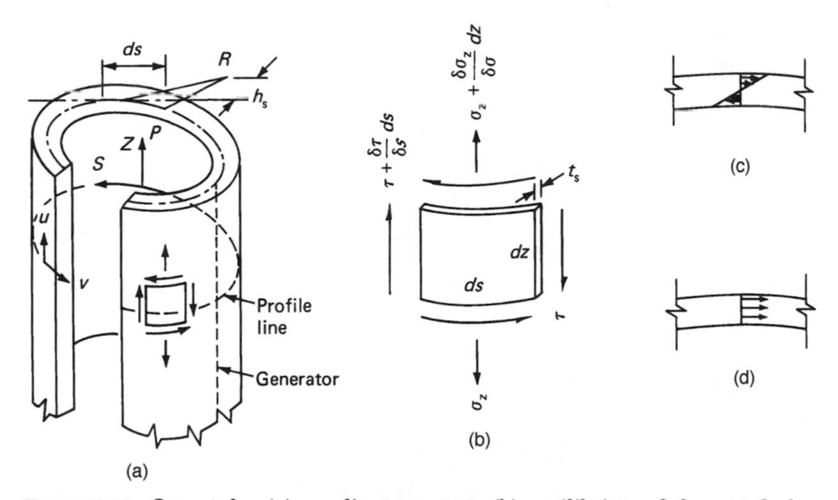


**Figure 10.43** Open tube: (a) coordinate system; (b) equilibrium of element $dz\,ds$; (c) St Venant's torsion shear stresses; (d) constrained (warping) torsion shear stresses.

taken at the base, and any generator is taken as the origin for the curvilinear coordinate $s$. Let $\theta$ be the angle of rotation of the profile at a distance $z$ from the base. This rotation is in the $xy$ plane and is measured with respect to any arbitrary center of rotation $R$.

Consider the displacements of any point $p$ on the middle surface of the tube. The transverse displacement $v$ in the direction of the tangent to the profile line is given by

$$v_{z,s} = \theta_z h_s \tag{10.45}$$

where $h_s$ is the perpendicular distance from the tangent at $p$ to the center of rotation $R$ (Fig. 10.43). If $u_{z,s}$ is the longitudinal displacement along the generator, then considering the displacements at $p$, of an element $dz\,ds$ (Fig. 10.43b), lying on the middle surface, the condition of zero shear strain is given by the relation

$$\gamma = \frac{\partial u}{\partial s} + \frac{\partial v}{\partial z} = 0$$

$$\frac{\partial u}{\partial s} + h_s \frac{d\theta}{dz} = 0 \tag{10.46}$$

Integrating,

$$u_{z,s} = -\int^{s} h_s\, ds \frac{d\theta}{dz} \tag{10.47}$$

The integral $s$ is taken along the profile from an arbitrary point to the point $p$ for which the longitudinal displacement is required. The product $h_s\, ds$ is equal to twice the area of the elementary triangle whose base and height are equal to $ds$ and $h_s$, respectively, and is usually given the symbol $d\omega$.

$$u_{z,s} = -\int^{s} \frac{d\theta}{dz} d\omega \tag{10.48}$$

$$= -\frac{d\theta}{dz} \omega_s \tag{10.49}$$

Since the displacement $u_{z,s}$ changes along the distance $z$, the strain $\varepsilon_z$ is given by

$$\varepsilon_z = \frac{\partial u}{\partial z} = -\frac{d^2\theta}{dz^2} \omega_s$$

Hence, corresponding stress $\sigma_{z,s}$ is

$$\sigma_{z,s} = \frac{E}{1-\mu^2} \varepsilon_z$$

$$= \frac{E}{1-\mu^2} \frac{d^2\theta}{dz^2} \omega_s \tag{10.50}$$

The origin of the coordinate $s$ can now be found from the condition that there is no applied vertidal load on the tube, i.e.,

$$\int_0^s \sigma_{z,s}\, t_s\, ds = 0 \tag{10.51}$$

where $t_s$ is the thickness of the tube.

The longitudinal stresses $\sigma_{z,s}$ are accompanied by shear stresses and are found from consideration of equilibrium of an element $t\, ds\, dz$ in the $z$ direction (Fig. 10.43b)

$$t_s \frac{\partial \sigma}{\partial z} + \frac{\partial t_s \tau}{\partial s} = 0 \tag{10.52}$$

$$\tau = \frac{E}{1-\mu^2}\frac{d^3\theta}{dz^3}\int_0^s \omega_s\, ds \tag{10.53}$$

Using the condition of zero external shear forces, it may be deduced that the origin of the arbitrary center of rotation $R$ is at the shear center. This determines completely the total stress distribution in terms of the derivatives of $\theta$.

The torque $M_1$ carried by the membrane shear stresses (Fig. 10.43d) which accompany the longitudinal stresses is given by

$$M_1 = \int_0^s \tau t_s h_s\, ds = -\frac{E}{1-\mu^2}\frac{d^3\theta}{dz^3}\int_0^s \omega_s^2 t_s\, ds \tag{10.54}$$

The quantity

$$\int_0^s \omega_s^2 t_s\, ds$$

is a structural constant, the so-called warping moment of inertia, and is usually denoted by $I_\omega$. Hence

$$M_1 = -\frac{-E}{1-\mu^2} I_\omega \frac{d^3\theta}{dz^3} \tag{10.55}$$

The torque $M_2$ carried by St Venant shear stresses (Fig. 10.43c) is given by

$$M_2 = GJ\frac{d\theta}{dz}$$

where $GJ$ is the St Venant torsional rigidity of the section.

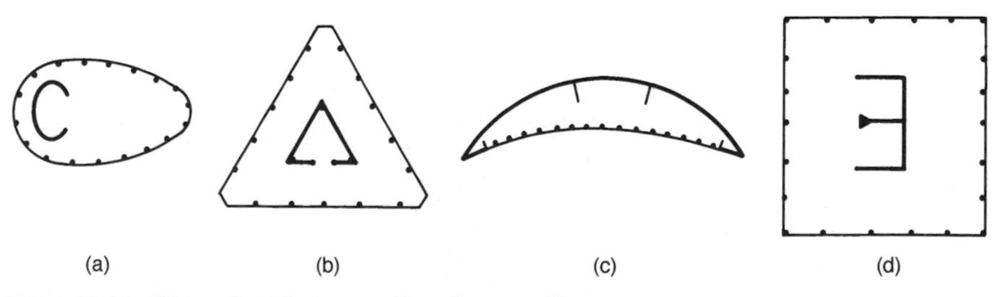


**Figure 10.44** Examples of open section shear walls.

The total torque $M = M_1 + M_2$ is then given by

$$M = \frac{-EI_\omega}{1-\mu^2}\frac{d^3\theta}{dz^3} + GJ\frac{d\theta}{dz} \tag{10.56}$$

Differentiating with respect to $z$, Eq. (10.56) becomes

$$\frac{EI_\omega}{1-\mu^2}\frac{d^4\theta}{dz^4} - GJ\frac{d^2\theta}{dz^2} = m_z \tag{10.57}$$

Using notation

$$k = l\sqrt{\frac{(1-\mu^2)GJ}{EI_\omega}}$$

Equation (10.57) can be written

$$\frac{d^4\theta}{dz^4} - \frac{k^2}{l^2}\frac{d^2\theta}{dz^2} = m_z \tag{10.58}$$

Equation (10.58) is the governing differential equation of constrained torsion. It can be used in the analysis of open section shear walls illustrated in Fig. 10.44.

**Longitudinal stresses and bimoment.** Consider the relation between the longitudinal stresses and warping,

$$\sigma_{z,s} = -\frac{E}{1-\mu^2}\frac{d^2\theta}{dz^2}\omega_s$$

Multiplying both sides of this equation by $\omega_s t_s$ and integrating over the whole profile gives

$$\int \sigma_{z,s}\omega_s t_s\, ds = -\frac{E}{1-\mu^2}\frac{d^2\theta}{dz^2}\int \omega_s^2 t_s\, ds$$

and since

$$\int \omega_s^2 t_s\, ds = I_\omega$$

$$\int \sigma_{z,s}\omega_s t_s\, ds = -\frac{E}{1-\mu^2}\frac{d^2\theta}{dz^2}I_\omega \tag{10.59}$$

The quantity

$$\int \sigma_{z,s}\omega_s t_s\, ds$$

is a generalized force called the bimoment and is represented by $B_z$. Thus

$$B_z = -\frac{EI_\omega}{1-\mu^2}\frac{d^2\theta}{dz^2}$$

From Eq. (10.50)

$$\frac{E}{1-\mu^2}\frac{d^2\theta}{dz^2} = \frac{\sigma_{z,s}}{\omega_s}$$

$$\therefore\ \sigma_{z,s} = \frac{B_z\omega_s}{I_\omega} \tag{10.60}$$

Hence the magnitude of bimoment at $z$ and the distribution of the sectorial area over the profile completely determine the longitudinal stresses.

**Solution of the differential equation.** Using the notation $f_{(z)} = m(1-\mu^2)/EI_\omega$, Eq. (10.42) can be written as

$$\frac{d^4\theta}{dz^4} - \frac{k^2}{l^2}\frac{d^2\theta}{dz^2} - f_{(z)} = 0 \tag{10.61}$$

The solution is of the form,

$$\theta_{(z)} = C_1 + C_2 z + C_3 \sinh\frac{k}{l}z + C_4 \cosh\frac{k}{l}z + \bar{\theta}_z \tag{10.62}$$

Differentiating Eq. (10.62) and using Eqs (10.56) and (10.57), equations for the displacements $\theta_z$ and $\theta'_z$ and the two forces $B_z$ and $M_z$ can be written thus,

$$\theta_z = C_1 + C_2 z + C_3 \sinh\frac{k}{l}z + C_4 \cosh\frac{k}{l}z + \bar{\theta}_z$$

$$\theta'_z = C_2 + C_3\frac{k}{l}\cosh\frac{k}{l}z + C_4\frac{k}{l}\sinh\frac{k}{l}z + \bar{\theta}'_z$$

$$B_z = -GJ\left[C_3 \sinh\frac{k}{l}z + C_4 \cosh\frac{k}{l}z + \frac{l^2}{k^2}\bar{\theta}''_z\right] \tag{10.63}$$

$$M_z = GJ\left[C_2 + \bar{\theta}'_z - \frac{l^2}{k^2}\bar{\theta}'''_z\right]$$

The constants $C_1$, $C_2$, $C_3$, and $C_4$ can be determined from the two boundary conditions at each end. However, calculations are greatly simplified if instead of the arbitrary constants $C_1$, $C_2$, $C_3$, and $C_4$, displacement and force boundary conditions, in terms of $\theta$, $\theta'$, $b$, and $M$, are used in Eq. (10.63). If $\theta_0$, $\theta'_0$, $B_0$, and $M_0$ are the two sets of displacements and forces at the section $z = 0$, and if there are no applied forces, the constants $C_1$, $C_2$, $C_3$, and $C_4$ from Eq. (10.47) are given by

$$\begin{aligned} C_1 &= \theta_0 + \frac{1}{GJ} B_0 \\ C_2 &= \frac{1}{GJ} M_0 \\ C_3 &= \frac{l}{k} \theta'_0 - \frac{l}{k}\frac{1}{GJ} M_0 \\ C_4 &= \frac{-1}{GJ} B_0 \end{aligned} \tag{10.64}$$

Substituting these in Eq. (10.63) and writing in matrix form the general equations for the four quantities, $\theta$, $\theta'$, $B$, and $M$ will be of the form

$$\begin{bmatrix} \theta_z \\ \theta'_z \\ \dfrac{B_z}{GJ} \\ \dfrac{M_z}{GJ} \end{bmatrix} = \begin{bmatrix} 1 & \dfrac{l}{k}\sinh\dfrac{k}{l}z & 1-\cosh\dfrac{k}{l}z & z-\dfrac{l}{k}\sinh\dfrac{k}{l}z \\ 0 & \cosh\dfrac{k}{l}z & \dfrac{-k}{l}\sinh\dfrac{k}{l}z & 1-\cosh\dfrac{k}{l}z \\ 0 & \dfrac{-l}{k}\sinh\dfrac{k}{l}z & \cosh\dfrac{k}{l}z & \dfrac{l}{k}\sinh\dfrac{k}{l}z \\ 0 & 0 & 0 & 1 \end{bmatrix} \begin{bmatrix} \theta_0 \\ \theta'_0 \\ \dfrac{B_0}{GJ} \\ \dfrac{M_0}{GJ} \end{bmatrix} \tag{10.65}$$

or in matrix notation, $Z_s = G_0 Z_0$, where $Z_s$ is the action matrix, $G_0$ the distribution matrix, and $Z_0$ the initial boundary restraint matrix. If, in addition to the boundary restraints, concentrated forces and displacements are applied at any section $z = t$, then, using the principle of superposition, the expressions for the actions at any section $z(t \leq z \leq d)$ will be of the form

$$Z_{(z)} = G_{0(z)} Z_0 + G_{(z-t)} Z_t \tag{10.66}$$

Where $G_{(z-t)}$ is of the same form as $G_{0(z)}$, except that the argument $(z - t)$ replaces $z$, and $z_t$ refers to the restraint matrix at $z = t$. The solution represented by Eq. (10.66) can easily be extended to other loading cases, such as several loads applied at various sections and distributed loads, by simple superposition and integration, respectively.

**Boundary conditions.** The horizontal loading on the core is replaced by a statically equivalent system of loads parallel to the $x$ axis and acting along the shear center axis and a uniform twisting moment. The simple beam theory is used to analyze the effects of loads through the shear center axis.

The effect of the uniformly distributed twisting moment $m_z$ is accounted for by considering the restraint vector at $z = t$ as $[0, 0, 0, m_z]$ and integrating the last column of the distribution matrix $G_{z-t}$ of Eq. (10.66) between the limits 0 and $l$.

Assuming the core to be completely rigid at the base, the boundary conditions at the ends of the core are

$$\begin{aligned} \text{Bottom:}\quad \theta_0 = 0 \quad \text{and} \quad \theta_0' = 0 \\ \text{Top:}\quad B_l = 0 \quad \text{and} \quad M_l = 0 \end{aligned} \tag{10.67}$$

Using these boundary conditions in the general Eqs (10.66), the expressions for the four quantities $\theta_z$, $\theta_z'$, $B_z$, and $M_z$ are written. The two quantities $\theta_z$ and $B_z$ which are a measure of deflection and stresses will be

$$\begin{aligned} \theta_z = \frac{-m}{GJ\cosh k}\left[\frac{-l^2}{k^2} - \frac{l^2}{k}\sinh k + z\left(l - \frac{z}{2}\right)\cosh k + \frac{l^2}{k^2}\cosh\frac{k}{l}z \right. \\ \left. + \frac{l^2}{k^2}\cosh\frac{k}{l}z + \frac{l^2}{k}\sinh\frac{k}{l}(l-z)\right] \end{aligned} \tag{10.68}$$

$$B_z = \frac{-ml^2}{k^2\cosh k}\left[\cosh k - \cosh\frac{k}{l}z - \frac{k\sinh k}{l}(l-z)\right] \tag{10.69}$$

Equations (10.68) and (10.69) are the basic equations applicable in the analysis of a wide variety of complex-shaped shear walls. Using these equations, a back-of-the-envelope type of calculations may be performed to get an idea of the maximum rotation at the top, and the maximum axial and shear stresses at the base of shear wall structures subjected to torsion.

### 10.4.10 Torsion analysis of shear wall structures: worked examples

The theory described previously, may be used to determine the stresses and rotations in shear wall structures subjected to torsion. To demonstrate the method, three buildings consisting of: (i) a single core; (ii) twin cores; and (iii) a randomly distributed shear walls will be considered.

**Example 1:** As a first example consider a 15-story building consisting of a rather large core as shown in Fig. 10.45. The core is singly symmetric, therefore its shear center lies on the axis of symmetry. To keep the derivation of shear center location some what general we will consider the core with arbitrary dimensions as in Fig. 10.46. The procedure for calculating the location O of the shear center, the principal sectorial coordinates and the warping moment of inertia $I_\omega$ are as follows.

1. Construct the $\omega_p$ diagram (Fig. 10.46b) which is the diagram of sectorial areas with respect to an arbitrary pole *P*. Using the $\omega_p$ and $x$ diagrams for the section, Fig. 10.46b and 10.46c, calculate the product of inertia of the $\omega_p$ diagram about the *Y* axis $I_{\omega_p x}$ using

$$I_{\omega_p x} = \int^A \omega_p x \, dA$$

in which $dA = t\,ds$, the area of a segment of the profile of thickness $t$ and length $ds$. The integral in Eq. (10.60) may be evaluated simply for straight-sided sections by using the product integrals given in Table 10.3.

3. Calculate $I_{yy}$, the second moment of area of the cross-section about the axis of symmetry.
4. As a final step, calculate the distance $\alpha_y$ of the shear center from *p* along the axis of symmetry, using

$$\alpha_x = \frac{I_{\omega_p x}}{I_{yy}}$$

5. The values the principal sectorial coordinates $\omega_p$ can be found by sweeping the ray *AP* around the profile and taking twice the values of the swept areas.
6. Sectorial moment of inertia $I_\omega$. This geometric property expresses the core's warping resistance. It is analogous to the moment of inertia in bending, $I_{xx}$ and $I_{yy}$.

   The sectorial moment of inertia is derived from the principal sectorial coordinate distribution using

$$I_\omega = \int^A \omega^2 \, dA = \int_0^s \omega^2 t \, ds$$

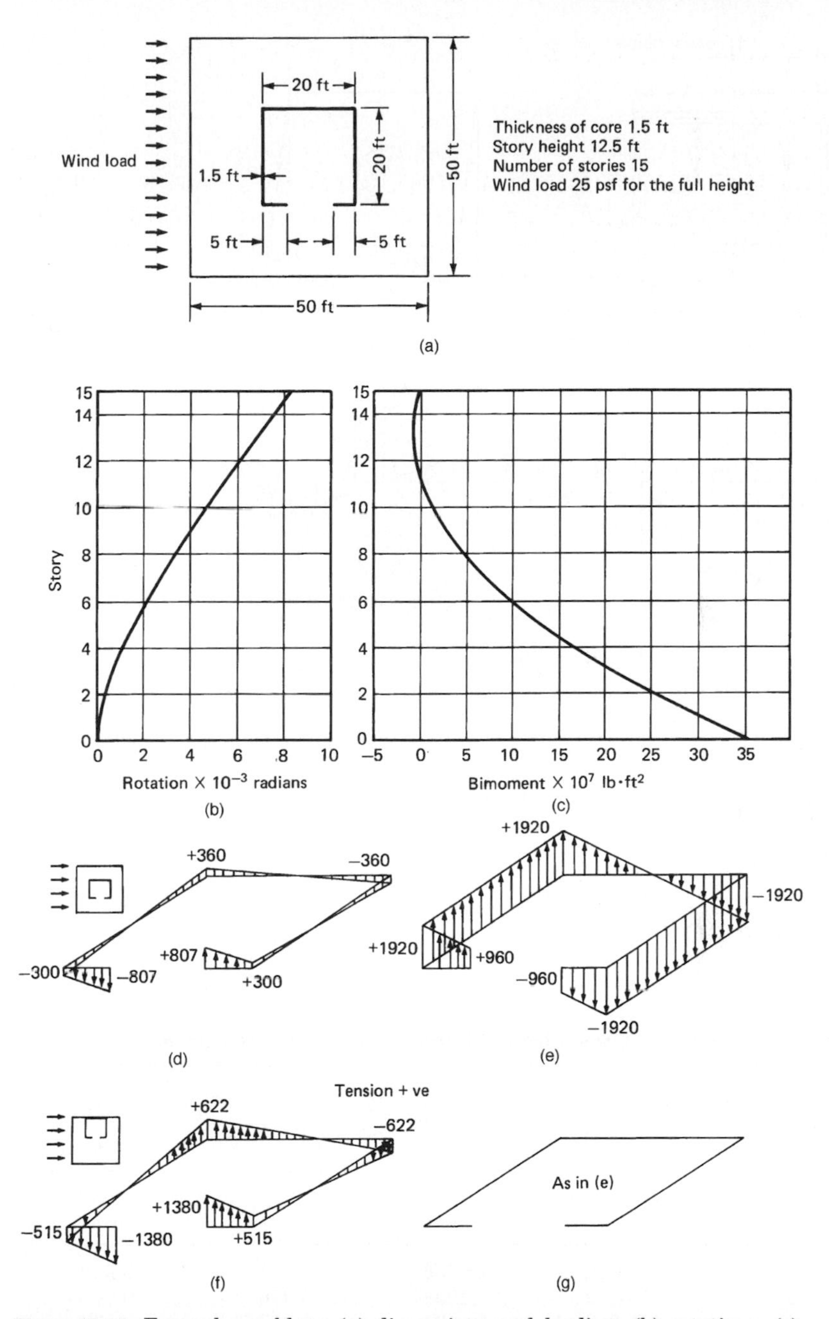


**Figure 10.45** Example problem: (a) dimensions and loading; (b) rotations; (c) bimoment variation; (d) warping stresses (core at center), psi; (e) bending stresses, psi; (f) warping stresses (offset core), psi; (g) bending stresses.

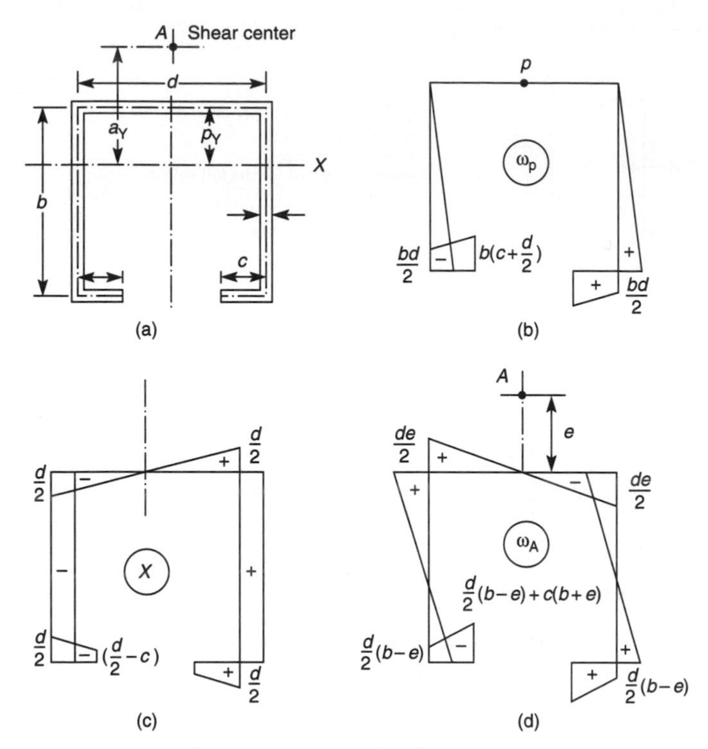


**Figure 10.46** (a) Core dimensions; (b) diagram of sectorial area $\omega$ with respect to arbitrary point $p$; (c) diagram of $x$ coordinate; (d) principal sectorial areas of $\omega$.

**TABLE 10.6 Example 3: Comparison of Bending Stresses at the Base**

| Wall no. | Section modulus in.$^3$ | Bending moment from comp. analysis kip-in | Extreme bending stresses (ksi) | |
|---|---|---|---|---|
| | | | Comp. analysis | Warping theory |
| 1 | 294,912 | 65,238 | 0.221 | 0.206 |
| 2 | 18,432 | 1038 | 0.056 | 0.04 |
| 3 | 165,888 | 64,191 | 0.387 | 0.360 |
| 4 | 73,728 | 6950 | 0.094 | 0.078 |
| 5 | 73,728 | 6950 | 0.094 | 0.078 |

For the assumed dimensions of the core shown in Fig. 10.46, the results for $\omega_s$ and $I_\omega$ are as indicated in Fig. 10.46b through d and as calculated in the following equation.

$$I_\omega = \int_0^s \omega^2 t\, ds = 2t\left[\frac{d^3 e^2}{24} + \frac{d^2 e^3}{12} + \frac{d^2(b-e)^3}{12}\right.$$
$$+ \frac{c}{6}\left\{\frac{d}{2}(b-e)\left[\frac{3d}{2}(b-e) + c(b+e)\right] + \left[\frac{d}{2}(b-e) + c(b+e)\right]\right.$$
$$\left.\left.\times\left[\frac{3d}{2}(b-e) + 2c(b+e)\right\}\right] \qquad (10.70)$$

For the example problem shown in Fig. 10.45 the stresses at the base due to bending and twisting are compared for two cases. In the first case the core is located centrally, and in the second case it is offset with respect to the plan dimensions. The numerical calculation of the various sectorial properties are not given since these are given with excruciating details in the other two examples.

A comparison of stresses for the example problem indicates that the warping stresses are of the same magnitude as the bending stresses. Observe that the warping stresses exceed the bending stress for the second case where the core is offset from the center of plan dimensions. The deflection at the building top, due to torsion bears a similar relation to the bending deflection, as will be demonstrated in the next example. All these examples, although simplified to make the numerical work less cumbersome, demonstrate the importance of considering warping stresses and deflection in practical building structures.

**Example 2.** Consider a 25-story, 300 ft (91.44 m) building consisting of two cores as shown in Fig. 10.47a,b. To keep the analysis simple assume that the resistance to lateral loads and torque is provided solely by the core. The building is subjected to a uniform wind load of 25 psf (1.197 kN/m$^2$) in the $x$ direction.

It is required to determine the maximum deflection and rotation at the top, and the vertical stresses at the base due to bending and twisting. An elastic modulus $E = 3600$ ksi (24,822 MPa) and a shear modulus $G = 1565$ ksi (10,791 MPa) are assumed for the concrete properties. The procedure is first described and then illustrated numerically.

Step 1. *Determine the sectorial properties.* For the given structure, by inspection, the location of shear center $O$ is determined at a point mid-way between the two cores. The $\omega$ diagram is related to the shear center $O$ as its pole and a point of zero warping deflection as an origin. In a symmetrical section, as in the example problem, the intersection of the axis of symmetry with the profile at $D$ defines a point of antisymmetrical behavior, and hence of zero warping deflection: therefore it may be used as the origin.

Values of $\omega$ are found from first principles, by sweeping the ray $OD$ around the profile and taking twice the values of the swept areas. For the example problem, the principle sectorial coordinate diagram is shown in Fig. 10.47c.

Step 2. *Determine the sectorial moment of inertia $I_\omega$ from Eq.* (10.42)

$$I_\omega = \int^A \omega^2 \, dA = \int^s \omega^2 t \, ds$$

Using the $\omega_s$ diagram (Fig. 10.47c) and the product integral table (Table 10.3), the value for $I_\omega$ is evaluated as shown in the worksheets.

Step 3. *Torsion constant $J$ for the core is determined from Eq.* (10.43). For one core, $J = \frac{1}{3}\sum bt^3 = \frac{1}{3} \times 1^3(3 \times 7.5 + 1 \times 19) = 13{,}834\ \text{ft}^4$, $J$ for two cores $= 13.834 \times 2 = 27.671\ \text{ft}^4$.

Step 4. *Determine eccentricity e of the line of action of wind resultant*

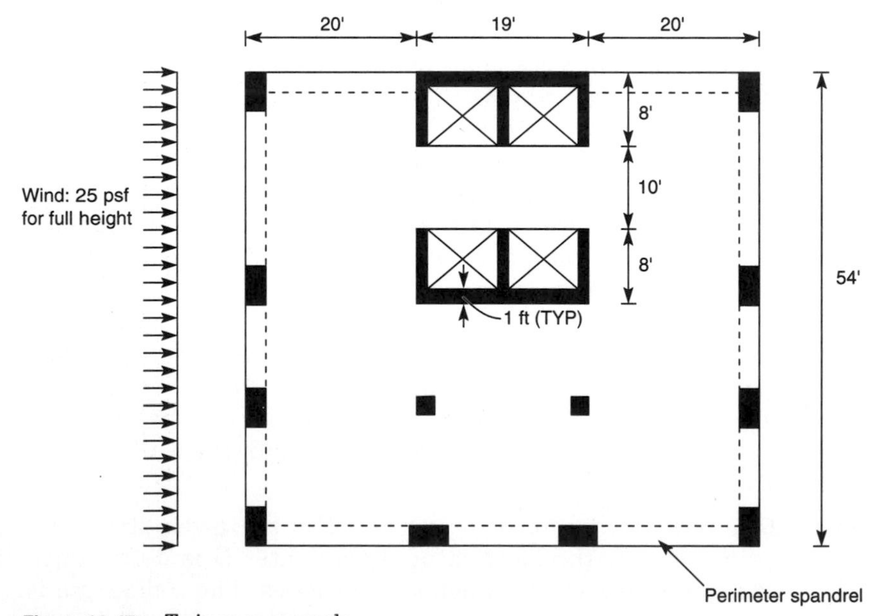


**Figure 10.47a** Twin-core example.

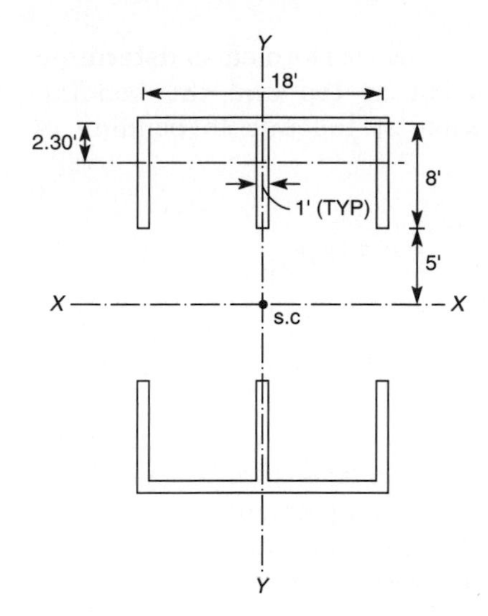


**Figure 10.47b** Core properties.

*from the shear center*. The resultant wind force per unit height of the building is equal to $25 \times 54 = 1.35$ kip/ft (1.83 kN/m) acting at 13.5 ft (4.12 m) to the south of shear center. Therefore the eccentricity, *e*, from the shear center is 13.5 ft (4.12 m). Since the external torque is the product of the horizontal loading and its eccentricity, the torsion due to wind is $1.35 \times 13.5 = 18.225$ kft/ft (24.70 kNm) per unit height, anticlockwise.

Step 5. *Determine bending deflection at the top and stresses at*

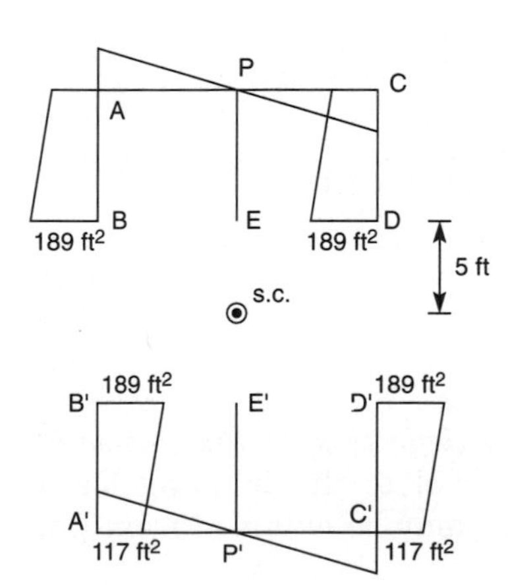


**Figure 10.47c** $\omega_s$ diagram (sectorial coordinates).

*the base.* A bending analysis is now performed to determine the maximum lateral deflection at top and the bending stresses at the base. Deflection at top due to bending is calculated as follows.

$$\Delta_{y(\max)} = \frac{wl^4}{8EI_{yy}} = 0.737 \text{ ft}$$

The deflection is $\frac{1}{406}$ of the height, as compared to the generally accepted limit of $\frac{1}{400}$ and therefore is acceptable. Observe that the building is very flexible in the $Y$ direction because the moment of inertia of the core $I_{xx}$ is about one-sixth of $I_{yy}$, indicating that supplemental bracing is required in the $y$ direction. One solution is to add primeter rigid frames as indicated in Fig. 10.47a. However, we continue the problem with the assumption made earlier, namely, that the core resists all the lateral loads.

$$\text{Maximum bending stress} = \frac{MC}{I} = \frac{60{,}750}{1788} \times 9.5$$

$$= 322.78 \text{ ksf}$$

$$= 2.24 \text{ ksi } (16.78 \text{ MPa})$$

The bending stress diagram is given in Fig. 10.47d.

Step 6. *Determine the parameter k using Eq.* (10.26)

$$k = l\sqrt{\frac{GJ}{EI_\omega}}$$

$$= 300\sqrt{\frac{1565 \times 144 \times 28}{92{,}7180 \times 2.3}}$$

$$= 1.08$$

Step 7. *Determine the rotation and total deflection at the corner of top floor.* The rotation at any level of the building for a uniformly distributed torque "$m$" may be obtained from Eq. (10.69).

At the top, $Z = l$. Substituting $z = l$ in Eq. (10.69) we get

$$\theta = \frac{-ml^2}{GJ\cosh k}\left[-\frac{l}{k^2} - \frac{l}{k}\sinh k + \frac{1}{2}\cosh k + \frac{1}{k^2}\cosh k\right] \quad (10.71)$$

Substituting for the various parameters we get $\theta_H = 0.0196$ radians, anticlockwise. Therefore, the additional deflection at the south-east corner $c$ of top floor due to torsion

$$\Delta_t = \theta_H \times \text{distance of } c \text{ from shear center}$$
$$= 0.196 \times 50.10 = 0.9821 \text{ ft } (0.30 \text{ m})$$

The total deflection at $c$ due to bending and torsion $= \Delta_b + \Delta_t = 1.47 + 0.9821 = 2.45$ ft (0.75 m). This value represents $\frac{1}{122}$ of the building height, an unacceptably large value, confirming the earlier observation that the building is too flexible in the $y$ direction.

Step 8. *Determine bimoments and warping stresses.* The warping stresses $\sigma_\omega$ at the base are determined from the bimoment $B$ at that level. The bimoment is obtained for a uniformly distributed torque from Eq. (10.69). Then, at any point on the section where the principal sectorial coordinate is $\omega_s$,

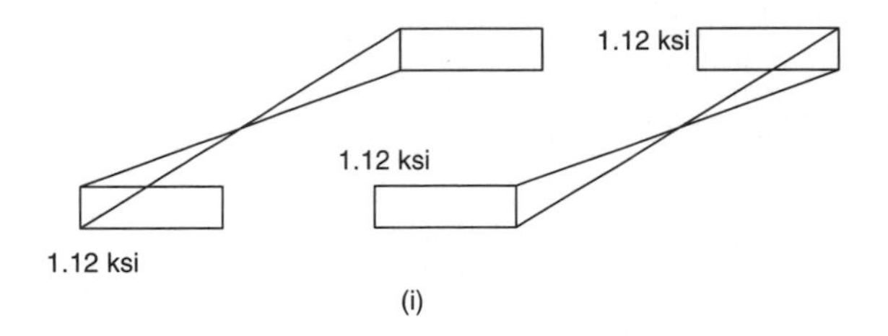


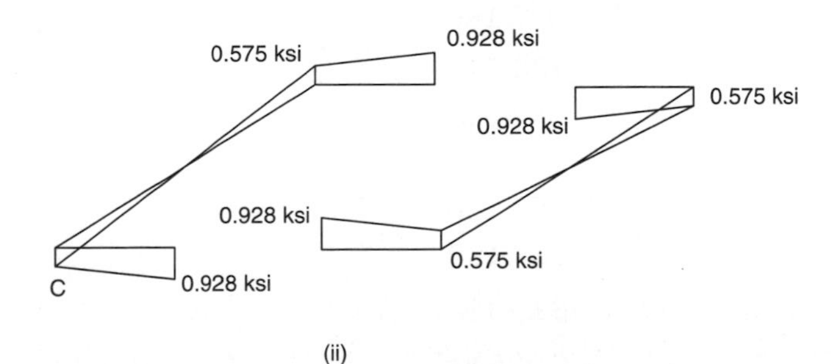


**Figure 10.47d** Comparison of stresses: (a) bending stress $\sigma_b$; (b) warping stress $\sigma_\omega$.

the vertical warping stress is obtained from Eq. (10.44). The total axial stresses due to horizontal loading is obtained by combining the warping stresses with the bending stresses.

The vertical stresses at the base due warping are determined from the bimoment at the base. This is given by Eq. (10.69)

$$B_z = -\frac{ml^2}{k \cosh k}\left[\cosh k - \cosh\frac{k_z}{l} - k \sinh\frac{k}{l}(l - z)\right] \quad (10.72)$$

At the base

$$z = 0, \quad B_0 = -\frac{ml^2}{k^2 \cosh k}[\cosh k - l - k \sinh k]$$

$$k = 1.08, \quad \sinh k = 1.3025, \quad \cosh k = 1.642 \quad (10.73)$$

$$\sinh 0 = 0, \quad \cosh 0 = 1$$

Substituting the above values

$$\begin{aligned} B_0 &= \frac{-ml^2}{1.08^2 \times 1642}[1.642 - 1 - 1.08 \times 1.3025] \\ &= \frac{-ml^2}{2.504} \\ &= -\frac{18.225 \times 300^2}{2.504} \\ &= 656{,}100 \text{ kip/ft}^2 \end{aligned}$$

Warping stresses are given by $\sigma_\omega = \frac{B_0}{I_\omega}\omega_s$. At $D$,

$$\begin{aligned} \sigma_\omega &= \frac{656{,}100}{927{,}180} \times 189 \\ &= 133.74 \text{ ksf} \\ &= 0.928 \text{ ksi} \end{aligned}$$

At $C$,

$$\begin{aligned} \sigma_\omega &= \frac{656{,}100}{927{,}180} \times 117 \\ &= 82.79 \text{ ksf} \\ &= 0.575 \text{ ksi} \end{aligned}$$

Figure 10.47d shows a comparison of bending and warping stresses. The importance of warping torsion is obvious.

## Example 2: Worksheet

### Core bending properties

$$\text{Area of one core} = (1+8)1 + (8-0.5) \times 1 \times 3$$
$$= 41.5 \text{ ft}^2$$

Determine c.g.

$$3 \times 7.5 \times 1\left(\frac{7.5}{2} + 0.5\right) = 41.5\,\bar{y}$$

$$\bar{y} = \frac{95.6}{41.5} = 2.30 \text{ ft}$$

$$I_{yy} = 19 \times \frac{1^3}{12} + 3\left(\frac{1 \times 7.5^3}{12}\right) + 19 \times 1 \times 2.30^2 + 3.75 \times 1 \times 1.96^2$$
$$= 294 \text{ ft}^4$$

$$I_{xx} \text{ for two cores} = 2 \times 294 = 588 \text{ ft}^4$$

$$I_{xx} = 1 \times \frac{1^3}{12} + 7.5 \times 1 \times 9^2 \times 2 + 3 \times 7.5 \times \frac{1^3}{12}$$
$$= 571.58 + 1215 + 1.87$$
$$= 1788.45 \text{ ft}^4 \quad \text{for one core}$$

$$I_{xx} \text{ for two cores} = 2 \times 1788.45 = 3577 \text{ ft}^4$$

### Core torsion properties

The shear center is located at the center of two cores. Using points $P$ and $P'$ as the principal poles, the $\omega_s$ is drawn by sweeping the radii $OP$ and $OP'$ around the profile

$\omega_s$ Diagram (sectorial coordinates)

#### Sectorial moment of inertia $I_\omega$

This is obtained by using the product integral table as follows:

| Segement | $\omega_s$ | $\int \omega_s^2\, ds$ |
|---|---|---|
| *PA, PC, P'A', P'C'* | 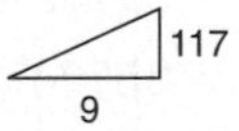  | $\frac{1}{3} \times 9 \times 117 \times 1 = 41{,}067 \text{ ft}^6$ |
| *AB, CD, A'B', C'D'* | 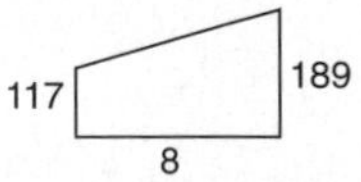  | $\frac{8}{6}[117(2 \times 117 + 189) + 189(2 \times 189 \times 117)]$ $= 190{,}728 \text{ ft}^6$ |

Total for half of one core $= 41{,}067 + 190{,}728 = 231{,}795 \text{ ft}^6$

Verify the integral $\int^{A} \omega^2 \, ds$ for segment $AB$ by treating the area as a combination of a rectangle and a triangle thus:

117 ⓐ ⓑ 72 117 8

$$\int \omega^2 \, ds = 117 \times 17 \times 8 + \tfrac{1}{3} \times 8 \times 72^2 + 2 \times \tfrac{1}{2} \times 117 \times 82 \times 8$$

$$= a^2 + b^2 + 2ab$$

$$= 109{,}512 + 13{,}824 + 67{,}392 = 190{,}728 \text{ ft}^6$$

$$I_\omega \text{ for two cores} = 4 \times 231{,}795 = 927{,}180 \text{ ft}^6$$

### St Venant torsion constant *GJ*

$$J \text{ for one core} = \sum \frac{mt^3}{3} = \frac{1}{3} \times 1^3(3 \times 7.5 + 1 \times 19) = 13{,}834 \text{ ft}^4$$

$$J \text{ for two core} = 2 \times 13{,}834 = 27.67 \text{ ft}^4$$

$$G = \frac{E}{2(1+\mu)} = \frac{E}{2.3} \qquad E = 57{,}000\sqrt{4000} = 3600 \text{ ksi} = 518{,}400 \text{ ksf}$$

$$G = \frac{518{,}400}{2(1+0.15)} \qquad \mu \text{ for concrete} = 0.15$$

$$= 225{,}391 \text{ ksf}$$

$$GJ = 225{,}391 \times 27.67$$

$$= 6{,}236{,}569 \text{ kft}^2$$

$$k = l\sqrt{\frac{GJ}{EI_\omega}}$$

$$= 300\sqrt{\frac{27.67}{927{,}180 \times 2.3}} = 1.08$$

### Bending analysis

#### Deflection

The building deflection at top is given by the relation

$$\Delta_b = \frac{wl^4}{8EI} \qquad w = 25 \times 54 = 1.35 \text{ kip/ft}$$

$$l = 300 \text{ ft}$$

$$= \frac{1.35 \times 300^4}{8 \times 518{,}400 \times 3677} \qquad E = 518{,}400 \text{ ksi}$$

$$I = 3577 \text{ ft}^4$$

$$= 0.737 \text{ ft}$$

$$\frac{\Delta_b}{l} = \frac{0.737}{3.00} = \frac{1}{406}$$

### Bending stresses

$$\text{Maximum bending moment at the base} = \frac{wl^2}{2}$$

$$= \frac{1.35 \times 300^2}{2}$$

$$= 60{,}750 \text{ kip ft}$$

$$\text{Bending stress } \sigma_b = \frac{MC}{I_{xx}}$$

$$= \frac{60{,}750 \times 9.5}{3577}$$

$$= 161.35 \text{ ksf}$$

$$= 1.120 \text{ ksi}$$

### Torsion analysis

#### Rotation

The rotation $\theta_l$ at the top, for an open section subjected to a uniform torque of $m$ is given by

$$\theta_l = -\frac{m}{GJ\cosh k}\left[-\frac{l^2}{k^2} - \frac{l^2}{k^2}\sinh k + l\left(l - \frac{l}{2}\right)\cosh k + \frac{l^2}{k^2}\cosh k + \frac{l^2}{k}\sinh\frac{k}{l}(l - l)\right]$$

Substituting

$$m = 1.35 \times 13.5 = 18.225 \text{ kft/ft}$$

$$l = 300 \text{ ft}$$

$$k = 1.08, \quad \sinh k = 1.3025, \quad \cosh k = 1.642$$

$$\sinh 0 = 0, \quad \cosh 0 = 1$$

$$\theta_l = 0.0196 \text{ radians.}$$

The distance $R$ of shear centre from the building corner

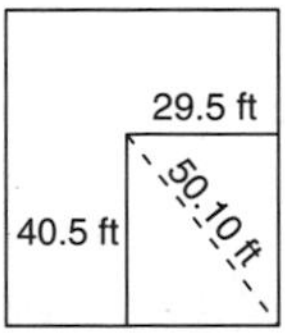

$$R = \sqrt{29.5^2 + 40.5^2} = 50.10 \text{ ft}$$

$$\text{Deflection due to torsion } \Delta_t = \theta_l \times R$$

$$= 0.0196 \times 50.10$$

$$= 0.9821 \text{ ft}$$

$$\text{Total deflection due to bending and torsion} = \Delta_b + \Delta_t$$

$$= 0.737 + 0.9821 = 1.72 \text{ ft}$$

The deflection index $\frac{\Delta}{l} = \frac{1.72}{300} = \frac{1}{174}$, an unacceptably large value indicating serious deficiency in the lateral load resisting system.

### Stresses due to torsion

The bimoment $B_z$ due to a uniformly applied torque $m$ is given by the relation

$$B_z = \frac{-ml^2}{k^2 \cosh k}\left[\cosh k - \cosh\frac{kz}{l} - k \sinh\frac{k}{l}(l - z)\right]$$

At $z = 0$, the bimoment $B_0$ at the base is given by

$$B_0 = \frac{-ml^2}{k^2 \cosh k}[\cosh k - 1 - k \sinh k]$$

Substituting the values for $m$, $l$, $k$, $\sinh k$ and $\cosh k$ as before,

$$B_0 = \frac{-ml^2}{2.504} \qquad m = 1.35 \times 13.5 = 18.225 \text{ kft/ft}$$

$$l = 300 \text{ ft}$$

$$B_0 = \frac{18.225 \times 300^2}{2.504}$$

$$= 656{,}100 \text{ kft}^2$$

Warping stress at $D$,

$$\sigma_{\omega(D)} = \frac{B_0}{I_\omega}\,\omega_s$$

$$= \frac{656{,}100 \times 189}{927{,}180}$$

$$= 133.74 \text{ ksf}$$

$$= 0.928 \text{ ksi}$$

Warping stress at $C$,

$$\sigma_{\omega(C)} = \frac{656{,}100 \times 117}{927{,}180}$$

$$= 82.79 \text{ ksf}$$

$$= 0.575 \text{ ksi}$$

As noted previously, the analogy between the warping torsion and bending may be used to get an approximate idea of the rotation and stresses in core structures. Using this approach, for the example problem we get the rotation at top

$$\theta_l = \frac{ml^4}{8EI_\omega} = \frac{13.5 \times 300^4}{8 \times 518{,}400 \times 927{,}180}$$
$$= 0.0285 \text{ radians}$$

as compared to the value 0.0196 radians obtained by the more accurate analysis.

The biomoment $B_0$, at the base is given by $B_0 = \frac{ml^2}{2}$ as compared to the value of $\frac{ml^2}{2.504}$ obtained before which is pretty good for a five-minute back-of-the-envelope type calculation.

**Example 3:** The warping theory described with reference to single and twin-core structures may also be used to determine the stresses and rotations in a shear wall structure consisting of randomly distributed planar shear walls. To demonstrate the method of analysis, a 15-story building is analyzed, first by using the warping theory, and then by a commercially available three-dimensional computer program. The result for torsion which includes the rotations of the core, and the bending stresses at the base are compared and shown to give almost identical results. The warping analysis is given in the work sheet. The high-lights of the analysis are as follows.

The building Fig. 10.48, consists of three walls, $W_1$, $W_2$, and $W_3$ in the transverse direction, and two walls $W_4$ and $W_5$ in the longitudinal direction. The uniform wind load of 25 psf (1.197 kN/m$^2$) gives a resultant of $25 \times 70 = 1.75$ kip/ft (2.37 kNm) acting at the center of gravity of the plan in the $Y$ direction. For the given structure, by inspection, the location of the shear center $O$ is determined to be on a line midway between the walls $W_4$ and $W_5$. The distance $x$ of the shear center $O$ from wall $W_1$, is obtained from the relation

$$\bar{x} = \frac{\sum I_{xx} x}{\sum I_{xx}}$$

Next the eccentricity $e$ which is the distance from the line of action of wind resultant from the shear center is determined to be 14 ft (4.27 m). The external torque, the product of the horizontal load and its eccentricity is evaluated to be equal to $1.75 \times 14 = 24.5$ kft/ft

(10.12 kNm/m). To keep the comparison simple, both the computer and warping analyses are performed for only a uniformly applied torsion. For the example problem, the center of gravity of each wall defines a point of antisymmetrical behavior, and hence of zero warping deflection. Therefore, these points are used as the origins for determining the values of sectorial coordinate $\omega$. A building analysis

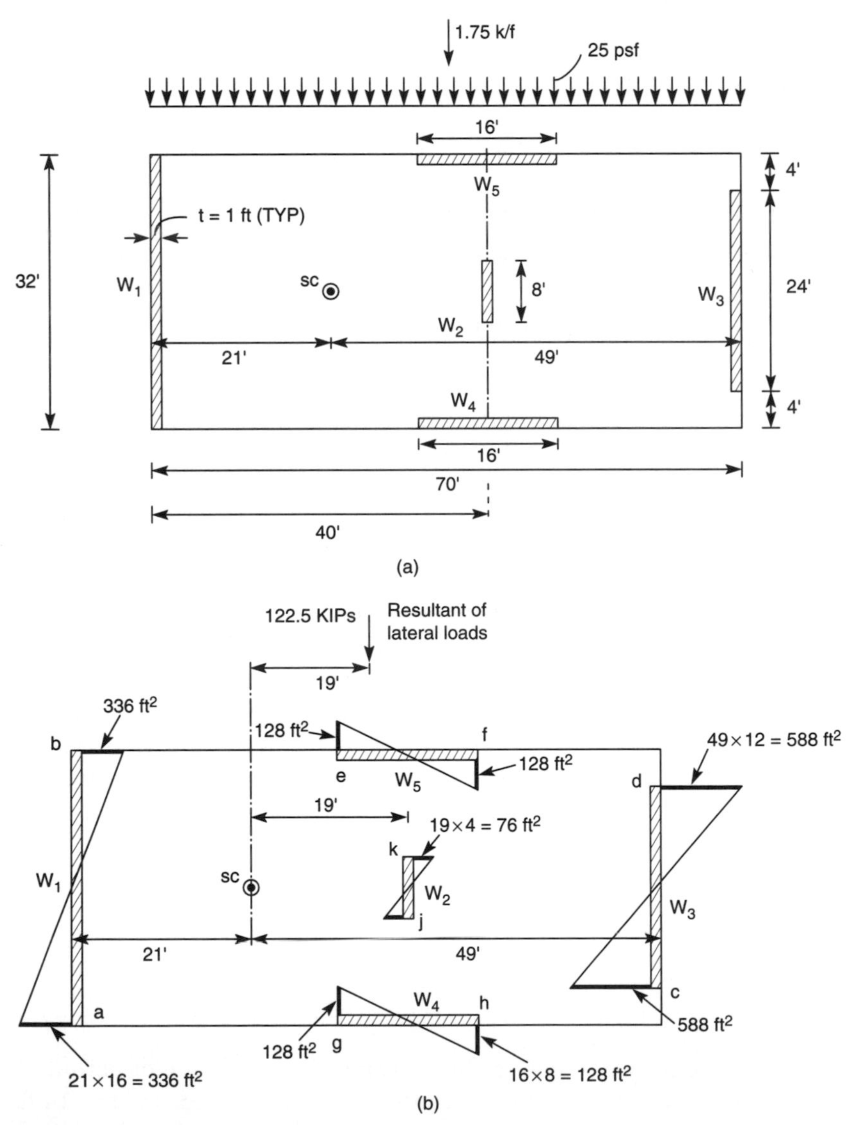


**Figure 10.48** Randomly distributed shear walls, torsion example: (a) plan; (b) rotation comparison.

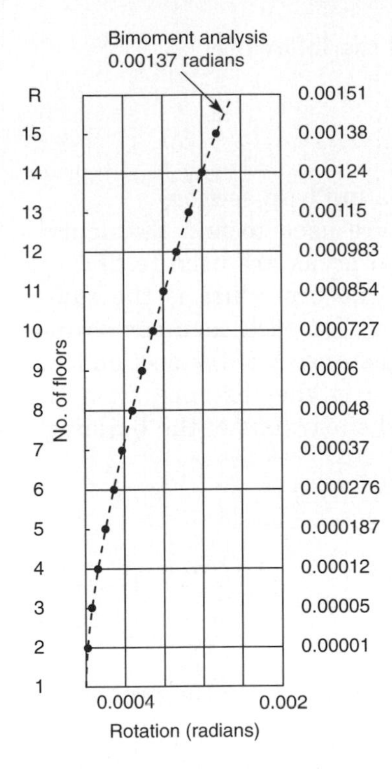


Rotation comparison: etabs version 6 versus warping analysis
(c)

**Figure 10.48** *(Continued)*.

is then performed to determine the rotation at the building top and the bending stresses at the base, as shown in the worksheets.

In the computer analysis the building is analyzed with five walls assuming the slab to be infinitely rigid in its own plane. In the final part of the work sheet, the results of rotations and bending stresses from the two analyses are compared. From the close correlation between the two, the applicability of warping analysis to practical planar shear walls is obvious.

#### Example 3: Worksheet

### Location of shear center

By inspection, the location of shear center s.c. is determined to be on the common neutral axis $x$-$x$. The location of the other shear axis is given by the relation

$$e = \frac{\sum I_x x}{\sum I_x}$$

Notice the similarity between this and the following:

$$\bar{y} = \frac{\sum Ay}{\sum A}$$

which is used to find the neutral axis of a built-up section.

Just as the areas of individual parts are used to find the neutral axis, the moments of inertia of individual areas are used to find the shear axis of the building (Fig. 10.48). The procedure is the same; select a reference axis ($y$-$y$), determine $I_x$ for each member section (about its own neutral axis $x$-$x$) and the distance $x$ of the section from the reference axis ($y$-$y$). The resultant $e$ is the distance from the chosen reference axis ($y$-$y$) to the parallel shear axis of the building.

$$e = \frac{I_{x_1}x_1 + I_{x_2}x_2 + I_{x_3}x_3 + I_{x_4}x_4}{I_{x_1} + I_{x_2} + I_{x_3} + I_{x_4}}$$

or:

$$e = \frac{\sum I_x x}{\sum I_x} \tag{10.74}$$

Calculate the location $\bar{x}_1$ of the shear center s.c. from the centre line of wall $W_1$.

$$\sum I_x = 1 \times \frac{32^3}{12} + 1 \times \frac{8^3}{12} + 1 \times \frac{24^3}{12}$$

$$= 2730.67 + 42.67 + 1152$$

$$= 3925.34 \text{ ft}^4$$

$$\bar{x}_1 = \frac{\sum I_x x}{\sum I_x}$$

$$= (2736.67 \times 0 + 42.67 \times 40 + 1152 \times 70)/3925.24$$

$$= \frac{82{,}346.8}{3925.24} = 20.979 \text{ ft} \quad \text{Use 21 ft}$$

Verify location of s.c. from the center line of east wall $W_3$.

$$\bar{x}_2 = (1152 \times 0 + 42.67 \times 30 + 2730.67 \times 70)/3925.24$$

$$= \frac{192{,}427}{3925.24} = 49 \text{ ft}$$

The eccentricity $e$ of the line of action of wind resultant from the shear center

$$e = 35.21 = 14 \text{ ft}$$

Torsional moment $m$ per foot height of the building

$$m = 1.75 \times 14 = 24.5 \text{ kf/f}$$

### Torsion properties

Using the centers of each wall as the principal poles, the sectorial coordinate diagram representing the warping coordinates for the composite building is drawn by sweeping the radius vector passing through the shear center. The resulting $\omega_s$ diagram is shown in Fig. 10.48b.

The warping moment of inertia for the building is calculated as before by using the product integral table.

$$\begin{aligned} I_\omega &= \tfrac{2}{3} \times 16 \times 336^2 + \tfrac{2}{3} \times 4 \times 76^2 + \tfrac{2}{3} \times 12 \times 588^2 \\ &\quad + \tfrac{2}{3} \times 8 \times 128^2 + \tfrac{2}{3} \times 8 \times 128^2 \\ &= 4{,}160{,}341 \text{ ft}^6 \end{aligned}$$

The St Venant's torsion $J$ is calculated from the relation

$$\begin{aligned} J &= \sum^{n} bt^3 \\ &= 1^3(32 + 8 + 24 + 16 + 16) \\ &= 96 \text{ ft}^4 \\ GJ &= 225{,}360 \times 96 \\ &= 216{,}345{,}60 \times \text{ kip/ft}^2 \\ k &= l\sqrt{\frac{GJ}{EI_\omega}}, \quad \frac{G}{E} = \frac{1}{2.3} \\ &= 180\sqrt{\frac{96}{4{,}160{,}341 \times 2.3}} = 0.57 \end{aligned}$$

### Torsional rotation

The rotation $\theta_l$ at top due to a uniformly distributed torque of $m$ per unit height is given by

$$\theta_l = \frac{-ml^2}{GJ \cosh k}\left[\frac{\cosh k - 1}{k^2} + \frac{\cosh k}{2} - \frac{\sinh k}{k}\right]$$

Substituting

$$k = 0.57, \quad \sinh k = \sinh 0.57 = 0.601$$

$$\cosh k = \cosh 0.57 = 1.167, \quad GJ = 216{,}345{,}60 \text{ kip/ft}^2$$

and

$$m = 1.75 \times 14 = 24.5 \text{ kf/f}$$

$$\theta_l = 0.00137 \text{ radians.}$$

### Bending stresses due to torsion

The bimoment $B_0$ at the base is given by

$$B_0 = -\frac{ml^2}{k^2 \cosh k}[\cosh k - 1 - k \sinh k]$$

Substituting

$$k = 0.57, \quad \sinh k = \sinh 0.57 = 0.601$$

$$\cosh k = \cosh 0.57 = 1.167, \quad m = 1.75(35 - 21) = 24.5 \text{ kft/ft}$$

$$l = 15 \text{ stories @ } 12 \text{ ft} = 180 \text{ ft}$$

$$B_0 = -\frac{24.5 \times 180^2}{0.57^2 \times 1.167}[1.167 - 1 - 0.57 + 0.601]$$

$$= 367{,}529 \text{ kip/ft}^2$$

The bending stresses at the extremities of walls are as follows.

Wall 1

$$\sigma_\omega \text{ at } a, b = \frac{367{,}529 \times 336}{4{,}160{,}341} = 29.68 \text{ kip/ft}^2 = 0.206 \text{ ksi}$$

Wall 3

$$\sigma_\omega \text{ at } c, d = \frac{367{,}529 \times 588}{4{,}160{,}341} = 51.941 \text{ kip/ft}^2 = 0.360 \text{ ksi}$$

Wall 5

$$\sigma_\omega \text{ at } e, f = \frac{367{,}524 \times 128}{4{,}160{,}341} = 11.30 \text{ kip/ft}^2 = 0.078 \text{ ksi}$$

Wall 4

$\sigma_\omega$ at $g, h$—similar to wall 5 at $e, f$.

Wall 2

$$\sigma_\omega \text{ at } j, k = \frac{367{,}529 \times 76}{4{,}160{,}341} = 671 \text{ kip/ft}^2 = 0.046 \text{ ksi}$$

### 10.4.11 Torsion analysis of steel braced core: worked example torsional

The torsional behavior of braced cores in steel buildings can be predicted with sufficient accuracy by using the nonuniform torsion theory given in the preceding section. For example, consider the typical floor plan of a 30-story steel building as shown in Fig. 10.49a. Assume that wind resistance is provided by X-bracing of the core all around except between two corridor columns as shown in Fig. 10.49a. Without the luxury of bracing all around its perimeter, the core loses much of its torsional resistance and behaves more like an open core as shown in Fig. 10.49b. In analyzing the core it will be assumed initially that the warping constraint of beam $AB$ is negligible. These effects are considered later. In the analytical model, the braces can be considered as equivalent solid walls subjected only to shear forces. The normal forces can be assumed to be sustained by the columns alone, acting as vertical ribs. The idealized analytical model is shown in Fig. 10.49c. The differential equation of nonuniform torsion is of the form

$$\frac{d^4\theta}{dz^4} - \frac{k^2}{l^2}\frac{d^2\theta}{dz^2} = m_z \frac{1-\mu^2}{EI_\omega} \tag{10.75}$$

where $\theta$ = the rotation of the core at section $z$ in the $xy$ plane
$k$ = the nondimensional parameter given by

$$k = l\sqrt{GJ/EI_\omega}$$

$l$ = the height of the core
$m_z$ = the intensity at $z$ of the applied torque
$\mu$ = the Poisson's ratio
$E$ = the modulus of elasticity
$I_\omega$ = the warping moment of inertia
$J$ = the St Venant's torsional moment of inertia of the braced core

For the case of no applied external forces, Eq. (10.75) has a solution of the form:

$$\theta_z = \theta_0 + \theta_0' \frac{l}{k}\sinh\frac{k}{l}z + \frac{B_0}{GJ}\left(1 - \cosh\frac{k}{l}z\right) + \frac{M_0}{GJ}\left(z - \frac{l}{k}\sinh\frac{k}{l}z\right) \tag{10.76}$$

where $\theta_0$, $\theta_0'$, $B_0$ and $M_0$ are the rotation, warping, bimoment, and torque at the section $z = 0$.

In a high-rise core, the boundary conditions at the foundation level are $\theta_0 = 0$ and $\theta_0' = 0$. At the top, since there is no applied torque or bimoment, $M_l = 0$ and $B_l = 0$. Using these conditions in Eq. (10.76), we could write the expressions for the rotation $\theta_z$ and bimoment $B_z$ at any section $z$. For the particular case of uniformly distributed

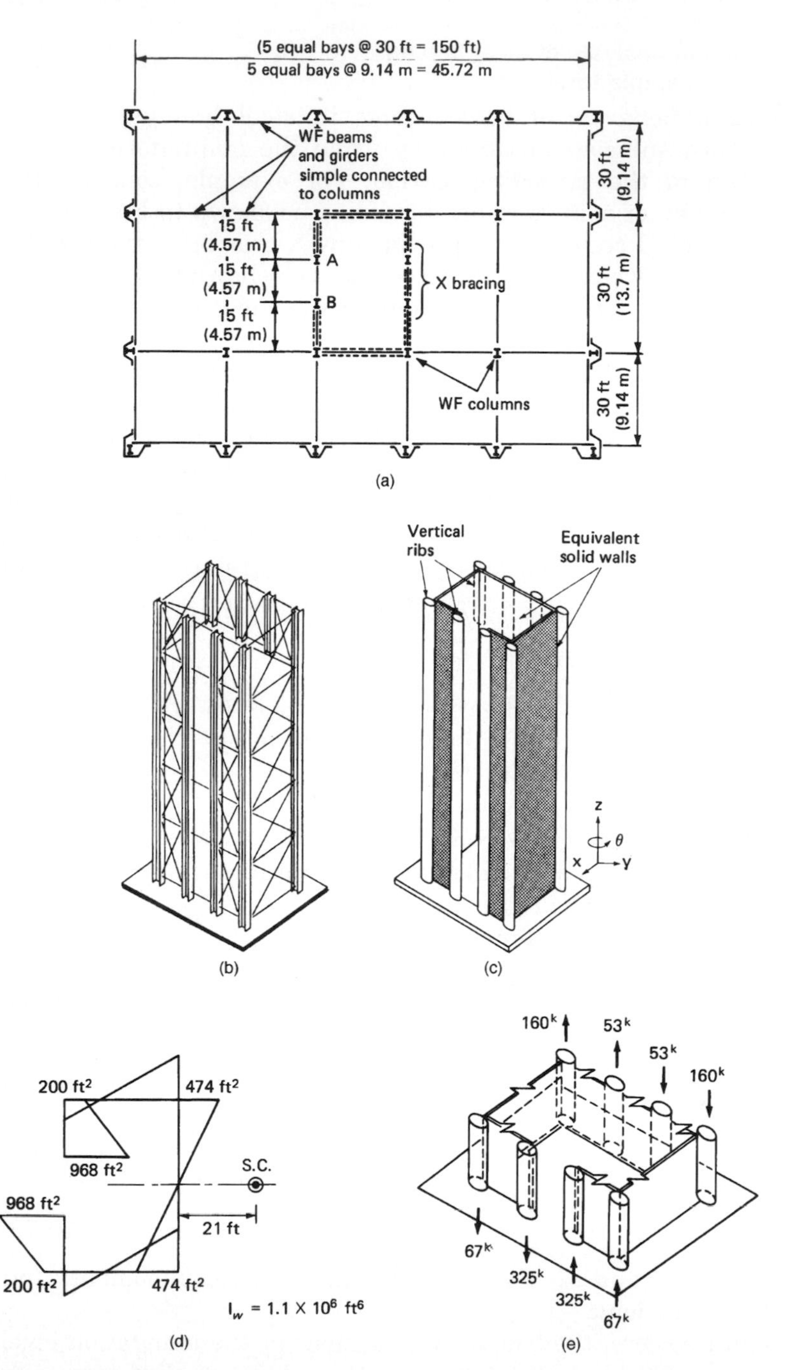


**Figure 10.49** Example problem: (a) typical floor plan; (b) asymmetrical bracing resulting in an open section core; (c) analytical model for torsion analysis; (d) diagram of principal sectorial coordinates; (e) axial forces in core columns due to bending; (f) axial force in core columns due to bending; (g) core warping constrained by floor beams.

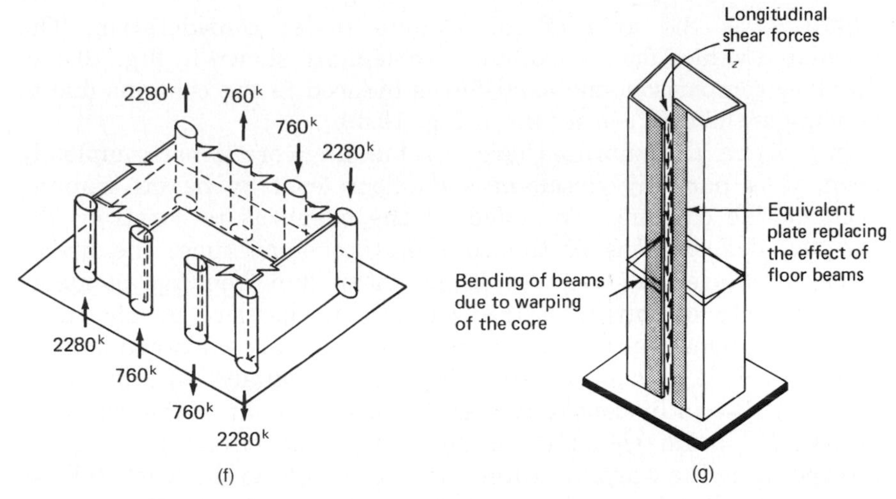


**Figure 10.49** (*Continued*).

twisting moment $m_z$, these expressions can be shown to be:

$$\theta_z = -\frac{m}{GJ\cosh k}\left[-\frac{l^2}{k^2} - \frac{l^2}{k}\sinh k + z\left(l - \frac{z}{2}\right)\cosh k + \frac{l^2}{k^2}\cosh\frac{k}{l}z + \frac{l^2}{k}\sinh\frac{k}{l}(l-z)\right] \quad (10.77)$$

$$B_z = -\frac{ml^2}{k^2\cosh k}\left[\cosh k - \cosh\frac{k}{l}z - k\sin\frac{k}{l}(l-z)\right] \quad (10.78)$$

Equations (10.77) and (10.78) are basic equations applicable to a wide variety of braced cores. These equations, together with the sectorial properties of the core, completely define the core rotations and axial stresses in the columns.

To show the order of the magnitude of axial forces induced in the columns, the results of torsion and bending analysis for the building shown in Fig. 10.49a are presented. A uniform wind load of 20 psf (97.64 kg/m$^2$) is assumed to act on the $x$ face of the building for the full height. The diagram of principal sectorial coordinates and the sectorial moment of inertia for the core are given in Fig. 10.49d.

Conservatively, it will be assumed that the St Venant's torsional stiffness of the core is small compared to the warping stiffness. The parameter $k$ is equal to zero, giving simple expressions for rotation and bimoment in the core. The bimoment $B_0$ at the base will be $ml^2/2$ giving a value of $5.5 \times 10^6$ kip/ft$^2$. The axial forces in the columns are calculated from the relation:

$$P_c = \frac{B_0\omega_c^2 A_c}{I_\omega} \quad (10.79)$$

where $A_c$ is the area of the column under consideration. The calculated values for the example problem are shown in Fig. 10.49e. To allow comparison, the axial forces induced in the columns due to bending of the core are shown in Fig. 10.49.

In practice, the unbraced face of a building core is not completely open; it is partially closed at each floor level by beams framing between the columns. The effect of the beams is to constrain the warping deformations of the core. At the same time, the beams themselves are subjected to large shear and bending moments in having to comply with the warping of the core as shown in Fig. 10.49g. One approach to the problem, which is explained later in this chapter, is to employ a stiffness method of analysis for the beam and core system by considering each story segment of the core as a thin-walled beam element and by adding the effect of beams by incorporating the warping stiffness of the beam into the total stiffness of the core. Herein a different method analogous to the continuous connection technique, which can be carried out without the aid of a computer, is presented. Briefly, the procedure consists of replacing the effect of individual beams by a continuous plate equivalent in mechanical properties to the connecting beams.

The structure is rendered statically determinate by introducing an imaginary cut along a line bisecting the imaginary plate. The equilibrium equation is written for the core in terms of external transverse forces and equivalent shear forces applied along the cut edges of the plate. The compatibility condition for the relative displacement at the cut section leads to the differential equation for the core beam assembly. The detailed steps are as follows.

The equilibrium equation for nonuniform torsion of a thin-walled beam subjected to the action of external transverse loads and longitudinal shear forces applied along the edges can be shown to be

$$EI_\omega \frac{d^4\theta}{dz^4} - GJ\frac{d^2\theta}{dz^2} = m + \frac{dT}{dz}\Omega \tag{10.80}$$

where $m$ = the external torsional moment at $z$

$dT/dz$ = the derivative of the shear forces $T_z$ at section $z$

$\Omega$ = twice the area of the enclosed contour between the core and the beam

The vertical displacement $u_{z,s}$ which results from warping of the thin-walled beam is expressed according to the equation

$$u_{z,s} = -\frac{d\theta}{dz}\omega_s \tag{10.81}$$

where $d\theta/dz$ is the relative warping dependent upon the $z$ coordinate and $\omega_s$ is the sectorial area which depends on the location of the point $s$ on the contour.

The relative displacement at the cut section due to warping can be written thus

$$\delta_1 = -\frac{d\theta}{dz}(\omega_L - \omega_k)$$

or

$$\delta_1 = -\frac{\partial\theta}{\partial z}\Omega \tag{10.82}$$

The relative displacement $\delta_2$ at the cut due to the flexibility of the beam under the application of a shear force $T_z h$ is given by

$$\delta_2 = \frac{T_z h}{G}\left(\frac{\alpha^2 G}{12EI_b} + \frac{1.2}{A_b}\right) \tag{10.83}$$

where $h$ = the story height
$\alpha$ = the beam length
$I_b$ = moment of inertia of beam about the $y$ axis
$A_b$ = the area of the beam
$E$ and $G$ = the familiar material properties of the beam

For compatibility of displacement, we should have $\delta_1 + \delta_2 = 0$; i.e.,

$$\frac{d\theta}{dz}\Omega + \frac{T_z h}{G}\left(\frac{\alpha^2 G}{12EI_b} + \frac{1.2}{A_b}\right) = 0 \tag{10.84}$$

Differentiating with respect to $z$, we have

$$\frac{d^2\theta}{dz^2}\Omega + \frac{dT}{dz}\frac{h}{G}\left(\frac{\alpha^2 G}{12EI_b} + \frac{1.2}{A_b}\right) = 0 \tag{10.85}$$

Substituting for $dt/dz$ in the equilibrium equation and using the notation

$$J_b = \frac{\Omega}{\alpha h}\left\{\frac{\alpha^2 G}{12EI_b} + \frac{1.2}{A_b}\right\} \tag{10.86}$$

we get

$$EI_\omega \frac{d^4\theta}{dz^4} - G(J + J_b)\frac{d^2\theta}{dz^2} = m_z \tag{10.87}$$

This equation is identical to Eq. (10.75). Therefore, the two solutions given in Eqs (10.75) and (10.78) for the rotation and bimoment of the open core can also be used for the solution of the core and beam assembly. For this purpose, it is only necessary to replace

St Venant's torsional constant $J$ by the sum $J + J_b$ as given by Eq. (10.87). The parameter $k$ is now computed from the equation

$$k = l\sqrt{\frac{G(J - J_b)}{EI_\omega}} \tag{10.88}$$

The calculation of bimoment and axial forces follows the procedure outlined earlier.

For purposes of illustration, it is assumed in the problem that a W16 × 26 beam is moment-connected across the corridor columns. The values for $J_b$ and $k$ are found to be respectively equal to 55.38 $ft^4$ (0.478 $m^4$) and 19.5. Substituting the value of $k$ in Eq. (10.78), the bimoment at the base is found to be equal to 60 percent of the value obtained for the open core. The resulting values of the column loads are also 60 percent of the values shown in Fig. 10.49e.

### 10.4.12 Warping torsion constants for open sections

It is perhaps evident by now that although the concept of warping torsion is easy to assimilate, the calculation of sectorial properties are rather tedious. To alleviate this problem, formulas for the sectorial properties of open sections commonly used in shear wall structures, are given Table 10.7.

Let us verify the value of $I_\omega$ derived previously in Example 2, by using the formula given in the table (ref. no. 7 in Table 10.7)

$$I_\omega = \frac{h^2 t_1 t_2 b_1^3 b_2^3}{12(t_1 b_1^3 + t_2 b_2^3)}$$

$$h = 30\text{ ft}, \quad t_1 = t_2 = 2\text{ ft}, \quad b_1 = 20\text{ ft}, \quad b_2 = 10\text{ ft}$$

$$I_\omega = \frac{30^2 \times 2 \times 2 \times 10^3 \times 20^3}{12(2 \times 10^3 + 2 \times 20^3)} = 133{,}336\text{ ft}^6$$

This confirms the accuracy of the calculations done previously.

### 10.4.13 Computer analysis

**Modeling techniques.** The classical warping torsion theory for open sections described in the previous sections is useful for the analysis of buildings with uniform single cores and for buildings with a random distribution of planar shear walls with uniform properties. It is also helpful in understanding warping behavior and in getting a feel for the magnitude of axial forces resulting from torsion. However, the properties of cores and shear walls usually vary with the height of buildings with the walls and cores often interconnected to other

**TABLE 10.7 Torsion Constants for Open Sections**

| Cross-section reference no. | Constants |
|---|---|
| 1. Channel 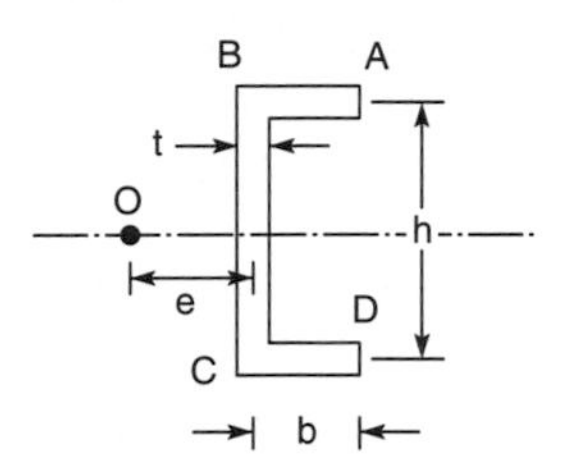  | $e = \dfrac{3b^2}{h + 6b}$<br>$J = \dfrac{t^3}{3}(h + 2b)$<br>$I_\omega = \dfrac{h^2b^2t}{12}\,\dfrac{2h + 3b}{h + 6b}$ |
| 2. *C* section | $e = b\dfrac{3h^2b + 6h^2b_1 - 8b_1^3}{h^2 + 6h^2b + 6h^2b_1 + 8h_1^3 - 12hb_1^2}$<br>$J = \dfrac{t^3}{2}(h + 2b + 2b_1)$<br>$I_\omega = l\left[\dfrac{h^2b^2}{2}\left(b_1 + \dfrac{b}{3} - e - \dfrac{2eb_1}{b} + \dfrac{2b_1^2}{k}\right) + \dfrac{h^2e^2}{2}\left(b + b_1 + \dfrac{h}{6} - \dfrac{2b_1^2}{h}\right) + \dfrac{2b_1^3}{3}(b + e)^2\right]$ |
| 3. Hat section 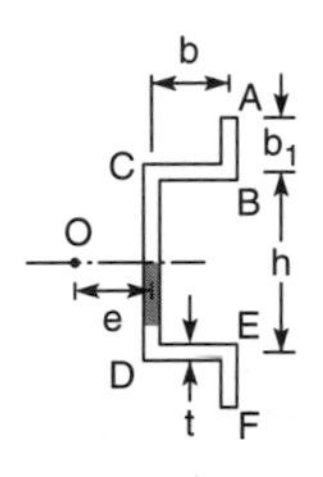  | $e = b\dfrac{3h^2b + 6h^2b_1 - 8b_1^2}{h^2 + 6h^2b + 6h^2b_1 + 8b_1^3 + 12hb_1^2}$<br>$J = \dfrac{t^3}{3}(h + 2b + 2b_1)$<br>$I_\omega = l\left[\dfrac{h^2b^2}{2}\left(b_1 + \dfrac{b}{3} - e - \dfrac{2eb_1}{b} - \dfrac{2b_1^2}{h}\right) + \dfrac{h^2c^2}{2}\left(b + b_1 + \dfrac{h}{6} + \dfrac{2b_1^2}{h}\right) + \dfrac{2b_1^2}{3}(b + e)^2\right]$ |
| 4. Twin channel with flanges inward 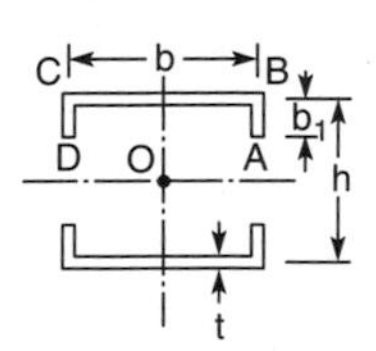  | $J = \dfrac{t^2}{3}(2b + 4b_1)$<br>$I_\omega = \dfrac{tb^2}{24}(8b_1^2 + 6h^2b_1 + h^2b + 12b_1^2h)$ |
| 5. Twin channel with flanges outward 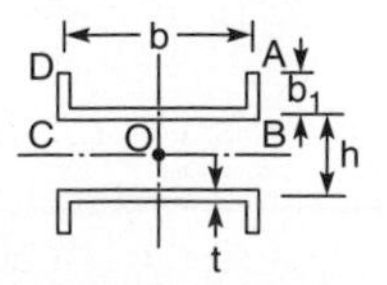  | $J = \dfrac{t^2}{3}(2h + 4b_1)$<br>$I_\omega = \dfrac{tb^2}{24}(8b_1^3 + 6h^2b_1 + h^2b - 12b_1^2h)$ |

**TABLE 10.7** (*Continued*).

| Cross-section reference no. | Constants |
|---|---|
| 6. Wide flanged beam with equal flanges 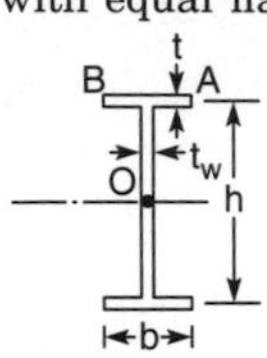  | $J = \frac{1}{3}(2t^3b + t_\omega^3h)$<br>$I_\omega = \frac{h^2tb^3}{24}$ |
| 7. Wide flanged beam with unequal flanges | $e = \frac{t_1b_1^2h}{t_1b_1^3 + t_2b_2^3}$<br>$J = \frac{1}{3}(t_1^3b_1 + t_2^3b_2 + t_w^3h)$<br>$I_\omega = \frac{h^2t_1t_2b_1^3b_2^3}{12(t_1b_1^3 + t_2b_2^3)}$ |
| 8. *Z* section 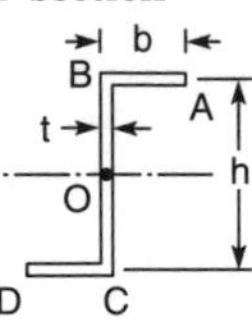  | $J = \frac{t^3}{3}(2b + h)$<br>$I_\omega = \frac{th^2b^3}{12}\left(\frac{b+h}{2b+h}\right)$ |
| 9. Segment of a circular tube 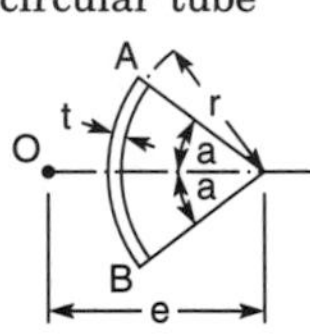  | $e = 2r\frac{\sin\alpha - \alpha\cos\alpha}{\alpha - \sin\alpha\cos\alpha}$<br>$J = \frac{2}{3}t^3r\alpha$<br>$I_\omega = \frac{2tr^5}{3}\left[\alpha^3 - 6\frac{(\sin\alpha - \alpha\cos\alpha)^2}{\alpha - \sin\alpha\cos\alpha}\right]$<br>For $\alpha = 45°$ $e = 1.06r$<br>$= 90°$ $e = 1.27r$<br>$= 180°$ $e = 2r$ |
| 10. | $e = 0.707ab^2\frac{3a - 2b}{2a^3 - (a-b)^3}$<br>$J = \frac{2}{3}t^3(a + b)$<br>$I_\omega = \frac{ta^4b^3}{6}\,\frac{3a + 2b}{2a^3 - (a-b)^2}$ |
| 11. | $J = \frac{1}{3}(4t^3b + t_w^3a)$<br>$I_\omega = \frac{a^2b^2t}{3}\cos^2\alpha$ |

lateral load-resisting elements such as moment frames. Such complex assemblies cannot be analyzed by the classical theory and it is necessary to revert to a stiffness matrix computer analysis, which can consider the total structure as an assemblage of discrete elements.

A significant aspect of commercial computer programs most often used in engineering practice for modeling of shear walls is that they do not require any knowledge of warping theory, nor do they require the calculation of the sectorial properties. Building systems with shear walls in practice are analyzed by using special panel elements that combine the versatility of the finite element method with the design requirement that the wall output be in terms of total moments and forces, instead of the usual finite element output of direct stresses, shear stresses and principal stress values.

A simple cantilever wall as shown in Fig. 10.50 may be modeled either as an assemblage of panel elements between the floors, or as a single column placed at the center of gravity of the wall. The single column is given the properties representing the wall axial area, shear area and moment of inertia. Rigid end zones on either side of the

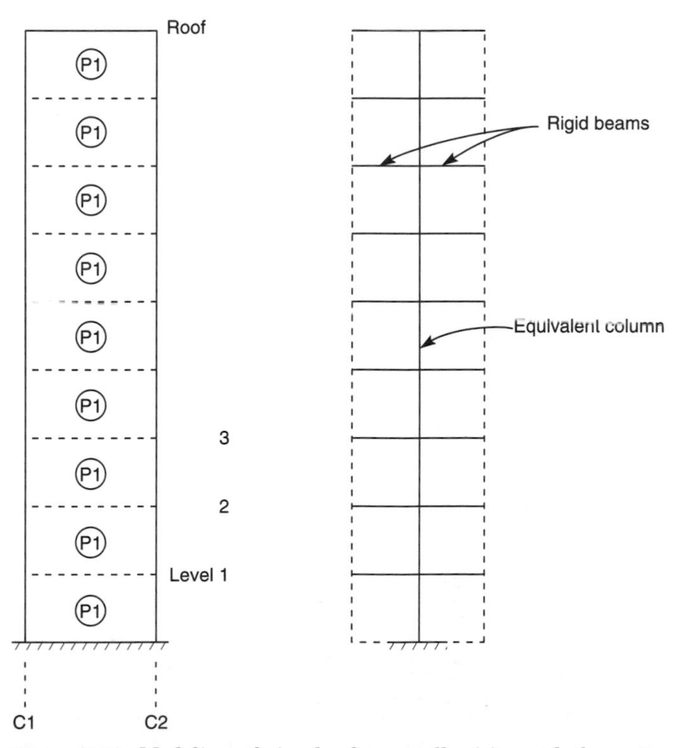


**Figure 10.50** Modeling of simple shear walls: (a) panel elements; (b) equivalent column.

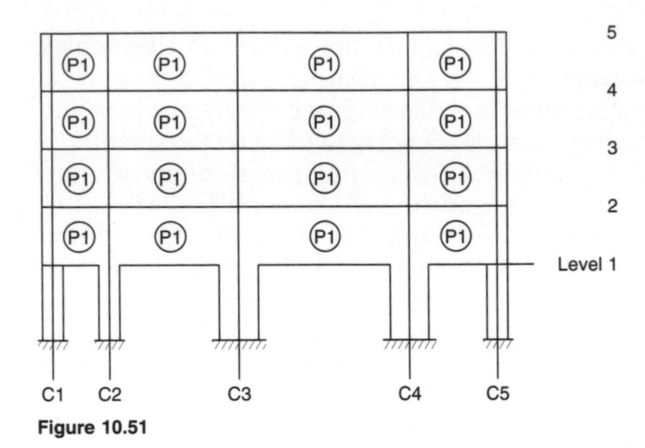


**Figure 10.51**

column are specified to capture the effects of finite dimensions of the wall on the stiffness of the system. In modeling the columns as panels, column lines $C_1$ and $C_2$ are used to define the extent of panels.

A discontinuous wall Fig. 10.51, in which the shear wall from

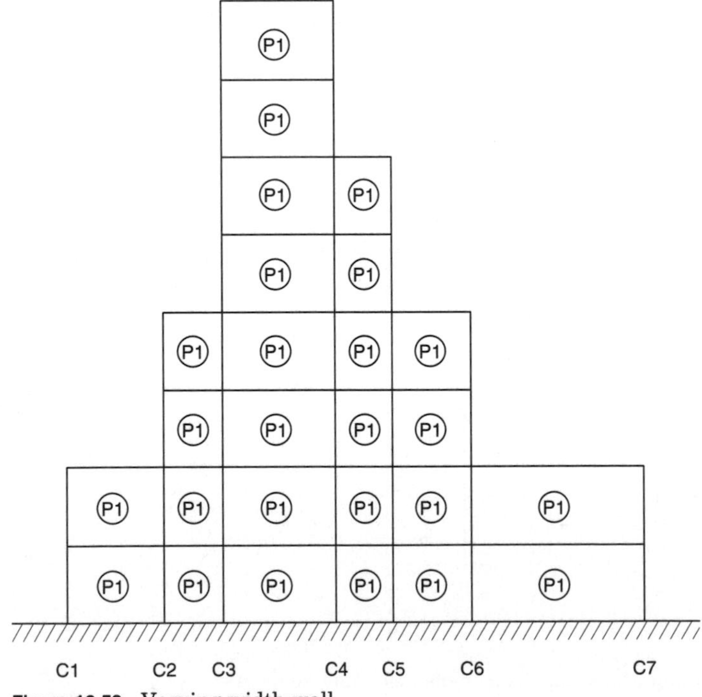


**Figure 10.52** Varying width wall.

above terminates on a series of columns below, requires four column lines $C_1$, $C_2$, $C_3$ and $C_4$, and three panels at each level.

When the width of a shear wall changes over the building height, as shown in Fig. 10.57, the model requires a column line corresponding to each width of wall. For example, in Fig. 10.52, seven columns lines are required to define the widths of the wall at various heights.

A wall with random openings may be modeled by using a combination of column lines and panel elements as shown in Fig. 10.53.

Modeling of three-dimensional shear walls such as *C, E,* and *L*-shapes is shown in Fig. 10.54 in which more than one panel element is used at each level.

## Single column model

As an alternative to defining three-dimensional walls with panel elements, it is possible to use a single column model with an extra, seventh degree-of-freedom per node, to represent warping. The principle advantage of the single warping-column model is its extremely concise form of representing warping. In this formulation, the seventh, warping, degree of freedom is taken to be $d\theta/dz$ to express the magnitude of warping, while $B$, the bimoment, is taken as the corresponding generalized force. In other words, just as we associate the rotations $\theta_x$ and $\theta_y$ with their corresponding moments $M_z$ and $M_y$, the rate of change of torsional rotation $\frac{d\theta}{dz}$ is associated with the bimoment $B$.

As part of his investigation into the torsional behavior of open section shear walls (ref. 20), the author developed a $14 \times 14$ stiffness

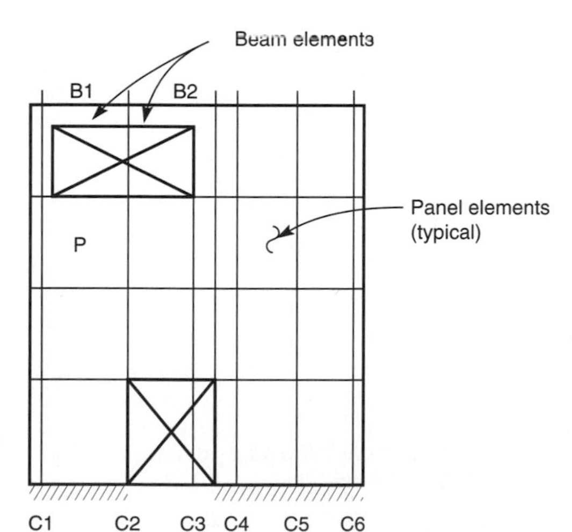


**Figure 10.53** Shear wall with random openings.

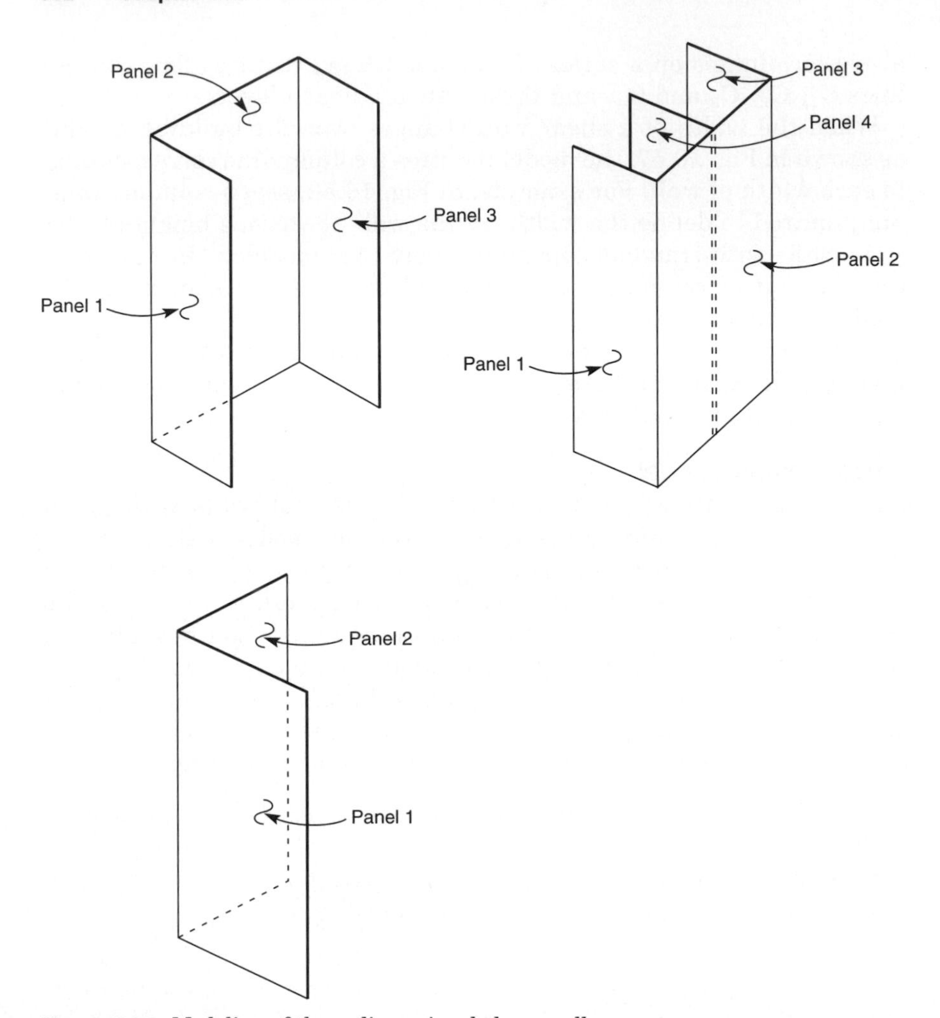


**Figure 10.54** Modeling of three-dimensional shear walls.

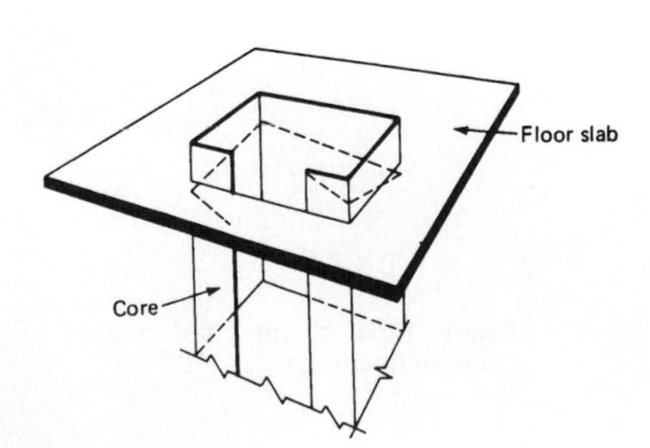


**Figure 10.55** Warping deformation of core.

matrix for defining the three-dimensional behavior of open sections. The purpose was to investigate the behavior of single and multiple open cores interconnected through rigid floor diaphragms. The investigation included the study of warping behavior of open sections as well as the "warping restraint" provided by the out-of-plane bending and twisting of the floor slab. A brief description of the methodology is given in the following sections.

**Warping stiffness of floor slab.** A floor slab surrounding core may be considered to have two distinct actions. One is to hold the cross-sectional shape of the core intact, by preventing its in-plane distortion, and the other is to restrain its longitudinal deformation due to warping. The in-plane action of the slab is tacitly taken into consideration in the torsion theory by assuming the profile of the core to be rigid. The out-of-planeto be rigid. The out-of-plane stiffness of the slab, the warping stiffness, is the subject of this discussion.

Figure 10.55 shows the warping deformation of the cone while Fig. 10.56 shows the corresponding deformation of the slab. Observe that the slab is under "torture" due to the out-of-plane bending and twisting indicating the likelihood of very large interacting forces. The warping of the cross-section which is the longitudinal displacement of the open core is given by the relation:

$$u_{z,s} = -\theta_z' \omega_s$$

The cross-sectional distortion $\theta_z'$ can be considered equivalent to a generalized displacement similar to, say, the rotation of the core about the $x$-axis. Just as the axial displacement due to $\theta_x$ varies across the cross-section as a function of the coordinate $y$, similarly the warping displacement varies across the profile in a manner similar to the warping coordinate.

Thus the distortion of the cross-section $\theta_z'$ can be considered as a generalized displacement. A unit value of $\theta_z'$, i.e., a "unit warping displacement" introduces up-and-down longitudinal displacements

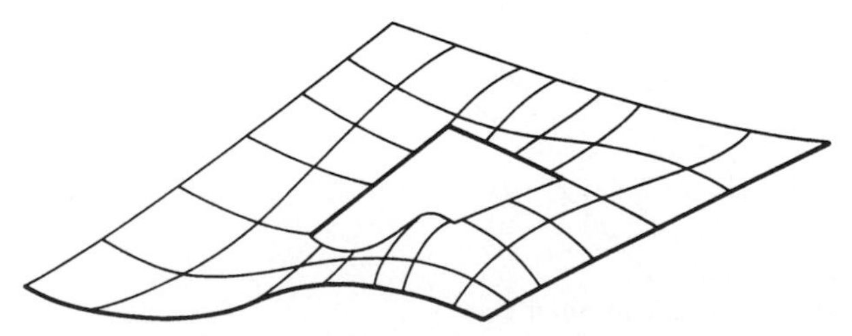

**Figure 10.56** Warping deformation of floor slab.

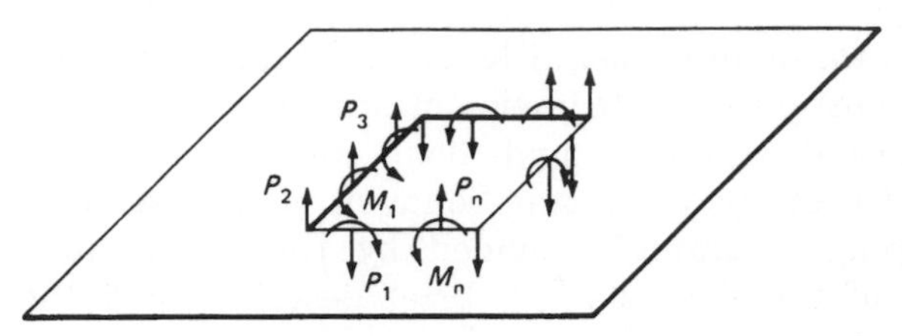


**Figure 10.57** Forces and moments due to warping of floor slab.

which vary across the cross-section in a manner similar to the warping coordinate.

When the core undergoes a warping deformation, the floor slab rigidly connected to the core is forced to bend and twist. The out-of-plane displacements of the slab and the warping displacement of the core must be the same along the boundary common to the core and the slab.

The displacement of the slab at its common boundary with the wall is, therefore, known from the warping displacement of the core. For a unit warping displacement of the core, the slab is displaced along the contact boundary in a manner similar to the warping coordinate diagram $\omega_s$ of the core. This displacement gives rise to continuous interactive forces consisting of distributed axial forces and moments at the inner edges of the slab. The evaluation of the warping stiffness of the slab, therefore, reduces to the determination of these interactive forces and moments which can be mathematically converted into a bimoment function, as will be seen shortly.

To study the slab-core interaction, consider the slab isolated from the core structure and subjected to an as yet unknown distribution of reactive forces and moments corresponding to the warping deformation of the core.

The resulting force system, which consists of concentrated forces

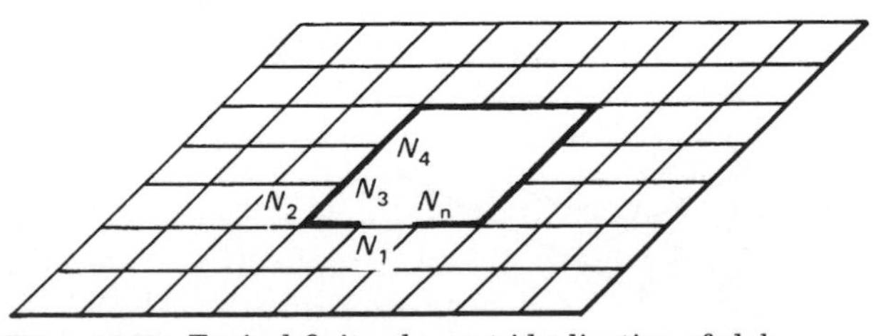


**Figure 10.58** Typical finite element idealization of slab.

$P_1, P_2, P_3, \ldots, P_n$, and moments $M_1, M_2, M_3, \ldots, M_n$ applied at the points $k = 1, 2, 3, \ldots, n$ of the cross-section, can be expressed as a bimoment by the relation

$$B_\omega = \sum_{k=1}^{n} P_k \omega_k + \sum_{k=1}^{n} M_k \frac{\partial \omega_k}{\partial_s} \tag{10.89}$$

where $B_\omega$ = the warping stiffness of floor slab

$\sum_{k=1}^{n} P_k \omega_k$ = the summation of the product of concentrated forces and the warping displacement

$\sum_{k=1}^{m} M_k \frac{\partial \omega_k}{\partial_s}$ = the summation of the concentrated transverse moments and the rate of change of warping function

The forces and moments are diagrammatically shown in Fig. (10.57). These are evaluated by using a finite element analysis.

**Finite element analysis.** A typical finite element idealization for the floor slab is shown in Fig. 10.58. To impose a unit $\theta_z'$ displacement at the inner boundary of the slab transverse displacements perpendicular to the plane of the slab and equal in magnitude and sense to the warping function are introduced at the nodes common to the core and the slab e.g., nodes $N_1, N_2, \ldots, N_n$ in Fig. 10.58. In addition to these vertical displacements, the slopes $\delta\omega/\delta x$ and $\delta\omega/\delta y$ are made equal in magnitude and sense to the slope of the $\omega_s$ diagram. Having thus given at the inner edge of the slab a displacement conforming to the warped outline of the core, the forces and transverse moments at the nodes are found from a finite element solution. The bimoment, which then corresponds to the required warping stiffness of the slab, is found from the relation given in Eq. (10.69). The effect of the floor system, which is mathematically equivalent to the bimoment, is incorporated into a stiffness type of analysis by adding the bimoment to the appropriate elements of the stiffness matrix of the core. A prerequisite for this operation is the derivation of stiffness coefficients for the open section. This is considered next.

**Twisting and warping stiffness of open sections.** The stiffness matrix for the open section is derived by solving the governing differential equation of torsion by imposing appropriate boundary conditions at each end. Defining a member as the segment of the core between two floors, the stiffness matrix for the restrained member is established first. Elements of this matrix are the values of restraint exerted at the ends of the member when unit displacements are imposed one at a time at each end. Concentrating on the torsion analysis only, the number of degrees of freedom at each end will be two, the rotation $\theta_z$ and the warping $\theta'_z$. Hence, the order of the restrained member stiffness matrix will be 4 by 4. Equation (10.49) is conveniently used for deriving the elements of the member stiffness matrix. This is reproduced here for convenience as Eq. (10.70).

$$\begin{bmatrix} \theta_z \\ \theta'_z \\ \dfrac{B_z}{GJ} \\ \dfrac{M_z}{GJ} \end{bmatrix} = \begin{bmatrix} 1 & \dfrac{l}{k}\sinh\dfrac{kz}{l} & 1-\cosh\dfrac{k}{l}z & z-\dfrac{l}{k}\sinh\dfrac{kz}{l} \\ 0 & \cosh\dfrac{kz}{l} & -\dfrac{k}{l}\sinh\dfrac{kz}{l} & 1-\cosh\dfrac{kz}{l} \\ 0 & -\dfrac{l}{k}\sinh\dfrac{kz}{l} & \cosh\dfrac{kz}{l} & \dfrac{l}{k}\sinh\dfrac{kz}{l} \\ 0 & 0 & 0 & 1 \end{bmatrix} \begin{bmatrix} \theta_0 \\ \theta'_0 \\ \dfrac{B_0}{GJ} \\ \dfrac{M_0}{GJ} \end{bmatrix} \tag{10.90}$$

Now, to find the elements of the first row of the stiffness matrix, which correspond to the torque and bimoment at each end of the beam required to produce a unit rotation of $\theta = 1$ at $z = 0$ while all the other three displacements are zero, it is necessary to introduce the appropriate boundary conditions in Eqs. (10.70). Thus

$$\begin{bmatrix} 0 \\ 0 \\ \dfrac{B_l}{GJ} \\ \dfrac{M_l}{GJ} \end{bmatrix} = \begin{bmatrix} 1 & \dfrac{l}{k}\sinh\dfrac{kz}{l} & 1-\cosh\dfrac{k}{l}z & z-\dfrac{l}{k}\sinh\dfrac{kz}{l} \\ 0 & \cosh\dfrac{kz}{l} & -\dfrac{k}{l}\sinh\dfrac{kz}{l} & 1-\cosh\dfrac{kz}{l} \\ 0 & -\dfrac{l}{k}\sinh\dfrac{kz}{l} & \cosh\dfrac{kz}{l} & \dfrac{l}{k}\sinh\dfrac{kz}{l} \\ 0 & 0 & 0 & 1 \end{bmatrix} \begin{bmatrix} 1 \\ 0 \\ \dfrac{B_0}{GJ} \\ \dfrac{M_0}{GJ} \end{bmatrix} \tag{10.91}$$

Solving these equations, the four forces $M_0$, $B_0$, $M_l$, and $B_l$ at the two ends are obtained. In the same manner, the remaining elements of the stiffness matrix are obtained by introducing the appropriate boundary conditions. These are shown diagrammatically in Fig. (10.59). Adapting the sign convention shown therein, the member stiffness matrix for the thin-walled beam subjected to torsion will

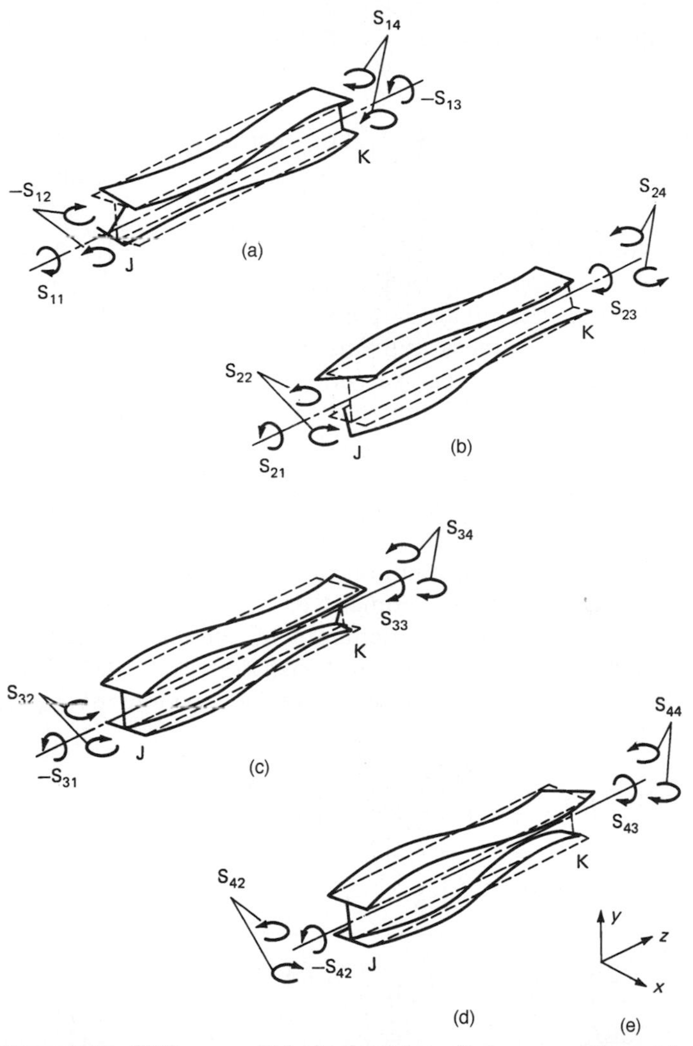


**Figure 10.59** Stiffness coefficients for thin-walled open sections: (a) unit rotation at $J$; (b) unit warping displacement at $J$; (c) unit rotation at $K$; (d) unit warping displacement at $K$; (e) coordinate axes.

take the form,

$$\left[\frac{GJ}{2+k\sinh k-2\cosh k}\right]$$

$$\times\begin{bmatrix} (1-\cosh k) & \frac{l}{k}(k\cosh k - \sin k) & -(1-\cosh k) & \frac{l}{k}(\sinh k-k) \\ \frac{-k}{l}\sinh k & -(1-\cosh k) & \frac{k}{l}\sinh k & -(1-\cosh k) \\ (1-\cosh k) & \frac{l}{k}(\sinh k-k) & -(1-\cosh h) & \frac{l}{k}(k\cosh k - \sinh k) \end{bmatrix} \tag{10.92}$$

To the best of the author's knowledge, he was the first to derive $\theta$, $\theta'$, $B$, and $M$ in the form of stiffness coefficients. Such a matrix is very convenient for including the effect of beams and slabs, and can be extended to a generalized stiffness analysis by treating open cores as special columns with seven degrees of freedom at each end.

When the torsional rigidity $GJ$ due to St Venant torsion becomes zero, the differential equation of constrained torsion will be analogous to the equation of bending of a beam. The elements of the [S] matrix (Eqns (10.72)] can be written directly from the beam stiffness matrix if $M$, $B$, $\theta$, $\theta'$, and $I_\omega$ are replaced by $P$, $M$, $w$, $w'$, and $I_{xx}$, where $P$ is the shear force, $M$ is the bending moment, $w$ is the displacement, $w'$ is the slope $dw/dz$, and $I_{xx}$ is the moment of inertia. Further, this analogy can be used to check the elements of the matrix [S] by writing the expanding forms for $\sinh k$ and $\cosh k$ in Eq. (10.72) and taking $k \to 0$ in the limit. For example, considering one element $S_{11}$ of Eq. (10.72) we get

$$S_{11}=\frac{GJk}{l(2+k\sinh k-2\cosh k)}\sinh k$$

when $k \leftarrow 0$,

$$S_{11}=\frac{EI_\omega k^2\left(\frac{k}{l}\right)\left(k+\frac{k^3}{6}+\frac{k^5}{120}\right)}{l^2\left[2+h\left(k+\frac{k^3}{6}+\frac{k^5}{120}\right)-2\left(1+\frac{k^2}{2}+\frac{k^4}{24}\right)\right.}$$

$$=\frac{12EI_\omega}{l^3}$$

which is of the same form as the element corresponding to the shear force in the well-known beam stiffness matrix.

**Stiffness method using warping-column model.** Building structures are generally analyzed as three-dimensional frames, with the members oriented in any direction and subjected to axial force, shear and moment in two orthogonal planes, and torsion about their linear axes. Therefore, a general beam or column element in the analysis of three-dimensional frames, must include forces in three directions and moments about three axes. Such a beam element with six displacements at each end is shown in Fig. 10.60. The stiffness matrix which is the relationship between the end forces and displacements is a $12 \times 12$ matrix, corresponding to six degrees of freedom at each end. The stiffness coefficients depicting the force-displacement relation for the three-dimensional beam element is found by combining the stiffness terms for axial deformation, bending about two axes, and torsion. The resulting $12 \times 12$ stiffness matrix is given in Fig. 10.60.

A non-planar shear wall such as an *I* or a *C* shaped wall is

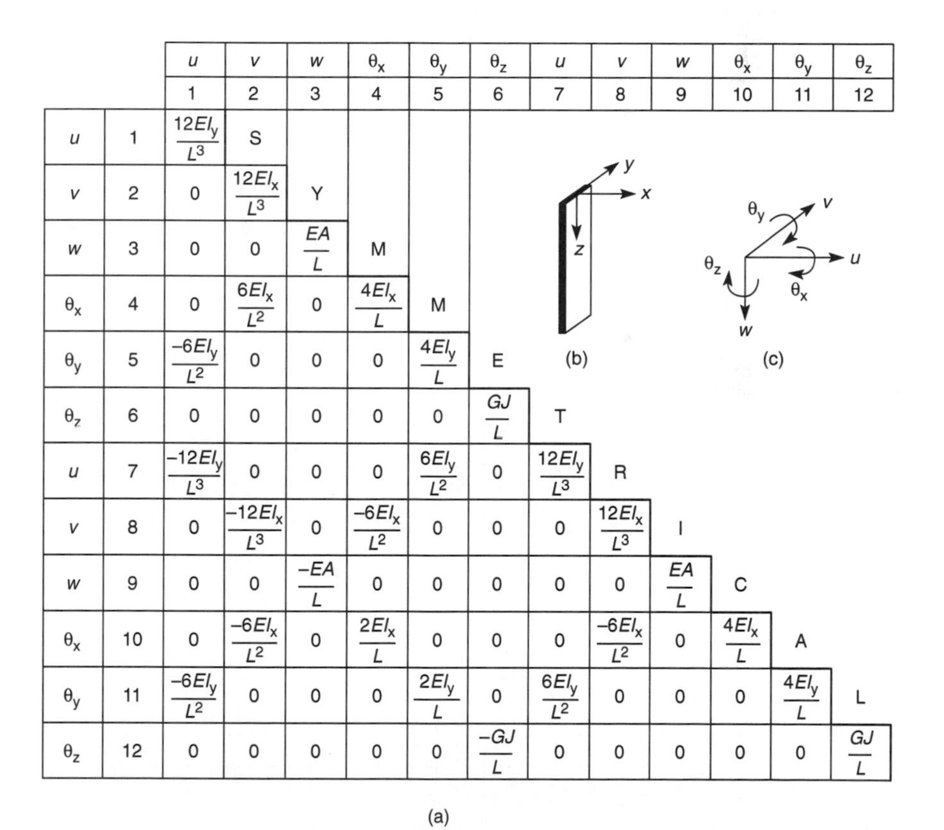

| | | $u$ | $v$ | $w$ | $\theta_x$ | $\theta_y$ | $\theta_z$ | $u$ | $v$ | $w$ | $\theta_x$ | $\theta_y$ | $\theta_z$ |
|---|---|---|---|---|---|---|---|---|---|---|---|---|---|
| | | 1 | 2 | 3 | 4 | 5 | 6 | 7 | 8 | 9 | 10 | 11 | 12 |
| $u$ | 1 | $\frac{12EI_y}{L^3}$ | S | | | | | | | | | | |
| $v$ | 2 | 0 | $\frac{12EI_x}{L^3}$ | Y | | | | | | | | | |
| $w$ | 3 | 0 | 0 | $\frac{EA}{L}$ | M | | | | | | | | |
| $\theta_x$ | 4 | 0 | $\frac{6EI_x}{L^2}$ | 0 | $\frac{4EI_x}{L}$ | M | | | | | | | |
| $\theta_y$ | 5 | $\frac{-6EI_y}{L^2}$ | 0 | 0 | 0 | $\frac{4EI_y}{L}$ | E | | | | | | |
| $\theta_z$ | 6 | 0 | 0 | 0 | 0 | 0 | $\frac{GJ}{L}$ | T | | | | | |
| $u$ | 7 | $\frac{-12EI_y}{L^3}$ | 0 | 0 | 0 | $\frac{6EI_y}{L^2}$ | 0 | $\frac{12EI_y}{L^3}$ | R | | | | |
| $v$ | 8 | 0 | $\frac{-12EI_x}{L^3}$ | 0 | $\frac{-6EI_x}{L^2}$ | 0 | 0 | 0 | $\frac{12EI_x}{L^3}$ | I | | | |
| $w$ | 9 | 0 | 0 | $\frac{-EA}{L}$ | 0 | 0 | 0 | 0 | 0 | $\frac{EA}{L}$ | C | | |
| $\theta_x$ | 10 | 0 | $\frac{-6EI_x}{L^2}$ | 0 | $\frac{2EI_x}{L}$ | 0 | 0 | 0 | $\frac{-6EI_x}{L^2}$ | 0 | $\frac{4EI_x}{L}$ | A | |
| $\theta_y$ | 11 | $\frac{-6EI_y}{L^2}$ | 0 | 0 | 0 | $\frac{2EI_y}{L}$ | 0 | $\frac{6EI_y}{L^2}$ | 0 | 0 | 0 | $\frac{4EI_y}{L}$ | L |
| $\theta_z$ | 12 | 0 | 0 | 0 | 0 | 0 | $\frac{-GJ}{L}$ | 0 | 0 | 0 | 0 | 0 | $\frac{GJ}{L}$ |



**Figure 10.60** (a) Twelve-by-twelve stiffness matrix for prismatic three-dimensional element; (b) coordinate axes; (c) positive sign convention.

modeled in a three-dimensional analysis as an assemblage of floor-to-floor panel elements connected along their edges. The continuous connection between the panels provides for the principal interaction, the vertical shear, along their connecting edges.

As an alternative technique, a three-dimensional wall may be represented in all its aspects of behavior including warping, by a warping column element, with seven degrees of freedom at each floor level. Its assigned properties would include the warping moment of inertia $I_\omega$, in addition to the familiar area $A$, to represent its resistance to axial load, inertias $I_x$ and $I_y$ to represent its resistance to bending about its principal axes, and $J$ to represent its St Venant's resistance to torsion. Such a single-column model, with an extra, seventh degree of freedom, as developed by the author in (ref. 20), is particularly suitable for open-section walls that are uniform over the height. The seventh, warping, degree of freedom is the parameter $d\theta/dz$ which expresses the magnitude of warping. It is used as the warping degree of freedom, while $B$, the bimoment, becomes the corresponding generalized force. Thus, with seven degrees of freedom per node, the column element (Fig. 10.61) has a 14 × 14 stiffness

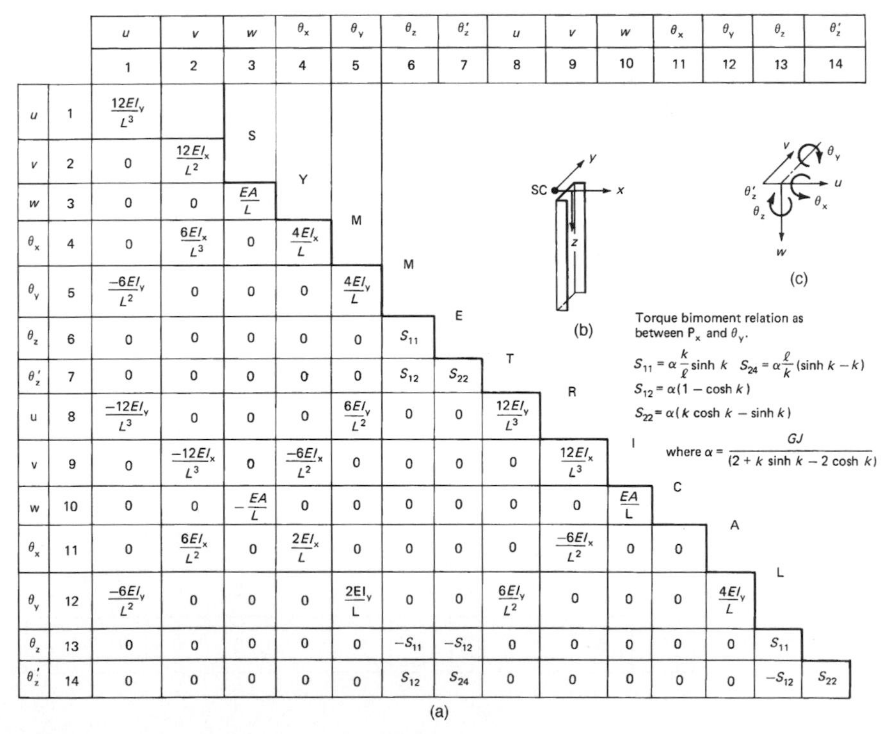


**Figure 10.61** (a) Fourteen-by-fourteen stiffness matrix for thin-walled open section; (b) coordinate axes; (c) positive sign convention.

matrix. A number of such story-height elements may be stacked vertically to represent a complete core. The interaction of slabs and beams at the floor levels may also be included in the stiffness matrix of the total structure by an appropriate combination of floor stiffness matrix with the stiffness matrix of the core, as given in the author's first publication (ref. 20). Engineers engaged in developing special-purpose computer programs may find this reference useful for including the additional warping degree of freedom for open-section shear wall buildings.

Chapter

# 11

# Structural Design

Thus far we have taken a world tour in Chap. 1, to witness structural schemes used in practice, have learned about steel, concrete and composite lateral and gravity systems in subsequent chapters, and finally in the previous chapter observed how these systems are analyzed, first for preliminary design using hand calculations, and then for final design using computer techniques. Our knowledge gained up to this point should enable us to conceptually design both the lateral and gravity systems. However, structural engineering does not stop here, at the conceptual level. Although the conceptual design is the most important step in developing an economical system, the concept has to be translated into buildable drawings by exercising judgement at each step of the design process. An important and inevitable step involves the so-called code check process in which the individual members of the system are verified to assure that they: (i) are not overstressed; and (ii) do not exceed the normally accepted deflection and stability criteria. Before the advent of computer-aided design, excruciating hand calculations supplemented by standard design tables were the only means of accomplishing this task. Today the code check process is invariably relegated to computer programs that not only analyze the building but more importantly check the individual members for stress, stability and deflection criteria.

It is imperative that engineers understand the codes and design specifications as both tools and constraints in design practice. Today's codes are usually a mutual concession between the past and the future knowledge of engineering, serving as a safety net for the young as well as the seasoned engineer. We will, in this chapter, discuss these to a limited extent to point out how they relate to the design of tall buildings. The specification will be the 1989 Specification for the Design Fabrication and Erection of Structural Steel for Buildings of AISC. The codes will be the 1994 Uniform Building Code (UBC) of the International Conference of Building Officials, and the ACI 318-89 (revised 1992).

## 11.1 Steel Design

The design procedure described in this section is based on the historic "Working Stress", the elastic approach. Safety is realized by limiting the permissible stresses to values considerably below the elastic limit of the steel. The design aspects as they apply to the column, beam and brace elements, and beam-to-column joints are described to give an overview as well as pertinent discussions of the code equations. Reference to applicable sections and equations for the AISC-ASD89 and UBC94 are indicated with the "AISC" and UBC prefix. For simplicity, all equations and descriptions presented in this section correspond to inch kip second units, by far the most prevalent in the U.S. practice.

### 11.1.1 Design load combinations

The design load combinations are required for determining the most severe combinations of the load conditions for which the structure needs to be designed. Generally, the structure is analyzed for a number of independent load conditions, and then the load combination multipliers are applied to obtain the design forces and moments for each load combination.

If a structure is subjected to dead $D$, and live load $L$ only, the stress check may need only one load combination, namely $D + L$ (AISC A4). However, in addition to the dead and live load the structure is invariably subjected to wind forces, and in seismic zones to earthquake forces. Considering wind from two mutually perpendicular directions $W_x$ and $W_y$ and noting that wind forces are reversible, the following nine load combinations may have to be investigated.

1. $D + L$
2. $0.75(D + L + W_x)$
3. $0.75(D + L - W_x)$

4. $0.75(D + L + W_y)$
5. $0.75(D + L - W_y)$
6. $0.75(D + \quad W_x)$
7. $0.75(D - \quad W_x)$
8. $0.75(D + \quad W_y)$
9. $0.75(D - \quad W_y)$

The 0.75 factor associated with wind load combinations is for the $33\frac{1}{3}$ percent increase in the stress allowed for short-term loads (AISC A5.2). By using this factor all stress ratios can be compared to an allowable stress ratio of 1.0, irrespective of whether the controlling load combination has contributions from short-term loads such as wind and seismic, or not.

If seismic design is based on the 94 UBC, the following load combinations may have to be considered:

1. $D + L$
2. $0.75(D + L + \sqrt{E_x^2 + E_y^2} + E_t)$
3. $0.75(D + L + \sqrt{E_x^2 + E_y^2} - E_t)$
4. $0.75(D + L - \sqrt{E_x^2 + E_y^2} + E_t)$
5. $0.75(D + L - \sqrt{E_x^2 + E_y^2} - E_t)$
6. $0.75(D + \quad \sqrt{E_x^2 + E_y^2} + E_t)$
7. $0.75(D + \quad \sqrt{E_x^2 + E_y^2} - E_t)$
8. $0.75(D - \quad \sqrt{E_x^2 + E_y^2} + E_t)$
9. $0.75(D - \quad \sqrt{E_x^2 + E_y^2} - E_t)$

Again the 0.75 factor is for the one-third increase in the allowable stresses, so that all stress ratios are compared against unity. In the above load combinations $E_x$, $E_y$, and $E_t$ are the $x$ and $y$ direction seismic loads and accidental torsion seismic load respectively.

### 11.1.2 Tension members

Although any type of steel cross-section may be used as a tension member, in practice the selection is usually influenced by the type connections at the ends.

The allowable stress in tension $F_t$ is based both on the yield criteria over the gross section and the fracture criteria based on effective net area. Thus

$$F_t = 0.6F_y \qquad \text{(yield criteria)}$$

$$F_t = 0.5F_u \qquad \text{(fracture criteria)}$$

For pin-connected members the allowable tension is given by

$$F_t = 0.45F_y \qquad \text{(across the pin hole)}$$

$$F_t = 0.6F_y \qquad \text{(across the body of eye bar)}$$

In the above equations, $F_y$ is the specified minimum yield stress of the type of steel being used, and $F_u$ is the minimum tensile strength, both in the units of kips per square inch.

A member selected for an axial tension $T$ thus requires a gross area

$$A_g \geqslant \frac{T}{0.6F_y}$$

Alternatively when applied to the effective net area

$$A_e \geqslant \frac{T}{0.5F_u}$$

$A_g$ is the gross area of the section and $A_e$ is the effective net area, both as defined in the AISC. If end connections are welded and the area not otherwise reduced, $A_g = A_e$. For bolted members, determination of the reduced cross section area $A_n$ is required by searching for different possible chains of holes. When a chain of holes is straight and perpendicular to the member axis

$$A_n = A_g - \phi N t$$

where $\phi$ = effective diameter of the hole = $\phi_b + \frac{1}{8}''$: ($\phi_b$ is the bolt diameter)
$N$ = number of bolts in the chain
$t$ = thickness of material being connected

Although the slenderness ratio $l/r$ for tension members is not critical, AISC recommends a nonmandatory limit of $l/r \leqslant 240$ for main members $l/r \leqslant 300$ for bracing and secondary members. Flexible members such as cables, round or square rods, or thin, wide bars are excluded from this requirement.

### 11.1.3 Members subject to bending

#### 11.1.3.1 Lateral stability

Consider a uniformly loaded continuous wide-flange beam as shown in Fig. 11.1. The beam segment between the points of contraflexure

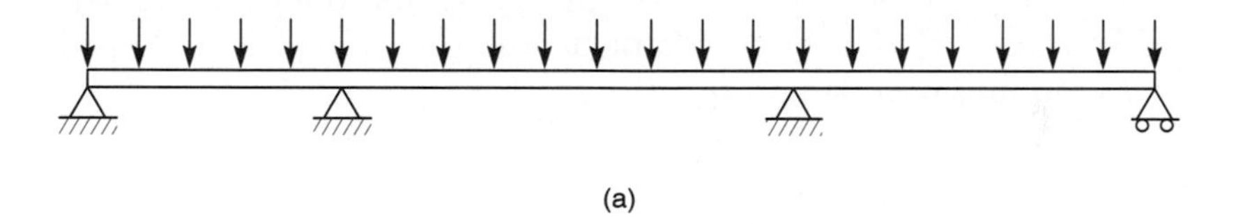

(a)

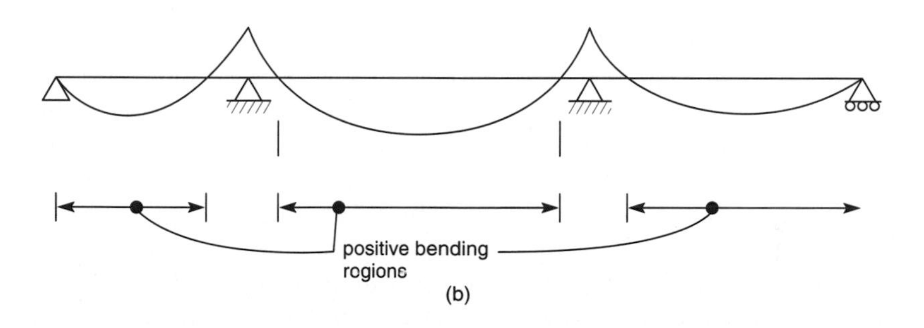


(b)

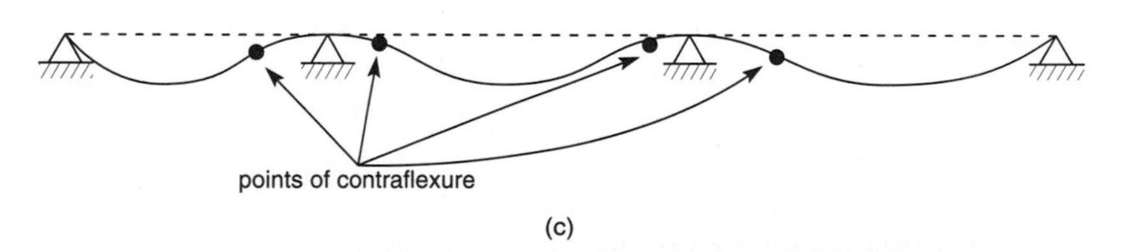


(c)

**Figure 11.1** Lateral buckling of beams: (a) continuous beam with uniformly distributed load; (b) bending moment diagram; (c) positive bending region between points of contraflexure.

is subjected to positive bending with the top portion in compression throughout this region, acting in a manner similar to a column. Unless there are closely spaced restraints, i.e. $l_y \leq L_c$, the compression portion of the beam has a tendency to buckle laterally at some value of critical moment, $M_{cr}$, analogous to the critical load, $P_{cr}$, at which the column would buckle. The mode of lateral buckling is in torsion, partly due St Venant's twisting, and partly due to warping torsion, the latter induced by the bending of beam flanges in opposite directions. As seen previously in Chap. 10, deep I-shaped open sections typically have large values for warping moment of inertia $I_\omega$. Consequently the buckling mode for such beams is dominated by flange bending. On the other hand, for shallow I-beams, the St Venant's torsion dominates the torsional response. These two considerations, the warping torsion, and the St Venant's torsion lead to the two AISC equations, discussed presently for determining the allowable stress $F_b$ for both compact and non-compact I-shaped sections with unbraced length $l_y > L_c$.

#### 11.1.3.2 Compact, semicompact and noncompact sections

There are two main categories of beams, compact and noncompact. Compact beams by virtue for their special controls on their geometry, are particularly stable. Semicompact beams satisfy all tests for compactness except one. All other beams are considered noncompact. Compact section criteria are based on the yield strength of steel, the type of cross section, the ratios of width to thickness of the elements of cross section, and the ratio $l/r_t$ of the length, $l$, between lateral supports of the compression flange to the weak radius of gyration, $r_t$ of that flange. Members, whether they are rolled shapes or shapes made from plates, are compact if they fulfil the criteria. To meet all criteria, the member must by symmetrical about its minor axis, meet certain stringent limits of $b_f/t_f$ and $h/t_w$ ratios, have unsupported length of compression flange less than $L_c$, and be bent about its major axis. Nonfulfilment of any of the criteria will degrade the member as noncompact. A beam is considered semi-compact when its $b_f/2t_f$ is in the range between the special compact limit of $65/\sqrt{F_y}$ and the limit $95/\sqrt{F_y}$. Linear adjustments are made in $F_b$ for both major and minor axis bending.

#### 11.1.3.3 Allowable bending stresses

To obtain the allowable bending stress the following criteria are used. For all I-sections, C-sections, T-sections, angles and double angles the allowable major direction bending stress is computed based on the compactness criteria and the laterally unbraced length, $l_y$. If $l_y$ is less than

$$\frac{76b_f}{\sqrt{F_y}} \text{ and } \frac{20{,}000}{\left(\frac{d}{A_f}\right)F_y} \qquad \text{(AISC F1-2)}$$

and the section is compact, the allowable major direction bending stress is taken as

$$F_{bx} = 0.66F_y \qquad \text{(AISC F1-1)}$$

If $l_y$ is less than the above limits and the section is non-compact,

$$F_{bx} = 0.60F_y \qquad \text{(AISC F1-5)}$$

If the ubraced length $l_y$ exceeds the above limits, then for both compact and non-compact sections

$$F_{bx} = \left[\frac{2}{3} - \frac{F_y\left(\frac{l}{r_T}\right)^2}{1530 \times 10^3 C_b}\right]F_y \leqslant 0.60F_y \qquad \text{(AISC F1-6)}$$

when $\sqrt{\frac{102 \times 10^3 C_b}{F_y}} \leqslant \frac{l}{r_T} \leqslant \sqrt{\frac{510 \times 10^3 C_b}{F_y}}$.

When $\frac{l}{r_T} \geqslant \sqrt{\frac{510 \times 10^3 C_b}{F_y}}$, $F_{bx} = \frac{170 \times 10^3 C_b}{\left(\frac{l}{r_T}\right)^2} \leqslant 0.60F_y$ (AISC 1-7)

For any value of $l/r_T$

$$F_{bx} = \frac{12 \times 10^3 C_b}{l_y\left(\frac{d}{A_f}\right)} \leqslant 0.6F_y \qquad \text{(AISC F1-8)}$$

In the above equations

$$C_b = 1.75 + 1.05\left(\frac{M_1}{M_2}\right) + 0.3\left(\frac{M_1}{M_2}\right)^2 \leqslant 2.3$$

$M_1$ and $M_2$ are end moments of the unbraced segment and $M_1$ is less than $M_2$. The ratio $\dfrac{M_1}{M_2}$ is positive for double curvature bending and negative for single curvature bending. Also, if any moment within the segment is greater than $M_2$, $C_b$ is taken as 1.0.

Figure 11.2 shows the curves for $F_{bx}$ for compact and noncompact sections. Note there are two curves. The first is based on the

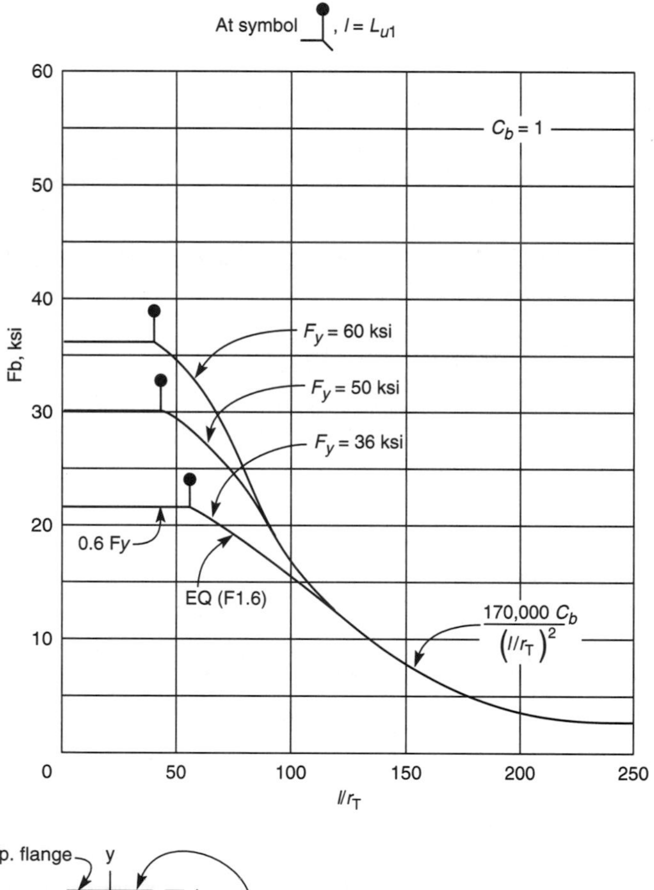


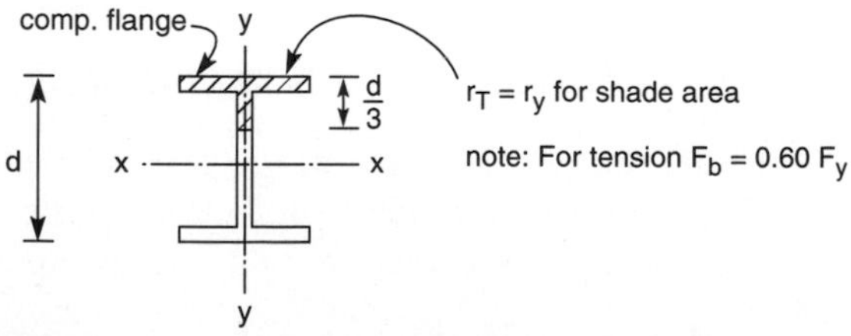


**Figure 11.2** Values of $F_b$ (compression) for compact and non-compact I-sections: (a) $F_b$ versus $l/r_t$; (b) $F_b$ versus $ld/A_f$.

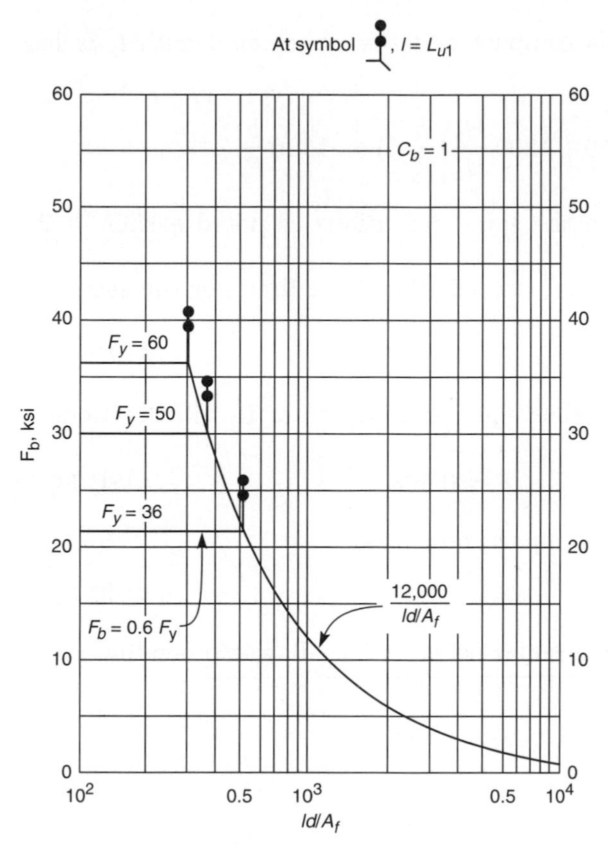


**Figure 11.2** (*Continued*).

assumption that buckling of the unbraced compression flange is initiated by warping torsion, and the second by St Venant's torsion.

The minor direction allowable bending stress, $F_{by}$ is taken as

$$F_{by} = 0.60F_y \qquad \text{(AISC F2-2)}$$

except in the case of compact I-sections, it is taken as

$$F_{by} = 0.75F_y \qquad \text{(AISC F2-1)}$$

This is because a compact I-shape bent about its mirror axis does not have to satisfy the criteria for unsupported length since the major axis stiffness provides a continuous lateral support. For box sections and rectangular tubes, the allowable bending stress in both the major and minor directions is taken as

$$F_b = 0.66F_y \qquad \text{(AISC F3-1)}$$

provided the section is compact and the unbraced length $l_y$ is less than the greater of

$$\left(1950 + 1200\frac{M_1}{M_2}\right)\frac{b}{F_y} \quad \text{or} \quad 1200\frac{b}{F_y}$$

where $M_1$ and $M_2$ have the same definitions as noted earlier in the formula for $C_b$.

If the unbraced length $l_y$ exceeds the above limits or the section is non-compact and

$$F_b = 0.60F_y \qquad \text{(AISC F3-3)}$$

For pipe sections the allowable bending stress in all directions is taken as

$$F_b = 0.66F_y \qquad \text{(AISC F3-1)}$$

provided the section is compact, otherwise

$$F_b = 0.60F_y \qquad \text{(AISC F3-3)}$$

For rectangular and circular sections the allowable bending stress in both major and minor directions is taken as

$$F_b = 0.66F_y$$

#### 11.1.3.4 Allowable shear stress

The allowable shear stress $F_v$ is taken as $0.40F_y$ (AISC F4-1). For very slender webs, where $\frac{h}{t_w} > \frac{380}{\sqrt{F_y}}$, a reduction in the allowable shear stress applies and must be separately investigated (AISC F4).

### 11.1.4 Members subjected to compression

#### 11.1.4.1 Buckling of columns

In structural design, a column is considered slender if its cross-sectional dimensions are small compared to its length. The degree of slenderness is measured in terms of the ratio $l/r$, where $l$ is the unsupported length of the column and $r$ is the radius of gyration. Whereas a stocky column fails by crushing or yielding, a slender column does so by buckling.

Before examining the AISC design equations, a review of column behavior is useful to understand the design parameters. Since the derivations of the column buckling formulas may be found in strength-of-material textbooks, the emphasis here is only on the column behavior as related to design.

Euler enunciated more than 200 years ago that a straight concentrically loaded pin-ended slender column fails by buckling at a critical load

$$P_c = \frac{\pi^2 EI}{l^2}$$

where $E$, $I$, and $l$ are the familiar notations for Young's modulus, moment of inertia, and the unsupported length of the column. Dividing $P_c$ by the cross-sectional area $A$ of the column, the expression for the critical load may be written in terms of the critical average stress $f_c$ on the gross section of the column

$$\frac{P_c}{A} = f_c = \frac{\pi^2 EI}{l^2 A}$$

Substituting $I = Ar^2$, where $r$ is the radius of gyration gives the stress form of Euler's equation

$$f_c = \frac{\pi^2 E}{(l/r)^2}$$

A plot of the critical stress versus the slenderness ratio, a so-called column curve, is shown in Fig. 11.3, illustrating the reduction in column strength as the slenderness increases. Stocky columns do not fail by buckling but do so by yielding or crushing of the material.

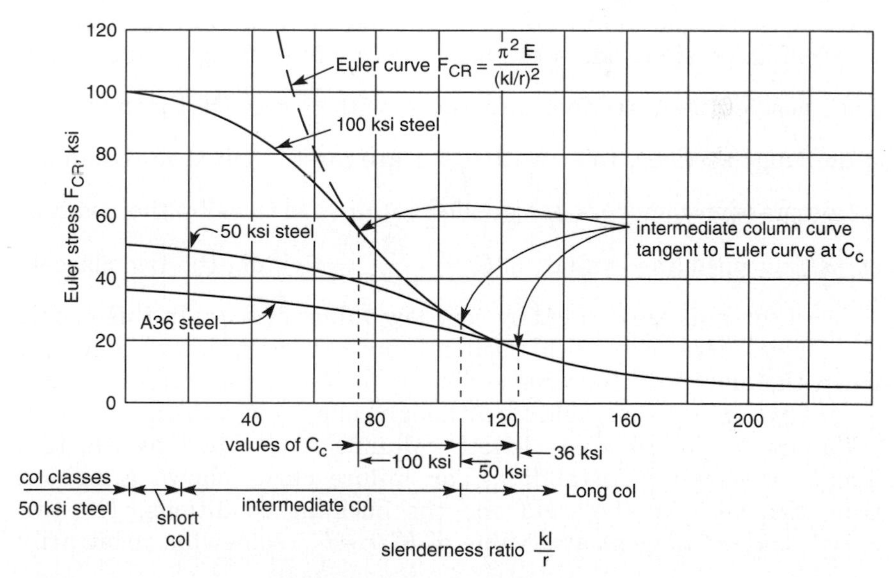


**Figure 11.3** Critical stress $\frac{P_{cr}}{A}$ versus $\frac{Kl}{r}$.

There is a limiting slenderness ratio below which failure occurs by crushing, while for larger values, the mode of failure is by buckling.

The expression for buckling load $P_c$ is for an idealized column supported by frictionless supports; a condition that exists rarely in practice. Building columns are connected to beams which restrain column rotation, thereby inducing end moments. Additionally, columns experience lateral deflections. Therefore, to determine the critical loads for practical cases, the idea of an effective length of column is used in design. The effective length is expressed as a product of actual length times a factor *K,* called the effective length factor. The critical load for practical cases is given by the relation

$$P_c = \frac{\pi^2 EI}{Kl^2}$$

**11.1.4.2 Column curves**

To understand the performance of compression members consider again the curves in Fig. 11.3 which show the failure stress verses the slenderness ratio, $Kl/r$, for three grades of steel. Two things are clear from the figure: the yield strength of steel is very significant for short columns, of decreasing significance through the intermediate range, and of no consequence in the performance of long columns. The most efficient use of the strength of steel is made by selecting columns in the intermediate range. To achieve large values of $I$ and $r$ for a given area $A$, a section that has the area distributed as far from its centroid as possible, offers the best choice, other things being equal. The most efficient sections are those with $\frac{r_x}{r_y} = 1$. Of the available wide-flange sections, those with $\frac{b}{d} \approx 1$ are most efficient for columns.

Compression members are divided into two classes by their values of $Kl/r$, with the value of $C_c = \sqrt{\frac{2\pi^2 E}{F_y}}$ dividing the two classes.

Short columns are defined by very low values of $Kl/r$. In this range, the Euler curve for critical load is approaching infinity. However, when the axial load becomes sufficient to cause yield stress, failure occurs by compression yielding although collapse is unlikely.

Failure for intermediate length columns is initiated by the tendency for buckling instability. The failure curve shows a smooth transition between the yield and the buckling conditions. The two curves become tangent at a value of $Kl/r = C_c$, somewhat arbitrarily chosen in the AISC specifications as $C_c = \sqrt{\frac{2\pi^2 E}{F_y}}$.

### 11.1.4.3 Allowable stresses

The allowable axial compressive stress value, $F_a$, for compact or non-compact sections, is evaluated as follows:

when $$\frac{Kl}{r} \leqslant C_c$$

$$F_a = \frac{\left[1.0 - \frac{\left(\frac{Kl}{r}\right)^2}{2C_c^2}\right]F_y}{\frac{5}{3} + \frac{3\left(\frac{Kl}{r}\right)}{8C_c} - \frac{\left(\frac{Kl}{r}\right)^3}{8C_c^3}} \qquad \text{(AISC E2-1)}$$

where $Kl/r$ is the larger of $K_x l_x/r_x$ and $K_y l_y/r_y$ and $C_c = \sqrt{(2\pi^2 E)/F_y}$. Otherwise, if $Kl/r > C_c$

$$F_a = \frac{12\pi^2 E}{23(Kl/r)^2} \qquad \text{(AISC E2-2)}$$

It is to be noted that for: (i) single angles $r_z$ is used in place of $r_x$ and $r_y$ and; (ii) for members in compression, $Kl/r$ must not be greater than 200 (AISC B7).

### 11.1.4.1 Stability of frames: effective length concept

Stability of frames is dealt in ASCE specification primarily by using the concept of effective length $Kl$. Frames are classified as braced and unbraced frames in their treatment of stability. As mentioned previously the basic column formulas work only for pin ended compression members with no lateral movement. Therefore an effective length factor $K$ is used to convert real cases to basic pin ended case. The term $Kl$ represents the distance between points of theoretical zero moments.

There are two unsupported lengths $l_x$ and $l_y$, corresponding to instability in the major and minor directions of the column. These are the lengths between the support points of the column in the corresponding directions.

Typically, in building design, all floor diaphragms are assumed to be lateral support points. Therefore, the unsupported length of a column is equal to the story height associated with the level. However, if a column is disconnected from any level, the unsupported length of the column is longer than the story height.

Therefore, in determining the values of $l_x$ and $l_y$ for the beam and column elements, the designer must recognize various aspects of the structure that have an effect on these lengths, such as member connectivity and diaphragm disconnections.

It should be noted that columns may have different unsupported lengths corresponding to the major and minor directions. For example, beams framing into columns in the column major and minor directions will give lateral support in both the directions. However, if a beam frames into only one direction of the column at a level where the column has been disconnected from the diaphragm, the beam gives lateral support only in that direction.

For beams any column, brace or wall support is generally assumed to be the location of the vertical support to the beam in the major direction as well as the lateral support to the beam in the minor direction. For brace elements, the unsupported length is generally assumed equal to the actual element length.

There are two K-factors, $K_x$ and $K_y$, associated with each column. These values correspond to instability associated with the major and minor directions of the column, respectively. The calculation of the K-factor in a particular direction involves the evaluation of the stiffness ratios, $G_{top}$ and $G_{bot}$ corresponding to the top and bottom support points of the column, in the direction under consideration where:

$$G_{top} = \frac{\dfrac{E_{ca}I_{ca}}{L_{ca}} + \dfrac{E_{cb}I_{cb}}{L_{cb}}}{\sum_{n=1}^{n_b} \dfrac{E_{gn}I_{gn}}{L_{gn}} \cos^2 \theta_n}$$

where $E_{ca}$ = modulus of elasticity of column above top lateral support point

$E_{cb}$ = modulus of elasticity of column below top lateral support point

$I_{ca}$ = moment of inertia of column above top lateral support point

$I_{cb}$ = moment of inertia of column below top lateral support point

$L_{ca}$ = unsupported length of column in direction under consideration above top lateral support point

$L_{cb}$ = unsupported length of column in direction under consideration below top lateral support point

$E_{gn}$ = modulus of elasticity of beam, $n$, at top lateral support point

$L_{gn}$ = major moment of inertia of beam, $n$, at top lateral support point

$n_b$ = number of beams that connect to the column at lateral support level

$\theta_n$ = angle between the column direction under consideration and the beam, $n$

For the K-factor calculation above, the unsupported lengths are generally based on full member lengths and do not consider any rigid end offsets.

The calculation for $G_{bot}$ is similar, as it corresponds to the bottom lateral support point. The column K-factor for the corresponding direction is then calculated by solving the following relationship for $\alpha$:

$$\frac{\alpha^2 G_{top} G_{bot} - 36}{6(G_{top} + G_{bot})} = \frac{\alpha}{\tan \alpha}$$

from which

$$K = \frac{\pi}{\alpha}$$

This relationship is the mathematical formulation for K-factor evaluation assuming the sidesway is uninhibited. The following are some important aspects associated with the column K-factor.

Cantilever beams and beams and columns having pin ends at the joint under consideration are excluded in the calculation of the stiffness $EI/L$ summations because they do not contribute to the rotational stiffness of the joint. A column or beam that has a pin at the far end from the joint under consideration will contribute only 50 percent of the calculated $EI/L$ value. If a pin release exists at a particular end of a column, the corresponding G-value is 10.0 in both directions. If there are no beams framing into a particular direction of a column, the associated G-value will be infinity. If the G-value at any one end of a column for a particular direction is infinity, the K-factor corresponding to the direction is equal to unity. If rotational releases exist at both ends of a column, the corresponding K-factors are equal to unity.

Observe that the above procedure for the calculation of K-factor can generate artificially high K-factors, under certain circumstances. For example, in Fig. 11.4 column line C2 has no beams framing in a direction parallel to the column minor direction. The $G_{top}$ and $G_{bot}$ values for this column are infinity. Such columns are considered to be laterally supported by the floor diaphragms with the column a K-factor of unity. Now consider the condition where the beams framing into a column are slightly inclined with the column major axis, as shown in Fig. 11.5 for column line C2. The small components of the beam stiffness in the column minor direction will generate small $G_{top}$ and $G_{bot}$ values for the column minor direction, resulting in a large minor direction K-factor. In general, such columns are laterally supported by the floor diaphragms in minor directions and should be assigned a K-factor of unity. For braced frames, the K-factors for the beam and brace elements are generally assumed to be unity.

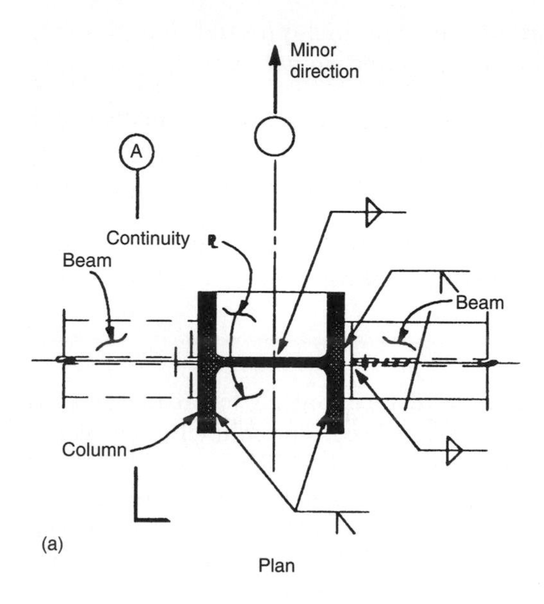


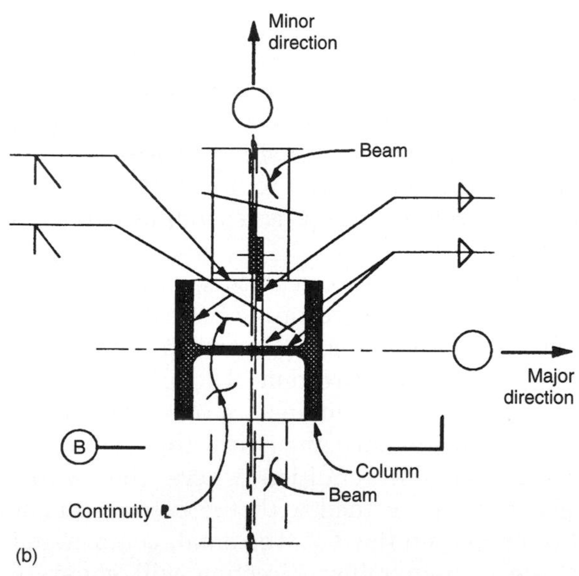


**Figure 11.4** Beams framing into columns in one direction: (a) beam framing into column flange; (b) beam framing into column web.

### 11.1.5 Members subjected to combined axial load and bending

#### 11.1.5.1 Secondary bending: $p$-$\Delta$ effects

Frame columns in buildings are in effect "beam-columns", i.e., they are subject to simultaneous bending caused by lateral loads, and axial compression due to gravity loads. Consider the column shown in Fig. 11.13a subjected to simultaneous action of axial load and moments at the ends. At any point, the total moment $M$ can be considered as a combination of the moment $M_0$ due to end moments plus the addition of the moment caused by $P$ acting at an eccentricity $y$

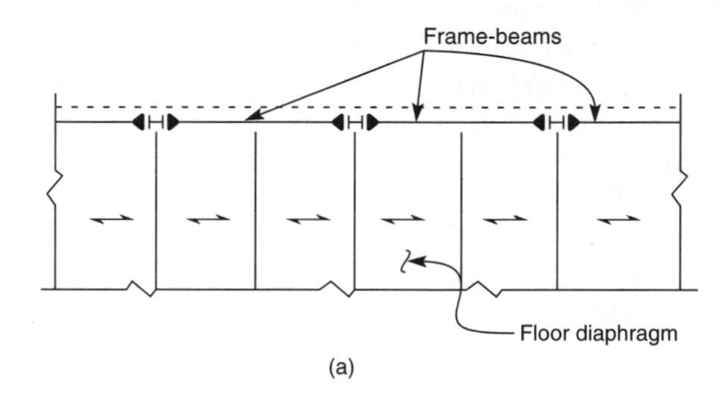


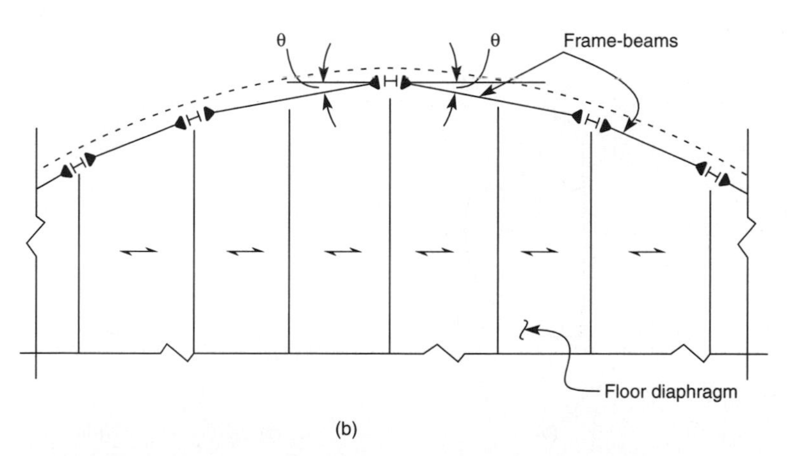


**Figure 1.5** Beam framing conditions for evaluation of effective length factor, $K$: (a) beams framing into columns without skew; (b) skewed beams framing into columns.

(Fig. 11.6b–d). Thus $M = M_0 + P_y$. Since the deflection is maximum at midheight, the secondary moment also reaches its maximum value at that height. A similar effect is caused when bending is produced by a lateral load as shown in Fig. 11.7. Since the deflection $y$ and hence the magnitude of the secondary moment is itself a function of the end moments, a differential equation formulation is required for determining the stresses in beam-columns. Simple cases of beam-columns subjected to end moments and concentrated loads, uniformly distributed loads, etc., have been solved by differential equation techniques. In a practical structure, such a closed-form solution is extremely complicated if not impossible. Therefore, various design standards such as the ACI code and AISC specifications give provisions for approximate evaluation of the slenderness effect. The method in essence requires that the moments obtained by a so-called first-order analysis be magnified by a moment magnification factor.

The direct addition of the maximum $p$-$\Delta$ moment to the maximum primary moment is valid only when the beam-column is subjected to equal moments at the ends subjecting the column to bend in a single curvature. For all other cases, it represents an upper bound, giving a moment magnification factor much larger than that in a real structure. If the two end moments are unequal but of the same sign, producing single curvature, the primary movement $M_0$ is certainly magnified but not to the same extent as when the moments are equal. If the end moments are of opposite sign, producing a reverse curvature in the column, the moment magnification effect will be very small. A moment magnification coefficient $C_m$ is

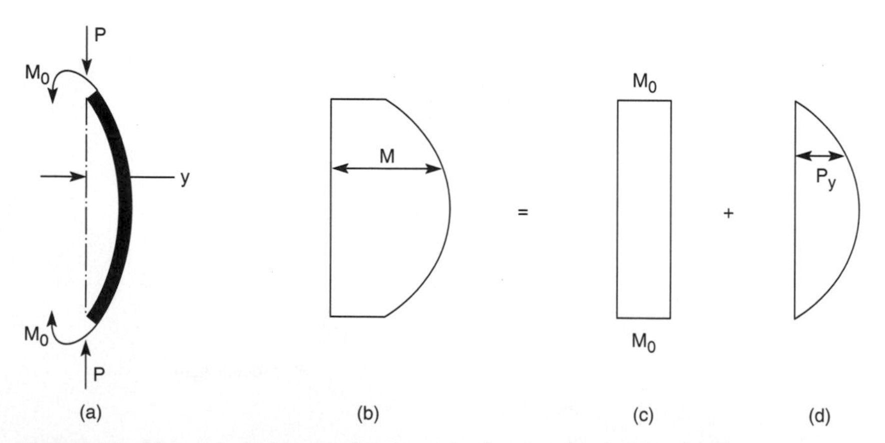


**Figure 11.6** Moments in beam-columns: (a) column subjected to simultaneous axial load and bending moments; (b) combined moment diagram; (c) moment diagram due to equal end moments $M_0$; (d) moment due to $P$-$\Delta$ effect.

therefore used to take into account the relative magnitude and sense of the two end moments. It is given by the expression

$$C_m = 0.6 + 0.4M_1/M_2 \tag{11.7}$$

In the above equation $M_1$ and $M_2$ represent the smaller and larger end moments, respectively. The ratio $M_1/M_2$ is defined as positive when the column bends in a single curvature and negative if the moments product reverse curvature. As can be expected, when $M_1 = M_2$ as in a column subjected to equal end moments, the value of $C_m$ becomes equal to 1.0. The above expression applies only to members braced against side sway. For columns which are part of the lateral resisting system, the maximum moment magnification occurs, i.e., $C_m = 1$ as illustrated in the following discussion.

Consider Fig. 11.8a, which shows the deflected shape of an unbraced portal frame subjected to the simultaneous action of gravity and lateral loads. Considering only the lateral loads, the deflection of the portal frame may be represented by solid lines as shown in Fig. 11.8a. The corresponding moments at the ends of a typical column are as shown in Fig. 11.8.

When axial load is imposed on the deflected shape of the frame, additional sway occurs in the frame as shown by dashed lines in Fig. 11.8a. This additional deflection imposes secondary moments in the column as shown in Fig. 11.8c. It is seen that both the primary and secondary moments are of the same sign and have maximum values at the same locations, namely at the two ends of the columns. They are, therefore, fully additive as shown in Fig. 11.8d, meaning that the value of $C_m = 1$ for unbraced frames.

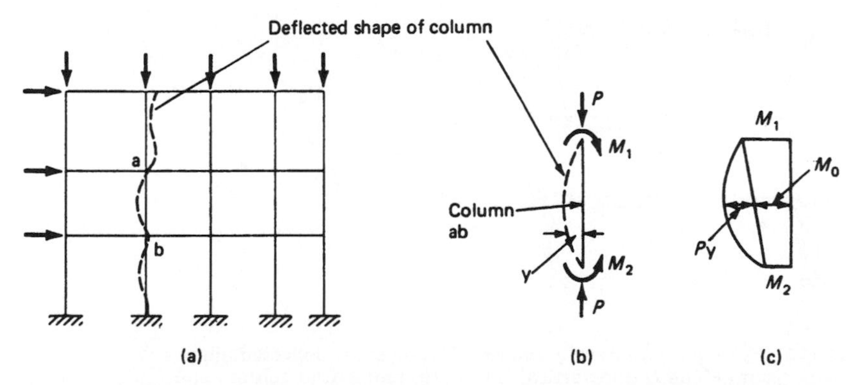


**Figure 11.7** Behavior of building column: (a) building frame showing deflected shape of column; (b) colunm subjected to the simultaneous action of axial loads and moments; (c) moment diagram due to end moments and $P$-$\Delta$ effect.

In American practice, both for steel and concrete buildings, the approach to the stability problem is to modify individual member design in a manner which approximately accounts for frame buckling effects. This is done by isolating a compression member together with its adjoining members at both ends and determining its critical load in terms of effective length factor $K$. The member is then analyzed as a beam-column by a simplified interaction equation which accounts for the moment magnification caused by the $p$-$\Delta$ effect. Instead of frame analysis for the $p$-$\Delta$ method, a member analysis is substituted.

We have seen earlier, using a total moment obtained by the direct addition of secondary and primary moments results in an overdesign if both these moments do not occur at the same location. The coefficient $C_m$ in the interaction equation prevents overdesign by reducing the design moment by taking into account the relative magnitude and sense of the moments occurring at the ends of columns.

Values of $C_m$ less than 1.0 increase $F_b$, offsetting the effects of axial load when the shape of elastic curve increases stability. Thus $C_m = 0.85$ when rotational restraints are present at the ends of the member. When there is no joint translation and where the shape of the curve is not affected by transverse loading, reverse curvature bending may reduce $C_m$ to as little as 0.4.

To prevent a dramatic increase in $F_b$ which can result in unsafe designs, an interaction equation which does not contain the term $C_m$, is also required to be satisfied.

The calculation of stress ratios in frame columns is essentially an exercise in the evaluation of stresses due to simultaneous axial and bending action.

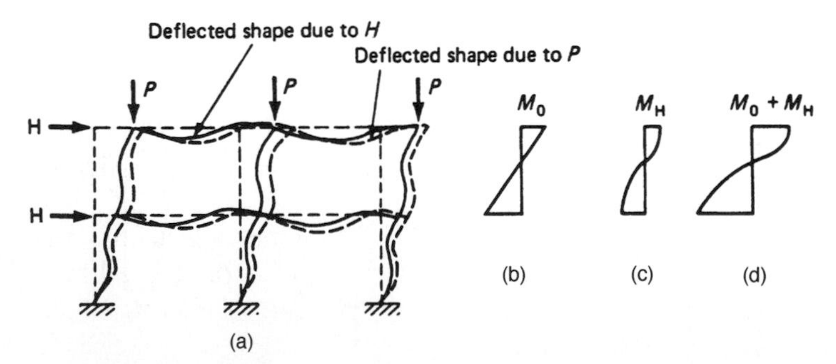


**Figure 11.8** $P$-$\Delta$ effect in laterally unbraced frames: (a) deflected shapes due to horizontal load $H$ and vertical load $P$; (b) moment at column ends due to horizontal load $H$; (c) moment at column ends due to axial loads $P$; (d) combined moment diagram due to $H$ and $P$. Maximum moment due to $H$ and $P$ occurs at the ends of columns resulting in $C_m = 1.0$.

### 11.1.5.2 Interaction equations

Prior to 1963, structural engineers could have made peace with the entire design process of beam-columns by using the formula

$$\frac{f_a}{F_a} + \frac{f_b}{F_b} \leqslant 1.0$$

Since then, engineers have had to deal with many seemingly formidable factors which have been added on to the above interaction equation. For example, the allowable bending stress $F_b$ now has a factor $(1 - f_a/F'_e)$ to account for the reduction in the bending capacity because of axial loads. The more the axial load in the column, the greater is the reduction of $F_b$. Reducing the allowable stress is mathematically equivalent to increasing the design moment for the $p$-$\Delta$ effects. $F'_c$ is the familiar Euler's stress divided by the same factor of safety, $\frac{23}{12}$, that governs the allowable stress of long columns.

Consideration of only uniaxial bending reduces the AISC equations to the less intimidating format as follows:

$$\frac{f_a}{F_a} + \frac{C_m f_b}{(1 - f_a/F'_e)F_b} \leqslant 1.0$$

$$\frac{f_a}{0.6F_y} + \frac{f_b}{F_b} \leqslant 1.0$$

where $f_a$ = axial stress in the column due to vertical loads
$F_a$ = allowable axial stress
$f_b$ = bending stress in the column
$F_b$ = allowable bending stress
$C_m$ = coefficient for modifying the actual bending moment to an equivalent moment diagram for purposes of evaluating secondary bending
$F'_e$ = Euler's stress divided by safety factor, $\frac{23}{12}$
$F_y$ = yield stress of column steel

As mentioned previously, a stress ratio greater than 1.0 indicates overstress requiring the redesign of the column.

For the general case of axial load plus biaxial bending, the interaction equations for calculating the stress ratios are as follows: If $f_a$ is compressive and $f_a/F_a > 0.15$, the compressive stress ratio $CR$ is given by the larger of $CR_{1a}$ and $CR_{1b}$, where

$$CR_{1a} = \frac{f_a}{F_a} + \frac{C_{mx} f_{bx}}{\left(1 - \frac{f_a}{F'_{ex}}\right)F_{bx}} + \frac{C_{my} f_{by}}{\left(1 - \frac{f_a}{F'_{ey}}\right)F_{by}} \qquad \text{(AISC H1-1)}$$

and

$$CR_{1b} = \frac{f_a}{0.60F_y} + \frac{f_{bx}}{F_{bx}} + \frac{f_{by}}{F_{by}} \qquad \text{(AISC H1-2)}$$

If $f_a/F_a \leqslant 0.15$, $CR = CR_2$, where

$$CR_2 = \frac{f_a}{F_a} + \frac{f_{bx}}{F_{bx}} + \frac{f_{by}}{F_{by}} \qquad \text{(AISC H1-3)}$$

where $C_{mx}$ and $C_{my}$ are coefficients representing distribution of moment along member length. Although, as mentioned previously, their value could be as low as 0.4, in practice they are conservatively assumed equal to 1.0 in most cases.

If $f_a$ is tensile or zero, the tensile stress ratio $TR$ is given by the larger of $TR_1$ and $TR_2$, where

$$TR_1 = \frac{f_a}{F_a} + \frac{f_{bx}}{F_{bx}} + \frac{f_{by}}{F_{by}} \qquad \text{(AISC H2-1)}$$

and

$$TR_2 = \frac{f_{bx}}{F_{bx}} + \frac{f_{by}}{F_{by}}$$

In the calculation of the tensile ratio, $TR$, the allowable bending stresses $F_{bx}$ and $F_{by}$ have a minimum value of $0.60F_y$. For circular sections a Square Root of Sum of the Squares (SRSS) combination is first made of the two bending components before adding the axial load component instead of the simple algebraic addition implied by the above formulas for $CR$ and $TR$.

#### 11.1.5.3 Direct analysis for *P*-Δ effects

It is important to realize that the moment magnification method using K-factors is an approximate method for evaluating the $P$-$\Delta$ effects. With the availability of computer programs which can directly account for $P$-$\Delta$ effects, it is not necessary to use the approximate moment magnification method for evaluating the effect of axial loads $P$ acting through the lateral deflection $\Delta$ of the structure. Analysis of structures by using programs with $P$-$\Delta$ capabilities is highly recommended for routine office use. With the results of the $P$-$\Delta$ analysis in hand, the engineer need not worry about calculating the effective length factor $K$ by using alignment charts or complicated equations. All columns whether they are gravity or frame columns can be designed by using the effective length factor $K = 1$. And moreover the $P$-$\Delta$ method is applicable to all types of construction—steel, concrete, or composite—as a general procedure. Although the procedure itself is not codified in the ACI and AISC, it is highly endorsed in their commentaries.

The SEAOC blue book gives a drift ratio value of $0.02/R_w$ as the threshold of lateral deformation beyond which the $P$-$\Delta$ effects become signficant. Consider for example, a building with a typical floor-to-floor height of 13′ 6″ with special moment resisting frame as the lateral system. With an $R_w = 12$, the limiting drift ratio for this building is approximately equal to 0.25″. $P$-$\Delta$ analysis is necessary in the lateral load analysis only when the drift ratio exceeds this value.

However, by using commercial programs it is easy to include $P$-$\Delta$ effects in a single solution without having to use iteration technique. Therefore, analysis of structures including $P$-$\Delta$ effects is highly recommended for office practice. It should be noted that while $P$-$\Delta$ effects are included in the programs, the effect of reduction of stiffness of columns due to axial loads, in general, is not accounted.

Observe that the columns of moment frames which are designed with $P$-$\Delta$ effects included need not have their bending stresses amplified by the term $(1 - f_a/F_e)$ in AISC Formula H1-1 or $\delta_s$ in ACI Formula 10-8, since these factors were intended to account for $P$-$\Delta$ effects.

It should be noted that $P$-$\Delta$ effects are potentially much more significant in lower seismic zones than in higher seismic zones, because the relative stiffness of lateral load resisting systems in higher seismic zones is required to be greater than those in lower seismic zones.

### 11.1.6 Calculation of stress ratios

First, for each load combination, the actual member stress components and corresponding stress allowables are calculated. Then, the elastic stress ratios are evaluated at critical locations under the influence of each of the design load combinations using the corresponding equations referred to earlier in this section. The controlling compression and/or tension stress ratio is calculated. A stress ratio greater than 1.0 indicates an over stress. Similarly, a shear stress ratio is separately calculated. For the general case, the member stresses calculated for each load combination are:

$$f_a = P/A$$

$$f_{bx} = M_x/S_x$$

$$f_{by} = M_y/S_y$$

$$f_{vx} = V_x/A_{vx}$$

$$f_{vy} = V_y/A_{vy}$$

For column, beam and brace elements the stress ratios are checked for combined axial force and biaxial bending effects. In general for columns and braces, the checks are evaluated at the top and bottom ends of the clear element length, while for the beams, these checks are calculated at several locations along the clear element length, and at each end.

### 11.1.7 Design of continuity plates

In a beam/column moment connection, a steel beam can frame into a column in the following ways:

1. The steel beam frames in a direction parallel to the column major direction, i.e., the beam frames into the column flange.
2. The steel beam frames in a direction parallel to the column minor direction, i.e., the beam frames into the column web.
3. The steel beam frames in a direction that is at an angle to both of the principal axes of the column, i.e., the beam frames partially into the column web and partially into the column flange.

To achieve a beam/column moment connection, continuity plates such as shown in Fig. 11.9 are usually welded to the column, in line with the top and bottom flanges of the beam, to transfer the compression and tension flange forces of the beam into the column. For connection conditions described by 2 and 3 above, the thickness of such plates is usually set equal to the flange thickness of the corresponding beam. However, for the connection condition described by 1 above, where the beam frames into the flange of the column, such continuity plates are not always needed. The requirement depends upon the magnitude of the beam-flange force and the properties of the column.

The continuity plate requirements are evaluated for moment frames for each of the beams that frame into column to determine the maximum continuity plate area. The continuity plate area required for a particular beam framing into a column is given by:

$$A_{cp} = \frac{P_{bf}}{F_{yc}} - t_{wc}(t_{fc} + 5k_c) \qquad \text{(AISC K1-9)}$$

If $A_{cp} \leq 0$, no continuity plates are required provided the following two conditions are also satisfied:

1. The depth of the column clear of the fillets, i.e., $d_c - 2k_c$ is less than or equal to:

$$\frac{4100 t_{wc}^3 \sqrt{F_{yc}}}{P_{bf}} \qquad \text{(AISC K1-8)}$$

2. The thickness of the column flange, $t_{fc}$, is greater than or equal to:

$$0.4\sqrt{\frac{P_{bf}}{F_{yc}}} \qquad \text{(AISC K1-1)}$$

where $P_{bf} = f_b A_{bf}$.

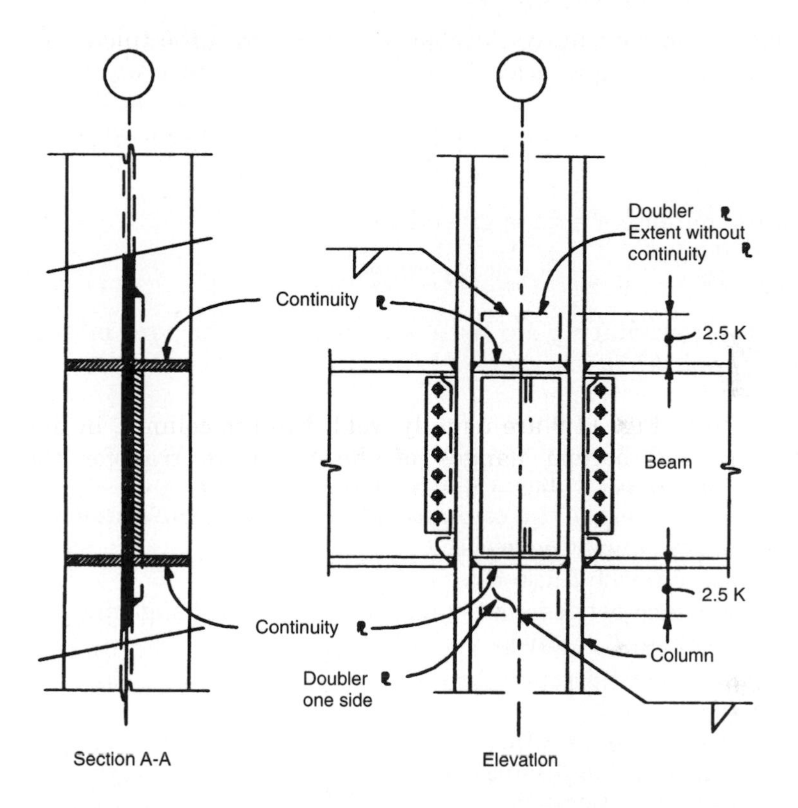


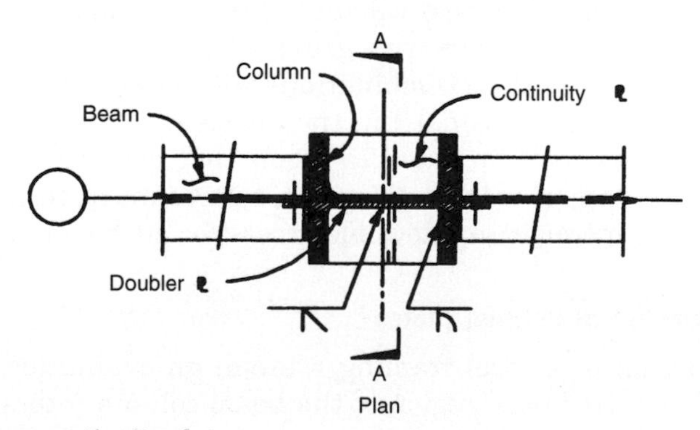


**Figure 11.9** Column continuity plates.

If continuity plates are required, they must satisfy a minimum area specification defined as follows:

1. The thickness of the stiffeners is at least $0.5fb$, or

$$t_{cp}^{min} = 0.5t_{fb} \qquad \text{(AISC K1.8.2)}$$

2. The width of the continuity plate on each side plus $\frac{1}{2}$ the thickness of the column web shall be less than $\frac{1}{3}$ of the beam flange width, or

$$b_{cp}^{min} = 2\left(\frac{b_{fp}}{3} - \frac{t_{wc}}{2}\right) \qquad \text{(AISC K1.8.1)}$$

so that the minimum area is given by:

$$A_{cp}^{min} = t_{cp}^{min} b_{cp}^{min}$$

Therefore, the continuity plate area required is either zero or the greater of $A_{cp}$ and $A_{cp}^{min}$.

where $A_{bf}$ = area of beam flange
$A_{cp}$ = required Continuity plate area
$F_{yb}$ = yield stress of beam material
$F_{yc}$ = yield stress of the column and continuity plate material
$t_{fb}$ = beam flange thickness
$t_{wc}$ = column web thickness
$k_c$ = distance between outer face of the column flange and web toe of its fillet
$d_c$ = column depth
$d_b$ = beam depth
$f_b$ = beam flange width
$t_{cb}$ = continuity plate thickness
$b_{cp}$ = continuity plate width
$f_b$ = bending stress calculated from the larger of $\frac{5}{3}$ of loading combinations with gravity loads only and $\frac{4}{3}$ times $\frac{4}{3}$ of the loading combinations with lateral loads (AISC K1.2). The additional $\frac{4}{3}$ in the above equation is included to account for the 0.75 factor that may be used in the lateral load combination to account in turn for the $\frac{1}{3}$ increase in allowable stress for such conditions

### 11.1.8 Design of doubler plates

In the design of a steel framing system, an evaluation of the shear forces that exist in the region of the beam column intersection known as the panel zone is of importance.

Although shear stresses seldom control the design of a beam or column, in a moment-resisting frame, the shear stress in the beam-column joint can be critical, especially when the column is subjected to major direction bending and the joint shear is resisted by the column web. When the column is subjected to minor axis bending, the joint shear is carried by column flanges. The resulting shear stresses are seldom critical, and therefore not generally investigated.

The high shear stresses in the panel zone, due to major direction bending of column, may require welding of additional plates to the column web. The requirement depends on the loading and the geometry of the beams that frame into the column. Beams framing either along the major direction of the column or at an angle, with components along the major direction may precipitate this requirement.

The shear force in the panel zone, is given by

$$V_p = P - V_c$$

or

$$V_p = \sum_{n=1}^{n_b} \frac{M_{bn}}{d_n - t_{f_n}} \cos\theta_n - V_c$$

The required web thickness to resist the shear force, $V_p$, is given by

$$t_r = \frac{V_p}{F_v d_c}$$

The extra thickness, or thickness of the doubler plate is given by

$$t_{dp} = t_r - t_{wc}$$

where $F_{yc}$ = yield stress of the column and doubler plate material
$t_r$ = required column web thickness
$t_{dp}$ = required doubler plate thickness
$t_{wc}$ = column web thickness
$V_p$ = panel zone shear
$V_c$ = column shear in column above
$P$ = beam flange forces
$n_b$ = number of beams connecting to column
$d_n$ = depth of $n$th beam connecting to column
$\theta_n$ = angle between $n$th beam and column major direction
$d_c$ = depth of column
$M_{bn}$ = calculated factored beam moment from the corresponding loading combination
$F_v = 0.40F_{yc}$

The largest calculated value of $t_{db}$ calculated for any of the load combinations based upon the factored beam moments controls the design.

### 11.1.9 Additional seismic requirements (UBC94)

This section highlights additional checks required for design based on the special seismic requirements of the Uniform Building Code, 1994, Section 2211. The special requirements are dependent on the type of framing used and are described below for each type of framing. The framing types described are

- Ordinary moment frames
- Special moment-resisting frames
- Braced frames
- Eccentrically braced frames
- Special concentrically braced frames

#### 11.1.9.1 Ordinary moment frames

The following additional checks are required for ordinary moment frames with $R_w = 6.0$

1. Whenever the axial stress, $f_a$, in columns due to the prescribed loading combinations exceeds $0.3F_y$, the following axial load combinations are checked with respect to the column axial load capacity (UBC 2211.5.1).

$$P_{DL} + 0.7P_{LL} \pm \tfrac{3}{8}R_w(\sqrt{P_X^2 + P_Y^2} + |P_T|)$$
$$0.85P_{DL} \pm \tfrac{3}{8}R_w(\sqrt{P_X^2 + P_Y^2} + |P_T|)$$

   where $P_{DL}$, $P_{LL}$, $P_X$, $P_Y$ and $P_T$ are axial loads due to dead loads, live loads, seismic loads in the $X$ direction, seismic loads in the $Y$ direction, and seismic loads due to accidental torsion, respectively.

   For the above loading combinations the column axial capacity in compression is taken as $1.7F_aA$ and the capacity in tension is taken as $F_yA$.

2. The beam connection shears are verified for the following additional loading combinations (UBC 2211.6)

$$V_{DL} + V_{LL} \pm \tfrac{3}{8}R_w(\sqrt{V_x^2 + V_y^2} + |V_T|)$$

   where $V_{DL}$, $V_{LL}$, $V_X$, $V_Y$ and $V_t$ are beam shears due to dead load, live load, seismic loads in the $X$ direction, seismic loads in the $Y$ direction, and seismic loads due to accidental torsion, respectively.

3. The continuity plates are designed for a beam flange force, $P_{bf} = A_{bf}F_y b$.

#### 11.1.9.2 Special moment resisting frames

For this framing system, the following additional checks are required.

1. The additional requirement of column axial strength as noted above for the ordinary moment frames also applies for the special moment resisting frames except the value for $R_w$ is assumed as 12.

2. The beam connection shears are calculated to allow for the development of the full plastic moment capacity of the beam (UBC 2211.7.1). Thus:

$$V = \frac{CM_{Pb}}{L} + V_{I,II,III}$$

where $V$ = shear force corresponding to the left and right of beam
$C$ = 0 if beam ends are pinned, or for cantilever beam
$C$ = 1 if one end of the beam is pinned
$C$ = 2 if no ends of the beam are pinned
$M_{Pb}$ = plastic moment capacity of the beam
$L$ = Clear length of the beam
$V_{I,II,III}$ = absolute maximum of the calculated factored beam shears at the corresponding beam ends from the vertical load combinations

3. The panel zone doubler plate is calculated to develop lesser of beam moments equal to 0.8 of the plastic moment capacity of the beam ($0.8 \sum M_{Pb}$), or beam moments due to gravity loads plus 1.85 times the seismic load. The capacity of the panel zone in resisting this shear is taken as (UBC 2211.7.2.1):

$$V_p = 0.55F_y d_c t_r \left(1 + \frac{3b_c t_{cf}^2}{d_b d_c t_r}\right)$$

giving the required panel zone thickness as

$$t_r = \frac{V_p}{0.55F_y d_c} - \frac{3b_c t_{cf}^2}{d_b d_c}$$

and the required doubler plate thickness as

$$t_{dp} = t_r - t_{wc}$$

where $b_c$ = width of column flange
$t_{cf}$ = thickness of column flange
$d_b$ = depth of deepest beam framing into the major direction of the column

4. Compact I-shaped beam sections are additionally checked for $b_f/2t_f$ to be less than $52/\sqrt{F_y}$. Compact I-shaped column sections are additionally checked for $b_f/2t_f$ to be less than the numbers given for plastic sections in the AISC. For example the limiting ratio of $\dfrac{b_f}{2t_f}$ for 50 ksi steel is 7.0. Compact box shaped column sections are additionally checked for $b_f/t_f$ and $d/t_w$ to be less than $110/\sqrt{F_y}$. If this criterion is not satisfied the designer must modify the section (UBC 2211.7.3).
5. For determining the need for continuity plates at joints due to tension transfer from the beam flanges, the force $P_{bf}$ is taken as $1.8A_{bf}F_{yb}$ (UBC 2211.7.4). For design of the continuity plate the beam flange force is taken as $A_{bf}F_{yb}$.
6. To review strong column weak beam criterion (UBC 2211.7.5) a beam/column plastic moment capacity ratio for every joint in the structure is required. For the major direction of column the capacity ratio is obtained as

$$R_{\text{maj}} = \sum_{n=1}^{nb} \frac{M_{pbn} \cos \theta_n}{m_{pcax} + M_{pcbx}}$$

For the minor direction the capacity ratio is obtained as

$$R_{\text{min}} = \sum_{n=1}^{nb} \frac{M_{pbn} \sin \theta_n}{M_{pcay} + M_{pcby}}$$

where $R_{\text{maj,min}}$ = plastic moment capacity ratios, in the major and minor directions of the column, respectively
$M_pb_n$ = plastic moment capacity of $n$th beam connecting to column
$\theta_n$ = angle between the $n$th beam and the column major direction.
$M_{pcax,y}$ = major and minor plastic moment capacities, reduced for axial force effects, of column above story level
$M_{pcbx,y}$ = major and minor plastic moment capacities, reduced for axial force effects, of column below story level
$nb$ = number of beams connecting to the column

The plastic moment capacity of the column is reduced for axial force effects as given by

$$M_{pc} = Z_c(F_{yc} - f_a)$$

where $Z_c$ = plastic modulus of column
$F_{yc}$ = yield stress of column material
$f_a$ = maximum axial stress in the column from any of the load combinations from gravity loads alone or $\frac{4}{3}$ of the loading combinations including lateral loads. As noted previously the $\frac{4}{3}$ factor is for the $\frac{1}{3}$ increase allowed for the load combinations

For the above calculations the column above is taken to be the same as the column below. It is assumed that the column splice is located at a minimum distance of 3′ 6″ above the story level.

7. Next, the laterally unsupported length of beams is checked to be less than $96r_y$ (UBC 2211.7.8).

#### 11.1.9.3 Braced frames

The additional checks for this framing system are as follows:

1. The additional requirement of column axial strength as noted previously for ordinary moment frames also applies to braced frames. However, this requirement is checked even when the axial force level does not exceed $0.3F_y$. The value for $R_w$ is assumed to be 6.
2. The maximum $L/r$ ratio of the braces is checked not to exceed $720/\sqrt{F_y}$ (UBC 2211.8.2.1).
3. The allowable compressive stress for braces is reduced by a factor $\beta$, where

$$\beta = 1/[1 + (Kl/r)/2C_c] \qquad \text{(UBC 2211.8.2.2)}$$

4. Bracing connection force is the smaller of the tensile strength of the brace ($F_yA$) and $\frac{3}{8}R_w$ times the prescribed seismic forces plus gravity loads (UBC 2211.8.3.1).
5. Chevron braces are designed for 1.5 times the specified loading combinations (UBC 2211.8.4.1). Note all braces are Chevron type unless a real column exists at each end of the brace either in the story above or below.

#### 11.1.9.4 Eccentrically braced frames (EBF)

The following additional checks are required for beams, columns and braces associated with eccentric braced frames configurations.

1. Compact I-sections are additionally checked for $\frac{b_f}{2t_f}$ to be less than $\frac{52}{\sqrt{F_y}}$. If this criteria is not satisfied the designer must modify the section property (UBC 2211.10.2).
2. The link beam strength in shear $V_s = 0.55F_y\,dt_w$ and moment $M_s = ZF_y$ are calculated. If $V_s \leqslant 2.2M_s/e$ the link beam strength is assumed to be governed by shear. If the above condition is not satisfied the link beam strength is assumed to be governed by flexure. When line beam strength is governed by shear, the axial and flexural properties (area, $A$ and section modulus, $S$) for use in the interaction equations are calculated based on beam flanges only (UBC 2211.10.3).
3. The link beam rotation, $\Theta$ (calculated as story drift times bay length divided by total lengths of link beams in the bay) relative to the rest of the beam at a total frame drift of $\frac{3}{8}R_w$ times the drift under seismic loads is checked to be less than the following values (UBC 2211.10.4):

$$\Theta \leqslant 0.060 \text{ where link clear length} \leqslant 1.6M_s/V_s$$

$$\Theta \leqslant 0.015 \text{ where link beam clear length} \geqslant 33.0M_s/V_s$$

and $\Theta \leqslant$ value interpolated between 0.060 and 0.015 as the link beam clear length varies from $1.6M_s/V_s$ to $3.0M_s/V_s$

The value of $R_w$ is assigned 12.0.

4. The link beam shear under the specified loading combinations is checked not to exceed $\frac{3}{4}$ times $0.8V_s$ (UBC 2211.10.5). The $\frac{3}{4}$ factor is to account for the 0.75 factor allowed for such combinations.
5. The brace strength in compression (computed as $P_{sc} = 1.7F_aA$) is required for design of the brace to beam connection (UBC 2211.10.6).
6. The link beam connection shear is calculated as equal to the link beam web shear capacity (UBC 2211.10.12).
7. The brace strength in compression (computed in 5 above) is checked to be at least 1.5 times the axial force corresponding to the controlling link beam strength (UBC 2211.10.13). The controlling link beam strength is either the shear strength, $V_s$, or the reduced flexural strength, $M_{rs}$, whichever produces the lower brace force. The value of $M_{rs}$ is taken as $M_{rs} = Z\,(F_y - f_a)$ (UBC 27211.10.3), where $f_a$ is the lower of the axial stress in the link beam corresponding to yielding of the link beam web in shear or the link beam flanges in flexure.

8. The beam strength in flexure reduced for axial loads of the beam outside of the link, is checked to be at least 1.5 times the moment corresponding to the controlling link beam strength (UBC 2211.10.13).

9. The column is checked not to become inelastic for gravity loads plus 1.25 times the column forces corresponding to the controlling link beam strength (UBC 2211.10.14). This condition is deemed satisfied by verifying the following two interaction equations:

$$\frac{P}{1.7AF_a} + \frac{M_x}{S_xF_y} + \frac{M_y}{S_yF_y} \leqslant 1.0 \text{ for compressive } P$$

and

$$\frac{P}{AF_y} + \frac{M_x}{S_xF_y} + \frac{M_y}{S_yF_y} \leqslant 1.0 \text{ for tensile } P$$

This condition does not govern if the columns meet the additional requirement of column axial strength as noted previously in Item #1 for braced frames.

10. Axial forces in the beams are included in checking of the beams (UBC 2211.10.17). Note that using a rigid diaphragm model in computer analyses will result in zero axial forces in the beams. Therefore in the computer model, some of the column lines from the diaphragm must be disconnected to allow for beams to carry the axial loads.

11. The laterally unsupported length of beam is checked to be less than $76b_f/\sqrt{F_y}$. If not, the section must be revised (2211.10.18).

12. The continuity plate requirements are checked for a beam flange force of $P_{bf} = A_{fb}F_{yb}$.

13. The doubler plate requirements are checked similar to the doubler plate checks for special moment resisting frames as discussed earlier.

#### 11.1.9.5 Special concentrically braced frames

For this framing system, the following additional checks are required.

1. The additional requirement of column axial strength as noted in Item #1 of ordinary moment frames also applies to special

concentrically braced frames. However, this requirement is checked even when the axial force level does not exceed $0.3F_y$. The value for $R_w$ is taken to be 9.

2. The maximum $Kl/r$ ratio of the braces is checked not to exceed $1000/\sqrt{F_y}$. If this check is not met, the section must be revised (UBC 211.9.2.1).
3. Bracing members are checked to be compact. The criteria used is as given in AISC Table B5.1, except for angles $b/t$ is limited to $52/F_y$, for box sections $b_f/t_f$ and $d_f t_w$ is limited to $110/\sqrt{F_y}$, for pipe sections $D/T$ is limited to $1300/\sqrt{F_y}$. If this criterion is not satisfied the designer must modify the section (UBC 2211.9.2.4).
4. Bracing connection force is the smaller of the tensile strength of the brace ($F_yA$) and $\frac{3}{8}R_w$ times the prescribed seismic forces plus gravity loads (UBC 2211.9.3.1).
5. Beams intersected by Chevron braces are checked to have a strength to support loads represented by the following loading combinations (UBC 2211.9.4.1):

$$1.2DL + 0.5LL + P_b$$

$$0.9DL + P_b$$

where $P_b$ is given by the difference of $F_yA$ for the tension brace and 0.3 times $(1.7F_aA)$ for the compression brace.

6. Compact I-shaped column sections are additionally checked for $b_f/2t_f$ to be less than the values given in the AISCS for plastic sections. Compact box-shaped column sections are additionally checked for $b_f/t_f$ and $d/t_w$ to be less than $110/\sqrt{F_y}$. If this criterion is not satisfied the designer should modify the section.

## 11.2 Concrete Design

The vocabulary of structural systems available for resisting lateral loads in concrete buildings is somewhat limited. Most common systems consist of shear walls and rigid frames, the latter acting either as a planar bent or as a tube, when arranged around the building perimeter. In this section the various aspects of design of beams, columns and shear walls based on the ACI 318-89 (revised 1992) and the UBC 1994 are described.

Since the two codes are similar, only the significant differences are identified, and for simplicity all descriptions and equations are given in inch-pound-second units.

### 11.2.1 Load combinations and $\phi$ factors

Strength design as covered by ACI 318-89, makes use of two sets of factors. One, called the load factor, provides for variable conditions that lead to possible overloads. The other, called the $\phi$-factor or strength reduction factor, provides for variable conditions that lead to possible understrengths. Different load factors are prescribed for dead, live, wind, and earthquake effects, and the factors vary with different combination of such effects. For example, it is permissible to apply a lower load factor to dead load than to live load, which is logical because the dead loads are usually known with greater precision than live loads.

Typically, building systems are subjected to vertical loads due to dead and live loads acting downwards. In addition to these loads, the building is subjected to lateral loads, resulting from wind or seismic forces. These loads which act along different directions are usually assumed to be in two mutually orthogonal and reversible directions. For a structure subjected to wind forces $W_x$ and $W_y$ from two mutually perpendicular and reversible directions the following loading combinations result

1. $1.4D + 1.7L$
2. $0.75(1.4D + 1.7L + 1.7W_x)$
3. $0.75(1.4D + 1.7L + 1.7W_y)$
4. $0.75(1.4D + 1.7L - 1.7W_x)$
5. $0.75(1.4D + 1.7L - 1.7W_y)$
6. $(0.9D + 1.3W_x)$
7. $(0.9D + 1.3W_y)$
8. $(0.9D - 1.3W_x)$
9. $(0.9D - 1.3W_y)$

The number of combinations for seismic conditions are the same although the load factors are different as given below.

1. $1.4D + 1.7L$
2. $1.05D + 1.275L + 1.4E_x$
3. $1.05D + 1.275L + 1.4E_y$
4. $1.05D + 1.275L - 1.4E_x$
5. $1.05D + 1.275L - 1.4E_y$
6. $0.9D + 1.4E_x$
7. $0.9D + 1.4E_y$
8. $0.9D - 1.4E_x$
9. $0.9D - 1.4E_y$

The $\phi$ factors prescribed in the codes are as follows:

Bending without axial load $\varphi = 0.90$

Axial tension, and axial tension with bending $\varphi = 90$

Axial compression, and axial compression with bending:

1. Members with spiral reinforcement $\varphi = 0.75$
2. Members with ties $\varphi = 0.70$

These may be increased linearly to 0.90 as explained shortly in the test.

Shear and torsion $\varphi = 0.85$

Bearing on concrete $\varphi = 0.70$

The following notations are used in the development of various design equations.

*Notation*

$A_{cv}$ Area used to determine shear stress, sq. in.
$A_g$ Gross section area, sq. in.
$A_h$ Area of shear reinforcement parallel to flexural reinforcement, sq. in.
$A_n$ Net cross-sectional area, sq in.
$A_s$ Total area of reinforcement, sq. in.
$A_{sc}$ Area of compression reinforcement, sq. in.
$A_{st}$ Area of tension reinforcement, sq. in.
$A_v$ Area of shear reinforcement perpendicular to flexural reinforcement, sq. in.
$A_{vd}$ Area of diagonal reinforcement, sq. in.

$a$ Depth of compression block, in.
$b$ Width of member, in.
$b_w$ Width of web (T-beam section), in.
$b_f$ Effective width of flange (T-beam section), in.
$c$ Depth to neutral axis, in.
$c_b$ Depth to neutral axis at balanced conditions, in.
$D$ Total depth of member, in.
$d$ Distance from compression face to tension reinforcement, in.
$d'$ Concrete cover to center of reinforcing, in.
$d_s$ Thickness of flange (T-beam section), in.
$E_c$ Modulus of elasticity of concrete, psi
$E_s$ Modulus of elasticity of reinforcement, assumed as 29,000,000 psi
$f'_c$ Specified compressive strength of concrete, psi
$f_y$ Specified yield strength of flexural reinforcement, psi
$f_{ys}$ Specified yield strength of shear reinforcement, psi
$L$ Length of member, in.
$M_b$ Nominal moment capacity at balanced strain conditions, lb in.
$M_o$ Nominal moment capacity with no axial load, lb in.
$M_n$ Factored moment, lb in.
$P_b$ Nominal axial load capacity at balanced strain conditions, lb
$P_{\max}$ Maximum axial load strength allowed, lb
$P_n$ Nominal load capacity, lb
$P_o$ Nominal axial load capacity at zero eccentricity, lb
$P_u$ Factored axial load, lb
$t$ Thickness of wall, in.
$V_c$ Nominal shear strength of concrete, lb
$V_m$ Nominal shear strength of masonry, lb
$V_s$ Nominal shear strength of shear reinforcement, lb
$V_u$ Factored shear force, lb
$\beta_1$ Factor for obtaining depth of compression block
$\varepsilon_c$ Strain in concrete
$\varepsilon_s$ Strain in reinforcing steel
$\varphi$ Strength reduction factor
$\rho_w$ $\dfrac{A_s}{bd}$

### 11.2.2 Beam design

The design of concrete beams consists of determining the required areas of steel for flexure and shear for the ultimate beam moments and shears. Only the design for major direction flexure and shear is considered in this section. Effects due to axial forces, minor direction bending and torsion are not investigated. The beam design consists of two phases; design for flexure, and design for shear.

#### 11.2.2.1 Design for flexure

In the design for flexure, the factored moments for each load combination are obtained by factoring the load conditions with the corresponding load factors. The beam section is then designed for the maximum positive $+M_u$ and maximum negative $-M_u$. Negative beam moments requiring top steel are always designed assuming the beam as a rectangular section. Positive beam moments requiring bottom steel may be designed by considering the beam as a rectangular section, or as a T-beam. In either case the usual practice is to assume that all sections are singly reinforced. In other words, the effect of any reinforcing in the compression zone of the beam section is conservatively neglected.

In designing for a factored negative moment, $-M_u$ (i.e. designing top steel) the depth of the compression block as shown in Fig. 11.10 is given by

$$a = d - \sqrt{d^2 + \frac{2M_u}{0.85 f'_c \varphi b}}$$

where $M_u$ is negative in the above equation. If $a > 0.75\beta_1 c_b$ a concrete compression over-stress results (ACI 10.3.3), where

$$\beta_1 = 0.85 - 0.05\left(\frac{f'_c - 4000}{1000}\right) \qquad \text{(ACI 10.2.7.3)}$$

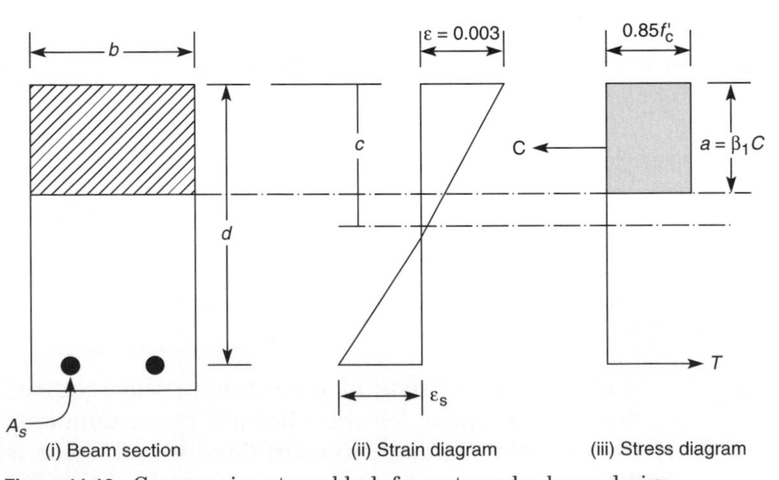


**Figure 11.10** Compression stress block for rectangular beam design.

with a maximum of 0.85 and a minimum of 0.65, and

$$c_b = \frac{87{,}000}{87{,}000 + f_y} d$$

The area of negative steel is then given by

$$-A_s = \frac{-M_u}{\varphi f_y\left(d - \dfrac{a}{2}\right)}$$

where the value of the capacity reduction factor $\varphi$ in the above equation is 0.90 (ACI 9.3.2.1). In designing for factored positive moment, $+M_u$, (i.e., designing bottom steel), the procedure for calculating the area of steel is exactly the same as above if the beam section is rectangular, i.e., no T-beam action is considered.

If the $T$-beam action is taken into account, the depth of the compression block is given by

$$a = d - \sqrt{d^2 - \frac{2M_u}{0.85 f'_c \varphi b_f}}$$

where $a < 0.75\beta_1 c_b$.

If $a < d_s$, the subsequent calculations for $A_s$ are exactly the same as for rectangular section design. If $a > d_s$, calculation for $A_s$ is in two parts. The first part is for balancing the compressive force from the flange, $C_f$, and the second is for balancing the compressive force from the web, $C_w$, as shown in Fig. 11.11.

$$C_f = 0.85 f'_c (b_f - b_w) d_s$$

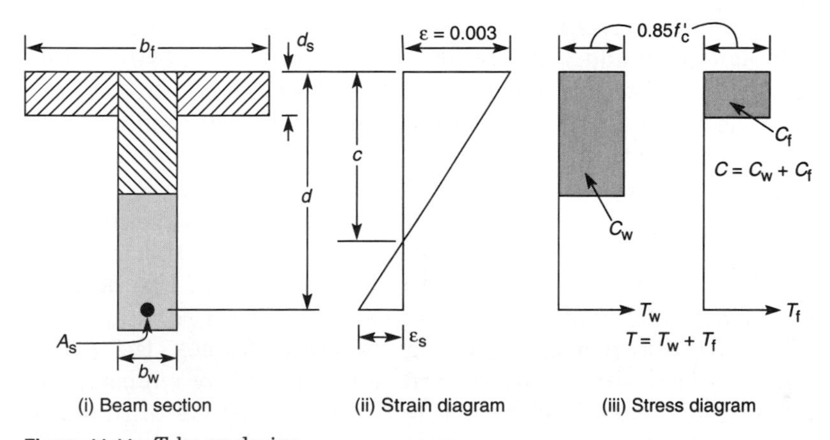


**Figure 11.11** $T$-beam design.

Therefore
$$A_{s_1} = \frac{C_f}{f_y}$$

and the portion of $M_u$ resisted by the flange is given by

$$M_{uf} = C_f\left(d - \frac{d_s}{2}\right)\varphi$$

Therefore, the balance of the moment, $M_u$ to be carried by the web is given by

$$M_{uw} = M_u - M_{uf}$$

The web is a rectangular section of dimensions $b_w$ and $d$, for which the depth of the compression block is recalculated as

$$a_1 = d - \sqrt{d^2 - \frac{2M_{uw}}{0.85f'_c\varphi b_w}}$$

where $a_1 \le 0.75\beta_1 c_b$ from which the second part of the reinforcing is calculated, giving

$$A_{s_2} = \frac{M_{uw}}{\varphi f_y\left(d - \frac{a_1}{2}\right)}$$

The total required reinforcing for the $T$-section is then given by

$$A_s = A_{s_1} + A_{s_2}$$

As before, the value for $\varphi$ is 0.90. The minimum flexural steel is given by

$$A_s(\min) = \frac{200b_w d}{f_y} \quad \text{(ACI 10.5.1 and 21.3.2.1)}$$

For special moment-resisting concrete frames, the beam design must satisfy the following additional conditions: (i) the beam flexural steel is limited to a maximum given by $A_s(\max) = 0.025bd$; (ii) at any support of the beam, the beam positive moment capacity must not be less than 50 percent of the beam negative moment capacity (ACI 21.3.2.2); (iii) the negative moment capacity at any location of the beam span must not be less than one-fourth of the negative moment capacity of any of the beam support. Similarly, the positive moment capacity must not be less than one-fourth of the positive moment capacity of any of the beam support (ACI 21.3.2.2).

For intermediate moment-resisting concrete frames the beam design must satisfy the following conditions: (i) at any support, the beam positive moment capacity must not be less than one-third of the beam negative moment capacity at that support (ACI 21.8.4.1); (ii) the negative moment capacity at any location of the beam span must not be less than one-fifth of the negative moment capacity at any beam support. Similarly, the positive moment capacity of the beam span must not be less than one-fifth of the positive moment capacity at any support (ACI 21.8.4.1.).

### 11.2.2.2 Design for shear

In the design of beam shear reinforcing of an ordinary moment-resisting concrete frame, the shear forces and moments for a particular load combination are obtained by factoring the analysis load condition with the corresponding load combination factors.

In the design of special and intermediate moment-resisting concrete frames, however, the shear force $V_u$ is calculated from the moment capacities of each end of the beam and the gravity shear forces. Therefore, for each load combination the gravity beam shear force is calculated using only the gravity load conditions to get unfactored $V_{D+L}$ for UBC code. However, for ACI code, the factored gravity loads are used to obtain $V_{D+L}$.

The design shear force $V_u$ is then given by

$$V_P + V_{D+L} \qquad \text{(ACI 21.3.4.1)}$$

where $V_P$ is the shear force obtained by applying the calculated ultimate moment capacities of the beam, acting in opposite directions, at the corresponding ends of the beams. Therefore, $V_P$ is the maximum of $V_{P_1}$ and $V_{P_2}$

Where

$$V_{P_1} = \frac{M_I^- + M_J^+}{L}$$

$$V_{P_2} = \frac{M_I^+ + M_J^-}{L}$$

where $M_I^-$ = moment capacity at end $I$, with top steel in tension, using a steel yield stress value of $\alpha f_y$ and no $\phi$ factors

$M_J^+$ = moment capacity at end $J$, with bottom steel in tension, using a steel yield stress value of $\alpha f_y$ and no $\phi$ factors

$M_I^+$ = moment capacity at end $I$, with bottom steel in tension, using a steel yield stress value of $\alpha f_y$ and no $\phi$ factors

$M_J^-$ = moment capacity at end $J$ with top steel in tension, using a steel yield stress value of $\alpha f_y$ and no $\phi$ factors

$L$ = clear span of beam

The overstrength factor $\alpha$ is always taken as 1.0 for intermediate moment-resisting frames. For special moment-resisting frames $\alpha$ is taken as 1.25. The moment capacities $M_I^-$, $M_J^+$, $M_I^+$ and $M_J^-$, are evaluated at actual support points for the beam. For intermediate moment resisting frames, an additional design shear force is calculated based on the specified load factors except the earthquake loads are doubled.

The allowable concrete shear capacity is given by

$$V_c = 2.0\sqrt{f'_c}\, b_w d$$

For special moment-resisting frames, $V_c$ is zero if $V_P > 0.5V_u$ (ACI 21.3.4.2). Given $V_u$ and $V_c$, the required shear reinforcing in area per unit length is calculated as

$$A_v = \frac{\frac{V_u}{\varphi} - V_c}{f_{ys} d} \qquad \text{(ACI 11.5.6.8)}$$

$$\left(\frac{V_u}{\varphi} - V_c\right) \text{ must not exceed } 8\sqrt{f'_c}\, bd$$

where $\varphi$, the strength reduction factor, is 0.85 (ACI 9.3.2.3). The maximum of all the calculated $A_v$ values, obtained from each load combination controls the design.

When $\frac{L}{d} > 5$ and if $\frac{V_u}{\varphi} > 0.5V_c$

$$A_{v_{\min}} = \frac{50b}{f_{ys}} \qquad \text{(ACI 11.5.5.3)}$$

$$A_{h_{\min}} = 0$$

and $\left(\frac{V_u}{\varphi} - V_c\right)$ must not exceed $8\sqrt{f'_c}\, bd$. When $\frac{L}{d} > 5$ and if $\frac{V_u}{\varphi} \leq 0.5V_c$ then

$$A_{v_{\min}} = A_{h_{\min}} = 0$$

For deep beams when $\frac{L}{d} \leq 5$

$$A_{v_{\min}} = 0.0015b \qquad \text{(ACI 11.8.9)}$$

$$A_{h_{\min}} = 0.0025b \qquad \text{(ACI 11.8.10)}$$

and $\left(\frac{V_u}{\varphi}\right)$ must not exceed $\frac{2}{3}\left(10 + \frac{L}{d}\right)\sqrt{f'_c}\, bd$ for $2 \leq L/d \leq 5$ (ACI 11.8.4). However, when $L/d$ is less than 2, $\frac{V_u}{\varphi}$ must not exceed $8\sqrt{f'_c}\, bd$. Where $\varphi$, the strength reduction factor is 0.85 (ACI 9.3.2.3) for non-seismic design and is taken as 0.60 (ACI 9.3.4.1) for seismic design. $L/d$ is the span to depth ratio. The maximum of all the calculated $A_v$ values, obtained from each load combination for each location controls the design.

**Shear reinforcement in coupling beams: seismic design.** When UBC94 seismic design is performed and $L/d < 4$ and $v_u > 4\sqrt{f'_c}$ for coupling beams, the area of one leg of diagonal shear reinforcement is calculated as

$$A_{vd} = \frac{V_u}{2f_y \sin \alpha}$$

where
$$\sin \alpha = \frac{0.80D}{\sqrt{L^2 + (0.80D)^2}}$$

### 11.2.2.3 Joint design of special moment-resisting frames

To ensure that the beam-column joint of special moment-resisting frames have adequate shear strength, first, a rational analysis of the beam column panel zone is performed to determine the shear forces generated in the joint. The next step is to check this against allowable shear stress.

The joint analysis is done in the major and the minor directions of the column. The procedure involves the following steps:

- Determination of panel zone design shear force.
- Determination of effective area of the joint.
- Verification of panel zone shear stress.

**Determination of panel zone shear force.** Consider the free body stress condition of a typical beam-column intersection showing the forces $P_u$, $V_u$, $M_u^L$ and $M_u^R$ (Fig. 11.12). The force $V_u^h$, the horizontal panel zone shear force, is to be calculated.

The forces $P_u$ and $V_u$ are the axial force and shear force, respectively, from the column framing into the top of the joint. The moment $M_u^L$ and $M_u^R$ are obtained from the beams framing into the joint. The joint shear force $V_u^h$ is calculated by resolving the moments into compression $C$ and tension $T$ forces. The location of $C$ or $T$ is determined by the direction of the moment using basic principles of ultimate strength theory. Noting that $T_L = C_L$ and $T_R = C_R$, $V_u^h = T_L + T_R - V_u$.

The moments and the $C$ and $T$ forces from beams that frame into the joint in a direction that is not parallel to the major or minor directions of the column are resolved along the direction that is being investigated.

In the design of special moment-resisting concrete frames, the evaluation of the design shear force is based upon the moment capacities (with reinforcing steel overstrength factor, $\alpha$, and no $\phi$ factors) of the beams framing into the joint (ACI 21.5.1.1). The $C$ and

$T$ forces are based upon these moment capacities. The column shear force $V_u$ is calculated from the beam moment capacities as follows:

$$V_u = \frac{M_u^L + M_u^R}{H}$$

It should be noted that the points of inflection shown on Fig. 11.12

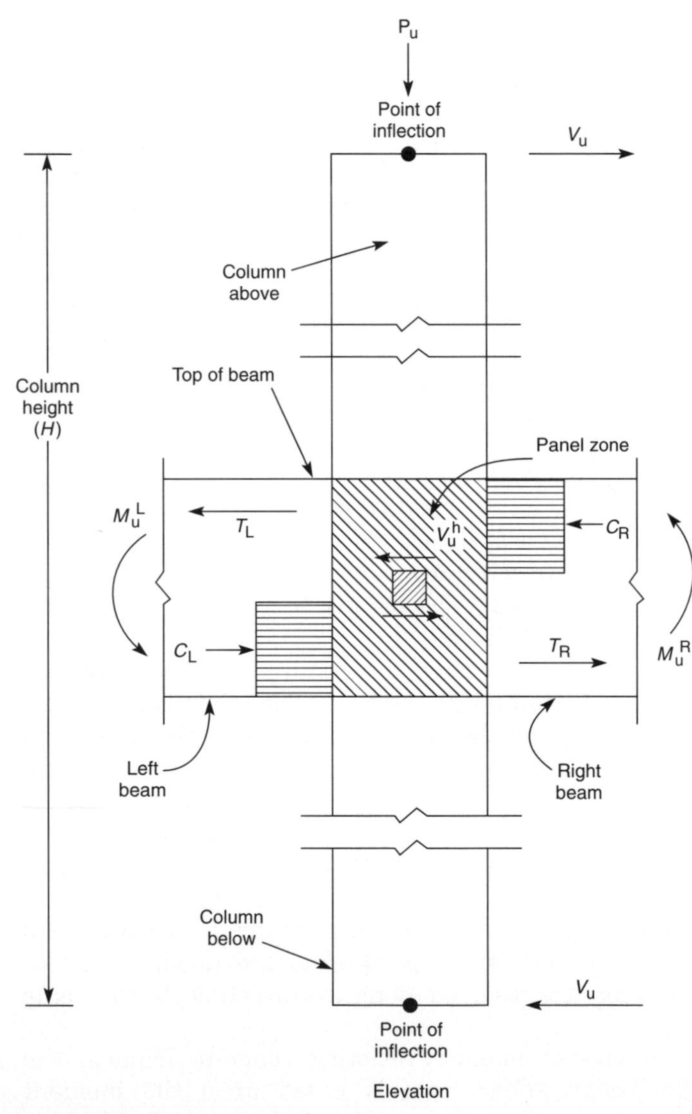


**Figure 11.12** Column panel shear forces.

are taken as midway between actual lateral support points for the columns.

The effects of load reversals, as illustrated in Cases 1 and 2 of Fig. 11.13 are investigated and the design is based upon the maximum of joint shears obtained from the two cases.

**Determine the effective area of joint.** The joint area that resists the shear forces is assumed always to be rectangular. The dimensions of the rectangle correspond to the major and minor dimensions of the column below the joint, except that if the beam framing into the joint is very narrow, the width of the joint is limited to the depth of the joint plus the width of the beam. The area of the joint is assumed not to exceed the area of the column below. It should be noted that if the beam frames into the joint eccentrically, the above assumptions may be unconservative.

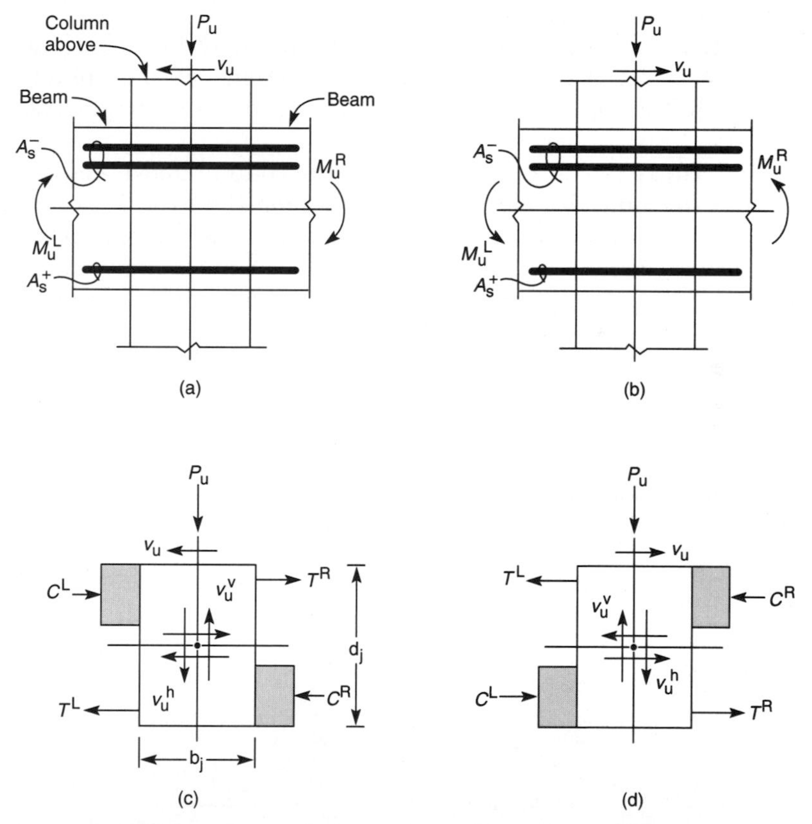


**Figure 11.13** Beam-column joint analysis: (a) forces and moments, case 1; (b) forces and moments, case 2; (c) resolved forces, case 1; (d) resolved forces, case 2.

**Check panel zone shear stress.** The panel zone shear stress is evaluated by dividing the shear force by the effective area of the joint and comparing it with the following allowable: (ACI 21.5.3).

For joints confined on all four faces $20\varphi\sqrt{f'_c}$

For joints confined on three faces or on two opposite faces $15\varphi\sqrt{f'_c}$

For all other joints $12\varphi\sqrt{f'_c}$

Where $\varphi = 0.85$

#### 11.2.2.4 Beam-column flexural capacity ratios

At a particular column direction, major or minor, the ACI and UBC codes require calculation of the ratio of the sum of the beam moment capacities to the sum of the column moment capacities (ACI 12.4.2.2). The capacities are calculated with no reinforcing overstrength factor, $\alpha$, and including $\phi$ factors.

The moment capacities of beams framing obliquely into the joint are resolved along the major and minor directions of the column before adding to the capacities of beams framing to the principle axes of the column.

The column capacity summation includes the column above and the column below the joint. For each load combination the axial force, $P_u$, is calculated from the analysis load conditions and the corresponding load combination factors. The moment capacity of the column under the influence of the corresponding axial load $P_u$ is then determined for the major and minor directions of the column, using a uniaxial column interaction diagram, similar to the one shown in Fig. 11.14

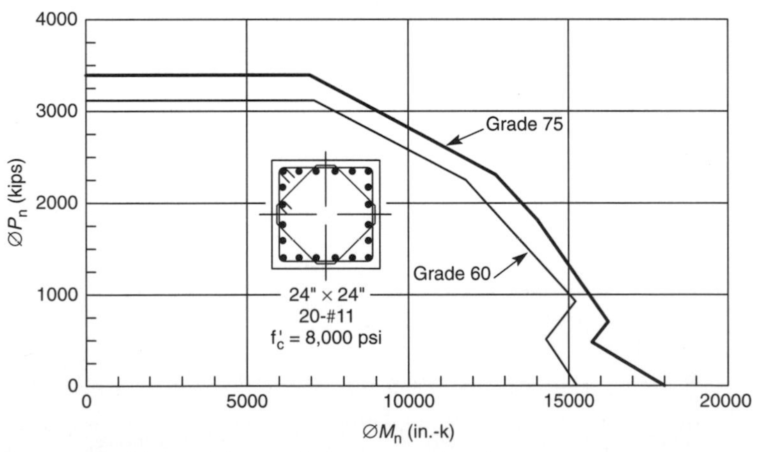


**Figure 11.14** Uniaxial interaction diagram for column.

for a $24 \times 24''$ column. The moment capacities of the two columns intersecting at a joint are added to give the capacity summation for the corresponding load combination. The minimum capacity summations obtained from all of the load combinations is used for verifying the beam/column capacity ratio.

### 11.2.3 Column design

Strength design for columns is a complete design since there are no further serviceability requirements to be satisfied. It consists of determining the critical load combinations and resulting factored loadings for each; selecting a column cross section satisfying all design assumptions for materials used under loads and moments equal to or larger than the factored loads and moments; and ensuring that slenderness effects do not reduce capacity.

Since the reduction in capacity for slenderness is dependent upon the cross-sectional properties, including reinforcement of the column selected, and the factored moments are also dependent upon the column gross cross-section, strength design for columns is in reality a two-stage review of a trial design. The manual calculations for such a review are formidable and become increasingly laborious if several trials are required.

The variety of cross-sectional shapes used for reinforced concrete columns is unlimited, but most columns are square or rectangular, because of practical requirements in layout of walls and partitions and ease of forming.

The design procedure for reinforced concrete columns subjected to baxial bending and axial loads involves the following steps. First, generate an axial load-bi-axial moment interaction surface diagram for the column as shown schematically in Fig. 11.15. Observe that the range of allowable reinforcement is 1 to 8 percent, except for special moment-resisting frames it is 1 to 6 percent. Then check the capacity of the column for the factored axial force and magnified biaxial bending moments obtained from each loading combination at each end of the column. Finally, design the column shear reinforcing.

#### 11.2.3.1 Generation of bi-axial interaction surfaces

With the general availability of computers it is no longer tedious to establish axial load-moment interaction diagrams for columns of various shapes with various percentages of reinforcement and concrete strengths. This is done by taking successive choices of neutral axis distance measured along one face of the column from the most heavily compressed corner, then calculating the axial force $P_n$ and the corresponding moments $M_{nx}$ and $M_{ny}$. Each sequence of calculations is repeated until the complete interaction diagram is obtained. The method is obviously impractical for hand calculations but the interactive steps are routinely performed by using computers which in most cases provide a graphical presentation of results.

The column capacity interaction volume is numerically described by a series of discrete points that are generated on the three-dimensional interaction failure surface. In addition to axial compression and biaxial bending, the general formulation allows for axial tension and biaxial bending considerations as shown in Fig. 11.15. The coordinates of these points are determined by rotating a plane of linear strain in three dimensions on the section of the column. The method is based on the principles of ultimate strength design, and is applicable for rectangular, square or circular, doubly symmetric

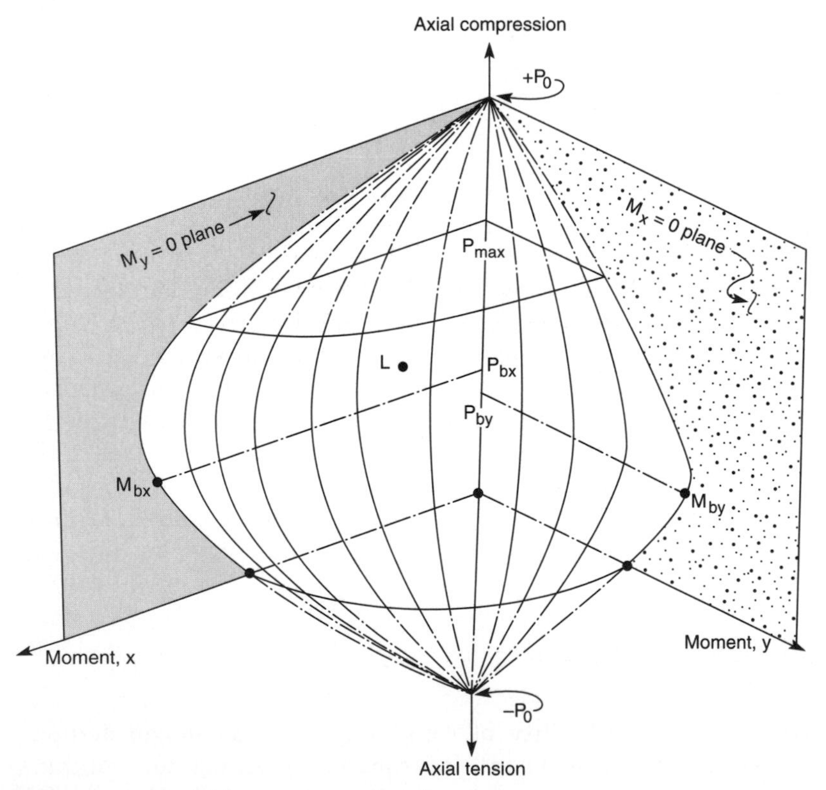


**Figure 11.15** Biaxial interaction surface for column.

column section. The linear strain diagram limits the maximum concrete strain, $\varepsilon_c$, at the extremity of the section, to 0.003. The stress in the steel is given by the product of the steel strain and the steel modules of elasticity, $\varepsilon_s E_s$, and is limited to the yield stress of the steel, $f_y$.

The concrete compression stress block is assumed to be rectangular, with a stress value of $0.85f'_c$ (see Fig. 11.16). The interaction procedure should account for the concrete area displaced by the reinforcement in the compression zone. The effects of the strength reduction factor, $\phi$, are included in the generation of the interaction surfaces. The maximum compressive axial load is limited to $P_{\max}$,

where $P_{\max} = 0.80P_0$ for columns with rectangular reinforcement patterns (i.e., tied columns)

$P_{\max} = 0.85P_0$ for columns with circular reinforcement patterns (i.e., columns with spiral reinforcing)

where

$$P_0 = \varphi_{\min}[0.85f'_c(A_g - A_{st}) + f_y A_{st}]$$

and

$$\varphi_{\min} = 0.70 \text{ for tied columns and}$$

$$\varphi_{\min} = 0.75 \text{ for spirally reinforced columns}$$

The values of $\phi$ used in the interaction diagram varies from $\phi_{\min}$ to 0.9 based on the axial load. For low values of axial load, $\phi$ is increased linearly from $\phi_{\min}$ to 0.9 as the axial load decreases from

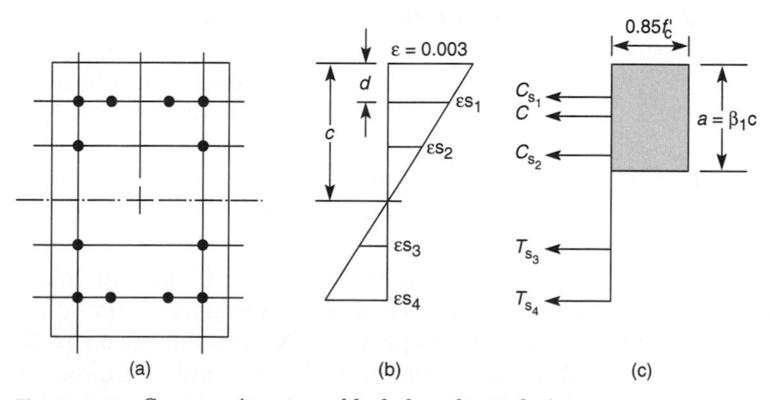


**Figure 11.16** Compression stress block for column design.

the smaller of $0.1f'A_gA$ or $P_b$ to zero. For axial tension, $\phi$ is always 0.9 (ACI 9.3.2.2.).

The column capacity is checked for each loading combination at the top and bottom of the column by verifying if the point, defined by the resulting axial load and biaxial moments, lies within the interaction volume. Before entering the interaction diagram to check the column capacity, the moment magnification factors are applied to the factored loads to obtain $P$, $M_x$ and $M_y$, so that:

$$P = P_u \quad \text{(no magnification)}$$

$$M_x = \delta_{bx}M_{uxb} + \delta_{sx}M_{uxs}$$

where

$$M_y = \delta_{by}M_{uyb} + \delta_{sy}M_{uys}$$

$P_u$ is the factored axial load; $M_{uxb}$ and $M_{uyb}$ are factored moments in the major and minor directions caused by gravity loads increased if necessary to satisfy minimum eccentricity requirements; $M_{uxs}$ and $M_{uys}$ are factored moments in the major and minor directions caused by lateral loads; and $\delta_{bx}$, $\delta_{by}$, $\delta_{sx}$ and $\delta_{sy}$ are moment magnification factors as explained presently. The point corresponding to $P$, $M_x$, $M_y$ is then placed in the interaction space as shown as point $L$ in Fig. 11.15. If the point lies within the interaction volume, the column capacity is adequate; if not, the column is over stressed.

#### 11.2.3.2 Determination of moment magnification factors

The moment magnification factors are different for braced frames and unbraced frames. They may also be different for major and minor directions of the column. The ACI code permits the use of axial load and end moments determined by a conventional frame analysis (which does not include the $P$-$\Delta$ effects) in the design of columns, provided the moments are magnified to take into account the $P$-$\Delta$ effects.

The ACI procedure for determining moment magnification factors in frame columns effectively braced against side sway is as follows.

1. Find the primary moments in the columns due to lateral loads including the effect of unsymmetry due to geometry and axial loads. In defining the member properties, due consideration is to be given for creep and cracking effects in beams and columns. In the absence of a more detailed analysis, the ACI code permits the

effective $EI$ to be determined by either of the following two relations;

$$EI = \frac{E_c I_g/5 + E_s I_s}{1+\beta}$$

$$EI = \frac{E_c I_g/2.5}{1+\beta}$$

where $E_c$ and $E_s$ = the moduli of elasticity of concrete and steel, respectively

$I_g$ and $I_s$ = the moments of inertia of the gross section of concrete and reinforcement about the axis of bending

$\beta$ = a ratio that takes into account the effect of creep and is equal to the ratio of maximum factored dead load moment to the maximum factored total load moment

2. From the lateral analysis obtain the moments $M_1$ and $M_2$ at the ends of the column and calculate the coefficient $C_m$ by the relation $C_m = 0.6 + 0.4M_1/M_2$.
3. Determine the effective length factor $K$ by using alignment charts. $K$ varies from 0.5 to 1.0 for braced columns.
4. Calculate the moment magnification factor $\delta$ by the relation:

$$\delta_b = \frac{C_m}{1 - P_u/\phi P_c} \geqslant 1.0$$

where $P_u$ = ultimate axial load on the column
$P_c$ = critical load
$\phi$ = capacity reduction faction

5. Obtain the magnified moment at the ends of the column by multiplying the results of the elastic analysis $M_{2b}$ by the moment magnification factor $\delta_b$.

$$M_c = \delta_b M_{2b}$$

6. Design the column for axial loads and magnified moment obtained in step 5 using column design charts or computer interaction diagrams.

The ACI procedure for designing columns in unbraced frames can be outlined as follows:

1. In a side sway mode, lateral instability can occur only if all the columns in a particular story buckle. Therefore, in calculating moment magnification use $\sum P_u$ and $\sum P_c$ for all the columns in

that story. For moments resulting from lateral loads, $C_m = 1$ because the maximum secondary moments occur at the same locations as the primary moments.

2. From the elastic analysis find the gravity moments occurring at the ends of the columns.
3. Calculate the coefficiient $C_m$ for individual buckling of columns by using the relation given previously for braced columns.
4. From $\sum P_u$ and $\sum P_c$ values of all columns in the particular story, calculate the moment magnification factor $\delta_s$ to reflect lateral drift resulting from lateral and unsymmetrical gravity loads

$$\delta_s = \frac{1}{1 - \sum P_u / \phi \sum P_c}$$

5. Obtain the design moment $M_c$ by multiplying the primary, i.e., the gravity moment $M_g$ and the secondary moments $M_s$ by their corresponding magnification factors thus:

$$M_c = \delta_b M_{2g} + \delta_s M_{2s}$$

The subscript 2 for $M_g$ and $M_s$ indicates that the larger of the two end moments is to be used in obtaining the design moment.

6. Design the column for the axial load and the amplified moments obtained in step 5.

Adjustments as outlined above are required to take into account the reduction in the strength of the structure caused by the effects of stability, the so-called $p$-$\Delta$ effect. The compensation for not including the $p$-$\Delta$ effect directly in the analysis is made by multiplying the results of first-order analysis by the moment magnifier.

A conventional first-order analysis assumes that the effect of deflections due to external loads does not alter the magnitude of internal moments and forces in the structure. However, in reality the axial loads $P$ acting through the deflection of the sturcture give rise to additional moment with a consequent increase in the deflection. The additional moment may be visualized as being caused by additional story shears called *sway forces* or $p$-$\Delta$ forces. The increase in deflection causes further increase in the moment and a consequent increase in the deflection. The process continues until either the structure comes to a stable equilibrium or, when the structure is very limber, it collapses. As mentioned earlier, until recently the only practical method of accounting for the $p$-$\Delta$ effect was to use the moment magnification method. However, with the availability of computer programs which can take into account directly the $p$-$\Delta$ effects. It is no longer necessary to calculate the effective length factors through alignment charts. All columns can be designed by taking the effective length factor $K$ and moment magnification factors equal to 1. Although the procedure itself is not described in the ACI, it is highly endorsed in its commentary.

If computer programs capable of performing second-order analysis are not available, the first-order analysis may be modified to include the $p$-$\Delta$ effect. In this method the lateral and vertical loads are applied to the structure and the relative first-order lateral displacements in each story are computed. The additional story shears caused by the $p$-$\Delta$ moments are computed from a knowledge of story displacement and the accumulated gravity loads at each story. The net story shear at any given level is obtained by the algebraic sum of story shears from the columns above and below the floor. The additional story shears are added to the applied loads and the structure is reanalyzed for the new set of lateral loads. Since these loads in general are larger than the first loading case, the resulting deflections are larger, requiring one or two more cycles of iteration for convergence. However, the following equation can be used to obtain directly the final second-order deflection from the first-order deflection thus

$$\Delta = \frac{1}{1 - \sum P\Delta_i / Hh}$$

where $\sum P$ = cumulative vertical load
$\Delta_i$ = the first-order story sway
$h$ = story height
$H$ = horizontal shear force

The above procedure can be outlined as follows:

1. From the first-order analysis determine $\Delta_i$ in each story.
2. Compute the second-order final deflection from the above equation.
3. Calculate the additional story shears caused by the $p$-$\Delta$ moments.
4. Perform another first-order analysis by subjecting the frame to the applied lateral loads plus story shears obtained in step 3.

Include gravity loads in steps 1 and 4 for unsymmetrical buildings. Step 4 gives the second-order moments and forces, which can be used in the design of members without having to resort to moment magnifier methods.

The designer, however, faces a particular problem in choosing a suitable value of flexural stiffness $EI$ under various loading conditions. The value of $EI$ should ideally reflect the amount of reinforcement, the extent of cracking, creep, reduction in stiffness due to axial loads and the inelastic behavior of steel and concrete. It is necessary to recognize the variation of stiffness along the entire length of each member by taking into account cracked and uncracked regions. Practical design considerations preclude the possibility of going into this detail for calculating the flexural stiffness of each and every member of a multistory building with thousands of members. Therefore, in practice, simplified methods are employed to compute $EI$.

#### 11.2.3.3 Additional seismic requirements

In seismic design using special moment-resisting frames, the column design is carried out for one other moment value about each axis separately. The moment is obtained by distributing to the top and bottom columns at a joint, a moment equal to $\frac{6}{5}$ths the sum of the moment capacities of the beams framing into the joint. In calculating the moment capacities of the beams for this purpose, no yield overstrength factors are used and $\phi$ values of 1.0 are used. A point of contraflexure at mid height of the columns is generally assumed to distribute this moment to top and bottom columns at a joint. The design for these moments is done in conjunction with the factored axial loads.

Detailing requirements for confinement reinforcement in seismic zones 3 and 4 are shown in Figs 11.17 and 11.18.

A summary of design criteria for seismic and non-seismic frames is given in Table 11.1.

#### 11.2.3.4 Design for shear

The shear reinforcing is designed for each loading combination in the major and minor directions of the column and involves the following steps

- Determination of factored forces acting on the section $P_u$ and $V_u$. Note that $P_u$ is needed for the calculation of $V_c$.
- Determination of shear force, $V_c$, that can be resisted by the concrete.
- Calculation of reinforcing steel required to carry the balance of shear force.

**Determination of factored forces.** In the design of column shear reinforcing of an ordinary moment resisting concrete frame, the column axial force, $P_u$ and the column shear force, $V_u$, in a particular direction are obtained by factoring the load conditions with the corresponding load combination factors.

In the design of intermediate moment-resisting concrete frames, the shear force $V_u$ in the column is also calculated from the moment capacities of the column associated with its factored axial force. For each load combination, the factored axial load $P_u$ is calculated, and then the moment capacity $M_u$ is obtained using the uniaxial interaction diagram. The shear force $V_u$ is calculated as:

$$V_u = \frac{\sum M_u}{H}$$

where $H$ = the clear height of the column and
$\sum M_u$ = the sum on top and bottom moment capacities.

Other values require to be checked for intermediate moment-resisting frames are based on the specific load factors except the

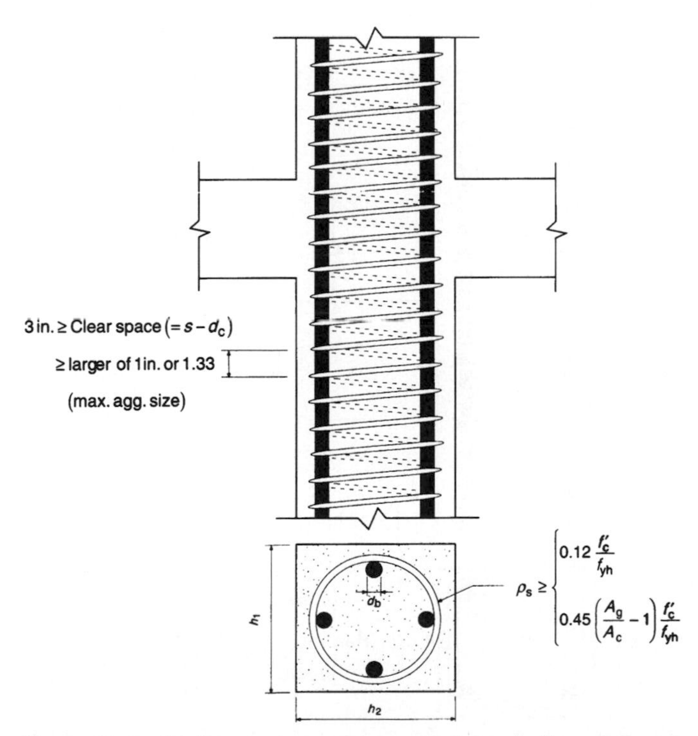


**Figure 11.17** Confinement requirements for spirally reinforced columns.

earthquake loads are doubled. In the design of special moment-resisting frames, the force $V_u$ in a particular direction is also calculated from the probable moment capacities of the beams that frame into the column (ACI 21.4.5.1). The probable moment capacities are based on the steel yield overstrength factor, $\alpha$ equal to 1.25, and $\phi$ of 1.0. To obtain the column shear from the beam moments, first calculate column shear for columns above and below a joint (bottom shear for column above the top shear for column below), assuming point of contraflexure in the columns at mid height. Since the top and bottom shear so computed could differ in value, an average of the two is used in design. The minimum $P_u$ from among the factored loads is used in conjunction with the above $V_u$. This completes the required design force set, $P_u$ and $V_u$.

**Determination of concrete shear capacity.** Given the design force set $P_u$ and $V_u$ the shear force carried by the concrete, $V_c$, is calculated as follows:

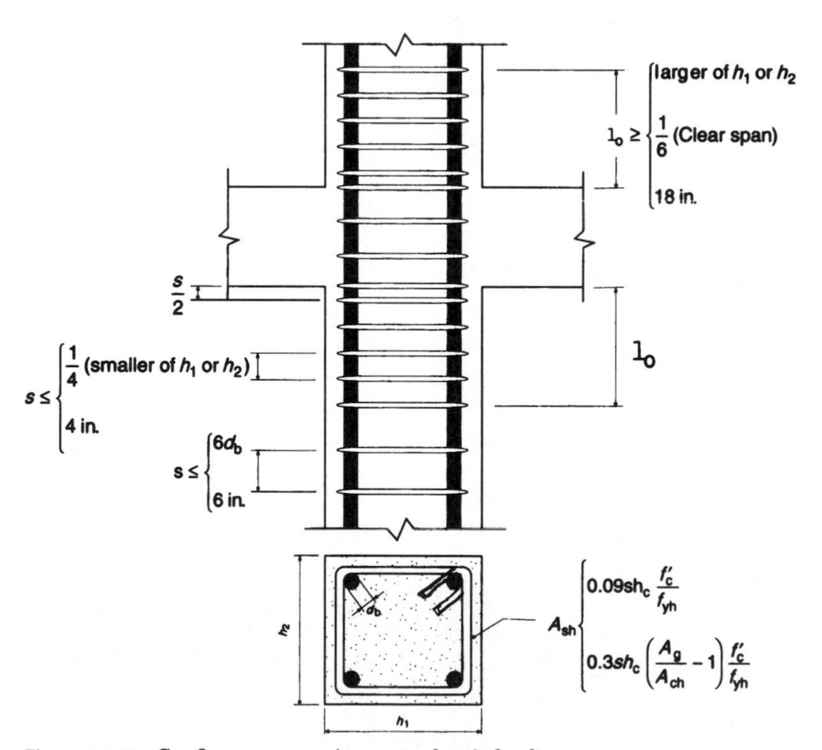


**Figure 11.18** Confinement requirements for tied columns.

**TABLE 11.1 Design Criteria for Seismic and Non-Seismic Frames**

| Type of checks/design | Ordinary moment-resisting frames (non-seismic) | Intermediate moment-resisting frames (seismic) | Special moment-resisting frames (seismic) |
|---|---|---|---|
| Column design flexure and axial loads | Ultimate load combinations $1\% < \rho < 8\%$ | Ultimate load combinations $1\% < \rho < 8\%$ | Ultimate load combinations $\frac{6}{5}$ beam capacity $\alpha = 1.0$ $1\% < \rho < 6\%$ |
| Column shears | Ultimate load combinations | Modified ultimate load combination (earthquake loads doubled) column capacity $\varphi = 1.0$ and $\alpha = 1.0$ | Ultimate load beam capacity combinations $\varphi = 1.0$ and $\alpha = 1.25$ |
| Beam design flexure | Ultimate load combinations | Ultimate load combinations | Ultimate load combinations $\rho_{max} \leqslant 0.025$ |
| Beam min. moment requirements | No requirement | $M^{+}_{u\text{END}} \geqslant \frac{1}{3}M^{-}_{u\text{END}}$ $M^{+}_{u\text{SPAN}} \geqslant \frac{1}{5}M^{+}_{u\text{END}}$ $M^{-}_{u\text{SPAN}} \geqslant \frac{1}{5}M^{-}_{u\text{END}}$ | $M^{+}_{u\text{END}} \geqslant \frac{1}{2}M^{-}_{u\text{END}}$ $M^{+}_{u\text{SPAN}} \geqslant \frac{1}{4}M^{+}_{u\text{END}}$ $M^{-}_{u\text{SPAN}} \geqslant \frac{1}{4}M^{-}_{u\text{END}}$ |
| Beam design shear | Ultimate load combinations | Modified ultimate load combinations (earthquake loads doubled). Beam capacity shear ($V_p$) with $\alpha = 1.0$ and $\varphi = 1.0$ plus $V_{D+L}$ | Beam capacity shear ($V_p$) with $\alpha = 1.25$ and $\varphi = 1.0$ plus $V_{D+L}$ |
| Joint design | No requirement | No requirement | Beam capacity with $\alpha = 1.25$ and $\varphi = 1.0$ |
| Beam/column ratios | No requirement | No requirement | Beam capacity $\alpha = 1.0$. Column capacity based on uniaxial capacity under axial loads from ultimate load combinations |

- If the column is subjected to axial compression, $P_u$ is positive (ACI 11.3.1.2)

$$V_c = 2\sqrt{f'_c}\left(1 + \frac{P_u}{2000A_g}\right)A_{cv}$$

where $V_c$ may not be greater than

$$V_c = 3.5\sqrt{f'_c}\sqrt{\left(1+\frac{P_u}{500A_g}\right)}A_{cv}$$

- If the column is subjected to axial tension, $P_u$ is negative (ACI 11.3.2.3)

$$V_c = 2\sqrt{f'_c}\left(1+\frac{P_u}{500A_g}\right)A_{cv} \geqslant 0$$

The term $\frac{P_u}{A_g}$ must have psi units.

For special moment-resisting frame concrete design, $V_c$ is considered zero if $P_u < 0.05f'_cA_g$ (ACI 21.4.5.2).

**Determination of shear reinforcement.** Given $V_u$ and $V_c$, the required shear reinforcing in area per unit length (e.g. square inches/foot) is given by

$$A_v = \frac{\frac{V_u}{\varphi} - V_c}{f_{ys}d}$$

where $\frac{V_u}{\phi} - V_c$ must not exceed $8\sqrt{f_c}A_{cv}$ (ACI 11.5.6.8) and $\phi$, the strength reduction factor, is 0.85 (ACI 9.3.2.3). The maximum of all the calculated $A_v$ values obtained from each load combination for the major and minor directions of the column controls the design.

### 11.2.4 Shear wall design

Prior to 1994 UBC, in the design of walls in seismic zones 3 and 4, special boundary zone reinforcement was required at the ends of walls when the gross section elastic stresses exceed $0.2f'_c$. This design philosophy assumed that the wall panel between the boundary elements has deteriorated by developing diagonal cracking from the repeated inelastic cycles typical of large earthquakes. This assumption therefore neglected the vertical load-carrying capacity of the panel with the result that the boundary elements are designed to resist the entire gravity loads tributary to the wall and the entire factored overturning moment caused by the seismic loads.

The current 1994 UBC allows the use of the entire wall cross-section to resist factored gravity loads and overturning moments due to seismic loads. New rules are used for determining when special confinement reinforcement is necessary at wall boundaries. Boundary zone confinement is not eliminated in the 1994 UBC. In fact, confinement of a larger portion of the wall may be required ensuring that those portions of the wall web, in which significant compressive strains develop, will be adequately confined to maintain vertical load stability.

The methodology for determining the need for confinement for the wall boundary reinforcement is based on basic engineering principles. Limitations on concrete compressive strains provide a basis to determine the need and extent of special confinement reinforcement. Basic strain in the concrete at yield level seismic load is limited to 0.003 which may be increased to 0.015 under the maximum seismic event. The reader is referred to the UBC for the specific formulas relating to the boundary elements.

The design of symmetric shear walls subjected to in-plane moments and shears as shown in Fig. 11.19 is considered in this section. Two designs are required;

(i) Design for overturning moment and axial load and
(ii) Design for shear.

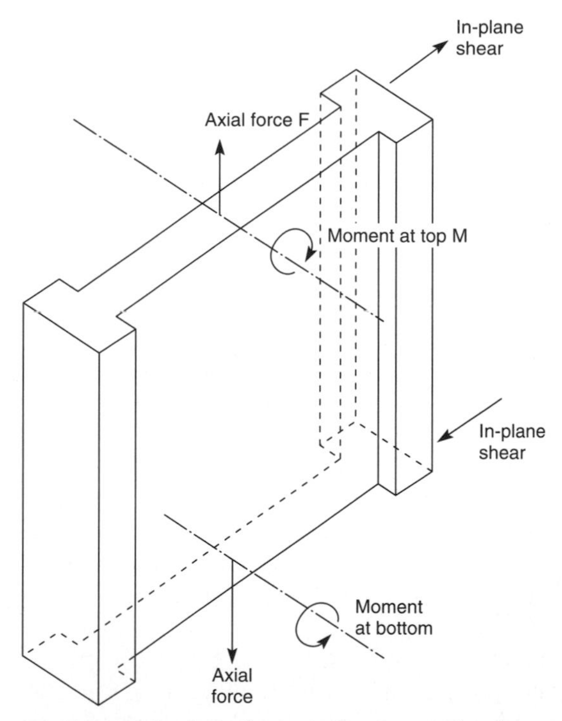


**Figure 11.19** In-plane forces and moments for shear wall design.

As explained in Chap. 10, a shear wall may be modeled using either column elements or an assemblage of panel elements with the major direction as the in-plane direction. When panel elements are used for modeling three-dimensional walls, it is imperative that the walls are modeled as an assemblage of planar segments rather than as a single unit. This is accomplished by treating each planar segment of three-dimensional wall as a discrete wall with vertical connectivity to invoke three-dimensional behavior.

#### 11.2.4.1 Design for overturning moment and axial load

Consider the wall shown in Fig. 11.20 subjected to a factored axial force $F$ and a factored overturning moment $M$. The design procedure

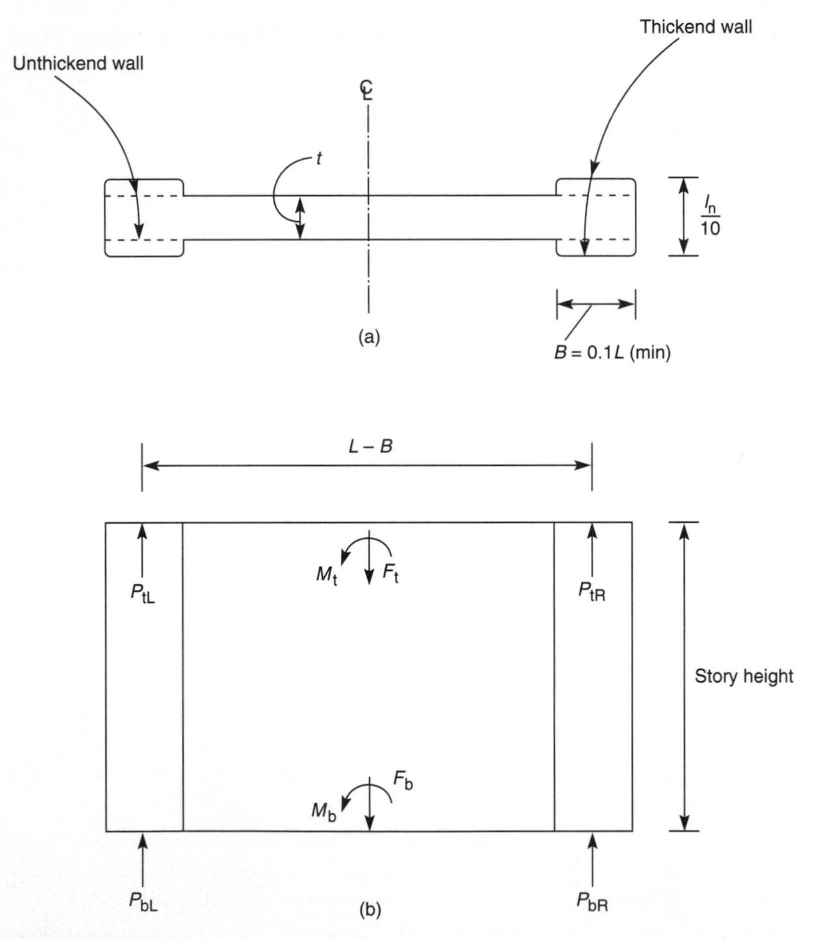


**Figure 11.20** Shear wall with thickened boundary elements: (a) wall section; (b) wall elevation.

starts by assuming an edge member of thickness $t$ and width $B_1$ at each end of the wall. The overturning moment and axial force are converted to an equivalent force-set $P_L$ and $P_R$ using the following relationship:

$$P_L = \frac{F}{2} + \frac{M}{ARM}$$

$$P_R = \frac{F}{2} - \frac{M}{ARM}$$

where $ARM = L - B_1$. For any given loading combination, the net values for $P_L$ or $P_R$ could be tension or compression. For dynamic loads, $P_L$ and $P_R$ are obtained by using modal combinations before combining with other loads. If any $P$ value is tension, the area of steel required for tension is calculated as

$$A_{st} = \frac{P}{\varphi f_y} \quad \text{where } \varphi = 0.90$$

If any $P$ value is compression, the compressive area of steel, $A_{sc}$, must satisfy the following relationship.

$$P = 0.80\varphi[0.85f'_c(A_g - A_{sc}) + f_y A_{sc}] \qquad \text{(ACI 10.3.5.2)}$$

where

$$A_g = tB_1 \quad \text{and} \quad \varphi = 0.70$$

from which

$$A_{sc} = \frac{\left(\dfrac{P}{0.80\varphi} - 0.85f'_f A_g\right)}{(f_y - 0.85f'_c)}$$

If the calculations show that the compressive steel area $A_{sc}$ is negative, no compressive steel is needed by the calculations. However, minimum reinforcement as required by the codes is provided. The maximum tensile and compressive reinforcing within the boundary zone $t$ times $B_1$, should be limited to some reasonable values to avoid overcrowding of reinforcement.

If $A_{st}$ and $A_{sc}$ are within reasonable limits, the design may be considered satisfactory. Othersie $B_1$ dimenison is incremented to $B_2$, and new values for $P_L$ and $P_R$ are calculated. The value of $ARM$ is changed to $(L - B_2)$, to obtain new values of $A_{st}$ and $A_{sc}$. This is continued until $A_{st}$ and $A_{sc}$ are within the allowed steel percentages.

If the value of width of the edge member $B$ increments to where it reaches a value equal to $L/2$, the dimensions of wall should be increased.

The tension design criteria should be satisfied before starting the compression design. This procedure maintains the tensile reinforcing as close to the outer edge of the wall as possible.

For walls with thickened edges the most common design procedure is to assume that the boundary element is limited to the $P_1$ times $t$ area, as shown in Fig. 11.20. This is an approximate but a convenient design method. Walls found to be over-stressed by using this method may very well be satisfactory if evaluated against a load-moment interaction curve generated for the entire wall. It should be noted, however, that some seismic codes require the use of edge members to resist the $P_L$ and $P_R$ values calculated.

#### 11.2.4.2 Design for shear

The shear reinforcing is designed for each of the factored loading combinations using the following steps. First, determine the factored forces acting on the section $P_u$, $M_u$ and $V_u$. Note that $P_u$ and $M_u$ are needed for the calculation of $V_c$. Then determine the shear force $V_c$, that can be resisted by the concrete. Finally, calculate the reinforcing steel required to carry the balance.

The ultimate design values for the wall axial force, $P_u$, the maximum moment along the member, $M_u$, and the wall shear force, $V_u$, are obtained by factoring the analysis load combinations with the corresponding load factors.

Next the shear force carried by the concrete, $V_c$, is calculated as follows (ACI 11.10.6):

$$V_c = 3.3\sqrt{f'_c}\,td + \frac{P_u d}{4L}$$

where $V_c$ may not be greater than

$$V_c = \left[0.6\sqrt{f'_c} + \frac{L\left(1.25\sqrt{f'_c} + 0.2\dfrac{P_u}{Lt}\right)}{\dfrac{M_u}{V_u} - \dfrac{L}{2}}\right] td$$

where $L$ and $t$ are the length and thickness of the wall, $d$ is taken as $0.8L$ and $P_u$ is negative for tension. The second equation does not apply if $\dfrac{M_u}{V_u} - \dfrac{L}{2}$ is negative.

Given $V_u$ and $V_c$, the required shear reinforcing in area per unit length (e.g. square inches per foot) is given by

$$A_v = \frac{\frac{V_u}{\varphi} - V_c}{f_{ys} d}$$

where

$$\frac{V_u}{\varphi} - V_c \text{ must not exceed } 8\sqrt{f'_c}\, td \qquad \text{(ACI 11.5.6.8)}$$

and

$$\frac{V_u}{\phi} \text{ must not exceed } 10\sqrt{f'_c}\, td \qquad \text{(ACI 11.10.3)}$$

where $\phi$, the strength reduction factor, is 0.85 (ACI 9.3.2.3). The maximum of all the calculated $A_v$ values obtained from each load combination controls the design.

The wall shear reinforcing requirements given above are based purely upon shear strength considerations. Other minimum shear steel requirements required by the code are not considered here.

#### 11.2.4.3 Additional seismic requirements

For shear design of walls subjected to seismic loads the following additional requirements should also be checked: the nominal shear strength of walls is limited to

$$V_n = \left(2\sqrt{f'_c} + \frac{A_v}{t} f_y\right) Lt$$

Since $V_u = \varphi V_n$, $A_v$ can be calculated as

$$A_v = \frac{\frac{V_u}{\varphi} - 2\sqrt{f'_c}\, Lt}{L f_y}$$

where $\varphi$ is 0.60 for shear (ACI 9.3.4.1)

Observe that a lower value of $A_v$ may be required based on ACI 21.6.4.3. Also in the satisfaction of ACI 21.6.4.6, use the more conservative assumption that the individual pier nominal shear strength does not exceed $8\sqrt{f'_c}\, Lt$.

### 11.2.4.4 Load-moment interaction diagram

When the geometry and the vertical reinforcing distribution for wall are known the capacity check of the section may be performed by generating a load-moment interaction curve for the entire wall section Fig. 11.21. As mentioned previously the interaction diagram is numerically described by a series of discrete points. In addition to axial compression and bending, the formulation for generating the interaction diagram should consider axial tension and bending. The formulation is based on the principles of ultimate strength design (ACI 10.3), with a linear strain diagram that limits the maximum concrete strain, $\varepsilon_c$, at the extremity of the section to 0.003.

The stress in the steel is given by the product of the steel strain and the steel modules of elasticity, $\varepsilon_s E_s$, and is limited to the yield stress of the steel, $F_y$. The concrete compression stress block is assumed to be rectangular, with a stress value of $0.85f'_c$ (see Fig. 11.2). The interaction should account for the concrete area displaced by the reinforcement in the compression zone.

The effects of the strength reduction factor, $\varphi$, are generally included in the generation of the interaction surfaces. The strength

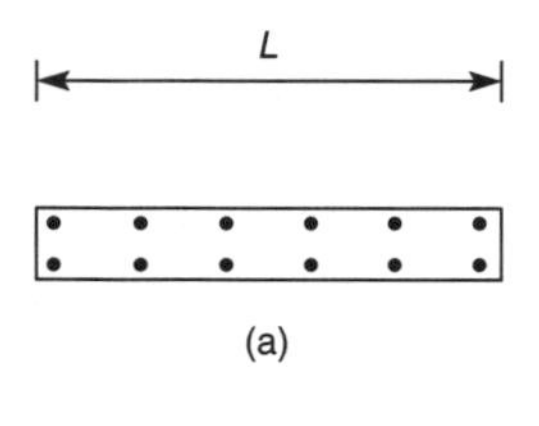


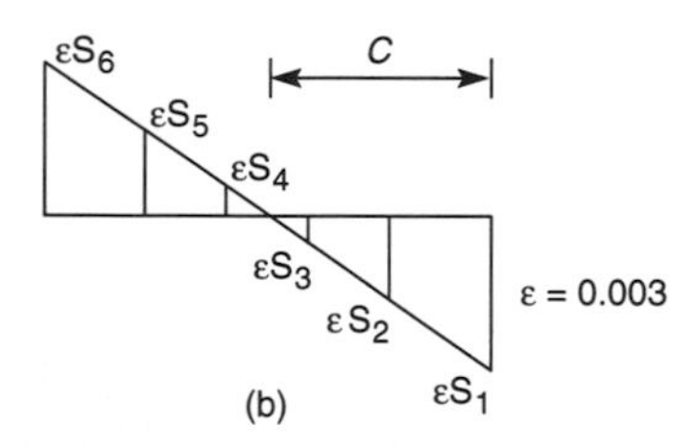


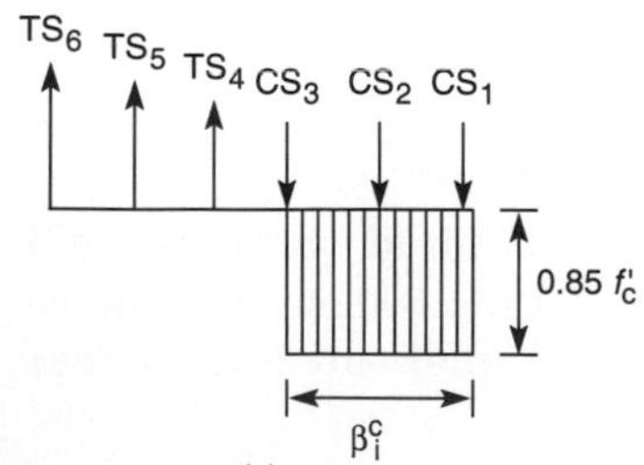


**Figure 11.21** Stress strain relationship in shear wall: (a) wall section; (b) strain diagram; (c) stress diagram.

reduction factor, $\varphi$, for high axial compression, with or without moment is assumed to be 0.70. For low values of axial load, $\phi$ is increased linearly from 0.70 to 0.90 as the axial force capacity, $\phi P_n$ decreases from the smaller of $0.10 f'_c A_g$ or $\varphi P_b$ to zero. In cases involving axial tension, $\phi$ is always 0.90 (ACI 9.3.2.2.). The maximum factored compressive axial load including the $\phi$ factor reduction is limited to $P_{max}$, where

$$P_{max} = 0.80\varphi[0.85 f'_c(A_g - A_s) + f_y A_s] \qquad \text{(ACI 10.3.5.2)}$$

After the moment interaction curve is generated, the wall capacity is checked for each loading combination. The first step is to determine the factored magnified moments and axial forces from the analysis load conditions and the specified load combination factors to give $P_u$ and $M_u$. The second step is to determine if the point, defined by the factored axial load and moment set, lies within the interaction curve. If the point lies within the interaction curve, the wall capacity is satisfactory. If not, a redesign of the wall is required.

### 11.2.5 Comments on seismic details

Reinforced concrete achieves ductility through careful limits on steel in tension and concrete in compression. Reinforced concrete beams with common proportions can possess ductility under monotonic loading even greater than common steel beams, in which local buckling is usually a limiting factor. However, providing stability of the resistance to reversed inelastic strains requires uncommon detailing. Thus, there is a wide range of reduction factors from elastic response to design response, depending on the detailing for stable and assured resistance. The essence of seismic detailing is to prevent premature shear failures in members and joints, buckling of compression bars, and concrete compression failures.

It is not sufficient to show strength capability, there must also be special details for the inelastic behavior of the system along with a verification of stability at deformations corresponding to maximum expected ground motion. Vertical loads must be supported even when maximum deformations are exceeded. The underlying principle of seismic design is that inelastic yielding is allowed in resisting seismic loads, as long as yielding does not impair the vertical load capacity of the structure.

**Joint shear.** Because the mechanism of shear failure in the joint is different than shear-flexure failure in beams, the nominal shear capacities are considerably higher than the values the designers in the non-seismic areas are accustomed to. For example, the 1994 UBC permits shear stress in planel zones to $20\sqrt{f'_c}$, $\sqrt{15f'_c}$, and $12\sqrt{f'_c}$ when the joint is confined on all four, three, and two faces of the column respectively. Compare this to the value of $4\sqrt{f'_c}$ allowed for punching shear. The relatively high values for joint shear may give the wrong impression that joint shear will not be a problem in sizing of columns. However, it should be noted that even with the very high shear stresses permitted, joint shear most often controls the size of frame-columns. Also note that shear reinforcement is required in the joints even though no increase in shear capacity is credited for its presence. Closely spaced ties are required to extend through the joint.

**Why strong column-weak beam?** The reason is to prevent story mechanism. This is accomplished by assuring that at each beam-column joint the flexural resistance of columns is substantially (20 percent by 1994 UBC) more than the flexural strength of beams. In calculating the nominal flexural strength of columns, the effect of column axial loads should be included.

**Why minimum positive reinforcement?** The reason for minimum positive moments at beam ends is because seismic loads are large in magnitude and reverse in direction. The bending moment and shear in a member, therefore, can be positive or negative at different points in time. Simple elastic analysis and load combinations cannot possibly give reliable results. Hence, the necessity to provide a minimum capacity for positive moments at the ends as well as negative moments at the midspans of beams.

To ensure that hinge develops in frame beams it is necessary to: (i) attain yeilding of reinforcement well before concrete fails in compression and; (ii) provide transverse reinforcement at close intervals to confine the concrete core within the longitudinal reinforcement.

Closely spaced ties enhance the ductility of concrete by allowing large compression strains to develop in concrete without spalling. The ties prevent buckling of longitudinal bars. A buckled or kinked bar has a tendency to fracture when the bar straightens in tension under load reversals. In seismic detailing it is necessary to use seismic hooks in the ties. This is because when concrete cover spalls, the hoops may themselves be exposed and lose their confining capacity. Therefore, ties must be anchored into the confined zone of concrete.

As stated many times in this book, a structure designed in accordance with seismic code provisions is expected to respond inelastically during strong earthquakes. The members must be designed and detailed with prior realization of the inevitability of inelastic response. The detailing provisions promote relatively benign ductile response rather than undesirable brittle response. This is achieved by ensuring that members in the critical regions of such mechanisms have inelastic energy-dissipation qualities. The energy dissipation should occur through yielding of reinforcement as opposted to the shearing or crushing failure of concrete. The critical region should be distributed uniformly throughout the structure rather than confined to selected fewer regions. Vertical elements which are expected to partake in energy dissipation should have proper confinement such that the vertical load-carrying capacity of the frame is not compromised. The provisions encourage formation of beam hinges rather than column hinges. To ensure adequate flexural ductility in critical regions of beams, the UBC prescribes an upper limit of 1.25 on the ratio of ultimate tensile strength to actual yield strength of reinforcement. Also, the amount by which the actual yield strength can exceed the minimum specified value is limited to 18,000 psi. Joints of frames must be designed for shears with the development of maximum beam moments assuming that the longitudinal reinforcement is stressed to 1.25 times the specified yield strength. This is to allow for the effects of strain hardening and for the fact that actual yield strengths usually exceed the specified minimum values.

Chapter

# 12

# Special Topics

The purpose of this chapter is twofold: (i) to examine in detail a number of special subjects which we have met in several ways in earlier chapters; and (ii) to touch briefly on certain topics which are unique to the design of tall buildings. Our consideration of special topics opens with a discussion of differential shortening of columns.

A column in a tall building undergoes axial shortening considerably more than its lower brethren, requiring special attention in its design. A related problem, by no means unique to tall buildings, but one that gets aggrevated to a greater extent, is the levelness of floors. Similarly, the problem of human response to transient vibration of floors is not unique to tall buildings but needs careful study because the cost of correcting the problem in a tall building with several floors is phenomenally more expensive than correcting the relatively fewer floors of a low-rise building. Next we will consider in some detail the behavior of panel zones and their effect on lateral deflections of buildings.

Next, we move on to the gray area of design of curtain wall systems that brings together several diversified disciplines. The section briefly presents the design and installation aspects of metal curtain walls, stone claddings, brick veneer, and glass fiber-reinforced concrete systems.

The next section is introductory in its presentation of mechanical devices which are used to increase the damping and thus reduce wind-induced sway acceleration and torsion-induced translation acceleration. Two devices, tuned mass damper and viscoelastic damper, are discussed. With the additional damping provided by these devices, the peak wind-induced resultant accelerations are designed to be within benchmark limit of 20 milli-g (one-fiftieth of the acceleration due to gravity) for a recurrence interval of once in 10 years.

The next section presents a brief discussion of design aspects of two types of foundations, the drilled pier or caisson foundation, and the mat foundation. The next section presents a discussion of seismic design of floor diaphragms for horizontal forces including the effect of plan irregularities commonly encountered in practice. Next, earthquake mitigation technologies that include seismic isolation and energy dissipation are discussed, followed by an overview of SAC guidelines for repair, modification, and design of welded steel moment frames. The next section gives unit structural quantities for several types of concrete floor systems and high rise steel and composite buildings. The final section gives an update on the 1997 UBC.

## 12.1 Differential Shortening of Columns

Columns in tall buildings are subjected to large axial displacements because they accumulate loads from a large number of floors and are also relatively long. A 60-story interior column in a steel building may shorten as much as 2 to 3 in. (50 to 76 mm) because of dead and live loads, while a concrete column may experience an additional 2 to 3 in. (50 to 76 mm) of shortening because of creep and shrinkage. If this shortening is not given due consideration, problems may develop in the performance of curtain walls and levelness of floor systems. Proper awareness of this problem is necessary on the part of the structural engineer, architect, and the curtain wall supplier to avoid lost time and money.

The maximum effect of column shortening is at the roof level reducing gradually towards the ground level. In a concrete frame the axial shortening may take several years to complete because of long-term effect of creep, although a major part of it occurs within the first few months of construction. There is very little the structural engineers can do to minimize frame shortening, but they should make the design team aware of the magnitude of frame shortening so that soft joints of appropriate widths are properly detailed between curtain wall joints to prevent load from being transferred into the building façade. Before fabrication of curtain wall connections, the in-place elevations of the structural frame should be measured and
,,in-place elevations of the structural frame should be measured and
provided to the curtain wall contractor. The fabrication should be based on these, rather than the theoretical elevations. There must be sufficient space at the joints between the panels to allow for the expected movement of the structure as well as thermal expansion and contraction of the panels themselves. Insufficient space may result in bowed curtain wall panels, or in extreme cases, the panels may even pop off the building.

A similar problem occurs when mechanical and plumbing lines are attached rigidly to the structure. Frame shortening may force the pipes to act as structural columns resulting in their distress. A general remedy is to make sure that nonstructural elements are not brought in to bear the vertical loads. Sufficient compensation should be provided during design and construction to make sure that nonstructural elements are separated from structural elements.

The axial loads in all columns of a tall building are seldom the same, giving rise to the problem of so-called differential shortening. The problem is more acute in a composite structure because slender steel columns are subjected to large axial loads during construction. Determining the magnitude of axial shortening in a composite system is complicated because many variables that contriute to the shortening of columns cannot be predicted with sufficient accuracy. The lower part of the column, which is encased in concrete, is continually undergoing creep, and because the age and strength of concrete keep changing, their effect on creep is difficult to predict with any precision. The steel column at any given period during construction is partly enclosed in concrete at lower floors, with the bare steel section projecting beyond the concreted levels by as many as 10 or 15 floors. Another factor difficult to predict is the gravity load redistribution due to continuity of beams attached to columns. If the building is founded on compressible material, foundation settlement is another factor that influences the relative changes in the elevations of the columns. The magnitude of load imbalance continually changes, making an accurate assessment of column shortening beyond the reach of day-to-day engineering practice. If all the variables are known, the prediction of differential shortening is no more complicated than a systematic evaluation of the $PL/AE$ equation.

The routine method of analysis of high-rise structures is usually performed for the full frame without taking into account the sequential nature of construction. This approach for a 60-story concrete building may result in a calculated axial shortening of about 3 in. (76 mm) of immediate axial shortening and a mind-boggling 3 to 5 in. (76 to 254 mm) of additional displacement when creep effects are included in the computation. Fortunately, the method of construction in concrete buildings more or less takes care of the immediate shortening, and to a limited extent the creep effects on lower-level columns. This is because buildings are constructed one floor at a time and since each floor is leveled at the time of its construction, the column shortening which has occurred prior to the construction of that floor is of no consequence. Also, the lower-story columns of tall buildings undergo considerably smaller creep and shrinkage because the load is applied incrementally over a 15 to 24-month construction period.

Creep is difficult to quantify because it is time-dependent. Initially the rate of creep is significant and diminishes as time progresses until it eventually reaches zero. Because of sustained loads the stress in concrete gradually gets transferred to the reinforcement with a simultaneous decrease in concrete stress.

Columns with different percentages of reinforcement and different volume-to-surface ratios creep and shrink differently. An increase in the percentage of reinforcement and volume-to-surface ratio reduces the strain due to creep and shrinkage under similar stresses. Differential shortening of columns induces moments in the connecting girders and spandrels resulting in gravity load transfers to adjacent columns. A column which has shortened less receives more load, thus compensating for the initial imbalance.

In this section we will not address the foreshortening of concrete columns because attempts to quantify the shrinkage and creep effects are considered beyond the scope of this work. Leaving the problem to more theoretical minds, we will proceed in this section to a brief discussion of differential shortening of steel columns. For this purpose a column isolated from the remainder of structural frame is studied as a cantilever. A closed-form solution for computing axial shortening of columns is presented with a numerical example to demonstrate the practicality of the method. The section concludes with suggested details for field adjustment of column heights.

### 12.1.1 Calculations

*Differential* rather than the *absolute* shortening of column is more significant. If all columns shorten by the same amount, the floors would still be level. Relative displacement between columns occurs because of the difference between the $P/A$ ratios of columns. If all columns in a building have the same slenderness ratio and are sized for gravity load requirement only, there will be no relative vertical movement between the columns. All columns will undergo the same displacement because the $P/A$ ratio is nearly constant for all columns. In a real building, this condition is seldom present. Usually the design of frame columns is governed by the combined gravity and lateral load, while non-frame columns are designed for gravity loads only.

For example, consider the tubular system used for buildings in the 50- to 80-story range. The system typically utilizes closely spaced exterior columns and widely spaced interior columns. Normally, in a steel system, high-strength steel columns up to 65 ksi are used in the interior of the building to collect gravity loads on large tributary areas resulting in a large $P/A$ ratio. The exterior columns, on the other hand, usually have a small $P/A$ ratio for two reasons; first, their tributary areas are small because of thier close spacing of usually 8 to 12 ft (2.44 to 3.66 m); second, the columns are sized from lateral displacement considerations, resulting in areas much in excess of those required from the strength consideration alone. Because of this imbalance in the gravity stress level, these two groups of columns undergo different axial shortenings; the interior columns shorten much more than the exterior columns.

A somewhat reversed condition occurs in buildings utilizing interior-braced core columns and widely spaced exterior columns; the exterior columns experience more axial shortening than the interior columns. The behavior of columns of buildings utilizing other structural systems, such as interacting core and exterior frames, tends to be somewhere in between these two limiting cases.

In all cases, it is relatively easy to evaluate the shortening of columns. The procedure requires a step-by-step manipulation of the basic $PL/AE$ equation as described below.

Consider a typical column of a 50-story building. Assume, for simplicity, that the variations in story heights, column areas, and the load increment at each floor are constant as shown in Fig. 12.1. The calculation of the axial shortening, in itself, at any floor is trivial. It is given by the summation equation:

$$\Delta_n = \frac{PL}{AE} \sum_{i=1}^{n} (NS + 1 - i) \qquad (12.1)$$

where $\Delta_n$ = the axial shortening at level $n$
$P$ = the load increment assumed constant at each level
$L$ = the story height assumed constant for the full height
$A$ = the column area assumed constant for the full height
$NS$ = the number of stories

Using this equation, the axial shortening for the simplified column (Fig. 12.1) works out as given in Table 12.1. Having obtained these values, the next step is to evaluate the column length correction $\Delta_c$ at different levels. This is given by the difference in the axial shortenings at the level under consideration and the floor immediately below it. For instance, from Table 11 this value at level 30 for the example column is given by $PL/AE$ $(1065 - 1044) = 21\,PL/AE$. The magnitude of this correction in a normally proportioned building is rather small, perhaps $\frac{1}{8}$ in. (3.17 mm) at the most. Instead of specifying this value as a correction at each level, in practice it is usual to lump these corrections for a few floors, for example, every tenth floor or so. The lumped correction at the level is simply the difference between the values of axial shortening at that level and at the lumped floor below it. For example, in Table 12.1 the lumped correction of 255 $PL/AE$ units at the 30th floor is equal to the column shortening at the 30th level less the shortening at the 20th level

$$\frac{PL}{AE}(1065 - 810) = 255\frac{PL}{AE} \text{ units}$$

Let us consider the practical case of a building column which has variations in story heights, load increments, and column areas as shown in Fig. 12.2. The summation equation for this general case takes the form:

$$\Delta_n = \frac{1}{E}\sum_{k=1}^{n}\frac{L_k}{A_k}\sum_{i=k}^{NS}P_i \tag{11.2}$$

Table 12.2 shows in a tabular form the assumed dimensions and loading conditions for the column and the computations for obtaining the column length shortening values. The last column of the table

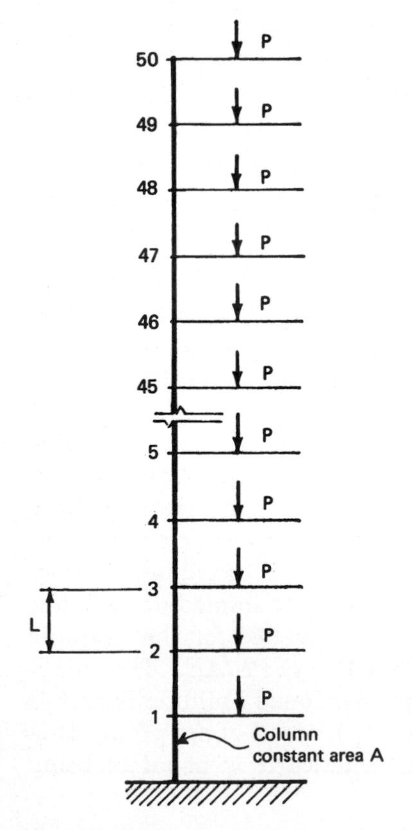


**Figure 12.1** Simplified column for calculating column shortening.

**TABLE 12.1 Axial Shortening Computations for Simplified Column**

| Level | Axial shortening | Column length correction at each level | Lumped column length correction |
|---|---|---|---|
| 50 | 1275 | 1 | 55 |
| 49 | 1274 | 2 | |
| 48 | 1272 | 3 | |
| 47 | 1269 | 4 | |
| 46 | 1265 | 5 | |
| 45 | 1260 | 6 | |
| 44 | 1245 | 7 | |
| 43 | 1247 | 8 | |
| 42 | 1239 | 9 | |
| 41 | 1230 | 10 | |
| 40 | 1220 | 11 | 155 |
| 39 | 1209 | 12 | |
| 38 | 1197 | 13 | |
| 37 | 1184 | 14 | |
| 36 | 1170 | 15 | |
| 35 | 1155 | 16 | |
| 34 | 1139 | 17 | |
| 33 | 1122 | 18 | |
| 32 | 1104 | 19 | |
| 31 | 1085 | 20 | |
| 30 | 1065 | 21 | 255 |
| 29 | 1044 | 22 | |
| 28 | 1022 | 23 | |
| 27 | 999 | 24 | |
| 26 | 975 | 25 | |
| 25 | 950 | 26 | |
| 24 | 924 | 27 | |
| 23 | 897 | 28 | |
| 22 | 869 | 29 | |
| 21 | 840 | 30 | |
| 20 | 810 | 31 | 355 |
| 19 | 779 | 32 | |
| 18 | 747 | 33 | |
| 17 | 714 | 34 | |
| 16 | 680 | 35 | |
| 15 | 645 | 36 | |
| 14 | 609 | 37 | |
| 13 | 572 | 38 | |
| 12 | 534 | 39 | |
| 11 | 495 | 40 | |
| 10 | 455 | 41 | 405 |
| 9 | 414 | 42 | |
| 8 | 372 | 43 | |
| 7 | 329 | 44 | |
| 6 | 285 | 45 | |
| 5 | 240 | 46 | |
| 4 | 194 | 47 | |
| 3 | 147 | 48 | |
| 2 | 99 | 49 | 50 |
| 1 | 50 | 50 | |

Note: All *values* are in terms of *PL*/*AE*.

shows the lumped corrections at levels 2, 10, 20, 30, 40, and the roof. Basically, these corrections represent the additional lengths over and above their theoretical lengths. For example, $\Delta_c = 1\frac{1}{4}$ in. (31.75 mm) at the tenth level means that the actual fabricated length of column should be $1\frac{1}{4}$ in. longer than the theoretical length. This overlength could be achieved by increasing the length of column in each tier by $\frac{1}{4}$ in. (6.35 mm) (ten stories equal five tiers, therefore, $\frac{1}{4}$ in. times 5 gives $1\frac{1}{4}$ in.). The fabricator may elect to increase the length in each story by $\frac{1}{8}$ in. (3.2 mm) instead of $\frac{1}{4}$ in. per tier to achieve the same $\Delta_c$ at the tenth floor.

The value of $\Delta_c = 2$ in. (50.8 mm) at the 20th floor means the overlength of columns between levels 1 and 20 should be 2 in. However, an overlength of $1\frac{1}{4}$ in. (31.75 mm) up to the tenth level has already been achieved by specifying $\Delta_c = 1\frac{1}{4}$ in. at the tenth level. Therefore, the increment between the 10th and 20th levels should be 2 in. less $1\frac{1}{4}$ in. $= \frac{3}{4}$ in. (19.0 mm). A correction table incorporating the above information for the example column is shown in Table 12.2.

### 12.1.2 Simplified approach

In a normally proportioned building the cross-sectional area of a column usually increases in a stepwise manner, from a minimum

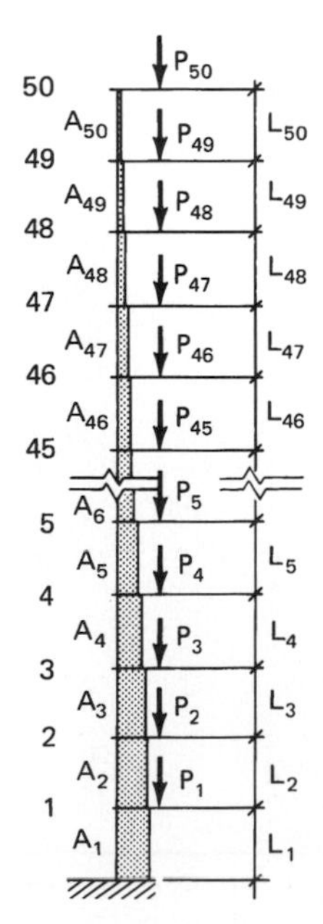


**Figure 12.2** Axial shortening computations for a practical column.

TABLE 12.2 **Axial Shortening Computations for Practical Column**

| Level | Accumulated load, kips | Column section | Story height, in. | Column shortening $\Delta n$, in. | Column length correction each level, in. | Lumped column length correction, in. | Column shortening from eq. (11.4), in. |
|---|---|---|---|---|---|---|---|
| 50 | 53 | W14 × 43 | 156 | 5.14 | 0.023 | 0.73 | 5.11 |
| 49 | 106 | 43 | 210 | 5.12 | 0.061 | | 5.08 |
| 48 | 159 | 53 | 168 | 5.05 | 0.051 | | 5.02 |
| 47 | 212 | 53 | 156 | 5.00 | 0.073 | | 4.95 |
| 46 | 265 | 68 | 156 | 4.93 | 0.071 | | 4.89 |
| 45 | 318 | 68 | 156 | 4.86 | 0.086 | | 4.82 |
| 44 | 371 | 84 | 156 | 4.77 | 0.081 | | 4.75 |
| 43 | 424 | 84 | 156 | 4.69 | 0.092 | | 4.67 |
| 42 | 477 | 95 | 156 | 4.60 | 0.092 | | 4.59 |
| 41 | 530 | 95 | 156 | 4.51 | 0.102 | | 4.50 |
| 40 | 583 | 111 | 156 | 4.41 | 0.09 | 1.02 | 4.42 |
| 39 | 636 | 111 | 156 | 4.32 | 0.105 | | 4.33 |
| 38 | 689 | 127 | 156 | 4.21 | 0.09 | | 4.24 |
| 37 | 742 | 127 | 156 | 4.12 | 0.107 | | 4.15 |
| 36 | 795 | 142 | 156 | 4.01 | 0.103 | | 4.04 |
| 35 | 848 | 142 | 156 | 3.91 | 0.109 | | 3.96 |
| 34 | 901 | 167 | 156 | 3.80 | 0.09 | | 3.86 |
| 33 | 954 | 167 | 156 | 3.71 | 0.105 | | 3.76 |
| 32 | 1007 | 176 | 156 | 3.62 | 0.105 | | 3.66 |
| 31 | 1060 | 176 | 156 | 3.50 | 0.110 | | 3.56 |
| 30 | 1113 | 202 | 156 | 3.39 | 0.101 | 1.07 | 3.46 |
| 29 | 1166 | 202 | 156 | 3.29 | 0.106 | | 3.36 |
| 28 | 1219 | 211 | 156 | 3.19 | 0.106 | | 3.26 |
| 27 | 1272 | 211 | 156 | 3.08 | 0.110 | | 3.16 |
| 26 | 1325 | 228 | 156 | 2.97 | 0.106 | | 3.12 |
| 25 | 1378 | 228 | 156 | 2.86 | 0.111 | | 2.95 |
| 24 | 1431 | 246 | 156 | 2.75 | 0.107 | | 2.85 |
| 23 | 1484 | 246 | 156 | 2.65 | 0.111 | | 2.74 |
| 22 | 1537 | 264 | 156 | 2.53 | 0.107 | | 2.64 |
| 21 | 1590 | 264 | 156 | 2.43 | 0.110 | | 2.53 |
| 20 | 1643 | 287 | 156 | 2.32 | 0.104 | 1.06 | 2.42 |
| 19 | 1696 | 287 | 156 | 2.20 | 0.108 | | 2.32 |
| 18 | 1749 | 314 | 156 | 2.10 | 0.101 | | 2.21 |
| 17 | 1802 | 314 | 156 | 2.00 | 0.105 | | 2.10 |
| 16 | 1855 | 314 | 156 | 1.90 | 0.108 | | 1.99 |
| 15 | 1908 | 314 | 156 | 1.79 | 0.111 | | 1.88 |
| 14 | 1961 | 342 | 156 | 1.68 | 0.104 | | 1.78 |
| 13 | 2014 | 342 | 156 | 1.58 | 0.107 | | 1.67 |
| 12 | 2067 | 370 | 156 | 1.47 | 0.101 | | 1.56 |
| 11 | 2120 | 370 | 156 | 1.37 | 0.104 | | 1.45 |
| 10 | 2173 | 370 | 156 | 1.26 | 0.107 | 0.53 | 1.34 |
| 9 | 2226 | 370 | 156 | 1.16 | 0.109 | | 1.23 |
| 8 | 2279 | 398 | 156 | 1.05 | 0.104 | | 1.12 |
| 7 | 2332 | 398 | 156 | 0.94 | 0.107 | | 1.01 |
| 6 | 2385 | 398 | 156 | 0.84 | 0.11 | | 0.89 |
| 5 | 2438 | 398 | 210 | 0.73 | 0.11 | | 0.78 |
| 4 | 2491 | 426 | 168 | 0.62 | 0.11 | | 0.67 |
| 3 | 2544 | 426 | 156 | 0.51 | 0.11 | | 0.56 |
| 2 | 2597 | 500 | 156 | 0.40 | 0.09 | | 0.45 |
| Mezzanine | 2650 | 500 | 240 | 0.31 | 0.15 | | 0.17 |
| 1 | 2770 | W14 × 500 | 240 | 0.16 | 0.16 | | 0.17 |

value at the roof to a maximum value at the base as shown in Fig. 12.3. The incremental steps are caused by the finite choice of column shapes. In tall buildings which merit column shortening investigations, the signficance of these incremental steps diminishes rather quickly as compared to a low-rise building column. Therefore, it is possible without losing meaningful accuracy, to express the load and cross-sectional properties by continuous mathematical expressions as indicated in Fig. 12.4. The gravity load distribution may be assumed to vary linearly throughout the height (Fig. 12.4). A similar linear assumption for the column area overestimates the actual column areas as shown by the dashed curve in Fig. 12.5. Although mathematically it is possible to derive an equation to fit the curve, the author proposes a modified linear variation as indicated in Fig. 12.5 in which the equivalent column area at the bottom is taken as $0.9 \times$ the actual area. This simplification leads to less formidable expressions for a closed-form solution without any meaningful loss in accuracy.

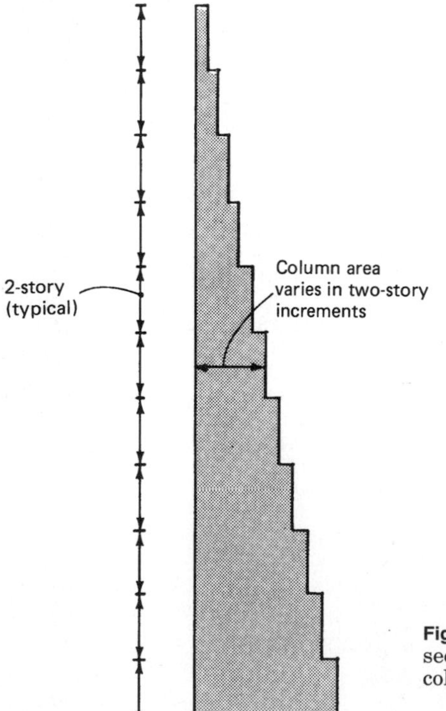


**Figure 12.3** Variation of cross-sectional area of a high-rise column.

**Derivation of closed-form solution**

The notations used in the derivation of closed form solution (Fig. 12.5a) are as follows:

$L$ = Height of the building (note previously in the long hand method, notion $L$ was used to denote story height)

$\Delta_z$ = Axial shortening at a height $x$ (also denoted as $z$), above foundation level

$A_t$ = column area at top

$A_b$ = Modified column area at bottom equal to 0.9 × actual area of column at bottom = $0.9 \times A_B$

$A_x$ = Area of column at height $x$ (also denoted as $z$), above foundation level

$\alpha$ = Rate of change of area of column

$P_t$ = Axial load at top

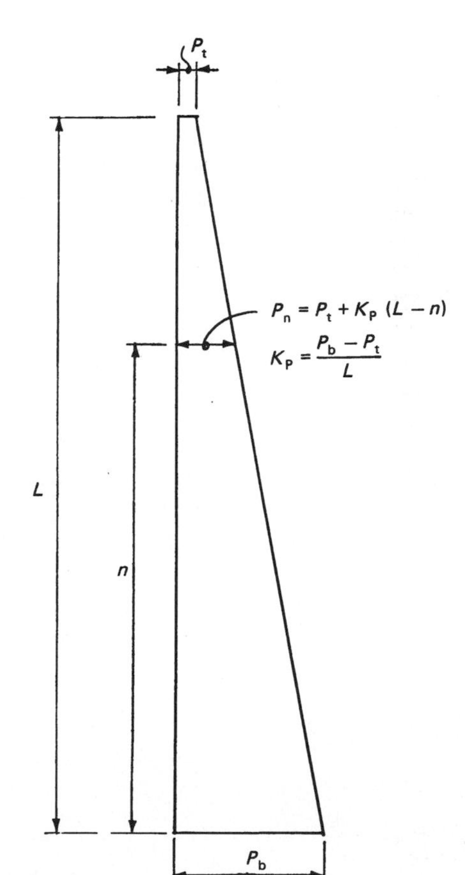


**Figure 12.4** Idealized gravity load distribution on a column.

$P_b$ = Axial load at bottom
$P_x$ = Axial load at height $x$ above foundation
$\beta$ = Rate of change of axial load
$\varepsilon_x$ = Axial strain at height $x$
$E$ = Modulus of elasticity

The area of column at height $z$ is given by

$$A_x = A_t \frac{x}{L} + A_b\left(1 - \frac{x}{L}\right)$$

$$= A_b - (A_b - A_t)\frac{x}{L}$$

$$= A_b - \alpha x$$

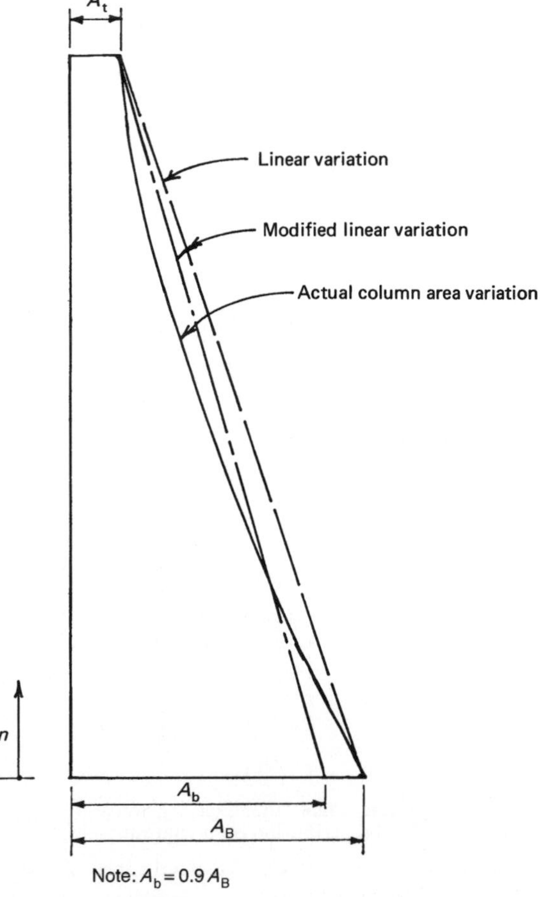


**Figure 12.5** Idealized column cross sectional areas.

where $$\alpha = \frac{A_b - A_t}{L}$$

The axial load at height $z$ above foundation is given by

$$P_x = P_t \frac{x}{L} + P_b\left(1 - \frac{x}{L}\right)$$

$$= P_b - \beta x$$

where $$\beta = \frac{P_t - P_t}{L}$$

The axial strain $$\varepsilon_x = \frac{Px}{A_x E}$$

Using vertical work: $$P_z^1 \, \Delta_z = \int_0^z P_x^1 \varepsilon_x \, dx$$

with $$P_z^1 = 1 = P_x^1,$$

$$\Delta_z = \frac{1}{E} \int_0^z \frac{P_b - \beta x}{A_b - \alpha x} \, dx$$

$$= \frac{P_b}{E} \int_0^z \frac{dx}{A_b - \alpha x} - \frac{\beta}{E} \int_0^z \frac{x \, dx}{A_b - \alpha x}$$

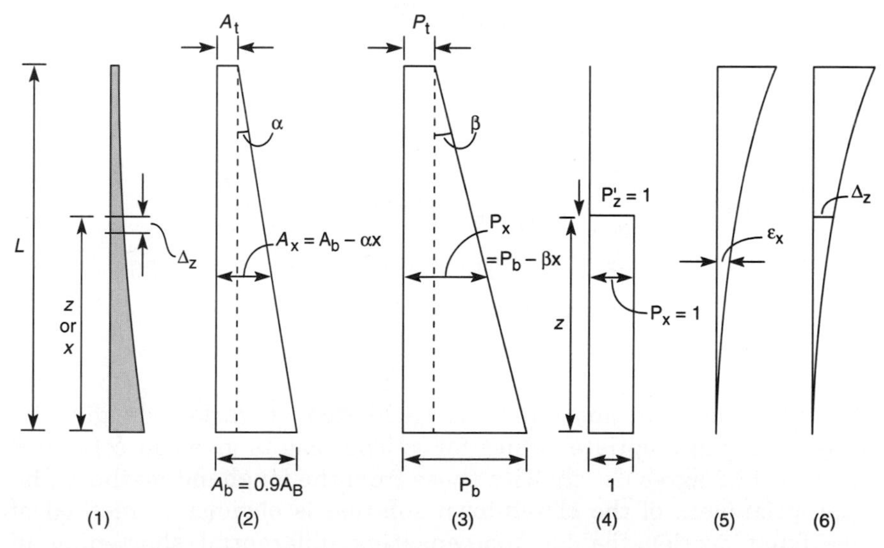


**Figure 12.5a** Axial shortening of columns: closed form solution: (1) axial shortening $\Delta_z$; (2) column area; (3) column axial load; (4) unit load at height $z$; (5) axial strain; (6) axial displacement.

Evaluating these integrals we get the following final expression for $\Delta_z$.

$$\Delta_z = \frac{P_b}{E}\left[-\frac{1}{\alpha}\ln\left(1-\frac{\alpha z}{A_b}\right)\right] - \frac{\beta}{E}\left[-\frac{1}{\alpha^2}\left\{\alpha z + A_b \ln\left(1-\frac{\alpha z}{A_b}\right)\right\}\right]$$

**Example problem.** Given:

Height of the building: $L = 682$ ft $= 8184$ in. (207.8 m)
Modulus of elasticity: $E = 29\,000$ ksi ($200 \times 10^3$ MPa)
Axial load at top: $P_t = 53$ kips (237.5 kN)
Area of column at top: $A_t = 12.48$ in.$^2$ (8052 mm$^2$)
Axial load at base: $P_b = 2770$ kips ($12.32 \times 10^3$ kN)
Actual column area at base: $A_B = 147$ in.$^2$ ($94.84 \times 10^3$ mm$^2$)
Reduced column area at base: $A_b = 0.9 \times 147 = 133.3$ in.$^2$ ($86.0 \times 10^3$ mm$^2$)

**Required.** Axial shortening of column at top.

**Solution** Since column shortening is calculated at top, $z = L$.

$$\alpha = \frac{A_b - A_t}{L} = \frac{133 - 12.48}{8184} = 0.014\,76 \text{ in.}^2/\text{in.}$$

$$\beta = \frac{P_b - P_t}{L} = \frac{2770 - 53}{8184} = 0.332 \text{ kip/in.}$$

$$\ln\left(1 - \frac{\alpha L}{A_b}\right) = \ln\left(1 - \frac{0.014\,76 \times 8184}{133.3}\right)$$
$$= \ln(0.093\,62)$$
$$= -2.368\,47$$

$$\Delta_L \text{ at top} = \frac{2770}{29\,000}\left\{-\frac{1}{0.014\,76} \times (-2.368\,47)\right\} - \frac{0.332}{29\,000}$$
$$\times \{-4590.15(0.014\,76 \times 8184 + 133.3 \times -2.368\,47\}$$
$$= 15.327 - 10.2$$
$$= 5.127$$

Similarly the axial shortening is calculated at various heights by substituting appropriate values for $z$. The results given in column 8 of Table 12.2 agree closely with those from the longhand method. The appropriateness of the closed-form solution is obvious. A method of specifying overlengths for compensating differential shortening of columns is shown in Table 12.3.

### 12.1.3 Column shortening verification during construction

Assuming that the engineer has appropriately corrected column lengths to compensate for axial shortening, it becomes somewhat difficult during steel erection to determine whether the variations in the top elevations of columns are within allowable limits. For example, let us say the actual variation between an interior and exterior column at the time of erection is 2 in. (50.8 mm) when the columns are erected halfway up a 40-story building. Let us say 1 in. (25.4 mm) out of this 2 in. is in excess of the allowable erection tolerance. It is not immediately clear how much of this excess 1 in. is due to the overlength allowed for column shortening, because the column has undergone partial shortening due to already existing dead and construction loads. Therefore, there is a need for a second set of of column shortening computations which gives the relative elevations of the columns "during erection." These values would indicate the amount by which the columns should be protruding above their theoretical location at the time of erection and, therefore, serve as a benchmark for checking the relative elevations of columns during construction.

To illustrate the above idea, let us consider again the overly simplified column of Fig. 12.1. The total overlength specified at level 30, for example, is 1065 *PL*/*AE* units (Table 12.1). When loads *P* are applied starting at level 1, the overlength correspondingly starts decreasing by a factor *PL*/*AE* units for load *P* at each level. When erection is at the 30th level, the total shortening at the level would be 855 *PL*/*AE* units. The residual overlength at this level will be 210 *PL*/*AE* units. Physically the residual overlength $\Delta_{R_n}$ at a given level represents the column shortening due to loads applied at and above that level as shown schematically in Fig. 12.6.

For the general case of column shown in Fig. 12.2, the residual overlength $\Delta_{R_m}$ at any level *n* works out to be

$$\Delta_{R_n} = \frac{1}{E} \sum_{i=n}^{NS} P_i \sum_{i=1}^{n} \frac{L_i}{A_i} \tag{12.5}$$

Values of $\Delta_{R_n}$ calculated by using eqn (12.5) are shown in Table 12.4.

### 12.1.4 Conclusions

Although the closed-form solution presented here is based on certain simplifying assumptions, the author believes that the accuracies obtained by the use of this simple method fully justify its application to normally proportioned tall buildings. Where the material and load distribution patterns are radically different from those shown in

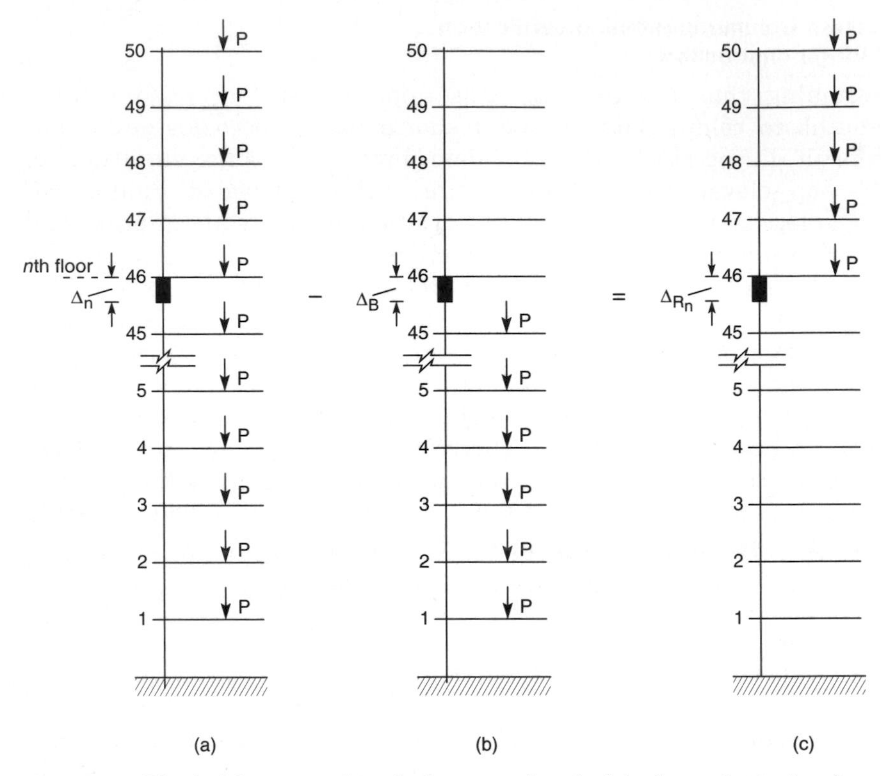


**Figure 12.6** Physical interpretation of column overlength: (a) column shortening due to loads at all floors; (b) column shortening due to loads below the $n$th level; (c) column residual overlength $\Delta_{R_n} = \Delta_n - \Delta_B$.

Fig. 12.3, the application of the more accurate long-hand procedure is recommended.

As mentioned previously, a steel frame does not have the luxury of built-in compensation of a cast-in-place concrete building because

**TABLE 12.3 Column Length Correction Table**

| Level | Column length shortening, in. (mm) | Correction to scheduled column lengths, in. (mm) |
|---|---|---|
| 50 | $5\frac{1}{8}$ (13.75) | 6 @ $\frac{1}{8} = \frac{3}{4}$ (19) |
| 40 | $4\frac{3}{8}$ (111) | 8 @ $\frac{1}{8} = 1$ (25.4) |
| 30 | $3\frac{3}{8}$ (85.7) | 8 @ $\frac{1}{8} = 1$ (25.4) |
| 20 | $2\frac{3}{8}$ (60.3) | 9 @ $\frac{1}{8} = 1\frac{1}{8}$ (28.6) |
| 10 | $1\frac{1}{4}$ (31.75) | 10 @ $\frac{1}{8} = 1\frac{1}{4}$ (31.75) |

**TABLE 12.4 Residual Overlength of Column During Construction**

| Level | Residual overlength from eqs (11.5), in. |
|---|---|
| 50 | 0.33 |
| 45 | 1.87 |
| 40 | 2.15 |
| 35 | 2.20 |
| 30 | 2.12 |
| 25 | 1.96 |
| 20 | 1.72 |
| 15 | 1.49 |
| 10 | 1.12 |
| 5 | 0.68 |

steel columns are prefabricated with predetermined lengths. Therefore, a suggested method for avoiding gross inaccuracies in predicting column length corrections would be to have the steel erector check the top elevation of columns at predetermined levels, say every fourth or sixth floor, and to make provisions for adjustments of column heights.

Some of the problems associated with differential shortening of columns can be eliminated in the fabrication stage. Certain columns can be made slightly longer than their nominal length to account for axial shortening. However, the uncertainties especially in composite structures are so many that some form of field adjustment is usually required.

For example, in steel buildings it may become necessary to place removable shims between columns and their foundations in order to compensate for differential shortening. When the erection of structural steel reaches predetermined levels, the columns are temporarily unloaded and steel shims removed. Another relatively simple method of adjusting column lengths which has been successfully used in composite construction is shown in Fig. 12.7.

## 12.2 Floor-Leveling Problems

Floor-leveling problems are of increasing importance because stronger building materials and more refined designs have resulted in lighter construction more prone to deflections than earlier heavier buildings. In framed buildings considerable trouble is encountered in trying to provide a level floor because of the many variable conditions that exist in practice. Concrete floors that are level at the time of construction may not be so at the time of occupancy. Many of the variable factors encountered are not mathematically determinable, even though an attempt is made by the engineer to compensate for such effects.

In steel buildings floor beams for the normal 30-42 ft spans are furnished with a predetermined camber, while in concrete buildings the camber is built in the form work. Usually the specified camber ranges from a minimum of $\frac{1}{2}$ in. (12.7 mm) to a maximum of 2.5 in. (63.5 mm). Cambers smaller than $\frac{1}{2}$ in. (12.7 mm) are difficult to achieve, while a camber of substantially greater than 2.5 in. (63.5 mm) will result in other serviceability problems for beams in the 30- to 42 ft (9.14- to 12.80 m) range. Cambers are specified anticipating that the loading of floors will overcome the camber, resulting in a level floor. This is not always the case because: (i) rolling and construction tolerance combined with the long-term effect of creep of concrete is enough to affect the final result up or down in both steel and concrete construction; (ii) most usually, camber is calculated as if the beam were pin-connected or completely fixed.

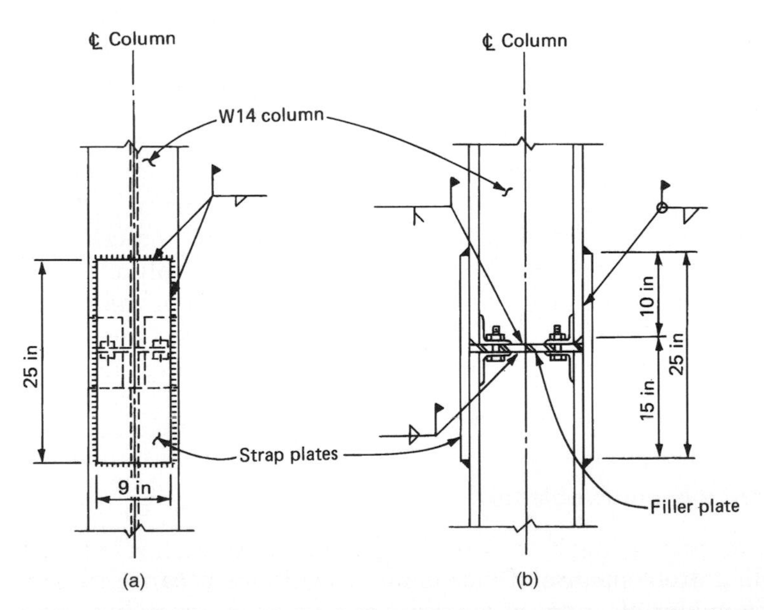


**Figure 12.7** Splice detail for field correction of column length: (a) elevation; (b) section.

Actual conditions vary. For example, even with simple connections, steel beams experience partial fixity. Depending upon the degree of fixity the final result could again be up or down; (iii) vertical members will shorten elastically during erection. The magnitude of elastic shortening between interior and exterior columns or between two adjacent exterior columns most usually is different, compounding floor-leveling problems.

Because of these variable factors, combined with the fact that none of these is mathematically determinable in the context of a practical design office, it makes it almost an accident if the floor turns out to be perfectly level. The problem comes to light at the time of interior finishing of the space when ceiling and partitions are being installed. One sure method of obtaining a level floor is to float the floor to remove the lumps and fill the low spots. Cement-based self-leveling underlayments are used for this purpose. In a floor built to commercially acceptable tolerance, the average fill over the entire floor area should not exceed $\frac{1}{2}$ in. (12.7 mm), which translates into an additional dead load of 6 psf (287.3 mm). Depending upon the type of construction, this additional load may represent an increase of 3 to 6 percent of the total working stress load. It is recommended that an allowance be made for this additional load in the design.

The most commonly specified tolerance for finished floor slab surfaces is $\frac{1}{8}$ in. (3.7 mm) in 10 ft (3.048 m), which is considered too stringent for most uses. The reasons for unlevelness are manyfold, including form work sagging, deflection of members due to dead and live loads, finishing irregularities, or errors in setting of steel beams or form work. As a result the as-built surface of the floor always exhibits bumps and dips.

In recognition of this problem, the American Concrete Institute has revised its "Standard Tolerances for Concrete Construction and Materials" (ACI 117). The standard includes floor finish tolerances based on two measuring methods: the F-number system and the straight-edge method. F numbers describe floor flatness. The larger the F number, the flatter the floor. An F-60 floor is roughly twice as flat as an F-30 floor.

## 12.3 Floor vibrations

Earlier chapters have dealt with the subject of vibration induced in a building by outside sources of vibrational loads such as wind and earthquakes. In addition to these, a building is subjected to a variety of vibrational loads that come from within. Although almost all loads except dead loads are nonstatic, internal sources of vibration that might be a cause of concern in an office or a residential building are the oscillating machinery, passage of vehicles, and various types of impact loads such as those caused by dancing, athletic activities, and even by pedestrian traffic. The trend in the design of floor framing systems of high rises is for long spans using structural systems of minimum weight. To this end, high-strength steel with lightweight concrete topping is routinely employed. With the use of lightweight concrete, most building codes allow for a reduction in the thickness of slab required for fire rating. This results in a further reduction in the mass and stiffness of the structural system, thereby increasing the period of the structure, which at times may approach the period of the source causing the vibration. Resonance may occur, causing large forces and amplitudes of vibration.

The performance of such structures can be greatly improved by adding nonstructural elements such as partitions and ceilings which contribute greatly to the damping of vibrations. Nonstructural elements may also add to the mass and stiffness to produce the desired degree of solidity. Although the essential requirement in establishing the adequacy of a floor system is its strength, large deflections can be objectional for several reasons: (i) excessive deflections and vibrations may give the user the negative impression that the building is not solid. In retail areas, for example, the china may rattle every time someone goes by, or mirrors in dressing rooms of clothing stores may shake, giving the customer the somewhat nebulous but real feeling that the structure is not solid. In extreme cases vibration may cause damage to the structure as a result of loosening of connections, brittle fracture of welds, etc. It is therefore important that the structure be able to absorb impact forces and vibrations without transmitting any humanly perceptible shaking or bouncing. Monolithic concrete buildings are more solid in this respect as compared to light-framed buildings with steel or precast concrete; (ii) excessive deflection may result in curvature or misalignments perceptible to the eye; (iii) large deflections may result in fracture of architectural elements such as plaster or masonry; (iv) large deflections may result in the transfer of load to nonstructural elements such as curtain wall frames.

It is difficult to establish a general criteria related to perception of vibrations. Feeling of bounciness varies from person to person, and what is objectionable to some may be barely noticeable to others. Among the criteria employed in the design of floor systems are limitations on the span-to-depth ratio and flexibility which normally lead to deeper sections than would be required from strength considerations alone. It is somewhat dubious that these limitations assure occupants' comfort.

Recognizing that there is no single scale by which the limit of tolerable deflection can be defined, the AISC specification does not specify any limit on the span-to-depth ratios for floor framing members. However, as a guide, the commentary on the specification recommends that the depth of fully stressed beams and girders in floors should not be less than $(F_y/800)$ times the span. If beams of lesser depth are used, it is recommended that the allowable bending stresses be decreased in the same ratio as the depth. Where human comfort is the criterion for limiting motion, the commentary recommends that the depth of steel beams supporting large open floor areas free of partitions and other sources of damping should not be less than one-twentieth of the span, to minimize perception of transient vibration due to pedestrian traffic.

Thus there is no clear-cut requirement on the flexibility to limit the perception of vibration by occupants. Flexibility limits are given, however, from other considerations such as fracture of architectural elements like plaster ceilings. The rule-of-thumb limitations are $\frac{1}{150}$ to $\frac{1}{180}$ of the span for visibly perceptible curvature and $\frac{1}{240}$ to $\frac{1}{360}$ of the span for curvature likely to result in fracture of applied ceiling finishes.

In the design of floor systems, fatigue damage due to transient vibrations is not a consideration because it is tacitly assumed that the number of cycles to which the floor system is subjected is well within the fatigue limitations. However, damage due to fatigue can be a cause of concern in floors subjected to aerobic exercise activities.

Human response is directly related to the characteristics of the vertical motion of the floor system. Users perceive floor vibrations more strongly when standing or sitting on the floor than when walking across it. Human response to vibration seems to be a factor for consideration in design only when a significant proportion of the users will be standing, walking slowly, or seated.

Most of the experiments done on human response to vibrations are related to the physical safety and performance abilities of physically conditioned young subjects in a vibrating environment such as the research supported by NASA and various defense agencies. Very little information is available on the comfort of humans subjected to unexpected vibrations during the course of their normal duties such as slowly walking across a floor or sitting at a desk. Comfort is a subjective human response and defies scientific quantification. Different people report the same vibrations to be perceptible, unpleasant, or even intolerable. A measure for human response to steady sinusoidal vibration taken from Ref. 60 is shown in Fig. 12.8. Although there is no simple physical characteristic of vibration that completely defines the human response, there is enough evidence to suggest that acceleration associated in the frequency range of 1 to 10 Hz is the preferable criteria. This is the range for normally encountered natural frequencies of floor beams. Investigations have shown that human susceptibility to building floor vibrations is

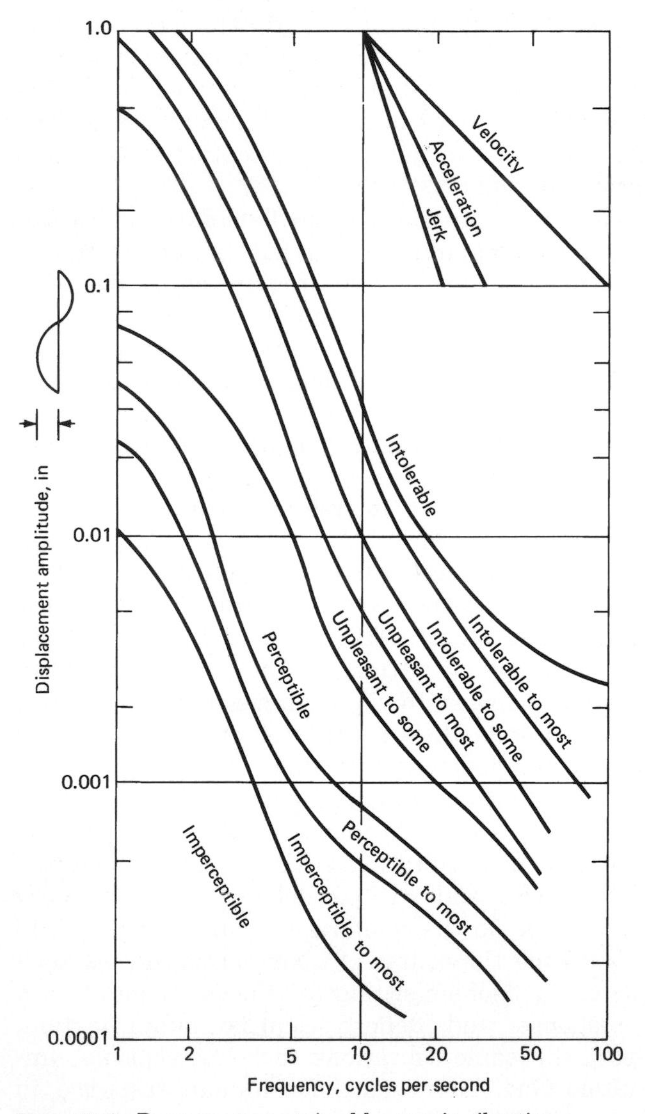


**Figure 12.8** Response to sustained harmonic vibration.

influenced by the rate at which the vibrations decay; people tend to be less sensitive to vibrations that decay rapidly. In fact, experiments have shown that people do not react to vibrations which persist for fewer than five cycles.

Human response ratings to a steady state of vibrations as originally documanted by two researchers, Richer and Meister, have been found to be too severe for the design of building floors subject to transient vibrations caused by human activity. Lenzen (Ref. 61) has modified the Richer and Meister rating scale by multipling the amplitude scale by 10 to account for the nonsteady state of vibrations. The modified curves which account implicitly for damping are shown in Fig. 12.9. In this figure the natural frequency $f$ is plotted on the horizontal scale and the amplitude $A_0$ is plotted on the vertical scale.

The natural frequency $f$ is given by the relation

$$f = 1.57\sqrt{\frac{EI_b g}{W_d l^4}}$$

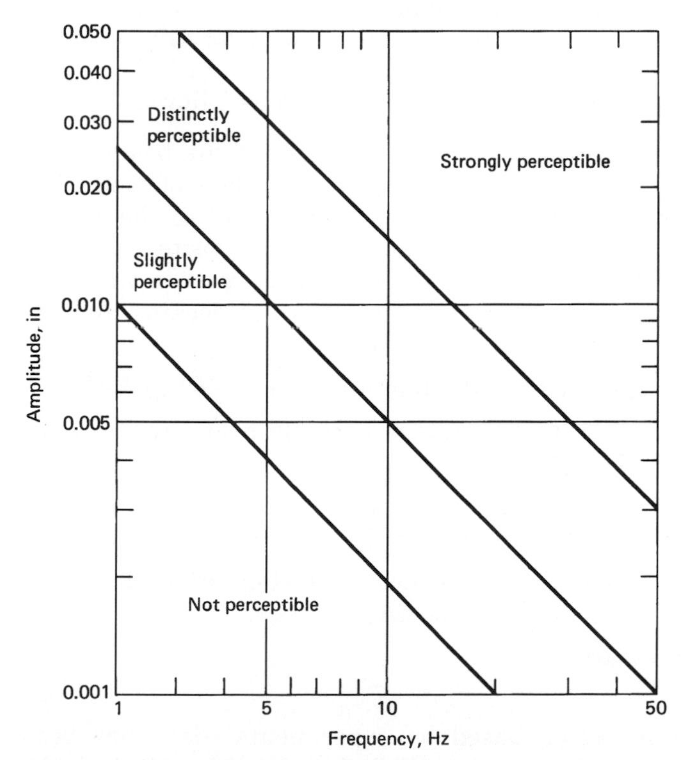


**Figure 12.9** Richer, Meister vibration criteria.

where $f$ = frequency in cycles per second
$E$ = the modulus of elasticity of the system in ksi
$I_b$ = transformed moment of inertia of the beam assuming full interaction with slab system in inches$^4$
$g$ = acceleration due to gravity, 386.4 in./s
$W_d$ = dead load tributary to beam in kips/in.
$l$ = effective span of beam, in inches

The design amplitude $A_0$ is obtained by modifying the initial amplitude of vibration of a simply supported beam subjected to the impact load of a 190 lb person executing a heel drop. The initial amplitude for the most common value of $E$ of 29,000 ksi is given by the relation

$$A_{0t} = (\text{DLF})_{\max} \times \frac{l^3}{80EI_b}$$

where $(\text{DLF})_{\max}$ is the maximum dynamic factor which can be obtained from a graph given in Ref. 62.

Since a floor system usually consists of a number of parallel beams, Ref. 62 suggests that the design amplitude be obtained by dividing the initial amplitude by a factor $N_{\text{eff}}$ to account for the action of multiple beams. Methods of estimating $N_{\text{eff}}$ are given in Ref. 62.

The design procedure can thus be summarized as follows:

1. Compute the transformed moment of inertia of the beam under investigation. Use full composite action regardless of method of construction and assume an effective width equal to the sum of half the distances to adjacent beams. For composite beams on metal deck use an effective slab depth that is equal in weight to the actual slab including concrete in valleys of decking and the weight of decking itself.
2. Compute the frequency from the relation $f = 1.57\sqrt{EI_b g/W_d l^4}$.
3. Compute the heel drop amplitude of a single beam by using the relation $A_{0t} = (\text{DLF})_{\max} \times l^3/80EI_b$.
4. Estimate the effective number of beams, $N_{\text{eff}}$ (Ref. 63) and compute the design amplitude by the relation $A_0 = A_{0t}/N_{\text{eff}}$.
5. Plot on the modified Richer–Meister scale (Fig. 12.9) the computed frequency $f$ and the amplitude $A_0$.
6. Redesign if necessary.

Another response rating based on experimental data has been developed by Wiss and Parmelee (Ref. 47). In thier method the response rating $R$ is given as a function of frequency, peak amplitude, and damping. Based upon the computed value of $R$, the expected human response is classified into one of the five following categories:

| | |
|---|---|
| 1. Imperceptible | $R < 1.5$ |
| 2. Barely perceptible | $1.5 < R < 2.5$ |
| 3. Distinctly perceptible | $2.5 < R < 3.5$ |
| 4. Strongly perceptible | $3.5 < R < 4.5$ |
| 5. Severe | $R > 4.5$ |

The response factor $R$ is given by

$$R = 5.08(FA_0/D^{0.217})^{0.265}$$

where $R$ = response rating
$F$ = frequency, in cycles per second
$A_0$ = Displacement in inches
$D$ = Damping ratio expressed as a ratio of actual damping to critical damping

The damping coefficient $D$, among other things, depends on the inherent characteristics of the floor, such as ceiling, duct work, flooring, furniture, and partitions. It should be noted that $D$ cannot be determined theoretically but can only be estimated in relation to existing floors and their contents. For a rough estimate, the Canadian Standards Association suggests the following values:

| | |
|---|---|
| Bare floors | $D = 0.03$ |
| Finished floor with ceiling, mechanical ducts, flooring and furniture | $D = 0.06$ |
| Finished floor with partitions | $D = 1.13$ |

Floor structures subjected to rhythmic activities such as dancing, aerobics, and other jumping exercises have been a source of annoyance to owners and eingineers alike. Unlike vibration problems encountered in office occupancies, the vibrations due to rhythmic activities are continuous. These vibrations can be greatly amplified when periodic forces are synchronized with the floor frequency, a condition called resonance. Unlike transient vibrations, continuous vibrations may not decay. The National Building Code of Canada (NBC) in its commentary recommends that floor frequencies less than 5 Hz should be avoided for light residential floors, schools, auditoriums, gymnasiums, and other similar occupancies. It recommends a frequency of 10 Hz or more for very repititive activities because of the possibility of getting resonance when the rhythmic beat is on every second cycle of vibration.

In a paper entitled "Vibration Criteria for Assembly Occupancies," Allen, Rainer, and Pernica have presented a procedure for designing floor structures subjected to rhythmic activities. Briefly the procedure is as follows:

1. Determine the density of occupancy based on type of activity. For example, if the floor area is 30 by 60 ft (9.15 by 18.3 m) and has an aerobic class of 50 people of average weight of 120 lb, the equivalent density of occupancy works out to be

$$\frac{50 \times 120}{30 \times 60} = 3.33 \text{ psf} \quad (159.6 \text{ Pa})$$

2. Choose an appropriate forcing frequency $f$ and a dynamic load factor $\alpha$. For aerobic exercises, the value of $f$ suggested in the paper is between 1.5 and 3 Hz, while the value for $\alpha$ is given as 1.5.
3. Choose an acceptable limiting acceleration ratio, $\alpha_0/g$ at the center of the floor. The suggested value for physical exercise activity is 0.05.
4. Determine the lowest acceptable fundamental frequency $f_0$ of the floor system by the relation:

$$f_0 \geqslant f\sqrt{1 + \frac{1.3}{\alpha_0/g}\,\frac{\alpha W_P}{W_t}}$$

where $w_p$ = weight per unit area of participants
$w_t$ = total weight per unit area of structure, participants, furniture, etc.

5. Determine the natural frequency $f_0$ of the floor structure. In addition to the weight of the floor structure itself, weights of participants and furniture, if any, are to be included in the computation of $f_0$.
6. The frequency $f_0$ should be greater than or equal to the frequency obtained in step 4. If not, the options are to stiffen the floor system, relocate the activity, convince the owner to accept a higher limiting acceleration by pointing out that no serious safety-related problems are known to have occurred for floors with frequencies higher than 6 Hz.

Increasing frequency of the floor system by increasing the stiffness is usually cost prohibitive. The most prudent course is to make building owners aware of vibration-related problems during the early design phase.

## 12.4 Panel Zone Effects

Structural engineers involved in the design of high-rise structures are confronted with many uncertainties when calculating lateral drifts. For example, they must decide the magnitude of appropriate wind loads and the limit of allowable lateral deflections and accelerations. Even assuming that these are well defined, another question that often comes up in modeling of building frames is whether or not one should consider the panel zones at the beam-column intersections as rigid.

The panel zone can be defined as that portion of the frame whose boundaries are within the rigid connection of two or more members with webs lying in a common plane. It is the entire assemblage of the joint at the intersection of moment-connected beams and columns. It could consist of just two orthogonal members as at the intersection of a roof girder and an exterior column, or it may consist of several members coming together as at an interior joint, or any other valid combination. In all these cases, the panel zone can be looked upon as a link for transferring loads from horizontal members to vertical members and vice versa. For example, consider the free-body diagram of a frame element consisting of an assemblage of two identical beams and columns with points of zero moments at the ends (Fig. 12.10). These zero moment ends are, in fact, representative of points of inflection in the members.

Consider the frame element subjected to lateral loads. It is easy to see that because of these loads the columns are subjected to horizontal shear forces and corresponding bending moments as shown in Fig. 12.10b. Equilibrium considerations result in vertical shear forces in the beams at the inflection points and corresponding bending moments in the beams. The panel zone thus acts as a device for transferring the moments and forces between columns and beams. In providing for this mechanism the panel zone itself is subjected to large shear stresses.

The presence of high shear forces in a panel zone is best explained with references to the connection shown in Fig. 12.11a. The bending moment in the beam can be considered to be carried as tensile forces in the top flange and compressive forces in the bottom flange and the shear stresses can be assumed to be carried by the web. In the panel zone, the tensile force in the top flange is carried into the web by horizontal shear forces and, by a similar action, is converted back into a tensile force in the outer flange of the column. The distribution of the actual state of stress in the panel zone is highly indeterminate, but a reasonable approximation can be obtained by assuming that the tensile stresses are reduced linearly from a maximum at the edge of the corner *B* or *D* to zero at the extenral corner. If members *AB* and *CD* are assumed as stiffeners, a distinct load path can be visualized for the compressive and tensile forces in the beam flange. Consideration of equilibrium of forces within the panel zone results in shear stress and a corresponding shear deformation as shown in Fig. 12.11d. It is this deformation that is of considerable interest in the calculation of drift of multistory buildings.

Before proceeding with a qualitative explanation of the behavior of panel zones and their influence on building drift, it is instructive to discuss some of the assumptions commonly made in the analysis of building frames. Prior to the availability of commercial analysis programs with built-in capability of treating panel zones as rigid joints, it was common practice to ignore their effects; the frame was usually modeled using actual properties along the centerlines of beams and columns.

If the size and number of joints in a frame were relatively large, an effort was made to include the effect of joint rigidity by artificially increasing the moments of inertia of beams and columns; the actual

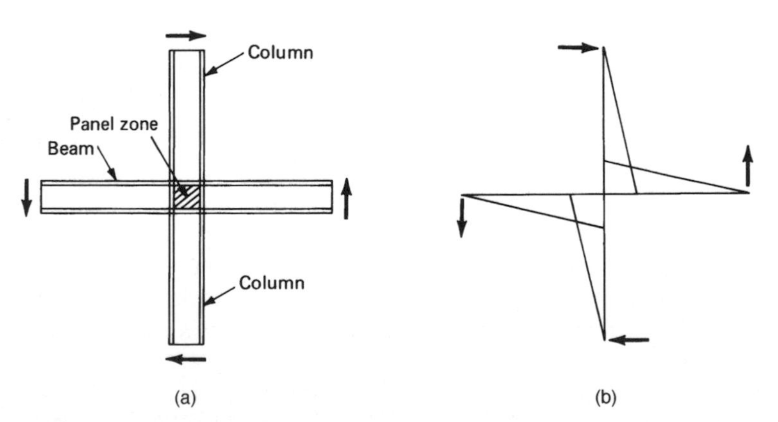


**Figure 12.10** Typical frame element: (a) free body diagram; (b) bending moments due to shear in beam and columns.

properties were usually multiplied by a square of the ratio of centerline dimensions to clear-span dimensions.

Nowadays, it is relatively easy to model the panel zone as a rigid element because of the availability of a large number of computer programs which include this feature. Flexibility of panel zones can also be considered in some of these programs, although somewhat awkwardly, by artificially decreasing the size of panel zones.

Computations of beam, column, and panel zone contributions to frame drift can be carried out by hand calculations by using virtual

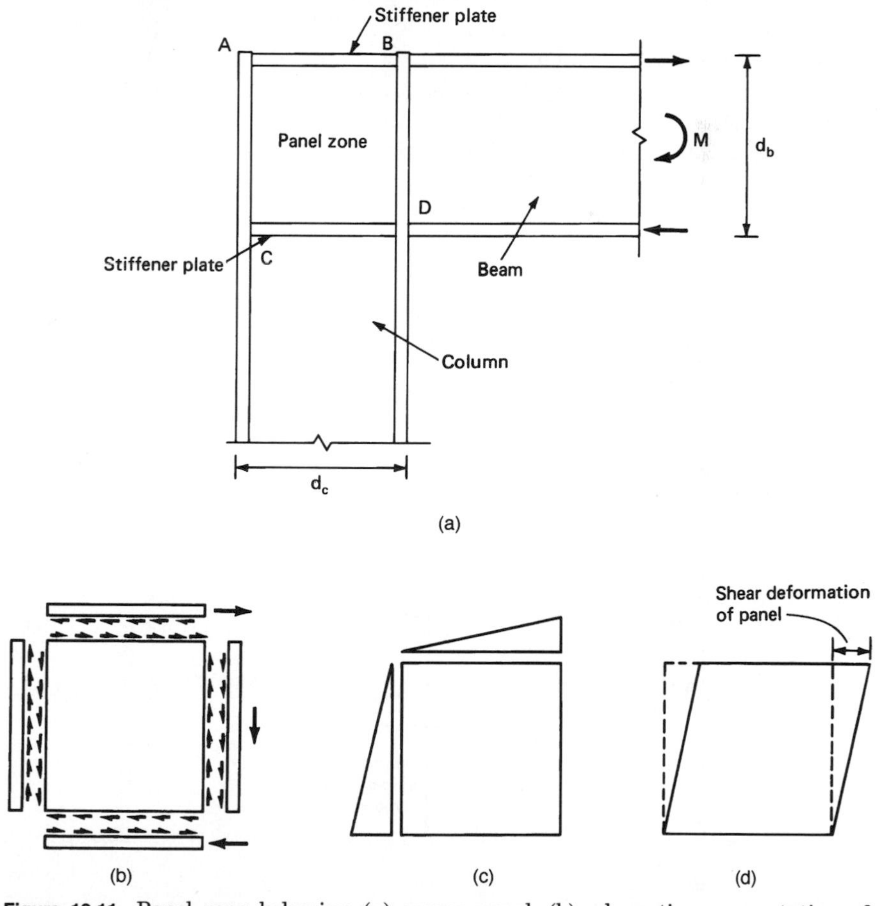


**Figure 12.11** Panel zone behavior: (a) corner panel; (b) schematic representation of shear forces in panel zone; (c) linear distribution of tensile stresses; (d) shear deformation of panel zone.

work method. For this purpose consider again the typical frame element subjected to horizontal shear forces $P_c$ and vertical shear forces $P_b$ at the inflection points (Fig. 12.12a).

The notations used in the development of the method are as

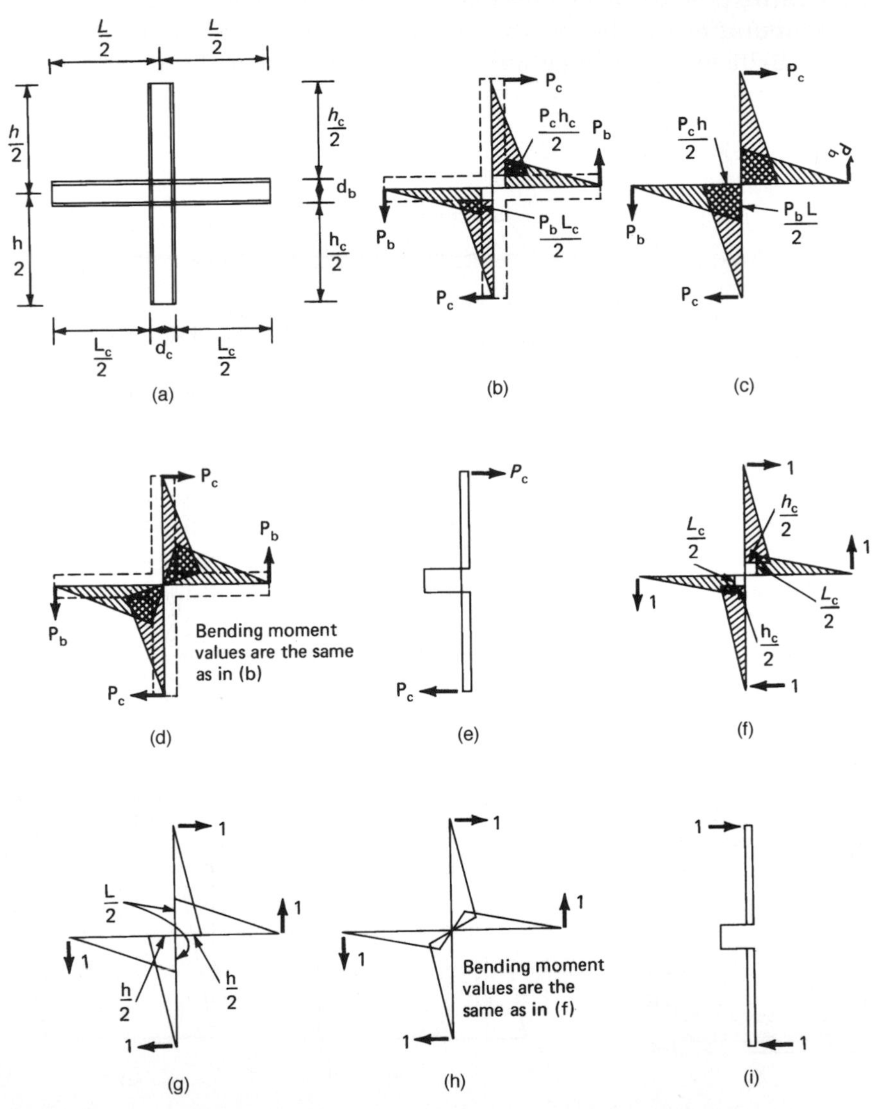


**Figure 12.12** Typical frame segment: (a) geometry; (b) bending moment diagram with rigid panel zone; (c) bending moment diagram without panel zone; (d) bending moment diagram with flexible panel zone; (e) shear force diagram; (f–i) unit load diagrams.

follows:

$d_b$ = depth of panel zone
$d_c$ = width of panel zone
$h_c$ = clear height of column
$L_c$ = clear span of beam
$L$ = center-to-center span of beam
$h$ = center-to-center height of column
$I_c$ = moment of inertia of column
$I_b$ = moment of inertia of beam
$E$ = modulus of elasticity
$G$ = shear modules
$\Delta_b$ = frame drift due to beam bending
$\Delta_c$ = frame drift due to column bending
$\Delta_p$ = frame drift due to panel zone shear deformation

The bending moment diagrams for the typical frame element can be obtained under three different assumptions.

1. The first assumption corresponds to ignoring the rigidity of panel zone; the bending moment diagrams for the external and unit loads can be assumed as shown in Figs 12.12c and 12.12g. The bending moments increase linearly from the point of contraflexure to the centerline of the joint. By integrating the moment diagrams shown in Figs 12.12c and 12.12g the column and beam bending contributions to the frame drift are given by:

$$\Delta_c = \frac{P_c h^3}{12EI_c}$$

$$\Delta_b = \frac{P_b L^3}{12EI_b}$$

2. In the second case, which corresponds to assuming that the panel zone is completely rigid, we get bending moment diagrams for external and unit loads as shown in Figs 12.12b and 12.12f. The bending moments increase linearly from the points of contraflexure but stop at the face of beams and columns. Integration of moment diagrams gives the expressions for $\Delta_c$ and $\Delta_b$ as follows:

$$\Delta_c = \frac{P_c h_c^3}{12EI_c}$$

$$\Delta_b = \frac{P_b L_c^3}{12EI_b}$$

3. The third assumption, which attempts to account for the flexibility of panel zones, results in bending moment and shear force diagrams for external and unit loads as shown in Figs 12.12d,h,e,i. Integration of bending moment and shear force diagrams leads to the following expressions in $\Delta_c$, $\Delta_b$, and $\Delta_p$.

$$\Delta_c = \frac{1}{12EI_c}(P_c h_c^3 + P_c h_c^2 d_b)$$

$$\Delta_b = \frac{1}{12EI_b}(P_b L_c^3 + P_b L_c^2 d_c)$$

$$\Delta_P = \frac{P_c h_c^2}{d_b t_w d_c}$$

The effect of panel zone continuity plates may be determined by performing a finite element analysis of a typical frame unit as shown in Fig. 12.13. A series of finite element analyses can be performed to

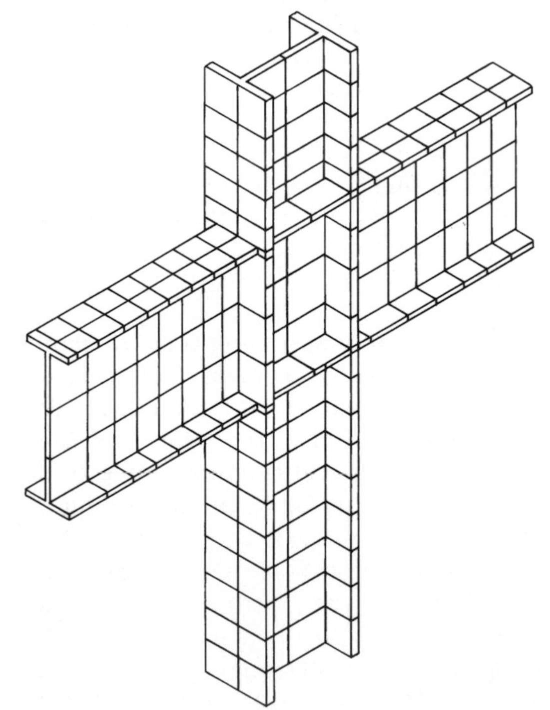

**Figure 12.13** Finite element idealization of typical frame unit.

relate the effect of panel zone to basic section properties of beam and columns of the typical unit. Halvorson (Ref. 64) indicates that for a typical 13 ft, 1 in. high by 15 ft long (4 by 4.57 m) unit consisting of W 36 × 300 columns and W 36 × 230 beams, that the frame stiffness is approximately 8, 15, or 22 percent stiffer than a stick element model depending upon whether no, AISE minimum-, or full-continuity plates are provided, respectively.

Using the virtual work expressions given above, or by performing a finite element analysis, it is relatively easy to compute the contribution of panel zone deformation to frame drift. The author recommends that before undertaking the analysis of large tubelike frames, representative frame elements, say at one-fourth, one-half, and three-fourths the height of the building be analyzed to get a feel for the contribution of panel zone deformation to frame drift. Armed with the results, it is relatively easy to modify the properties of beam and columns such that the overall behavior of the frame is properly represented in the model.

## 12.5 Cladding Systems

Cladding is an expensive part of the building, taking up as much as 10 to 20 percent of a building's initial cost. If the building façade is improperly designed or installed, repairs can cost many times more than the original cost because of difficulties of recladding a building that is already in use. Until the invention of structural steel building frame, about a hundred years ago, exterior building walls were load-bearing elements pierced by windows to provide light and ventilation. The walls served the dual purpose of providing support for the upper floors while simultaneously protecting the building against the elements. Today the only function of building exterior is to act as a filter system.

The performance of a cladding system depends largely on the effect of natural forces on the cladding material. The natural forces are sunlight, temperature, water, wind, and seismic forces, in addition to the ever present gravity forces. Sunlight, particularly in the form of ultraviolet rays, produces chemical changes which cause fading and degradation of materials. Temperature creates expansion and contraction of materials in addition to the temperature differential between interior and exterior of the cladding. Water, in the form of wind-driven rain, can penetrate through the small openings and appear on the interior face of curtain walls. If trapped within the wall in the form of vapor, it can cause serious damage. Wind, acting either inward or outward, creates stresses that require the structural properties of the framing members, panels, and glass to be determined for its maximum effects. Gravity loads of cladding systems cause deflections in horizontal load-carrying members. Although gravity load itself is a one-time load, the members supporting the cladding system are subjected to variable floor and roof live loads, requiring that connections of the cladding to the frame be designed to provide sufficient relative movement to ensure that displacements do not impose vertical loads on the cladding itself. In addition, concrete buildings are subjected to creep, compounding the problem of cladding design.

In the design of curtain walls there are three matters of chief concern: (i) structural integrity; (ii) provision for movement; and (iii) weathertightness. For structural integrity of the wall the stiffness rather than the requirement of strength is the primary concern, although anchorage failure due to inadequate strength should be given proper attention. Providing proper resistance to lateral loads is a routine procedure, especially in view of the fact that we now know more about the nature of wind loads than we did only a decade ago. The signficance of negative wind pressure or suction forces acting on the wall augmented by the internal building pressure caused by air leakage is well understood, as reflected in most building codes.

In designing a curtain wall it is important to provide for ample movement between the wall components themselves, relative movements between the components, and relative movement between the cladding and the building frame to which it is attached. Although the causes of movement are well known, it is not practical to predict the magnitude of the movements accurately. Since the movements manifest themselves at the joints of the cladding, the success of a cladding system lies in the proper detailing of the joints. Provision must be made to accommodate both the vertical and horizontal movement in the plane of the wall. Other design considerations such as weather-tightness, moisture control, thermal insulation, and sound transmissions are areas normally outside the sphere of structural engineering and therefore will not be discussed here.

Common types of curtain wall failures are: (i) structural failure of anchorages resulting in other wall elements disengaging from the building, creating serious hazard to the pedestrians; (ii) glass breakage, which in many instances has been a safety hazard; (iii) excessive air infiltration, which prevents proper conditioning of the interior space; and (iv) accumulation of condensation, causing significant damage to interior finishes.

Failure of cladding can arise because of differential vertical and lateral movement between building columns. Thermal movement, long-term creep of a concrete frame, and foundation settlements can also contribute to the failure. Structural engineers normally do not have a direct involvement in the design of building cladding systems. Typically architect selects the cladding material for appearance and specifies requirements for the weatherproofing, performance, and durability of the cladding. From a knowledge of the weight of the cladding and designated connections points, the engineer designs the structure to carry the anticipated loads. The anticipated movements of the structural frame should be indicated on the drawings in order for the curtain wall supplier to properly detail its connection to the structure. The effect of floor deflections due to the weight of cladding and live load, the magnitude of anticipated building drift, column shortening, foundation settlement, and thermal effects should all be taken into consideration.

The design of cladding systems in North America is traditionally by the architects. Many structural engineering firms specifically exclude this service by inserting a "no involvement with cladding" clause in their service contract. In spite of these attempts to keep out of conflict, if the building cladding turns out to be less than adequate, be it for watertightness, breakage, or any number of other problems, the structural engineer will be drawn into the dispute sooner or later. Therefore, it is important to communicate to the design team the expected behavior of the building frame both under gravity and lateral loads.

Building cladding is one place where a number of independent designs come together. The selection of cladding material for appearance and details for weatherproofing and performance characteristics are in the domain of the architect. The design of the structure for the weight of cladding is the responsibility of the structural engineer. It is likely that the exact weight and connection points of the cladding may not be known at the time the building frame is designed but may have to be conservatively assumed. Short- and long-term deflections of the structure due to gravity loads, column shortening due to creep and shrinkage in concrete and composite structures, expansion and contraction due to temperature, and lateral deflection of the building, especially the interstory drift, are some of the pieces of information required by the cladding supplier. The engineer should have a good understanding of tolerance required both horizontally and vertically for proper installation of curtain walls. In a core-braced steel frame it is likely that the perimeter spandrel is designed as a simply supported beam for gravity loads and hence somewhat limber as compared to a wind-resisting moment-connected spandrel. Even when the gravity loads are heavy, typical simple connections at beam ends have very little stiffness to prevent the rotation of spandrel beams. This is, perhaps, the single most common cause for cladding problems.

The basic components that make up the exterior envelope consist of vision panels and spandrel panels. While vision panels are invariably of glass, the spectrum of materials available for spandrel panels is an ever increasing one. To name a few, spandrel panels can be made of aluminum, steel, glass, masonry, precast concrete, or glass fiber-reinforced concrete. To reduce costs, many designs use a laminated thin-gauge metal sheet bonded to a metal honeycomb backup. Typically framing employed for the mullions is aluminum because of the ease of producing complex custom shapes by extrusion.

The next sections discuss the design and installation aspects of exterior cladding systems. The proper design, installation, and operation of cladding in buildings is a culmination of the coordination activities among architects, engineers, cladding manufacturers, glazing contractors, building officials, and, of course, building owners. The following sections briefly explain how the glass commonly used in a building is made and discusses the strength and testing aspects of glass. Other cladding systems such as metal curtain wall, stone cladding, brick veneer systems, and glass fiber-reinforced concrete systems are also briefly discussed.

### 12.5.1 Glass

Just as floor slabs are designed to carry dead load and occupancy loads, and roof decks to carry dead loads, snow loads and uplift forces due to wind, so must windows and spandrel panels be designed to withstand the lateral loads due to wind. Glass panels in curtain walls are structural elements insofar as they resist wind pressure or suction and transfer wind forces to the building frame. In today's high-rise architecture, glass is employed in more imaginative ways than ever before. At present, the only way to construct a building 50- or 100-stories high is to frame it on a skeleton of steel or concrete and to cloak it with a skin of relatively lightweight cladding. Glass has always been the most suitable material in the eyes of modern architects as witnessed by scores of tall buildings built within the last two decades.

To counteract the fundamental drawback of glass, namely the inordinate amount of heat loss and gain in its pure state, the glass industry has invented a whole series of products intended to make glass buildings economical. There are many tinted glasses designed to reduce the blinding effect of the sun; there are tinted-glass sandwiches, double-glazed panels designed to reduce the blinding effect and to increase the insulating value of the glass wall. All these types of glass units may be used not only as vision panels but also as spandrel panels. There are reflective glasses coated with silver or gold mirror films and innumerable shading devices that reduce glare as well as heat gain and thus air-conditioning loads.

In laying the groundwork for a discussion of glass design, it may be beneficial to explain briefly how different types of glass commonly used in building are made.

Glass is an amorphous, organic, transparent, or translucent substance made of a mixture of silicates, borates, or phosphates and cooled from a liquid to a solid state. Archeological evidence from pre-Roman times indicates that glass making originated in the Near East, probably in the third millennium before Christ. Sand, flint, and quartz are the major sources of silica for glass manufacturing. Typical glass batches include, in addition to sand and other raw materials, up to 50 percent of broken glass of related composition, called waste. This waste promotes melting and homogenization of glass. Impurities, normally in the form of iron traces, cause the glass to be green or brown. To achieve a clear substance magnesium oxide is added to counteract the effect of iron traces. Glass can be colored by dissolving in it certain oxides and sulfides.

Carefully measured ingredients are mixed and allowed to undergo initial fusion before being subjected to full heat. In modern glass plants, the glass is melted in large tank furnaces heated by gas, oil, or electricity. Window glass, in use since the first century, was originally made by blowing hollow cylinders that were slit and flattened into sheets. In modern processes molten glass can be directly drawn into sheets. Glass thus made is not entirely uniform in thickness because of the nature of the process by which it is made. The variations in thickness distort the appearance of objects viewed through panes of glass. The traditional method of overcoming such defects has been to grind and polish the glass.

The modern procedure of making "plate glass," first introduced in France in 1668, consists of rolling the glass continuously between double rollers. After the rough sheet has been annealed, both sides of the glass sheet are finished continuously and simultaneously. *Annealing* is a process in which the glass is reheated to a temperature high enough to relieve internal stresses and then slowly cooled to avoid introducing new stresses.

Grinding and polishing are now being replaced by the more economical "float-glass" process. In this process, flat surfaces are formed on both sides by floating a continuous sheet of glass on a bath of molten tin. The temperature is high enough to allow the surface imperfections to be removed by fluid flow of glass. The temperature is gradually lowered as the glass moves along the tin bath and it passes through a long annealing oven at the end.

Annealed glass which has been reheated to a temperature near its softening point and forced to cool rapidly under carefully controlled conditions is described as "heat-treated" glass. Very often stresses are introduced intentionally in glass to impart strength to it. The objective is to introduce a surface compression because glass always breaks as a result of tensile stresses that generally originate across an infinitesimal surface scratch. Compression on the surface increases the amount of tensile stresses that can be endured before breakage occurs. One of the oldest methods of introducing surface compression is called "thermal tempering." It consists of heating the glass almost to the softening point and then cooling it rapidly with an air blast or by plunging it into a liquid bath. This process rapidly hardens the surface, and the subsequent contraction of the slower-cooling interior portions of the glass pulls the surface into compression. Surface compressions approaching 35 ksi (241.32 MPa) can be obtained in thick pieces by this method. Heat-treated glasses are classified as either "fully tempered" or "heat strengthened" according to the magnitude of compressive stresses induced during heat treatment. Federal specification DD-G-1403B calls for a minimum surface compression of 10 ksi (69 MPa) for a minimum edge compression of 9.7 ksi (66.8 MPa) for the glass to be classified as fully tempered glass. The corresponding minimum requirements for heat-strengthened glass are 3.5 and 5.5 ksi (24.1 and 38 MPa). Below this level the glass is classified as annealed glass.

The fracture characteristics of heat-strengthened glass vary widely. Annealed glass fractures at about 3.5 ksi (24.1 MPa) while fully tempered glass does so at 10 ksi (69 MPa). The characteristic feature of tempered glass is that the glass fractures into small relatively harmless fragments, reducing the likelihood of injury to people. Therefore, many building regulations require the use of tempered glass for skylights, overhead glazing, sloped glazing, and other safety glazing applications.

"Laminated glass" units are made by bonding together two or more lights of glass with an elastomer interlayer. The interlayer is commonly a plastic film of 0.030 in. (0.76 mm) thickness. The laminated glass unit can be made either with annealed, heat-strengthened, or fully tempered glass sheets in any combination. The interlayer does not possess the strength or the stiffness necessary to render the composite unit as strong as an equivalent monolithic light of same thickness. The strength of a laminated unit is taken as 60 percent of the strength of a monolithic light of equal thickness. The actual strength could vary anywhere from 25 percent to full strength of an equivalent monolithic light depending upon the capacity of the interlayer to transmit horizontal shear loads.

"Insulating glass unit" consists of two lights of monolithic glass separated by a spacer and sealed around the perimeter. The sealed air space acts as a layer of insulation, greatly improving the heat-resisting properties of the unit. The edge seal may be fused glass or may be composed of elastomeric sealants and silicones capable of providing a moisture seal around the air space for the normal life of a building. The spacer contains a dessicant to absorb any moisture that may cause the fogging of the glass. Because the air space within an insulating unit is sealed, any pressure applied onto one face is effectively transferred to the other light, making the two lights share the external pressure.

"Wire glass," made by introducing wire mesh into the molten glass before it passes between the rollers, is used to prevent glass from shattering if it is struck. The presence of mesh does not increase the resistance of the light to breakage; it simply holds the pieces together should breakage occur. Building codes commonly consider the strength of wired glass as 50 percent the strength of an annealed monolithic light of the same thickness. Instead of casting a wire mesh within the glass light, an elastomeric film can be applied to the surface to improve the resistance to falling from the frame after breakage.

The strength of glass panels cannot be determined by the classical method of plate analysis because such an analysis is valid only for deflections considerably smaller than the plate thickness. Glass panels deflect many times their thickness and as a result develop membrane stresses which add significantly to their strength and stiffness. Moreover, glass is a brittle material exhibiting no observable yield strength. As such its behavior under loads is best described by evaluating the strength under full-scale test results. Test results show a wide variation of strength for the same size and support conditions of glass panels. This is because the mechanism of glass failure is complex and is highly sensitive to different characteristics of flaws such as their size, orientation, and severity. The only practical approach is to evaluate the effects using an appropriate statistical model. To incorporate the scatter of statistical analysis, a so-called design factor is correlated with probability of failure at full design load. The glass industry uses the normal distribution as the standard model in recommending glass thickness for various design conditions. The published charts of recommended thickness are based on an expected glass breakage probability of 8 lights per 1000, resulting in a design factor of 2.5.

It should be noted that from the statistical point of view, it is virtually impossible to design a glass light without some probability of its failure under design load. Therefore, in conducting a full-scale mock-up test of curtain walls, glass breakage is not considered as a cause for test failure. Glass breakage may very well occur, since the test loads are usually much larger than the design load of glass. If breakage occurs, broken glass is replaced with plywood in order to complete mock-up tests. However, if the test indicates that premature failure of glass is due to inadequate stiffness of the glass supporting system, stricter deflection control should be imposed in the supporting system.

In designing cladding systems with double-glazed insulating sandwiches, care should be taken to prevent the buildup of ultraviolet light, which is known to play havoc with the strip of sealants used on the sides and edges.

Tempered or semitempered glass commonly used in spandrel areas appears to be susceptible to breakage from even minute inclusions of nickel sulfide. Tempered glass is used in these areas because of the large amount of heat that builds up in the space behind the spandrel between the ceiling and floor above. Heat-strengthened glass which is not fully tempered and does not have the same mechanical and thermal endurance qualities is considered much less susceptible to breakage from infiltration of nickel and carbon sulfide.

The performance of the glass supporting members such as mullions may have a bearing on the actual load-carrying capacity of the window if they are signficantly more flexible than the support provided in strength tests. Moreover, relatively large lateral deflection may occur under design loads, resulting in in-plane movements and a tendency for the glass to slip out of its retaining frame.

The most commonly recommended deflection limitation for the glass supporting system is 1:175 of the span. However, stiffer supports may be required for proper weathertightness, durability of sealants, or appearance.

Glass design from a structural point of view consists of selecting an appropriate thickness for a given area and design pressure from charts based on tests conducted by the glass industry.

### 12.5.2 Metal curtain wall

This is one of the most popular methods of cladding the exterior of a building. In simple terms it consists of a metal framework in which metal, glass, and other surface material are housed. Although aluminum curtain walls of today appear to present endless variety of design, the majority of these designs can be classified into the following five generally recognized systems.

1. Stick system.
2. Unit system.
3. Unit and mullion system.
4. Panel system.
5. Column and spandrel cover system.

In the stick system the vertical mullions are usually attached first to the structural frame with anchors followed by installation of the window sill and head section. Spandrel panels and vision glass are attached within the mullion framework to complete the system. In the unit system the curtain wall is preassembled into large units complete with spandrel panels and sometimes also with glass panels. The units may be one, two, or sometimes three stories in height. Typical units are designed to snap in place for a sequential interlocking installation.

The unit and mullion system, as the name implies, is a system that attempts to capture the advantages of both the stick and the unit systems. Mullions are installed first, followed by placement of preassembled frame units between them. The panel system is similar in principle to the unit system, the main difference being that the panels do not consist of vertical and horizontal mullions but are integral units formed from sheet metal or laminated aluminum honeycomb panels. Unlike the other systems, the panel system does not have to follow the rigid discipline of vertical and horizontal grid pattern and can be made to represent a wider range of architectural design flexibility.

The column and spandrel cover system can be thought of as a response of the cladding design to the tube system of lateral bracing consisting of closely spaced columns and deep spandrels. Advantage is taken of the close spacing of columns by spanning the glazing units directly between the column covers. The system can be engineered to clearly express the structural skeleton and permits a wide latitude of aesthetic expression for column and spandrels.

### 12.5.3 Stone cladding

Stone provides a distinctive alternative to glass and metal curtain walls. Distinctive pinks, reds, grays, and blacks of polished granite and the white, buff, and green of marble along with the black and blue of slate are appearing on the urban skyline. Their popularity has increased in such an extent that major glass and aluminum curtain wall manufacturers have adapted their systems to accept stone inserts. Angles, straps, anchors, and wire ties are common methods of stone erection subject to individual design considerations. New methods have been developed to cut natural stone more precisely and quickly. Because stone is a product of nature with inherent variations in physical properties, proper selection of material and its preparation as a cladding system requires careful evaluation. To this end the C-18 Committee on Natural Building Stones of the American Society of Testing and Materials has developed standards for a variety of building stones.

Stone can be finished in a variety of ways. It can be polished, honed, rubbed, flame or thermal finished, or may be used without any finish to retain the natural sawn appearance. Not all ranges of finishes are applicable to all types of stones. The wide range of finish choices applicable for dense stones such as granite diminishes to a limited choice for soft stones such as limestone and sandstone.

There are four primary types of stone veneer attachment to the building exterior. In the first two, the stone is installed piece-by-piece on site, while in the other two types it is fabricated offsite and installed as an integral panel. In the conventional piece-by-piece system, the stone is installed by using relieving angles, strap anchors, dowels, mortar or other mechanical systems. In the standard method which is perhaps the oldest, the stone is attached to a masonry backup using portland cement motor spots. In the mechanical system the stone panel is retained in place by angles fitted into slots cut into the sides of the stone. It is important to note that noncorrosive metals must be protected from electrolytic reaction.

In the prefabricated system, panels are attached off the job site to a concrete back-up system. The stone is lowered, finished-face down, into prepared forms. Stainless steel anchors are placed into precision-drilled holes on the back of the stone. The location and number of anchors used are functions of the spanning capability of the stone. A bond breaker such as a liquid applied membrane or a sheet of polyethylene is laid on the back of the stone prior to concrete pour. The purpose of bond breaker is to accommodate differential movement between the stone and the concrete backup. After the concrete is cured, the panels are trucked to the job site, hoisted into position, and secured to the structural frame.

Another method of prefabrication is to fasten stone pieces directly onto steel trusses. Yet another is to fasten a fiberglass mesh portland cement backing board to the truss and attaching the stone by using the mortar method.

A system popular in certain parts of North America consists of mullions with precut stone veneer in the non-vision areas.

### 12.5.4 Brick veneer systems

Many problems that occur in a brick façade stem from the expansion of the brick due to moisture and direct sunlight, effects of freezing and thawing, and problems arising from insufficient firing of brick. Brick cladding attached to concrete frame buildings is subjected to serious stress buildups because concrete tends to creep with time while masonry has a tendency to swell. To alleviate cladding problems it is necessary to provide enough shelf angles for support and sufficient horizontal expansion joints in the wall to allow the two components—concrete and brick—to move independently.

Adequate drainage should be provided to assure that water is not trapped behind the façade of buildings in areas where winters are harsh. Otherwise water freezes and expands, dislodging the brick. A similar problem occurs if brick is supported on untreated steel structure. When steel rusts it flakes and expands, pushing out the brick. Proper coating of steel to prevent rust is necessary. The cladding joints should be detailed in such a way as not to jostle each other under building movements.

### 12.5.5 Glass fiber-reinforced concrete cladding

Glass fiber-reinforced concrete (GFRC) cladding is a composite material consisting of a portland cement base with glass fibers randomly dispersed throughout the material. The fibers add to the tensile and impact strengths, making possible the production of strong yet lightweight architectural panels. The panels can be made much thinner than a conventional precast system thus resulting in a lightweight system. The lightweight GFRC cladding can provide significant savings by minimizing structural framing and foundation costs for multistory construction especially in areas with poor supporting soil.

### 12.5.6 Curtain wall mock-up tests

In high-rise construction it is a normal practice to test a full-scale mock-up of cladding to evaluate the performance of cladding against the various environmental elements. Structural framing, glass, sealants, gaskets, and anchorage devices representing the actual job site conditions. In most cases, the cladding materials for the mock-up are supplied by the individual job supplier under the direction of the curtain wall subcontractor. The height of mock-up may range from one to four stories, typically forming one side of a four-sided roofed test chamber.

Three performance characteristics are commonly investigated in the mock-up tests: resistance to air infiltration, resistance to water penetration, and sturctural performance. Additionally other tests to measure heat and sound transmission characteristics may also be conducted. Traditionally structural engineers have very little input for the tests except for the one that measures the structural characteristics. Therefore, in this section a brief description of the test for structural performance is given. The reader is referred to other publications, such as *Aluminum Curtain Wall Design Guide Manual,* for a description of tests not covered in this section.

The governing factor in the structural design of framing members and penals is usually the stiffness rather than strength. Analytically it is impossible to account for all the interactions that exist in a curtain wall which is comprised of many interdependent elements such as fastenings, anchors, nonrigid joints, seals, and gaskets. Physical testing is often the most reliable means of verifying the performance. Useful information can be obtained by loading the specimens to failure to get an insight into the ultimate capacity of the system and to identify its weak spots. Structural testing is conducted by the static method using an air chamber, subjecting the test specimen to both positive and negative pressures.

Air is pumped into the chamber to simulate outward-acting, i.e., negative wind loads, and is sucked out of the chamber to simulate inward-acting, i.e., positive wind loads. A monometer is used to measure the air pressure differential between the outside and inside of the pressure chamber. In the dynamic water test a propeller is used to create air turbulence to simulate wind storms.

Typically, the structural test consists of positive and negative uniform static design load and a proof load of 150 percent of the design load. Seismic load test is conducted by subjecting the wall to a racking displacement in the plane of the wall equal to the calculated floor-to-floor lateral displacement, and for proof load twice that displacement.

## 12.6 Mechanical Damping Systems

The exact dynamic behavior of tall buildings is impossible to predict with any great certainty because of the complicated nature of wind and the uncertainty in the evaluation of building stiffness. However, this much is well understood: designing a tall building to meet a given drift criterion under equivalent static forces will not automatically preclude creature-comfort problems.

The intrinsic stiffness of buildings can be increased up to a point beyond which it becomes prohibitively costly to do so. Although computer programs are capable of analytically determining the dynamic characteristics of buildings, it should be remembered that the damping characteristics cannot be estimated closer than say, 30 percent until after the building is constructed. From the fundamental dynamic equations, it is well known that the wind-induced building response is inversely proportional to the square root of total damping comprised of aerodynamic plus structural damping. To reduce the response by 50 percent, we have to increase the structural damping by about four times. Because the inherent damping of a building under wind loads is in the range of 0.5 to 1.5 percent, it is impractical to increase this value to achieve a meaningful reduction in response.

Installation of external damping systems offers an effective way of increasing the comfort of occupants. There are two types of externally installed damping systems. One called "passive viscoelastic damper" was first used in the twin towers of the World Trade Center in New York. A more recent application has been in the 76-story building called Columbia Center in Seattle. Viscoelastic dampers used in this building consist of steel plates coated with a polymer compound. The plates are sandwiched between a system of relatively stationary plates. As the building sways under the action of wind loads, the steel plates which are attached to structural members are subjected alternately to compression and tension. The viscoelastic polymer is in turn subjected to shearing deformations absorbing the strain energy created in the structural members. The dissipation of energy into heat reduces the building sway.

In another system called the "tuned mass damper" or TMD, the building oscillations are controlled in high winds by placing near the top of the building a large mass that oscillates in a direction counter to the direction of building deflection. This system is used on two buildings—the City Corp Center in New York and the John Hancock Tower in Boston. The TMD in the Hancock Tower consists of a huge concrete block weighing about 2 percent of the building weight. It rests on a smooth concrete surface with a large spring connected to one side of the concrete block, while a piston similar to a shock absorber is connected to the other side. The building accelerations are continuously monitored and when accelerations exceed 3 milli-g, indicating the possibility of a heavy windstorm, the oil pumps which levitate the block above the concrete surface are automatically activated. The mass of concrete is thus free to move back and forth.

The device is said to be tuned because during the installation of the damper the mass of concrete and the stiffness of spring are adjusted to the same frequency as that of the building. As the building sways, the mass begins to move in the opposite direction, creating a force that opposes or dampens the motion of the building.

## 12.7 Foundations

The structural design of a skyscraper foundation is primarily determined by loads transmitted by its many floors to the ground on which the building stands. To keep its balance in high windstorms and earthquakes, its foundation requires special consideration because the lateral loads which must be delivered to the soil are rather large. Where load-bearing rock or stable soils such as compact glacial tills are encountered at reasonable depth, as in Dallas with limestone with a bearing pressure of 50 tons per square foot ($47.88 \times 10^2$ kPa), Chicago with hard pan at 20 to 40 tons per square foot ($19.15 \times 10^2$ to $38.3 \times 10^2$ kPa), the foundation may be directly carried down to the load-bearing strata. This is accomplished by utilizing deep basements, caissons, or piles to carry the column loads down through poor soils to compact materials. The primary objective of a foundation system is to provide reasonable flexibility and freedom in architectural layout; it should be able to accommodate large variations in column loadings and spacings without adversely affecting the structural system due to differential settlements.

Many principal cities of the world are fortunate to be underlain by incompressible bedrock at shallow depths, but certain others rest on thick deposits of compressible soil. The soils underlying downtown Houston, for example, are primarily clays that are susceptible to significant volume changes due to changes in applied loads. The loads on such compressible soils must be controlled to keep settlements to acceptable limits.

Usually this is done by excavating a weight of soil equal to a significant portion of the gross weight of the structure. The net allowable pressure that the soil can be subjected to is dependent upon the physical characteristics of the soil. Where soil conditions are poor, a weight of the soil equal to the weight of the building may have to be excavated to result in what is commonly known as fully compensated foundation. Construction of deep foundations may create a serious menace to many older neighboring buildings in many ways. If the water table is high, installation of pumps may be required to reduce the water pressure during the construction of basements and may even require a permanent dewatering system. Depending upon the nature of subsoil conditions, the water table under the adjoining facilities can be lowered, creating an adverse effect on neighboring buildings. Another effect to be kept in mind is the settlement of nearby structures from the weight of the new building.

For buildings in seismic zones, in addition to the stiffness and load distribution, it is important to consider the rigitity of the foundation. During earthquakes, the building displacements are increased by the angular rotation of the foundation due to rocking action. The effect is an increase in the natural period of vibration of the building.

Loads resulting at the foundation level due to wind or earthquake must be delivered ultimately to the soil. The vertical component due to overturning effects is resisted by the soil in a manner similar to the effects of gravity loads. The lateral component is resisted by: (i) shear resistance of piles or piers; (ii) axial loads in batter piles; (iii) shear along the base of the structure; (iv) lateral resistance of soil pressure acting against foundation walls, piers, etc. Depending upon the type of foundation, one or more of the above may play a predominant role in resisting the lateral component.

Much engineering judgement is required to reach a sound conclusion on the allowable movements that can be safely tolerated in a tall building. A number of factors need to be taken into account. These are

1. Type of framing emoloyed for the building.
2. Magnitude of total as well as differential movement.
3. Rate at which the predicted movement takes place.
4. Type of movement whether the deformation of the soil causes tilting or vertical displacement of the building.

Every city has its own particular characteristics in regard to design and construction of foundations for tall buildings which are characterized by the local geology and groundwater conditions. Their choice for a particular project is primarily influenced by economic and soil conditions, and even under identical conditions can vary in different geographical locations. In this section a brief description of two types, namely, the pile and mat foundations, is given, highlighting their practical aspects.

### 12.7.1 Pile foundation

Pile foundation using either driven piles or drilled piers (also called caissons) are finding more and more application in tall building foundation systems. Driven piles usually consist of prestressed precast piles, or steel pipes with pipe, box, or steel H sections. Drilled piers may consist of either straight shafts or may have bells or underreams at the bottom. The number of different pile and caisson types in use is continually changing with the development of pile-driving and earth-drilling equipment.

Driven piles can be satisfactorily founded in nearly all types of soil conditions. When soils overlying the foundation stratum are soft, normally no problem is encountered in driving the piles. If variations occur in the level of the bearing stratum, it will be necessary to use different lengths of piles over the site. A bearing type of pile or pier receives its principal vertical support from a soil or rock layer at the bottom of the pier, while a friction-type pier receives its vertical support from skin resistance developed along the shaft. A combination pier, as the name implies, provides resistance from a combination of bearing at the bottom and friction along the shaft. The function of a foundation is to transfer axial loads, lateral loads, and bending moments to the soil or rock surrounding and supporting it.

The design of a pier consists of two steps: (i) determination of pier size, based on allowable bearing and skin friction if any, of the foundation material; and (ii) design of the concrete pier itself as a compression member. Piers that cannot be designed in plain concrete with practical dimensions can be designed in reinforced concrete in accordance with the provisions of the applicable codes, such as the ACI code. When tall buildings are constructed with deep basements, the earth pressure on the basement walls may be sufficient to resist the lateral loads from the superstructure. However, the necessary resistance must be provided by the piers when there is no basement, when the depth of basement walls below the surface is too shallow, or when the lateral movements associated with the mobilization of adequate earth pressure are too large to be tolerated. In such cases it is necessary to design the piers for lateral forces at the top, axial forces from gravity loads and overturning, and concentrated moments at the top. One method of evaluating lateral response of piers is to use the theory of beam on elastic foundation by considering the lateral reaction of the soil as an equivalent lateral elastic spring.

The effect of higher concentration of gravity loading over the plan area of a tall building often necessitates use of piles in large groups. In comparison to the stresses in the soil produced by a single pile, the influence of a group of piles extends to a significantly greater distance both laterally and vertically. The resultant effect on both ultimate resistance to failure and overall settlement are significantly different than the summation of individual pile contributions. Because of group action, the ultimate resistance is less while the overall settlement is more.

Oftentimes the engineers and architects are challenged to create a floating effect for the building. This is usually achieved by not bringing the façade right to the ground and by using glass sively on the ground to create an open feeling in the lobby. A ¸exten¸sively on the ground to create an open feeling in the lobby. A structural system which uses a heavily braced core and a nominal moment frame on the perimeter presents itself as a solution, the core resisting most of the overturning moment and shear while the perimeter frame provides the torsional resistance. Because of the limited width of the core, strong uplift forces are created in the core columns due to lateral loads. A similar situation develops in the corner columns of exterior-braced tube structures. One of the methods of overcoming the uplift forces is to literally anchor the columns into bedrock. A system of post-tensioned, high-strength anchors about 30 ft or so are driven into bedrock and high-strength grout is injected at the tips to anchor the post-tensioned steel into the rock. A concrete pier constructed below the foundation is secured to the rock by the post-tensioned anchors. Anchor bolts for steel columns cast in the pier transfer the tensile forces from columns to the pier. Another method of securing the columns is to thread the post-tensioned anchors directly through the base plate assemblies of the column.

Figure 12.14 shows the plan and cross-section of a foundation system for a corner column of an X-braced tube building. The spread footing founded on limestone resists the compressive forces while the belled pier under the spread footing is designed to resist uplift forces. To guard against the failure of rock due to horizontal fissures a series of rock bolts are installed around the perimeter of spread footing.

### 12.7.2 Mat foundation

**General considerations.** The absence of high bearing and side friction capacities of stratum at a reasonable depth beneath the footprint of the building precludes the use of piles or deep underreamed footings. In such circumstances, mat foundations are routinely used under tall buildings, particularly when the soil conditions result in conventional footings or piles occupying most of the footprint of the building. Although it may be possible to construct a multitude of individual or combined footings under each vertical load-bearing element, mat foundations are preferred because of the tendency of the mat to equalize the foundation settlements. Because of continuity, mat foundations have the capacity to bridge across local weak spots in substratums. Mat foundations are predominantly used in two instances: (i) whenever the underlying load-bearing stratum consists of soft, compressible material with low bearing capacity; and (ii) as a

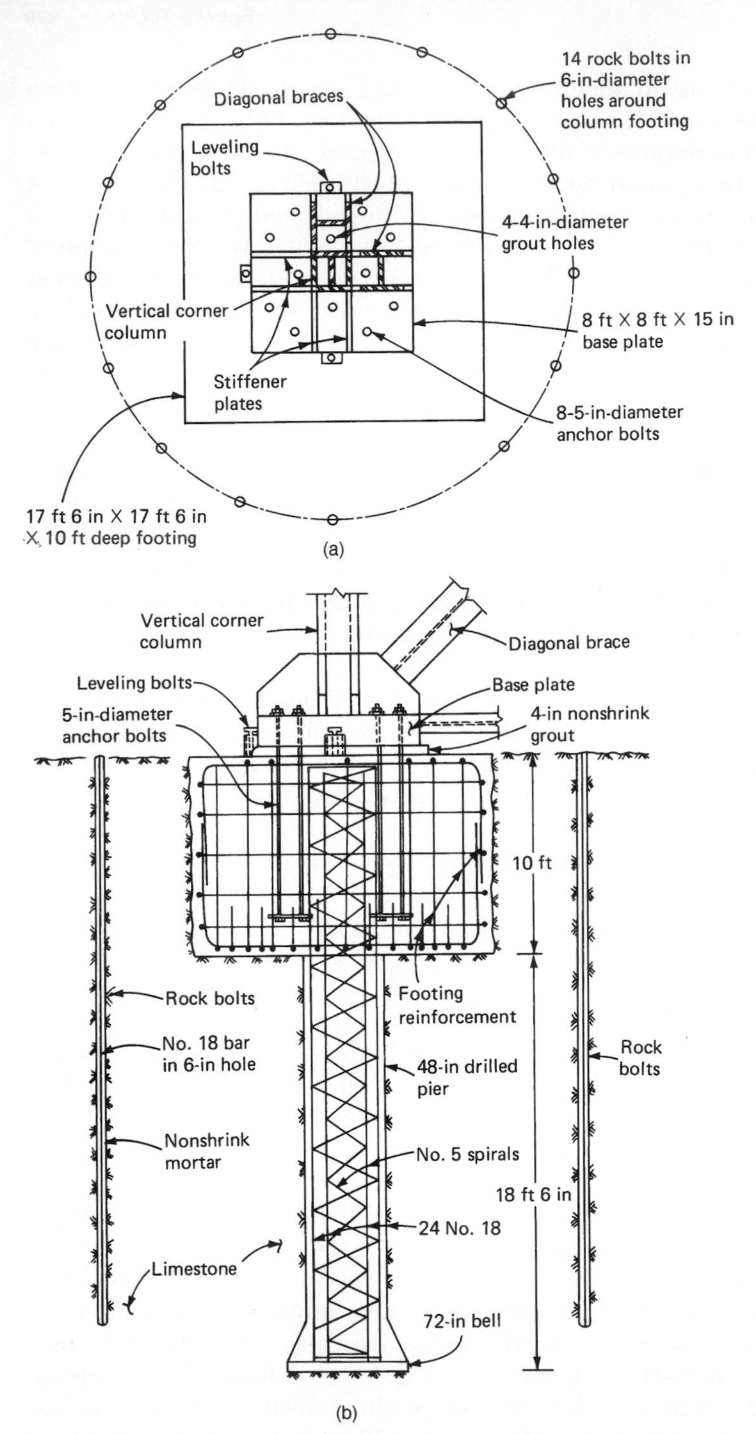


**Figure 12.14** Foundation system for a corner column for an *X*-braced tube building: (a) plan; (b) section.

giant pile cap to distribute the building load to a cluster of piles placed under the footprint of the building.

Mat foundations are ideal when the superstructure load is delivered to the foundation through a series of vertical elements resulting in a more or less uniform bearing pressure. It may not be a good solution when high concentrations of loading occur over limited plan area. For example, in a core-supported structure carrying most of the building load, if not the entire load, it is uneconomical to spread the load over the entire footprint of the building because this would involve construction of exceptionally thick and heavily reinforced mat. A more direct solution is to use driven piles or drilled caissons directly under the core.

The plan dimensions of the mat are determined such that the mat contact pressure does not exceed the allowable bearing capacity prescribed by the geotechnical consultant. Typically three types of allowable pressures are to be recognized: (i) net sustained pressure under sustained gravity loads; (ii) gross pressure under total design gravity loads; and (iii) gross pressure under both gravity and lateral loads.

In arriving at the net sustained pressure the loads to be considered on the mat area should consist of

1. Gravity load due to the weight of the structural frame.
2. Weight of curtain wall, cooling tower, and other mechanical equipment.
3. An allowance for actual ceiling construction including air conditioning duct work, lights sprinklers, and fireproofing.
4. Probable weight of partition based on single and multitenant layouts.
5. Probable sustained live load.
6. Loads applied to the mat from backfill, slab, pavings, etc.
7. Weight of mat.
8. Weight of soil removed from grade to the bottom of the mat.

This last item accounts for the reduction in overburden pressure and therefore is subtracted in calculating the net sustained pressure.

In calculating the sustained pressure on mats, typically less than the code prescribed values are used for items 4 and 5, requiring engineering judgement in their estimation. In the opinion of the author a total of 20 psf (958 Pa) for these items appears to be adequate. A limit on sustained pressure is basically a limit on the settlement of the mat. In practical cases of mat design it is not uncommon to have the calculated sustained pressure under isolated regions of mat somewhat larger than the prescribed limits. This situation should be reviewed with the geotechnical engineer and usually is of no concern as long as the overstress is limited to a small portion of the mat.

The gross pressure on the mat is equivalent to the loads obtained from items 1 through 7. The weight of the soil removed from grade to the bottom of the mat is not subtracted from the total load because the gross pressure is of concern. Also, in calculating the weight of partition and live loads, the code-specified values are used.

The transitory nature of lateral loads is recognized in mat design by allowing a temporary overstress on the soil. This concept is similar to the 33 percent increase in stresses allowed in most building codes for wind and seismic loads. From an academic point of view, the ideal thickness for a mat is the one that is just right from punching shear considerations while at the same time minimum reinforcement of $0.002A_c$ provided for temperature works just right from flexural considerations. However, in practice it is found that it is more economical to construct pedestals or provide shear reinforcement in the mat rather than to increase the basic mat thickness.

In detailing the flexural reinforcement there appear to be two schools of thought. One school maintains that it is more economical to limit the largest bar size to a #11 bar which can be lap-spliced. However, this limitation may force the use of as many as four layers of reinforcement both at top and bottom of mat. The other school promotes the use of #14 and #18 bars with mechanical tension splices. This requires fewer bars, resulting in cost savings in the placement of reinforcement. The choice is, of course, a matter of economy as perceived by the contractor.

**Analysis.** A vast majority of soil-structure interaction takes place under sustained gravity loads. Although the interaction is complicated by the nonlinear and time-dependent behavior of soils, it is convenient for analytical purposes to represent the soil as an equivalent elastic spring. This concept was first proposed by Winkler in 1867 and hence the name Winkler spring. He proposed that the force and vertical displacement relationship of the soil be expressed in terms of a constant $K$ called the modulus of subgrade reaction. It is easy to incorporate the effect of the soil by simply including a spring with a stiffness factor in terms of force per unit length beneath each reaction. However, it should be remembered that the modulus is not a fundamental property of the soil. It depends on many things, including the size of the loaded area and the length of time it is loaded. Consequently, the modulus of subgrade reaction used for calculating the spring constants must be consistent with the type and duration of loading applied to the mat.

Prior to the availability of finite element programs, mat analysis used to be undertaken by using a grid analysis by treating the mat as an assemblage of linear elements. The grid members are assigned equivalent properties of a rectangular mat section tributary to the grid. The magnitude of the Winkler's spring constant at each grid intersection is calculated on the basis of tributary area of the joint.

The preferred method for analyzing mats under tall buildings is to use a finite element computer program. With the availability of computers, analytical solutions for complex mats are no longer cumbersome; engineers can incorporate the following complexities into the solution with a minimum of effort:

1. Varying subgrade modulus.
2. Mats of complex shapes.
3. Mats with nonuniform thickness.
4. Mats subjected to arbitrary loads due to axial loads and moments.
5. Soil-structure interaction in cases where the rigidity of the structure significantly affects the mat behavior.

As in other finite element idealizations, the mathematical model for the mat consists of an assemblage of discretized elements interconnected at the nodes. It is usual practice to use rectangular or square elements instead of triangular elements because of the superiority of the former in solving plate-bending problems. The element normally employed is a plate-bending element with 12 degrees of freedom for three generalized displacements at each node. The reaction of the soil is modeled as a series of independent elastic springs located at each node in the computer model. The behavior of the soil tributary to each node is mathematically represented as a Winkler spring at each node. There is no continuity between the springs other than through the mat. Also, the springs because of their very nature can only resist compression loads although computationally it is not possible to impose this restriction in a linear elastic analysis. Therefore, it is necessary to review the spring reactions for any possible tensile support reactions. Should this have occurred in the analysis, it is necessary to set the spring constant to zero at these nodes and to perform a new analysis. This iterative procedure is carried out until the analysis shows no tensile forces in springs.

In modeling the mat as an assemblage of finite elements, the following key factors should be considered: (i) grid lines that delineate the mat into finite elements should encompass the boundaries of the slab, as well as all openings. They should also occur between elements with changes in thicknesses. Skew boundaries of mat not parallel to the orthogonal grid lines may be approximated by steps that closely resemble the skewed boundary; (ii) grid lines should intersect preferably at the location of all columns. Minor deviations are permissible without loss of meaningful accuracy; (iii) a finer grid should be used to define regions subjected to severe displacement gradients. This can be achieved by inserting additional grid lines adjacent to major columns and shear walls.

Although it is possible to construct an analytical model consisting of both the mat and superstructure, practical budgetary and time considerations preclude use of such complex analyses in everyday practice. Admittedly the trend, with the availability of computers and general analysis programs, is certainly toward this end. However, the current practice of accounting for superstructure interaction is to simulate the stiffness of superstructure by incorporating artificially stiff elements in mat analysis. Although the procedure is approximate, it has the advantage of being simple and yet capable of capturing the essential stiffness contribution of the superstructure.

The complex soil-structure interaction can be accounted for in the design by the following iterative procedure. Initially the pressure distribution under the mat is calculated on the assumption of a rigid mat. The geotechnical engineer uses this value to obtain the deformation and hence the modulus of subgrade reaction at various points under the footprint of the mat. Under uniform pressure the soil generally shows greater deformations at the center than at the edges of the mat. The modulus of subgrade reaction, which is a function of the displacement of the soil, therefore has higher values at the edges than at the center. The finite element mat analysis is performed using the varying moduli of subgrade reaction at different regions of the mat. A new set of values for contact pressures is obtained and processed by the geotechnical engineer to obtain a new set of values of soil displacements and hence the moduli of subgrade reaction. The process is repeated until the deflections predicted by the mat finite element analysis and the settlement predicted by the soil deflection due to consolidation and recompression of soil stratum converge to a desirable degree.

Two examples are presented following to give the reader a feel for the physical behavior of mats. The first consists of a mat for a

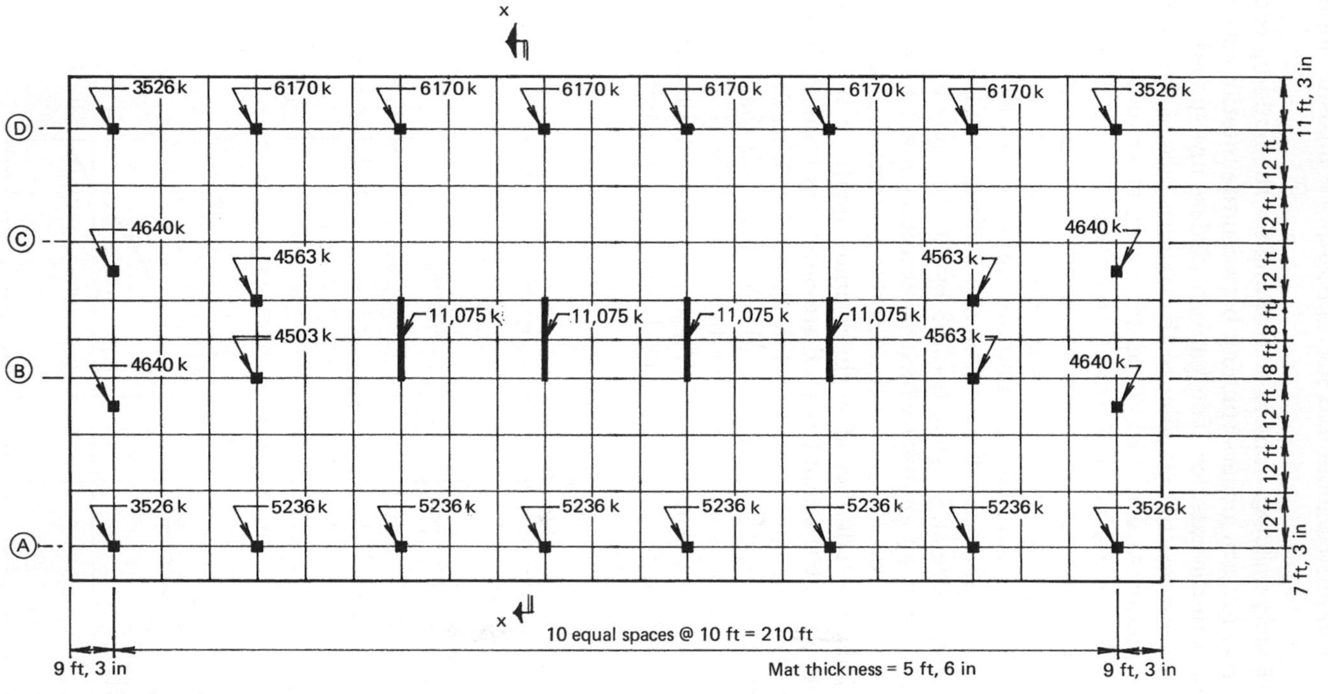


**Figure 12.15** Foundation mat for a 25-story building; finite element idealization and column ultimate loads.

25-story concrete office building and the second example highlights the behavior of an octagonal mat for an 85-story composite building.

**Mat for a 25-story building.** The floor framing for the building consists of a system of haunch girders running between the interior core walls and columns to the exterior. The haunch girders are spaced at 30 ft (9 m) on centers and run parallel to the narrow face of the building. Skip joists spaced at 6 ft (1.81 m) center-to-center span between the haunch girders. A 4 in. (101.6 mm) thick concrete slab spanning between the skip joists completes the floor framing system. Lightweight concrete is used for floor framing members while normal weight concrete is employed for columns and shear walls.

Shown in Fig. 12.15 is a finite element idealization of the mat. The typical element size of 12 by 10 ft (3.63 by 3.03 m) may appear to be rather coarse, but an analysis which used a finer mesh in which the typical element was 3 by 2.5 ft (0.9 by 0.76 m) showed results identical to that obtained for the coarse mesh. The calculated ultimate loads at the top of the mat are shown in Fig. 12.15. It may be noted that the finite element idealization is chosen in such a manner that the location of almost all columns, with the exception of four exterior columns on the narrow face, coincide with the intersection of the finite element mesh. The loads at these locations are applied directly at the nodes. The loads on the four exterior columns

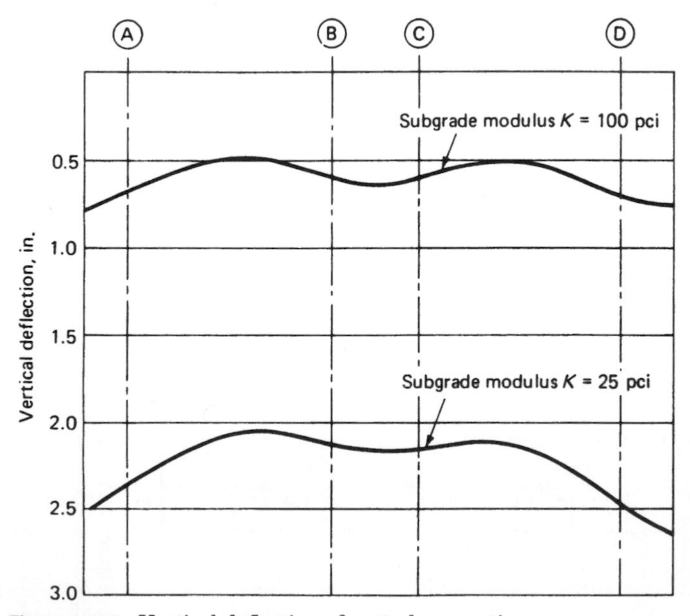


**Figure 12.16** Vertical deflection of mat along section $x$–$x$.

are, however, divided into two equal loads and applied at the two nodes nearest to the column. The resulting discrepancy in the analytical results has very little impact, if any, on the settlement behavior and the selection of reinforcement for the mat.

Assuming a value of 100 lb/in.$^3$ (743 kg/mm$^3$) for the subgrade modules, the spring constant at a typical interior node may be shown to be equal to 1728 kips/in. ($196 \times 10^3$ N/m). Figure 12.17 shows the mat deflection comparison for two values of subgrade reaction, namely 100 and 25 lb/in.$^3$ (743 and 185.75 kg/mm$^3$). As can be expected, the mat experiences a larger deflection when supported on relatively softer springs (Fig. 12.16). The variation of curvature, which is a measure of bending moments in the mat, is relatively constant for the two cases. This can be verified further by comparing the bending-moment diagrams shown in Fig. 12.17. Also shown in this figure are the bending moments obtained by assuming the mat as a continuous beam supported by three rows of supports corresponding to exterior columns and interior shear walls, and subjected to the reaction of the soil acting vertically upward. The results for the example mat appear to indicate that mat reinforcement selected on the basis of any of the three analyses will result in adequate design.

**Mat for an 85-story building.** Figure 12.18 shows a finite element idealization of a mat for a proposed 85-story composite building, in Houston, Texas. Noted that diagonal boundaries of the mat are approximated in a stepwise pattern using rectangular finite elements. To achieve economy, the thickness of the mat was varied; a thicker mat was proposed under the columns where the loads and thus the bending of the mat were expected to be severe. A relatively

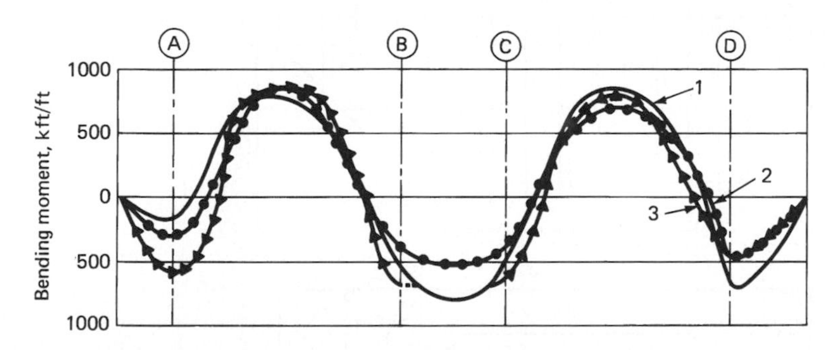


1. Modulus of subgrade reaction K = 100 pci
2. Modulus of subgrade reation K = 25 pci
3. Continuous beam analysis

**Figure 12.17** Bending moment variation along section $x$–$x$.

thin mat section was proposed for the center of the mat. The appropriateness of choosing two mat thicknesses can be appreciated by studying the pressure contours in Fig. 12.19a and b. The pressure contours plotted in these figures were obtained from computer analyses for two different loading conditions—gravity loads acting alone and gravity loads combined with wind loads. No uplift due to wind loads was evident.

## 12.8 Seismic Design of Diaphragms

### 12.8.1 Introduction

Why is it that the design of floor and roof diaphragms is much more significant in seismic design than in wind design? The answer lies in the magnitude of, and the manner in which the loads are originated.

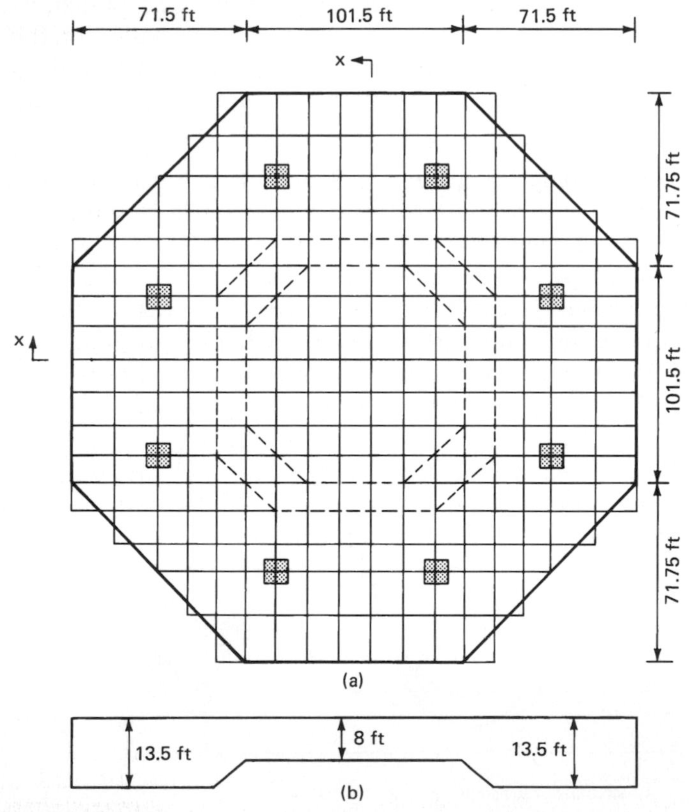


**Figure 12.18** Foundation mat for a proposed 85-story office building: (a) finite element idealization; (b) cross-section $x$-$x$ through mat.

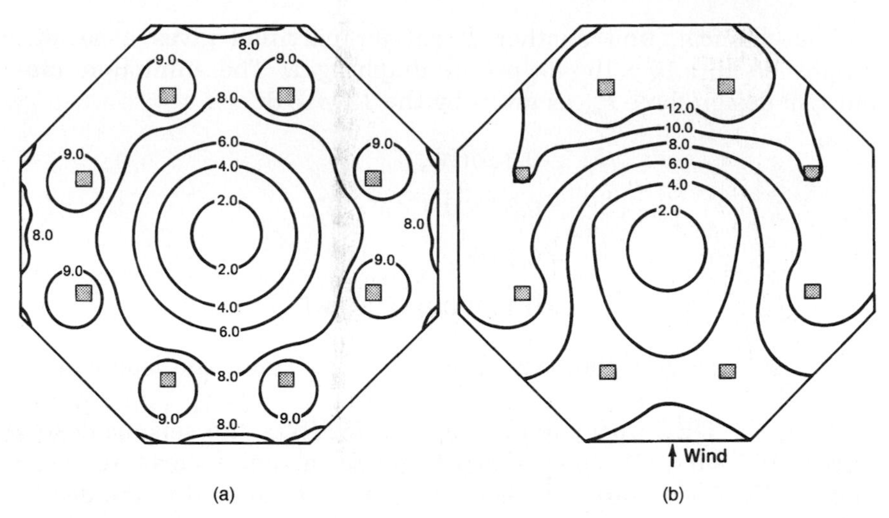


**Figure 12.19** Contact pressure contour under the mat (ksf): (a) dead loads plus live loads; (b) dead plus live plus wind loads (note $K = 100\,\text{lb/in.}^3$).

Wind is an externally applied load, and although it acts in a dynamic fashion, its basic effect is similar to a statically applied load. Wind buffeting the exterior curtain wall is successively resisted by the glass, the mullions, the floor slab, and finally by the lateral load resisting elements such as shear walls and frames. To get a feel for the difference in the magnitude of wind and seismic loads consider a building $200 \times 100$ ft, with a 14 ft floor-to-floor height, subjected to a uniform wind load of 30 psf. The wind load tributary to the floor in the short direction $= 30 \times 14 \times 200/1000 = 84$ kip. The load per ft of diaphragm $= 420$ plf. Now consider the seismic design of building in zone 4 using an Importance factor $I = 1.0$. Assume the building is of steel construction, with a floor construction consisting of 3 in. deep metal deck and $3\frac{1}{4}$ in. lightweight concrete topping. The seismic dead load of the floor system including an allowance of 10 psf for ceiling

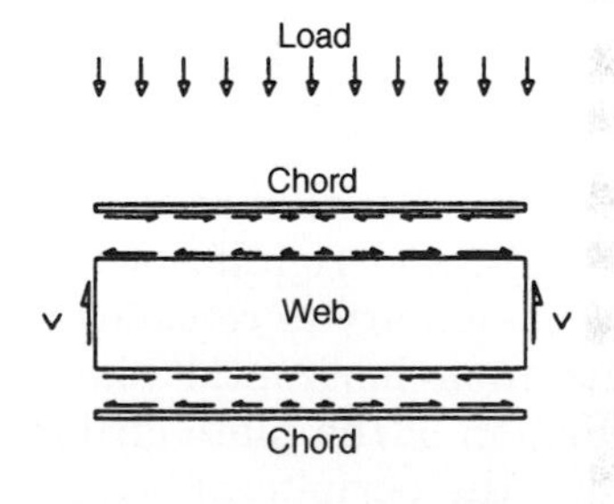


**Figure 12.20** Diaphragm as a beam: free-body diagram.

and mechanical, and another 10 psf for partition gives a seismic weight of $46 + 10 + 10 = 66$ psf of diaphragm. The minimum diaphragm design force $F_{px}$ as given by the 1994 UBC is:

$$\begin{aligned} F_{px} &= 0.35ZIW_{px} \\ &= 0.35 \times 0.4 \times 1 \times W_{px} \\ &= 0.14W_{px} \\ &= 0.14 \times 66 = 9.24 \text{ psf} \end{aligned}$$

$F_{px}$ per ft $= 9.24 \times 100 = 924$ lb which is 2.2 times the load caused by wind.

Using the UBC upper limit of $F_{px} = 0.75ZIW_{px}$, the seismic design force $= 0.75 \times 0.4 \times 1 \times 66 = 19.8$ psf giving a unit shear of 1980 pounds. This is almost a five-fold increase as compared to the design value due wind.

Observe that the comparison just made is between the design values. In reality the actual seismic loads are several times larger than the design loads and could be as much as $\frac{3}{8}R_w$ times, or even larger than the design load. For the example building assuming $R_w = 12$, the load could be 4.5 times the design value or as much as 8910 pounds per ft of diaphragm. This value compared to 420 plf for wind gives an idea of the vast difference in the order of magnitude of diaphragm forces produced during a windstorm and in a large seismic event.

### 12.8.2 Diaphragm behavior

Buildings are composed of vertical and horizontal structural elements which resist lateral forces. Horizontal forces on a structure produced by seismic ground motion originate at the centroid of the mass of the building elements and are proportional to the masses of these elements. These forces include inertia forces originating from the weight of the diaphragm and the elements attached there to, as well as forces that are required to be transferred to vertical resisting elements because of offsets or changes of stiffness in vertical resisting elements above and below the diaphragm.

Horizontal forces at any floor or roof level are distributed to the vertical resisting elements by using the strength and rigidity of the floor or roof deck to act as a diaphragm. It is customary to consider a diaphragm analogous to a plate girder laid in a horizontal plane where the floor or roof deck performs the function of the plate girder web, the beams function as web stiffeners, and the peripheral beams or integral reinforcement function as flanges (Fig. 12.20).

The diaphragm chord can take on many forms such as a line of edge beams connected to the floor, or reinforcing in the edge of a slab or reinforcing in a spandrel (Fig. 12.21). Boundary members at edges of diaphragms must be designed to resist direct tensile or compressive chord stresses, including adequate splices at points of discontinuity. For instance, in a steel frame building the spandrel beams acting as a diaphragm flange component require a splice design at the columns for the tensile and compressive stresses induced by diaphragm action. The fundamental requirements for the chord are the continuity of the chord and the connection with the slab.

To continue the similarity with the beam, an opening in the floor for a stair, elevator or a sklylight may weaken the floor just as a hole in the web for a mechanical duct weakens the beam. Similarly a break in the edge of the floor may weaken the diaphragm just as a notch in a flange weakens the beam. In each case the diaphragm should be detailed such that all stresses around the openings are developed into the diaphragm.

Another beam analogy applicable to diaphragms is the rigidity of the diaphragm compared to the walls or frame that provide lateral support, and transmit the lateral forces to the ground. A metal-deck roof is relatively flexible compared to concrete walls while a concrete floor is relatively rigid compared to steel moment frames.

Yet another beam characteristic is continuity over intermediate supports. Consider, for example, a three-bay building with three spans and four supports. If the diaphragm is relatively rigid, the chords may be designed like flanges of a beam continuous over the intermediate supports. On the other hand if the diaphragm is flexible, it may be designed as a simple beam spanning between walls with no consideration of continuity. However, in the latter case, it should remembered that the diaphragm really is continuous, that its

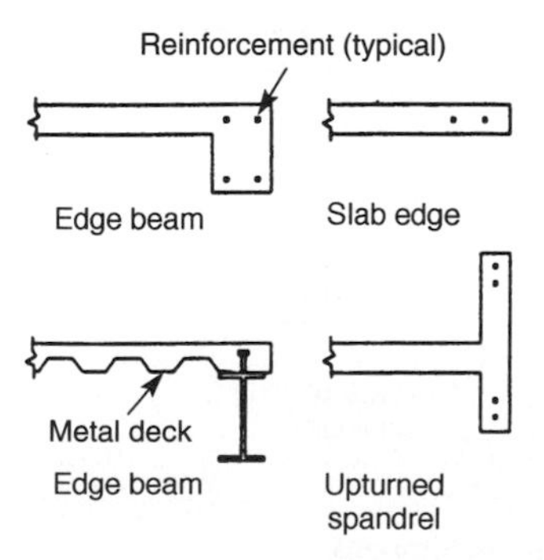


**Figure 12.21** Chord sections.

continuity is simply being neglected. The consequence of the neglect may well be some damage where adjacent spans meet.

Another essential consideration is that the connection between the beam and its support must be adequate for the transfer of diaphragm shear to the shear wall. As part of the diaphragm design, it may be necessary to add a collector or drag strut to collect the diaphragm shear and drag it into the vertical subsystems. This technique has the desirable effect of precluding a concentration of stress in the diaphragm alongside the vertical subsystem. The collector, however, becomes a critical element: it must be continuous, developing its required capacity across any interrupting elements such as cross beams, and there must be an adequate connection to transfer the collector force into vertical system.

In buildings the effect of flexible diaphragms on the normal walls must be considered. Figure 12.22 shows the deflected shape of the normal wall of a two-story building. In the building on the left (the more usual case), the floor is stiffer than the roof, and the wall is bent out of its plane, with the maximum movement near the upper floor. In the building on the right (an unusual case), the roof is stiffer, and the bending can be more severe at the floor level.

A common plan irregularity is an odd shape: a single isolated odd shape or one that results from offsetting each module in a group of buildings. Shown in Fig. 12.23 are examples of plan shapes with re-entrant corners. If the tensile capacity provided at the re-entrant corners at the points marked "X" is not sufficient, a local concentration of damage may occur, and this could lead to loss of support for beams or slabs.

Another common irregularity is an inset, usually for an entrance

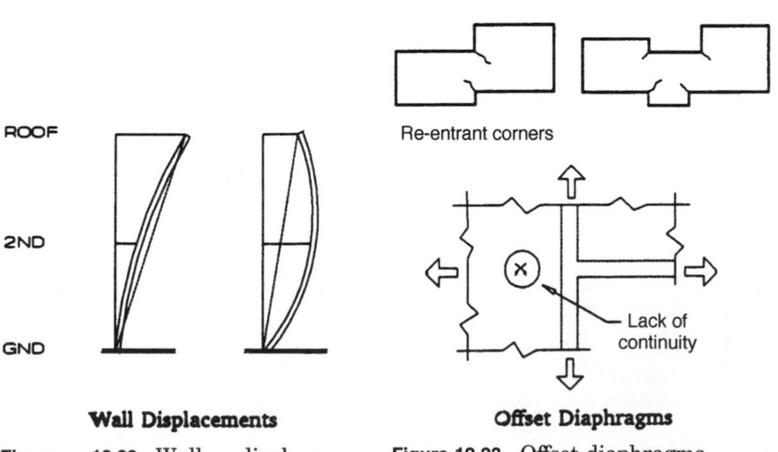


**Figure 12.22** Wall displacements.

**Figure 12.23** Offset diaphragms.

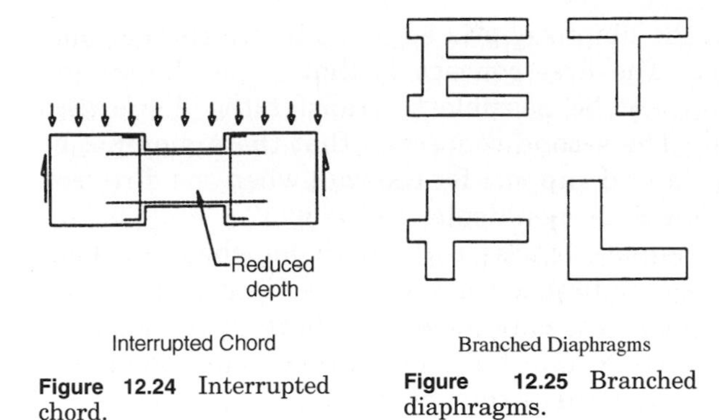


**Figure 12.24** Interrupted chord.

**Figure 12.25** Branched diaphragms.

or loading bay resulting in reduced chord depth (Fig. 12.24). Using the beam analogy, the inset is like a notch in the beam flange requiring development of chord forces on either side of notch.

Buildings of E-, T-, X-, and L-shaped plans (Fig. 12.25) are also troublesome. The branches of these shapes have modes of vibration that the designer may not have considered in a simple design for north–south and east–west forces. Large tensile forces can develop in the re-entrant corners. Often this condition is aggravated by holes in the diaphragm made for elevators and stairs.

The L- and T-shaped buildings should have the flange, i.e., the chord stresses, developed through or into the heel of L or T. This is analogous to a girder with a deep haunch in which the girder flange forces are developed into the haunch.

Small openings have not been a life-safety issue in buildings, but large openings as shown in Figure 12.26 can have a disastrous effect if not properly accounted for in design.

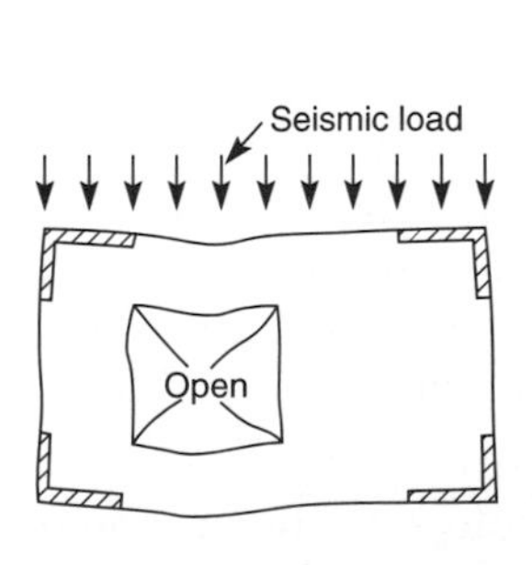


Diaphragm Opening

**Figure 12.26** Diaphragm opening.

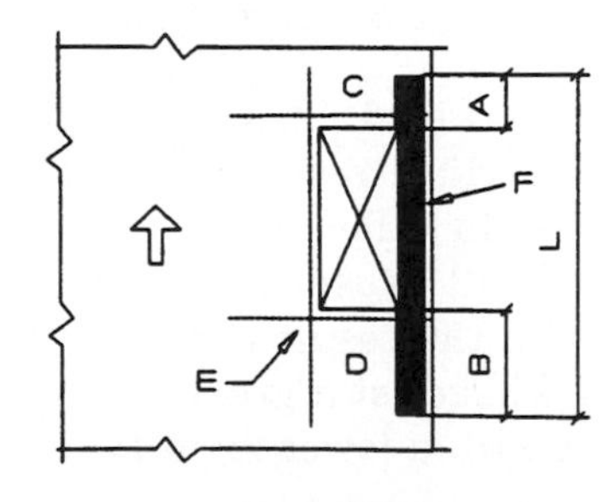


Diaphragm
Opening at wall

**Figure 12.27** Diaphragm opening.

When an opening in the diaphragm is adjacent to a wall or frame, there are two concerns. The first concern is that if the diaphragm opening is long, it may not be possible to transfer the diaphragm shear to the shear wall. The second concern is that there may not be enough floor to provide lateral support for the wall when the direction of the earthquake loading is perpendicular to the wall.

In Fig. 12.27, the opening reduces the length for shear transfer from $L$ to $A + B$. Stresses are higher than otherwise in regions $C$ and $D$; moreover, a region such as $C$ may have so little rigidity compared to $D$ that it is relatively ineffective for shear transfer. Reinforcing is necessary at the re-entrant corners such as $E$.

For earthquake loads in the other direction (i.e., perpendicular to the wall), the wall is a normal wall. It should be designed as a beam to span the length of the opening $F$, to collect out-of-plane forces from above and below and deliver them through suitable connections into and diaphragm at the ends of the opening.

### 12.8.3 Rigid, semi-rigid and flexible diaphragms

The total shear, which includes the forces contributed through the diaphragm as well as the forces contributed from the vertical resisting elements above the diaphragm, at any level are distributed to various vertical elements of the lateral force-resisting system such as shear walls or moment-resisting frames, in proportion to their rigidities considering the rigidity of the diaphragm.

The effect of diaphragm stiffness on the distribution of lateral forces is schematically illustrated in Fig. 12.28. For this purpose, diaphragms are classified into three groups of flexibilities relative to the flexibilities of the walls. These are: (i) rigid; (ii) semi-rigid; and (iii) flexible, diaphragms. No diaphragm is actually infinitely rigid and no diaphragm capable of carrying a load is infinitely flexible.

1. A rigid diaphragm (Fig. 12.28a) is assumed to distribute horizontal forces to the vertical resisting elements in proportion to their relative rigidities. In other words, under symmetrical loading a rigid diaphragm will cause each vertical element to deflect an equal amount with the result that a vertical element with a high relative rigidity will resist a greater proportion of the lateral force than an element with a lower rigidity.
2. A flexible diaphragm (Fig. 12.28c) is analogous to a shear deflecting continuous beam or series of beams spanning between supports. The supports are considered non-yielding, as the relative stiffness of the vertical resisting elements compared to that of

the diaphragm is large. Thus a flexible diaphragm is considered to distribute the lateral forces to the vertical resisting elements on a tributary load basis. A flexible diaphragm is considered incapable of distributing torsional stresses resulting from eccentricity of masses.

3. Semi-rigid diaphragms are those which have significant deflection under load but which also have sufficient stiffness to distribute a portion of their load to vertical elements in proportion to the rigidities of the vertical resisting elements. The action is analogous to a continuous beam of appreciable stiffness on yielding supports (Fig. 12.28b). The support reactions are dependent on

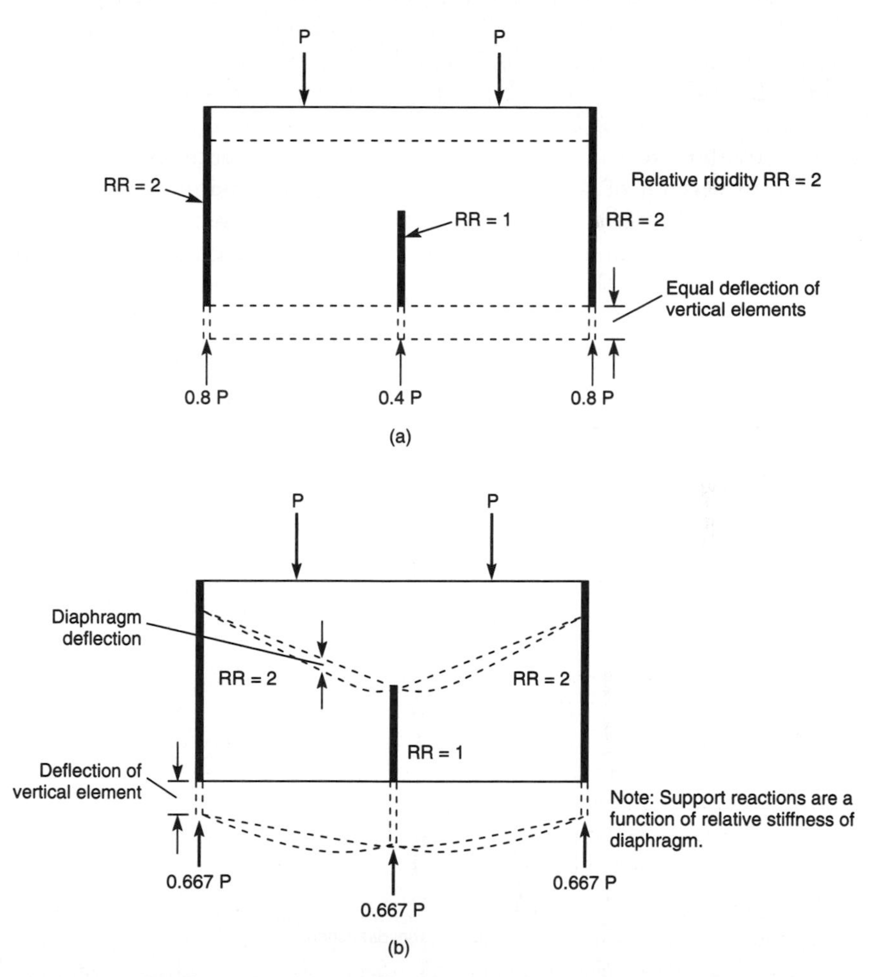


**Figure 12.28** Rigid, semi-rigid and flexible diaphragms: (a) rigid diaphragm; (b) semi-rigid diaphragm; (c) flexible diaphragm.

the relative stiffness of both diaphragm and vertical elements. A rigorous analysis is sometimes very time consuming and frequently unjustified. In such cases a design based on reasonable limits may be used; however, the calculations must reasonably bracket the likely range of reactions and deflections.

A torsional moment is generated whenever the center of gravity, *CG,* of the lateral forces fails to coincide with the center of rigidity, *CR,* of the vertical resisting elements, providing the diaphragm is sufficiently rigid to transfer torsion. The magnitude of the torsional moment that is required to be distributed to the vertical resisting elements by a diaphragm is determined by the larger of the following: (i) the sum of the moments created by the physical eccentricity of the translational forces at the level of the diaphragm from the center of rigidity of the resisting elements ($M = Fe$, where $e$ = distance between *CG* and *CR*; or (ii) the sum of the moments created by an "accidental" torsion of 5 percent. This is an arbitrary code requirement equivalent to the story shear acting with an eccentricity of not less than 5 percent of the maximum building dimension at that level.

The torsional shears are combined with the direct, translational shears. However, when the torsional shears are opposite in direction to the direct shears, the lateral forces are not decreased.

### 12.8.4 Metal deck diaphragms

Bare metal deck can be used as a diaphragm when the individual panels are properly welded to each other and to the supporting

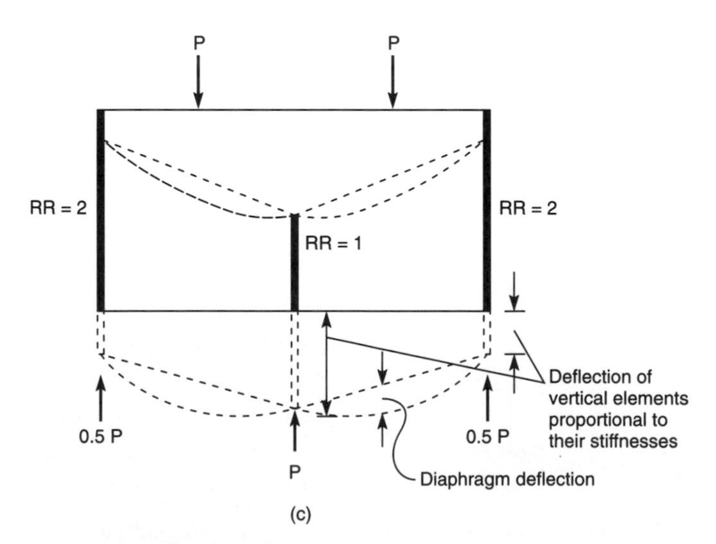


**Figure 12.28** *(Continued).*

framing. The strength of the diaphragm depends on the profile and gauge of the deck and the layout and size of the welds. Allowable values of metal deck diaphragms are usually obtained from approved data developed by the industry from test and analytical work.

Metal deck used in floors has a concrete fill. In some cases the strength of the diaphragm is considered to be that of the metal deck acting by itself, with the concrete functioning only as a topping that produces a level floor and covers conduit laid on the deck. In these cases the concrete can have a stiffening effect that makes the capacity of the system greater than that of the bare deck. In other cases, the metal deck is considered to act only as a form, and the diaphragm is treated as a reinforced concrete diaphragm.

Concrete-filled metal decks generally make excellent diaphragms

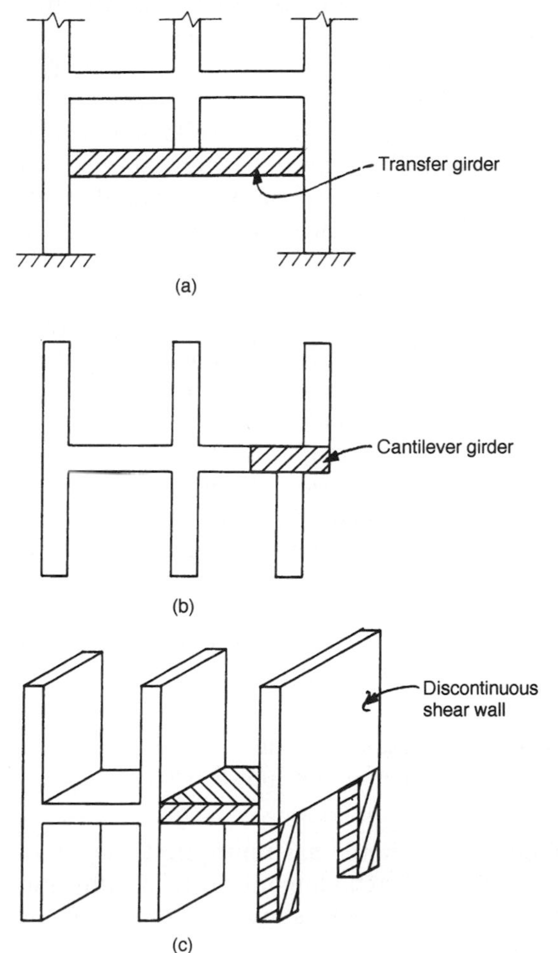


**Figure 12.29** Irregularities in force transfers: (a) transfer girder supporting discontinuous column; (b) cantilever girder supporting discontinuous column; (c) shear force transfer through diaphragm due to discontinuous shear wall.

and usually are not a problem as long as the basic requirements for chords, collectors, and reinforcing around openings are met. However it is necessary to check for conditions that can weaken the diaphragm: troughs, gutters, electrical raceways and recesses for architectural purposes which can have the effect of reducing the concrete diaphragm to the bare deck diaphragm.

### 12.8.5 Design criteria

The UBC 94 requires that the diaphragm at each level be designed to span horizontally between the lateral load resisting elements. In addition to the forces $F_{px}$, the diaphragm should transfer the shears from discontinuous shear walls and frames. A question often comes up while designing diaphragms for discontinuous walls; it is whether these loads should be increased by the magnification factor $\frac{3}{8}R_w$.

The 1994 UBC code language appears to be vague but the requirement is implicit in the 1996 SEAOC Blue Book. According to this, diaphragms supporting discontinuous shear walls can cause concentrations of inealstic behavior nullifying the assumption used in the selection of $R_w$, namely the inelastic behavior is reasonably well distributed throughout the entire system. Since the irregularities and discontinuities create concentration of inelastic behavior, these elements, meaning the transfer girder in Fig. 12.29a, cantilever girder in Fig. 12.29b, and the columns and diaphragm in Fig. 12.29c, should be designed for an $R_w$ value less than those used for the entire structural system. Multiplying the code forces by $\frac{3}{8}R_w$ to account for the discontinuities appears to be a reasonable approach. A better approach, as mentioned many times throughout this work, is to eliminate the irregularities and discontinuities altogether. It is not always easy but engineers should try to convince the architect and owners that seismic design is a shared responsibility. They should bring to their attention the poor performance of such systems in past earthquakes.

The 1994 UBC gives the following special provisions for diaphragm design of buildings in zones 3 and 4 with plan irregularities.

1. The connection of drag and collectors to the vertical elements should be designed without considering the one-third increase in allowable stresses.
2. The diaphragm wings should be designed assuming that they may move independently of each other. Each element of the diaphragm including the chords, drags and collectors should be designed for the more severe of the two conditions:

(a) projecting wings move in the same direction
(b) projecting wings move in opposite directions.

## 12.9 Earthquake Hazard Mitigation Technology

Earthquake mitigation technology includes seismic isolation, energy dissipation, ductility, damping systems, and other technologies, which strive to reasonably protect buildings and nonstructural components, building content, and functional capacity from earthquake damage. The potential advantage of this technology lies on its effectiveness in protecting nonstructural components and building contents thereby improving the capability of the facility to function immediately after the earthquake.

Two main concepts used in achieving the above concept are: (i) seismic isolation and; (ii) energy dissipation. The idea in seismic isolation is to shift the fundamental period of vibration of the building away from the predominant period of earthquake-induced ground motion to reduce the forces transmitted into the structure. Isolation provides a means of limiting the earthquake entering the structure by decoupling the building's base from its superstructure. The energy dissipation concept, on the other hand, allows seismic energy into the building. However, by incorporating damping mechanisms into the lateral load resisting system of structure, earthquake energy is dissipated as the structure sways back and forth due to seismic loading. A brief description of each of these concepts follows.

### 12.9.1 Seismic base isolation

#### 12.9.1.1 Introduction

The demands of seismic loads far exceed the traditional capacity of members normally associated with building structures. To build a structure to resist seismic loads in the same manner as it would resist gravity or wind loads, results in unacceptably expensive and architecturally unmanagebale buildings. The general public may be unaware, but the premise of seismic design is to reduce the likelihood of total collapse by dissipating the seismic energy by allowing the structural members to undergo permanent deformation.

Earthquakes, when they occur, cause high accelerations in stiff buildings and large interstory drifts in flexible structures. These two factors impose difficulties in the design of essential facilities such as hospitals, communication and emergency centers, etc. which must remain operational immediately after an earthquake when needed most. Although theoretically it may be possible to build-in enough strength in these structures to retain their functionality by resisting the seismic forces elastically by brute force, a relatively new technique termed "seismic isolation" is finding more and more application in the design of essential facility. With this technique, the building is detached or isolated from the ground in such a way that only a very small portion of seismic ground motions is transmitted up through the building (Fig. 12.30). In other words, although the ground underneath it may vibrate violently, the building itself would remain relatively stable. This results in a significant reduction in floor accelerations and interstory drifts, thereby providing protection to the building contents and components. Therefore, seismic isolation maximizes the possibility that the essential facility will remain operational after a major seismic event.

In simple terms the seismic isolation consists of mounting the building on seismic isolators which have large flexibility in the horizontal plane. At the same time dampers may be introduced to reduce the amplitude of motion caused by the earthquake.

Although each earthquake is unique, it can be stated in general that earthquake ground motions result in greater acceleration response in a structure at shorter periods than at longer periods. A seismic isolation system exploits this phenomenon by shifting the fundamental period of the building from the more force-vulnerable shorter periods to the less force-vulnerable longer periods.

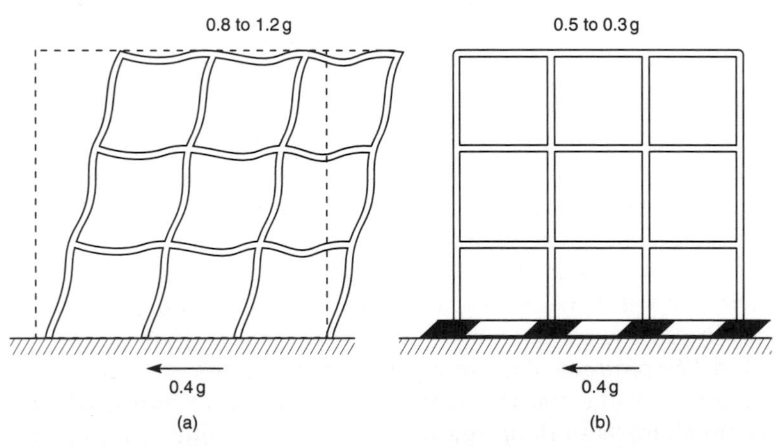


**Figure 12.30** Response of a fixed-base and base-isolated building: (a) fixed-base; (b) base-isolated. Base isolation typically reduces the maximum roof acceleration response of low rise buildings from maximum of 0.8 g to 1.2 g to maximum of 0.15 g to 0.3 g.

The principle of seismic isolation is to introduce flexibility in the basic structure in the horizontal plane, while at the same time adding damping elements to restrict the resulting motions. A practical base isolation system should consist of:

1. A flexible mounting to increase the period of vibration of the building sufficiently to reduce forces in the structure above.
2. A damper or energy dissipater to reduce the relative deflections between the building and the ground to a practical level.
3. A method of providing rigidity to control the behavior under minor earthquakes and wind loads.

Flexibility can be introduced at the base of the building by many possible devices such as elastomeric bearing, rollers, sliding plates, cable suspension, sleeved piles, rocking foundations, etc. Substantial reductions in acceleration with increase in period and consequent reduction in base shear are possible, the degree of reduction depending on the initial fixed-base period and shape of the response curve. However, the decrease in base shear comes at a price; the flexibility introduced at the base will give rise to large relative displacements across the flexible mount. Hence the necessity of providing additional damping at the level of isolators. This can be provided through hysteric energy dissipation of mechanical devices which use the plastic deformation of either lead or mild steel to achieve high damping. New seismic isolation devices such as high density rubber HDR, lead rubber bearing LRB, friction pendulum system, FPS, and viscous damping devices, VDD are some of the other devices for achieving high damping.

While a flexible mounting is required to isolate the building from seismic loads, its flexibility under frequently occurring wind and minor earth tremors is undesirable because the building motions may be perceived by the building occupants and create discomfort. Therefore the devices introduced at the base must be stiff enough at these loads, such that the building response is as if it is on a fixed base. Lead rubber bearings and other mechanical energy dissipaters provide the desired rigidity at low loads while providing flexibility at seismic loads.

The plane of isolation, meaning the level at which the flexible mountings are introduced, is usually a trade-off between the amount of additional construction, access to the building, utility connections, access to the isolators for inspection, etc. Generally one isolator per column is used. Occasionally when the column loads are high, of the order of several thousands of kips, more than one isolator may be used at the column location. For shear walls, one or more isolators are used at each end, and if the wall is long, isolators are placed along its length, the spacing depending upon the spanning ability of the wall between the isolators.

#### 12.9.1.2 Salient features

1. Access for inspection and replacement of bearings should be provided at bearing locations.
2. Stub walls or columns to function as backup systems should be provided to support the building in the event of isolator failure.
3. A diaphragm capable of delivering lateral loads uniformly to each bearing is preferable. If the shear distribution is unequal the bearings should be arranged such that larger bearings are under stiffer elements.
4. A moat to allow free-movement for the maximum predicted horizontal displacement must be provided around the building (Figs 12.31 and 12.32).
5. The isolator must be free to deform horizontally in shear and must be capable of transferring maximum seismic forces between the super substructure and the foundation.
6. The bearings should be tested to ensure that they have lateral stiffness properties that are both predictable and repeatable. The tests should show that over a wide range of shear strains, the effective horizontal stiffness and area of the hysteresis loop are in agreement with those used in the design.

When earthquakes occur, the elastomeric bearings used for base isolation are subjected to large horizontal displacements, as much as 15 in. or greater in a 10-story steel-framed building. They must therefore be designed to carry the vertical loads safely at these displacements.

#### 12.9.1.3 SEAOC Blue Book (1996) requirements

General requirements for seismic isolated structures can be found in Section 150 of the 1996, sixth edition Blue Book which is the publication "Recommended Lateral Force Requirements and Commentary" published by the Seismology Committee of Structural Engineers Association of California. Although simple procedures such as an equivalent static procedure and a response spectrum analysis are permissible under the regulations, the restrictions are so many that in all but very simple and regular buildings, it is mandatory to perform response spectrum and time-history analysis.

As an approximate first-step solution it is possible to use a linear analysis program to analyze a base-isolated building. This may be accomplished by adding an extra level to the analytical model to capture the effect of isolators. The overall shear stiffness of this additional soft story is made equivalent to that of the seismic isolation system. To represent the hysteretic damping an equivalent viscous damping may be assumed for the isolator. The properties of the equivalent columns are manipulated to represent the equivalent linear properties of the isolators. The input response spectrum is

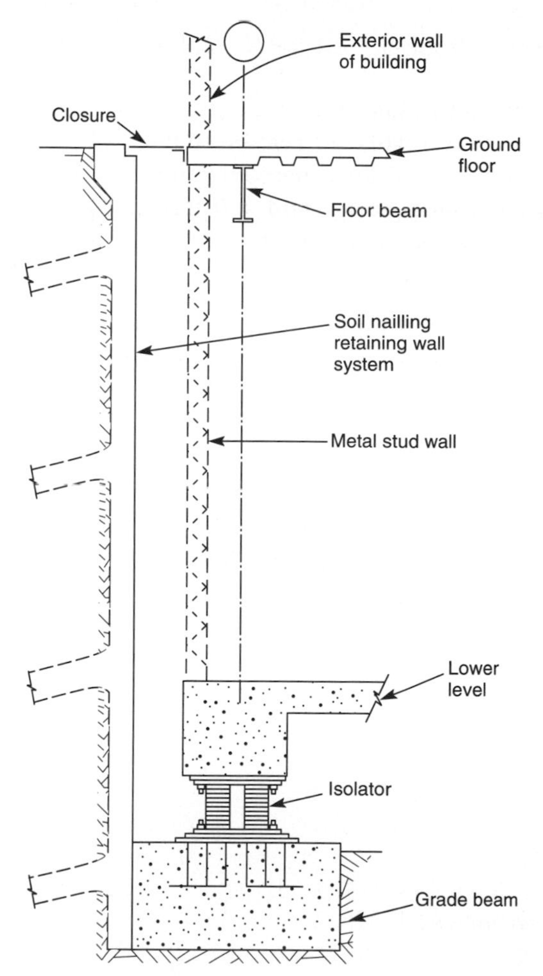


**Figure 12.31** Moat around base isolated building.

modified to include the effect of hysteric damping provided by the isolators.

The design displacement $D$ for the isolation system is given by the relation

$$D = \frac{10ZNS_1T_1}{B}$$

where $D$ = isolator displacement in inches
$Z$ = seismic Zone coefficient (0.4 for zone 4, 0.3 for zone 3, 0.2 for zone ZB)
$N$ = near field coefficient varying from a minimum of 1 to a maximum of 1.5 depending on: (i) site proximity to active fault; and (ii) the magnitude of maximum capable earthquake
$S_1$ = soil type coefficient (1.0 for $S_1$, 1.4 for $S_2$, 2.3 for type $S_3$) and 2.7 for $S_4$)
$T_1$ = isolated building period in seconds
$B$ = damping coefficient of the isolated system (0.8 for 2 percent or less damping, 1.0 for 5 percent damping, 1.2 for 10 percent, 1.5 for 20 percent, and 1.7 percent for 30 percent damping, 1.9 for 40 percent and 2.0 for 50 percent or greater damping)

The total design displacement $D_T$ should include the effects of torsion and shall be less than 1.1$D$. The reader is referred to the Blue Book

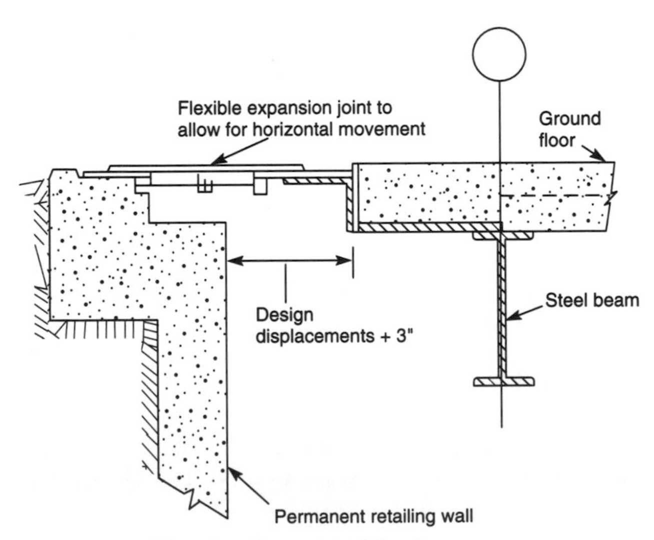


**Figure 12.32** Moat detail at ground level.

for additional requirement related to the calculation of total displacement.

The period of the isolated building is calculated based on the minimum effective stiffness of the isolation system as follows:

$$T_1 = 2\pi \sqrt{\frac{W_I}{K_{\min} g}}$$

where $W_I$ = total seismic dead load of the structure above the isolation interface

$g$ = gravity constant (386.4 in./sec$^2$)

$K_{\min}$ = minimum effective lateral stiffness of the system, determined by test, kips/in.

The reason for basing the period on the minimum effective stiffness of the isolation system is to ensure that the longest period which results in the maximum displacement is obtained from the formula.

The Blue Book specifies minimum lateral forces in terms of the design displacement. Two levels of forces are prescribed, one for the element at or below the plane of isolation, $V_b$, and the other for elements above the isolators, $V_s$.

For working stress design of elements at or below the isolation surface the design shear force, $V_b$, is given by

$$V_b = \frac{K_{\max} D}{1.5}$$

where $K_{\max}$ = maximum effective stiffness of the isolation system, as determined by tests, kips/in., and

$D$ = Design displacement, inches. For working-stress design of structure above the isolation plane, the design shear, $V_s$ is given by

$$V_s = \frac{K_{\max} D}{R_{wI}}$$

where $K_{\max}$ and $D$ are as defined earlier, and $R_{wI}$ = numerical coefficient related to the type of lateral force resisting system above the isolation interface. Representative values of $R_{wI}$ are, 1.8 for ordinary moment resisting steel frames and 3.0 for special moment resisting frames. The value of $V_s$ calculated above shall not be less than:

1. The 1994 UBC design base shear calculated for a fixed-base structure of the same weight $W_I$ and a period $T_1$, equal to the isolated period of the isolated structure.
2. The base shear corresponding to the design wind shear.

3. The force required to fully activate the isolation system corresponding to the friction level of the sliding system, or the ultimate capacity of the wind-resisting system.

The basic shear $V_s$ is distributed over the height of the structure above the isolation base according to the equation

$$F_x = \frac{V_s w_x h_x}{\sum_{i=1}^{n} w_i h_i}$$

The stresses in each structural element is calculated for the effect of lateral force $F_x$ applied at each level above the base.

The 1996 Blue Book requirements for base isolated structures permit, as for fixed-based structures, three methods of analysis: the equivalent static, the response spectrum analysis, and the time-history analysis. While for the fixed-based structures, the first two are the most common, they are not so for base isolation structures because the limitations imposed on these analyses preclude their use in all but very simple structures.

Several modeling approaches may be used for the analysis of seismic isolated buildings. One approach uses the familiar three-dimensional elastic modeling techniques with a linearized representation of isolation system by using an "effective stiffness" factor for modeling the isolators.

The second approach attempts to capture the non-linear behavior of the isolators. For this purpose, the elastic behavior of the structure above the base isolator, is condensed into an equivalent stiffness matrix. The reduced three-dimensional model is then coupled to the isolators through the common degrees-of-freedom. The isolators themselves incorporate the non-linear hysteretic characteristics. The combined model consisting of the reduced elastic model of the building above the base isolators, and the non-linear isolators are analyzed to determine the deformation and force in the isolators as well as those in the building above. An iterative procedure is used to delete the isolators with uplift forces.

It should be noted that the above procedure is greatly simplified with the availability of computer programs that include three-dimensional elastic modeling of the superstructure and three-dimensional nonlinear modeling of the isolators.

The time-history analysis consists of solving the dynamic equations of motion to obtain the time-wise response of the building. Only material nonlinearity is generally considered, with the elements taking excursions beyond the elastic range being modeled as non-linear elements. The analysis is performed with at least three matched pairs of horizontal time-history components. Design parameters of interest which include member forces, connection forces, interstory drift, isolator displacements, and overturning forces are calculated for each time-history analysis pair.

Figure 12.33 shows photographs of base-isolators used for a hospital building in Los Angeles, California.

### 12.9.2 Energy dissipation

Unlike base isolation, which involves a period shift of the structure from the predominant period of earthquake motion, passive energy dissipation allows earthquake energy into the building. Through appropriate configuration of the lateral resisting system, earthquake energy is directed towards energy dissipation devices located within the lateral resisting elements to intercept this energy. There are four general types of energy dissipation systems: (i) metallic; (ii) friction; (iii) viscoelastic; and (iv) viscous and semi-viscous systems. The common feature of all these devices is the transformation of earthquake energy into heat, which is dissipated into the structure.

#### 12.9.2.1

Metallic systems—In these systems, energy dissipation is achieved through yielding of metal parts. These can be fabricated from steel, lead or special shape memory alloys. Lead extruders force the material through an orifice within a cylinder for energy dissipation. The shape memory alloys have potential capabilities of dissipating energy without incurring damage, as in the case of steel dampers when they yield. These systems are referred to as amplitude dependent systems since the amount of energy dissipated, which is hysteretic in nature, is usually proportional to force and displacement. Their force-displacement characteristics are configured to be highly nonlinear to take full advantage of their energy dissipation potential. The devices are most often located within structural lateral-load-resisting elements such as braced frames.

#### 12.9.2.2

Friction systems—In this system the friction surfaces are clamped with prestressing bolts (Fig. 12.34). The characteristic feature of this system is that almost perfect rectangular hysteretic behavior is exhibited. These systems are referred to as displacement dependent systems since the amount of energy dissipated is proportional to displacement. Contact surfaces used are lead bronze against stainless steel, or teflon against stainless steel.

(a)

(b)

**Figure 12.33** Base-isolation of Martin Luther King Jr. Hospital (Structural engineers, John A. Martin & Asso., Los Angeles): (a) base-isolater; (b) base-isolater adjacent to basement wall; (c) structural steel framing above isolater. (Photos: courtesy of Dr. Farzad Naeim)

(c)

**Figure 12.33** *(Continued)*.

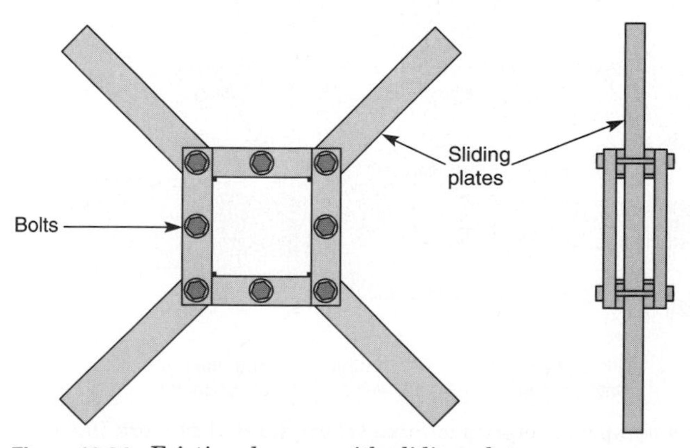


**Figure 12.34** Friction damper with sliding plates.

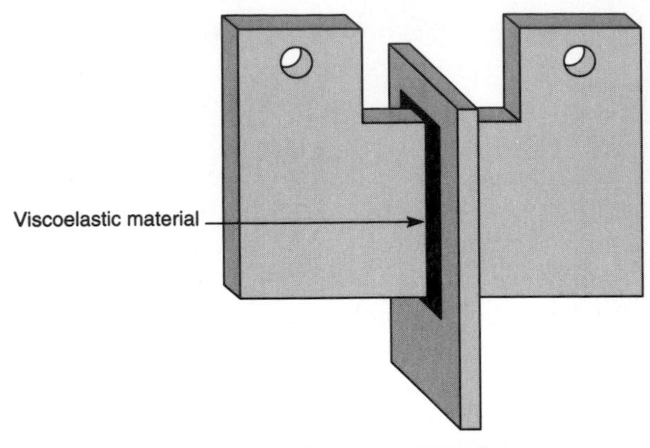


**Figure 12.35** Viscoelastic damper: stacked plates separated by inert polymer materials.

### 12.9.2.3

Viscoelastic systems—These systems use materials similar to elastomer. The materials are usually bonded to steel, and dissipate energy when sheared, similar to elastomeric bearings (Fig. 12.35). The materials also exhibit restoring force capabilities. Stiffness properties of some viscoelastic materials are temperature and frequency dependent. These variations should be taken into consideration in the design of these systems.

### 12.9.2.4

Viscous and semi-viscous fluid systems—These systems dissipate energy by forcing a fluid through an orifice, and in that respect are similar to the shock absorbers of an automobile (Fig. 12.36). The

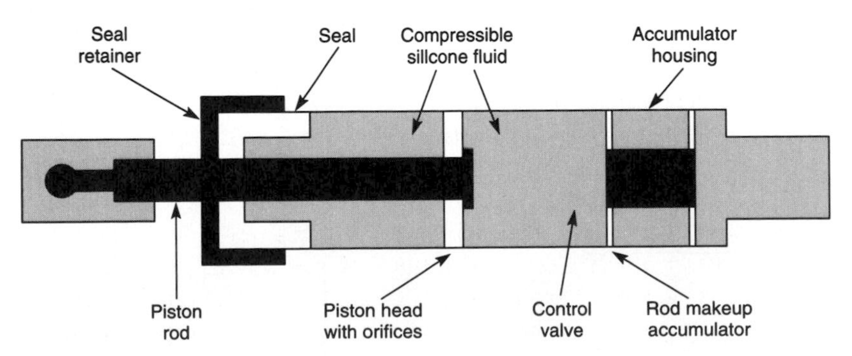


**Figure 12.36** Fluid viscous dampers are pistons in metal cylinders that work like shock absorbers.

fluids used are usually of high viscosity, such as silicone. The orifices are specially designed for optimum performance. The unique feature of these devices is that their damping characteristics and hence the amount of energy dissipated can be made proportional to the velocity.

Fluid viscous damping reduces stress and deflection because the force from the damping is completely out of phase with stresses due to seismic loadings. This is because the damping force varies with stroking velocity. Other types of damping such as yielding elements, friction devices, plastic hinges, and visco-elastic elastomers do not vary their output with velocity; hence they can increase column stress while reducing deflection. To understand the out of phase response of fluid viscous dampers, consider a building shaking laterally back and forth during a seismic event. The stress in a lateral load-resisting element such as a frame column is at a maximum when the building has deflected a maximum amount from its normal position. This is also the point at which the building reverses direction to move back in the opposite direction. If a fluid viscous damper is added to the building, damping force will drop to zero at this point of maximum deflection. This is because the damper stroking velocity goes to zero as the building reverses direction. As the building moves back in the opposite direction, maximum damper force occurs at maximum velocity, which occurs when the building goes through its normal, upright position. This is also the point where the stresses in the lateral load resisting elements are at a minimum. This out of phase response is the most desirable feature of fluid viscous damping.

## 12.10 Welded Moment Connections Subjected to Large Inelastic Demands

### 12.10.1 Introduction

Traditionally, structural engineers have calculated the seismic demand in moment frames by sizing the members for strength and drift using either code equivalent static or scaled dynamic forces, and then developing the strength of members. Since 1988, developing the strength has been accomplished by prescriptive means, taking for granted that the prescribed connections would be strong and behave in a nearly plastic manner. It was assumed that the girder would yield in bending, or the panel zone in shear, producing the plastic rotations necessary to dissipate the energy generated during a large earthquake. The perceived advantage of this connection in a seismic steel structure was that the base metal and weld metal ductility would be adequate to sustain inelastic structure behavior.

The seismic events known as the Northridge earthquake in Los Angeles, and the Kobe earthquake in Japan shook-up mother earth, the buildings resting upon her, and the very philosophy behind the seismic resistant design of rigid steel frame.

Consider the events in the City of Los Angeles located in seismic Zone 4, which is considered by experts to be the most seismically active zone in the country. The city is bounded on the east by the San Andreas Fault and the region is interlaced with other earthquake faults running through, adjacent to and under the city.

The 1994 Northridge earthquake caused considerable damage to Los Angeles buildings and structures, including welded steel moment frame buildings located in San Fernando Valley and West Los Angeles areas. Experts expect a massive earthquake on one of the faults under city within the next 30 years, and several earthquake similar in intensity to Northridge earthquake during the same period. It is believed that the damage to welded steel moment frame buildings could expose occupants of these buildings to a potential life-safety risk in future earthquakes.

The widespread damage to steel moment-resisting frames surprised most design professionals and academicians because this type of frame was considered by many to be the best earthquake-resisting system.

The damage to steel moment-resisting frames wreaked by the Northridge earthquake raised many serious questions for the earthquake engineering community, including

- Why did the welded beam-to-column connection fare so poorly?
- What plastic hinge rotations are likely in the UBC compliant steel moment frames?
- How can earthquake-damage connections be repaired?
- What are appropriate standards for the design of new steel moment-resisting frames?
- What are appropriate strategies for retrofitting vulnerable steel moment-resisting frames?
- How can the steel beam-to-column connection be improved?

In an attempt to answer these questions, a joint venture of design professionals and researchers was formed. Known by the acronym SAC, the joint venture partners are the Structural Engineers

Association of California, the applied Technology Council and the California Universities for research in Earthquake Engineering. In its interm guidelines dated August 1995, the SAC Venture has banished the use of welded steel moment connections for use in the construction of new buildings intended to resist ground shaking through inelastic behavior.

This is because contrary to the intended behavior, in many connections brittle fractures initiated at very low level of plastic demand, and in some instances, while the structure remained elastic. Typically fractures initiated near, or at, the complete joint penetration weld between the column flange and the beam bottom flange. Once initiated, these fractures progressed along a member of different paths, depending upon the individual joint conditions. Initially, the initiation of fractures was attributed to a number of factors including; notch effects created by the backing bar commonly left in place after welding of the joint, substandard welding with excessive porosity and slag inclusions as well as incomplete fusion; and potentially, pre-earthquake fractures resulting from initial shrinkage of the highly restrained weld during cool-down. However, it is now certain that these were not the only causes of fractures. The other contributing factors include; lamellar weakness in the through-thickness direction of structural steel shapes, non-uniform distribution of stress across the girder flange at the connection to the column, and large secondary stresses induced at the joint by kinking of column flanges resulting from panel zone shear deformation.

During the Northridge earthquake, once fractures initiated in beam-column joints, they progressed along different paths. In some instances, the fractures initiated but did not grow. In other instances, they progressed completely through the thickness of the weld. In some cases, the weld remained intact but a portion of the column flange remained bonded to the beam flange pulling free from the remainder of the column. This fracture pattern sometimes termed a "nugget" or "divot" failure was perhaps the most threatening in that a number of fractures progressed completely through the column flange. The fractures were along a near horizontal plane aligning with the beam lower flange, and in some instances fracturing the entire cross section of the column. Despite the obvious local loss of strength, many damaged buildings in the Northridge Earthquake did not display characteristic signs of structural damage, such as permanent drifts, or excessive damage to architectural elements. None of the buildings collapsed, although many did, in Japan, after the Kobe earthquake.

In reaction to the widespread damage discovered following the

Northridge earthquake, supported by results of tests on beam-column assemblages performed at the university of Austin, Texas, it was concluded that the prequalified connection specified in the UBC-94, was fundamentally flawed and was banished for new construction. Until then, a girder to column connection was deemed satisfactory for resisting seismic loads, if the beam flanges were butt welded to column by full penetration welds, and the beam web attached to column with a shear tab using slip-critical high strength bolts, welds or both. As of September 1994, it has become the responsibility of the engineers to prove either by cyclic test or by calculation, that the proposed connection can sustain inelastic rotation and develop the strength of the girder in flexure.

The damage to framing elements observed after the Northridge earthquake have been categorized by the SAC Joint Venture into five categories: Weld (W); Girder (G); Columns (C); Panel Zone (P) or Shear tab (S) category.

Based on limited testing, as well as observed behavior of actual buildings in the Northridge earthquake the SAC joint venture has issued recommendations for the repair of damaged elements to restore a building to its pre-earthquake condition. The repair techniques include:

1. Replacement of portion of column and beam section.
2. Replacement of connection elements.
3. Replacement of connection weld.
4. Repair to any of the aforementioned components.

Alteration of connection to improve its earthquake performance and that of the structure as a whole typically involves substantial changes to the connections geometry and capacity. Connection modification is intended to permit inelastic frame behavior. Therefore, the connection is proportioned to accommodate plastic deformation through the development of plastic hinges at predetermined locations in the girder. The criteria for modifications is to provide sufficient strength in the connection through the use of cover plates, haunches, side plates, etc., to force the plastic hinge to form away from the column face. As an alternate to strengthening the connection, the desired condition of formation of hinge away from the column may be attained through local weakening of the beam section. In either case, the plastic rotation of modified connections should be capable of developing a minimum capacity of the order of 0.025 to

0.030 radian, for a moment frame system and somewhat less when braced frames or other elements are introduced into the moment frames to control its lateral deformation.

There appears to be several potential details that can be used to modify girder-column joints in existing welded moment frames structures.

The basic idea in modifying the connection is to move the formation on hinges away from the column face to reduce through-thickness stresses in the column flanges. Connections relying solely on through-thickness strength are unreliable as demonstrated by the Northridge earthquake. Therefore, a high amount of structural redundancy is recommended for frames which rely solely on through-thickness strength of column flange.

An overview of the report released by SAC committee (FEMA 267) is given in the following figures reproduced from the report. The reader should, however, ultimately rely on the complete wording of the report, which is not repeated here.

### 12.10.2 Overview of SAC's guidelines

Figure 12.37 illustrates elements of welded moment frame that experienced structural damage in the Northridge Earthquake. These

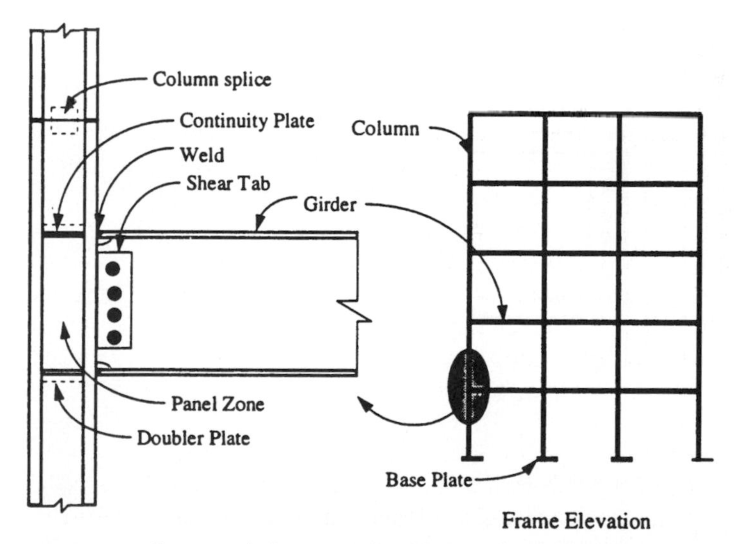


**Figure 12.37** Structural elements of welded steel moment frame.

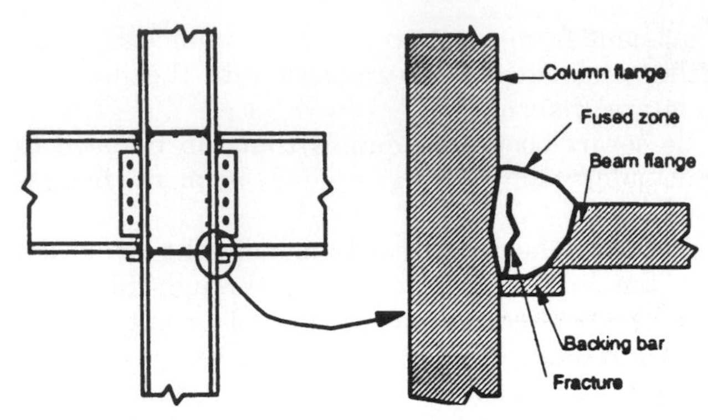


**Figure 12.38** Typical zone of fracture initiation in beam-column connection.

elements included girders, columns, column panel zones including girder flange continuity plates and column web doubler plates, welds of beam to column flanges and shear tabs which connect girder webs to column flanges.

Observed damages in welds, column, girder, panel zones and shear tabs are shown in Figs 12.38–12.43. Recommended post-earthquake repairs are shown in Figs 12.44–12.52.

It should be noted that by doing the repair the essential behavior of the connection is not altered as a result of the repair.

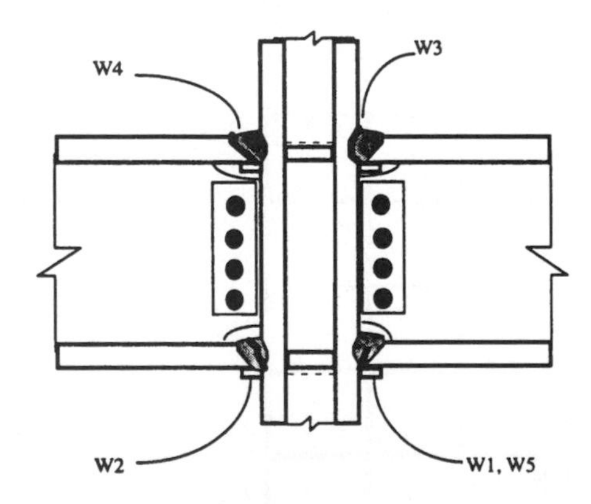


W1 : Weld root indications
W1a: Incipient indications–depth <3/16" or $T_f/4$; width $<b_f/4$
W1b: Root indications larger than that for W1a
W2 : Crack through weld metal thickness
W3 : Fracture at column interface
W5 : UT detectable indication–non-rejectable

**Figure 12.39** Types of weld damage.

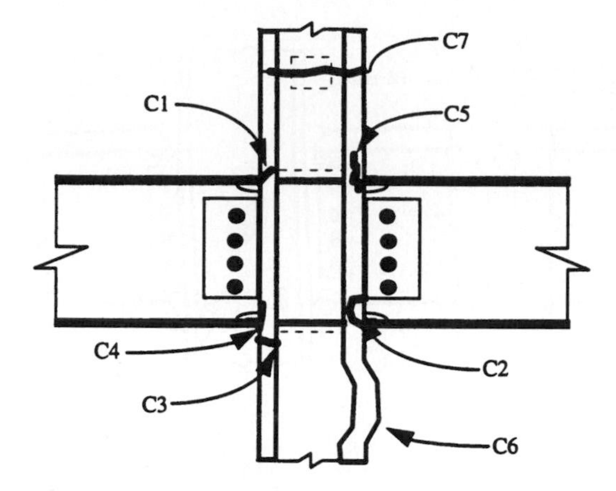


C1 : Incipient flange crack
C2 : Flange tear out or divot
C3 : Full or partial flange crack outside HAZ
C4 : Full or partial flange crack in HAZ
C5 : Lamellar flange tearing
C6 : Buckled flange
C7 : Column Splice Failure

**Figure 12.40** Types of column damage.

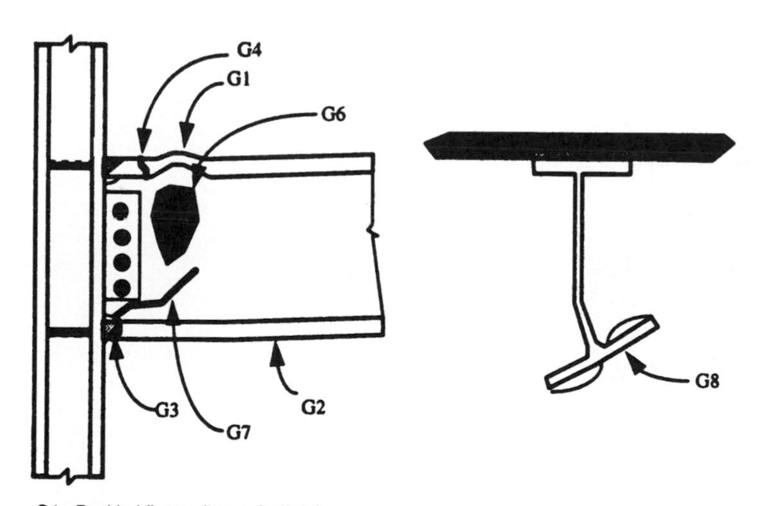


G1 : Buckled flange (top or bottom)
G2 : Yielded flange (top or bottom)
G3 : Flange fracture in HAZ (top or bottom)
G4 : Flange fracture outside HAZ (top or bottom)
G5 : Flange fracture top and bottom
G6 : Yielded or buckling of web
G7 : Fracture of web
G8 : Lateral torsional buckling of section

**Figure 12.41** Types of girder damage.

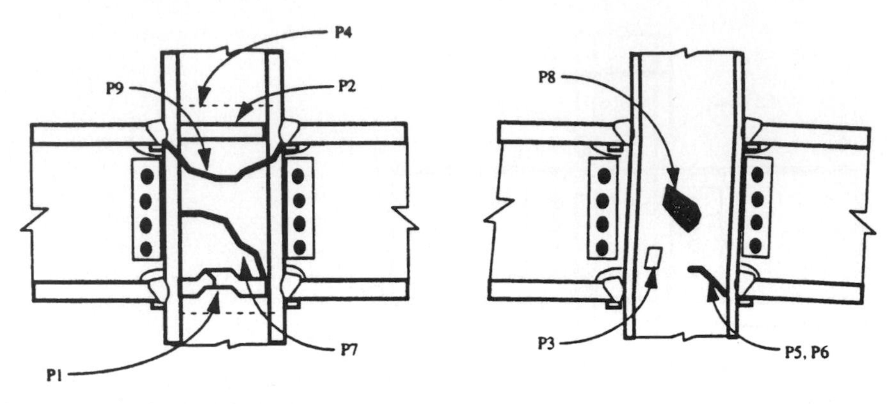


P1 : fracture, buckle or yield of continuity plate
P2 : Fracture in continuity plate welds
P3 : Yielding or ductile deformation of web
P4 : Fracture of doubler plate welds
P5 : Partial depth fracture to doubler plate
P6 : Partial depth fracture in web
P7 : Full or near full depth fracture in web or doubler
P8 : Web buckling
P9 : Severed Column

**Figure 12.42** Types of panel zone damage.

In contrast, connection modification refers to alternation made to the connection to improve the performance of the structure as a whole. The basic intent of connection modification is to enhance the expected performance of the building in future earthquakes by

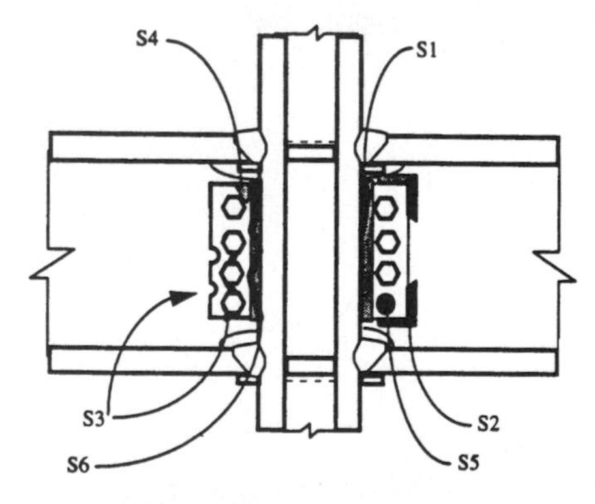


S1 : Partial crack at weld to column
S1a : girder flanges sound
S1b : girder flange cracked
S2 : Fracture of supplemental weld
S2a : girder flanges sound
S2b : girder flange cracked
S3 : Fracture through tab at bolts or severe distortion
S4 : Yielding or buckling of tab
S5 : Loose, damaged or missing bolts
S6 : Full length fracture of weld to column

**Figure 12.43** Types of shear tab damage.

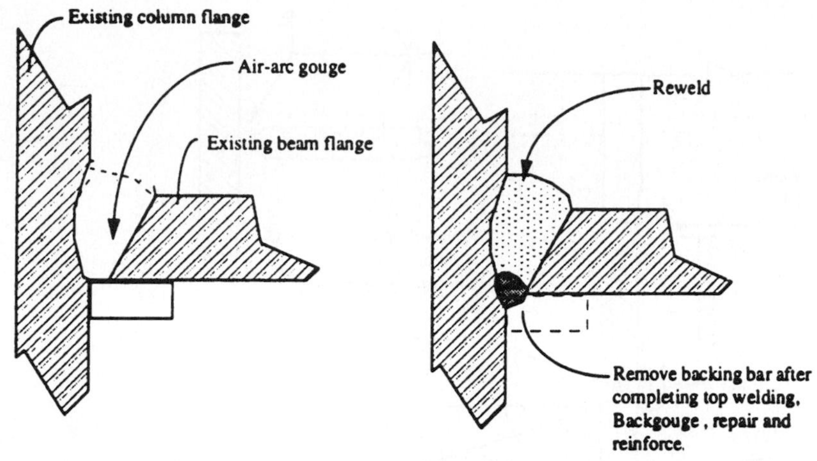


**Figure 12.44** Backgouge and re-weld repair.

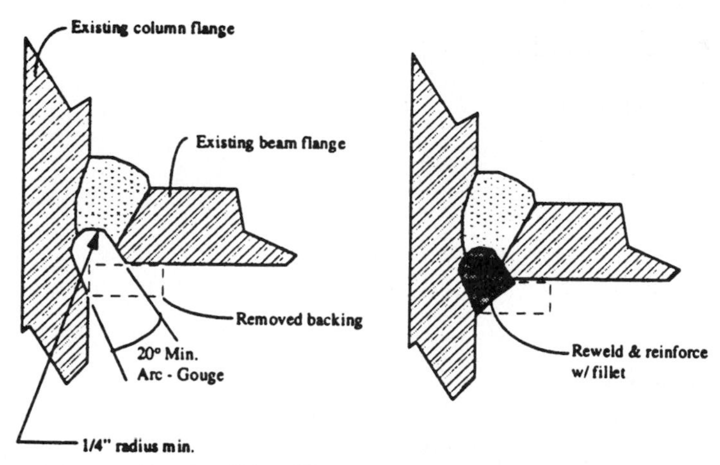


**Figure 12.45** Repair of Type $W_1$ root weld defect.

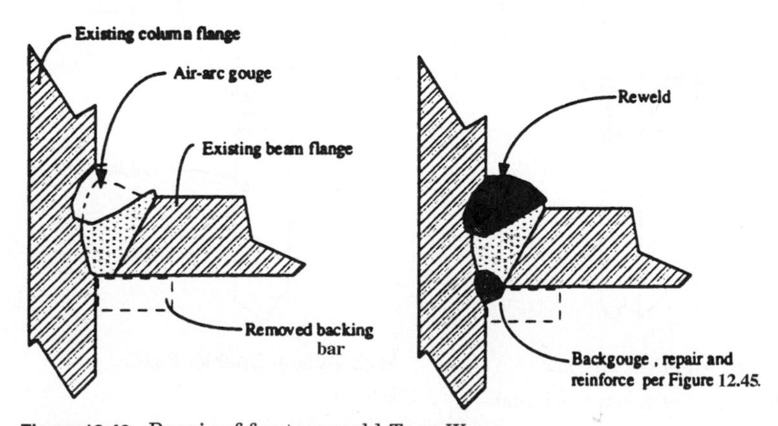


**Figure 12.46** Repair of fracture weld Type $W_1$.

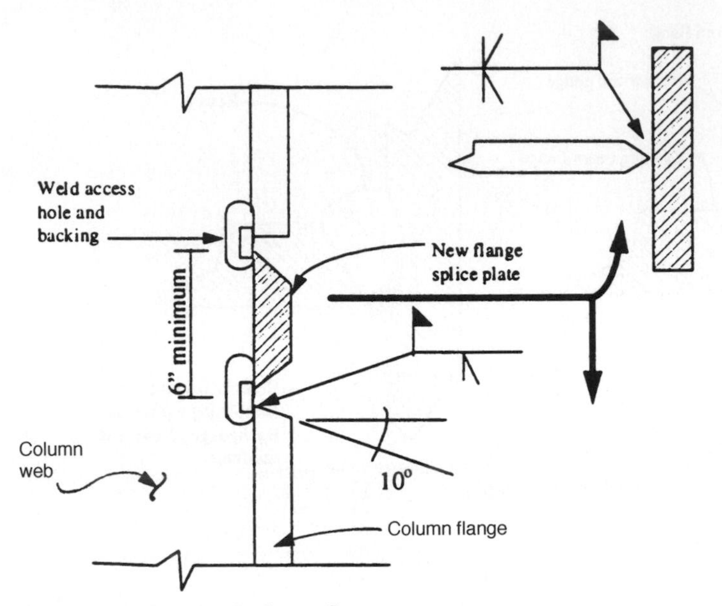


**Figure 12.47** Repair of column flange.

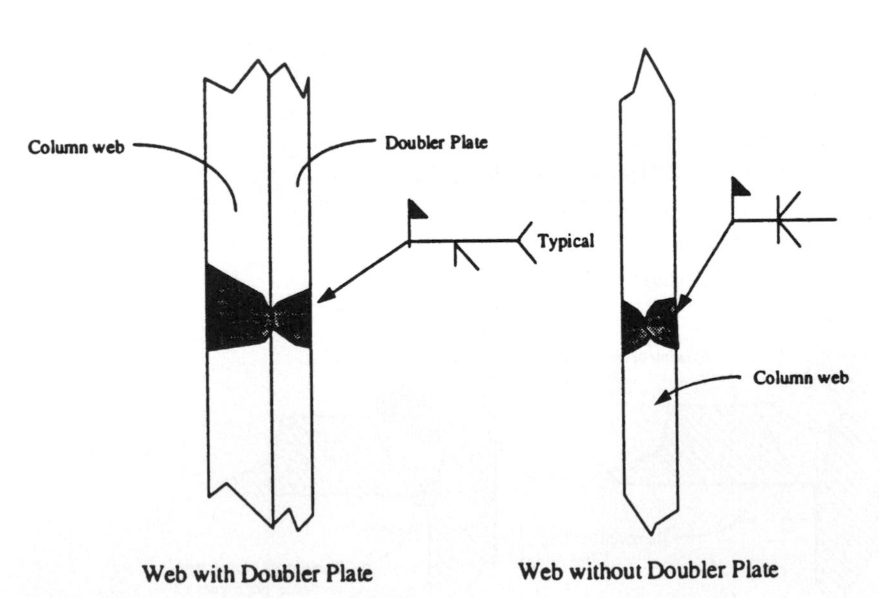


**Figure 12.48** Repair of column web plate.

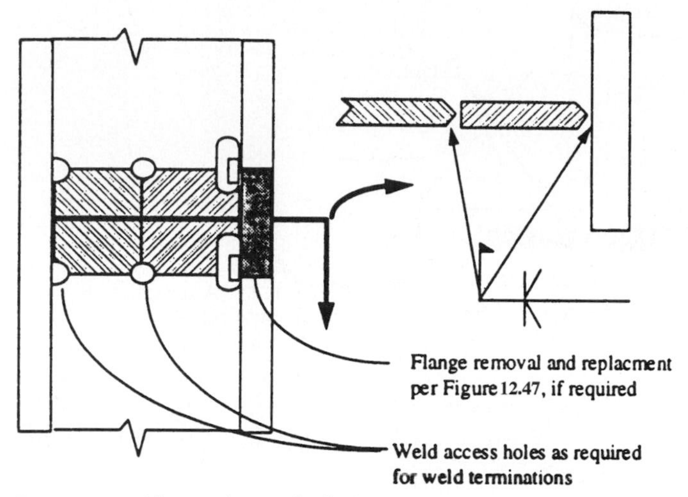


**Figure 12.49** Alternative method of web repair of columns without doubler plates.

making the connections sufficiently strong such that inelastic hinge forms in the girder, and not at the column-girder connection (Fig. 12.53). Recommended methods of reinforcing the connections to achieve this goal are shown in Figs 12.52–12.58 and 12.20–12.26.

As in connection modifications, the intent in new construction is to force the plastic hinge away from the face of the column to a

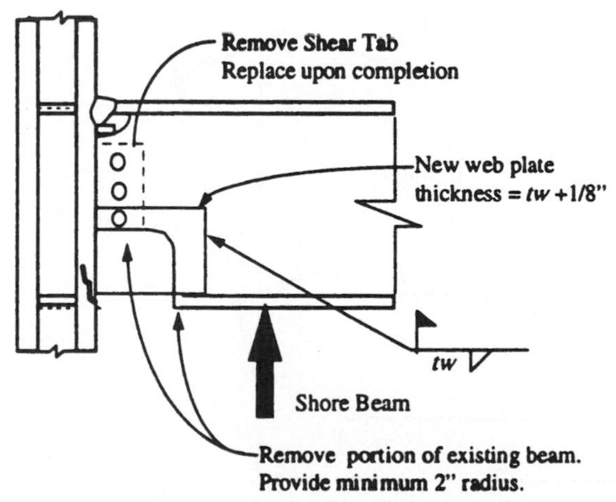


**Figure 12.50** Temporary removal of beam section for repair access.

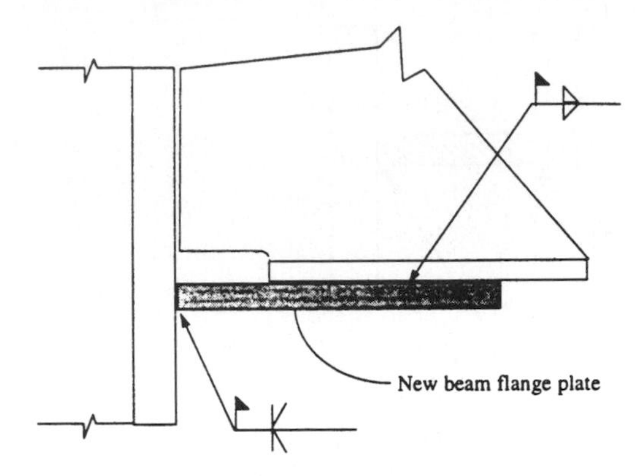


**Figure 12.51** Alternative beam-flange plate replacement.

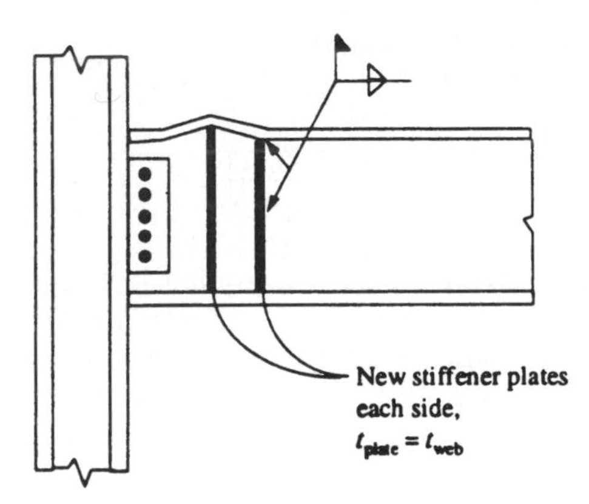


**Figure 12.52** New stiffners attachment at buckled girder flange.

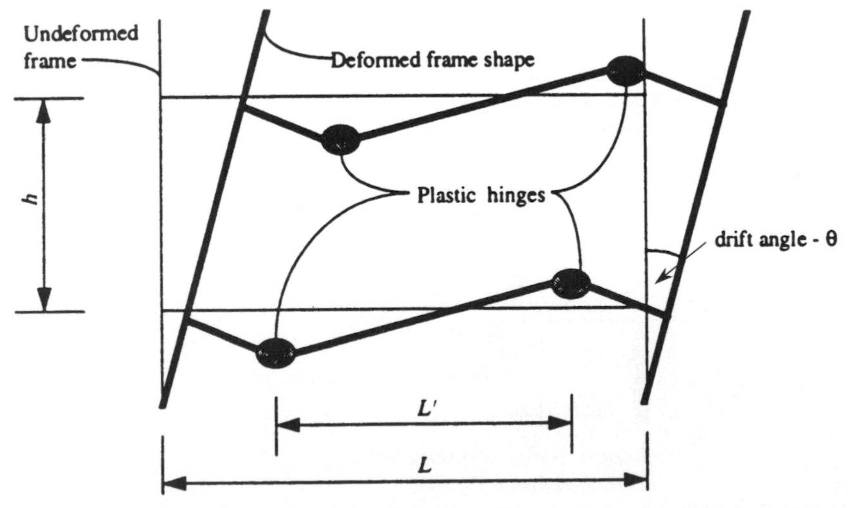


**Figure 12.53** Desired plastic frame behavior: hinges form in girders, not in columns.

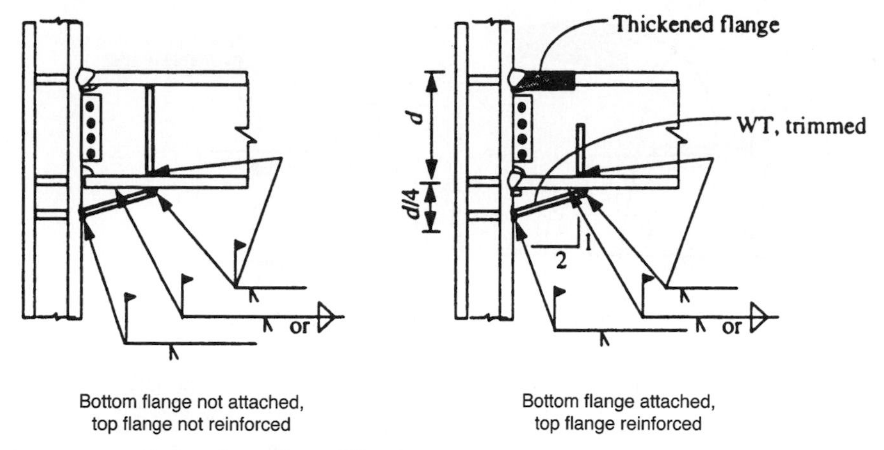


**Figure 12.54** Bottom haunch modification.

predetermined location within the girder span. Although connection design by calculations alone is acceptable in certain restricted conditions, the SAC guidelines strongly recommends cyclic seismic testing of connections that replicate as closely as practical the anticipated conditions in field.

Several configurations of moment frame connections, which promote plastic hinging in the beams rather than in the columns, are offered for consideration in new construction. The reader is referred to SAC's interim report for further details.

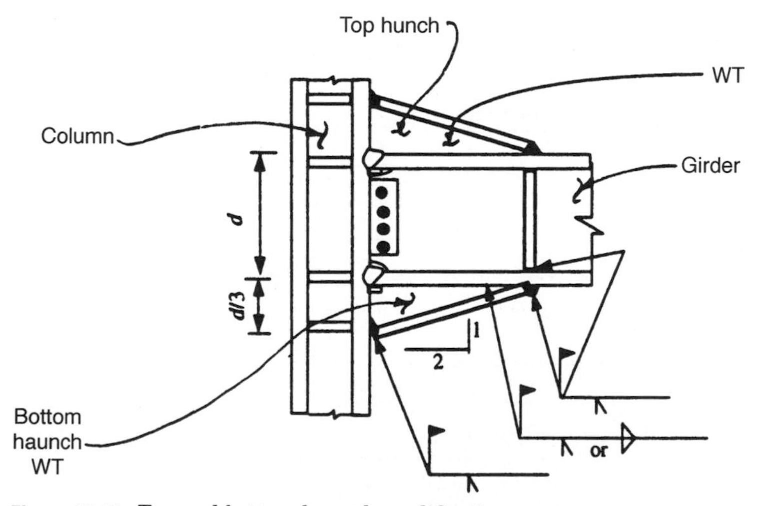


**Figure 12.55** Top and bottom haunch modification.

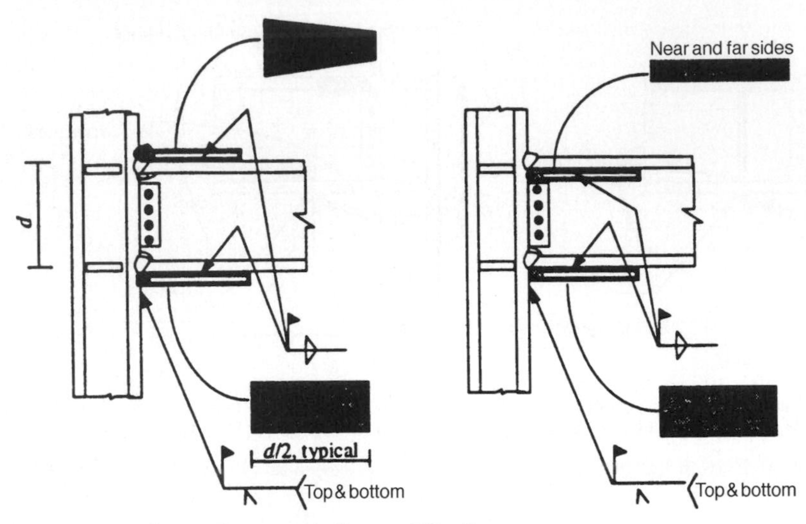


**Figure 12.56** Cover plate connection modifications.

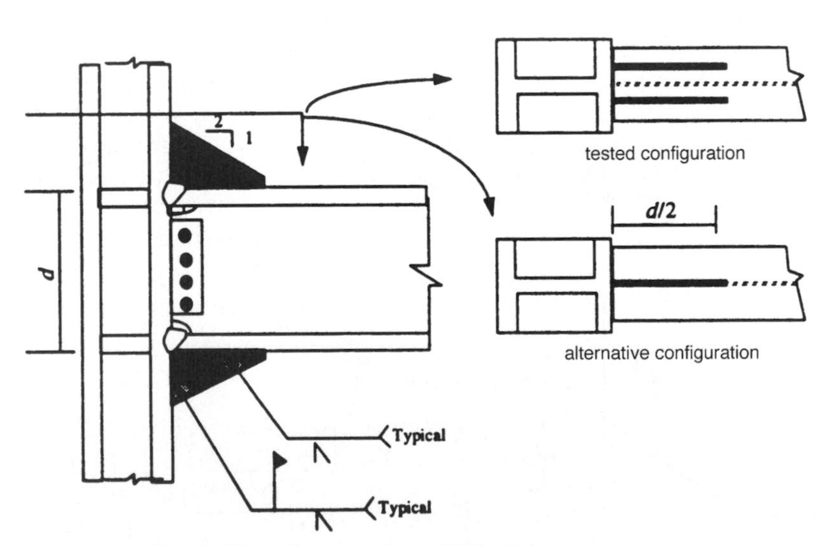


**Figure 12.57** Upstanding rib connection modifications.

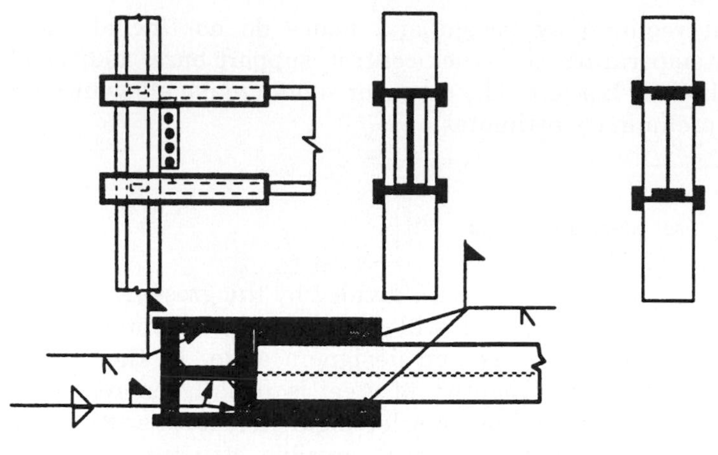


**Figure 12.58** Side plate connection modifications.

## 12.11 Unit Structural Quantities

### 12.11.1 Introduction

Quantities are physical items of construction to which unit costs are applied to arrive at a total construction cost. These are relatively easy to obtain once complete working drawings and specifications have been prepared. Prior to this point, however, the estimator or engineer must use "conceptual estimating" to determine approximate quantities. Conceptual estimates require considerable judgement in addition to unit quantities to correct so-called average unit quantities to reflect complexity of construction operations, expected time required for construction etc.

Typically, units of structural quantities are dimensional based on linear feet, square feet, or cubic feet. These result in unit quantities such as pounds per linear feet, plf, pounds per square foot, psf, etc.

### 12.11.2 Concrete floor framing quantities

Reinforcement and concrete unit quantities for various concrete floor-framing systems are shown in Figs 12.59–12.64. Live loads shown in the figures are working loads, and range from a typical office live load of 50 psf to a maximum of 200 psf appropriate for heavily loaded warehouse floors. The rebar quantities shown are for reinforcement required by design and hence do not include bars required for temperature and crack control, support bars, additional lengths required for laps etc. The engineer should make allowance for these in the preliminary estimates.

### 12.11.3 Structural steel quantities

The total quantity of structural steel divided by the gross area of the building has always been, and will always be an item of great interest to building developers and designers alike. In the United States practice, this unit quantity of steel is usually expressed in terms of pounds per square foot. In selecting a structural system, the usual practice is to look into several possible structural options. Assuming all preliminary schemes meet the basic architectural requirements, the deciding factor most often is the unit quantity of the system. Although unit quantity of steel may not address other

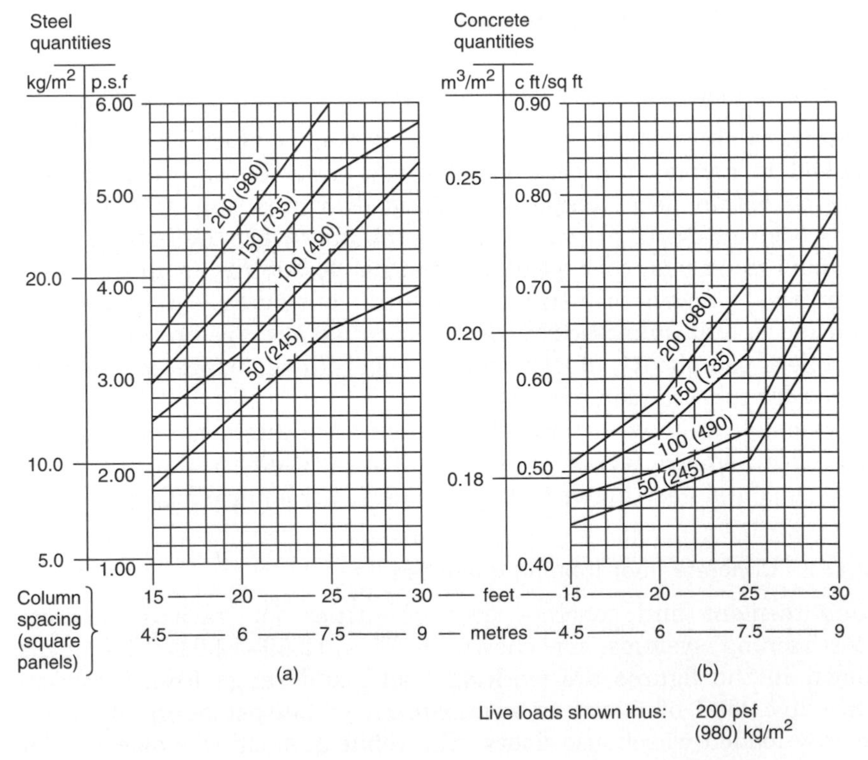


**Figure 12.59** One-way solid slab, unit quantities: (a) reinforcement; (b) concrete.

factors relevant to the construction cost (such as the unit price of steel for the particular scheme, whether or not the scheme has compositing of vertical systems, the cost of fireproofing, and the cost of borrowing money for the period of construction), most often, at least in the preliminary scheme selection stage, it is the unit quantity that pushes a particular scheme to the forefront.

Prior to the advent of tubular and mega frame systems, most of the buildings were designed using braced or moment frames as lateral load-resisting systems. Consequently, their poundage is very heavy as compared to present-day schemes.

Leaving aside the pre-1960 buildings, it is of interest for conceptual estimating purposes, to assemble the unit structural steel quantities for buildings that have been built within the last three decades. Figure 12.65 shows the general trend in the increase of unit quantity of steel as the building height is increased.

Four distinct regions *A*, *B*, *C* and *D* are shown in the figure. Region *A* is for buildings up to 30 stories, *B* for buildings between 30

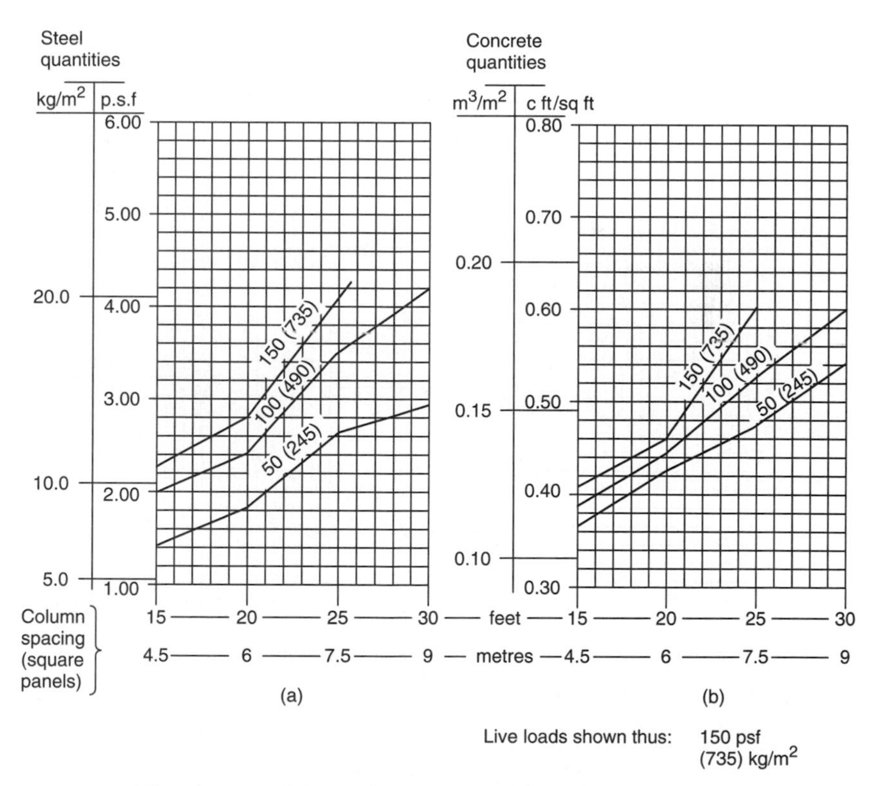


**Figure 12.60** One-way pan joist, unit quantities: (a) reinforcement; (b) concrete.

and 50 stories, *C* is for buildings 50 to 70 stories while *D* is for high-rises in excess of 70 stories. The use of this figure is best explained with respect to the following examples.

**Example 1:** A proposed ten-story building in wind-controlled, low-seismic risk area. The engineer is asked to come up with unit quantity of structural steel for purposes of a conceptual cost estimate.

For a ten-story building it is seen from Fig. 12.65 that the lower and upper bounds for the unit quantity are 10.5 and 15 psf respectively, which is a rather wide range. The engineer now has to make some judgment calls, depending on what is known about the building at this stage of the game. Some relevant questions are: (i) what are the typical spans for the floor framing? (ii) Are there any undue restrictions for beam and girder depths? (iii) What is the likely lateral framing system? If none of these questions can be answered with any great certainty the engineer must make some judgment calls. For the current example, the designer decides that the subject building is an

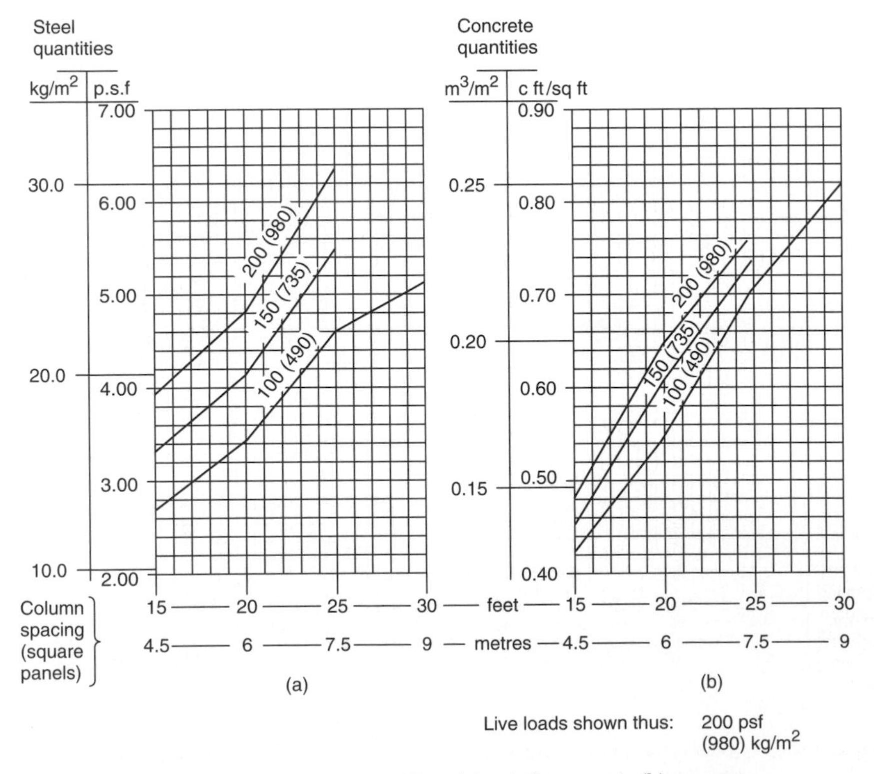


**Figure 12.61** Two-way slab unit quantities: (a) reinforcement; (b) concrete.

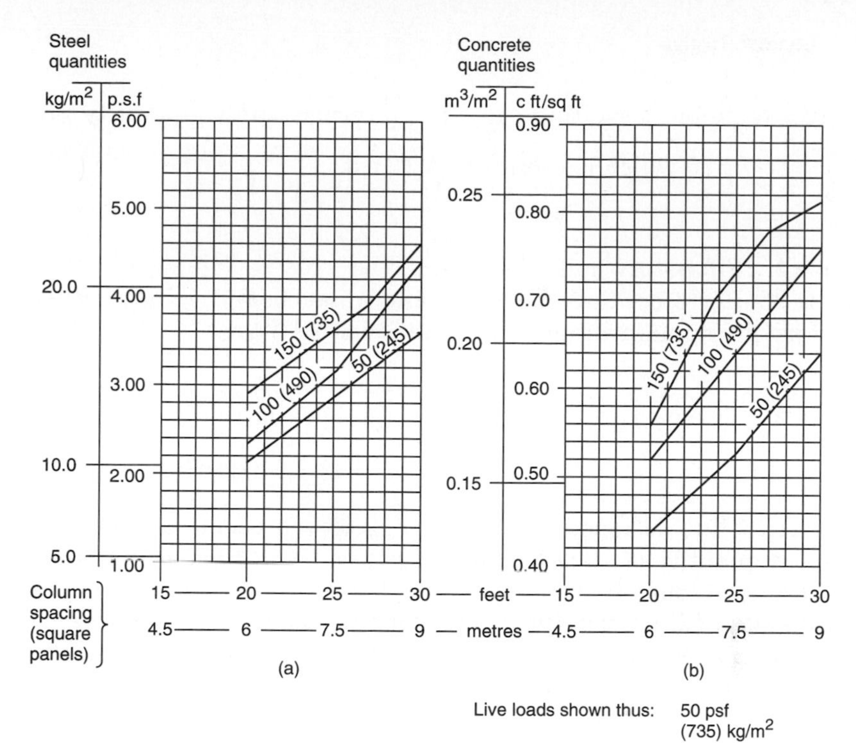


**Figure 12.62** Waffle slab, unit quantities; (a) reinforcement; (b) concrete.

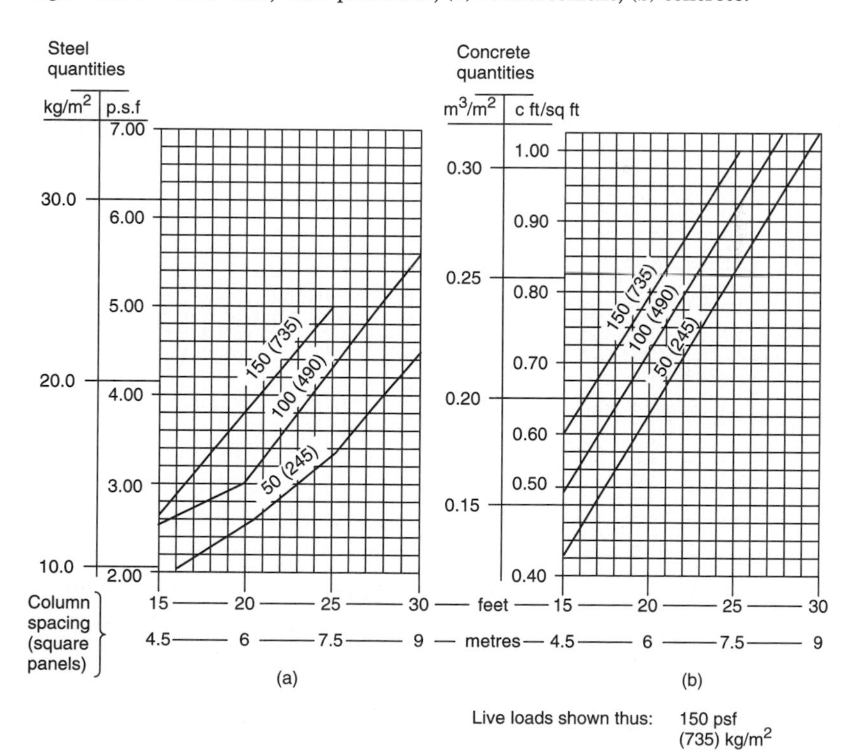


**Figure 12.63** Flat plate, unit quantities: (a) reinforcement; (b) concrete.

average building resulting in an average unit quantity of $\frac{10.5 + 15}{2} = 12.75$ which is rounded off 13 psf.

**Example 2:** Fifty-story building.
Using a similar procedure the lower and upper bound values are 21 and 32 psf giving an average unit quantity of 26.5 psf.

**Example 3:** Twenty-two story building: Design controlled by seismic loads.

The straight line designed as $S$ in Fig. 12.65 represents an approximate unit quantity of steel for buildings located in high-risk seismic zones (zones 3 and 4). Observe that this line stops at 25 stories implying that design of taller buildings, in general, is controlled by wind loads. The cut-off level of 25 stories with building periods in the

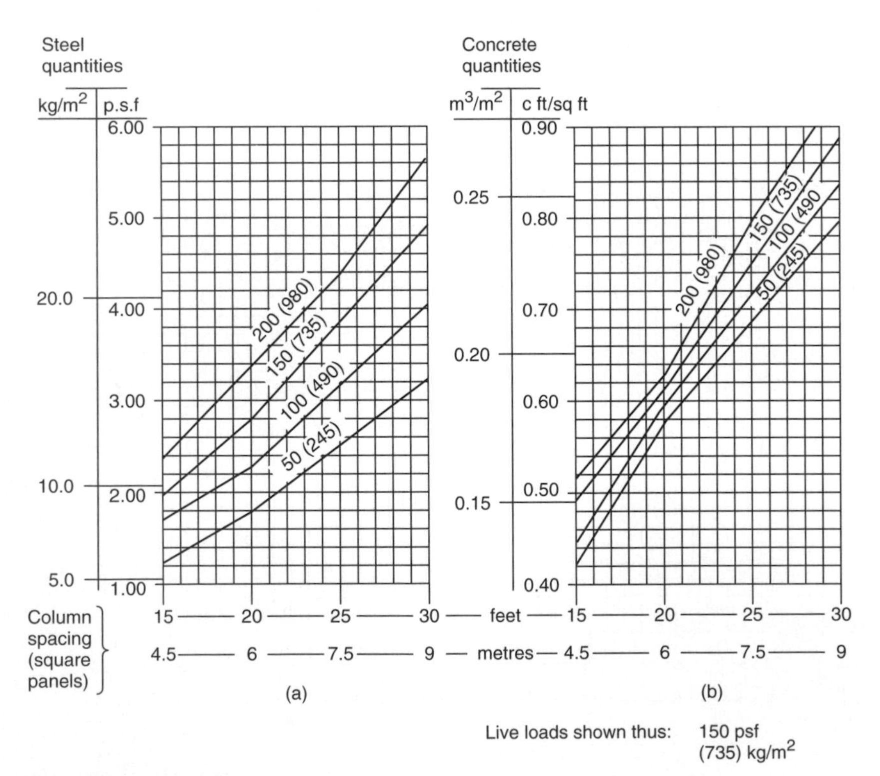


**Figure 12.64** Flat slab, unit quantities: (a) reinforcement; (b) concrete.

range of 2–3 sec is considered the threshold where the wind design requirements exceed those of seismic loads. For the example building, the unit quantity of steel is given by the straight line $S$ as 22 psf.

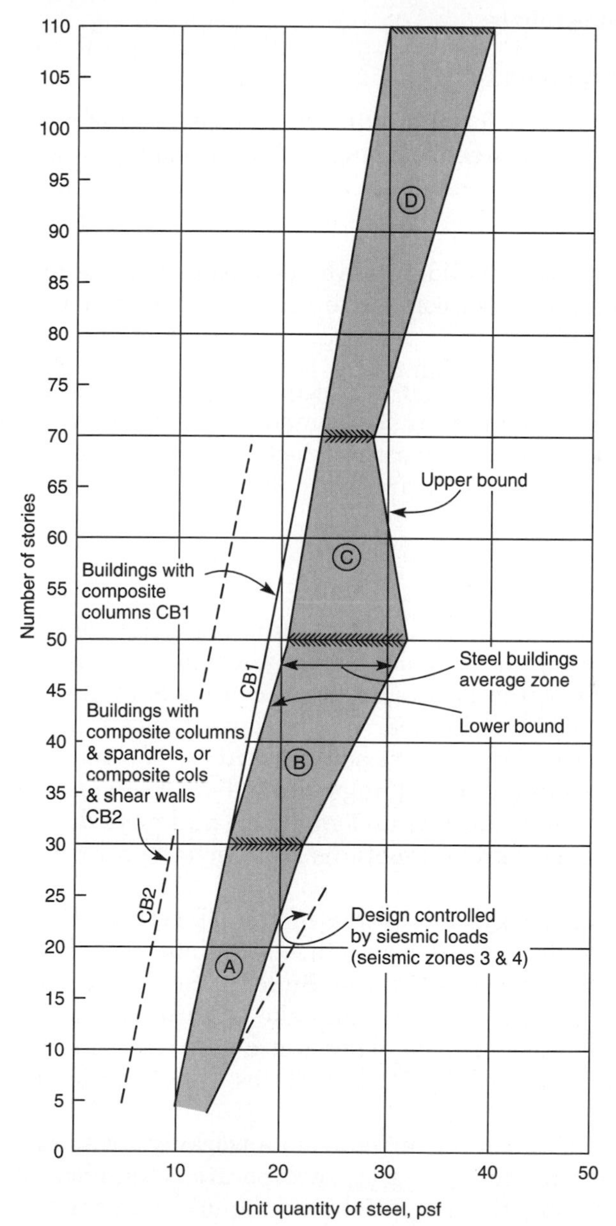


**Figure 12.65** Structural steel unit quantities.

**Example 4:** Now consider the same building in a seismically low-risk area such as Houston, Texas. We go to the region designated as an average zone on the graph. From the graph the unit weight is seen to vary from a low of 13.5 psf to a maximum of 19.5 psf. As in the previous case we use an average unit quantity of 16.5 psf for the preliminary conceptual estimate.

**Example 5:** 70-story composite building.

The line designated as *CB* 1 is for approximate unit quantity of steel for buildings using composite steel columns as lateral load resisting frame columns. As a preliminary estimate a unit quantity of 23 psf is appropriate for the building.

**Example 6:** 50-story composite building with composite columns and composite girders (buildings with composite columns and composite shear walls similar).

From the graph noted as *CB* 2, we get a unit quantity of 13.5 psf for this building. Note that design rules for composite construction in seismic zones 3 and 4 are not well established in North America. Therefore, graphs *CB* 1 and *CB* 2 are valid for buildings in low seismic risk zones only.

## 12.12 1997 UBC Update

The seismic provisions in the 1997 Uniform Building Code contain significant changes affecting the seismic resistant design of buildings. For buildings in California, the most significant change is related to the amplification of forces in areas near major active faults. This is done through the introduction of near-source factors, $N_a$ and $N_v$. These factors affect the design of buildings throughout the spectral range, but particularly affect mid-rise and high-rise buildings due to their responses to long period ground motions for which the near field effect is most pronounced.

Developed by the Seismology Committee of the Structural Engineers Association of Southern California (SEAOC), these provisions have paralleled a similar approach by the Building Seismic Safety Council (BSSC) for the 1997 National Earthquake Hazards Reduction Program (NEHRP) provisions. The provisions are expected to serve as a source document for the entire United States beginning in the year 2000.

The code provisions have been converted from a working stress to a strength design basis. The changes also incorporate a number of important lessons from Northridge and Kobe earthquakes. In general it is intended to provide parity with previous requirements, except for longer period buildings in near-field locations and for systems with poor redundancy.

### Overview of code changes

1. The adoption of ASCE-7 load factors for strength-based load combinations. In addition, working stress load combinations are maintained as an alternative.
2. The $R$-factor which is a numerical coefficient representative of inherent overstrength and global ductility capacity of lateral force-resisting system has been adjusted to provide a strength level base shear.
3. The incorporation of a Redundancy/Reliability Factor $\rho$, which is intended to encourage redundant lateral force resisting systems by penalizing non-redundant ones through higher lateral force requirements.
4. The incorporation of near-source factors $N_a$ and $N_v$ in Seismic Zone 4 that are intended to recognize the amplified ground motions that occur at close distances to the fault. The $N_a$ and $N_v$ factors address near field effects through the use of spectral value maps which are being developed by BSSC based on new USGS maps. The maps are not yet completed at this time (September 1997) in a format suitable for design applications but are expected to be ready in time for code adaptation by different agencies.

   The design base shear, as determined in the 1994 UBC, is a function of an assumed level of ground motion. In Seismic Zone 4, this level of ground motion has been taken as being effective peak ground acceleration EPA of 0.4 g. While no formal relationship exists between the EPA and the peak ground acceleration PGA, typically the EPA is taken at about two-thirds of the peak ground acceleration (PGA).

   Strong motion measurements in recent large earthquakes, such as the Northridge and Kobe events, showed that ground motions are significantly greater near the earthquake source. These events had near-source ground motions that greatly exceeded the EPA level for Zone 4 in the 1994 UBC. It has also been observed that the amplification of long-period ground motions is also greater with less competent site soil conditions.

   These near-source factors apply only to Seismic Zone 4 because it is believed that the near-source effect is only significant for large earthquakes. Research of the ground motions from Northridge, Kobe, and other events has indicated that the amount of near-source effect is greater at the long periods than the short periods.

The two near-source factors result in a greater amplification of the ground motions for long periods than for those short periods.

5. The adoption of a new set of soil profile categories which are used in combination with Seismic Zone Factors $Z$ and near-source factors $N_a$ and $N_v$, to provide site-dependent ground motion coefficients $C_a$ and $C_v$. These numerical coefficients define ground motion response within the acceleration and velocity-controlled ranges of the spectrum. The design response spectrum differs from the current one in two ways: (i) the constant velocity portion is now defined by $1/T$, as opposed to $1/T^{2/3}$, causing it to drop more rapidly in that range; and (ii) the plateau in the constant acceleration domain varies with $C_a$ rather than being a constant value for all soil profiles.
6. Substantial revisions have been made to the lateral force requirements for elements of structures, nonstructural components and equipment supported by structures. These provisions are believed to represent more accurately the lateral forces on elements by recognizing varying diaphragm accelerations, component amplification, component response modification, and ground motion response.
7. An important concept in the code change is the use of *elastic response parameters* to define unreduced forces and displacements with $R = 1$ for calculations involving drift and deformation compatibility. In addition, the parameter $E_m$ has been introduced to represent the maximum earthquake force that can be developed in the structure for use in addressing non-ductile conditions, similar to the $3R_w/8$ parameter in the 1994 UBC.
8. A new set of soil profile types has been introduced and is based on the average soil properties for the upper 100 ft of the soil profile. These are designated as $S_A$ through $S_E$. There is a sixth profile type $S_F$ for soils vulnerable to ground failure during seismic events, highly organic soils, or very thick, soft deposits of clay, which require a site-specific evaluation.
9. Introduction of Redundancy/Relaibility Factor $\rho$

   This factor represents the first attempt to codify a characteristic that has long been recognized as desirable in seismic force resisting systems. Previous editions of the Blue Book Commentary have recommended making lateral systems as redundant as possible; however, nothing was included in the provisions to encourage redundancy or any guidance given as to what constitutes a redundant structure.

The Redundancy Factor, $\rho$, varies between 1.0 for a structure with sufficient redundancy to 1.5 or a structure lacking in redundancy. The factor is given by the following formula:

$$\rho = 2 - 20/[r_{\max}(A_B)^{1/2}]$$

where $r_{\max}$ is the maximum element-story shear ratio, defined as the ratio of the design story shear in the most heavily loaded single element to the total design story shear. $A_B$ is the area of the base of the ground floor of the building.

The intent of defining $r_{\max}$ in terms of the element story in *the most heavily loaded single element* at each level is to account for torsional irregularities in the system. It is defined separately in the provisions for shear walls, braced frames and moment frames, and includes elements existing at or below the two-thirds height of the building.

For special moment-resisting frames, the provisions require that the number of bays of frames be increased such that the redundancy factor does not exceed 1.25. This is intended to address the damage to steel frame buildings in the Northridge earthquake, particularly those lacking redundancy. Moment frame design is often controlled by drift, and since the redundancy factor does not affect the drift calculation, current design practice would not be affected by the required increase in lateral forces. The 1.25 value will result in a minimum number of lateral force resisting elements for special moment frame buildings.

In accordance with the 1997 UBC, the base shear is calculated as follows:

$$V = \frac{C_v I}{RT} W \quad \text{and need not exceed} \quad V = \frac{2.5 C_a I}{R} W$$

The first formula defines the constant velocity range, and the second the constant acceleration range. In addition, the long period minimum value is set as:

$$V = 0.11\, C_a I W$$

with the additional Zone 4 minimum of: $V = \dfrac{0.8\, Z N_v}{R} W$

The Redundancy Factor, $\rho$, is applied to the base shear to obtain the

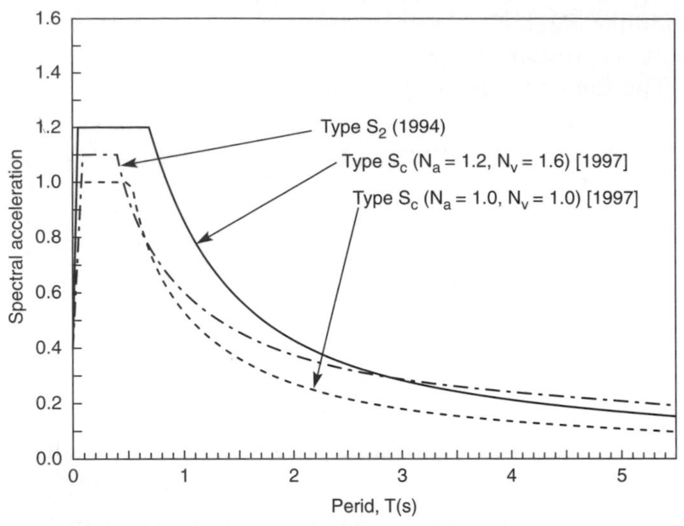


**Figure 12.66** Comparison of unreduced response spectra for zone 4: 1994 UBC-$S_2$ soil and 1997 UBC-$S_c$ soil.
*Note:* long period minimum and zone 4 minimum are not shown.

earthquake load, $E$, which is included in the basic load combinations with a 1.0 factor since the base shear is set at the *strength* level.

$$E = \rho E_h + E_v$$

$$E_m = \Omega_0 E_h$$

where $E$ = the earthquake load resulting from the combination of the horizontal component $E_h$, and the vertical component $E_v$

$E_h$ = the earthquake load due to base shear $V$

$E_m$ = the estimated maximum earthquake force that can be developed in the structure

$E_v$ = the load effect resulting from the vertical component of earthquake ground motion, and may be taken as zero for Allowable Stress Design

$\Omega_0$ = the seismic force amplification factor that is required to account for structural overstrength

$\rho$ = Reliability/Redundancy Factor as defined earlier

A comparison of unreduced response spectra for zone 4 between 1994 UBC soil type $S_2$ and 1997 UBC soil type $S_c$ is given in Fig. 12.66.

# Selected References

1. Farzad Naeim, Editor, "The Seismic Design Handbook." New York, Van Nostrand Reinhold, 1989.
2. Uniform Building Code, 1991, 1994 & 1997.
3. American National Standards Institute (ANSI) A58.1, 1982.
4. "Structuring Tall Buildings," *Progressive Architecture,* December 1980.
5. J. E. Cermak, "Applications of Fluid Mechanics to Wind Engineering," *Journal of Fluids Engineering,* March 1975.
6. "Wind Effects on High Rise Buildings," Symposium Proceedings, March 1970, Northwestern University, Illinois.
7. T. Tschanz, "Measurement of Total Dynamic Loads Using Elastic Models With High Natural Frequencies," in Timothy J. Reinhold (ed.), *Wind Tunnel Modeling for Civil Engineering Applications,* Cambridge University Press, 1982.
8. "Tall Buildings," Conference Proceedings, 1984, Singapore.
9. Robert A. Coleman, *Structural System Designs,* Englewood Cliffs, N.J., Prentice-Hall, 1983.
10. *Post-Tension Manual.* Prestressed Concrete Institute.
11. T. Y. Lin and S. D. Stotesbury, *Structural Concepts and Systems for Architects and Engineers.* New York, John Wiley.
12. James R. Libby, *Modern Pre-Stressed Concrete.* New York, Van Nostrand Reinhold, 1984.
13. T. Y. Lin and S. D. Stotesbury, *Design of Prestressed Concrete Structure.* New York, John Wiley.
14. George Winter and Arthur H. Nilson, *Design of Concrete Structures.* New York, McGraw-Hill.
15. Frederick S. Merritt, *Building Design and Construction Handbook.* New York, McGraw-Hill.
16. Gaylord and Gaylord, *Design of Steel Structures.* New York, McGraw-Hill.
17. N. M. Newmark and E. Rosenblueth, *Fundamentals of Earthquake Engineering.* Englewood Cliffs, N.J., Prentice-Hall, 1971.
18. Coull and Stafford-Smith, *Tall Buildings with Particular Reference to Shear Wall Structures.* New York, Pergamon Press, 1967.
19. Monoru Wakabayashi, *Design of Earthquake Resistant Buildings,* New York, McGraw-Hill, 1986.
20. Bungale S. Taranath, *Structural Analysis & Design of Tall Buildings.* New York, McGraw-Hill, 1988.
21. Kenneth Leet, *Reinforced Concrete Design.* New York, McGraw-Hill.
22. John M. Biggs, *Introduction to Structural Dynamics.* New York, McGraw-Hill, 1964.
23. Gaylord and Gaylord, *Structural Engineering Handbook.* New York, McGraw-Hill.
24. Sol. E. Cooper with Andrew C. Chen, "Designing Steel Structures—Methods and Cases," Englewood Cliffs, N.J., Prentice-Hall, 1985.
25. John Karlberg, *Premliminary Design for Post-tensioned Structures, Structural Engineering Practice, Vol. 2.* New York, Marcel Dekker, Inc., 1983.
26. Wolfgang Schueller, *High Rise Building Structures*, 1976.
27. Zbirohowski-Koscia, *Thin-Walled Beams.* Crosby Lockwood, 1967.
28. Milo S. Ketchum (ed.), *Structural Engineering Practice.* Vol. nos. 1 & 2, 1982; vol. no. 4, 1982–83; vol. nos. 1, 2, & 3, 1983; vol. no. 4, 1983–84. New York, Marcel Dekker.

29. *National Building Code of Canada and Its Supplement.*
30. N. W. Murray, *Introduction to the Theory of Thin-Walled Structures.* New York, Oxford University Press, 1984.
31. Timothy J. Reinhold (ed.), *Wind Tunnel Modeling for Civil Engineering Applications.* Cambridge University Press, 1982.
32. M. B. Kanchi, *Matrix Methods of Structural Analysis.* New York, Wiley Eastern Limited.
33. Mario Paz, *Structural Dynamics.* New York, Van Nostrand Reinhold, 1985.
34. J. R. Choudhury, "Analysis of Plain and Spatial Systems of Interconnected Shear Walls," Ph.D. Thesis, University of Southampton, 1968.
35. V. Z. Vlasov, *Thin-Walled Elastic Beams.* Washington, D.C., National Science Foundation.
36. Bungale S. Taranath, "Torsional Behavior of Open-Section Shear Wall Structures," Ph.D. Thesis, University of Southampton.
37. Aslam Qadeer, "Interaction of Floor Slabs and Shear Walls," Ph.D. Thesis, University of Southampton, 1968.
38. Bungale S. Taranath, "A New Look at Composite High-Rise Construction," Our World in Concrete and Structures, Singapore, 1983.
39. Bungale S. Taranath *et al.*, "A Practical Computer Method of Analysis for Complex Shear Wall Structure," ASCE 8th Conference in Electronic Computation, 1983.
40. Bungale S. Taranath, "Composite Design of First City Tower," *The Structural Engineer,* 1982.
41. Bungale S. Tanarath, "Differential Shortening of Columns in High-Rise Buildings," *Jounal of Torsteel*, 1981.
42. Bungale S. Tanarath, "Analysis of Interconnected Open Section Wall Structures," *ASCE Journal*, 1986.
43. Bungale S. Tanarath, "The Effect of Warping on Interconnected Shear Wall Flat Plate Structures," *Proceedings of the Institution of Civil Engineers*, 1976.
44. Bungale S. Tanarath, "Torsion Analysis of Braced Multi-Storey Buildings," *The Structural Engineer*, London, 1975.
45. Bungale S. Tanarath, "Optimum Belt Truss Locations for High-Rise Buildings," *AISC*, vol. II, 1974.
46. Kenneth H. Lenzen, "Vibrations of Steel Joist–Concrete Slab Floors," *AISC Engineering Journal,* July 1966.
47. J. F. Wiss and R. H. Parmalee, "Human Perception of Transient Vibrations," *Journal of Structural Division ASCE*, vol. 100, April 1974, p. 773–783.
48. Thomas M. Murray, "Design to Prevent Floor Vibrations," *Engineering Journal–American Institute of Steel Construction,* third quarter, 1975.
49. R. Halvorson and N. Isyumov, "Comparison of Predicted and Measured Dynamic Behavior of Allied Bank Plaza," in N. Isyumov and T. Tschanz (eds): *Building Motion in Wind*, New York, ASCE, 1982.
50. Bryan Stafford Smith and Alex Coull "Tall Building Structures, Analysis and Design," John Wiley & Sons, Inc. 1991.
51. NEHRP, Recommended Provisions for the Development of Seismic Regulations for New Buildings, Part 1, Provisions, Part 2, Commentary (FEMA 222 and 223, 1991 Edition)
52. Guide to Application of the NEHRP Recommended Provisions in Earthquake-resistant Building Design (FEMA 140, Sept. 1990).
53. NEHRP Handbook for the Seismic Evaluation of Existing Buildings (FEMA 178, June 1992).
54. NEHRP Handbook for Seismic Rehabilitation of Existing Buildings (FEMA 172, June 1992).
55. Vision 2000: A framework for performance-based engineering of buildings, by Ronald O. Hamburger, Anthony B. Court, & Jeffrey R. Sonlager, SEAOC proceedings 1995 convention.
56. Alexander Coull and J. R. Choudhury, "Analysis of Coupled Shear Walls," *ACI Journal,* September, 1967.
57. ETABS Version 6, Steeler, *Stress Check of Steel Frames*; Conker, *Design of Concrete Frames by Computers and Structures, Inc.*, January 1995.

58. *A Rational System for Earthquake Risk Management* by M. Russel Nester and Allen R. Porush.
59. *Seismic Design Guidelines for Upgrading Existing Buildings:* Department of Army, Navy, & Air Force, PB 89-220453.
60. Building Code Requirements for Structural Concrete (ACI 318-95) and Commentary (ACI 318R-95). American Concrete Institute.
61. Manual of Steel Construction, Allowable Stess Design, Ninth Edition, American Institute of Steel Construction, 1989.
62. Seismology Committee, Structural Engineers Association of California, Recommended Lateral Force Requirements and Commentary (Blue Book), 1996, Sixth Edition.

Appendix

# Conversion Factors: U.S. Customary Units to SI Metric Units

### Lengths or displacements

1 ft = 0.3048 m
1 in = 25.40 mm
1 mile (U.S. statute) = 1609.3 m (1.609 km)
1 mile (international nautical) = 1852.0 m (1.852 km)

### Area

1 $in^2$ = 645.2 $mm^2$
1 $ft^2$ = 0.0929 $m^2$

### Section modulus or volume

1 $in^3 = 16.39 \times 10^3$ $mm^3$
1 $ft^3$ = 0.0283 $m^3$

### Moment of inertia

1 $in^4 = 0.4162 \times 10^6$ $mm^4$
1 $ft^4$ = 0.008631 $m^4$

## Loads

1 lb = 4.448 N
1 kip (1000 lb) = 4.448 kN
1 lb/ft = 14.59 N/m
1 kip/ft = 14.59 kN/m
1 lb/in = 1.7513 N/m

## Bending moment or torque

1 lb · in = 0.11298 Nm
1 lb · ft = 1.3558 Nm
1 kip · ft = 1.3558 kNm

## Stress, modulus of elasticity, and surface loads

1 $lb/in^2$ = 0.006895 MPa
1 $kip/in^2$ = 6.895 MPa
1 $lb/ft^2$ = 0.0479 $kN/m^2$
1 $kip/ft^2$ = 47.88 $kN/m^2$

## Unit weight, density

1 $lb/ft^3$ = 16.03 $kg/m^3$
1 lb mass = 0.45359 kg
1 ton mass = 907.18 kg

# Index


Acceleration
  building top floors: 2, 39, 81:
    Devenport's criteria for, 246
Accelerograph, 267, 269
Across-wind, 80
Along-wind, 121
Aeronautical wind tunnel, 125, 222
  (*See also* Wind tunnel engineering)
AISC (American Institute of Steel
    Construction) design specification
    and commentary, 578
ANSI Standard A58.1 (American
    National Standard Minimum Design
    Loads for Building and Other
    structures), 136
ASCE Standard 7 (1993), 119, 131, 147–
    149
ASCE Standard 7 (1995), 179–205
  building classification for wind, 151
  example problems, 162, 166, 168–178,
    194–197, 199–205
  exposure category constants, 154, 185
  external pressure coefficients: 156, 157
  cladding design, 156, 159, 167
  gust response factors, 164
  gust response sensitivity study, 178–
    180, 197–199
  importance factor for wind, 152, 182
  internal pressure coefficient, 158, 161
  overview, 148, 181
  pressure profile factor, 171
  resonance factor, 173
  structure size factor, 174
  surface coefficient factor, 172
  surface friction factor, 172
  velocity exposure coefficient, 152, 155,
    184, 185
ASCE Standard 7 (1995), (*Cont.*):
  wind pressure for bracing design, 186
  wind speed-up $K_z$ factor, 187
  wind map, 153, 183
Anemometer, 117

Blue Book, 304
BOCA (Building Officials and Code
    Administrators), International, wind
    loads, code related to, 130, 131
Boundary layer, atmospheric, 116
Brace(ing):
  Chevron, 78, 91, 573
  eccentric, 95, 99,
Braced core 67,
Braced frame, 398, 421
  types of, 423–426,
Braced tube, 91
Buckling load, Rayleigh–Ritz method,
    526
Bundled (cellular, modular) tube, 398,

Caisson foundation, 23, 24, 28, 29, 30, 52
Cantilever method, 705, 706, 708, 710,
    712
Cellular tube, 528
  two-celled tube, 490
  (*See also* Bundled tube)
Cladding, 126, 221, 913–924
    brick veneer, 923
  design pressure comparison with
    overall loads, 229
Column:
  composite, 36, 38, 39, 54, 57, 58, 61,
    62, 63, 66, 67, 531–575
Composite beams, 664–687
  design example, 682
Composite columns, 15, 16, 17, 701–703
  interaction diagram, 703
Composite (hybrid, mixed) construction,
    16
Composite haunched girders:
  description of, 687–689
Composite metal deck-slab: 661–664
  SDI specifications, 662–664
Composite trusses, 689
  framing plan, elevation and details,
    690

Concrete design
beam design, 849–855
comments on seismic details, 877
column bi-axial bending, 859–862
load combinations, 847–848
panel zone shear forces, 855–858
$P\Delta$ effects (see also steel design), 862–866
seismic requirements, 866–870
shear wall, design 870–877
strength reduction factors, 847, 848
Connection:
AISC Types 1, 2 and 3, 399, 405, 406
double-angle, 41
end plate, 403
seat angle with top plate, 414
semi-rigid, 408, 410
top and bottom clip angle, 412, 413
unstiffened seated beam, 399–405
welded, beam-to-column, 404
Coupled shear walls:
continuous medium method, 724–732
CQC combination, 332, 333
Creature comfort, 106
(*See also* Acceleration, Devenport's criteria for; Sway)
Critical damping, 279
(*See also* Damping)
Curtain wall: 920, 921
Cyclones, 112

Damping, 39, 279, 280
Coulomb, 279
critical, 279
devices, mechanical, to reduce sway, 13
friction, 124, 279
hysteresis, 280
internal, 124
mechanical, 924–926
radiation, 280
representative values, 81,
viscous, external and internal, 279
Deflection:
frame: bending component, 418–419
equivalent cantilever method, 719
example portal, 718
shear component, 419–420, 716
Density, 40
Diagonally braced tube, (frame), 10, 49
Diaphragm, 289, 290
flexible, 300, 944
rigid, 944
Deflection (*Cont.*):
seismic design of, 938–949
semirigid, 944
Differential shortening of columns, 12, 27, 39, 882–897
Drift, 39, 40, 77, 126, 420, (Deflection; Lateral drift; Sway)
Drilled pier, (*See* Foundation; Pile)
Ductile frame, 10
Ductility, 26, 274, 283, 289, 291, 394

Earthquake: (*see also* seismic design)
accidental eccentricity, 300
attenuation, 275, 278
behavior of tall buildings during, 276–281
elastic rebound theory, 261
El Centro ground acceleration record, 269
El Centro ground displacement record, 270
El Centro ground velocity record, 270
epicenter of, 278
focus of, 266
ground motion specification, 275
influence of soil, 278
intensity of, 268
Mercalli scale, 269
magnitude, 268
Northridge, 270, 271, 417
recent earthquakes, 262
Richter scale, 268
Earthquake hazard mitigation technology
friction dampers, 957
metallic dampers, 957
passive energy dissipation, 957–961
SEAOC Blue Book requirements, 952–957
seismic base isolation, 949–961
viscoelastic damper, 960
viscous and semi-viscous dampers, 960
Eccentric brace, 398, 424, 435, 530
analysis and design of, 438
behavior of, 436–438
deflection of, 439
ductility of, 436
elevation and detail of, 437
types of, 437
Effective width of slab, 503
El Centro earthquake,
(*see* Earthquake)

Fastest-mile wind velocity, 119
Floor levelness, problems with, 897–899
Foundation: 926–938
  for corner column, X-braced building, 930
  pile, drilled, underreamed, 927–929
Framed tube, 398, 466–480, 722
  behavior of, 467–471
  column transfer in, 479
    axial forces due to, 479
    plate girder, 478
      shoring of, 478
    transfer girder, 478
  shear transfer through diaphragm, 480
Free-form tubes, 469
  schematic plans, 471

Glass: 916–920
  annealing of, 917
  float, 917
  heat-strengthened, 918
  heat-treated, 918
  insulating, 99
  laminated, 918
  plate, 917
  tempered, 918
  wire, 919
Glass fiber-reinforced concrete, 73, 923
Gradient height, 115, 116
Gradient velocity or wind speed, 107
Gravity systems:
  composite, 661–703
    beams, 664–687
    columns, 701–703
    haunch girders, 687–689
    metal deck, 661, 662
    stub girders, 689–701
    trusses, 689
  in concrete, 585–660
  ACI coefficients, 594
  ACI minimum thickness, two-way slabs, 611
    band beam, 590, 591
  design examples:
    one-way slab and beams, 592
    T-beam, 600
    two-way slab, 610
    flat plate, 586
    flat slab, 587
    haunch girder, 591
    one-way joist, 588, 589
    skip joist, 588, 590
    waffle, 587, 588
Gravity systems (*Cont.*):
  for steel buildings, 577–584
    columns, 583, 584
    metal deck, 580
    open-web joist, 580
    wide-flange beams, 582

Haunch girder, concrete, 29, 34
Hurricanes, 111, 112, 113, 114
Hysteretic behavior; 291, 292

Interior brace:
  for full depth of building, 529,

Lateral bracing:
  composite construction, 531–575
    composite columns, 535–539
    composite diagonals, 539
    composite girder, 534, 535
    composite shear walls, 539, 540, 541–545
    composite shear wall-frame interaction, 545, 546
    composite slabs, 534
    composite tubes, 546–549
    example projects, 550–575
    high efficiency structure, 573–575
    mixed floor system, 572
    vertically mixed system, 549
  concrete construction, 501–530
    column and slab, 502–505
    core structure, 509
    coupled shear walls, 506
    exterior diagonal tube, 523–524
    flat slab: and walls, 505
      walls and columns, 505
  frame tubes, 522
  perimeter tube, 507, 508
  rigid frame, 506
  rigid frame-haunch girder, 508
  shear wall-frame, 510–522
Loads:
  dead, 579
  live, 579
Lumping techniques, 732–736
Mat foundation, 24, 29, 30, 38, 40, 78, 82, 94
Modal analysis, hand calculation, 335–346
Motion perception, 3, 221, 258
Motion sickness, 105

National Building Code of Canada (NBC 1990), 119, 131, 205–221
wind loads, 205–221
design example, 215–221

Optimum location:
for single outrigger, 457–459
for two outriggers, 459–466
Outrigger and belt truss system, 398, 445–466, 527, 528
behavior, 448
example projects, 20, 23, 39, 43, 59, 463–466
optimum location: for single outrigger, 457–459
for two outriggers, 459, 466
tied contilever model, 446
Outrigger truss, 31, 36, 38

*p*-Δ effect, 276, 302
Panel zone, 721, 907–913
Partial computer model, 736–737
Perception of motion, 13, 40
Pile (pier), 38, 40, 66, 73, 93, 94,
Portal frame:
deflection comparison, 715
Portal method, 705, 706, 709, 713
comparison with computer analysis, 715
Posttension steel, 616
Posttension systems, 613–660
beam and slab system, 25, 28
building examples, 623–625
cracking problems, 625
design steps, 618
load balance technique in, 620, 628, 629
practical considerations, 622, 623
preliminary design, 626–660
beam, 645–649
continuous span, 634–635
flat plate, 650–660
one-way slab, 637–645
simple spans, 631–634
secondary moments, 620, 627
strength design, 660
span-to-depth ratios, 619
temporary pour-strip, 624, 626
Preliminary hand calculations, 705
Pressure:
wind:
internal, 129, 230
Prestressed concrete, (*see also* posttension systems) 613–660
advantages of, 613, 614
creep in, 625
losses in, 616
materials, 616, 617
methods of prestressing, 614
building examples, 623–625
Probability analysis, wind load, 119

Redundancy, 293, 304
Resonance:
of building, 124, 279
Response spectrum, 81, 321–326
Return period, 119
Reynolds number, 125
Richter scale, 76,
Rigid frame, unbraced, moment, 31, 398, 415
response of, 416
interaction with cross braces, 440–444

SEAOC Vision 2000, 370–374
St. Venant's torsion, 739
Seismic design: (*see also* Earthquake)
irregularities, 286–287, 304, 305, 307
response spectrum method, 315
dynamic approach, 281, 309–346
dynamic approach UBC overview, 331–335
static approach, 281, 282, 293
UBC method, 293–309
UBC design example, 305
uncertainties, 272–275
seismic separation, 281
Seismograph, 262, 267, 268
conceptual model, 268
Seismic Vulnerability Study, 346–374
FEMA-178 Method, 353–360
Tri-services manual, 360–370
capacity spectrum method, 365–370
pushover analysis, 365
Self-limiting deformation, 405
Semirigid connection, 398
Shear studs (connectors) for composite beams, 664, 666, 667, 671, 672, 673, 674, 679
AISC minimum requirements, 681
AISC placement criteria, 681
AISC shear reduction factors, 670, 672
description of, 666
for full composite action, 676, 677
maximum spacing, 670

Shear studs (connectors) for composite beams, (*Cont.*)
for moment due to concentrated loads, 679
for partial composite action, 680, 685
recommended installation sequence, 681
in stub girders, 699
Shear wall-frame interaction, 29
Slurry wall, 52, 93
Special Moment Resisting Frame (SMRF, ductile frame), 53, 75, 82, 83
SRSS combination, 332, 335, 336, 337, 341, 344, 345, 346
Staggered truss, 398, 426, 428
behavior, 429
column design in, 433
conceptual model, 431, 432
floor design in, 429–433
load path in, 432
plan and perspective, 428
in semicircular building, 430
span-to-depth ratio, 434
truss design in, 434
Standard Building Code (SBC), wind loads, 130, 136
Steel design:
braced frames, 843
compact, semicompact and noncompact sections, 818–822
compression members, 822–825
continuity plates, 836–838
doubler plates, 838–840
eccentrically braced frames (EBF), 843–845
effective length concept, 825–829
lateral stability, 817
load combinations, 814–815
ordinary moment frames, 840, 848
$p$-$\Delta$ effects (*see also* concrete design), 829–835
special concentrically braced frames, 845, 846
Special Moment Resisting Frame (SMRF), 841–843
tension members, 815
UBC (1994) seismic requirements, 840–846
Stress relieving of welded joints, 77
Strouhal number, 124, 125
Stub girder, 689–701

Time-history analysis, 269, 326–331, 334
Top-down construction, 52
Tornado, 111, 113, 114
Torsion:
analogy with bending, 739
bimoment, 747, 748, 763, 779
circular shaft, 741
fourteen-by-fourteen matrix, 810
I-section, 748–754
sectorial coordinate, 754–756
sectorial moment of inertia, 758, 759, 760, 781
shear center, 744, 756, 757, 759
shear flow, 743, 745, 746
St. Venant, 747
steel braced core, 791–796
terminology, 739
torsion constants, 758, 797, 798
twelve-by-twelve matrix, 809
Vlasov theory, 764–771
warping (warping stiffness), 746, 747, 761–771, 805–811
worked examples
Truss, 69
Tube, 5, 9, 18
bundled, 5, 487–495, 525
circular, 470
free-form, 471
irregular shaped, 476
partial framed, 65
perimeter, 25, 31, 86
rectangular, axial stresses in, 470
shear lag in, 469, 471, 472, 474, 475
triangular, 470
trussed, 398, 480–487
Tuned mass damper, 925
Turbulence, wind, 114
Typhoons, 111, 112

Ultimate high-efficiency structures, 495–499
floor plan, 498
Khan's concept, 498
Uniform Building Code (UBC), 11, 130, 136–147
Update (1997), 982–986
Unit structural quantities, 970–982
variation with height, 114, 115

Velocity gradient, 115
Vibration:
in floor system: 899–906
NBC design recommendation, 905

Vibration (*Cont.*):
  Richer-Meister rating, 903
  Wiss and Parmelee rating, 904
Virendeel system, 17, 23, 26, 43, 53, 54, 55, 57, 58, 62, 63, 70, 82
Viscosity, 115
Viscoelastic damper, 925
Viscous damping (*see* Damping)
Vlasov's torsion theory (*see* Torsion)
Vortex shedding, 80, 114, 121, 123, 124, 125, 221

Waffle system, 83
Warping (*see* Torsion)
Welded moment connections, SAC guidelines, 961–970
Wind:
  across-wind response (*see* Vortex shedding), 121, 122
  dynamic action of, 114
  easterlies, 109
  fluctuations of, 110
  gustiness, 110, 125
  local, 110, 129,
  prevailing, 109
  probabilistic analysis of, 114, 119
  seasonal, 110
  six components of, 121
  trade wind, 108, 109, 110
  turbulence, due to, 116, 117, 118, 122
  westerlies, 108, 110
Wind tunnel engineering, 221–260
  aeroelastic study, 39, 232–246
  comparison with code winds, 259
  flexible model, 240–246
  model requirements, 235
  rigid models, 225–232, 236, 239
  when to undertake, 234
  development of, 221–222
  description, 222
  field measurement comparison, 256–258
  five-component force balance model, 39, 250–253
  high-frequency force balance model, 246–253
  objectives, 224–225
  pedestrian wind studies, 253–256
  overall design loads, 230
  requirements, 223

## ABOUT THE AUTHOR

Dr. Bungale S. Taranath, Ph.D., SE, FI. Struct E, is a senior project manager with the largest consulting firm in the United States, John A. Martin and Associates located in Los Angeles, California. He has extensive experience in the design of concrete, steel, and composite tall buildings and has served as principal-in-charge on many notable high-rise buildings. He has held positions as a senior project engineer in Chicago, Illinois and as vice president and principal-in-charge with two consulting firms in Houston, Texas. Dr. Taranath is a fellow of the Institution of Structural Engineers, London, England; a member of the American Society of Civil Engineers and the American Concrete Institute; and a registered structural and professional engineer in several states. He has conducted research into the behavior of tall buildings and shear wall structures and is the author of a number of published papers on torsion analysis and multistory construction projects including *Structural Analysis & Design of Tall Buildings,* published by McGraw-Hill in 1988.